D0221428

MATHEMATICS FOR SCIENTISTS AND ENGINEERS

MATHEMATICS FOR SCIENTISTS AND ENGINEERS

HAROLD COHEN
California State University, Los Angeles

PRENTICE HALL, Englewood Cliffs, New Jersey 07632

Library of Congress Cataloging-in-Publication Data

Cohen, Harold, (date)
Mathematics for scientists and engineers/by Harold Cohen.
p. cm.
Includes index.
ISBN 0-13-563156-4
1. Mathematics. I. Title.
QA37.2.C63 1992
510—dc20

Acquisitions editor: Timothy Bozik
Editorial/production supervision and
interior design: Maria McColligan
Copyeditor: Barbara Zeiders
Cover design: Bruce Kenselaar
Prepress buyer: Paula Massenaro
Manufacturing buyer: Lori Bulwin
Supplements editor: Alison Munoz

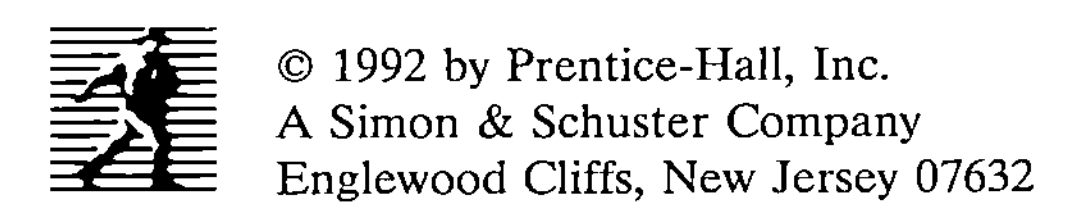

Printed in the United States of America

10 9 8 7 6 5 4 3 2 1

ISBN 0-13-563156-4

Prentice-Hall International (UK) Limited, *London*
Prentice-Hall of Australia Pty. Limited, *Sydney*
Prentice-Hall Canada Inc., *Toronto*
Prentice-Hall Hispanoamericana, S.A., *Mexico*
Prentice-Hall of India Private Limited, *New Delhi*
Prentice-Hall of Japan, Inc., *Tokyo*
Simon & Schuster Asia Pte. Ltd., *Singapore*
Editora Prentice-Hall do Brasil, Ltda., *Rio de Janiero*

This book is dedicated to my children, Lisa and David, and to my parents. It is also dedicated to those students who have taken the Mathematical Methods of Physics course at California State University, Los Angeles. They helped in its creation and refinement.

CONTENTS

CHAPTER 12 INTEGRAL EQUATIONS 624

CHAPTER 13 NUMERICAL METHODS 668

PREFACE

This text is a mathematics book. The author's training and discipline of interest is physics, and the book is the result of teaching a course entitled "Mathematical Methods of Physics" offered by the Department of Physics and Astronomy at California State University, Los Angeles.

However, the book is intended to serve students in many disciplines. As such, specific examples tend to be purely mathematical in nature. When avoidable, they do not involve content specific to physics. When examples involving a knowledge of physics are presented, attempts have been made to keep the physics at an elementary level. In those instances where an example involves a nonelementary physics concept, the accompanying discussion describing the physics is designed to give the reader sufficient background to understand the mathematical treatment of the problem. The author hopes that he has succeeded in this endeavor.

The reason for this mathematical approach evolved from the author's experience teaching the above-mentioned course. By keeping the physics content of the course elementary and minimal, word spread around campus that the course was appropriate for students in other disciplines. As a result, the course currently attracts students outside physics, primarily from our mathematics and chemistry programs. It was the goal of the author to create an applied mathematics textbook for a broad audience, including engineers and people in several fields of science, not just physics. If someone is using this text in an applied mathematics course for students in a particular discipline, it is intended that the instructor will include additional examples and problems from that discipline.

Harold Cohen

MATHEMATICS FOR SCIENTISTS AND ENGINEERS

CHAPTER 1

VECTOR ANALYSIS

1.1 Vector Arithmetic

Many quantities and equations can be described compactly using vectors. In introductory physics and engineering courses, one is usually introduced to vector descriptions in three-dimensional space. The **i**, **j**, and **k** vectors, called the *Cartesian basis vectors*, are parallel to the x, y, and z axes (see Figure 1.1). They form a mutually orthogonal or perpendicular set of basis vectors in three-dimensional space. They are said to be *normalized*, which means that they are of length 1. Then any vector in the space can be written as a linear combination of **i**, **j**, and **k**. That is,

$$\mathbf{V} = V_x\mathbf{i} + V_y\mathbf{j} + V_z\mathbf{k} \tag{1.1}$$

The scalar quantities, V_x, V_y, and V_z are called the x, y, and z *components* of the vector **V**.

Arithmetic operations can be performed on vectors similar to the way they are performed on scalar parameters. One is usually introduced to methods of vector addition and subtraction in introductory courses. The reader is referred to an introductory textbook for these operations.

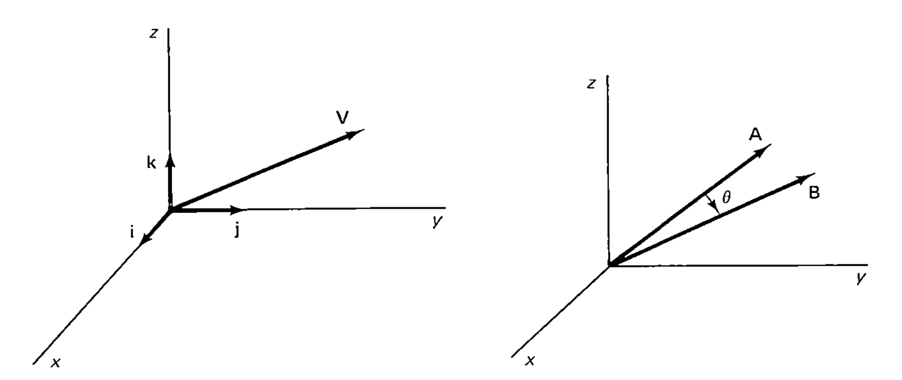

Figure 1.1 Vector **V** and Cartesian basis vectors.

Figure 1.2 Angle between two vectors.

Scalar and vector products of two vectors

Two types of vector multiplication are defined in three-dimensional space. The *dot* (or *scalar*) *product* of **A** and **B** yields a scalar quantity. In terms of the angle θ between the vectors (Figure 1.2), the dot product can be written

$$\mathbf{A} \cdot \mathbf{B} = |\mathbf{A}|\,|\mathbf{B}|\cos\theta = \mathbf{B} \cdot \mathbf{A} \tag{1.2}$$

If **A** and **B** are orthogonal, $\theta = \pi/2$ and $\mathbf{A} \cdot \mathbf{B} = 0$. If **A** is parallel to **B**, $\theta = 0$ and $\mathbf{A} \cdot \mathbf{B} = |\mathbf{A}|\,|\mathbf{B}|$. Thus the Cartesian basis vectors satisfy

$$\text{(a)}\quad \mathbf{i} \cdot \mathbf{i} = \mathbf{j} \cdot \mathbf{j} = \mathbf{k} \cdot \mathbf{k} = 1 \qquad \text{(b)}\quad \mathbf{i} \cdot \mathbf{j} = \mathbf{i} \cdot \mathbf{k} = \mathbf{j} \cdot \mathbf{k} = 0 \tag{1.3}$$

With these prescriptions, the dot product of two vectors is

$$\begin{aligned}\mathbf{A} \cdot \mathbf{B} &= (A_x\mathbf{i} + A_y\mathbf{j} + A_z\mathbf{k}) \cdot (B_x\mathbf{i} + B_y\mathbf{j} + B_z\mathbf{k}) \\ &= (A_xB_x\mathbf{i} \cdot \mathbf{i} + A_yB_y\mathbf{j} \cdot \mathbf{j} + A_zB_z\mathbf{k} \cdot \mathbf{k}) + (A_xB_y + A_yB_x)\mathbf{i} \cdot \mathbf{j} \\ &+ (A_xB_z + A_zB_x)\mathbf{i} \cdot \mathbf{k} + (A_yB_z + A_zB_y)\mathbf{j} \cdot \mathbf{k} = (A_xB_x + A_yB_y + A_zB_z)\end{aligned} \tag{1.4}$$

Instead of labeling the components x, y, and z, we can denote them by 1, 2, and 3. Then the dot product can be written as

$$\mathbf{A} \cdot \mathbf{B} = A_1B_1 + A_2B_2 + A_3B_3 = \sum_{l=1}^{3} A_lB_l \tag{1.5a}$$

From this description it is easy to see that the dot product can be defined in a space of arbitrary dimension, N, as

$$\mathbf{A} \cdot \mathbf{B} = \sum_{l=1}^{N} A_lB_l \tag{1.5b}$$

The *cross* or *vector-product* is denoted by $\mathbf{A} \times \mathbf{B}$. This product is only defined in three-dimensional space and the result is a vector. The magnitude of $\mathbf{A} \times \mathbf{B}$ is expressed in terms of the angle between the vectors as

$$|\mathbf{A} \times \mathbf{B}| = |\mathbf{A}|\,|\mathbf{B}|\sin\theta \tag{1.6}$$

Thus if **A** and **B** are parallel or antiparallel, $\theta = 0$ or π and $\mathbf{A} \times \mathbf{B} = 0$. If **A** and **B** are perpendicular, $\theta = \pi/2$ and $|\mathbf{A} \times \mathbf{B}| = |\mathbf{A}|\,|\mathbf{B}|$.

The direction of the cross product of **A** and **B** is perpendicular to the plane containing **A** and **B**, and is defined by a *right-hand rule*. A common description of this right-hand rule is illustrated in Figure 1.3. One orients one's right hand with its fingers in the direction of the first vector in the product (**A** in this discussion) such that when the right hand is closed, the fingers rotate into the second vector (**B** in this case). When the hand is closed, the thumb will point in the direction of $\mathbf{A} \times \mathbf{B}$.

From Figure 1.3, it is noted that $\mathbf{B} \times \mathbf{A}$ is oppositely directed to $\mathbf{A} \times \mathbf{B}$. Since, by equation 1.6, they are equal in magnitude,

$$\mathbf{A} \times \mathbf{B} = -\mathbf{B} \times \mathbf{A} \tag{1.7}$$

Applying either the right-hand rule, equation 1.6, or equation 1.7 to the Cartesian basis vectors, it is straightforward to deduce that

$$\begin{aligned}&\text{(a)}\quad \mathbf{i} \times \mathbf{j} = -\mathbf{j} \times \mathbf{i} = \mathbf{k} \qquad &&\text{(b)}\quad \mathbf{j} \times \mathbf{k} = -\mathbf{k} \times \mathbf{j} = \mathbf{i} \\ &\text{(c)}\quad \mathbf{k} \times \mathbf{i} = -\mathbf{i} \times \mathbf{k} = \mathbf{j} \qquad &&\text{(d)}\quad \mathbf{i} \times \mathbf{i} = \mathbf{j} \times \mathbf{j} = \mathbf{k} \times \mathbf{k} = 0\end{aligned} \tag{1.8}$$

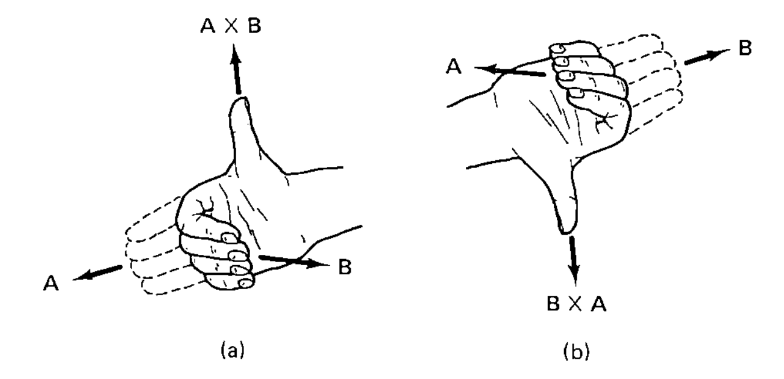

Figure 1.3 Right-hand rule describing the directions of **A** × **B** and **B** × **A**.

Therefore,

$$\begin{aligned}\mathbf{A}\times\mathbf{B} &= (A_x\mathbf{i}+A_y\mathbf{j}+A_z\mathbf{k})\times(B_x\mathbf{i}+B_y\mathbf{j}+B_z\mathbf{k})\\ &= (A_xB_y-A_yB_x)\mathbf{i}\times\mathbf{j}+(A_yB_z-A_zB_y)\mathbf{j}\times\mathbf{k}+(A_zB_x-A_xB_z)\mathbf{k}\times\mathbf{i}\\ &\quad + A_xB_x\mathbf{i}\times\mathbf{i}+A_yB_y\mathbf{j}\times\mathbf{j}+A_zB_z\mathbf{k}\times\mathbf{k}\end{aligned} \tag{1.9a}$$

Using the fact that the cross product of a vector with itself is zero, we obtain

$$\mathbf{A}\times\mathbf{B} = (A_yB_z-A_zB_y)\mathbf{i}+(A_zB_x-A_xB_z)\mathbf{j}+(A_xB_y-A_yB_x)\mathbf{k} \tag{1.9b}$$

Even though **A** × **B** has x, y, and z components, the cross product has a different vector character than **A** or **B**. Consider describing **A** and **B** by the bases $\mathbf{i}'=-\mathbf{i}$, $\mathbf{j}'=-\mathbf{j}$, and $\mathbf{k}'=-\mathbf{k}$. Such an inversion results in a coordinate system that is called left-handed (see Figure 1.4). Using the right-hand rule described in Figure 1.3 gives

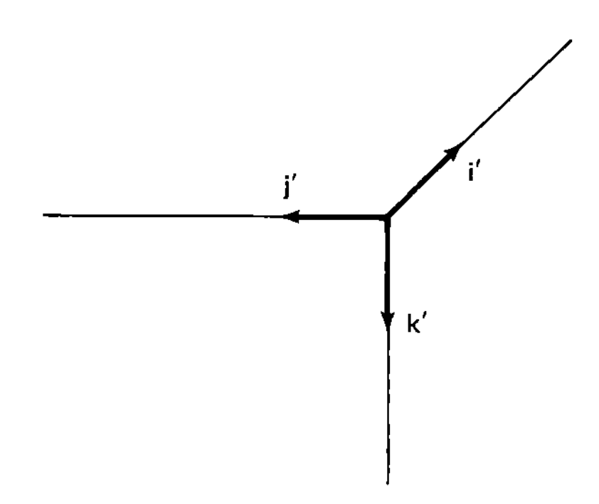

Figure 1.4 Basis vectors of a left-handed coordinate system.

$$\text{(a)}\quad \mathbf{i}'\times\mathbf{j}'=-\mathbf{k}' \qquad \text{(b)}\quad \mathbf{j}'\times\mathbf{k}'=-\mathbf{i}' \qquad \text{(c)}\quad \mathbf{k}'\times\mathbf{i}'=-\mathbf{j}' \tag{1.10}$$

However, performing the same manipulations on $\mathbf{i}'$, $\mathbf{j}'$, and $\mathbf{k}'$ using the left hand, we obtain

$$\text{(a)}\quad \mathbf{i}'\times\mathbf{j}'=\mathbf{k}' \qquad \text{(b)}\quad \mathbf{j}'\times\mathbf{k}'=\mathbf{i}' \qquad \text{(c)}\quad \mathbf{k}'\times\mathbf{i}'=\mathbf{j}' \tag{1.11}$$

Upon inversion, the vectors **A** and **B** in a right-handed coordinate system become −**A** and −**B** in a left-handed system. However, when inverting the coordinate frame, $\mathbf{A}\times\mathbf{B}\rightarrow(-\mathbf{A})\times(-\mathbf{B})=\mathbf{A}\times\mathbf{B}$; that is, the cross product does not change sign upon inversion. Vectors such as **A** and **B** that change sign upon inversion are called *real* or *true vectors*. A vector such as the cross product, that does not change sign under inversion, is referred to as a *pseudovector*. If **A** is a real vector and **B** is a pseudovector, $\mathbf{A} \times \mathbf{B} \to (-\mathbf{A}) \times (+\mathbf{B}) = -\mathbf{A} \times \mathbf{B}$. Since $\mathbf{A} \times \mathbf{B}$ changes sign, it is a real vector.

Another method of evaluating the vector product is achieved by an arithmetic operation on an array of numbers called *taking the determinant*. We discuss the evaluation of determinants in greater detail in Chapter 8. For now we restrict ourselves to the evaluation of 2×2 and 3×3 arrays.

A 2×2 determinant is evaluated by multiplying along the diagonals as shown in equation 1.12a. The products of the elements along the left-to-right diagonals have positive signs associated with them. The products along the right-to-left diagonals are associated with negative signs. These products are added with their appropriate signs. The multiplication is indicated by the diagonal lines.

$$\begin{vmatrix} a_{11} & a_{12} \\ a_{21} & a_{22} \end{vmatrix} = a_{11}a_{22} - a_{12}a_{21} \tag{1.12a}$$

Consider the 3×3 array $\begin{vmatrix} a_{11} & a_{12} & a_{13} \\ a_{21} & a_{22} & a_{23} \\ a_{31} & a_{32} & a_{33} \end{vmatrix}$

The evaluation of this 3×3 determinant is accomplished in a way similar to the evaluation of the 2×2 array using the following operations:

1. Copy the first and second rows below the third.
2. Multiply the elements along the diagonal lines (as shown below) until all rows have been exhausted.
3. Add the products along each diagonal together with positive signs if the diagonals slant from left to right, with negative signs if the diagonals slant from right to left.

For the 3×3 array, these operations are indicated by the diagonal lines.

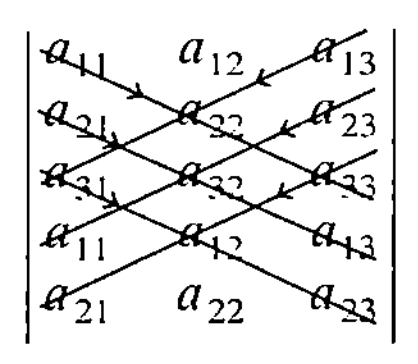

Because there is no sixth row in the array above, all rows have been accessed after multiplication along three left-to-right diagonals. Similarly, it is only possible to multiply along three right-to-left diagonals. The result of the multiplications along the diagonals with the appropriate signs is

$$\begin{vmatrix} a_{11} & a_{12} & a_{13} \\ a_{21} & a_{22} & a_{23} \\ a_{31} & a_{32} & a_{33} \end{vmatrix} = \tag{1.12b}$$

$$(a_{11}a_{22}a_{33} + a_{21}a_{32}a_{13} + a_{31}a_{12}a_{23}) - (a_{13}a_{22}a_{31} + a_{23}a_{32}a_{11} + a_{12}a_{21}a_{33})$$

With this prescription, the vector product can be written as the determinant

$$\mathbf{A} \times \mathbf{B} = \begin{vmatrix} \mathbf{i} & \mathbf{j} & \mathbf{k} \\ A_x & A_y & A_z \\ B_x & B_y & B_z \end{vmatrix} \tag{1.13}$$

Example 1.1

To illustrate these manipulations, we will determine the cross product of

$$\text{(a)} \quad \mathbf{A} \equiv 3\mathbf{i} + 2\mathbf{j} \qquad \text{(b)} \quad \mathbf{B} \equiv 3\mathbf{j} + 2\mathbf{k} \tag{1.14}$$

By direct multiplication, referring to equations 1.8, we obtain

$$\mathbf{A} \times \mathbf{B} = (3\mathbf{i} + 2\mathbf{j}) \times (3\mathbf{j} + 2\mathbf{k}) = 9\mathbf{i} \times \mathbf{j} + 6\mathbf{i} \times \mathbf{k} + 6\mathbf{j} \times \mathbf{j} + 4\mathbf{j} \times \mathbf{k} = 4\mathbf{i} - 6\mathbf{j} + 9\mathbf{k} \tag{1.15}$$

To evaluate this cross product by the determinant process, we refer to equation 1.12b to obtain

$$\mathbf{A} \times \mathbf{B} = \begin{vmatrix} \mathbf{i} & \mathbf{j} & \mathbf{k} \\ 3 & 2 & 0 \\ 0 & 3 & 2 \end{vmatrix} = (4\mathbf{i} + 9\mathbf{k} + 0\mathbf{j}) - (0\mathbf{k} + 0\mathbf{i} + 6\mathbf{j}) = 4\mathbf{i} - 6\mathbf{j} + 9\mathbf{k} \quad \square \tag{1.16}$$

Levi–Civita three-index symbol

To define another method of describing the cross product, we use the 1, 2, 3 notation for the components of a vector. The Levi–Civita three-index symbol is denoted by ε_{rst}. The subscripts r, s, and t can have the values 1, 2, 3. ε_{rst} is defined by the following properties:

1. If any two indices have the same numerical value, the ε symbol is zero: for example, $\varepsilon_{121} = 0$.
2. $\varepsilon_{123} \equiv +1$.
3. If any two indices are interchanged, the symbol is the negative of its value before the interchange. For example, $\varepsilon_{123} = +1$, so $\varepsilon_{213} = -1$. Similarly, $\varepsilon_{321} = -1$, so $\varepsilon_{312} = +1$. In the first example, 2 and 1 are interchanged; in the second, 3 and 1 have been interchanged. In the third example, 2 and 3 were interchanged, followed by the interchange of 1 and 3.

With this definition, it is straightforward to show that

$$(\mathbf{A} \times \mathbf{B})_r = \sum_{s,t=1}^{3} \varepsilon_{rst} A_s B_t \tag{1.17}$$

As an example, choosing $r = 1$,

$$(\mathbf{A} \times \mathbf{B})_1 = \varepsilon_{123} A_2 B_3 + \varepsilon_{132} A_3 B_2 \tag{1.18a}$$

In this expansion, all terms with $s = 1$ and $t = 1$ have been omitted. By property 1, since $r = 1$, the ε symbol will be zero for s and/or t also equal to 1. Using $\varepsilon_{123} = +1$ and $\varepsilon_{132} = -1$ gives

$$(\mathbf{A} \times \mathbf{B})_1 = A_2 B_3 - A_3 B_2 \tag{1.18b}$$

which is identical to the other determinations of the first component of $\mathbf{A} \times \mathbf{B}$.

Scalar and vector triple products

Triple products are the products of three vectors. With three vectors **A**, **B**, and **C**, two types of triple products can be formed. The scalar triple product is of the form $\mathbf{A} \cdot (\mathbf{B} \times \mathbf{C})$, and $\mathbf{A} \times (\mathbf{B} \times \mathbf{C})$ is an example of the vector triple product. Using the description of the cross product of two vectors in terms of the Levi–Civita symbol, the scalar triple product can be written

$$\mathbf{A} \cdot (\mathbf{B} \times \mathbf{C}) = \sum_{r=1}^{3} A_r (\mathbf{B} \times \mathbf{C})_r = \sum_{rst=1}^{3} \varepsilon_{rst} A_r B_s C_t \tag{1.19a}$$

Since A_r, B_s, and C_t are scalar quantities, the order in which they are multiplied is immaterial. Thus

$$\mathbf{A} \cdot (\mathbf{B} \times \mathbf{C}) = \sum_{rst=1}^{3} \varepsilon_{rst} C_t A_r B_s \tag{1.19b}$$

Using the third property of the Levi–Civita symbol,

$$\varepsilon_{rst} = -\varepsilon_{rts} = +\varepsilon_{trs} \tag{1.20}$$

$$\Rightarrow \quad \mathbf{A} \cdot (\mathbf{B} \times \mathbf{C}) = \sum_{rst=1}^{3} \varepsilon_{trs} C_t A_r B_s \tag{1.21}$$

But

$$\sum_{rs=1}^{3} \varepsilon_{trs} A_r B_s = (\mathbf{A} \times \mathbf{B})_t \tag{1.22}$$

$$\Rightarrow \quad \mathbf{A} \cdot (\mathbf{B} \times \mathbf{C}) = \sum_{t=1}^{3} C_t (A \times B)_t = \mathbf{C} \cdot (\mathbf{A} \times \mathbf{B}) = (\mathbf{A} \times \mathbf{B}) \cdot \mathbf{C} \tag{1.23}$$

That is, if the order in which the vectors are written is left unchanged, the dot and cross operators can be interchanged in the scalar triple product.

If $\mathbf{B} = \mathbf{A}$ in equation 1.23,

$$\mathbf{A} \cdot (\mathbf{A} \times \mathbf{C}) = (\mathbf{A} \times \mathbf{A}) \cdot \mathbf{C} = 0 \tag{1.24}$$

Since $\mathbf{A} \cdot (\mathbf{A} \times \mathbf{C}) = 0$, $\mathbf{A} \times \mathbf{C}$ is orthogonal to $\mathbf{A}$, as stated earlier; that is, $\mathbf{A} \times \mathbf{C}$ is perpendicular to the plane containing $\mathbf{A}$ and $\mathbf{C}$.

Writing

$$\mathbf{B} \times \mathbf{C} = \mathbf{i}(\mathbf{B} \times \mathbf{C})_x + \mathbf{j}(\mathbf{B} \times \mathbf{C})_y + \mathbf{k}(\mathbf{B} \times \mathbf{C})_z \tag{1.25a}$$

the scalar triple product can be written

$$\mathbf{A} \cdot (\mathbf{B} \times \mathbf{C}) = A_x(\mathbf{B} \times \mathbf{C})_x + A_y(\mathbf{B} \times \mathbf{C})_y + A_z(\mathbf{B} \times \mathbf{C})_z \tag{1.25b}$$

We see that $\mathbf{A} \cdot (\mathbf{B} \times \mathbf{C})$ has a similar form to $\mathbf{B} \times \mathbf{C}$ in that if in the expansion of $\mathbf{B} \times \mathbf{C}$, we replace the basis vectors $\mathbf{i}, \mathbf{j}$ and $\mathbf{k}$ by the corresponding components of $\mathbf{A}$, we obtain $\mathbf{A} \cdot (\mathbf{B} \times \mathbf{C})$. Therefore, since

$$\mathbf{B} \times \mathbf{C} = \begin{vmatrix} \mathbf{i} & \mathbf{j} & \mathbf{k} \\ B_x & B_y & B_z \\ C_x & C_y & C_z \end{vmatrix} \tag{1.13}$$

the scalar triple product can be written

$$\mathbf{A} \cdot (\mathbf{B} \times \mathbf{C}) = \begin{vmatrix} A_x & A_y & A_z \\ B_x & B_y & B_z \\ C_x & C_y & C_z \end{vmatrix} \tag{1.26}$$

Recall that if the coordinate system is inverted to a left-handed frame, a true vector will change sign and a pseudovector will not. Because it has no dependence on direction, a scalar should be independent of whether the basis vectors are **i**, **j**, and **k** or −**i**, −**j**, and −**k**. That is, a scalar should not change sign under inversion. However, under inversion, the scalar triple product of three true vectors becomes

$$\mathbf{A}\cdot(\mathbf{B}\times\mathbf{C}) \to (-\mathbf{A})\cdot[(-\mathbf{B})\times(-\mathbf{C})] = -\mathbf{A}\cdot(\mathbf{B}\times\mathbf{C}) \tag{1.27a}$$

Therefore, the scalar triple product of three real vectors is not a true scalar. As such, it is called a *pseudoscalar*. Similarly, if, for example, **A** and **B** are true vectors and **C** is a pseudovector,

$$\mathbf{A}\cdot(\mathbf{B}\times\mathbf{C}) \to \mathbf{A}\cdot(\mathbf{B}\times\mathbf{C}) \tag{1.27b}$$

and so is a true scalar.

In writing the scalar triple product, it is not necessary to place the parentheses around the cross product. The form $\mathbf{A}\cdot\mathbf{B}\times\mathbf{C}$ is unambiguous since $(\mathbf{A}\cdot\mathbf{B})\times\mathbf{C}$ is not defined. However, when writing the vector triple product, it is important where one places the parentheses. Both $\mathbf{A}\times(\mathbf{B}\times\mathbf{C})$ and $(\mathbf{A}\times\mathbf{B})\times\mathbf{C}$ are defined and they are not the same. Consider

$$[\mathbf{A}\times(\mathbf{B}\times\mathbf{C})]_l = \sum_{rn=1}^{3} \varepsilon_{lrn} A_r (\mathbf{B}\times\mathbf{C})_n = \sum_{nst=1}^{3} \varepsilon_{lrn}\varepsilon_{nst} A_r B_s C_t \tag{1.28}$$

Consider the term

$$\sum_{n=1}^{3} \varepsilon_{lrn}\varepsilon_{nst} = \varepsilon_{lr1}\varepsilon_{1st} + \varepsilon_{lr2}\varepsilon_{2st} + \varepsilon_{lr3}\varepsilon_{3st} \tag{1.29}$$

$\varepsilon_{lr1}\varepsilon_{1st}$, for example, is nonzero only when l, r, s, and $t \neq 1$. In addition, the ε symbol is zero if l and r have the same value, or if s and t have the same value. Therefore, the possible choices for l and r and s and t that yield nonzero results are

$$\text{(a)}\quad l = 2, r = 3 \quad\text{or}\quad l = 3, r = 2 \qquad \text{(b)}\quad s = 2, t = 3 \quad\text{or}\quad s = 3, t = 2, \tag{1.30}$$

An identical analysis for any value of n leads to the fact that for $\varepsilon_{lrn}\varepsilon_{nst}$ to be nonzero, we must have

$$\text{(a)}\quad l = s \quad\text{and}\quad r = t \quad\text{or}\quad \text{(b)}\quad l = t \quad\text{and}\quad r = s \tag{1.31}$$

and none of these indices have the same value as n.

The Kronecker symbol, δ_{pq}, is defined such that

$$\delta_{pq} = \begin{cases} 0 & p \neq q \\ 1 & p = q \end{cases} \tag{1.32}$$

Thus we can write

$$\text{(a)}\quad \varepsilon_{lr1}\varepsilon_{1st} = a_1\delta_{ls}\delta_{rt} + b_1\delta_{lt}\delta_{rs} \qquad \text{(b)}\quad \varepsilon_{lr2}\varepsilon_{2st} = a_2\delta_{ls}\delta_{rt} + b_2\delta_{lt}\delta_{rs}$$

$$\text{(c)}\quad \varepsilon_{lr3}\varepsilon_{lst} = a_3\delta_{ls}\delta_{rt} + b_3\delta_{lt}\delta_{rs} \tag{1.33}$$

Summing these three terms and defining $(a_1 + a_2 + a_3) \equiv \alpha$ and $(b_1 + b_2 + b_3) \equiv \beta$, we obtain

$$\sum_{n=1}^{3} \varepsilon_{lrn}\varepsilon_{nst} = \alpha\delta_{ls}\delta_{rt} + \beta\delta_{lt}\delta_{rs} \tag{1.34}$$

To determine α and β, we note that l, r, s, and t are free indices. That is, they are not summed over. We can therefore assign any value we choose to each of them. For example, if we choose $l = s = 1$ and $r = t = 2$, we obtain, from equation 1.34,

$$\sum_{n=1}^{3} \varepsilon_{12n}\varepsilon_{n12} = \varepsilon_{123}\varepsilon_{312} = \alpha\delta_{11}\delta_{22} + \beta\delta_{12}\delta_{21} = \alpha \tag{1.35}$$

where we have used the fact that $\varepsilon_{n12} = \varepsilon_{n21} = 0$ for $n = 1$ and 2. Using $\varepsilon_{123} = \varepsilon_{312} = +1$, we obtain $\alpha = +1$. It is equally straightforward to show that $\beta = -1$ (see Problem 3). Thus

$$\sum_{n=1}^{3} \varepsilon_{lrn}\varepsilon_{nst} = \delta_{ls}\delta_{rt} - \delta_{lt}\delta_{rs} \tag{1.36}$$

With this result, equation 1.28 becomes

$$[\mathbf{A} \times (\mathbf{B} \times \mathbf{C})]_l = \sum_{rst=1}^{3} (\delta_{ls}\delta_{rt} - \delta_{lt}\delta_{rs}) A_r B_s C_t \tag{1.37}$$

From the first summation, consider the term $\sum_{rt=1}^{3} \delta_{rt} A_r C_t$. From the definition of the Kronecker δ-symbol, the only nonzero terms are those for which $r = t$. Thus

$$\sum_{t=1}^{3} \delta_{rt} C_t = C_r \tag{1.38}$$

$$\Rightarrow \quad \sum_{rt=1}^{3} \delta_{rt} A_r C_t = \sum_{r=1}^{3} A_r C_r = \mathbf{A} \cdot \mathbf{C} \tag{1.39}$$

By an identical argument, all terms in the first sum over s are zero except the term for which $s = l$. Thus

$$\sum_{s=1}^{3} \delta_{ls} B_s = B_l \tag{1.40}$$

Therefore, the first sum becomes $B_l(\mathbf{A} \cdot \mathbf{C})$. Using identical analyses, the second sum can be shown to be $C_l(\mathbf{A} \cdot \mathbf{B})$ (see Problem 5). Thus

$$[\mathbf{A} \times (\mathbf{B} \times \mathbf{C})]_l = B_l(\mathbf{A} \cdot \mathbf{C}) - C_l(\mathbf{A} \cdot \mathbf{B}) \tag{1.41a}$$

or writing this in vector form instead of in terms of components, we have

$$[\mathbf{A} \times (\mathbf{B} \times \mathbf{C})] = \mathbf{B}(\mathbf{A} \cdot \mathbf{C}) - \mathbf{C}(\mathbf{A} \cdot \mathbf{B}) \tag{1.41b}$$

Example 1.2

As an example, the scalar triple product of

$$\text{(a)}\quad \mathbf{A} \equiv \mathbf{i} + 2\mathbf{j} \qquad \text{(b)}\quad \mathbf{B} \equiv 2\mathbf{j} + \mathbf{k} \qquad \text{(c)}\quad \mathbf{C} \equiv 2\mathbf{i} + \mathbf{k} \tag{1.42}$$

is found using direct multiplication to be

$$\mathbf{A} \cdot \mathbf{B} \times \mathbf{C} = (\mathbf{i} + 2\mathbf{k}) \cdot [(2\mathbf{j} + \mathbf{k}) \times (2\mathbf{i} + \mathbf{k})] = (\mathbf{i} + 2\mathbf{k}) \cdot (2\mathbf{i} + 2\mathbf{j} - 4\mathbf{k}) = -6 \tag{1.43a}$$

Using the determinant form of the scalar triple product (equation 1.26) yields

$$\mathbf{A} \cdot \mathbf{B} \times \mathbf{C} = \begin{vmatrix} 1 & 0 & 2 \\ 0 & 2 & 1 \\ 2 & 0 & 1 \end{vmatrix} = -6 \tag{1.43b}$$

The vector triple product, found by direct multiplication, is

$$\begin{aligned} \mathbf{A} \times (\mathbf{B} \times \mathbf{C}) &= (\mathbf{i} + 2\mathbf{k}) \times [(2\mathbf{j} + \mathbf{k}) \times (2\mathbf{i} + \mathbf{k})] = (\mathbf{i} + 2\mathbf{k}) \times (2\mathbf{i} + 2\mathbf{j} - 4\mathbf{k}) \\ &= -4\mathbf{i} + 8\mathbf{j} + 2\mathbf{k} \end{aligned} \tag{1.44a}$$

The same result is obtained using equation 1.41b.

$$\begin{aligned} \mathbf{A} \times (\mathbf{B} \times \mathbf{C}) &= (2\mathbf{j} + \mathbf{k})[(\mathbf{i} + 2\mathbf{k}) \cdot (2\mathbf{i} + \mathbf{k})] - (2\mathbf{i} + \mathbf{k})[(\mathbf{i} + 2\mathbf{k}) \cdot (2\mathbf{j} + \mathbf{k})] \\ &= 4(2\mathbf{j} + \mathbf{k}) - 2(2\mathbf{i} + \mathbf{k}) = -4\mathbf{i} + 8\mathbf{j} + 2\mathbf{k} \quad \square \end{aligned} \tag{1.44b}$$

Simulated vector division

Consider the equation $A = zB$, where A and B are known parameters and z is unknown. If A and B are scalars, a solution for z is easily obtained by division. But if A and B are vectors, the ratio $\mathbf{A}/\mathbf{B}$ is not defined. The equation $\mathbf{A} = z\mathbf{B}$ implies that **A** and **B** are parallel or antiparallel. We cannot simply express z as $|\mathbf{A}|/|\mathbf{B}|$ since this results in a positive value for z. It is possible for z to be negative (when **A** and **B** are antiparallel).

To solve for z, we must simulate vector division. Since we can only divide by a scalar, we must make the multiplier of z a scalar before we divide. A scalar quantity can be made from a vector by a dot product. Since **A** and **B** are parallel or antiparallel, $\mathbf{A} \cdot \mathbf{B} \neq 0$. Thus we can take the dot product of both sides of

$$\mathbf{A} = z\mathbf{B} \tag{1.45}$$

with either **A** or **B**. Then

$$\text{(a)} \quad \mathbf{A} \cdot \mathbf{B} = zB^2 \qquad \text{or} \qquad \text{(b)} \quad A^2 = z\mathbf{A} \cdot \mathbf{B} \tag{1.46}$$

and we can then solve for z as

$$\text{(a)} \quad z = \frac{\mathbf{A} \cdot \mathbf{B}}{B^2} \qquad \text{or} \qquad \text{(b)} \quad z = \frac{A^2}{\mathbf{A} \cdot \mathbf{B}} \tag{1.47}$$

1.2 Geometry of Lines and Planes

To give the reader some experience with geometric manipulation of vectors, we consider a small set of examples in which we determine some of the properties of lines and planes. Two points are required to determine a line and three noncollinear points are needed to define a plane. Since the coordinates of the points that determine the line or plane being studied are defined relative to a coordinate axis system, these points can also be defined by vectors from the origin to the points (see Figures 1.5). Thus a line is said to be defined by two vectors, and a plane by three vectors.

Properties of a plane

The general form for the equation of a plane is

$$Ax + By + Cz + D = 0 \tag{1.48a}$$

A, B, C, and D are constants, and x, y, and z are the coordinates of any point lying in the plane. The equation of a particular plane requires determining the values of the

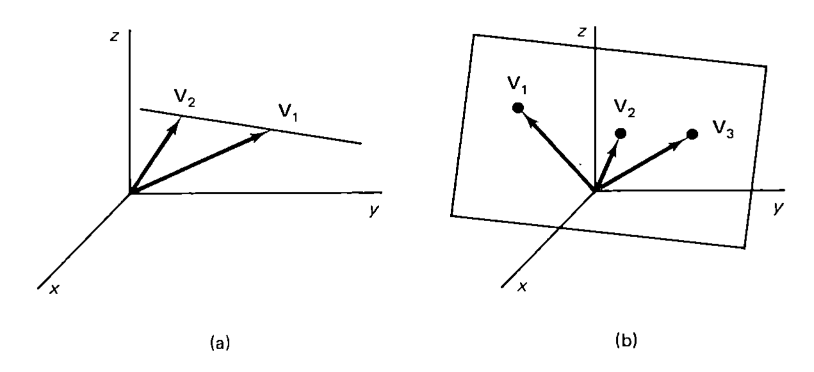

Figure 1.5 (a) Vectors defining a line; (b) vectors defining a plane.

constants. If we were sure that D were not zero, we could divide equation 1.48a by D to obtain

$$A'x + B'y + C'z + 1 = 0 \tag{1.48b}$$

where $A' = A/D$, $B' = B/D$, and $C' = C/D$. Then substitution of the coordinates of the three known points into this equation would allow us to solve three simultaneous equations for A', B', and C'. D will be zero if the plane passes through the origin. One cannot tell a priori if D is nonzero just from a knowledge of three points in the plane. In addition, such a method is not an analysis illustrating the use of vectors. For that reason, such an approach will not be considered.

Let vectors $\mathbf{V}_1$, $\mathbf{V}_2$, and $\mathbf{V}_3$ define the plane. Then the vectors $\mathbf{V}_1 - \mathbf{V}_2$ and $\mathbf{V}_2 - \mathbf{V}_3$ lie in the plane. Thus the cross product of these two vectors will be perpendicular or normal to the plane. If this cross product is divided by its magnitude, the resulting vector will be the unit (length 1) normal vector to the plane. That is, the unit normal to the plane is

$$\mathbf{n} = \frac{(\mathbf{V}_1 - \mathbf{V}_2) \times (\mathbf{V}_2 - \mathbf{V}_3)}{|(\mathbf{V}_1 - \mathbf{V}_2) \times (\mathbf{V}_2 - \mathbf{V}_3)|} \tag{1.49}$$

As will soon become clear, it it not necessary to normalize the vector to unit length to determine the equation of the plane. However, for other properties of the plane, the unit length normal vector will be required, so it is introduced here.

Let a general point in the plane have coordinates x, y, and z. Then the vector

$$\mathbf{r} = x\mathbf{i} + y\mathbf{j} + z\mathbf{k} \tag{1.50}$$

defines a point in the plane. Therefore, the vector $\mathbf{r} - \mathbf{V}_1$ is a vector lying in the plane. (Obviously, $\mathbf{V}_2$ or $\mathbf{V}_3$ can be used in place of $\mathbf{V}_1$.) Thus $\mathbf{r} - \mathbf{V}_1$ is perpendicular to the normal to the plane, so

$$\mathbf{n} \cdot (\mathbf{r} - \mathbf{V}_1) = 0 \tag{1.51}$$

As mentioned above, the magnitude of the normal vector at this point is immaterial since this dot product is zero.

The equation $\mathbf{n} \cdot (\mathbf{r} - \mathbf{V}_1) = 0$ is a scalar equation for an arbitrary point in the plane. It is therefore the equation that describes the plane.

Referring to Figure 1.6, from an analysis of the vectors that define the plane, one can find the perpendicular distance from the origin to the plane. Viewing the plane

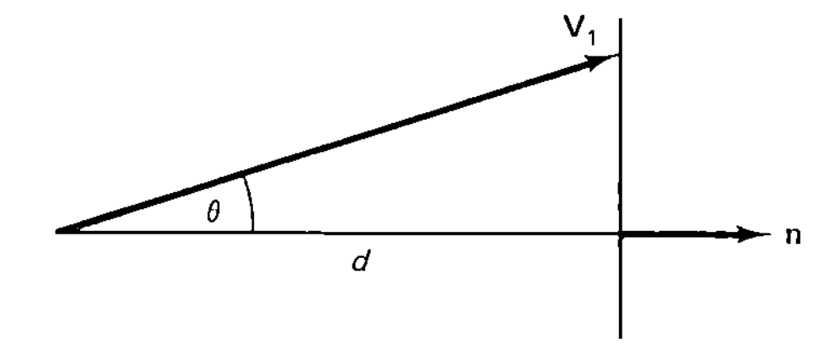

Figure 1.6 Edgewise view of plane describing perpendicular distance from origin to plane.

edgewise, it is evident that since **n** has unit length

$$d = |\mathbf{V}_1|\,|\mathbf{n}|\cos\theta = \mathbf{V}_1 \cdot \mathbf{n} \tag{1.52}$$

(Obviously, $\mathbf{V}_2$ or $\mathbf{V}_3$ could also be used to find d.) It is clear that in determining the perpendicular distance from the origin to the plane, we must use the normal vector of unit length.

Example 1.3

As an illustrative example, we will determine the equation of the plane defined by the points (3, 5, 2), (1, 0, 0), and (3, − 1, 0), and its perpendicular distance from the origin. To find the equation of the plane, we first find the normal vector, **n**. From the points given, the three known vectors defining the plane are

$$\text{(a)}\quad \mathbf{V}_1 = 3\mathbf{i} + 5\mathbf{j} + 2\mathbf{k} \qquad \text{(b)}\quad \mathbf{V}_2 = \mathbf{i} \qquad \text{(c)}\quad \mathbf{V}_3 = 3\mathbf{i} - \mathbf{j} \tag{1.53}$$

Therefore, $\mathbf{V}_1 - \mathbf{V}_2 = (2\mathbf{i} + 5\mathbf{j} + 2\mathbf{k})$ and $\mathbf{V}_2 - \mathbf{V}_3 = -2\mathbf{i} + \mathbf{j}$ lie in the plane. Thus the unit normal is

$$\mathbf{n} = \frac{(2\mathbf{i} + 5\mathbf{j} + 2\mathbf{k}) \times (-2\mathbf{i} + \mathbf{j})}{|(2\mathbf{i} + 5\mathbf{j} + 2\mathbf{k}) \times (-2\mathbf{i} + \mathbf{j})|} = \frac{(-\mathbf{i} - 2\mathbf{j} + 6\mathbf{k})}{\sqrt{41}} \tag{1.54}$$

A general vector lying in the plane is

$$\mathbf{r} - \mathbf{V}_1 = (x - 3)\mathbf{i} + (y - 5)\mathbf{j} + (z - 2)\mathbf{k} \tag{1.55}$$

so the equation of the plane is

$$\text{(a)}\quad \mathbf{n} \cdot (\mathbf{r} - \mathbf{V}_1) = 0 = -(x - 3) - 2(y - 5) + 6(z - 2) \quad \text{or} \quad \text{(b)}\quad x + 2y - 6z - 1 = 0 \tag{1.56}$$

As a check that the correct equation of the plane has been determined, we must ascertain that all three points defining the plane satisfy its equation. It is straightforward to verify that this is so for the current example.

The perpendicular distance from the origin to the plane is determined by

$$d = \mathbf{n} \cdot \mathbf{V}_1 = \frac{(-\mathbf{i} - 2\mathbf{j} + 6\mathbf{k}) \cdot (3\mathbf{i} + 5\mathbf{j} + 2\mathbf{k})}{\sqrt{41}} = \frac{-1}{\sqrt{41}} \tag{1.57}$$

The negative sign is unimportant in the determination of the perpendicular distance of the plane from the origin. It simply means that the normal vector we have determined is directed opposite to the normal vector shown in Figure 1.6. That is, it points toward the origin rather than away from it. Therefore, the negative sign can be ignored and the perpendicular distance from the origin to the plane is $1/\sqrt{41}$. In general, the perpendicular distance from the origin to the plane should be expressed as

$$d = |\mathbf{V}_1 \cdot \mathbf{n}| \quad \square \tag{1.58}$$

Properties of a line

Two points, and therefore two vectors, are required to define a line. The vectors $\mathbf{V}_1$, $\mathbf{V}_2$, and the line form a plane with $\mathbf{V}_1$ and $\mathbf{V}_2$ lying in the plane. We can thus find the normal to that plane by taking the cross product of $\mathbf{V}_1$ and $\mathbf{V}_2$. This normal is also a normal to the line. Obviously, there are an infinite number of vectors that are normal to a line. They lie in a plane that is perpendicular to the line. Two such vectors are illustrated in Figure 1.7. The unit normal

$$\mathbf{n}_1 = \frac{\mathbf{V}_1 \times \mathbf{V}_2}{|\mathbf{V}_1 \times \mathbf{V}_2|} \tag{1.59}$$

is perpendicular to the plane containing $\mathbf{V}_1$, $\mathbf{V}_2$, and the line.

Referring to Figure 1.7, $\mathbf{V}_1 - \mathbf{V}_2$ is a vector along the line. A second vector can be found that is perpendicular to both the line and to $\mathbf{n}_1$ from

$$\mathbf{n}_2 = \frac{\mathbf{n}_1 \times (\mathbf{V}_1 - \mathbf{V}_2)}{|\mathbf{n}_1 \times (\mathbf{V}_1 - \mathbf{V}_2)|} \tag{1.60}$$

Since $\mathbf{n}_2$ is perpendicular to $\mathbf{n}_1$, it must lie in the plane formed by $\mathbf{V}_1$, $\mathbf{V}_2$, and the line.

Let

$$\mathbf{r} = x\mathbf{i} + y\mathbf{j} + z\mathbf{k} \tag{1.50}$$

define an arbitrary point on the line. Since $\mathbf{r} - \mathbf{V}_1$ lies along the line, and thus lies in the plane, it is perpendicular to $\mathbf{n}_1$. Therefore,

$$\mathbf{n}_1 \cdot (\mathbf{r} - \mathbf{V}_1) = 0 \tag{1.61a}$$

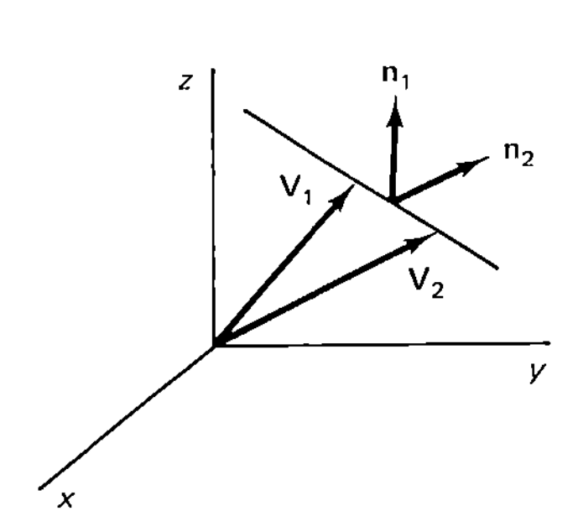

Figure 1.7 Two vectors normal to a line.

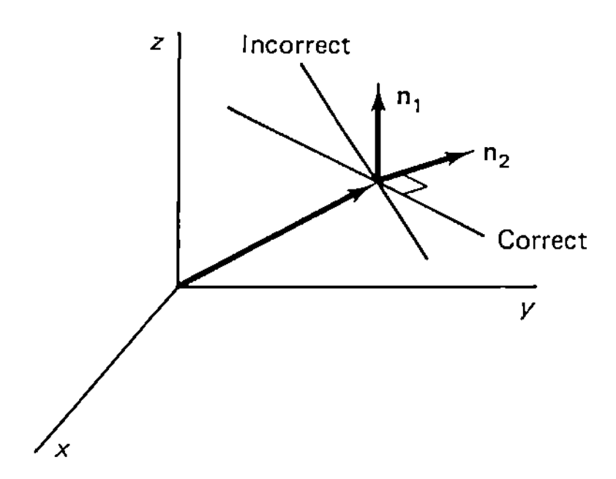

Figure 1.8 Two lines containing one point that are perpendicular to a normal vector.

There are clearly an infinite number of lines that lie in the plane and contain the point (x, y, z). Thus there are an infinite number of lines in the plane that are perpendicular to $\mathbf{n}_1$. Figure 1.8 shows two such lines. However, only the line we are seeking is also perpendicular to $\mathbf{n}_2$. Thus to determine the line uniquely, we must also require that

$$\mathbf{n}_2 \cdot (\mathbf{r} - \mathbf{V}_1) = 0 \tag{1.61b}$$

Therefore, in three-dimensional space, two equations are required to define a line.

Another argument to convince oneself of the need for two equations is to note that one equation is needed to specify the plane in which the lines lies. But since a plane

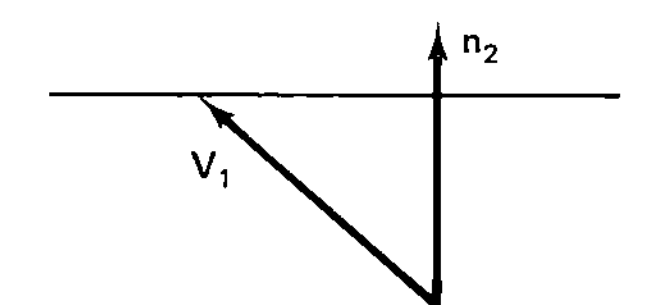

Figure 1.9 Perpendicular distance from the origin to the line.

contains an infinite number of lines, a second equation is required to specify the particular line in that plane.

We can also find the perpendicular distance from the origin to the line. Referring to Figure 1.9, this distance is

$$d = \big| |\mathbf{V}_1| \; |\mathbf{n}_2| \cos\theta \big| = |\mathbf{V}_1 \cdot \mathbf{n}_2| \tag{1.62}$$

Example 1.4

As an illustrative example, let us find the equations of a line defined by the points (2, 1, − 1) and (1, 0, 3), and its perpendicular distance from the origin. The normal vectors are

$$\mathbf{n}_1 = \frac{(2\mathbf{i} + \mathbf{j} - \mathbf{k}) \times (\mathbf{i} + 3\mathbf{k})}{|(2\mathbf{i} + \mathbf{j} - \mathbf{k}) \times (\mathbf{i} + 3\mathbf{k})|} = \frac{3\mathbf{i} - 7\mathbf{j} - \mathbf{k}}{\sqrt{59}} \tag{1.63a}$$

$$\mathbf{n}_2 = \frac{\mathbf{n}_1 \times (\mathbf{V}_1 - \mathbf{V}_2)}{|\mathbf{n}_1 \times (\mathbf{V}_1 - \mathbf{V}_2)|} = \frac{29\mathbf{i} + 11\mathbf{j} + 10\mathbf{k}}{\sqrt{1061}} \tag{1.63b}$$

Thus, using equations 1.61, we obtain

$$\text{(a)} \quad \mathbf{n}_1 \cdot (\mathbf{r} - \mathbf{V}_1) = 3x - 7y - z = 0 \qquad \text{(b)} \quad \mathbf{n}_2 \cdot (\mathbf{r} - \mathbf{V}_1) = 29x + 11y + 10z - 59 = 0 \tag{1.64}$$

These are the two equations that describe the line.

The perpendicular distance from the origin to the line is

$$d = \mathbf{n}_2 \cdot \mathbf{V}_1 = \frac{59}{\sqrt{1061}} \quad \square \tag{1.65}$$

Analysis of two parallel planes

A plane is defined by the points P_1, P_2, P_3, and equivalently by the vectors $\mathbf{V}_1, \mathbf{V}_2, \mathbf{V}_3$. Let us find the equation of a plane that is parallel to the plane containing these points and is A units farther from the origin.

Since the planes are parallel, the normal to the plane closer to the origin is the same as the normal to the more distant plane, the equation of which we are trying to find. Since the three vectors to points in the closer plane are known, the normal to the two planes is found as discussed above.

$$\mathbf{n}_1 = \mathbf{n}_2 = \frac{(\mathbf{V}_1 - \mathbf{V}_2) \times (\mathbf{V}_1 - \mathbf{V}_3)}{|(\mathbf{V}_1 - \mathbf{V}_2) \times (\mathbf{V}_1 - \mathbf{V}_3)|} \equiv \mathbf{n} \tag{1.49}$$

Of course, one must determine that the normal is directed away from the origin (toward

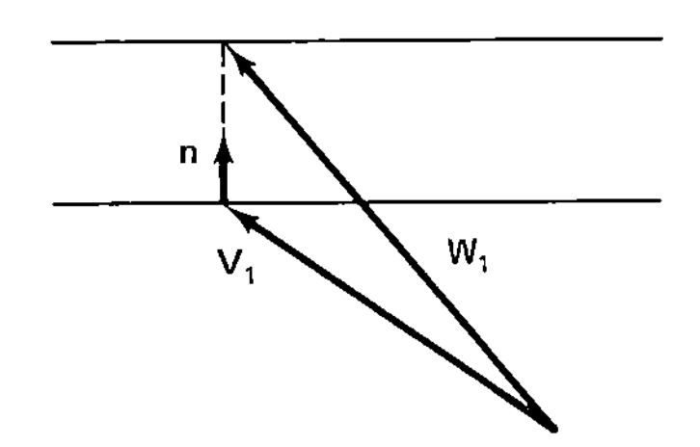

Figure 1.10 Two parallel planes, viewed edge on, A units apart.

the more distant plane). This can be ascertained by first determining the perpendicular distance from the origin to the closer plane. If that is positive, the normal points in the correct direction. If not, the normal found is the negative of the correct one.

Let the vectors $\mathbf{W}_1$, $\mathbf{W}_2$, and $\mathbf{W}_3$ define points in the more distant plane. Looking at the planes edge on as in Figure 1.10, we see that three points in the more distant plane can be found by

$$\text{(a)}\quad \mathbf{W}_1 = \mathbf{V}_1 + A\mathbf{n} \qquad \text{(b)}\quad \mathbf{W}_2 = \mathbf{V}_2 + A\mathbf{n} \qquad \text{(c)}\quad \mathbf{W}_3 = \mathbf{V}_3 + A\mathbf{n} \tag{1.66}$$

Now that three points in this more distant plane are known, we can proceed as before to find the equation of this plane. We have already determined the perpendicular distance from the origin to the closer plane, so the perpendicular distance to the second plane, being A units larger, is known.

Example 1.5

To illustrate, consider a plane containing the points (5, 1, 1), (3, 4, 5), and (1, 1, 1). We will determine the equation of a plane that is 5 units closer to the origin, and the perpendicular distance of both planes.

The vectors to the three points are

$$\text{(a)}\quad \mathbf{V}_1 = 5\mathbf{i} + \mathbf{j} + \mathbf{k} \qquad \text{(b)}\quad \mathbf{V}_2 = 3\mathbf{i} + 4\mathbf{j} + 5\mathbf{k} \qquad \text{(c)}\quad \mathbf{V}_3 = \mathbf{i} + \mathbf{j} + \mathbf{k} \tag{1.67}$$

Referring to equation 1.49, we find the normal to the first plane (and thus both planes) to be

$$\mathbf{n} = \frac{(2\mathbf{i} - 3\mathbf{j} - 4\mathbf{k}) \times (4\mathbf{i})}{|(2\mathbf{i} - 3\mathbf{j} - 4\mathbf{k}) \times (4\mathbf{i})|} = \frac{12\mathbf{k} - 16\mathbf{j}}{|12\mathbf{k} - 16\mathbf{j}|} = \frac{3\mathbf{k} - 4\mathbf{j}}{5} \tag{1.68}$$

To determine whether $\mathbf{n}$ points toward or away from the origin, we note that

$$d = \mathbf{n} \cdot \mathbf{V}_1 = -\tfrac{1}{5} \tag{1.69}$$

The negative value of d indicates that $\mathbf{n}$ points toward the origin. Thus

$$\mathbf{n}' = -\mathbf{n} = \tfrac{1}{5}(4\mathbf{j} - 3\mathbf{k}) \tag{1.70}$$

will point away from the origin as in Figure 1.6.

We note that since the first plane is $\frac{1}{5}$ unit from the origin, a plane that is 5 units closer to the origin will lie on the opposite side of the origin from the first. To determine points in the second plane, we define

$$\text{(a)}\quad \mathbf{W}_1 = \mathbf{V}_1 - 5\mathbf{n}' = 5\mathbf{i} - 3\mathbf{j} + 4\mathbf{k} \qquad \text{(b)}\quad \mathbf{W}_2 = \mathbf{V}_2 - 5\mathbf{n}' = 3\mathbf{i} + 8\mathbf{k}$$
$$\text{(c)}\quad \mathbf{W}_3 = \mathbf{V}_3 - 5\mathbf{n}' = \mathbf{i} - 3\mathbf{j} + 4\mathbf{k} \tag{1.71}$$

Thus the equation of this second plane is

$$\mathbf{n} \cdot (\mathbf{r} - \mathbf{W}_1) = 4(y + 3) - 3(z - 4) = 4y - 3z + 24 = 0 \tag{1.72}$$

The distance of the second plane from the origin is 5 units less than the distance of the first plane from the origin, that is, $d_2 = \frac{24}{5}$ units from the origin, on the opposite side of the origin from the first plane. □

1.3 Non-Cartesian Coordinate Systems

Many problems in which there is a high degree of geometric symmetry can be simplified when expressed in a non-cartesian coordinate system that reflects the symmetry of the problem. We will illustrate this in computing the surface area and volume of a sphere and a cylinder.

Spherical coordinates

In Cartesian coordinates, points are located by specifying the values of x, y, and z. The three spherical coordinates r, θ, and ϕ locate an arbitrary point as illustrated in Figure 1.11. The coordinate r is the length of the line drawn from the origin to the point in question. θ is the angle that this radial line makes with the z-axis and is called the *polar angle*. The projection of the radial line into the x-y plane is of length $r \sin \theta$. It makes an angle ϕ with the x-axis, which is called the *azimuthal angle*.

The location of a point using spherical coordinates is specified by particular values of r, θ, and ϕ. All points with a common value of r lie on a spherical shell of radius r. Any point on that shell can be accessed by specifying a value of θ in the range $[0, \pi]$ and a value of $\phi \in [0, 2\pi]$. To access any spherical shell, r can be varied between 0 and ∞. This specifies the ranges of the three spherical coordinates.

In Cartesian coordinates, the basis vectors are defined as vectors of length 1, in a direction of increasing value of the corresponding coordinate. $\mathbf{i}$, for example, points in the direction of increasing x. Analogously, the unit vectors in spherical coordinates point in the direction of increasing value of corresponding coordinate parameter, as shown in Figure 1.12. The spherical unit vectors are denoted by $\mathbf{r}_0$, $\boldsymbol{\theta}_0$, and $\boldsymbol{\phi}_0$.

$\mathbf{r}_0$ is directed outward from the origin toward the point, along the radial line. If a circle of radius r containing the z-axis is drawn through the point, $\boldsymbol{\theta}_0$ will be directed along the tangent to that circle at the point in the direction of increasing θ. $\boldsymbol{\phi}_0$ lies in the x-y plane tangent to a circle, the radius of which is $r \sin \theta$, the length of the x-y projection of the radial line, and points in the direction of increasing ϕ.

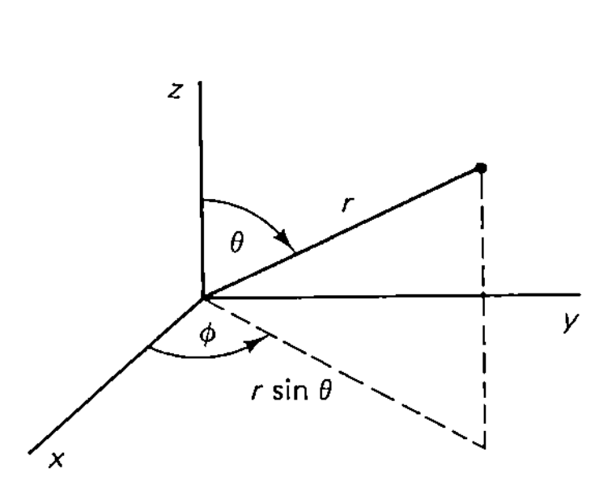

Figure 1.11 Specification of a point using spherical coordinates.

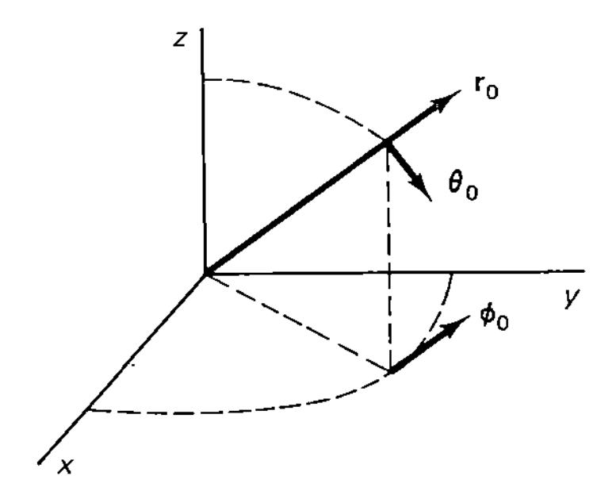

Figure 1.12 Spherical basis vectors.

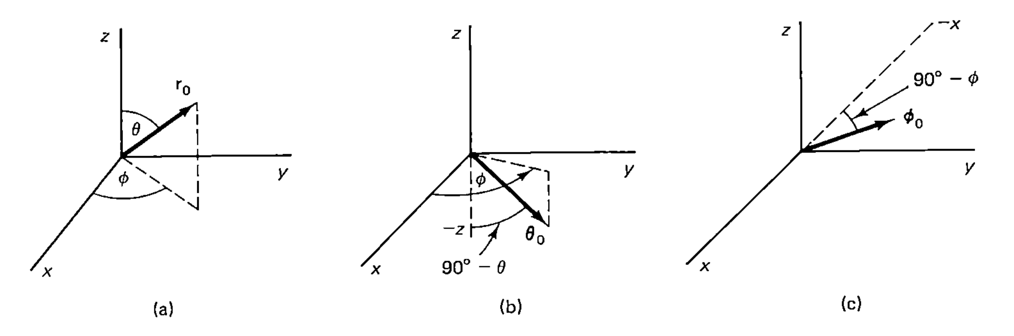

Figure 1.13 Spherical basis vectors displaced to the origin.

We can obtain relations between the spherical and Cartesian basis vectors. These relations will be useful in determining how $\mathbf{r}_0$, $\boldsymbol{\theta}_0$, and $\boldsymbol{\phi}_0$ change in direction when moving from point to point. Figures 1.13 show $\mathbf{r}_0$, $\boldsymbol{\theta}_0$, $\boldsymbol{\phi}_0$ referred to the origin. From these we see that

$$\mathbf{r}_0 = \sin\theta\cos\phi\mathbf{i} + \sin\theta\sin\phi\mathbf{j} + \cos\theta\mathbf{k} \tag{1.73a}$$

$$\boldsymbol{\theta}_0 = \cos\theta\cos\phi\mathbf{i} + \cos\theta\sin\phi\mathbf{j} - \sin\theta\mathbf{k} \tag{1.73b}$$

$$\boldsymbol{\phi}_0 = -\sin\phi\mathbf{i} + \cos\phi\mathbf{j} \tag{1.73c}$$

From equations 1.73 it is straightforward to show that the spherical basis vectors are orthonormal,

$$\text{(a)}\quad \mathbf{r}_0 \cdot \mathbf{r}_0 = \boldsymbol{\theta}_0 \cdot \boldsymbol{\theta}_0 = \boldsymbol{\phi}_0 \cdot \boldsymbol{\phi}_0 = 1 \quad \text{(b)}\quad \mathbf{r}_0 \cdot \boldsymbol{\theta}_0 = \mathbf{r}_0 \cdot \boldsymbol{\phi}_0 = \boldsymbol{\theta}_0 \cdot \boldsymbol{\phi}_0 = 0 \tag{1.74}$$

and form a right-handed system

$$\mathbf{r}_0 \times \boldsymbol{\theta}_0 = \boldsymbol{\phi}_0, \qquad \boldsymbol{\theta}_0 \times \boldsymbol{\phi}_0 = \mathbf{r}_0, \qquad \boldsymbol{\phi}_0 \times \mathbf{r}_0 = \boldsymbol{\theta}_0 \tag{1.75}$$

The infinitesmal changes in the spherical basis vectors can be determined by taking the differentials of equations 1.73. Since **i**, **j**, and **k** are constant vectors, only the differentials with respect to angles are taken. Thus

$$d\mathbf{r}_0 = (\cos\theta\cos\phi\mathbf{i} + \cos\theta\sin\phi\mathbf{j} - \sin\theta\mathbf{k})\,d\theta + (-\sin\theta\sin\phi\mathbf{i} + \sin\theta\cos\phi\mathbf{j})\,d\phi \tag{1.76}$$

Using equations 1.73b and 1.73c, and referring to Problem 18 we obtain

$$\text{(a)}\quad d\mathbf{r}_0 = \boldsymbol{\theta}_0\,d\theta + \sin\theta\boldsymbol{\phi}_0\,d\phi \qquad \text{(b)}\quad d\boldsymbol{\theta}_0 = -\mathbf{r}_0\,d\theta + \boldsymbol{\phi}_0\cos\theta\,d\phi$$

$$\text{(c)}\quad d\boldsymbol{\phi}_0 = -(\mathbf{r}_0\sin\theta + \boldsymbol{\theta}_0\cos\theta)\,d\phi \tag{1.77}$$

The position vector is expressed in Cartesian coordinates as $\mathbf{r} = x\mathbf{i} + y\mathbf{j} + z\mathbf{k}$. In spherical coordinates, $\mathbf{r} = r\mathbf{r}_0$. Equating the two expressions,

$$r\mathbf{r}_0 = x\mathbf{i} + y\mathbf{j} + z\mathbf{k} \tag{1.78}$$

and using equation 1.73a, it is straightforward to derive equations relating x, y, and z to r, θ, and ϕ. One obtains

$$\text{(a)}\quad x = r\sin\theta\cos\phi \qquad \text{(b)}\quad y = r\sin\theta\sin\phi \qquad \text{(c)}\quad z = r\cos\theta \tag{1.79}$$

These equations can be inverted to obtain r, θ, and ϕ in terms of x, y, and z. We obtain

$$\text{(a)}\ r = \sqrt{x^2 + y^2 + z^2} \qquad \text{(b)}\ \cos\theta = \frac{z}{\sqrt{x^2 + y^2 + z^2}} \qquad \text{(c)}\ \tan\phi = \frac{y}{x} \qquad (1.80)$$

Cylindrical coordinates

The cylindrical coordinate system is another commonly used non-Cartesian set of coordinates. In Figure 1.14 we illustrate the location of a point using cylindrical coordinates. ρ is the length of a line from the origin to the projection of the point into the x-y plane. ϕ is the angle that this radial line makes with the x-axis, and z is the Cartesian coordinate z, the perpendicular distance from the point to the x-y plane.

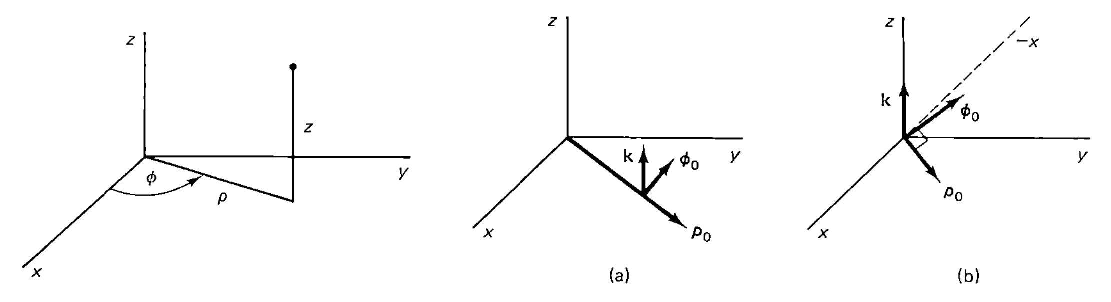

Figure 1.14 Location of a point using cylindrical coordinates.

Figure 1.15 (a) Cylindrical basis vectors; (b) cylindrical basis vectors displaced to the origin.

The equations connecting the Cartesian and cylindrical coordinates, are easily obtained. Referring to Figure 1.14, we have

$$\text{(a)}\quad x = \rho\cos\phi \qquad \text{(b)}\quad y = \rho\sin\phi \qquad \text{(c)}\quad z = z \qquad (1.81)$$

Referring to Figure 1.14, it is evident that $\rho = r\sin\theta$, the projection of the spherical r coordinate into the x-y plane. Therefore, the cylindrical coordinate ϕ is identical to the spherical azimuthal angle ϕ.

As is done in the Cartesian and spherical representations, cylindrical basis vectors are defined to be of unit length in the direction of increasing corresponding coordinate. The cylindrical bases are depicted in Figure 1.15a. Referring to Figure 1.15, it is easy to deduce the relations between $\boldsymbol{\rho}_0, \boldsymbol{\phi}_0, \mathbf{k}$ and $\mathbf{i}, \mathbf{j}, \mathbf{k}$. They are

$$\text{(a)}\quad \boldsymbol{\rho}_0 = \cos\phi\mathbf{i} + \sin\phi\mathbf{j} \qquad \text{(b)}\quad \boldsymbol{\phi}_0 = -\sin\phi\mathbf{i} + \cos\phi\mathbf{j} \qquad \text{(c)}\quad \mathbf{k} = \mathbf{k} \qquad (1.82)$$

As with the other two coordinate systems we have studied, these basis vectors form an orthonormal right-handed system. That is, using equations 1.82, it follows that

$$\text{(a)}\quad \boldsymbol{\rho}_0\cdot\boldsymbol{\rho}_0 = \boldsymbol{\phi}_0\cdot\boldsymbol{\phi}_0 = \mathbf{k}\cdot\mathbf{k} = 1 \qquad \text{(b)}\quad \boldsymbol{\rho}_0\cdot\boldsymbol{\phi}_0 = \boldsymbol{\rho}_0\cdot\mathbf{k} = \boldsymbol{\phi}_0\cdot\mathbf{k} = 0 \qquad (1.83)$$

$$\text{(a)}\quad \boldsymbol{\rho}_0\times\boldsymbol{\phi}_0 = \mathbf{k} \qquad \text{(b)}\quad \boldsymbol{\phi}_0\times\mathbf{k} = \boldsymbol{\rho}_0 \qquad \text{(c)}\quad \mathbf{k}\times\boldsymbol{\rho}_0 = \boldsymbol{\phi}_0 \qquad (1.84)$$

The variation of the basis vectors can be found from equations 1.82. We obtain

$$\text{(a)}\quad d\boldsymbol{\rho}_0 = \boldsymbol{\phi}_0\,d\phi \qquad \text{(b)}\quad d\boldsymbol{\phi}_0 = -\boldsymbol{\rho}_0\,d\phi \qquad \text{(c)}\quad d\mathbf{k} = 0 \qquad (1.85)$$

Referring to Figure 1.14, the position vector can be written in terms of the cylindrical coordinates as

$$\mathbf{r} = \rho\boldsymbol{\rho}_0 + z\mathbf{k} \qquad (1.86)$$

Comparing this to $\mathbf{r} = x\mathbf{i} + y\mathbf{j} + z\mathbf{k}$ leads to

$$\text{(a)} \quad x = \rho \cos \phi \qquad \text{(b)} \quad y = \rho \sin \phi \qquad \text{(c)} \quad z = z \tag{1.81}$$

which were deduced earlier. These equations can be inverted easily to obtain

$$\text{(a)} \quad \rho = \sqrt{x^2 + y^2} \qquad \text{(b)} \quad \tan \phi = \frac{y}{x} \qquad \text{(c)} \quad z = z \tag{1.87}$$

Variation of a vector

As mentioned above, one advantage of using Cartesian bases, and expressing the non-Cartesian basis vectors in terms of $\mathbf{i}, \mathbf{j}, \mathbf{k}$, is that when evaluating integrals or derivatives of vector functions, the Cartesian bases are constant. Thus the variation in the vector function

$$\mathbf{G} = \mathbf{G}_x\mathbf{i} + \mathbf{G}_y\mathbf{j} + \mathbf{G}_z\mathbf{k} \tag{1.88a}$$

arises entirely from the variation of G_x, G_y, and G_z. If $\mathbf{G}$ is expressed in spherical coordinates, for example, then

$$\mathbf{G} = \mathbf{G}_r\mathbf{r}_0 + G_\theta \boldsymbol{\theta}_0 + G_\phi \boldsymbol{\phi}_0 \tag{1.88b}$$

and the variation of $\mathbf{G}$ arises not only from the change in G_r, G_θ, and G_ϕ, but also from changes in $\mathbf{r}_0$, $\boldsymbol{\theta}_0$, and $\boldsymbol{\phi}_0$. For example,

$$\frac{d\mathbf{G}}{dt} = \frac{dG_x}{dt}\mathbf{i} + \frac{dG_y}{dt}\mathbf{j} + \frac{dG_z}{dt}\mathbf{k} \tag{1.89a}$$

in Cartesian coordinates. If $\mathbf{G}$ is expressed in spherical coordinates,

$$\frac{d\mathbf{G}}{dt} = \frac{dG_r}{dt}\mathbf{r}_0 + G_r\frac{d\mathbf{r}_0}{dt} + \frac{dG_\theta}{dt}\boldsymbol{\theta}_0 + G_\theta\frac{d\boldsymbol{\theta}_0}{dt} + \frac{dG_\phi}{dt}\boldsymbol{\phi}_0 + G_\phi\frac{d\boldsymbol{\phi}_0}{dt} \tag{1.89b}$$

Line, area, and volume elements

Often, one must evaluate integrals over a curve, a surface, or throughout a volume of space. It is therefore important to deduce the differential line, surface, and volume element in the coordinate system in which one is working.

A *differential line element* is defined as a segment of a curve with an associated direction. In Cartesian coordinates, the differential line element is

$$d\mathbf{l} = dx\,\mathbf{i} + dy\,\mathbf{j} + dz\,\mathbf{k} \tag{1.90a}$$

To obtain an expression for $d\mathbf{l}$ in spherical coordinates, we note that since $\mathbf{r} = x\mathbf{i} + y\mathbf{j} + z\mathbf{k}$, $d\mathbf{l} = d\mathbf{r}$. In spherical coordinates, $\mathbf{r} = r\mathbf{r}_0$. Thus, referring to equation 1.77a, we have

$$d\mathbf{l} = (dr)\mathbf{r}_0 + r(d\mathbf{r}_0) = dr\,\mathbf{r}_0 + r\,d\theta\,\boldsymbol{\theta}_0 + r\sin\theta\,d\phi\,\boldsymbol{\phi}_0 \tag{1.90b}$$

Similarly, in cylindrical coordinates, $\mathbf{r} = \rho\boldsymbol{\rho}_0 + z\mathbf{k}$. Therefore,

$$d\mathbf{l} = (d\rho)\boldsymbol{\rho}_0 + \rho(d\boldsymbol{\rho}_0) + dz\,\mathbf{k} = d\rho\,\boldsymbol{\rho}_0 + \rho\,d\phi\,\boldsymbol{\phi}_0 + dz\,\mathbf{k} \tag{1.90c}$$

In the Cartesian system, the differential surface element is an infinitesmal rectangle in the plane in which the integration is being performed. If this plane is parallel to the x-y plane, for example, we will denote the surface element as dS_{xy}. As can be seen from Figure 1.16,

$$\text{(a)} \quad dS_{xy} = dx\,dy \qquad \text{(b)} \quad dS_{yz} = dy\,dz \qquad \text{(c)} \quad dS_{xz} = dx\,dz \tag{1.91}$$

A *differential volume element* in Cartesian coordinates is an infinitesmal rectangular parallelepiped, the sides of which are of lengths dx, dy, and dz. The volume of such

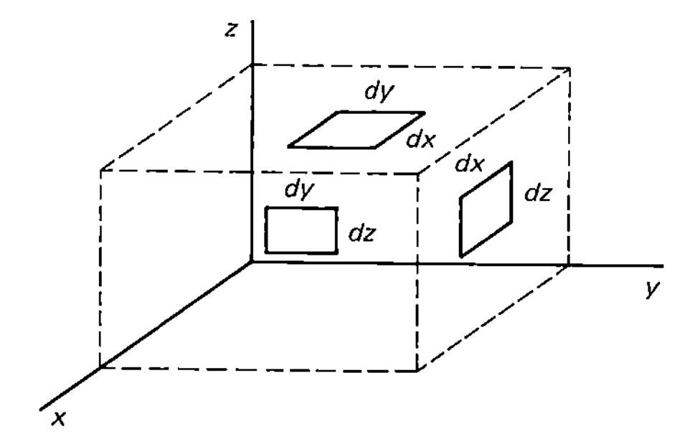

Figure 1.16 Differential surface elements in planes parallel to the x-y, x-z, and y-z planes.

a parallelepiped is

$$dV = dx\,dy\,dz \tag{1.92}$$

The *area element* can also be described in terms of cross products. We will discuss this in Cartesian coordinates, and for simplicity, we will consider a differential area in the x-y plane.

It is customary to associate a direction with an area element by defining a unit vector normal to the surface of the area element. Then

$$d\mathbf{S} = \mathbf{n}\,dS \tag{1.93}$$

For example, $\pm\mathbf{k}$ are the two unit normals to the x-y plane. Therefore, depending on how the direction is defined, the vector area element in a plane parallel to the x-y plane is

$$d\mathbf{S}_{xy} = \pm\mathbf{k}\,dx\,dy \tag{1.94}$$

The direction of the normal is defined by a right-hand rule. For example, the surface with area $dx\,dz$ is taken to be enclosed by line elements $\pm dx\,\mathbf{i}$ and $\pm dz\,\mathbf{k}$ one possible choice of which is shown in Figure 1.17. With the directions of the line elements defined, curling the fingers of the right hand in the direction indicated by the line elements points the thumb in the direction of the normal to the surface.

Referring to Figure 1.18, the area element in the x-y plane is bounded by the vector line elements $d\mathbf{l}_1 = dx\,\mathbf{i}$ and $d\mathbf{l}_2 = dy\,\mathbf{j}$. Therefore, this area element can also be

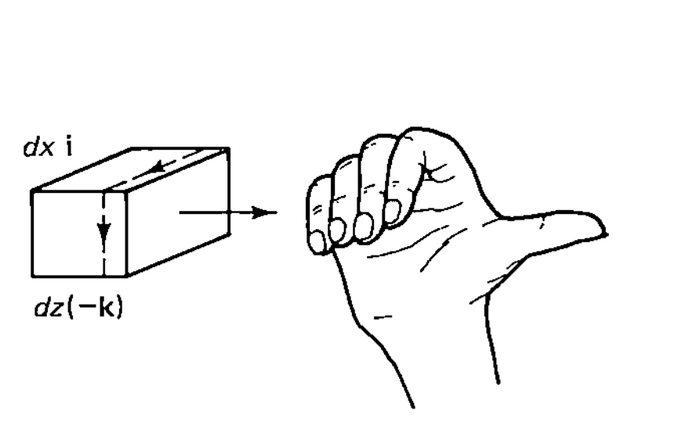

Figure 1.17 Right-hand rule for direction of normal to a surface.

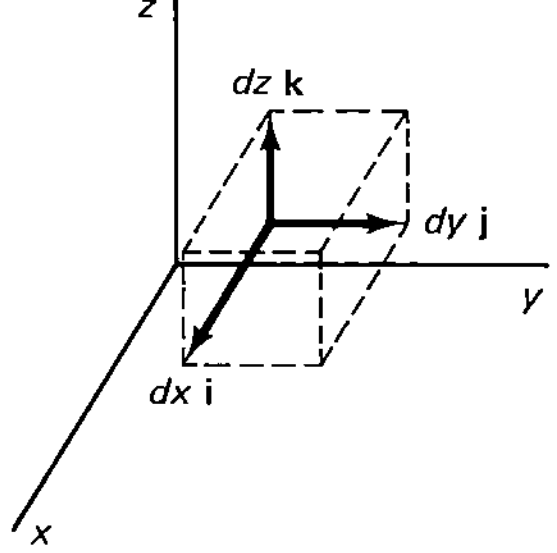

Figure 1.18 Differential line elements defining area and volume elements in Cartesian coordinates.

written

$$d\mathbf{S}_{xy} = d\mathbf{l}_1 \times d\mathbf{l}_2 \tag{1.95a}$$

With $d\mathbf{l}_1 = dx\,\mathbf{i}$, $d\mathbf{l}_2 = dy\,\mathbf{j}$, and $d\mathbf{l}_3 = dz\,\mathbf{k}$, it follows that

$$\text{(b)}\quad d\mathbf{S}_{yz} = d\mathbf{l}_2 \times d\mathbf{l}_3 \qquad \text{(c)}\quad d\mathbf{S}_{zx} = d\mathbf{l}_3 \times d\mathbf{l}_1 \tag{1.95}$$

This result can easily be generalized. If $d\mathbf{l}_i$ is a differential line element forming one side of a surface element, and $d\mathbf{l}_j$ is the differential line element that forms the second side of that surface element, then

$$d\mathbf{S}_{ij} = d\mathbf{l}_i \times d\mathbf{l}_j \tag{1.96}$$

where i, j cyclically take on the values 1, 2, and 3.

The volume element in Cartesian coordinates can be written in terms of the scalar triple product of vector line elements. For example, from $d\mathbf{S}_{yz} = d\mathbf{l}_2 \times d\mathbf{l}_3 = dy\,dz\,\mathbf{i}$, and $d\mathbf{l}_1 = dx\,\mathbf{i}$, it is easy to see that the volume element of equation 1.92 is obtained by

$$dV = d\mathbf{l}_1 \cdot d\mathbf{l}_2 \times d\mathbf{l}_3 \tag{1.97a}$$

Since the scalar triple product is invariant under the interchange of the dot and cross operations, this can also be written

$$dV = d\mathbf{l}_1 \times d\mathbf{l}_2 \cdot d\mathbf{l}_3 \tag{1.97b}$$

Using similar analysis, we will now show that the area and volume elements in spherical and cylindrical coordinates can be written in the same form as equations 1.96 and 1.97. That is, an area element is the cross product of two line elements, and the volume element is the scalar triple product of three line elements.

Figure 1.19 depicts the sides of a differential area element on the surface of a spherical shell of radius R. As can be seen, the side $dl_2 = R\,d\theta$. Since the radius of the circle that traces out dl_3 is the length of the x-y projection of the radial line, $dl_3 = R\sin\theta\,d\phi$. Thus the area element on the shell is

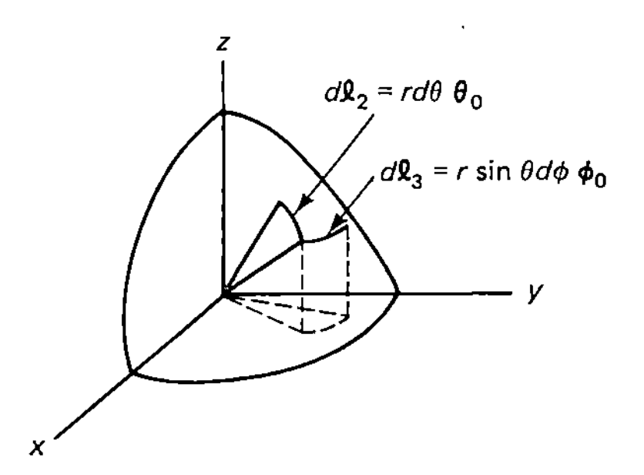

Figure 1.19 Differential area element on the surface of a sphere.

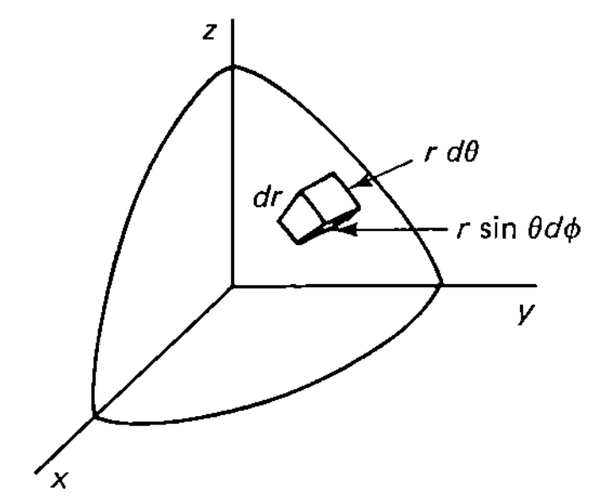

Figure 1.20 Differential volume element of a sphere.

$$dS_{r=R} = R^2 \sin\theta\, d\theta\, d\phi \equiv R^2\, d\Omega \tag{1.98}$$

$d\Omega$ is called the *differential solid angle*. The direction perpendicular to the spherical shell is along the unit vector $\mathbf{r}_0$.

In this representation it is easy to find the area of a spherical shell of radius R by integrating θ from 0 to π, and ϕ from 0 to 2π. The result is

$$S_{\text{shell}} = R^2 \int_0^{2\pi} d\phi \int_0^{\pi} \sin\theta \, d\theta = 4\pi R^2 \tag{1.99}$$

Such a computation using Cartesian coordinates is quite unwieldy.

Referring to Figure 1.19, the vector line elements that bound the area element are

$$\text{(a)} \quad d\mathbf{l}_2 = R\, d\theta \, \boldsymbol{\theta}_0 \qquad \text{(b)} \quad d\mathbf{l}_3 = R \sin\theta \, d\phi \, \boldsymbol{\phi}_0 \tag{1.100}$$

Therefore, the vector area element on the surface of the spherical shell is

$$d\mathbf{S}_{r=R} = d\mathbf{l}_2 \times d\mathbf{l}_3 = R^2 \sin\theta \, d\theta \, d\phi \, \boldsymbol{\theta}_0 \times \boldsymbol{\phi}_0 = R^2 \, d\Omega \, \mathbf{r}_0 \tag{1.100c}$$

in agreement with equation 1.98.

The differential volume element is formed by taking a segment between two infinitesmally separated spherical shells. Each shell has a differential area element

$$dS = r^2 \, d\Omega \tag{1.100d}$$

and the area elements are separated by a distance dr as shown in Figure 1.20. Thus

$$dV = r^2 \, d\Omega \, dr = r^2 \, dr \sin\theta \, d\theta \, d\phi \tag{1.101}$$

As just deduced in equation 1.100d, the area element on the surface of a spherical shell is $d\mathbf{S} = r^2 \, d\Omega \, \mathbf{r}_0 = d\mathbf{l}_2 \times d\mathbf{l}_3$. Since $d\mathbf{l}_1 = dr \, \mathbf{r}_0$,

$$dV = d\mathbf{l}_1 \cdot d\mathbf{l}_2 \times d\mathbf{l}_3 = r^2 \, dr \, d\Omega \tag{1.102}$$

$$\Rightarrow \quad V = \int_0^R r^2 \, dr \int_0^{2\pi} d\phi \int_0^{\pi} \sin\theta \, d\theta = \tfrac{4}{3}\pi R^3 \tag{1.103}$$

for a sphere at radius R. It is much more difficult to evaluate this volume using Cartesian coordinates.

Referring to Figure 1.21, an identical type of analysis for the area and volume elements in cylindrical coordinates leads to the following results: The area element on the curved and top surfaces of a cylinder of radius P are

$$\text{(a)} \quad dS_{\text{curved}} = P \, d\phi \, dz \qquad \text{(b)} \quad dS_{\text{top}} = \rho \, d\rho \, d\phi \tag{1.104}$$

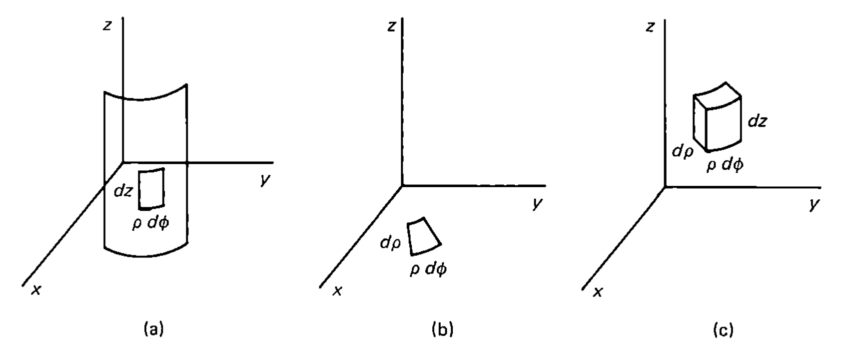

Figure 1.21 Area and volume elements of a cylinder.

Thus the total surface area of a cylinder of radius P and of length L is given by

$$S = S_{\text{curved}} + 2S_{\text{top}} = \int_0^L dz \int_0^{2\pi} P\,d\phi + 2\int_0^P \rho\,d\rho \int_0^{2\pi} d\phi = 2\pi PL + 2\pi P^2 \tag{1.105}$$

The volume element of a cylinder is

$$dV = \rho\,d\rho\,d\phi\,dz \tag{1.106}$$

$$\Rightarrow \quad V = \int_0^{2\pi} d\phi \int_0^P \rho\,d\rho \int_0^L dz = \pi P^2 L \tag{1.107}$$

1.4 Operations with ∇

Vector functions are vectors, the components of which are functions of the coordinate variables. That is, a vector function in Cartesian coordinates will have the general form

$$\mathbf{G}(x, y, z) = G_x(x, y, z)\mathbf{i} + G_y(x, y, z)\mathbf{j} + G_z(x, y, z)\mathbf{k} \tag{1.108}$$

A simple example of such a vector function is the position vector

$$\mathbf{r} = x\mathbf{i} + y\mathbf{j} + z\mathbf{k} = r\mathbf{r}_0.$$

In an analogous way, it is possible to define a vector differential operator ∇. In Cartesian coordinates, it is written

$$\nabla = \mathbf{i}\frac{\partial}{\partial x} + \mathbf{j}\frac{\partial}{\partial y} + \mathbf{k}\frac{\partial}{\partial z} \tag{1.109}$$

Since this operator (referred to as "del" or gradient) is a vector operator, it can operate "multiplicatively" in three different ways.

For a scalar function $F(x, y, z)$,

$$\nabla F = \mathbf{i}\frac{\partial F}{\partial x} + \mathbf{j}\frac{\partial F}{\partial y} + \mathbf{k}\frac{\partial F}{\partial z} = \text{gradient of } F \text{ (grad } F) \tag{1.110}$$

is a vector function.

Consider $\nabla F \cdot d\mathbf{r}$, where in Cartesian coordinates,

$$d\mathbf{r} = dx\,\mathbf{i} + dy\,\mathbf{j} + dz\,\mathbf{k} \tag{1.111}$$

Then

$$\nabla F \cdot d\mathbf{r} = \frac{\partial F}{\partial x}dx + \frac{\partial F}{\partial y}dy + \frac{\partial F}{\partial z}dz = dF \tag{1.112}$$

To obtain a more graphical description of the gradient operation, consider an arbitrary scalar function $F(x, y, z)$. The constraint $F = C_1$ = constant defines a surface in three-dimensional space. $F = C_2$ = constant defines a second surface. Let the two surfaces be differentially separated. By this is meant

$$F(x, y, z) = C_1 \tag{1.113a}$$

$$F(x + dx, y + dy, z + dz) = C_2 = F(x, y, z) + dF = C_1 + dF \tag{1.113b}$$

From Figure 1.22 we see that F differs from $F + dF$ by a component that lies along the normal to one of the surfaces and a component that lies along the tangent to that

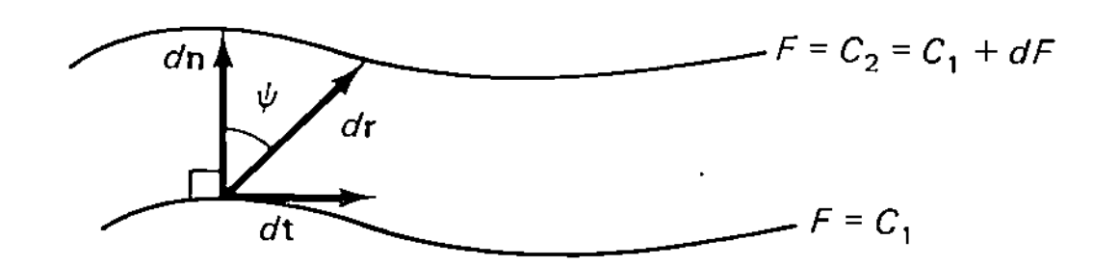

Figure 1.22 Two differentially separated surfaces.

surface. That is,

$$dF = \frac{\partial F}{\partial n}\, dn = \frac{\partial F}{\partial t}\, dt \tag{1.114}$$

But the tangents are along surfaces that are defined by constant values of F. That is, F does not vary in the tangential direction, so

$$\text{(a)} \quad \frac{\partial F}{\partial t} = 0 \quad \Rightarrow \quad \text{(b)} \quad dF = \frac{\partial F}{\partial n}\, dn \tag{1.115}$$

Referring to Figure 1.22, we see that

$$dn = dr \cos \psi = \mathbf{n} \cdot d\mathbf{r} \tag{1.116}$$

Therefore,

$$dF = \frac{\partial F}{\partial n} \mathbf{n} \cdot d\mathbf{r} \tag{1.117}$$

Comparing this to

$$dF = \nabla F \cdot d\mathbf{r} \tag{1.112}$$

$$\Rightarrow \quad \nabla F = \frac{\partial F}{\partial n} \mathbf{n} \tag{1.118}$$

That is, the gradient of a scalar function is directed along the normal to the surface of constant F at that point. For that reason, the gradient is also referred to as the normal derivative.

Since ∇ is a vector operator, it can operate "multiplicatively" on vector functions in two ways. One can "dot" the operator into a vector function or one can take the "cross product" with a vector function.

Let $\mathbf{G}(x, y, z)$ be a vector function. The vector current $\rho\mathbf{v}$ of a fluid (of density ρ, flowing with a velocity $\mathbf{v}$), and the electric field, are two examples of vector functions of physical significance. From the definition of the dot product,

$$\nabla \cdot \mathbf{G} = \frac{\partial G_x}{\partial x} + \frac{\partial G_y}{\partial y} + \frac{\partial G_z}{\partial z} \equiv \text{divergence of } \mathbf{G} = (\text{div}\,\mathbf{G}) \tag{1.119}$$

The cross product of ∇ with $\mathbf{G}$ is

$$\nabla \times \mathbf{G} = \left(\frac{\partial G_z}{\partial y} - \frac{\partial G_y}{\partial z}\right)\mathbf{i} + \left(\frac{\partial G_x}{\partial z} - \frac{\partial G_z}{\partial x}\right)\mathbf{j} + \left(\frac{\partial G_y}{\partial x} - \frac{\partial G_x}{\partial y}\right)\mathbf{k}$$

$$\equiv \text{curl of } \mathbf{G} = (\text{curl}\,\mathbf{G}) \tag{1.120}$$

To get a physical sense of the divergence of a vector, consider a Cartesian volume element and a vector function $\mathbf{G}(x, y, z)$ as shown in Figure 1.23. In Figure 1.23, the components G_x and G_z are not shown.

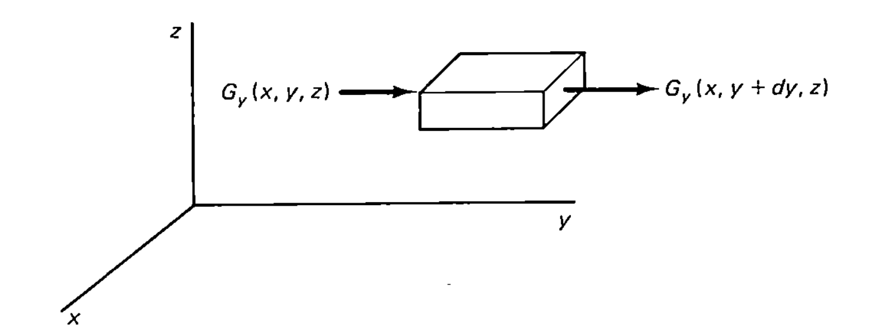

Figure 1.23 Cartesian volume element and vector field lines in y-direction passing through it.

The net change in G_y between the two sides of the differential volume element is

$$G_y(x, y + dy, z) - G_y(x, y, z) = \frac{\partial G_y}{\partial y} dy \tag{1.121}$$

The area element in the y-direction is $dx\,dz$. Thus the change in G_y multiplied by the area element in the y-direction is

$$\frac{\partial G_y}{\partial y} dx\,dy\,dz = dV \cdot (y \text{ part of } \nabla \cdot \mathbf{G}) \tag{1.122}$$

Combining this with similar contributions from the x and z directions, we obtain

$$\nabla \cdot \mathbf{G}\, dV = \text{scalar product of the net change in } G \text{ and the corresponding area element} \tag{1.123}$$

Therefore, $\nabla \cdot \mathbf{G}$ is the change per unit length in the vector $\mathbf{G}$ through a region of space.

For example, the electric field between the plates of an ideal capacitor is constant. Let the volume under consideration be a cube in the space between the plates. We will use the concept, attributed to Faraday, that the number of field lines we draw crossing a unit area is a measure of the strength of the field. Therefore, since the field is constant throughout the region between the plates, the number of field lines that are drawn emerging from the right-hand side of the cubical volume in Figure 1.24 is the same as the number that are drawn entering the left-hand side of that cubic volume. Since the difference in the number of lines leaving and the number entering is zero, the divergence of the electric field at every point inside the cube is zero.

Now consider a volume of space surrounding a positive point charge as in Figure 1.25. Since the point charge is the source of the electric field and there is no negative charge to act as a sink of field lines, no lines enter the volume, but a nonzero number of

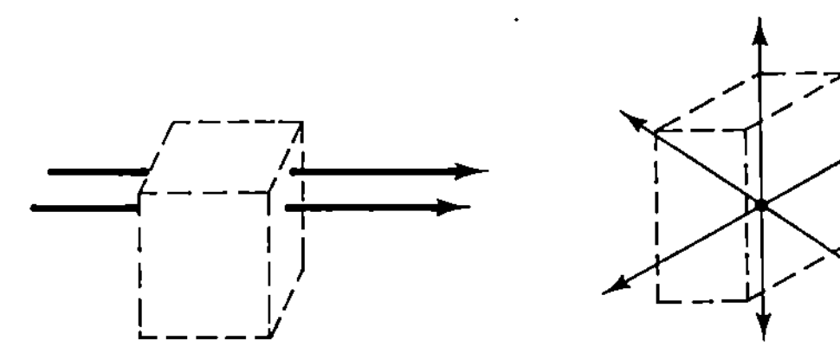

Figure 1.24 Constant electric field through a cubical volume.

Figure 1.25 Electric field due to a point charge through a cubical volume.

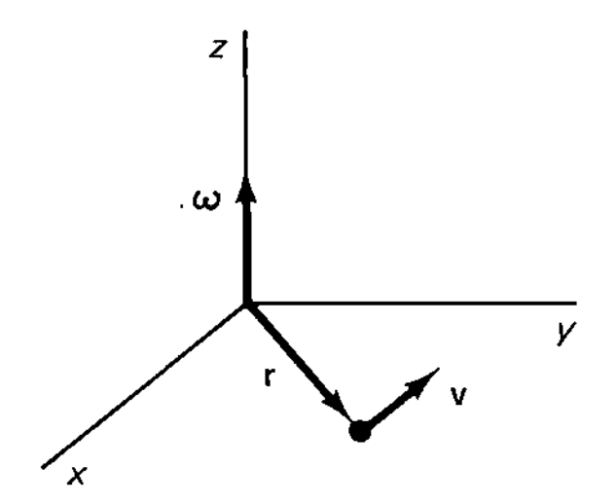

Figure 1.26 Particle describing circular motion in the x-y plane.

lines leave the volume. For this configuration, the divergence of the electric field is nonzero.

From these two examples we can infer that if $\nabla \cdot \mathbf{G} \neq 0$, there must be a source or sink (terminus) of the vector field. Conversely, if $\nabla \cdot \mathbf{G} = 0$, we can conclude that there is no source or sink of the field.

To obtain a sense of the meaning of the curl of a vector, consider a particle moving in a circle. For simplicity, we will constrain the particle to the x-y plane (or equivalently, we will define the particle's plane of motion as the x-y plane). Its angular velocity vector will point along the z-axis. Referring to Figure 1.26 and using the right-hand rule for the cross product, the particle's tangential velocity $\mathbf{v}$ can be written as $\mathbf{v} = \boldsymbol{\omega} \times \mathbf{r}$.

Consider

$$\nabla \times \mathbf{v} = \mathbf{i}\left(\frac{\partial v_z}{\partial y} - \frac{\partial v_y}{\partial z}\right) + \mathbf{j}\left(\frac{\partial v_x}{\partial z} - \frac{\partial v_z}{\partial x}\right) + \mathbf{k}\left(\frac{\partial v_y}{\partial x} - \frac{\partial v_x}{\partial y}\right) \tag{1.124}$$

where

$$\text{(a)} \quad \mathbf{v} = \omega\mathbf{k} \times (x\mathbf{i} + y\mathbf{j} + z\mathbf{k}) = \omega(x\mathbf{j} - y\mathbf{i}) \quad \Rightarrow \quad \text{(b)} \quad \nabla \times \mathbf{v} = 2\omega\mathbf{k} = 2\boldsymbol{\omega} \tag{1.125}$$

That is, $\nabla \times \mathbf{v}$ is a measure of the rate at which the particle rotates. From this example, we develop the idea that the curl of a vector is a measure of how much rotation or curvature the vector undergoes at a point. A vector $\mathbf{G}$, for which $\nabla \times \mathbf{G} = 0$, then, is said to be *irrotational*.

Integral theorems

We will now develop integral theorems involving the gradient of a scalar function, the divergence, and the curl of a vector function.

Consider the quantity involving the vector function $\mathbf{G}$,

$$\frac{1}{dV} \sum_{\substack{\text{all} \\ \text{faces}}} \mathbf{G} \cdot \mathbf{n}\, dS$$

where dV is the volume of the infinitesimal rectangular parallelepiped of Figure 1.27. The sum is over all the faces of the parallelepiped, and dS is the area element of each of its faces. For example, the two faces of the parallelepiped that are parallel to the x-z plane have normal vectors $+\mathbf{j}$ at $y + dy$ and $-\mathbf{j}$ at y. Thus the contribution to the above sum from this pair of faces is

$$\begin{aligned} \frac{1}{dV} \sum_{\substack{y \\ \text{faces}}} \mathbf{G} \cdot \mathbf{n}\, dS &= \frac{1}{dx\, dy\, dz}\left[G_y(x, y + dy, z) - G_y(x, y, z)\right] dx\, dz \\ &= \frac{\partial G_y}{\partial y} = y\text{-term in } \nabla \cdot \mathbf{G} \end{aligned} \tag{1.126}$$

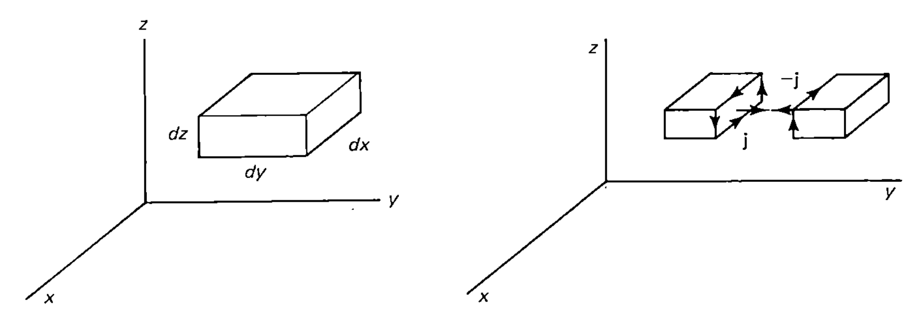

Figure 1.27 Infinitesimal rectangular parallelepiped.

Figure 1.28 Two adjacent infinitesimal parallelepipeds.

Adding analogous contributions from the pairs of faces parallel to the other two coordinate planes, we obtain

$$\frac{1}{dV} \sum_{\substack{\text{all}\\ \text{faces}}} \mathbf{G} \cdot \mathbf{n}\, dS = \nabla \cdot \mathbf{G} \tag{1.127a}$$

Henceforth we will drop the "all faces" indicator; the summation symbol will imply that.

Consider two adjacent infinitesimal parallelepipeds. As shown in Figure 1.28, the normals to adjoining faces are of opposite sign ($\pm\mathbf{j}$ in the figure). If the two parallelepipeds are in contact with one another, the x-z faces in the figure that are touching would have the same areas with oppositely directed normals. Thus, when adding contributions of a function, evaluated at points on these faces in contact, the resultant would be zero.

If we now combine an infinite number of these infinitesimal parallelpipeds to make a macroscopic volume, the contributions from the internal surfaces of the parallelepipeds will cancel. Therefore, the resulting sum will include only contributions from unpaired surfaces, which comprise the outer surface of the macroscopic volume. That is, the *divergence* or *Gauss' theorem* is

$$\int \nabla \cdot \mathbf{G}\, dV = \sum \int \mathbf{G} \cdot \mathbf{n}\, dS \tag{1.127b}$$

and the sum is over all faces of the outer surface of the volume defined by the left-hand-side integral.

From this point on, the sum over all surfaces of the enclosed volume will be implied by the surface integral, and the summation symbol will not be explicitly written. That is, we will express Gauss's theorem as

$$\int \nabla \cdot \mathbf{G}\, dV = \int \mathbf{G} \cdot \mathbf{n}\, dS \tag{1.128}$$

We now apply this analysis to the quantity

$$\frac{1}{dV} \sum F\mathbf{n}\, dS$$

where F is a scalar function. As before, the contribution from the pair of faces parallel to the x-z plane is

$$\frac{1}{dx\,dy\,dz}[F(x, y + dy, z) - F(x, y, z)]\mathbf{j}\, dx\,dz = \mathbf{j}\frac{\partial F}{\partial y} = \mathbf{j}(\nabla F)_y \tag{1.129}$$

or, adding the contributions from all faces of the parallelepiped, we obtain

$$\nabla F = \frac{1}{dV} \sum F \mathbf{n}\, dS \tag{1.130a}$$

Therefore, by adding the contributions from an infinite number of infinitesimal parallelepipeds, the result for a macroscopic volume is

$$\int \nabla F\, dV = \int F \mathbf{n}\, dS \tag{1.130b}$$

Similarly, for the vector function $\mathbf{G}$, consider $\frac{1}{dV} \sum \mathbf{G} \times \mathbf{n} dS$. Taking the contribution from the faces parallel to the x-z plane, we obtain

$$\frac{1}{dV}[\mathbf{G}(x, y + dy, z) - \mathbf{G}(x, y, z)] \times \mathbf{j}\, dx\, dz$$

$$= \frac{1}{dx\, dy\, dz}\Big\{\mathbf{k}[G_x(x, y + dy, z) - G_x(x, y, z)] - \mathbf{i}[G_z(x, y + dy, z) - G_z(x, y, z)]\Big\}\, dx\, dz$$

$$= \mathbf{k}\frac{\partial G_x}{\partial y} - \mathbf{i}\frac{\partial G_z}{\partial y} = \frac{\partial \mathbf{G}}{\partial y} \times \mathbf{j} \tag{1.131a}$$

By identical analysis, the contribution from the pairs of faces parallel to the y-z and x-y planes are

$$\text{(b)}\quad \mathbf{j}\frac{\partial G_z}{\partial x} - \mathbf{i}\frac{\partial G_y}{\partial x} = \frac{\partial \mathbf{G}}{\partial y} \times \mathbf{i} \qquad \text{(c)}\quad \frac{\partial G_y}{\partial z}\mathbf{i} - \frac{\partial G_x}{\partial z}\mathbf{j} = \frac{\partial \mathbf{G}}{\partial z} \times \mathbf{k} \tag{1.131}$$

respectively. Combining equations 1.131, we obtain

$$\frac{1}{dV} \sum \mathbf{G} \times \mathbf{n}\, dS = \mathbf{i}\left(\frac{\partial G_y}{\partial z} - \frac{\partial G_z}{\partial y}\right) + \mathbf{j}\left(\frac{\partial G_z}{\partial x} - \frac{\partial G_x}{\partial z}\right) + \mathbf{k}\left(\frac{\partial G_x}{\partial y} - \frac{\partial G_v}{\partial x}\right) = -\nabla \times \mathbf{G} \tag{1.132a}$$

Therefore, over a macroscopic volume

$$\int \nabla \times \mathbf{G}\, dV = \int \mathbf{n} \times \mathbf{G}\, dS \tag{1.132b}$$

The negative sign has been absorbed into the interchange of factors in the cross product in $\mathbf{n}$ and $\mathbf{G}$.

Comparing equations 1.128, 1.130b, and 1.132b, we see that if Q represents either a vector or a scalar function, and if $\nabla * Q$ represents either the divergence or curl of the vector function or the gradient of the scalar function, all three of these integral theorems are of the form

$$\int \nabla * Q\, dV = \int \mathbf{n} * Q\, dS \tag{1.133}$$

As noted above, the integral over the entire surface enclosing the volume implies the sum over all faces.

From equation 1.130a,

$$\nabla F = \frac{1}{dV} \sum F \mathbf{n}\, dS \tag{1.130a}$$

where dS is the area element of the surface with normal $\mathbf{n}$. As such, we will denote this

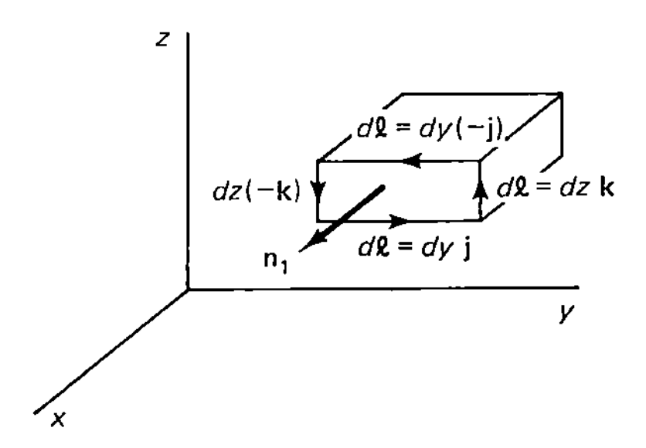

Figure 1.29 Rectangular parallelepiped with loop enclosing an area element parallel to the y-z plane.

area element by dS_n to distinguish it from another area element we will introduce shortly.

Referring to the rectangular parallelepiped of Figure 1.29, let $\mathbf{n}_1$ be the normal to one of the faces. For example, the surface outlined by the line elements depicted in figure 1.29 is parallel to the y-z plane, and using the right-hand rule illustrated in Figure 1.17, the normal to this face is $\mathbf{n}_1 = +\mathbf{i}$.

Referring to equation 1.130a, consider the quantity

$$\mathbf{n}_1 \times \nabla F = \frac{1}{dV} \sum F \mathbf{n}_1 \times \mathbf{n}\, dS_n \tag{1.134}$$

with $\mathbf{n}_1 = +\mathbf{i}$ as in Figure 1.29. Because of the factor $\mathbf{n}_1 \times \mathbf{n}$, when summing over the six faces of the rectangular parallelepiped, the contributions from the two faces with normals $\pm\mathbf{n}_1 = \pm\mathbf{i}$ will be zero. That is, the sum will only contain terms from those faces for with normals $\mathbf{n} = \pm\mathbf{j}$, and $\mathbf{n} = \pm\mathbf{k}$. We consider the quantity $(1/dV)\mathbf{n}_1 \times \mathbf{n}\, dS_n$ over each of these four faces.

For the right-hand face of the parallelepiped, for which $\mathbf{n} = \mathbf{j}$, and along which $d\mathbf{l} = dz\,\mathbf{k}$,

$$\frac{1}{dV}\mathbf{n}_1 \times \mathbf{n}\, dS_n = \mathbf{i} \times \mathbf{j}\frac{dx\,dz}{dx\,dy\,dz} = \frac{dz\,\mathbf{k}}{dy\,dz} \tag{1.135}$$

$dz\,\mathbf{k} = d\mathbf{l}$ along this right-hand face, and $dy\,dz$ is the area element of the face enclosed by the loop that has normal $\mathbf{n}_1$. We designate this area element as

$$dS_{n_1} = dy\,dz \tag{1.136}$$

Thus along this right-hand face,

$$\frac{1}{dV}\mathbf{n}_1 \times \mathbf{n}\, dS_n = \frac{d\mathbf{l}}{dS_{n_1}} \tag{1.137a}$$

Similarly, along the top face of the rectangular parallelepiped, $\mathbf{n} = \mathbf{k}$, $d\mathbf{l} = dy(-\mathbf{j})$, and $dS_n = dx\,dy$. Thus

$$\frac{1}{dV}\mathbf{n}_1 \times \mathbf{n}\, dS_n = \mathbf{i} \times \mathbf{k}\frac{dx\,dy}{dx\,dy\,dz} = \frac{dy(-\mathbf{j})}{dy\,dz} = \frac{d\mathbf{l}}{dS_{n_1}} \tag{1.137b}$$

Along the left-hand face, $\mathbf{n} = -\mathbf{j}$, $d\mathbf{l} = dz(-\mathbf{k})$, and $dS_n = dx\,dz$. Thus

$$\frac{1}{dV}\mathbf{n}_1 \times \mathbf{n}\, dS_n = \frac{dz(-\mathbf{k})}{dy\,dz} = \frac{d\mathbf{l}}{dS_{n_1}} \tag{1.137c}$$

with an identical result for the bottom face. That is, in general,

$$\frac{1}{dV}\mathbf{n}_1 \times \mathbf{n}\, dS_n = \frac{d\mathbf{l}}{dS_{n_1}} \tag{1.138}$$

Therefore, equation 1.130a becomes

$$\mathbf{n}_1 \times \nabla F = \sum_{\substack{\text{around}\\ \text{loop}}} F\frac{d\mathbf{l}}{dS_{n_1}} \tag{1.139a}$$

The sum around the loop is a sum over differential line elements. Since dS_{n_1} is the same for all segments of the loop, it can be removed from the sum and equation 1.139a can be written

$$\mathbf{n}_1 \times \nabla F\, dS_{n_1} = \sum_{\substack{\text{around}\\ \text{loop}}} F\, d\mathbf{l} \tag{1.139b}$$

We view dS_{n_1} as the area element of the surface enclosed by the loop, and $\mathbf{n}_1$ as the normal to that surface as defined by the right-hand rule illustrated in Figure 1.17.

Adding contributions from an infinite number of these infinitesimal parallelepipeds to make a macroscopic surface, and recognizing that cancellations will occur between neighboring faces of adjacent parallelepipeds, we obtain

$$\int_{\substack{\text{enclosed}\\ \text{surface}}} \mathbf{n}_1 \times \nabla F\, dS_{n_1} = \oint F\, d\mathbf{l} \tag{1.140}$$

The only faces of the parallelepipeds for which there will be no cancellations are those for which there are no adjacent faces. These unpaired faces make up the outer surface of the macroscopic volume formed by the infinite aggregate of microscopic parallelepipeds. Thus dS_{n_1} is an area element of that outer surface, $\mathbf{n}_1$ is the normal to that surface, and $d\mathbf{l}$ is a differential line element along the loop that defines the enclosed surface. The notation $\oint$ means integration around the closed loop.

From equation 1.132a,

$$\nabla \times \mathbf{G} = \frac{1}{dV}\sum \mathbf{n} \times \mathbf{G}\, dS_n \tag{1.132a}$$

Referring again to the rectangular parallelepiped of Figure 1.27 or 1.29,

$$\mathbf{n}_1 \cdot \nabla \times \mathbf{G} = \frac{1}{dV}\sum \mathbf{n}_1 \cdot \mathbf{n} \times \mathbf{G}\, dS_n \tag{1.141a}$$

Since the dot and cross operations can be interchanged, this can be written

$$\mathbf{n}_1 \cdot \nabla \times \mathbf{G} = \frac{1}{dV}\sum (\mathbf{n}_1 \times \mathbf{n}) \cdot \mathbf{G}\, dS_n \tag{1.141b}$$

which now contains the quantity $(1/dV)\mathbf{n}_1 \times \mathbf{n}\, dS_n$. Using equation 1.138, equation 1.141b becomes

$$\mathbf{n}_1 \cdot \nabla \times \mathbf{G} = \frac{1}{dS_{n_1}}\sum d\mathbf{l} \cdot \mathbf{G} \tag{1.142}$$

or for a macroscopic volume

$$\int_{\substack{\text{enclosed}\\ \text{surface}}} \mathbf{n}_1 \cdot \nabla \times \mathbf{G}\, dS_{n_1} = \oint \mathbf{G} \cdot d\mathbf{l} \tag{1.143}$$

This important theorem is known as *Stokes's theorem* or the *curl theorem*.

Finally, consider $(\mathbf{n}_1 \times \nabla) \times \mathbf{G}$. From the analysis of the vector triple product, this can be shown quite straightforwardly to be

$$(\mathbf{n}_1 \times \nabla) \times \mathbf{G} = \nabla(\mathbf{n}_1 \cdot \mathbf{G}) - \mathbf{n}_1(\nabla \cdot \mathbf{G}) \tag{1.144}$$

Using equation 1.127a for the divergence of a vector function, and equation 1.130a for the gradient of a scalar, equation 1.144 can be written

$$(\mathbf{n}_1 \times \nabla) \times \mathbf{G} = \frac{1}{dV} \sum [(\mathbf{n}_1 \cdot \mathbf{G})\mathbf{n} - \mathbf{n}_1(\mathbf{G} \cdot \mathbf{n})]\, dS_n \tag{1.145}$$

Referring to equation 1.41b for the vector triple product, we obtain

$$(\mathbf{n}_1 \cdot \mathbf{G})\mathbf{n} - \mathbf{n}_1(\mathbf{G} \cdot \mathbf{n}) = -\mathbf{G} \times (\mathbf{n}_1 \times \mathbf{n}) = (\mathbf{n}_1 \times \mathbf{n}) \times \mathbf{G} \tag{1.146}$$

$$\Rightarrow \quad (\mathbf{n}_1 \times \nabla) \times \mathbf{G} = \frac{1}{dV} \sum (\mathbf{n}_1 \times \mathbf{n}) \times \mathbf{G}\, dS_n \tag{1.147}$$

As before, this also contains the quantity $(1/dV)\mathbf{n}_1 \times \mathbf{n}\, dS_n$. Therefore, using equation 1.138, we obtain

$$\int (\mathbf{n}_1 \times \nabla) \times \mathbf{G}\, dS_{n_1} = \oint d\mathbf{l} \times \mathbf{G} \tag{1.148}$$

Since $\mathbf{n}_1$ and dS_{n_1} are the only normal and area element involved in these integral theorems, dispensing with these subscripts will not create any confusion. We will do so from now on.

In summary, we have deduced

$$\int \nabla \cdot \mathbf{G}\, dV = \int \mathbf{G} \cdot \mathbf{n}\, dS \qquad \text{(Gauss's theorem)} \tag{1.128}$$

$$\int \mathbf{n} \cdot \nabla \times \mathbf{G}\, dS = \oint \mathbf{G} \cdot d\mathbf{l} \qquad \text{(Stokes's theorem)} \tag{1.143}$$

$$\int \nabla F\, dV = \int F\mathbf{n}\, dS \quad (1.130\text{b}) \qquad \int \nabla \times \mathbf{G}\, dV = \int \mathbf{n} \times \mathbf{G}\, dS \quad (1.132\text{b})$$

$$\int \mathbf{n} \times \nabla F\, dS = \oint F\, d\mathbf{l} \quad (1.140) \qquad \int (\mathbf{n} \times \nabla) \times \mathbf{G}\, dS = \oint d\mathbf{l} \times \mathbf{G} \quad (1.148)$$

In equations 1.140, 1.143, and 1.148, $\mathbf{n}$ is the normal vector to the surface enclosed by the loop. The direction of the normal is obtained using the right-hand rule defined earlier (see Figure 1.17).

Although these theorems were derived using a rectangular infinitesimal parallelepiped, the macroscopic volume and surface obtained by combining an infinite number of the infinitesimal parallelepipeds were not specified. Thus these theorems are valid for volumes and surfaces of any shape.

Gauss's theorem and Stokes's theorem have many important applications to scientific problems, while the other four theorems do not. For that reason, we will only investigate these two further.

Example 1.6

As an example, we will verify the divergence theorem or Gauss's theorem for the vector

$$\mathbf{G} = 3xy\mathbf{i} + x^2\mathbf{j} + xyz\mathbf{k} \tag{1.149}$$

over a cube of sides 1 with one corner situated at the origin as shown in Figure 1.30. The

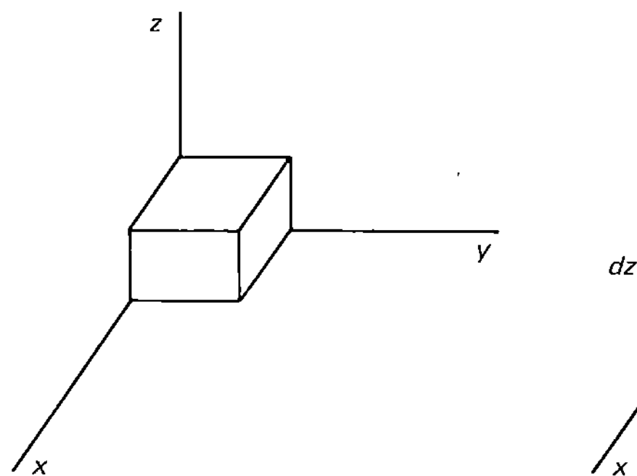

Figure 1.30 Cube of side 1, with one corner at the origin.

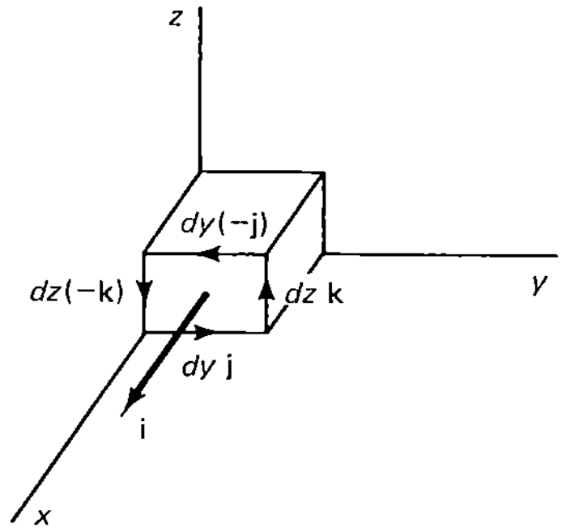

Figure 1.31 Face of a cube of side 1, enclosed by a square loop.

divergence of the vector field of equation 1.149 is

$$\nabla \cdot \mathbf{G} = \frac{\partial}{\partial x}(3xy) + \frac{\partial}{\partial y}(x^2) + \frac{\partial}{\partial z}(xyz) = y(x+3) \tag{1.150}$$

$$\Rightarrow \quad \int \nabla \cdot \mathbf{G}\, dV = \int_0^1 dx \int_0^1 dy \int_0^1 dz\, y(x+3) = \tfrac{7}{4} \tag{1.151a}$$

Now consider the surface integral

$$\int \mathbf{G} \cdot \mathbf{n}\, dS = \int_0^1 \int_0^1 [G_x(1, y, z) - G_x(0\ y, z)]\, dy\, dz$$

$$+ \int_0^1 \int_0^1 [G_y(x, 1, z) - G_y(x, 0, z)]\, dx\, dz + \int_0^1 \int_0^1 [G_z(x, y, 1) - G_z(x, y, 0)]\, dx\, dy$$

$$= \int_0^1 dz \int_0^1 dy(3y) + \int_0^1 dz \int_0^1 dx(x^2 - x^2) + \int_0^1 dx \int_0^1 dy(xy) = \tfrac{7}{4} \tag{1.151b}$$

as expected. □

Example 1.7

We will now verify the curl theorem or Stokes's theorem for the same vector **G**, over one face of the same cube shown in Figure 1.30. We will choose the face that is parallel to the y-z plane, and at $x = 1$. The loop that encloses that face is shown in Figure 1.31.

Using the right-hand rule, the normal to the face enclosed by the loop in Figure 1.31 is **i**. The Stokes integral over the face enclosed by that loop is

$$\int (\nabla \times \mathbf{G}) \cdot \mathbf{n}\, dS = \int (\nabla \times \mathbf{G})_x\, dy\, dz = \int_0^1 \int_0^1 dy\, dz \left(\frac{\partial G_z}{\partial y} - \frac{\partial G_y}{\partial z} \right) = \int_0^1 \int_0^1 dy\, dz(xz) = \tfrac{1}{2} \tag{1.152a}$$

The Stokes integral around the loop is

$$\oint \mathbf{G} \cdot d\mathbf{l} = \int_0^1 G_y(1, y, 0)\, dy + \int_0^1 G_z(1, 1, z)\, dz \tag{1.152b}$$

$$+ \int_0^1 - G_y(1, y, 1)\, dy + \int_0^1 - G_z(1, 0, z)\, dz = \int_0^1 dy + \int_0^1 z\, dz - \int_0^1 dy - \int_0^1 0\, dz = \tfrac{1}{2}$$

which verifies Stokes's theorem.

In the third and fourth integrals, the negative signs arise from the fact that $d\mathbf{l}_3 = -\mathbf{j}\,dy$ and $d\mathbf{l}_4 = -\mathbf{k}\,dz$. The limits on these integrals must be taken in the positive sense (from the smaller to the larger value) since the direction of the line element is accounted for by the sign of the unit vector defining the direction of the line element. □

A most important application of Gauss's and Stokes's theorems and the vector identities introduced here is to Maxwell's equations for the electric and magnetic fields, and the charges and currents that create them. Maxwell's equations in a vacuum (the absence of a material medium) are

$$\text{(a)}\quad \nabla \cdot \mathbf{E} = \frac{\rho}{\varepsilon_0} \qquad \text{(b)}\quad \nabla \cdot \mathbf{B} = 0 \qquad \text{(c)}\quad \nabla \times \mathbf{E} = -\frac{\partial \mathbf{B}}{\partial t}$$

$$\text{(d)}\quad \nabla \times \mathbf{B} = \mu_0 \mathbf{J} + \mu_0 \varepsilon_0 \frac{\partial \mathbf{E}}{\partial t} \tag{1.153}$$

where ρ is the volume charge density (charge per unit volume), and $\mathbf{J}$ is the real current density (current per unit area). That is,

$$\text{(a)}\quad \rho = \frac{dq_{\text{real}}}{dV} \qquad \text{(b)}\quad I = \frac{dq_{\text{real}}}{dt} = \int \mathbf{J} \cdot \mathbf{n}\,dS \tag{1.154}$$

Gauss's law for the electric and magnetic fields are obtained by applying Gauss's theorem to equations 1.153a and 1.153b. They are

$$\text{(a)}\quad \int \nabla \cdot \mathbf{E}\,dV = \int \mathbf{E} \cdot \mathbf{n}\,dS = \frac{1}{\varepsilon_0}\int \rho\,dV = \frac{q_{\text{real}}}{\varepsilon_0} \qquad \text{(b)}\quad \int \nabla \cdot \mathbf{B}\,dV = \int \mathbf{B} \cdot \mathbf{n}\,dS = 0 \tag{1.155}$$

Equation 1.153b for the magnetic field is a mathematical expression of the fact that isolated magnetic charges (magnetic monopoles) do not exist. If there were magnetic monopoles, one could be surrounded with a volume like in Figure 1.25 and $\nabla \cdot \mathbf{B}$ would be nonzero through that volume. Thus since $\nabla \cdot \mathbf{B} = 0$ at every point in space, the minimal configuration for magnetic poles is dipole. This means that for every north pole there must be a south pole; or for every source of magnetic field, there is a sink of magnetic field. The dipole is the simplest configuration for the magnetic field such that the difference between the number of field lines entering and leaving an arbitrary volume will be zero. If magnetic monopoles are found to exist, equation 1.153b will have to be modified to include a magnetic monopole density (analogous to equation 1.153a, which contains the electric monopole or charge density).

Faraday's law is obtained from equation 1.153c using Stokes's theorem. Integrating the left-hand side of this equation, we obtain an expression for the induced electromotive force (emf):

$$\int \nabla \times \mathbf{E} \cdot \mathbf{n}\,dS = \int \mathbf{E} \cdot d\mathbf{l} = -\Delta V \tag{1.156a}$$

$$\Rightarrow \quad \Delta V = \int \frac{\partial \mathbf{B}}{\partial t} \cdot \mathbf{n}\,dS = \frac{\partial}{\partial t}\int \mathbf{B} \cdot \mathbf{n}\,dS = \frac{\partial \Phi_B}{\partial t} \tag{1.156b}$$

where Φ_B is the magnetic flux, defined by

$$\Phi_B \equiv \int \mathbf{B} \cdot \mathbf{n}\,dS \tag{1.157}$$

Φ_B is a measure of the number of magnetic field lines crossing the area specified in its integral definition.

Ampère's law is obtained by applying Stokes's theorem to equation 1.153d:

$$\int \nabla \times \mathbf{B} \cdot \mathbf{n}\, dS = \oint \mathbf{B} \cdot d\mathbf{l} = \mu_0 \int \mathbf{J} \cdot \mathbf{n}\, dS + \mu_0 \varepsilon_0 \int \frac{\partial \mathbf{E}}{\partial t} \cdot \mathbf{n}\, dS \tag{1.158a}$$

$$\Rightarrow \quad \oint \mathbf{B} \cdot d\mathbf{l} = \mu_0 I + \mu_0 \varepsilon_0 \frac{\partial \Phi_E}{\partial t} \tag{1.158b}$$

Φ_E is the electric flux, defined for the electric field in the same way that the magnetic flux is defined for magnetic field lines. That is,

$$\Phi_E = \int \mathbf{E} \cdot \mathbf{n}\, dS \tag{1.159}$$

is a measure of the number of electric field lines crossing area S.

Second operations with ∇

By operating on a scalar or vector function with the ∇ operator, we have formed other scalar and vector functions. We can apply the ∇ operator again, to these resulting functions to obtain other scalar and vector functions.

Since ∇F is a vector function, we can take the divergence or the curl of it. The divergence

$$\nabla \cdot \nabla F \equiv \nabla^2 F = \frac{\partial^2 F}{\partial x^2} + \frac{\partial^2 F}{\partial y^2} + \frac{\partial^2 F}{\partial z^2} \tag{1.160}$$

is an important operation on a scalar function that occurs frequently in scientific problems. Many phenomena are described by a second-order partial differential equation involving the ∇^2 operator applied to a physically meaningful function.

For example, Poisson's equation for the electrostatic potential and Schrödinger's equation for the quantum wave function of a nonrelativistic system are

$$\text{(a)} \quad \nabla^2 V = -\frac{\rho}{\varepsilon_0} \qquad \text{(b)} \quad -\frac{\hbar^2}{2m}\nabla^2 \Psi + V\Psi = i\hbar \frac{\partial \psi}{\partial t} = E\Psi \tag{1.161}$$

As will be seen when solving Problem 28,

$$\nabla \times \nabla F = 0 \tag{1.162}$$

for any scalar function F. That is, ∇F is irrotational. For that reason, if $\nabla \times \mathbf{G} = 0$ ($\mathbf{G}$ is irrotational), it can be written as $\mathbf{G} = \nabla F$ (the gradient of a scalar function).

For example, in an electromagnetic problem, if the charge does not move, the current is zero, and the electric field will not vary with time. From equation 1.153d, zero current and a static electric field leads to $\nabla \times \mathbf{B} = 0$. In addition, $\nabla \cdot \mathbf{B} = 0$ always. These two conditions imply that $\mathbf{B} = 0$. Physically, this situation produces no magnetic field. With $\mathbf{B} = 0$, equations 1.153a and 1.153c become $\nabla \times \mathbf{E} = 0$ and $\nabla \cdot \mathbf{E} = \rho/\varepsilon_0$. Since $\mathbf{E}$ is irrotational under these static constraints, we can write $\mathbf{E}$ as the gradient of a scalar function. That is, $\nabla \times \mathbf{E} = 0$ implies that $\mathbf{E}$ can be expressed in terms of the electrostatic potential as

$$\text{(a)} \quad \mathbf{E} = -\nabla V \quad \Rightarrow \quad \text{(b)} \quad \nabla \cdot \mathbf{E} = \nabla \cdot (-\nabla V) = -\nabla^2 V = \frac{\rho}{\varepsilon_0} \tag{1.163}$$

which is Poisson's equation.

Since $\nabla \cdot \mathbf{G}$ is a scalar function, the only way to apply a second operation with ∇ is to take the gradient of $\nabla \cdot \mathbf{G}$. This has very little application in scientific problems and has no properties that are of particular scientific interest. For this reason, it will not be discussed further.

Since $\nabla \times \mathbf{G}$ is a vector, we can take the divergence and the curl of it. A second part of Problem 28 is to show that

$$\nabla \cdot \nabla \times \mathbf{G} = 0 \tag{1.164}$$

for any vector function $\mathbf{G}$. Thus any divergenceless vector can be written as the curl of a vector function.

For example, the magnetic field is divergenceless (equation 1.153b). Thus the magnetic vector potential $\mathbf{A}$ can be introduced from this as

$$\mathbf{B} = \nabla \times \mathbf{A} \tag{1.165}$$

Recall that the curl of the gradient of any scalar function is zero. Thus we can add the gradient of any scalar function to $\mathbf{A}$ without affecting the magnetic field $\mathbf{B}$, which is the measurable quantity. That is, if

$$\mathbf{A}' = \mathbf{A} + \nabla\chi \tag{1.166}$$

then

$$\mathbf{B}' = \nabla \times \mathbf{A}' = \nabla \times \mathbf{A} + \nabla \times \nabla\chi = \nabla \times \mathbf{A} = \mathbf{B} \tag{1.167}$$

This transformation of $\mathbf{A}$, which leaves $\mathbf{B}$ unchanged, is called a gauge transformation, and $\mathbf{B}$ is said to be invariant under this gauge transformation. The scalar function can be chosen in any way that will simplify the problem at hand.

1.5 Generalized Coordinates

We have thus far discussed vector algebra and vector calculus in Cartesian, spherical, and cylindrical coordinates. Using these as guides, we can describe vector operations in a general orthogonal coordinate system.

As we did when describing spherical and cylindrical coordinates, we begin by defining the generalized coordinates we will use for locating a point. They will be denoted by

$$\mathbf{u} = (u_1, u_2, u_3) \tag{1.168}$$

As with Cartesian, spherical and cylindrical systems, there will be transformation equations relating the generalized coordinates to the Cartesian coordinates, for example.

In the Cartesian frame, the vector displacement between two points is given by

$$d\mathbf{l} = dx\,\mathbf{i} + dy\,\mathbf{j} + dz\,\mathbf{k} \tag{1.169}$$

In spherical coordinates, for example, the analogous displacement is given by

$$d\mathbf{l} = dr\,\mathbf{r}_0 + r\,d\theta\,\boldsymbol{\theta}_0 + r\sin\theta\,d\phi\,\boldsymbol{\phi}_0 \tag{1.170}$$

Note that in spherical coordinates, the multipliers of the infinitesimal coordinate differences dr, $d\theta$, and $d\phi$ are functions of the coordinates. This is true in any coordinate system. For Cartesian coordinates the multipliers are obviously 1. Therefore, in a generalized system, we must include such functional multipliers.

We denote these multipliers by $h_i(\mathbf{u})$ $(i = 1, 2, 3)$. They are also referred to as scaling functions. $\mathbf{v}_i$ $(i = 1, 2, 3)$ will denote the unit basis vectors in the generalized frame. Since the coordinate system is right-handed and orthonormal,

$$\text{(a)}\quad \mathbf{v}_1 \times \mathbf{v}_2 = \mathbf{v}_3 \qquad \text{(b)}\quad \mathbf{v}_2 \times \mathbf{v}_3 = \mathbf{v}_1 \qquad \text{(c)}\quad \mathbf{v}_3 \times \mathbf{v}_1 = \mathbf{v}_2 \qquad \text{(d)}\quad \mathbf{v}_i \cdot \mathbf{v}_j = \delta_{ij} \tag{1.171}$$

Then

$$d\mathbf{l} = h_1\, du_1\, \mathbf{v}_1 + h_2\, du_2\, \mathbf{v}_2 + h_3\, du_3\, \mathbf{v}_3 \tag{1.172}$$

Specifically, for spherical coordinates,

$$\text{(a)}\quad h_1 = 1 \qquad \text{(b)}\quad h_2 = r \qquad \text{(c)}\quad h_3 = r\sin\theta \tag{1.173}$$

and, of course,

$$\text{(a)}\quad \mathbf{v}_1 = \mathbf{r}_0 \qquad \text{(b)}\quad \mathbf{v}_2 = \boldsymbol{\theta}_0 \qquad \text{(c)}\quad \mathbf{v}_3 = \boldsymbol{\phi}_0 \tag{1.174}$$

Once the scaling functions h_i and the basis vectors $\mathbf{v}_i$ are specified, the operation of ∇ on functions of generalized coordinates can be derived.

To deduce the gradient of a scalar function of the generalized coordinates, let $F = F(u_1, u_2, u_3)$. Then, by the chain rule,

$$\nabla F = \frac{\partial F}{\partial x}\mathbf{i} + \frac{\partial F}{\partial y}\mathbf{j} + \frac{\partial F}{\partial z}\mathbf{k} = \sum_{l=1}^{3}\left(\frac{\partial F}{\partial u_l}\frac{\partial u_l}{\partial x}\mathbf{i} + \frac{\partial F}{\partial u_l}\frac{\partial u_l}{\partial y}\mathbf{j} + \frac{\partial F}{\partial u_l}\frac{\partial u_l}{\partial z}\mathbf{k}\right) = \sum_{l=1}^{3}\frac{\partial F}{\partial u_l}\nabla u_l \tag{1.175}$$

In spherical coordinates, for example, since $h_r = 1$,

$$\nabla r = \frac{\partial r}{\partial x}\mathbf{i} + \frac{\partial r}{\partial y}\mathbf{j} + \frac{\partial r}{\partial z}\mathbf{k} = \frac{(x\mathbf{i} + y\mathbf{j} + z\mathbf{k})}{r} = \frac{\mathbf{r}}{r} = \frac{1}{h_r}\mathbf{r}_0 \tag{1.176}$$

For $\cos\theta = z/r$, we obtain

$$\text{(a)}\quad \frac{\partial\theta}{\partial x} = \frac{zx}{r^3\sin\theta} = \frac{\cos\theta\cos\phi}{r} \qquad \text{(b)}\quad \frac{\partial\theta}{\partial y} = \frac{zy}{r^3\sin\theta} = \frac{\cos\theta\sin\phi}{r}$$

$$\text{(c)}\quad \frac{\partial\theta}{\partial z} = \frac{z^2}{r^3\sin\theta} - \frac{1}{r\sin\theta} = -\frac{\sin\theta}{r} \tag{1.177}$$

Therefore, referring to equation 1.73b, we have

$$\nabla\theta = \frac{1}{r}(\cos\theta\cos\phi\mathbf{i} + \cos\theta\sin\theta\mathbf{j} - \sin\theta\mathbf{k}) = \frac{1}{r}\boldsymbol{\theta}_0 = \frac{1}{h_\theta}\boldsymbol{\theta}_0 \tag{1.178}$$

From $\tan\phi = y/x$,

$$\text{(a)}\quad \frac{\partial\phi}{\partial x} = -\frac{y}{x^2\sec^2\phi} = -\frac{\sin\phi}{r\sin\theta} \qquad \text{(b)}\quad \frac{\partial\phi}{\partial y} = \frac{1}{x\sec^2\phi} = \frac{\cos\phi}{r\sin\theta}$$

$$\text{(c)}\quad \frac{\partial\phi}{\partial z} = 0 \tag{1.179}$$

Therefore, referring to equation 1.73c yields

$$\nabla\phi = \frac{-\sin\phi\mathbf{i} + \cos\phi\mathbf{j}}{r\sin\theta} = \frac{1}{r\sin\theta}\boldsymbol{\phi}_0 = \frac{1}{h_\phi}\boldsymbol{\phi}_0 \tag{1.180}$$

Substituting equation 1.176b, 1.178, and 1.180 into equation 1.175, the gradient in spherical coordinates is

$$\nabla F = \frac{1}{h_r}\frac{\partial F}{\partial r}\mathbf{r}_0 + \frac{1}{h_\theta}\frac{\partial F}{\partial \theta}\boldsymbol{\theta}_0 + \frac{1}{h_\phi}\frac{\partial F}{\partial \phi}\boldsymbol{\phi}_0 \tag{1.181}$$

We note that ∇r is along $\mathbf{r}_0$, $\nabla\theta$ is along $\boldsymbol{\theta}_0$, and $\nabla\phi$ is in the direction $\boldsymbol{\phi}_0$. This is expected since it was demonstrated earlier that ∇f is along a normal to the surface defined by $f =$ constant. The surface defined by $r =$ constant is a spherical surface. Then ∇r must be along the normal to that sphere, $\mathbf{r}_0$. The surface defined by constant θ is the slanted surface of a cone, the normal to which is $\boldsymbol{\theta}_0$ (see Figure 1.33). $\nabla\theta$ should therefore be directed along $\boldsymbol{\theta}_0$. $\nabla\phi$ must be in the direction of the plane containing the z-axis, making an angle ϕ to the x-axis. $\boldsymbol{\phi}_0$ is the normal to such a plane.

As the reader will show in Problems 35 and 36, the gradient of a scalar function in cylindrical coordinates is

$$\nabla F = \frac{1}{h_\rho}\frac{\partial F}{\partial \rho}\boldsymbol{\rho}_0 + \frac{1}{h_\phi}\frac{\partial F}{\partial \phi}\boldsymbol{\phi}_0 + \frac{1}{h_z}\frac{\partial F}{\partial z}\mathbf{k} \tag{1.182}$$

Generalizing the results expressed in equations 1.181 and 1.182, the gradient can be written

$$\nabla F = \frac{1}{h_1}\frac{\partial F}{\partial u_1}\mathbf{v}_1 + \frac{1}{h_2}\frac{\partial F}{\partial u_2}\mathbf{v}_2 + \frac{1}{h_3}\frac{\partial F}{\partial u_3}\mathbf{v}_3 \tag{1.183}$$

To deduce the divergence of

$$\mathbf{G}(u_1, u_2, u_3) \equiv G_1\mathbf{v}_1 + G_2\mathbf{v}_2 + G_3\mathbf{v}_3 \tag{1.184}$$

we note from equation 1.183 that by setting $F = u_1$, then $F = u_2$, then $F = u_3$, we obtain

$$\nabla u_l = \frac{\mathbf{v}_l}{h_l} \qquad l = 1, 2, 3 \tag{1.185}$$

$$\Rightarrow \quad \nabla u_l \times \nabla u_m = \frac{1}{h_l h_m}\mathbf{v}_l \times \mathbf{v}_m = \varepsilon_{lmn}\frac{1}{h_l h_m}\mathbf{v}_n \tag{1.186}$$

where ε_{lmn}, the Levi–Civita symbol, guarantees that $l \neq m \neq n$. Thus

$$\nabla \cdot (\nabla u_l \times \nabla u_m) = \varepsilon_{lmn}\nabla \cdot \left(\frac{1}{h_l h_m}\mathbf{v}_n\right) \tag{1.187}$$

Using identity (c) of Problem 32, and the result that $\nabla \times \nabla F = 0$, we obtain

$$\nabla \cdot (\nabla u_l \times \nabla u_m) = \nabla u_m \cdot (\nabla \times \nabla u_l) - \nabla u_l \cdot (\nabla \times \nabla u_m) = 0 \tag{1.188}$$

$$\Rightarrow \quad \nabla \cdot \left(\frac{1}{h_l h_m}\mathbf{v}_n\right) = 0 \qquad l \neq m \neq n \tag{1.189}$$

Writing $\mathbf{G}$ as

$$\mathbf{G} = (h_2 h_3 G_1)\frac{\mathbf{v}_1}{h_2 h_3} + (h_3 h_1 G_2)\frac{\mathbf{v}_2}{h_3 h_1} + (h_1 h_2 G_3)\frac{\mathbf{v}_3}{h_1 h_2} \tag{1.190}$$

the divergence of $\mathbf{G}$ is

$$\nabla \cdot \mathbf{G} = \nabla \cdot \left[(h_2 h_3 G_1)\frac{\mathbf{v}_1}{h_2 h_3}\right] + \nabla \cdot \left[(h_3 h_1 G_2)\frac{\mathbf{v}_2}{h_3 h_1}\right] + \nabla \cdot \left[(h_1 h_2 G_3)\frac{\mathbf{v}_3}{h_1 h_2}\right] \tag{1.191}$$

We use identity (a) of Problem 32 to evaluate the divergence of the product of each scalar function $h_l h_m G_n$ with the vector $\mathbf{v}_n/h_l h_m$. For example,

$$\nabla \cdot \left[(h_2 h_3 G_1)\frac{\mathbf{v}_1}{h_2 h_3}\right] = (h_2 h_3 G_1)\nabla \cdot \left[\frac{\mathbf{v}_1}{h_2 h_3}\right] + \frac{\mathbf{v}_1}{h_2 h_3} \cdot \nabla(h_2 h_3 G_1) \tag{1.192a}$$

But it was shown in equation 1.189 that the first term is zero. We refer to equation 1.183 for the first component of the gradient and obtain

$$\nabla \cdot \left[(h_2 h_3 G_1)\frac{\mathbf{v}_1}{h_2 h_3}\right] = \frac{\mathbf{v}_1}{h_2 h_3} \cdot \nabla(h_2 h_3 G_1) = \frac{1}{h_1 h_2 h_3}\frac{\partial}{\partial u_1}(h_2 h_3 G_1) \tag{1.192b}$$

With analogous results for the remaining two terms, the divergence of a vector function in generalized coordinates is

$$\nabla \cdot \mathbf{G} = \frac{1}{h_1 h_2 h_3}\left[\frac{\partial}{\partial u_1}(h_2 h_3 G_1) + \frac{\partial}{\partial u_2}(h_1 h_3 G_2) + \frac{\partial}{\partial u_3}(h_1 h_2 G_3)\right] \tag{1.193}$$

From equations 1.183 and 1.193, it is straightforward to obtain

$$\nabla^2 F = \frac{1}{h_1 h_2 h_3}\left[\frac{\partial}{\partial u_1}\left(\frac{h_2 h_3}{h_1}\frac{\partial F}{\partial u_1}\right) + \frac{\partial}{\partial u_2}\left(\frac{h_3 h_1}{h_2}\frac{\partial F}{\partial u_2}\right) + \frac{\partial}{\partial u_3}\left(\frac{h_1 h_2}{h_3}\frac{\partial F}{\partial u_3}\right)\right] \tag{1.194}$$

To express the curl of $\mathbf{G}(u_1, u_2, u_3)$ in generalized coordinates, we note that

$$\nabla \times \nabla u_l = \nabla \times \left(\frac{\mathbf{v}_l}{h_l}\right) = 0 \tag{1.195}$$

Therefore, if we write

$$\mathbf{G} = h_1 G_1 \frac{\mathbf{v}_1}{h_1} + h_2 G_2 \frac{\mathbf{v}_2}{h_2} + h_3 G_3 \frac{\mathbf{v}_3}{h_3} \tag{1.196}$$

the curl of $\mathbf{G}$ becomes

$$\nabla \times \mathbf{G} = \nabla \times \left(h_1 G_1 \frac{\mathbf{v}_1}{h_1}\right) + \nabla \times \left(h_2 G_2 \frac{\mathbf{v}_2}{h_2}\right) + \nabla \times \left(h_3 G_3 \frac{\mathbf{v}_3}{h_3}\right) \tag{1.197}$$

Identity (b) of Problem 32 is used to evaluate each of these terms. For example,

$$\nabla \times \left(h_1 G_1 \frac{\mathbf{v}_1}{h_1}\right) = h_1 G_1 \nabla \times \left(\frac{\mathbf{v}_1}{h_1}\right) + \nabla(h_1 G_1) \times \left(\frac{\mathbf{v}_1}{h_1}\right) \tag{1.198}$$

As shown in equation 1.195, the first term is zero. Thus, using equation 1.183 for the appropriate components of $\nabla(h_1 G_1)$, we obtain

$$\nabla \times \left(h_1 G_1 \frac{\mathbf{v}_1}{h_1}\right) = -\frac{1}{h_1 h_2}\frac{\partial}{\partial u_2}(h_1 G_1)\mathbf{v}_3 + \frac{1}{h_1 h_3}\frac{\partial}{\partial u_3}(h_1 G_1)\mathbf{v}_2 \tag{1.199}$$

With analogous expressions for $\nabla \times [h_2 G_2(\mathbf{v}_2/h_2)]$ and $\nabla \times [h_3 G_3(\mathbf{v}_3/h_3)]$. It is then straightforward to determine that these results can be expressed compactly in the form of a determinant:

$$\nabla \times \mathbf{G} = \frac{1}{h_1 h_2 h_3}\begin{vmatrix} h_1 \mathbf{v}_1 & h_2 \mathbf{v}_2 & h_3 \mathbf{v}_3 \\ \dfrac{\partial}{\partial u_1} & \dfrac{\partial}{\partial u_2} & \dfrac{\partial}{\partial u_3} \\ h_1 G_1 & h_2 G_2 & h_3 G_3 \end{vmatrix} \tag{1.200}$$

For example, in spherical coordinates, with $h_1 = 1$, $h_2 = r$, and $h_3 = r \sin\theta$, we have

$$\nabla F = \frac{\partial F}{\partial r}\mathbf{r}_0 + \frac{1}{r}\frac{\partial F}{\partial \theta}\boldsymbol{\theta}_0 + \frac{1}{r\sin\theta}\frac{\partial F}{\partial \phi}\boldsymbol{\phi}_0 \tag{1.201a}$$

$$\nabla \cdot \mathbf{G} = \frac{1}{r^2\sin\theta}\left[\frac{\partial}{\partial r}(r^2\sin\theta G_r) + \frac{\partial}{\partial \theta}(r\sin\theta G_\theta) + \frac{\partial}{\partial \phi}(rG_\phi)\right] \tag{1.201b}$$

$$\nabla^2 F = \frac{1}{r^2}\frac{\partial}{\partial r}\left(r^2\frac{\partial F}{\partial r}\right) + \frac{1}{r^2\sin\theta}\frac{\partial}{\partial\theta}\left(\sin\theta\frac{\partial F}{\partial\theta}\right) + \frac{1}{r^2\sin^2\theta}\frac{\partial^2 F}{\partial\phi^2}$$

$$= \frac{1}{r}\frac{\partial^2}{\partial r^2}(rF) + \frac{1}{r^2\sin\theta}\frac{\partial}{\partial\theta}\left(\sin\theta\frac{\partial F}{\partial\theta}\right) + \frac{1}{r^2\sin^2\theta}\frac{\partial^2 F}{\partial\phi^2} \tag{1.201c}$$

$$\nabla \times \mathbf{G} = \frac{1}{r^2\sin\theta}\begin{vmatrix} \mathbf{r}_0 & r\boldsymbol{\theta}_0 & r\sin\theta\boldsymbol{\phi}_0 \\ \frac{\partial}{\partial r} & \frac{\partial}{\partial\theta} & \frac{\partial}{\partial\phi} \\ G_r & rG_\theta & r\sin\theta G_\phi \end{vmatrix} = \frac{1}{r\sin\theta}\left(\frac{\partial}{\partial\theta}(\sin\theta G_\phi) - \frac{\partial G_\theta}{\partial\phi}\right)\mathbf{r}_0$$

$$+ \frac{1}{r}\left(\frac{1}{\sin\theta}\frac{\partial G_r}{\partial\phi} - \frac{\partial}{\partial r}(rG_\phi)\right)\boldsymbol{\theta}_0 + \frac{1}{r}\left(\frac{\partial}{\partial r}(rG_\theta) - \frac{\partial G_r}{\partial\theta}\right)\boldsymbol{\phi}_0 \tag{1.201d}$$

The area and volume elements in a generalized coordinate frame are straightforward to obtain. Figure 1.32 is a representation of a generalized infinitesimal volume element, analogous to the infinitesimal spherical volume element in Figure 1.20. From Figure 1.32 we see that

$$dV = h_1 h_2 h_3\, du_1\, du_2\, du_3 \tag{1.202}$$

$$dS_{u_1=\text{const.}} = h_2 h_3\, du_2\, du_3 \tag{1.203}$$

with $\mathbf{v}_1$ the normal to the surface defined by u_1 = constant. The volume element in Figure 1.32 is drawn like a Cartesian volume element but, in fact, represents the infinitesimal volume element in a general, orthogonal, right-handed system.

For example, in spherical coordinates,

$$dV = 1 \cdot r \cdot r\sin\theta\, dr\, d\theta\, d\phi = r^2\, dr\, d\Omega \tag{1.204}$$

$$dS_{r=R=\text{const.}} = R \cdot R\sin\theta\, d\theta\, d\phi = R^2\, d\Omega \tag{1.205a}$$

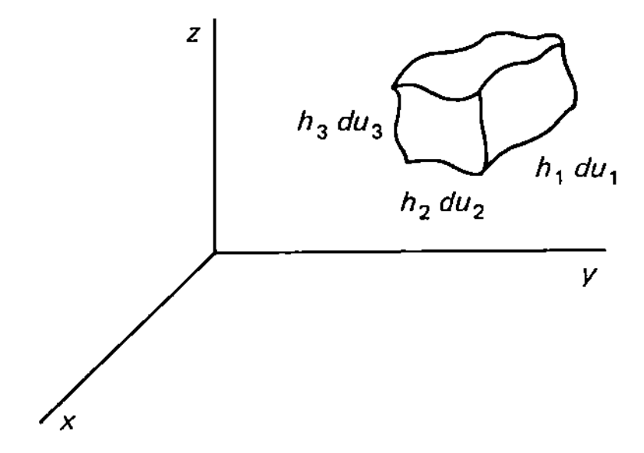

Figure 1.32 Differential volume element described in generalized curvilinear coordinates.

$\theta = \theta_1 =$ constant defines the surface of a cone (as in Figure 1.33). The area element on this surface is

$$dS_{\theta=\theta_1=\text{const.}} = h_1 h_3\, du_1\, du_3 = 1 \cdot r \sin\theta_1\, dr\, d\phi \tag{1.205b}$$

Jacobian of a transformation

The transformation of area and volume elements from one coordinate frame to another is achieved from the equations relating the coordinates of the two systems. As an illustrative example, we will determine the area and volume elements in cylindrical coordinates from the transformation equations between Cartesian and cylindrical coordinates: equations 1.81.

Referring to Figure 1.21b, the line elements in the x-y plane for constant angle ϕ and constant radius ρ are

$$\text{(a)}\quad d\mathbf{l}_1 = d\rho\, \boldsymbol{\rho}_0 \qquad \text{(b)}\quad d\mathbf{l}_2 = \rho\, d\phi\, \boldsymbol{\phi}_0 \tag{1.206}$$

Then

$$d\mathbf{S} = d\mathbf{l}_1 \times d\mathbf{l}_2 \tag{1.207}$$

In terms of the Cartesian coordinates, these differential line elements are

$$\text{(a)}\quad d\mathbf{l}_1 = dx\,\mathbf{i} + dy\,\mathbf{j} \qquad \text{(b)}\quad d\mathbf{l}_2 = dx\,\mathbf{i} + dy\,\mathbf{j} \tag{1.208}$$

($dz = 0$ for line elements that lie in the x-y plane.) Referring to equation 1.206a, $\phi =$ constant along $d\mathbf{l}_1$. Thus

$$\text{(a)}\quad dx = \frac{\partial x}{\partial \rho}\, d\rho \qquad \text{(b)}\quad dy = \frac{\partial y}{\partial \rho}\, d\rho \tag{1.209}$$

Similarly, referring to equation 1.206b, $\rho =$ constant along $d\mathbf{l}_2$. Thus

$$\text{(c)}\quad dx = \frac{\partial x}{\partial \phi}\, d\phi \qquad \text{(d)}\quad dy = \frac{\partial y}{\partial \phi}\, d\phi \tag{1.209}$$

Therefore,

$$\text{(a)}\quad d\mathbf{l}_1 = \left(\mathbf{i}\frac{\partial x}{\partial \rho} + \mathbf{j}\frac{\partial y}{\partial \rho}\right) d\rho \qquad \text{(b)}\quad d\mathbf{l}_2 = \left(\mathbf{i}\frac{\partial x}{\partial \phi} + \mathbf{j}\frac{\partial y}{\partial \phi}\right) d\phi \tag{1.210}$$

Thus

$$d\mathbf{S} = d\mathbf{l}_1 \times d\mathbf{l}_2 = \left(\frac{\partial x}{\partial \rho}\frac{\partial y}{\partial \phi} - \frac{\partial x}{\partial \phi}\frac{\partial y}{\partial \rho}\right)\mathbf{k}\, d\rho\, d\phi \tag{1.211}$$

As can be seen by comparison with equation 1.12a, equation 1.211 can be written as the 2×2 determinant

$$d\mathbf{S}_{xy} = \mathbf{k}\begin{vmatrix} \dfrac{\partial x}{\partial \rho} & \dfrac{\partial y}{\partial \rho} \\ \dfrac{\partial x}{\partial \phi} & \dfrac{\partial v}{\partial \phi} \end{vmatrix} d\rho\, d\phi \tag{1.212a}$$

This determinant for the transformation from Cartesian to cylindrical coordinates is called the Jacobian determinant for the transformation of the area element $d\mathbf{S}_{xy}$.

The result can also be obtained from the determinental description of the cross product of equation 1.13.

$$d\mathbf{S}_{xy} = d\mathbf{l}_1 \times d\mathbf{l}_2 = \begin{vmatrix} \mathbf{i} & \mathbf{j} & \mathbf{k} \\ \dfrac{\partial x}{\partial \rho} & \dfrac{\partial y}{\partial \rho} & 0 \\ \dfrac{\partial x}{\partial \phi} & \dfrac{\partial y}{\partial \phi} & 0 \end{vmatrix} d\rho\, d\phi \tag{1.212b}$$

Using the expansion of a 3×3 determinant given in equation 1.12b, it is straightforward to demonstrate that this is the same as the 2×2 determinant of equation 1.212a.

We can compute the area element along the curved side of a cylinder by specifying the line elements for $\rho = P =$ constant. Referring to Figure 1.21a, we have

$$d\mathbf{l}_2 = dx\,\mathbf{i} + dy\,\mathbf{j} \tag{1.208b}$$

with the variation of x and y arising from the variation of ϕ. Therefore, as discussed above,

$$\text{(a)}\quad dx = \frac{\partial x}{\partial \phi}\, d\phi \qquad \text{(b)}\quad dy = \frac{\partial y}{\partial \phi}\, d\phi \tag{1.209}$$

Thus, with $d\mathbf{l}_3 = dz\,\mathbf{k}$,

$$d\mathbf{S}_{\rho=P} = d\mathbf{l}_2 \times d\mathbf{l}_3 = \left(\frac{\partial x}{\partial \phi}\mathbf{i} + \frac{\partial y}{\partial \phi}\mathbf{j}\right) \times \mathbf{k}\, d\phi\, dz = \left(-\frac{\partial x}{\partial \phi}\mathbf{j} + \frac{\partial y}{\partial \phi}\mathbf{i}\right) d\phi\, dz \tag{1.213}$$

Using the determinant expression for the cross product, with $\partial z/\partial \phi = 0$, this can be written as

$$d\mathbf{S}_{\rho=P} = \begin{vmatrix} \mathbf{i} & \mathbf{j} & \mathbf{k} \\ \dfrac{\partial x}{\partial \phi} & \dfrac{\partial y}{\partial \phi} & 0 \\ \dfrac{\partial x}{\partial z} & \dfrac{\partial y}{\partial z} & \dfrac{\partial z}{\partial z} \end{vmatrix} d\phi\, dz = \begin{vmatrix} 1 & \mathbf{j} & \mathbf{k} \\ \dfrac{\partial x}{\partial \phi} & \dfrac{\partial y}{\partial \phi} & 0 \\ 0 & 0 & 1 \end{vmatrix} = \begin{vmatrix} \mathbf{i} & \mathbf{j} \\ \dfrac{\partial x}{\partial \phi} & \dfrac{\partial y}{\partial \phi} \end{vmatrix} \tag{1.214}$$

This area element has a normal that changes direction from point to point. Therefore, unlike the determinant of equation 1.212b, it cannot be written as single Cartesian basis vector multiplying a scalar 2×2 determinant containing partial derivatives. Usually, when one refers to a Jacobian determinant, one is talking about a scalar determinant. We see by this example that the Jacobian determinant for an area element is not necessarily a scalar.

As an example of the transformation of the volume element, we will consider the transformation between Cartesian and spherical coordinates. From equation 1.97,

$$dV = d\mathbf{l}_1 \cdot (d\mathbf{l}_2 \times d\mathbf{l}_3) \tag{1.97a}$$

Referring to Figure 1.20, the line element along the radial direction expressed in the two coordinate systems is

$$d\mathbf{l}_1 = dr\,\mathbf{r}_0 = dx\,\mathbf{i} + dy\,\mathbf{j} + dz\,\mathbf{k} \tag{1.215}$$

In the radial direction, θ and ϕ are constant. Therefore, x, y, and z vary only with respect to r. Thus, along $d\mathbf{l}_1$,

$$\text{(a)} \quad dx = \frac{\partial x}{\partial r}\,dr \qquad \text{(b)} \quad dy = \frac{\partial y}{\partial r}\,dr \qquad \text{(c)} \quad dz = \frac{\partial z}{\partial r}\,dr \tag{1.216}$$

Along $d\mathbf{l}_2$, r and ϕ are constant, so

$$d\mathbf{l}_2 = r\,d\theta\,\boldsymbol{\theta}_0 = dx\,\mathbf{i} + dy\,\mathbf{j} + dz\,\mathbf{k} \tag{1.217}$$

Therefore, since only θ is variable,

$$\text{(a)} \quad dx = \frac{\partial x}{\partial \theta}\,d\theta \qquad \text{(b)} \quad dy = \frac{\partial y}{\partial \theta}\,d\theta \qquad \text{(c)} \quad dz = \frac{\partial z}{\partial \theta}\,d\theta \tag{1.218}$$

Similarly, along $d\mathbf{l}_3$, where r and θ are constant,

$$d\mathbf{l}_3 = r\sin\theta\,d\phi\,\boldsymbol{\phi}_0 = dx\,\mathbf{i} + dy\,\mathbf{j} + dz\,\mathbf{k} \tag{1.219}$$

with

$$\text{(a)} \quad dx = \frac{\partial x}{\partial \phi}\,d\phi \qquad \text{(b)} \quad dy = \frac{\partial y}{\partial \phi}\,d\phi \qquad \text{(c)} \quad dz = \frac{\partial z}{\partial \phi}\,d\phi \tag{1.220}$$

Therefore,

$$\text{(a)} \quad d\mathbf{l}_1 = \left(\frac{\partial x}{\partial r}\mathbf{i} + \frac{\partial y}{\partial r}\mathbf{j} + \frac{\partial z}{\partial r}\mathbf{k}\right) dr \qquad \text{(b)} \quad d\mathbf{l}_2 = \left(\frac{\partial x}{\partial \theta}\,\mathbf{i} + \frac{\partial y}{\partial \theta}\,\mathbf{j} + \frac{\partial z}{\partial \theta}\,\mathbf{k}\right) d\theta$$

$$\text{(c)} \quad d\mathbf{l}_3 = \left(\frac{\partial x}{\partial \phi}\mathbf{i} + \frac{\partial y}{\partial \phi}\mathbf{j} + \frac{\partial z}{\partial \phi}\mathbf{k}\right) d\phi \tag{1.221}$$

Using the determinant description of the scalar triple product of equation 1.26, we obtain

$$dV = d\mathbf{l}_1 \cdot (d\mathbf{l}_2 \times d\mathbf{l}_3) = \begin{vmatrix} \dfrac{\partial x}{\partial r} & \dfrac{\partial y}{\partial r} & \dfrac{\partial z}{\partial r} \\ \dfrac{\partial x}{\partial \theta} & \dfrac{\partial y}{\partial \theta} & \dfrac{\partial z}{\partial \theta} \\ \dfrac{\partial x}{\partial \phi} & \dfrac{\partial y}{\partial \phi} & \dfrac{\partial z}{\partial \phi} \end{vmatrix} dr\,d\theta\,d\phi \tag{1.222}$$

This determinant is the Jacobian determinant for the transformation of the volume element from Cartesian to spherical coordinates.

With these examples as guides, we can generalize the transformation of area and volume elements from Cartesian to any curvilinear coordinate system.

Referring to Figure 1.32, the line elements in generalized coordinates are

$$\text{(a)} \quad d\mathbf{l}_1 = h_1\,du_1\,\mathbf{v}_1 \qquad \text{(b)} \quad d\mathbf{l}_2 = h_2\,du_2\,\mathbf{v}_2 \qquad \text{(c)} \quad d\mathbf{l}_3 = h_3\,du_3\,\mathbf{v}_3 \tag{1.223}$$

Along $d\mathbf{l}_1$, u_2, and u_3 are constant. Therefore, since x, y, and z vary only with u_1,

$$d\mathbf{l}_1 = dx\,\mathbf{i} + dy\,\mathbf{j} + dz\,\mathbf{k} = \left(\frac{\partial x}{\partial u_1}\mathbf{i} + \frac{\partial y}{\partial u_1}\mathbf{j} + \frac{\partial z}{\partial u_1}\mathbf{k}\right) du_1 \tag{1.224a}$$

By identical argument, only u_2 varies along $d\mathbf{l}_2$ and along $d\mathbf{l}_3$, only u_3 is variable. Thus

$$\text{(b)} \quad d\mathbf{l}_2 = \left(\frac{\partial x}{\partial u_2}\mathbf{i} + \frac{\partial y}{\partial u_2}\mathbf{j} + \frac{\partial z}{\partial u_2}\mathbf{k}\right) du_2 \qquad \text{(c)} \quad d\mathbf{l}_3 = \left(\frac{\partial x}{\partial u_3}\mathbf{i} + \frac{\partial y}{\partial u_3}\mathbf{j} + \frac{\partial z}{\partial u_3}\mathbf{k}\right) du_3 \tag{1.224}$$

Therefore, for example,

$$d\mathbf{S}_{u_1=\text{const.}} = d\mathbf{l}_2 \times d\mathbf{l}_3 = \begin{vmatrix} \mathbf{i} & \mathbf{j} & \mathbf{k} \\ \dfrac{\partial x}{\partial u_2} & \dfrac{\partial y}{\partial u_2} & \dfrac{\partial z}{\partial u_3} \\ \dfrac{\partial x}{\partial u_3} & \dfrac{\partial y}{\partial u_3} & \dfrac{\partial z}{\partial u_3} \end{vmatrix} du_2\, du_3 \tag{1.225}$$

and from equations 1.26,

$$dV = d\mathbf{l}_1 \cdot d\mathbf{l}_2 \times d\mathbf{l}_3 = \begin{vmatrix} \dfrac{\partial x}{\partial u_1} & \dfrac{\partial y}{\partial u_1} & \dfrac{\partial z}{\partial u_1} \\ \dfrac{\partial x}{\partial u_2} & \dfrac{\partial y}{\partial u_2} & \dfrac{\partial z}{\partial u_2} \\ \dfrac{\partial x}{\partial u_3} & \dfrac{\partial y}{\partial u_3} & \dfrac{\partial z}{\partial u_3} \end{vmatrix} du_1, du_2, du_3 \tag{1.226}$$

The determinants in equations 1.225 and 1.226 are the Jacobians of the transformations from Cartesian to any generalized coordinate system for an area element and volume element.

The Jacobian of a transformation from one generalized coordinate system to another is described in terms of the scaling functions h_i, and the derivatives of the transformation equations. Let a transformation be made from coordinates (u_1, u_2, u_3) with scaling functions h_1, h_2, h_3 to a system specified by coordinates (t_1, t_2, t_3). On the surface defined by t_3 = constant, the Jacobian for the transformation of the area element dS_{12} is

$$J_{12} = \begin{vmatrix} \mathbf{v}_1 & \mathbf{v}_2 & \mathbf{v}_3 \\ h_1 \dfrac{\partial u_1}{\partial t_1} & h_2 \dfrac{\partial u_2}{\partial t_1} & h_3 \dfrac{\partial u_3}{\partial t_1} \\ h_1 \dfrac{\partial u_1}{\partial t_2} & h_2 \dfrac{\partial u_2}{\partial t_2} & h_3 \dfrac{\partial u_3}{\partial t_2} \end{vmatrix} \tag{1.227}$$

The Jacobian for the transformation of the volume element is

$$J = \begin{vmatrix} h_1 \dfrac{\partial u_1}{\partial t_1} & h_2 \dfrac{\partial u_2}{\partial t_1} & h_3 \dfrac{\partial u_3}{\partial t_1} \\ h_1 \dfrac{\partial u_1}{\partial t_2} & h_2 \dfrac{\partial u_2}{\partial t_2} & h_3 \dfrac{\partial u_3}{\partial t_2} \\ h_1 \dfrac{\partial u_1}{\partial t_3} & h_2 \dfrac{\partial u_2}{\partial t_3} & h_3 \dfrac{\partial u_3}{\partial t_3} \end{vmatrix} \tag{1.228}$$

These expressions are obtained from the determinental description of the cross product, and the scalar triple product, using the same analysis that yields the Jacobians for transformations from Cartesian to a non-Cartesian coordinate system.

Example 1.8

As an example, we will evaluate the Jacobian of the transformation of the volume element from spherical to cylindrical coordinates. With

$$\text{(a)}\quad h_r = 1 \qquad \text{(b)}\quad h_\theta = r = \sqrt{\rho^2 + z^2} \qquad \text{(c)}\quad h_\phi = r \sin\theta = \rho \tag{1.229}$$

the Jacobian of the transformation of the volume element from spherical to cylindrical coordinates is

$$J = \begin{vmatrix} \dfrac{\partial r}{\partial \rho} & r\dfrac{\partial \theta}{\partial \rho} & r\sin\theta\dfrac{\partial \phi}{\partial \rho} \\ \dfrac{\partial r}{\partial \phi} & r\dfrac{\partial \theta}{\partial \phi} & r\sin\theta\dfrac{\partial \phi}{\partial \phi} \\ \dfrac{\partial r}{\partial z} & r\dfrac{\partial \theta}{\partial z} & r\sin\theta\dfrac{\partial \phi}{\partial z} \end{vmatrix} = \begin{vmatrix} \dfrac{\rho}{r} & \dfrac{z}{r} & 0 \\ 0 & 0 & \rho \\ \dfrac{z}{r} & -\dfrac{\rho}{r} & 0 \end{vmatrix} = \rho \tag{1.230}$$

Therefore, as expected,

$$dV = \rho\, d\rho\, d\phi\, dz \tag{1.231}$$

PROBLEMS

1. Show that 3×3 determinant multiplication can be achieved by copying the first and second columns after column 3, and multiplying along diagonals as described in the text, until all columns are exhausted.

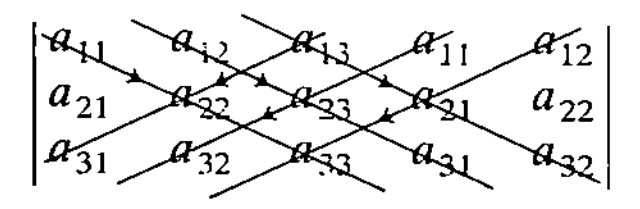

2. Use the definition of the cross product in terms of the Levi–Civita symbol to prove that $\mathbf{A} \times \mathbf{A} = 0$ for any vector $\mathbf{A}$.
3. (a) Prove that $\beta = -1$ in the identity
   $$\sum_{n}^{3} \varepsilon_{lrn}\varepsilon_{nst} = \alpha\delta_{ls}\delta_{rt} + \beta\delta_{lt}\delta_{rs}$$
   (b) From this result and the proof that $\alpha = 1$ (equation 1.35) prove that $\sum_{klm} \varepsilon_{jlk}\varepsilon_{klm}\varepsilon_{mnp} = 2\varepsilon_{jpn}$
4. Prove that $\sum_{rst} \delta_{lt}\delta_{rs} A B_r B_s C_t = C_l(\mathbf{A} \cdot \mathbf{B})$
5. Obtain the reduction for $(\mathbf{A} \times \mathbf{B}) \times \mathbf{C}$ analogous to the reduction
   $$\mathbf{A} \times (\mathbf{B} \times \mathbf{C}) = \mathbf{B}(\mathbf{A} \cdot \mathbf{C}) - \mathbf{C}(\mathbf{A} \cdot \mathbf{B})$$
6. (a) Using the description of the cross product in terms of the Levi–Civita symbol, show that
   $$(\mathbf{A} \times \mathbf{B}) \times (\mathbf{C} \times \mathbf{D}) = \mathbf{C}(\mathbf{A} \times \mathbf{B} \cdot \mathbf{D}) - \mathbf{D}(\mathbf{A} \times \mathbf{B} \cdot \mathbf{C})$$
   (b) Is $(\mathbf{A} \times \mathbf{B}) \times (\mathbf{C} \times \mathbf{D})$ a true vector or a pseudovector if:
   (1) $\mathbf{A}$ and $\mathbf{C}$ are real vectors, $\mathbf{B}$ and $\mathbf{D}$ are pseudovectors?
   (2) $\mathbf{A}$, $\mathbf{B}$, and $\mathbf{C}$ are real vectors, $\mathbf{D}$ is a pseudovector?
   (3) All four vectors are real vectors?
7. Show that
   $$(\mathbf{A} \times \mathbf{B}) \cdot (\mathbf{C} \times \mathbf{D}) = (\mathbf{A} \cdot \mathbf{C})(\mathbf{B} \cdot \mathbf{D}) - (\mathbf{A} \cdot \mathbf{D})(\mathbf{B} \cdot \mathbf{C})$$
   (a) Using equation 1.41b for the vector product of three vectors.
   (b) Using the Levi–Civita description of the cross product.
   (c) Is $(\mathbf{A} \times \mathbf{B}) \cdot (\mathbf{C} \times \mathbf{D})$ a true scalar or a pseudoscalar if
   (1) $\mathbf{A}$ and $\mathbf{C}$ are real vectors, $\mathbf{B}$ and $\mathbf{D}$ are pseudovectors?
   (2) $\mathbf{A}$, $\mathbf{B}$, and $\mathbf{C}$ are real vectors, $\mathbf{D}$ is a pseudovector?
   (3) All four vectors are pseudovectors?
8. Prove the cyclic identity
   $$\mathbf{A} \times (\mathbf{B} \times \mathbf{C}) + \mathbf{B} \times (\mathbf{C} \times \mathbf{A}) + \mathbf{C} \times (\mathbf{A} \times \mathbf{B}) = 0$$
   (a) Using equation 1.41b.
   (b) Using the description of the cross product in terms of the Levi–Civita three-index symbol.
9. The vector $\mathbf{D}$ is defined in terms of the vectors $\mathbf{A}$ and $\mathbf{B}$ as
   $$\mathbf{D} = \alpha\mathbf{A} + \beta\mathbf{B}$$
   The angle between $\mathbf{A}$ and $\mathbf{D}$ is θ, and $\mathbf{A}$ and $\mathbf{B}$ are orthogonal.
   (a) Find α and β in terms of the angle θ and the lengths of $\mathbf{A}$, $\mathbf{B}$, and $\mathbf{D}$.
   (b) Use this result to deduce the trigonometric form of the Pythagorean theorem.
10. A plane is defined by the points $(2, 1, 0)$, $(1, 2, -3)$, and $(5, 0, 2)$. Find the equation of this plane and its perpendicular distance from the origin.
11. A plane is specified by the equation $2x - y + 5z - 4 = 0$. Find the perpendicular distance of the plane from the origin.
12. A plane contains the point $(1, 1, 1)$ and is specified by the equation $x + 4y + 3z = D$. Find the perpendicular distance from the origin to this plane.
13. A line contains the points $(1, 1, 1)$ and $(1, 2, 3)$. Find the equations that specify this line and the perpendicular distance from the origin to the line.

14. Two parallel planes are separated by a distance of 5 units. The plane that is more distant from the origin contains the points $(5, 5, 0)$, $(0, 5, 5)$, and $(5, 0, 5)$. Find the equation of each plane and the perpendicular distance from the origin to each plane.

15. In three-dimensional space, there are eight octants. They are defined by:
(a) $x, y, z \geqslant 0$ (b) $x \leqslant 0,\ y, z \geqslant 0$
(c) $x, y \leqslant 0,\ z \geqslant 0$ (d) $x, z \leqslant 0,\ y \geqslant 0$
(e) $x, y, z \leqslant 0$ (f) $x \geqslant 0,\ y, z \leqslant 0$
(g) $x, y \geqslant 0,\ z \leqslant 0$ (h) $x, z \geqslant 0,\ y \leqslant 0$
For each octant, identify the range of the spherical angles θ and ϕ.

16. For the vector function $\mathbf{G} = r^2\mathbf{r}$, evaluate $\int \cos\theta \mathbf{G}\, dS$ over the surface of a sphere of radius R. (Note that this integral is *not* $\int \mathbf{G} \cdot \mathbf{n}\, dS$, which is a scalar. This integral is a vector.)

17. The vector field $\mathbf{G}$ is given by $\mathbf{G} = y\mathbf{i} - x\mathbf{j} + z\mathbf{k}$. Identify the spherical components of $\mathbf{G}$; that is, G_r, G_θ, and G_ϕ.

18. Use equations 1.73 to show that

$$d\boldsymbol{\theta}_0 = -\mathbf{r}_0\, d\theta + \boldsymbol{\phi}_0 \cos\theta\, d\phi$$

$$d\boldsymbol{\phi} = -(\mathbf{r}_0 \sin\theta + \boldsymbol{\theta}_0 \cos\theta)\, d\phi$$

19. Find the area element $dS_{yz} = dy\, dz$ in spherical and cylindrical coordinates.

20. A sphere of radius 2 meters has a 30° conic section removed from it (Figure 1.33). Find the area of the remaining spherical surface (do not include the inside surface of the conic "hole") and the volume of the remaining solid.

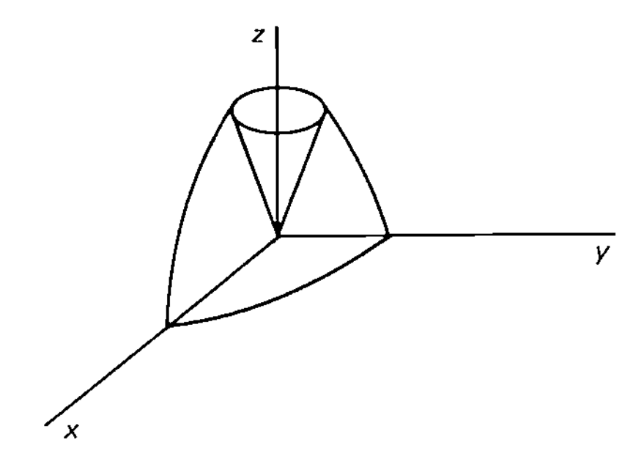

Figure 1.33 Sphere with a conic section removed.

21. A particle's position is described in spherical coordinates. The particle is moving in a circle of radius R. The circle is in a plane parallel to the x-y plane. That is, the particle moves along the rim of the circle on the top of the cone of Figure 1.33. Therefore, the particle's position at any instant is defined by $\theta = \theta_0 =$ constant, $r \sin\theta_0 = R$, $\phi = \phi(t)$. By differentiating $\mathbf{r} = r\mathbf{r}_0$:
(a) Prove that the velocity of the particle is tangent to the circle and that its speed is given by $|\mathbf{v}| = R\omega$.
(b) If $\theta = \pi/2$, show that the particle's acceleration is entirely centripetal and has a magnitude $|\mathbf{a}| = v^2/R = R\omega^2$.

22. A vector field $\mathbf{G}$ is given by $\mathbf{G} = (\mathbf{r} + r\boldsymbol{\theta}_0)F(r)$. ($r = |\mathbf{r}|$, and $\mathbf{r}$ is the radius vector in spherical coordinates. $\boldsymbol{\theta}_0$ is the unit θ vector). Determine the function $F(r)$, if any, that will make $\mathbf{G}$ irrotational.

23. An ellipsoid of revolution is defined by the equation

$$\frac{x^2}{a^2} + \frac{y^2}{b^2} + \frac{z^2}{c^2} = R^2$$

where a, b, and c are constants. Analogous to a spherical surface, the surface of the ellipsoid is defined by $R =$ constant. Find the unit normal to this surface at the point $(1, 4, 4)$ for $c = 2b = 4a$.

24. For the vector field $\mathbf{G}$ of Problem 17, find the radial component of $\nabla \times \mathbf{G}$ expressed in spherical coordinates, that is, $(\nabla \times \mathbf{G})_r$.

25. Prove the validity of the divergence theorem for the vector function

$$\mathbf{G} = yze^x\mathbf{i} + xze^y\mathbf{j} + xye^z\mathbf{k}$$

through the volume of a cube of side 2, with one corner at the origin (Figure 1.34).

26. Prove the validity of Stokes's theorem for the vector field of Problem 25 over the face of the cube in Figure 1.34 defined by $x = 2$.

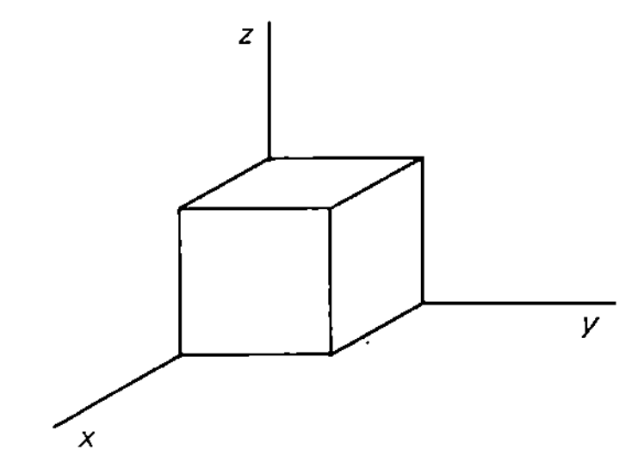

Figure 1.34 Cube of side 2.

27. Prove that $\oint \mathbf{r}F(r) \cdot d\mathbf{l} = 0$ around any arbitrary closed loop for any scalar function $F(r)$.

28. Use the description of the cross product in terms of the Levi–Civita symbol to prove that:
(a) $\nabla \times \nabla F = 0$ for arbitrary F.
(b) $\nabla \cdot \nabla \times \mathbf{G} = 0$ for arbitrary $\mathbf{G}$.

29. If V is the volume of an arbitrary closed region of space and $\mathbf{r}$ is the position vector $\mathbf{r} = x\mathbf{i} + y\mathbf{j} + z\mathbf{k}$, prove that $\int \mathbf{r} \cdot \mathbf{n}\, dS = 3V$

30. Prove that:

(a) $\int \nabla \times \mathbf{G} \cdot \mathbf{n}\, dS = 0$ over the entire surface of an arbitrary closed object.

(b) $\int \nabla F \cdot d\mathbf{l} = 0$ around an arbitrary closed loop enclosing an area that has no holes in it.

31. Show that in spherical coordinates, the operator $\mathbf{r} \times \nabla$ is

$$\mathbf{r} \times \nabla = \boldsymbol{\phi}_0 \frac{\partial}{\partial \theta} - \boldsymbol{\theta}_0 \frac{1}{\sin\theta} \frac{\partial}{\partial \phi}$$

(*Hint*: Consider $\mathbf{r} \times \nabla \Psi$ in spherical coordinates for an arbitrary scalar function Ψ.)

32. Prove the following identities:

(a) $\nabla \cdot (F\mathbf{G}) = \mathbf{G} \cdot \nabla F + F\nabla \cdot \mathbf{G}$

(b) $\nabla \times (F\mathbf{G}) = F\nabla \times \mathbf{G} + (\nabla F) \times \mathbf{G}$

(c) $\nabla \cdot (\mathbf{G} \times \mathbf{H}) = \mathbf{H} \cdot (\nabla \times \mathbf{G}) - \mathbf{G} \cdot (\nabla \times \mathbf{H})$

(d) $\nabla(\mathbf{G} \cdot \mathbf{H}) = \mathbf{G} \times (\nabla \times \mathbf{H}) + \mathbf{H} \times (\nabla \times \mathbf{G}) + (\mathbf{G} \cdot \nabla)\mathbf{H} + (\mathbf{H} \cdot \nabla)\mathbf{G}$
where, for example

$$(\mathbf{G} \cdot \nabla)\mathbf{H} = G_x \frac{\partial \mathbf{H}}{\partial x} + G_y \frac{\partial \mathbf{H}}{\partial y} + G_z \frac{\partial \mathbf{H}}{\partial z}$$

(e) $\nabla \times (\mathbf{G} \times \mathbf{H}) = (\mathbf{H} \cdot \nabla)\mathbf{G} - (\mathbf{G} \cdot \nabla)\mathbf{H} + \mathbf{G}(\nabla \cdot \mathbf{H}) - \mathbf{H}(\nabla \cdot \mathbf{G})$

33. $\mathbf{A}$ is the magnetic vector potential and $\mathbf{B}$ is the magnetic field obtained from it. If the electric field is static (does not depend on time) and if there are no free currents, show that

$$\nabla \cdot (\mathbf{A} \times \mathbf{B}) = \mathbf{B} \cdot \mathbf{B} = B^2$$

34. In a source-free space ($\rho = 0$, $\mathbf{J} = 0$), show that the electric and magnetic fields satisfy

$$\nabla^2 \mathbf{E} - \frac{1}{c^2} \frac{\partial^2 \mathbf{E}}{\partial t^2} = 0 \quad \text{and} \quad \nabla^2 \mathbf{B} - \frac{1}{c^2} \frac{\partial^2 \mathbf{B}}{\partial t^2} = 0$$

These equations form the basis for the electromagnetic theory of light.

35. Identify the generalized functions h_ρ, h_ϕ, and h_z of cylindrical coordinates.

36. Find ∇F, $\nabla \cdot \mathbf{G}$, $\nabla^2 F$, and $\nabla \times \mathbf{G}$ in cylindrical coordinates.

37. Find the Jacobian for the transformation of the volume element from Cartesian to cylindrical coordinates.

38. Parabolic coordinates are defined in terms of spherical and Cartesian coordinates as follows:

$$\eta = r + r\cos\theta = \sqrt{x^2 + y^2 + z^2} + z,$$

$$\xi = r - r\cos\theta = \sqrt{x^2 + y^2 + z^2} - z, \quad \text{and}$$

$$\phi = \phi = \tan^{-1}(y/x).$$

This represents a parabola rotated about the z-axis (Figure 1.35). Find the Jacobian of the transformation for the volume element:

(a) From Cartesian to parabolic coordinates (b) From spherical to parabolic coordinates

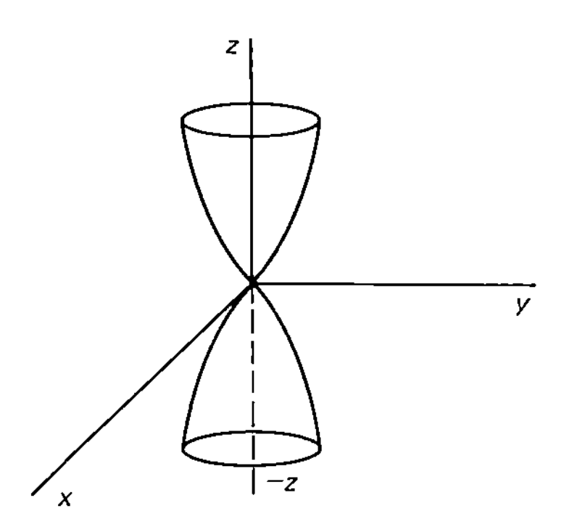

Figure 1.35 Paraboloid of revolution.

CHAPTER 2

COMPLEX ANALYSIS

One of the most useful branches of mathematics for solving problems in the applied sciences is that of complex variables. Among the many problems that are amenable to solution by complex analysis are the evaluation of many types of integrals, and the solution to Laplace's equation (see equation 1.161a with $\rho = 0$) for certain geometric configurations.

2.1 Complex Numbers

A *complex number* is a prescription for locating a point in the two-dimensional plane. That is, instead of specifying the position of a point in the x-y plane by the vector $\mathbf{z} = x\mathbf{i} + y\mathbf{j}$, a point in the complex plane is defined by the complex number

$$z = x + iy \tag{2.1}$$

with $i^2 = -1$.

Rather than labeling the axes in terms of the x and y basis vectors, the x and y axes are renamed the real and imaginary axes as shown in Figure 2.1, and a complex number is defined by the addition of a real part (x in equation 2.1) and i multiplied by an imaginary part (y in equation 2.1). That is, the imaginary part of z, denoted by Im z, is y, not iy.

The *complex conjugate* of a complex number, denoted by z^*, is defined by replacing i by $-i$ everywhere in the complex number. For example,

$$\text{(a)} \quad z = i^i \quad \Rightarrow \quad \text{(b)} \quad z^* = (-i)^{(-i)} \tag{2.2}$$

Referring to Figure 2.1, it is evident from the Pythagorean theorem that

$$|z| = \sqrt{x^2 + y^2} \tag{2.3a}$$

which is the length of the line from the origin to the point. $|z|$ is called the length, the magnitude, or the modulus of the complex number. It is a simple arithmetic operation to show that

$$|z| = \sqrt{zz^*} \tag{2.3b}$$

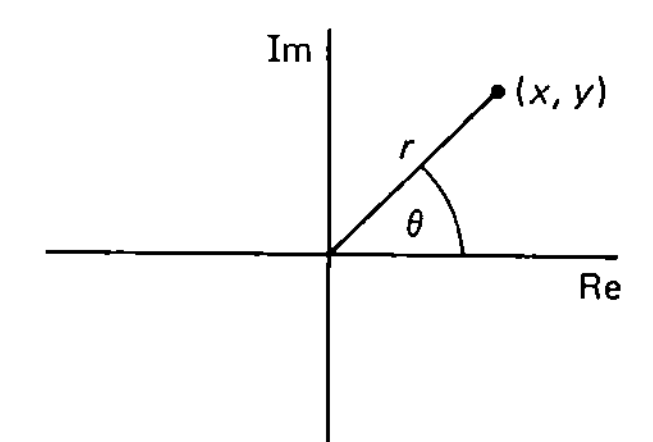

Figure 2.1 Location of a point in the complex plane.

Complex numbers can be represented in polar coordinates as well as Cartesian coordinates. Writing $x = r\cos\theta$ and $y = r\sin\theta$, the *trigonometric form* of $z = x + iy$ is

$$z = r(\cos\theta + i\sin\theta) \tag{2.4}$$

As seen in Figure 2.1, θ is defined as the angle between the x-axis and the line drawn from the origin to the point. r is the length of that line (i.e., $r = |z|$). Henceforth, when appropriate to the discussion, we will also use this line to represent the corresponding complex number pictorially. From equation 2.1 or 2.4, we see that

$$zz^* = r^2 \tag{2.5}$$

With r = constant, the differential of equation 2.4 is

$$dz = r(-\sin\theta + i\cos\theta)\,d\theta = r(i^2\sin\theta + i\cos\theta)\,d\theta = ir(\cos\theta + i\sin\theta)\,d\theta = iz\,d\theta \tag{2.6}$$

Integrating this leads straightforwardly to the *expontential form* of a complex number,

$$z = re^{i\theta} \tag{2.7}$$

$$\Rightarrow \quad z^* = re^{-i\theta} \tag{2.8}$$

Then, again

$$zz^* = r^2 \tag{2.5}$$

By setting $r = 1$, we obtain the identity

$$\cos\theta + i\sin\theta = e^{i\theta} \tag{2.9}$$

Example 2.1

For example,

$$i^i = e^{i\log i} \tag{2.10}$$

We can express i in trigonometric and exponential form by writing

$$i = \cos\left(\frac{\pi}{2}\right) + i\sin\left(\frac{\pi}{2}\right) = e^{i(\pi/2)} \tag{2.11}$$

$$\Rightarrow \quad e^{i\log i} = e^{i\log[e^{i(\pi/2)}]} = e^{i[i(\pi/2)]} = e^{-\pi/2} \tag{2.12}$$

which is real. □

Just as a vector is zero only if each of its components is zero, a complex number can be zero only if its real part is zero and separately its imaginary part is zero. If two complex numbers are equal, then

$$z_1 - z_2 = (x_1 - x_2) + i(y_1 - y_2) = 0 \tag{2.13}$$

Since the real and imaginary parts of equation 2.13 are separately zero,

$$\text{(a)} \quad x_1 = x_2 \qquad \text{(b)} \quad y_1 = y_2 \tag{2.14}$$

That is, if two complex numbers are equal, their real parts are equal, and separately, their imaginary parts are equal.

Using this, we can derive *deMoivre's theorem*. Starting with

$$e^{in\theta} = \left(e^{i\theta}\right)^n \tag{2.15}$$

we can use the identity expressed in equation 2.9 to write

$$\cos(n\theta) + i\sin(n\theta) = (\cos\theta + i\sin\theta)^n \tag{2.16}$$

Expanding the right-hand side of equation 2.16 by the binomial expansion, and separately equating real and imaginary parts, leads to identities for $\cos(n\theta)$ and $\sin(n\theta)$.

Example 2.2

For example,

$$\cos(2\theta) + i\sin(2\theta) = \cos^2\theta + 2i\cos\theta\sin\theta - \sin^2\theta \tag{2.17}$$

$$\Rightarrow \quad \text{(a)} \quad \cos(2\theta) = \cos^2\theta - \sin^2\theta \qquad \text{(b)} \quad \sin(2\theta) = 2\sin\theta\cos\theta \quad \square \tag{2.18}$$

The hyperbolic functions are defined in terms of the sines, cosines, and other trigonometric functions of imaginary angles. Using equation 2.9 gives

$$\text{(a)} \quad \cos\theta = \frac{\left(e^{i\theta} + e^{-i\theta}\right)}{2} \qquad \text{(b)} \quad \sin\theta = \frac{\left(e^{i\theta} - e^{-i\theta}\right)}{2i} \tag{2.19}$$

If $\theta = iw$, with w real,

$$\cos(iw) = \frac{\left(e^{i(iw)} + e^{-i(iw)}\right)}{2} = \frac{\left(e^{w} + e^{-w}\right)}{2} \equiv \cosh(w) \tag{2.20a}$$

$$\sin(iw) = \frac{\left(e^{i(iw)} - e^{-i(iw)}\right)}{2i} = i\frac{\left(e^{w} - e^{-w}\right)}{2} \equiv i\sinh(w) \tag{2.20b}$$

Thus $\cosh(w)$ and $\sinh(w)$ are real.

From $\sin^2(iw) + \cos^2(iw) = 1$, we obtain the Pythagorean theorem in terms of the hyperbolic cosh and sinh functions:

$$\cosh^2 w - \sinh^2 w = 1 \tag{2.21}$$

Other hyperbolic functions are defined in analogy with the trigonometric functions. For example,

$$\tanh w = \frac{\sinh(w)}{\cosh(w)} \tag{2.22}$$

Complex numbers and AC electrical circuits

In analyzing complicated dc electrical circuits containing resistors, one tries to find the equivalent resistance of simple series and parallel subcircuits, until the complicated circuit has been reduced to a single equivalent resistance. However, ac circuits that contain capacitors and inductors cannot be treated as straightforwardly because in these devices, the current through the device and the potential difference across the device are not in phase.

To describe what that means, consider an ac current through a resistor. When the current through the resistor is a maximum, the potential difference across it is a maximum, and when one of these quantities is a minimum, the other is a minimum. The relation between the two, of course, defines resistance by *Ohm's law*, $V = IR$. Saying that the current and potential difference are both maximum (or both minimum) at the same instant is stating that the two quantities are in phase with each other for a resistor.

In a *capacitor*, a potential difference will appear across the device only after charge has been deposited on its plates. Maximum current flow occurs when the capacitor is uncharged and thus does not present an emf that opposes current flow. Thus the potential difference across the capacitor is a minimum (zero) when the current in the capacitive circuit is a maximum. When the potential difference across the capacitor is a maximum, equal to and opposing the potential difference across the charging device, the capacitor is fully charged and the current has dropped to zero. Therefore, when an uncharged capacitor is first connected in a circuit, current must flow in the circuit before a potential difference appears across the capacitor. We say, then, that potential difference lags behind the current in a capacitive circuit.

In an *inductor*, when current is initially sent through the coil, a potential (an induced back emf) appears across the device that opposes this increased current flow and so, initially, no current flows in the circuit. As the current through the device increases to a maximum, the potential difference across the device drops to zero. Therefore, the potential difference appears across the inductor before the current flows in the circuit containing it. That is, in a circuit containing an inductor, the potential difference across the device leads the current through the circuit containing the inductor.

Thus, in a circuit containing a capacitor or an inductor, the current through the device and potential difference across the device are not in phase with one another. For each of these devices, in an ideal situation (no resistance in the circuit containing the device), the phase differences between current and potential difference is 90°. That is, when one electrical quantity is a maximum, the other is zero. In Figure 2.2a we display the sinusoidal current through a purely inductive or purely capacitive circuit. Figure 2.2b and c show representative sinusoidal potential differences across these devices. As can be seen, at $t = 0$, V_L has a maximum positive value. The function $\sin \theta$ has a positive maximum when $\theta = 90°$. Similarly, V_C has a negative maximum value at this instant, and $\sin \theta$ is maximum and negative when $\theta = -90°$.

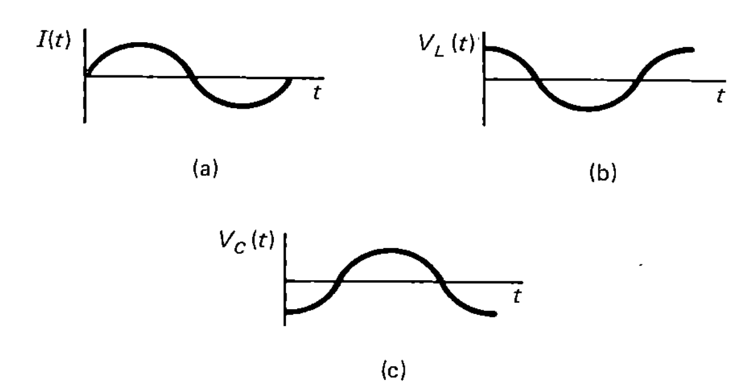

Figure 2.2 (a) Sinusoidal current through a capacitive or inductive circuit; (b) sinusoidal potential difference across an inductor; (c) sinusoidal potential difference across a capacitor.

Figure 2.3a is a phase diagram illustrating the relative phase between the current and potential difference in an ideal resistor, capacitor, and inductor. The current is taken to have a phase angle of 0° and is thus real. As such it is represented along the real axis. Then the potential difference across the resistor is in phase with the current and thus also has a phase angle of 0°. The potential difference across the capacitor is 90° behind the current, and the potential difference across the inductor leads the current through it by 90°. Mnemonics often employed to remember these facts are

$ELI \Rightarrow$ emf (E) in an inductor (L) appears before current (I)

$ICE \Rightarrow$ current (I) in a capacitor (C) appears before emf (E)

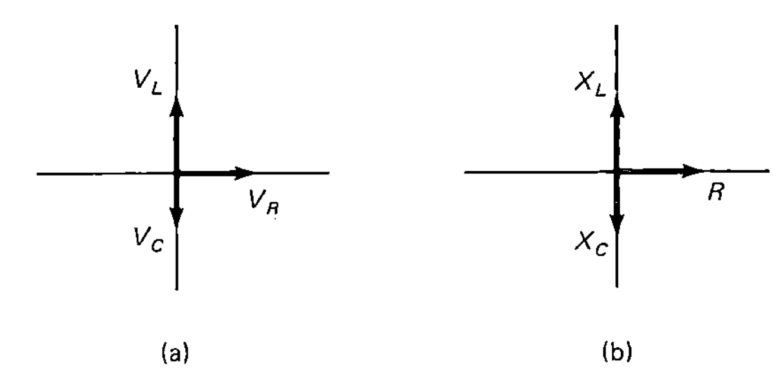

Figure 2.3 (a) Phase diagram for the potential difference across the three devices in an ac circuit; (b) phase diagram for the potential difference divided by the current for the three devices in an ac circuit.

An Ohm's law exists for capacitors and inductors in ac circuits. If the angular frequency of the ac generator is $\omega = 2\pi f$, the reactances (which are the ac equivalent of resistances for these devices) are defined such that

$$\text{(a)} \quad V_L = IX_L \quad \text{where} \quad \text{(b)} \quad X_L = \omega L \tag{2.23}$$

$$\text{(a)} \quad V_C = IX_C \quad \text{with} \quad \text{(b)} \quad X_C = \frac{1}{\omega C} \tag{2.24}$$

Thus the orientation of the lines drawn in Figure 2.3a represents the phase of the potential difference across each device relative to an arbitrary 0° phase chosen for the current through the device. The lengths of the lines represents the magnitude of each potential difference.

Figure 2.3b is the phase diagram equivalent to Figure 2.3a but described in terms of reactances instead of potential differences. Since the current has been chosen to be a real number, dividing the potential differences by the current does not affect the various phases. Therefore, the lengths of the lines can represent the magnitudes of the reactances in the circuit.

A complex number called the *impedance* of the circuit is defined by

$$Z = R + i(X_L - X_C) \tag{2.25}$$

By defining the complex impedance this way, the complex number Z carries the phases implicit in the diagram of Figure 2.3b.

If the complex impedances are now treated the same way resistors in a multiresistor dc circuit are handled, the ac circuit can be analyzed the same way as a dc resistive circuit. That is, for two devices in series

$$\text{(a)} \quad Z_{\text{net}} = Z_1 + Z_2 \quad \text{and} \quad \text{(b)} \quad \frac{1}{Z_{\text{net}}} = \frac{1}{Z_1} + \frac{1}{Z_2} \tag{2.26}$$

for two devices in parallel.

Example 2.3

As an example, consider the circuit shown in Figure 2.4. We will assume that the circuit draws a current of 2 A from the generator, and that the generator operates at 60 Hz. With $\omega = 2\pi f = 377$ Hz, the reactances are $X_L = \omega L = 11.3\ \Omega$ and $X_C = 1/\omega C = 13.3\ \Omega$.

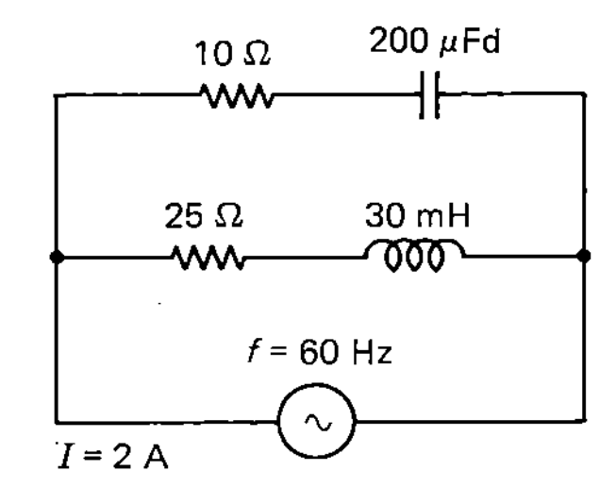

Figure 2.4 AC circuit to be analyzed using complex numbers.

Therefore,

$$\text{(a)}\quad Z_1 = R_1 + iX_C = (10 - 13.3i)\ \Omega \qquad \text{(b)}\quad Z_2 = R_2 + iX_L = (25 + 11.3i)\ \Omega \tag{2.27}$$

Treating the impedances of each branch of the circuit as if they were dc resistances, the net impedance of the circuit is obtained by

$$\frac{1}{Z_{\text{net}}} = \frac{1}{10 - 13.3i} + \frac{1}{25 + 11.3i} = \frac{10 + 13.3i}{10^2 + 13.3^2} + \frac{25 - 11.3i}{25^2 + 11.3^2} = 0.06933 + 0.03302i \tag{2.28a}$$

$$\Rightarrow \quad Z_{\text{net}} = \frac{1}{0.06933 + 0.03302i} = (11.76 - 5.60i)\ \Omega \tag{2.28b}$$

Therefore, for a real current of 2 A,

$$V_{\text{net}} = IZ_{\text{net}} = (23.52 - 11.20i)\ \text{volts} \tag{2.29}$$

Because the current has been chosen to be real, the phase of the voltage is determined by the phase of the complex impedance. Since the imaginary part of Z_{net} is negative, the imaginary part of V_{net} is negative. That is, the voltage lags behind the current in this circuit. This is the relative phase relation for a circuit containing a resistor and a capacitor. For that reason, such a circuit is said to be a *capacitive circuit*.

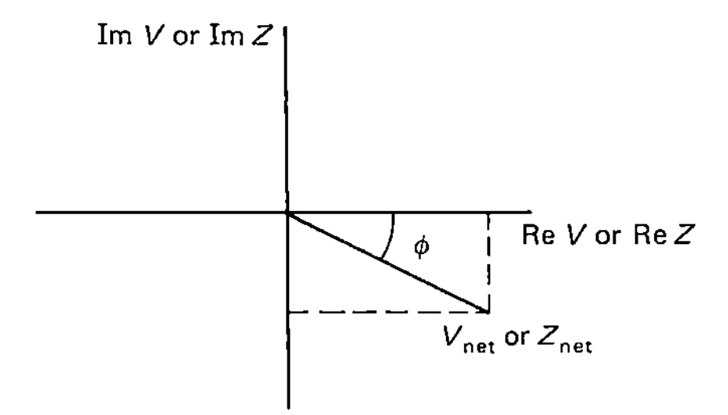

Figure 2.5 Resulting phase diagram for circuit of Figure 2.4.

It is straightforward to find the phase angle between voltage and current. Figure 2.5 is a phase diagram for the net impedance or potential difference for the circuit of Example 2.3. From this phase diagram,

$$\tan\phi = \frac{X_{\text{net}}}{R_{\text{net}}} = -\frac{5.64}{11.79} \tag{2.30}$$

Thus the voltage lags behind the current by 25.57°.

Referring to equation 2.28b, the entire circuit could be replaced by an equivalent circuit containing one 11.76-Ω resistor and one capacitor of reactance 5.60 Ω. Such a capacitor in a circuit with this generator would then have a capacitance of 470.66 μFd. □

If the imaginary part of Z_{net} were positive, the voltage would lead the current and the circuit would be termed inductive. The equivalent circuit would then contain a single resistor and one inductor. The inductance of the equivalent inductor could then be found straightforwardly from the resultant inductive reactance.

2.2 Functions of a Complex Variable

Analyticity of a function

If the real and imaginary parts of z are variable, a function $F(x, y)$ can be treated as a function of a complex variable $F(z)$. Then calculus operations can be performed on it.

By definition, the derivative $F'(z) = dF/dz$ is

$$\frac{dF}{dz} \equiv \lim_{\Delta z \to 0}\left(\frac{F(x + \Delta x, y + \Delta y) - F(x, y)}{\Delta x + i\,\Delta y}\right) \tag{2.31}$$

where $\Delta z \to 0$ means $\Delta x \to 0$ and $\Delta y \to 0$. Since $F(z)$ is complex, we write it as

$$F(z) = F(x, y) = U(x, y) + iV(x, y) \tag{2.32}$$

Then

$$\frac{dF}{dz} = \lim_{\substack{\Delta x \to 0 \\ \Delta y \to 0}} \left(\frac{\Delta U + i\,\Delta V}{\Delta x + i\,\Delta y} \right) \tag{2.33}$$

Since the limit with respect to two variables must be taken, there are an infinite number of ways to approach this limit. For example, we could take $\Delta x \to 0$ first then take $\Delta y \to 0$. Setting $\Delta x = 0$, equation 2.33 becomes

$$\frac{dF}{dz} = \lim_{\Delta y \to 0} \left(\frac{\Delta U + i\,\Delta V}{i\,\Delta y} \right) = -i\frac{\partial U}{\partial y} + \frac{\partial V}{\partial y} \tag{2.34a}$$

If Δy is set to 0 first, then $\Delta x \to 0$, we obtain

$$\frac{dF}{dz} = \lim_{\Delta x \to 0} \left(\frac{\Delta u + i\,\Delta V}{\Delta x} \right) = \frac{\partial U}{\partial x} + i\frac{\partial V}{\partial x} \tag{2.34b}$$

Unless the limit is independent of the path along which Δz approaches 0, the derivative is not defined. Therefore, if $F'(z)$ exists, the derivative as expressed in equation 2.34a must be the same as that given in equation 2.34b. A function for which the derivative at a point is independent of how one approaches the limit is said to be analytic at that point. Equating the two expressions above for the derivative, we obtain a condition for the analyticity of a function at a specified point.

$$-i\frac{\partial U}{\partial y} + \frac{\partial V}{\partial y} = \frac{\partial U}{\partial x} + i\frac{\partial V}{\partial x} \tag{2.35}$$

Equating the real parts and separately the imaginary parts of equation 2.35, analyticity at a point requires that both

$$\text{(a)} \quad \frac{\partial U}{\partial x} = \frac{\partial V}{\partial y} \qquad \text{and} \qquad \text{(b)} \quad \frac{\partial U}{\partial y} = -\frac{\partial V}{\partial x} \tag{2.36}$$

be satisfied at that point. These are called the *Cauchy–Riemann* (CR) *conditions*.

If one of the CR conditions is not satisfied at a point, the function is not analytic (even if the other CR condition is satisfied at the point). If a function is not analytic at a point, it is said to have a singularity at that point.

If $F(z)$ is analytic, its derivative with respect to z is obtained from either one of equations 2.34.

Example 2.4

Consider

$$F(z) = e^z = e^{(x+iy)} = e^x(\cos y + i \sin y) \tag{2.37}$$

$$\Rightarrow \quad \text{(a)} \quad U(x, y) = e^x \cos y \qquad \text{(b)} \quad V(x, y) = e^x \sin y \tag{2.38}$$

Applying the CR conditions to U and V, we obtain

$$\text{(a)} \quad \frac{\partial U}{\partial x} = e^x \cos y = \frac{\partial V}{\partial y} \qquad \text{(b)} \quad \frac{\partial V}{\partial x} = e^x \sin y = -\frac{\partial U}{\partial y} \tag{2.39}$$

Since both CR conditions are satisfied, e^z is analytic for all finite values of z. Such a function that is analytic at all finite z is called an entire function.

The derivative of e^z, using equation 2.34a for example, is

$$\frac{dF}{dz} = \frac{\partial U}{\partial x} + i\frac{\partial V}{\partial x} = e^x \cos y + ie^x \sin y = e^z \quad \square \tag{2.40}$$

We note that even though z is actually the sum of two variables, dF/dz is the derivative one would obtain treating z as a single variable. This result is true in general for analytic functions.

Example 2.5

To illustrate the fact that a function can be analytic at some points and not be analytic at others, consider

$$F(z) = \frac{1}{z} = \frac{1}{x + iy} = \frac{x - iy}{x^2 + y^2} \tag{2.41}$$

$$\Rightarrow \quad \text{(a)} \quad U = \frac{x}{x^2 + y^2} \qquad \text{(b)} \quad V = -\frac{y}{x^2 + y^2} \tag{2.42}$$

$$\Rightarrow \quad \text{(a)} \quad \frac{\partial U}{\partial x} = \frac{y^2 - x^2}{(x^2 + y^2)^2} = \frac{\partial V}{\partial y} \qquad \text{(b)} \quad \frac{\partial U}{\partial y} = -\frac{2xy}{(x^2 + y^2)^2} = -\frac{\partial V}{\partial x} \tag{2.43}$$

It appears that the CR conditions are satisfied for all x, y. However, at $x = y = 0$, these expressions become infinite. Thus, at $z = 0$, we cannot say that the CR conditions are valid. $F(z)$ is, therefore, analytic everywhere except at $z = 0$. $\square$

Example 2.6

Consider

$$F(z) = z^* = x - iy \tag{2.44}$$

$$\Rightarrow \quad \text{(a)} \quad U = x \qquad \text{(b)} \quad V = -y \tag{2.45}$$

$$\Rightarrow \quad \text{(a)} \quad \frac{\partial U}{\partial y} = 0 = -\frac{\partial V}{\partial x} \quad \text{but} \quad \text{(b)} \quad \frac{\partial U}{\partial x} = 1 \qquad \text{(c)} \quad \frac{\partial V}{\partial y} = -1 \tag{2.46}$$

Since both CR conditions are not satisfied for any value of z, $F(z) = z^*$ is not analytic anywhere. $\square$

Let $F(z)$ be a function that is analytic at and around the point $z_0 = x_0 + iy_0$. Then the CR conditions hold in the region around z_0.

$$\text{(a)} \quad \frac{\partial U}{\partial x} = \frac{\partial V}{\partial y} \qquad \text{(b)} \quad \frac{\partial U}{\partial y} = -\frac{\partial V}{\partial x} \tag{2.36}$$

If we differentiate the first CR equation with respect to x and the second with respect to y, then add the two equations, we find that U satisfies

$$\frac{\partial^2 U}{\partial x^2} + \frac{\partial^2 U}{\partial y^2} = 0 = \nabla^2 U \tag{2.47a}$$

which is *Laplace's equation* (Poisson's equation with $\rho = 0$) in two dimensions. Similarly, differentiating equation 2.36a with respect to y, and equation 2.36b with respect to x and subtracting, we see that V also satisfies the two-dimensional Laplace equation

$$\frac{\partial^2 V}{\partial x^2} + \frac{\partial^2 V}{\partial y^2} = 0 = \nabla^2 V \tag{2.47b}$$

Thus, satisfying the CR conditions means that U and V also satisfy the two-dimensional Laplace equation. Such functions are called harmonic functions.

If U and V are harmonic functions, then $F = U + iV$ must also satisfy Laplace's equation if it is analytic in the region around z_0. This idea will be useful for problems in electrostatics for which the geometric symmetry allows us to treat the problem in two dimensions.

2.3 Evaluation of Integrals: Part I

Integrals of analytic functions

Consider the integral of a function around a closed path in the complex plane. The path over which the function is integrated is called the *contour*. Let the function be analytic at every point inside and on the contour. If the contour is closed as in Figure 2.6, we denote the integral by $\oint F(z)\, dz$.

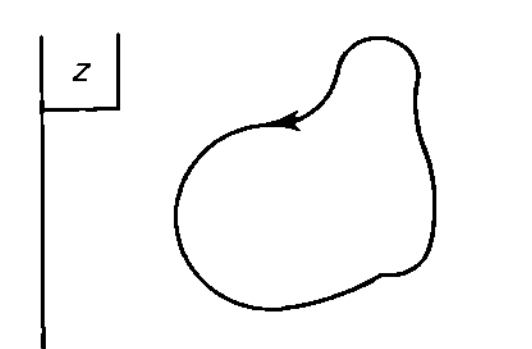

Figure 2.6 Closed-loop contour.

In the two-dimensional plane, the positive real axis is defined as 0°, and increasing angle is obtained by rotating a radius line in the counterclockwise direction. Therefore, we will associate a positive value to an integral if the contour is traversed in the counterclockwise direction. Rotating the radius line in the clockwise direction results in decreasing the angle (increasing negative angle). As such, a negative value will be associated with an integral over a contour that is traversed in the clockwise direction.

Writing $F(z)$ and dz in terms of their real and imaginary parts gives

$$\oint F(z)\, dz = \oint (U + iV)(dx + i\, dy) = \oint (U\, dx - V\, dy) + i\oint (V\, dx + U\, dy) \tag{2.48}$$

Recall that Stokes's theorem is

$$\int \nabla \times \mathbf{G} \cdot \mathbf{n}\, dS = \oint \mathbf{G} \cdot d\mathbf{l} \tag{1.143}$$

Since the complex z-plane is the x-y plane, $dS = dx\, dy$, $\mathbf{n} = \mathbf{k}$, and Stokes's theorem becomes

$$\int (\nabla \times \mathbf{G})_3\, dx\, dy = \oint \mathbf{G} \cdot (dx\, \mathbf{i} + dy\, \mathbf{j}) \tag{2.49a}$$

$$\Rightarrow \quad \int \left(\frac{\partial G_y}{\partial x} - \frac{\partial G_x}{\partial y} \right) dx\, dy = \oint (G_x\, dx + G_y\, dy) \tag{2.49b}$$

If we let $G_x = U$ and $G_y = -V$, equation 2.49b becomes

$$\oint (U\, dx - V\, dy) = \int \left(-\frac{\partial V}{\partial x} - \frac{\partial U}{\partial y} \right) dx\, dy \tag{2.50}$$

Since we have specified that $F(z)$ is analytic at every point inside and on the contour, the CR conditions are satisfied by U and V. It is essential that $F(z)$ be analytic at all points inside the contour so that the CR conditions hold at every point inside the area enclosed by the contour. Then

$$\frac{\partial V}{\partial x} = -\frac{\partial U}{\partial y} \tag{2.36b}$$

everywhere inside and on the contour, and the integrand of the integral on the right-hand side of equation 2.50 is zero. Thus, if $F(z)$ is analytic inside and on the contour, the real part of equation 2.48 is zero.

If we now let $G_x = V$ and $G_y = U$ and apply the same analysis, we find that the imaginary part of equation 2.48 is also zero. Thus we have shown that the integral of $F(z)$ is zero around a closed contour surrounding a region of analyticity of $F(z)$.

Consider such a region of analyticity of $F(z)$ and let a closed contour defining that region be made up of two pieces labeled C_1 and C_2 (Figure 2.7). Since the two pieces form a closed contour, and all points inside and on C_1 and C_2 are points of analyticity of $F(z)$,

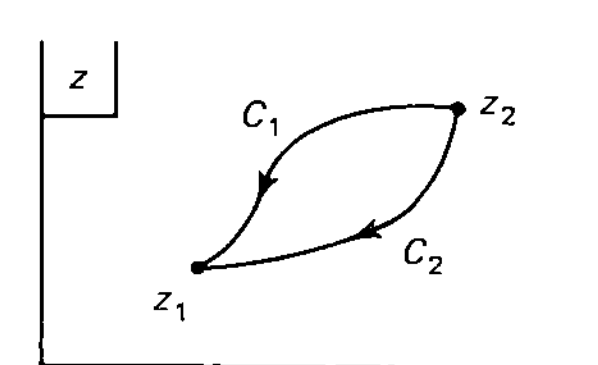

Figure 2.7 Closed contour constructed of two different segments.

$$\text{(a)} \quad \oint F(z)\,dz = 0 = \int_{\substack{z_1 \\ C_1}}^{z_2} F(z)\,dz - \int_{\substack{z_1 \\ C_2}}^{z_2} F(z)\,dz \quad \Rightarrow \quad \text{(b)} \quad \int_{\substack{z_1 \\ C_1}}^{z_2} F(z)\,dz = \int_{\substack{z_1 \\ C_2}}^{z_2} F(z)\,dz \tag{2.51}$$

That is, the integral is independent of the path taken from z_1 to z_2.

An equivalent way of describing this is to say that the integral is unchanged if the contour C_1 is deformed into the contour C_2 as long as all the points encountered in the deformation are points of analyticity of the integrand.

Example 2.7

As an example, consider the integral

$$I \equiv \int_0^{1+i} z^2\,dz \tag{2.52}$$

over the three contours shown in Figure 2.8. For the contour of Figure 2.8a, $z = x + i0$ along the first part of the path and $z = 1 + iy$ along the second part of the path. Therefore,

$$\int_{C_a} z^2\,dz = \int_0^1 x^2\,dx + \int_0^1 (1 + iy)^2 i\,dy = \tfrac{1}{3}(1 + i)^3 \tag{2.53a}$$

For the contour of Figure 2.8b, $z = 0 + iy$ along the first segment and $z = x + i$ along the

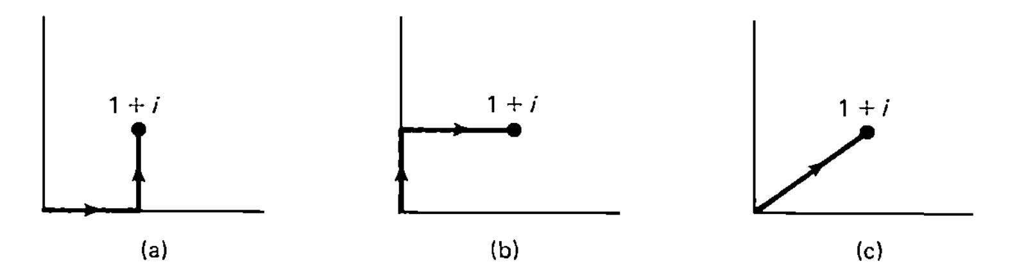

Figure 2.8 Three contours to illustrate that the integral of an analytic function is independent of path.

second piece. Thus

$$\int_{C_b} z^2\,dz = \int_0^1 (iy)^2 i\,dy + \int_0^1 (x+i)^2\,dx = \tfrac{1}{3}(1+i)^3 \tag{2.53b}$$

Along the third contour, $x = y$, so $z = x(1 + i)$. Therefore,

$$\int_{C_c} z^2\,dz = \int_0^1 x^2(1+i)^2\,dx(1+i) = \tfrac{1}{3}(1+i)^3 \quad \square \tag{2.53c}$$

Integrals of functions containing singularities

There are two types of singular points that an integrable function can have. One is called a *pole*, the other a *branch point*. We will consider functions with pole singularities first.

$F(z)$ is said to have an Nth-*order pole* at the point z_0 if there exists an integer N such that

$$\text{(a)} \quad \lim_{z \to z_0} F(z) = \infty \qquad \text{while} \qquad \text{(b)} \quad \lim_{z \to z_0} (z - z_0)^N F(z) \equiv L(z_0) \tag{2.54}$$

where $L(z_0)$ is a finite, nonzero number, which, of course depends on z_0. $F(z)$ is then said to have a pole of order N at z_0.

Example 2.8

For example,

$$F(z) = \frac{\sin z}{(z - 3)} \tag{2.55}$$

is infinite at $z = 3$, while

$$\lim_{z \to 3} (z - 3)F(z) = \sin(3) \tag{2.56}$$

is a nonzero finite number. Then $\sin z/(z - 3)$ has a first-order pole at $z = 3$. □

A first-order pole is also referred to as a *simple pole*. The function $\tan z$ has an infinite number of simple poles, one at each odd multiple of $\pi/2$.

Example 2.9

The function $(\sin z)/z^3 \to 0/0$ when $z \to 0$. For such a case, it is not clear if the function has a pole, and if so, what the order of the pole is. One way of determining this for $(\sin z)/z^3$ is to look at the behavior of the numerator near the pole. When z is very small, $\sin z \simeq z$. Thus in the vicinity of the pole, $(\sin z)/z^3 \simeq 1/z^2$, so this function, when multiplied by z^2, will be finite and nonzero at $z = 0$. That is, $(\sin z)/z^3$ has a second-order pole at $z = 0$. □

The function $e^{1/z}$ becomes infinite when $z \to 0$ from the positive real axis. However, there is no integer, N, such that $z^N e^{1/z}$ becomes finite at $z = 0$. Therefore, this singularity is not a pole. We also note that the limiting value of $e^{1/z}$ is not unique. For example, when $z \to 0$ from the negative real axis, $e^{1/z} \to 0$, and when this limit is approached from the positive or negative imaginary axis, $e^{1/z} \to e^{\pm i\infty}$.

A singularity that cannot be removed by multiplying the function by some power of $(z - z_0)$ is called a *nonremovable* or *essential singularity*. Functions with essential singularities are generally nonintegrable. For that reason we will not consider them further.

Cauchy's residue theorem for functions with simple poles

Cauchy's theorem provides a method for evaluating an integral of a function around a closed contour that encircles a pole of the function. We will investigate simple ($N = 1$) poles first.

The function that is left after the pole has been removed is called the *residue* of the pole. For a simple pole, this is the finite nonzero number

$$\lim_{z \to z_0} (z - z_0)F(z) \equiv R(z_0) \tag{2.57}$$

From this definition, then, when z is very close, but not quite equal to z_0, we can approximate $F(z)$ by

$$F(z) \simeq \frac{R(z_0)}{(z - z_0)} \tag{2.58}$$

This approximation becomes more exact the closer z is to z_0.

Suppose $F(z)$ has a simple pole at z_0. Consider the integral of this function around the contour shown in Figure 2.9a, which encloses the pole.

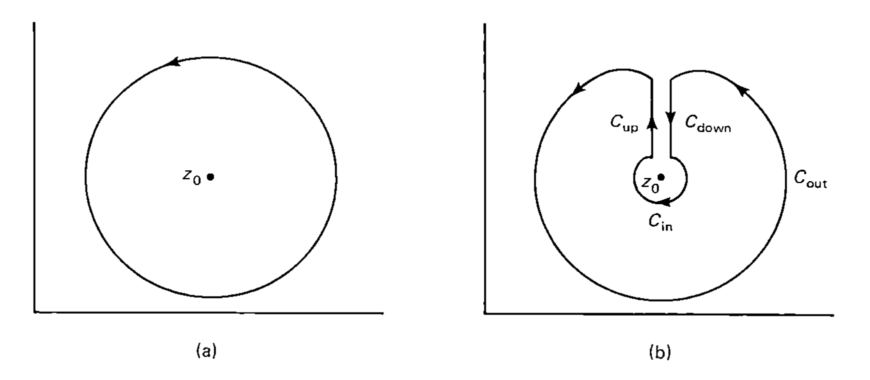

Figure 2.9 (a) Contour encircling a simple pole of a function; (b) contour that excludes the simple pole.

To analyze the integral around the contour that encloses the pole, we will consider the contour of Figure 2.9b. Note that this contour is constructed so that the point z_0 is outside the contour. Thus the integral around the contour of Figure 2.9b is around a closed contour that encloses only points of analyticity of $F(z)$. Thus, around the contour of Figure 2.9b,

$$\oint F(z)\, dz = 0 \tag{2.59a}$$

Writing this in terms of the four parts of the contour shown in Figure 2.9b yields

$$\oint F(z)\, dz = 0 = \int_{C_{out}} F(z)\, dz + \int_{C_{down}} F(z)\, dz + \int_{C_{in}} F(z)\, dz + \int_{C_{up}} F(z)\, dz \tag{2.59b}$$

We will take the gap between C_{up} and C_{down} to be very small, eventually going to the limit of a zero-width gap. In that limit, the integral around the outer contour of Figure 2.9b will equal the integral around the contour of Figure 2.9a, the integral we are analyzing.

Since there are no singularities between C_{up} and C_{down} and since these parts of the contour are traversed in opposite directions,

$$\int_{C_{down}} F(z)\,dz + \int_{C_{up}} F(z)\,dz = 0 \tag{2.60}$$

Equivalently, based on our analysis of the integral represented in Figure 2.7, C_{up} can be deformed into C_{down}, so they are actually the same path, traversed in opposite directions.

Note that the outer contour is traversed in a counterclockwise direction and the inner contour in a clockwise direction. Denoting the clockwise direction by "c" and the counterclockwise direction by "cc", the contour integral becomes

$$\oint F(z)\,Dz = \oint_{C_{out}}^{cc} F(z)\,dz + \oint_{C_{in}}^{c} F(z)\,dz = \oint_{C_{out}}^{cc} F(z)\,dz - \oint_{C_{in}}^{cc} F(z)\,dz = 0 \tag{2.61}$$

We are trying to evaluate the integral around the outer contour. To do so, we note that the inner contour can be shrunk to as small a circle as we wish as long as we do not collapse it to a point at z_0. Thus we take the inner contour to be a circle centered on z_0, with infinitesimal radius ρ (Figure 2.10).

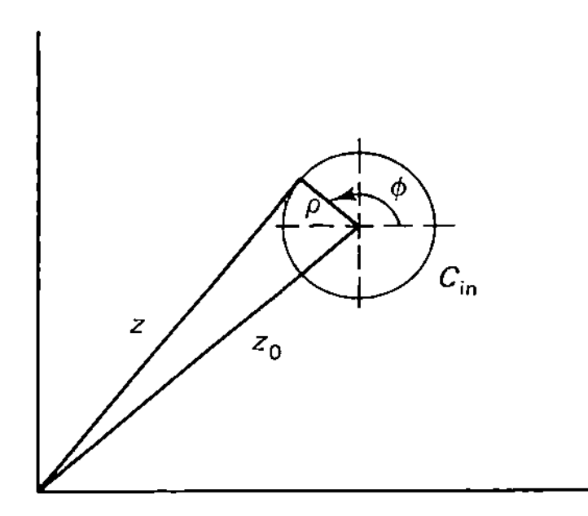

Figure 2.10 Small circle around the simple pole at z_0.

A point on this small circle is of the form

$$\text{(a)}\quad z = z_0 + \rho e^{i\phi} \qquad \Rightarrow \qquad \text{(b)}\quad dz = i\rho e^{i\phi}\,d\phi \tag{2.62}$$

Therefore,

$$\oint_{C_{in}}^{cc} F(z)\,dz = \int_0^{2\pi} F(z_0 + \rho e^{i\phi}) i\rho e^{i\phi}\,d\phi \tag{2.63}$$

Since ρ is very small, every point z on the inner contour is very close to z_0. Thus we can approximate the function by

$$F(z) \simeq \frac{R(z_0)}{z - z_0} = \frac{R(z_0)}{\rho e^{i\phi}} \tag{2.64}$$

This becomes exact when $\rho \to 0$. Thus

$$\lim_{\rho \to 0} \oint_{C_{in}}^{cc} F(z)\,dz = R(z_0) \int_0^{2\pi} i\,d\theta = 2\pi i R(z_0) \tag{2.65}$$

Therefore,

$$\oint_{C_{out}}^{cc} F(z)\,dz = 2\pi i R(z_0) \tag{2.66a}$$

This is *Cauchy's integral theorem* for a function with one simple pole.

If the integral around the outer contour had been taken in the clockwise direction, the integral around the inner contour would also be taken in the clockwise direction. Then we would obtain

$$\oint_{C_{\text{out}}}^{c} F(z)\,dz = \oint_{C_{\text{in}}}^{c} F(z)\,dz = \lim_{\rho\to 0}\int_{2\pi}^{0} F(z_0 + \rho e^{i\phi}) i e^{i\phi}\,d\phi = -2\pi i R(z_0) \qquad (2.66b)$$

A simpler derivation of this result is achieved by recognizing that there are no singularities of $F(z)$ between the outer contour and z_0. Thus the integral around the outer contour is the same as that around the infinitesimal circle centered at z_0. The latter integral was shown above to be $2\pi i R(z_0)$. Thus, equating the two integrals yields Cauchy's theorem as expressed in equations 2.66.

Example 2.10

As an example, we use Cauchy's residue theorem to evaluate $\oint e^z \sin z/(z - 2i)\,dz$ around circles of radius 5 and 1 centered at $z = 0$. The contours and the pole are shown in Figure 2.11. The integrand has a simple pole at $z = 2i$, the residue of which is

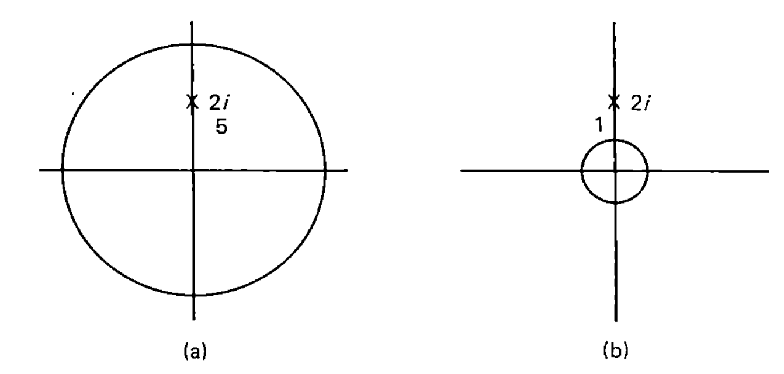

Figure 2.11 Contours for evaluating $\oint[e^z \sin z/(z - 2i)]\,dz$.

$$R(2i) = \lim_{z\to 2i}(z - 2i)\frac{e^z \sin z}{(z - 2i)} = e^{2i}\sin 2i = ie^{2i}\sinh(2) \qquad (2.67)$$

For an origin-centered circle of radius 5, the pole is inside the contour. Therefore,

$$\oint \frac{e^z \sin z}{(z - 2i)}\,dz = 2\pi i R(2i) = 2\pi i\left[ie^{2i}\sinh(2)\right] = -2\pi e^{2i}\sinh(2) \qquad (2.68a)$$

For an origin-centered circle of radius 1, the pole is outside the contour. Thus

$$\oint \frac{e^z \sin z}{(z - 2i)}\,dz = 0 \quad \square \qquad (2.68b)$$

As discussed above, the value of an integral such as the one in Example 2.10 depends only on whether the pole is inside or outside the contour, not on the shape of the contour. Therefore, in this example, if the integral were evaluated around a square of side 8, centered at $(-4, -4)$, as shown in Figure 2.12a, the result would be $-2\pi e^{2i}\sinh(2)$, identical to the result of Example 2.10, part (a). Similarly, the equilateral triangle of Figure 2.12b does not enclose the pole. Thus the result is zero, the same as that of Example 2.10, part (b). Thus the square of Figure 2.12a could be deformed into the circle of Figure 2.11a without encountering any singularities. Similarly, the triangle of Figure 2.12b is deformable into the small circle of Figure 2.11b.

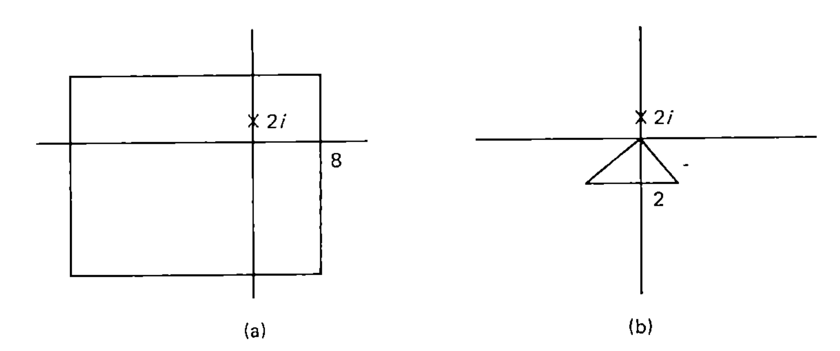

Figure 2.12 Equivalent contours for evaluating $\oint [e^z \sin z/(z - 2i)]\, dz$.

If the integrand has several simple poles, a contour can be constructed to exclude all of them in the same way that one was constructed to exclude the single simple pole in Figure 2.9b. Such a contour for three simple poles is shown in Figure 2.13b. The analysis of the integration around each small contour surrounding a pole is independent of the existence of the other poles. It therefore proceeds exactly as the analysis for one simple pole. Each pole contributes $2\pi i R(\text{pole})$. Thus Cauchy's theorem for a function containing many simple poles is

$$\oint^{cc} F(z)\, dz = 2\pi i \sum \text{residues} \tag{2.69a}$$

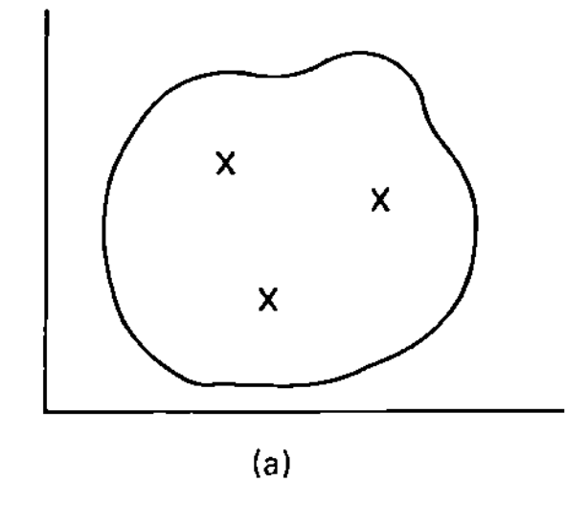

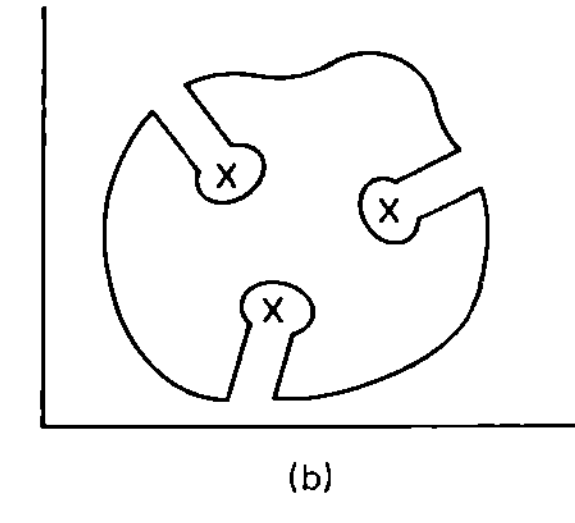

Figure 2.13 (a) Contour encircling three simple poles; (b) contour excluding three simple poles.

If the integral is taken in the clockwise direction, using an analysis identical to the one that leads to equation 2.66b, Cauchy's theorem is

$$\oint^{c} F(z)\, dz = -2\pi i \sum \text{residues} \tag{2.69b}$$

Henceforth, we omit the designation of the direction on the integral sign unless it is necessary for clarity. It will be understood that unless otherwise specified, the contour will be traversed in the counterclockwise direction.

Example 2.11

As an example of the evaluation of an integral of a function with more than one pole, consider

$$I \equiv \oint \frac{e^z}{(z^2 + 1)}\, dz \tag{2.70}$$

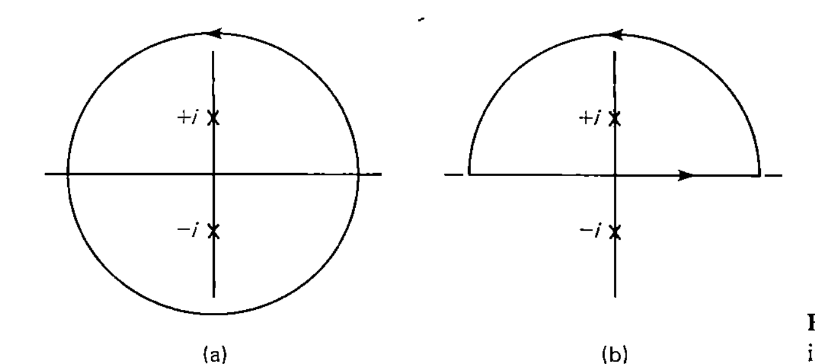

Figure 2.14 Contours for two examples involving $\oint [e^z/(z^2+1)]\,dz$.

around an origin-centered circle of radius 2 as shown in Figure 2.14a. The integrand has two simple poles, at $z = \pm i$, both of which are inside the contour. Were this integral to be evaluated around the semicircle of Figure 2.12b, there would only be a contribution from the pole at $z = +i$. Writing the integrand as $e^z/[(z+i)(z-i)]$, Cauchy's theorem for the integral around contour 2.14a yields

$$\oint \frac{e^z}{(z^2+1)}\,dz = 2\pi i[R(i) + R(-i)] \tag{2.71}$$

with

$$R(i) = \lim_{z\to i}(z-i)\frac{e^z}{(z+i)(z-i)} = \frac{e^i}{2i} \tag{2.72a}$$

$$R(-i) = \lim_{z\to -i}(z+i)\frac{e^z}{(z+i)(z-i)} = \frac{e^{-i}}{-2i} \tag{2.72b}$$

Therefore, around the contour of Figure 2.14a,

$$\oint \frac{e^z}{(z^2+1)}\,dz = 2\pi i\frac{(e^i - e^{-i})}{2i} = 2\pi i\sin(1) \tag{2.73}$$

The integral around the contour of Figure 2.14b is

$$\oint \frac{e^z}{(z^2+1)}\,dz = 2\pi i R(i) = \pi e^i \quad \square \tag{2.74}$$

Cauchy's theorem for higher-order poles

Consider the function

$$F(z') = \frac{f(z')}{(z'-z)} \tag{2.75}$$

and let $f(z')$ be an entire function. Then $F(z)$ has one simple pole at $z' = z$. We consider

$$\oint F(z')\,dz' = \oint \frac{f(z')}{(z'-z)}\,dz' \tag{2.76}$$

By Cauchy's theorem, since $F(z')$ has a pole at $z' = z$,

$$\oint F(z')\,dz' = 2\pi i R(z) \tag{2.77}$$

with

$$R(z) = \lim_{z' \to z} (z' - z) F(z') = \lim_{z' \to z} (z' - z) \frac{f(z')}{(z' - z)} = f(z) \tag{2.78}$$

Since $f(z)$ is analytic for all points in and on a contour C, we obtain the integral representation for $f(z)$:

$$f(z) = \frac{1}{2\pi i} \oint \frac{f(z')}{(z' - z)} \, dz' \tag{2.79}$$

We can use this result to develop integral representations of the derivatives of $f(z)$. Since all the z-dependence of the integral is in the denominator of the integrand,

$$\text{(a)} \quad \frac{df}{dz} = \frac{1}{2\pi i} \oint \frac{f(z')}{(z' - z)^2} \, dz' \qquad \text{(b)} \quad \frac{d^2 f}{dz^2} = \frac{2}{2\pi i} \oint \frac{f(z')}{(z' - z)^3} \, dz'$$

or, in general,

$$\text{(c)} \quad \frac{d^n f}{dz^n} = \frac{n!}{2\pi i} \oint \frac{f(z')}{(z' - z)^{n+1}} \, dz' \tag{2.80}$$

Equation 2.80c gives a prescription for evaluating integrals of a particular form in which the integrand can have a higher-than-first-order pole.

Example 2.12

For example, we can evaluate

$$I \equiv \oint \frac{z e^z}{(z - 3)^3} \, dz \tag{2.81}$$

around an origin-centered circle of radius 5 (Figure 2.15). Comparing equations 2.80c and 2.81, we identify $f(z)$ as ze^z. Then, from equations 2.80,

$$\text{(a)} \quad \left. \frac{d^2 f}{dz^2} \right)_{z=3} = \frac{2!}{2\pi i} \oint \frac{f(z)}{(z - 3)^3} \, dz \quad \Rightarrow \quad \text{(b)} \quad \oint \frac{z e^z}{(z - 3)^3} \, dz = \left. \frac{2\pi i}{2!} \frac{d^2}{dz^2} (z e^z) \right)_{z=3}$$

$$= 5\pi i e^3 \quad \square \tag{2.82}$$

If there are several (two or more) poles inside the contour, a very straightforward extension of equation 2.80c can be derived. For example, consider an integral of the form

$$I \equiv \oint \frac{g(z)}{(z - z_0)^{M+1} (z - z_1)^{N+1}} \, dz \tag{2.83}$$

where we will assume that $g(z)$ is analytic everywhere inside and on the contour, and z_0 and z_1 are inside the contour. The sum of the residues will contain contributions from

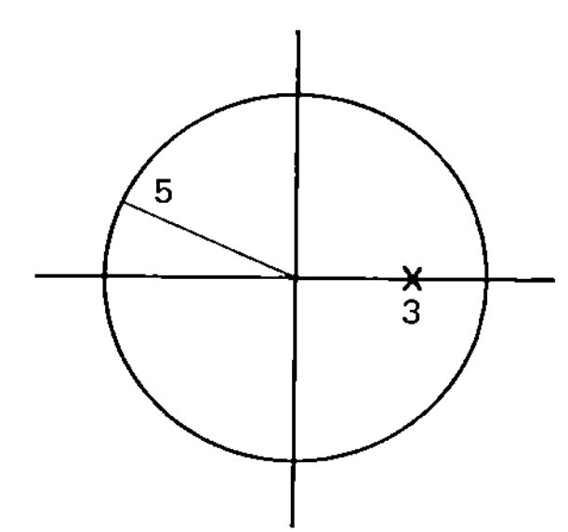

Figure 2.15 Contour for $\oint [ze^z/(z - 3)^3] \, dz$.

both poles. The part of the integrand that is analytic at z_0 is

$$f(z) = \frac{g(z)}{(z - z_1)^{N+1}} \tag{2.84}$$

and not just $g(z)$. Thus the contribution from the $M + 1$ order pole at z_0 is

$$\frac{2\pi i}{M!} \frac{d^M}{dz^M} \left(\frac{g(z)}{(z - z_1)^{N+1}} \right)_{z=z_0} \tag{2.85}$$

There is an analogous contribution from the $N + 1$ order pole at z_1. Thus

$$\oint \frac{g(z)}{(z - z_0)^{M+1}(z - z_1)^{N+1}} dz$$

$$= 2\pi i \left[\frac{1}{M!} \frac{d^M}{dz^M} \left(\frac{g(z)}{(z - z_1)^{N+1}} \right)_{z=z_0} + \frac{1}{N!} \frac{d^N}{dz^N} \left(\frac{g(z)}{(z - z_0)^{M+1}} \right)_{z=z_1} \right] \tag{2.86}$$

It is very straightforward (but quite cumbersome) to generalize the expression for an integral like that of equation 2.86 to one with an arbitrary number of poles. It will be left to the interested reader to do so.

Example 2.13

As an example, we use equation 2.86 to evaluate

$$I \equiv \oint_{\substack{\text{unit}\\\text{circle}}} \frac{e^z}{(z - \frac{1}{2})^2 z^3 (z - \frac{3}{2})} dz \tag{2.87}$$

The unit circle is an origin centered circle of radius 1. Since $z = \frac{3}{2}$ is outside the unit circle, there is no contribution to the integral from this pole. Therefore,

$$\oint_{\substack{\text{unit}\\\text{circle}}} \frac{e^z}{(z - \frac{1}{2})^2 z^3 (z - \frac{3}{2})} dz = 2\pi i \left[\frac{1}{1!} \frac{d}{dz} \left(\frac{e^z}{z^3 (z - \frac{3}{2})} \right)_{z=1/2} \right.$$

$$\left. + \frac{1}{2!} \frac{d^2}{dz^2} \left(\frac{e^z}{(z - \frac{1}{2})^2 (z - \frac{3}{2})} \right)_{z=0} \right] = 2\pi i \left(32 e^{1/2} - \frac{5641}{108} \right) \quad \square \tag{2.88}$$

It will be shown later that the residue of a pole of order $N + 1$ at z_0 is

$$R(z_0) = \frac{1}{N!} \frac{d^N}{dz^N} \left((z - z_0)^{N+1} F(z) \right)_{z=z_0} \tag{2.89}$$

For

$$F(z) = \frac{f(z)}{(z - z_0)^{N+1}} \tag{2.75}$$

this residue becomes

$$R(z_0) = \frac{1}{N!} \frac{d^N}{dz^N} f(z) \bigg|_{z=z_0} \equiv \frac{1}{N!} f^{(N)}(z_0) \tag{2.90}$$

Thus for higher-order poles, the integrals of equations 2.80c and 2.86 are consistent with

Cauchy's theorem of equations 2.69, namely,

$$\oint \frac{f(z)}{(z-z_0)^{M+1}(z-z_1)^{N+1}\cdots(z-z_k)^{P+1}}\,dz = 2\pi i \sum \text{residues} \tag{2.91}$$

2.4 Series Representations of Functions

Taylor series for an analytic function

Cauchy's theorem can be used to derive a series representation for an analytic function $F(z)$ in the region around $z = z_0$. Let $F(z')$ be analytic at all points inside and on a specified contour in the z'-plane. We will take the contour to be a circle centered at z_0 and take z to be an arbitrary point inside this circle (Figure 2.16). Since $F(z')$ is analytic at $z' = z$,

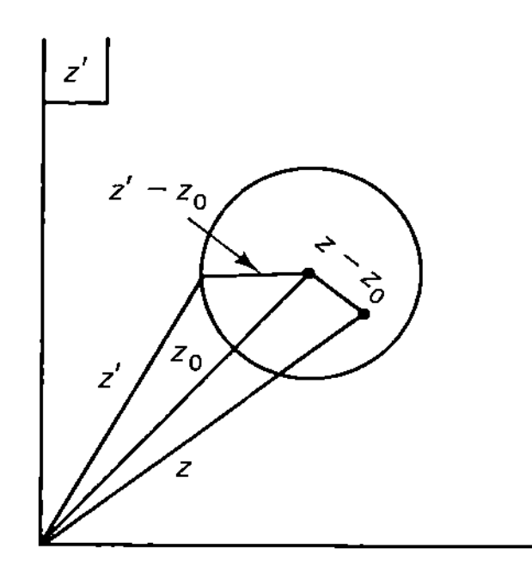

Figure 2.16 Contour and selected points in the region around z_0 for deriving the Taylor series for an analytic function.

$$F(z) = \frac{1}{2\pi i}\oint \frac{F(z')}{(z'-z)}\,dz' \tag{2.79}$$

We define

$$h \equiv z - z_0 \tag{2.92}$$

$$\Rightarrow \quad F(z) = F(z_0 + h) = \frac{1}{2\pi i}\oint \frac{F(z')}{(z'-z_0-h)}\,dz' \tag{2.93}$$

Writing

$$z' - z_0 - h = (z'-z_0)\left[1 - \frac{h}{(z'-z_0)}\right] \tag{2.94}$$

we note that for all points z' on the contour

$$\text{(a)} \quad |z'-z_0| > |h| \quad \Rightarrow \quad \text{(b)} \quad \left|\frac{h}{(z'-z_0)}\right| < 1 \tag{2.95}$$

In an elementary algebra course, the reader has probably learned that by a process called synthetic division, one can write

$$\frac{1}{(1-x)} = 1 + x + x^2 + x^3 + \cdots = \sum_{n=0}^{\infty} x^n \tag{2.96}$$

for $|x| < 1$. Thus

$$\frac{1}{(z'-z_0-h)} = \frac{1}{(z'-z_0)}\left[1 + \frac{h}{(z'-z_0)} + \frac{h^2}{(z'-z_0)^2} + \cdots\right] = \sum_{n=0}^{\infty} \frac{h^n}{(z'-z_0)^{n+1}} \tag{2.97}$$

Therefore,

$$F(z_0 + h) = \frac{1}{2\pi i} \sum_{n=0}^{\infty} h^n \oint \frac{F(z')}{(z' - z_0)^{n+1}} dz' \tag{2.98}$$

Since $F(z')$ is analytic for all points inside and on the contour,

$$\oint \frac{F(z')}{(z' - z_0)^{n+1}} dz' = \frac{2\pi i}{n!} \frac{d^n}{dz^n} F(z) \Bigg)_{z=z_0} \equiv \frac{2\pi i}{n!} F^{(n)}(z_0) \tag{2.99}$$

With $h = z - z_0$,

$$F(z) = \sum_{n=0}^{\infty} \frac{F^{(n)}(z_0)}{n!} (z - z_0)^n \tag{2.100}$$

This expansion of an analytic function is referred to as the *Taylor series expansion* of $F(z)$ around the point z_0. If $z_0 = 0$, the series is called the *MacLaurin series.*

To say that the Taylor series is a valid representation of $F(z)$ means that at a particular value of z, the value one obtains for the series is the same as the value obtained for the function. Another way of stating this is to say that the series on the right-hand side of equation 2.100 converges to the function on the left-hand side of equation 2.100 at the point z.

Recall that the contour used to develop the Taylor series is a circle centered at z_0, which does not enclose any singularities of $F(z)$. Referring to Figure 2.17, let z_1 be the singularity of $F(z)$ that is closest to z_0. Since the function is analytic everywhere inside the contour, z_1 must be outside the contour. Recall that the integral of equation 2.93 is unchanged if we enlarge the circle, as long as we do not cross a singularity of $F(z')$. Therefore, we can enlarge the circle until it just encounters z_1 without affecting the development of the Taylor series. This largest circle for which the series converges is called the *circle of convergence*, and its radius, $|z_1 - z_0|$, is called the *radius of convergence* of the Taylor series.

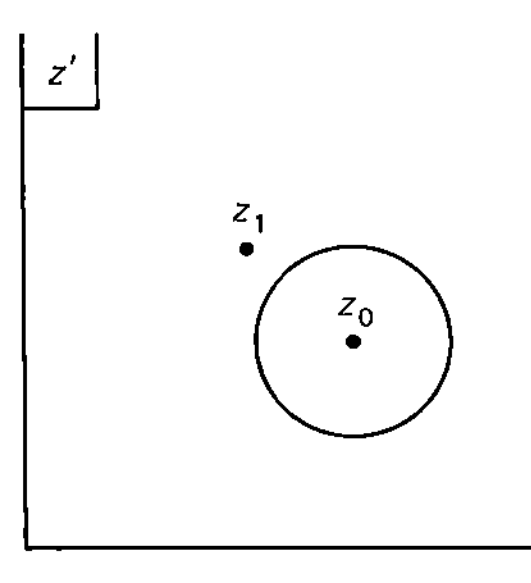

Figure 2.17 Contour used to obtain series representation of a function analytic at z_0 with singularity of the function at z_1.

Obviously, $1/(1 - x)$ has a singularity at $x = 1$. Therefore, a series expansion around $x = 0$ is valid inside and on a circle with a radius that is less than 1. That is why the series description of $1/(1 - x)$ in equation 2.96 is valid only for $|x| < 1$.

Since an entire function is analytic at all finite values of z, it has an infinite radius of convergence. That is, its Taylor series converges to the function at any finite value of z.

Example 2.14

As an example, we derive the Taylor series expansion of e^z around $z = z_0$. The series is

$$F(z) = e^z = F(z_0) + F'(z_0)(z - z_0) + \frac{1}{2!} F''(z_0)(z - z_0)^2 + \cdots \tag{2.101}$$

Since every derivative of e^z is e^z, $F^{(n)}(z_0) = e^{z_0}$ and the Taylor series is

$$\text{(a)} \quad e^z = e^{z_0} \sum_{n=0}^{\infty} \frac{(z - z_0)^n}{n!} \quad \Rightarrow \quad \text{(b)} \quad e^{(z-z_0)} = \sum_{n=0}^{\infty} \frac{(z - z_0)^n}{n!} \tag{2.102}$$

Letting $x = z - z_0$, we obtain

$$e^x = \sum_{n=0}^{\infty} \frac{x^n}{n!} \tag{2.102c}$$

which is the MacLaurin series for e^x. □

Example 2.15

As another example, we deduce the MacLaurin series for $\sin z$.

$$\text{(a)} \quad F(z) = \sin z \quad \Rightarrow \quad \text{(b)} \quad F(0) = 0 \tag{2.103}$$

Then

$$\begin{aligned} &\text{(a)} F'(z) = \cos z && \Rightarrow \quad \text{(b)} \quad F'(0) = 1 \\ &\text{(c)} \quad F''(z) = -\sin z && \Rightarrow \quad \text{(d)} \quad F''(0) = 0 \end{aligned} \tag{2.104}$$

and so on. Therefore,

$$\sin z = z - \frac{z^3}{3!} + \frac{z^5}{5!} - \cdots = \sum_{n=0}^{\infty} (-1)^n \frac{z^{2n+1}}{(2n+1)!} \quad \square \tag{2.105}$$

L'Hospital's rule

L'Hospital's rule is a prescription for determining the limit of a function at a point z_0, when the function evaluated at z_0 takes the form $0/0$ or ∞/∞. This is studied in a first calculus course. We derive the rule using the Taylor series representation just discussed.

If a the limit of a function at z_0 is to be $0/0$ or ∞/∞, it must be of the form

$$F(z) = \frac{P(z)}{Q(z)} \tag{2.106}$$

such that $P(z)$ and $Q(z)$ are both 0, or are both infinite at z_0.

If the limiting value of $F(z)$ is of the form $0/0$, that means that $P(z)$ and $Q(z)$ are both 0 at $z = z_0$. If P and Q are analytic at z_0, they can be expanded in a Taylor series around the point. Thus

$$P(z) = P(z_0) + P'(z_0)(z - z_0) + \frac{1}{2!}P''(z_0)(z - z_0)^2 + \cdots \tag{2.107}$$

$$Q(z) = Q(z_0) + Q'(z_0)(z - z_0) + \frac{1}{2!}Q''(z_0)(z - z_0)^2 + \cdots \tag{2.108}$$

However, $P(z_0) = Q(z_0) = 0$. Therefore, if $P'(z_0) \neq 0$ and $Q'(z_0) \neq 0$,

$$F(z_0) = \lim_{z \to z_0} \frac{P'(z_0)(z - z_0) + (1/2!)P''(z_0)(z - z_0)^2 + \cdots}{Q'(z_0)(z - z_0) + (1/2!)Q''(z_0)(z - z_0)^2 + \cdots} \tag{2.109}$$

Canceling one factor of $(z - z_0)$ from numerator and denominator, the limit then becomes

$$\lim_{z \to z_0} F(z) = \frac{P'(z_0)}{Q'(z_0)} \tag{2.110}$$

If the first derivatives of P and Q are also zero, we write

$$P(z) = \frac{1}{2!}P''(z_0)(z - z_0)^2 + \frac{1}{3!}P'''(z_0)(z - z_0)^3 + \cdots \tag{2.111a}$$

$$Q(z) = \frac{1}{2!}Q''(z_0)(z - z_0)^2 + \frac{1}{3!}Q'''(z_0)(z - z_0)^3 + \cdots \tag{2.111b}$$

After canceling $(z - z_0)^2$ from numerator and denominator,

$$\lim_{z \to z_0} F(z) = \frac{P''(z_0)}{Q''(z_0)} \tag{2.112}$$

Extending this analysis to any order, if the first $n - 1$ derivatives of $P(z)$ and of $Q(z)$ are zero, then

$$\lim_{z \to z_0} F(z) = \frac{P^{(n)}(z_0)}{Q^{(n)}(z_0)} \tag{2.113}$$

If $P(z_0) = Q(z_0) = \infty$, we write

$$F(z) = \frac{P(z)}{Q(z)} = \frac{1/Q(z)}{1/P(z)} \equiv \frac{S(z)}{R(z)} \tag{2.114}$$

where $R(z_0) = S(z_0) = 0$. We will first consider the case where P and Q have first-order infinities; that is, $R'(z_0)$ and $S'(z_0)$ are not zero. Then using equation 2.110, we have

$$\lim_{z \to z_0} F(z) = \frac{S'(z_0)}{R'(z_0)} = \lim_{z \to z_0} \frac{Q'/Q^2}{P'/P^2} = \lim_{z \to z_0} \frac{P^2(z)}{Q^2(z)} \frac{Q'(z)}{P'(z)} \tag{2.115}$$

Using the property that the limit of a product is equal to the product of limits, this becomes

$$\lim_{z \to z_0} \frac{P(z)}{Q(z)} = \left(\lim_{z \to z_0} \frac{P(z)}{Q(z)} \right)^2 \left(\lim_{z \to z_0} \frac{Q'(z)}{P'(z)} \right) \tag{2.116}$$

$$\Rightarrow \quad \lim_{z \to z_0} \frac{P(z)}{Q(z)} = \lim_{z \to z_0} \frac{P'(z)}{Q'(z)} = \lim_{z \to z_0} F(z) \tag{2.117}$$

Analysis for higher-order infinities leads to results identical to equation 2.113.

Laurent expansion of a nonanalytic function

If $F(z')$ has a singularity at $z' = z_0$, the Taylor series of equation 2.100 is not a valid representation of the function. This is evident from the fact that equation 2.100 does not contain any singular structure at z_0. To develop a series representation for $F(z)$ (called the *Laurent series*), which contains the singularity at z_0, we again start with the Cauchy integral representation of $F(z)$,

$$F(z) = \frac{1}{2\pi i} \oint \frac{F(z')}{(z' - z)} dz' \tag{2.79}$$

Since $F(z')$ is singular at z_0, the contour we use must exclude this point. The appropriate contour is shown in Figure 2.18.

As with the development of the Taylor series, we define

$$z - z_0 \equiv h \tag{2.92}$$

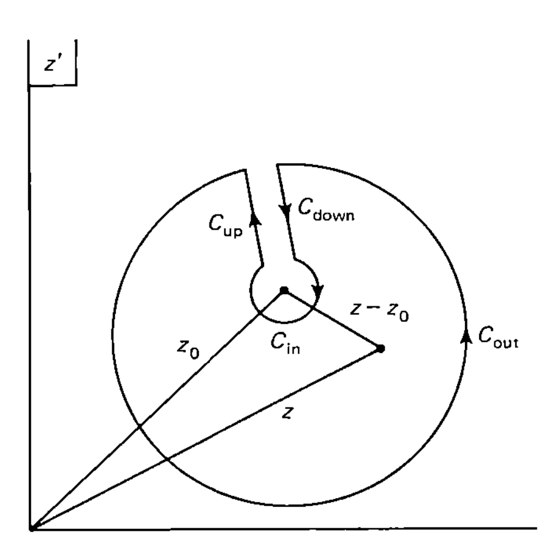

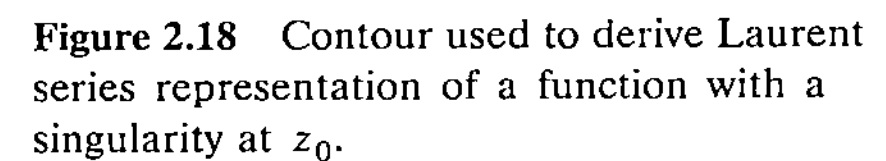

Figure 2.18 Contour used to derive Laurent series representation of a function with a singularity at z_0.

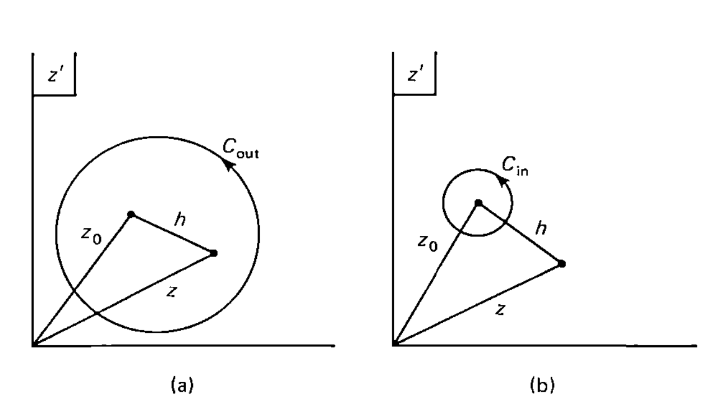

Figure 2.19 (a) Outer and (b) inner pieces of the contour used to derive the Laurent series representation of a function.

so that

$$F(z_0 + h) = \frac{1}{2\pi i}\oint \frac{F(z')}{(z' - z_0 - h)}\, dz' \tag{2.118}$$

To exclude the singularity at z_0, the contour is divided into four parts. As discussed in developing Cauchy's theorem, the integrals along C_{up} and C_{down} cancel. After reversing the direction of traversal of the inner contour so that all integrals are taken in the counterclockwise direction, equation 2.118 becomes

$$F(z_0 + h) = \frac{1}{2\pi i}\left[\oint_{C_{out}} \frac{F(z')}{(z' - z_0 - h)}\, dz' - \oint_{C_{in}} \frac{F(z')}{(z' - z_0 - h)}\, dz'\right] \tag{2.119}$$

All points on the outer contour satisfy $|h| < |z' - z_0|$ as depicted in Figure 2.19a. We see in Figure 2.19b that all points on the inner contour satisfy $|h| > |z' - z_0|$. To expand $1/(z' - z_0 - h)$ as in equation 2.97, this denominator must be manipulated so that the terms in the expansion have a magnitude less than 1. Thus for points on the outer contour, the expansion of $1/(z' - z_0 - h)$ is the one made to develop the Taylor series expansion of a nonsingular function: namely,

$$\frac{1}{(z' - z_0 - h)} = \frac{1}{(z' - z_0)}\left[\frac{1}{1 - h/(z' - z_0)}\right] = \sum_{n=0}^{\infty} \frac{h^n}{(z' - z_0)^{n+1}} \tag{2.97}$$

On the inner contour, since $|h| > |z' - z_0|$, we must make the expansion in terms of $(z' - z_0)/h$. Thus we write

$$\frac{1}{(z' - z_0 - h)} = -\frac{1}{h}\left[\frac{1}{1 - (z' - z_0)/h}\right] = -\sum_{n=0}^{\infty} \frac{(z' - z_0)^n}{h^{n+1}} \tag{2.120}$$

Equation 2.119 then becomes

$$F(z_0 + h) = F(z) =$$

$$\frac{1}{2\pi i}\left[\sum_{n=0}^{\infty} h^n \oint_{C_{out}} \frac{F(z')}{(z' - z_0)^{n+1}}\, dz' + \sum_{n=0}^{\infty} \frac{1}{h^{n+1}} \oint_{C_{in}} F(z')(z' - z_0)^n\, dz'\right] \tag{2.121}$$

In the second sum of equation 2.121, let $n = -n' - 1$. This second sum then becomes

$$\sum_{n'=-\infty}^{-1} h^{n'} \oint_{C_{\text{in}}} \frac{F(z')}{(z' - z_0)^{n'+1}} dz' = \sum_{n=-\infty}^{-1} h^{n} \oint_{C_{\text{in}}} \frac{F(z')}{(z' - z_0)^{n+1}} dz' \tag{2.122}$$

where the unnecessary prime on n' has been ignored. Then equation 2.121 becomes

$$F(z) = \frac{1}{2\pi i}\left[\sum_{n=0}^{\infty} h^n \oint_{C_{\text{out}}} \frac{F(z')}{(z' - z_0)^{n+1}} dz' + \sum_{n=-\infty}^{-1} h^n \oint_{C_{\text{in}}} \frac{F(z')}{(z' - z_0)^{n+1}} dz'\right] \tag{2.123}$$

Since the singularities of both $F(z')$ and $1/(z' - z_0)$ are at $z' = z_0$, there are no singularities of the integrands of equation 2.123 between C_{out} and C_{in}. Thus

$$\oint_{C_{\text{out}}} \frac{F(z')}{(z' - z_0)^{n+1}} dz' = \oint_{C_{\text{in}}} \frac{F(z')}{(z' - z_0)^{n+1}} dz' \tag{2.124}$$

That is, C_{in} can be expanded to become C_{out} and the two integrals can be combined to yield

$$F(z) = \frac{1}{2\pi i} \sum_{n=-\infty}^{\infty} h^n \oint \frac{F(z')}{(z' - z_0)^{n+1}} dz' \tag{2.125}$$

Because $F(z')$ is not analytic at z_0, the integral of equation 2.125 is not a multiple of the nth derivative of $F(z)$ at z_0. That is,

$$\oint \frac{F(z')}{(z' - z_0)^{n+1}} dz' \neq \left.\frac{2\pi i}{n!} \frac{d^n}{dz^n} F(z)\right)_{z=z_0} \tag{2.126}$$

However, this integral is a parameter that depends on z_0. We define

$$a_n(z_0) \equiv \frac{1}{2\pi i} \oint \frac{F(z')}{(z' - z_0)^{n+1}} dz' \tag{2.127}$$

$$\Rightarrow \quad F(z) = \sum_{n=-\infty}^{\infty} a_n(z_0) h^n = \sum_{n=-\infty}^{\infty} a_n(z_0)(z - z_0)^n \tag{2.128}$$

This is the Laurent series expansion of $F(z)$. As with the Taylor series, the contour around which the integral defining $a_n(z_0)$ is taken can be enlarged until the next singularity of $F(z)$ is encountered. Denoting the next singularity of $F(z)$ as z_1 (see Figure 2.17), the circle of convergence is centered at z_0 and has a radius of convergence $|z_1 - z_0|$.

Returning to the definition of $a_n(z_0)$ in equation 2.127, suppose that $F(z')$ is analytic at z_0. Then, for $n \geqslant 0$, $n! a_n(z_0)$ is the nth derivative $F^{(n)}(z_0)$ (see equation 2.80c). For $n \leqslant -1$,

$$a_n(z_0) = \frac{1}{2\pi i} \oint F(z')(z' - z_0)^{|n|-1} dz' \tag{2.129}$$

Since both $F(z')$ and $(z' - z_0)^{|n|-1}$ are analytic for this range of n-values, this integral is zero. Therefore, if $F(z)$ is analytic, all the terms in the Laurent sum with $n \leqslant -1$ are zero, and the Laurent series reduces to the Taylor series.

Consider the Laurent series when the singularity of $F(z)$ is a pole at z_0 of order M. Then

$$\lim_{z \to z_0} (z - z_0)^M F(z) = \text{a finite, nonzero number} \tag{2.130}$$

Thus, for any integer $K \geqslant 0$, $(z - z_0)^{M+K}F(z)$ is analytic at z_0. From equation 2.127, the $-M - L$ coefficient is

$$a_{-(M+L)}(z_0) = \frac{1}{2\pi i}\oint F(z')(z' - z_0)^{M+L-1}\, dz' \qquad (2.131)$$

If $L \geqslant 1$, $F(z')(z' - z_0)^{M+L-1}$ is analytic inside and on the contour, and $a_{-(M+L)} = 0$. That is, if $F(z)$ has an Mth-order pole at z_0,

$$a_{-(M+1)} = a_{-(M+2)} = a_{-(M+3)} = \cdots = 0 \qquad (2.132)$$

$$\Rightarrow \quad F(z) = \frac{a_{-M}}{(z - z_0)^M} + \frac{a_{-(M-1)}}{(z - z_0)^{M-1}} + \cdots + a_0 + a_1(z - z_0) + \cdots \qquad (2.133)$$

We note that the part of this series that contains nonnegative powers of $z - z_0$ is an analytic function. That is,

$$F(z) = \sum_{n=1}^{M} \frac{a_{-n}}{(z - z_0)^n} + \text{analytic part} \qquad (2.134)$$

Consider the integral of a function with an Mth-order pole around a contour enclosing the pole (Figure 2.20).

$$\oint F(z)\, dz = \sum_{n=1}^{M} a_{-n} \oint \frac{dz}{(z - z_0)^n} + \oint (\text{analytic part})\, dz \qquad (2.135)$$

Obviously, the integral of the analytic function around a closed contour is zero. We choose the contour to be a small circle of radius ρ, centered at z_0. (We will eventually take the limit $\rho = 0$.) Then for any point on the small circle,

$$\text{(a)} \quad z = z_0 + \rho e^{i\phi} \quad \Rightarrow \quad \text{(b)} \quad dz = i\rho e^{i\phi}\, d\phi \qquad (2.136)$$

$$\Rightarrow \quad a_{-n} \oint \frac{dz}{(z - z_0)^n} = a_{-n} \int_0^{2\pi} \frac{i\rho e^{i\phi}}{\rho^n e^{in\phi}}\, d\phi = a_{-l} \int_0^{2\pi} i\rho^{-n+1} e^{-i(n-1)\phi}\, d\phi \qquad (2.137)$$

But

$$\int_0^{2\pi} e^{-i(n-1)\phi}\, d\phi = \begin{cases} 0 & n \neq 1 \\ 2\pi & n = 1 \end{cases} \qquad (2.138)$$

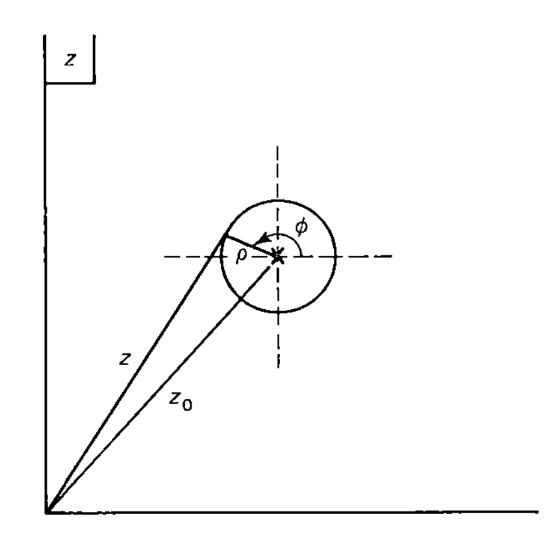

Figure 2.20 Small circle around the Mth-order pole of a function.

Therefore, in terms of the Kronecker δ-symbol,

$$a_{-n}\oint\frac{dz}{(z-z_0)^n} = a_{-n}2\pi i\delta_{n1} \tag{2.139}$$

Thus, equation 2.135 becomes

$$\oint F(z)\,dz = 2\pi i a_{-1}(z_0) \tag{2.140}$$

which is independent of the order of the pole.

If the pole is a simple pole, equation 2.133 is

$$F(z) = \frac{a_{-1}(z_0)}{(z-z_0)} + \text{analytic part} \tag{2.141}$$

Then

$$\lim_{z\to z_0}(z-z_0)F(z) = a_{-1}(z_0) + \lim_{z\to z_0}(z-z_0)\cdot\text{analytic part} = a_{-1}(z_0) \tag{2.142}$$

That is, $a_{-1}(z_0)$ satisfies the original definition of the residue of a function with a simple pole. For that reason, $a_{-1}(z_0)$ is called the residue for a pole of any order. It is evident that determining this residue is essential for evaluating the integral of a function with a pole.

Methods of determining the residue

We now present three straightforward approaches for determining $a_{-1}(z_0)$.

I) Direct Laurent expansion

From equations 2.128, $a_{-1}(z_0)$ is the coefficient of the $1/(z-z_0)$ term in the Laurent expansion. Therefore, by determining the Laurent series for the function, we can identify the residue directly from the expansion.

Example 2.16

As an example, we find the residue of

$$F(z) = \frac{e^z}{(z-1)^4} \tag{2.143}$$

which has a fourth-order pole at $z = 1$. As determined in equation 2.102a, the expansion of e^z around $z = 1$ is

$$e^z = e\sum_{n=0}^{\infty}\frac{(z-1)^n}{n!} \tag{2.144}$$

Therefore, the Laurent series for $F(z)$ is

$$F(z) = \frac{e^z}{(z-1)^4} = e\left[\frac{1}{(z-1)^4} + \frac{1}{(z-1)^3} + \frac{1}{2!}\frac{1}{(z-1)^2} + \frac{1}{3!}\frac{1}{(z-1)} + \frac{1}{4!} + \cdots\right] \tag{2.145}$$

from which the coefficient of $1/(z-1)$ is

$$a_{-1}(1) = \frac{e}{3!} \quad \square \tag{2.146}$$

II) Derivative method

If $F(z)$ has a pole of order M, its Laurent expansion is

$$F(z) = \frac{a_{-M}}{(z-z_0)^M} + \frac{a_{-(M-1)}}{(z-z_0)^{M-1}} + \cdots + \frac{a_{-1}}{(z-z_0)} + \text{analytic part} \tag{2.133}$$

Then

$$\begin{gathered}(z-z_0)^M F(z) = \\ a_{-M} + a_{-(M-1)}(z-z_0) + \cdots + a_{-1}(z-z_0)^{M-1} + (z-z_0)^M \cdot \text{analytic part}\end{gathered} \tag{2.147}$$

We note that $a_{-1}(z_0)$ is multiplied by $(z-z_0)^{M-1}$. By taking $M-1$ derivatives of $(z-z_0)^M F(z)$, all terms containing powers of $(z-z_0)$ lower than $M-1$ will be zero. Then

$$\begin{gathered}\frac{d^{M-1}}{dz^{M-1}}\left[(z-z_0)^M F(z)\right] = \\ (M-1)!a_{-1} + \frac{M!}{1!}a_0(z-z_0) + \frac{(M+1)!}{2!}a_1(z-z_0)^2 + \cdots\end{gathered} \tag{2.148}$$

Setting $z = z_0$, all terms except the one containing a_{-1} are zero. Therefore,

$$a_{-1}(z_0) = \frac{1}{(M-1)!}\frac{d^{M-1}}{dz^{M-1}}\left[(z-z_0)^M F(z)\right]_{z=z_0} \tag{2.149}$$

If $F(z)$ has a pole of order M, it can be written as

$$F(z) \simeq \frac{f(z)}{(z-z_0)^M} \tag{2.150}$$

in the region near z_0. Then equation 2.149 becomes

$$a_{-1}(z_0) = \frac{1}{(M-1)!}\frac{d^{M-1}}{dz^{M-1}}f(z)\Big)_{z=z_0} \tag{2.151}$$

Referring to equation 2.80c, if $f(z)$ is analytic, we found from Cauchy's theorem that

$$\oint \frac{f(z)}{(z-z_0)^M}\,dz = \frac{2\pi i}{(M-1)!}\frac{d^{M-1}}{dz^{M-1}}f(z)\Big)_{z=z_0} \tag{2.152}$$

Therefore, we see that

$$\oint \frac{f(z)}{(z-z_0)^M}\,dz = 2\pi i a_{-1}(z_0) = 2\pi i \cdot \text{residue} \tag{2.153}$$

as mentioned earlier.

Example 2.17

Using

$$F(z) = \frac{e^z}{(z-1)^4} \tag{2.143}$$

$$\Rightarrow \quad (z-1)^4 F(z) = e^z \tag{2.154}$$

$$\Rightarrow \quad a_{-1}(1) = \frac{1}{3!}\frac{d^3}{dz^3}e^z\Big)_{z=1} = \frac{e}{3!} \tag{2.155}$$

III) Ratio method

Often, when $F(z)$ has an Mth-order pole, it can be written as the ratio of two analytic functions,

$$F(z) = \frac{P(z)}{Q(z)} \tag{2.156}$$

and it is an Mth-order zero of $Q(z)$ that gives rise to the Mth-order pole of $F(z)$. That is, the Taylor series for $Q(z)$ is of the form

$$Q(z) = \frac{Q^{(M)}(z_0)}{M!}(z - z_0)^M + \text{higher-order terms in } (z - z_0) \tag{2.157a}$$

For example, if $F(z)$ has a simple pole, then

$$Q(z) = Q'(z_0)(z - z_0) + \frac{1}{2!}Q''(z_0)(z - z_0)^2 \tag{2.157b}$$

$$\Rightarrow \quad F(z) = \frac{P(z)}{Q'(z_0)(z - z_0) + (1/2!)Q''(z_0)(z - z_0)^2 + \cdots} \tag{2.158}$$

Then

$$a_{-1}(z_0) = \lim_{z \to z_0} (z - z_0)\left[\frac{P(z)}{Q'(z_0)(z - z_0) + (1/2!)Q''(z_0)(z - z_0)^2 + \cdots}\right] = \frac{P(z_0)}{Q'(z_0)} \tag{2.159}$$

For a simple pole, this result can also be deduced quite simply using either the Laurent series expansion or the derivative method discussed above. In addition, for a simple pole, equation 2.159 can also be obtained straightforwardly using L'Hospital's rule. For a simple pole, the residue is

$$a_{-1}(z_0) = \lim_{z \to z_0} \frac{(z - z_0)P(z)}{Q(z)} \tag{2.160}$$

which takes the form $0/0$. From L'Hospital's rule,

$$a_{-1}(z_0) = \lim_{z \to z_0} \frac{\frac{d}{dz}[(z - z_0)P(z)]}{\frac{d}{dz}Q(z)} = \frac{[(z - z_0)P'(z) + P(z)]}{Q'(z)} = \frac{P(z_0)}{Q'(z_0)} \tag{2.161}$$

This ratio method becomes cumbersome rather quickly. As the reader will show in Problem 13, if the pole is of second order, the residue is

$$a_{-1}(z_0) = \frac{2[3P'(z_0)Q''(z_0) - Q'''(z_0)P(z_0)]}{3[Q''(z_0)]^2} \tag{2.162}$$

Example 2.18

As an example, $\tan z$ has simple poles at all odd multiples of $\pi/2$. Thus, to apply this technique to find the residue at $z = \pi/2$, we write $\tan z = \sin z/\cos z$ and identify $P(z) = \sin z$, and $Q(z) = \cos z$. Therefore, equation 2.159 yields

$$a_{-1}\left(\frac{\pi}{2}\right) = \lim_{z \to \pi/2} (z - \pi/2)\tan z = \frac{P(\pi/2)}{Q'(\pi/2)} = \frac{\sin(\pi/2)}{-\sin(\pi/2)} = -1 \quad \square \tag{2.163}$$

Example 2.19

The function $e^z/\sin^2 z$ has second-order poles at integer multiples of π. With $P(z) = e^z$ and $Q(z) = \sin^2 z$, equation 2.162 yields

$$a_{-1}(2\pi) = \left[\frac{2}{3}e^z\left(\frac{\cos(2z) + 2\sin(2z)}{2\cos^2(2z)}\right)\right]_{z=2\pi} = \frac{e^{2\pi}}{3} \quad \square \tag{2.164}$$

If the function has two or more poles, the techniques presented here are easily applied.

Example 2.20

For example,

$$F(z) = \frac{e^z}{z(z+2)^2} \tag{2.165}$$

has a simple pole at $z = 0$ and a second-order pole at $z = -2$. Therefore, at $z = -2$, the analytic part of $F(z)$ is e^z/z and it is this function that will be expanded around $z = -2$ (if the direct Laurent expansion method is used), or will be differentiated (if the derivative method is used). If the ratio method is used, we can either identify e^z/z as $P(z)$ and $(z+2)^2$ as $Q(z)$ or take $P(z)$ to be e^z and $z(z+2)^2$ as $Q(z)$.

Laurent expansion approach: Using equation 2.102a with $z_0 = -2$,

$$e^z = e^{-2}\sum_{n=0}^{\infty}\frac{(z+2)^n}{n!} \tag{2.166}$$

$$\frac{1}{z} = -\frac{1}{2-(z+2)} = -\frac{1}{2}\frac{1}{(1-\frac{1}{2}(z+2))} = -\frac{1}{2}\sum_{n=0}^{\infty}\frac{(z+2)^n}{2^n} \tag{2.167}$$

Therefore,

$$\begin{aligned}\frac{e^z}{z(z+2)^2} &= -\frac{e^{-2}}{2}\frac{\left(1+\frac{3}{2}(z+2)+\frac{5}{2}(z+2)^2+\cdots\right)}{(z+2)^2}\\ &= -\frac{e^{-2}}{2}\left(\frac{1}{(z+2)^2}+\frac{3}{2}\frac{1}{(z+2)}+\frac{5}{2}+\cdots\right)\end{aligned} \tag{2.168}$$

$$\Rightarrow \quad a_{-1}(-2) = -\tfrac{3}{4}e^{-2} \tag{2.169}$$

Derivative method: From

$$(z+2)^2F(z) = \frac{e^z}{z} \tag{2.170}$$

we obtain

$$a_{-1}(-2) = \frac{1}{1!}\frac{d}{dz}\left(\frac{e^z}{z}\right)_{z=-2} = -\tfrac{3}{4}e^{-2} \tag{2.171}$$

Identifying the function as the ratio of analytic functions: We take

$$\begin{aligned}&\text{(a)}\ \ P(z) = e^z && \text{(b)}\ \ Q(z) = z(z+2)^2\\ \Rightarrow\ &\text{(c)}\ \ P'(-2) = e^{-2} && \text{(d)}\ \ P''(-2) = e^{-2} \qquad \text{(e)}\ \ Q'(-2) = 0\\ &\text{(f)}\ \ Q''(-2) = -4 && \text{(g)}\ \ Q'''(-2) = 6\end{aligned} \tag{2.172}$$

Thus, from equation 2.162, the residue is

$$a_{-1}(-2) = -\tfrac{3}{4}e^{-2} \quad \square \tag{2.173}$$

2.5 Evaluation of Integrals: Part II

Integrals over the entire real axis

The results just developed can now be applied to the evaluation of certain types of integrals that arise in scientific problems. One such class of integrals is of the form

$$I = \int_{-\infty}^{\infty} F(x)\, dx \tag{2.174}$$

If this integral is to exist, $F(x)$ must approach 0 at ∞ at a rapid enough rate (the meaning of which will be determined later). For the moment, we assume that the rate at which $F(x)$ approaches 0 is fast enough to apply the techniques developed previously. We also assume that $F(x)$ does not have any singularities on the x-axis.

Consider the integral $\oint F(z)\, dz$ around the contour shown in Figure 2.21. The contour is comprised of the real axis from $-\infty$ to ∞, and a semicircle of very large (eventually infinite) radius. On the real axis, $z = x$. Thus

$$\oint F(z)\, dz = \int_{-\infty}^{\infty} F(x)\, dx + \int_{C_\infty} F(z)\, dz \tag{2.175}$$

Points on the infinite semicircle are of the form $Re^{i\phi}$, and since that part of the contour is a semicircle, R = constant. For these points, $dz = iRe^{i\phi}\, d\phi$. Thus for infinite radius,

$$\int_{C_\infty} F(z)\, dz = \lim_{R\to\infty} \int_0^{\infty} F(Re^{i\phi}) Rie^{i\phi}\, d\phi \tag{2.176}$$

$e^{i\phi}$ is a complex number, the magnitude of which is 1 and therefore cannot affect the behavior of the integrand when $R \to \infty$. Thus the behavior of the integral for infinite argument is determined by $RF(R)$. If

$$\lim_{R\to\infty} RF(R) = 0 \tag{2.177}$$

the integral on the infinite semicircle is zero, and

$$\oint F(z)\, dz = \int_{-\infty}^{\infty} F(x)\, dx \tag{2.178}$$

Equation 2.177 is what is meant by the condition "fast enough." For example, if $F(z)$ behaves like $1/z^p$ for large z, then $F(z)$ approaches zero when $z \to \infty$ at a fast enough rate if $p > 1$.

Since we can use Cauchy's theorem to determine the integral around the closed contour of Figure 2.21, the integral of equation 2.174 can be evaluated as

$$\int_{-\infty}^{\infty} F(x)\, dx = 2\pi i \sum R_{+1/2} \tag{2.179}$$

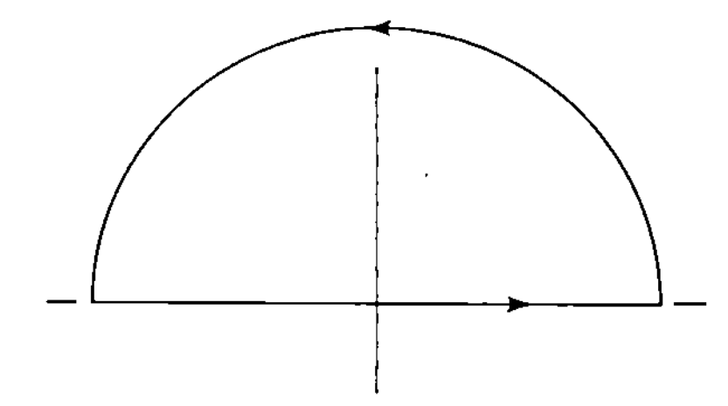

Figure 2.21 Possible contour for evaluating $\int_{-\infty}^{\infty} F(x)\, dx$.

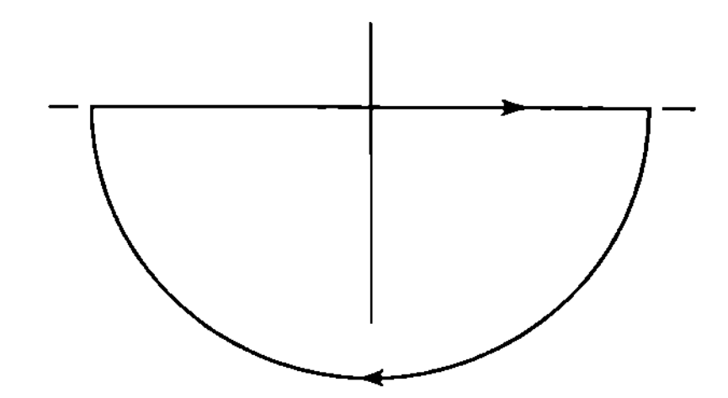

Figure 2.22 Possible contour for evaluating $\int_{-\infty}^{\infty} F(x)\, dx$.

$\Sigma R_{+1/2}$ denotes the sum of the residues in the upper half-plane, the half-plane enclosed by the contour of Figure 2.21.

We also note in Figure 2.21 that by closing the contour in the upper half-plane, integration along the real axis from $-\infty$ to $+\infty$ forces the contour to be traversed in the counterclockwise direction.

Many integrals of functions that approach zero fast enough in either half-plane ($1/z^N$, for example) can be evaluated just as easily by closing the contour in the lower half-plane, as shown in Figure 2.22. But for this contour, integrating along the real axis from $-\infty$ to $+\infty$ causes the integration around the contour to be in the clockwise direction. Thus, for such problems, closing in the lower half-plane would yield

$$\int_{-\infty}^{\infty} F(x)\,dx = -2\pi i \sum R_{-1/2} \tag{2.180}$$

Example 2.21

As an example, consider

$$I = \int_{-\infty}^{\infty} \frac{dx}{(1+x^2)} \tag{2.181}$$

We begin by considering

$$\oint \frac{dz}{(1+z^2)} = \int_{-\infty}^{\infty} \frac{dx}{(1+x^2)} + \int_{C_\infty} \frac{dz}{(1+z^2)} \tag{2.182}$$

Figure 2.23 shows the contour closed in the upper half-plane, along with the two poles of $1/(1+z^2)$. On the infinite semicircle

$$\text{(a)}\quad \frac{1}{(1+z^2)} \to \frac{1}{z^2} \to 0 \qquad \text{and} \qquad \text{(b)}\quad \lim_{R\to\infty} R\frac{1}{(1+R^2)} = 0 \tag{2.183}$$

Thus $1/(1+z^2)$ approaches zero at infinity "fast enough."

Because only the pole at $z = +i$ is inside the contour,

$$\int_{-\infty}^{\infty} \frac{dx}{(1+x^2)} = 2\pi i R(+i) \tag{2.184}$$

Evaluating

$$R(+i) = (z-i)\frac{1}{(z+i)(z-i)}\bigg)_{z=i} = \frac{1}{2i} \tag{2.185}$$

$$\Rightarrow \quad \int_{-\infty}^{\infty} \frac{dx}{(1+x^2)} = \pi \tag{2.186}$$

Were we to close in the lower half-plane as shown in Figure 2.24, the integral would be

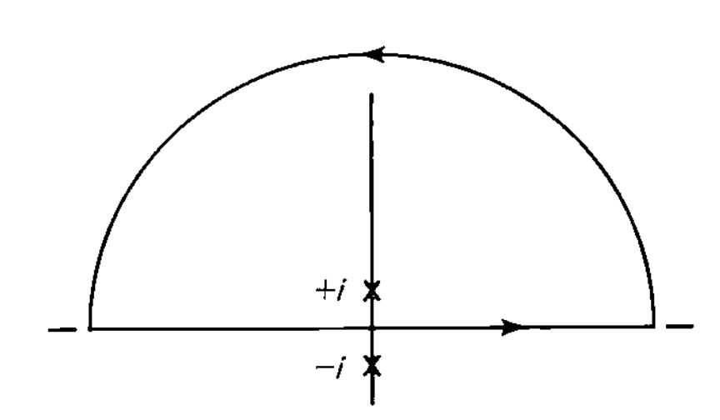

Figure 2.23 Possible contour for evaluating $\int_{-\infty}^{\infty} dx/(1+x^2)$.

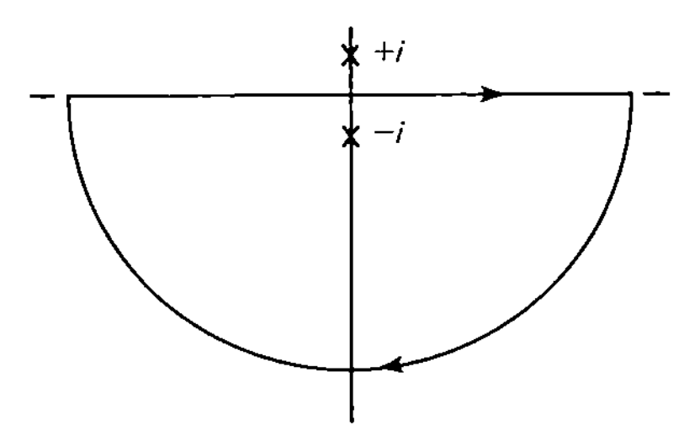

Figure 2.24 Possible contour for evaluating $\int_{-\infty}^{\infty} dx/(1+x^2)$.

evaluated as

$$\int_{-\infty}^{\infty} \frac{dx}{(1+x^2)} = -2\pi i R(-i) \tag{2.187}$$

The residue is

$$R(-i) = -\frac{1}{2i} \tag{2.188}$$

$$\Rightarrow \quad \int_{-\infty}^{\infty} \frac{dx}{(1+x^2)} = \pi \quad \square \tag{2.186}$$

Consider applying this approach to $\int_0^\infty F(x)\,dx$. Since the lower limit is 0, the contour does not encompass the entire real axis. Therefore, we cannot close the contour in one of the half-planes, and thus cannot apply Cauchy's theorem to evaluate the integral. However, if $F(x)$ is an even function that approaches zero at infinity "fast enough," we can write

$$\int_0^{\infty} F(x)\,dx = \tfrac{1}{2}\int_{-\infty}^{\infty} F(x)\,dx \tag{2.189}$$

and use Cauchy's theorem to evaluate the integral. That is, for even functions that approach zero at infinity "fast enough,"

$$\text{(a)} \quad \int_0^{\infty} F(x)\,dx = i\pi \sum R_{+1/2} \qquad \text{(b)} \quad \int_0^{\infty} F(x)\,dx = -i\pi \sum R_{-1/2} \tag{2.190}$$

Fourier integrals

There are certain types of integrals that can be evaluated using Cauchy's theorem, for which one must close the contour in only one of the half-planes. One such integral, called a *Fourier integral*, is of the form

$$I = \int_{-\infty}^{\infty} F(x) e^{ikx}\,dx \tag{2.191}$$

where $F(z)$ in general has poles in both half-planes, but not on the x-axis.

To demonstrate that for a Fourier integral, one is not free to choose the half-plane in which to close the contour, we will take $k > 0$, and attempt to evaluate the Fourier integral by closing the contour in the lower half-plane as in Figure 2.22.

Taking the integrals over the two parts of the contour, we have

$$\oint F(z) e^{ikz}\,dz = \int_{-\infty}^{\infty} F(x) e^{ikx}\,dx + \int_{C_{-\infty}} F(z) e^{ikz}\,dz \tag{2.192}$$

On the infinite semicircle, $z = Re^{i\phi}$, with ϕ ranging from 2π to π. That is,

$$\int_{C_{-\infty}} F(z) e^{ikz}\,dz = \int_{2\pi}^{\pi} F(Re^{i\phi}) e^{ikRe^{i\phi}} iRe^{i\phi}\,d\phi \tag{2.193}$$

But

$$e^{ikRe^{i\phi}} = e^{ikR\cos\phi} e^{-kR\sin\phi} \tag{2.194}$$

For $\pi < \phi < 2\pi$, $\sin\phi < 0$. Thus in the lower half-plane,

$$e^{-kR\sin\phi} = e^{+kR|\sin\phi|} \tag{2.195}$$

When $R \to \infty$, this exponential becomes infinite for $k > 0$. Therefore, the integral around the semicircle in the lower half-plane is infinite.

With $k > 0$, consider closing the contour in the upper half-plane, as in Figure 2.21. For such a contour, ϕ ranges from 0 to π on the infinite semicircle. On that part of the contour, $\sin \phi > 0$ and

$$e^{-kR\sin\phi} \to 0 \qquad k > 0 \tag{2.196}$$

when $R \to \infty$. So for positive k, the contour must be closed in the upper half-plane, and

$$\oint F(z)e^{ikz}\,dz = \int_{-\infty}^{\infty} F(x)e^{ikx}\,dx = 2\pi i \sum R_{+1/2} \qquad k > 0 \tag{2.197a}$$

Similarly, if k is positive, we use the same reasoning in analyzing $\oint F(z)e^{-ikz}\,dz$ in order to evaluate $\int_{-\infty}^{\infty} F(x)e^{-ikx}\,dx$. We find that we must close the contour in the lower half-plane for this negative exponent. The contour is then traversed in the clockwise direction, so

$$\oint F(z)e^{-ikz}\,dz = \int_{-\infty}^{\infty} F(x)e^{-ikx}\,dx = -2\pi i \sum R_{-1/2} \qquad k > 0 \tag{2.197b}$$

Example 2.22

As an example, we evaluate

$$I = \int_{-\infty}^{\infty} \frac{e^{2ix}}{(1+x^2)}\,dx \tag{2.198}$$

Since $k = 2$ in the exponent is positive, we must close the contour in the upper half-plane, thereby enclosing the pole at $z = +i$. Using equation 2.197a,

$$\int_{-\infty}^{\infty} \frac{e^{2ix}}{(1+x^2)}\,dx = 2\pi i R(+i) = \pi e^{-2} \tag{2.199}$$

which is real. From this we note that

$$\int_{-\infty}^{\infty} \frac{e^{2ix}}{(1+x^2)}\,dx = \int_{-\infty}^{\infty} \frac{\cos 2x}{(1+x^2)}\,dx + i\int_{-\infty}^{\infty} \frac{\sin 2x}{(1+x^2)}\,dx = \pi e^{-2} \tag{2.200}$$

Equating real and imaginary parts separately, we obtain

$$\text{(a)} \quad \int_{-\infty}^{\infty} \frac{\cos 2x}{(1+x^2)}\,dx = \pi e^{-2} \qquad \text{(b)} \quad \int_{-\infty}^{\infty} \frac{\sin 2x}{(1+x^2)}\,dx = 0 \quad \square \tag{2.201}$$

The reader is cautioned that this approach cannot be used to evaluate any integral involving an exponential integrand. It is only applicable to an integrand containing an exponential function that has a linear exponent.

Example 2.23

For example, consider

$$\oint F(z)e^{ikz^2}\,dz = \int_{-\infty}^{\infty} F(x)e^{ikx^2}\,dx + \int_{C_\infty} F(z)e^{ikz^2}\,dz \tag{2.202}$$

with $k > 0$. On the infinite semicircle, with $z = Re^{i\phi}$,

$$e^{ikz^2} = e^{ikR^2\cos(2\phi)}e^{-kR^2\sin(2\phi)} \tag{2.203}$$

In the upper half-plane ϕ ranges from 0 to π. For $0 < \phi < \pi/2$, $\sin(2\phi) > 0$, so the exponential $\to 0$ when $R \to \infty$. But for $\pi/2 < \phi < \pi$, $\sin(2\phi) < 0$, and the exponential is infinite for infinite R. Thus the integral over half the infinite semicircle will be infinite. A similar argument would apply if we tried to close in the lower half-plane. Thus the integral of equation 2.202, if it exists, cannot be evaluated using Cauchy's theorem. □

Integrals involving sines and cosines

Another type of integral that can be evaluated using Cauchy's theorem is of the form

$$I = \int_0^{2\pi} F(\cos\theta, \sin\theta)\, d\theta \tag{2.204}$$

To evaluate this, we write

$$\text{(a)}\quad \cos\theta = \frac{(e^{i\theta} + e^{-i\theta})}{2} \qquad \text{(b)}\quad \sin\theta = \frac{(e^{i\theta} - e^{-i\theta})}{2i} \tag{2.19}$$

We then make the substitution

$$\text{(a)}\quad z = e^{i\theta} \qquad \Rightarrow \qquad \text{(b)}\quad dz = ie^{i\theta}\, d\theta = iz\, d\theta \tag{2.205}$$

Since $|z| = |e^{i\theta}| = 1$, with $0 \leqslant \theta \leqslant 2\pi$, these points lie on the unit circle. Thus the integral of equation 2.204 becomes

$$I = \oint_{\substack{\text{unit}\\ \text{circle}}} F\left(\frac{(z + z^{-1})}{2}, \frac{(z - z^{-1})}{2i}\right) \frac{dz}{iz} \tag{2.206}$$

In this form, the integral is evaluated by Cauchy's theorem as

$$\oint_{\substack{\text{unit}\\ \text{circle}}} F\left(\frac{(z + z^{-1})}{2}, \frac{(z - z^{-1})}{2i}\right) dz = 2\pi i \sum R(\text{inside unit circle}) \tag{2.207}$$

Example 2.24

As an example, consider

$$I = \int_0^{2\pi} \frac{\cos\theta}{(20 + 12\sin\theta)}\, d\theta \tag{2.208}$$

Writing

$$\text{(a)}\quad \cos\theta = \frac{z + z^{-1}}{2} \qquad \text{(b)}\quad \sin\theta = \frac{z - z^{-1}}{2i} \tag{2.209}$$

$$\Rightarrow \quad \int_0^{2\pi} \frac{\cos\theta}{(20 + 12\sin\theta)}\, d\theta = \frac{1}{12} \oint_{\substack{\text{unit}\\ \text{circle}}} \frac{z^2 + 1}{\left[z\left(z^2 + \frac{10}{3}iz - 1\right)\right]}\, dz \tag{2.210}$$

This integrand has poles at $z = 0, -3i, -i/3$. Clearly, two of these lie inside the unit circle, and one is outside. Therefore,

$$\int_0^{2\pi} \frac{\cos\theta}{(20 + 12\sin\theta)}\, d\theta = 2\pi i\left[R(0) + R\left(-\frac{i}{3}\right)\right] \tag{2.211}$$

The reader can easily verify that $R(0) = -\frac{1}{12}$ and $R(-i/3) = \frac{1}{12}$. Therefore, $I = 0$. This can easily be verified by integrating the function directly.

$$\int_0^{2\pi} \frac{\cos\theta}{20 + 12\sin\theta}\, d\theta = \frac{1}{12}\int_0^{2\pi} \frac{d(12\sin\theta)}{20 + 12\sin\theta} = \frac{1}{12}\log(20 + 12\sin\theta)\Big|_0^{2\pi} = 0 \quad \square \tag{2.212}$$

Integrals of the form

$$(a)\quad I = \int_{-\infty}^{\infty} F(x)\cos(kx)\,dx \qquad (b)\quad I = \int_{-\infty}^{\infty} F(x)\sin(kx)\,dx \tag{2.213}$$

can be easily put in forms we have already investigated. Writing

$$(a)\quad \cos(kx) = \frac{(e^{ikx} + e^{-ikx})}{2} \qquad (b)\quad \sin(kx) = \frac{(e^{ikx} - e^{-ikx})}{2i} \tag{2.214}$$

the integrals of equations 2.213 take the form

$$I = \text{constant} \cdot \int_{-\infty}^{\infty} F(x)(e^{ikx} \pm e^{-ikx})\,dx \tag{2.215}$$

We have already discussed the evaluation of $\int_{-\infty}^{\infty} F(x)e^{\pm ikx}\,dx$. Therefore, we can evaluate the integrals of equations 2.213 by applying Cauchy's theorem to two exponential integrals: one for a positive exponent, and a second for a negative exponent.

If $F(x)$ is real or if it is purely imaginary, the integrals of equations 2.213 can be evaluated by applying Cauchy's theorem for a single exponential integral. For example, if $F(x)$ is real,

$$\int_{-\infty}^{\infty} F(x)\cos(kx)\,dx = \int_{-\infty}^{\infty} F(x)\mathrm{Re}(e^{ikx})\,dx = \mathrm{Re}\int_{-\infty}^{\infty} F(x)e^{ikx}\,dx \tag{2.216a}$$

$$\int_{-\infty}^{\infty} F(x)\sin(kx)\,dx = \int_{-\infty}^{\infty} F(x)\mathrm{Im}(e^{ikx})\,dx = \mathrm{Im}\int_{-\infty}^{\infty} F(x)e^{ikx}\,dx \tag{2.216b}$$

If $F(x)$ is purely imaginary, it can be written as $F(x) = iG(x)$, with $G(x)$ real. Then

$$\int_{-\infty}^{\infty} F(x)\cos(kx)\,dx = i\int_{-\infty}^{\infty} G(x)\mathrm{Re}(e^{ikx})\,dx = i\,\mathrm{Re}\int_{-\infty}^{\infty} G(x)e^{ikx}\,dx \tag{2.216c}$$

$$\int_{-\infty}^{\infty} F(x)\sin(kx)\,dx = i\int_{-\infty}^{\infty} G(x)\mathrm{Im}(e^{ikx})\,dx = i\,\mathrm{Im}\int_{-\infty}^{\infty} G(x)e^{ikx}\,dx \tag{2.216d}$$

2.6 Multivalued Functions, Branch Points, and Cuts

A branch point singularity occurs when a function is not single-valued. To see what this means, consider the polar form of z:

$$z(r,\theta) = re^{i\theta} \tag{2.217}$$

Increasing θ by 2π,

$$z(r,\theta + 2\pi) = re^{i(\theta+2\pi)} = re^{i\theta} = z(r,\theta) \tag{2.218}$$

That is, when θ is increased by 2π, z returns to its original value.

If a function $F(z)$ satisfies

$$F[z(r,\theta + 2\pi)] = F[z(r,\theta)] \tag{2.219}$$

then $F(z)$ is said to be single-valued. A function that does not satisfy equation 2.219 is said to be multivalued.

One type of function that arises in many scientific problems is

$$F(z) = z^n \tag{2.220}$$

Then

$$\text{(a)} \quad F[z(r,\theta)] = r^n e^{in\theta} \qquad \text{(b)} \quad F[z(r,\theta + 2\pi)] = r^n e^{in\theta} e^{2\pi in} \tag{2.221}$$

If n is an integer, $e^{2\pi in} = 1$ and F has the same value when the argument of z is $\theta + 2\pi$ as it has when the argument of z is θ. That is, when n is an integer, F is a single-valued function. If n is not an integer, $e^{in2\pi} \neq 1$ and F is multivalued.

Example 2.25

For example, if $n = \frac{1}{3}$,

$$e^{2\pi i/3} = \cos\left(\frac{2\pi}{3}\right) + i\sin\left(\frac{2\pi}{3}\right) = \frac{(-1 + i\sqrt{3})}{2} \tag{2.222}$$

Thus

$$F[z(r,\theta)] = [z(r,\theta)]^{1/3} = r^{1/3}e^{i\theta/3} \tag{2.223a}$$

$$\begin{aligned} F[z(r,\theta + 2\pi)] &= [z(r,\theta + 2\pi)]^{1/3} = r^{1/3}e^{i\theta/3}\frac{(-1 + i\sqrt{3})}{2} \\ &= F[z(r,\theta)]\frac{(-1 + i\sqrt{3})}{2} \end{aligned} \tag{2.223b}$$

If θ is increased by another 2π (4π in all),

$$F[z(r,\theta + 4\pi)] = r^{1/3}e^{i\theta/3}\frac{(-1 - i\sqrt{3})}{2} = F[z(r,\theta)]\frac{(-1 - i\sqrt{3})}{2} \tag{2.223c}$$

Since $e^{6\pi i/3} = 1$, a third increase of θ by 2π (6π in all) yields

$$F[z(r,\theta + 6\pi)] = F[z(r,\theta)] \tag{2.223d}$$

Since at one value of z, there are three possible values of F, F is a triple-valued function. In general, if N is an integer, $z^{1/N}$ is an N-valued function. □

In the z-plane, increasing θ by 2π brings z back to its original value. However, consider a plane in which the axes are labeled not by the real and imaginary parts of z but by the real and imaginary parts of a multivalued function F. When θ is increased by 2π, F does not return to its original value, even though it is evaluated at the same value of z. To keep track of this multivaluedness of F, we conjecture not one complex F-plane but a multiple layer of many complex F-planes. When one increases θ by 2π, F returns to a point on an adjacent complex plane. The planes are called Riemann sheets of the function, and a multivalued function is, therefore, also an analytic function in a multisheeted set of complex planes. For example, $F(z) = z^{1/3}$ is a three-sheeted function since it has three different values at a given value of z.

For the function $z^{1/3}$, with $z = re^{i\theta}$, we note that this multivaluedness occurs for any value of $r > 0$. When $r = 0$, $z = 0$. The point where the multivaluedness begins ($z = 0$ for $z^{1/3}$) is called the branch point or the point of accumulation. The multivaluedness occurs for any value of θ. Therefore, there exists a line, starting at the branch point, and running to ∞ at an arbitrary value of θ, for which the function has different values on the two sides of the line. For example, Figure 2.25 shows the values of $z^{1/3}$ on two sides of a line in the complex F-plane that runs from 0 to ∞ at an arbitrarily chosen θ. On the upper side of the line, $F(z)$ has the value

$$F[z(r,\theta)] = r^{1/3}e^{i\theta/3} \tag{2.224a}$$

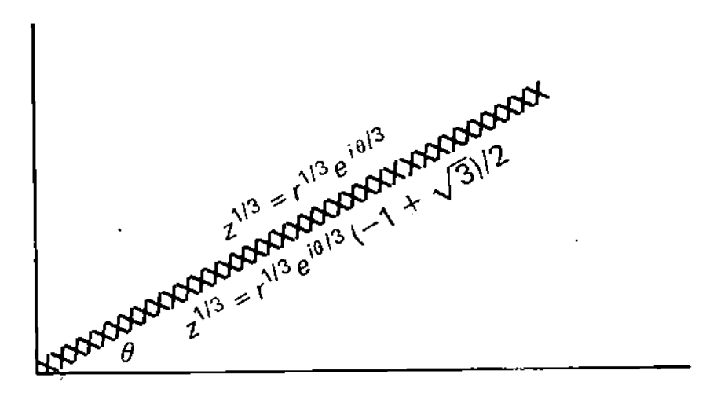

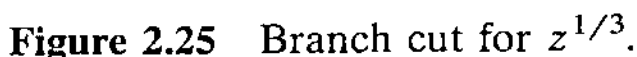
Figure 2.25 Branch cut for $z^{1/3}$.

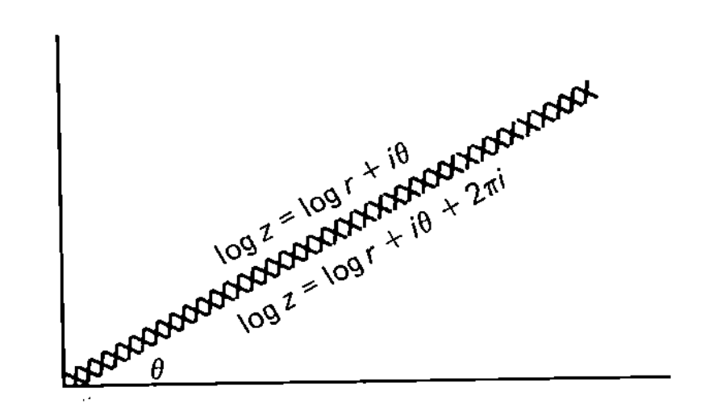

Figure 2.26 Branch cut for $\log z$.

while on the lower side of the line, F has the value

$$F[z(r,\theta+2\pi)] = r^{1/3}e^{i\theta/3}\frac{(-1+i\sqrt{3})}{2} \tag{2.224b}$$

If we take the difference between the functional values on two sides of a line, we get a nonzero value. That is the function is discontinuous across the line. The line is sometimes called a *line of discontinuity*, or more commonly, it is called a *branch line*, a *branch cut*, or just a *cut*. It is envisioned as cutting the complex F-plane to give access to the next sheet.

In addition to the fractional power of z, another multivalued function that arises frequently in scientific problems is the logarithmic function

$$F[z(r,\theta)] = \log(z) = \log r + i\theta \tag{2.225a}$$

Increasing θ by 2π yields

$$F[z(r,\theta+2\pi)] = \log r + i\theta + i2\pi \tag{2.225b}$$

It is easy to see that no matter how many times θ is increased by 2π, $\log z$ will never return to the value of equation 2.225a. Thus $\log z$ has an infinite number of values, or is an infinite-sheeted function.

We note here that we are denoting the natural logarithm, $\log_e$, by "log," although we are not explicitly using the subscript. If it occurs that we must discuss the common logarithm, it will be denoted by $\log_{10}$.

Like the fractional root function, the logarithm function has a branch point at the value of z for which the argument of the logarithm is zero. As discussed earlier, the cut can be chosen to run in any direction from the branch point to ∞. A typical cut, and the values of $\log z$ above and below the cut, are shown in Figure 2.26.

Consider the analytic structure of

$$F(z) = \sqrt{z} \tag{2.226}$$

On the first sheet,

$$F[z(r,\theta)] = r^{1/2}e^{i\theta/2} \tag{2.227a}$$

Increasing θ by 2π we obtain

$$F[z(r,\theta+2\pi)] = r^{1/2}e^{i\theta/2}e^{2\pi i/2} = -F[z(r,\theta)] \tag{2.227b}$$

A second increase of 2π yields

$$F[z(r,\theta+4\pi)] = r^{1/2}e^{i\theta/2}e^{4\pi i/2} = F[z(r,\theta)] \tag{2.227c}$$

So, as expected, $\sqrt{z}$ is a double-valued function.

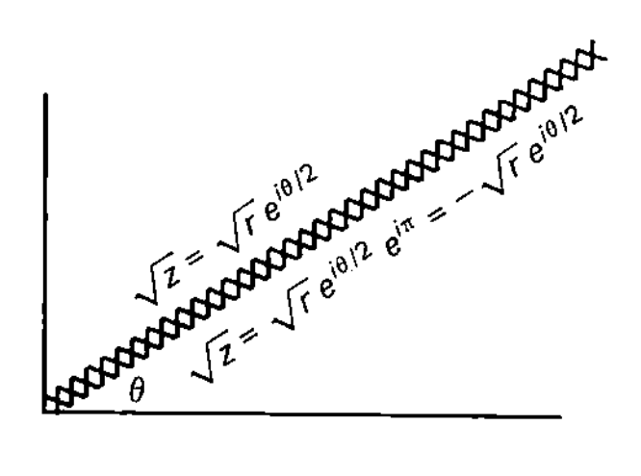

Figure 2.27 Branch cut for $\sqrt{z}$.

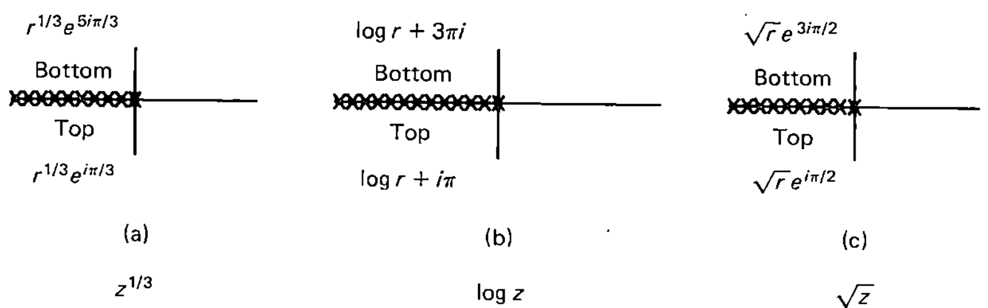

Figure 2.28 Cuts for (a) $z^{1/3}$; (b) $\log z$; and (c) $\sqrt{z}$ oriented along the negative real axis.

For a cut running from 0 to ∞ at an arbitrary angle, the function is $+r^{1/2}e^{i\theta/2}$ above the cut and $-r^{1/2}e^{i\theta/2}$ below the cut, as shown in Figure 2.27. We are using the term the *top of the cut* or *above the cut* to mean that side of the cut at which we would start, so that by moving in the counterclockwise direction (the direction of increasing angle), we would eventually arrive on the other side of the cut, which we term the *bottom of the cut* or *below the cut*. For example, for the three multivalued functions considered thus far, if the cut were oriented at an angle $\theta = \pi$ in Figure 2.25, 2.26, or 2.27, those cuts would appear as shown in Figure 2.28. In Figure 2.28, then, *above the cut* would be a point on the underside of the negative real axis, in the third quadrant.

2.7 Evaluation of Integrals: Part III

We can use the properties of multivalued functions to evaluate integrals using Cauchy's theorem. Consider a function $F(z)$ that has a branch point at $z = z_0$ and an associated cut running to ∞ at some angle to the real axis (Figure 2.29). In most cases the cut will be taken along the real axis. Let $F(z)$ have the functional value $F_{\text{above}}(z)$ (sometimes denoted by F^+) just above the cut, and $F_{\text{below}}(z)$ (or F^-) just below the cut. At any

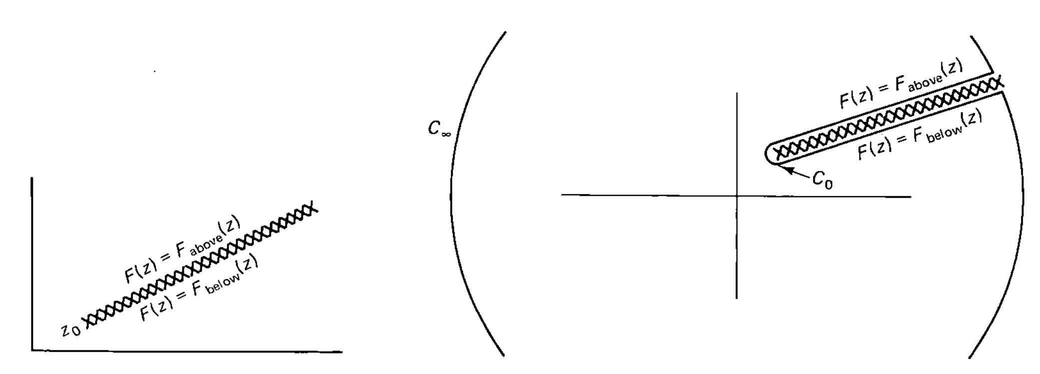

Figure 2.29 Branch structure of a function $F(z)$.

Figure 2.30 Contour for the integral of a multivalued function.

value of z such that $|z| > |z_0|$, these functional values are different. That is,

$$F_{\text{above}}(z) - F_{\text{below}}(z) \equiv \Delta(z) \neq 0 \qquad |z| > |z_0| \tag{2.228}$$

$\Delta(z)$ is called the *discontinuity* of the function across the cut.

Consider $\oint F(z)\,dz$ around the contour shown in Figure 2.30. Then

$$\oint F(z)\,dz = \int_{z_0}^{\infty} F_{\text{above}}(z)\,dz + \int_{C_\infty} F(z)\,dz + \int_{\infty}^{z_0} F_{\text{below}}(z)\,dz + \int_{C_0} F(z)\,dz \tag{2.229}$$

where, as in earlier discussions, C_∞ is the circular contour of infinite radius. C_0 is an infinitesimal circle (of radius zero) that goes around the branch point. For now, we will assume that the integrals over these segments of the contour are zero. Then, after inverting the limits on the integral below the cut, we obtain

$$\oint F(z)\,dz = \int_{z_0}^{\infty} [F_{\text{above}}(z) - F_{\text{below}}(z)]\,dz = \int_{z_0}^{\infty} \Delta(z)\,dz \tag{2.230}$$

From Cauchy's theorem, the contour integral is related to the sum of the residues of the enclosed poles. Therefore,

$$\int_{z_0}^{\infty} \Delta(z)\,dz = 2\pi i \sum [\text{residues of } F(z)] \tag{2.231}$$

Thus, to evaluate the integral of a function $\Delta(z)$, we look for a function $F(z)$ that has a branch point at the lower limit of the integral and an associated cut that can be taken to ∞, poles that are those of $\Delta(z)$, and a discontinuity that is proportional to $\Delta(z)$. When such a function is determined, its integral can be evaluated from a determination of the residues.

Example 2.26

To illustrate, consider

$$I = \int_0^{\infty} \frac{1}{(1 + x^3)}\,dx \tag{2.232}$$

Since the integrand is not even,

$$I \neq \frac{1}{2}\int_{-\infty}^{\infty} \frac{1}{(1 + x^3)}\,dx \tag{2.233}$$

To evaluate this integral using the properties of a multivalued function, we note that the lower limit on the integral is zero, the integrand has poles at the three values of $(-1)^{1/3}$, and the numerator of the integrand is a constant. Thus we seek a multivalued function that has a branch point at $z = 0$, three poles at the three cube roots of -1, and a discontinuity across the cut that is proportional to $1/(1 + z^3)$. Since we have determined that the discontinuity across the logarithm cut is a constant, the function that has these properties is

$$F(z) = \frac{\log z}{(1 + z^3)} \tag{2.234}$$

Therefore, we consider the integral

$$I = \oint \frac{\log z}{(1 + z^3)}\,dz \tag{2.235}$$

The poles of $F(z)$ are at

$$z = (-1)^{1/3} = \left(e^{i(\pi + 2k\pi)}\right)^{1/3} \qquad k = 0, 1, 2 \tag{2.236}$$

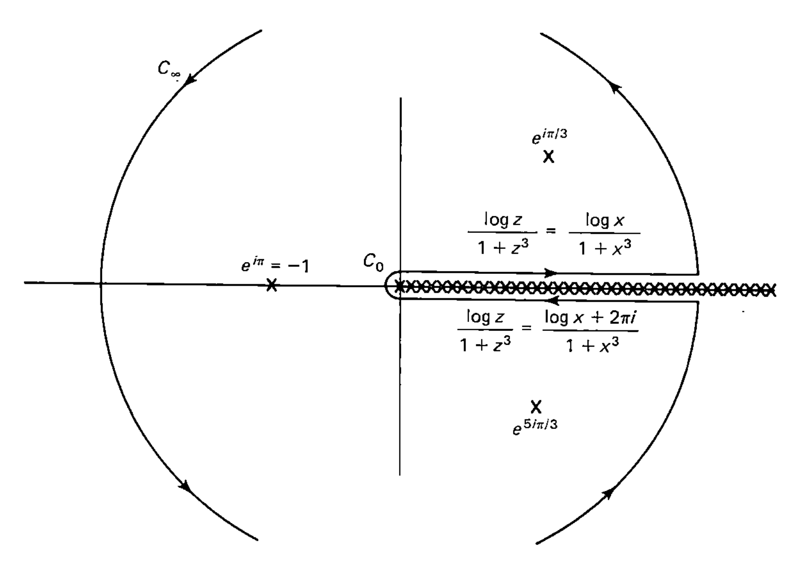

Figure 2.31 Pac-Man contour for the integral of equation 2.235.

That is, the positions of the poles are

$$\text{(a)} \quad z_0 = e^{i\pi/3} \qquad \text{(b)} \quad z_1 = e^{i\pi} = -1 \qquad \text{(c)} \quad z_2 = e^{5i\pi/3} \tag{2.237}$$

We can take the cut associated with the logarithmic branch point at $z = 0$ to run ∞ in any direction we choose. Since the integral we are trying to evaluate runs along the positive real axis in the z-plane from 0 to ∞, we take the cut along the positive real axis from 0 to ∞.

The structure of $F(z)$ and the contour we choose for the contour integral is shown in Figure 2.31. We dub this the "*Pac-Man*" *contour*. Writing the integral along different pieces of the contour, we have

$$\oint \frac{\log z}{(1+z^3)}\, dz = \int_0^\infty \frac{\log x}{(1+x^3)}\, dx + \int_{C_\infty} \frac{\log z}{(1+z^3)}\, dz + \int_\infty^0 \frac{\log x + 2\pi i}{(1+x^3)}\, dx + \int_{C_0} \frac{\log z}{(1+z^3)}\, dz \tag{2.238}$$

On the infinite circle, where $z = Re^{i\theta}$,

$$\frac{\log z}{(1+z^3)}\, dz = \frac{\log R + i\theta}{R^3 e^{3i\theta}} iRe^{i\theta}\, d\theta \to \frac{\log R}{R^2} \to 0 \tag{2.239}$$

Therefore, the integral over C_∞ is zero.

One C_0, $z = \rho e^{i\phi}$, with $\rho \to 0$ and $\phi \in [0, 2\pi]$. Then

$$\frac{\log z}{(1+z^3)}\, dz = \frac{\log \rho + i\phi}{(1+\rho^3 e^{3i\phi})} i\rho e^{i\phi}\, d\phi \tag{2.240}$$

As $\rho \to 0$, this goes to zero due to the overall multiplicative factor of ρ, and the property that

$$\lim_{\rho \to 0} \rho \log \rho = 0 \tag{2.241}$$

Thus the integral around C_0 is also zero. Combining these results with an inversion of the limits on the second semi-infinite integral yields

$$\oint \frac{\log z}{(1+z^3)}\, dz = \int_0^\infty \frac{\log x}{(1+x^3)}\, dx - \int_0^\infty \frac{\log x + 2\pi i}{(1+x^3)}\, dx = -2\pi i \int_0^\infty \frac{1}{(1+x^3)}\, dx \tag{2.242}$$

which is a constant multiple of the integral that we wish to evaluate.

Since the contour of Figure 2.31 encloses the three poles of the integrand of equation 2.235, Cauchy's theorem yields

$$\oint \frac{\log z}{(1+z^3)}\,dz = 2\pi i \sum \text{residues} = -2\pi i \int_0^\infty \frac{1}{(1+x^3)}\,dx \tag{2.243}$$

$$\Rightarrow \quad \int_0^\infty \frac{1}{(1+x^3)}\,dx = -\left[R(e^{i\pi/3}) + R(e^{i\pi}) + R(e^{5i\pi/3})\right] \tag{2.244}$$

where the residues are residues of the integrand of equation 2.235:

$$F(z) = \frac{\log z}{(1+z^3)} \tag{2.234}$$

For example,

$$R(e^{i\pi}) = \lim_{z\to e^{i\pi}} (z - e^{i\pi}) \frac{\log z}{(z - e^{i\pi/3})(z - e^{i\pi})(z - e^{5i\pi/3})} = \frac{i\pi}{3} \tag{2.245}$$

From this we find that the integral (equation 2.232) is $-2\pi/3\sqrt{3}$. □

We again offer a note of caution. In the z-plane, $e^{5i\pi/3}$ has the same value as $e^{-i\pi/3}$. But $\log e^{5i\pi/3} \neq \log e^{-i\pi/3}$. Starting at the top of the cut ($\theta = 0$), we cannot get to this point by rotating in the clockwise direction by $-\pi/3$ since doing so would require that we cross the cut. We must traverse points in the counterclockwise direction, increasing θ by $5\pi/3$. Therefore, this complex number must be kept in the form $e^{5i\pi/3}$.

As discussed above, the reason that the contour integral of equation 2.235 yields an evaluation of the original integral is that the discontinuity across the cut of the logarithmic function (the part of the integrand of equation 2.235 that contains the cut structure) is a constant ($-2\pi i$).

Example 2.27

Consider the integral

$$I = \int_1^\infty \frac{1}{(1+x^3)}\,dx \tag{2.246}$$

This integral is almost the same as that of equation 2.234 except that the lower limit is 1 instead of 0. Taking a clue from the evaluation of the integral of Example 2.26, we look for a function that has a constant discontinuity (the logarithm) and a branch point at $z = 1$ and three poles at the three roots of -1. From our previous analysis, it is clear that

$$F(z) = \frac{\log(z-1)}{(1+z^3)} \tag{2.247}$$

has the required properties. If $F(z)$ is integrated over a Pac-Man contour with the cut starting at $z = 1$ and extending to $+\infty$, along the real axis, as in Figure 2.32, we obtain

$$\begin{aligned}\oint \frac{\log(z-1)}{(1+z^3)}\,dz &= \int_1^\infty \frac{\log(x-1)}{(1+x^3)}\,dx + \int_{C_\infty} \frac{\log(z-1)}{(1+z^3)}\,dz \\ &\quad + \int_\infty^1 \frac{\log(x-1) + 2\pi i}{(1+x^3)}\,dx + \int_{C_0} \frac{\log(z-1)}{(1+z^3)}\,dz\end{aligned} \tag{2.248a}$$

Again, the integrals around C_∞ and C_0 are zero, and

$$\oint \frac{\log(z-1)}{(1+z^3)}\,dz = -2\pi i \int_1^\infty \frac{1}{(1+x^3)}\,dx = 2\pi i \sum \text{residues}\left(\frac{\log(z-1)}{(1+z^3)}\right) \tag{2.248b}$$

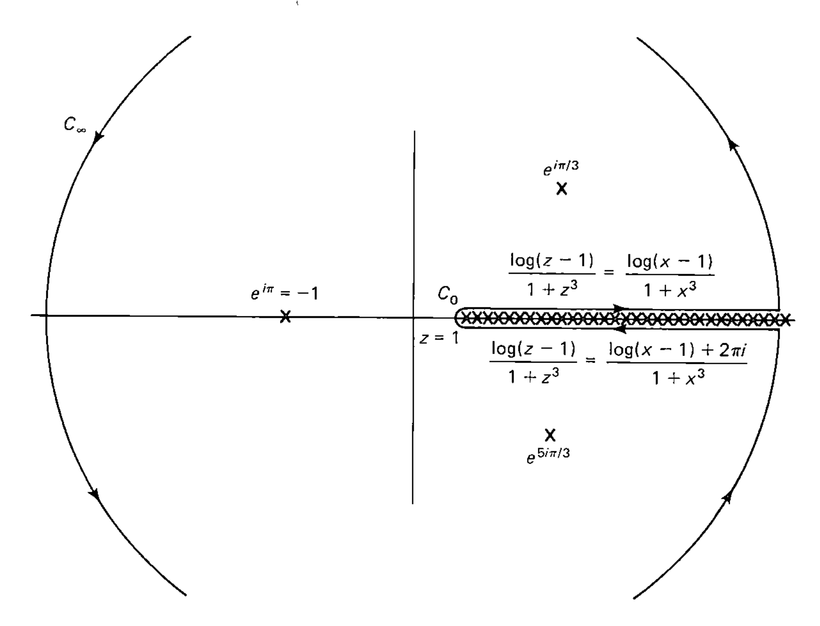

Figure 2.32 Contour for evaluating $\int_1^\infty dx/(1+x^3)$.

As noted in Example 2.26, $e^{5i\pi/3}$ cannot be expressed as $e^{-i\pi/3}$ because of the logarithm. The result is

$$\int_1^\infty \frac{dx}{(1+x^3)} = \frac{\pi}{\sqrt{3}} - \frac{1}{3}\log 2 \quad \square \tag{2.249}$$

Example 2.28

Consider the integral

$$I = \int_0^\infty \frac{\sqrt{x}}{(1+x^3)}\,dx \tag{2.250}$$

Clearly, to evaluate this integral using Cauchy's theorem, we must consider a function that has three poles at the three cube roots of -1, a branch point at $z = 0$, and a discontinuity across the associated cut that is proportional to $\sqrt{x}$. From our previous discussions, the appropriate function is

$$F(z) = \frac{\sqrt{z}}{(1+z^3)} \tag{2.251}$$

The singularity structure and the Pac-Man contour are shown in Figure 2.33. Then

$$\oint \frac{\sqrt{z}}{(1+z^3)}\,dz = \int_0^\infty \frac{\sqrt{x}}{(1+x^3)}\,dx + \int_{C_\infty} \frac{\sqrt{z}}{(1+z^3)}\,dz + \int_\infty^0 \frac{-\sqrt{x}}{(1+x^3)}\,dx + \int_{C_0} \frac{\sqrt{z}}{(1+z^3)}\,dz \tag{2.252}$$

using the same arguments presented in Example 2.26, it is straightforward to show that the integrals over C_0 and C_∞ are zero. Thus, changing the negative sign upon inversion of the limits, we obtain

$$\oint \frac{\sqrt{z}}{(1+z^3)}\,dz = 2\int_0^\infty \frac{\sqrt{x}}{(1+x^3)}\,dx = 2\pi i \sum \text{residues}\left(\frac{\sqrt{z}}{(1+z^3)}\right) \tag{2.253}$$

Evaluating these residues, we obtain

$$\int_0^\infty \frac{\sqrt{x}}{(1+x^3)}\,dx = \frac{\pi}{3} \quad \square \tag{2.254}$$

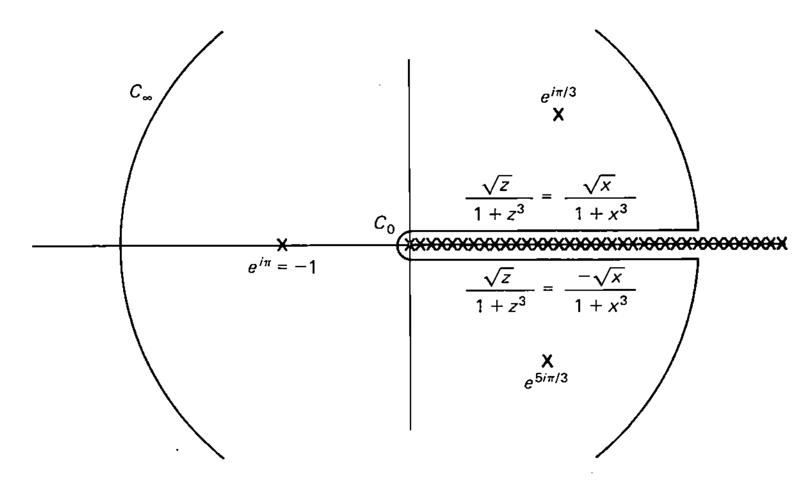

Figure 2.33 Contour for evaluating $\int_0^\infty [\sqrt{x}/(1+x^3)]\,dx$.

We note that Cauchy's theorem cannot be used to evaluate

$$I = \int_1^\infty \frac{\sqrt{x}}{(1+x^3)}\,dx \tag{2.255}$$

$\sqrt{z}$ has a branch point at $z = 0$, and there is no function that has a branch point at the lower limit, $z = 1$, which also has a discontinuity across the associated cut that is proportional to $\sqrt{z}$.

Multiple branch points

If a function has only one branch point, the cut extends from the branch point to ∞. If the function has multiple branch points, it may be possible to choose the cut to run between finite points. For example, consider the function

$$F(z) = \sqrt{z^2 - a^2} = \sqrt{(z-a)}\sqrt{(z+a)} \tag{2.256}$$

$\sqrt{(z-a)}$ has a branch point at $z = a$ and $\sqrt{(z+a)}$ has a branch point at $z = -a$. Separately, each cut runs to ∞ in some direction. Two possible choices for the cut structure are shown in Figures 2.34.

We first consider the cut structure shown in Figure 2.34a. Note that the region above the positive real axis (first quadrant) is the top of the right-hand cut, and the region underneath the negative real axis (third quadrant) is the top of the left-hand cut for this function. In the region to the right of a along the positive real axis, then, $\sqrt{z+a}$ is analytic. Therefore, in that region $\sqrt{z^2 - a^2}$ is multivalued because of the multivaluedness of $\sqrt{z-a}$. Similarly, in the region to the left of $-a$ along the negative real axis, $\sqrt{z^2 - a^2}$ is multivalued because $\sqrt{z+a}$ is multivalued.

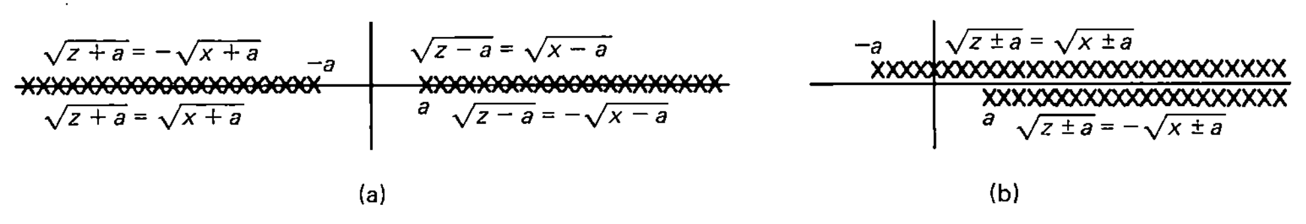

Figure 2.34 Two possible cut structures for the function $\sqrt{z^2 - a^2}$.

Example 2.29

To illustrate that the assignment of the values of $\sqrt{z \pm a}$ as shown in Figure 2.34a are correct, we consider the integral

$$I = \int_2^{\infty} \frac{1}{\left(\sqrt{x^2 - 4}\,(1 + x^2)\right)}\, dx \tag{2.257}$$

As will be seen, we cannot use Cauchy's theorem to evaluate this, but only with the sign assignment shown in Figure 2.34a will we obtain a consistent result for this integral.

To attempt to evaluate the integral of equation 2.257, we should consider the contour of Figure 2.35 for the integral

$$I = \oint \frac{1}{\left(\sqrt{z^2 - 4}\,(1 + z^2)\right)}\, dz \tag{2.258}$$

Taking the integral along various pieces of the contour, we have

$$\begin{aligned}\oint \frac{1}{\left(\sqrt{z^2 - 4}\,(1 + z^2)\right)}\, dz &= \int_2^{\infty} \frac{1}{\left(\sqrt{x^2 - 4}\,(1 + x^2)\right)}\, dx + \int_{-\infty}^{-2} \frac{1}{\left(-\sqrt{x^2 - 4}\,(1 + x^2)\right)}\, dx \\ &\quad + \int_{-2}^{-\infty} \frac{1}{\left(\sqrt{x^2 - 4}\,(1 + x^2)\right)}\, dx + \int_{\infty}^{2} \frac{1}{\left(-\sqrt{x^2 - 4}\,(1 + x^2)\right)}\, dx \\ &= 2\pi i[R(+i) + R(-i)] \end{aligned} \tag{2.259}$$

The integrals on the infinite semicircles and infinitesmal circles around the branch points have been omitted since they are zero.

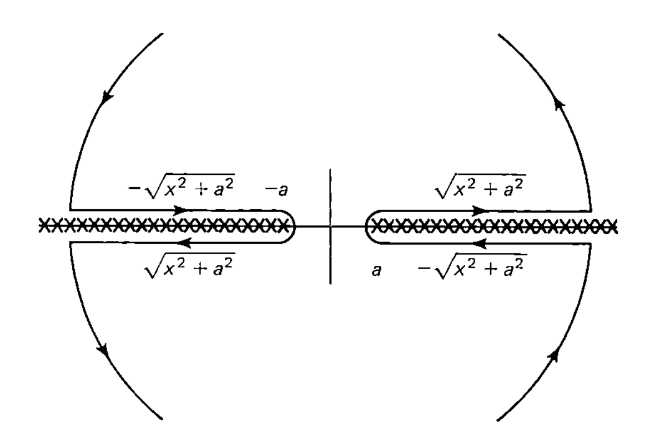

Figure 2.35 Contour for $\oint 1/[\sqrt{z^2 - 4}\,(1 + z^2)]\, dz$.

By inverting limits and substituting $x = -x$ in the integrals over negative x-values, we can see that the four integrals add to zero. Since the poles are not in regions above or below a cut, $R(\pm i)$ can be evaluated by simply setting $z = \pm i$ in $\sqrt{z^2 - 4}$. It is straightforward to show that

$$\text{(a)} \quad R(\pm i) = \pm \frac{1}{2i\sqrt{-5}} \qquad \Rightarrow \qquad \text{(b)} \quad R(+i) + R(-i) = 0 \tag{2.260}$$

consistent with the sum of the integrals. Had we taken any other sign assignment for the values of the function above and below the left hand cut, we would not have obtained the correct result of zero. □

Example 2.30

The integral

$$I = \int_2^{\infty} \frac{x}{\sqrt{x^2 - 4}\,(1 + x^2)}\, dx \tag{2.261}$$

can be evaluated using the structure of Figure 2.34a by considering

$$I = \oint \frac{1}{\left(\sqrt{z^2 - 4}\,(z - i)\right)}\, dz \tag{2.262}$$

around the contour of Figure 2.35. Writing the integral in terms of integrals over the various segments of the contour gives

$$\oint \frac{1}{\left(\sqrt{z^2-4}\,(z-i)\right)}\, dz = \int_2^{\infty} \frac{1}{\left(\sqrt{x^2-4}\,(x-i)\right)}\, dx + \int_{-\infty}^{-2} \frac{1}{\left(-\sqrt{x^2-4}\,(x-i)\right)}\, dx$$

$$+ \int_{-2}^{-\infty} \frac{1}{\left(\sqrt{x^2-4}\,(x-i)\right)}\, dx + \int_{\infty}^{2} \frac{1}{\left(-\sqrt{x^2-4}\,(x-i)\right)}\, dx$$

$$= 2\int_2^{\infty} \frac{1}{\left(\sqrt{x^2-4}\,(x-i)\right)}\, dx + 2\int_{-2}^{-\infty} \frac{1}{\left(\sqrt{x^2-4}\,(x-i)\right)}\, dx \tag{2.263}$$

where, again, the integrals around the infinite semicircles and the small contours around the branch points, which are zero, have been omitted.

By substituting $x = -x$ in the integral from -2 to $-\infty$, and noting that the integrand of the contour integral has one pole at $z = i$, we obtain

$$\oint \frac{1}{\left(\sqrt{z^2-4}\,(z-i)\right)}\, dz = 2\int_2^{\infty} \frac{1}{\left(\sqrt{x^2-4}\,(x-i)\right)}\, dx + 2\int_2^{\infty} \frac{1}{\left(\sqrt{x^2-4}\,(x+i)\right)}\, dx$$

$$= 4\int_2^{\infty} \frac{x}{\left(\sqrt{x^2-4}\,(1+x^2)\right)}\, dx = 2\pi i R(+i) \tag{2.264}$$

The residue of the pole at $z = i$ is $1/i\sqrt{5}$. Therefore,

$$\int_2^{\infty} \frac{x}{\sqrt{x^2-4}\,(1+x^2)}\, dx = \frac{\pi}{2\sqrt{5}} \tag{2.265}$$

The integral can also be evaluated directly using the substitution $u^2 = x^2 - 4$, so the reader can verify this result. □

We now investigate the structure of $\sqrt{z^2 - a^2}$ shown in Figure 2.34b in more detail. The cut associated with the branch point at $z = +a$ is shown in Figure 2.36. Since this branch point and cut are associated with the factor $\sqrt{z - a}$, the factor $\sqrt{z + a}$ is analytic and continuous across this cut. Thus the value of the function $\sqrt{z^2 - a^2}$ is that shown in Figure 2.36. By an identical argument, the values of the function on the two sides of the cut associated with the $z = -a$ branch point are those shown in Figure 2.37.

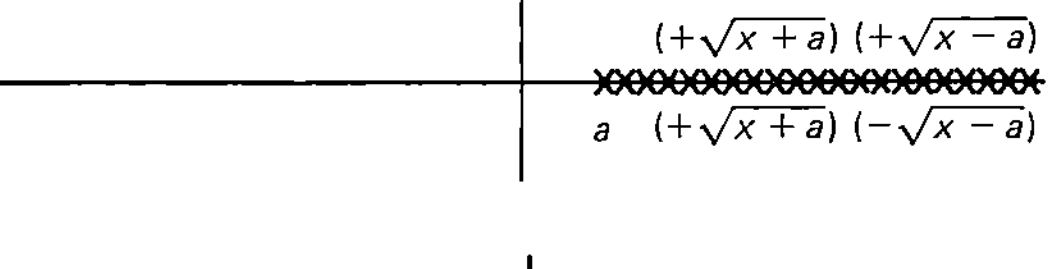

Figure 2.36 Values of factors of $\sqrt{z^2 - a^2}$ on the top and bottom of cut associated with branch point at $z = +a$.

−a
(+√x + a) (+√x − a)
(−√x + a) (+√x − a)

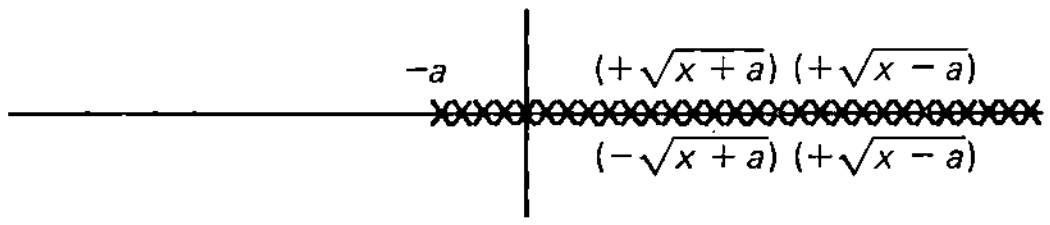

Figure 2.37 Values of factors of $\sqrt{z^2 - a^2}$ on the top and bottom of cut associated with branch point at $z = -a$.

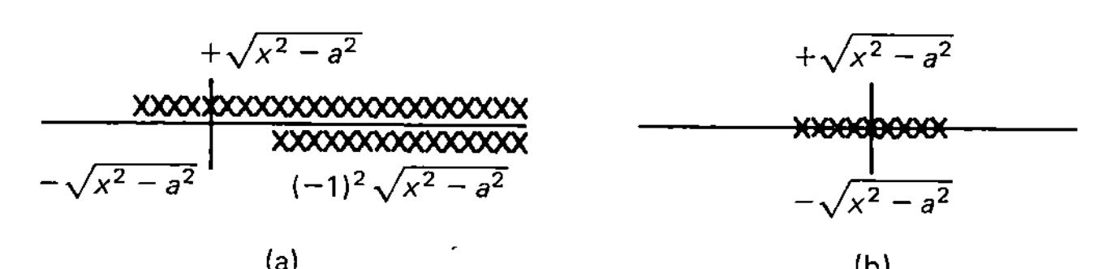

Figure 2.38 Equivalent cut structures for $\sqrt{z^2-a^2}$.

Consider the cut structure shown in Figure 2.38a. In the region for $x < -a$, there is no cut, and thus there is no discontinuity of the function across the real axis. In the region $-a \leqslant x \leqslant a$, there is only one cut associated with the branch point at $z = -a$. Across this cut, the function changes sign because $\sqrt{x+a}$ changes sign while $\sqrt{x-a}$ does not. For the region $x > a$, both factors change sign across the real axis. Thus the product does not change sign, so the function has no discontinuity in this region. That is, the function is analytic for $x > a$. Combining the cuts results in the structure shown in Figure 2.38b.

Example 2.31

The cut structure of Figure 2.38b is appropriate for evaluating the integral

$$I = \int_{-1}^{1} \frac{1}{\left(\sqrt{1-x^2}\,(1+x^2)\right)}\,dx \tag{2.266}$$

To do so, we consider

$$I = \oint \frac{1}{\left(\sqrt{1-z^2}\,(1+z^2)\right)}\,dz \tag{2.267}$$

around the contour shown in Figure 2.39. We separate contributions from the various parts of the contour, writing

$$\oint \frac{1}{\left(\sqrt{1-z^2}\,(1+z^2)\right)}\,dz = \int_{-1}^{1} \frac{1}{\left(\sqrt{1-x^2}\,(1+x^2)\right)}\,dx + \int_{1}^{-1} \frac{1}{\left(-\sqrt{1-x^2}\,(1+x^2)\right)}\,dx$$
$$+ \int_{-1}^{-\infty} \frac{1}{\left(\sqrt{1-x^2}\,(1+x^2)\right)}\,dx + \int_{-\infty}^{-1} \frac{1}{\left(\sqrt{1-x^2}\,(1+x^2)\right)}\,dx \tag{2.268}$$

From the functional dependence, it is clear that the integrals around the infinite part of the

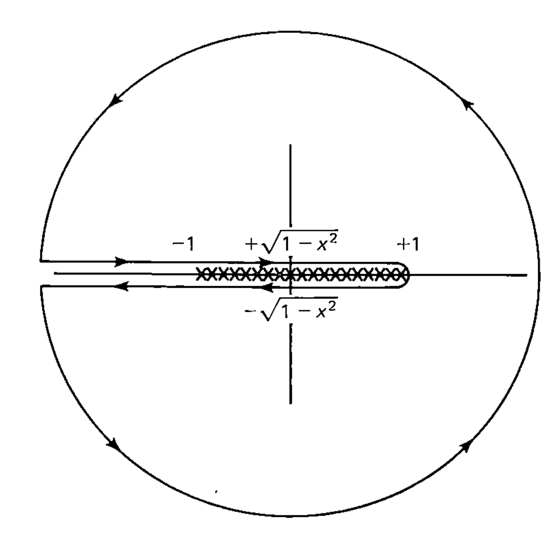

Figure 2.39 Contour of evaluating $\int_{-1}^{1} 1/[\sqrt{1-x^2}\,(1+x^2)]\,dx$.

contour and the infinitesmal circles around $z = \pm 1$ are zero, and have been omitted. For this cut structure, the function is continuous across the real axis for $x < -1$, so

$$\int_{-1}^{-\infty} \frac{1}{\left(\sqrt{1-x^2}\,(1+x^2)\right)}\,dx + \int_{-\infty}^{-1} \frac{1}{\left(\sqrt{1-x^2}\,(1+x^2)\right)}\,dx = 0 \tag{2.269}$$

We also note that the negative sign of the second integral on the right-hand side of equation 2.268 becomes positive when the limits are inverted. These manipulations lead to

$$\oint \frac{1}{\left(\sqrt{1-z^2}\,(1+z^2)\right)}\,dz = 2\int_{-1}^{1} \frac{1}{\left(\sqrt{1-x^2}\,(1+x^2)\right)}\,dx \tag{2.270}$$

$$\Rightarrow \quad \int_{-1}^{1} \frac{1}{\left(\sqrt{1-x^2}\,(1+x^2)\right)}\,dx = i\pi[R(+i) + R(-i)] \tag{2.271}$$

Since $\pm i$ are above and below the cut, respectively, they must be written as

$$\text{(a)} \quad i = e^{i\pi/2} \qquad \text{(b)} \quad -i = e^{3i\pi/2} \tag{2.272}$$

That is, because the real part of $\pm i$ is in the region containing the cut, $-i$ cannot be written as $e^{-i\pi/2}$. Then

$$R(+i) = \lim_{z \to e^{i\pi/2}} (z - e^{i\pi/2}) \frac{1}{\sqrt{1-z^2}\,(z - e^{i\pi/2})(z - e^{3i\pi/2})} \tag{2.273}$$

The factor $e^{i\pi/2} - e^{3i\pi/2}$ is not under a square root. Therefore, it can be evaluated as $i - (-i) = 2i$. Since the pole at $e^{i\pi/2}$ is above the cut and $e^{3i\pi/2}$ is below the cut,

$$\text{(a)} \quad \sqrt{1-z^2}\Big|_{z=e^{i\pi/2}} = +\sqrt{2} \qquad \text{(b)} \quad \sqrt{1-z^2}\Big|_{z=e^{3i\pi/2}} = -\sqrt{2} \tag{2.274}$$

Thus

$$\text{(a)} \quad R(+i) = \frac{1}{2i\sqrt{2}} \qquad \text{(b)} \quad R(-i) = \frac{1}{2i\sqrt{2}} \tag{2.275}$$

$$\Rightarrow \quad \int_{-1}^{1} \frac{1}{\left(\sqrt{1-x^2}\,(1+x^2)\right)}\,dx = \frac{\pi}{\sqrt{2}} \quad \square \tag{2.276}$$

2.8 Evaluation of Integrals: Part IV

Cauchy principal value integrals

In our discussion up to now, we have only considered situations for which the poles were inside or outside the contour. The integral of a function is essentially a sum of an infinite number of infinitesmal terms. If the pole is on the contour, one of those terms is infinite, so the integral is infinite.

Consider the integral of a function that has a simple pole that appears to be on the contour (Figure 2.40). In light of the discussion above, the integral around this contour is infinite. To obtain a finite result, we will have to traverse a slightly different contour, one that avoids the pole. There are two possible ways to avoid the pole of Figure 2.40.

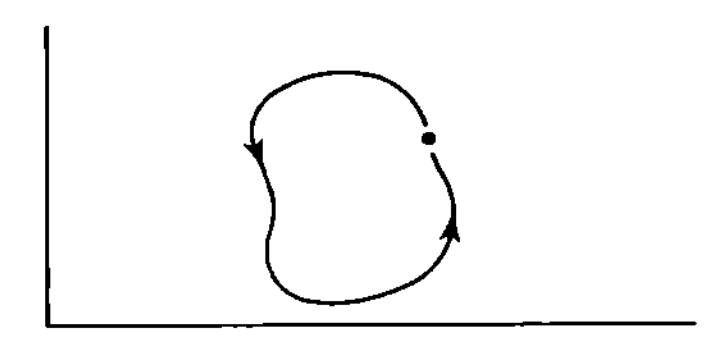

Figure 2.40 Contour that encounters a pole.

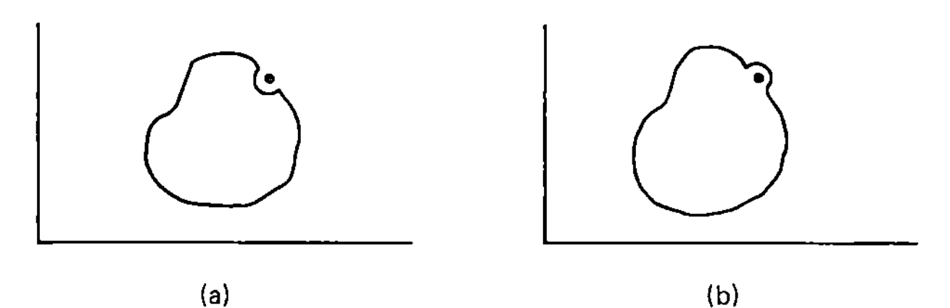

Figure 2.41 Contours that avoid a pole.

They are shown in Figures 2.41. That is, we can go around the pole in such a way that the pole is either outside the contour (Figure 2.41a) or inside the contour (Figure 2.41b). The integrals around these two contours are different and they are both finite.

An integral that one typically encounters in scientific problems is one over an infinite contour along the real axis with a pole on the real axis. It is of the form

$$I = \int_{-\infty}^{\infty} \frac{f(x)}{(x - x_0)}\, dx \tag{2.277}$$

with x_0 real. We assume that $f(x)/(x - x_0)$ approaches zero "fast enough" so that when evaluating the integral using Cauchy's theorem, the contour can be closed in at least one of the half-planes.

The pole at x_0 can be avoided in either one of the two ways shown in Figures 2.42. For this treatment, we will close our contour in the upper half-plane. The results we obtain will be identical to those obtained had we closed in the lower half-plane.

Let each of the contours C_+ and C_- in Figures 2.42 be a semicircle with small radius ρ. We will eventually take $\rho \to 0$. For the closed contour of Figure 2.42a,

$$\oint \frac{f(z)}{(z - x_0)}\, dz = \int_{-\infty}^{x_0-\rho} \frac{f(x)}{(x - x_0)}\, dx + \int_{C_+} \frac{f(z)}{(z - x_0)}\, dz + \int_{x_0+\rho}^{\infty} \frac{f(x)}{(x - x_0)}\, dx \tag{2.278}$$

where the integral around the infinite semi-circle has been omitted since it is zero.

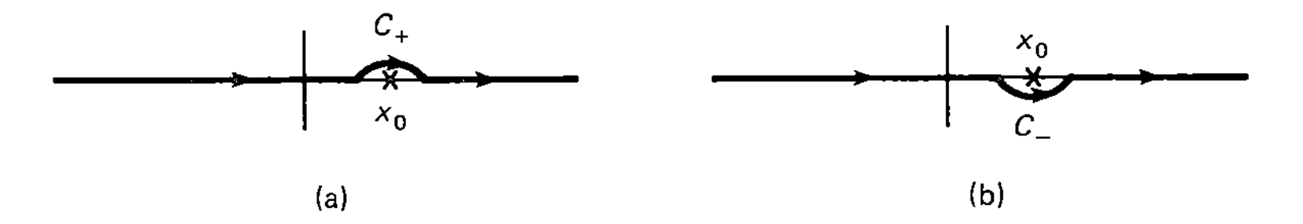

Figure 2.42 Two contours that avoid the pole of $f(x)/(x - x_0)$.

Using Cauchy's theorem, equation 2.278 becomes

$$\oint \frac{f(z)}{(z - x_0)}\, dz = \int_{-\infty}^{x_0-\rho} \frac{f(x)}{(x - x_0)}\, dx + \int_{C_+} \frac{f(z)}{(z - x_0)}\, dz + \int_{x_0+\rho}^{\infty} \frac{f(x)}{(x - x_0)}\, dx$$

$$= 2\pi i \sum R_{+1/2}\left(\frac{f(z)}{(z - x_0)}\right) \tag{2.279}$$

Because the contour of Figure 2.42a, closed in the upper half-plane, does not include the pole at x_0, the residue of the integrand at x_0 is not included in the sum of residues.

On C_+,

$$z = x_0 + \rho e^{i\phi} \tag{2.280}$$

and because the contour along the real axis is traversed from $-\infty$ to $+\infty$, ϕ ranges from

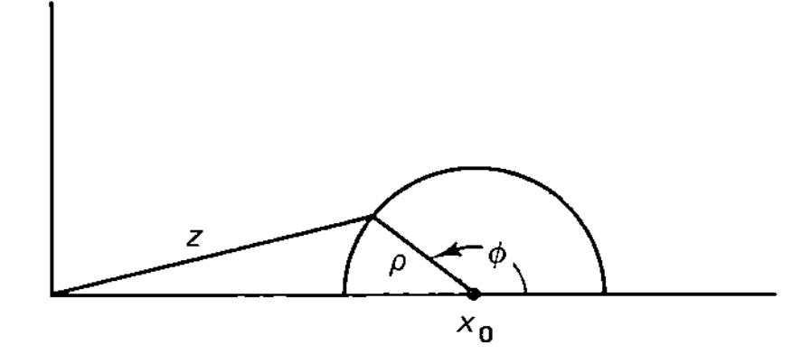

Figure 2.43 Small semicircular contour C_+ that avoids the pole at x_0.

π to 0 (see Figure 2.43). Thus

$$\int_{C_+} \frac{f(z)}{(z - x_0)}\, dz = \int_{\pi}^{0} \frac{f(x_0 + \rho e^{i\phi})}{\rho e^{i\phi}} i\rho e^{i\phi}\, d\phi \tag{2.281}$$

In the limit when $\rho \to 0$, $f(x_0 + \rho e^{i\phi})$ becomes independent of ϕ and this integral becomes

$$\lim_{\rho \to 0} \int_{C_+} \frac{f(z)}{(z - x_0)}\, dz = f(x_0) \int_{\pi}^{0} i\, d\phi = -i\pi f(x_0) \tag{2.282}$$

Thus equation 2.279 becomes

$$\lim_{\rho \to \infty} \left[\int_{-\infty}^{x_0 - \rho} \frac{f(x)}{(x - x_0)}\, dx + \int_{x_0 + \rho}^{\infty} \frac{f(x)}{(x - x_0)}\, dx \right] - i\pi f(x_0) = 2\pi i \sum R_{+1/2} \tag{2.283}$$

Note that the two integrals in the brackets are integrals just up to and just beyond the pole, but contain no information about how to go around the pole. Therefore, they are independent of our choice of paths C_+ or C_-. The sum of these two integrals, in the limit of zero radius ρ, is called the *Cauchy principal value integral*. It is denoted by

$$\lim_{\rho \to 0} \left[\int_{-\infty}^{x_0 - \rho} \frac{f(x)}{(x - x_0)}\, dx + \int_{x_0 + \rho}^{\infty} \frac{f(x)}{(x - x_0)}\, dx \right] \equiv P \int_{-\infty}^{\infty} \frac{f(x)}{(x - x_0)}\, dx \tag{2.284}$$

We will also denote the principal value integral as

$$P \int_{-\infty}^{\infty} \frac{f(x)}{(x - x_0)}\, dx = \int_{-\infty}^{\infty} \frac{f(x)}{(x - x_0)_P}\, dx \tag{2.285}$$

With this result, equation 2.283 becomes

$$P \int_{-\infty}^{\infty} \frac{f(x)}{(x - x_0)}\, dx = i\pi f(x_0) + 2\pi i \sum R_{+1/2} \tag{2.286}$$

To illustrate that the principal value integral is independent of the choice of the small contour used to avoid the pole, consider the contour of Figure 2.42b, again closing in the upper half-plane. Since the integral over the infinite semicircle is zero, it will be omitted. Then

$$\oint \frac{f(z)}{(z - x_0)}\, dz = P \int_{-\infty}^{\infty} \frac{f(x)}{(x - x_0)}\, dx + \int_{C_-} \frac{f(z)}{(z - x_0)}\, dz \tag{2.287}$$

By closing in the upper half-plane, the contour of Figure 2.42b encloses the pole at x_0 inside the contour. Therefore, the sum of residues includes the residue of the pole at x_0. This residue is

$$R(x_0) = \lim_{z \to x_0} (z - x_0) \frac{f(z)}{(z - x_0)} = f(x_0) \tag{2.288}$$

and Cauchy's theorem yields

$$P\int_{-\infty}^{\infty} \frac{f(x)}{(x - x_0)}\, dx + \int_{C_-} \frac{f(z)}{(z - x_0)}\, dz = 2\pi i \sum R_{+1/2} + 2\pi i f(x_0) \tag{2.289}$$

Referring to Figure 2.42b, a point on C_- has the form $z = x_0 + \rho e^{i\phi}$. Traversing the real axis from $-\infty$ to ∞ requires ϕ to range from π to 2π on C_-. Thus, as $\rho \to 0$,

$$\int_{C_-} \frac{f(z)}{(z - x_0)}\, dz \to f(x_0) \int_{\pi}^{2\pi} \frac{i\rho e^{i\phi}}{\rho e^{i\phi}}\, d\phi = i\pi f(x_0) \tag{2.290}$$

and equation 2.289 becomes

$$P\int_{-\infty}^{\infty} \frac{f(x)}{(x - x_0)}\, dx = i\pi f(x_0) + 2\pi i \sum R_{+1/2} \tag{2.291}$$

which is identical to equation 2.286. Therefore, as stated, the principal value integral is independent of how we avoid the pole.

Displacement of the pole

Feynman invented a technique for dealing with the avoidance of a pole on the contour. Suppose we want to evaluate the integral of equation 2.278, avoiding the pole around C_+ as in Figure 2.42a; that is, we choose the contour to go slightly above the pole. Equivalently, we can say that we choose the pole to be slightly below the contour. Feynman's idea is to move the pole slightly below the contour rather than moving the contour slightly above the pole. This is illustrated in Figures 2.44. That is, in Figure 2.44b, the pole has been given an infinitesmal negative imaginary part, so the pole is now at $z = x_0 - i\varepsilon$, where ε has an infinitesmal positive value.

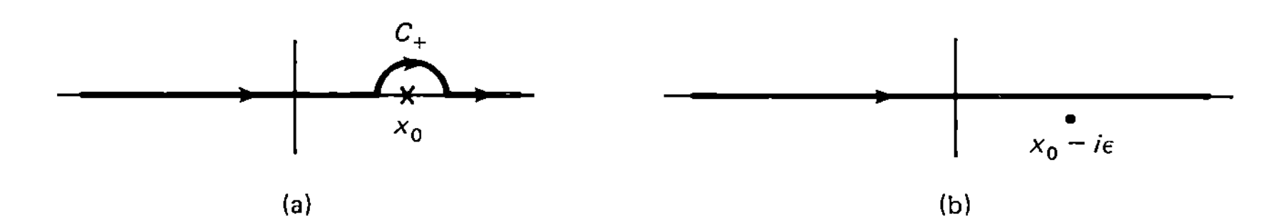

Figure 2.44 Equivalence of distorting the contour slightly above the pole and displacing the pole slightly below the contour.

Similarly, avoiding the pole by going around contour C_-, which is slightly below the pole as in Figure 2.42b, is equivalent to raising the pole an infinitesmal amount above the contour as shown in Figure 2.45. Thus, in Figure 2.45b the pole is at $z = x_0 + i\varepsilon$.

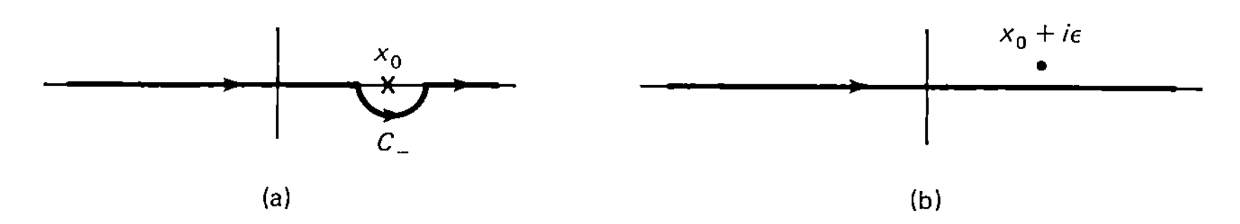

Figure 2.45 Equivalence of distorting the contour slightly below the pole and displacing the pole slightly above the contour.

Since displacing the pole to $z = x_0 - i\varepsilon$, as in Figure 2.44b, is equivalent to avoiding the pole around the contour C_+,

$$I = \oint \frac{f(z)}{(z - x_0 + i\varepsilon)}\, dz$$

must yield the same result as that obtained from the analysis of equation 2.278. That is, omitting the integral on the infinite semicircle, which is zero, the integral around the contour closed in the upper half-plane, yields

$$\oint \frac{f(z)}{(z - x_0 + i\varepsilon)}\, dz = \int_{-\infty}^{\infty} \frac{f(x)}{(x - x_0 - i\varepsilon)}\, dx = 2\pi i \sum R_{+1/2} \tag{2.292}$$

Comparing this to equation 2.283, we obtain

$$\int_{-\infty}^{\infty} \frac{f(x)}{(x - x_0 + i\varepsilon)}\, dx = P\int_{-\infty}^{\infty} \frac{f(x)}{(x - x_0)}\, dx - i\pi f(x_0) \tag{2.293a}$$

Displacing the pole to $z = x_0 + i\varepsilon$ yields the same result as that obtained using the contour C_-. Therefore,

$$\int_{-\infty}^{\infty} \frac{f(x)}{(x - x_0 - i\varepsilon)}\, dx = P\int_{-\infty}^{\infty} \frac{f(x)}{(x - x_0)}\, dx + i\pi f(x_0) \tag{2.293b}$$

Dirac δ-symbol

The Dirac δ-symbol is similar in concept to the Kronecker δ-symbol. However, it depends on continuous variables rather than discrete indices and is defined only under an integral. It is defined such that for any continuous function $G(x)$,

$$\int_a^b G(y)\delta(x - y)\, dy = \begin{cases} G(x) & a < x < b \\ 0 & x \text{ outside this range} \end{cases} \tag{2.294}$$

Strictly speaking, $\delta(x - y)$ is not a function and is not defined except under an integral as specified in equation 2.294. There is a whole class of entities like the Dirac δ-function called distributions, which are defined only under integrals. However, when expressed as a function, the *Dirac δ-symbol* (also called the *Dirac δ-function*) has the values

$$\delta(x - y) = \begin{cases} 0 & \text{for } x \neq y \\ \infty & \text{for } x = y \end{cases} \tag{2.295}$$

To get a sense of the functional description of the δ-symbol, let ε be an infinitesmally small positive number and $F(y)$ be an arbitrary function. Then $y > y - \varepsilon$ and $y < y + \varepsilon$. Therefore, from the definition of $\delta(x - y)$ expressed in equation 2.294,

$$\int_{-\infty}^{x-\varepsilon} F(y)\delta(x - y)\, dy = \int_{x+\varepsilon}^{\infty} F(y)\delta(x - y)\, dy = 0 \tag{2.296a}$$

For both of these integrals, the ranges of integration are such that $y \neq x$. Since $F(y)$ is arbitrary, the integrals can be zero only if

$$\delta(y - x) = 0 \qquad y \neq x \tag{2.296b}$$

Therefore, with ε small enough, we have deduced the values of $\delta(x - y)$ for all values of y except $y = x$. Again referring to the definition of equation 2.294, we have

$$\int_{x-\varepsilon}^{x+\varepsilon} \delta(x - y)\, dy = 1 \tag{2.297a}$$

Figure 2.46 Dirac δ-symbol as a function.

Thus $\delta(x - y)$ as a function is of zero width such that the area under the function is 1. As such, the functional height of $\delta(x - y)$ must be infinite (so that $0 \cdot \infty = 1$). That is,

$$\delta(x - y) = \infty \qquad y = x \tag{2.297b}$$

Therefore, as a function, $\delta(x - y)$ is described in Figure 2.46.

Equations 2.293 can be rewritten in terms of the Dirac δ-function as

$$\int_{-\infty}^{\infty} \frac{f(x)}{(x + x_0 + i\varepsilon)}\, dx = P\int_{-\infty}^{\infty} \frac{f(x)}{(x - x_0)}\, dx - i\pi\int_{-\infty}^{\infty} f(x)\delta(x - x_0)\, dx \tag{2.298a}$$

$$\int_{-\infty}^{\infty} \frac{f(x)}{(x - x_0 - i\varepsilon)}\, dx = P\int_{-\infty}^{\infty} \frac{f(x)}{(x - x_0)}\, dx + i\pi\int_{-\infty}^{\infty} f(x)\delta(x - x_0)\, dx \tag{2.298b}$$

These equations have been derived for any arbitrary function $f(x)$. Thus the integrands must be equal. That is,

$$\frac{1}{(x - x_0 + i\varepsilon)} = \frac{1}{(x - x_0)_P} - i\pi\delta(x - x_0) \tag{2.299a}$$

$$\frac{1}{(x - x_0 - i\varepsilon)} = \frac{1}{(x - x_0)_P} + i\pi\delta(x - x_0) \tag{2.299b}$$

Example 2.32

To illustrate how we can use the Feynman idea to evaluate principal value integrals, consider the integral

$$I = P\int_{-\infty}^{\infty} \frac{1}{(x^2 + x_1^2)(x - x_0)}\, dx \tag{2.300}$$

with x_0 and x_1 real. This integral can be evaluated by applying equation 2.286 directly. The integrand has a pole on the real axis at $z = x_0$ and poles at $z = \pm ix_1$. Closing in the upper half-plane, with one pole at $z = ix_1$, equation 2.286 yields

$$P\int_{-\infty}^{\infty} \frac{1}{[(x^2 + x_1^2)(x - x_0)]}\, dx = 2\pi i R(ix_1) + \frac{i\pi}{(x_0^2 + x_1^2)} \tag{2.301}$$

The residue is evaluated straightforwardly as

$$R(ix_1) = \frac{1}{2ix_1(ix_1 - x_0)} \tag{2.302}$$

and equation 2.301 becomes

$$P\int_{-\infty}^{\infty} \frac{1}{[(x^2 + x_1^2)(x - x_0)]}\, dx = 2\pi i \frac{1}{2ix_1(ix_1 - x_0)} + i\pi \frac{1}{(x_0^2 + x_1^2)} = \frac{-\pi x_0}{x_1(x_0^2 + x_1^2)} \tag{2.303}$$

which is real, as it must be.

Using the Feynman idea, we write the distribution

$$\frac{1}{(x - x_0)_P} = Re\left(\frac{1}{(x - x_0 \pm i\varepsilon)}\right) \tag{2.304}$$

$$\Rightarrow \quad P\int_{-\infty}^{\infty} \frac{1}{[(x^2 + x_1^2)(x - x_0)]}\, dx = Re\int_{-\infty}^{\infty} \frac{1}{[(x^2 + x_1^2)(x - x_0 \pm i\varepsilon)]}\, dx \tag{2.305}$$

By choosing $+i\varepsilon$ and closing in the upper half-plane (since the integral over the infinite semicircle is zero), we do not enclose the pole at $x_0 - i\varepsilon$ inside the contour. In that way the residue at $x_0 - i\varepsilon$ does not have to be calculated, saving a little arithmetic. Then

$$\oint \frac{1}{[(z^2 + x_1^2)(z - x_0 + i\varepsilon)]}\, dz = \int_{-\infty}^{\infty} \frac{1}{[(x^2 + x_1^2)(x - x_0 + i\varepsilon)]}\, dx \tag{2.306}$$

with the contour and poles shown in Figure 2.47.

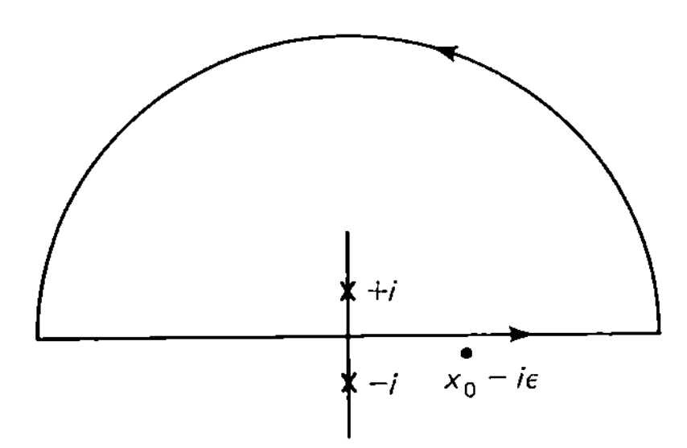

Figure 2.47 Contour for evaluating $\int_{-\infty}^{\infty} 1/[(z^2 + x_1^2)(z - x_0 + i\varepsilon)]\, dz$.

Evaluating the contour integral by Cauchy's theorem, we obtain

$$\int_{-\infty}^{\infty} \frac{1}{[(x^2 + x_1^2)(x - x_0 + i\varepsilon)]}\, dx = 2\pi i R(ix_1) = \frac{\pi}{x_1(ix_1 - x_0 + i\varepsilon)} \tag{2.307a}$$

The denominator has an imaginary part $x_1 + \varepsilon$, and since ε is infinitesmal, it is therefore negligible compared to x_1. Thus

$$\int_{-\infty}^{\infty} \frac{1}{[(x^2 + x_1^2)(x - x_0 + i\varepsilon)]}\, dx = \frac{\pi}{x_1(ix_1 - x_0)} \tag{2.307b}$$

$$\Rightarrow \quad P\int_{-\infty}^{\infty} \frac{1}{[(x^2 + x_1^2)(x - x_0)]}\, dx = Re\left(\frac{\pi}{x_1(ix_1 - x_0)}\right) = \frac{-\pi x_0}{x_1(x_1^2 + x_0^2)} \quad \square \tag{2.308}$$

Using this example as a guide, in general, if a function $G(x)$ is real, then

$$P\int_{-\infty}^{\infty} \frac{G(x)}{(x - x_0)}\, dx = Re\int_{-\infty}^{\infty} \frac{G(x)}{(x - x_0 \pm i\varepsilon)}\, dx \tag{2.309}$$

The integral $\int_{-\infty}^{\infty}[G(x)/(x - x_0 \pm i\varepsilon)]\, dx$ can be evaluated by Cauchy's theorem using the methods described above. To minimize the amount of calculation required, the sign of $i\varepsilon$ should be taken to place this pole outside the contour.

If $G(x)$ is complex, an equation similar to equation 2.309 can be written to evaluate

$$I = P\int_{-\infty}^{\infty} \frac{G(x)}{(x - x_0)}\, dx \tag{2.310}$$

except that when $G(x)$ is complex, two integrals must be evaluated instead of one.

Writing $G(x) = G_1(x) + iG_2(x)$, then

$$\begin{aligned} P\int_{-\infty}^{\infty} \frac{G(x)}{(x - x_0)}\, dx &= P\int_{-\infty}^{\infty} \frac{G_1(x)}{(x - x_0)}\, dx + iP\int_{-\infty}^{\infty} \frac{G_2(x)}{(x - x_0)}\, dx \\ &= Re\int_{-\infty}^{\infty} \frac{G_1(x)}{(x - x_0 \pm i\varepsilon)}\, dx + iRe\int_{-\infty}^{\infty} \frac{G_2(x)}{(x - x_0 \pm i\varepsilon)}\, dx \end{aligned} \tag{2.311}$$

Since $G_1(x)$ and $G_2(x)$ are real functions, each integral can be evaluated using Cauchy's theorem as described above.

Another approach is to use equations 2.299 to write

$$\frac{1}{(x - x_0)_P} = \frac{1}{2}\left(\frac{1}{(x - x_0 + i\varepsilon)} + \frac{1}{(x - x_0 - i\varepsilon)}\right) \tag{2.312}$$

Then, for an arbitrary complex function $G(x)$,

$$P\int_{-\infty}^{\infty} \frac{G(x)}{(x - x_0)}\, dx = \frac{1}{2}\left[\int_{-\infty}^{\infty} \frac{G(x)}{(x - x_0 + i\varepsilon)}\, dx + \int_{-\infty}^{\infty} \frac{G(x)}{(x - x_0 - i\varepsilon)}\, dx\right] \tag{2.313}$$

Each integral is then evaluated using Cauchy's theorem.

Pole on the cut

The approach used to evaluate an integral for an integrand that has a pole somewhere on a cut is no more complicated than methods we have developed up to now. Consider an integral around a Pac-Man contour of a function that has a cut along the real axis, and a pole on the real axis. As an example, we illustrate how to evaluate

$$I = \oint \frac{f(z)}{(z - x_0)}\, dz \tag{2.314}$$

where $f(z)$ has a branch point at $z = a$ and an associated cut running to ∞ along the real axis, and $x_0 > a$ is real and on the cut. We consider the contour shown in Figure 2.48. We assume that the function is such that the integrals around the infinite circle and the infinitesmal contour around the branch point are zero.

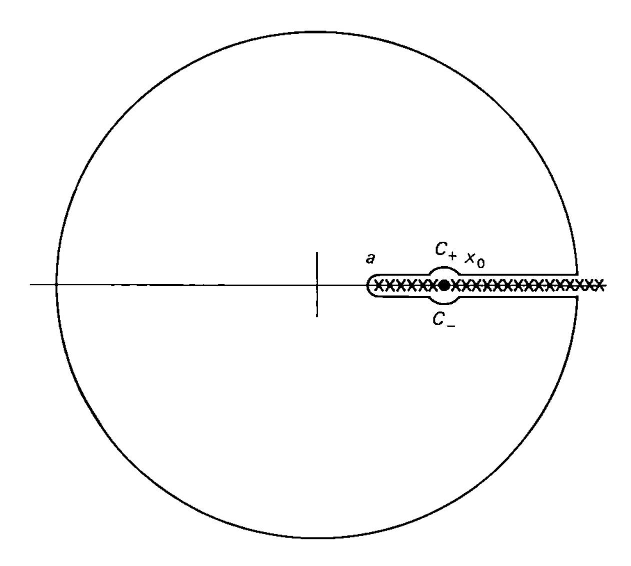

Figure 2.48 Contour for evaluating an integral of a function that has a pole on a cut.

Taking small semicircles around the pole, we have

$$\oint \frac{f(z)}{(z-x_0)}\,dz = \int_a^{x_0-\rho} \frac{f[z(x,0)]}{(x-x_0)}\,dx + \int_{C_+} \frac{f_{\text{above}}(z)}{(z-x_0)}\,dz + \int_{x_0+\rho}^{\infty} \frac{f[z(x,0)]}{(x-x_0)}\,dx$$

$$+ \int_{\infty}^{x_0+\rho} \frac{f[z(x,2\pi)]}{(x-x_0)}\,dx + \int_{C_-} \frac{f_{\text{below}}(z)}{(z-x_0)}\,dz + \int_{x_0-\rho}^{a} \frac{f[z(x,2\pi)]}{(x-x_0)}\,dx$$

$$= 2\pi i \sum \text{residues off real axis} \tag{2.315}$$

Inverting the limits of the fourth and sixth integrals and taking $\rho \to 0$, equation 2.315 becomes

$$\oint \frac{f(z)}{(z-x_0)}\,dz = P\int_a^{\infty} \frac{f[z(x,0)] - f[z(x,2\pi)]}{(x-x_0)}\,dx + \int_{C_+} \frac{f_{\text{above}}(z)}{(z-x_0)}\,dz$$

$$+ \int_{C_-} \frac{f_{\text{below}}(z)}{(z-x_0)}\,dz = 2\pi i \sum R(\text{off the real axis}) \tag{2.316}$$

The quantity

$$\Delta(x) = f[z(x,0)] - f[z(x,2\pi)] \tag{2.317}$$

is the discontinuity of $f(z)$ across the cut.

To evaluate the integrals of equation 2.316, over $C_\pm$, we note that on these small semicircles,

$$z = x_0 + \rho e^{i\phi} \tag{2.280}$$

Because of the direction of transversal of the contour, ϕ ranges from π to 0 on C_+. On C_-, the range of ϕ is from 2π to π. Therefore, the integrals of equation 2.316 over C_+ become

$$\int_{\pi}^{0} \frac{f_{\text{above}}[z(x_0,0)]}{\rho e^{i\phi}} i\rho e^{i\phi}\,d\phi + \int_{2\pi}^{\pi} \frac{f_{\text{below}}[z(x_0,2\pi)]}{\rho e^{i\phi}} i\rho e^{i\phi}\,d\phi$$

$$= -\pi\left[f_{\text{above}}[z(x_0,0)] + f_{\text{below}}[z(x_0,2\pi)]\right] \tag{2.318}$$

Therefore, equation 2.316 becomes

$$P\int_a^{\infty} \frac{\Delta(x)}{(x-x_0)}\,dx - i\pi[f(x_0,0) + f(x_0,2\pi)] = 2\pi i \sum R(\text{off the real axis}) \tag{2.319}$$

Example 2.33

To illustrate this approach, consider the evaluation of

$$I = P\int_0^{\infty} \frac{1}{(1-x^3)}\,dx \tag{2.320}$$

The integrand has poles at $(+1)^{1/3} = e^{i2\pi k/3}$ $k = 0, 1, 2$. Thus

$$\text{(a)}\quad z_0 = e^0 = 1 \qquad \text{(b)}\quad z_1 = e^{2\pi i/3} \qquad \text{(c)}\quad z_2 = e^{4\pi i/3} \tag{2.321}$$

This integral can be evaluated by direct integration or by displacing the pole an infinitesimal distance from the real axis and applying Cauchy's theorem. Writing

$$P\int_0^\infty \frac{1}{(1-x^3)}\,dx = Re\int_0^\infty \frac{1}{(1-x\pm i\varepsilon)(x^2+x+1)}\,dx \tag{2.322a}$$

or

$$P\int_0^\infty \frac{1}{(1-x^3)}\,dx = \frac{1}{2}\left[\int_0^\infty \frac{1}{(1-x+i\varepsilon)(x^2+x+1)}\,dx + \int_0^\infty \frac{1}{(1-x-i\varepsilon)(x^2+x+1)}\,dx\right] \tag{2.322b}$$

and using methods developed earlier, we obtain

$$P\int_0^\infty \frac{1}{(1-x^3)}\,dx = \frac{\pi}{3\sqrt{3}} \tag{2.323}$$

The integral of equation 2.320 can also be evaluated by considering a function that has a branch point at $z = 0$, three poles at the values given in equation 2.321, and a discontinuity across the cut that is proportional to $1/(1-x^3)$. Clearly, the function that has these properties is

$$F(z) = \frac{\log z}{(1-z^3)} \tag{2.324}$$

Therefore, by evaluating

$$I = \oint \frac{\log z}{(1-z^3)}\,dz \tag{2.325}$$

around the contour of Figure 2.49 we must evaluate an integral containing a pole on a cut.

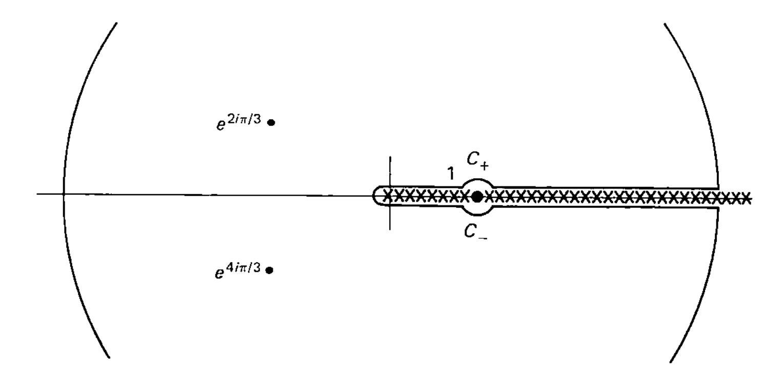

Figure 2.49 Contour and singularity structure for evaluating $\oint \log z/(1-z^3)\,dz$.

Separating out the various pieces of the contour, and recognizing that the integrals around the infinite circle and infinitesimal contour around the branch point are zero, the integral of equation 2.325 can be written

$$\oint \frac{\log z}{(1-z^3)}\,dz = P\int_0^\infty \frac{\log x}{(1-x^3)}\,dx + \int_{C_+} \frac{\log z}{(1-z^3)}\,dz + P\int_\infty^0 \frac{\log x + 2\pi i}{(1-x^3)}\,dx + \int_{C_-} \frac{\log z}{(1-z^3)}\,dz \tag{2.326}$$

Inverting the limits on the second semi-infinite integral and applying Cauchy's theorem, we obtain

$$- 2\pi i P\int_0^{\infty} \frac{1}{(1 - x^3)}\, dx + \int_{C_+} \frac{\log z}{(1 - z^3)}\, dz + \int_{C_-} \frac{\log z}{(1 - z^3)}\, dz$$

$$= 2\pi i\left[R(e^{2\pi i/3}) + R(e^{4\pi i/3})\right] \tag{2.327}$$

On C_+, ϕ ranges from π to 0, and on C_-, ϕ ranges from 2π to π. Points on $C_\pm$ are of the form

$$z = 1 + \rho e^{i\phi} \tag{2.328}$$

Because of the cut, 1 must be written

$$\text{(a)}\quad 1 = e^{i0} \text{ on } C_+ \qquad \text{(b)}\quad 1 = e^{2\pi i} \text{ on } C_- \tag{2.329}$$

Therefore,

$$\int_{C_+} \frac{\log z}{(1 - z^3)}\, dz = \log(e^{i0})\int_{\pi}^{0} \frac{i\rho e^{i\phi}}{-3\rho e^{i\phi}}\, d\phi = 0 \tag{2.330a}$$

$$\int_{C_-} \frac{\log z}{(1 - z^3)}\, dz = \log(e^{2\pi i})\int_{2\pi}^{\pi} \frac{i\rho e^{i\phi}}{-3\rho e^{i\phi}}\, d\phi = -\frac{2\pi^2}{3} \tag{2.330b}$$

Equation 2.327 then becomes

$$-2\pi i P\int_0^{\infty} \frac{1}{(1 - x^3)}\, dx - \frac{2\pi^2}{3} = 2\pi i\left[R(e^{(2/3)i\pi}) + R(e^{(4/3)i\pi})\right] \tag{2.331}$$

Straightforwardly,

$$\text{(a)}\quad R(e^{2\pi i/3}) = \frac{4\pi}{3\sqrt{3}}\frac{1}{(3 - i\sqrt{3})} \qquad \text{(b)}\quad R(e^{4\pi i/3}) = -\frac{8\pi}{3\sqrt{3}}\frac{1}{(3 + i\sqrt{3})} \tag{2.332}$$

$$\Rightarrow \quad -2\pi i P\int_0^{\infty} \frac{1}{(1 - x^3)}\, dx - \frac{2\pi^2}{3} = -\frac{2\pi^2 i}{3\sqrt{3}} - \frac{2\pi^2}{3} \tag{2.333}$$

$$\Rightarrow \quad P\int_0^{\infty} \frac{1}{(1 - x^3)}\, dx = \frac{\pi}{3\sqrt{3}} \qquad \square \tag{2.323}$$

2.9 Conformal Mapping

Definition

A *mapping* is a process of substituting one complex variable for another, in the form

$$z' = f(z) \tag{2.334}$$

Suppose a point z_0 in the z-plane maps or transforms into a point z_0' in the z'-plane. Let C_1 and C_2 be two curves that pass through the point z_0, and let the points on these curves in the region around z_0 map into curves C_1' and C_2' in the region around z_0'.

As displayed in Figure 2.50a we construct tangent lines to C_1 and C_2 at the point z_0 and call the angle between these tangents α. Referring to Figure 2.50b the angle subtended by the tangents to C_1' and C_2' is denoted by α'. The mapping is said to be *conformal* if $\alpha = \alpha'$; that is, the mapping is angle-preserving.

If the function generating the transformation is analytic at z_0, the mapping of z_0 into z_0' is conformal. To demonstrate this, let the function of equation 2.334 be analytic. Then from the definition of analyticity, $dz'/dz)_{z_0}$ is independent of the path along which z_0 is approached.

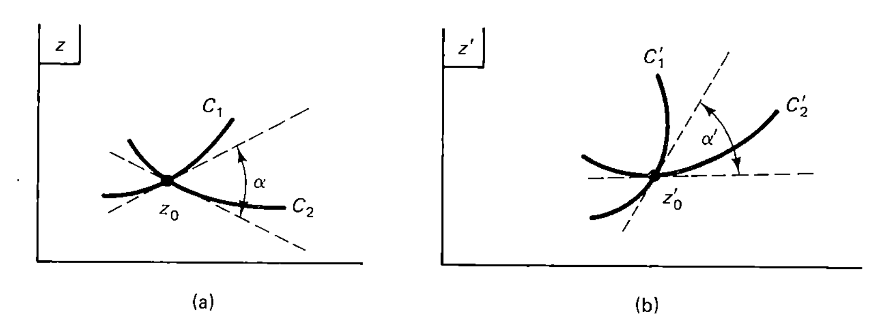

Figure 2.50 Angle formed by the tangents to two curves passing through a specified point.

A curve in the z-plane is defined by a parameter λ. For example, the points on the circumference of a circle are parametrized by the constant radius λ by

$$F(z) = zz^* = \lambda^2 \tag{2.335}$$

Let the curve C' in the z'-plane be parametrized by λ as

$$B(z') = \lambda \tag{2.336a}$$

This will define those points in the z'-plane that lie on C'. Those points were mapped from the curve C in the z-plane by the mapping $z' = f(z)$. Therefore,

$$B[f(z)] = \lambda = A(z) \tag{2.336b}$$

defines the points on C. That is, although the constraining functions A and B are different, the curve C and its image C' are parametrized by the same parameter λ.

Let the curves C_1 and C_2 of Figure 2.50a, and their images C_1' and C_2' of Figure 2.50b, by parameterized by λ_1 and λ_2, respectively. We define four complex numbers:

$$\begin{aligned} &(\text{a}) \quad \left.\frac{dz}{d\lambda_1}\right)_{z_0} \equiv \rho_1 e^{i\theta_1} \qquad (\text{b}) \quad \left.\frac{dz}{d\lambda_2}\right)_{z_0} \equiv \rho_2 e^{i\theta_2} \\ &(\text{c}) \quad \left.\frac{dz'}{d\lambda_1}\right)_{z_0} \equiv \rho_1' e^{i\theta_1'} \qquad (\text{d}) \quad \left.\frac{dz'}{d\lambda_2}\right)_{z_0} \equiv \rho_2' e^{i\theta_2'} \end{aligned} \tag{2.337}$$

Since the mapping is via an analytic function, $dz'/dz)_{z_0}$ is independent of whether z_0 is approached along C_1 or along C_2. If the path taken to z_0 is C_1 (and thus C_1' is taken to z_0'), the mapping of the tangent to C_1 at z_0 is

$$\left.\frac{dz'}{dz}\right)_{z_0} = \frac{dz'}{d\lambda_1}\frac{d\lambda_1}{dz} = \frac{dz'/d\lambda_1)_{z_0'}}{dz/d\lambda_1)_{z_0}} = \frac{\rho_1'}{\rho_1}e^{i(\theta_1'-\theta_1)} \tag{2.338a}$$

If z_0 is approached along C_2, the transformation of the tangent to C_2 at z_0 is

$$\left.\frac{dz'}{dz}\right)_{z_0} = \frac{dz'}{d\lambda_2}\frac{d\lambda_2}{dz} = \frac{dz'/d\lambda_2)_{z_0'}}{dz/d\lambda_2)_{z_0}} = \frac{\rho_2'}{\rho_2}e^{i(\theta_2'-\theta_2)} \tag{2.338b}$$

Since the transformation function $f(z)$ is analytic, the derivative is independent of path. Therefore,

$$\frac{\rho_1'}{\rho_1}e^{i(\theta_1'-\theta_1)} = \frac{\rho_2'}{\rho_2}e^{i(\theta_2'-\theta_2)} \tag{2.339}$$

We take the logarithm of equation 2.339 and separately equate the real and imaginary parts. By equating the real parts, we obtain

$$\text{(a)} \quad \log \frac{\rho_1'}{\rho_1} = \log \frac{\rho_2'}{\rho_2} \quad \Rightarrow \quad \text{(b)} \quad \frac{\rho_1'}{\rho_1} = \frac{\rho_2'}{\rho_2} \tag{2.340}$$

From the equality of the imaginary parts

$$\text{(a)} \quad \theta_1' - \theta_1 = \theta_2' - \theta_2 \quad \Rightarrow \quad \text{(b)} \quad \theta_1' - \theta_2' = \theta_1 - \theta_2 \tag{2.341}$$

We note here that $dz'/dz)_{z_0}$ cannot be zero. If it were, ρ_1'/ρ_1 and ρ_2'/ρ_2 would be zero. Then we could not take the logarithm of equation 2.338b and could not make any definitive statement about the relation between the angles.

Referring to Figure 2.50a we see that since, for example, $dz/d\lambda_1$ defines the tangent to the curve C_1 in the z-plane, θ_1 is the angle between that tangent and the real axis. Analogously, θ_2 is the angular orientation of the tangent to C_2. Analogous meaning for θ_1' and θ_2' is obtained from Figure 2.50b. Therefore,

$$\text{(a)} \quad \theta_1 - \theta_2 = \alpha \qquad \text{(b)} \quad \theta_1' - \theta_2' = \alpha' \tag{2.342}$$

From equation 2.341b, then

$$\alpha' = \alpha \tag{2.343}$$

There is no restriction on how the magnitudes of the complex quantities ρ_1 and ρ_2 change under a conformal mapping, but angles between the tangents to curves remain unchanged by such a mapping.

Applications

In certain three-dimensional problems, the geometric symmetry is such that the analysis is independent of one of the coordinates. Thus the problem becomes two-dimensional. For example, the electric field due to a uniform infinitely long line of charge is symmetric in the dimension along the line of charge. This is illustrated in Figure 2.51a. The electric field is the same in magnitude and direction at points P_1 and P_2 which are the same distance from the line. The three-dimensional problem of Figure 2.51a can be analyzed as the two-dimensional one shown in Figure 2.51b, which is the view of Figure 2.51a looking along the line of charge.

In some of these two-dimensional problems, it is possible to map a complicated geometry into a simple one, making the problem in the new geometry straightforward. The technique will be illustrated with some examples in electrostatics. In particular, we consider problems in which we determine the electrostatic potential in a two-dimensional space in regions in which the charge density ρ is zero. In such a problem, Maxwell's equations lead to Laplace's equation,

$$\nabla^2 V(x, y) = 0 \tag{2.344}$$

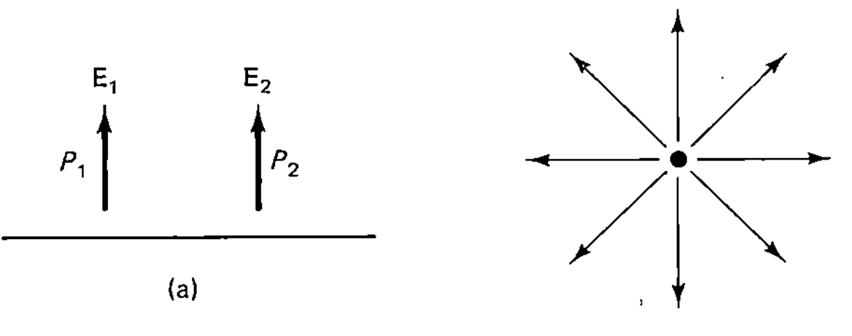

Figure 2.51 Electric field due to an infinite line of charge (a) at two points equidistant from the line and (b) viewed along the line of charge.

Recall it was shown earlier (equation 2.47) that if a function is analytic, it satisfies Laplace's equation in two dimensions. Thus we expect the potential $V(x, y)$ to be an analytic function of the complex variable z. One of the properties of a conformal transformation via an analytic function is that equation 2.344 is valid in the transformed complex plane.

This fact can be verified in a rather straightforward way. Let V be the electrostatic potential for a problem that can be treated in two dimensions. We then consider $V(x, y)$ as a function of the complex variable. That is,

$$V(x, y) = V(z) \tag{2.345}$$

Let $z' = f(z)$ be a mapping by an analytic function in the region around z_0 such that $df/dz)_{z_0} \neq 0$. Because $f(z)$ is an analytic function, $dz'/dz)_{z_0}$ exists, and therefore the inverse mapping

$$z = f^{-1}(z') \equiv g(z') \tag{2.345}$$

is analytic in the region around z_0' (the image of z_0). Differentiating $z' = f(z)$, we have

$$\text{(a)} \quad 1 = \frac{df}{dz}\frac{dz}{dz'} \qquad \Rightarrow \qquad \text{(b)} \quad \left.\frac{dz}{dz'}\right)_{z_0'} = \frac{1}{df/dz)_{z_0}} \tag{2.346}$$

Since $df/dz)_{z_0}$ exists and is nonzero, $dz/dz')_{z_0'}$ exists, indicating that the inverse transformation is analytic. Thus if a potential $V(z)$ is analytic, its image, $V(g(z')) = V(z')$ is also analytic, and will also satisfy Laplace's equation in the primed variables.

The conditions at the geometric boundaries of the problem that are imposed on most physically significant functions (such as the electrostatic potential) will usually be of one of the following types:

$$\text{(a)} \quad V(z_{\text{boundary}}) = C_1 = \text{constant} \qquad \text{(b)} \quad \left.\frac{dV}{dn}\right|_{z_{\text{boundary}}} = C_2 = \text{constant}$$

$$\text{(c)} \quad V(z_{\text{boundary}}) + \text{constant} \cdot \left.\frac{dV}{dn}\right|_{z_{\text{boundary}}} = C_3 = \text{constant} \tag{2.347}$$

They are known as (a) Dirichlet, (b) Neumann, and (c) Gauss or mixed boundary conditions, respectively. The point on the boundary maps into its image by

$$\text{(a)} \quad z'_{\text{boundary}} = f(z_{\text{boundary}}) \qquad \Rightarrow \qquad \text{(b)} \quad z_{\text{boundary}} = f^{-1}(z'_{\text{boundary}}) = g(z'_{\text{boundary}}) \tag{2.348}$$

Therefore, the three boundary conditions can be written

$$\begin{aligned} &\text{(a)} \quad V\left[g(z'_{\text{boundary}})\right] = C_1 \qquad \text{(b)} \quad \left.\frac{dV[g(z')]}{dn}\right|_{z'_{\text{boundary}}} = C_2 \\ &\text{(c)} \quad V\left[g(z'_{\text{boundary}})\right] + \text{constant} \cdot \left.\frac{dV[g(z')]}{dn}\right|_{z'_{\text{boundary}}} = C_3 \end{aligned} \tag{2.349}$$

That is, the value(s) of the potential at the boundaries remain unchanged by the mapping. Then in the z'-plane, in terms of x' and y', V will also satisfy the Laplace equation.

$$\nabla^2 V(x', y') = 0 \tag{2.350}$$

subject to the same boundary conditions as those that apply in the z-plane.

The goal is to find a conformal transformation that makes a complicated geometry simple.

Example 2.34

Consider an infinite conducting material that has a wedge-shaped space cut out of it (Figure 2.52). Charge is distributed throughout the conductor such that the conducting material is at a potential V_0. The angle of the wedge is β. We wish to find the potential at all points inside the wedged space. Points on the edge of the wedge along the real axis are of the form

$$z = re^{i0} = r \tag{2.351a}$$

The points on the slanted edge are of the form

$$z = re^{i\beta} \tag{2.351b}$$

Consider the transformation

$$z' = z^{\pi/\beta} \tag{2.352}$$

From equation 2.351a, the points on the edge of the wedge along the real axis map into

$$z' = \left(re^{i0}\right)^{\pi/\beta} = r^{\pi/\beta} \tag{2.353}$$

These points are real, so

$$\text{(a)} \quad x' = r^{\pi/\beta} \qquad \text{(b)} \quad y' = 0 \tag{2.354}$$

Since $0 \leqslant r \leqslant \infty$, the points described in equation 2.351a map into the positive real axis in the z'-plane.

From equation 2.351b, the points on the slanted edge map into

$$z' = \left(re^{i\beta}\right)^{\pi/\beta} = r^{\pi/\beta}e^{i\pi} = -r^{\pi/\beta} \tag{2.355}$$

Since this is real and negative, and in the z-plane, $0 \leqslant r \leqslant \infty$, these points are mapped into the negative real axis of the z'-plane (Figure 2.53).

A point in the space inside the wedge is of the form $z = re^{i\theta}$. Then such a point is mapped into

$$z' = \left(re^{i\theta}\right)^{\pi/\beta} = r^{\pi/\beta}e^{i(\theta\pi/\beta)} \tag{2.356}$$

Since $0 < \theta/\beta < 1$, the argument of z' for points mapped from inside the wedge are between 0 and π. Thus these points map into the upper half z'-plane.

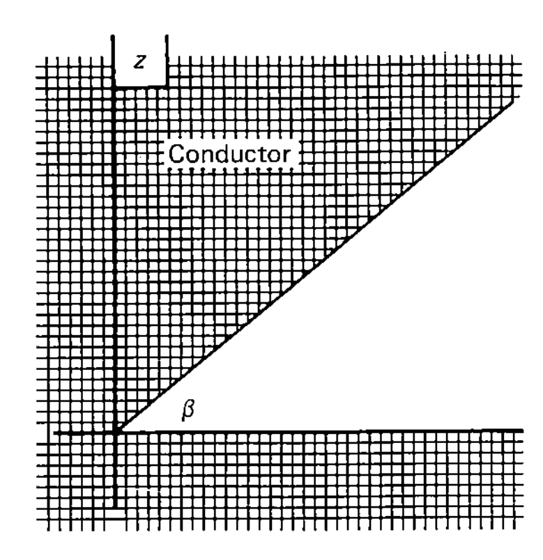

Figure 2.52 Wedge-shaped space cut from a conducting material.

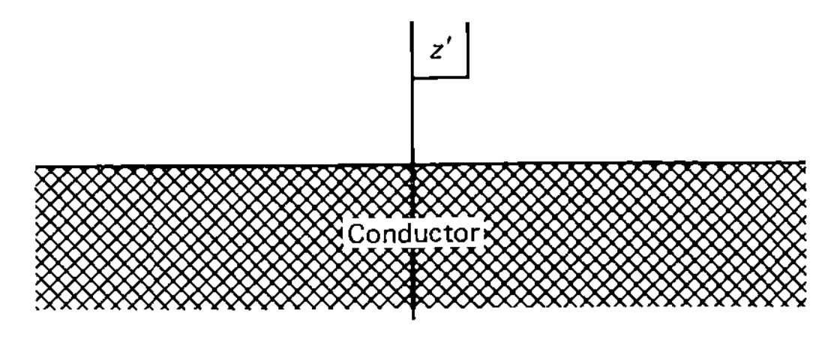

Figure 2.53 Geometry into which the wedge of Figure 2.52 is mapped by $z' = z^{\pi/\beta}$.

From the symmetry of the problem in the z'-plane, it is obvious that the potential cannot depend on x'. That is,

$$V(x', y') = V(y') \tag{2.357}$$

Since V is analytic (it satisfies the Laplace equation in two dimensions), it can be expanded in a MacLaurin series:

$$V(y') = \sum_{n=0}^{\infty} a_n y'^n \tag{2.358}$$

Laplace's equation for a potential that only depends on y' becomes

$$\frac{d^2V}{dy'^2} = 0 \tag{2.359}$$

The only way this can be satisfied by the series representation of $V(y')$ for all y' is if $a_2 = a_3 = \cdots = 0$. Therefore, the series reduces to

$$V(y') = a_0 + a_1 y' \tag{2.360}$$

This is the result obtained when equation 2.359 is integrated twice.

From Gauss's law, the electric field for a flat plate with a surface charge density σ is the constant

$$E = \frac{\sigma}{\varepsilon_0} = -\frac{dV}{dy'} \tag{2.361}$$

Applying this, and the boundary condition that $V = V_0$ at any point on the surface of the conductor ($y' = 0$), we obtain

$$V(y') = V_0 - Ey' = V_0 - \frac{\sigma}{\varepsilon_0} y' \tag{2.362}$$

To complete the problem, we must now transform back to the z-plane. Since $z' = x' + iy'$, we have $y' = \text{Im } z'$. Using equation 2.356 yields

$$y' = \text{Im } z' = \text{Im}(r^{\pi/\beta} e^{i(\pi\theta/\beta)}) = r^{\pi/\beta} \sin\left(\frac{\theta\pi}{\beta}\right) \tag{2.363}$$

Therefore, the potential in terms of r, and θ is

$$V(r, \theta) = V_0 - \frac{\sigma}{\varepsilon_0} r^{\pi/\beta} \sin\left(\frac{\theta\pi}{\beta}\right) \tag{2.364}$$

This can also be expressed in Cartesian coordinates by using

$$\text{(a)} \quad r = \sqrt{x^2 + y^2} \qquad \text{(b)} \quad \theta = \tan^{-1} \frac{y}{x} \tag{2.365}$$

By setting $V(r, \theta) = \text{constant}$ in equation 2.364, we obtain the equation for the equipotential surfaces of this configuration in the z-plane. They are

$$r^{\pi/\beta} \sin\left(\frac{\theta\pi}{\beta}\right) = \text{constant} \quad \square \tag{2.366}$$

Example 2.35

As a second example, consider the problem of a wedge shaped space at an angle β, with an insulating strip separating two sections of the conductor (Figure 2.54). Because of the insulating strip, the potentials of the two sections of conductor can be different values.

For this problem we cannot use the transformation $z' = z^{\pi/\beta}$. We have seen that this mapping will transform the geometry of Figure 2.54 into a flat plate. But the insulating strip allows the potential on the slanted part of the wedge to be different than the potential on

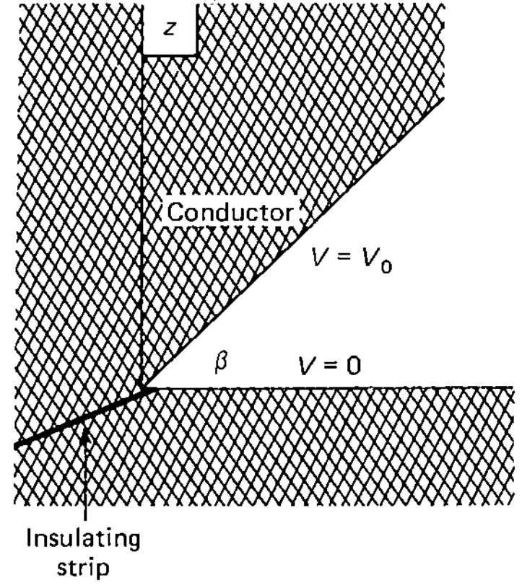

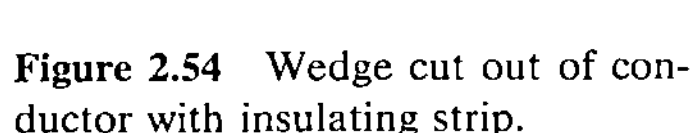

Figure 2.54 Wedge cut out of conductor with insulating strip.

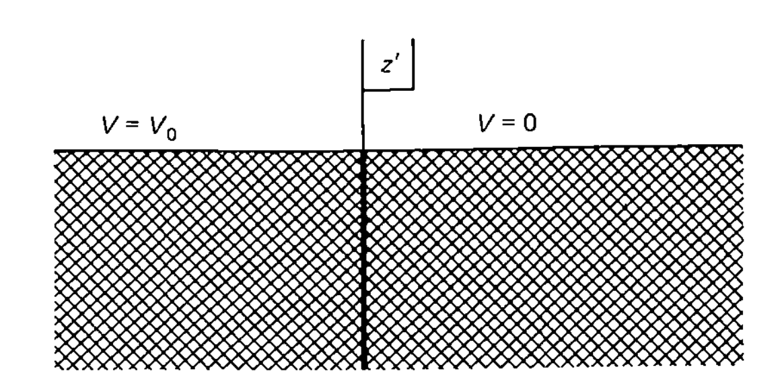

Figure 2.55 Mapping of wedge with insulating strip under the mapping $z' = z^{\pi/\beta}$.

the flat part of the wedge. Thus in the z' plane, the part of the plate along the negative real axis would be at a different potential than the part along the positive real axis, as illustrated in Figure 2.55. Therefore, the symmetry in the x' variable that yields straightforward results in Example 2.34 does not exist for this geometry.

Consider the transformation

$$z' = \log z \tag{2.367}$$

Points on the flat part of the wedge, $z = re^{i0} = r$, map into

$$z' = \log r = x' \tag{2.368}$$

Since $0 \leqslant r \leqslant \infty$, this region maps into $-\infty \leqslant x' \leqslant \infty$. That is, these points map into the entire real axis in the z'-plane.

Points on the slanted edge are of the form $z = re^{i\beta}$, so

$$z' = \log r + i\beta = x' + i\beta \tag{2.369}$$

Since the imaginary part of each point is constant, these points form a flat surface parallel to the real z'-axis and displaced from it by a distance β. Since $x' = \log r$ and $0 \leqslant r \leqslant \infty$, this surface also runs from $-\infty$ to $+\infty$.

A point in the space inside the wedge is of the form $z = re^{i\theta}$ with $0 < \theta < \beta$. These map into points $z' = \log r + i\theta$ which have real parts ranging from $-\infty$ to $+\infty$, and imaginary parts that are between 0 and β. These are the points between the plates in Figure 2.56.

Thus the mapping of equation 2.367 transforms the geometry of Figure 2.54 into an infinite parallel plate capacitor with a plate separation β, and a potential difference V_0. For

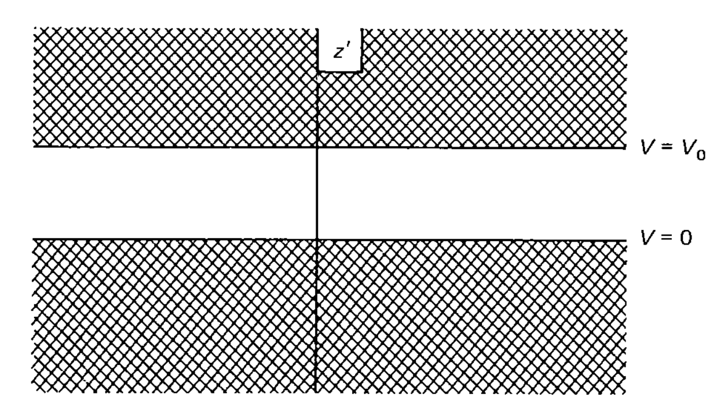

Figure 2.56 Mapping of wedge with insulating strip of Figure 2.54 under mapping $z' = \log z$.

such a capacitor, the constant electric field between the plates is

$$E = -\frac{\Delta V}{d} = -\frac{V_0}{\beta} \tag{2.370}$$

As in the previous example,

$$V = a_0 + a_1 y' \tag{2.371}$$

The boundary condition $V(y' = 0) = 0$, and

$$E = -\frac{dV}{dy'} = -\frac{V_0}{\beta} = -a_1 \tag{2.372}$$

$$\Rightarrow \quad V = \frac{V_0}{\beta} y' \tag{2.373}$$

As before, we must express the result in the z-plane by using the inverse mapping. From

$$y' = \operatorname{Im} z' = \operatorname{Im}(\log r + i\theta) = \theta = \tan^{-1}\left(\frac{y}{x}\right) \tag{2.374}$$

$$\Rightarrow \quad \text{(a)} \quad V(r, \theta) = \frac{V_0}{\beta}\theta \qquad \text{(b)} \quad V(x, y) = \frac{V_0}{\beta}\tan^{-1}\left(\frac{y}{x}\right) \tag{2.375}$$

The equipotential surfaces are found by setting $V =$ constant. V will be constant when y/x is constant. So the equipotential surfaces in the z-plane are straight lines specified by $y = cx$. □

2.10 Singularities of Functions Defined by Integrals

In several instances, functions are defined by integrals of other functions. Two such examples are the gamma function (a generalization of $N!$) and the error function.

$$\Gamma(z) = \int_0^\infty x^z e^{-x}\, dx \tag{2.376}$$

$$\operatorname{erf}(z) = \frac{2}{\sqrt{\pi}} \int_0^z e^{-x^2}\, dx \tag{2.377}$$

These and other integral defined functions are discussed in detail in Chapter 6.

In many cases a result must be left in the form of an integral because the integral cannot be evaluated in terms of elementary functions. In this section we investigate the singularity structure of functions of the form

$$F(z) = \int_A^B G(z, z')\, dz' \tag{2.378}$$

The singularity structure of $F(z)$ is deduced from the singularities of $G(z, z')$ and the contour from A to B. If $G(z, z')$ has no singularities in the z'-plane, then the sum of an infinite number of analytic terms is analytic, for any z. Thus $F(z)$ is analytic at all z, and the integral of equation 2.378 will be independent of the path taken between A and B. Therefore, if $F(z)$ is singular, $G(z, z')$ must have at least one singularity. We take that singularity of $G(z, z')$ to be at z_0'. If z_0' does not depend on z, it is called a fixed singularity. If z_0' varies with z, it is called a movable singularity. As long as the contour does not encounter the singularity, the sum is finite and $F(z)$ is analytic.

Example 2.36

For example, the integrand of

$$F(z) = \int_{-1}^{1} \frac{z}{(z' - i)}\, dz' \tag{2.379}$$

has a fixed singularity at $z' = i$, and the singularity is not on the contour. Therefore,

$$F(z) = z \log\left(\frac{(1-i)}{-(1+i)}\right) = z \log\left[\frac{e^{-i\pi/4}}{e^{-3i\pi/4}}\right] = iz\frac{\pi}{2} \tag{2.380}$$

which is analytic at all finite z. □

If z_0' depends on z, the singularity will move around in the z'-plane when z varies. If it moves toward the contour, and if $G(z, z')$ is analytic at all other points z', the contour can be deformed away from the singularity for most points along the contour. That is, the integral of equation 2.378 will be the same whether integrated over the contour of Figure 2.57a or over the contour of Figure 2.57b. The only points on the contour that cannot be deformed away from $z_0'(z)$ are the endpoints. Thus if z takes on values such that $z_0'(z)$ encounters one of the endpoints, $F(z)$ will be singular at that value of z.

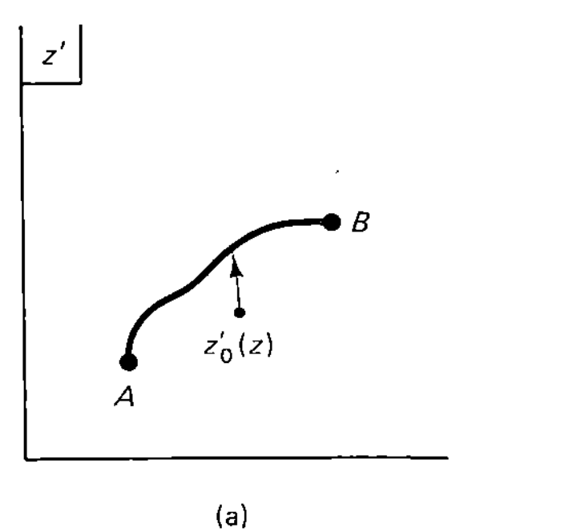

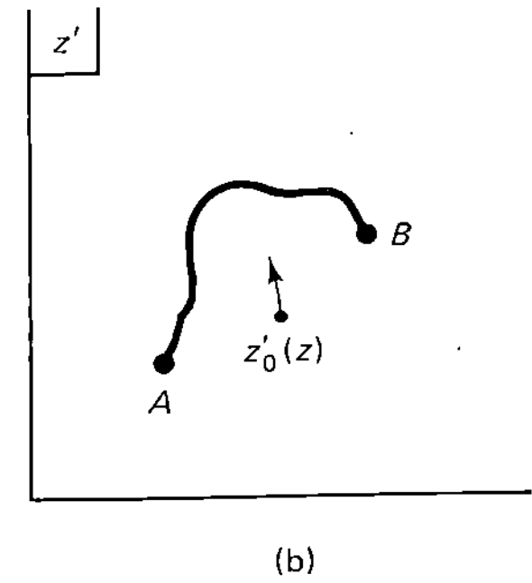

Figure 2.57 (a) Movement of a singularity toward the contour and (b) deformation of the contour away from the singularity.

Example 2.37

As an example, consider

$$\text{(a)} \quad F(z) = \int_A^B \frac{1}{(z' - z^2)}\, dz' \quad \Rightarrow \quad \text{(b)} \quad G(z, z') = \frac{1}{(z' - z^2)} \tag{2.381}$$

Clearly, $G(z, z')$ has a simple pole at $z_0' = z^2$. Thus when

$$\text{(a)} \quad z_0'(z) = z^2 = A \qquad \text{(b)} \quad z_0'(z) = z^2 = B, \tag{2.382}$$

$F(z)$ will be singular. This intergral, of course, can be evaluated directly.

$$F(z) = \log\left(\frac{(B - z^2)}{(A - z^2)}\right) \tag{2.383}$$

It is obvious that $F(z)$ has branch singularities at $z = \pm\sqrt{A}$ and at $z = \pm\sqrt{B}$, as specified. □

The types of singularities of $F(z)$ that arise from the singularities of $G(z, z')$ coinciding with the endpoints of the contour are called *endpoint singularities*. Now consider a function $G(z, z')$ that has singularities at more than one point. If $G(z, z')$ has singularities at $z_0'(z)$ and $z_1'(z)$, then either the singularities will lie on the same side of the contour, or they will be on opposite sides of the contour as illustrated in Figure 2.57.

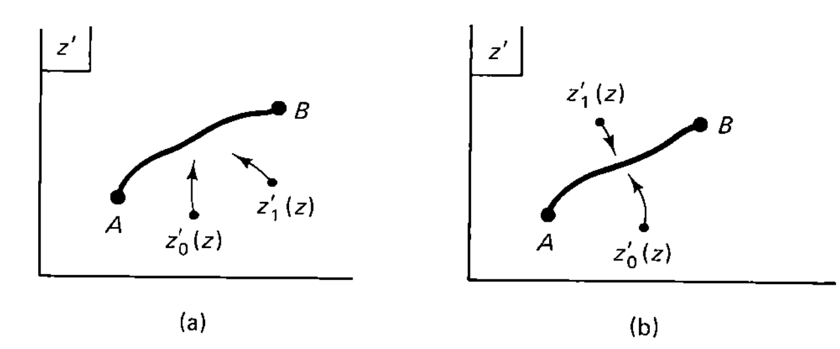

Figure 2.58 (a) Two singularities that cannot pinch the contour; (b) two singularities that can pinch the contour.

When z is varied, and these singularities move around in the z'-plane, both singularities can approach the contour. In Figure 2.58a the two singularities approach the contour from the same side. The contour can be deformed away, and $F(z)$ will not be singular (except for possible endpoint coincidences). However, if z_0' and z_1' are on opposite sides of the contour, they can approach the same point on the contour, and it cannot be deformed away from the singularity. In that case, the two singularities pinch the contour and $F(z)$ is singular at the value of z that causes the pinch. As expected, this type of a singularity of $F(z)$ is called a *pinch singularity*.

Example 2.38

An integral similar to one that occurs in a quantum scattering problem is

$$F(x) = \int_{-b}^{b} \frac{1}{\left[(x'-x)^2 - m^2 + i\varepsilon\right]\left[(x'+x)^2 - m^2 + i\varepsilon\right]}\, dx' \qquad (2.384)$$

where m is the mass of a particle involved in the scattering, and ε is the Feynman ε. The integrand has singularities as shown in Figure 2.59. Since ε will eventually be 0, endpoint singularities of $F(x)$ arise from

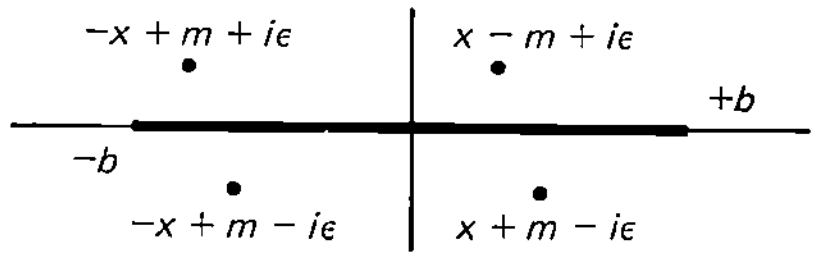

Figure 2.59 Singularities and contour for the integral of equation 2.384.

$$\text{(a)}\quad x - m = \pm b \quad \Rightarrow \quad x = m \pm b \qquad \text{(b)}\quad x + m = \pm b \quad \Rightarrow \quad x = -m \pm b$$

$$(2.385)$$

$$\text{(c)}\quad -x - m = \pm b \quad \Rightarrow \quad x = m \pm b \qquad \text{(d)}\quad -x + m = \pm b \quad \Rightarrow \quad x = -m \pm b$$

Equation 2.385a is the same as equation 2.385c, and equations 2.385b and 2.385d are identical. Thus there are four endpoint singularities of $F(x)$, at

$$\text{(a)}\quad x = m + b \quad \text{(b)}\quad x = m - b \quad \text{(c)}\quad x = -m + b \quad \text{(d)}\quad x = -m - b \qquad (2.386)$$

We refer to Figure 2.59 to investigate pinch singularities. We see that when $\varepsilon \to 0$, $-x + m - i\varepsilon$ coincides with $x + m - i\varepsilon$ at $x = 0$. But there is no pinch of the contour since they approach the contour from the same side. Thus $F(x)$ is not singular at $x = 0$. For the same reason, there is no pinch singularity when $-x - m$ coincides with $x - m$ at $x = 0$.

Pinch singularities of $F(x)$ will arise from singularities of the integrand that are on opposite sides of the contour. For $m > 0$, $-x - m$ cannot equal $-x + m$. Similarly, $x - m$ and $x + m$ cannot be equal. Thus, pinch singularities of $F(x)$ do not arise from a pinch by these poles. However, the contour can be pinched when

$$\text{(a)}\quad -x - m = x + m \quad \Rightarrow \quad \text{(b)}\quad x = -m \qquad (2.387)$$

$$\text{(a)}\quad -x + m = x - m \quad \Rightarrow \quad \text{(b)}\quad x = +m \qquad (2.388)$$

These, of course, will pinch the contour only if $-b < m < b$. So, for $-b < m < b$, $F(x)$ will have pinch singularities at $x = \pm m$. In the actual scattering problems, $b = \infty$, so there are never endpoint singularities, and there are always pinch singularities at $x = \pm m$. □

2.11 Dispersion Relations

A *dispersion relation* is an integral representation of a function. It describes the real part of the function in terms of an integral involving the imaginary part of that function, and expresses the imaginary part of the function as an integral involving its real part. The designation "dispersion relations" came from the work of Kramers and Kronig, who first derived a pair of integrals that related the real (dispersive) and imaginary (absorptive) parts of the complex index of refraction for electromagnetic waves in material media.

Consider a function $F(z)$ that may have poles at $z_0, z_1, \ldots, z_N$ inside a contour C, with corresponding residues $R_0, R_1, \ldots, R_N$. Then by Cauchy's theorem,

$$\frac{1}{2\pi i}\oint \frac{F(z')}{(z'-z)}\,dz' = F(z) + \sum_{i=1}^{N} R_i \tag{2.389}$$

A contour that is appropriate for a large number of scientific problems runs along the real axis from $-\infty$ to $+\infty$, and closes in one of the half planes, as shown in Figure 2.60.

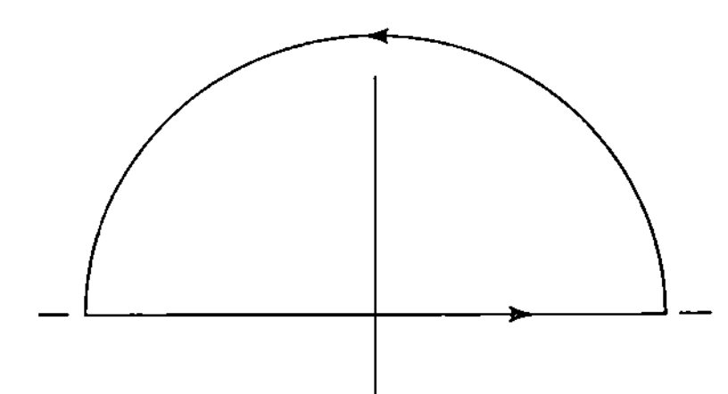

Figure 2.60 Contour appropriate for deducing dispersion relations.

For simplicity, we will assume that $F(z)$ is analytic inside and on such a contour, and approaches zero at infinity "fast enough." Then all residues are zero, and Cauchy's theorem becomes

$$\frac{1}{2\pi i}\oint \frac{F(z')}{(z'-z)}\,dz' = \begin{cases} F(z) & \operatorname{Im} z > 0 \\ 0 & \operatorname{Im} z < 0 \end{cases} \tag{2.390}$$

Taking $\operatorname{Im} z \geqslant 0$, we obtain

$$F(z) = \frac{1}{2\pi i}\oint \frac{F(z')}{(z'-z)}\,dz' \tag{2.391}$$

Using the integral around the contour of Figure 2.60, let z approach the real axis from the upper half-plane as $z = x + i\varepsilon$. Then, taking the integral on the infinite semicircle to be zero, equation 2.391 becomes

$$F(x) = \frac{1}{2\pi i}\int_{-\infty}^{\infty} \frac{F(x')}{(x'-x-i\varepsilon)}\,dx' \tag{2.392}$$

Using equations 2.299, we have

$$F(x) = \frac{1}{2\pi i}\int_{-\infty}^{\infty} F(x')\left(\frac{1}{(x'-x)_P} + i\pi\delta(x'-x)\right)dx'$$

$$= \frac{1}{2\pi i} P\int_{-\infty}^{\infty} \frac{F(x')}{(x'-x)}\,dx' + \frac{1}{2}F(x) \tag{2.393a}$$

$$\Rightarrow \quad F(x) = \frac{1}{i\pi} P\int_{-\infty}^{\infty} \frac{F(x')}{(x'-x)}\,dx' \tag{2.393b}$$

In many problems in which dispersion relations are applicable, one part of the function (e.g., the imaginary part) may be known, perhaps in analytic form or numerically from experimental measurements. Dispersion relations are used to determine the unknown part of the function. To put equation 2.393b into a more convenient form for such a determination, we write

$$F(x) = \operatorname{Re} F(x) + i \operatorname{Im} F(x) \tag{2.394}$$

Substituting this into equation 2.393b, and equating real parts and separately imaginary parts, we obtain

$$\text{(a)} \quad \operatorname{Re} F(x) = \frac{1}{\pi} P \int_{-\infty}^{\infty} \frac{\operatorname{Im} F(x')}{(x'-x)}\, dx' \quad \text{(b)} \quad \operatorname{Im} F(x) = -\frac{1}{\pi} P \int_{-\infty}^{\infty} \frac{\operatorname{Re} F(x')}{(x'-x)}\, dx' \tag{2.395}$$

Often, x represents a physical parameter such as the kinetic energy of a particle, that does not take on negative values. Then one part of $F(x)$, Im $F(x)$ for example, would be known only at positive x-values. Therefore, half the region of integration of equation 2.395a is inaccessible physically. Since it is not possible to know Im $F(x)$ over the entire range of x, dispersion relations expressed in the form of equations 2.395 would not be useful for the analysis of such a problem.

Reflection symmetry

However, some scientifically meaningful functions satisfy symmetry conditions that allow us to express the integrals in equations 2.395 over just the positive half (or just the negative half) of the range of integration. One such symmetry, called the *reflection symmetry*, specifies that F is even under reflection if

$$\text{(a)} \quad F^*(-z^*) = +F(z) \quad \text{while} \quad \text{(b)} \quad F^*(-z^*) = -F(z) \tag{2.396}$$

means that F is odd under reflection.

Let $F(z)$ be even under this reflection. We write equation 2.393b as

$$F(x) = \frac{1}{i\pi} P \left[\int_{-\infty}^{0} \frac{F(x')}{(x'-x)}\, dx' + \int_{0}^{\infty} \frac{F(x')}{(x'-x)}\, dx' \right] \tag{2.397}$$

On the real axis, $z'^* = z' = x'$. Referring to equation 2.396a, on the real axis, a function even under this reflection satisfies

$$F(-x') = F^*(x') \tag{2.398}$$

In the first integral, with $F(x')$ even, substituting $x' = -x'$ yields

$$F(x) = \frac{1}{i\pi} P \left[\int_{\infty}^{0} \frac{F^*(x')}{(x'+x)}\, dx' + \int_{0}^{\infty} \frac{F(x')}{(x'-x)}\, dx' \right] \tag{2.399}$$

After inverting the limits on the first integral and taking the real and imaginary parts of this, we obtain

$$\begin{aligned} &\text{(a)} \quad \operatorname{Re} F(x) = \frac{2}{\pi} P \int_{0}^{\infty} x' \frac{\operatorname{Im} F(x')}{(x'^2 - x^2)}\, dx' \\ &\text{(b)} \quad \operatorname{Im} F(x) = -\frac{2}{\pi} x P \int_{0}^{\infty} \frac{\operatorname{Re} F(x')}{(x'^2 - x^2)}\, dx' \end{aligned} \tag{2.400}$$

Example 2.39

As an example, we will determine the function that has the following properties:

1. It is even under the reflection described in equation 2.396a.
2. The real part of the function is $1/(x^2 + \alpha^2)$ with α = constant.

Since Re $F(x)$ is known, we substitute this into equation 2.400b to obtain

$$\text{Im } F(x) = -\frac{2}{\pi} x P \int_0^\infty \frac{1}{(x'^2 + \alpha^2)(x'^2 - x^2)} dx' \tag{2.401}$$

At this point, we present some techniques for evaluating this integral. Writing

$$\frac{1}{(x'^2 + \alpha^2)(x'^2 - x^2)} = \frac{1}{(x^2 + \alpha^2)} \left[\frac{1}{(x'^2 - x^2)} - \frac{1}{(x'^2 + \alpha^2)} \right] \tag{2.402}$$

$$\Rightarrow \quad \text{Im } F(x) = -\frac{2}{\pi} \frac{1}{(x^2 + \alpha^2)} P \int_0^\infty \left[\frac{1}{(x'^2 - x^2)} - \frac{1}{(x'^2 + \alpha^2)} \right] dx' \tag{2.403}$$

To evaluate the first integral, we use the identity

$$\frac{1}{(x'^2 - x^2)_P} = \text{Re} \left[\frac{1}{(x'^2 - x^2 - i\varepsilon)} \right] \tag{2.404}$$

The first integral then becomes

$$P \int_0^\infty \frac{1}{(x'^2 - x^2)} dx' = \text{Re} \int_0^\infty \frac{1}{(x'^2 - x^2 - i\varepsilon)} dx' = \frac{1}{2x} \text{Re} \log \left(\frac{(x' - x - i\varepsilon)}{(x' + x + i\varepsilon)} \right)_0^\infty \tag{2.405}$$

We take the cut of the logarithm to be along the positive x' axis and $x > 0$. Therefore, when $x' = 0, x' - x - i\varepsilon = -x - i\varepsilon$ is just below the negative real axis. Since there is no cut in this region, we can write

$$x' - x - i\varepsilon|_{x'=0} = xe^{i\pi} \tag{2.406a}$$

Since $x + i\varepsilon$ has a small positive imaginary part, and a positive real part

$$x' + x + i\varepsilon|_{x'=0} = xe^{i0} = x \tag{2.406b}$$

When $x' \to \infty$, x can be ignored. Then $x' - i\varepsilon$ is just below the cut along the positive real axis, and can thus be written

$$x' - x - i\varepsilon|_{x' \to \infty} = x'e^{2\pi i} \tag{2.406c}$$

Similarly, when $x' \to \infty$, $x' + i\varepsilon$ is just above the cut and therefore,

$$x' + x + i\varepsilon|_{x' \to \infty} = x'e^{i0} = x' \tag{2.406d}$$

Therefore,

$$\log \left(\frac{x' - x - i\varepsilon}{x' + x + i\varepsilon} \right)_0^\infty = \log \left(\frac{x'}{x'} \right)_{x' \to \infty} + 2\pi i - \log \left(\frac{x}{x} \right) - i\pi = i\pi \tag{2.407}$$

from which

$$P \int_0^\infty \frac{1}{(x'^2 - x^2)} dx = \text{Re}(i\pi) = 0 \tag{2.408}$$

The second integral is easily evaluated with a substitution $x' = \alpha \tan \theta$. The result is

$$\int_0^\infty \frac{1}{(x'^2 + \alpha^2)} dx' = \frac{\pi}{2\alpha} \tag{2.409}$$

From these results, we find that

$$\text{Im } F(x) = \frac{x}{\alpha(x^2 + \alpha^2)} \tag{2.410}$$

$$\Rightarrow \quad F(x) = \left(1 + i\frac{x}{\alpha}\right) \frac{1}{(x^2 + \alpha^2)} \quad \square \tag{2.411}$$

Subtracted dispersion relations

There are some problems that are amenable to dispersion relations, for which the function has a specified value at some point. Such a situation can arise from a boundary condition, as in the potential problems we solved using conformal maps. Or they might come about in problems for which an initial value of the function is specified. The dispersion relations of equations 2.395 or 2.400 do not require the function to have a specified value at a specific point. Such a condition must be designed into the dispersion relations.

Let the function $G(x)$ be required to have the value $G(x_0)$ at the point x_0. In equations 2.395, let

$$F(x) = \frac{[G(x) - G(x_0)]}{(x - x_0)} \tag{2.412}$$

At $x = x_0$,

$$F(x_0) = G'(x_0) \tag{2.413}$$

which we assume is well defined. Then equations 2.395 become

$$\mathrm{Re}\left[\frac{[G(x) - G(x_0)]}{(x - x_0)}\right] = \frac{1}{\pi} P \int_{-\infty}^{\infty} \frac{\mathrm{Im}[G(x') - G(x_0)]}{(x' - x_0)(x' - x)} \, dx' \tag{2.414a}$$

$$\mathrm{Im}\left[\frac{[G(x) - G(x_0)]}{(x - x_0)}\right] = -\frac{1}{\pi} P \int_{-\infty}^{\infty} \frac{\mathrm{Re}[G(x') - G(x_0)]}{(x' - x_0)(x' - x)} \, dx' \tag{2.414b}$$

from which

$$\mathrm{Re}\, G(x) = \mathrm{Re}\, G(x_0) + \frac{x - x_0}{\pi} P \int_{-\infty}^{\infty} \frac{\mathrm{Im}[G(x') - G(x_0)]}{(x' - x_0)(x' - x)} \, dx' \tag{2.415a}$$

$$\mathrm{Im}\, G(x) = \mathrm{Im}\, G(x_0) - \frac{x - x_0}{\pi} P \int_{-\infty}^{\infty} \frac{\mathrm{Re}[G(x') - G(x_0)]}{(x' - x_0)(x' - x)} \, dx' \tag{2.415b}$$

This is an example of what is called a once-subtracted or singly subtracted dispersion relation. If $G(z)$ is even under the reflection described above, once subtracted dispersion relations, integrated over $[0, \infty]$ can be obtained straightforwardly.

We note that the denominators of equations 2.415 contain two factors of x', whereas the denominators of the integrands in the unsubtracted dispersion relations of equations 2.395 contain only one factor of x'. Thus the integrands of the subtracted dispersion integrals approach zero faster at infinity that do those in the unsubtracted dispersion integrals. If $F(x)$ does not approach zero "fast enough," one cannot describe $F(x)$ by an unsubtracted dispersion relation. However, it may still be possible to describe $F(x)$ by a subtracted dispersion relation because of the additional factor of x' in the denominator of equations 2.415.

The analysis that leads to dispersion relations can be applied to functions that have branch points as well as poles. A scattering amplitude is a function that describes the quantum scattering of two or more particles. An amplitude will have a branch point in the complex energy variable at the energy where a two-particle system has a resonance, or at the mass of a particle that is created out of the collision.

For a function that has a branch point at $z = M$, for which the associated cut is taken along the positive real axis, Cauchy's theorem is written

$$F(z) = \frac{1}{2\pi i} \int_{M}^{\infty} \frac{F(z')}{(z' - z)} \, dz' \tag{2.416}$$

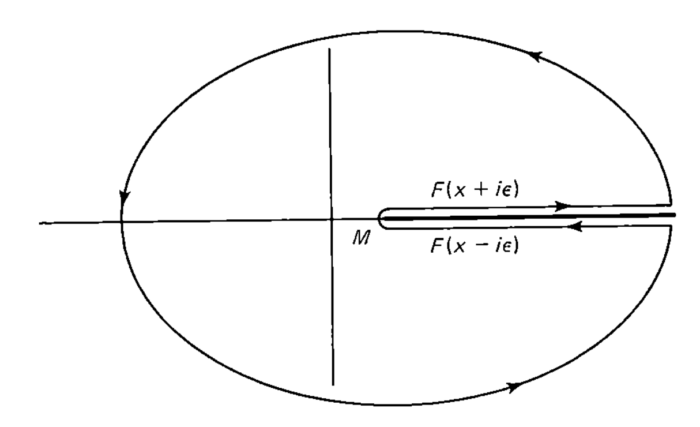

Figure 2.61 Branch structure of a scattering amplitude and contour for its dispersion integral representation.

around the Pac-Man contour of Figure 2.61. Explicitly writing out the contributions from each of the pieces of the contour yields

$$\begin{aligned} F(z) &= \frac{1}{2\pi i}\int_M^\infty \frac{F(x'+i\varepsilon)}{(x'-z)}\,dx' + \int_\infty^M \frac{F(x'-i\varepsilon)}{(x'-z)}\,dx' \\ &= \frac{1}{2\pi i}\int_M^\infty \frac{F(x'+i\varepsilon)-F(x'-i\varepsilon)}{(x'-z)}\,dx' = \frac{1}{2\pi i}\int_M^\infty \frac{\Delta(x')}{(x'-z)}\,dx' \end{aligned} \tag{2.417}$$

where the contributions from the infinite circle and contour around the branch point are assumed to be zero and have been omitted. $\Delta(x')$ is the discontinuity across the cut.

Schwarz reflection principle

Many functions involved in quantum scattering processes satisfy a reflection principle called the *Schwarz reflection principle*. Let $F(z)$ have the following properties:

1. $F(z)$ is analytic at and in a region around a real point x, and thus is analytic in a region that includes a part of the real axis.
2. $F(z)$ is real for real values of z in the region described in property 1.

In the situation we are considering, the cut is chosen along the real axis. Therefore, the region of analyticity required in property 1 includes a point on the real axis at some value of $x < M$. For values of z around x, $F(z)$ has a Taylor expansion

$$F(z) = \sum_{n=0}^{\infty} \frac{F^{(n)}(x)}{n!}(z-x)^n \tag{2.418}$$

Referring to property 2, since $F(z=x)$ is real, every derivative $F^{(n)}(x)$ is real. Thus,

$$F^*(z) = \sum_{n=0}^{\infty} \frac{F^{(n)}(x)}{n!}(z^*-x)^n = F(z^*) \tag{2.419}$$

This is the Schwarz reflection principle.

For a function that satisfies the Schwarz reflection principle

$$F(x'-i\varepsilon) = F[(x'+i\varepsilon)^*] = F^*(x'+i\varepsilon) \tag{2.420}$$

Thus, the discontinuity across the cut is

$$\Delta(x') = F(x'+i\varepsilon) - F(x'-i\varepsilon) = F(x'+i\varepsilon) - F^*(x'+i\varepsilon) = 2i\,\text{Im}\,F(x'+i\varepsilon) \tag{2.421}$$

and equation 2.417 becomes

$$F(z) = \frac{1}{\pi}\int_M^\infty \frac{\operatorname{Im} F(x' + i\varepsilon)}{(x' - z)}\,dx' \tag{2.422}$$

$\operatorname{Im} F(x' + i\varepsilon)$ indicates that we are evaluating $\operatorname{Im} F(x')$ above the cut. We will drop the $i\varepsilon$ in what follows, and understand that the point x' on the cut is approached from above.

The quantum scattering amplitude $F(E)$ satisfies a condition referred to as the *optical theorem*:

$$\sigma(E) = \frac{4\pi}{E}\operatorname{Im} F(E) \tag{2.423}$$

where $\sigma(E)$, called the *scattering cross section*, is one of the parameters that is measurable in a scattering experiment. Therefore, if z in equation 2.422 is a complex energy variable, $\operatorname{Im} F(x')$ is, in principle, known from the cross section $\sigma(x')$. Thus the complete amplitude $F(x)$ can be determined from an equation such as equation 2.422.

Using the identity of equation 2.299a with $z = x + i\varepsilon$,

$$\frac{1}{(x' - x - i\varepsilon)} = \frac{1}{(x' - x)_P} + i\pi\delta(x' - x) \tag{2.299a}$$

$$\Rightarrow \quad F(x + i\varepsilon) = \frac{1}{\pi}P\int_M^\infty \frac{\operatorname{Im} F(x')}{(x' - x)}\,dx' + i\operatorname{Im} F(x) \tag{2.424}$$

Writing

$$F(x) = \operatorname{Re} F(x) + i\operatorname{Im} F(x) \tag{2.425}$$

and equating the imaginary parts of equation 2.424 yields the identity $\operatorname{Im} F(x) = \operatorname{Im} F(x)$. Equating the real parts of equation 2.424, we obtain

$$\operatorname{Re} F(x) = \frac{1}{\pi}P\int_M^\infty \frac{\operatorname{Im} F(x')}{(x' - x)}\,dx' \tag{2.426}$$

A once-subtracted form of equation 2.422 can also be obtained straightforwardly. Let $G(x)$ have the value $G(x_0)$ at $x_0 < M$. Then substituting

$$F(z) = \frac{[G(z) - G(x_0)]}{(z - x_0)} \tag{2.412}$$

into equation 2.422 yields

$$\frac{[G(z) - G(x_0)]}{(z - x_0)} = \frac{1}{\pi}\int_M^\infty \frac{\operatorname{Im}[G(x') - G(x_0)]}{(x' - x_0)(x' - z)} \tag{2.427}$$

Since F, and therefore G, is to be real on the real axis for $x_0 < M$,

$$\operatorname{Im} G(x_0) = 0 \tag{2.428}$$

$$\Rightarrow \quad G(z) = G(x_0) + \frac{(z - x_0)}{\pi}\int_M^\infty \frac{\operatorname{Im} G(x')}{(x' - z)(x' - x_0)}\,dx' \tag{2.429}$$

Letting $z = x + i\varepsilon$ and using equation 2.299a, equation 2.429 becomes

$$G(x + i\varepsilon) = G(x_0) + \frac{(x - x_0)}{\pi}P\int_M^\infty \frac{\operatorname{Im} G(x')}{(x' - x_0)(x' - x)}\,dx' + i\operatorname{Im} G(x) \tag{2.430}$$

As before, since $G(x_0)$ must be real in order that G satisfy the Schwarz reflection condition, equating the imaginary part leads to the identity $\text{Im}\, G(x) = \text{Im}\, G(x)$. Equality of the real part of equation 2.430 yields

$$\text{Re}\, G(x) = G(x_0) + \frac{(x - x_0)}{\pi} P \int_M^\infty \frac{\text{Im}\, G(x')}{(x' - x_0)(x' - x)}\, dx' \qquad (2.431)$$

Like the dispersion relations for functions without branch points, we see that the denominator of the integrand of equation 2.429 has two factors involving x', whereas the integrand of equation 2.422 has only one factor containing x'. Again, this additional factor in the denominator could make the integrand of the subtracted dispersion relation approach zero at infinity "fast enough" in cases where the integrand of the unsubtracted dispersion integral does not.

Example 2.40

As an example, we determine the function with the following properties:

1. $F(z) \to 0$ for $|z| \to 0$.
2. $F(z)$ satisfies the Schwarz reflection principle.
3. $F(z)$ is analytic on the real axis for $x > 1$ and for $-1 < x$.
4. $F(3) = 2$.
5. $\text{Im}\, F(x) = x$ for $-1 \leqslant x \leqslant 1$.

The cut structure implied by properties 3 and 5 is shown in Figure 2.62. By properties 2 and 3 $\text{Im}\, F(x) = 0$ for $x > 1$ and for $x < -1$. Consider

$$\frac{[F(z) - F(3)]}{z - 3} = \frac{1}{2\pi i} \oint \frac{[F(z') - F(3)]}{(z' - 3)(z' - z)}\, dz' \qquad (2.432)$$

around the contour shown in Figure 2.61. Omitting the contributions from the infinite circle and the contour around the branch points gives

$$\frac{[F(z) - F(3)]}{(z - 3)} = \frac{1}{2\pi i} \left[\int_{-1}^{1} \frac{[F(x' + i\varepsilon) - F(3)]}{(x' - 3)(x' - z)}\, dx' + \int_{1}^{-1} \frac{[F(x' - i\varepsilon) - F(3)]}{(x' - 3)(x' - z)}\, dx' \right] \qquad (2.433)$$

Since $F(x)$ is analytic on the real axis for $x < -1$, the integral from -1 to $-\infty$ cancels the

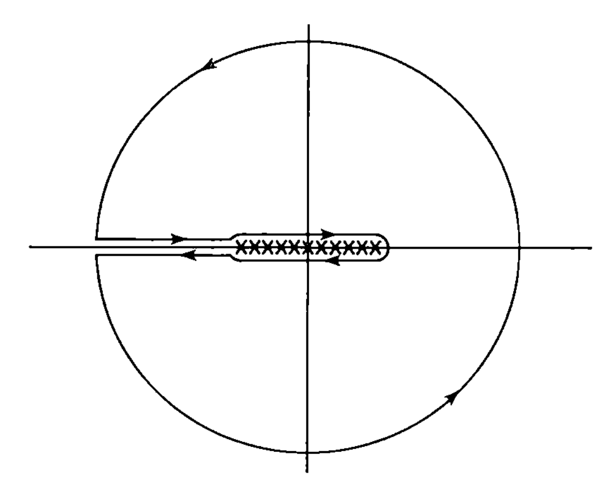

Figure 2.62 Cut structure and contour for Example 2.40.

integral from $-\infty$ to -1. Therefore,

$$F(z) = F(3) + \frac{(z-3)}{2\pi i}\int_{-1}^{1}\frac{[F(x'+i\varepsilon) - F(x'-i\varepsilon)]}{(x'-3)(x'-z)}\,dx'$$

$$= 2 + \frac{(z-3)}{2\pi i}\int_{-1}^{1}\frac{2i\,\mathrm{Im}\,F(x')}{(x'-3)(x'-z)}\,dz' = 2 + \frac{(z-3)}{\pi}\int_{-1}^{1}\frac{x'}{(x'-3)(x'-z)}\,dx'$$

$$= 2 + \frac{1}{\pi}\left(z\log\left(\frac{(z-1)}{(z+1)}\right) + 3\log 2\right) \quad \square \tag{2.434}$$

Integral representation of the Dirac δ-function

The dispersion relations of equations 2.395 or equations 2.400 can be used to derive integral representations of the Dirac δ-function. Equation 2.395b is an expression for $\mathrm{Im}\,F(x)$. Substituting it into the integral of equation 2.395a, we obtain

$$\mathrm{Re}\,F(x) = -\frac{1}{\pi^2}\int_{-\infty}^{\infty}\frac{dx'}{(x'-x)_P}\int_{-\infty}^{\infty}\frac{\mathrm{Re}\,F(x'')}{(x''-x')_P}\,dx'' \tag{2.435}$$

Since these integrals must be finite if any of this discussion is to yield meaningful results, the order of integration of the double integral can be interchanged, and equation 2.435 can be written

$$\mathrm{Re}\,F(x) = -\frac{1}{\pi^2}\int_{-\infty}^{\infty}\mathrm{Re}\,F(x'')\,dx''\int_{-\infty}^{\infty}\frac{1}{(x'-x)_P(x''-x')_P}\,dx' \tag{2.436}$$

Since this has been derived without specifying $F(x)$, $\mathrm{Re}\,F(x)$ is an arbitrary function. Comparing equation 2.436 to the definition of the Dirac δ-function given in equation 2.295 allows us to identify

$$\delta(x-x'') = -\frac{1}{\pi^2}\int_{-\infty}^{\infty}\frac{1}{(x'-x)_P(x''-x')_P}\,dx' \tag{2.437}$$

We also note that the integral of equation 2.437 has the properties stated for the Dirac δ-function. Note that this integral can be evaluated by writing

$$\int_{-\infty}^{\infty}\frac{1}{(x'-x)_P(x''-x')_P}\,dx' = \frac{1}{(x-x'')}\int_{-\infty}^{\infty}\left[\frac{1}{(x'-x)_P} - \frac{1}{(x'-x'')_P}\right]dx' \tag{2.438}$$

$$\Rightarrow \quad \delta(x-x'') = \frac{1}{(x-x'')}\log\left[\frac{(x'-x)}{(x'-x'')}\right]_{x'=-\infty}^{x'=\infty} = \begin{cases} 0 & x \neq x'' \\ \infty & x = x'' \end{cases} \tag{2.439}$$

This is the description of the Dirac δ-symbol when expressed as a function.

PROBLEMS

1. Express the following complex numbers in exponential form. From that, determine the real and imaginary part of each number.
   (a) $(1 - 2i)^{-i}$
   (b) $(i^i)^i$
   (c) $(i)^{(i^i)}$
   (d) $\log(3 + 4i)$
   (e) $\log(-10)$
   (f) $\sin(1 + i)$
2. Use deMoivre's theorem to derive identities for $\sin(3\theta)$ and $\cos(3\theta)$ in terms of $\sin\theta$ and $\cos\theta$.
3. Use identities between different forms of a complex number to prove that

$$\sin(u + v) = \sin u \cos v + \cos u \sin v$$

$$\cos(u + v) = \cos u \cos v - \sin u \sin v$$

4. (a) Show that both circuits in Figures 2.63 are purely resistive if $\omega^2 = 1/LC$.
   (b) The circuit of Figure 2.63a can be replaced by a 20-Ω resistor in series with a 50-mH inductor. What is the frequency of the generator, and what is the capacitance of the capacitor in the original circuit?

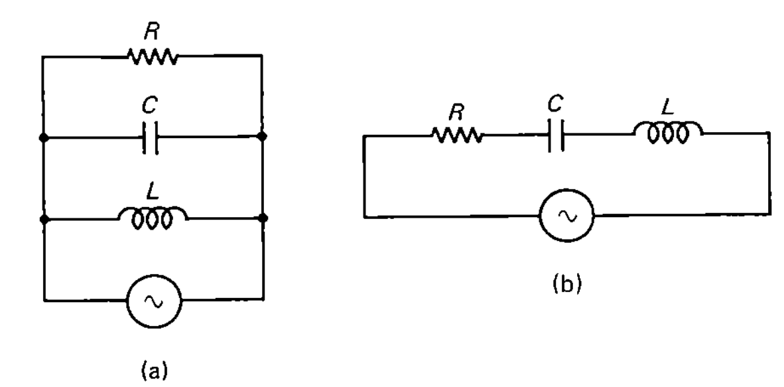

Figure 2.63 AC circuit with all components (a) in parallel and (b) in series.

5. (a) Prove that $\sin(z^2)$ is an entire function and determine $d/dz \sin(z^2)$
   (b) Identify the real and imaginary parts of $\sin(z^2)$ and show that they are harmonic functions.
6. Prove that $|z|^2$ is analytic at only one point and determine that point.
7. Evaluate

$$\int_0^{1+i\pi} e^z \, dz$$

along the three paths shown in Figure 2.64.

8. Evaluate

$$\oint \frac{z^2 \cos(z^2)}{(z - 1 + i)} dz$$

   (a) Around an origin-centered circle of radius $\frac{1}{2}$
   (b) Around an origin-centered circle of radius $\frac{3}{2}$
   (c) Around a square centered at the origin of side 5
9. For the three contours of Problem 8, evaluate

$$\oint \frac{ze^z}{(z - 1)(z + 2i)} dz$$

10. For the three contours of Problem 8, evaluate

$$\oint \frac{ze^z}{(z - 1 - i)^3} dz$$

11. Evaluate

$$\oint \frac{ze^z}{(z - 1)^3(z + 2i)^2} dz$$

around an origin-centered circle of radius:
   (a) $\frac{1}{2}$
   (b) $\frac{3}{2}$
   (c) $\frac{5}{2}$

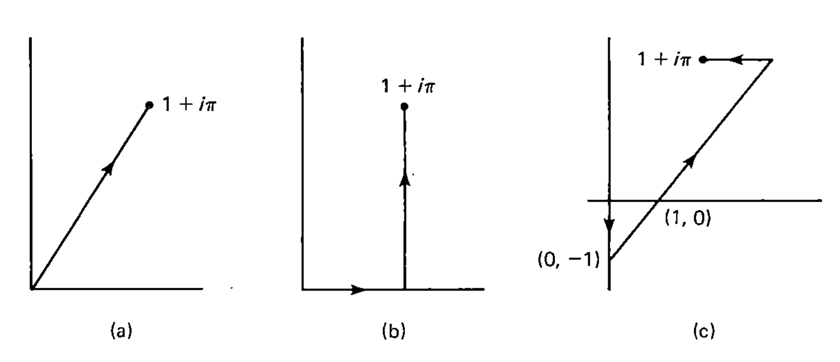

Figure 2.64 Three paths for evaluating $\int_0^{1+i\pi} e^z \, dz$.

12. Find the first three nonzero terms in the Taylor series for $F(z) = e^{z \sin z}$ around:
(a) $z = \pi$
(b) $z = 0$
[*Note:* It is important first to demonstrate that $F(z)$ is analytic in the region around the point of expansion, so that a Taylor expansion exists around that point.]

13. A function $F(z)$ can be written as the ratio of two analytic functions:

$$F(z) = \frac{P(z)}{Q(z)}$$

(a) $F(z)$ has a second-order pole at $z = z_0$ because $P(z_0) \neq 0$ and $Q(z_0) = Q'(z_0) = 0$. Thus the Taylor series for $Q(z)$ is

$$Q(z) = \frac{1}{2!}Q''(z_0)(z - z_0)^2 + \cdots$$

Prove that the residue of the pole of $F(z)$ at z_0 is
$a_{-1}(z_0) =$

$$\frac{2}{3[Q''(z_0)]^2}[3P'(z_0)Q''(z_0) - Q'''(z_0)P(z_0)]$$

(b) If $P(z_0) = 0$ and $P'(z_0) \neq 0$, the pole of $P(z)/Q(z)$ is actually a first-order pole. Show that the residue of such a function is correctly given by the expression obtained in part (a).

14. Find the residue of the pole at $z = \pi$ of the function

$$G(z) = \frac{e^{z \sin z}}{z(z - \pi)^2}$$

(a) By a direct Laurent expansion
(b) By the derivative method
(c) By recognizing that $G(z)$ is the ratio of two functions that are analytic at $z = \pi$

Evaluate the integrals of Problems 15 through 25 by Cauchy's theorem only.

15. $$\int_{-\infty}^{\infty} \frac{x^2}{(1 + x^4)}\, dx$$

16. $$\int_0^{\infty} \frac{x^2}{(1 + x^4)}\, dx$$

17. $$\int_{-\infty}^{\infty} \frac{e^{-2ix}}{(1 + x^4)}\, dx$$

From this, determine

$$\int_{-\infty}^{\infty} \frac{\cos(2x)}{(1 + x^4)}\, dx$$

and

$$\int_{-\infty}^{\infty} \frac{\sin(2x)}{(1 + x^4)}\, dx$$

18. $$\int_0^{\infty} \frac{\cos(3x)}{(1 + x^2)(4 + x^2)}\, dx$$

19. $$\int_{-\infty}^{\infty} \frac{\sin(3x)}{(x - i)^2(1 + x^2)}\, dx$$

20. $$\int_0^{2\pi} \sin^4 \theta \, d\theta$$

21. $$\int_0^{\pi} \frac{1}{(2 + \cos \theta)}\, d\theta$$

22. $$\int_2^{\infty} \frac{1}{x(x + 4)^3}\, dx$$

23. Show that

$$\int_0^{\infty} \frac{x^{1/N}}{(1 + x^2)}\, dx = \pi \frac{\sin(\pi/2N)}{\sin(\pi/N)} = \frac{\pi}{2} \sec\left(\frac{\pi}{2N}\right)$$

for $N > 1$

24. $$\int_{-\infty}^{-2} \frac{1}{(x^3 - 8)}\, dx$$

25. Evaluate

$$\int_{-\infty}^{\infty} \frac{e^{cx}}{(e^x + 1)}\, dx \qquad \text{for } 0 < c < 1$$

Hint: Consider the contour of Figure 2.65 for

$$\lim_{L \to \infty} \oint \frac{e^{cz}}{(e^z + 1)}\, dz$$

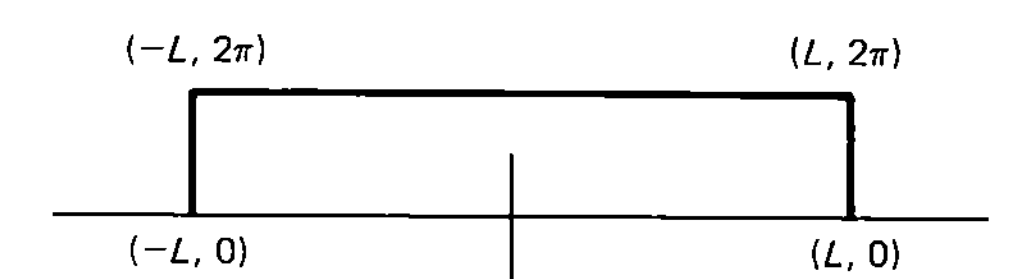

Figure 2.65 Contour for Problem 25.

26. Determine the analytic structure of

$$F(z) = \frac{z^{1/4}}{(1 + z^2)}$$

Identify all poles and branch points. For any branch point, let the associated cut run from the branch point to ∞ at an arbitrary angle θ (see Figure 2.25, for example) and deduce the values of the function on the two sides of the cut. Then specify these functional values when the cut is taken along the
(a) Positive real axis
(b) Negative real axis
(c) Positive imaginary axis

27. Using the analysis of Problem 26, apply Cauchy's theorem to evaluate

$$\int_{-\infty}^{0} \frac{x^{1/4}}{(1 + x^2)}\, dx$$

28. Show that the function $\log[(z + 1)/(z - 1)]$ can be assigned the cut structure shown in Figure 2.66.

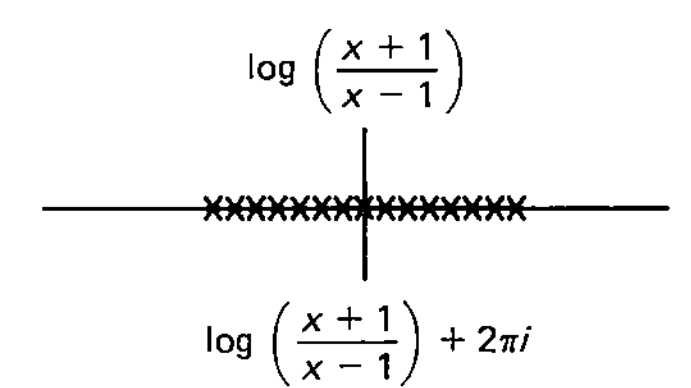

Figure 2.66 Possible cut structure for $\log[(z+1)/(z-1)]$.

29. Show that for any complex function $F(x)$

$$P\int_{-\infty}^{\infty} \frac{F(x)}{(x-x_0)}\, dx =$$

$$\frac{1}{2}\left[\int_{-\infty}^{\infty} \frac{F(x)}{(x-x_0-i\varepsilon)}\, dx + \int_{-\infty}^{\infty} \frac{F(x)}{(x-x_0+i\varepsilon)}\, dx\right]$$

30. Evaluate

$$P\int_{-\infty}^{\infty} \frac{e^{-3ix}}{(x-2)(1+x^2)}\, dx$$

(a) By integrating around the pole as in Figure 2.44a
(b) By integrating around the pole as in Figure 2.45a
(c) By displacing the pole slightly off the real axis

31. Use the results of Problem 30 to evaluate

$$P\int_{-\infty}^{\infty} \frac{\sin(3x)}{(x-2)(1+x^2)}\, dx$$

32. Evaluate

$$P\int_{0}^{\infty} \frac{\sqrt{x}}{(x-3)(x^2+4)}\, dx$$

33. A semi-infinite slot of width $\pi/2$ is cut out of an infinite conducting block as shown in Figure 2.67.
(a) What is the geometry of this configuration in the z'-plane under the mapping $z' = \sin^2(z)$?

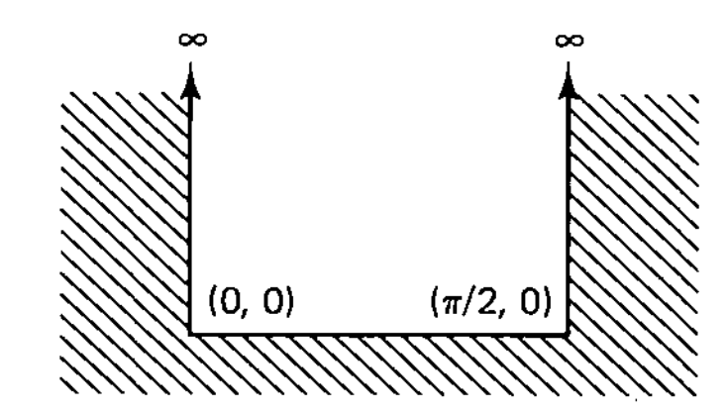

Figure 2.67 Slot cut in infinite conducting medium.

(b) If charge distributed throughout the conductor keeps the conductor at a potential V_0, find the potential in the vacuum region.
(c) Deduce the equation for the equipotential surfaces for this geometry.
(d) Find the electric field in the vacuum region. (For static charge distributions, $\mathbf{E} = -\nabla V$.)

34. Find the geometry in the z'-plane of a pair of semi-infinite parallel plates in the z-plane separated by $2d$ (see Figure 2.68) under the mapping

$$z' = \log\left(\cosh \frac{\pi z}{2d}\right)$$

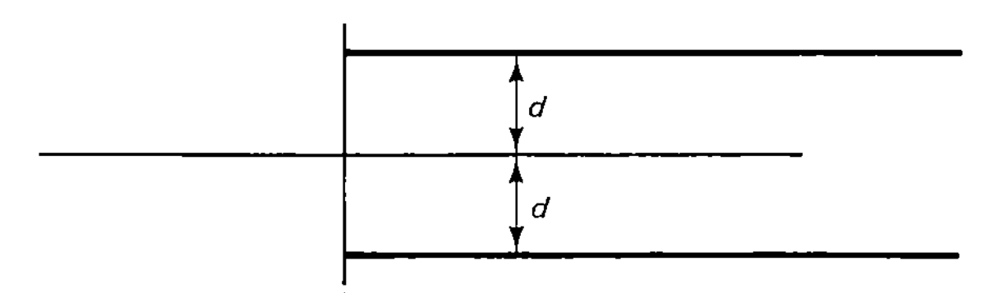

Figure 2.68 Semi-infinite parallel plates.

35. The functions $F(z)$ and $G(z)$ are defined by

$$F(z) \equiv \int_{-5}^{5} \frac{\log(z' - \frac{1}{3}z)}{(z'+z+1)}\, dz'$$

and

$$G(z) \equiv \int_{-5}^{5} \frac{\log(z' - \frac{1}{3}z)}{(z'-z-1)}\, dz'$$

Find the position of all singularities of $F(z)$ and $G(z)$ and identify each singularity as either an end-point coincidence or a pinch singularity. (*Hint:* Begin by specifying the sign of $\operatorname{Im} z = y$. For example, take $y > 0$.)

36. The complex function $F(x)$ has the following properties:

1. $F(x) \to 0$ as $x \to \infty$
2. $F(\frac{1}{2}) = 2$
3. $\operatorname{Im} F(x) = 1/(1+x^2)$, for $x \leqslant 0$
4. $F^*(x) = F(-x)$

Find $F(x)$.

37. Find the function that has the following properties:

1. $F(z)$ satisfies the Schwarz reflection principle.
2. $F(z) \to 0$ when $|z| \to 0$.
3. $F(z)$ is analytic on the real axis for $x < 1$.
4. $\operatorname{Im} F(x) = 1/x$ for $x \geqslant 1$.
5. $F(0) = 1$.

CHAPTER 3

INFINITE SERIES

In Chapter 2 we developed the methods for determining the Taylor and Lauent series representations of known functions. That is, from a known function and its derivatives, we determined an infinite series that was a valid representation of the function. In this chapter we discuss the properties of series, and develop methods for determining the function that a series represents: that is, the function to which the series converges.

Some of the more commonly used MacLaurin series representations of elementary functions that we discussed in Chapter 2 are listed below.

$$e^z = 1 + z + \frac{z^2}{2!} + \frac{z^3}{3!} + \cdots = \sum_{n=0}^{\infty} \frac{z^n}{n!} \tag{3.1}$$

$$\sin z = z - \frac{z^3}{3!} + \frac{z^5}{5!} - \cdots = \sum_{n=0}^{\infty} (-1)^n \frac{z^{2n+1}}{(2n+1)!} \tag{3.2}$$

$$\cos z = 1 - \frac{z^2}{2!} + \frac{z^4}{4!} - \cdots = \sum_{n=0}^{\infty} (-1)^n \frac{z^{2n}}{(2n)!} \tag{3.3}$$

$$\log(1+z) = z - \frac{z^2}{2} + \frac{z^3}{3} - \cdots = \sum_{n=1}^{\infty} (-1)^{n+1} \frac{z^n}{n} \tag{3.4}$$

Not all series of interest are infinite sums. For example, the binomial series is

$$(1+z)^N = \sum_{n=0}^{N} \frac{N!}{n!(N-n)!} z^n \tag{3.5}$$

The geometric series is the finite sum

$$g_N(z) = 1 + z + z^2 + \cdots + z^N = \sum_{n=0}^{N} z^n \tag{3.6a}$$

To obtain a functional expression for the geometric series, consider

$$zg_N(z) = z + z^2 + z^3 + \cdots + z^N + z^{N+1}$$

$$= -1 + 1 + z + z^2 + \cdots + z^N + z^{N+1} = -1 + g_N(z) + z^{N+1} \tag{3.6b}$$

Therefore,

$$g_N(z) = \frac{(1 - z^{N+1})}{(1 - z)} \tag{3.7}$$

In Chapter 2 we determined that an infinite series representation may only be valid over a certain range of the argument of the function, which depends on the singularity structure of a function. That range of convergence may or may not be infinite. As we discussed in Chapter 2, e^z, $\sin z$, and $\cos z$ are entire functions, which means that the series of equations 3.1, 3.2, and 3.3 converge for all finite values of z. That is, the radius of convergence is infinite. $\log(1 + z)$ has a branch singularity at $z = -1$. Therefore, the series of equation 3.4 converges to $\log(1 + z)$ for $|z| < 1$. One important aspect of the study of infinite series is an investigation of the convergence properties of a series.

3.1 Convergence Tests

The most general form of a series is a sum of terms, each of which is a function of the independent variable (or variables). Consider the series involving a single independent variable, z.

$$S(z) = \sigma_0(z) + \sigma_1(z) + \sigma_2(z) + \cdots + \sigma_N(z) + \cdots \tag{3.8}$$

If the sum terminates after a finite number of terms, it can always be summed with sufficient patience. If each term in the finite series is finite for a specified value of z, the function to which the finite series sums can be determined (as in equation 3.7, for example).

If the series contains an infinite number of terms, the series may not be finite at a specified value of z, even if each term is finite at that value of z. Such a series is said to diverge at z.

We define the Nth partial sum of the infinite series

$$\text{(a)} \quad S(z) = \sum_{n=0}^{\infty} \sigma_n(z) \qquad \text{to be} \qquad \text{(b)} \quad S_N(z) \equiv \sum_{n=0}^{N-1} \sigma_n(z) \tag{3.9}$$

If

$$\lim_{N\to\infty} S_N(z) = S(z) \neq \infty \tag{3.10}$$

the series is said to *converge* to $S(z)$ at the specified value of z. One condition necessary for convergence comes from this requirement. The condition for convergence implied by equation 3.10 can be written as

$$\lim_{N\to\infty} [S(z) - S_N(z)] = \lim_{N\to\infty} \left[\sum_{n=0}^{\infty} \sigma_n(z) - \sum_{n=0}^{N-1} \sigma_n(z) \right] = 0 \tag{3.11}$$

Using a nonrigorous argument, we see that as N approaches ∞, the second sum subtracts all the terms except the "last term" or $\sigma_\infty(z)$. That is, equation 3.11b requires that if a series is to converge, for large enough index n, the terms must be decreasing as n increases, and

$$\lim_{n\to\infty} \sigma_n(z) = 0 \tag{3.12}$$

This requirement is a necessary condition for convergence, but it is not sufficient. If the terms of a series satisfy equation 3.12, the series may still diverge at a specified value of z. Thus additional tests must be developed to determine whether a series converges.

Cauchy root test

We want to determine if the series

$$S(z) = \sum_{n=0}^{\infty} \sigma_n(z) \tag{3.9a}$$

is convergent. The nth root of $\sigma_n(z)$ is defined as

$$r_n = [\sigma_n(z)]^{1/n} \tag{3.13}$$

The convergence properties of $S(z)$ are determined as follows. If

$$r \equiv \lim_{N\to\infty} r_N \begin{cases} < 1 & \text{series converges} \\ = 1 & \text{convergence indeterminate} \\ > 1 & \text{series diverges} \end{cases} \tag{3.14}$$

To prove this, we divide the series into two parts, a finite sum, and an infinite sum. For some large but finite N, we write

$$S(z) = \sum_{n=0}^{N-1} \sigma_n(z) + \sum_{n=N}^{\infty} \sigma_n(z) \equiv S_1(z) + S_2(z) \tag{3.15}$$

Assuming that all terms $\sigma_n(z)$ are finite, as mentioned above, the convergence of $S(z)$ is determined by the convergence of S_2. From equation 3.13 we can write each term as

$$\sigma_n(z) = (r_n)^n \tag{3.16}$$

By choosing N very large (almost infinite), we can approximate r_n by its limiting value,

$$r_n \simeq r \tag{3.17}$$

Therefore, $S_2(z)$ becomes

$$S_2(z) \simeq r^N + r^{N+1} + \cdots = r^N(1 + r + r^2 + r^3 + \cdots) = r^N \sum_{n=0}^{\infty} r^n \tag{3.18}$$

Referring to equation 3.7, we see that the series $\sum_{n=0}^{\infty} r^n$ is the limiting value of the finite geometric series $g_N(r)$. From equation 3.7 we see that if $r < 1$, the series converges to $1/(1-r)$. If $r > 1$, the series diverges, and if $r = 1$, the convergence of the series is indeterminate. These are the convergence conditions expressed in equation 3.14.

Example 3.1

As an example, consider

$$S = \sum_{n=1}^{\infty} \frac{1}{n^n} \tag{3.19}$$

Then

$$\text{(a)} \quad r_n = \left(\frac{1}{n^n}\right)^{1/n} = \frac{1}{n} \quad \Rightarrow \quad \text{(b)} \quad r = \lim_{n\to\infty} \frac{1}{n} = 0 \tag{3.20}$$

Therefore, $\sum_{n=1}^{\infty} 1/n^n$ converges. □

Example 3.2

As a second example, consider

$$S = \sum_{n=1}^{\infty} \frac{1}{n^2} \tag{3.21}$$

It will be demonstrated later, using other tests, that this series converges. However,

$$\text{(a)} \quad r_n = \left(\frac{1}{n^2}\right)^{1/n} = n^{-2/n} \quad \Rightarrow \quad \text{(b)} \quad \log(r_n) = -\frac{2}{n}\log(n) \tag{3.22}$$

Therefore,

$$\text{(a)} \quad \lim_{n \Rightarrow \infty} r = \log_{n \Rightarrow \infty}(r_n) = 0 \quad \Rightarrow \quad \text{(b)} \quad r = \lim_{n \to \infty} r_n = 1 \tag{3.23}$$

Thus the convergence of this series is indeterminate using the root test. □

Comparison test

As with the root test, we first divide $S(z)$ into a finite sum $S_1(z)$, and an infinite series $S_2(z)$, as in equation 3.15. As before, if the series $S(z)$ is to converge, the series

$$S_2(z) = \sum_{n=N}^{\infty} \sigma_n(z) \tag{3.24}$$

(with N large) must converge. Let the infinite part of the series

$$\text{(a)} \quad T(z) = \sum_{n=0}^{\infty} \tau_n(z) \quad \text{be} \quad \text{(b)} \quad T_2(z) = \sum_{n=N}^{\infty} \tau_n(z) \tag{3.25}$$

and assume that the series for $T(z)$ is known to converge. As with the series for $S_2(z)$, we will take $\tau_n(z) \geqslant 0$ for all $n \geqslant N$. If

$$\tau_n(z) \geqslant \sigma_n(z) \tag{3.26a}$$

for all $n \geqslant N$, the series representation for S_2 is convergent.

The comparison test can also be used to determine if the series is divergent. If the series

$$T_2(z) = \sum_{n=N}^{\infty} \tau_n(z) \tag{3.25b}$$

is known to be a divergent series, and if

$$\sigma_n(z) \geqslant \tau_n(z) \tag{3.26b}$$

for all $n \geqslant N$, then the series $\sum_{n=0}^{\infty} \sigma_n(z)$ must also be a divergent series.

The proof of the comparison test for convergence or divergence is almost axiomatic. If the series representation of $T_2(z)$ converges, and if $\tau_n(z) \geqslant \sigma_n(z)$ for all $n \geqslant N$, then

$$\sum_{n=N}^{\infty} \tau_n(z) = \text{finite value} \geqslant \sum_{n=N}^{\infty} \sigma_n(z) \tag{3.27a}$$

That is, the series representation of $S_2(z)$ sums to a finite value; and so it converges.

By an identical argument, if the infinite sum denoted by $T_2(z)$ diverges, and if $\tau_n(z) \leqslant \sigma_n(z)$ for all $n \geqslant N$, then

$$\sum_{n=N}^{\infty} \tau_n(z) = \infty \leqslant \sum_{n=N}^{\infty} \sigma_n(z) \tag{3.27b}$$

so the series representation of $S_2(z)$ is infinite (i.e., divergent).

Example 3.3

As an example, let us test the convergence of the series

$$S(z) = \sum_{n=0}^{\infty} \frac{z^n}{(n!)^2} \tag{3.28}$$

The clue to deciding what known convergent series to compare this to is that the denominator contains $n!$. Of the six series listed at the beginning of the chapter, only the series representation

$$T(z) = e^z = \sum_{n=0}^{\infty} \frac{z^n}{n!} \tag{3.29}$$

has terms that contain $1/n!$. Comparing

$$\text{(a)} \quad \tau_n(z) = \frac{z^n}{n!} \qquad \text{and} \qquad \text{(b)} \quad \sigma_n(z) = \frac{z^n}{(n!)^2} \tag{3.30}$$

we see that $\tau_n(z) \geqslant \sigma_n(z)$ for all values of n. (Note that in this example, we do not have to separate out a finite number of lower-index terms that might violate the inequality of equation 3.26a.) Since the series for $T(z)$ converges to e^z, for any value of z, we are assured that the series representation of $S(z)$ converges to a finite value for any z. □

Example 3.4

To illustrate divergence using the comparison test, consider

$$S = \sum_{n=1}^{\infty} \frac{1}{\sqrt{n}} \tag{3.31}$$

Referring to equation 3.4, we note that

$$\lim_{z \to -1} \log(1+z) = \infty = -\sum_{n=1}^{\infty} \frac{1}{n} \tag{3.32}$$

That is, the series

$$T = \sum_{n=1}^{\infty} \frac{1}{n} \tag{3.33}$$

is divergent. Since

$$\frac{1}{\sqrt{n}} \geqslant \frac{1}{n} \tag{3.34}$$

for all n, the original series satisfies

$$\sum_{n=1}^{\infty} \frac{1}{\sqrt{n}} \geqslant \sum_{n=1}^{\infty} \frac{1}{n} = \infty \tag{3.35}$$

Thus the series of equation 3.31 is divergent. □

Cauchy ratio test

A major disadvantage to the comparison test is that it requires knowledge of a series of known convergence or divergence that can be compared term by term to the series being tested. Often, such a known series is not available. The Cauchy ratio test is one of the most commonly used methods for testing the convergence or divergence of a series. It does not require knowledge of the convergence properties of another series.

To test the convergence of the series

$$S(z) = \sum_{n=0}^{\infty} \sigma_n(z) \tag{3.9a}$$

we define the ratio

$$\rho_N(z) \equiv \left| \frac{\sigma_{N+1}(z)}{\sigma_N(z)} \right| \tag{3.36}$$

for large N. $\rho_N(z)$ is called the *Cauchy ratio*. We note that $\rho_N(z)$ is defined as an absolute value. Therefore, the terms in the series being tested do not have to be positive. The *Cauchy ratio test* specifies that if

$$\lim_{N\to\infty} \rho_N(z) \text{ is} \begin{cases} > 1 & \text{series diverges} \\ = 1 & \text{convergence indeterminant} \\ < 1 & \text{series converges} \end{cases} \tag{3.37}$$

at the specified value of z.

We note the similarity in the convergence conditions of the ratio test to those of the root test. The proof of equation 3.37 is similar to the proof of the convergence criteria of the root test.

Referring to equation 3.7, the geometric series is

$$g_N(\rho) = \sum_{n=0}^{N} \rho^n = \frac{1-\rho^{N+1}}{1-\rho} \tag{3.7}$$

Then as discussed for the root test,

$$\lim_{N\to\infty} g_N(\rho) = \begin{cases} \infty & \text{if } \rho > 1 \\ \text{indeterminate value} & \text{if } \rho = 1 \\ \dfrac{1}{1-\rho} & \text{if } \rho < 1 \end{cases} \tag{3.38}$$

If some of the terms in $S(z)$ are negative, $S(z)$ is smaller than the series obtained by summing the absolute values of terms. Such a series is called an *absolute series*.

$$S_{\text{abs}}(z) \equiv \sum_{n=0}^{\infty} |\sigma_n(z)| \tag{3.39}$$

Therefore, if the absolute series converges, the comparison test assures us that the original series is convergent. The Cauchy ratio test provides a determination of the convergence of the absolute series. Thus we will assume that for all n, $\sigma_n(z) \geqslant 0$.

We again divide the series into a finite sum and an infinite series, with N large. If $S_2(z)$ is to converge, its terms must satisfy

$$\sigma_{n+1}(z) < \sigma_n(z) \tag{3.40}$$

If N is large enough, ρ_N will be very close to its limiting value ρ. Thus from the definition of $\rho_N \simeq \rho$, we can write the terms in the series for $S_2(z)$ as

$$\text{(a)} \quad \sigma_{N+1}(z) = \rho\sigma_N(z) \qquad \text{(b)} \quad \sigma_{N+2}(z) = \rho\sigma_{N+1}(z) = \rho^2\sigma_N(z)$$

or

$$\text{(c)} \quad \sigma_{N+k}(z) = \rho^k\sigma_N(z) \qquad k \geqslant 1, \tag{3.41}$$

Therefore, S_2 can be written

$$S_2 = \sum_{n=N}^{\infty} \sigma_n(z) = \sigma_N(z) \sum_{k=0}^{\infty} \rho^k = \sigma_N(z) \frac{1 - \rho^{N+1}}{1 - \rho} \tag{3.42}$$

Comparing this to equation 3.38, we see that $\sum_{k=0}^{\infty} \rho^k$ is divergent for $\rho > 1$, of indeterminant convergence if $\rho = 1$, and convergent for $\rho < 1$. Thus $S_2(z)$ satisfies the conditions stated in equation 3.37.

Example 3.5

To illustrate, consider

$$S(z) = \sum_{n=0}^{\infty} \frac{z^n}{(n!)^2} \tag{3.28}$$

Using the comparison test, we demonstrated that this series was convergent for all values of z. The Cauchy ratio for this series is

$$\rho_N = \left| \frac{z^{N+1}/[(N+1)!]^2}{z^N/(N!)^2} \right| = |z| \frac{1}{(N+1)^2} \xrightarrow[N \to \infty]{} 0 \tag{3.43}$$

Since $\lim_{N \to \infty} \rho_N < 1$, the series of equation 3.28 converges. □

Example 3.6

In Example 3.2, we were unable to determine the convergence of

$$S = \sum_{n=1}^{\infty} \frac{1}{n^2} \tag{3.21}$$

using the root test.

For this series, the Cauchy ratio is

$$\rho_N = \frac{1/(N+1)^2}{1/N^2} = \left(\frac{N}{N+1} \right)^2 \xrightarrow[N \to \infty]{} 1 \tag{3.44}$$

Therefore, the convergence of the series of equation 3.21 is also indeterminate using the Cauchy ratio test. □

Note that in our test of the series of equation 3.28, we determined that the series converges for all values of z. In addition to being able to ascertain whether a series converges or not, if the terms in the series depend on a variable, it may be possible to determine the radius of convergence of such a series using the Cauchy ratio test.

Example 3.7

As an example, consider the series

$$S(z) = \sum_{n=1}^{\infty} (-1)^{n+1} \frac{1}{n} \left(\frac{z}{2} \right)^n \tag{3.45}$$

The Cauchy ratio is,

$$\rho = \lim_{N \to \infty} \left| \frac{\left(\frac{1}{2}z\right)^{N+1}}{\left(\frac{1}{2}z\right)^N} \frac{N}{N+1} \right| = \left| \frac{z}{2} \right| \tag{3.46}$$

To ensure convergence, this ratio must be less than 1. That is, the series of equation 3.45 converges for

$$\text{(a)} \quad \left|\frac{z}{2}\right| < 1 \qquad \text{or} \qquad \text{(b)} \quad |z| < 2 \tag{3.47}$$

Comparing equation 3.45 to equation 3.4, we see that for $|z| < 2$, the series of equation 3.45 converges to $\log(1 + \frac{1}{2}z)$. Since the function has a branch point at $z = -2$, the radius of convergence of its MacLaurin series is 2. □

Cauchy integral test

In both the root test, which is rarely used, and the widely used Cauchy ratio test, there remains an indeterminacy when the root, or the ratio is 1 for infinite value of the index. The Cauchy integral test may yield information about the convergence of series that cannot be obtained from any of the other tests we have discussed.

We begin by again dividing $S(z)$ into a finite sum and an infinite series.

$$S(z) = \sum_{n=0}^{N-1} \sigma_n(z) + \sum_{n=N}^{\infty} \sigma_n(z) \equiv S_1(z) + S_2(z) \tag{3.15}$$

for large N, and we investigate the convergence of $S_2(z)$. Since successive values of n differ by 1, we change nothing by multiplying the terms in S_2 by $\Delta n = (n + 1) - n$. Then

$$S_2(z) = \sum_{n=N}^{\infty} \sigma_n(z)\, \Delta n \tag{3.48}$$

Since N has been chosen large, the terms in the series $S_2(z)$ must be decreasing as n increases if there is any chance that S_2 converges. In Figure 3.1 we illustrate typical values of $\sigma_n(z)$ for a fixed value of z, at various values of n. The dashed curve is the envelope of the points defined by $\sigma_n(z)$. We point out that each $\sigma_n(z)$ has been drawn having a positive value. With this, we are indicating that the Cauchy integral test is applicable only to an absolute series. (At least all the terms in $S_2(z)$ must be positive to within an overall negative sign.)

Since the rectangles are of width Δn, the area under the rectangles is

$$A_{\text{rectangles}} = \sum_{n=N+1}^{\infty} \sigma_n(z)\, \Delta n = S_2(z) - \sigma_N(z) \tag{3.49}$$

The envelope of the points is obtained by treating n as a continuous variable. That is, the envelope for a particular value of z, is specified by the function $\sigma(n, z)$. The area

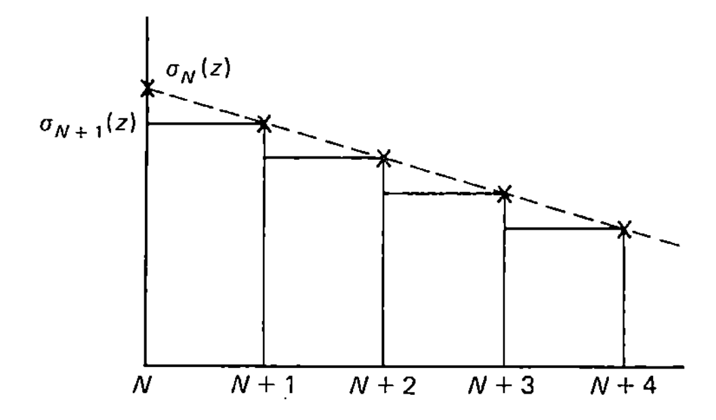

Figure 3.1 Graphical representation of $\sigma_n(z)$ and its envelope.

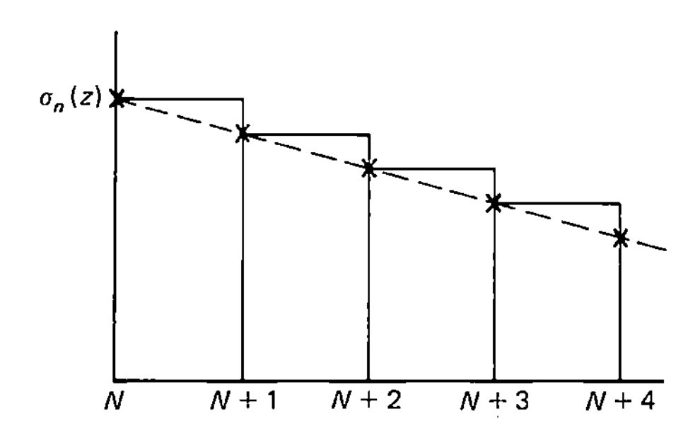

Figure 3.2 Graphical representation of $\sigma_n(z)$ and its envelope.

under this envelope is

$$A_{\text{envelope}} = \int_N^\infty \sigma(n, z)\, dn \tag{3.50}$$

This integral is referred to as the Cauchy envelope integral. It is evident from Figure 3.1 that

$$A_{\text{envelope}} > A_{\text{rectangles}} \tag{3.51}$$

$$\Rightarrow \quad \int_N^\infty \sigma(n, z)\, dn > S_2(z) - \sigma_N(z). \tag{3.52}$$

Since $\sigma_N(z)$ is finite, $S_2(z)$ will converge if $\int_N^\infty \sigma(n, z)\, dn$ is finite.

If the area under the envelope is infinite, equation 3.52 becomes $S_2(z) < \infty$. This does not mean that $S_2(z)$ must be finite. $S_2(z)$ could be a lower-order infinity. However, if the envelope integral is infinite, this ensures that the series $S_2(z)$ diverges.

To see this, consider the representation of $\sigma_n(z)$ as shown in Figure 3.2. We note that the rectangles in Figure 3.1 have been constructed so that each rectangle drawn between n and $n + 1$ is of height $\sigma_{n+1}(z)$. Those of Figure 3.2 are constructed such that the height of each rectangle represents the $\sigma_n(z)$ from n to $n + 1$. Thus each rectangle in Figure 3.2 is taller than the corresponding rectangle of Figure 3.1. The area under the rectangles of Figure 3.2 is

$$A_{\text{rectangles}} = \sum_{n=N}^{\infty} \sigma_n(z) = S_2(z) \tag{3.53}$$

while the area under the envelope is, as before,

$$A_{\text{envelope}} = \int_N^\infty \sigma(n, z)\, dn \tag{3.50}$$

In the representation of Figure 3.2,

$$A_{\text{rectangles}} > A_{\text{envelope}} \tag{3.54}$$

$$\Rightarrow \quad S_2(z) > \int_N^\infty \sigma(n, z)\, dn \tag{3.55}$$

If the envelope integral is infinite, $S_2(z)$ must be infinite; that is, S_2 is divergent. Therefore, convergence or divergence is completely determined by whether the envelope integral is finite or infinite.

Example 3.8

Using the Cauchy integral test, we are able to test the convergence of the constant series

$$S = \sum_{n=1}^{\infty} \frac{1}{n^r} \tag{3.56}$$

for all possible values of r. The Cauchy envelope integral is

$$\int_N^\infty \frac{dn}{n^r} = \left\{ \begin{array}{ll} -\dfrac{1}{(r-1)n^{r-1}} & \text{for} \quad r \neq 1 \\ \log n & \text{for} \quad r = 1 \end{array} \right\}_N^\infty \tag{3.57}$$

For any finite value of N, no matter how large, the integrated expression will be finite. Therefore, the finiteness of the Cauchy envelope integral is determined by the infinite upper limit. For that reason, many texts will not include the lower limit in determining the envelope integral.

Evaluating this integral at ∞, we see from equation 3.57 that

$$\int_N^\infty \frac{dn}{n^r} \text{ is} \begin{cases} \text{infinite} & \text{for } r \leqslant 1 \\ \text{finite} & \text{for } r > 1 \end{cases} \tag{3.58}$$

$$\Rightarrow \quad S = \sum_{n=1}^{\infty} \frac{1}{n^r} \tag{3.56}$$

converges for $r > 1$ and diverges for $r \leqslant 1$. Thus, for example, $\sum_{n=1}^{\infty} 1/n^2$ converges, a fact that could not be ascertained by other convergence tests. □

The major drawback to the integral test is that sometimes the envelope integrals cannot be evaluated at all, or can only be evaluated in such a way that it is impossible to determine the convergence or divergence of the series being tested.

Example 3.9

For example, referring to equation 3.4, $\sum_{n=1}^{\infty} z^n/n$ converges to $-\log(1 + z)$ for $|z| < 1$. To ascertain this by the integral test requires evaluation of

$$\int_N^\infty \frac{z^n}{n}\, dn = \int_N^\infty \frac{e^{n \log z}}{n}\, dn = \sum_{k=0}^{\infty} \frac{(\log z)^k}{k!} \int_N^\infty n^{k-1}\, dn = \left[\log n + \sum_{k=1}^{\infty} \frac{(\log z)^k}{k!} \frac{n^k}{k}\right]_N^\infty \tag{3.59}$$

Clearly, it is difficult at best to demonstrate convergence for $|z| < 1$ from this expression. The Cauchy ratio test or the root test yields convergence for $|z| < 1$ straightforwardly. □

We see from these discussions, then, that no single test is best for all series. The preferred test depends on the individual series being investigated. However, the ratio and integral tests are among the most commonly used approaches for testing the convergence of an absolute series.

Alternating series

An *alternating series* is one in which the sign of each term is the negative of the sign of the preceding term. The series for $\sin z$, $\cos z$, and $\log(1 + z)$ of equations 3.2, 3.3, and 3.4 are examples of alternating series.

Since all the terms in an absolute series are positive (except for a possible overall multiplication of the series by -1),

$$\sum_n^{\infty} \sigma_n(z) < \sum_n^{\infty} |\sigma_n(z)| \tag{3.60}$$

The sum involving the absolute values of the individual terms is called the *absolute series*. From equation 3.60, we see that the convergence requirements for an alternating series are less stringent than for an absolute series. For example, it is possible for the alternating series to converge, while the corresponding absolute series is divergent. If the alternating series converges but the absolute series is divergent, the convergence is said to be *conditional*. If the absolute series converges, the alternating series must converge and the convergence is termed *absolute convergence*.

The condition for the convergence of an alternating series is very simple. The necessary and sufficient condition for the alternating series

$$S(z) = \sum_{n=0}^{\infty} (-1)^n |\sigma_n(z)| \tag{3.61}$$

to be convergent is

$$\lim_{n \to \infty} |\sigma_n(z)| = 0 \tag{3.62}$$

The proof of this again requires that we divide the series into a finite sum $S_1(z)$ and an infinite series $S_2(z)$. N is chosen to be large enough that

$$|\sigma_{n+1}(z)| < |\sigma_n(z)| \tag{3.63}$$

for all $n \geqslant N$. We also require that N be even. Then, for an alternating series, $S_2(z)$ can be written

$$\begin{aligned} S_2(z) &= \sum_{n=N}^{\infty} (-1)^n |\sigma_n(z)| \\ &= |\sigma_N(z)| - (|\sigma_{N+1}(z)| - |\sigma_{N+2}(z)|) - (|\sigma_{N+3}(z)| - |\sigma_{N+4}(z)|) - \cdots \end{aligned} \tag{3.64}$$

N is chosen to be an even number, so that the first term in the S_2 series is positive. Because the terms are decreasing with increasing index, each of the terms in the parentheses is positive. Therefore, this sum is less than the first term. That is,

$$S_2(z) < |\sigma_N(z)| \tag{3.65}$$

Thus if $|\sigma_N(z)|$ is finite for arbitrarily large N, $S_2(z)$ converges. That is, $S_2(z)$ converges if

$$\lim_{N \to \infty} \sigma_N(z) = \text{finite number} \tag{3.66}$$

But convergence, in general, requires that this limit be zero. Therefore, the necessary and sufficient condition that an alternating series converge is

$$\lim_{N \to \infty} \sigma_N(z) = 0 \tag{3.67}$$

Equation 3.65 provides an upper bound for $S_2(z)$ for an alternating series. Using a similar analysis, we can also obtain a lower bound of $S_2(z)$.

Since the terms of S_2 are decreasing with increasing index, we can write $S_2(z)$ as

$$S_2(z) = (|\sigma_N(z)| - |\sigma_{N+1}(z)|) + (|\sigma_{N+2}(z)| - |\sigma_{N+3}(z)|) + \cdots \tag{3.68}$$

Each of the parenthetical differences is positive, so this sum is larger than the terms in the first parentheses. That is,

$$S_2(z) > |\sigma_N(z)| - |\sigma_{N+1}(z)| \tag{3.69}$$

Example 3.10

For example, since

$$\lim_{n \to \infty} \frac{1}{n^r} = 0 \tag{3.70}$$

for $r > 0$, the series

$$S = \sum_{n=1}^{\infty} \frac{(-1)^n}{n^r} \tag{3.71}$$

converges for all $r > 0$. For $r = 1$, using equation 3.4, we see that this series converges to $-\log(2)$. However, using Cauchy integral test, it was shown that the absolute series

$$S_{\text{abs}} = \sum_{n=1}^{\infty} \frac{1}{n^r} \tag{3.72}$$

converges only for $r > 1$. Therefore, the alternating series

$$S = \sum_{n=1}^{\infty} \frac{(-1)^n}{n^r} \tag{3.71}$$

diverges for $r < 0$, is conditionally convergent for $0 \leqslant r \leqslant 1$, and is absolutely convergent for $r > 1$. □

Rearrangement of a series. There is one caveat that must be noted in dealing with an alternating series. If the series is conditionally convergent, rearranging the order of the terms in the series can change the value to which the series converges. If the series is absolutely convergent, such a rearrangement will have no effect on the value of the series.

Example 3.11

To illustrate, consider the series

$$S = \sum_{n=1}^{\infty} \frac{(-1)^{n+1}}{n} \tag{3.73}$$

From equation 3.4, with $z = 1$, we see that this alternating series converges to log(2). Using the Cauchy integral test, it was demonstrated earlier that the absolute series diverges. Therefore, the convergence of the alternating series is conditional. Writing this series out term by term, we have

$$\text{(a)}\quad S = 1 - \frac{1}{2} + \frac{1}{3} - \frac{1}{4} + \cdots \qquad \Rightarrow \qquad \text{(b)}\quad \tfrac{1}{2}S = \frac{1}{2} - \frac{1}{4} + \frac{1}{6} - \cdots \tag{3.74}$$

Adding equations 3.74, we obtain

$$\begin{aligned} \tfrac{3}{2}S &= (1+0) + \left(-\frac{1}{2} + \frac{1}{2}\right) + \left(\frac{1}{3} + 0\right) + \left(-\frac{1}{4} - \frac{1}{4}\right) + \left(\frac{1}{5} + 0\right) \\ &\quad + \left(-\frac{1}{6} + \frac{1}{6}\right) + \left(\frac{1}{7} + 0\right) + \left(-\frac{1}{8} - \frac{1}{8}\right) + \left(\frac{1}{9} + 0\right) + \cdots \\ &= 1 + \frac{1}{3} - \frac{1}{2} + \frac{1}{5} + \frac{1}{7} - \frac{1}{4} + \frac{1}{9} + \cdots \end{aligned} \tag{3.75}$$

We note that the series of equation 3.75 for $\frac{3}{2}S$ has the same terms as the series of equation 3.74a for S, except that the terms appear in different order. That is, the rearranged series of equation 3.75 converges to $\frac{3}{2}$ log(2). □

Complex series

If a series is complex, the analysis is identical to that for a real series, except that one must analyze two series instead of one. That is, if $\sigma_n(z) = u_n(z) + iv_n(z)$, then

$$S(z) = \sum_{n=0}^{\infty} \sigma_n(z) \equiv U(z) + iV(z) \tag{3.76}$$

is convergent only if the series for $U(z)$ is convergent, and separately the series for $V(z)$ is convergent. If either of these real series diverges, the series for $S(z)$ diverges. If one of the series is absolutely convergent and the other is conditionally convergent, the convergence of the series for $S(z)$ is conditional.

Uniform convergence

Let

$$S(z) = \sum_{n=0}^{\infty} \sigma_n(z) \tag{3.9a}$$

be a convergent series, and, as before, we express $S(z)$ as

$$S(z) = \sum_{n=0}^{N-1} \sigma_n(z) + \sum_{n=N}^{\infty} \sigma_n(z) = S_1(z) + S_2(z) \tag{3.15}$$

We can interpret $S_2(z)$ as the error incurred in approximating $S(z)$ by the finite sum $S_1(z)$.

As noted earlier, the convergence properties of $S(z)$ are determined by the convergence of $S_2(z)$. Thus the convergence properties of $S(z)$ would be unaffected by ignoring the finite sum $S_1(z)$.

$S(z)$ is said to be *uniformly convergent* in the interval $z \in [a, b]$ if for any arbitrarily small positive number ε, it is possible to find an integer N_0 independent of z, such that for all $z \in [a, b]$, the error $S_2(z)$ satisfies

$$|S_2(z)| < \varepsilon \tag{3.77}$$

for all $N \geqslant N_0$.

Example 3.12

The series

$$S(z) = z + z(z-1) + z^2(z-1) + \cdots + z^{n-1}(z-1) + \cdots$$

$$= z + z^2 - z + z^3 - z^2 + \cdots + z^n - z^{n-1} + \cdots = z + \sum_{n=0}^{\infty} z^n(z-1) \tag{3.78a}$$

is an example of a series that is convergent in the open interval $z \in (0, 1)$ but is not uniformly convergent in this interval. To see this, we note from equation 3.78a that the partial sum is

$$S_N(z) = z + \sum_{n=0}^{N-1} z^n(z-1) = z^N \tag{3.78b}$$

Therefore, since $z < 1$,

$$\lim_{N\to\infty} S_N(z) = 0 \tag{3.79}$$

That is, the series of equation 3.78a converges to zero. The error in truncating the infinite series by the partial sum

$$S(z) \simeq S_{N_0}(z) \simeq z + \sum_{n=0}^{N_0-1} z^n(z-1) \tag{3.80}$$

is

$$|S(z) - S_{N_0}(z)| = z^{N_0} - 0 = z^{N_0} \tag{3.81}$$

Thus this series is uniformly convergent if we can find an integer N_0, independent of z, such that

$$z^{N_0} < \varepsilon \tag{3.82}$$

Such an integer would satisfy

$$N_0 > \frac{\log \varepsilon}{\log z} \tag{3.83}$$

The maximum value of N_0 is obtained from the smallest value of $\log z$, which approaches zero as $z \to 1$. Thus since the maximum N_0 is infinite and there is no finite value of N_0 that is independent of z, for which question 3.82 is satisfied, the convergence of the series of equation 3.78a is not uniform. □

Example 3.13

The geometric series

$$g(z) = \sum_{n=0}^{\infty} z^n \tag{3.6a}$$

is an example of a series that converges uniformly in the interval $z \in [0, \frac{1}{2}]$. We note that in this range of z,

$$g(z) = \lim_{N\to\infty} g_N(z) = \lim_{N\to\infty} \frac{1 - z^N}{1 - z} = \frac{1}{1 - z} \tag{3.84}$$

Therefore, the error in approximating $g(z)$ by $g_N(z)$ is

$$|S_2(z)| = |g(z) - g_{N_0}(z)| = \frac{z^{N_0}}{1 - z} \tag{3.85}$$

To specify that this error be less than ε, we require that

$$N_0 > \frac{\log[\varepsilon(1 - z)]}{\log z} \tag{3.86}$$

If a finite value of N_0 can be found that is greater than the maximum value of the right-hand side of equation 3.86, N_0 will be independent of z, and the geometric series will be uniformly convergent for all $z \in [0, \frac{1}{2}]$. We note that this maximum value is achieved for $z = \frac{1}{2}$. Therefore

$$N_0 > \left[\frac{\log[\varepsilon(1 - z)]}{\log z}\right]_{\max} = \frac{\log(\frac{1}{2}\varepsilon)}{\log(\frac{1}{2})} = 1 + \frac{\log \varepsilon}{\log(2)} \tag{3.87}$$

Since a finite value of N_0 can be found that is independent of z, the geometric series converges uniformly in the specified range of z.

For example,

$$S_2(z) = \sum_{n=N}^{\infty} z^n \tag{3.88}$$

will be less than 0.001 for all $z \in [0, \frac{1}{2}]$, for all

$$N \geqslant N_0 > 1 - \frac{\log(0.001)}{\log(2)} = 10.97 \tag{3.89}$$

That is,

$$\sum_{n=N}^{\infty} z^n < 0.001 \tag{3.90}$$

for all $z \in [0, \frac{1}{2}]$ for any $N \geqslant 11$. □

Weierstrass *M*-test

A widely used test of uniform convergence of an infinite series is the Weierstrass *M*-test. For the remainder series

$$S_2(z) = \sum_{n=N}^{\infty} \sigma_n(z) \tag{3.91}$$

let $\{M_n\}$ be an infinite set of positive numbers such that

$$M_n \geqslant |\sigma_n(z)| \tag{3.92}$$

for every $z \in [a, b]$. If the series

$$M_2 \equiv \sum_{n=N}^{\infty} M_n \tag{3.93}$$

converges, then

$$S_2(z) = \sum_{n=N}^{\infty} \sigma_n(z) \tag{3.91}$$

and thus $S(z)$ converges uniformly in the interval $[a, b]$.

The proof of this is almost axiomatic. If $M_n \geqslant |\sigma_n(z)| \geqslant \sigma_n(z)$, then

$$\sum_{n=N}^{\infty} M_n \geqslant \sum_{n=N}^{\infty} |\sigma_n(z)| \geqslant \left| \sum_{n=N}^{\infty} \sigma_n(z) \right| \tag{3.94}$$

But since $\sum_{n=N}^{\infty} M_n$ converges,

$$\lim_{n \to \infty} M_n = 0 \tag{3.95}$$

Thus, by a suitably large choice of N, $\sum_{n=N}^{\infty} M_n$ can be made as small as required. That is, we can always find a value of N such that

$$M_2 = \sum_{n=N}^{\infty} M_n < \varepsilon \tag{3.96}$$

where ε is arbitrarily small. Thus

$$|S_2(z)| = \left| \sum_{n=N}^{\infty} \sigma_n(z) \right| = \left| S(z) - \sum_{n=0}^{N-1} \sigma_n(z) \right| \leqslant \sum_{n=N}^{\infty} M_n < \varepsilon \tag{3.97}$$

Example 3.14

As an example, consider the series

$$S(z) = \sum_{n=1}^{\infty} \frac{e^{-nz}}{n^2} \tag{3.98}$$

In the range $z \in [0, \infty]$, $e^{-nz} \leqslant 1$ for all n. Thus, if we take

$$M_n = \frac{1}{n^2} \tag{3.99}$$

we have

$$\sigma_n(z) = \frac{e^{-nz}}{n^2} \leqslant M_n \tag{3.100}$$

As was demonstrated in Example 3.8, $\sum_{n=1}^{\infty} 1/n^2$ converges. Therefore,

$$S(z) = \sum_{n=1}^{\infty} \frac{e^{-nz}}{n^2} \tag{3.98}$$

converges uniformly in the range $z \in [0, \infty]$. □

The importance of the study of uniform convergence is embodied in the following three theorems:

Theorem 3.1. Let $\{\sigma_n(z)\}$ be an infinite set of functions that are continuous over the interval $[a, b]$. If the series $\sum_{n=0}^{\infty} \sigma_n(z)$ is uniformly convergent for all $z \in [a, b]$, then the function $S(z)$ to which the series converges is continuous over $z \in [a, b]$.

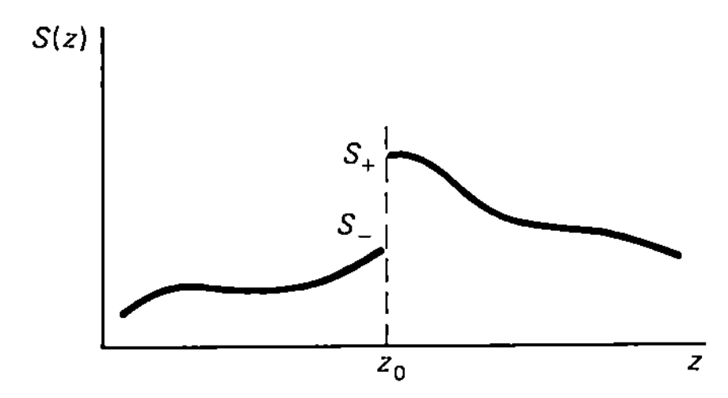

Figure 3.3 Discontinuous function.

A function $S(z)$ is continuous if the limit

$$\lim_{h \to 0} S(z+h) = S(z) \quad (3.101)$$

is independent of how h approaches zero. An illustration of a function that is discontinuous at a point z_0 is shown in Figure 3.3. Referring to the figure if h approaches zero from a positive value, $S(z_0 + h) \to S_+$. If z_0 is approached from negative values of h, $S(z_0 + h) \to S$. Therefore, if $S(z)$ is continuous at z, and ε is an arbitrarily small positive number, one can always find a value of δ small enough that $S(z)$ satisfies

$$|S(z+\delta) - S(z-\delta)| < \varepsilon \quad (3.102)$$

If

$$S(z) = \sum_{n=0}^{N-1} \sigma_n(z) + \sum_{n=N}^{\infty} \sigma_n(z) = S_1(z) + S_2(z) \quad (3.15)$$

is uniformly convergent, then N can be chosen such that

$$|S_2(z)| < \varepsilon \quad (3.76)$$

for an arbitrarily small ε. Therefore,

$$\begin{aligned} |S(z+\delta) - S(z-\delta)| &= |S_1(z+\delta) - S_1(z-\delta) + S_2(z+\delta) - S_2(z-\delta)| \\ &\leqslant |S_1(z+\delta) - S_1(z-\delta)| + |S_2(z+\delta) - S_2(z-\delta)| \end{aligned} \quad (3.103)$$

Since $S_1(z)$ is a finite sum of continuous functions $\sigma_n(z)$, $S_1(z)$ is continuous. Therefore,

$$|S_1(z+\delta) - S_1(z-\delta)| < \varepsilon \quad (3.104)$$

Because $S_2(z)$ satisfies the condition for a uniformly convergent series expressed in equation 3.76,

$$|S_2(z+\delta) - S_2(z-\delta)| < |S_2(z+\delta)| + |S_2(z-\delta)| < 2\varepsilon \quad (3.105)$$

Thus

$$|S(z+\delta) - S(z-\delta)| < 3\varepsilon \equiv \varepsilon' \quad (3.106)$$

where ε' is an arbitrarily small positive number. Since equation 3.106 is valid for all values of z for which $S(z)$ is uniformly convergent, $S(z)$ is continuous over this range of z.

Theorem 3.2. If

$$S(z) = \sum_{n=0}^{\infty} \sigma_n(z) \quad (3.9a)$$

is uniformly convergent over the interval $z \in [a, b]$, and if α and β are both within this interval, then

$$I = \int_{\alpha}^{\beta} \sum_{n=0}^{\infty} \sigma_n(z)\, dz \tag{3.107a}$$

can be integrated term by term and the sum of the integrals will converge to

$$I = \int_{\alpha}^{\beta} S(z)\, dz \tag{3.107b}$$

To prove this, consider

$$\int_{\alpha}^{\beta} [S(z) - S_1(z)]\, dz = \int_{\alpha}^{\beta} S_2(z)\, dz \tag{3.108}$$

Since an integral is an infinite sum of infinitesmal terms

$$\left| \int_{\alpha}^{\beta} S_2(z)\, dz \right| \leqslant \int_{\alpha}^{\beta} |S_2(z)|\, dz \tag{3.109}$$

$$\Rightarrow \quad \left| \int_{\alpha}^{\beta} [S(z) - S_1(z)]\, dz \right| \leqslant \int_{\alpha}^{\beta} |S_2(z)|\, dz \tag{3.110}$$

But since $S(z)$ converges uniformly, $S_2(z)$ satisfies

$$|S_2(z)| < \varepsilon \tag{3.76}$$

and equation 3.110 becomes

$$\left| \int_{\alpha}^{\beta} [S(z) - S_1(z)]\, dz \right| < \int_{\alpha}^{\beta} \varepsilon\, dz = \varepsilon(\beta - \alpha) \equiv \varepsilon' \tag{3.111}$$

Because ε' is an arbitrarily small number,

$$\int_{\alpha}^{\beta} S(z)\, dz \simeq \int_{\alpha}^{\beta} S_1(z)\, dz \tag{3.112}$$

to any degree of accuracy. But $S_1(z)$ is a finite sum. Therefore, it can be integrated term by term. That is,

$$\int_{\alpha}^{\beta} S_1(z)\, dz = \sum_{n=0}^{N-1} \int_{\alpha}^{\beta} \sigma_n(z)\, dz \tag{3.113}$$

When $N \to \infty$,

$$\int_{\alpha}^{\beta} S_1(z)\, dz \to \int_{\alpha}^{\beta} S(z)\, dz = \sum_{n=0}^{\infty} \int_{\alpha}^{\beta} \sigma_n(z)\, dz \tag{3.114}$$

Theorem 3.3. If

$$T(z) = \sum_{n=0}^{\infty} \tau_n(z) \tag{3.115}$$

converges for all $z \in [a, b]$ and if

$$S(z) = \sum_{n=0}^{\infty} \frac{d\tau_n(z)}{dz} \tag{3.116}$$

converges uniformly over $[a, b]$, then

$$S(z) = \frac{dT(z)}{dz} \tag{3.117}$$

The proof of this is quite straightforward. Since

$$S(z) = \sum_{n=0}^{\infty} \frac{d\tau_n(z)}{dz} \tag{3.116}$$

converges uniformly, then by Theorem 3.2, this series can be integrated term by term. Thus

$$\int S(z)\,dz = \sum_{n=0}^{\infty} \int \frac{d\tau_n(z)}{dz}\,dz = \sum_{n=0}^{\infty} \tau_n(z) + \sum_{n=0}^{\infty} c_n = T(z) + C \tag{3.118}$$

The derivative of this equation yields the desired result:

$$S(z) = \frac{dT(z)}{dz} \tag{3.117}$$

3.2 Arithmetic Combinations of Power Series

A Taylor series of the form

$$S(z) = \sum_{n=0}^{\infty} \alpha_n (z - z_0)^n \tag{3.119}$$

is also referred to as a power series. Often, z_0 will be zero, and the power series becomes a MacLaurin series. Such is the case in the examples of recognizable series representations expressed in equations 3.1 through 3.7. In the following discussions, we take $z_0 = 0$.

Addition and subtraction

Let

$$\text{(a)}\quad S(z) = \sum_{n=0}^{\infty} \sigma_n(z) = \sum_{n=0}^{\infty} \alpha_n z^n \qquad \text{(b)}\quad T(z) = \sum_{n=0}^{\infty} \tau_n(z) = \sum_{n=0}^{\infty} \beta_n z^n \tag{3.120}$$

be two convergent series. The sum or difference in these functions is then

$$S(z) \pm T(z) = \sum_{n=0}^{\infty} (\alpha_n z^n \pm \beta_n z^n) = \sum_{n=0}^{\infty} (\alpha_n \pm \beta_n) z^n \equiv \sum_{n=0}^{\infty} \gamma_n z^n \tag{3.120c}$$

That is, the sum or difference of two power series is a power series, and the coefficient of each power of z is the sum or difference of the coefficients of the same power of z in the two constituent series.

Multiplication

The product of two series is

$$S(z) \cdot T(z) \equiv U(z) = \left(\sum_{n=0}^{\infty} \alpha_n z^n \right) \left(\sum_{k=0}^{\infty} \beta_k z^k \right) = \alpha_0\beta_0 + (\alpha_1\beta_0 + \alpha_0\beta_1) z$$
$$+ (\alpha_2\beta_0 + \alpha_1\beta_1 + \alpha_0\beta_2) z^2 + (\alpha_3\beta_0 + \alpha_2\beta_1 + \alpha_1\beta_2 + \alpha_0\beta_3) z^3 + \cdots \tag{3.121}$$

We define the coefficients of the power series for $U(z)$ to be μ_n. Then

$$\text{(a)} \quad U(z) = \sum_{n=0}^{\infty} \mu_n z^n \quad \Rightarrow \quad \text{(b)} \quad \mu_n = \sum_{k=0}^{n} \alpha_k \beta_{n-k} \tag{3.122}$$

Division

If $T(z) \neq 0$ for a specified value of z, we define

$$\text{(a)} \quad V(z) \equiv \frac{S(z)}{T(z)} \equiv \sum_{n=0}^{\infty} \nu_n z^n \quad \Rightarrow \quad \text{(b)} \quad S(z) = T(z)V(z) \tag{3.123}$$

Referring to equation 3.122b, from the series form of this we have

$$\text{(a)} \quad \sum_{n=0}^{\infty} \alpha_n z^n = \left(\sum_{l=0}^{\infty} \beta_l z^l \right)\left(\sum_{k=0}^{\infty} \nu_k z^k \right) \quad \Rightarrow \quad \text{(b)} \quad \alpha_n = \sum_{k=0}^{n} \nu_k \beta_{n-k} \tag{3.124}$$

Thus

$$\text{(a)} \quad \alpha_0 = \nu_0 \beta_0 \quad \Rightarrow \quad \text{(b)} \quad \nu_0 = \frac{\alpha_0}{\beta_0} \tag{3.125}$$

For $n = 1$ and $n = 2$, we obtain

$$\text{(c)} \quad \alpha_1 = \nu_0 \beta_1 + \nu_1 \beta_0 \qquad \text{(d)} \quad \alpha_2 = \nu_0 \beta_2 + \nu_1 \beta_1 + \nu_2 \beta_0 \tag{3.125}$$

Since ν_0 is known from equation 3.125b, we can solve equation 3.125c for ν_1. From this, equation 3.125d yields a value for ν_2. By this process, all coefficients for the ratio series can, in principle, be obtained. Of course, one cannot continue this process indefinitely, so one must be able to deduce a general expression for the coefficients ν_n, or the series for $V(z)$ must be approximated as a finite sum.

3.3 Summing a Power Series

To sum a convergent series means to find the function (if the terms in the series depend on a variable) or the value (if they do not) to which the series converges. To do this, we compare the series we are trying to sum to various known series such as those described in equations 3.1 through 3.7. Often, the series we are trying to sum will not be in a recognizable form, and it will be necessary to manipulate it so that it can be compared with one of the known series. There is no set prescription for the type and method of manipulation, so that much of our description of the manipulation of unknown series will be by example.

Algebraic manipulation

Example 3.15

Consider the series

$$\begin{aligned} S &= \sum_{n=0}^{\infty} \frac{1}{(2n+1)(2n+2)} = \sum_{n=0}^{\infty} \left(\frac{1}{(2n+1)} - \frac{1}{(2n+2)} \right) \\ &= \left(1 - \frac{1}{2}\right) + \left(\frac{1}{3} - \frac{1}{4}\right) + \left(\frac{1}{5} - \frac{1}{6}\right) + \cdots = \sum_{n=1}^{\infty} \frac{(-1)^{n+1}}{n} \end{aligned} \tag{3.126}$$

The key to recognizing which series this should be compared to is the existence of the factor $1/n$ contained in each term, like the logarithm series. Comparing this to equation 3.4, we see that $S = \log(2) = 0.6931$. □

Example 3.16

Consider the series

$$S(z) = \sum_{n=1}^{\infty} \frac{z^{2n+1}}{4^n n!} = \frac{z^3}{4(1!)} + \frac{z^5}{4^2(2!)} + \frac{z^7}{4^3(3!)} + \cdots \tag{3.127}$$

Even though all the powers of z are odd, (or can all be made even by factoring out one factor of z) each term in this series contains $1/n!$ and n takes on all values. This suggests a comparison of $S(z)$ with the exponential series of equation 3.1 rather than the sine series of equation 3.2, the terms of which contain $1/n!$ for odd values of n, or the cosine series in which n takes on only even values. Then

$$S(z) = z\left(\frac{(\frac{1}{2}z)^2}{1!} + \frac{(\frac{1}{2}z)^4}{2!} + \frac{(\frac{1}{2}z)^6}{3!} + \cdots\right) = z\left(-1 + \sum_{n=0}^{\infty} \frac{(\frac{1}{2}z)^{2n}}{n!}\right)$$

$$= z(-1 + e^{(z/2)^2}) \quad \square \tag{3.128}$$

Noting the factor $1/n$ in each term of the series of Example 3.15 and the $1/n!$ for all n-values in the series of Example 3.16 are the types of clues to look for in deciding the direction in which the manipulation should proceed.

Manipulation by calculus operations

If the series we are trying to sum depends on a variable, we can perform calculus operations on the series in an attempt to put it into a recognizable form. If the terms of the series are constant, it may be possible to sum the series using calculus operations by considering a similar series in which the terms depend on a variable. We may then perform calculus operations to manipulate this variable-dependent series until it is in a recognizable form. We would then evaluate the variable at a specified value to obtain the series we are summing.

Example 3.17

For example, consider the series

$$S = \sum_{n=1}^{\infty} \frac{1}{n(n+1)3^n} \tag{3.129}$$

The fact that the terms contain factors of n in the denominator but not $n!$ indicates that we investigate a manipulation of the logarithm series. Since our series is not alternating, we begin with the series for $\log(1 - z)$ instead of $\log(1 + z)$.

Consider

$$-\log(1 - z) = \sum_{n=1}^{\infty} \frac{z^n}{n} \tag{3.130}$$

for $|z| < 1$. In each term we can obtain a denominator containing $n(n + 1)$ from a term z^n/n by integrating such a term once. Thus we consider

$$-\int \log(1 - z)\, dz = \sum_{n=1}^{\infty} \frac{z^{n+1}}{n(n+1)} + C = (1 - z)[\log(1 - z) - 1] \tag{3.131}$$

To evaluate the constant of integration, we note that there is at least one factor of z in every term in the sum. Therefore, the sum is zero for $z = 0$. Setting $z = 0$, we obtain $C = -1$. Therefore,

$$\sum_{n=1}^{\infty} \frac{z^{n+1}}{n(n+1)} = (1 - z)[\log(1 - z) - 1] + 1 \tag{3.132}$$

Setting $z = \frac{1}{3}$ we obtain

$$\sum_{n=1}^{\infty} \frac{1}{n(n+1)3^{n+1}} = \frac{1}{3}\sum_{n=1}^{\infty} \frac{1}{n(n+1)3^{n}} = \frac{1}{3}S = \frac{2}{3}\left(\log\left(\frac{2}{3}\right) - 1\right) + 1 \tag{3.133a}$$

$$\Rightarrow \quad S = 1 + 2\log(\tfrac{2}{3}) \quad \square \tag{3.133b}$$

Example 3.18

As a second example, we sum the series

$$S = \sum_{n=0}^{\infty} \frac{3^n}{(n+3)(n+4)n!} \tag{3.134}$$

The existence of $n!$ in the denominator and the fact that n takes on all values indicate that we investigate a series involving e^z.

$$e^z = \sum_{n=0}^{\infty} \frac{z^n}{n!} \tag{3.1}$$

We also note that we can obtain factors $(n + 3)(n + 4)$ in the denominator of a term by integrating z^{n+2} twice. To obtain terms involving z^{n+2} from terms containing z^n, we consider the series

$$z^2e^z = \sum_{n=0}^{\infty} \frac{z^{n+2}}{n!} \tag{3.135}$$

Integrating this once, we obtain

$$\int z^2e^z\,dz = \sum_{n=0}^{\infty} \frac{z^{n+3}}{(n+3)n!} + C_1 = e^z(z^2 - 2z + 2) \tag{3.136}$$

As before, we evaluate C_1 by setting $z = 0$. Since the series is zero for $z = 0$, we obtain $C_1 = 2$. Therefore,

$$\sum_{n=0}^{\infty} \frac{z^{n+3}}{(n+3)n!} = e^z(z^2 - 2z + 2) - 2 \tag{3.137}$$

A second integration of equation 3.135 yields

$$\sum_{n=0}^{\infty} \frac{z^{n+4}}{(n+3)(n+4)n!} + C_2 = e^z(z^2 - 4z + 6) - 2z \tag{3.138}$$

Setting $z = 0$, yields $C_2 = 6$. Thus

$$\sum_{n=0}^{\infty} \frac{z^{n+4}}{(n+3)(n+4)n!} = e^z(z^2 - 4z + 6) - 2z - 6 \tag{3.139}$$

By bringing a factor of z^4 outside the sum, then setting $z = 3$, we obtain an expression for the series of equation 3.134.

$$3^4\sum_{n=0}^{\infty} \frac{3^n}{(n+3)(n+4)n!} = 3^4S = 3e^3 - 12 \tag{3.140a}$$

$$\Rightarrow \quad S = \sum_{n=0}^{\infty} \frac{3^n}{(n+3)(n+4)n!} = \frac{3e^3 - 12}{3^4} = 0.5958 \quad \square \tag{3.140b}$$

In Examples 3.17 and 3.18 additional factors involving the summation index appeared in the denominator of the terms. These were generated by integrating the terms of the series of a known function. If additional factors containing the summation index appear in the numerator, these can be obtained by differentiation of the terms of the known series.

Example 3.19

As an example, consider

$$S = \sum_{n=0}^{\infty} \frac{(n+1)(n+2)}{3^n n!} \tag{3.141}$$

The existence of $n!$ in the denominator of each term indicates that we manipulate a series involving e^z. The factors $(n+1)(n+2)$ in the numerator suggest two derivatives of the series we will manipulate. To obtain factors $(n+2)$ and $(n+1)$, we must differentiate z^{n+2} twice. For those reasons, we begin with the series

$$z^2 e^z = \sum_{n=0}^{\infty} \frac{z^{n+2}}{n!} \tag{3.135}$$

Then

$$\frac{d^2}{dz^2}(z^2 e^z) = \sum_{n=0}^{\infty} \frac{(n+2)(n+1)z^n}{n!} \tag{3.142}$$

If this is now evaluated at $z = \frac{1}{3}$ we obtain a value for the series of equation 3.141. That is,

$$\frac{d^2}{dz^2}(z^2 e^z)_{z=1/3} = \sum_{n=0}^{\infty} \frac{(n+1)(n+2)}{3^n n!} = S = \frac{31}{9} e^{1/3} \quad \square \tag{3.143}$$

Example 3.20

The series

$$S = \sum_{n=1}^{\infty} \frac{(n+1)}{n 3^n} \tag{3.144}$$

is evaluated by considering

$$\frac{d}{dz}[z \log(1-z)]_{z=1/3} = -\sum_{n=1}^{\infty} \frac{(n+1)}{n3^n} = -S = \log(\tfrac{2}{3}) - \tfrac{1}{2} \tag{3.145a}$$

$$\Rightarrow \quad S = \tfrac{1}{2} - \log(\tfrac{2}{3}) \quad \square \tag{3.145b}$$

Combinations of integrals and derivatives might be required to sum a series.

Example 3.21

For example, to evaluate

$$S = \sum_{n=0}^{\infty} \frac{(n+3)}{(n+2)2^n n!} \tag{3.146}$$

we must consider

$$I = \int \frac{1}{z} \frac{d}{dz}(z^3 e^z)\, dz \tag{3.147}$$

We determine the constant of integration, and set $z = \frac{1}{2}$. These operations yield

$$S = 4 - e^{1/2} \quad \square \tag{3.148}$$

Example 3.22

Consider the series

$$S \equiv \sum_{n=0}^{\infty} (-1)^n \frac{\pi^{2n-1}(2n+3)}{16^n (2n)!(n+1)} \tag{3.149}$$

The factors $1/(2n)!$ and $(-1)^n$ suggest that we manipulate the series

$$\cos x = \sum_{n=0}^{\infty} (-1)^n \frac{x^{2n}}{(2n)!} \tag{3.3}$$

To produce the factor $(n+1)$ in the denominator, we note that

$$\int x^{2n+1}\, dx = \frac{x^{2n+2}}{2(n+1)} \tag{3.150}$$

Thus we consider

$$\int x \cos x\, dx = \sum_{n=0}^{\infty} (-1)^n \frac{1}{2(n+1)} \frac{x^{2n+2}}{(2n)!} + C = x \sin x + \cos x \tag{3.151}$$

At $x = 0$, each term in the series is zero. Thus

$$C = 1 \tag{3.152}$$

$$\Rightarrow \quad x \sin x + \cos x - 1 = \frac{1}{2} \sum_{n=0}^{\infty} (-1)^n \frac{1}{(n+1)} \frac{x^{2n+2}}{(2n)!} \tag{3.153}$$

The factor $(2n+3)$ in the numerator is obtained from

$$\begin{aligned} \frac{d}{dx}[x(x \sin x + \cos x - 1)] &= \frac{1}{2} \sum_{n=0}^{\infty} (-1)^n \frac{(2n+3)}{(n+1)} \frac{x^{2n+2}}{(2n)!} \\ &= x \sin x + (x^2+1) \cos x - 1 \end{aligned} \tag{3.154}$$

Setting $x = \pi/4$, we obtain

$$\frac{\pi}{4\sqrt{2}} + \left(\frac{\pi^2}{16} + 1\right) - 1 = \frac{\pi^3}{32} \sum_{n=0}^{\infty} (-1)^n \frac{\pi^{2n-1}(2n+3)}{16^n (2n)!(n+1)} = \frac{\pi^3}{32} S \tag{3.155}$$

$$\Rightarrow \quad S = \frac{32}{\pi^3}\left[\frac{\pi}{4\sqrt{2}} + \left(\frac{\pi^2}{16} + 1\right) - 1\right] = 0.721037 \quad \square \tag{3.156}$$

Example 3.23

As a final example, we evaluate the finite sum

$$S = \sum_{n=1}^{N} \frac{N!}{(n-1)!(N-n)!} \tag{3.157}$$

by manipulating the binomial series of equation 3.5. Consider

$$\frac{d}{dz}(1+z)^N = N(1+z)^{N-1} = \sum_{n=1}^{N} \frac{nN!}{n!(N-n)!} z^{n-1} = \sum_{n=1}^{N} \frac{N!}{(n-1)!(N-n)!} z^{n-1} \tag{3.158}$$

Since the $n = 0$ term in the binomial series is independent of z, its derivative is zero. Therefore, the sum of equation 3.150 starts at $n = 1$. If we set $z = 1$, we obtain

$$\sum_{n=1}^{N} \frac{N!}{(n-1)!(N-n)!} = N 2^{N-1} \quad \square \tag{3.159}$$

Estimating the value of an absolute series

If all attempts for summing a series exactly are unsuccessful, one should then try to obtain an approximate value of the sum. We first consider approximating an absolute series (in which all terms are positive except for a possible overall negative sign).

As we have done in deducing convergence tests, we divide the infinite series into a finite sum and infinite series.

$$S(z) = \sum_{n=0}^{N-1} \sigma_n(z) + \sum_{n=N}^{\infty} \sigma_n(z) \equiv S_1(z) + S_2(z) \tag{3.15}$$

Perhaps the simplest approximation to the series is to take N large enough that S_2 can be ignored. If the series converges fairly rapidly, it may be well approximated by taking a reasonably small number of terms for S_1.

Example 3.24

As an example, consider the series of equation 3.140b:

$$S = \sum_{n=0}^{\infty} \frac{3^n}{(n+3)(n+4)n!} = \frac{3e^3 - 12}{3^4} = 0.5958 \tag{3.140b}$$

If we estimate this series by the nine-term partial sum, we obtain

$$\sum_{n=0}^{8} \frac{3^n}{(n+3)(n+4)n!} = \frac{1}{12} + \frac{3}{20} + \frac{9}{60} + \frac{27}{252} + \frac{81}{1344} + \frac{243}{8640} + \frac{729}{64{,}800} + \frac{2187}{554{,}400} + \frac{6561}{5{,}322{,}240} \simeq 0.5953 \tag{3.160}$$

which is a very reasonable approximation to the value of the infinite series. □

The Cauchy integral test was developed for determining the convergence of an absolute series. If the Cauchy integral can be evaluated, it is possible to use it to estimate S_2 and therefore to estimate the value of an absolute series. Combining the results of

$$\int_N^{\infty} \sigma(n, z)\, dn > S_2(z) - \sigma_N(z) \tag{3.52}$$

$$S_2(z) > \int_N^{\infty} \sigma(n, z)\, dn \tag{3.55}$$

provides upper and lower bounds on $S_2(z)$ in terms of the Cauchy envelope integral. That is,

$$S_2(z) > \int_N^{\infty} \sigma(n, z)\, dn > S_2(z) - \sigma_N(z) \tag{3.161a}$$

$$\Rightarrow \quad \int_N^{\infty} \sigma(n, z)\, dn + \sigma_N(z) > S_2(z) > \int_N^{\infty} \sigma(n, z)\, dn \tag{3.161b}$$

If N is large, so that $\sigma_N(z)$ can be considered small, then equation 3.161b provides a small range of allowed values for S_2. A reasonable estimate of S_2 is obtained by taking the average of the upper and lower bounds. That is, we can approximate S_2 by

$$S_2(z) \simeq \tfrac{1}{2}\sigma_N(z) + \int_N^{\infty} \sigma(n, z)\, dn \tag{3.162}$$

Example 3.25

For example, we determined earlier that

$$S = \sum_{n=0}^{\infty} \frac{1}{(2n+1)(2n+2)} = \log(2) = 0.6931 \tag{3.126a}$$

Writing S as

$$S = \sum_{n=0}^{4} \frac{1}{(2n+1)(2n+2)} + S_2 \tag{3.163}$$

we estimate S_2 to be

$$S_2 \simeq \int_5^\infty \frac{dn}{(2n+1)(2n+2)} + \frac{1}{2}\sigma_5 = \frac{1}{2}\log\left(\frac{12}{11}\right) + \frac{1}{2}\frac{1}{(2\cdot 5+1)(2\cdot 5+2)} = 0.0473 \tag{3.164}$$

Therefore,

$$S \simeq \sum_{n=0}^{4} \frac{1}{(2n+1)(2n+2)} + 0.0473 = 0.6929 \tag{3.165}$$

This is a better estimate of the value of the series than the one obtained by approximating S by the partial sum

$$S \simeq \sum_{n=0}^{4} \frac{1}{(2n+1)(2n+2)} = 0.6456 \quad \square \tag{3.166}$$

Estimating the value of an alternating series

If the series to be estimated is an alternating series, we can combine

$$S_2(z) < |\sigma_N(z)| \tag{3.65}$$

$$S_2(z) > |\sigma_N(z)| - |\sigma_{N+1}(z)| \tag{3.69}$$

to obtain bounds on $S_2(z)$. With N even, then

$$|\sigma_N(z)| > S_2(z) > |\sigma_N(z)| - |\sigma_{N+1}(z)| \tag{3.167}$$

As we did with the bounds used to estimate S_2 for an absolute series, we approximate S_2 for an alternating series by the average of the upper and lower bounds described in equation 3.167. That is,

$$S_2 \simeq |\sigma_N(z)| - \tfrac{1}{2}|\sigma_{N+1}(z)| \tag{3.168}$$

Example 3.26

As an example, consider

$$S = \sum_{n=0}^{\infty} \frac{(-1)^n}{(2n+1)!}4^n = \frac{1}{2}\sum_{n=0}^{\infty} \frac{(-1)^n}{(2n+1)!}2^{2n+1} = \frac{\sin(2)}{2} = 0.4546 \tag{3.169}$$

Since we must begin the S_2 series with an even index term, we take

$$S_1 = \frac{1}{2}\sum_{n=0}^{1} \frac{(-1)^n}{(2n+1)!}2^{n+1} = \frac{1}{2}\left(2 - \frac{2^3}{3!}\right) = \frac{1}{3} \tag{3.170a}$$

and approximate S_2 by

$$S_2 \simeq |\sigma_2| - \tfrac{1}{2}|\sigma_3| = \frac{1}{2}\left(\frac{2^5}{5!} - \frac{1}{2}\frac{2^7}{7!}\right) = 0.1270 \tag{3.170b}$$

Therefore, our estimate of the value of the series is

$$S \simeq 0.3333 + 0.1270 = 0.4603 \tag{3.171}$$

This is a reasonably good estimate of S, particularly considering the fact that we have taken very few terms for S_1. $\square$

Padé approximants

If the series to be estimated is a power series, an approximate function called the *Padé approximant* can sometimes yield a better estimate of the series than truncation of the original series, or approximating S_2 as described above. It can be particularly useful if the series is slowly converging. Let

$$S(z) = \sum_{n=0}^{\infty} \alpha_n z^n \tag{3.172}$$

be the series we are trying to estimate.

One way to represent a function is as an infinitely continued fraction. A typical infinitely continued fraction of a function that depends on z is

$$S(z) = A_0 + \cfrac{A_1 z}{B_0 + \cfrac{B_1 z}{C_0 + \cfrac{C_1 z}{D_0 + \cfrac{D_1 z}{\ddots}}}} \tag{3.173}$$

If the continuation of the fractions is terminated after a finite number of terms, the result can be written as the ratio of two polynomials. For example, if the process is terminated after the four terms shown in equation 3.173, the expression can be rationalized to

$$S(z) \simeq A_0 + \cfrac{A_1 z}{B_0 + \cfrac{B_1 z}{C_0 + \cfrac{C_1 z}{D_0 + D_1 z}}} = \frac{p_0 + p_1 z + p_2 z^2}{q_0 + q_1 z + q_2 z^2} \equiv \frac{P_2(z)}{Q_2(z)} \tag{3.174}$$

where the p_i and q_i are combinations of the A's, B's, C's and D's of equation 3.173. By extension, then, in principle, the infinitely continued fraction can be written as the ratio of two infinite series

$$S(z) = \frac{P(z)}{Q(z)} \tag{3.175}$$

where

$$\text{(a)} \quad P(z) = \sum_{n=0}^{\infty} p_n z^n \qquad \text{(b)} \quad Q(z) = \sum_{n=0}^{\infty} q_n z^n \tag{3.176}$$

where the p_i and q_i are combinations of the coefficients of the infinitely continued fraction.

The coefficients in the $P(z)$ and $Q(z)$ series are determined by requiring that the coefficient of each power of z in $P(z)/Q(z)$ is the same as the coefficient of that power of z in $S(z)$. This is most easily accomplished by rewriting equation 3.175 in series form as

$$\left[\sum_{n=0}^{\infty} \alpha_n z^n\right]\left[\sum_{m=0}^{\infty} q_m z^m\right] = \sum_{k=0}^{\infty} p_k z^k \tag{3.177}$$

then multiplying the $S(z)$ and $Q(z)$ series term by term, and equating the coefficients of corresponding powers of z.

In practice, it is impossible to carry out an infinite number of such processes. The Padé approximant is defined by truncating the continued fraction after a finite number of terms. As indicated in the example of equation 3.174, if the fraction is truncated after $2N$ (an even number of) terms, the result is a ratio $P_N(z)/Q_N(z)$ where $P_N(z)$ and $Q_N(z)$ are two polynomials each of order N. This Padé approximant is called the Nth diagonal approximant and is denoted by $S^{[N,N]}(z)$. The terminology *diagonal* refers to the fact that the numerator and denominator polynomials are of the same order. It is of the form

$$S^{[N,N]}(z) = \frac{P_N(z)}{Q_N(z)} = \frac{p_0 + p_1 z + \cdots + p_N z^N}{q_0 + q_1 z + \cdots + q_N z^N} \tag{3.178}$$

The generalization to the nondiagonal Padé approximant is an obvious extension of the description of the diagonal approximant. The $[M, N]$ Padé approximant is defined by

$$S^{[M,N]}(z) = \frac{P_M(z)}{Q_N(z)} = \frac{p_0 + p_1 z + \cdots + p_M z^M}{q_0 + q_1 z + \cdots + q_N z^N} \tag{3.179a}$$

If $q_0 \neq 0$, the p_m and q_n coefficients can be divided by q_0 and renamed, so that $S^{[M,N]}(z)$ can be written

$$S^{[M,N]}(z) = \frac{p_0 + p_1 z + \cdots + p_M z^M}{1 + q_1 z + \cdots + q_N z^N} \tag{3.179b}$$

Equivalently, there is no loss of generality in setting $q_0 = 1$.

The criterion for determining the coefficients is to require $S^{[M,N]}(z)$ be the same as the partial sum $S_{M+N}(z)$ for all powers of z up to z^{M+N}. That is,

$$\Delta(z) \equiv S_{M+N}(z) - S^{[M,N]}(z) \tag{3.180}$$

contains no terms in z of power less than $M + N + 1$. This is denoted by

$$S_{M+N}(z) \cdot Q_N(z) - P_M(z) = O(z^{M+N+1}) \tag{3.181}$$

The symbol $O(z^{M+N+1})$ denotes an unspecified polynomial the lowest-order term of which is z^{M+N+1}. This condition will yield enough equations to determine the $M + N - 1$ coefficients in the two polynomials.

Example 3.27

As an example, we approximate

$$S(z) = -\log(1 - z) = z + \frac{z^2}{2} + \frac{z^3}{3} + \frac{z^4}{4} + \cdots \tag{3.182}$$

by the [2, 2] Padé approximant,

$$S^{[2,2]}(z) = \frac{p_0 + p_1 z + p_2 z^2}{1 + q_1 z + q_2 z^2} \tag{3.183}$$

To determine the p's and q's, we require that

$$\Delta(z) = \left(z + \frac{z^2}{2} + \frac{z^3}{3} + \frac{z^4}{4}\right)(1 + q_1 z + q_2 z^2) - (p_0 + p_1 z + p_2 z^2) = O(z^5) \tag{3.184}$$

Thus

$$\begin{aligned} &-p_0 + (1 - p_1)z + \left(\tfrac{1}{2} + q_1 - p_2\right)z^2 \\ &\quad + \left(\frac{1}{3} + \frac{1}{2}q_1 + q_2\right)z^3 + \left(\frac{1}{4} + \frac{1}{3}q_1 + \frac{1}{2}q_2\right)z^4 + O(z^5) = O(z^5) \end{aligned} \tag{3.185}$$

which is satisfied by setting the coefficients of each of the terms up to z^4 to 0. This yields

$$\text{(a)}\quad p_0 = 0 \quad \text{(b)}\quad p_1 = 1 \quad \text{(c)}\quad p_2 = -\tfrac{1}{2} \quad \text{(d)}\quad q_1 = -1 \quad \text{(e)}\quad q_2 = \tfrac{1}{6} \tag{3.186}$$

$$\Rightarrow \quad S^{[2,2]}(z) = \frac{z - \frac{1}{2}z^2}{1 - z + \frac{1}{6}z^2} \simeq -\log(1 - z) \tag{3.187}$$

For example, $-\log(1 - 0.5) = 0.693147$. The fourth partial sum and the [2, 2] Padé approximant have the values

$$\text{(a)}\quad S_4(0.5) = \sum_{n=1}^{4} \frac{z^n}{n}\bigg|_{z=0.5} = 0.682292 \qquad \text{(b)}\quad S^{[2,2]}(0.5) = 0.692308 \quad \square \tag{3.188}$$

3.4 Other Processes Using Series

Estimating integrals

When a definite integral cannot be evaluated exactly, it may be possible to obtain an accurate approximation using a series expansion of the integrand.

Example 3.28

As an example, since

$$I(x) = \int \sin(x^2)\, dx \tag{3.189a}$$

cannot be evaluated in terms of elementary functions, an exact value of

$$I(0.5) - I(0) = \int_0^{0.5} \sin(x^2)\, dx \tag{3.189b}$$

cannot be obtained. However, by expanding the sine function in a MacLaurin series, and integrating the power series term by term, we obtain a series expression for the integral. We can obtain an accurate estimate of the integral by truncating the series after a finite number of terms. The number of terms one retains, of course, depends on the degree of accuracy required by the user.

For this example, if five-decimal-place accuracy is required, consider

$$\int_0^{0.5} \sin x^2\, dx = \int_0^{0.5} \left(x^2 - \frac{x^6}{3!} + \frac{x^{10}}{5!} + \cdots \right) dx = 0.04167 - 0.00019 + 3.7 \times 10^{-7} \tag{3.190}$$

It is clear that we can terminate this expansion after two terms to achieve five-decimal-place accuracy. Since the sine function is an entire function, the series expansion will be valid for any limits. However, a significantly larger range of integration would probably require more than two terms to obtain five-decimal-place accuracy. $\square$

Example 3.29

An application of this method of approximation arises in the Debye theory of specific heat. Debye showed that the contribution from the vibration of lattice ions to the specific heat of a material is given by the expression

$$C_v = 9R\left(\frac{T}{\Theta_D}\right)^3 \int_0^{\Theta_D/T} \frac{x^4 e^x}{(e^x - 1)^2}\, dx \tag{3.191}$$

where T is the absolute temperature of the material, and Θ_D is a constant called the Debye temperature. It is a parameter that is specific to the lattice structure of the material being studied. Since this integral cannot be evaluated exactly, we must resort to approximation.

The Stefan–Boltzmann law states that at low temperatures, the thermal energy of a substance is given by

$$E = AT^4 \tag{3.192}$$

where A is a constant and T is the absolute temperature of the material. Therefore, at low temperatures, the specific heat is

$$C_v = \frac{\partial E}{\partial T} = 4AT^3 \tag{3.193}$$

That is, the specific heat varies as the cube of the absolute temperature.

Referring to the Debye expression, if T is small compared to Θ_D, the upper limit of the integral is large. For large values of x, the integrand can be approximated by $x^4 e^{-x}$, which is small. Thus very little error is introduced by taking the upper limit of the integral to be ∞ for small T. Then

$$C_v \simeq 9R\left(\frac{T}{\Theta_D}\right)^3 \int_0^\infty \frac{x^4 e^x}{(e^x - 1)^2}\, dx = \text{constant} \cdot T^3 \tag{3.194}$$

in agreement with the Stefan–Boltzmann law.

At high (room) temperature, $T \gg \Theta_D$. Therefore, the limits of integration constrain x to a narrow range of values. As such, an approximation of the Debye integral can be obtained by expanding the integrand in a MacLaurin series, integrating term by term, and truncating the series after a desired accuracy is achieved. The first nonzero term in the expansion of the integrand is

$$\frac{x^4 e^x}{(e^x - 1)^2} \simeq \frac{x^4(1 + x)}{\left(x + \frac{1}{2}x^2\right)^2} \simeq x^2 \tag{3.195}$$

Thus the lowest-order expression for the Debye specific heat at high temperature is the expected Dulong–Petit law,

$$C_v \simeq 9R\left(\frac{T}{\Theta_D}\right)^3 \int_0^{\Theta_D/T} x^2\, dx = 3R \qquad \square \tag{3.196}$$

It is sometimes possible to use the series expansion of one function to obtain a series representation of another.

Example 3.30

To illustrate, consider the infinite geometric series in x^2.

$$\frac{1}{1 - x^2} = \sum_{n=0}^{\infty} x^{2n} \qquad |x| < 1 \tag{3.197}$$

Integrating this, we obtain

$$\int \frac{dx}{1 - x^2} = \frac{1}{2}\log\left(\frac{1 + x}{1 - x}\right) = \sum_{n=0}^{\infty} \frac{x^{2n+1}}{2n + 1} + C \tag{3.198}$$

By setting $x = 0$, we obtain $C = 0$. Thus, using the geometric series in x^2 of equation 3.197, we have determined the MacLaurin series for

$$\log\left(\frac{1 + x}{1 - x}\right) = 2\sum_{n=0}^{\infty} \frac{x^{2n+1}}{(2n + 1)} \qquad \square \tag{3.199}$$

3.5 Bernoulli and Euler Series

Bernoulli numbers

The *Bernoulli numbers*, B_n, are defined from coefficients in the series expansion of

$$F(z) \equiv \frac{z}{(e^z - 1)} = \sum_{n=0}^{\infty} B_n \frac{z^n}{n!} \tag{3.200}$$

Using

$$e^z - 1 = \sum_{k=1}^{\infty} \frac{z^k}{k!} = z \sum_{k=0}^{\infty} \frac{z^k}{(k+1)!} \tag{3.201}$$

equation 3.200 becomes

$$1 = \sum_{k,n=0}^{\infty} B_n \frac{z^{n+k}}{n!(k+1)!} \tag{3.202}$$

By equating the coefficients of z^0 on each side of equation 3.202, we obtain

$$B_0 = 1 \tag{3.203}$$

The equality of the coefficients of z^1 with $B_0 = 1$ yields

$$\text{(a)} \quad \frac{B_0}{(0!)(2!)} + \frac{B_1}{(1!)(1!)} = 0 \quad \Rightarrow \quad \text{(b)} \quad B_1 = -\tfrac{1}{2} \tag{3.204}$$

Similarly, the equating the coefficients of z^2, we find that

$$B_2 = \tfrac{1}{6} \tag{3.205}$$

and from the equality of all remaining odd powers of z, we obtain

$$B_3 = B_5 = \cdots = 0 \tag{3.206}$$

That is, all odd-index Bernoulli numbers are zero except B_1.

The Bernoulli numbers satisfy a somewhat unusual recurrence relation. Consider the binomial expansion of the quantity

$$(B+1)^n - B^n = \sum_{k=0}^{n} \frac{n!}{k!(n-k)!} B^k - B^n = 0 \tag{3.207}$$

If each power of B is then replaced by the Bernoulli number of corresponding index,

$$B^k \to B_k \tag{3.208}$$

we obtain the recurrence relation that determines a Bernoulli number of a specified index from the values of Bernoulli numbers of lower index. Since the leading term in $(B+1)^n$ is B^n and this is subtracted away in the difference $(B+1)^n - B^n$, in order to find B_N using this recurrence relation, we must choose $n = N + 1$.

Example 3.31

As an example, we determine B_3 using equation 3.207. As such, we must set $n = 4$. Then

$$(B+1)^4 - B^4 = 0 \tag{3.209}$$

$$\Rightarrow \quad B^4 + 4B^3 + 6B^2 + 4B^1 + B^0 - B^4 = 0 \tag{3.210}$$

Setting $B^k \to B_k$, equation 3.210 becomes

$$4B_3 + 6B_2 + 4B_1 + B_0 = 0 \tag{3.211}$$

With the values of B_0, B_1, and B_2 given in equations 3.203 through 3.205, we obtain $B_3 = 0$. □

Both the recurrence relation and the process of equating coefficients of like powers of x in equation 3.202 become cumbersome very quickly. To develop another method for computing the Bernoulli numbers, we return to the defining series

$$F(z) \equiv \frac{z}{(e^z - 1)} = \sum_{n=0}^{\infty} B_n \frac{z^n}{n!} \tag{3.200}$$

This series representation is a MacLaurin expansion in a region of analyticity of $F(z)$, which has poles at those values of z for which $e^z = 1$, except at $z = 0$. That is, the poles are at $z = 2\pi i k$ with $k = \pm 1, \pm 2, \ldots$. These are depicted in Figure 3.4. As we will demonstrate when determining the residues, these are simple poles.

The MacLaurin series is only a valid representation of $F(z)$ in a region that excludes these poles. From the position of the poles, it is evident that the radius of convergence of the MacLaurin series is 2π. We initially take the region of convergence to be a small circle of radius $< 2\pi$ as shown in Figure 3.5.

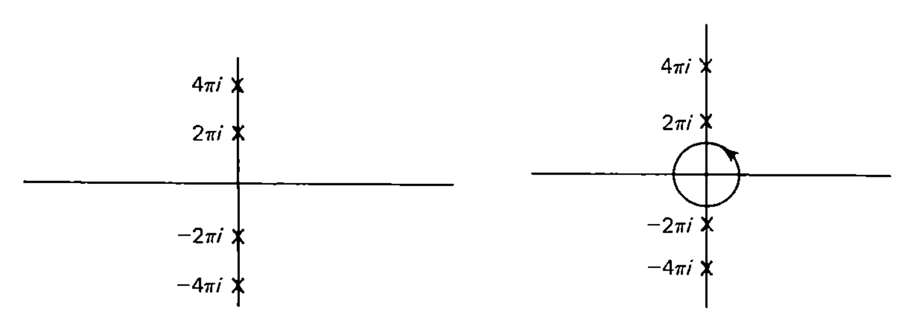

Figure 3.4 Poles of $z/(e^z - 1)$.

Figure 3.5 Small circular region for the MacLaurin expansion of the Bernoulli series.

In Chapter 2 we developed the Taylor series representation of a function in its region of analyticity as

$$F(z) = \sum_{n=0}^{\infty} \frac{F^{(n)}(z_0)}{n!}(z - z_0)^n \tag{2.100}$$

It was shown using Cauchy's theorem that the nth derivative of $F(z_0)$ is

$$F^{(n)}(z_0) = \frac{n!}{2\pi i}\oint \frac{F(z)}{(z - z_0)^{n+1}}\, dz \tag{2.80c}$$

where the contour encloses z_0 and none of the poles of $F(z)$. Comparing equation 2.100 to equation 3.200, we note that for the series defining the Bernoulli numbers, $z_0 = 0$, and that $F^{(n)}(z_0 = 0) = B_n$. Thus, for the Bernoulli series, equation 2.80c becomes

$$B_n = \frac{n!}{2\pi i}\oint \frac{z}{(e^z - 1)} \frac{1}{z^{n+1}}\, dz \tag{3.212}$$

Since $z/(e^z - 1)$ is analytic at the origin, the integrand of equation 3.212 has one pole of order $n + 1$ at the origin arising from $1/z^{n+1}$. The contour must enclose only that pole. We therefore choose the small circle of Figure 3.5, traversed in the counterclockwise direction, as our contour.

Since there are no singularities of the integrand along the real axis except at $z = 0$, and since

$$\lim_{x\to\infty} \frac{x}{(e^x - 1)} \frac{1}{x^{n+1}} = 0 \tag{3.213}$$

the circle of Figure 3.5 can be stretched into the oblong contour of Figure 3.6 without changing the value of the integral for B_n. Consider the integrand of equation 3.212 on an infinite circle. With $z = R\cos\theta + iR\sin\theta$, on the part of the circle where $\cos\theta > 0$,

$$\lim_{z\to\infty} \frac{z}{(e^z - 1)} \frac{1}{z^{n+1}} \sim R^{-n} e^{-R\cos\theta} \to 0 \tag{3.214a}$$

On the part of the circle where $\cos\theta < 0$,

$$\lim_{z\to\infty} \frac{z}{(e^z - 1)} \frac{1}{z^{n+1}} \sim \frac{R^{-n}}{(e^{-R|\cos\theta|} - 1)} \sim R^{-n} \to 0 \tag{3.214b}$$

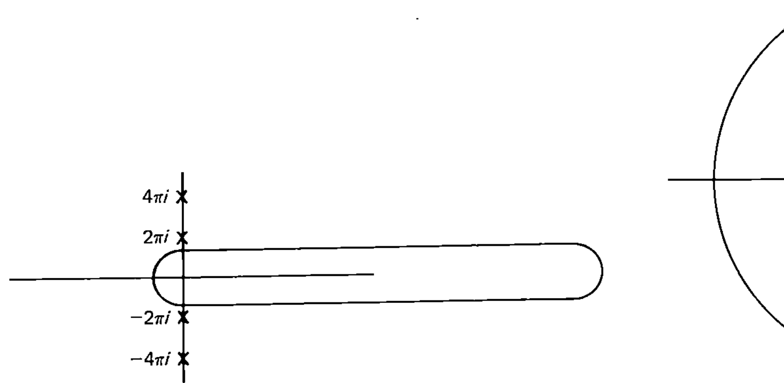

Figure 3.6 Equivalent contour for computing B_n.

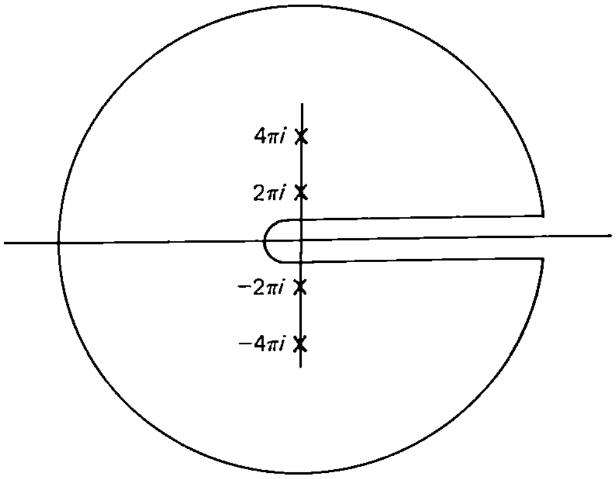

Figure 3.7 Equivalent contour for computing B_n.

Therefore, the integrand is zero not only at infinity on the real axis, but at any point on an infinitely large circle. As such, nothing is changed by removing the section of the contour on the real axis at infinity (one point) where the integrand is zero and replacing it with another contour (the infinitely large circle) along which the integrand is zero. That is, the computation of B_n is unchanged by taking the integral around the contour shown in Figure 3.7. With this contour, the pole of the integrand at $z = 0$ is excluded, and all the poles of $z/(e^z - 1)$ on the imaginary axis are enclosed by the contour. Note also that by "breaking" the contour at $x = \infty$ and substituting the infinite circle, the contour of Figure 3.7 is being traversed in a clockwise direction. Therefore, by Cauchy's theorem,

$$B_n = \frac{n!}{2\pi i} \oint \frac{1}{(e^z - 1)} \frac{1}{z^n}\, dz = -n! \sum_{k=-\infty}^{\infty} R(2\pi ik) \qquad k \neq 0 \tag{3.215}$$

We can explicitly exclude the $k = 0$ term by writing the sum in equation 3.215 as

$$\sum_{k=-\infty}^{\infty} R(2\pi ik) = \sum_{k=-\infty}^{-1} R(2\pi ik) + \sum_{k=1}^{\infty} R(2\pi ik) = \sum_{k=1}^{\infty} [R(2\pi ik) + R(-2\pi ik)] \tag{3.216}$$

The kth residue (evaluated by using L'Hospital's rule, for example) is

$$R(2\pi ik) = \lim_{z\to 2\pi ik} \left[\frac{(z - 2\pi ik)}{(e^z - 1) z^n} \right] = \frac{1}{(2\pi ik)^n} \tag{3.217}$$

As noted earlier, since the residue, obtained by simply removing the singularity (without taking derivatives), is finite and nonzero, the poles shown in Figure 3.4 are simple poles. With these residues, equation 3.215 becomes

$$B_n = -\frac{n!}{(2\pi i)^n}\sum_{k=1}^{\infty}\left[\frac{1}{(-k)^n}+\frac{1}{k^n}\right] = -\frac{n!}{(2\pi i)^n}\left[1+(-1)^n\right]\sum_{k=1}^{\infty}\frac{1}{k^n} \tag{3.218}$$

The series in equation 3.218 is called the *Riemann zeta function* and is denoted by

$$\zeta(n) \equiv \sum_{k=1}^{\infty}\frac{1}{k^n} \tag{3.219}$$

We note that if $n = 1$, the Riemann zeta series diverges. But for $n = 1$, the factor $[1 + (-1)^n]$ in equation 3.218 is zero. Thus B_1 is indeterminate by this calculation. For all $n > 1$, the Riemann zeta series converges. Therefore, from this computation we see that all odd-index Bernoulli numbers are zero except B_1, and that for all even values of n,

$$B_n = -\frac{2(n!)}{(2\pi i)^n}\zeta(n) \tag{3.220}$$

To explicitly specify that n is even, we let $n' = 2n$. Then dropping the prime, equation 3.220 becomes

$$B_{2n} = -\frac{2(2n)!}{(2\pi)^{2n}(i)^{2n}}\zeta(2n) = (-1)^{n+1}\frac{2(2n)!}{(2\pi)^{2n}}\zeta(2n) \tag{3.221}$$

From this, we can easily find expressions for even powers of π in terms of the Riemann zeta function. For example, for $n = 1$, with $B_2 = \frac{1}{6}$ we obtain

$$\text{(a)}\quad B_2 = \frac{4}{4\pi^2}\zeta(2) \quad\Rightarrow\quad \text{(b)}\quad \pi^2 = 6\sum_{k=1}^{\infty}\frac{1}{k^2} \tag{3.222}$$

Alternatively, with the value of B_{2n}, one can sum the Riemann zeta series exactly. That is,

$$\zeta(2n) = \sum_{k=1}^{\infty}\frac{1}{k^{2n}} = \frac{1}{2}(-1)^{n+1}(2\pi)^{2n}\frac{B_{2n}}{(2n)!} \tag{3.323}$$

Table 3.1 lists the Bernoulli numbers and the Bernoulli numbers divided by $n!$ for $n \leqslant 40$.

Convergence of the Bernoulli series

Since all odd-index Bernoulli numbers are zero (except for B_1), $[B_{2n}/(2n)!]z^{2n}$ and $[B_{2n+2}/(2n+2)!]z^{2n+2}$ are two successive terms in the Bernoulli series. Using the Cauchy ratio test, we see that the Bernoulli series converges for

$$|z|^2 \lim_{n\to\infty}\left|\frac{B_{2n+2}/(2n+2)!}{B_{2n}/(2n)!}\right| < 1 \tag{3.224}$$

From equation 3.221 we see that

$$\frac{B_{2n+2}/(2n+2)!}{B_{2n}/(2n)!} = -\frac{1}{4\pi^2}\frac{\zeta(2n+2)}{\zeta(2n)} \tag{3.225}$$

Referring to equation 3.219, we note that the series for $\zeta(2n)$ is an absolute series, the

TABLE 3.1 BERNOULLI NUMBERS AND BERNOULLI NUMBERS DIVIDED BY THE FACTORIAL OF THE INDEX

n	B_n	$\frac{B_n}{n!}$
0	1.00000	1.00000
1	−0.50000	−0.50000
2	0.16667	8.33333×10^{-2}
4	-3.33333×10^{-2}	-1.38889×10^{-3}
6	2.38095×10^{-2}	3.30688×10^{-5}
8	-3.33333×10^{-2}	-8.26720×10^{-7}
10	7.57576×10^{-2}	2.08768×10^{-8}
12	-2.53114×10^{-1}	-5.28419×10^{-10}
14	1.16667	1.33825×10^{-11}
16	−7.09216	-3.38968×10^{-13}
18	5.49712×10^{1}	8.58606×10^{-15}
20	-5.29124×10^{2}	-2.17487×10^{-16}
22	6.19212×10^{3}	5.50090×10^{-18}
24	-8.65803×10^{4}	-1.39545×10^{-19}
26	1.42552×10^{6}	3.53471×10^{-21}
28	-2.72982×10^{7}	-8.95352×10^{-23}
30	6.01581×10^{8}	2.26795×10^{-24}
32	-1.51163×10^{10}	-5.74479×10^{-26}
34	4.29615×10^{11}	1.45517×10^{-27}
36	-1.37117×10^{13}	-3.68600×10^{-29}
38	4.88332×10^{14}	9.33673×10^{-31}
40	-1.92966×10^{16}	-2.36502×10^{-32}

first term of which is 1. Thus $\zeta(2n) \geqslant 1$ for all n. As $n \to \infty$, all terms with $k > 1$ become infinitesmally small compared to the $k = 1$ term. Thus

$$\lim_{n\to\infty} \zeta(2n+2) = \lim_{n\to\infty} \zeta(2n) = 1 \tag{3.226}$$

$$\Rightarrow \quad \lim_{n\to\infty} \left| \frac{B_{2n+2}/(2n+2)!}{B_{2n}/(2n)!} \right| = \frac{1}{4\pi^2} \tag{3.227}$$

and from equation 3.224, we obtain the fact that the Bernoulli series converges for $|z| < 2\pi$. This had been determined earlier from the analyticity of the function $F(z)$ to which the Bernoulli series sums.

We can determine an approximate value for the Riemann zeta function, using the method developed for estimating the value of an absolute series in terms of the Cauchy envelope integral. Writing

$$\zeta(n) = \sum_{k=1}^{N-1} \frac{1}{k^n} + S_2 \tag{3.228}$$

S_2 is then estimated to be

$$S_2 \simeq \frac{1}{2}\sigma_N + \int_N^\infty \sigma(k)\, dk = \frac{1}{2N^n} + \int_N^\infty \frac{dk}{k^n} = \frac{1}{2N^n} + \frac{1}{(n-1)N^{(n-1)}} \tag{3.229}$$

This result can then be used to provide an accurate estimate of the Bernoulli numbers.

Example 3.32

For example, from equation 3.221,

$$B_6 = (-1)^4 \frac{2(6!)}{(2\pi)^6} \zeta(6) \tag{3.230}$$

Writing

$$\zeta(6) = \sum_{k=1}^{2} \frac{1}{k^6} + S_2 \tag{3.231}$$

we find that

$$\text{(a)} \quad S_1 = \sum_{k=1}^{2} \frac{1}{k^6} = 1.01563 \qquad \text{(b)} \quad S_2 \simeq \frac{1}{2 \cdot 3^6} + \frac{1}{5 \cdot 3^5} = 1.50892 \times 10^{-3} \tag{3.232}$$

$$\Rightarrow \quad \zeta(6) \simeq 1.01713 \tag{3.233}$$

$$\Rightarrow \quad B_6 \simeq 2.38046 \times 10^{-2} \tag{3.234}$$

Had we taken more terms for the series S_1, the estimate of B_6 would have been more accurate. □

Bernoulli functions

Like the Bernoulli numbers, the Bernoulli functions, denoted by $\beta_n(s)$, are defined from the coefficients of a series. The defining series is

$$\frac{ze^{sz}}{(e^z - 1)} = \sum_{n=0}^{\infty} \beta_n(s) \frac{z^n}{n!} \tag{3.235}$$

Setting $s = 0$, and referring to equation 3.200, equation 3.235 becomes

$$\frac{z}{(e^z - 1)} = \sum_{n=0}^{\infty} \beta_n(0) \frac{z^n}{n!} \tag{3.236}$$

$$\Rightarrow \quad \beta_n(0) = B_n \tag{3.237}$$

Consider

$$\frac{\partial}{\partial s}\left(\frac{ze^{sz}}{(e^z - 1)}\right) = \frac{z^2 e^{sz}}{(e^z - 1)} = \sum_{n=0}^{\infty} \beta_n'(s) \frac{z^n}{n!} \tag{3.238a}$$

But

$$\frac{z^2 e^{sz}}{(e^z - 1)} = z\left(\frac{ze^{sz}}{(e^z - 1)}\right) = \sum_{n=0}^{\infty} \beta_n(s) \frac{z^{n+1}}{n!} \tag{3.238b}$$

Substituting $n' = n + 1$ in equation 3.238b and dropping the prime, we equate the series in equations 3.238a and 3.238b to obtain

$$\sum_{n=1}^{\infty} \beta_{n-1}(s) \frac{z^n}{(n-1)!} = \sum_{n=0}^{\infty} \beta_n'(s) \frac{z^n}{n!} \tag{3.239}$$

Equating the coefficients of like powers of z, we obtain

$$\text{(a)} \quad \beta_0'(s) = 0 \qquad n = 0 \qquad \text{(b)} \quad \frac{\beta_n'(s)}{n!} = \frac{\beta_{n-1}(s)}{(n-1)!} \qquad n > 0 \tag{3.240}$$

From equation 3.240a, we obtain

$$\beta_0(s) = \text{constant} \tag{3.241}$$

But

$$\text{(a)} \quad \beta_0(0) = B_0 = 1 \qquad \Rightarrow \qquad \text{(b)} \quad \beta_0(s) = 1 \tag{3.242}$$

From equation 3.240b, we have

$$\beta_n'(s) = n\beta_{n-1}(s) \tag{3.243}$$

For example,

$$\text{(a)} \quad \beta_1'(s) = \beta_0 = 1 \qquad \Rightarrow \qquad \text{(b)} \quad \beta_1(s) = s + c \tag{3.244}$$

Then

$$\text{(a)} \quad \beta_1(0) = B_1 = -\tfrac{1}{2} \qquad \Rightarrow \qquad \text{(b)} \quad \beta_1(s) = s - \tfrac{1}{2} \tag{3.245}$$

If we now set $s = 1$ in equation 3.235, we obtain

$$\frac{ze^z}{(e^z - 1)} = \sum_{n=0}^{\infty} \beta_n(1) \frac{z^n}{n!} \tag{3.246a}$$

Dividing numerator and denominator by e^z, this becomes

$$\frac{-z}{(e^{-z} - 1)} = \sum_{n=0}^{\infty} \beta_n(1) \frac{z^n}{n!} \tag{3.246b}$$

Comparing this to equation 3.200, we see that the function on the left-hand side of equation 3.246b is the Bernoulli series in terms of $-z$. Therefore,

$$\sum_{n=0}^{\infty} \beta_n(1) \frac{z^n}{n!} = \sum_{n=0}^{\infty} B_n \frac{(-z)^n}{n!} = \sum_{n=0}^{\infty} (-1)^n \beta_n(0) \frac{z^n}{n!} \tag{3.247}$$

Equating the coefficients of like powers of z, we find

$$\beta_n(1) = (-1)^n \beta_n(0) = (-1)^n B_n \tag{3.248}$$

Euler – MacLaurin integration formula

The Euler–MacLaurin formula for estimating the value of the integral

$$I = \int_a^b f(y)\, dy \tag{3.249a}$$

is developed in terms of the Bernoulli functions and the Bernoulli numbers. The formula is applicable to integrals with limits [0, 1]. For any values of $[a, b]$, the integral can be put in the form

$$I = \int_0^1 F(x)\, dx \tag{3.249b}$$

The transformations

$$\text{(a)} \quad x = \frac{y - a}{b - a} \quad a, b \text{ finite} \qquad \text{(b)} \quad x = \frac{a}{y} \quad \begin{array}{l} a \neq 0, \\ b = \infty \end{array}$$

$$\text{(c)} \quad x = \frac{2}{\pi} \tan^{-1} y \quad \begin{array}{l} a = 0, \\ b = \infty \end{array} \qquad \text{(d)} \quad x = \frac{(\pi/2) - \tan^{-1} y}{\pi} \quad \begin{array}{l} a = -\infty, \\ b = \infty \end{array} \tag{3.250}$$

map $y \in [a, b]$ into $x \in [0, 1]$. Thus there is no loss of generality in considering an integral over the range [0, 1].

Since

$$\beta_1'(x) = B_0 = 1 \tag{3.244a}$$

equation 3.249b can be written

$$I = \int_0^1 \beta_1'(x) F(x)\, dx \tag{3.251}$$

Integrating this by parts yields

$$I = F(x)\beta_1(x)\Big|_0^1 - \int_0^1 \beta_1(x)F'(x)\,dx \tag{3.252}$$

With

$$\text{(a)}\quad \beta_1(1) = -\beta_1(0) = -B_1 = \tfrac{1}{2} \qquad \text{(b)}\quad \beta_1(x) = \tfrac{1}{2}\beta_2'(x) \tag{3.253}$$

the integral becomes

$$I = \tfrac{1}{2}[F(1) + F(0)] - \tfrac{1}{2}\int_0^1 F'(x)\beta_2'(x)\,dx \tag{3.254}$$

Integrating by parts again, and using

$$\text{(a)}\quad \beta_2(1) = \beta_2(0) = B_2 = \tfrac{1}{6} \qquad \text{(b)}\quad \beta_2(x) = \tfrac{1}{3}\beta_3'(x) \tag{3.255}$$

we have

$$I = \tfrac{1}{2}[F(1) + F(0)] - \tfrac{1}{12}[F'(1) - F'(0)] - \frac{1}{3!}\int_0^1 F''(x)\beta_3'(x)\,dx \tag{3.256}$$

Continuing this process, we obtain

$$\begin{aligned} I = \tfrac{1}{2}[F(1) + F(0)] &- \sum_{n=1}^{N} \frac{B_{2n}}{(2n)!}\left[F^{(2n-1)}(1) - F^{(2n-1)}(0)\right] \\ &+ \frac{1}{(2N)!}\int_0^1 F^{(2N)}(x)\beta_{2N}(x)\,dx \end{aligned} \tag{3.257}$$

where the kth derivative of $F(x)$ is denoted by $F^{(k)}(x)$ as introduced in Chapter 2.

Referring again to Table 3.1, we note that even though B_n increases with increasing index, $B_n/n!$ decreases with increasing n. Therefore, if $F(x)$ is such that $F^{(n)}(0)$ and $F^{(n)}(1)$ decrease, or at least do not increase very rapidly with increasing n, the series of equation 3.257 will converge at a reasonable rate. Then, ignoring the remainder integral, after a few terms, the finite sum will yield an acceptable estimate of $\int_0^1 F(x)\,dx$.

Example 3.33

As an example, consider

$$\int_0^1 e^x\,dx = e - 1 = 1.71828 \tag{3.258}$$

Using the Euler–MacLaurin formula, we can approximate this integral by

$$\int_0^1 e^x\,dx \simeq \frac{1}{2}(e^1 + e^0) - \frac{1}{2!}B_2(e^1 - e^0) - \frac{1}{4!}B_4(e^1 - e^0) = 1.71834 \quad \square \tag{3.259}$$

Euler numbers

The Euler numbers are defined by the series for $\sec z$.

$$\sec z = \frac{1}{\cos z} \equiv \sum_{n=0}^{\infty} (-1)^n E_{2n}\frac{z^{2n}}{(2n)!} \tag{3.260}$$

for $z \neq$ an odd multiple of $\pi/2$.

TABLE 3.2 EULER NUMBERS AND EULER NUMBERS DIVIDED BY THE FACTORIAL OF THE INDEX

n	E_n	$\frac{E_n}{n!}$
0	1.00000	1.00000
2	−1.00000	−0.50000
4	5.00000	2.08333×10^{-1}
6	-6.10000×10^{1}	-8.47222×10^{-2}
8	1.38500×10^{3}	3.43503×10^{-2}
10	-5.05210×10^{4}	-1.39222×10^{-2}
12	2.70277×10^{6}	5.64250×10^{-3}
14	-1.99361×10^{8}	-2.28682×10^{-3}
16	1.93915×10^{10}	9.26813×10^{-4}
18	-2.40488×10^{12}	-3.75623×10^{-4}
20	3.70371×10^{14}	1.52234×10^{-4}
22	-6.93489×10^{16}	-6.16983×10^{-5}
24	1.55145×10^{19}	2.50054×10^{-5}
26	-4.08707×10^{21}	-1.01343×10^{-5}
28	1.25226×10^{24}	4.10727×10^{-6}
30	-4.41544×10^{26}	-1.66462×10^{-6}
32	1.77519×10^{29}	6.74643×10^{-7}
34	-8.07233×10^{31}	-2.73423×10^{-7}
36	4.12221×10^{34}	1.10814×10^{-7}
38	-2.34896×10^{37}	-4.49112×10^{-8}
40	1.48512×10^{40}	1.82018×10^{-8}

Multiplying the cosine series by the series on the right side of equation 3.260, we obtain

$$1 = \sum_{k,n=0}^{\infty} (-1)^{n+k} E_{2n} \frac{z^{2n+2k}}{(2n)!(2k)!} \tag{3.261}$$

Equating the coefficients of like powers of z, we obtain values for the Euler numbers. Table 3.2 lists values of the Euler numbers E_n and $E_n/n!$ for $n \leqslant 40$.

The Euler numbers satisfy a recurrence relation similar to the one satisfied by the Bernoulli numbers. If

$$(E+1)^k + (E-1)^k = 0 \tag{3.262}$$

is expanded in a binomial series, and

$$E^k \to E_k \tag{3.263}$$

one can find a particular Euler number from the values of the Euler numbers of lower index. From this, for example, one finds that all Euler numbers of odd index are zero.

As with the Bernoulli numbers, determining the Euler numbers using the defining secant series, or the recurrence relation of equation 3.262 quickly becomes a cumbersome process as the value of the index gets larger. Recalling the analysis that leads to equation 3.221 for the Bernoulli numbers, we note that equation 3.260 is a MacLaurin expansion of $\sec z$, and will converge in a region that excludes the poles of $\sec z$ at odd integer multiples of $\pi/2$. Figure 3.8 shows these poles of $\sec z$ and a small circle enclosing a region of convergence of the Euler series.

Again using equations 2.80c, and 2.100, the Euler numbers can be written in terms of the Cauchy integral as

$$F^{(2n)}(0) = (-1)^n E_{2n} = \frac{(2n)!}{2\pi i} \oint \frac{\sec z}{z^{n+1}}\, dz \tag{3.264}$$

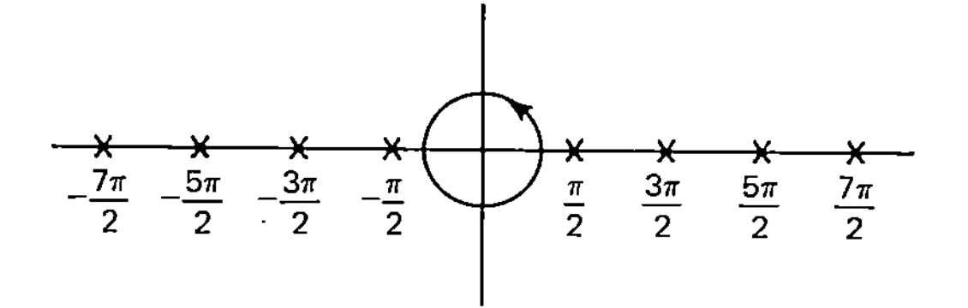

Figure 3.8 Region of convergence of the MacLaurin series expansion of $\sec z$ and the poles of $\sec z$.

From

$$\sec z = \frac{1}{\cos z} = \frac{2}{e^{iz} + e^{-iz}} \tag{3.265}$$

we note that at $z = \pm i\infty$, $\sec z \to 0$. Therefore, in an analysis similar to that used for the Bernoulli numbers, the contour of Figure 3.8 can be stretched out along either the positive or negative imaginary axis to infinity. We will choose the positive imaginary axis as shown in Figure 3.9

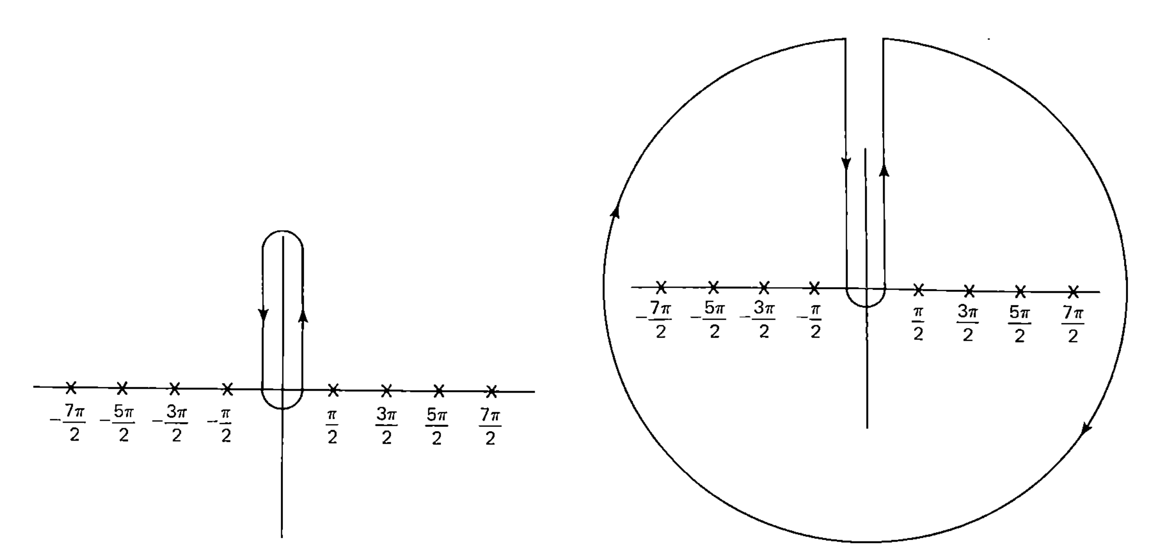

Figure 3.9 Equivalent contour for computing the Euler numbers.

Figure 3.10 Equivalent contour for computing the Euler numbers.

On an infinite circle, with $z = R\cos\phi + iR\sin\phi$, and $R \to \infty$, equation 3.265 becomes

$$\sec z = \frac{2}{e^{iR\cos\phi}e^{-R\sin\phi} + e^{-iR\cos\phi}e^{R\sin\phi}} \tag{3.266}$$

Regardless of the sign of $\sin\phi$, when $R \to \infty$, $\sec z \to 0$. Thus, as was done in the computation of the Bernoulli numbers, we can "break" the contour of Figure 3.9 at $z = i\infty$ and replace it with the infinite circle. That is, the contour of Figure 3.9 yields the same result as that shown in Figure 3.10. Note that the evolution of this contour has resulted in a clockwise traversal of the contour, as was the case in the analysis for the

Bernoulli numbers. Therefore,

$$(-1)^n E_{2n} = \frac{(2n)!}{2\pi i}\oint \frac{\sec z}{z^{2n+1}}\,dz = -(2n)!\sum_{k=-\infty}^{\infty} R\left(k\frac{\pi}{2}\right) \qquad k \text{ odd} \tag{3.267a}$$

$$\Rightarrow \quad E_{2n} = (-1)^{n+1}(2n)!\sum_{k=1}^{\infty}\left[R\left(k\frac{\pi}{2}\right) + R\left(-k\frac{\pi}{2}\right)\right] \qquad k \text{ odd} \tag{3.267b}$$

Using L'Hospital's rule,

$$R\left(k\frac{\pi}{2}\right) = \lim_{z\to k\pi/2}\frac{[z - k(\pi/2)]}{z^{2n+1}\cos z} = -\frac{1}{[k(\pi/2)]^{2n+1}}\frac{1}{\sin[k(\pi/2)]} \tag{3.268}$$

We note that the residue is an even function of k. Therefore,

$$R\left(-k\frac{\pi}{2}\right) = R\left(k\frac{\pi}{2}\right) \tag{3.269}$$

$$\Rightarrow \quad E_{2n} = (-1)^n 2(2n)!\sum_{k=1}^{\infty}\frac{1}{[k(\pi/2)]^{2n+1}}\frac{1}{\sin[k(\pi/2)]} \qquad k \text{ odd} \tag{3.270}$$

To explicitly express that k is odd, we set $k' = 2k + 1$, then drop the primes, to obtain

$$E_{2n} = \frac{4(-1)^n(2n)!2^{2n}}{\pi^{2n+1}}\sum_{k=0}^{\infty}\frac{(-1)^k}{(2k+1)^{2n+1}} \tag{3.271}$$

where we have used $\sin[(2k + 1)\pi/2] = \cos(k\pi) = (-1)^k$.

The alternating series in equation 3.271, which is similar in structure to the Riemann zeta function, is denoted by $S(n)$. We see that since

$$\lim_{k\to\infty}\frac{1}{(2k+1)^{(2n+1)}} = 0 \tag{3.272}$$

$S(n)$ converges for all $n \geqslant 0$. The convergence for $n = 0$ is conditional, the convergence for $n > 0$ is absolute.

Using an independent determination to find a specific Euler number, such as the recurrence relation of equation 3.262, equation 3.271 can be used to sum the alternating series defined in equation 3.271 [or express an odd power of π in terms of the alternating series $S(n)$].

Example 3.34

For example, with $n = 0$, and $E_0 = 1$, equation 3.271 becomes

$$\pi = 4S(0) = 4\sum_{k=0}^{\infty}\frac{(-1)^k}{(2k+1)} \tag{3.273}$$

Setting $n = 1$, with $E_2 = -1$, we find

$$\sum_{k=0}^{\infty}\frac{(-1)^k}{(2k+1)^3} = \frac{\pi^3}{32} \quad \square \tag{3.274}$$

Convergence of the Euler series

Referring to equation 3.260, the Euler series is written as an alternating series. However, from equation 3.271, we note that E_{2n} contains the factor $(-1)^n$. Therefore, equation 3.260 becomes

$$\sec z = 4 \sum_{n=0}^{\infty} \frac{2^{2n}}{\pi^{2n+1}} S(n) z^{2n} \tag{3.275}$$

which is an absolute series. From the Cauchy ratio test, the Euler series converges for

$$|z|^2 \lim_{n \to \infty} \frac{2^{2n+2}\pi^{2n+3}}{2^{2n}/\pi^{2n+1}} \frac{S(n+1)}{S(n)} = \frac{4}{\pi^2} |z|^2 \lim_{n \to \infty} \frac{S(n+1)}{S(n)} < 1 \tag{3.276}$$

Using the same analysis of $S(n)$ that was used for $\zeta(n)$, we obtain

$$\lim_{n \to \infty} S(n) = 1 \tag{3.277}$$

In addition, since $S(n)$ is an alternating series, it is straightforward to see that $S(n)$ is less than its first term. That is,

$$S(n) \leqslant 1 \tag{3.278}$$

for any n. Thus, from equation 3.277, we see that the Euler series converges for $|z| < \pi/2$, a fact we had deduced from the singularity structure of $\sec z$.

Using equation 3.168, we can obtain an estimate of the value of $S(n)$ in terms of the upper and lower bounds of the remainder series. This estimate is

$$\begin{aligned} S(n) &= S_1(n) + S_2(n) \\ &\simeq \sum_{k=1}^{N} \frac{(-1)^k}{(2k+1)^{(2n+1)}} + \frac{1}{(2N+1)^{(2n+1)}} - \frac{1}{2} \frac{1}{(2N+3)^{(2n+1)}} \end{aligned} \tag{3.279}$$

For N large enough, the estimate of $S_2(n)$ will be a small correction to the finite sum. From this, accurate values of E_{2n} can be determined.

PROBLEMS

1. Using the fact that $\sum_{n=1}^{\infty} 1/n^2$ converges, show that the series

$$S(z) = \sum_{n=1}^{\infty} \frac{\cos(nz)}{n^2}$$

converges uniformly for all real values of z.

2. Use the definition of uniform convergence to prove that

$$S(x) = \sum_{n=1}^{\infty} \frac{1}{(x+n)(x+n+1)}$$

converges uniformly for $x \in [0, \infty]$.

3. Use the Cauchy root test to show that

$$S(x) = \sum_{n=0}^{\infty} e^{-nx}$$

converges for all $x > 0$.

4. Determine the range of convergence of

$$S(x) = \sum_{n=1}^{\infty} ne^{-nx}$$

(a) Using the Cauchy ratio test
(b) Using the Cauchy integral test

5. Use the Cauchy ratio test to determine the range of convergence of

$$S(x) = \sum_{n=1}^{\infty} \frac{nx^n}{\sqrt{n!}}$$

6. Show that the series

$$S = \sum_{n=0}^{\infty} (-1)^n \frac{n!}{(2n)!}$$

converges.

7. Use the Cauchy ratio test to determine the range of x (if any) for which the series

$$S(x) = \sum_{n=0}^{\infty} (-1)^n \frac{n!}{(2n)!} x^{2n}$$

converges absolutely.

8. Does the series

$$S = \sum_{n=1}^{\infty} \frac{\log n}{n}$$

converge?

Sum the series of Problems 9 and 10 by performing algebraic manipulations to put each series in the form of one of the series of equations 3.1 through 3.7.

9.

$$S = \sum_{n=1}^{\infty} \frac{1}{(n+1)(n+2)(n+3)}$$

10.

$$S(x) = \sum_{n=0}^{\infty} \frac{x^n}{(n+1)(n+2)(n+3)(n+4)}$$

Sum the series of Problems 11 through 19 by performing calculus manipulation to put each series in the form of one of the series of equations 3.1 through 3.7.

11.

$$S(x) = \sum_{n=1}^{\infty} ne^{-nx}$$

12.

$$S = \sum_{n=0}^{\infty} (-1)^n \frac{(2n+2)}{(2n+3)!} \left(\frac{\pi}{2}\right)^{2n}$$

(*Hint*: Consider the series for $x \sin x$.)

13.

$$S = \sum_{n=0}^{N} \frac{N!(n+2)(n+1)}{n!(N-n)!3^n}$$

14.

$$S = \sum_{n=0}^{\infty} (-1)^n \frac{2^n}{(n+1)(2n+1)!}$$

(*Hint*: Consider the series for $\sin \sqrt{x}$.)

15.

$$S(x) = \sum_{n=0}^{\infty} \frac{x^n}{(n+1)(n+2)}$$

16.

$$S = \sum_{n=0}^{\infty} \frac{2^n}{n!} \frac{(n+1)}{(n+3)}$$

17.

$$S = \sum_{n=0}^{5} \frac{(n+1)}{n!(5-n)!(n+2)}$$

Do not sum this by explicitly evaluating each term individually and then adding the terms.

18. For the two series

$$S_a = \sum_{n=1}^{\infty} \frac{(2n+3)}{n} \left(\frac{4}{9}\right)^n \quad \text{and}$$

$$S_b = \sum_{n=1}^{\infty} (-1)^n \frac{(2n+1)}{n} \left(\frac{4}{9}\right)^n$$

(a) Show that each converges.
(b) Sum each series by considering one of the series given in equations 3.1 through 3.7 as a function of z^2.

19. Sum the series

$$S = \sum_{n=0}^{\infty} (-1)^n \frac{1}{4^n(2n+1)!} \frac{(n+2)}{(2n+3)}$$

20. Develop the [2, 2] Padé approximant to

$$S(x) = \sum_{n=1}^{\infty} \frac{x^n}{n^2}$$

21. A power series is given by

$$S(x) = \sum_{n=0}^{\infty} \sigma_n x^n$$

Find a general expression for the Padé approximants $S^{[1,1]}(x)$ and $S^{[2,2]}(x)$ in terms of the appropriate σ_n.

Estimate the value of the series of Problems 22 and 23 by
(1) The [1, 1] Padé approximant
(2) The [2, 2] Padé approximant
(3) Dividing each series into a finite sum up to $n = 3$, and a second sum from 4 to ∞, and estimating the second sum.

For all three methods, compare your result to the exact value of the series.

22.

$$S = \sum_{n=1}^{\infty} n^2 e^{-n/2}$$

(*Hint*: Consider a series in e^{-x}.)

23.

$$S = \sum_{n=0}^{\infty} \frac{(-1)^n}{4^n n!}$$

24. Without using the Euler–MacLaurin integration formula, estimate the value of

$$\int_0^{0.5} e^{-\sqrt{x}}\, dx$$

accurately to three decimal places.

25. Find the first nonzero correction to the room-temperature Debye specific heat (the Dulong–Petit law, $C_v = 3R$) from the Debye integral with $T \gg \Theta_D$.

26. **(a)** Using the MacLaurin series representations for $\sin x$ and $\cos x$, determine the first three nonzero terms for the MacLaurin series expansion of $\tan x$.

(b) From these first terms in the $\tan x$ series, determine the first three nonzero terms in the MacLaurin series for $\sec^2 x$ and for $\log(\cos x)$.

27. Obtain the values of the Bernoulli numbers B_4 and B_6 from the recurrence relation

$$(B+1)^n - B^n = 0$$

28. **(a)** Determine an expression for π^4 in terms of the Riemann zeta function $\zeta(4)$.

(b) Express $\zeta(4)$ as

$$\zeta(4) = \sum_{k=1}^{4} \frac{1}{k^4} + S_2$$

and estimate the value of S_2 to obtain an approximate value of π.

29. Starting with the identity

$$\cot x = \frac{\cos x}{\sin x} = i\left[\frac{e^{ix} + e^{-ix}}{e^{ix} - e^{-ix}}\right]$$

show that

$$\pi z \cot \pi z = 1 - 2 \sum_{n=1}^{\infty} z^{2n} \zeta(2n)$$

(*Hint*: Recall the series for generating the Bernoulli numbers.)

30. Determine the Bernoulli functions $\beta_2(s)$ and $\beta_3(s)$.

31. Prove that

$$\beta_n^{(k)}(s) \equiv \frac{d^k \beta_n}{ds^k} = \frac{n!}{(n-k)!}\beta_{n-k}(s) \qquad n \geqslant k$$

and

$$\beta_n^{(k)}(s) = 0 \qquad n < k$$

32. **(a)** Determine an expression for π^5 in terms of the appropriate series

$$S(n) = \sum_{k=0}^{\infty} \frac{(-1)^k}{(2k+1)^{2n+1}}$$

(b) Express $S(2)$ as

$$S(2) = \sum_{k=0}^{3} \frac{(-1)^k}{(2k+1)^{2n+1}} + S_2$$

and estimate S_2 to obtain an approximate value of π.

33. Evaluate the first three nonzero terms in the Euler–MacLaurin estimate of

$$\int_0^1 e^{-x^2}\, dx$$

CHAPTER 4

FOURIER SERIES AND INTEGRAL TRANSFORMS

4.1 Fourier Series

Periodicity

Many physical phenomena are described by periodic functions. A function is periodic in time if the kinematic and dynamic variables that describe a physical system have the same values at the end of a fixed time interval (the period) that they had at the beginning of that interval.

An example of such a system is a harmonic oscillator such as a pendulum. The values of the position, velocity, and acceleration of the pendulum at an instant $t + T$ are the same as the values of these quantities at time t. The time interval T is the period.

Spatial periodicity occurs if the kinematic and dynamic variables repeat themselves after a fixed spatial interval. Wave motion exhibits spatial as well as temporal periodicity.

In general, if at point x, at some instant t, $Q(x, t)$ is the value of a kinematic or dynamic variable (such as the velocity of a point in the medium supporting a wave), spatial periodicity and periodicity in time are characterized by

$$\text{(a)}\quad Q(x+\lambda, t) = Q(x, t) \qquad \text{(b)}\quad Q(x, t+T) = Q(x, t) \tag{4.1}$$

In Chapters 2 and 3 we studied several properties of series representations of functions. These studies focused primarily on power series. A periodic function could be described as a power series. For example, we could write

$$\text{(a)}\quad Q(x,t) = \sum_{n=0}^{\infty} a_n(x)t^n \quad \text{with} \quad \text{(b)}\quad a_n(x) = \sum_{m=0}^{\infty} b_{nm}x^m \tag{4.2}$$

However, if $Q(x, t)$ is periodic in x and/or t, the information about the periodicity is completely obscured by the power series representation. As an example,

$$\sin(\omega t) = \sum_{n=0}^{\infty} (-1)^n \frac{(\omega t)^{2n+1}}{(2n+1)!} \tag{3.2}$$

is a function that is periodic in time with a period $2\pi/\omega$. The periodicity of the function is not at all evident from this representation.

For simplicity, we will begin by considering a periodic function of one variable, t. That is, we will consider the function $Q(t)$ periodic with period

$$T = \frac{2\pi}{\omega} \tag{4.3}$$

Although we refer to T as if it were a temporal period, it represents the interval over which a function of any variable repeats itself. That is, T can represent a spatial period (wavelength), for example.

Fourier sine-cosine and exponential series

To make the periodicity of a function more transparent, let $Q(t)$ be a periodic function with period T given in equation 4.3. We can write $Q(t)$ as a sum of functions

$$Q(t) = \sum_{l=0}^{\infty} q_l(t) \tag{4.4}$$

and require that each $q_l(t)$ be periodic with a period T. Since $\sin(\omega t)$ and $\cos(\omega t)$ have the desired periodicity, if we express $q_l(t)$ as

$$q_l(t) = a_l[\cos(\omega t)]^l + b_l[\sin(\omega t)]^l \tag{4.5a}$$

the periodicity of $Q(t)$ is expressed in a transparent way; that is,

$$Q(t) = \sum_{l=0}^{\infty} \{a_l[\cos(\omega t)]^l + b_l[\sin(\omega t)]^l\} \tag{4.5b}$$

If we express $q_l(t)$ as

$$q_l(t) = A_l \cos(l\omega t) + B_l \sin(l\omega t) \tag{4.6a}$$

which also has the required periodicity, then

$$Q(t) = \sum_{l=0}^{\infty} [A_l \cos(l\omega t) + B_l \sin(l\omega t)] = A_0 + \sum_{l=1}^{\omega} [A_l \cos(l\omega t) + B_l \sin(l\omega t)] \tag{4.6b}$$

Using deMoivre's theorem, we can write

$$[\sin(\omega t)]^l = \sum_{m=0}^{\infty} [\alpha_{lm}^s \sin(m\omega t) + \beta_{lm}^s \cos(m\omega t)] \tag{4.7a}$$

and

$$[\cos(\omega t)]^l = \sum_{m=0}^{\infty} [\alpha_{lm}^c \sin(m\omega t) + \beta_{lm}^c \cos(m\omega t)] \tag{4.7b}$$

Substituting equations 4.7 into equation 4.5b and renaming the coefficients, we see that the series of equation 4.5b can be written in the form expressed in equation 4.6b. That is, they are not different series. Thus a general form for a series representation of a periodic function is

$$Q(t) = A_0 + \sum_{l=1}^{\infty} [A_l \cos(l\omega t) + B_l \sin(l\omega t)] \tag{4.6b}$$

This is called the *Fourier sine–cosine series*.

A second, equivalent form of the Fourier series can be obtained by writing the sine and cosine in their exponential forms

$$\text{(a)}\quad \cos(l\omega t) = \frac{e^{il\omega t} + e^{-il\omega t}}{2} \qquad \text{(b)}\quad \sin(l\omega t) = \frac{e^{il\omega t} - e^{-il\omega t}}{2i} \tag{2.19}$$

Then equation 4.6b becomes

$$Q(t) = C_0 + \sum_{l=1}^{\infty} \left[C_l e^{il\omega t} + C_{-l} e^{-il\omega t}\right] = \sum_{l=-\infty}^{\infty} C_l e^{il\omega t} \tag{4.8}$$

Equation 4.8 is the Fourier exponential series for $Q(t)$.

The coefficients of the exponential and the sine–cosine series are related by

$$\begin{aligned} &\text{(a)}\quad C_0 = A_0 \qquad \text{(b)}\quad C_l = \tfrac{1}{2}(A_l - iB_l) \qquad l > 0 \\ &\text{(c)}\quad C_{-l} = \tfrac{1}{2}(A_l + iB_l) \qquad l > 0 \end{aligned} \tag{4.9}$$

$$\begin{aligned} \Rightarrow\quad &\text{(a)}\quad A_0 = C_0 \qquad \text{(b)}\quad A_l = C_l + C_{-l} \qquad l > 0 \\ &\text{(c)}\quad B_l = i(C_l - C_{-l}) \qquad l > 0 \end{aligned} \tag{4.10}$$

If $Q(t)$ is real, the Fourier sine–cosine series of equation 4.6b must be real. In that case A_l and B_l will be real, and equations 4.9b and 4.9c yield $C_l = C^*_{-l}$. If $Q(t)$ is purely imaginary, A_l and B_l will be imaginary leading to $C_l = -C^*_{-l}$.

The Fourier series will be defined if the value of each coefficient can be determined. To obtain these coefficients, consider

$$I = \int_{t_0}^{t_0+T} Q(t) e^{-ik\omega t}\, dt \tag{4.11}$$

where t_0 is an arbitrary initial value of t, and T is the period given in equation 4.3. Using the Fourier exponential series expression for $Q(t)$ of equation 4.8, this integral becomes

$$\int_{t_0}^{t_0+T} Q(t) e^{-ik\omega t}\, dt = \sum_{l=-\infty}^{\infty} C_l \int_{t_0}^{t_0+T} e^{i(l-k)\omega t}\, dt \tag{4.12}$$

The integral under the sum is

$$\text{(a)}\quad \int_{t_0}^{t_0+T} e^{i(l-k)\omega t}\, dt = \begin{cases} 0 & k \neq l \\ T & k = l \end{cases} \quad \Rightarrow \quad \text{(b)}\quad \int_{t_0}^{t_0+T} e^{i(l-k)\omega t}\, dt = T\delta_{kl} \tag{4.13}$$

where we have used the fact that

$$e^{i(l-k)\omega T} = e^{i(l-k)2\pi} = 1 \tag{4.14}$$

for all k and l.

Equation 4.13b is an expression of the fact that the exponential functions $e^{il\omega t}$ form a set of functions that are mutually orthogonal to one another just as the basis vectors discussed in Chapter 1 form a set of mutually orthogonal vectors. Integration over one period for the exponential functions is equivalent to the dot product for vectors. Using this orthogonality, equation 4.11 becomes

$$C_l = \frac{1}{T} \int_{t_0}^{t_0+T} Q(t) e^{-il\omega t}\, dt \tag{4.15}$$

Thus we can obtain the coefficients of the Fourier exponential series by evaluating integrals of known functions. From these, we can determine the coefficients of the

Fourier sine–cosine series. For example,

$$A_0 = \frac{1}{T}\int_{t_0}^{t_0+T} Q(t)\,dt \tag{4.16a}$$

which is the time average of $Q(t)$ averaged over one period. For $l > 0$

$$A_l = C_l + C_{-l} = \frac{1}{T}\int_{t_0}^{t_0+T} Q(t)(e^{-il\omega t} + e^{il\omega t})\,dt = \frac{2}{T}\int_{t_0}^{t_0+T} Q(t)\cos(l\omega t)\,dt \tag{4.16b}$$

$$B_l = i(C_l - C_{-l}) = \frac{i}{T}\int_{t_0}^{t_0+T} Q(t)(e^{-il\omega t} - e^{il\omega t})\,dt = \frac{2}{T}\int_{t_0}^{t_0+T} Q(t)\sin(l\omega t)\,dt \tag{4.16c}$$

Consider the Fourier exponential series

$$Q(t) = \sum_{l=-\infty}^{\infty} C_l e^{il\omega t} \tag{4.8}$$

with

$$C_l = \frac{1}{T}\int_{t_0}^{t_0+T} Q(t')e^{-il\omega t'}\,dt' \tag{4.15}$$

Substituting equation 4.14 into equation 4.8, we have

$$Q(t) = \sum_{l=-\infty}^{\infty} e^{il\omega t}\frac{1}{T}\int_{t_0}^{t_0+T} Q(t')e^{-il\omega t'}\,dt' = \int_{t_0}^{t_0+T} Q(t')\left[\frac{1}{T}\sum_{l=-\infty}^{\infty} e^{il\omega(t-t')}\right]dt' \tag{4.17}$$

Since the Dirac δ-function is defined by

$$Q(t) = \int_{t_0}^{t_0+T} Q(t')\delta(t-t')\,dt' \qquad t_0 \leqslant t \leqslant t_0 + T \tag{4.18}$$

for an arbitrary function $Q(t)$, we see that

$$\delta(t-t') = \frac{1}{T}\sum_{l=-\infty}^{\infty} e^{il\omega(t-t')} \tag{4.19}$$

This is a series representation of the Dirac δ-function. In chapter 6 it will be demonstrated that a representation of the Dirac δ-function can be obtained in terms of sets of other mutually orthogonal functions. Equation 4.17 is known as the closure property in terms of the infinite set of mutually orthogonal exponential functions $\{e^{il\omega t}\}$.

Fourier series for even and odd functions

Let $Q(t) = Q(-t)$. That is $Q(t)$ is an even function. In terms of the Fourier exponential series, this equality becomes

$$\sum_{l=-\infty}^{\infty} C_l e^{il\omega t} = \sum_{l=-\infty}^{\infty} C_l e^{-il\omega t} \tag{4.20a}$$

In the sum on the right-hand side of equation 4.20a, we set $l' = -l$ and drop the prime. Then equation 4.20a becomes

$$\sum_{l=-\infty}^{\infty} C_l e^{il\omega t} = \sum_{l=-\infty}^{\infty} C_{-l} e^{il\omega t} \tag{4.20b}$$

Since the exponential functions are orthogonal as expressed by equation 4.13b, we can equate the coefficients of corresponding exponential terms to obtain

$$C_l = C_{-l} \tag{4.21}$$

for $Q(t)$ even.

We write equation 4.8 as

$$Q(t) = \sum_{l=-\infty}^{-1} C_l e^{il\omega t} + C_0 + \sum_{l=1}^{\infty} C_l e^{il\omega t} \tag{4.22}$$

and replace l by $-l$ in the first sum. With equation 4.21, we obtain

$$Q(t) = C_0 + \sum_{l=1}^{\infty} \left(C_l e^{il\omega t} + C_{-l} e^{-il\omega t}\right) = C_0 + \sum_{l=1}^{\infty} C_l \left(e^{il\omega t} + e^{-il\omega t}\right)$$

$$= C_0 + 2\sum_{l=1}^{\infty} C_l \cos(l\omega t) \tag{4.23}$$

If $C_l = C_{-l}$, then from equation 4.10b, $2C_l = A_l$ and equation 4.23 becomes

$$Q(t) = A_0 + \sum_{l=1}^{\infty} A_l \cos(l\omega t) \tag{4.24}$$

Note that equation 4.24 is the same as the Fourier sine–cosine series of equation 4.6b with the sine coefficients zero. That is, if $Q(t)$ is even, we can express it as a Fourier cosine series, with no sine terms in the series. This is to be expected since the cosine is an even function, the sine an odd function.

Similarly, if $Q(t)$ is odd, $Q(t) = -Q(-t)$ and equation 4.8 becomes

$$\sum_{l=-\infty}^{\infty} C_l e^{il\omega t} = -\sum_{l=-\infty}^{\infty} C_l e^{-il\omega t} \tag{4.25a}$$

Again, we replace l by $-l$ in the right-hand sum, and equation 4.25a becomes

$$\sum_{l=-\infty}^{\infty} C_l e^{il\omega t} = -\sum_{l=-\infty}^{\infty} C_{-l} e^{il\omega t} \tag{4.25b}$$

The orthogonality of the exponential function yields

$$C_l = -C_{-l} \tag{4.26}$$

if $Q(t)$ is odd. Thus $C_0 = -C_0 = 0$.

Again, we divide the exponential series into two sums and replace l by $-l$ in the sum from $-\infty$ to -1, to obtain

$$Q(t) = C_0 + \sum_{l=1}^{\infty} \left(C_l e^{il\omega t} + C_{-l} e^{il\omega t}\right) = \sum_{l=1}^{\infty} C_l \left(e^{il\omega t} - e^{-il\omega t}\right)$$

$$= \sum_{l=1}^{\infty} 2iC_l \sin(l\omega t) \tag{4.27}$$

With $C_l = -C_{-l}$ for an odd function, equation 4.10c yields $2iC_l = B_l$ and equation 4.27

becomes

$$Q(t) = \sum_{l=1}^{\infty} B_l \sin(l\omega t) \tag{4.28}$$

which is the Fourier sine series. As expected, an odd function has a Fourier series containing terms involving only the odd parity sine functions.

In summary, we see that if a function exhibits a specific parity, we can express it as a Fourier cosine series (if it is even) or a Fourier sine series (if it is odd).

Parseval's theorem

Let $Q(t)$ and $R(t)$ be two periodic functions with the same periodicity, T. The Fourier exponential series of each is

$$\text{(a)} \quad Q(t) = \sum_{l=-\infty}^{\infty} C_l e^{il\omega t} \qquad \text{(b)} \quad R(t) = \sum_{m=-\infty}^{\infty} \Gamma_m e^{im\omega t} \tag{4.29}$$

Consider the integral

$$\langle QR \rangle \equiv \frac{1}{T}\int_{t_0}^{t_0+T} Q(t)R(t)\,dt \tag{4.30}$$

which is the product of $Q(t)$ and $R(t)$ averaged over one period. The notation $\langle F \rangle$ is often used to indicate an average value of a quantum mechanical quantity F.

Substituting equations 4.29 into equation 4.30, we find

$$\langle QR \rangle = \frac{1}{T}\sum_{l,m=-\infty}^{\infty} C_l \Gamma_m \int_{t_0}^{t_0+T} e^{i(l+m)\omega t}\,dt \tag{4.31}$$

Using the orthogonality of the exponential function as described in equation 4.13, equation 4.31 becomes

$$\frac{1}{T}\int_{t_0}^{t_0+T} Q(t)R(t)\,dt = \sum_{l=-\infty}^{\infty} C_l \Gamma_{-l} = C_0\Gamma_0 + \sum_{l=1}^{\infty} (C_l\Gamma_{-l} + C_{-l}\Gamma_l) \tag{4.32}$$

This is called the *generalized Parseval theorem*.

Originally. Parseval's theorem was derived for $Q(t) = R(t)$. When $Q(t) = R(t)$, equation 4.32 becomes

$$\frac{1}{T}\int_{t_0}^{t_0+T} [Q(t)]^2\,dt = C_0^2 + 2\sum_{l=1}^{\infty} C_l C_{-l} \tag{4.33}$$

Using equations 4.9, Parseval's theorem can be expressed in terms of the coefficients of the Fourier sine–cosine series as

$$\frac{1}{T}\int_{t_0}^{t_0+T} Q(t)R(t)\,dt = A_0\alpha_0 + \frac{1}{2}\sum_{l=1}^{\infty} (A_l\alpha_l + B_l\beta_l) \tag{4.34}$$

where α_l and β_l are the coefficients of the Fourier sine–cosine series of $R(t)$. They are related to Γ_l in the same way that A_l and B_l are related to C_l as specified in equations 4.9. When $R(t) = Q(t)$, Parseval's theorem, in terms of the sine–cosine series coefficients becomes

$$\frac{1}{T}\int_{t_0}^{t_0+T} [Q(t)]^2\,dt = A_0^2 + \frac{1}{2}\sum_{l=1}^{\infty} (A_l^2 + B_l^2) \tag{4.35}$$

Square wave

Example 4.1

As an example, consider the Fourier series for the square wave defined by

$$Q(t) = \begin{cases} +Y & 0 \leqslant t \leqslant \frac{T}{2} \\ -Y & \frac{T}{2} \leqslant t \leqslant T \end{cases} \tag{4.36a}$$

which is pictured in Figure 4.1.

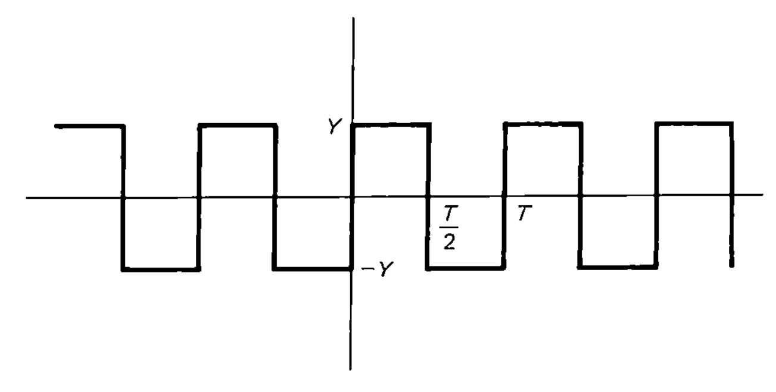

Figure 4.1 Periodic square wave.

The description of a periodic function as indicated, for example, in equation 4.36a, is nonrigorous and somewhat incomplete. As seen in Figure 4.1, the function actually satisfies

$$Q(t) = \begin{cases} +Y & nT \leqslant t \leqslant \left(n + \frac{1}{2}\right)T \\ -Y & \left(n + \frac{1}{2}\right)T \leqslant t \leqslant nT \end{cases} \qquad n = 0, \pm 1, \pm 2, \ldots \tag{4.36b}$$

However, it is customary to describe a periodic function over one period the way the square wave is described in equation 4.36a, even though what is actually meant is the description of equation 4.36b. We will continue to describe periodic functions in the somewhat incomplete form such as equation 4.36a, but the reader should be aware that such a description actually represents the function for all $t \in [-\infty, \infty]$.

The coefficients of the Fourier exponential series for the square wave are

$$C_l = \frac{1}{T}\left[\int_0^{T/2} (+Y)e^{-il\omega t}\, dt + \int_{T/2}^{T} (-Y)e^{-il\omega t}\, dt\right] \tag{4.37}$$

which yields

$$\text{(a)}\quad C_0 = 0 \quad \text{(b)}\quad C_l = 0 \quad l > 0, \quad \text{even} \quad \text{(c)}\quad C_l = \frac{2Y}{il\pi} \quad l > 0, \quad \text{odd} \tag{4.38}$$

Substituting these values for C_l into equation 4.8, we obtain the Fourier exponential series of the function of equation 4.36a:

$$Q(t) = \frac{2Y}{i\pi}\left[\left(\frac{e^{i\omega t}}{1} + \frac{e^{-i\omega t}}{-1}\right) + \left(\frac{e^{3i\omega t}}{3} + \frac{e^{-3i\omega t}}{-3}\right) + \cdots\right]$$

$$= \frac{2Y}{i\pi}2i\left[\frac{\sin(\omega t)}{1} + \frac{\sin(3\omega t)}{3} + \cdots\right] = \frac{4Y}{\pi}\sum_{l=0}^{\infty}\frac{\sin[(2l+1)\omega t]}{(2l+1)} \tag{4.39}$$

Therefore, the Fourier series for the function of equation 4.36a is a sine series. This is expected since the square wave depicted in Figure 4.1 is an odd function.

If we apply Parseval's theorem to the function of equation 4.36a using the coefficients of the exponential series in equations 4.38, equation 4.33 becomes

$$\frac{1}{T}\int_{t_0}^{t_0+T}[Q(t)]^2\,dt = \frac{8Y^2}{\pi^2}\sum_{\substack{l=1\\ l\text{ odd}}}^{\infty}\frac{1}{l^2} \tag{4.40}$$

To ensure that l is odd, we make the substitution $l = 2k + 1$, and equation 4.40 becomes

$$\frac{1}{T}\int_{t_0}^{t_0+T}[Q(t)]^2\,dt = \frac{8Y^2}{\pi^2}\sum_{k=0}^{\infty}\frac{1}{(2k+1)^2} \tag{4.41a}$$

But referring to Figure 4.1, we have

$$\frac{1}{T}\int_{t_0}^{t_0+T}[Q(t)]^2\,dt = \frac{1}{T}\left[\int_{t_0}^{t_0+T/2}(+Y)^2\,dt + \int_{t_0+T/2}^{t_0+T}(-Y)^2\,dt\right] = Y^2 \tag{4.41b}$$

Thus, from Parseval's theorem, we obtain an exact value for the series:

$$\sum_{k=0}^{\infty}\frac{1}{(2k+1)^2} = \frac{\pi^2}{8} \quad \square \tag{4.42}$$

Sawtooth or linear wave

Example 4.2

As another example, consider the waveform

$$Q(t) = \alpha t \qquad -\pi \leqslant t \leqslant \pi \tag{4.43}$$

which is depicted in Figure 4.2. In this example, $t_0 = -\pi$ and $T = 2\pi$. Thus $\omega = 1$.

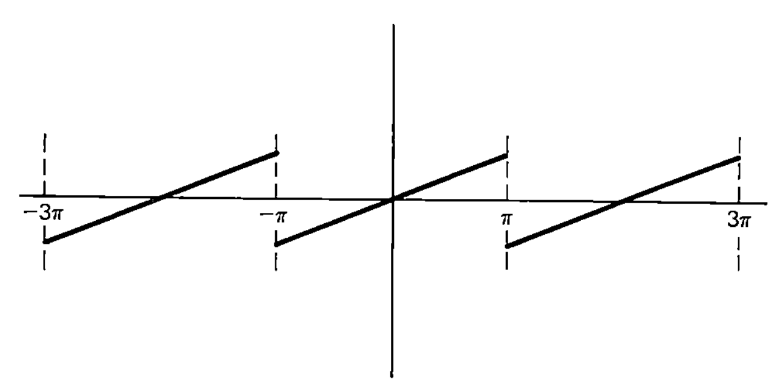

Figure 4.2 Periodic linear function.

It is evident that the function of equation 4.43 is odd, so its Fourier series must be a sine series.

$$\alpha t = \sum_{l=1}^{\infty} B_l \sin(l\omega t) \tag{4.44}$$

with

$$B_l = \frac{1}{\pi}\int_{-\pi}^{\pi}\alpha t \sin(l\omega t)\,dt = -\frac{2\alpha}{l}\cos(l\pi) = (-1)^{l+1}\frac{2\alpha}{l} \tag{4.45}$$

Therefore,

$$\text{(a)}\quad \alpha t = 2\alpha\sum_{l=1}^{\infty}(-1)^{l+1}\frac{\sin(lt)}{l} \quad\Rightarrow\quad \text{(b)}\quad t = 2\sum_{l=1}^{\infty}(-1)^{l+1}\frac{\sin(lt)}{l} \tag{4.46}$$

If we set $t = \pi/2$, we obtain

$$\frac{\pi}{2} = 2\sum_{l=1}^{\infty}(-1)^{l+1}\frac{\sin(l\pi/2)}{l} \tag{4.47}$$

If l is even, $\sin(l\pi/2) = 0$. Thus if we set $l = 2k + 1$ to ensure that l is odd, $\sin[(k + \frac{1}{2})\pi] = (-1)^k$ and equation 4.47 becomes

$$\pi = 4 \sum_{k=0}^{\infty} (-1)^k \frac{1}{(2k+1)} \tag{3.273}$$

This result was obtained in Chapter 3 using properties of the Euler numbers. □

Obtaining the Fourier series of one function from that of another

As we did when discussing power series, we can determine the Fourier series for a function from the Fourier series representation of another function.

Example 4.3

To illustrate, we determine the Fourier series representation of a parabolic wave defined by

$$Q(t) = t^2 \qquad -\pi \leqslant t \leqslant \pi \tag{4.48}$$

from the Fourier series for the linear wave. Starting with

$$t = 2 \sum_{l=1}^{\infty} (-1)^{l+1} \frac{\sin(lt)}{l} \tag{4.46b}$$

the periodic function of equation 4.48 is obtained by an integral.

$$t^2 = 2\int t\,dt = 4 \sum_{l=1}^{\infty} (-1)^{l+1} \int \frac{\sin(lt)}{l}\,dt + K = 4 \sum_{l=1}^{\infty} \frac{(-1)^l}{l^2} \cos(lt) + K \tag{4.49a}$$

Setting $t = 0$, we obtain a value for the constant K. The Fourier series for the parabolic wave is, then

$$t^2 = 4 \sum_{l=1}^{\infty} (-1)^l \left[\frac{\cos(lt) - 1}{l^2} \right] \quad \Box \tag{4.49b}$$

It is important when integrating one function to obtain the Fourier series representation of another function, that care be taken at points of discontinuity if they exist. If t_1 is a point of discontinuity, then

$$\int Q(t)\,dt \bigg|_{t=t_1} = \frac{1}{2} \lim_{\varepsilon \to 0} \left[\int Q(t)\,dt \bigg|_{t=t_1-\varepsilon} + \int Q(t)\,dt \bigg|_{t=t_1+\varepsilon} \right] \tag{4.50}$$

Approximation of a Fourier series

A technical application of Fourier series occurs in the construction of wave generators. The reader has probably learned in an introductory physics course that by rotating a coil of wire in a constant magnetic field, one can produce a sinusoidal wave of a specified frequency. From the Fourier sine–cosine series representations of functions, we see that by combining such sine and cosine waves with appropriate coefficients, waveforms of shapes other than sine and cosine waves can be generated.

For example, one can create a square wave such as is defined by equation 4.36a by adding an infinite number of sine waves of different frequencies together with coefficients given by the Fourier series of equation 4.39. Of course, in practice, it is not possible to include an infinite number of terms. Thus, to construct a device that produces square waves, one adds together a large, but finite number of sinusoidal waves, obtaining an approximation to a square wave. If the Fourier series converges, using a finite number of terms in the Fourier series will generate an approximation to the desired pattern. The number of terms that must be used will depend on the degree of precision required.

To construct a square-wave generator that will produce a pattern approximating that shown in Figure 4.1, one must build a device that will superimpose several sine waves with the coefficients of the Fourier series of equation 4.39. In Figure 4.3 we illustrate the superposition of 3, 10, 15 and 100 terms in the Fourier series for the square wave.

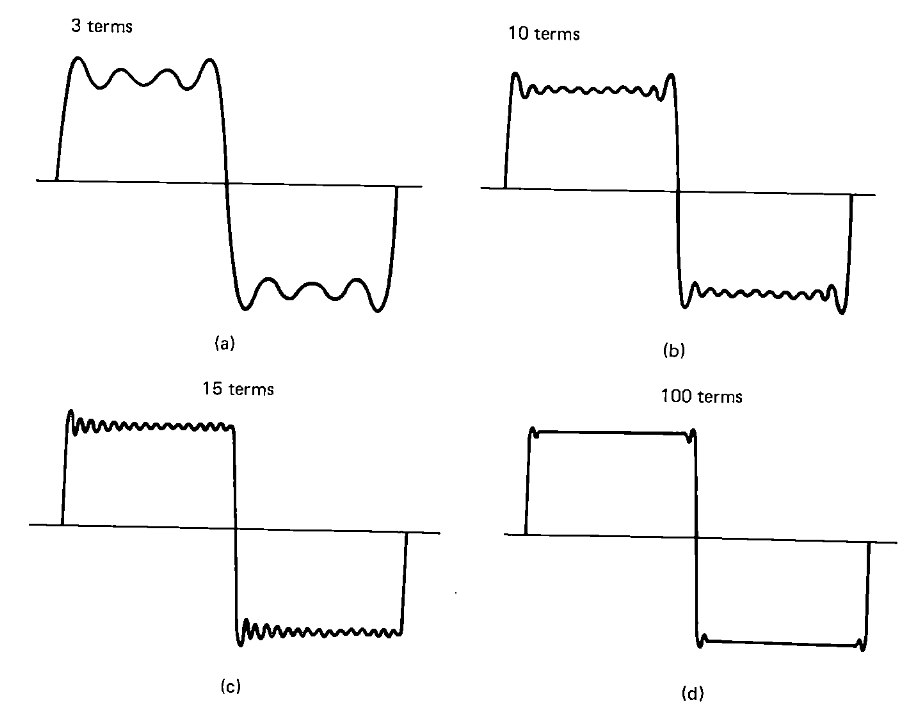

Figure 4.3 Partial sum approximations of the Fourier series for a square wave.

Note that by taking a large number of terms, the truncated Fourier series approximates the square wave quite well. However, as seen in Figure 4.3, there is a small spike on either side of the point of discontinuity. As the number of terms increases, this spike becomes narrower, but it will never disappear. Its height is approximately 18% of the amplitude Y. The existence of this spike for any partial sum approximation of the Fourier series is known as the *Gibbs phenomenon*. For a further explanation, the reader is referred to the text by J. Mathews and R. L. Walker, *Mathematical Methods of Physics*, 2nd ed. (Menlo Park, Calif.: W. A. Benjamin, Inc., 1970).

Convergence of the Fourier series and the Dirichlet theorem

In order that a Fourier series be well approximated by a partial Fourier sum, the Fourier series must converge. The convergence of a power series was tested by investigating the properties of the terms in the series.

The test of convergence of the Fourier series is applied to the function being represented by the series. If $Q(t)$ satisfies the following conditions:

1. $Q(t)$ has a finite number of point discontinuities over one period. This means that any discontinuity of $Q(t)$ exists at a point t_1 rather than over an extended region.

2. $Q(t)$ has a finite number of maxima and minima over one period.

3. $\int_{t_0}^{t_0+T} |Q(t)|\, dt$ is finite.

then the Fourier series converges to $Q(t)$ if t is not a point of discontinuity. The series converges to $\frac{1}{2} \lim_{\varepsilon \to 0} [Q(t - \varepsilon) + Q(t + \varepsilon)]$ if t is a point of discontinuity. This applies to both the sine–cosine series of equations 4.7 (with coefficients described in equations 4.10), and the Fourier exponential series of equation 4.8, (with coefficients found using equations 4.15 or 4.16). This is known as the Dirichlet theorem for the convergence of the Fourier series. For its somewhat involved proof, the reader is referred to the text by M. R. Spiegel, *Fourier Analysis with Applications to Boundary Value Problems* in Schaum's Outline Series (New York: McGraw-Hill Book Company, 1973).

The Dirichlet theorem is sufficient to guarantee convergence of the Fourier series. That is, if it is satisfied, the Fourier series converges to the specified function. However, if it is not satisfied, that does not guarantee that the Fourier series diverges. The series may still converge as specified, but convergence is not assured.

It is straightforward to see that approximating the Fourier sine–cosine series by the Nth partial sum

$$Q(t) \simeq A_0 + \sum_{l=1}^{N} (A_l \cos(l\omega t) + B_l \sin(l\omega t)) \tag{4.51a}$$

is the same as approximating the Fourier exponential series by

$$\sum_{l=-\infty}^{\infty} C_l e^{il\omega t} \simeq \sum_{l=-N}^{N} C_l e^{il\omega t} \tag{4.51b}$$

We estimate the error involved in such an approximation by defining the mean square error to be

$$\begin{aligned} E_N \equiv \frac{1}{T} \int_{t_0}^{t_0+T} \Bigg| A_0 &+ \sum_{l=1}^{\infty} [A_l \cos(l\omega t) + B_l \sin(l\omega t)] \\ &- A_0 - \sum_{l=1}^{N} [A_l \cos(l\omega t) + B_l \sin(l\omega t)] \Bigg|^2 dt \end{aligned} \tag{4.52a}$$

which is the same as

$$\begin{aligned} E_N &= \frac{1}{T} \int_{t_0}^{t_0+T} \left| \sum_{l=-\infty}^{\infty} C_l e^{il\omega t} - \sum_{l=-N}^{N} C_l e^{il\omega t} \right|^2 dt \\ &= \frac{1}{T} \int_{t_0}^{t_0+T} \left| \sum_{l=N+1}^{\infty} C_l e^{il\omega t} + \sum_{l=-\infty}^{-(N+1)} C_l e^{il\omega t} \right|^2 dt \\ &= \frac{1}{T} \int_{t_0}^{t_0+T} \left| \sum_{l=N+1}^{\infty} \left(C_l e^{il\omega t} + C_{-l} e^{-il\omega t}\right) \right|^2 dt \\ &= \frac{1}{T} \int_{t_0}^{t_0+T} \left[\sum_{l=N+1}^{\infty} \left(C_l e^{il\omega t} + C_{-l} e^{-il\omega t}\right) \right] \left[\sum_{n=N+1}^{\infty} \left(C_n^* e^{-in\omega t} - C_{-n}^* e^{-in\omega t}\right) \right] dt \end{aligned} \tag{4.52b}$$

Since both $l > 0$ and $n > 0$, the orthogonality of the exponential functions over one period yields

$$E_N = \sum_{l=N+1}^{\infty} \left[|C_l|^2 + |C_{-l}|^2 \right] \tag{4.53}$$

If the Fourier series converges, the mean square error $\to 0$ as $N \to \infty$. To see this, consider

$$C_l = \frac{1}{T} \int_{t_0}^{t_0+T} Q(t) e^{-il\omega t}\, dt \qquad l \geqslant N+1 \tag{4.15}$$

Let $Q(t)$ have a point discontinuity at a point t_1 that is not t_0 or $t_0 + T$. Then

$$C_l = \frac{1}{T} \lim_{\varepsilon \to 0} \left[\int_{t_0}^{t_1-\varepsilon} Q(t) e^{-il\omega t}\, dt + \int_{t_1+\varepsilon}^{t_0+T} Q(t) e^{-il\omega t}\, dt \right] \tag{4.54}$$

Integrating by parts, we obtain

$$\begin{aligned} C_l = \frac{1}{T} \lim_{\varepsilon \to 0} \Bigg[& \frac{Q(t) e^{-il\omega t}}{-il\omega} \Bigg|_{t_0}^{t_1-\varepsilon} + \frac{Q(t) e^{-il\omega t}}{-il\omega} \Bigg|_{t_1+\varepsilon}^{t_0+T} \\ & - \frac{1}{(-il\omega)} \int_{t_0}^{t_1-\varepsilon} Q'(t) e^{il\omega t}\, dt - \frac{1}{(-il\omega)} \int_{t_1+\varepsilon}^{t_0+T} Q'(t) e^{-il\omega t}\, dt \Bigg] \end{aligned} \tag{4.55a}$$

When $Q(t)$ has a discontinuity at t_1, $Q'(t)$ has an integrable singularity at t_1. Thus equation 4.54 becomes

$$C_l = \frac{e^{-il\omega t_1}}{(-il\omega T)} \lim_{\varepsilon \to 0} \left[Q(t_1 - \varepsilon) - Q(t_1 + \varepsilon) \right] - \frac{1}{(-il\omega T)} \int_{t_0}^{t_0+T} Q'(t) e^{-il\omega t}\, dt \tag{4.55b}$$

where we have used the fact that $Q(t_0)e^{-il\omega t_0} = Q(t_0 + T)e^{-il\omega(t_0+T)}$. A second integration by parts would yield an integrated term containing a factor $1/(-il\omega)^2 T$. For large enough l, this term will be small compared to the first term of equation 4.55b, which has a $1/l$ dependence. Thus, for large l, C_l can be approximated by

$$C_l \simeq \frac{e^{-il\omega t_1}}{(-2\pi i l)} \lim_{\varepsilon \to 0} \left[Q(t_1 - \varepsilon) - Q(t_1 + \varepsilon) \right] \tag{4.56}$$

Defining

$$M_0 \equiv \frac{\left| \lim_{\varepsilon \to 0} \left[Q(t_1 - \varepsilon) - Q(t_1 + \varepsilon) \right] \right|}{2\pi} \tag{4.57}$$

$$\Rightarrow \quad |C_l| = |C_{-l}| \simeq \frac{M_0}{l} \tag{4.58}$$

$$\Rightarrow \quad E_N \simeq 2M_0 \sum_{l=N+1}^{\infty} \frac{1}{l^2} \tag{4.59a}$$

which is convergent.

The series of equation 4.59a will be recognized as the infinite part of the Riemann zeta function $\zeta(2)$ (see equation 3.219). For sufficiently large N, the sum is dominated by the first term. Thus with $N + 1 \simeq N$,

$$E_N \simeq \frac{2M_0}{N^2} \tag{4.59b}$$

Therefore, the error is zero when $N \to \infty$.

If $Q(t)$ is continuous over a period but $Q'(t)$ has a discontinuity at t_1, the integrated term in equation 4.55b is zero and C_l becomes

$$C_l = -\frac{1}{(-il\omega T)} \lim_{\varepsilon \to 0} \left[\int_{t_0}^{t_1-\varepsilon} Q'(t) e^{-il\omega t}\, dt + \int_{t_1+\varepsilon}^{t_0+T} Q'(t) e^{-il\omega t}\, dt \right] \tag{4.60}$$

Integration by parts yields

$$C_l = \frac{1}{T} \lim_{\varepsilon \to 0} \left[\frac{Q'(t)e^{-il\omega t}}{(-il\omega)^2} \Bigg|_{t_0}^{t_1-\varepsilon} + \frac{Q'(t)e^{-il\omega t}}{(-il\omega)^2} \Bigg|_{t_1+\varepsilon}^{t_0+T} \right] - \frac{1}{(-il\omega)^2 T} \int_{t_0}^{t_0+T} Q''(t) e^{-il\omega t}\, dt$$

$$\simeq \frac{e^{-il\omega t_1}}{(-il\omega)^2 T} \lim_{\varepsilon \to 0} \left[Q'(t_1 - \varepsilon) - Q'(t_1 + \varepsilon) \right] \tag{4.61}$$

for large l. By an appropriate definition of M_1, equation 4.61 yields

$$|C_l| = |C_{-l}| \simeq \frac{M_1}{l^2} \tag{4.62}$$

$$\Rightarrow \quad E_N \simeq 2M_1 \sum_{l=N+1}^{\infty} \frac{1}{l^4} \simeq \frac{2M_1}{N^4} \tag{4.63}$$

The series of equation 4.63 is the infinite part of the Riemann zeta function $\zeta(4)$. It approaches zero faster at infinite N than does the infinite part of $\zeta(2)$.

By extending this analysis, we can conclude that if the first $(k - 1)$ derivatives of $Q(t)$ are continuous over one period and the kth derivative has a finite number of point discontinuities, the mean square error obtained by approximating the Fourier series by the Nth partial sum is given by

$$E_N \simeq 2M_k \sum_{l=N+1}^{\infty} \frac{1}{(l^2)^{k+1}} \simeq \frac{2M_k}{N^{2k+2}} \tag{4.64}$$

Fourier series in more than one variable

The Fourier series representation of a function that is periodic in two variables is a simple extension of the single variable Fourier series. We first determine the Fourier series in terms of one of the variables. For example, we can write

$$Q(x, t) = \sum_{l=-\infty}^{\infty} C_l(x) e^{il\omega t} \tag{4.65}$$

If $Q(x, t)$ is periodic in x as well, with periodicity λ, each $C_l(x)$ must reflect this periodicity. That is,

$$C_l(x + \lambda) = C_l(x) \tag{4.66}$$

Thus, as long as the Dirichlet theorem is satisfied for each $C_l(x)$, every C_l will have a Fourier series representation,

$$C_l(x) = \sum_{m=-\infty}^{\infty} F_{lm} e^{imkx} \tag{4.67}$$

The periodicity of $C_l(x)$ requires that

$$k = \frac{2\pi}{\lambda} \tag{4.68}$$

k is called the wave number if x is a spatial variable. Equation 4.65 then becomes

$$Q(x, t) = \sum_{l, m = -\infty}^{\infty} F_{lm} e^{i(mkx + l\omega t)} \tag{4.69}$$

where

$$F_{lm} = \frac{1}{\lambda} \int_{x_0}^{x_0+\lambda} C_l(x) e^{-imkx} \, dx = \frac{1}{\lambda T} \int_{x_0}^{x_0+\lambda} dx \int_{t_0}^{t_0+T} Q(x, t) e^{-i(mkx + l\omega t)} \, dt \tag{4.70}$$

When a function has periodicity in the three spatial variables x, y, and z, and in the time variable, the Fourier expansion is an obvious extension of equation 4.69. If we denote the wave numbers for the spatial variables as k_i, $i = 1, 2, 3$, and the corresponding wavelengths as λ_i, then

$$Q(\mathbf{r}, t) = \sum_{lmpq = -\infty}^{\infty} G_{lmpq} e^{i(mk_1x + pk_2y + qk_3z + l\omega t)} \tag{4.71}$$

$$\Rightarrow \quad \lambda_1 \lambda_2 \lambda_3 T G_{lmpq} = \int_{x_0}^{x_0+\lambda_1} \int_{y_0}^{y_0+\lambda_2} \int_{z_0}^{z_0+\lambda_3} \int_{t_0}^{t_0+T} Q(\mathbf{r}, t) e^{-i(l\omega t + mk_1x + pk_2y + qk_3z)} \, d^3r \, dt \tag{4.72}$$

4.2 Fourier Transforms

Beginning with

$$Q(t) = \sum_{l=-\infty}^{\infty} C_l e^{il\omega t} \tag{4.8}$$

we take t_0 to be $-T/2$. Thus, from equation 4.15, the Fourier coefficient is

$$C_l = \frac{1}{T} \int_{-T/2}^{T/2} Q(t) e^{-il\omega t} \, dt \tag{4.15}$$

Since the indices in equation 4.8 differ by 1, nothing is changed by multiplying each term in the series by $\Delta l = (l + 1) - l = 1$. In addition, we multiply each term by the number 1 in the form $\omega T / 2\pi$. That is, we write equation 4.8 as

$$Q(t) = \sum_{l=-\infty}^{\infty} \frac{T}{2\pi} C_l e^{il\omega t} \Delta(l\omega) \tag{4.73}$$

If T is infinite, $Q(t)$ is a nonperiodic function. But $T \to \infty$ requires that $\omega \to 0$ so that $\omega T = 2\pi$ is maintained. When $\omega \to 0$, $\Delta(l\omega) \to 0$. That is, $\Delta(l\omega)$ becomes a differential, and $l\omega$ behaves like a continuous variable. Therefore, we can then express C_l as a function of the product $l\omega$, and equation 4.73 becomes

$$\text{(a)} \quad Q(t) = \frac{1}{2\pi} \int_{-\infty}^{\infty} [TC(l\omega)] e^{i(l\omega)t} \, d(l\omega) \quad \Rightarrow \quad \text{(b)} \quad TC(l\omega) = \int_{-\infty}^{\infty} Q(t) e^{-i(l\omega)t} \, dt \tag{4.74}$$

It is now natural to define the product $l\omega$ as a new variable Ω, so equations 4.74 are written

$$\text{(a)} \quad Q(t) = \frac{1}{2\pi} \int_{-\infty}^{\infty} [TC(\Omega)] e^{i\Omega t} \, d\Omega \qquad \text{(b)} \quad TC(\Omega) = \int_{-\infty}^{\infty} Q(t) e^{-i\Omega t} \, dt \tag{4.75}$$

We note here that the factor of $1/2\pi$ multiplies the integral for $Q(t)$ but no such factor appears for the integral of $TC(\Omega)$. By redefining the function $TC(\Omega)$, one can specify how the factor of 2π is to multiply each integral. For example, if

$$P(\Omega) \equiv TC(\Omega) \tag{4.76}$$

equations 4.75 are written

$$\text{(a)} \quad Q(t) = \frac{1}{2\pi}\int_{-\infty}^{\infty} P(\Omega)e^{i\Omega t}\,d\Omega \qquad \text{(b)} \quad P(\Omega) = \int_{-\infty}^{\infty} Q(t)e^{-i\Omega t}\,dt \tag{4.77}$$

Or, one can make the factor of 2π appear symmetrically in both integral representations. Defining

$$P(\Omega) \equiv \frac{TC(\Omega)}{\sqrt{2\pi}} \tag{4.78}$$

$$\Rightarrow \quad \text{(a)} \quad Q(t) = \frac{1}{\sqrt{2\pi}}\int_{-\infty}^{\infty} P(\Omega)e^{i\Omega t}\,d\Omega \qquad \text{(b)} \quad P(\Omega) = \frac{1}{\sqrt{2\pi}}\int_{-\infty}^{\infty} Q(t)e^{-i\Omega t}\,dt \tag{4.79}$$

It is easy to define an integral pair with factors of 2π in any desired combination that is compatible with equations 4.75. Equations 4.77 and equations 4.79 are the two most commonly used forms of the pair of Fourier transforms. We will restrict our discussion to the pair of integrals for which the factor of $\sqrt{2\pi}$ appears symmetrically, namely equations 4.79.

Each function, $Q(t)$ and $P(\Omega)$, is called the Fourier transform or the Fourier inverse of the other. They are also sometimes denoted by

$$\text{(a)} \quad P(\Omega) = F.T.[Q(t)] \qquad \text{(b)} \quad Q(t) = F.T.[P(\Omega)] \tag{4.80}$$

Properties of Fourier transforms

1. Consider

$$|P(\Omega)| = \left|\int_{-\infty}^{\infty} Q(t)e^{-i\Omega t}\frac{dt}{\sqrt{2\pi}}\right| \leqslant \int_{-\infty}^{\infty} |Q(t)e^{-i\Omega t}|\frac{dt}{\sqrt{2\pi}} = \int_{-\infty}^{\infty} |Q(t)|\frac{dt}{\sqrt{2\pi}} \tag{4.81}$$

Therefore, $P(\Omega)$ is finite if $Q(t)$ is integrable. By identical analysis with the inverse transform, it is straightforward to deduce that $Q(t)$ is finite if $P(\Omega)$ is integrable.

2. Let $Q(t)$ be real, and define $P(\Omega) \equiv U(\Omega) + iV(\Omega)$. Then

$$\begin{aligned} U(\Omega) + iV(\Omega) &= \int_{-\infty}^{\infty} Q(t)e^{-i\Omega t}\frac{dt}{\sqrt{2\pi}} \\ &= \int_{-\infty}^{\infty} Q(t)\cos(\Omega t)\frac{dt}{\sqrt{2\pi}} - i\int_{-\infty}^{\infty} Q(t)\sin(\Omega t)\frac{dt}{\sqrt{2\pi}} \end{aligned} \tag{4.82}$$

Therefore,

$$\text{(a)} \quad U(\Omega) = \int_{-\infty}^{\infty} Q(t)\cos(\Omega t)\frac{dt}{\sqrt{2\pi}} \qquad \text{(b)} \quad V(\Omega) = -\int_{-\infty}^{\infty} Q(t)\sin(\Omega t)\frac{dt}{\sqrt{2\pi}} \tag{4.83}$$

$U(\Omega)$ is called the *Fourier cosine inverse* of $Q(t)$ and $-V(\Omega)$ is the *Fourier sine inverse* of $Q(t)$.

Since all Ω-dependence in equations 4.83 occurs in $\cos(\Omega t)$ and $\sin(\Omega t)$, it immediately follows that $U(\Omega) = \mathrm{Re}[P(\Omega)]$ is an even function, and $V(\Omega) = \mathrm{Im}[P(\Omega)]$ is odd. Therefore,

$$\begin{aligned} &\text{(a)} \quad U(\Omega) = U(-\Omega) \qquad \text{(b)} \quad V(\Omega) = -V(-\Omega) \\ &\Rightarrow \qquad \text{(c)} \quad P(-\Omega) = U(\Omega) - iV(\Omega) = P^*(\Omega) \end{aligned} \tag{4.84}$$

if $Q(t)$ is real. This is referred to as a *reflection principle.*

Not only does the reality of $Q(t)$ lead to equation 4.84c, but conversely, if equation 4.84c is satisfied, this guarantees that $Q(t)$ is real. That is, the reality of $Q(t)$ is a necessary and sufficient condition to guarantee that $P(\Omega)$ satisfies the reflection principle of equation 4.84c.

To demonstrate this, consider

$$Q(t) = \int_{-\infty}^{\infty} P(\Omega)e^{i\Omega t}\frac{d\Omega}{\sqrt{2\pi}} \tag{4.85a}$$

Taking the complex conjugate of equation 4.85a, we obtain

$$Q^*(t) = \int_{-\infty}^{\infty} P^*(\Omega)e^{-i\Omega t}\frac{d\Omega}{\sqrt{2\pi}} = \int_{-\infty}^{\infty} P(-\Omega)e^{-i\Omega t}\frac{d\Omega}{\sqrt{2\pi}} \tag{4.85b}$$

Replacing $\Omega \to -\Omega$ in the second integral, we obtain

$$Q^*(t) = \int_{-\infty}^{\infty} P(\Omega)e^{i\Omega t}\frac{d\Omega}{\sqrt{2\pi}} = Q(t) \tag{4.86}$$

Thus $Q(t)$ is real.

3. If $Q(t)$ is imaginary, it can be written as

$$Q(t) = iq(t) = \int_{-\infty}^{\infty} P(\Omega)e^{i\Omega t}\frac{d\Omega}{\sqrt{2\pi}} \tag{4.87}$$

$$\Rightarrow \quad q(t) = \int_{-\infty}^{\infty} -iP(\Omega)e^{i\Omega t}\frac{d\Omega}{\sqrt{2\pi}} = \int_{-\infty}^{\infty} p(\Omega)e^{i\Omega t}\frac{d\Omega}{\sqrt{2\pi}} \tag{4.88}$$

where $p(\Omega) \equiv -iP(\Omega)$. Since $q(t)$ is real, $p(-\Omega) = p^*(\Omega)$, or

$$-P(-\Omega) = P^*(\Omega) \tag{4.89}$$

Following the analysis of property 2, it is straightforward to show that $-P(-\Omega) = P^*(\Omega)$ ensures that $Q(t)$ is imaginary.

4. If $Q(t) = 0$, $P(\Omega) = 0$ and $P(\Omega) = 0 \Rightarrow Q(t) = 0$. From this it follows immediately that if

$$\text{(a)} \quad F.T.[Q(t)] = F.T.[R(t)] \quad \Rightarrow \quad \text{(b)} \quad Q(t) = R(t) \tag{4.90}$$

That is,

$$\text{(a)} \quad F.T.[Q(t) - R(t)] = 0 \quad \Rightarrow \quad \text{(b)} \quad Q(t) - R(t) = 0 \tag{4.91}$$

5. Consider

$$Q(t) = \int_{-\infty}^{\infty} P(\Omega) e^{i\Omega t} \frac{d\Omega}{\sqrt{2\pi}} = \int_{-\infty}^{\infty} P(\Omega)\cos(\Omega t) \frac{d\Omega}{\sqrt{2\pi}} + i\int_{-\infty}^{\infty} P(\Omega)\sin(\Omega t) \frac{d\Omega}{\sqrt{2\pi}} \tag{4.92a}$$

$$Q(-t) = \int_{-\infty}^{\infty} P(\Omega) e^{-i\Omega t} \frac{d\Omega}{\sqrt{2\pi}} = \int_{-\infty}^{\infty} P(\Omega)\cos(\Omega t) \frac{d\Omega}{\sqrt{2\pi}} - i\int_{-\infty}^{\infty} P(\Omega)\sin(\Omega t) \frac{d\Omega}{\sqrt{2\pi}} \tag{4.92b}$$

If $Q(t)$ is an even function, then $Q(t) = Q(-t)$. Therefore, separately equating the real and imaginary parts of equations 4.92, we obtain

$$\text{(a)} \quad Q(t) = \int_{-\infty}^{\infty} P(\Omega)\cos(\Omega t) \frac{d\Omega}{\sqrt{2\pi}} \qquad \text{(b)} \quad \int_{-\infty}^{\infty} P(\Omega)\sin(\Omega t) \frac{d\Omega}{\sqrt{2\pi}} = 0 \tag{4.93}$$

In order that an integral over symmetric limits be zero, the integrand must be odd. That is, if

$$\int_{-L}^{L} F(t)\, dt = G(t)\Big|_{-L}^{L} = G(L) - G(-L) = 0 \tag{4.94}$$

$G(t)$ must be even, which implies that $F(t)$ must be odd. Therefore, if $Q(t)$ is even, the integrand of equation 4.93b must be an odd function of Ω. Since $\sin(\Omega t)$ is odd, $P(\Omega)$ must be an even function of Ω. Thus $Q(t)$ being an even function implies that $P(\Omega)$ is even.

If $Q(t)$ is odd, then $Q(t) = -Q(-t)$. Then

$$\text{(a)} \quad Q(t) = \int_{-\infty}^{\infty} P(\Omega)\sin(\Omega t) \frac{d\Omega}{\sqrt{2\pi}} \qquad \text{(b)} \quad \int_{-\infty}^{\infty} P(\Omega)\cos(\Omega t) \frac{d\Omega}{\sqrt{2\pi}} = 0 \tag{4.95}$$

Because $\cos(\Omega t)$ is an even function of Ω, $P(\Omega)$ must be odd in order to satisfy equation 4.95b.

In summary, then, if $Q(t)$ and $P(\Omega)$ are Fourier inverses of one another, and if one of these functions has a specified parity, the other function has the same parity.

6. Let α and β be constants, and consider

$$\text{F.T.}[\alpha Q_1(t) + \beta Q_2(t)] = \int_{-\infty}^{\infty} [\alpha Q_1(t) + \beta Q_2(t)] e^{-i\Omega t} \frac{dt}{\sqrt{2\pi}}$$
$$= \alpha\int_{-\infty}^{\infty} Q_1(t) e^{-i\Omega t} \frac{dt}{\sqrt{2\pi}} + \beta\int_{-\infty}^{\infty} Q_2(t) e^{-i\Omega t} \frac{dt}{\sqrt{2\pi}} = \alpha\text{F.T.}[Q_1(t)] + \beta\text{F.T.}[Q_2(t)] \tag{4.96}$$

7. With α constant, consider

$$\text{F.T.}[Q(\alpha t)] = \int_{-\infty}^{\infty} Q(\alpha t) e^{-i\Omega t} \frac{dt}{\sqrt{2\pi}} \tag{4.97}$$

Taking $\alpha > 0$, let $t' = \alpha t$. Since $\alpha > 0$, $t \in [-\infty, \infty] \Rightarrow t' \in [-\infty, \infty]$. Therefore,

$$\text{F.T.}[Q(\alpha t)] = \frac{1}{\alpha}\int_{-\infty}^{\infty} Q(t') e^{-i\Omega t'/\alpha} \frac{dt'}{\sqrt{2\pi}} = \frac{1}{\alpha} P\left(\frac{\Omega}{\alpha}\right) = \frac{1}{|\alpha|} P\left(\frac{\Omega}{\alpha}\right) \tag{4.98a}$$

If $\alpha < 0$, then $t' = \alpha t = -|\alpha|t$, and

$$\text{F.T.}[Q(\alpha t)] = \frac{1}{|\alpha|}\int_{-\infty}^{\infty} Q(t')e^{-i(\Omega/\alpha)t'}\frac{dt'}{\sqrt{2\pi}} = \frac{1}{|\alpha|}P\left(\frac{\Omega}{\alpha}\right) \tag{4.98b}$$

Therefore, for any $\alpha \neq 0$,

$$\text{F.T.}[Q(\alpha t)] = \frac{1}{|\alpha|}P\left(\frac{\Omega}{\alpha}\right) \tag{4.99}$$

8. Consider

$$\text{F.T.}[Q(t - t_0)] = \int_{-\infty}^{\infty} Q(t - t_0)e^{-i\Omega t}\frac{dt}{\sqrt{2\pi}} \tag{4.100}$$

Let $t' = t - t_0$. Then

$$\text{F.T.}[Q(t - t_0)] = e^{-i\Omega t_0}\int_{-\infty}^{\infty} Q(t')e^{-i\Omega t'}\frac{dt'}{\sqrt{2\pi}} = e^{-i\Omega t_0}\text{F.T.}[Q(t)] \tag{4.101}$$

That is, a displacement in the t-variable by an amount t_0 results in a multiplication of the Fourier transform of $Q(t)$ by $e^{-i\Omega t_0}$.

9. Consider

$$\text{F.T.}\left[\frac{dQ}{dt}\right] = \int_{-\infty}^{\infty} \frac{dQ}{dt}e^{-i\Omega t}\frac{dt}{\sqrt{2\pi}} \tag{4.102}$$

Integration by parts yields

$$\text{F.T.}\left[\frac{dQ}{dt}\right] = Q(t)e^{-i\Omega t}\Big|_{-\infty}^{\infty} + i\Omega\int_{-\infty}^{\infty} Q(t)e^{-i\Omega t}\frac{dt}{\sqrt{2\pi}} \tag{4.103}$$

As discussed in Chapter 2, in order for integrals of this type to exist, $Q(\pm\infty)$ must be zero. When t is large, the factor $e^{-i\Omega t}$ oscillates rapidly for small variations of t but is still a number of magnitude 1. Therefore, at $\pm\infty$, the integrated term of equation 4.103 is zero. The remaining integral is the Fourier transform of $Q(t)$. Thus

$$\text{F.T.}\left[\frac{dQ}{dt}\right] = i\Omega\text{F.T.}[Q(t)] \tag{4.104a}$$

Using equation 4.104a, the second derivative can be written

$$\text{F.T.}\left[\frac{d^2Q}{dt^2}\right] = i\Omega\text{F.T.}\left[\frac{dQ}{dt}\right] = (i\Omega)^2\text{F.T.}[Q(t)] \tag{4.104b}$$

It is easy to see that for any n,

$$\text{F.T.}\left[\frac{d^nQ}{dt^n}\right] = (i\Omega)^n\text{F.T.}[Q(t)] \tag{4.104c}$$

10. Consider the Fourier transform of the Dirac δ-symbol.

$$\text{F.T.}[\delta(t - t_0)] = \int_{-\infty}^{\infty} \delta(t - t_0)e^{-i\Omega t}\frac{dt}{\sqrt{2\pi}} = \frac{1}{\sqrt{2\pi}}e^{-i\Omega t_0} \tag{4.105}$$

Therefore, the δ-symbol is the Fourier inverse of $(1/\sqrt{2\pi})e^{-i\Omega t_0}$. That is,

$$\delta(t - t_0) = \frac{1}{2\pi}\int_{-\infty}^{\infty} e^{i\Omega(t-t_0)}\,d\Omega \tag{4.106a}$$

By replacing $\Omega \to -\Omega$, the integral representation of the δ-symbol can also be written

$$\delta(t - t_0) = \frac{1}{2\pi}\int_{-\infty}^{\infty} e^{-i\Omega(t-t_0)}\, d\Omega = \frac{1}{2\pi}\int_{-\infty}^{\infty} e^{i\Omega[-(t-t_0)]}\, d\Omega = \delta[-(t - t_0)] \qquad (4.106b)$$

Thus we see that the δ-function is an even function.

This result can be obtained in another way. For the Fourier inverse pairs

$$\text{(a)}\quad Q(t) = \int_{-\infty}^{\infty} P(\Omega)e^{i\Omega t}\frac{d\Omega}{\sqrt{2\pi}} \qquad \text{(b)}\quad P(\Omega) = \int_{-\infty}^{\infty} Q(t)e^{-i\Omega t}\frac{dt}{\sqrt{2\pi}} \qquad (4.79)$$

we substitute equation 4.79b into equation 4.79a. This yields

$$Q(t) = \int_{-\infty}^{\infty} e^{i\Omega t}\frac{d\Omega}{\sqrt{2\pi}}\int_{-\infty}^{\infty} Q(t')e^{-i\Omega t'}\frac{dt'}{\sqrt{2\pi}} = \int_{-\infty}^{\infty} Q(t')\, dt'\frac{1}{2\pi}\int_{-\infty}^{\infty} e^{i\Omega(t-t')}\, d\Omega \qquad (4.107)$$

Since the Dirac δ-symbol is defined such that

$$Q(t) = \int_{-\infty}^{\infty} Q(t')\delta(t - t')\, dt' \qquad (4.108)$$

for any $Q(t)$, we see by comparison of equations 4.107 and 4.108 that

$$\delta(t - t') = \frac{1}{2\pi}\int_{-\infty}^{\infty} e^{i\Omega(t-t')}\, d\Omega \qquad (4.106a)$$

11. Defining the open integral

$$\int Q(t')\, dt' \equiv R(t') \qquad (4.109)$$

$$\Rightarrow \quad \text{F.T.}\left[\int_{t_0}^{t} Q(t')\, dt'\right] = \text{F.T.}[R(t) - R(t_0)] = \text{F.T.}[R(t)] - \text{F.T.}[R(t_0)] \qquad (4.110)$$

Using equation 4.104a, we obtain

$$\text{F.T.}[R(t)] = \frac{1}{i\Omega}\text{F.T.}\left[\frac{dR}{dt}\right] = \frac{1}{i\Omega}\text{F.T.}[Q(t)] \qquad (4.111)$$

$$\text{F.T.}[R(t_0)] = R(t_0)\int_{-\infty}^{\infty} e^{-i\Omega t}\frac{dt}{\sqrt{2\pi}} = \sqrt{2\pi}\, R(t_0)\delta(\Omega) \qquad (4.112)$$

$$\Rightarrow \quad \text{F.T.}\left[\int_{t_0}^{t} Q(t')\, dt'\right] = \frac{1}{i\Omega}\text{F.T.}[Q(t)] - \sqrt{2\pi}\, R(t_0)\delta(\Omega) \qquad (4.113)$$

which contains a δ-function singularity.

12. The convolution or *Faltung* (which is German for *folding*) of two functions over the interval $[-\infty, \infty]$ is defined as

$$Q_{12}(t) = \int_{-\infty}^{\infty} Q_1(t - t')Q_2(t')\, dt' \qquad (4.114)$$

The Fourier transform of $Q_{12}(t)$ is

$$P_{12}(\Omega) = \int_{-\infty}^{\infty} Q_{12}(t)e^{-i\Omega t}\frac{dt}{\sqrt{2\pi}} = \int_{-\infty}^{\infty}\int_{-\infty}^{\infty} Q_1(t - t')Q_2(t')\, dt'\, e^{-i\Omega t}\frac{dt}{\sqrt{2\pi}} \qquad (4.115)$$

If we transform the t variable to t'' by $t = t'' + t'$, we obtain

$$P_{12}(\Omega) = \int_{-\infty}^{\infty}\int_{-\infty}^{\infty} Q_1(t'')Q_2(t')\, dt'\, e^{-i\Omega t''}e^{-i\Omega t'}\frac{dt''}{\sqrt{2\pi}} \tag{4.116}$$

Since all t'' dependence resides in Q_1 and one exponential factor, and all t' dependence appears in Q_2 and a second exponential, this substitution yields a product of independent integrals. That is, equation 4.116 is

$$P_{12}(\Omega) = \sqrt{2\pi}\int_{-\infty}^{\infty} Q_1(t'')e^{-i\Omega t''}\frac{dt''}{\sqrt{2\pi}}\int_{-\infty}^{\infty} Q_2(t')e^{-i\Omega t'}\frac{dt'}{\sqrt{2\pi}} = \sqrt{2\pi}P_1(\Omega)P_2(\Omega) \tag{4.117a}$$

$$\Rightarrow \quad \text{F.T.}[Q_{12}(t)] = \sqrt{2\pi}\,\text{F.T.}[Q_1(t)]\text{F.T.}[Q_2(t)] \tag{4.117b}$$

13. Consider

$$I_{12} = \int_{-\infty}^{\infty} Q_1(t)Q_2^*(t)\, dt \tag{4.118}$$

Writing $Q_1(t)$ and $Q_2(t)$ in terms of Fourier transforms, this becomes

$$\begin{aligned} I_{12} &= \frac{1}{2\pi}\int_{-\infty}^{\infty} dt \int_{-\infty}^{\infty} P_1(\Omega)e^{i\Omega t}\, d\Omega \int_{-\infty}^{\infty} P_2^*(\Omega')e^{-i\Omega' t}\, d\Omega' \\ &= \int_{-\infty}^{\infty} P_1(\Omega)\, d\Omega \int_{-\infty}^{\infty} P_2^*(\Omega')\frac{1}{2\pi}\int_{-\infty}^{\infty} dt\, e^{i(\Omega-\Omega')t} \end{aligned} \tag{4.119}$$

Referring to equation 4.106b, we see that

$$\frac{1}{2\pi}\int_{-\infty}^{\infty} dt\, e^{i(\Omega-\Omega')t} = \delta(\Omega - \Omega') \tag{4.120}$$

Therefore, equation 4.119 is

$$\begin{aligned} \int_{-\infty}^{\infty} Q_1(t)Q_2^*(t)\, dt &= \int_{-\infty}^{\infty} P_1(\Omega)\, d\Omega \int_{-\infty}^{\infty} P_2^*(\Omega')\, d\Omega'\delta(\Omega - \Omega') \\ &= \int_{-\infty}^{\infty} P_1(\Omega)\, P_2^*(\Omega)\, d\Omega \end{aligned} \tag{4.121}$$

This is another theorem called Parseval's theorem. It is the Parseval theorem for Fourier transforms.

Evaluating Fourier integrals

The Fourier transform integrals of equations 4.79 are one of the types of integrals we evaluated in Chapter 2 using Cauchy's theorem. Many Fourier integrals can be evaluated by this approach. When using Cauchy's theorem, one must be careful to close the contour in the correct half-plane. For example, when determining $P(\Omega)$ from

$$P(\Omega) = \frac{1}{\sqrt{2\pi}}\int_{-\infty}^{\infty} Q(t)e^{-i\Omega t}\, dt \tag{4.79b}$$

Ω can have both positive and negative values. When $\Omega > 0$, the contour must be closed in the lower half-plane, which requires that the contour be transversed in the clockwise direction. When $\Omega < 0$, the contour must be closed in the upper half-plane and is transversed in the counterclockwise direction.

Example 4.4

For example, let us find

$$P(\Omega) = \text{F.T.}\left[\frac{1}{t^2+\alpha^2}\right] = \int_{-\infty}^{\infty} \frac{1}{t^2+\alpha^2} e^{-i\Omega t} \frac{dt}{\sqrt{2\pi}} \qquad \alpha \text{ real} \tag{4.122}$$

The integrand has poles at $t = \pm i\alpha$. Since the denominator contains α^2, its sign cannot be specified. As such, we will take α to be positive. Then, for $\Omega > 0$, we close in the lower half-plane as shown in Figure 4.4. For $\Omega < 0$, the contour must be closed in the upper half-plane (Figure 4.5). This yields

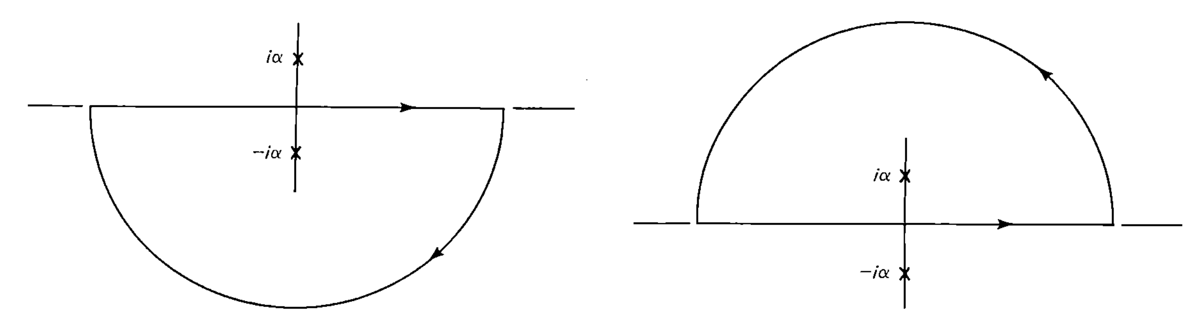

Figure 4.4 Contour for $\Omega > 0$.

Figure 4.5 Contour for $\Omega < 0$.

$$\text{(a)} \quad P(\Omega) = -2\pi i R(-i\alpha) = \frac{1}{\alpha}\sqrt{\frac{\pi}{2}}\, e^{-\alpha\Omega} \qquad \Omega > 0$$

$$\text{(b)} \quad P(\Omega) = 2\pi i R(+i\alpha) = \frac{1}{\alpha}\sqrt{\frac{\pi}{2}}\, e^{\alpha\Omega} \qquad \Omega < 0 \tag{4.123}$$

$$\Rightarrow \quad \text{(c)} \quad P(\Omega) = \frac{1}{\alpha}\sqrt{\frac{\pi}{2}}\, e^{\alpha|\Omega|} \quad \square$$

One very important application of Fourier transforms arises in quantum theory. The behavior of a quantum mechanical system is described in terms of probabilities, which are determined from a function called a wave function or probability amplitude. If $\psi(z)$ is the wave function of a nonrelativistic one-dimensional quantum system, determined from Schrödinger's equation, the probability that the particle exists in a small region of width dz around the point z is given by

$$\rho(z)\, dz = |\psi(z)|^2\, dz \tag{4.124}$$

It is also possible to describe the behavior of the system in terms of its momentum p. The wave function in the momentum description, $\phi(p)$, is related to the coordinate wave function by the Fourier transform. That is, $\psi(z)$ and $\phi(p)$ are Fourier inverses of one another. They are related by

$$\text{(a)} \quad \phi(p) = \int_{-\infty}^{\infty} \psi(z) e^{-i(p/\hbar)z} \frac{dz}{\sqrt{2\pi}} \qquad \text{(b)} \quad \psi(z) = \int_{-\infty}^{\infty} \phi(p) e^{i(p/\hbar)z} \frac{dp}{\sqrt{2\pi}} \tag{4.125}$$

where $\hbar$ is Planck's constant divided by 2π.

Example 4.5

For example, using the coordinate representation of the Schrödinger equation, one can determine the coordinate space wave functions for the one-dimensional harmonic oscillator (e.g., a mass m on a spring of constant k that has a natural angular frequency $\omega_0 = \sqrt{k/m}$).

The coordinate space wave function for the lowest–energy (ground) state of this system is

$$\psi_0(z) = \beta^{1/2}\pi^{-1/4}e^{-\beta^2 z^2/2} \tag{4.126}$$

where $\beta^2 = m\omega_0/\hbar$. The exponential containing the square of the variable is called the *Gaussian exponential function*.

The momentum space wave function for this state is

$$\phi(p) = \beta^{1/2}\pi^{-1/4}\int_{-\infty}^{\infty} e^{-\beta^2 z^2/2}\, e^{-i(p/\hbar)z}\frac{dz}{\sqrt{2\pi}} \tag{4.127}$$

This can be evaluated by completing the square in the exponential.

$$\frac{\beta^2 z^2}{2} + i\frac{p}{\hbar}z = \frac{\beta^2}{2}\left[\left(z + \frac{ip}{\hbar\beta^2}\right)^2 + \frac{p^2}{\hbar^2\beta^4}\right] \tag{4.128}$$

Then equation 4.127 becomes

$$\phi(p) = \left(\frac{\beta}{2}\right)^{1/2}\pi^{-3/4}e^{-(p^2/2\hbar^2\beta^2)}\int_{-\infty}^{\infty} e^{-\beta^2(z+iP/\hbar\beta^2)^2/2}\,dz \tag{4.129}$$

Substituting

$$x \equiv \frac{\beta}{\sqrt{2}}\left(z + \frac{ip}{\hbar\beta^2}\right) \tag{4.130}$$

$$\Rightarrow \quad \phi(p) = \beta^{-1/2}\pi^{-3/4}e^{-(p^2/2\hbar^2\beta^2)}\int_{-\infty}^{\infty} e^{-x^2}\,dx \tag{4.131}$$

To evaluate the integral of equation 4.131, we note that the integrand is an even function. Since a definite integral is unchanged by renaming the variable of integration,

$$I = \int_{-\infty}^{\infty} e^{-x^2}\,dx = 2\int_0^{\infty} e^{-x^2}\,dx = 2\int_0^{\infty} e^{-y^2}\,dy \tag{4.132}$$

Therefore, the product of these integrals is

$$I^2 = 4\int_0^{\infty}\int_0^{\infty} e^{-(x^2+y^2)}\,dx\,dy \tag{4.133}$$

Figure 4.6 shows the paths of integration for this double integral in the x-y plane. Recognizing that $x^2 + y^2$ is the square of the distance from the origin to the point (x, y) and $dx\,dy$ is the area element containing the point, we can express this integral in terms of circular polar coordinates (ρ, ϕ). With $\rho \in [0, \infty]$ and $\phi \in [0, \pi/2]$,

$$\text{(a)} \quad I^2 = 4\int_0^{\pi/2} d\phi \int_0^{\infty} e^{-\rho^2}\rho\,d\rho = \pi \qquad \Rightarrow \qquad \text{(b)} \quad I = \sqrt{\pi} \tag{4.134}$$

Thus

$$\phi(p) = \pi^{-1/4}\beta^{-1/2}e^{-(p^2/2\hbar^2\beta^2)} \qquad \square \tag{4.135}$$

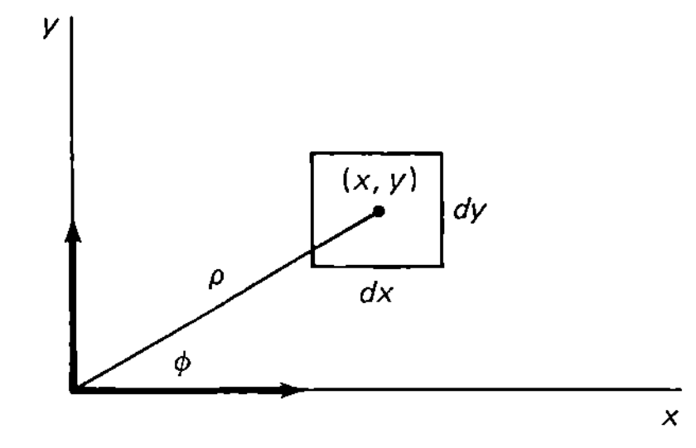

Figure 4.6 Paths of integration for $\int_0^{\infty}\int_0^{\infty} e^{-(x^2+y^2)}\,dx\,dy$.

TABLE 4.1 A SHORT TABLE OF SELECTED FOURIER TRANSFORMS

	$Q(t) = \int_{-\infty}^{\infty} P(\Omega) e^{i\Omega t} \frac{d\Omega}{\sqrt{2\pi}}$		$P(\Omega) = \int_{-\infty}^{\infty} Q(t) e^{-i\Omega t} \frac{dt}{\sqrt{2\pi}}$
1.	$\frac{1}{t^2+\alpha^2}$		$\frac{1}{\alpha}\sqrt{\frac{\pi}{2}}\, e^{-\alpha\Omega}$ $\quad \Omega > 0$ $\frac{1}{\alpha}\sqrt{\frac{\pi}{2}}\, e^{\alpha\Omega}$ $\quad \Omega < 0$
2.	$e^{-\alpha t^2}$		$\frac{1}{\sqrt{2\alpha}} e^{-\Omega^2/4\alpha}$
3.	$\frac{1}{(\alpha + it)^n}$	$\alpha > 0$ $n > 0$ and integer	$\frac{\sqrt{2\pi}\,\Omega^{n-1}}{(n-1)!} e^{-\alpha\Omega}$ $\quad \Omega < 0$ 0 $\quad \Omega > 0$
4.	$\frac{1}{(\alpha - it)^n}$	$\alpha > 0$ $n > 0$ and integer	0 $\quad \Omega < 0$ $-\frac{\sqrt{2\pi}\,(-\Omega)^{n-1}}{(n-1)!} e^{\alpha\Omega}$ $\quad \Omega > 0$
5.	$\frac{e^{-\lambda t}}{\alpha + e^{-t}}$	$\alpha > 0$ $0 < \lambda < 1$	$\sqrt{\frac{\pi}{2}}\, \alpha^{\lambda - 1 - i\Omega} \csc[\pi(\lambda - i\Omega)]$

Note that both the coordinate and momentum space wave functions depend on their respective variables through the Gaussian exponential function. That is, the Fourier transform of a Gaussian exponential is a Gaussian exponential.

These examples are guides as to how a large number of Fourier transform integrals are evaluated. Table 4.1 is a small sample of Fourier transform pairs.

Fourier transforms in higher dimensions

Consider a function of two independent variables, $Q(t_1, t_2)$. The Fourier transform of this function in the two variables can be developed for each variable separately. For example, since t_1 and t_2 are independent, we can fix t_2 and write a one-dimensional transform pair in t_1, Ω_1.

$$\text{(a)}\quad Q(t_1, t_2) = \int_{-\infty}^{\infty} R(\Omega_1, t_2) e^{i\Omega_1 t_1} \frac{d\Omega_1}{\sqrt{2\pi}} \qquad \text{(b)}\quad R(\Omega_1, t_2) = \int_{-\infty}^{\infty} Q(t_1, t_2) e^{-i\Omega_1 t_1} \frac{dt_1}{\sqrt{2\pi}} \tag{4.136}$$

We now write a transform pair in the t_2 variable.

$$\begin{aligned} &\text{(a)}\quad R(\Omega_1, t_2) = \int_{-\infty}^{\infty} P(\Omega_1, \Omega_2) e^{i\Omega_2 t_2} \frac{d\Omega_2}{\sqrt{2\pi}} \\ &\text{(b)}\quad P(\Omega_1, \Omega_2) = \int_{-\infty}^{\infty} R(\Omega_1, t_2) e^{-i\Omega_2 t_2} \frac{dt_2}{\sqrt{2\pi}} \end{aligned} \tag{4.137}$$

Combining equations 4.137 with equations 4.136, we obtain

$$\begin{aligned} &\text{(a)}\quad Q(t_1, t_2) = \int_{-\infty}^{\infty} P(\Omega_1, \Omega_2) e^{i(\Omega_1 t_1 + \Omega_2 t_2)} \frac{d\Omega_1\, d\Omega_2}{2\pi} \\ &\text{(b)}\quad P(\Omega_1, \Omega_2) = \int_{-\infty}^{\infty} Q(t_1, t_2) e^{-i(\Omega_1 t_1 + \Omega_2 t_2)} \frac{dt_1\, dt_2}{2\pi} \end{aligned} \tag{4.138}$$

We see from this example that it is straightforward to describe Fourier transforms of functions of more than one variable. For each variable for which a transform is generated, there is an integration from $-\infty$ to ∞, an exponential factor $e^{\pm i\Omega_n t_n}$, and a factor of $1/\sqrt{2\pi}$.

For example, if Q is a function of the three Cartesian coordinates and its Fourier inverse is a function of three wave numbers

$$\text{(a)} \quad Q(x, y, z) \equiv Q(\mathbf{r}) \qquad \text{(b)} \quad P(k_x, k_y, k_z) \equiv P(\mathbf{k}) \tag{4.139}$$

then

$$Q(\mathbf{r}) = \int_{-\infty}^{\infty}\int_{-\infty}^{\infty}\int_{-\infty}^{\infty} P(\mathbf{k}) e^{i\mathbf{k}\cdot\mathbf{r}} \frac{d^3k}{(2\pi)^{3/2}} \tag{4.140a}$$

$$P(\mathbf{k}) = \int_{-\infty}^{\infty}\int_{-\infty}^{\infty}\int_{-\infty}^{\infty} Q(\mathbf{r}) e^{-i\mathbf{k}\cdot\mathbf{r}} \frac{d^3r}{(2\pi)^{3/2}} \tag{4.140b}$$

4.3 Laplace Transforms

It was shown above that for the Fourier transform to exist, $\int_{-\infty}^{\infty} |Q(t)|\, dt$ has to be finite. This requires that $Q(\pm\infty) = 0$. But if $Q(t)$ has the properties

$$\text{(a)} \quad Q(t) = 0 \quad \text{for } t < 0 \qquad \text{(b)} \quad \lim_{t\to\infty} Q(t)e^{-ct} = 0 \quad \mathrm{Re}(c) > 0 \tag{4.141}$$

we can deduce an integral transform called the Laplace transform for $Q(t)$. We note that property 2 does not require $Q(\infty)$ to be zero or even finite. It only requires that $Q(t)$ approach its limit at infinity at a rate slower than e^{ct}.

We define

$$F(t) \equiv \begin{cases} Q(t)e^{-ct} & t > 0 \\ 0 & t < 0 \end{cases} \qquad \text{with } c \text{ real and } > 0 \tag{4.142}$$

The Heaviside step function is defined as

$$\Theta(t) = \begin{cases} 1 & t > 0 \\ 0 & t < 0 \end{cases} \tag{4.143}$$

Then $F(t)$ can be written as

$$F(t) = \Theta(t)Q(t)e^{-ct} \tag{4.144}$$

The Fourier transform of $F(t)$ is

$$P(\Omega) = \int_{-\infty}^{\infty} \Theta(t)Q(t)e^{-(c+i\Omega)t} \frac{dt}{\sqrt{2\pi}} = \int_0^{\infty} Q(t)e^{-(c+i\Omega)t} \frac{dt}{\sqrt{2\pi}} \tag{4.145}$$

Thus

$$F(t) = Q(t)e^{-ct} = \int_{-\infty}^{\infty} P(\Omega)e^{i\Omega t} \frac{d\Omega}{\sqrt{2\pi}} \qquad t > 0 \tag{4.146a}$$

$$\Rightarrow \quad Q(t) = \int_{-\infty}^{\infty} P(\Omega)e^{(c+i\Omega)t} \frac{d\Omega}{\sqrt{2\pi}} \qquad t > 0 \tag{4.146b}$$

It is natural to make the change of variable

$$s \equiv c + i\Omega \tag{4.147}$$

so the integral transforms become

$$\text{(a)} \quad P(s) = \int_0^\infty Q(t) e^{-st} \frac{dt}{\sqrt{2\pi}} \qquad \text{(b)} \quad Q(t) = \int_{c-i\infty}^{c+i\infty} P(s) e^{st} \frac{ds}{i\sqrt{2\pi}} \tag{4.148}$$

To shift the position of the factors of 2π, we define

$$G(s) = \sqrt{2\pi}\, P(s) \tag{4.149}$$

$$\Rightarrow \quad \text{(a)} \quad G(s) = \int_0^\infty Q(t) e^{-st}\, dt \qquad \text{(b)} \quad Q(t) = \frac{1}{2\pi i} \int_{c-i\infty}^{c+i\infty} G(s) e^{st}\, ds \tag{4.150}$$

The functions defined by equations 4.150 are known as the *Laplace inverses* of each other. They are also denoted by L.T., just as F.T. is used to denote Fourier transforms. In some of the literature, the Laplace transform refers only to equation 4.150a, and the integral of equation 4.150b is called the *inverse Laplace transform*.

Properties of Laplace transforms

1. The Laplace transform of a linear combination of functions satisfies

$$\text{L.T.}[\alpha Q_1(t) + \beta Q_2(t)] = \alpha\, \text{L.T.}[Q_1(t)] + \beta\, \text{L.T.}[Q_2(t)] \tag{4.151}$$

where α and β are constants.

2. Consider

$$\text{L.T.}\left[\frac{dQ}{dt}\right] = \int_0^\infty e^{-st} \frac{dQ}{dt}\, dt = e^{-st} Q(t)\Big|_0^\infty + s \int_0^\infty e^{-st} Q(t)\, dt \tag{4.152}$$

Since $\text{Re}(s) = c > 0$, the integrated term at the upper limit $= 0$. Thus the integrated term is $-Q(0)$. The remaining integral is L.T.$[Q(t)]$. Therefore,

$$\text{L.T.}\left[\frac{dQ}{dt}\right] = s\, \text{L.T.}[Q(t)] - Q(0) \tag{4.153}$$

Applying this to the second derivative, we obtain

$$\text{L.T.}\left[\frac{d^2Q}{dt^2}\right] = s\, \text{L.T.}\left[\frac{dQ}{dt}\right] - Q'(0) = s^2\, \text{L.T.}[Q(t)] - sQ(0) - Q'(0) \tag{4.154}$$

It is straightforward to deduce an expression for the nth derivative from this. The result is

$$\begin{aligned} \text{L.T.}\left[\frac{d^nQ}{dt^n}\right] &= s^n\, \text{L.T.}[Q(t)] - s^{n-1}Q(0) - s^{n-2}Q'(0) - \cdots \\ &= s^n\, \text{L.T.}[Q(t)] - \sum_{l=1}^{n} s^{n-l} Q^{(l-1)}(0) \end{aligned} \tag{4.155}$$

where, as introduced earlier, $Q^{(k)}(0)$ is the kth derivative of $Q(t)$ at $t = 0$.

3. With the definition

$$R(t') \equiv \int Q(t')\, dt' \tag{4.109}$$

consider

$$\text{L.T.}\left[\int_0^t Q(t')\, dt'\right] = \text{L.T.}[R(t) - R(0)] \tag{4.156}$$

Using equation 4.153 and the fact that $R(0)$ is a constant, these transforms are

$$\text{L.T.}[R(t)] = \frac{\text{L.T.}[R'(t)] + R(0)}{s} = \frac{\text{L.T.}[Q(t)] + R(0)}{s} \tag{4.157a}$$

$$\text{L.T.}[R(0)] = R(0)\int_0^\infty e^{-st}\,dt = \frac{R(0)}{s} \tag{4.157b}$$

$$\Rightarrow \quad \text{L.T.}\left[\int_0^t Q(t')\,dt'\right] = \frac{\text{L.T.}[Q(t)]}{s} \tag{4.158}$$

4. Consider the Laplace transform of the Dirac δ-function

$$\text{L.T.}[\delta(t - t_0)] = \int_0^\infty \delta(t - t_0)e^{-st}\,dt = \begin{cases} e^{-st_0} & t_0 > 0 \\ 0 & t_0 < 0 \end{cases} \tag{4.159}$$

Taking the inverse transform, we obtain another integral representation of the δ-symbol.

$$\delta(t - t_0) = \frac{1}{2\pi i}\int_{c-i\infty}^{c+i\infty} e^{s(t-t_0)}\,dt \qquad t_0 > 0 \tag{4.160}$$

5. The Laplace transform of the Heaviside step function is

$$\text{L.T.}[\Theta(t - t_0)] = \int_0^\infty \Theta(t - t_0)e^{-st}\,dt = \int_{t_0}^\infty e^{-st}\,dt = \frac{e^{-st_0}}{s} \tag{4.161a}$$

$$\Rightarrow \quad \Theta(t - t_0) = \frac{1}{2\pi i}\int_{c-i\infty}^{c+i\infty} e^{s(t-t_0)}\frac{ds}{s} \tag{4.161b}$$

This is an integral representation of $\Theta(t - t_0)$. Taking the derivative of $\Theta(t - t_0)$ as expressed in equation 4.161b, and comparing the result to equation 4.160, we see that

$$\frac{d}{dt}\Theta(t - t_0) = \frac{1}{2\pi i}\int_{c-i\infty}^{c+i\infty} e^{s(t-t_0)}\,ds = \delta(t - t_0) \tag{4.162}$$

Evaluation of Laplace integrals

Obtaining an expression for $G(s) = \text{L.T.}[Q(t)]$ is often a straightforward evaluation of an integral. In most cases, to determine $Q(t)$ from $G(s)$, the integral of equation 4.150b is evaluated using Cauchy's theorem. To do so, a closed contour must be constructed that contains the path from $c - i\infty$ to $c + i\infty$ in the complex s-plane as shown in Figure 4.7.

Example 4.6

To illustrate, consider

$$\text{L.T.}[\cosh(\alpha t)] = \int_0^\infty e^{-st}\frac{(e^{\alpha t} + e^{-\alpha t})}{2}\,dt = \frac{1}{2}\left[\frac{e^{-(s-\alpha)t}}{-(s - \alpha)}\Bigg|_0^\infty + \frac{e^{-(s+\alpha)t}}{-(s + \alpha)}\Bigg|_0^\infty\right] \tag{4.163}$$

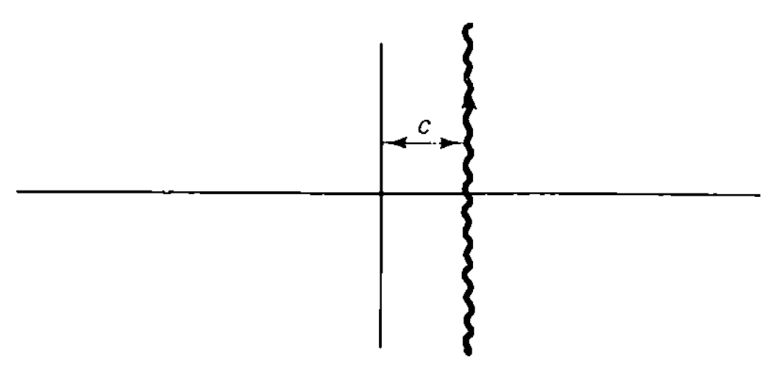

Figure 4.7 Contour in complex s-plane for evaluating Laplace transform integral to determine $Q(t)$.

If $\text{Re}(s - \alpha) > 0$, the first integrated term is zero at ∞. Thus the first term, $1/(s - \alpha)$, is finite. Similarly, the second term is finite if $\text{Re}(s + \alpha) > 0$. Thus L.T.[$\cosh(\alpha t)$] will be finite if $\text{Re}(s) > \pm\text{Re}(\alpha)$ or $\text{Re}(s) > |\text{Re}(\alpha)|$. If $\text{Re}(s) \leqslant |\text{Re}(\alpha)|$, L.T.[$\cosh(\alpha t)$] $= \infty$.

Taking $\text{Re}(s) > |\text{Re}(\alpha)|$ we have

$$\text{L.T.}[\cosh(\alpha t)] = \frac{s}{s^2 - \alpha^2} \tag{4.164}$$

$$\Rightarrow \quad \cosh(\alpha t) = \frac{1}{2\pi i}\int_{c-i\infty}^{c+i\infty} \frac{s}{s^2 - \alpha^2} e^{st}\, ds = \frac{1}{2\pi i}\int_{c-i\infty}^{c+i\infty} \frac{1}{2}\left(\frac{1}{s-\alpha} + \frac{1}{s+\alpha}\right) e^{st}\, ds \tag{4.165}$$

For the following discussion, we will take α to be real and positive. Thus the condition $\text{Re}(s) > |\text{Re}(\alpha)|$ becomes $c > \alpha$.

Consider

$$I_1 = \frac{1}{2\pi i}\int_{c-i\infty}^{c+i\infty} \frac{1}{s-\alpha} e^{st}\, ds \tag{4.166}$$

As a first step to evaluating this integral, we redefine the variable of integration as $s' = s - c$. This transforms the contour shown in Figure 4.8a into the contour of Figure 4.8b and shifts the position of the pole to $-(c - \alpha)$. Then

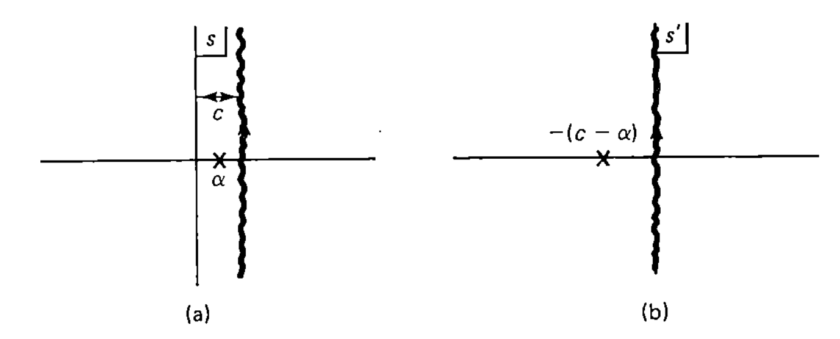

Figure 4.8 (a) Contour in the s-plane; (b) contour in the s'-plane.

$$I_1 = \frac{1}{2\pi i} e^{ct}\int_{-i\infty}^{+i\infty} \frac{1}{s' + (c - \alpha)} e^{s't}\, ds' \tag{4.167}$$

With $c - \alpha > 0$, this integrand has a simple pole on the negative real s'-axis. This is shown in Figure 4.8b.

The integral of equation 4.167 can be evaluated by Cauchy's theorem by closing the contour in either the left half-plane or the right half-plane as shown in Figures 4.9.

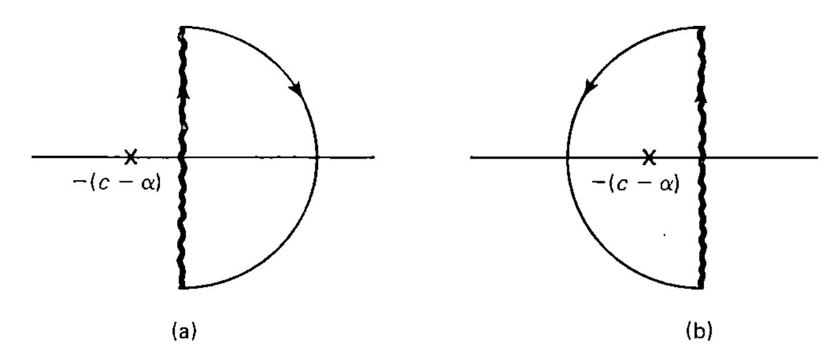

Figure 4.9 (a) Contour closed in right half-plane; (b) contour closed in left half-plane.

Since s' is a complex variable, $e^{s't}$ can be written

$$e^{s't} = e^{|s'|t(\cos\theta + i\sin\theta)} \tag{4.168}$$

If we close in the right half-plane as in Figure 4.9a, the integrand is analytic in the region inside the contour. Therefore, the integral around the closed contour is zero. But in the right half-plane, $\cos\theta > 0$ and

$$\lim_{|s'|\to\infty} e^{|s'|\cos\theta} \to \infty \tag{4.169}$$

Thus, on the infinite semicircle, the integrand becomes infinite and we cannot relate the closed contour integral to I_1.

By closing in the left half-plane as in Figure 4.9b, we enclose the pole at $-(c-\alpha)$. In the left half-plane, $\cos\theta < 0$. Thus on the infinite semicircle,

$$\lim_{|s'|\to\infty} e^{|s'|\cos\theta} \to 0 \tag{4.170}$$

Therefore, we must close the contour in the left half-plane. Since the contour is traversed in the counterclockwise direction, this results in

$$\oint \frac{1}{s' + (c-\alpha)} e^{s't}\, ds' = 2\pi i R[-(c-\alpha)] = \int_{-i\infty}^{+i\infty} \frac{1}{s' + (c-\alpha)} e^{s't}\, ds' = 2\pi i e^{-(c-\alpha)t} \tag{4.171}$$

$$\Rightarrow \quad I_1 = e^{\alpha t} \tag{4.172a}$$

By identical analysis, the second integral is evaluated by closing in the left half-plane, which straightforwardly results in

$$I_2 \equiv \frac{1}{2\pi i} \int_{c-i\infty}^{c+i\infty} \frac{1}{s+\alpha} e^{st}\, ds = e^{-\alpha t} \tag{4.172b}$$

Thus we obtain the expected result

$$\cosh(\alpha t) = \tfrac{1}{2}(e^{\alpha t} + e^{-\alpha t}) \quad \square \tag{4.173}$$

The Laplace transform integral

$$Q(t) = \frac{1}{2\pi i} \int_{c-i\infty}^{c+i\infty} G(s) e^{st}\, ds \tag{4.150b}$$

is known as the *Bromwich integral*. As noted in Example 4.6, we see that if the contour is closed in the right half-plane, the factor $e^{st} \to \infty$ on the infinite part of the contour. Thus for any Bromwich integral, the contour must be closed in the left half s-plane in order to evaluate such an integral by Cauchy's theorem. This contour, shown in Figure 4.10, is called the *Bromwich contour*.

In Example 4.6 the Bromwich integral is evaluated for a function $G(s)$ that contains only pole singularities. It is sometimes possible to evaluate the Bromwich integral when the integrand has a branch point. In such a case, the Bromwich contour must be modified.

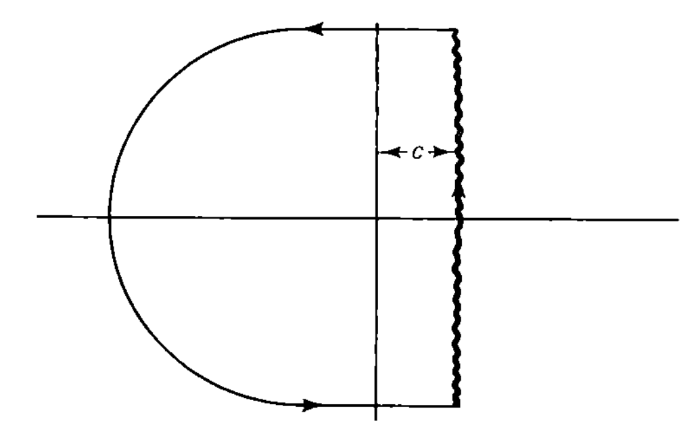

Figure 4.10 Bromwich contour for evaluating the inverse Laplace transform.

Example 4.7

To illustrate, consider

$$Q(t) = \frac{1}{2\pi i}\int_{c-i\infty}^{c+i\infty} \log(s+\beta)e^{st}\,ds \tag{4.174}$$

with β real and positive. $\log(s+\beta)$ has a branch point at $s = -\beta$, and we can take the associated cut in any direction that is necessary. If the cut were taken to run to $+\infty$ along the positive real axis, the part of the Bromwich contour from $c - i\infty$ to $c + i\infty$ would run over the cut, which is not allowed. Since the contour must be closed in the left half-plane, and since the contour cannot run over a cut, the contour and cut structure required to evaluate this integral are those shown in Figure 4.11.

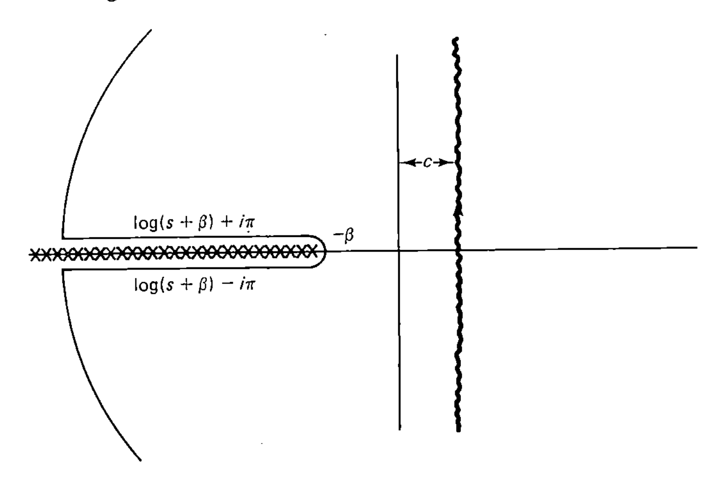

Figure 4.11 Modified Bromwich contour for L.T.[$\log(s+\beta)$].

The integrand of equation 4.174 has no poles. Therefore,

$$\frac{1}{2\pi i}\oint \log(s+\beta)e^{st}\,ds = 0 = \frac{1}{2\pi i}\left[\int_{c-i\infty}^{c+i\infty} \log(s+\beta)e^{st}\,ds + \int_{C_{+\infty}} \log(s+\beta)e^{st}\,ds \right.$$

$$\left. + \int_{-\infty}^{-\beta}[\log(s+\beta)+i\pi]e^{st}\,ds + \int_{-\beta}^{-\infty}[\log(s+\beta)-i\pi]e^{st}\,ds + \int_{C_{-\infty}} \log(s+\beta)e^{st}\,ds\right] \tag{4.175}$$

Since $e^{st} \to 0$ on the infinite arcs in the left half-plane,

$$\int_{C_{\pm\infty}} \log(s+\beta)e^{st}\,ds = 0 \tag{4.176a}$$

and, since $t > 0$,

$$\begin{aligned}\int_{-\infty}^{-\beta}(\log(s+\beta)+i\pi)e^{st}\,ds &+ \int_{-\beta}^{-\infty}(\log(s+\beta)-i\pi)e^{st}\,ds \\ &= -2\pi i\int_{-\beta}^{-\infty} e^{st}\,ds = 2\pi i\frac{e^{-\beta t}}{t}\end{aligned} \tag{4.176b}$$

$$\Rightarrow \quad Q(t) = \frac{1}{2\pi i}\int_{c-i\infty}^{c+i\infty} \log(s+\beta)e^{st}\,ds = \frac{e^{-\beta t}}{t} \quad \square \tag{4.177}$$

TABLE 4.2 SELECTED LIST OF LAPLACE TRANSFORMS

$Q(t) = \frac{1}{2\pi i}\int_{c-i\infty}^{c+i\infty} G(S)e^{st}\,ds$	$G(s) = \int_0^\infty Q(t)e^{-st}\,dt$
1. $\Theta(t - t_0)$	$\frac{e^{-st_0}}{s}$
2. $\delta(t - t_0)$	e^{-st_0}
3. t^n $\quad n = \text{integer} \geqslant 0$	$\frac{n!}{s^{n+1}}$
4. $\cosh(\alpha t)$ $\quad \alpha$ real $< \text{Re}(s)$	$\frac{s}{s^2 - \alpha^2}$
5. $\sinh(\alpha t)$ $\quad \alpha$ real $< \text{Re}(s)$	$\frac{\alpha}{s^2 - \alpha^2}$
6. $e^{-\beta t}\sin(\omega t)$ $\quad \omega$ real	$\frac{\omega}{s^2 + 2\beta s + \omega_0^2}$ $\quad \omega_0^2 = \omega^2 + \beta^2$
7. $e^{-\beta t}\cos(\omega t)$ $\quad \omega$ real	$\frac{\beta + s}{s^2 + 2\beta s + \omega_0^2}$ $\quad \omega_0^2 = \omega^2 + \beta$
8. $\frac{e^{-\beta t}}{t}$ $\quad \beta$ real, > 0	$\log(s + \beta)$
9. $e^{-\alpha t}$ $\quad \alpha$ complex	$\frac{1}{s + \alpha}$
10. $\frac{\alpha e^{-\alpha t} - \beta e^{-\beta t}}{(\alpha - \beta)}$	$\frac{s}{(s + \alpha)(s + \beta)}$

These examples illustrate methods of evaluating inverse Laplace integrals. With techniques such as these, one can generate tables of Laplace transforms. Table 4.2 is a small sampling of Laplace transform pairs. Tables containing large numbers of Laplace transforms are presented elsewhere [see, e.g., the text by M. R. Spiegel, *Laplace Transforms* in the Schaum Outline Series (New York: McGraw-Hill Book Company, 1964)]. The reader is referred to these for more complete lists of these transforms.

Laplace transform of a periodic function

If $Q(t)$ is periodic, then

$$Q(t + nT) = Q(t) \tag{4.178}$$

for any integer n. We can then write equation 4.150a as

$$\begin{aligned} G(s) &= \int_0^\infty Q(t)e^{-st}\,dt = \int_0^T Q(t)e^{-st}\,dt + \int_T^{2T} Q(t)e^{-st}\,dt + \int_{2T}^{3T} Q(t)e^{-st}\,dt + \cdots \\ &= \sum_{n=0}^{\infty} \int_{nT}^{(n+1)T} Q(t)e^{-st}\,dt \end{aligned} \tag{4.179}$$

For $nT \leqslant t \leqslant (n + 1)T$, we make the substitution $t' = t - nT$. Then equation 4.179 becomes

$$G(s) = \sum_{n=0}^{\infty} \int_0^T Q(t' + nT)e^{-st'}e^{-nsT}\,dt' \tag{4.180}$$

Using equation 4.178 and the fact that e^{-nsT} is independent of t', equation 4.180 becomes

$$G(s) = \left[\int_0^T Q(t')e^{-st'}\,dt'\right] \sum_{n=0}^{\infty} e^{-nsT} = \left[\int_0^T Q(t')e^{-st'}\,dt'\right] \sum_{n=0}^{\infty} (e^{-sT})^n \tag{4.181}$$

Since $\text{Re}(s) > 0$, $e^{-sT} < 1$. Therefore, the sum in equation 4.181 is the infinite geometric series

$$\sum_{n=0}^{\infty} (e^{-sT})^n = \frac{1}{(1 - e^{-sT})} \tag{4.182}$$

Defining

$$\Phi(s) \equiv \int_0^T Q(t')e^{-st'}\,dt' \tag{4.183}$$

$$\Rightarrow \quad G(s) = \text{L.T.}[Q(t)] = \frac{\Phi(s)}{(1 - e^{-sT})} \tag{4.184}$$

$$\Rightarrow \quad Q(t) = \text{L.T.}[G(s)] = \frac{1}{2\pi i}\int_{c-i\infty}^{c+i\infty} \frac{\Phi(s)}{(1 - e^{-sT})} e^{st}\,ds \tag{4.185}$$

This integrand has simple poles at those values of s for which $e^{-sT} = 1$. These values are

$$s = \frac{2\pi in}{T} = in\omega \qquad n = 0, \pm 1, \pm 2, \ldots \tag{4.186}$$

From the discussion above, we know that we must use the Bromwich contour to evaluate the integral of equation 4.185 by Cauchy's theorem. In doing so, we enclose all the poles of the integrand. The contour and the positions of the poles are shown in Figure 4.12. Therefore, equation 4.185 becomes

$$Q(t) = \frac{1}{2\pi i}\int_{c-i\infty}^{c+i\infty} \frac{\Phi(s)}{(1 - e^{-st})} e^{st}\,ds = 2\pi i \sum_{n=-\infty}^{\infty} R\left(\frac{2\pi in}{T}\right) \tag{4.187}$$

where

$$R\left(\frac{2\pi in}{T}\right) = \frac{1}{2\pi i} \lim_{s \to 2\pi in/T} \left(s - \frac{2\pi in}{T}\right) \frac{\Phi(s)}{(1 - e^{-sT})} e^{st} \tag{4.188}$$

Since both numerator and denominator approach zero, a simple way to evaluate this

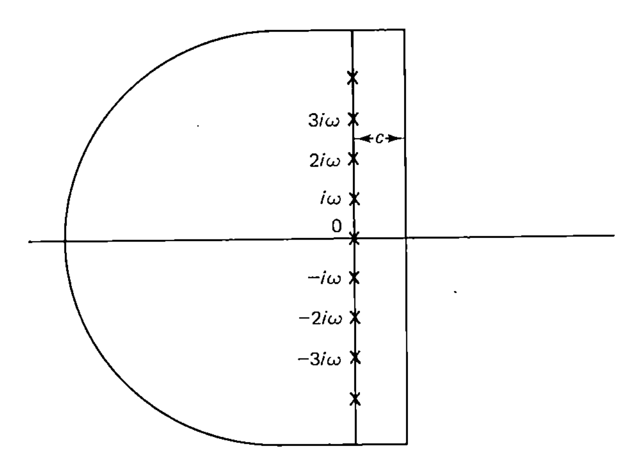

Figure 4.12 Bromwich contour and poles for evaluating Laplace transform integral of a periodic function.

limit is by L'Hospital's rule. We obtain

$$R\left(\frac{2\pi in}{T}\right) = \frac{\Phi(2\pi in/T)e^{(2\pi in/T)t}}{T} = \frac{\Phi(in\omega)e^{in\omega t}}{T} \tag{4.189}$$

$$\Rightarrow \quad Q(t) = \sum_{n=-\infty}^{\infty} \frac{\Phi(in\omega)}{T} e^{in\omega t} \tag{4.190}$$

This is the Fourier exponential series. From equation 4.182b, the coefficients of the series are

$$\frac{\Phi(in\omega)}{T} = \frac{1}{T}\int_0^T Q(t)e^{-in\omega t}\, dt \tag{4.191}$$

Comparing this to equation 4.15 with $t_0 = 0$, we see that equation 4.191 is the expression for C_n that we obtained when developing the Fourier exponential series.

4.4 Other Integral Transforms

Other integral transforms have been developed for various purposes. They have limited use and thus their properties and applications will not be developed in this book. The reader interested in a more detailed discussion of them is referred to a two-volume series edited by A. Erdelyi, *Table of Integral Transforms-The Bateman Project* (New York: McGraw-Hill Book Company, 1954). For completeness, we merely introduce the pairs of integral transforms here.

Hankel transforms

In Chapter 6 we introduce a function called the *Bessel function* which arises as part of a solution of a differential equation that describes certain types of physical problems. The Bessel function of order ν is denoted by $J_\nu(x)$. The *Hankel transforms* (also called the *Fourier–Bessel transforms*) are

$$\text{(a)} \quad P_\nu(\Omega) = \int_0^\infty Q(t)J_\nu(\Omega t)t\, dt \qquad \text{(b)} \quad Q(t) = \int_0^\infty P_\nu(\Omega)J_\nu(\Omega t)\Omega\, d\Omega \tag{4.192}$$

Hilbert transforms

The *Hilbert transforms* are defined in terms of Cauchy principal value integrals. They are

$$\text{(a)} \quad P(\Omega) = \frac{1}{\pi}P\int_{-\infty}^\infty \frac{Q(t)}{(t-\Omega)}\, dt \qquad \text{(b)} \quad Q(t) = \frac{1}{\pi}P\int_0^\infty \frac{P(\Omega)}{(\Omega - t)}\, d\Omega \tag{4.193}$$

Comparing these to equations 2.395, we see that the Hilbert transform pairs are the real and imaginary parts of a function that can be described by a dispersion relation. Thus some of the properties of Hilbert transforms have been discussed in Section 2.5.

Mellin transforms

Finally, the *Mellin transform* integrals are

$$\text{(a)} \quad G(\alpha) = \int_0^\infty x^{\alpha-1}Q(x)\, dx \qquad \text{(b)} \quad Q(x) = \frac{1}{2\pi i}\int_{-i\infty}^{i\infty} x^{-\alpha}G(\alpha)\, d\alpha \tag{4.194}$$

Consider the Fourier transforms of equations 4.79:

$$\text{(a)} \quad Q(t) = \int_{-\infty}^\infty P(\Omega)e^{i\Omega t}\frac{d\Omega}{\sqrt{2\pi}} \qquad \text{(b)} \quad P(\Omega) = \int_{-\infty}^\infty Q(t)e^{-i\Omega t}\frac{dt}{\sqrt{2\pi}} \tag{4.79}$$

Making the substitutions

$$\text{(a)} \quad x = e^t \qquad \text{(b)} \quad \alpha = -i\Omega \tag{4.195}$$

we obtain

$$\text{(a)} \quad Q(x) = \int_{-i\infty}^{i\infty} P(\alpha)x^{-\alpha}\frac{d\alpha}{i\sqrt{2\pi}} \qquad \text{(b)} \quad P(\alpha) = \int_0^{\infty} Q(x)x^{\alpha-1}\frac{dx}{\sqrt{2\pi}} \tag{4.196}$$

Defining $G(\alpha) = \sqrt{2\pi}P(\alpha)$, equations 4.196 become

$$\text{(a)} \quad Q(x) = \frac{1}{2\pi i}\int_{-i\infty}^{i\infty} G(\alpha)x^{-\alpha}\,d\alpha \qquad \text{(b)} \quad G(\alpha) = \int_0^{\infty} Q(x)x^{\alpha-1}\,dx \tag{4.197}$$

That is, the Mellin transforms are a different form of the Fourier transforms.

PROBLEMS

1. Show that the set of functions $\{\sin(l\omega t), \cos(l\omega t)\}$ form a mutually orthogonal set of functions. This is done by evaluating

$$\int_{t_0}^{t_0+T} \sin(l\omega t)\cos(m\omega t)\,dt$$

$$\int_{t_0}^{t_0+T} \sin(l\omega t)\sin(m\omega t)\,dt$$

$$\int_{t_0}^{t_0+T} \cos(l\omega t)\cos(m\omega t)\,dt$$

for l and m integers. Note that we need only concern ourselves with l and $m \geqslant 1$ since the sine–cosine series is summed over integers $\geqslant 1$.

2. If $Q(t)$ has periodicity T, and if $Q^{(k)}(t) \equiv d^kQ/dt^k$ exists at the point t, show that $Q^{(k)}(t)$ has the same periodicity T.

3. Prove that for any integer n, the coefficients defined by

$$\Gamma_l \equiv \int_{t_0+nT}^{t_0+(n+1)T} Q(t)e^{-il\omega t}\,dt$$

are identical to the coefficients of the Fourier exponential series representation of $Q(t)$ given in equation 4.15. This illustrates the fact that the coefficients of this series can be determined by integration over any single period.

4. The function $Q(t)$ is defined by

$$Q(t) = \sin^2(\pi t)$$

(a) What is the natural period of $Q(t)$?

(b) Determine the Fourier exponential and Fourier sine–cosine series for $Q(t)$ over its natural period.

(c) Apply Parseval's theorem to $\sin^2(\pi t)$ to evaluate

$$\int_0^1 \sin^4(\pi t)\,dt$$

5. **(a)** Determine the natural period of

$$Q(t) = \sin^3\!\left(\frac{\pi}{2}t\right)$$

(b) Determine both the Fourier exponential and Fourier sine–cosine series for $Q(t)$.

6. Determine both the Fourier exponential and sine–cosine series for the following functions.

(a) $Q(t) = t, \quad 0 \leqslant t \leqslant \pi$

(b) $Q(t) = t, \quad -\pi/2 \leqslant t \leqslant \pi/2$

(c) $Q(t) = \begin{cases} +Y & 0 \leqslant t \leqslant \frac{1}{4}T \\ 0 & \frac{1}{4}T \leqslant t \leqslant \frac{3}{4}T \\ -Y & \frac{3}{4}T \leqslant t \leqslant T \end{cases}$

7. **(a)** Determine the Fourier series representation of

$$Q(t) \equiv \begin{cases} \sin^2(2t) & 0 \leqslant t \leqslant \dfrac{\pi}{2} \\ 0 & \dfrac{\pi}{2} \leqslant t \leqslant \pi \end{cases}$$

(b) Apply Parseval's theorem for the Fourier series to show that

$$\pi^2 = \frac{256}{3}\sum_{k=0}^{\infty}\frac{1}{(2k+1)^2\left(4-(2k+1)^2\right)^2}$$

8. From the Fourier series for

$$Q(t) = t^2 \qquad -\pi \leqslant t \leqslant \pi$$

given in equation 4.49b, determine a series expression for $Q(\pi/2)$ to obtain the series description of π^2 given in equation 4.42.

9. From the Fourier series for

$$Q(t) = t^2 \qquad -\pi \leqslant t \leqslant \pi$$

given in equation 4.49b, determine the Fourier sine–cosine series for

$$R(t) = t^3 \qquad -\pi \leqslant t \leqslant \pi$$

(*Note*: To express the series as a Fourier sine–cosine series, all t-dependence must be in the form of $\sin(l\omega t)$ and $\cos(l\omega t)$.)

10. (a) Determine the Fourier series for

$$Q(T) = \cosh^2(\alpha t) \qquad \alpha \text{ real}, \quad -\frac{\pi}{4} \leqslant t \leqslant \frac{\pi}{4}$$

(b) From the result of part (a), determine the Fourier series for

$$R(t) = \cosh(\alpha t)\sinh(\alpha t)$$

$$\alpha \text{ real}, \quad -\frac{\pi}{4} \leqslant t \leqslant \frac{\pi}{4}$$

11. (a) In the text it was shown that if a function is real, the coefficients of its Fourier exponential series must satisfy $C_l = C^*_{-l}$. Show that if $C_l = C^*_{-l}$, the Fourier series must be real.

(b) In the text it was shown that if a function is imaginary, the coefficients of its Fourier exponential series must satisfy $C_l = -C^*_{-l}$. Show that if $C_l = -C^*_{-l}$, the Fourier series must be imaginary.

12. Apply Parseval's theorem for the Fourier series to the periodic sawtooth function of equation 4.43 to show that

$$\zeta(2) \equiv \sum_{l=1}^{\infty} \frac{1}{l^2} = \frac{\pi^2}{6}$$

which is a result given in equation 3.219, obtained by analysis of Bernoulli numbers.

13. A quantum particle is confined to a one-dimensional box that extends from $x = -L$ to $x = +L$. Its coordinate-space wave function in the nth quantum state is found to be

$$\psi_n(x) = \begin{cases} \dfrac{1}{\sqrt{2}} \cos\left(\dfrac{n\pi x}{2L}\right) & n \text{ odd} \\[2ex] \dfrac{1}{\sqrt{2}} \sin\left(\dfrac{n\pi x}{2L}\right) & n \text{ even} \end{cases}$$

with $\psi_n(x) = 0$, $x < -L$, and $x > L$. Find the momentum space wave function of this particle in the nth quantum state.

14. (a) Find the Fourier transform of the exponential decay function

$$Q(t) = \Theta(t)e^{-\Gamma t} \qquad \Gamma \text{ is real and positive}$$

$\Theta(t)$ is the Heaviside step function described in equation 4.143.

(b) Find the Fourier transform of the displacement function of a damped harmonic oscillator, given by

$$y(t) = \Theta(t)e^{-\Gamma t}\sin(\omega_0 t) \qquad \Gamma \text{ is real and positive}$$

$\Theta(t)$ is the Heaviside step function described in equation 4.143.

15. Find the inverse Fourier transform of

$$P(\Omega) = e^{-\Gamma|\Omega|}$$

with Γ real and positive.

16. A periodic function $Q(t)$ has a Fourier exponential series given by

$$Q(t) = \sum_{l=-\infty}^{\infty} C_l e^{il\omega t}$$

with

$$C_l = \frac{1}{T}\int_{t_0}^{t_0+T} Q(t)e^{-il\omega t}\,dt$$

What is the Fourier transform of $Q(t)$?

17. Determine the Fourier transform of the first derivative of the Dirac δ-function $d\delta(t - t_0)/dt$, where t_0 is finite.

18. Use the convolution theorem for Fourier transforms to evaluate the integral

$$Q(t) \equiv \int_{-\infty}^{\infty} e^{-(t-t')^2/2}\, e^{-t'^2/2}\,dt'$$

19. In Cartesian coordinates, the three-dimensional Dirac δ-function is defined as $\delta(\mathbf{r} - \mathbf{r}_0) = \delta(x - x_0)\delta(y - y_0)\delta(z - z_0)$. Deduce the integral representation of $\delta(\mathbf{r} - \mathbf{r}_0)$ that is the extension to three dimensions of the one-dimensional representation of equations 4.106.

20. $Q(t)$ is an Mth-order polynomial defined by

$$Q(t) = \sum_{l=0}^{M} \alpha_l t^l$$

Show that the Laplace transform of $Q(t)$ is given by

$$\text{L.T.}[Q(t)] = \sum_{l=0}^{M} \frac{l!\alpha_l}{s^{l+1}}$$

21. Prove the convolution theorem for Laplace transforms

$$\text{L.T.}\left[\int_0^t Q(t - t')R(t')\,dt'\right] = \text{L.T.}[Q(t)]\text{L.T.}[R(t)]$$

22. $\Theta(t)$ is the Heaviside step function defined in equation 4.143. Determine the Laplace transform of the following functions.
    (a) $Q(t) = \Theta(t)\sin(\alpha t)$, α is real
    (b) $Q(t) = \Theta(t)\cosh^2(\alpha t)$, α is real and positive
    What other conditions (if any) must be imposed on α?

23. Derive the integral representation of $\delta(t - t_0)$ as shown in equation 4.160 by substituting the expression of $G(s) = \text{L.T.}[Q(t)]$ given in equation 4.150a, into the integrand for $Q(t)$ in equation 4.150b.

24. Using Cauchy's residue theorem, evaluate the Bromwich integral to find the (inverse) Laplace transform of the following functions.
    (a) $G(s) = \dfrac{1}{s^{n+1}}$
    (b) $G(s) = \dfrac{1}{s^2 + \alpha^2}$, α real
    (c) $G(s) = \dfrac{s}{(s^2 + \alpha^2)^2}$, α real
    (d) $G(s) = \dfrac{1}{(s + \alpha)^n}$, α real and positive
    (e) $G(s) = \dfrac{e^{-\beta s}}{(s^2 + \alpha^2)^2}$, α, β real
    What additional restriction(s) (if any) must be imposed on β?
    (f) $G(s) = \dfrac{1}{\sqrt{s} + \beta}$

CHAPTER 5

ORDINARY DIFFERENTIAL EQUATIONS

In general, ordinary linear differential equations contain terms that involve derivatives of a function that depends on one independent variable. Thus all the derivatives that appear in the differential equation are total derivatives. By contrast, partial differential equations contain partial derivatives of a dependent function of two or more independent variables.

The analysis of many scientific phenomena lead to a description in terms of an ordinary linear differential equation. A classic example of this is the description of the motion of a particle using Newton's second law. The net force acting on the particle will, in general, be time dependent and depend on the position and velocity of the particle. Then Newton's second law for the motion of the particle in one dimension is

$$F_{\text{net}}\left(y, \frac{dy}{dt}, t\right) = m\frac{d^2y}{dt^2} \tag{5.1a}$$

The most general form of an ordinary linear differential equation of order N is

$$F\left(x, y, \frac{dy}{dx}, \frac{d^2y}{dx^2}, \ldots, \frac{d^Ny}{dx^N}\right) = 0 \tag{5.1b}$$

where the order of a differential equation is the order of the highest derivative in the equation.

5.1 First-Order Differential Equations

The general description of a first-order ordinary differential equation is

$$\text{(a)} \quad M(x, y)\frac{dy}{dx} + N(x, y) = 0 \quad \Rightarrow \quad \text{(b)} \quad M(x, y)\, dy + N(x, y)\, dx = 0 \tag{5.2}$$

Finding a solution to this equation means determining the unknown function y in terms of x.

If equation 5.2b can be manipulated so that it is in the form

$$Q(y)\, dy + R(x)\, dx = 0 \tag{5.3}$$

then equation 5.2a is said to be separable, and equation 5.3 is its separated form. Integration of equation 5.3 yields a solution in the form

$$F(x, y) = C \tag{5.4}$$

That is, it is not necessary to express the solution as

$$y = f(x) \tag{5.5}$$

The goal of the discussion that follows is to separate the differential equation. Therefore, even if the integrals of $Q(x)$ and $R(x)$ cannot be evaluated in terms of elementary functions, we will assume that the differential equation has been solved if it can be separated.

To evaluate the constant, one must have information about the value of y at one specified value of x. For a first-order equation, such information is called an *initial condition* (if x represents a time variable) or a *boundary condition* (if x represents a spatial variable).

We will focus first on methods of separating these first-order equations.

M and *N* are products of two functions

The differential equation in the form of equation 5.2b is most easily separated if both $M(x, y)$ and $N(x, y)$ are products of two functions, one that depends only on x, and one that is dependent on y. If

$$\text{(a)} \quad M(x, y) = M_1(x)M_2(y) \qquad \text{(b)} \quad N(x, y) = N_1(x)N_2(y) \tag{5.6}$$

it is straightforward to rewrite equation 5.2b in the separated form

$$\frac{M_2(y)}{N_2(y)}\, dy + \frac{M_1(x)}{N_1(x)}\, dx = 0 \tag{5.7}$$

Substitution of variable

Let z be some combination of x and y such that M and N depend on x and y as

$$\text{(a)} \quad M(x, y) = M[z(x, y)] \qquad \text{(b)} \quad N(x, y) = N[z(x, y)] \tag{5.8}$$

With

$$dz = \frac{\partial z}{\partial x}\, dx + \frac{\partial z}{\partial y}\, dy \tag{5.9}$$

the differential equation can be written in the form

$$M(z)\, dz + \left[N(z)\frac{\partial z}{\partial y} - M(z)\frac{\partial z}{\partial x}\right] dx = 0 \tag{5.10a}$$

by substituting for dy from equation 5.9. Similarly, substituting for dx instead of dy, the differential equation can be written

$$N(z)\, dz + \left[M(z)\frac{\partial z}{\partial x} - N(z)\frac{\partial z}{\partial y}\right] dy = 0 \tag{5.10b}$$

Equation 5.10a can be separated if the combination in brackets can be written as a product of two functions, one that depends only on x, and one that is dependent on z. That is,

$$N(z)\frac{\partial z}{\partial y} - M(z)\frac{\partial z}{\partial x} = A(x)B(z) \tag{5.11}$$

$$\Rightarrow \quad \frac{M(z)}{B(z)}\, dz + A(x)\, dx = 0 \tag{5.12}$$

Similarly, if

$$M(z)\frac{\partial z}{\partial x} - N(z)\frac{\partial z}{\partial y} = C(y)D(z) \tag{5.13}$$

equation 5.10b is separable in terms of y and z in the form

$$\frac{N(z)}{D(z)}\,dz + C(y)\,dy = 0 \tag{5.14}$$

For example, if x and y occur in both M and N in the form

$$\text{(a)}\quad M(x, y) = M(\alpha x + \beta y) \qquad \text{(b)}\quad N(x, y) = N(\alpha x + \beta y) \tag{5.15}$$

where α and β are constants, equation 5.2b can be separated. Setting

$$z = \alpha x + \beta y \tag{5.16}$$

$$\Rightarrow \quad \frac{M(z)}{\beta N(z) + \alpha M(z)}\,dz + dx = 0 \tag{5.17a}$$

The equation can also be separated in terms of y and z as

$$\frac{N(z)}{\alpha M(z) - \beta N(z)}\,dz + dy = 0 \tag{5.17b}$$

Integration of equations 5.17 yield solutions in the form

$$\text{(a)}\quad G(x, z) = G(x, \alpha x + \beta y) = \text{const.} \qquad \text{(b)}\quad H(y, z) = H(y, \alpha x + \beta y) = \text{const.} \tag{5.18}$$

Example 5.1

As an example, consider

$$\text{(a)}\quad (x - 4y + 2)^2\frac{dy}{dx} - (x - 4y + 5) = 0 \qquad \text{with} \qquad \text{(b)}\quad y(5) = 0 \tag{5.19}$$

The substitution

$$z = x - 4y \tag{5.20}$$

$$\Rightarrow \quad dy + \frac{(z + 5)}{(z^2 - 16)}\,dz = 0 \tag{5.21}$$

The integral of this yields

$$y + \tfrac{9}{8}\log(z - 4) - \tfrac{1}{8}\log(z + 4) = C \tag{5.22}$$

Substituting for z, the solution is

$$y + \tfrac{9}{8}\log(x - 4y - 4) - \tfrac{1}{8}\log(x - 4y + 4) = C \tag{5.23}$$

From $y(5) = 0$, we obtain the solution

$$y + \tfrac{9}{8}\log(x - 4y - 4) - \tfrac{1}{8}\log(x - 4y + 4) = -\tfrac{1}{8}\log(9) \quad \square \tag{5.24}$$

If M and N depend on x and y as some power of y/x,

$$\text{(a)}\quad M(x, y) = M\left(\left(\frac{y}{x}\right)^p\right) \qquad \text{(b)}\quad N(x, y) = N\left(\left(\frac{y}{x}\right)^p\right) \tag{5.25}$$

one has some flexibility in the choice of substitution variable. For example,

$$z = \left(\frac{y}{x}\right)^p \tag{5.26}$$

$$\Rightarrow \quad dy = z^{1/p}\,dx + \frac{x}{p} z^{(1/p-1)}\,dz \tag{5.27}$$

yields a separated equation for any value of p, in the form

$$\frac{z^{(1/p-1)}M(z)}{p[N(z) + z^{1/p}M(z)]}\,dz + \frac{dx}{x} = 0 \tag{5.28}$$

Other possible choices (and perhaps ones that will make the integrals easier to evaluate) would be to make a substitution

$$\text{(a)} \quad z = \frac{y}{x} \qquad \text{or} \qquad \text{(b)} \quad z = \frac{x}{y} \tag{5.29}$$

For example, if we were to choose the substitution $z = y/x$, the differential equation would be separated as

$$\frac{M(z)}{zM(z) + N(z)}\,dz + \frac{dx}{x} = 0 \tag{5.30}$$

Example 5.2

As an example, consider

$$\left[4 - 3\left(\frac{y}{x}\right)^2\right] dy + \left(\frac{y}{x}\right)^4 dx = 0 \tag{5.31}$$

If we consider M and N as functions of y/x, the substitution

$$z = \frac{y}{x} \tag{5.29a}$$

$$\Rightarrow \quad \frac{(4 - 3z^2)}{z(z^3 - 3z^2 + 4)}\,dz + \frac{dx}{x} = \frac{(4 - 3z^2)}{z(z+1)(z-2)^2}\,dz + \frac{dx}{x} = 0 \tag{5.32}$$

Taking the M and N to be functions of $(y/x)^2$, the equation separates in the form

$$\frac{(4 - 3z)}{2(z^{5/2} - 3z^2 + 4z)}\,dz + \frac{dx}{x} = 0 \tag{5.33}$$

which is difficult to integrate. The integral of equation 5.32 is

$$\log y + \log x - \tfrac{1}{9}\log(y + x) - \tfrac{8}{9}\log(y - 2x) - \frac{4x}{3(y - 2x)} = C \quad \square \tag{5.34}$$

One approach to integrating equation 5.33 is to make the substitution $z = u^2$. The resulting equation will be in the same form as the differential equation expressed in equation 5.32. This example illustrates the fact that when the differential equation depends on some power of the ratio y/x, there is a flexibility in choosing the substitution variable, but that one substitution may be preferred over other choices.

Example 5.3

As a second example, consider

$$\text{(a)} \quad \left(\frac{x}{y}\right)^2 dy + (1 + e^{y/x})\,dx = 0 \qquad \text{with} \qquad \text{(b)} \quad y(1) = 0 \tag{5.35}$$

We note here that both $z = y/x$ and $z = x/y$ seem to be reasonable choices for a substitution variable. However, the exponential in equation 5.35a occurs as $e^{y/x}$. It is invariably easier to evaluate integrals involving e^z than to evaluate integrals involving $e^{1/z}$. Thus it is preferable to make the substitution $z = y/x$, which leads to

$$\frac{dx}{x} + (z + z^2 + z^2 e^z)^{-1}\, dz = 0 \tag{5.36}$$

$$\Rightarrow \quad \log(x) + \tfrac{1}{2}z^2 + \tfrac{1}{3}z^3 + [z^2 - 2z + 2]e^z$$
$$= \log(x) + \tfrac{1}{2}\left(\frac{y}{x}\right)^2 + \tfrac{1}{3}\left(\frac{y}{x}\right)^3 + \left[\left(\frac{y}{x}\right)^2 - 2\left(\frac{y}{x}\right) + 2\right]e^{y/x} = C \tag{5.37}$$

From $y(1) = 0$, we obtain $C = 2$. □

If equation 5.11 or 5.13 cannot be satisfied, the differential equation cannot be separated by substitution. To illustrate this, let M and N depend on x and y in the form

$$z = \alpha x^p + \beta y^q \tag{5.38}$$

If the differential equation is separable in x and z, then equation 5.11 becomes

$$q\beta y^{q-1}N(z) + p\alpha x^{p-1}M(z) = A(x)B(z) \tag{5.39}$$

This can only be satisfied by taking $p = q = 1$. That is, only the linear sum of x and y of equation 5.16 is amenable to separation by substitution.

Similarly, as illustrated in Problem 5, a combination of the form

$$z = x^p y^q \tag{5.40}$$

is separable by substitution only if $p = -q$.

Differential equations involving homogeneous functions

A function $F(x, y)$ is said to be *homogeneous of order p* if

$$F(\alpha x, \alpha y) = \alpha^p F(x, y) \tag{5.41}$$

If $M(x, y)$ and $N(x, y)$ are homogeneous functions of the same order p, the differential equation can be separated.

Writing M and N as

$$M(x, y) = M\left(x \cdot 1, x \cdot \frac{y}{x}\right) = x^p M\left(1, \frac{y}{x}\right) = x^p M\left(\frac{y}{x}\right) \tag{5.42a}$$

$$N(x, y) = N\left(x \cdot 1, x \cdot \frac{y}{x}\right) = x^p N\left(1, \frac{y}{x}\right) = x^p N\left(\frac{y}{x}\right) \tag{5.42b}$$

$$\Rightarrow \quad M\left(\frac{y}{x}\right) dy + N\left(\frac{y}{x}\right) dx = 0 \tag{5.43}$$

In this form the equation is separable by a substitution

$$z = \left(\frac{y}{x}\right)^q \tag{5.44}$$

As indicated in Example 5.2, q will usually be chosen to be 1.

Example 5.4

As an example, consider

$$2xy\, dy - (x^2 + y^2)\, dx = 0 \tag{5.45}$$

Since $2xy$ and $(x^2 + y^2)$ are homogeneous of order 2, equation 5.45 is written as

$$x^2\left[2\left(\frac{y}{x}\right) dy - \left(1 + \left(\frac{y}{x}\right)^2\right) dx\right] = 0 \tag{5.46}$$

After cancellation of x^2 and the substitution $z = y/x$, equation 5.46 is separated in the form

$$\frac{2z}{(z^2 - 1)}\,dz + \frac{dx}{x} = 0 \tag{5.47}$$

The solution to this equation (without initial/boundary conditions) is

$$\text{(a)}\quad x(z^2 - 1) = C \qquad \Rightarrow \qquad \text{(b)}\quad y^2 - x^2 = Cx \quad \square \tag{5.48}$$

Exact equations

As can be seen from our discussion thus far, the solution to a first-order differential equation will be of the form

$$F(x, y) = C \tag{5.4}$$

$$\Rightarrow \quad dF = \frac{\partial F}{\partial y}\,dy + \frac{\partial F}{\partial x}\,dx = 0 \tag{5.49}$$

This will be the same as

$$M(x, y)\,dy + N(x, y)\,dx = 0 \tag{5.2b}$$

if

$$\text{(a)}\quad M(x, y) = \frac{\partial F}{\partial y} \qquad \text{(b)}\quad N(x, y) = \frac{\partial F}{\partial x} \tag{5.50}$$

If $M(x, y)$ and $N(x, y)$ can be expressed as partial derivatives of one function, the differential equation of is said to be *exact*. Then, finding the solution to the differential equation becomes the task of determining the function $F(x, y)$.

To ascertain whether an equation is exact, let us assume that M and N satisfy equations 5.50. We then consider the following partial derivatives of these equations:

$$\text{(a)}\quad \frac{\partial M}{\partial x} = \frac{\partial^2 F}{\partial x\,\partial y} \qquad \text{(b)}\quad \frac{\partial N}{\partial y} = \frac{\partial^2 F}{\partial y\,\partial x} \tag{5.51}$$

If $F(x, y)$ is continuous and infinitely differentiable, its second derivative is independent of the order in which the partial derivatives are taken. This is satisfied by most functions encountered in scientific problems. Then the functions M and N of an exact equation satisfy

$$\frac{\partial M}{\partial x} = \frac{\partial N}{\partial y} \tag{5.52}$$

If equation 5.52, is satisfied, the differential equation is exact.

To find $F(x, y)$, we begin with one of the partial derivatives of equations 5.50. For example, setting

$$M(x, y) = \frac{\partial F}{\partial y} \tag{5.50a}$$

$$\Rightarrow \quad F(x, y) = \int M(x, y)\,\partial y + G(x) \tag{5.53}$$

The notation of partial integration in equation 5.53 means that the integration is to be performed on the explicit y dependence of M, treating x as a constant. Since x is treated as a constant, the *constant of integration* can depend on x. It is obvious that the function $F(x, y)$ of equation 5.53 satisfies equation 5.50a.

To determine $G(x)$, we impose

$$N(x, y) = \frac{\partial F}{\partial x} \tag{5.50b}$$

on $F(x, y)$ as given in equation 5.53. We obtain

$$N(x, y) = \frac{\partial}{\partial x} \int M(x, y)\, \partial y + \frac{dG}{dx} \tag{5.54a}$$

$$\Rightarrow \quad G(x) = \int \left[N(x, y) + \frac{\partial}{\partial x} \int M(x, y)\, \partial y \right] dx \tag{5.54b}$$

Thus with equation 5.53, the solution to the differential equation is

$$F(x, y) = \int M(x, y)\, \partial y + \int \left[N(x, y) - \frac{\partial}{\partial x} \int M(x, y)\, \partial y \right] dx = C \tag{5.55a}$$

If we begin with equation 5.50b, an analogous development yields the solution

$$F(x, y) = \int N(x, y)\, \partial x + \int \left[M(x, y) - \frac{\partial}{\partial y} \int N(x, y)\, \partial x \right] dy = C \tag{5.55b}$$

which, of course, must be the same solution expressed in equation 5.55a.

Since dx in the second integral of equation 5.55a is an exact differential rather than a partial differential, the integral can be evaluated only if the integrand is a function of x and contains no y-dependence. Similarly, dy in the corresponding integral in equation 5.55b is an exact differential, requiring the function in brackets to be a function of y only. It is easy to see that since M and N satisfy the condition for exactness specified in equation 5.52, such requirements are satisfied.

Example 5.5

To illustrate, consider

$$\left(\frac{1}{y} - \frac{x}{y^2} \right) dy + \left(2x + \frac{1}{y} \right) dx = 0 \tag{5.56}$$

The first step in the process is to determine if this equation is exact. With

$$\text{(a)} \quad M(x, y) = \left(\frac{1}{y} - \frac{x}{y^2} \right) \quad \Rightarrow \quad \text{(b)} \quad \frac{\partial M}{\partial x} = -\frac{1}{y^2} \tag{5.57}$$

Similarly

$$\text{(a)} \quad N(x, y) = \left(2x + \frac{1}{y} \right) \quad \Rightarrow \quad \text{(b)} \quad \frac{\partial N}{\partial y} = -\frac{1}{y^2} \tag{5.58}$$

Therefore, the differential equation is exact.

The solution is found by first determining

$$\text{(a)} \quad \int M(x, y)\, \partial y = \log y + \frac{x}{y} \quad \Rightarrow \quad \text{(b)} \quad \frac{\partial}{\partial x} \int M(x, y)\, \partial y = \frac{1}{y} \tag{5.59}$$

Then

$$N(x, y) - \frac{\partial}{\partial x}\int M(x, y)\,\partial y = 2x \tag{5.59c}$$

$$\Rightarrow \quad G(x) = \int\left[N(x, y) - \frac{\partial}{\partial x}\int M(x, y)\,\partial y\right] dx = x^2 \tag{5.59d}$$

$$\Rightarrow \quad F(x, y) = \log y + \frac{x}{y} + x^2 = C \quad \square \tag{5.60}$$

Integrating factor for nonexact differential equations

If the differential equation is not exact, then

$$\frac{\partial M}{\partial x} \neq \frac{\partial N}{\partial y} \tag{5.61}$$

Since

$$M(x, y)\, dy + N(x, y)\, dx = 0 \tag{5.2b}$$

is not changed by multiplication by an overall factor $\mu(x, y)$, the solution to

$$\mu(x, y) M(x, y)\, dy + \mu(x, y) N(x, y)\, dx = 0 \tag{5.62}$$

is the same as the solution to the original equation.

For some differential equations, it is possible to find a $\mu(x, y)$ such that equation 5.62 is exact. Such a function is called an *integrating factor*. Equation 5.62 is exact if

$$\text{(a)} \quad \frac{\partial}{\partial x}(\mu M) = \frac{\partial}{\partial y}(\mu N) \qquad \Rightarrow \qquad \text{(b)} \quad \mu\left(\frac{\partial M}{\partial x} - \frac{\partial N}{\partial y}\right) = N\frac{\partial \mu}{\partial y} - M\frac{\partial \mu}{\partial x} \tag{5.63}$$

To see the conditions for which an integrating factor can be found, let $\mu(x, y)$ depend on x and y in the combination

$$\text{(a)} \quad z = z(x, y) \qquad \Rightarrow \qquad \text{(b)} \quad \mu(x, y) = \mu[z(x, y)] \tag{5.64}$$

Then equation 5.63b becomes

$$\mu(z)\left(\frac{\partial M}{\partial x} - \frac{\partial N}{\partial y}\right) = N\frac{d\mu}{dz}\frac{\partial z}{\partial y} - M\frac{d\mu}{dz}\frac{\partial z}{\partial x} \tag{5.65}$$

$$\Rightarrow \quad \frac{d\mu(z)}{\mu(z)} = \frac{(\partial M/\partial x - \partial N/\partial y)}{(N\,\partial z/\partial y - M\,\partial z/\partial x)}\, dz \tag{5.66}$$

Defining

$$\frac{(\partial M/\partial x - \partial N/\partial y)}{(N\,\partial z/\partial y - M\,\partial z/\partial x)} \equiv T(x, y) \tag{5.67}$$

equation 5.66 can be integrated if $T(x, y)$ can be written as

$$T(x, y) = T(z) \tag{5.68}$$

That is, if $T(x, y)$ depends on x and y in the same combination that μ does, then

$$\mu(z) = e^{\int T(z)\, dz} \tag{5.69}$$

Once $\mu(x, y)$ is determined,

$$\mu M\, dy + \mu N\, dx = 0 \tag{5.62}$$

is exact and the solution to the differential equation is

$$F(x, y) = \int \mu M \, \partial y + \int \left[\mu N - \frac{\partial}{\partial x} \int \mu M \, \partial y \right] dx = C \tag{5.70a}$$

or

$$F(x, y) = \int \mu N \, \partial x + \int \left[\mu M - \frac{\partial}{\partial y} \int \mu N \, \partial x \right] dy = C \tag{5.70b}$$

In order to determine $T(x, y) = T(z)$ from equation 5.67, the combination $z(x, y)$ must be known beforehand. Clearly, the functional form of $z(x, y)$ is not obvious. As such, this technique will only be useful if, by judicious trial and error, one can determine the necessary $z(x, y)$.

Example 5.6

As an example, we will determine the integrating factor $\mu(x, y)$ that makes

$$x(x + y) \, dy + y(3x + y) \, dx = 0 \tag{5.71}$$

exact. Note that this equation is homogeneous and can be solved using the method discussed earlier for such differential equations. However, we will obtain a solution by determining an integrating factor that depends on x and y in the combination

$$z = xy(2x + y) \tag{5.72}$$

Evaluating

$$\frac{(\partial M/\partial x - \partial N/\partial y)}{(N\partial z/\partial y - M\partial z/\partial x)} = T(x, y) \tag{5.67}$$

$$\Rightarrow \quad T(x, y) = -\frac{1}{xy(2x + y)} = -\frac{1}{z} \tag{5.73}$$

$$\Rightarrow \quad \mu(z) = e^{-\int dz/z} = \frac{1}{z} = \frac{1}{xy(2x + y)} \tag{5.74}$$

Using either of equations 5.70, we obtain

$$\frac{x^2 y}{(2x + y)} = C \quad \square \tag{5.75}$$

A simpler form of this analysis occurs if an integrating factor can be found that depends on just one of the variables. If $\mu = \mu(x)$, then $z = x$, and equation 5.66 becomes

$$\frac{d\mu(x)}{\mu(x)} = \frac{(\partial M/\partial x - \partial N/\partial y)}{-M} dx \tag{5.76}$$

One can therefore ascertain if a single-variable integrating factor will exist that depends only on x by determining if

$$\frac{(\partial M/\partial x - \partial N/\partial y)}{-M} = \text{a function of } x \text{ only} \equiv T(x) \tag{5.77}$$

Similarly, if $\mu = \mu(y)$, then $z = y$ and equation 5.66 becomes

$$\frac{d\mu(y)}{\mu(y)} = \frac{(\partial M/\partial x - \partial N/\partial y)}{N} dy \tag{5.78}$$

One can thus determine whether a single-variable integrating factor depending on y exists by determining if

$$\frac{(\partial M/\partial x - \partial N/\partial y)}{N} = \text{a function of } y \text{ only} \equiv T(y) \tag{5.79}$$

If equation 5.77 or 5.79 is satisfied, then

$$\text{(a)} \quad \mu(x) = e^{\int T(x)\,dx} \qquad \text{or} \qquad \text{(b)} \quad \mu(y) = e^{\int T(y)\,dy} \tag{5.80}$$

As a note of caution, it is important to be aware that the functions M and N could satisfy

$$\text{(a)} \quad \frac{(\partial M/\partial x - \partial N/\partial y)}{-M} = T(y) \qquad \text{or} \qquad \text{(b)} \quad \frac{(\partial M/\partial x - \partial N/\partial y)}{N} = T(x) \tag{5.81}$$

Simply because these combinations involving $M(x, y)$ and $N(x, y)$ are each functions of a single variable does not guarantee that a single-variable integrating factor exists. $(\partial M/\partial x - \partial N/\partial y)/(-M)$ must depend only on x if a single-variable integrating factor $\mu(x)$ is to exist, and $(\partial M/\partial x - \partial N/\partial y)/N$ must be a function only of y if $\mu(y)$ is to exist.

Example 5.7

As an example, let us find a single-variable integrating factor that will make

$$\cos x\,dy - y(\sin x + 3)\,dx = 0 \tag{5.82}$$

exact. Since

$$\frac{\partial M}{\partial x} - \frac{\partial N}{\partial y} = 3 \tag{5.83}$$

we see that the differential equation is not exact. Because

$$\frac{(\partial M/\partial x - \partial N/\partial y)}{N} = -\frac{3}{y(\sin x + 3)} \tag{5.84a}$$

contains x-dependence, a single-variable integrating factor that depends on y does not exist. But

$$\frac{(\partial M/\partial x - \partial N/\partial y)}{M} = \frac{3}{\cos x} \tag{5.84b}$$

is independent of y. Thus a single-variable integrating factor that depends only on x does exist. It is

$$\text{(a)} \quad \mu(x) = e^{-\int (3/\cos x)\,dx} = (\sec x + \tan x)^{-3} \tag{5.85}$$

The solution to the differential equation, then, is found by:

$$\text{(a)} \quad \int \mu M\,\partial y = y\frac{\cos x}{(\sec x + \tan x)^3} \qquad \text{(b)} \quad \frac{\partial}{\partial x}\int \mu M\,\partial y = -y\frac{(\sin x + 3)}{(\sec x + \tan x)^3} = \mu N$$

$$\text{(c)} \quad \int \left[\mu N - \frac{\partial}{\partial x}\int \mu M\,\partial y\right] dx = \int [\mu N - \mu N]\,dx = 0 \tag{5.86}$$

$$\Rightarrow \quad y\frac{\cos x}{(\sec x + \tan x)^3} = y\frac{\cos^4 x}{(1 + \sin x)^3} = C \quad \square \tag{5.87}$$

Bernoulli's differential equation

The general form of the Bernoulli differential equation of index n is

$$\frac{dy}{dx} + P(x)y = Q(x)y^n \tag{5.88}$$

where n can have any value.

1. For $n = 0$ the equation has the form

$$dy + (yP(x) - Q(x))\,dx = 0 \tag{5.89}$$

We note that since

$$\text{(a)}\quad M(x, y) = 1 \qquad \text{(b)}\quad N(x, y) = yP(x) - Q(x) \tag{5.90}$$

$$\Rightarrow \quad \frac{(\partial M/\partial x - \partial N/\partial y)}{-M} = P(x) \tag{5.91}$$

Therefore, a single-variable integrating factor $\mu(x)$ exists. It is

$$\mu(x) = e^{\int P(x)\,dx} \tag{5.92}$$

Following the steps outlined in equations 5.70, the solution is

$$ye^{\int P(x)\,dx} - \int Q(x)e^{\int P(x)\,dx}\,dx = C \tag{5.93}$$

2. For $n = 1$, Bernoulli's equation is

$$dy + y[P(x) + Q(x)]\,dx = 0 \tag{5.94}$$

which is separable. The solution is

$$y = Ce^{\int [P(x) - Q(x)]\,dx} \tag{5.95}$$

3. For $n \neq 0, 1$, we write equation 5.88 in the form

$$y^{-n}\frac{dy}{dx} + y^{1-n}P(x) = Q(x) \tag{5.96}$$

The substitution

$$v = y^{1-n} \tag{5.97}$$

$$\Rightarrow \quad \frac{1}{(1-n)}\frac{dv}{dx} + vP(x) = Q(x) \tag{5.98}$$

which is the zero-index Bernoulli equation in the variable v. The solution for v is obtained using equation 5.93. The result is

$$v(x) = Ce^{(n-1)\int P(x)\,dx} + (1-n)e^{(n-1)\int P(x)\,dx}\int dx\,Q(x)e^{(1-n)\int P(x)\,dx} \tag{5.99}$$

5.2 Higher-Order Differential Equations with Constant Coefficients

Heaviside operator theory

A general method of solution, called the *Heaviside operator method*, can be developed to solve

$$\sum_{n=0}^{N} a_n \frac{d^n y}{dx^n} = F(x) \tag{5.100}$$

if each of the coefficients a_n is a constant. To achieve this, the Heaviside differential operator, D, is defined as

$$D \equiv \frac{d}{dx} \tag{5.101}$$

such that

$$D^k y \equiv \frac{d^k y}{dx^k} \tag{5.102}$$

The inverse operator, then, is an integral operator defined by

$$D^{-1}F(x) \equiv \int F(x)\, dx + C \tag{5.103}$$

The differential equation, written in terms of the differential operator, becomes

$$\sum_{n=0}^{N} a_n D^n y(x) = F(x) \tag{5.104}$$

If z is an algebraic parameter, we define the polynomial function

$$P(z) = \sum_{n=0}^{N} a_n z^n \tag{5.105a}$$

Thus if z is replaced by D, the polynomial function becomes a polynomial operator

$$P(D) = \sum_{n=0}^{N} a_n D^n \tag{5.105b}$$

and the differential equation can be written

$$P(D)y(x) = F(x) \tag{5.106}$$

If the differential equation has a solution, an inverse operator, denoted by $P^{-1}(D)$ must exist such that

$$P(D)P^{-1}(D) = P^{-1}(D)P(D) = 1 \tag{5.107}$$

With this definition of $P^{-1}(D)$, the solution to equation 5.106 can be written formally as

$$y(x) = P^{-1}(D)F(x) + y_0(x) \tag{5.108}$$

It is straightforward to see that $y(x)$ will be the solution to equation 5.106 by requiring that

$$P(D)y_0(x) = 0 \tag{5.109}$$

That is, to every solution for which $F(x) \neq 0$, we must add a function $y_0(x)$ that satisfies the homogeneous equation. The part of the solution obtained from $P^{-1}(D)F(x)$ is called the particular solution. Thus the complete solution to the differential equation, as noted in equation 5.108, is the sum of the homogeneous and particular solutions

$$y(x) = y_0(x) + y_p(x) \tag{5.110}$$

Solution to the homogeneous equation

Consider the polynomial as a function of the algebraic quantity z. Let the roots of the polynomial

$$P(z) = \sum_{n=0}^{N} a_n z^n \tag{5.111}$$

be $z_1, z_2, \ldots, z_N$. Since the coefficient of D^N cannot be zero for a differential equation of order N, $a_N \neq 0$. Thus the equation can be divided by a_N or, by renaming the remaining coefficients, a_N can be taken to be 1. Then $P(z)$ can be written

$$P(z) = (z - z_1)(z - z_2)(z - z_3) \cdots (z - z_N) \tag{5.112a}$$

$$\Rightarrow \quad P(D) = (D - z_1)(D - z_2)(D - z_3) \cdots (D - z_N) \tag{5.112b}$$

$$\Rightarrow \quad (D - z_1)(D - z_2)(D - z_3) \cdots (D - z_N) y_0(x) = 0 \tag{5.113}$$

We will first consider the homogeneous equation for which all of the roots of the polynomial are distinct. We define

$$y_1(x) \equiv (D - z_2)(D - z_3) \cdots (D - z_N) y_0(x) \tag{5.114}$$

Then the differential equation can be written

$$\text{(a)} \quad (D - z_1) y_1(x) = 0 \quad \Rightarrow \quad \text{(b)} \quad \frac{dy_1}{dx} = z_1 y_1 \tag{5.115}$$

$$\Rightarrow \quad y_1(x) = A_1 e^{z_1 x} \tag{5.116}$$

Therefore, from equation 5.114,

$$(D - z_2)(D - z_3) \cdots (D - z_N) y_0 = A_1 e^{z_1 x} \tag{5.117}$$

Defining

$$y_2(x) \equiv (D - z_3) \cdots (D - z_N) y_0(x) \tag{5.118}$$

equation 5.117 becomes

$$\text{(a)} \quad (D - z_2) y_2(x) = A_1 e^{z_1 x} \quad \Rightarrow \quad \text{(b)} \quad \frac{dy_2}{dx} - z_2 y_2 = A_1 e^{z_1 x} \tag{5.119}$$

This is Bernoulli's equation of index $n = 0$. Writing it in the form

$$dy_2 - (z_2 y_2 + A_1 e^{z_1 x})\, dx = 0 \tag{5.119c}$$

and comparing it to equation 5.89, we see that

$$\text{(a)} \quad P(x) = -z_2 \qquad \text{(b)} \quad Q(x) = A_1 e^{z_1 x} \tag{5.120}$$

Referring to equation 5.93, the solution is

$$y_2 e^{-z_2 x} - \int A_1 e^{(z_1 - z_2)x}\, dx = A_2 \tag{5.121}$$

Since we are considering the solution for nonrepeated roots, $z_1 \neq z_2$. Therefore, the exponent under the integral is nonzero, and equation 5.121 becomes

$$y_2 = \frac{A_1}{(z_1 - z_2)} e^{z_1 x} + A_2 e^{z_2 x} \tag{5.122}$$

We continue this process, defining

$$y_3 = (D - z_4)(D - z_5) \cdots (D - z_N) y_0 \tag{5.123}$$

so that the differential equation for y_3 is, again, a Bernoulli equation of index 0.

$$\frac{dy_3}{dx} - z_3 y_3 = \frac{A_1}{(z_1 - z_2)} e^{z_1 x} + A_2 e^{z_2 x} \tag{5.124}$$

Renaming the coefficients of the exponential functions, the solution to equation 5.124 with z_1, z_2, and z_3 distinct, is

$$y_3 = B_1 e^{z_1 x} + B_2 e^{z_2 x} + B_3 e^{z_3 x} \tag{5.125}$$

Continuing this process until all factors of $(D - z_i)$ are exhausted, the solution to the homogeneous equation with all the roots distinct becomes

$$y_0 = \sum_{n=1}^{N} B_n e^{z_n x} \tag{5.126}$$

The constants B_n are undetermined. They must be fixed by initial or boundary conditions.

We next obtain the solution to the homogeneous equation when the polynomial operator has repeated roots. For simplicity, we will assume that just one of the roots is repeated l times, and designate this root z_N. Then the polynomial operator of equation 5.112b becomes

$$P(D) = (D - z_1)(D - z_2) \cdots (D - z_{N-l})(D - z_N)^l \tag{5.127}$$

Defining

$$y_{N-l} \equiv (D - z_N)^l y_0 \tag{5.128a}$$

the homogeneous differential equation becomes

$$(D - z_1)(D - z_2) \cdots (D - z_{N-l}) y_{N-l} = 0 \tag{5.128b}$$

Since none of the roots of the operator of equation 5.128b is a multiple root, repeating our previous analysis we obtain the solution

$$y_{N-l} = (D - z_N)^l y_0 = \sum_{n=1}^{N-l} B_n e^{z_n x} \tag{5.129}$$

As before, we solve this one factor at a time. Defining

$$y_{N-l+1} \equiv (D - z_N)^{l-1} y_0 \tag{5.130}$$

we again obtain a zero-index Bernoulli equation

$$(D - z_N) y_{N-l+1} = \sum_{n=1}^{N-l} B_n e^{z_n x} \tag{5.131}$$

$$\Rightarrow \quad y_{N-l+1} e^{-z_N x} - \int \sum_{n=1}^{N-l} B_n e^{(z_n - z_N)x}\, dx = A_0 \tag{5.132a}$$

Since $z_n \neq z_N$ for any n,

$$y_{N-l+1} = (D - z_N)^{l-1} y_0 = A_0 e^{z_N x} + \sum_{n=1}^{N-l} B_n e^{z_n x} \tag{5.132b}$$

where, as was done above, constants have been renamed, using the old symbol to designate the new constant. For example, B_n in equation 5.132b is $1/(z_n - z_N)$ multiplied by B_n of equation 5.132a. In the analysis that follows, when convenient, we will rename constants of integration without further comment.

Defining

$$\text{(a)}\quad y_{N-l+2} \equiv (D - z_N)^{l-2} y_0$$

$$\Rightarrow \quad \text{(b)}\quad (D - z_N) y_{N-l+2} = A_0 e^{z_N x} + \sum_{n=1}^{N-l} B_n e^{z_n x} \tag{5.133}$$

This has solution $$y_{N-l+2} e^{-z_N x} - \int \left[A_0 e^{z_N x} + \sum_{n=1}^{N-l} B_n e^{z_n x} \right] e^{-z_N x}\, dx = A_1 \tag{5.134a}$$

$$\Rightarrow \quad y_{N-l+2} = \sum_{n=1}^{N-l} B_n e^{z_n x} + (A_0 x + A_1) e^{z_N x} \tag{5.134b}$$

This process is continued until the solution for y_0 is obtained. The solution is

$$y_0 = \sum_{n=1}^{N-l} B_n e^{z_n x} + \left(C_1 x^{l-1} + C_2 x^{l-2} + \cdots + C_l \right) e^{z_N x} \tag{5.135}$$

We note that if the Nth root is not repeated, it has a multiplicity $l = 1$. In that case, the multiple of $e^{z_N x}$ is a polynomial of order zero (a constant), and the solution expressed in equation 5.135 is

$$y_0 = \sum_{n=1}^{N-1} B_n e^{z_n x} + C_1 e^{z_N x} \equiv \sum_{n=1}^{N} B_n e^{z_n x} \tag{5.136}$$

which is the solution to the homogeneous equation with all roots distinct (equation 5.126).

In summary, then, for every root z_k of the polynomial, the homogeneous solution contains a multiple of an exponential factor $e^{z_k x}$. That multiple is a polynomial in x, the order of which is one less than the multiplicity of z_k.

Therefore, the problem of finding the solution to the homogeneous differential equation is reduced to the problem of finding the roots of an algebraic polynomial and determining the multiplicity of each root.

Example 5.8

As an example, consider the differential equation

$$\begin{aligned} &\frac{d^6 y_0}{dx^6} - 4\frac{d^5 y_0}{dx^5} - 2\frac{d^4 y_0}{dx^4} + 16\frac{d^3 y_0}{dx^3} + 5\frac{d^2 y_0}{dx^2} - 20\frac{dy_0}{dx} - 12 y_0 \\ &= (D^6 - 4D^5 - 2D^4 + 16D^3 + 5D^2 - 20D - 12) y_0 = 0 \end{aligned} \tag{5.137}$$

This extremely unwieldy polynomial operator has been constructed from

$$P(D) = (D-2)^2 (D+1)^3 (D-3) \tag{5.138}$$

Written as an algebraic polynomial, $P(z)$ obviously has roots $z = 2$ (repeated twice), $z = -1$ (repeated three times), and $z = 3$ (repeated once or nonrepeated). Therefore, the solution to equation 5.137b is

$$y_0 = (B_0 + B_1 x) e^{2x} + \left(B_2 + B_3 x + B_4 x^2 \right) e^{-x} + B_5 e^{3x} \quad \square \tag{5.139}$$

Example 5.9

Consider a second example in which one of the roots of the polynomial is zero.

$$(D+2)^4 D^2 (D-1) y = 0 \tag{5.140}$$

The operator polynomial, written in terms of an algebraic variable, has roots $z = -2$ (quadruple root), $z = 0$ (double root), and $z = 1$ (nonrepeated root). Therefore, the solution is

$$y = \left(B_0 + B_1 x + B_2 x^2 + B_3 x^3\right)e^{-2x} + \left(B_4 + B_5 x\right) + B_6 e^x \tag{5.141}$$

That is, the root $z = 0$ contributes a polynomial of order 1 multiplied by $e^0 = 1$. □

In general, for an Nth-order homogeneous differential equation, there are N constants of integration, which must be determined by applying N initial or boundary conditions. These will be expressed as values of the solution and/or its derivatives at different selected values of the independent variable. Initial conditions will be in the form of values of the solution and its first $N - 1$ derivatives at one value of the independent time variable (which is designated as the initial instant). Boundary conditions will be given as N values of the solution y, at N different values of the independent spatial variable.

If z_k is an l-fold repeated root (including $l = 1$), the factor $(D - z_k)^l$ in the polynomial operator gives rise to l terms in the solution of the form $Bx^m e^{z_k x}$ with $m \leqslant l - 1$.

Consider

$$D\left[e^{\alpha x}F(x)\right] = \alpha e^{\alpha x}F(x) + e^{\alpha x}DF(x) = e^{\alpha x}(D + \alpha)F(x) \tag{5.142a}$$

$$D^2\left[e^{\alpha x}F(x)\right] = \alpha^2 e^{\alpha x}F(x) + 2\alpha e^{\alpha x}DF(x) + e^{\alpha x}D^2F(x) = e^{\alpha x}(D + \alpha)^2 F(x) \tag{5.142b}$$

$$\Rightarrow \quad D^l\left[e^{\alpha x}F(x)\right] = e^{\alpha x}(D + \alpha)^l F(x) \tag{5.143}$$

That is, the factor $e^{\alpha x}$ can be "moved through" to the left of the operator (so it is no longer operated on) by replacing D by $D + \alpha$ in the operator.

Therefore,

$$(D - z_k)^l\left[x^m e^{z_k x}\right] = e^{z_k x}(D + z_k - z_k)^l x^m = e^{\alpha x}D^l x^m \tag{5.144}$$

But $D^l x^m = 0$ since $m \leqslant l - 1$. Thus each term of the form $x^m e^{z_k x}$ is independently a solution to the homogeneous differential equation.

Returning to Example 5.9, the sum of terms of equation 5.141,

$$y_1(x) = \left(B_0 + B_1 x + B_2 x^2 + B_3 x^3\right)e^{-2x} \tag{5.145}$$

arise from inverting the factor $(D + 2)^4$ in the operator of equation 5.140. For each term in this part of the solution,

$$P(D)B_m x^m e^{-2x} = D^2(D - 1)(D + 2)^4\left[B_m x^m e^{-2x}\right] = D^2(D - 1)\left[B_m e^{-2x}D^4 x^m\right] \tag{5.146}$$

Since $m = 0$, 1, 2, or 3, $D^4 x^m = 0$. Thus, $B_3 x^3 e^{-2x}$ is, by itself, a solution to the homogeneous equation. That is, when the polynomial operator operates on $B_3 x^3 e^{-2x}$, we obtain zero, independent of the other six terms in the solution expressed in equation 5.141.

Thus as noted above, each term in this part of the solution satisfies the homogeneous differential equation, independent of the existence of the other terms in the solution. This result is valid for each term that multiplies an undetermined constant in equation 5.141. With this example as a guide, we see that for a homogeneous equation of order N, there are N independent solutions multiplied by N undetermined coefficients.

Digressing temporarily, we note that the result expressed in equation 5.144 provides a method for evaluating certain integrals containing an exponential integrand. Consider

$$I = \int e^{\alpha x} G(x)\, dx \tag{5.147}$$

Since the operator D^{-1} is the integral operator, equation 5.147 can be written

$$I = D^{-1}[e^{\alpha x} G(x)] = e^{\alpha x}(D + \alpha)^{-1} G(x) \tag{5.148}$$

If, for example, $G(x) = x^k$, the integral becomes

$$I = \int x^k e^{\alpha x}\, dx = e^{\alpha x}(D + \alpha)^{-1} x^k = \frac{e^{\alpha x}}{\alpha} \sum_{l=0}^{\infty} (-1)^l \left(\frac{D}{\alpha}\right)^l x^k \tag{5.149a}$$

Since the lth derivative of x^k is zero for $l > k$, the sum can be terminated at $l = k$. Therefore,

$$\int x^k e^{\alpha x}\, dx = \frac{e^{\alpha x}}{\alpha} \sum_{l=0}^{k} (-1)^l \left(\frac{D}{\alpha}\right)^l x^k \tag{5.149b}$$

For $l \leqslant k$,

$$D^l x^k = \frac{k!}{(k-l)!} x^{k-l} \tag{5.150}$$

$$\Rightarrow \quad \int x^k e^{\alpha x}\, dx = \frac{e^{\alpha x}}{\alpha} \sum_{l=0}^{k} \frac{(-1)^l}{\alpha^l} \frac{k!}{(k-l)!} x^{k-l} \tag{5.151}$$

Solution to the inhomogeneous equation

To solve equation 5.106 with $F(x) \neq 0$, we proceed as we did when solving the homogeneous differential equation. We find the roots of the operator polynomial and write the differential equation as

$$(D - z_1)(D - z_2) \cdots (D - z_N) y = F(x) \tag{5.152}$$

As before, we define

$$y_1 \equiv (D - z_2) \cdots (D - z_N) y \tag{5.153a}$$

$$\Rightarrow \quad (D - z_1) y_1 = F(x) \tag{5.153b}$$

which is a Bernoulli equation of index 0. Referring to equation 5.93, the solution is

$$y_1 e^{-z_1 x} - \int F(x) e^{-z_1 x}\, dx = A_1 \tag{5.154}$$

Assuming the integral in equation 5.154 can be evaluated, $y_1(x)$ is, in principle, determined.

Defining

$$y_2(x) \equiv (D - z_3) \cdots (D - z_N) y \tag{5.155}$$

equation 5.153a becomes

$$(D - z_2) y_2 = y_1(x) \tag{5.156}$$

with $y_1(x)$ known, this is a Bernoulli equation of index 0, with solution

$$y_2 e^{-z_2 x} - \int y_1(x) e^{-z_2 x}\, dx = A_2 \tag{5.157}$$

This procedure is continued until a solution for $y(x)$ is obtained.

Example 5.10

As an example, consider

$$(D^2 - 4)y = \cosh x \tag{5.158}$$

Defining

$$\text{(a)} \quad y_1 \equiv (D - 2)y \Rightarrow \quad \text{(b)} \quad (D + 2)y_1 = \cosh x \tag{5.159}$$

which is a Bernoulli equation of index 0. With $z_1 = -2$, the solution is

$$\text{(a)} \quad y_1 e^{2x} - \int e^{2x} \cosh x\, dx = A_1$$

$$\Rightarrow \quad \text{(b)} \quad y_1(x) = (D - 2)y = A_1 e^{-2x} + \tfrac{1}{2}\left(\frac{e^x}{3} + e^{-x}\right) \tag{5.160}$$

With $z_2 = 2$, the solution to equation 5.160b, which is the solution to the original differential equation, is

$$y(x) = -\tfrac{1}{4}A_1 e^{-2x} A_2 e^{2x} - \tfrac{1}{3}\cosh x \equiv B_1 e^{-2x} + B_2 e^{2x} - \tfrac{1}{3}\cosh x \tag{5.161}$$

Since the roots of $z^2 - 4$ are $z = \pm 2$, the homogeneous solution to equation 5.158 is

$$y_0 = B_1 e^{2x} + B_2 e^{-2x} \tag{5.162}$$

Therefore, the first two terms of equation 5.161b duplicate the solution to the homogeneous equation. That is, by retaining the constants of integration, the solution to the inhomogeneous equation includes the homogeneous solution as well. Therefore, equation 5.161b is the complete solution to the differential equation. The part that does not duplicate the homogeneous solution, $-\frac{1}{3}\cosh x$, is the particular solution. □

In general, when solving the inhomogeneous equation, if the undetermined constants of integration are taken to be nonzero, all or part of the homogeneous solution will be obtained along with the particular solution. By setting undetermined constants to zero, we will obtain just the particular solution.

Analysis with specific inhomogeneous terms

As noted above, the general approach to determining the particular solution to an Nth-order equation with constant coefficients requires the solution of N zero-index Bernoulli equations. When the inhomogeneous term $F(x)$ is one of a few special forms, the particular solution can be obtained by simpler methods that avoid solving several Bernoulli equations. Those specific forms are

$$\text{(a)} \quad F(x) = e^{\alpha x} \qquad \text{(b)} \quad F(x) = \sin(\beta x) \text{ or } \cos(\beta x)$$

$$\text{(c)} \quad F(x) = \sinh(\gamma x) \text{ or } \cosh(\gamma x) \qquad \text{(d)} \quad F(x) = \text{polynomial in } x \tag{5.163}$$

Of course, since the sine, cosine, sinh, and cosh functions can be written in terms of exponentials, the solution with these inhomogeneous terms can be obtained using the same method we will develop for the solution involving the exponential function. But for some equations involving these functions, a different and faster approach is possible, so they will be treated separately.

As noted above, terms containing undetermined constants will duplicate part or all of the homogeneous solution. For each of the inhomogeneous functions of equations

5.163, we will be developing a method of finding the particular solution only. Thus, unless noted otherwise, we will set the terms containing undetermined constants to zero, recognizing that those terms will be included when the homogeneous solution is added to the particular solution.

Formally, the particular solution is of the form

$$y(x) = P^{-1}(D)F(x) \tag{5.164}$$

If we consider the algebraic analog of $P(D)$, the function $P^{-1}(z)$ can be expanded in a MacLaurin series as long as $P(0) \neq 0$. If $P(0) = 0$, $P(z)$ can be written as $z^l R(z)$, with $R(0) \neq 0$. In that case,

$$P^{-1}(D) = D^{-l}R^{-1}(D) \tag{5.165}$$

The operator D^{-l} is the l-fold integration operator. That is, ignoring constants of integration,

$$D^{-l}F(x) = \underbrace{\int dx \int dx \cdots \int F(x)\, dx}_{\longleftarrow\; l \text{ times} \;\longrightarrow} \equiv G(x) \tag{5.166}$$

Thus if $P(0) = 0$, the solution to the differential equation is

$$y(x) = R^{-1}(D)D^{-l}F(x) = R^{-1}(D)G(x) \tag{5.167}$$

Therefore, once the methods are developed for $P(0) \neq 0$, the solution with $P(0) = 0$ can be obtained straightforwardly.

Unless an operator $P(D)$ is an integer power of D, the inverse operator is not expressible in terms of integrals. Thus, evaluating $P^{-1}(D)F(x)$ becomes the problem of determining the effect of the inverse operator on a specific function $F(x)$, rather than one of expressing $P^{-1}(D)$ in terms of known operators (integrals and derivatives).

Taking $P(0) \neq 0$, we expand $P^{-1}(z)$ in a MacLaurin series,

$$P^{-1}(z) = \sum_{n=0}^{\infty} a_n z^n \tag{5.168}$$

Therefore, the particular solution to the differential equation can be written

$$y(x) = \sum_{n=0}^{\infty} a_n D^n F(x) \tag{5.169}$$

Exponential inhomogeneous term

With $F(x) = e^{\alpha x}$, equation 5.169 is

$$y(x) = \sum_{n=0}^{\infty} a_n D^n e^{\alpha x} \tag{5.170}$$

We will first find the solution for $P(\alpha) \neq 0$. Since

$$D^n e^{\alpha x} = \alpha^n e^{\alpha x} \tag{5.171}$$

$$\Rightarrow \quad y(x) = \sum_{n=0}^{\infty} a_n \alpha^n e^{\alpha x} = \left[\sum_{n=0}^{\infty} a_n \alpha^n\right] e^{\alpha x} \tag{5.172a}$$

Comparing this sum to equation 5.168, we see that

$$y(x) = P^{-1}(\alpha)e^{\alpha x} = \frac{e^{\alpha x}}{P(\alpha)} \tag{5.172b}$$

That is, for $F(x) = e^{\alpha x}$ and $P(\alpha) \neq 0$, the particular solution is found simply by replacing D in the polynomial operator by α and algebraically solving for $y(x)$.

Example 5.11

As an example, we will "solve" (i.e., write down the solution to)

$$(D^3 + 3D - 5)(D^2 - 4)y = e^{3x} \tag{5.173}$$

$$\Rightarrow \quad P(3) = (27 + 9 - 5)(9 - 4) = 155 \neq 0 \tag{5.174}$$

$$\Rightarrow \quad y(x) = \frac{e^{3x}}{155} \quad \square \tag{5.175}$$

If $P(\alpha) = 0$, then α is an l-fold root of $P(z)$. That is,

$$P(z) = (z - \alpha)^l R(z) \tag{5.176}$$

where $R(\alpha) \neq 0$. Therefore, the differential equation can be written

$$R(D)(D - \alpha)^l y = e^{\alpha x} \tag{5.177a}$$

$$\Rightarrow \quad (D - \alpha)^l y = \frac{e^{\alpha x}}{R(\alpha)} \tag{5.177b}$$

We define

$$y_1 \equiv (D - \alpha)^{l-1} y \tag{5.178}$$

The zero index Bernoulli equation and its solution are

$$\text{(a)} \quad Dy_1 - \alpha y_1 = \frac{e^{\alpha x}}{R(\alpha)} \qquad \text{(b)} \quad y_1 = \left(\frac{x}{R(\alpha)} + A_0\right)e^{\alpha x} = (D - \alpha)^{l-1} y \tag{5.179}$$

Defining

$$y_2 \equiv (D - \alpha)^{l-2} y \tag{5.180}$$

the resulting zero-index Bernoulli equation and its solution are

$$\text{(a)} \quad Dy_2 - \alpha y_2 = \left(\frac{x}{R(\alpha)} + A_0\right)e^{\alpha x} \qquad \text{(b)} \quad y_2 = \left(\frac{x^2}{2!R(\alpha)} + A_0 x + A_1\right)e^{\alpha x} \tag{5.181}$$

Continuing this process until all factors of $(D - \alpha)$ are exhausted, the solution to the differential equation (with constants of integration renamed) is

$$y(x) = \left(\frac{x^l}{l!R(\alpha)} + B_0 x^{l-1} + B_1 x^{l-2} + \cdots + B_{l-1}\right)e^{\alpha x} \tag{5.182}$$

As mentioned above, when the polynomial operator has an l-fold repeated root α, terms involving the undetermined constants $B_0, B_1, \ldots, B_{l-1}$ duplicate terms in the homogeneous solution. Therefore, the particular solution with $F(x) = e^{\alpha x}$ and α an l-fold root of the polynomial is

$$y(x) = P^{-1}(D)e^{\alpha x} = \frac{x^l e^{\alpha x}}{l!R(\alpha)} \tag{5.183}$$

Note that if α is not a root of $P(z)$, then $l = 0$, and $R(\alpha) = P(\alpha)$. In that case, equation 5.183 becomes the solution expressed in equation 5.172b. Therefore, equation 5.183 is a general description of the particular solution to this differential equation whether α is a root of $P(z)$ or not.

Example 5.12

As an example, consider

$$(D+4)^3(D-2)^2(D+1)y = 9e^{-4x} \tag{5.184}$$

Since $\alpha = -4$ is a triple root of $P(z)$,

$$R(D) = (D-2)^2(D+1) \tag{5.185}$$

$$\Rightarrow \quad y(x) = \frac{9x^3e^{-4x}}{3!\left[(-6)^2(-3)\right]} = -\frac{x^3e^{-4x}}{72} \quad \square \tag{5.186}$$

As a straightforward extension of this analysis, if

$$F(x) = Ae^{\alpha x} + Be^{\beta x} \tag{5.187}$$

the particular solution to the differential equation is

$$y = AP^{-1}(D)e^{\alpha x} + BP^{-1}(D)e^{\beta x} \tag{5.188}$$

Each term is treated separately, so when the inhomogeneous term is a sum of exponentials, the particular solution is determined in the same way that we obtained the particular solution for a single exponential.

Sine or cosine inhomogeneous term

Since $\cos(\beta x)$ and $\sin(\beta x)$ can be written as sums of exponential functions, we can evaluate $P^{-1}(D)\cos(\beta x)$ and $P^{-1}(D)\sin(\beta x)$ using the results of the preceding section: namely,

$$P^{-1}(D)e^{\alpha x} = \frac{x^l}{l!}\frac{e^{\alpha x}}{R(\alpha)} \tag{5.183}$$

Both $\cos(\beta x)$ and $\sin(\beta x)$ can be written as a single expression

$$\frac{e^{i\beta x} + \sigma^2 e^{-i\beta x}}{2\sigma} = \begin{cases} \cos(\beta x) & \sigma = 1 \\ \sin(\beta x) & \sigma = i \end{cases} \tag{5.189}$$

Writing

$$P(D) = (D - i\beta)^{l_1}(D + i\beta)^{l_2}R(D) \tag{5.190}$$

such that $R(\pm i\beta) \neq 0$, and referring to equation 5.183, we obtain

$$P^{-1}(D)\left(\frac{e^{i\beta x} + \sigma^2 e^{-i\beta x}}{2\sigma}\right) = \frac{1}{2\sigma}\left[\frac{x^{l_1}}{l_1!}\frac{e^{i\beta x}}{(2i\beta)^{l_2}R(i\beta)} + \sigma^2\frac{x^{l_2}}{l_2!}\frac{e^{-i\beta x}}{(-2i\beta)^{l_1}R(-i\beta)}\right] \tag{5.191}$$

In many situations, $l_1 = l_2 \equiv l$. That is, $P(D)$ can be written

$$P(D) = (D^2 + \beta^2)^l R(D) \tag{5.192}$$

$$\Rightarrow \quad P^{-1}(D)\left(\frac{e^{i\beta x} + \sigma^2 e^{-i\beta x}}{2\sigma}\right) = \frac{1}{2\sigma}\frac{x^l}{l!}\left[\frac{e^{i\beta x}}{(2i\beta)^l R(i\beta)} + \sigma^2\frac{e^{-i\beta x}}{(-2i\beta)^l R(-i\beta)}\right] \tag{5.193}$$

If the constant multipliers of the various powers of D in an operator are all real (the a_n in equation 5.111), the polynomial operator is real. If $P(D)$, and therefore $R(D)$, is real, we note that the second term on the right-hand side of equation 5.193 is the complex conjugate of the first term. Thus

$$P^{-1}(D)\left(\frac{e^{i\beta x}+\sigma^2 e^{-i\beta x}}{2\sigma}\right)=\frac{1}{2\sigma}\frac{x^l}{l!}\left[\frac{e^{i\beta x}}{(2i\beta)^l R(i\beta)}+\sigma^2\left(\frac{e^{-i\beta x}}{(2i\beta)^l R(i\beta)}\right)^*\right] \tag{5.194}$$

Then, for $\sigma = 1$, and $\sigma = i$ respectively,

$$P^{-1}(D)\cos(\beta x)=\frac{x^l}{l!}\,\mathrm{Re}\left[\frac{e^{i\beta x}}{(2i\beta)^l R(i\beta)}\right] \tag{5.195a}$$

$$P^{-1}(D)\sin(\beta x)=\frac{x^l}{l!}\,\mathrm{Im}\left[\frac{e^{i\beta x}}{(2i\beta)^l R(i\beta)}\right] \tag{5.195b}$$

When $P(D)$ is real, these results can be achieved by operating on one exponential rather than on a sum of two exponentials. With $P(D)$ real,

$$y=P^{-1}(D)\cos(\beta x)=P^{-1}(D)\mathrm{Re}(e^{i\beta x})=\mathrm{Re}(P^{-1}(D)e^{i\beta x}) \tag{5.196a}$$

Allowing for the possibility that $P(-\beta^2)$ could be zero, we write

$$y=\mathrm{Re}\left[(D-i\beta)^l(D+i\beta)^l R^{-1}(D)e^{i\beta x}\right]=\mathrm{Re}\left[(D-i\beta)^l\frac{e^{i\beta x}}{(2i\beta)^l R(i\beta)}\right]$$

$$=\frac{x^l}{l!}\mathrm{Re}\left[\frac{e^{i\beta x}}{(2i\beta)^l R(i\beta)}\right] \tag{5.196b}$$

which is identical to the result expressed in equation 5.195a.

Similarly, if $P(D)$ is real,

$$y=P^{-1}(D)\sin(\beta x)=P^{-1}(D)\mathrm{Im}(e^{i\beta x})=\mathrm{Im}(P^{-1}(D)e^{i\beta x}) \tag{5.197a}$$

Following the same steps that lead to equation 5.196b, this becomes

$$y=\frac{x^l}{l!}\,\mathrm{Im}\left[\frac{e^{i\beta x}}{(2i\beta)^l R(i\beta)}\right] \tag{5.197b}$$

which is the result expressed in equation 5.195b.

By an identical analysis, if all coefficients are imaginary, the operator $P(D)$ is imaginary and can be written

$$P(D)=iQ(D) \tag{5.198}$$

where $Q(D)$ is real. Thus the development leading to equations 5.194 and 5.195 are unchanged, with the exception that there is an overall multiplication of each expression by $-i$.

Example 5.13

As an example, consider the differential equation

$$(D^3+3D-1)y=\sin(2x) \tag{5.199}$$

Replacing the D operator with $2i$, we see that $D^3 + 3D - 1 \to -(1 + 2i) \neq 0$. Therefore, $l = 0$, and $R(D) = P(D)$ is real. Referring to equation 5.197b, the particular solution to the differential equation is

$$y = -\text{Im}\left(\frac{e^{2ix}}{1+2i}\right) = \tfrac{2}{5}\cos 2x - \tfrac{1}{5}\sin(2x) \quad \square \tag{5.200}$$

With $F(x) = \cos(\beta x)$ or $F(x) = \sin(\beta x)$, a simpler approach exists when the polynomial operator depends on D only as D^2. As will be seen, the method of solution depends on whether or not $P(-\beta)^2 = 0$. We will first consider $P(-\beta^2) \neq 0$.

If $P(D) = P(D^2)$, then

$$y = P^{-1}(D^2)\cos(\beta x) = \sum_{n=0}^{\infty} a_n(D^2)^n \cos(\beta x) \tag{5.201a}$$

$$y = P^{-1}(D^2)\sin(\beta x) = \sum_{n=0}^{\infty} a_n(D^2)^n \sin(\beta x) \tag{5.201b}$$

Using

$$\text{(a)} \quad D^2\cos(\beta x) = -\beta^2\cos(\beta x) \qquad \text{(b)} \quad D^2\sin(\beta x) = -\beta^2\sin(\beta x) \tag{5.202}$$

$$\Rightarrow \quad y = \sum_{n=0}^{\infty} a_n(-\beta^2)^n \cos(\beta x) = P^{-1}(-\beta^2)\cos(\beta x) \tag{5.203a}$$

$$y = \sum_{n=0}^{\infty} a_n(-\beta^2)^n \sin(\beta x) = P^{-1}(-\beta^2)\sin(\beta x) \tag{5.203b}$$

That is, if $P(D) = P(D^2)$ and $P(-\beta^2) \neq 0$, the particular solution is obtained by replacing D^2 by $-\beta^2$ in the polynomial operator and dividing by the number $P(-\beta^2)$.

Example 5.14

As an example, the operator $(D^4 - 1)(D^2 + 2)$ is not zero when D^2 is replaced by -9. Therefore, we can write down the particular solution to

$$(D^4 - 1)(D^2 + 2)y = \cos(3x) \tag{5.204a}$$

by replacing D^2 by -9 and dividing by the resulting number. The result is

$$y(x) = -\frac{\cos(3x)}{560} \quad \square \tag{5.204b}$$

If $P(-\beta^2) = 0$, then $-\beta^2$ is an l-fold root of $P(D^2)$. With $R(D^2)$ defined by

$$P(D^2) \equiv (D^2 + \beta^2)^l R(D^2) \tag{5.205}$$

with $R(-\beta^2) \neq 0$, the differential equation can be reduced to

$$\text{(a)} \quad (D^2 + \beta^2)^l y = \frac{\cos(\beta x)}{R(-\beta^2)} \quad \text{or} \quad \text{(b)} \quad (D^2 + \beta^2)^l y = \frac{\sin(\beta x)}{R(-\beta^2)} \tag{5.206}$$

Consider

$$(D^2 + \beta^2)^l y = (D + i\beta)^l (D - i\beta)^l y = \frac{1}{2\sigma}\frac{(e^{i\beta x} + \sigma^2 e^{-i\beta x})}{R(-\beta^2)} \tag{5.207}$$

We have already shown that

$$\text{(a)} \quad (D + i\beta)^{-l} e^{i\beta x} = \frac{e^{i\beta x}}{(2i\beta)^l} \qquad \text{(b)} \quad (D - i\beta)^{-l} e^{i\beta x} = \frac{x^l}{l!} e^{i\beta x} \tag{5.208}$$

with analogous results when these operators operate on $e^{-i\beta x}$. Thus the solution to equation 5.207 is

$$y = (D + i\beta)^{-l}(D - i\beta)^{-l}\frac{1}{2\sigma}\frac{(e^{i\beta x} + \sigma^2 e^{-i\beta x})}{R(-\beta^2)}$$

$$= \frac{x^l}{l!}\frac{1}{2\sigma R(-\beta^2)}\left[\frac{e^{i\beta x}}{(2i\beta)^l} + \sigma^2\frac{e^{-i\beta x}}{(-2i\beta)^l}\right] \tag{5.209}$$

Therefore, with $\sigma = 1$, and $\sigma = i$ respectively,

$$y = (D^2 + \beta^2)^{-l}R^{-1}(D)\cos(\beta x) = \frac{x^l}{l!}\frac{1}{R(-\beta^2)}\operatorname{Re}\left[\frac{e^{i\beta x}}{(2i\beta)^l}\right] \tag{5.210a}$$

$$y = (D^2 + \beta^2)^{-l}R^{-1}(D)\sin(\beta x) = \frac{x^l}{l!}\frac{1}{R(-\beta^2)}\operatorname{Im}\left[\frac{e^{i\beta x}}{(2i\beta)^l}\right] \tag{5.210b}$$

If l is even, $i^l = (-1)^{l/2}$ and equation 5.210a becomes

$$(D^2 + \beta^2)^{-l}R^{-1}(D)\cos(\beta x) = \frac{x^l}{l!}\frac{1}{R(-\beta^2)}(-1)^{l/2}\frac{\cos(\beta x)}{(2\beta)^l} \tag{5.211a}$$

If l is odd, $i^l = i(-1)^{(l-1)/2}$ and equation 5.210a becomes

$$(D^2 + \beta^2)^{-l}R^{-1}(D)\cos(\beta x) = \frac{x^l}{l!}\frac{1}{R(-\beta^2)}(-1)^{(l-1)/2}\frac{\sin(\beta x)}{(2\beta)^l} \tag{5.211b}$$

Similarly, equation 5.210b becomes

$$(D^2 + \beta^2)^{-l}R^{-1}(D)\sin(\beta x) = \frac{x^l}{l!}\frac{1}{R(-\beta^2)}(-1)^{l/2}\frac{\sin(\beta x)}{(2\beta)^l} \tag{5.211c}$$

when l is even, and

$$(D^2 + \beta^2)^{-l}R^{-1}(D)\sin(\beta x) = \frac{x^l}{l!}\frac{1}{R(-\beta^2)}(-1)^{(l+1)/2}\frac{\cos(\beta x)}{(2\beta)^l} \tag{5.211d}$$

when l is odd.

Example 5.15

It is therefore straightforward to determine (write down) the particular solution to

$$(D^2 + 4)^3(D^2 - 1)^2 y = \cos(2x) \tag{5.212}$$

Since $-\beta^2 = -4$, it is evident that $R(D^2) = (D^2 - 1)^2$ and $l = 3$. Thus, from equation 5.211b,

$$y = \frac{x^3}{3!}\frac{1}{(-4-1)^2}(-1)^{(1/2)(3-1)}\frac{\sin(2x)}{(2\cdot 2)} = -x^3\frac{\sin(2x)}{600} \quad \square \tag{5.213}$$

When $P(D)$ contains terms in odd as well as even powers of D, a technique called rationalizing the operator allows us to use the properties of $P^{-1}(D^2)$ operating on $\cos(\beta x)$ or $\sin(\beta x)$. If the operator is very unweildy, it may be simpler to express sines and cosines as exponentials and use the techniques we have already developed. However, when convenient, rationalizing the operator is a useful technique. The method will be illustrated by example.

Example 5.16

Formally, the particular solution to

$$\text{(a)}\quad (D^2 + D - 1)y = \sin(2x) \qquad \text{is} \qquad \text{(b)}\quad y = (D^2 + D - 1)^{-1}\sin(2x) \tag{5.214}$$

D^2 and -1 are terms containing even powers of D. The presence of D does not permit us to simply replace D^2 by -4 in equation 5.214b. However, the operator

$$P(-D)P^{-1}(-D) = (D^2 - D - 1)(D^2 - D - 1)^{-1} = 1 \tag{5.215}$$

is the identity operator. Therefore, if we operate on y in equation 5.214b with this identity operator, it will not affect the solution to the differential equation. Thus

$$y = (D^2 - D - 1)(D^2 - D - 1)^{-1}(D^2 + D - 1)^{-1}\sin(2x) \tag{5.216}$$

The product of the two inverse operators is

$$(D^2 - 1 - D)^{-1}(D^2 - 1 + D)^{-1} = \left((D^2 - 1)^2 - D^2\right)^{-1} \tag{5.217}$$

$$\Rightarrow \quad y = (D^2 - 1 - D)\left((D^2 - 1)^2 - D^2\right)^{-1}\sin(2x) \tag{5.218}$$

Since $[(D^2 - 1)^2 - D^2]^{-1}$ contains terms in D^2 only, D^2 can be replaced by -4 in this operator, and equation 5.218 becomes

$$y = (D^2 - 1 - D)\frac{\sin(2x)}{29} \tag{5.219}$$

The operations remaining are differentiations of $\sin(2x)$, which can be performed straightforwardly. The particular solution then is

$$y = -\left(\frac{5\sin(2x) + 2\cos(2x)}{29}\right) \quad \square \tag{5.220}$$

In general, if $P(D)$ contains odd as well as even powers of D, the particular solution, by rationalizing the operator, is

$$y = P(-D)P^{-1}(-D)P^{-1}(D)\begin{cases}\sin(\beta x)\\ \cos(\beta x)\end{cases} \tag{5.221}$$

$P^{-1}(-D)P^{-1}(D)$ is the rationalized inverse operator and depends on D^2. Operating with $P^{-1}(-D)P^{-1}(D)$ on $\sin(\beta x)$ or $\cos(\beta x)$, is achieved by replacing D^2 by $-\beta^2$. The remaining derivatives specified by $P(-D)$ on $\sin(\beta x)$ and/or $\cos(\beta x)$ are straightforward.

Sinh or cosh inhomogeneous term

As might be expected, since hyperbolic sine and hyperbolic cosine are closely related to the sine and cosine and the exponential functions, many of the techniques presented above are easily modified for $P^{-1}(D)\cosh(\gamma x)$ and $P^{-1}(D)\sinh(\gamma x)$.

One approach is to write the inhomogeneous term as a sum or difference of exponential functions

$$\frac{e^{\gamma x} + \tau e^{-\gamma x}}{2} = \begin{cases}\cosh(\gamma x) & \tau = 1\\ \sinh(\gamma x) & \tau = -1\end{cases} \tag{5.222}$$

and obtain the solution in the form

$$y = P^{-1}(D)\left(\frac{e^{\gamma x} + \tau e^{-\gamma x}}{2}\right) \tag{5.223}$$

Accounting for the possibility that $P(\gamma) = 0$ and/or $P(-\gamma) = 0$, we write $P(D)$ as

$$P(D) = (D + \gamma)^{l_1}(D - \gamma)^{l_2}R(D) \tag{5.224}$$

with $R(\pm\gamma) \neq 0$. Then

$$\begin{aligned} P^{-1}(D)\left(\frac{e^{\gamma x} + \tau e^{-\gamma x}}{2}\right) &= \frac{1}{2}\left[\frac{(D-\gamma)^{l_2}e^{\gamma x}}{(2\gamma)^{l_1}R(\gamma)} + \tau\frac{(D+\gamma)^{l_1}e^{-\gamma x}}{(-2\gamma)^{l_2}R(-\gamma)}\right] \\ &= \frac{1}{2}\left[\frac{x^{l_2}}{l_2!}\frac{e^{\gamma x}}{(2\gamma)^{l_1}R(\gamma)} + \tau\frac{x^{l_1}}{l_1!}\frac{e^{-\gamma x}}{(-2\gamma)^{l_2}R(-\gamma)}\right] \end{aligned} \tag{5.225}$$

Example 5.17

For example, consider

$$(D^2 - D - 1)y = \cosh(2x) \tag{5.226}$$

Replacing D by ± 2, we see that $P(2) = 1 \neq 0$, and $P(-2) = 5 \neq 0$. Therefore, $l_1 = l_2 = 0$ and $R(D) = P(D)$. With $\tau = +1$, the solution from equation 5.225, is

$$y = \frac{1}{2}\left(\frac{e^{2x}}{1} + \frac{e^{-2x}}{5}\right) \tag{5.227}$$

Writing

$$\text{(a)}\quad e^{\gamma x} = \cosh(\gamma x) + \sinh(\gamma x) \qquad \text{(b)}\quad e^{-\gamma x} = \cosh(\gamma x) - \sinh(\gamma x) \tag{5.228}$$

the solution can also be expressed as

$$y = \tfrac{3}{5}\cosh(2x) + \tfrac{2}{5}\sinh(2x) \quad \square \tag{5.229}$$

Example 5.18

To illustrate the approach when $P(\gamma) = 0$ and/or $P(-\gamma) = 0$, consider

$$(D^2 - 1)^2(D + 1)(D^2 - 2)y = \sinh(x) \tag{5.230}$$

Since $(D^2 - 1) \to 0$ when D is replaced by ± 1, and $(D + 1) \to 0$ when D is replaced by -1, we see that $R(D) = (D^2 - 2)$, $l_1 = 2$, and $l_2 = 3$. With $\tau = -1$, the solution, from equation 5.225, is

$$y = \frac{1}{2}\left[\frac{x^2}{2!}\frac{e^x}{2^2(-1)} - \frac{x^3}{3!}\frac{e^{-x}}{(-2)^3(-1)}\right] = -\frac{1}{16}\left[x^2e^x - \frac{x^3e^{-x}}{6}\right] \quad \square \tag{5.231}$$

As with the sine and cosine, if the polynomial operator depends only on even powers of D, one can essentially write down the particular solution. Like the technique developed for the sine or cosine inhomogeneous terms, the method depends on whether or not $P(\gamma^2) = 0$. We will first consider $P(\gamma^2) \neq 0$.

If $P(D) = P(D^2)$, then

$$y = P^{-1}(D^2)\cosh(\gamma x) = \sum_{n=0}^{\infty} a_n (D^2)^n \cosh(\gamma x) \tag{5.232a}$$

$$y = P^{-1}(D^2)\sinh(\gamma x) = \sum_{n=0}^{\infty} a_n (D^2)^n \sinh(\gamma x) \tag{5.232b}$$

But

$$\text{(a)} \quad D^2 \cosh(\gamma x) = \gamma^2 \cosh(\gamma x) \qquad \text{(b)} \quad D^2 \sinh(\gamma x) = \gamma^2 \sinh(\gamma x) \tag{5.233}$$

$$\Rightarrow \quad y = \sum_{n=0}^{\infty} a_n (\gamma^2)^n \cosh(\gamma x) = P^{-1}(\gamma^2)\cosh(\gamma x) \tag{5.234a}$$

$$y = \sum_{n=0}^{\infty} a_n (\gamma^2)^n \sinh(\gamma x) = P^{-1}(\gamma^2)\sinh(\gamma x) \tag{5.234b}$$

for $P(\gamma^2) \neq 0$.

Example 5.19

As an example, the operator $(D^6 - 9)(D^4 + 1)$ is not zero when D^2 is replaced by 4. Therefore, replacing D^2 by 4 and dividing by the resulting number, the particular solution to

$$\text{(a)} \quad (D^6 - 9)(D^4 + 1)y = \sinh(2x) \qquad \text{is} \qquad \text{(b)} \quad y(x) = \frac{\sinh(2x)}{935} \quad \square \tag{5.235}$$

If $P(\gamma^2) = 0$, then γ^2 is an l-fold root of $P(D^2)$. We then define $R(D^2)$ by

$$P(D^2) \equiv (D^2 - \gamma^2)^l R(D^2) \tag{5.236}$$

such that $R(\gamma^2) \neq 0$. Comparing this to equation 5.224, we see that $l_1 = l_2 = l$. Therefore, using equation 5.225, the solution is

$$y = P^{-1}(D)\left(\frac{e^{\gamma x} + \tau e^{-\gamma x}}{2}\right) = \frac{x^l}{2(2\gamma)^l l! R(\gamma^2)}\left(e^{\gamma x} + (-1)^l \tau e^{-\gamma x}\right) \tag{5.237}$$

Example 5.20

For example,

$$(D^2 - 1)^4 (D^2 + 2)y = \cosh(x) \tag{5.238}$$

Clearly, since $\gamma = 1$, $l = 4$ and $R(D^2) = (D^2 + 2)$. Thus, with $\tau = +1$, the particular solution is

$$y = \frac{x^4}{2(2 \cdot 1)^4 4! 3}\left(e^x + (-1)^4(+1)e^{-x}\right) = \frac{x^4 \cosh(x)}{1152} \quad \square \tag{5.239}$$

As with the treatment involving sine or cosine, when $P(D)$ contains terms in odd as well as even powers of D, rationalizing the operator can yield a simplified method of solution. Again the method will be illustrated by example.

Example 5.21

The particular solution to

$$\text{(a)} \quad (D^2 + 2D - 1)y = \sinh(2x) \qquad \text{is} \qquad \text{(b)} \quad y = (D^2 + 2D - 1)^{-1}\sinh(2x) \tag{5.240}$$

As before, the operator is rationalized by multiplication by the identity operator,

$$P(-D)P^{-1}(-D) = (D^2 - 2D - 1)(D^2 - 2D - 1)^{-1} \tag{5.241}$$

$$\Rightarrow \quad y = (D^2 - 1 - 2D)\left((D^2 - 1)^2 - 4D^2\right)^{-1}\sinh(2x) \tag{5.242a}$$

Replacing D^2 by 4 in $[(D^2 - 1)^2 - 4D^2]^{-1}$, equation 5.242a becomes

$$y = (D^2 - 1 - 2D)\frac{\sinh(2x)}{-7} \tag{5.242b}$$

All that remains is to evaluate the indicated derivatives of $\sinh(2x)$. The solution is

$$y = \left(\frac{3\sinh(2x) - 4\cosh(2x)}{-7}\right) \quad \square \tag{5.243}$$

Polynomial inhomogeneous term

Let $F(x)$ be a polynomial of order N. We write it as

$$F(x) = \sum_{k=0}^{N} \phi_k x^k \tag{5.244}$$

If $P(0) = 0$,

$$P^{-1}(D)F(x) = R^{-1}(D)D^{-l}F(x) = R^{-1}(D)G(x) \tag{5.167}$$

with $R(0) \neq 0$. Since D^{-l} represents an l-fold integral, $G(x)$ is a polynomial of order $N + l$. Of course, if $P(0) \neq 0$, then $G(x) = F(x)$, $R(D) = P(D)$, and $l = 0$.

Formally, the solution is written as

$$y = R^{-1}(D)G(x) = \sum_{n=0}^{\infty} a_n D^n G(x) \tag{5.245}$$

Since $G(x)$ is a polynomial of order $N + l$, operating on $G(x)$ with derivatives of order higher than D^{N+l} will yield zero. Therefore,

$$\sum_{n=0}^{\infty} a_n D^n G(x) = \sum_{n=0}^{N+l} a_n D^n G(x) \tag{5.246}$$

That is, finding the particular solution becomes the task of evaluating the derivatives of various powers of x.

Example 5.22

The particular solution to

(a) $(D^2 - 3D + 2)y = 4x^3 + 2x - 1$ is (b) $y = (D^2 - 3D + 2)^{-1}(4x^3 + 2x - 1)$

(5.247)

Since $P(0) \neq 0$, we can expand $P^{-1}(D)$ in a MacLaurin series by writing

$$(D^2 - 3D + 2)^{-1} = \frac{1}{2}\left(1 + \frac{(D^2 - 3D)}{2}\right)^{-1} \tag{5.248}$$

and expanding this in the geometric power series in $(D^2 - 3D)/2$. We obtain

$$\frac{1}{2}\left(1 + \frac{(D^2 - 3D)}{2}\right)^{-1} = \frac{1}{2}\left(1 - \left[\frac{(D^2 - 3D)}{2}\right] + \left[\frac{(D^2 - 3D)}{2}\right]^2 - \cdots\right) \tag{5.249}$$

Since $F(x)$ is a third-order polynomial, it is only necessary to retain terms in this expansion up to D^3. That is, we write equation 5.249 as

$$\frac{1}{2}\left(1+\frac{(D^2-3D)}{2}\right)^{-1}=\frac{1}{2}\left(1+\frac{3}{2}D+\frac{7}{4}D^2+\frac{15}{8}D^3+O(D^4)\right) \quad (5.250)$$

Therefore, the particular solution is

$$y=\tfrac{1}{2}\left[1+\tfrac{3}{2}D+\tfrac{7}{4}D^2+\tfrac{15}{8}D^3+O(D^4)\right](4x^3+2x-1)=2x^3+9x^2+22x+\tfrac{47}{2} \quad \square \quad (5.251)$$

Example 5.23

As a second example, consider

$$(D^3-4D)y=x^2 \quad (5.252a)$$

For this example, $P(0)=0$. Therefore, the particular solution is

$$y=(D^2-4)^{-1}D^{-1}x^2=(D^2-4)^{-1}\frac{x^3}{3} \quad (5.252b)$$

Expanding the operator as

$$(D^2-4)^{-1}=-\frac{1}{4}\left(1-\frac{D^2}{4}\right)^{-1}=-\frac{1}{4}\left(1+\frac{D^2}{4}+\frac{D^4}{16}+\cdots\right) \quad (5.253)$$

and ignoring terms of order D^4 and higher, we obtain the solution

$$y=-\frac{1}{4}\left(\frac{x^3}{3}+\frac{x}{2}\right) \quad \square \quad (5.254)$$

Combinations of specified inhomogeneous terms

Now that we have developed techniques for dealing individually with each of the inhomogeneous terms specified in equations 5.163, we investigate methods for determining the particular solution when the inhomogeneous term is a combination of these functions.

If these functions are combined by addition (or subtraction),

$$\text{(a)} \quad P(D)y=F_1(x)\pm F_2(x) \quad \Rightarrow \quad \text{(b)} \quad y=P^{-1}(D)F_1(x)\pm P^{-1}(D)F_2(x) \quad (5.255)$$

where each $F(x)$ is one of the four functions listed in equations 5.163. Each term is evaluated by one of the techniques already developed.

If $F(x)$ is a product of several (two or more) of these functions, it is also possible to use the approaches that have already been developed to determine a particular solution. For example, with $G(x)$ one of these specified functions, consider

$$P(D)y=e^{\alpha x}G(x) \quad (5.256)$$

Referring to equation 5.144, this has the particular solution

$$y=P^{-1}(D)[e^{\alpha x}G(x)]=\sum_{l=0}^{\infty}a_lD^l[e^{\alpha x}G(x)]=e^{\alpha x}\sum_{l=0}^{\infty}a_l(D+\alpha)^lG(x)$$
$$=e^{\alpha x}P^{-1}(D+\alpha)G(x) \quad (5.257)$$

Since $G(x)$ is one of the functions specified in equations 5.163, the techniques for evaluating $P^{-1}(D + \alpha)G(x)$ have already been established.

If $F(x)$ is of the form $\cos(\beta x)G(x)$, $\sin(\beta x)G(x)$, $\cosh(\gamma x)G(x)$, or $\sinh(\gamma x)G(x)$, these can also be written as a sum of exponentials multiplying $G(x)$. Thus the solution is the sum of two terms of the form expressed in equation 5.257.

However, if $P(D)$ and $G(x)$ are real, the solution to

$$\text{(a)}\quad P(D)y = \cos(\beta x)G(x) \qquad \text{(b)}\quad P(D)y = \sin(\beta x)G(x) \tag{5.258}$$

can be found more readily as

$$\text{(c)}\ \ y = \operatorname{Re}\left[e^{i\beta x}P^{-1}(D + i\beta)G(x)\right] \ \ \text{(d)}\ \ y = \operatorname{Im}\left[e^{i\beta x}P^{-1}(D + i\beta)G(x)\right] \tag{5.258}$$

If $P(D)$ and/or $G(x)$ are purely imaginary, the imaginary quantity can be written as a real quantity multiplied by i, and a similar result can be developed.

Example 5.24

We will illustrate the method of approach when $F(x)$ is a product of two or more of these special functions by example. The particular solution to

$$\text{(a)}\quad (D^2 + 4D + 5)y = x^2\cos(x)e^{-2x} \qquad \text{(b)}\quad y = (D^2 + 4D + 5)^{-1}(x^2\cos(x)e^{-2x}) \tag{5.259}$$

The solution is obtained by moving exponentials to the left of the operator and expressing $\cos(x)$ as $\operatorname{Re}(e^{ix})$. We obtain

$$\begin{aligned} y &= e^{-2x}\left[(D-2)^2 + 4(D-2) + 5\right]^{-1}\left[x^2\cos(x)\right] \\ &= e^{-2x}(D^2+1)^{-1}\left[x^2\cos(x)\right] = e^{-2x}(D^2+1)^{-1}\operatorname{Re}(x^2e^{ix}) \end{aligned} \tag{5.260a}$$

Since $(D^2 + 1)^{-1}$ is real, this can be written

$$\begin{aligned} y &= e^{-2x}\operatorname{Re}\left[(D^2+1)^{-1}(x^2e^{ix})\right] = e^{-2x}\operatorname{Re}\left[e^{ix}\left((D+i)^2+1\right)^{-1}x^2\right] \\ &= e^{-2x}\operatorname{Re}\left[e^{ix}(D+2i)^{-1}D^{-1}x^2\right] = e^{-2x}\operatorname{Re}\left[e^{ix}(D+2i)^{-1}\frac{x^3}{3}\right] \\ &= e^{-2x}\operatorname{Re}\left[\frac{e^{ix}}{6i}\left(1+\frac{D}{2i}\right)^{-1}x^3\right] \end{aligned} \tag{5.260b}$$

Expanding the inverse operator, this becomes

$$\begin{aligned} y &= e^{-2x}\operatorname{Re}\left[\frac{e^{ix}}{2i}\left\{\frac{x^3}{3} - \frac{x}{2} + i\left(\frac{x^2}{2} - \frac{1}{4}\right)\right\}\right] \\ &= \frac{e^{-2x}}{2}\left[\left(\frac{x^3}{3} - \frac{x}{2}\right)\sin x + \frac{1}{2}\left(x^2 - \frac{1}{2}\right)\cos x\right] \quad \square \end{aligned} \tag{5.261}$$

Assuming the form of the solution

For some of the inhomogeneous terms specified in equations 5.163, finding the particular solution is sometimes simplified by using our knowledge of the form of the solution.

When $F(x) = e^{\alpha x}$ the particular solution will be of the form

$$y = A\frac{x^l}{l!}e^{\alpha x} \tag{5.262}$$

where α is an l-fold root of the operator polynomial. Writing $P(D) = (D - \alpha)^l R(D)$, substitution of this assumed form of the solution into the differential equation $P(D)y = e^{\alpha x}$ yields

$$A = \frac{1}{R(\alpha)} \tag{5.263}$$

This is identical to the constant obtained in equation 5.183. Thus, for the exponential inhomogeneous term, no simplification is obtained by assuming the form of the solution and determining a constant coefficient from the differential equation.

If $F(x) = \cos(\beta x)$ or $F(x) = \sin(\beta x)$, and if $P(D)$ is not in the form $P(D^2)$, and if $P(\pm i\beta) \neq 0$, we have seen that the particular solution is a linear combination of $\cos(\beta x)$ and $\sin(\beta x)$. Therefore, writing the solution as

$$y = A\cos(\beta x) + B\sin(\beta x) \tag{5.264}$$

we can determine A and B by substituting the assumed solution into the differential equation and separately equating the coefficients of $\sin(\beta x)$ and $\cos(\beta x)$.

Example 5.25

As an example, consider

$$(D^2 - D + 2)y = 2\sin(4x) \tag{5.265}$$

Noting that $(\pm 4i)^2 \mp 4i + 2 \neq 0$, we know that the solution will be of the form

$$y = A\cos(4x) + B\sin(4x) \tag{5.266}$$

Substituting this solution into the differential equation, we obtain

$$(-14A - 4B)\cos(4x) + (4A - 14B)\sin(4x) = 2\sin(4x) \tag{5.267}$$

Thus, by separately equating the coefficients of $\cos(4x)$ and $\sin(4x)$, we obtain

$$\text{(a)}\quad 14A + 4B = 0 \qquad \text{(b)}\quad 4A - 14B = 2 \tag{5.268}$$

$$\Rightarrow \quad y = \frac{2\cos(4x) - 7\sin(4x)}{53} \quad \square \tag{5.269}$$

If $P(\pm i\beta) = 0$, A and B become polynomials in x. In that case, a solution of the form of equation 5.264 leads to coupled differential equations for the functions A and B. For that reason, if $P(\pm i\beta) = 0$, rather than assuming the form of the solution, it is often less work to use the method of solution described in equation 5.191, for example.

If $F(x) = \cosh(\gamma x)$ or $F(x) = \sinh(\gamma x)$, and $P(\pm\gamma) \neq 0$, the particular solution will be of the form

$$y = A\cosh(\gamma x) + B\sinh(\gamma x) \tag{5.270}$$

where A and B are constants. A and B are determined by substituting this solution into the differential equation and separately equating the coefficients of $\sinh(\gamma x)$ and $\cosh(\gamma x)$.

Example 5.26

As an example, consider

$$(D^2 + 2D - 3)y = 4\cosh(2x) \tag{5.271}$$

Since $(\pm 2)^2 + 2(\pm 2) - 3 \neq 0$, the particular solution will be of the form

$$y = A\cosh(2x) + B\sinh(2x) \tag{5.272}$$

Substituting this solution into the differential equation, we obtain

$$(A + 4B)\cosh(2x) + (4A + B)\sinh(2x) = 4\cosh(2x) \tag{5.273}$$

$$\Rightarrow \quad \text{(a)} \quad A + 4B = 4 \qquad \text{(b)} \quad 4A + B = 0 \tag{5.274}$$

$$\Rightarrow \quad y = \frac{16\sinh(2x) - 4\cosh(2x)}{15} \quad \square \tag{5.275}$$

As with equations involving sine and cosine inhomogeneous terms, when γ is a root of the polynomial operator, the undetermined coefficients become polynomials in x and one is led to coupled differential equations for these coefficients. Therefore, if $P(\pm\gamma) = 0$, it is often less work to use the method of solution described in equation 5.225.

If $F(x)$ is a polynomial of order N, and if $P(z)$ has an l-fold root at $z = 0$, the particular solution is a polynomial of order $N + l$. Writing

$$y = \sum_{k=0}^{N+l} A_k x^k \tag{5.276}$$

the coefficients A_k will be determined by substituting of the assumed solution into the differential equation, and equating the coefficient of each power of x on the two sides of the equation.

Example 5.27

For example, we will find the particular solution to

$$(D^3 - 3D)y = 3x^2 \tag{5.277}$$

Since $z = 0$ is a root of $P(z)$ of multiplicity 1, the particular solution is a third-order polynomial

$$y = \sum_{k=0}^{3} A_k x^k = A_3 x^3 + A_2 x^2 + A_1 x + A_0 \tag{5.278}$$

Substituting this into the differential equation yields

$$-9A_3 x^2 - 6A_2 x + (6A_3 - 3A_1) = 3x^2 \tag{5.279}$$

$$\Rightarrow \quad \text{(a)} \quad A_3 = -\tfrac{1}{3} \quad \text{(b)} \quad A_2 = 0 \quad \text{(c)} \quad A_1 = -\tfrac{2}{3} \quad \text{(d)} \quad A_0 \quad \text{indeterminate} \tag{5.280}$$

Since A_0 is indeterminate, the term containing it is a duplication of part of the homogeneous solution and thus is not part of the particular solution. Thus the particular solution is

$$y = -\frac{(x^3 + 2x)}{3} \quad \square \tag{5.281}$$

If $F(x)$ is a product of two or more of these functions, and one of the factors in the product is an exponential, the particular solution to

$$P(D)y = e^{\alpha x}G(x) \tag{5.282}$$

will be of the form

$$y = e^{\alpha x} y_1(x) \tag{5.283}$$

Substituting this solution into equation 5.282 yields the differential equation for y_1

$$P(D + \alpha)y_1 = G(x) \tag{5.284}$$

When $G(x)$ is one of the forms specified in equations 5.163, one can assume the form of solution for y_1 and evaluate the constants from equation 5.284.

Example 5.28

For example,

$$(D^2 - 2)y = 3xe^{-x} \tag{5.285}$$

has a solution of the form

$$y = e^{-x}y_1(x) \tag{5.286}$$

$$\Rightarrow \quad \left((D-1)^2 - 2\right)y_1 = (D^2 - 2D - 1)y_1 = 3x \tag{5.287}$$

Noting that the polynomial operator of equation 5.287 is not zero when D is replaced by 0, we can assume a first-order polynomial solution for y_1. With

$$y_1 = A_1x + A_0 \tag{5.288}$$

equation 5.286 becomes

$$-A_1x - (2A_1 + A_0) = 3x \tag{5.289}$$

$$\Rightarrow \quad \text{(a)} \quad A_1 = -3 \qquad \text{(b)} \quad A_0 = 6 \tag{5.290}$$

$$\Rightarrow \quad y = (-3x + 6)e^{-x} \quad \square \tag{5.291}$$

Solution using integral transforms

Differential equations with constant coefficients can be reduced to a problem of evaluating an integral transform. In Chapter 4, only the Fourier and Laplace transforms were discussed in detail. For that reason, we develop the methods of solution in terms of these integral transforms.

We begin with the differential equation in the form

$$P(D)y(x) = \sum_{l=0}^{N} a_l D^l y(x) = F(x) \tag{5.106}$$

Solution using Fourier transforms

In some problems no initial or boundary conditions are specified. That is, it is implied that the instant the system began operating, or a region where the physical parameters have specific values, are not important to the study of that system. For example, in an *RLC* series ac circuit, the steady-state solution for the time dependence of the charge on the capacitor or the current through the resistor and inductor is unaffected by the instant when the ac generator was switched on. In such problems the system can be analyzed using Fourier transforms. As will be seen, the Fourier solution will be independent of undetermined constants. Since there will be no initial conditions to determine such constants, the Fourier solution will be the particular solution.

We define the Fourier inverse of $y(x)$ to be $Y(k)$, and the Fourier inverse of $F(x)$ to be $G(k)$. As shown in Chapter 4,

$$\text{F.T.}\left[D^l y(x)\right] = (ik)^l \text{F.T.}\left[y(x)\right] = (ik)^l Y(k) \tag{4.104c}$$

Taking the Fourier transform of the differential equation and using the distributive property of integral transforms, we obtain

$$\text{F.T.}\left[\sum_{l=0}^{N} a_l D^l y(x)\right] = \sum_{l=0}^{N} a_l (ik)^l Y(k) = P(ik)Y(k) = \text{F.T.}[F(x)] = G(k) \tag{5.292}$$

$$\Rightarrow \quad Y(k) = \frac{G(k)}{P(ik)} \tag{5.293}$$

and the particular solution to the differential equation is reduced to the task of evaluating the Fourier integral

$$y(x) = \int_{-\infty}^{\infty} Y(k) e^{-ikx} \frac{dk}{\sqrt{2\pi}} = \int_{-\infty}^{\infty} \frac{G(k)}{P(ik)} e^{-ikx} \frac{dk}{\sqrt{2\pi}} \tag{5.294}$$

Note that if we try to solve the homogeneous equation using Fourier transform methods, $F(x) = 0$ requires $G(k) = 0$, which leads to $Y(k) = 0$. From equation 5.294, then, the only solution to the homogeneous equation by Fourier transform methods is the trivial one.

Example 5.29

Consider a series circuit consisting of a resistor, capacitor, and inductor driven by a sinusoidal AC generator that has an angular frequency ω_0 (see Figure 5.1). All transient effects that arise when the generator is initially turned on, decay exponentially to negligible values. In most circuits, this decay is fairly rapid. For this example, all transient effects will be ignored and only the steady-state quantities will be considered.

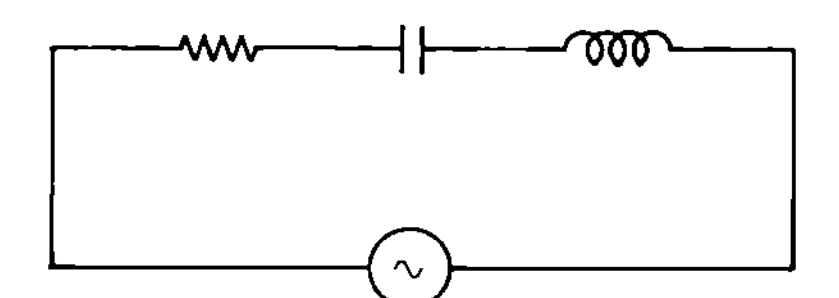

Figure 5.1 *RLC* series circuit.

The steady-state current through the resistor and inductor, and the charge on the capacitor, satisfy

$$IR + \frac{1}{C}Q + L\frac{dI}{dt} = V_0 \sin(\omega_0 t) \tag{5.295a}$$

With $I(t) = dQ/dt$, the differential equation for the charge on the capacitor is

$$L\frac{d^2Q}{dt^2} + R\frac{dQ}{dt} + \frac{1}{C}Q = V_0 \sin(\omega_0 t) \tag{5.295b}$$

Using the Heaviside operator method, the particular solution to this equation is

$$Q(t) = V_0\left(LD^2 + RD + \frac{1}{C}\right)^{-1} \sin(\omega_0 t) = V_0 \,\text{Im}\left[\left(LD^2 + RD + \frac{1}{C}\right)^{-1} e^{i\omega_0 t}\right] \tag{5.296}$$

$$= V_0 \,\text{Im}\left[\frac{e^{i\omega_0 t}}{-L\omega_0^2 + iR\omega_0 + 1/C}\right] = V_0\left[\frac{(-L\omega_0^2 + 1/C)\sin(\omega_0 t) - R\omega_0 \cos(\omega_0 t)}{R^2\omega_0^2 + (-L\omega_0^2 + 1/C)^2}\right]$$

To solve this differential equation using the Fourier transform, we define $P(\Omega) \equiv$ F.T.$[Q(t)]$ and take the Fourier transform of equation 5.295b. Using

$$\text{F.T.}\left[D^l Q(t)\right] = (i\Omega)^l P(\Omega) \tag{4.104c}$$

we obtain

$$\left(-L^2\Omega^2 + iR\Omega + \frac{1}{C}\right)P(\Omega) = V_0 \text{ F.T.}[\sin(\omega_0 t)] \tag{5.297}$$

where

$$\text{F.T.}[\sin(\omega_0 t)] = \frac{1}{2i}\int_{-\infty}^{\infty}[e^{-i(\Omega-\omega_0)t} - e^{-i(\Omega+\omega_0)t}]\frac{dt}{\sqrt{2\pi}} \tag{5.298a}$$

Referring to the integral representation of the Dirac δ-symbol of equations 4.106, we obtain

$$\text{F.T.}[\sin(\omega_0 t)] = \frac{\sqrt{2\pi}V_0}{2i}[\delta(\omega_0 - \Omega) - \delta(\omega_0 + \Omega)] \tag{5.298b}$$

Therefore,

$$P(\Omega) = \frac{\sqrt{2\pi}V_0}{2i}\frac{[\delta(\omega_0 - \Omega) - \delta(\omega_0 + \Omega)]}{[-L\Omega^2 + iR\Omega + 1/C]} \tag{5.299}$$

$$\Rightarrow \quad Q(t) = \frac{\sqrt{2\pi}V_0}{2i}\int_{-\infty}^{\infty}\frac{[\delta(\omega_0 - \Omega) - \delta(\omega_0 + \Omega)]}{[-L\Omega^2 + iR\Omega + 1/C]}e^{i\Omega t}\frac{d\Omega}{\sqrt{2\pi}}$$

$$= \frac{V_0}{2i}\left[\frac{e^{i\omega_0 t}}{(-L\omega_0^2 - iR\omega_0 + 1/C)} - \frac{e^{-i\omega_0 t}}{(-L\omega_0^2 - iR\omega_0 + 1/C)}\right] \tag{5.296}$$

$$= V_0 \text{ Im}\left[\frac{e^{i\omega_0 t}}{(-L\omega_0^2 + iR\omega_0 + 1/C)}\right] = V_0\left[\frac{(-L\omega_0^2 + 1/C)\sin(\omega_0 t) - R\omega_0\cos(\omega_0 t)}{(R^2\omega_0^2 + (-L\omega_0^2 + 1/C)^2)}\right]$$

which is identical to the solution obtained by Heaviside methods. □

Solution using Laplace transforms

If $y(x)$ and $F(x)$ in equation 5.106 are specified to be zero for negative values of x, the solution to the differential equation can be obtained using Laplace transforms. We define

$$\text{(a)} \quad Y(s) \equiv \text{L.T.}[y(x)] \quad \Rightarrow \quad \text{(b)} \quad G(s) = \text{L.T.}[F(x)] \tag{5.300}$$

and apply the Laplace transform to the differential equation. We have deduced that

$$\text{L.T.}[D^l y(x)] = s^l Y(s) - \sum_{k=1}^{l} s^{l-k-1}y^{(k-1)}(0) \tag{4.155}$$

Therefore, the Laplace transform of

$$P(D)y = F(x) \tag{5.106}$$

leads to

$$P(s)Y(s) - \sum_{l=1}^{N} a_l \sum_{k=1}^{l} s^{l-k-1}y^{(k-1)}(0) = G(s) \tag{5.301a}$$

$$\Rightarrow \quad Y(s) = \frac{1}{P(s)}\left[G(s) + \sum_{l=1}^{N} a_l \sum_{k=1}^{l} s^{l-k-1}y^{(k-1)}(0)\right] \tag{5.301b}$$

The solution can then be found by evaluating the Bromwich integral,

$$y(x) = \frac{1}{2\pi i}\int_{c-i\infty}^{c+i\infty} \frac{1}{P(s)}\left[G(s) + \sum_{l=1}^{N} a_l \sum_{k=1}^{l} s^{l-k-1} y^{(k-1)}(0)\right] e^{sx}\, ds \tag{5.302}$$

or using a table of Laplace transforms.

We note that when using Laplace transforms, the constants $y^{(k)}(0)$ occur naturally in the solution. If the independent variable, x, is time, and $x = 0$ is the initial instant for the system, the values of the constants $y^{(k)}(0)$ specify the initial conditions of the system. That is, when the Laplace transforms are used to solve such a problem, the initial conditions arise naturally in the solution.

The homogeneous solution is obtained by setting $G(s) = 0$, and we see that a nontrivial solution to the homogeneous equation is found using Laplace transforms. From equation 5.302, we see that the particular solution is

$$y_p(x) = \frac{1}{2\pi i}\int_{c-i\infty}^{c+i\infty} \frac{G(s)}{P(s)} e^{sx}\, ds \tag{5.303a}$$

and thus the homogeneous solution is

$$y_0(x) = \frac{1}{2\pi i}\sum_{l=1}^{N} a_l \sum_{k=1}^{l} y^{(k-1)}(0) \int_{c-i\infty}^{c+i\infty} \frac{s^{l-k-1}}{P(s)} e^{st}\, ds \tag{5.303b}$$

Therefore, the solution represented in equation 5.302 is the complete solution to the differential equation.

Example 5.30

To illustrate, we will consider the problem of a harmonic oscillator with a natural frequency ω_0 that is driven and damped by forces

$$\text{(a)}\quad F_{\text{driving}} = F_0 \cos(\omega_0 t) \qquad \text{(b)}\quad F_{\text{damping}} = -\lambda \frac{dy}{dt} \tag{5.304}$$

The differential equation for the displacement of the oscillator from equilibrium is

$$F_{\text{driving}} + F_{\text{damping}} + F_{\text{Hookes}} = m\frac{d^2y}{dt^2} = F_0\cos(\omega_0 t) - \lambda\frac{dy}{dt} - m\omega_0^2 y \tag{5.305a}$$

$$\Rightarrow \quad \left(mD^2 + \lambda D + m\omega_0^2\right)y = F_0\cos(\omega_0 t) \tag{5.305b}$$

For arithmetic simplicity, we will assume that $\lambda^2 \ll m^2\omega_0^2$.

We will solve this differential equation using Laplace transform methods subject to the initial conditions

$$\text{(a)}\quad y(0) = 0 \qquad \text{(b)}\quad y'(0) = v_{\max} = \omega_0 y_{\max} \tag{5.306}$$

Using equation 4.155, and the initial conditions, we have

$$\text{L.T.}[Dy] = sY(s) - y(0) = sY(s) \tag{5.307a}$$

$$\text{L.T.}[D^2y] = s^2Y(s) - sy(0) - y'(0) = s^2Y(s) - \omega_0 y_{\max} \tag{5.307b}$$

We must also evaluate

$$\text{L.T.}[\cos(\omega_0 t)] = \text{Re}\int_0^\infty e^{-st}e^{i\omega_0 t}\, dt = \frac{s}{\left(s^2 + \omega_0^2\right)} \tag{5.307c}$$

This can also be obtained from a table of Laplace transforms (e.g., item 7 of Table 4.2 with $\beta = 0$). Therefore, the Laplace transform of the differential equation yields

$$Y(s) = \frac{\omega_0 y_{\max}}{\left(s^2 + (\lambda/m)s + \omega_0^2\right)} + \frac{F_0}{m} \frac{s}{\left(s^2 + \omega_0^2\right)} \frac{1}{\left(s^2 + (\lambda/m)s + \omega_0^2\right)} \tag{5.308}$$

Writing

$$\frac{s}{\left(s^2 + \omega_0^2\right)} \frac{1}{\left(s^2 + (\lambda/m)s + \omega_0^2\right)} = \frac{m}{\lambda}\left[\frac{1}{\left(s^2 + \omega_0^2\right)} - \frac{1}{\left(s^2 + (\lambda/m)s + \omega_0^2\right)}\right] \tag{5.309}$$

$$\Rightarrow \quad Y(s) = \frac{\left(\omega_0 y_{\max} - F_0/\lambda\right)}{\left(s^2 + (\lambda/m)s + \omega_0^2\right)} + \frac{F_0}{\lambda} \frac{1}{\left(s^2 + \omega_0^2\right)} \tag{5.310}$$

We will first determine $y(t)$ by evaluating the Bromwich integral. The first term of equation 5.310 has two simple poles at

$$s = -\frac{\lambda}{2m} \pm \tfrac{1}{2}\sqrt{\frac{\lambda^2}{m^2} - 4\omega_0^2} \tag{5.311a}$$

Using the approximation $\lambda^2 \ll m^2\omega_0^2$, we ignore λ^2/m^2 under the square root. Thus the poles of the first term in equation 5.310 are at

$$s \simeq -\frac{\lambda}{2m} \pm i\omega_0 \tag{5.311b}$$

The second term has simple poles at

$$s = \pm i\omega_0 \tag{5.312}$$

These poles, along with the Bromwich contour, are shown in Figure 5.2. Therefore,

$$\begin{aligned} y(t) &= \frac{1}{2\pi i}\int_{c-i\infty}^{c+i\infty} Y(s)e^{st}\,ds \\ &= R\left(-\frac{\lambda}{2m} + i\omega_0\right) + R\left(-\frac{\lambda}{2m} - i\omega_0\right) + R(i\omega_0) + R(-i\omega_0) \end{aligned} \tag{5.313a}$$

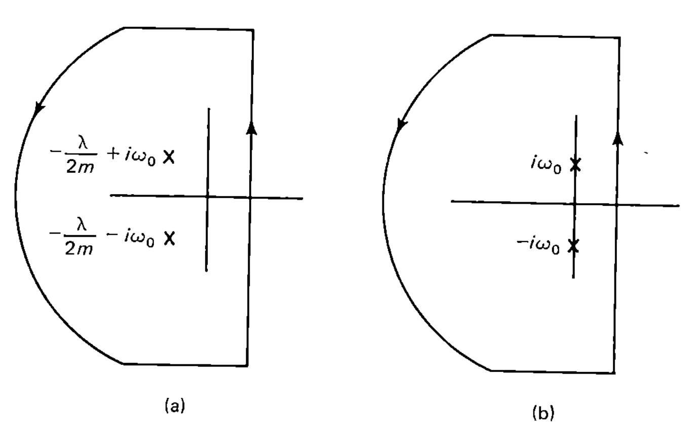

Figure 5.2 Poles and Bromwich contour for a damped harmonic oscillator driven by a force $F_0 \cos(\omega_0 t)$.

The evaluation of these residues is straightforward. The result is

$$y(t) = \left[\left(\omega_0 y_{\max} - \frac{F_0}{\lambda}\right)\frac{e^{-\lambda t/2m}}{\omega_0} + \frac{F_0}{\omega_0 \lambda}\right]\sin(\omega_0 t) \tag{5.313b}$$

The inverse Laplace transform can also be deduced from Laplace transform tables. Referring to Table 4.2, we note from item 6 that with $\omega^2 = \omega_0^2 - \beta^2$,

$$\frac{1}{\omega}\,\text{L.T.}\left[e^{-\beta t}\sin(\omega t)\right] = \frac{1}{\left(s^2 + 2\beta s + \omega_0^2\right)} \tag{5.314}$$

Referring to equation 5.310, we see that $\beta = \lambda/2m \ll \omega_0$. Thus $\omega \simeq \omega_0$. Therefore,

$$\frac{1}{\omega_0}\,\text{L.T.}\left[e^{-\lambda t/2m}\sin(\omega_0 t)\right] = \frac{1}{\left(s^2 + (\lambda/m)s + \omega_0^2\right)} \tag{5.315a}$$

Also, using item 6 of Table 4.2, with $\beta = 0$,

$$\frac{1}{\omega_0}\,\text{L.T.}[\sin(\omega_0 t)] = \frac{1}{\left(s^2 + \omega_0^2\right)} \tag{5.315b}$$

Therefore,

$$\begin{aligned} Y(s) &= \frac{(\omega_0 y_{\max} - F_0/\lambda)}{\left(s^2 + (\lambda/m)s + \omega_0^2\right)} + \frac{F_0}{\lambda}\frac{1}{\left(s^2 + \omega_0^2\right)} \\ &= \left(\omega_0 y_{\max} - \frac{F_0}{\lambda}\right)\frac{1}{\omega_0}\,\text{L.T.}\left[e^{-\lambda t/2m}\sin(\omega_0 t)\right] + \frac{F_0}{\lambda}\frac{1}{\omega_0}\,\text{L.T.}[\sin(\omega_0 t)] \end{aligned} \tag{5.316a}$$

$$\Rightarrow \quad y(t) = \text{L.T.}[Y(s)] = \left[\left(\omega_0 y_{\max} - \frac{F_0}{\lambda}\right)\frac{e^{-\lambda t/2m}}{\omega_0} + \frac{F_0}{\omega_0\lambda}\right]\sin(\omega_0 t) \tag{5.316b}$$

This is the result obtained by evaluating the Bromwich integral. □

Variation of parameters

The method of variation of parameters can sometimes yield a solution when the other methods already studied are not applicable. To apply the technique, one must be able to determine all independent solutions to the homogeneous equation. For an Nth-order differential equation, we denote these independent homogeneous solutions by $y_k(x)$, $1 \leqslant k \leqslant N$. If the coefficient of each power of D in the polynomial is constant, these independent solutions will be a product of some power of x and an exponential, such as was shown in equation 5.135. In general, the complete homogeneous solution will be

$$y_0(x) = \sum_{k=1}^{N} A_k y_k(x) \tag{5.317}$$

To apply the method of variation of parameters, it is assumed that the particular solution has the same form as the homogeneous solution but the constants A_k are replaced by functions $A_k(x)$. That is,

$$y_p(x) = \sum_{k=1}^{N} A_k(x) y_k(x) \tag{5.318}$$

The task of determining the particular solution is that of determining the N functions $A_k(x)$. There is only one constraint that can be imposed on these functions; $y_p(x)$ must satisfy the differential equation. Therefore, $N-1$ additional conditions must be specified in order that all $A_k(x)$ be determined, and these additional constraints can be imposed somewhat arbitrarily.

In order to form the differential equation, derivatives of y_p up to $D^N y_p$ must be evaluated. At each step, the sum of terms containing the first derivatives of the $A_k(x)$ is set to zero. These are the additional constraints imposed on the $A_k(x)$.

For example, the first derivative of $y_p(x)$ is

$$Dy_p = \sum_{k=1}^{N} A_k(x)[Dy_k(x)] + \sum_{k=1}^{N} [DA_k(x)]\, y_k(x) \tag{5.319}$$

Setting

$$\sum_{k=1}^{N} [DA_k(x)]\, y_k(x) = 0 \tag{5.320}$$

imposes one condition on the $A_k(x)$. With this condition, the first derivative becomes

$$Dy_p = \sum_{k=1}^{N} A_k(x)[Dy_k(x)] \tag{5.321}$$

Therefore,

$$D^2 y_p = \sum_{k=1}^{N} A_k(x)[D^2 y_k(x)] + \sum_{k=1}^{N} [DA_k(x)][Dy_k(x)] \tag{5.322}$$

A second constraint for the $A_k(x)$ is obtained by again setting the sum of terms containing first derivatives of $A_k(x)$ to zero,

$$\sum_{k=1}^{N} [DA_k(x)][Dy_k(x)] = 0 \tag{5.323}$$

$$\Rightarrow \quad D^2 y_p = \sum_{k=1}^{N} A_k(x)[D^2 y_k(x)] \tag{5.324}$$

This process is continued until $N-1$ equations containing first derivatives of the $A_k(x)$ are obtained. They are of the form

$$\sum_{k=1}^{N} [DA_k(x)][D^m y_k(x)] = 0 \qquad 0 \leqslant m \leqslant N-2 \tag{5.325}$$

From this process we obtain

$$D^{N-1} y_p(x) = \sum_{k=1}^{N} A_k(x)[D^{N-1} y_k(x)] \tag{5.326}$$

$$\Rightarrow \quad D^N y_p(x) = \sum_{k=1}^{N} A_k(x)[D^N y_k(x)] + \sum_{k=1}^{N} [DA_k(x)][D^{N-1} y_k(x)] \tag{5.327}$$

Substituting this Nth derivative and all lower-order derivatives of y_p (in the form expressed in equations 5.321, 5.324, and so on), into the differential equation, we obtain the necessary Nth equation needed to specify the N functions $A_k(x)$. The differential equation plus the $N-1$ equations in the form of equation 5.325 are a set of N coupled first-order differential equations in $A_k(x)$.

Example 5.31

As an example we will solve the equation

$$(D^3 + D)y = \sec(x) \tag{5.328}$$

$\sec(x)$ is not one of the inhomogeneous functions specified in equations 5.163. Thus the techniques developed for those inhomogeneous terms cannot be applied to this differential equation.

The roots of the algebraic polynomial are $z = 0, \pm i$. Therefore, the homogeneous solution is

$$y_0 = A_1 e^{ix} + A_2 e^{-ix} + A_3 \equiv A\sin(x) + B\cos(x) + C \tag{5.329}$$

From this, we take the particular solution to be

$$y_p(x) = A(x)\sin(x) + B(x)\cos(x) + C(x) \tag{5.330}$$

Therefore,

$$Dy_p = A(x)\cos(x) - B(x)\sin(x) + [A'(x)\sin(x) + B'(x)\cos(x) + C'(x)] \tag{5.331}$$

Setting the sum of terms involving $A'(x)$, $B'(x)$ and $C'(x)$ to zero, we have

$$A'(x)\sin(x) + B'(x)\cos(x) + C'(x) = 0 \tag{5.332}$$

This is one constraint on these functions. Then the first derivative of y_p becomes

$$Dy_p = A(x)\cos(x) - B(x)\sin(x) \tag{5.333}$$

$$\Rightarrow \quad D^2 y_p = -A(x)\sin(x) - B(x)\cos(x) + [A'(x)\cos(x) - B'(x)\sin(x)] \tag{5.334}$$

Setting

$$A'(x)\cos(x) - B'(x)\sin(x) = 0 \tag{5.335}$$

$$\Rightarrow \quad D^2 y_p = -A(x)\sin(x) - B(x)\cos(x) \tag{5.336}$$

$$\Rightarrow \quad D^3 y_p = -A(x)\cos(x) + B(x)\sin(x) - [A'(x)\sin(x) + B'(x)\cos(x)] \tag{5.337}$$

Combining equations 5.333 and 5.337, the differential equation becomes

$$(D^3 + D)y_p = A'(x)\sin(x) + B'(x)\cos(x) = -\sec(x) \tag{5.338}$$

We now have a set of three coupled first-order differential equations (equations 5.332, 5.335, and 5.338) involving $A(x)$, $B(x)$, and $C(x)$.

From equation 5.335

$$A'(x) = B'(x)\tan(x) \tag{5.339}$$

Therefore, equation 5.338 becomes

$$\text{(a)} \quad B'(x) = -1 \quad \Rightarrow \quad \text{(b)} \quad B(x) = -x + b \tag{5.340}$$

With this result, equation 5.339 is

$$\text{(a)} \quad A'(x) = -\tan(x) \quad \Rightarrow \quad \text{(b)} \quad A(x) = \log[\cos(x)] + a \tag{5.341}$$

With these expressions for $A'(x)$ and $B'(x)$, equation 5.332 yields

$$\text{(a)} \quad C'(x) = \sec(x) \quad \Rightarrow \quad \text{(b)} \quad C(x) = \log[\sec(x) + \tan(x)] + c \tag{5.342}$$

Therefore,

$$y_p = \log[\cos(x)]\sin(x) - x\cos(x) + \log[\sec(x) + \tan(x)] + a\sin(x) + b\cos(x) + c \tag{5.343}$$

We note that the last three terms duplicate the homogeneous solution. If the homogeneous solution is separately added to the particular solution, these terms can be ignored. This is accomplished by setting the constants of integration in equations 5.340b, 5.341b, and 5.342b to zero. □

Finally, we note that in the development of the method of variation of parameters, it was never necessary to require that the coefficients of the various powers of D be constants, although the example presented was an equation of this type. Therefore, if the solution to the homogeneous equation can be found, the method of variation of parameters is applicable in solving differential equations in which the coefficients of D are functions of x.

5.3 Higher-Order Differential Equations with Nonconstant Coefficients Series or Frobenius's method

The *method of Frobenius*, or *solution in series*, is a widely used method for determining the solution to a differential equation in which the coefficients of the various powers of the D operator are functions of the independent variable. Since the description of the method for an equation of general order can be rather unwieldy, we restrict our discussion to second-order linear differential equations. Many scientific problems are described by such a differential equation.

The most general form of a second-order linear differential equation is

$$[D^2 + A(x)D + B(x)]y = F(x) \tag{5.344}$$

The complete solution to this equation will be the sum of the homogeneous and particular solutions. The method for determining the particular solution by series methods is a simple extension of the approach to solving the homogeneous equation. Therefore, we concentrate on determination of the homogeneous solution and discuss the solution to the nonhomogeneous equation by series methods at the appropriate point in this development.

In addition to the series method, the particular solution may also be determinable by the method of variation of parameters once the homogeneous solution is found. Green's function methods described in Chapter 11 are another widely used approach for finding the particular solution.

Fuchs's theorem

As discussed earlier, a linear homogeneous differential equation of order N will have N independent solutions. Thus, for a second-order equation, there will be two independent solutions, and the complete solution will be of the form

$$y(x) = c_1 y_1(x) + c_2 y_2(x) \tag{5.345}$$

Fuchs's theorem provides a prescription that specifies when $y_1(x)$ and $y_2(x)$ have series representations. To apply Fuchs's theorem, we first define what is meant by the singularity structure of a homogeneous differential equation.

1. If $A(x)$ and $B(x)$, as defined in equation 5.344, are both analytic at and in a region surrounding a point x_0, then x_0 is called an *ordinary point*. In such a case, both $A(x)$ and $B(x)$ can be expanded in Taylor series about the point x_0.

$$\text{(a)} \quad A(x) = \sum_{k=0}^{\infty} a_k(x - x_0)^k \qquad \text{(b)} \quad B(x) = \sum_{k=0}^{\infty} b_k(x - x_0)^k \tag{5.346}$$

2. If $A(x)$ and/or $B(x)$ is singular at x_0, the differential equation is said to have a *singularity* at x_0. The singularity is called a *regular singularity* if both $(x - x_0)A(x)$ and $(x - x_0)^2B(x)$ are analytic at and around x_0. Thus $A(x)$ can have at most a simple pole, and the singularity of $B(x)$ can be a pole of order no higher than 2. Of course, one of the functions could be analytic, or $B(x)$ could have a simple pole and the equation would still have a singularity.

   If $A(x)$ has a simple pole, its Laurent series representation is

$$A(x) = \frac{a_{-1}}{(x - x_0)} + \sum_{k=0}^{\infty} a_k(x - x_0)^k \tag{5.347a}$$

   If $B(x)$ has a second-order pole, its Laurent series is

$$B(x) = \frac{b_{-2}}{(x - x_0)^2} + \frac{b_{-1}}{(x - x_0)} + \sum_{k=0}^{\infty} b_k(x - x_0)^k \tag{5.347b}$$

3. If the singularity structure of $A(x)$ and/or $B(x)$ at x_0 is such that x_0 is neither an ordinary point nor a regular singularity, then the singularity of the differential equation is called an *irregular singularity*.

Fuchs's theorem specifies the circumstances under which we can expect series representations for the two independent solutions to a second-order differential equation.

1. If x_0 is an ordinary point, Taylor series representations about x_0 exist for both independent solutions.
2. If x_0 is a regular singularity of the differential equation, at least one of the two independent solutions will have a Taylor series expansion about x_0. It is possible that both solutions will have Taylor series representations, but we can only be assured of finding one series solution. We will discuss the conditions for which only one or both solutions have series representations later.
3. If x_0 is an irregular singularity, we cannot be assured that either independent solution will have a Taylor expansion about x_0. A Taylor series representation about x_0 for one or both solutions is not ruled out, but there is no assurance that either has a series expansion.

If a series solution is to exist, the approach is to assume a series form for the solution:

$$y(x) = \sum_{l=0}^{\infty} c_l(x - x_0)^{l+s} \tag{5.348}$$

The solution will be determined when the parameter s (which is called the *index* of the solution) and all the coefficients c_l are determined.

x_0 is an ordinary point

For an ordinary point, both solutions have Taylor series expansions around x_0. Therefore, all terms in the series for $y(x)$ will contain nonnegative integer powers of $(x - x_0)$. If $s = 0$, all powers of $(x - x_0)$ will be nonnegative. The index can also have values other than zero, but it will always be possible to duplicate the solution for $s \neq 0$ by taking $s = 0$ and adjusting the coefficients c_l. If x_0 is an ordinary point, the index s can be chosen to be zero. It is not necessary to do so, but zero will always be one possible value of s.

For example, if the leading power of $(x - x_0)$ in the solution is $(x - x_0)^3$ one way this will be achieved is by selecting $s = 3$ as the value of the index, with $c_0 \neq 0$. Another way to obtain a leading term $(x - x_0)^3$ is to take $s = 0$, and require $c_0 = c_1 = c_2 = 0$.

As will be seen in subsequent examples, s can sometimes be a negative integer. The solution will then require that the first several coefficients be zero, so that all terms contain nonnegative powers of $(x - x_0)$. That is, if one chooses $s < 0$, then $c_0 = c_1 = \cdots = c_{|s|-1} = 0$, so that

$$y = \sum_{l=0}^{\infty} c_l (x - x_0)^{l-|s|} = \sum_{l=|s|}^{\infty} c_l (x - x_0)^{l-|s|} \tag{5.349}$$

which contains only nonnegative powers of $(x - x_0)$.

With the series representations of $A(x)$ and $B(x)$ given in equations 5.346, the homogeneous differential equation

$$\left[D^2 + A(x) D + B(x)\right] y = 0 \tag{5.350}$$

in series form is

$$\begin{aligned} &\sum_{l=0}^{\infty} c_l (l + s)(l + s - 1)(x - x_0)^{l+s-2} \\ &\quad + (a_0 + a_1(x - x_0) + \cdots) \sum_{l=0}^{\infty} c_l (l + s)(x - x_0)^{l+s-1} \\ &\quad + (b_0 + b_1(x - x_0) + \cdots) \sum_{l=0}^{\infty} c_l (x - x_0)^{l+s} = 0 \end{aligned} \tag{5.351a}$$

$$\begin{aligned} \Rightarrow \quad &\sum_{l=0}^{\infty} \left[c_l (l + s)(l + s - 1)(x - x_0)^{l+s-2} + c_l (l + s) \sum_{k=0}^{\infty} a_k (x - x_0)^{l+s+k-1} \right. \\ &\quad \left. + c_l \sum_{k=0}^{\infty} b_k (x - x_0)^{l+s+k} \right] = 0 \end{aligned} \tag{5.351b}$$

If equation 5.351b is to be valid for all values of x, the coefficient of each power of $(x - x_0)$ on the left-hand side must equal the coefficient of the same power of $(x - x_0)$ on the right-hand side of the equation: namely, zero.

If the differential equation of equations 5.344 is inhomogeneous, and if $F(x)$ can be expanded in a Taylor series, the method of solution is a straightforward extension of the approach described above. Writing

$$F(x) = \sum_{n=0}^{\infty} f_n (x - x_0)^n \tag{5.352}$$

$$\begin{aligned} \Rightarrow \quad &\sum_{l=0}^{\infty} \left[c_l (l + s)(l + s - 1)(x - x_0)^{l+s-2} + c_l (l + s) \sum_{k=0}^{\infty} a_k (x - x_0)^{l+s+k-1} \right. \\ &\quad \left. + c_l \sum_{k=0}^{\infty} b_k (x - x_0)^{l+s+k} \right] = \sum_{n=0}^{\infty} f_n (x - x_0)^n \end{aligned} \tag{5.353}$$

As with the homogeneous equation, coefficients of like powers of $(x - x_0)$ are equated. Therefore, in obtaining the solutions to the inhomogeneous equation, the coefficients of the various powers of $(x - x_0)$ on the left side of equation 5.353 are equated to the coefficients f_n on the right side of this equation. Thus the solution for $F(x) \neq 0$ by series methods is a straightforward extension of the method of solution for the homogeneous equation.

Returning to the homogeneous equation, the lowest power of $(x - x_0)$ in equation 5.351b arises from the $l = 0$ term in the first sum. It is

$$T_0 = c_0 s(s-1)(x-x_0)^{s-2} \tag{5.354a}$$

The term involving $(x - x_0)^{s-1}$ arises from the $l = 1$ term of the first sum and the $l = k = 0$ term in the second sum. That term is

$$T_1 = (c_1(s+1)s + c_0 a_0 s)(x-x_0)^{s-1} \tag{5.354b}$$

To determine the terms involving $(x - x_0)$ to powers higher than $s - 1$, consider the terms multiplying $(x - x_0)^{r+s}$ with $r \geqslant 0$. They will arise from a combination of terms from all three sums. The contribution to the $(x - x_0)^{r+s}$ term from the first sum of equation 5.353 arises from $l - 2 = r$. $r \geqslant 0$ requires $l \geqslant 2$. This term is

$$U_r = c_{r+2}(r+s+2)(r+s+1)(x-x_0)^{r+s} \tag{5.354c}$$

The $(x - x_0)^{r+s}$ term in the second sum arises for $l + k - 1 = r$, with $l + k \geqslant 1$. The contributions to this term come from the terms with

$$\text{(a)} \quad l = r+1, \quad k = 0 \qquad \text{(b)} \quad l = r, \quad k = 1 \cdots \qquad \text{(c)} \quad l = 0, \quad k = r+1 \tag{5.355}$$

That is, the $(x - x_0)^{r+s}$ term from the second sum is

$$V_r = \left[\sum_{k=0}^{r+1} c_{r-k+1} a_k (r - k + 1 + s)\right](x-x_0)^{r+s} \tag{5.356}$$

By identical analysis, it is straightforward to see that the contribution to the $(x - x_0)^{r+s}$ term from the third sum is

$$W_r = \left[\sum_{k=0}^{r} c_{r-k} b_k\right](x-x_0)^{r+s} \tag{5.357}$$

Therefore, the $(x - x_0)^{r+s}$ term in equation 5.351b is

$$T_r = \left[c_{r+2}(r+s+2)(r+s+1) + \sum_{k=0}^{r+1} c_{r-k+1} a_k (r+s-k+1) + \sum_{k=0}^{r} c_{r-k} b_k\right](x-x_0)^{r+s} \tag{5.358}$$

The coefficient of each power of $(x - x_0)$ must be zero. Thus, from equations 5.354 and 5.358, we obtain

$$\text{(a)} \quad c_0 s(s-1) = 0 \qquad \text{(b)} \quad [c_1(s+1) + a_0 c_0]s = 0 \tag{5.359}$$

$$c_{r+2} = \frac{-\sum_{k=0}^{r+1} c_{r-k+1} a_k (r+s-k+1) - \sum_{k=0}^{r} c_{r-k} b_k}{(r+s+2)(r+s+1)} \tag{5.360a}$$

Equations 5.359a and 5.359b determine the allowed values of the index s. As such, they are called the *indicial equations*. Note that one solution, $s = 0$, satisfies both equations for any value of c_0 and c_1. With $s = 0$, equation 5.360a becomes

$$c_{r+2} = \frac{-\sum_{k=0}^{r+1} c_{r-k+1} a_k (r-k+1) - \sum_{k=0}^{r} c_{r-k} b_k}{(r+2)(r+1)} \tag{5.360b}$$

with $r \geqslant 0$. For $r = 0$, we see that c_2 is determined in terms of c_0 and c_1. Setting $r = 1$, c_3 is a combination of c_1 and c_2 and therefore is a combination of c_0 and c_1. From this type of recurrent determination of the coefficients, we see that all coefficients $c_2, c_3, \ldots$ are combinations of c_0 and c_1. Equation 5.360a (for any value of the index) or equation 5.360b (specifically for $s = 0$) is called the *recurrence relation* for the coefficients.

From the discussion above, we see that when x_0 is an ordinary point, $s = 0$ is one possible solution for the index that satisfies both indicial equations for arbitrary values of c_0 and c_1. From the recurrence relation all coefficients are expressible in terms of c_0 and c_1. Therefore, the series solution is in the form

$$y = c_0[\text{series in } (x - x_0)] + c_1[\text{series in } (x - x_0)] \tag{5.361}$$

Because the coefficients in each of the series are found from the recurrence relation, each of the series separately must satisfy the differential equation. That is, these series are the two required independent solutions to the differential equation, and both can be obtained by choosing $s = 0$ as a solution to the indicial equations.

From equations 5.359 we see that there are other values of the index that satisfy the indicial equations. However, using them will produce solutions that duplicate one or both of the independent solutions found for $s = 0$.

Example 5.32

As an example, consider Airy's equation

$$(D^2 - x)y = 0 \tag{5.362}$$

Comparing this to equation 5.350, we see that $x_0 = 0$ and

$$\text{(a)} \quad A(x) = 0 \qquad \text{(b)} \quad B(x) = -x \tag{5.363}$$

We express the solution for y as the MacLaurin series

$$y = \sum_{l=0}^{\infty} c_l x^{l+s} \tag{5.364}$$

Substituting this into equation 5.362, the series form of the Airy differential equation is

$$\sum_{l=0}^{\infty} c_l (l+s)(l+s-1) x^{l+s-2} - \sum_{l=0}^{\infty} c_l x^{l+s+1} = 0 \tag{5.365a}$$

We note that the three lowest powers of x in the first sum of equation 5.365a, arising from $l = 0$, 1, and 2, are x^{s-2}, x^{s-1}, and x^s. The lowest power of x in the second sum is x^{s+1}. Thus there are no terms in the second sum that will combine with the $l = 0$, 1, and 2 terms from the first sum. We therefore write equation 5.365a as

$$c_0 s(s-1)x^{s-2} + c_1(s+1)sx^{s-1} + c_2(s+2)(s+1)x^s + \sum_{l=3}^{\infty} c_l(l+s)(l+s-1)x^{l+s-2} - \sum_{l=0}^{\infty} c_l x^{l+s+1} = 0 \tag{5.365b}$$

Equating the coefficients of x^{s-2}, x^{s-1}, and x^s to zero, we obtain

$$\text{(a)} \quad c_0 s(s-1) = 0 \qquad \text{(b)} \quad c_1(s+1)s = 0 \qquad \text{(c)} \quad c_2(s+1)(s+2) = 0 \tag{5.366}$$

These are the indicial equations for the Airy equation. Since each of the first three terms in equation 5.365b is zero, the differential equation, in series form, is

$$\sum_{l=3}^{\infty} c_l(l+s)(l+s-1)x^{l+s-2} - \sum_{l=0}^{\infty} c_l x^{l+s+1} = 0 \tag{5.367}$$

In order to adjust the powers of x to be the same in both sums, we define a new summation index in the first sum, l', such that $l + s - 2 = l' + s + 1$, or $l = l' + 3$. Substituting $l' + 3$ for l in the first sum, then dropping the prime, equation 5.367 becomes

$$\sum_{l=0}^{\infty} [c_{l+3}(l+s+3)(l+s+2) - c_l]x^{l+s+1} = 0 \tag{5.368}$$

It is now straightforward to set the coefficient of each power of x to zero to obtain the recurrence relation

$$c_{l+3} = \frac{c_l}{(l+s+3)(l+s+2)} \tag{5.369}$$

Referring to equations 5.366, one set of solutions to the indicial equations is $c_0 = c_1 = c_2 = 0$. With this choice, we see from the recurrence relation that $c_3 = c_4 = c_5 = 0$, from which $c_6 = c_7 = c_8 = 0$, and so on. That is, by choosing $c_0 = c_1 = c_2 = 0$, we obtain the trivial solution, $y = 0$. Thus to obtain a nontrivial solution, we must use solutions to the indicial equations in which at least one of the coefficients is nonzero.

For example, equation 5.366a will be nonzero with $c_0 \neq 0$ if $s = 0$ or $s = 1$. Equation 5.366b can be satisfied with $c_1 \neq 0$ by choosing $s = 0$ or $s = -1$. Similarly, equation 5.366c is satisfied with $c_2 \neq 0$ if $s = -1$ or $s = -2$. As noted above in general, one possible choice is $s = 0$. With this choice, both c_0 and c_1 are arbitrary but from equation 5.366c, c_2 must be zero. If we choose $s = -1$, both c_1 and c_2 can be nonzero, and c_0 must be zero. If $s = 1$ or $s = -2$, only one coefficient can be nonzero, and two coefficients must be zero.

Choosing $s = 0$, the recurrence relation of equation 5.369 becomes

$$c_{l+3} = \frac{c_l}{(l+3)(l+2)} \tag{5.370}$$

$$\text{(a)} \quad \Rightarrow \quad c_3 = \frac{c_0}{3 \cdot 2} = 1 \cdot \frac{c_0}{3!} \qquad \text{(b)} \quad c_6 = \frac{c_3}{6 \cdot 5} = \frac{c_0}{6 \cdot 5 \cdot 3 \cdot 2} = 4 \cdot 1 \cdot \frac{c_0}{6!}$$

$$\text{(c)} \quad c_9 = \frac{c_6}{9 \cdot 8} = 7 \cdot 4 \cdot 1 \cdot \frac{c_0}{9!} \tag{5.371}$$

$$\Rightarrow \quad c_{3N} = (3N-2)(3N-5)(3N-8) \cdots \frac{c_0}{(3N)!} \qquad N = 1, 2, \ldots \tag{5.372}$$

Similarly,

$$\text{(a)} \quad c_4 = \frac{c_1}{4 \cdot 3} = 2 \cdot \frac{c_1}{4!} \qquad \text{(b)} \quad c_7 = \frac{c_4}{7 \cdot 6} = 5 \cdot 2 \cdot \frac{c_1}{7!}$$

$$\text{(c)} \quad c_{10} = \frac{c_7}{10 \cdot 9} = 8 \cdot 5 \cdot 2 \cdot \frac{c_1}{10!} \tag{5.373}$$

$$\Rightarrow \quad c_{(3N+1)} = (3N-1)(3N-4)(3N-7) \cdots \frac{c_1}{(3N+1)!} \qquad N = 1, 2, \ldots \tag{5.374}$$

When $s = 0$, this requires that $c_2 = 0$. Therefore,

$$c_5 = c_8 = \cdots = 0 \tag{5.375}$$

$$\Rightarrow \quad c_{(3N+2)} = 0 \qquad N = 1, 2, \ldots \tag{5.376}$$

Therefore, from equations 5.372 and 5.374, the solution to Airy's equation is

$$\begin{aligned} y = c_0\left[1 + \frac{x^3}{3 \cdot 2} + \frac{x^6}{6 \cdot 5 \cdot 3 \cdot 2} + \frac{x^9}{9 \cdot 8 \cdot 6 \cdot 5 \cdot 3 \cdot 2} + \cdots\right] \\ + c_1\left[x + \frac{x^4}{4 \cdot 3} + \frac{x^7}{7 \cdot 6 \cdot 4 \cdot 3} + \frac{x^{10}}{10 \cdot 9 \cdot 7 \cdot 6 \cdot 4 \cdot 3} + \cdots\right] \end{aligned} \tag{5.377}$$

If s had been chosen to be -1, c_0 would have to be zero and c_1 and c_2 would be arbitrary. With $s = -1$ and $c_0 = 0$, the recurrence relation is

$$c_{l+3} = \frac{c_l}{(l+2)(l+1)} \tag{5.378}$$

$$\Rightarrow \quad c_3 = c_6 = c_9 = \cdots = c_{3N} = 0 \qquad N = 0, 1, 2, \ldots \tag{5.379}$$

From the recurrence relation, it is straightforward to obtain the solution for $s = -1$. It is

$$\begin{aligned} y = c_1 x^{-1}\left[x + \frac{x^4}{3 \cdot 2} + \frac{x^7}{6 \cdot 5 \cdot 3 \cdot 2} + \frac{x^{10}}{9 \cdot 8 \cdot 6 \cdot 5 \cdot 3 \cdot 2} + \cdots\right] \\ + c_2 x^{-1}\left[x^2 + \frac{x^5}{4 \cdot 3} + \frac{x^8}{7 \cdot 6 \cdot 4 \cdot 3} + \frac{x^{11}}{10 \cdot 9 \cdot 7 \cdot 6 \cdot 4 \cdot 3} + \cdots\right] \end{aligned} \tag{5.380}$$

Comparing this to equation 5.377, we see that this is the same solution obtained with $s = 0$. What is called c_0 in equation 5.377 is designated c_1 in equation 5.380, and c_2 in equation 5.380 is c_1 in equation 5.377.

As will be shown in Problem 33, by choosing $s = 1$, or $s = -2$, only one solution is obtained and that solution is a duplication of one of the two independent series solutions obtained above. Note that neither independent solution contains singularities. □

Using this example as a guide, we note that when x_0 is an ordinary point, if a single choice of the index will yield two independent solutions to the equations, any other valid choice of index will duplicate one or both independent solutions. $s = 0$ will always be one of these possible choices. If a choice of index is made that yields only one independent solution, the second independent solution is found by making a second choice of the index consistent with the indicial equations.

The series representations are valid solutions to the differential equation within the range of convergence of the series. From the recurrence relation for the coefficients of the Airy solution, we see that two successive terms in each independent series differ by three powers of x. Thus the Cauchy ratio for either Airy series is

$$\rho_N = \left|\frac{c_{N+3}x^{N+3}}{c_N x^N}\right| = |x^3|\left|\frac{c_{N+3}}{c_N}\right| \tag{5.381}$$

From the recurrence relation,

$$\frac{c_{N+3}}{c_N} = \frac{1}{(N+s+3)(N+s+2)} \tag{5.382}$$

$$\Rightarrow \quad \lim_{N\to\infty} \rho_N = |x^3| \lim_{N\to\infty} \frac{1}{(N+s+3)(N+s+2)} = |x^3| \cdot 0 < 1 \tag{5.383}$$

Thus the Airy series solutions converge for all finite values of x.

x_0 is a regular singular point

When x_0 is a regular singularity, the approach is identical to that used for ordinary points. However, at least one solution will contain a singularity at x_0. For that reason, s does not have to be an integer. That is, the solution could contain a branch point. For example, if one value of the index were $s = \frac{1}{2}$, that solution would contain a square root branch point at x_0.

If x_0 is a regular singularity, $(x - x_0)A(x)$ and $(x - x_0)^2B(x)$ are analytic at x_0. Writing the differential equation as

$$\left\{(x - x_0)^2 D^2 + (x - x_0)[(x - x_0)A(x)D] + (x - x_0)^2 B(x)\right\}y = 0 \tag{5.384}$$

we define the analytic functions

$$\text{(a)}\quad P(x) \equiv (x - x_0)A(x) \qquad \text{(b)}\quad Q(x) \equiv (x - x_0)^2 B(x) \tag{5.385}$$

$$\Rightarrow \quad \left[(x - x_0)^2 D^2 + (x - x_0)P(x)D + Q(x)\right]y = 0 \tag{5.386}$$

$P(x)$ and $Q(x)$ can now be expanded in Taylor series;

$$\text{(a)}\quad P(x) = \sum_{k=0}^{\infty} p_k (x - x_0)^k \qquad \text{(b)}\quad Q(x) = \sum_{k=0}^{\infty} q_k (x - x_0)^k \tag{5.387}$$

Assuming the series solution

$$y = \sum_{l=0}^{\infty} c_l (x - x_0)^{l+s} \tag{5.348}$$

the series form of the differential equation is

$$\begin{aligned}\sum_{l=0}^{\infty} c_l(l + s)(l + s - 1)(x - x_0)^{l+s} + \sum_{\substack{l=0\\k=0}}^{\infty} c_l p_k (l + s)(x - x_0)^{l+s+k}& \\ + \sum_{\substack{l=0\\k=0}}^{\infty} c_l q_k (x - x_0)^{l+s+k} = 0&\end{aligned} \tag{5.388}$$

The lowest power of $(x - x_0)$ in each sum is $(x - x_0)^s$ arising from $l = 0$ and $k = 0$ terms. From this we obtain the indicial equation

$$c_0[s(s - 1) + p_0 s + q_0] = 0 \tag{5.389}$$

$$\Rightarrow \quad s_{1,2} = \tfrac{1}{2}\left[(1 - p_0) \pm \sqrt{(1 - p_0)^2 - 4q_0}\right] \tag{5.390}$$

The indicial equation is obtained by setting the coefficient of $(x - x_0)^s$ to zero. Thus the recurrence relation is found by setting the coefficient of $(x - x_0)^{r+s}$ to zero for $r \geqslant 1$. In the first sum of equation 5.388, this coefficient is $c_r(r + s)(r + s - 1)$. In the second and third sums, the coefficient of $(x - x_0)^{r+s}$ arises from terms in which $l + k = r$. These are obtained for

$$\text{(a)}\quad l = r, \quad k = 0 \qquad \text{(b)}\quad l = r - 1, \quad k = 1 \cdots \qquad \text{(c)}\quad l = 0, \quad k = r \tag{5.391}$$

That is, the coefficient of $(x - x_0)^{r+s}$ in the second sum is $\sum_{k=0}^{r} c_{r-k} p_k (r - k + s)$ and in the third sum, this coefficient is $\sum_{k=0}^{r} c_{r-k}q_k$. Therefore, equation 5.388 can be rewritten

$$\sum_{r=1}^{\infty} \left\{c_r(r+s)(r+s-1) + \sum_{k=0}^{r} c_{r-k}[p_k(r-k+s) + q_k]\right\}(x-x_0)^{r+s} = 0 \tag{5.392}$$

$$\Rightarrow \quad c_r(r+s)(r+s-1) + \sum_{k=0}^{r} c_{r-k}[p_k(r-k+s) + q_k] = 0 \tag{5.393a}$$

Note that the $k = 0$ term in the sum contains c_r. That is, equation 5.393a can be written

$$c_r(r+s)(r+s-1) + \sum_{k=0}^{r} c_{r-k}[p_k(r-k+s) + q_k]$$

$$= c_r[(r+s)(r+s-1) + p_0(r+s) + q_0] + \sum_{k=1}^{r} c_{r-k}[p_k(r-k+s) + q_k] = 0 \tag{5.393b}$$

Therefore, the recurrence relation is

$$c_r = \frac{-\sum_{k=1}^{r} c_{r-k}[p_k(r-k+s) + q_k]}{(r+s)(r+s-1) + p_0(r+s) + q_0} \tag{5.394}$$

Since the indicial equation is quadratic, there will be two values for the index, and therefore, two series solutions, one for each value of s found in equation 5.390. If $s_1 = s_2$, of course, only one series solution exists.

If s_1 and s_2 differ by a positive integer, let

$$s_1 - s_2 \equiv N \geqslant 1 \tag{5.395}$$

From equation 5.390,

$$s_1 - s_2 = \sqrt{(1-p_0)^2 - 4q_0} \tag{5.396}$$

$$\Rightarrow \quad \text{(a)} \quad s_1 = \tfrac{1}{2}(1-p_0) + \frac{N}{2} \qquad \text{(b)} \quad s_2 = \tfrac{1}{2}(1-p_0) - \frac{N}{2} \tag{5.397}$$

For $s = s_1$, the recurrence relation of equation 5.394 is

$$c_r = \frac{-\sum_{k=1}^{r} c_{r-k}[p_k(r-k+s_1) + q_k]}{(r+s_1)(r+s_1-1) + p_0(r+s_1) + q_0} \tag{5.398a}$$

With $s = s_2 = s_1 - N$, the recurrence relation becomes

$$c_r = \frac{-\sum_{k=1}^{r} c_{r-k}[p_k(r-k+s_1-N) + q_k]}{(r+s_1-N)(r+s_1-N-1) + p_0(r+s_1-N) + q_0} \tag{5.398b}$$

Since N is a positive integer, one possible value of r is N. Then

$$c_N = \frac{-\sum_{k=1}^{N} c_{N-k}[p_k(-k+s_1)+q_k]}{s_1(s_1-1)+p_0 s_1+q_0} \tag{5.399}$$

Because s_1 satisfies the indicial equation, the denominator of equation 5.399 is zero. If the numerator is not zero, the series contains coefficients that are infinite. In that case, a convergent series solution with $s = s_2$ does not exist.

One way the numerator of equation 5.399 will be zero is if each $c_{N-k} = 0$, $1 \leqslant k \leqslant N-1$. Then the series solution for $s = s_2$ will be of the form

$$\begin{aligned} y &= (x-x_0)^{s_2}\left[c_N(x-x_0)^N + c_{N+1}(x-x_0)^{N+1} + \cdots\right] \\ &= (x-x_0)^{s_2+N}\left[c_N + c_{N+1}(x-x_0) + \cdots\right] = (x-x_0)^{s_1}\left[c_N + c_{N+1}(x-x_0) + \cdots\right] \end{aligned} \tag{5.400}$$

This is identical in form to the series solution for $s = s_1$. Thus, if the numerator is zero because the coefficients c_{N-k} are zero, the series solution for $s = s_2$ is a duplication of the series solution for $s = s_1$.

The numerator of the recurrence relation in equation 5.399 will also be zero if $p_k(-k+s_1)+q_k = 0$ for each k. In that case, c_N is arbitrary and a second series solution will exist even when the indices differ by an integer. Therefore, as stated in Fuchs's theorem, when x_0 is a regular singularity, one series solution is guaranteed, and it may be possible to obtain both solutions in series.

Method of reduction of order and the second solution

When only one series solution can be found, the second solution to the differential equation can be obtained by a technique developed by d'Alembert, called the method of reduction of order.

Let $y_1(x)$ be a known series solution to the differential equation. We are assured by Fuchs's theorem of finding this. Let the second solution we are trying to determine be denoted by $y_2(x)$. We define a function $v(x)$ such that

$$y_2(x) \equiv v(x)y_1(x) \tag{5.401}$$

Since y_2 must satisfy the differential equation, substituting this second solution into equation 5.386 yields

$$\begin{aligned} &(x-x_0)^2 y_1(D^2v) + 2(x-x_0)^2(Dy_1)(Dv) + (x-x_0)P(x)y_1(Dv) \\ &\quad + v\left[(x-x_0)^2D^2y_1 + (x-x_0)P(x)Dy_1 + Q(x)y_1\right] = 0 \end{aligned} \tag{5.402}$$

Because y_1 satisfies the differential equation, the term in the brackets multiplying v is zero and the differential equation for $v(x)$ becomes

$$(x-x_0)^2 y_1 D^2 v + \left[2(x-x_0)^2 Dy_1 + (x-x_0)P(x)y_1\right]Dv = 0 \tag{5.403}$$

With $v' = Dv$, this is seen to be a separable first-order differential equation for v':

$$\frac{dv'}{v'} + \left[2\frac{dy_1}{y_1} + \frac{P(x)}{(x-x_0)}\,dx\right] = 0 \tag{5.404}$$

$$\Rightarrow \quad v' = \frac{1}{(y_1)^2}e^{-\int P(x)/(x-x_0)\,dx} \tag{5.405a}$$

Since we will be multiplying y_2 by a constant to obtain the complete solution, we are taking the constants of integration in this development to be zero. Integrating equation 5.405a we obtain

$$v(x) = \int dx \frac{1}{[y_1(x)]^2} e^{-\int P(x)/(x-x_0)\,dx} \tag{5.405b}$$

$$\Rightarrow \quad y_2(x) = y_1(x) \int \left[\frac{1}{[y_1(x)]^2} e^{-\int P(x)/(x-x_0)\,dx} \right] dx \tag{5.406}$$

Wronskian determination of the second solution

This second solution can be obtained in another way. Let y_1 be the solution obtainable as a series and y_2 be the second solution we are seeking. Each satisfies the differential equation. That is,

$$D^2 y_1 + \frac{P(x)}{(x-x_0)} D y_1 + \frac{Q(x)}{(x-x_0)^2} y_1 = 0 \tag{5.407a}$$

$$D^2 y_2 + \frac{P(x)}{(x-x_0)} D y_2 + \frac{Q(x)}{(x-x_0)^2} y_2 = 0 \tag{5.407b}$$

Multiplying equation 5.407a by y_2 and equation 5.407b by y_1 and subtracting, we obtain

$$\frac{(y_1 y_2'' - y_2 y_1'')}{(y_1 y_2' - y_2 y_1')} = -\frac{P(x)}{(x-x_0)} \tag{5.408}$$

The denominator, $y_1 y_2' - y_2 y_1'$ is called the *Wronskian* of y_1 and y_2. It can be written as a 2×2 determinant,

$$W_{12} = \begin{vmatrix} y_1 & y_2 \\ y_1' & y_2' \end{vmatrix} \tag{5.409}$$

We note that

$$\frac{dW_{12}}{dx} = y_1 y_2'' + y_1' y_2' - y_2' y_1' - y_2 y_1'' = y_1 y_2'' - y_2 y_1'' \tag{5.410}$$

which is the numerator of the right-hand side of equation 5.408. Thus, equation 5.408 can be written

$$\frac{1}{W_{12}} \frac{dW_{12}}{dx} = -\frac{P(x)}{(x-x_0)} \tag{5.411}$$

$$\Rightarrow \quad W_{12}(x) = W_0 e^{-\int P(x)/(x-x_0)\,dx} \tag{5.412}$$

Since $P(x)$ is analytic at x_0, it can be expanded in a Taylor series about x_0. Thus

$$\int \frac{P(x)}{(x-x_0)} dx = p_0 \log(x-x_0) + p_1(x-x_0) + p_2(x-x_0)^2 + \cdots \tag{5.413}$$

$$\Rightarrow \quad W_{12}(x) = \frac{W_0}{(x-x_0)^{p_0}} e^{-[p_1(x-x_0)+p_2(x-x_0)^2+\cdots]} \tag{5.414}$$

We note that the exponential is nonzero for all finite x. In addition, $1/(x-x_0)^{p_0} \neq 0$ for any finite x (except $x = x_0$ with $p_0 < 0$). Thus, for $x \neq x_0$, if $W_{12}(x) = 0$, then $W_0 = 0$ and $W_{12}(x) = 0$ for all x. Conversely, if there is one value of x (other than x_0) for which $W_{12}(x) \neq 0$, then $W_0 \neq 0$, and $W_{12}(x)$ will be nonzero for all finite x.

If $W_{12} = 0$, then

$$\text{(a)} \quad y_1 y_2' = y_2 y_1' \qquad \Rightarrow \qquad \text{(b)} \quad \frac{y_1'}{y_1} = \frac{y_2'}{y_2} \tag{5.415}$$

$$\Rightarrow \quad y_2 = A y_1 \tag{5.416}$$

where A is constant. That is, if $W_{12} = 0$, y_1 and y_2 are not independent. Conversely, if $W_{12} \neq 0$, y_2 is not a multiple of y_1 and is therefore independent of y_1.

With $W_{12} = y_1 y_2' - y_2 y_1' \neq 0$, equation 5.412 can be written

$$y_2' - \frac{y_1'}{y_1} y_2 = \frac{W_0}{y_1} e^{-\int P(x)/(x-x_0)\,dx} \tag{5.417}$$

This is a Bernoulli equation of index zero. From equation 5.93, its solution, in terms of the known y_1, is

$$\frac{y_2}{y_1} - W_0 \int \frac{dx}{y_1^2} e^{-\int P(x)/(x-x_0)\,dx} = C \tag{5.418a}$$

$$\Rightarrow \quad y_2(x) = W_0 y_1(x) \int \frac{dx}{y_1^2} e^{-\int P(x)/(x-x_0)\,dx} + C y_1(x) \tag{5.418b}$$

Since the complete solution to the differential equation is obtained by adding a constant multiple of y_2 to y_1, there is no loss in generality in taking $C = 0$, $W_0 = 1$. In that case, the second solution expressed in equation 5.418b is identical to that obtained in equation 5.406.

If a second series solution does not exist when s_1 and s_2 differ by a positive integer or zero, the second solution for the smaller root $s_2 = s_1 - N$ is obtained from equation 5.406. Writing

$$\text{(a)} \quad P(x) = p_0 + p_1(x - x_0) + \cdots \qquad \text{(b)} \quad y_1(x) = (x - x_0)^{s_1}[c_0 + c_1(x - x_0) + \cdots] \tag{5.419}$$

the integrand of equation 5.406 is

$$\frac{1}{(x - x_0)^{2s_1}[c_0 + c_1(x - x_0) + \cdots]^2} e^{-[p_0 \log(x - x_0) + p_1(x - x_0) + \cdots]} \tag{5.420}$$

$$= \frac{1}{(x - x_0)^{2s_1 + p_0}} \frac{1}{[c_0 + c_1(x - x_0) + \cdots]^2} e^{-[p_1(x - x_0) + \cdots]} \equiv \frac{1}{(x - x_0)^{2s_1 + p_0}} U(x)$$

Since $c_0 \neq 0$ in y_1, $U(x)$ is analytic at x_0, and thus has a Taylor expansion about x_0. Therefore, the second solution is

$$y_2(x) = (x - x_0)^{s_1}[c_0 + c_1(x - x_0) + \cdots] \int \frac{dx}{(x - x_0)^{2s_1 + p_0}} [U_0 + U_1(x - x_0) + \cdots] \tag{5.421}$$

Referring to equation 5.397a, $2s_1 + p_0 = N + 1$. Thus the integral in equation 5.421 is

$$\int dx\left[\frac{U_0}{(x-x_0)^{N+1}} + \frac{U_1}{(x-x_0)^N} + \cdots + \frac{U_N}{(x-x_0)} + \cdots\right]$$
$$= -\frac{U_0}{N}\frac{1}{(x-x_0)^N} - \frac{U_1}{(N-1)}\frac{1}{(x-x_0)^{N-1}} + \cdots + U_N\log(x-x_0) + \cdots \tag{5.422}$$

If the roots are equal, $N = 0$ and the leading term in equation 5.422 is $U_0 \log(x - x_0)$. Multiplying equation 5.422 by $y_1(x)$, with $s_2 = s_1 - N$ the second solution becomes

$$\begin{aligned} y_2(x) &= U_N y_1(x)\log(x-x_0) \\ &\quad - \left[\frac{U_0}{N}(x-x_0)^{-N} + \frac{U_1}{(N-1)}(x-x_0)^{-N+1} + \cdots\right]\sum_{l=0}^{\infty} c_l(x-x_0)^{l+s_1} \\ &= U_N y_1(x)\log(x-x_0) \\ &\quad - \left[\frac{U_0}{N} + \frac{U_1}{(N-1)}(x-x_0) + \cdots\right](x-x_0)^{s_1-N}\sum_{l=0}^{\infty} c_l(x-x_0)^l \\ &= U_N y_1(x)\log(x-x_0) + (x-x_0)^{s_2}\sum_{l=0}^{\infty}\gamma_l(x-x_0)^l \end{aligned} \tag{5.423}$$

We note that if $U_N = 0$, the logarithmic term is absent.

Example 5.33

To illustrate this development, we consider three examples. The first is

$$\left[2x^2D^2 - xD + (1 + x)\right]y = 0 \tag{5.424}$$

which has a regular singularity at $x = 0$. Writing the solution in series form,

$$y = \sum_{l=0}^{\infty} c_l x^{l+s} \tag{5.425}$$

the series form of the differential equation is

$$\begin{aligned} &\sum_{l=0}^{\infty}\left[2c_l(l+s)(l+s-1)x^{l+s} - c_l(l+s)x^{l+s} + c_l x^{l+s} + c_l x^{l+s+1}\right] \\ &\quad = \sum_{l=0}^{\infty} c_l[(l+s)(2l+2s-3)+1]x^{l+s} + \sum_{l=0}^{\infty} c_l x^{l+s+1} = 0 \end{aligned} \tag{5.426}$$

The lowest power of x in the first sum is x^s, whereas the lowest power of x in the second sum is x^{s+1}. Therefore, we write equation 5.426 in the form

$$c_0[s(2s-3)+1]x^s + \sum_{l=1}^{\infty} c_l[(l+s)(2l+2s-3)+1]x^{l+s} + \sum_{l=0}^{\infty} c_l x^{l+s+1} = 0 \tag{5.427a}$$

Replacing l by $l + 1$ in the first sum, and combining the two sums, this becomes

$$c_0[s(2s-3)+1]x^s + \sum_{l=0}^{\infty}\left[c_{l+1}(l+s)(2l+2s+1) + c_l\right]x^{l+s+1} = 0 \tag{5.427b}$$

The indicial equation is obtained by setting the coefficient of x^s to zero:

$$c_0(2s^2 - 3s + 1) = 0 \tag{5.428}$$

The recurrence relation results from equating the coefficients of all higher powers of x to zero. Therefore,

$$c_{l+1} = \frac{-c_l}{(l + s)(2l + 2s + 1)} \tag{5.429}$$

We note from equation 5.429 that every coefficient is a multiple of c_0. If we satisfy the indicial equation by setting $c_0 = 0$, we obtain the trivial solution. Therefore, the nontrivial solutions arise by choosing s such that

$$2s^2 - 3s + 1 = 0 \tag{5.430}$$

Thus series solutions exist for $s = 1$, and $s = \frac{1}{2}$. Since these roots do not differ by an integer, we expect two independent series solutions, one for each value of s.

For $s = 1$,

$$c_{l+1} = \frac{-c_l}{(l + 1)(2l + 3)} \tag{5.431}$$

$$\Rightarrow \quad y_1 = c_0\left[x - \frac{x^2}{3} + \frac{x^3}{30} - \frac{x^4}{630} + \cdots\right] \tag{5.432}$$

For $s = \frac{1}{2}$ we obtain the solution

$$y_2 = c_0' x^{1/2}\left[1 - x + \frac{x^2}{6} - \frac{x^3}{90} + \cdots\right] \tag{5.433}$$

Note that y_2 and therefore the complete solution has a square root branch point at $x = 0$, reflecting the singularity of the differential equation.

The range of convergence of each of these series is determined by the behavior of the Cauchy ratio

$$\rho_N = \left|\frac{c_{N+1}x^{N+1}}{c_N x^N}\right| = |x|\left|\frac{c_{N+1}}{c_N}\right| \tag{5.434}$$

at large N. From the recurrence relation of equation 5.429,

$$\rho_N = |x|\frac{1}{(N + s)(2N + 2s + 1)} \to \frac{|x|}{2N^2} \to |x| \cdot 0 < 1 \tag{5.435}$$

That is, the series solutions converge for all finite x. □

Example 5.34

As an example of a differential equation for which the values of the index differ by an integer, consider

$$(xD^2 - 1)y = 0 \tag{5.436}$$

which has a regular singularity at $x = 0$. The series solution

$$y = \sum_{l=0}^{\infty} c_l x^{l+s} \tag{5.437}$$

$$\Rightarrow \quad \sum_{l=0}^{\infty} \left[c_l(l + s)(l + s - 1)x^{l+s-1} - c_l x^{l+s}\right] = 0 \tag{5.438}$$

There is one indicial equation

$$c_0 s(s - 1) = 0 \tag{5.439}$$

which has solutions $c_0 = 0$, $s = 0$, or $s = 1$. The recurrence relation is

$$c_{l+1} = \frac{c_l}{(l+s+1)(l+s)} \tag{5.440}$$

We note from the recurrence relation that to obtain the nontrivial solution, c_0 cannot be zero. Thus the solution to the indicial equation must be obtained by taking $s_1 = 1$ and $s_2 = 0$, which differ by an integer.

For $s = 1$,

$$\text{(a)} \quad c_{l+1} = \frac{c_l}{(l+2)(l+1)} \quad \Rightarrow \quad \text{(b)} \quad c_l = \frac{c_0}{(l+1)!\,l!} \tag{5.441}$$

$$\Rightarrow \quad y_{s=1} = c_0 \sum_{l=0}^{\infty} \frac{x^{l+1}}{(l+1)!\,l!} \tag{5.442}$$

For $s = 0$,

$$c_{l+1} = \frac{c_l}{l(l+1)} \tag{5.443}$$

If $c_0 \neq 0$, then for $l = 0$, we obtain $c_1 = \infty$, and a second solution for $s = 0$ does not exist. $c_1 = \infty$ can be avoided by taking $c_0 = 0$. Then c_1 is arbitrary and using equation 5.443, we generate $c_2, c_3, \ldots$ in terms of c_1. We obtain

$$\text{(a)} \quad c_2 = \frac{c_1}{2 \cdot 1} \qquad \text{(b)} \quad c_3 = \frac{c_2}{3 \cdot 2 \cdot 2 \cdot 1} = \frac{c_1}{3! \cdot 2!} \qquad \text{(c)} \quad c_4 = \frac{c_1}{4! \cdot 3!} \tag{5.444}$$

$$\Rightarrow \quad c_l = \frac{c_1}{l! \cdot (l-1)!} \tag{5.445}$$

Then, with $s = 0$ and $c_0 = 0$,

$$y_{s=0} = c_1 \sum_{l=1}^{\infty} \frac{x^l}{l!\,(l-1)!} \tag{5.446a}$$

Replacing l by $l - 1$, this becomes

$$y_{s=0} = c_1 \sum_{l=0}^{\infty} \frac{x^{l+1}}{l!\,(l+1)!} \tag{5.446b}$$

which is identical to the series for $s = 1$ expressed in equation 5.442. Thus, only one solution exists in series form.

To obtain the second solution, we use equation 5.406. Since there is no term in the differential equation in Dy, $P(x) = 0$. Therefore,

$$y_2 = \left[\frac{1}{c_0} \sum_{k=0}^{\infty} \frac{x^{k+1}}{(k+1)!\,k!}\right] \int \left[\sum_{l=0}^{\infty} \frac{x^{l+1}}{(l+1)!\,l!}\right]^{-2} dx \tag{5.447}$$

Using the recurrence relation of equation 5.440, the Cauchy ratio is

$$\rho_N = |x| \frac{1}{(N+s+2)(N+s+1)} \rightarrow |x| \cdot 0 < 1 \tag{5.448}$$

Therefore, the one series solution converges for all finite x. □

Example 5.35

Finally, we present an example for which two independent series solutions exist even though the two indices differ by an integer. We consider the differential equation

$$\left[x^2D^2 + xD + \left(x^2 - \tfrac{1}{4}\right)\right] y = 0 \tag{5.449}$$

which is known as the Bessel equation of half-order.

From the series solution

$$y = \sum_{l=0}^{\infty} c_l x^{l+s} \tag{5.450}$$

we obtain Bessel's differential equation in series form.

$$\sum_{l=0}^{\infty} \left[c_l\left((l+s)^2 - \tfrac{1}{4}\right)x^{l+s} + c_l x^{l+s+2} \right] = 0 \tag{5.451}$$

The indicial equations arise from the $l = 0$ and $l = 1$ terms in the first sum. They are

$$\text{(a)} \quad c_0\left(s^2 - \tfrac{1}{4}\right) = 0 \qquad \text{(b)} \quad c_1\left[(s+1)^2 - \tfrac{1}{4}\right] = 0 \tag{5.452}$$

If we choose to satisfy the first indicial equation by setting $s = \frac{1}{2}$, we must take $c_1 = 0$. However, we note that by selecting $s = -\frac{1}{2}$, both c_0 and c_1 are arbitrary, and we know from experience that if two series solutions exist, this choice will yield those two solutions simultaneously. With this choice, we have two indices that differ by $N = 1$. In the discussion following equation 5.399, we noted that if the indices differ by an integer, a second independent series solution will exist if the terms in the numerator of the recurrence relation, $p_k(-k + s_1) + q_k = 0$ for each k. For the half-order Bessel equation, since $r = N = 1$, the only value k can have is $k = 1$. Referring to equation 5.449, we see that

$$\text{(a)} \quad p_k = \delta_{k0} \qquad \text{(b)} \quad q_k = \delta_{k2} - \tfrac{1}{4}\delta_{k0} \tag{5.453}$$

where δ_{kn} is the Kronecker δ-symbol. Specifically, for $k = 1$, $p_1 = q_1 = 0$, so the numerator of the recurrence relation of equation 5.399 is zero, and both c_0 and c_1 are arbitrary, and two independent series solutions exist for the half-order Bessel equation.

Choosing $c_1 = 0$, equation 5.452a requires that $s^2 = \frac{1}{4}$ or $s = \pm\frac{1}{2}$. Then

$$\text{(a)} \quad c_{l+2} = \frac{-c_l}{\left(l + \frac{5}{2}\right)^2 - \frac{1}{4}} \quad s = \tfrac{1}{2} \qquad \text{(b)} \quad c_{l+2} = \frac{-c_l}{\left(l + \frac{3}{2}\right)^2 - \frac{1}{4}} \quad s = -\tfrac{1}{2} \tag{5.454}$$

From these recurrence relations, we obtain the solutions

$$y_{1/2} = c_0 x^{1/2}\left[1 - \frac{x^2}{6} + \frac{x^4}{120} - \cdots\right] = c_0 x^{1/2}\left[1 - \frac{x^2}{3!} + \frac{x^4}{5!} - \cdots\right] = c_0 \frac{\sin x}{\sqrt{x}} \tag{5.455a}$$

$$y_{-1/2} = c_0' x^{-1/2}\left[1 - \frac{x^2}{2} + \frac{x^4}{24} - \cdots\right] = c_0' x^{-1/2}\left[1 - \frac{x^2}{2!} + \frac{x^4}{4!} - \cdots\right] = c_0' \frac{\cos x}{\sqrt{x}} \tag{5.455b}$$

We note that both solutions contain square root branch points at $x = 0$, reflecting the singularity structure of the Bessel equation. We see that the solution for $s = \frac{1}{2}$ is finite at $x = 0$, while the solution for $s = -\frac{1}{2}$ is infinite at $x = 0$.

From our discussions in Chapter 3, we know that both $\sin x$ and $\cos x$ converge for all finite values of x. Therefore, both Bessel half-order series converge for all finite x. □

Euler's equation

If the functions $P(x)$ and $Q(x)$ defined in equations 5.385 are constants p_0 and q_0, the differential equation becomes

$$\left[(x - x_0)^2 D^2 + p_0(x - x_0)D + q_0\right] y = 0 \tag{5.456a}$$

which is known as the Euler equation. In series form, this differential equation is

$$\sum_{l=0}^{\infty} c_l\left[(l+s)(l+s-1) + p_0(l+s) + q_0\right](x - x_0)^{l+s} = 0 \tag{5.456b}$$

We note that terms involving coefficients with different subscripts do not combine in this equation. Therefore, each term in the series must separately be zero. That is, either

$$\text{(a)} \quad c_l = 0 \qquad \text{or} \qquad \text{(b)} \quad (l+s)(l+s-1) + p_0(l+s) + q_0 = 0 \tag{5.457}$$

Thus there is no recurrence relation for the coefficients of the Euler equation.

Note that l appears in the combination $l + s$ in exponent of $(x - x_0)$ and in the indicial equation, equation 5.457b. We define this combination $l + s \equiv m$, and equation 5.457b is

$$m(m-1) + mp_0 + q_0 = 0 \tag{5.458}$$

Since there are only two values of m that satisfy this quadratic equation, there are only two nonzero coefficients. We denote them by $c_{\pm}$. That is, the series solution to the Euler equation reduces to two terms,

$$y = c_+(x - x_0)^{m_+} + c_-(x - x_0)^{m_-} \tag{5.459}$$

where

$$m_{\pm} = \frac{(1 - p_0) \pm \sqrt{(1 - p_0)^2 - 4q_0}}{2} \tag{5.460}$$

If $(1 - p_0)^2 = 4q_0$, then $m_+ = m_-$ and this approach yields only one solution. The second solution is obtained using the method of reduction of order. With $m_+ = m_- \equiv m$, and $y_1 = (x - x_0)^m$, the second solution, from equation 5.406, becomes

$$y_2 = (x - x_0)^m \int \frac{dx}{(x - x_0)^{2m}} e^{-p_0 \log(x - x_0)} = (x - x_0)^m \int \frac{dx}{(x - x_0)^{2m + p_0}} \tag{5.461}$$

But from equation 5.460, when $(1 - p_0)^2 = 4q_0$, $2m + p_0 = 1$. Therefore, the second solution is

$$y_2 = (x - x_0)^m \log(x - x_0) \tag{5.462}$$

Thus, when there is only one index, the complete solution to the Euler equation is

$$y = c_1(x - x_0)^m + c_2(x - x_0)^m \log(x - x_0) \tag{5.463}$$

Equations involving multiple-term recurrence relations

All of the examples presented thus far have involved differential equations for which the recurrence relations have been of the form

$$c_{l+k} = c_l F(l, s) \tag{5.464}$$

where k is a positive integer and F is some function of l and s. For example, for the Airy equation,

$$c_{l+3} = \frac{1}{(l + s + 3)(l + s + 2)} c_l \tag{5.465}$$

This recurrence relation involves only two coefficients, c_{l+3} and c_l.

Example 5.36

To illustrate the analysis of equations in which the recurrence relation involves more than two coefficients, consider the equation

$$\left[3xD^2 + (1 + x^2)D + (1 - x)\right]y = 0 \tag{5.466}$$

which has a regular singularity at $x = 0$. Substituting the series

$$y = \sum_{l=0}^{\infty} c_l x^{l+s} \tag{5.467}$$

into the differential equation yields

$$\sum_{l=0}^{\infty} [3c_l(l+s)(l+s-1)x^{l+s-1} + c_l(l+s)x^{l+s-1} + c_l(l+s)x^{l+s+1} + c_l x^{l+s}$$

$$-c_l x^{l+s+1}] = \sum_{l=0}^{\infty} [c_l(l+s)(3l+3s-2)x^{l+s-1} + c_l x^{l+s} + c_l(l+s-1)x^{l+s+1}] = 0 \tag{5.468}$$

We express the terms involving x^{s-1} and x^s explicitly, writing equation 5.468 as

$$c_0 s(3s-2)x^{s-1} + [c_1(s+1)(3s+1) + c_0]x^s$$

$$+ \sum_{l=2}^{\infty} c_l(l+s)(3l+3s-2)x^{l+s-1} + \sum_{l=1}^{\infty} c_l x^{l+s} + \sum_{l=0}^{\infty} c_l(l+s-1)x^{l+s+1} = 0 \tag{5.469}$$

Setting the coefficient of x^{s-1} to zero, the indicial equation is

$$c_0 s(3s-2) = 0 \tag{5.470}$$

With $c_0 \neq 0$, the allowed values of the index are $s = 0, \frac{2}{3}$. By setting the coefficient of x^s to zero, we obtain

$$c_1 = \frac{-c_0}{(s+1)(3s+1)} \tag{5.471}$$

This is not an indicial equation, but is more in the form of a recurrence relation.

To obtain the recurrence relation, we substitute $l' = l - 2$ in the first sum of equation 5.469 and $l' = l - 1$ in the second sum. Then, dropping the prime and setting the coefficients of x^{s-1} and x^s to zero, the differential equation in series form becomes

$$\sum_{l=0}^{\infty} [c_{l+2}(l+s+2)(3l+3s+4) + c_{l+1} + c_l(l+s-1)]x^{l+s+1} = 0 \tag{5.472}$$

from which we obtain the three-term recurrence relation

$$c_{l+2}(l+s+2)(3l+3s+4) + c_{l+1} + c_l(l+s-1) = 0 \tag{5.473}$$

For $s = 0$, the recurrence relation is

$$c_{l+2}(l+2)(3l+4) + c_{l+1} + c_l(l-1) = 0 \tag{5.474}$$

We note that if we set $l = -1$ and take $c_{-1} = 0$, we obtain the expression for c_1 given in equation 5.471. Therefore, for $l = 0$, with $c_1 = -c_0$ from equation 5.471,

$$\text{(a)} \quad 8c_2 + c_1 - c_0 = 8c_2 - 2c_0 = 0 \quad \Rightarrow \quad \text{(b)} \quad c_2 = \frac{c_0}{4} \tag{5.475}$$

For $l = 1$, with $c_1 = -c_0$,

$$\text{(c)} \quad 21c_3 + c_2 = 0 \quad \Rightarrow \quad \text{(d)} \quad c_3 = -\frac{c_2}{21} = \frac{-c_0}{84} \tag{5.475}$$

For $l = 2$, we obtain

$$c_4 = \frac{-5c_0}{840} \tag{5.475e}$$

and so on. Therefore, the series solution for $s = 0$ is

$$y_{s=0} = c_0\left[1 + \frac{x^2}{4} - \frac{x^3}{84} - \frac{5x^4}{840} + \cdots\right] \tag{5.476}$$

For $s = \frac{2}{3}$, the recurrence relation is

$$c_{l+2}(3l + 8)(l + 2) + c_{l+1} + c_l(l - \tfrac{1}{3}) = 0 \tag{5.477}$$

and from equation 5.471,

$$\begin{aligned} &\text{(a)} \quad c_1 = -\frac{c_0}{5} \quad \Rightarrow \quad \text{(b)} \quad c_2 = \frac{c_0}{30} \\ \Rightarrow \quad &\text{(c)} \quad c_3 = \frac{c_0}{330} \quad \Rightarrow \quad \text{(d)} \quad c_4 = -\frac{c_0}{1540} \end{aligned} \tag{5.478}$$

Therefore, the solution for $s = \frac{2}{3}$ is

$$y_{s=2/3} = c_0 x^{2/3}\left[1 - \tfrac{1}{5}x + \tfrac{1}{30}x^2 + \tfrac{1}{330}x^3 - \tfrac{1}{1540}x^4 + \cdots\right] \tag{5.479}$$

which has a fractional power branch point at $x = 0$.

To determine the range of convergence of the series solutions, the Cauchy ratio must be less than 1.

$$\rho_N = \left|\frac{c_{N+1}x^{N+1}}{c_N x^N}\right| = |x|\left|\frac{c_{N+1}}{c_N}\right| < 1 \tag{5.480}$$

From the recurrence relation of equation 5.473,

$$\frac{c_{N+2}}{c_{N+1}}(N + s + 2)(3N + 3s + 4) + 1 + \frac{c_N}{c_{N+1}}(N + s - 1) = 0 \tag{5.481}$$

For large N, this becomes

$$\frac{c_{N+2}}{c_{N+1}} + \frac{1}{N}\frac{c_N}{c_{N+1}} + \frac{1}{3N^2} = 0 \tag{5.482}$$

But for large N, the Cauchy ratio can be written as

$$\left|\frac{c_{N+2}}{c_{N+1}}\right| = \left|\frac{c_{N+1}}{c_N}\right| = \frac{\rho_N}{|x|} \tag{5.483}$$

Thus equation 5.480 can be written in terms of ρ_N as

$$\frac{\rho_N}{|x|} + \frac{1}{3N}\frac{|x|}{\rho_N} + \frac{1}{3N^2} = 0 \tag{5.484a}$$

For large N, the solution to this quadratic equation for $\rho_N/\,|x|$ yields

$$\text{(b)} \quad \lim_{N\to\infty}\frac{\rho_N}{|x|} = \lim_{N\to\infty}\frac{1}{\sqrt{-3N}} = 0 \quad \Rightarrow \quad \text{(c)} \quad \lim_{N\to\infty}\rho_N = |x| \cdot 0 < 1 \tag{5.484}$$

So the series converges for all finite values of x. □

PROBLEMS

1. Separate the differential equation

$$6(x + 2y)\,dy + \tfrac{1}{2}(x^2 + 4xy + 4y^2 + 9)\,dx = 0$$

and find its solution subject to $y(0) = 2$.

2. Solve the differential equation

$$e^{\sqrt{x/y}}\,dy + \frac{y}{x}\left[\frac{y}{x} - e^{\sqrt{x/y}}\right]dx = 0$$

subject to $y(1) = 1$. In choosing the substitution to be made, use the fact that it is often easier to integrate functions involving e^z rather than functions containing $e^{\sqrt{z}}$, $e^{1/z}$, or $e^{1/\sqrt{z}}$.

3. (a) Show that if x and y appear in the functions $M(x, y)$ and $N(x, y)$ in the combination

$$z = \frac{(1 + \alpha y)}{x}$$

that the first-order differential equation

$$M(x, y)\,dy + N(x, y)\,dx = 0$$

is separable in x and z, and is also separable in y and z.

(b) Show that if x and y appear in the functions $M(x, y)$ and $N(x, y)$ in the combination

$$z = x^p(1 + \alpha y)$$

that the first-order differential equation

$$M(x, y)\,dy + N(x, y)\,dx = 0$$

is separable in x and z, and is also separable in y and z only for $p = -1$.

4. Find the solution to the differential equation

$$\left[\frac{(1 + 2y)}{x}\right]dy - \left[3 + \frac{(1 + 2y)^2}{x^2}\right]dx = 0$$

subject to $y(1) = 0$.

5. Show that if $M(x, y)$ and $N(x, y)$ depend on x and y in the combination

$$z = x^p y^q$$

the differential equation

$$M(x, y)\,dy + N(x, y)\,dx = 0$$

is separable in x and z, and is also separable in y and z only if $p = -q$.

6. Show that the differential equation

$$xy^2\frac{dy}{dx} - (x^3 + y^3) = 0$$

is homogeneous of order 3, and determine its solution subject to $y(1) = 3$.

7. Determine the order of homogeneity of the following differential equations, and find the solution of each equation subject to the initial (boundary) condition specified.

(a) $x\,dy - [y + \sqrt{x^2 + y^2}]\,dx = 0$, with $y(1) = 0$

(b) $2x^{1/2}\,dy + 3x^{-1/2}y\,dx = 0$, with $y(1) = 1$

(c) $x\cos\left(\frac{y}{x}\right)\frac{dy}{dx} + \left[x - y\cos\left(\frac{y}{x}\right)\right] = 0$, with $y(1) = \frac{\pi}{2}$

8. Show that the differential equation

$$(3x^2y - 4y^3 + 1)\,dy + 3x(x + y^2)\,dx = 0$$

is exact, and find its solution subject to $y(0) = -1$.

9. Show that the differential equation

$$\left[\frac{1}{y} - \frac{x}{y\sqrt{x^2 + y^2}}\right]\frac{dy}{dx} + \frac{1}{\sqrt{x^2 + y^2}} = 0$$

is exact and determine its solution subject to $y(1) = 0$.

10. (a) Show that the differential equation

$$M(x, y)\,dy + N(x, y)\,dx = 0$$

is exact if $M(x, y)$ and $N(x, y)$ can be written

$$M(x, y) = A(x) + B(y) \quad \text{and}$$

$$N(x, y) = y\frac{dA}{dx} + C(x)$$

(b) Determine the solution to such an equation in terms of the functions $A(x)$, $B(y)$, and $C(x)$.

(c) Apply this approach to find the solution to the differential equation

$$\left(x^3 + \frac{1}{y}\right)dy + \left(3x^2y + \frac{1}{x^2}\right)dx = 0$$

subject to $y(1) = 1$.

11. Show that the differential equation

$$2x^{1/2}\,dy + 3x^{-1/2}y\,dx = 0$$

can be made exact by multiplication by an integrating factor $\mu(z)$ with $z = xy$. Find that integrating factor and solve the equation.

12. Show that the differential equation

$$(ye^x - 2)\,dy + (x^2e^y - 2)\,dx = 0$$

can be made exact by multiplication by an integrating factor that depends on $z = x + y$. Find that integrating factor and solve the equation subject to $y(0) = 0$.

13. Find the single-variable integrating factor that makes each of the equations below exact. Solve the differential equation subject to the specified initial (boundary) condition.

(a) $xy^2\dfrac{dy}{dx} - (x^3 + y^3) = 0$, with $y(1) = 1$

(b) $(2x - ye^y)\,dy + y\,dx = 0$, with $y(0) = 0$

(c) $(ye^{-x} + 1)\,dy + (y + e^{-x})\,dx = 0$, with $y(0) = 1$

(d) $y^2 \sin(2x)\,dx - (y\cos^2 x + 1)\,dy = 0$, with $y(0) = 1$

14. For each of the differential equations below, determine if a single-variable integrating factor exists that makes the equation exact.

(a) $3xy^2\,dx + (y^2 \tan x + 3x^2 y)\,dy = 0$

(b) $2(x + y^2\cos^2 x)\,dx + y(\sin(2x) - 2x)\,dy = 0$

15. Determine the solution to each of the following homogeneous differential equations by the Heaviside operator method.

(a) $(D^2 - 3D + 1)y = 0$

(b) $(3D^3 - 2D^2 + 2D)y = 0$

(c) $(D^3 - 2D^2 + D)y = 0$

(d) $(D^3 - 2D^2 + D)^2 y = 0$

(e) $(D^2 + \omega^2)y = 0$

16. Find the particular solution to each of the following inhomogeneous differential equations using Heaviside operator methods by inverting the operator polynomial.

(a) $(D^3 - D + 3)y = e^{-2x}$

(b) $(D^4 - 4D^3 + 6D^2 - 4D + 1)y = e^{-x}$

(c) $(D^4 - 4D^3 + 6D^2 - 4D + 1)y = e^x$

(d) $(D^2 + D - 6)^2 y = e^{3x} + e^{-2x}$

(e) $(D^2 - D - 6)^2 y = e^{3x} + e^{-2x}$

17. Find the particular solution to each of the following differential equations by the Heaviside operator method by noting that the polynomial operator is real.

(a) $(D^2 - D - 1)y = \sin(3x)$

(b) $(D^3 - D - 1)y = \cos(3x)$

18. Find the particular solution to each of the following differential equations using the Heaviside operator method by writing the inhomogeneous trigonometric functions in terms of exponentials.

(a) $(D^2 + iD + 2i + 1)y = \sin x$

(b) $(D^2 + iD + 2)y = \cos(2x)$

19. Find the particular solution to each of the following differential equations by Heaviside operator techniques, using the properties of a polynomial operator that contains only even powers of the differential operator.

(a) $(D^4 + 3D^2 + 2)y = \cos(2x)$

(b) $(D^4 - 5D^2 - 36)y = \sin(3x)$

(c) $(D^4 - 5D^2 - 36)y = \cos(2x)$

(d) $(D^5 - D)y = 2\sin(2x)$

20. Find the particular solution to each of the following differential equations by rationalizing the polynomial operator.

(a) $(D^3 - D + 1)y = \cos x$

(b) $(D^2 + iD - 2)y = \sin(2x)$

(c) $(D^2 + D + 1 - i)y = \cos x$

21. Determine the particular solution to each of the following differential equations using Heaviside operator methods by expressing the inhomogeneous function in terms of exponentials.

(a) $(D^3 - 2D - 4)y = \cosh(2x)$

(b) $(D^4 - 4D^3 + 6D^2 - 4D + 1)y = \sinh(4x)$

(c) $(D^3 - 3D^2 + 3D - 1)y = \cosh x - \sinh x$

(d) $(D^3 - 3D^2 + 3D - 1)y = \cosh x + \sinh x$

22. Determine the particular solution to each of the following differential equations by Heaviside operator methods using the properties of a polynomial operator that contains only even powers of the differential operator.

(a) $(D^6 + 2D^2 + 1)y = \cosh(2x)$

(b) $(D^4 - 4D^2 + 3)y = \sinh x$

(c) $(D^3 - 9D)(D^2 - 2)y = 3\cosh(3x)$

23. Determine the particular solution to each of the following differential equations using Heaviside operator methods by rationalizing the polynomial operator.

(a) $(D^2 - D - 1)y = \cosh(2x)$

(b) $(D^2 - D)y = \sinh(3x)$

(c) $(D^3 - 1)y = \sinh\left(\dfrac{x}{2}\right)$

(d) $(D^2 + 3D - 4)y = \cosh(4x)$

(e) $(D^2 - 3D)y = 2\sinh(2x)$

24. Find the particular solution to each of the following differential equations using Heaviside methods by expanding the inverse operator.

(a) $(D^2 + 3D - 1)y = x^2$

(b) $(D^2 + 3D - 1)y = x^2 + x$

(c) $(D^3 - D)y = x^2$

(d) $(D^3 - 1)y = x^2$

(e) $(D^3 - D)y = x$

(f) $(D^5 - D^2)y = x + 1$

25. Using arguments involving Heaviside operator methods, explain (without obtaining the solutions) why the particular solution to

$$(D^5 + 5D^4 - D + 1)y = x$$

is the same as the particular solution to

$$(D^4 - 3D^3 + D^2 - D + 1)y = x$$

26. Find the particular solution to the following differential equations using Heaviside operator methods.

(a) $(D^2 - 3D + 2)y = xe^x\cosh(2x)$

(b) $(D^3 - 1)y = xe^x$

(c) $(D^3 - 1)y = e^x \sin(2x)$
(d) $(D^2 - 1)y = x \cosh x$

27. Find the particular solution to each of the following differential equations by Heaviside methods using the fact that the polynomial operator is real.
(a) $(D^2 - 2D + 2)y = xe^{2x} \cos x$
(b) $(D^2 - D + 3)y = x \sin x$
(c) $(D^2 - 1)y = x(2 \sin x + 3 \cos x)$

28. Find the particular solution to each of the following differential equations using Heaviside operator methods, by your recognition of the form of the solution.
(a) $(D^3 - 2D^2 + 1)y = 2 \sin(2x)$
(b) $(D^2 + D + 2)y = 3 \sin(2x) + 2 \cos(3x)$
(c) $(D^2 - 3D + 1)y = x^2$
(d) $(D^3 + 4D^2)y = 2x$
(e) $(D^3 + 1)y = 18x^2e^{-x}$

29. An oscillating mass m on a spring with a spring constant $k = m\omega_0^2$ is damped by a force $F_{\text{damping}} = -\lambda(dy/dt)$. It is driven by a force $F_{\text{driving}} = F_0e^{-\lambda|t|/m}$. Deduce the differential equation satisfied by this oscillator, and find its solution using Fourier transform methods. For arithmetic simplification, take $\omega_0^2 \gg \lambda^2/m^2$.

30. A series RLC circuit is driven by a DC battery that has a terminal potential difference V_0 (see Figure 5.3). The capacitor is initially uncharged. At an instant designated as $t = 0$, the switch S is closed. Deduce the differential equation for $Q(t)$, the charge on the capacitor at any instant. Find the solution to the differential equation by Laplace transform methods, recognizing that the inductor initially opposes current flow in the circuit. Therefore, $I(0) = 0$. For arithmetic simplification, take $R^2/L^2 \ll 1/LC$.
(a) Determine the solution for $Q(t)$ by evaluating the Bromwich integral.
(b) Determine the solution for $Q(t)$ by using Table 4.2, a table of Laplace transforms.

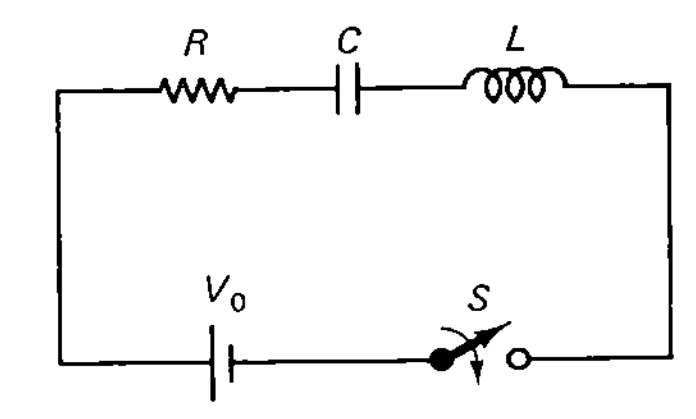

Figure 5.3 *RLC* series DC circuit.

31. A series RLC circuit contains no battery or AC generator (see Figure 5.4). The capacitor is placed in the circuit after having been given a charge Q_0. At an instant designated at $t = 0$, the switch S is closed. Deduce the differential equation for $Q(t)$, the charge on the capacitor at any instant. Recognizing that the inductor initially opposes any current flow in the circuit, so that $I(0) = 0$, solve the differential equation by Laplace transform methods.
(a) Determine the solution for $Q(t)$ by evaluating the Bromwich integral.
(b) Determine the solution for $Q(t)$ using Table 4.2, a table of Laplace transforms.

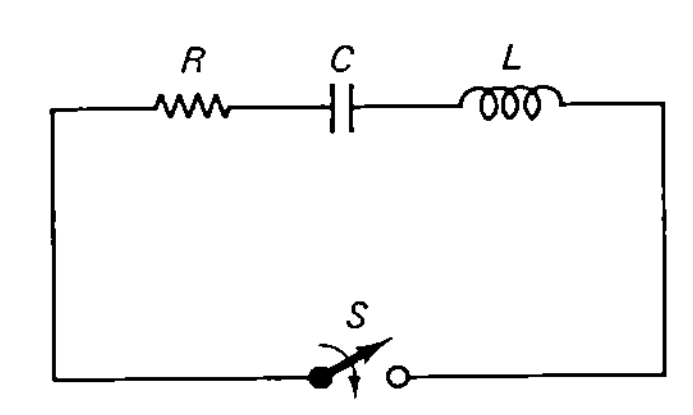

Figure 5.4 *RLC* series circuit.

32. Find the particular solution to each of the following differential equations using the method of variation of parameters.
(a) $(D^2 + 1)y = \tan x$
(b) $(D^2 + 4)y = 2 \cot(2x)$
(c) $(D^2 - 2D + 2)y = e^x \sec x \csc x$
(d) $(D + 1)^2y = \dfrac{e^{-x}}{x^2} \log x$
(e) $(D^2 + 2D - 8)y = e^{2x}\left[\dfrac{6}{x} - \dfrac{1}{x^2}\right]$

33. As was described in the text, $s = +1$ and $s = -2$ are possible values of the index for the Frobenius solution to the Airy equation (see equations 5.366). Show that the series solution obtained for each of these values of the index duplicates part of the complete solution to the Airy equation for $s = 0$ given in equation 5.377.

34. For each of the following differential equations:
(1) Identify which point(s), if any, are singular points. If there are any singular points, specify whether they are regular or irregular singularities.
(2) Determine the indicial equation(s) and all possible values of the index that will yield independent solutions. Obtain all independent solutions to the differential equation. If only one series solution exists, determine the second solution. If any point is a singularity of the differential equation, specify where that singularity appears in the solution.
(3) For each independent series solution, determine the range of convergence of the series.
(a) $(D^2 - x^2D - x)y = 0$
(b) $(D^2 + \omega^2x^2)y = 0$
(c) $(D^2 - 2xD + 2\lambda)y = 0$
This is the Hermite differential equation. The solution is denoted by $H_\lambda(x)$. Show that if $\lambda \to$

N, where N is a non-negative integer, the solution $H_N(x)$ is a polynomial of order N.

(d) $(3xD^2 + 2D + 1)y = 0$

(e) $(xD^2 + xD + 1)y = 0$

(f) $(xD^2 + D + x)y = 0$

This is the differential equation for the zero-order Bessel function.

(g) $[2xD^2 + (1 - 2x^2)D - (1 + 4x)]y = 0$

(h) $(x^3D^2 + xD + 1)y = 0$

35. Find the two independent series solutions to each of the following differential equations.

(a) $(2x^2D^2 + xD - 1)y = 0$

(b) $(x^2D^2 + xD + \frac{1}{4})y = 0$

(c) $(2xD^2 - 3D)y = 0$

CHAPTER 6

SPECIAL FUNCTIONS

There are certain functions that arise in the solution to many scientific problems that are designated as special functions. In this chapter we determine many of the properties of the more commonly occurring of these functions. The known properties of these functions are voluminous, and we do not attempt to develop all of them. Rather, in deriving those properties that are presented, it is the intent to point out several methods of manipulation that the reader can employ in dealing with these and similar special functions.

6.1 Functions Defined by Integrals

Gamma and beta functions

The most frequently used description of the Γ-function is its Euler integral representation

$$\Gamma(p) \equiv \int_0^\infty x^{p-1}e^{-x}\,dx \tag{6.1}$$

Another integral representation can be obtained from this by making the substitution $w = e^{-x}$. Then

$$\Gamma(p) = (-1)^p \int_0^1 [\log(w)]^{p-1}\,dw \tag{6.2}$$

Returning to the form of the Γ-function expressed in equation 6.1, a single integration by parts yields

$$\Gamma(p) = -x^{p-1}e^{-x}\big|_0^\infty + (p-1)\int_0^\infty x^{p-2}e^{-x}\,dx \tag{6.3a}$$

For $Re\,p > 1$, the integrated term is zero, and

$$\Gamma(p) = (p-1)\int_0^\infty x^{p-2}e^{-x}\,dx = (p-1)\Gamma(p-1) \tag{6.3b}$$

From equation 6.3b

$$\text{(a)}\quad \Gamma(p-1) = (p-2)\Gamma(p-2) \quad\Rightarrow\quad \text{(b)}\quad \Gamma(p) = (p-1)(p-2)\Gamma(p-2) \tag{6.4}$$

$$\Rightarrow \quad \Gamma(p) = (p-1)(p-2)(p-3)\cdots(p-k)\Gamma(p-k) \tag{6.5}$$

If p is a positive integer, N, this can be written

$$\Gamma(N) = (N-1)(N-2)(N-3)\cdots 2\cdot 1\cdot \Gamma(1) \tag{6.6}$$

Using the fact that

$$\Gamma(1) = \int_0^\infty e^{-x}\,dx = 1 \tag{6.7}$$

$$\Rightarrow \quad \Gamma(N) = (N-1)! \tag{6.8}$$

$$\Rightarrow \quad 0! = \Gamma(1) = 1 \tag{6.9}$$

For this reason, for any p, $\Gamma(p)$ is referred to as the *generalized factorial function*. That is, when one refers to the factorials of noninteger quantities, one is referring to the Γ-function.

For example,

$$(\tfrac{1}{2})! = \Gamma(\tfrac{3}{2}) = \int_0^\infty x^{1/2}e^{-x}\,dx \tag{6.10}$$

With the substitution $y = x^{1/2}$, this becomes

$$(\tfrac{1}{2})! = 2\int_0^\infty y^2 e^{-y^2}\,dy \tag{6.11a}$$

This is integrated once by parts, letting $u = y$, $dv = ye^{-y^2}\,dy$. The result is

$$(\tfrac{1}{2})! = \int_0^\infty e^{-y^2}\,dy = \frac{\sqrt{\pi}}{2} \tag{6.11b}$$

(see equations 4.132 through 4.134).

Using the integral representation, we next evaluate the Γ-function for negative integers and $\Gamma(0)$. To do this, consider

$$\Gamma(p)\Gamma(q) = \int_0^\infty\int_0^\infty u^{p-1}v^{q-1}e^{-u}e^{-v}\,du\,dv \tag{6.12}$$

We make the substitutions $x = u^{1/2}$ and $y = v^{1/2}$ to obtain

$$\Gamma(p)\Gamma(q) = 4\int_0^\infty\int_0^\infty x^{2p-1}e^{-x^2}y^{2q-1}e^{-y^2}\,dx\,dy \tag{6.13}$$

This double integral is an integral over the first quadrant in the x-y plane. Expressing the integrand in polar coordinates, with $x = \rho\cos\phi$, $y = \rho\sin\phi$, and $dA = dx\,dy = \rho\,d\rho\,d\phi$, this becomes

$$\Gamma(p)\Gamma(q) = 4\int_0^{\pi/2}\cos^{2p-1}\phi\,\sin^{2q-1}\phi\,d\phi\int_0^\infty \rho^{2p+2q-1}e^{-\rho^2}\,d\rho \tag{6.14}$$

If we let $t = \rho^2$,

$$\int_0^\infty \rho^{2p+2q-1} e^{-\rho^2} \, d\rho = \tfrac{1}{2}\int_0^\infty t^{p+q-1} e^{-t} \, dt = \tfrac{1}{2}\Gamma(p+q) \tag{6.15}$$

$$\Rightarrow \quad \frac{\Gamma(p)\Gamma(q)}{\Gamma(p+q)} = 2\int_0^{\pi/2} \cos^{2p-1}\phi \sin^{2q-1}\phi \, d\phi \equiv \beta(p,q) \tag{6.16}$$

As can be seen from the combination of Γ-functions, $\beta(p, q) = \beta(q, p)$. This symmetry can also be deduced from the integral definition by substituting $\phi' = \pi/2 - \phi$.

From the substitution $t = \cos^2\phi$ we obtain another representation of the β-function.

$$\beta(p,q) = \int_0^1 t^{p-1}(1-t)^{q-1} \, dt \tag{6.17}$$

Setting $t = x/(1+x)$ the β-function can also be written as

$$\beta(p,q) = \int_0^\infty \frac{x^{p-1}}{(1+x)^{p+q}} \, dx \tag{6.18}$$

To evaluate the Γ-function for zero or negative integers, we consider

$$\beta(p, 1-p) = \int_0^\infty \frac{x^{p-1}}{(1+x)} \, dx = \Gamma(p)\Gamma(1-p) \tag{6.19}$$

Unless p is an integer, this integrand, in the complex z-plane, has a fractional root point at $z = 0$ and a simple pole at $z = -1$. To evaluate the integral, we take the cut associated with the branch point to run from 0 to ∞, and consider

$$I \equiv \oint \frac{z^{p-1}}{(1+z)} \, dz \tag{6.20}$$

around the Pac-Man contour in Figure 6.1.

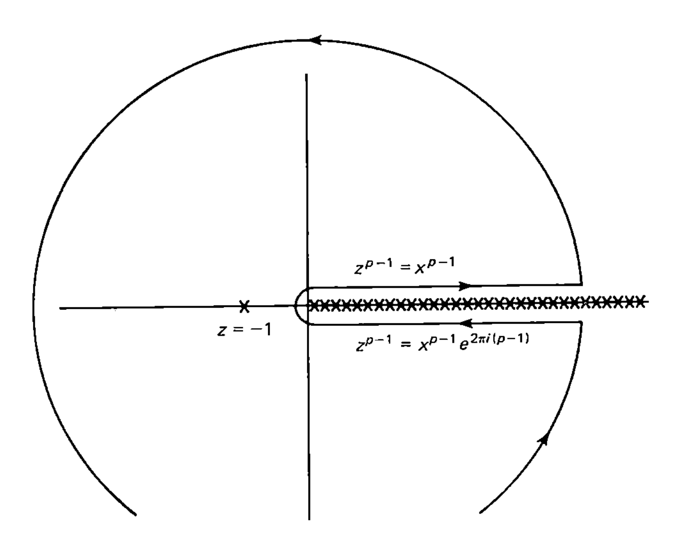

Figure 6.1 Contour for evaluating $\int_0^\infty [x^{p-1}/(1+x)] \, dx$.

Writing the integral over the various segments of the contour

$$I = \int_0^{\infty} \frac{x^{p-1}}{(1+x)}\, dx + \oint_{C_\infty} \frac{z^{p-1}}{(1+z)}\, dz + e^{2\pi i(p-1)} \int_{\infty}^{0} \frac{x^{p-1}}{(1+x)}\, dx$$

$$= (1 - e^{2\pi i p}) \int_0^{\infty} \frac{x^{p-1}}{(1+x)}\, dx + \oint_{C_\infty} \frac{z^{p-1}}{(1+z)}\, dz = 2\pi i R(-1) = 2\pi i e^{i\pi(p-1)} \quad (6.21)$$

On the infinite circle, $z = Re^{i\phi}$ and

$$\oint_{C_\infty} \frac{z^{p-1}}{(1+z)}\, dz \to i \int_0^{2\pi} R^{p-1} e^{i\phi(p-1)}\, d\phi \quad (6.22)$$

This will be zero in the limit of infinite R for $Re(p) < 1$. From equation 6.21, for these values of p, then,

$$\beta(p, 1-p) = \Gamma(p)\Gamma(1-p) = \int_0^{\infty} \frac{x^{p-1}}{(1+x)}\, dx = 2\pi i \frac{e^{i\pi(p-1)}}{(1 - e^{2\pi i p})}$$

$$= -\frac{2\pi i}{(e^{-i\pi p} - e^{i\pi p})} = \frac{\pi}{\sin \pi p} \quad (6.23)$$

If p is a negative integer $-N$ (N positive), this becomes

$$\Gamma(-N)\Gamma(1+N) = \infty \quad (6.24)$$

$$\Rightarrow \quad \Gamma(-N) = -(N+1)! = \infty \quad (6.25)$$

since $\Gamma(1+N)$ is finite. Similarly, if $p = 0$,

$$\Gamma(0)\Gamma(1) = \infty \quad (6.26)$$

$$\Rightarrow \quad \Gamma(0) = (-1)! = \infty \quad (6.27)$$

Figure 6.2 displays the dependence of the Γ-function on its argument.

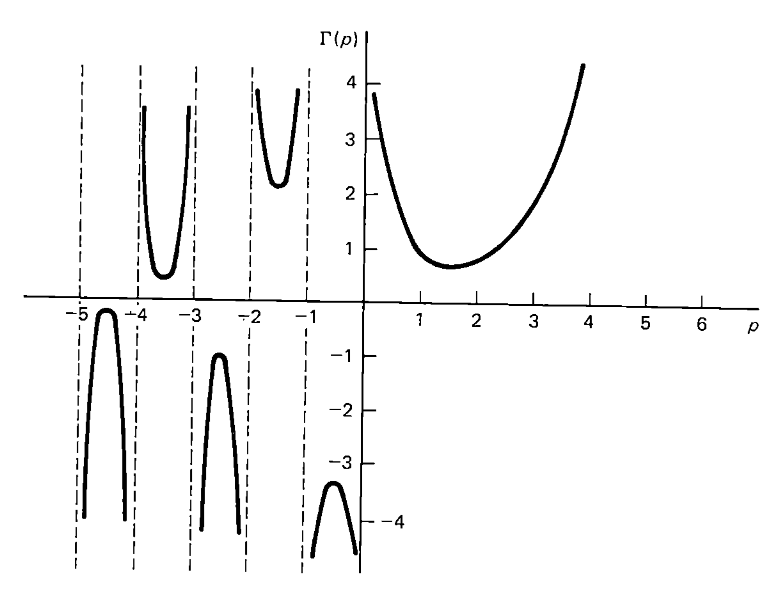

Figure 6.2 The Γ-function.

From equation 6.23, with $p = \frac{1}{2}$, we obtain

$$\text{(a)}\quad \Gamma(\tfrac{1}{2}) = \sqrt{\pi} \qquad \Rightarrow \qquad \text{(b)}\quad \Gamma(\tfrac{3}{2}) = \tfrac{1}{2}\Gamma(\tfrac{1}{2}) = \frac{\sqrt{\pi}}{2} = (\tfrac{1}{2})! \tag{6.28}$$

the result obtained in equation 6.11.

Again, using equation 6.23 with $p = q + \frac{1}{2}$, we obtain

$$\Gamma(\tfrac{1}{2} + q)\Gamma(\tfrac{1}{2} - q) = \frac{\pi}{\cos(\pi q)} \tag{6.29}$$

and also using $\Gamma(1 - p) = -p\Gamma(-p)$, we obtain

$$\Gamma(p)\Gamma(-p) = -\frac{\pi}{p \sin(\pi p)} \tag{6.30}$$

Another representation of the Γ-function can be deduced from

$$\Gamma(p) = \int_0^\infty x^{p-1} e^{-x}\, dx \tag{6.1}$$

Using

$$e^{-x} = \lim_{m \to \infty} \left(1 - \frac{x}{m}\right)^m \tag{6.31}$$

$$\Rightarrow \quad \Gamma(p) = \lim_{m \to \infty} \int_0^\infty x^{p-1}\left(1 - \frac{x}{m}\right)^m dx \tag{6.32}$$

Making the substitution, $y = x/m$, this becomes

$$\Gamma(p) = \lim_{m \to \infty} m^p \int_0^1 y^{p-1}(1 - y)^m\, dy = \lim_{m \to \infty} m^p \beta(p, m + 1)$$

$$= \lim_{m \to \infty} m^p \frac{\Gamma(p)\Gamma(m + 1)}{\Gamma(p + m + 1)} \tag{6.33a}$$

Writing $\Gamma(m + 1) = m!$, and $\Gamma(p + m + 1) = (p + m)(p + m - 1) \cdots (p + 2)(p + 1)p\Gamma(p)$, this becomes

$$\Gamma(p) = \lim_{m \to \infty} m^p \frac{m!}{p(p + 1) \cdots (p + m)} \tag{6.33b}$$

Equation 6.33b is called the *Euler limit representation* of the Γ-function.

Using the description of the β-function expressed in equation 6.17, consider

$$\beta(p, p) = \int_0^1 t^{p-1}(1 - t)^{p-1}\, dt = \frac{[\Gamma(p)]^2}{\Gamma(2p)} \tag{6.34}$$

Substituting $t = \frac{1}{2}(1 + u)$, this becomes

$$\frac{[\Gamma(p)]^2}{\Gamma(2p)} = \frac{1}{2^{2p-1}} \int_{-1}^1 (1 - u^2)^{p-1}\, du$$

$$= \frac{1}{2^{2p-1}}\left[\int_{-1}^0 (1 - u^2)^{p-1}\, du + \int_0^1 (1 - u^2)^{p-1}\, du\right] \tag{6.35a}$$

In the first integral, we replace u by $-u$ to obtain

$$\frac{[\Gamma(p)]^2}{\Gamma(2p)} = \frac{1}{2^{2p-2}} \int_0^1 (1 - u^2)^{p-1}\, du \tag{6.35b}$$

If we let $t = u^2$, and use $\Gamma(\frac{1}{2}) = \sqrt{\pi}$ this becomes

$$\frac{[\Gamma(p)]^2}{\Gamma(2p)} = \frac{1}{2^{2p-1}} \int_0^1 (1-t)^{p-1} t^{-1/2}\, dt = \frac{1}{2^{2p-1}} \beta\left(p, \frac{1}{2}\right)$$

$$= \frac{1}{2^{2p-1}} \frac{\Gamma(p)\Gamma(\frac{1}{2})}{\Gamma(p+\frac{1}{2})} \tag{6.36}$$

$$\Rightarrow \quad \frac{\Gamma(p)\Gamma(p+\frac{1}{2})}{\Gamma(2p)} = \frac{\sqrt{\pi}}{2^{2p-1}} \tag{6.37a}$$

To express this in a form we will find more useful later, let $p = q + \frac{1}{2}$. Then

$$\frac{\Gamma(q+1)\Gamma(q+\frac{1}{2})}{\Gamma(2q+1)} = \frac{\sqrt{\pi}}{2^{2q}} \tag{6.37b}$$

Equations 6.37 are two forms of an identity known as the *Legendre duplication formula*.

Taylor series expansion of the Γ-function and the polygamma functions

We can estimate the value of the Γ-function at small values of p by expanding $\Gamma(p+1)$ in a MacLaurin series, and retain only the first few terms in the expansion. [$\Gamma(p)$ cannot be expanded about $p = 0$ since $\Gamma(0) = \infty$.] To obtain the MacLaurin series requires the evaluation of the derivatives of $\Gamma(p+1)$. However, both the integral form presented in equation 6.1, and the description in terms of the Euler limit, as expressed in equation 6.33b, are very cumbersome to differentiate.

Starting with the limiting form in equation 6.33b, consider

$$\log[\Gamma(p+1)] = \lim_{m\to\infty} \log\left[m^{p+1} \frac{m!}{(p+1)(p+2)\cdots(p+m+1)}\right]$$

$$= \lim_{m\to\infty}\left[\log(m!) + (p+1)\log(m) - \sum_{k=0}^{m} \log(p+1+k)\right] \tag{6.38}$$

Therefore,

$$\psi(p+1) \equiv \frac{d}{dp}\log[\Gamma(p+1)] = \frac{\Gamma'(p+1)}{\Gamma(p+1)} = \lim_{m\to\infty}\left[\log(m) - \sum_{k=0}^{m} \frac{1}{(p+1+k)}\right] \tag{6.39}$$

$\psi(p+1)$ is called the *digamma function*.

When $m \to \infty$, both $\log(m)$ and the series are infinite, but their difference is finite. To see this, we write

$$\lim_{m\to\infty} \sum_{k=0}^{m} \frac{1}{(p+1+k)} = \frac{1}{(p+1)} + \lim_{m\to\infty} \sum_{k=1}^{m} \left[\frac{1}{(p+1+k)} - \frac{1}{k}\right] + \lim_{m\to\infty} \sum_{k=1}^{m} \frac{1}{k}$$

$$= \frac{1}{(p+1)} - (p+1) \lim_{m\to\infty} \sum_{k=1}^{m} \frac{1}{k(p+1+k)} + \lim_{m\to\infty} \sum_{k=1}^{m} \frac{1}{k} \tag{6.40}$$

We have shown using the Cauchy integral test, that $\sum_{k=1}^{\infty} 1/k^2$ converges (see Example 3.8). For $p + 1 > 0$, $1/k(p+1+k) < 1/k^2$. Therefore, by the comparison test, it is clear that $\sum_{k=1}^{\infty} 1/k(p+1+k)$ converges. Thus we can take m to ∞ for this series, and the digamma function becomes

$$\psi(p+1) = \lim_{m\to\infty}\left[\log(m) - \sum_{k=1}^{m}\frac{1}{k}\right] - \frac{1}{(p+1)} + (p+1)\sum_{k=1}^{\infty}\frac{1}{k(p+1+k)} \tag{6.41}$$

Using the Cauchy integral test, we have shown (see equation 3.161b) that when we write the series $S \equiv \lim_{m\to\infty}\sum_{k=1}^{m} 1/k$ as

$$S = \lim_{m\to\infty}\left[\sum_{k=1}^{N-1}\frac{1}{k} + \sum_{k=N}^{m}\frac{1}{k}\right] \equiv S_1 + S_2 \tag{6.42}$$

that S_2 is bounded by

$$\begin{aligned} &\text{(a)} \quad \lim_{m\to\infty}\int_N^m \frac{dk}{k} < S_2 < \frac{1}{N} + \lim_{m\to\infty}\int_N^m \frac{dk}{k} \\ \Rightarrow \quad &\text{(b)} \quad \lim_{m\to\infty}\log(m) < S_2 < \frac{1}{N} + \lim_{m\to\infty}\log(m) \end{aligned} \tag{6.43}$$

Therefore, S contains a logarithmic infinity, so

$$\lim_{m\to\infty}(\log(m) - S_1 - S_2) \neq \infty \tag{6.44}$$

The finite difference

$$\lim_{m\to\infty}\left[\sum_{k=1}^{m}\frac{1}{k} - \log(m)\right] \equiv \gamma = 0.5772156\ldots \tag{6.45}$$

is known as the *Euler–Mascheroni constant*. Using the estimate of S_2 expressed in equation 3.162,

$$S_2 \simeq \frac{1}{2}s_N + \lim_{m\to\infty}\int_N^m s(k)\,dk = \frac{1}{2N} + \lim_{m\to\infty}\int_N^m\frac{dk}{k} = \frac{1}{2N} + \lim_{m\to\infty}\log(m) - \log(N) \tag{6.46}$$

$$\Rightarrow \quad \gamma = \lim_{m\to\infty}[S_1 + S_2 - \log(m)] \simeq \sum_{k=1}^{N-1}\frac{1}{k} + \frac{1}{2N} - \log(N) \tag{6.47}$$

As an example, with $N = 500$, we obtain $\gamma \simeq 0.5772153$. With this definition of γ, the digamma function of equation 6.41 is

$$\psi(p+1) = -\gamma - \frac{1}{(p+1)} + (p+1)\sum_{k=1}^{\infty}\frac{1}{k(p+1+k)} \tag{6.48}$$

To generate the MacLaurin expansion of $\Gamma(p+1)$ we must determine the derivatives of the digamma function. Denoting the derivatives as $\psi^{(n)}(p+1)$, and using equation 6.39, we have

$$\psi'(p+1) = \psi^{(1)}(p+1) = \lim_{m\to\infty}\sum_{k=0}^{m}\frac{1}{(p+1+k)^2} \tag{6.49a}$$

Since the series is convergent, we can take m to be infinite, so

$$\psi^{(1)}(p+1) = \sum_{k=0}^{\infty} \frac{1}{(p+1+k)^2} \tag{6.49b}$$

The higher derivatives are easily obtained from this.

$$\psi^{(n)}(p+1) = -\sum_{k=0}^{\infty} \frac{(-1)^n n!}{(p+1+k)^{n+1}} \tag{6.50}$$

This nth derivative of $\log \Gamma(p+1)$ is known as the nth polygamma function.

With these derivatives established, the MacLaurin series for $\log \Gamma(p+1)$ becomes

$$\log \Gamma(p+1) = \log \Gamma(1) + \sum_{n=0}^{\infty} \frac{\psi^{(n)}(1)}{(n+1)!} p^{n+1} = \sum_{n=0}^{\infty} \frac{\psi^{(n)}(1)}{(n+1)!} p^{n+1} \tag{6.51}$$

Example 6.1

Therefore, for example,

$$\log \Gamma(1.05) \simeq \psi(1)(0.05) + \frac{\psi^{(1)}(1)}{2!}(0.05)^2 + \frac{\psi^{(2)}(1)}{3!}(0.05)^3 \tag{6.52}$$

$$\psi(1) = -\gamma - 1 + \sum_{k=1}^{\infty} \frac{1}{k(k+1)} = -\gamma - 1 + \sum_{k=1}^{\infty} \left[\frac{1}{k} - \frac{1}{(k+1)}\right] \tag{6.53}$$

But referring to equation 3.132, with $z = 1$, we obtain

$$\sum_{k=1}^{\infty} \left[\frac{1}{k} - \frac{1}{(k+1)}\right] = \left(1 - \frac{1}{2}\right) + \left(\frac{1}{2} - \frac{1}{3}\right) + \left(\frac{1}{3} - \frac{1}{4}\right) + \cdots = 1 \tag{6.54}$$

$$\Rightarrow \quad \psi(1) = -\gamma \simeq -0.57722 \tag{6.55a}$$

In addition, dividing the series representations of $\psi^{(1)}(1)$ and $\psi^{(2)}(2)$ into a finite sum S_1 and an infinite series S_2 and estimating S_2 by the average of its upper and lower bounds, we obtain

$$\text{(b)} \quad \psi^{(1)}(1) \simeq 1.64493 \qquad \text{(c)} \quad \psi^{(2)}(1) \simeq -2.40412 \tag{6.55}$$

$$\Rightarrow \quad \text{(a)} \quad \log \Gamma(1.05) \simeq -0.02685 \quad \Rightarrow \quad \text{(b)} \quad \Gamma(1.05) \simeq 0.97350 \quad \square \tag{6.56}$$

We note, parenthetically, that

$$\psi^{(1)}(1) = \psi'(1) = \sum_{k=0}^{\infty} \frac{1}{(k+1)^2} = \sum_{k=1}^{\infty} \frac{1}{k^2} = \zeta(2) \tag{6.57}$$

Many other properties of the polygamma functions that will not be presented here are described in the literature. For these additional properties, the reader is referred to other sources, such as the book edited by M. Arbramowitz and I. A. Stegun, *Handbook of Mathematical Functions* (Washington, D.C.: National Bureau of Standards, 1964).

Stirling's approximation for $\Gamma(N)$ for large N

The *Stirling approximation* is a formula that yields an approximate value for $\Gamma(N+1) = N!$ when N is very large. As we will see, N is not restricted to integer values, so the Stirling approximation yields estimates of the factorial function for any large number.

We begin with the expression for the Γ-function

$$\Gamma(N + 1) = N! = \int_0^\infty x^N e^{-x}\, dx \tag{6.1}$$

and make the substitution $y = x/N$. Then

$$N! = N^{N+1} \int_0^\infty y^N e^{-Ny}\, dy = N^{N+1} \int_0^\infty e^{-N[y - \log y]}\, dy \tag{6.58}$$

Since $y - \log y > 0$ for all values of y, $e^{-N[y-\log y]} \to 0$ very quickly for N very large. Thus the major contribution to the integral will come from the region around the minimum of $y - \log y$. Differentiating this function, we see that this minimum occurs at $y = 1$. Therefore, we expand $y - \log y$ around $y = 1$. With $\delta \equiv y - 1$,

$$y - \log y = (1 + \delta) - \log(1 + \delta) = (1 + \delta) - \left(\delta - \frac{\delta^2}{2} + \frac{\delta^3}{3} - \cdots\right)$$

$$= 1 + \frac{\delta^2}{2} + O(\delta^3) \tag{6.59}$$

Retaining only the first nonzero term in δ, the approximation becomes

$$N! \simeq N^{N+1} e^{-N} \int_{-1}^\infty e^{-N\delta^2/2}\, d\delta \tag{6.60a}$$

Since N is very large, the contribution to the integral from the range $\delta = -1$ to $\delta = -\infty$ is very small. Therefore, within the limits of the approximation, no significant error is introduced by extending the lower limit to $-\infty$. Therefore,

$$N! \simeq N^{N+1} e^{-N} \int_{-\infty}^\infty e^{-N\delta^2/2}\, d\delta \tag{6.60b}$$

With the substitution $u = \delta(N/2)^{1/2}$, this becomes

$$N! \simeq N^{N+1} e^{-N} \left(\frac{2}{N}\right)^{1/2} \int_{-\infty}^\infty e^{-u^2}\, du = \sqrt{2\pi N}\, N^N e^{-N} \tag{6.61}$$

Equation 6.61 is referred to as the *Stirling approximation* to $N!$. The fractional error incurred in this approximation is $\simeq 1/12N$.

Example 6.2

As an example, $5! = 120$. By the Stirling approximation,

$$5! \simeq \sqrt{10\pi}\, 5^5 e^{-5} = 118.02 \tag{6.62}$$

The fractional error is $1.98/120 = 0.0165$, which is essentially $1/(12 \cdot 5)$. □

Example 6.3

Stirling's approximation can also be used to approximate numbers such as $(10.05)! = \Gamma(11.05)$. The result is

$$(10.05)! \simeq \sqrt{2\pi \cdot 10.05}\quad 10.05^{10.05} e^{-10.05} = 4.048 \times 10^6 \quad \square \tag{6.63}$$

The factorial of a small noninteger number can also be approximated by the Stirling approximation.

Example 6.4

For example,

$$(0.05)! = \Gamma(1.05)$$

$$= \frac{(10.05)!}{10.05 \cdot 9.05 \cdot 8.05 \cdot 7.05 \cdot 6.05 \cdot 5.05 \cdot 4.05 \cdot 3.05 \cdot 2.05 \cdot 1.05} \simeq 0.96547 \qquad (6.64)$$

Comparing this to the result obtained by a MacLaurin expansion, we note that there is agreement to approximately 1%. To achieve a higher degree of precision, more terms in the MacLaurin series should be retained, or if the Stirling approximation is used, one should apply the Stirling approximation to a larger number, starting with (20.05)!, for example. □

The definition of the β-function and its relation to the Γ-function lends itself to the evaluation of integrals of the type

$$I = \int_0^{\pi/2} (\sin x)^r (\cos x)^s \, dx \qquad (6.65)$$

Example 6.5

As an example, consider

$$I = \int_0^{\pi/2} \frac{(\sin x)^{9/2}}{\sqrt{\cos x}} \, dx \qquad (6.66)$$

If $2p - 1 = \frac{9}{2}$ and $2q - 1 = -\frac{1}{2}$, the β-function is

$$\beta(\tfrac{11}{4}, \tfrac{1}{4}) = \frac{\Gamma(\frac{11}{4})\Gamma(\frac{1}{4})}{\Gamma(3)} = 2\int_0^{\pi/2} \frac{(\sin x)^{9/2}}{\sqrt{\cos x}} \, dx \qquad (6.67)$$

Using

$$\Gamma(\tfrac{11}{4}) = \tfrac{7}{4}\Gamma(\tfrac{7}{4}) = \tfrac{7}{4}\tfrac{3}{4}\Gamma(\tfrac{3}{4}) \qquad (6.68a)$$

$$\Rightarrow \quad \Gamma(\tfrac{3}{4})\Gamma(\tfrac{1}{4}) = \Gamma(p)\Gamma(1-p)|_{p=3/4} = \Gamma(p)\Gamma(1-p)|_{p=1/4} = \frac{\pi}{\sin(\pi/4)} = \pi\sqrt{2} \qquad (6.68b)$$

$$\Rightarrow \quad \int_0^{\pi/2} \frac{(\sin x)^{9/2}}{\sqrt{\cos x}} \, dx = \frac{21\pi\sqrt{2}}{64} \quad \square \qquad (6.69)$$

In this example, we were able to evaluate the integral exactly because of the choice of exponents of $\sin x$ and $\cos x$. However, with the Stirling approximation, integrals involving any exponents can be evaluated approximately.

Example 6.6

Consider

$$\int_0^{\pi/2} (\sin x)^{3/4} (\cos x)^{1/6} \, dx = \tfrac{1}{2}\beta(\tfrac{7}{8}, \tfrac{7}{12}) = \frac{1}{2}\frac{\Gamma(\frac{7}{8})\Gamma(\frac{7}{12})}{\Gamma(\frac{35}{24})} \qquad (6.70)$$

Using equation 6.5, we have

$$\Gamma(11\tfrac{7}{8}) = (10\tfrac{7}{8})(9\tfrac{7}{8})(8\tfrac{7}{8})(7\tfrac{7}{8})(6\tfrac{7}{8})(5\tfrac{7}{8})(4\tfrac{7}{8})(3\tfrac{7}{8})(2\tfrac{7}{8})(1\tfrac{7}{8})(\tfrac{7}{8})\Gamma(\tfrac{7}{8})$$

$$\simeq \sqrt{2\pi(10\tfrac{7}{8})}\, e^{-(10\frac{7}{8})}(10\tfrac{7}{8})^{(10\frac{7}{8})} = 2.91 \times 10^7 \qquad (6.71)$$

$$\Rightarrow \quad \Gamma(\tfrac{7}{8}) \simeq 1.081 \qquad (6.72a)$$

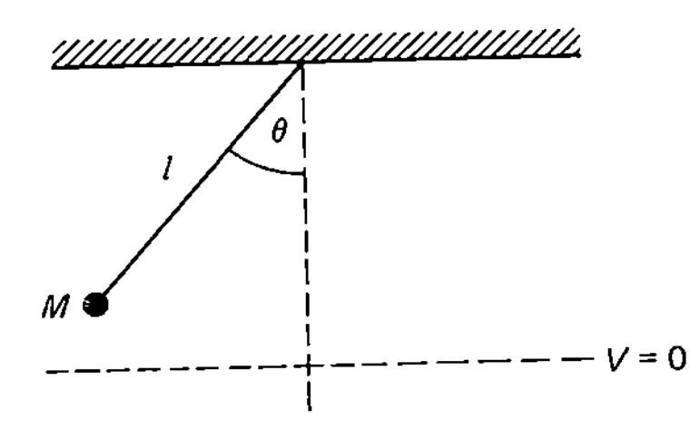

Figure 6.3 Motion of a planar pendulum.

With identical analyses, we obtain the estimates

$$\text{(b)}\quad \Gamma(\tfrac{7}{12}) \simeq 1.517 \qquad \text{(c)}\quad \Gamma(\tfrac{35}{24}) \simeq 0.8792 \tag{6.72}$$

$$\Rightarrow \quad \int_0^{\pi/2} (\sin x)^{3/4} (\cos x)^{1/6}\, dx \simeq 0.9326 \quad \square \tag{6.73}$$

Elliptic integral

Example 6.7

Consider the motion of a pendulum constructed of a mass M tied to a string of length l and caused to oscillate in a plane. In Figure 6.3 it is shown that for convenience, we define the zero level of potential energy to be the level of the bottom of the pendulum's arc. If the pendulum is pulled out to an angle θ_0 and released, conservation of energy yields

$$\text{(a)}\quad Mgl(1 - \cos\theta_0) = \tfrac{1}{2}Ml^2\dot{\theta}^2 + Mgl(1 - \cos\theta) \quad \Rightarrow \quad \text{(b)}\quad \dot{\theta} = \left(\frac{2g}{l}(\cos\theta - \cos\theta_0)\right)^{1/2} \tag{6.74}$$

where $\dot{\theta} = d\theta/dt$.

To determine the period of oscillation, we note that when the pendulum swings from $\theta = 0$ to $\theta = \theta_0$, it has completed $\frac{1}{4}$ of an oscillation. Therefore,

$$\int_0^{\theta_0} \frac{d\theta}{\sqrt{(\cos\theta - \cos\theta_0)}} = \left(\frac{2g}{l}\right)^{1/2} \int_0^{T/4} dt = \frac{1}{4}\left(\frac{2g}{l}\right)^{1/2} T \tag{6.75}$$

In the small-angle approximation, when $\sin\theta \simeq \theta$, $\cos\theta \simeq 1 - \theta^2/2$, equation 6.75 can be written

$$T \simeq 4\left(\frac{l}{g}\right)^{1/2} \int_0^{\theta_0} \frac{d\theta}{\sqrt{(\theta_0^2 - \theta^2)}} \tag{6.76}$$

which can be evaluated straightforwardly with the substitution $\theta = \theta_0 \sin x$. The result is the expected one,

$$T = 2\pi\left(\frac{l}{g}\right)^{1/2} \tag{6.77}$$

If θ_0 is large, of course, the small-angle approximation is invalid, and the period must be determined by evaluating the integral of equation 6.75. What is meant by large θ_0 depends on the degree of precision acceptable for the problem being considered. For example, for $\theta = 5^0 = \pi/36$, $\theta - \sin\theta \simeq 1.1 \times 10^{-4}$. If this level of precision is acceptable, 5^0 can be considered small.

With θ_0 arbitrary, we express this integral in its most common form by writing $\cos\theta = 1 - 2\sin^2(\theta/2)$. The period is, then,

$$T = 2\left(\frac{l}{g}\right)^{1/2} \int_0^{\theta_0} \left[\left\{\sin^2\left(\frac{\theta_0}{2}\right) - \sin^2\left(\frac{\theta}{2}\right)\right\}\right]^{-1/2} d\theta \tag{6.78}$$

If the substitution $\sin(\theta/2) = \sin(\theta_0/2)\sin\phi$ is made, and $\sin(\theta_0/2) \equiv k$, this becomes

$$T = 4\left(\frac{l}{g}\right)^{1/2} \int_0^{\pi/2} \frac{d\phi}{\sqrt{1 - k^2 \sin^2 \phi}} \equiv 4\left(\frac{l}{g}\right)^{1/2} K(k^2) \tag{6.79}$$

$K(k^2)$ is called the *complete elliptic integral of the first kind*. A second elliptic integral, called the *complete elliptic integral of the second kind*, is defined by

$$E(k^2) \equiv \int_0^{\pi/2} \sqrt{1 - k^2 \sin^2 \phi}\, d\phi \tag{6.80}$$

We note that

$$\text{(a)} \quad K(1) = \infty \qquad \text{(b)} \quad E(1) = 1 \qquad \text{(c)} \quad K(0) = E(0) = \frac{\pi}{2} \quad \square \tag{6.81}$$

Pochammer symbols

A series representation for the elliptic integrals can be developed starting with the binomial expansion of the integrand. Consider

$$(1 - x)^{-\alpha} = 1 + \alpha x + \alpha(\alpha + 1)\frac{x^2}{2!} + \cdots = \sum_{m=0}^{\infty} \frac{\Gamma(\alpha + m)}{\Gamma(\alpha)} \frac{x^m}{m!} \tag{6.82}$$

The *Pochammer symbol* is a shorthand notation for the coefficient of this expansion. It is defined by

$$\text{(a)} \quad (\alpha)_m \equiv (\alpha + m - 1)(\alpha + m - 2) \cdots (\alpha + 1)\alpha = \frac{\Gamma(\alpha + m)}{\Gamma(\alpha)} \quad \Rightarrow \quad \text{(b)} \quad (\alpha)_0 = 1 \tag{6.83}$$

Therefore, the binomial series can be written

$$(1 - x)^{-\alpha} = \sum_{m=0}^{\infty} (\alpha)_m \frac{x^m}{m!} \tag{6.84}$$

$$\Rightarrow \quad (1 - x)^{-1/2} = \sum_{m=0}^{\infty} \left(\frac{1}{2}\right)_m \frac{x^m}{m!} \tag{6.85}$$

We now analyze $\Gamma(m + \frac{1}{2})$ a little further. We write

$$\Gamma(m + \tfrac{1}{2}) = (m - \tfrac{1}{2})(m - \tfrac{3}{2}) \cdots \tfrac{3}{2}\tfrac{1}{2}\Gamma(\tfrac{1}{2}) = \frac{(2m - 1)(2m - 3) \cdots \cdot 3 \cdot 1}{2^m}\Gamma(\tfrac{1}{2}) \tag{6.86}$$

This numerator is similar to a factorial product, except that the factors differ by 2 instead of by 1. Such a product is referred to as a *double factorial*. That is

$$N!! \equiv N(N - 2)(N - 4) \cdots 5 \cdot 3 \cdot 1 \qquad N \text{ odd} \tag{6.87a}$$

$$N!! \equiv N(N - 2)(N - 4) \cdots 6 \cdot 4 \cdot 2 \qquad N \text{ even} \tag{6.87b}$$

$$\Rightarrow \quad \Gamma(m + \tfrac{1}{2}) = \frac{(2m - 1)!!}{2^m}\Gamma(\tfrac{1}{2}) = \frac{(2m - 1)!!}{2^m}\sqrt{\pi} \tag{6.88}$$

From Legendre's duplication formula expressed in equation 6.37b,

$$\Gamma(m + \tfrac{1}{2}) = \frac{\sqrt{\pi}}{2^{2m}} \frac{\Gamma(2m+1)}{\Gamma(m+1)} = \frac{\sqrt{\pi}}{2^{2m}} \frac{(2m)!}{m!} \tag{6.89}$$

Therefore, for an odd integer, $(2m - 1)$,

$$(2m-1)!! = \frac{1}{2^m} \frac{(2m)!}{m!} \tag{6.90}$$

For an even integer $(2m)$, the double factorial can be written

$$(2m)!! = (2m)(2m-2)(2m-4) \cdots (2 \cdot 3)(2 \cdot 2)(2 \cdot 1) = 2^m m! \tag{6.91}$$

Using this analysis, we can write

$$(1-x)^{-1/2} = \sum_{n=0}^{\infty} \frac{(2n)!}{2^{2n}(n!)^2} x^n = \frac{1}{\sqrt{n}} \sum_{n=0}^{\infty} \frac{\Gamma(n+\frac{1}{2})}{n!} x^n \tag{6.92}$$

$$\Rightarrow \quad K(k^2) = \sum_{n=0}^{\infty} \frac{\Gamma(n+\frac{1}{2})}{\sqrt{\pi}\, n!} k^{2n} \int_0^{\pi/2} \sin^{2n} \phi \, d\phi \tag{6.93}$$

From the definition of the β-function of equation 6.16,

$$\int_0^{\pi/2} \sin^{2n} \phi \, d\phi = \tfrac{1}{2}\beta(\tfrac{1}{2}, n + \tfrac{1}{2}) = \frac{1}{2} \frac{\Gamma(\frac{1}{2})\Gamma(n+\frac{1}{2})}{\Gamma(n+1)} = \frac{\sqrt{\pi}\,\Gamma(n+\frac{1}{2})}{2n!} \tag{6.94}$$

Therefore, the MacLaurin series for the elliptic integral of the first kind is

$$K(k^2) = \frac{1}{2} \sum_{n=0}^{\infty} \left[\frac{\Gamma(n+\frac{1}{2})}{n!}\right]^2 k^{2n} = \frac{\pi}{2} \sum_{n=0}^{\infty} \left[\left(\frac{1}{2}\right)_n\right]^2 \frac{k^{2n}}{(n!)^2} \tag{6.95}$$

The series representation for $E(k^2)$ is developed in the same way from the binomial expansion of $(1-x)^{1/2}$. This can be deduced from equations 6.82 or 6.84. We employ a slightly different approach to introduce a manipulation we will find useful later. We start with

$$(1-x)^{1/2} = (1-x)(1-x)^{-1/2} = (1-x) \sum_{n=0}^{\infty} \left(\frac{1}{2}\right)_n \frac{x^n}{n!}$$

$$= 1 + \sum_{n=1}^{\infty} \left[\frac{1}{(n+1)!}\left(\frac{1}{2}\right)_{n+1} - \frac{1}{n!}\left(\frac{1}{2}\right)_n\right] x^{n+1} = 1 - \frac{1}{2} \sum_{n=0}^{\infty} \left(\frac{1}{2}\right)_n \frac{x^{n+1}}{(n+1)!} \tag{6.96}$$

$$\Rightarrow \quad E(k^2) = \int_0^{\pi/2} d\phi - \frac{1}{2} \sum_{n=0}^{\infty} \left(\frac{1}{2}\right)_n \frac{k^{2n+2}}{(n+1)!} \int_0^{\pi/2} \sin^{2n+2} \phi \, d\phi \tag{6.97}$$

Using the description of the β-function expressed in equation 6.16 again, this becomes

$$\begin{aligned} E(k^2) &= \frac{\pi}{2} - \frac{1}{4} \sum_{n=0}^{\infty} \left[\frac{\Gamma(n+\frac{1}{2})}{(n+1)!}\right]^2 \left(n + \frac{1}{2}\right) k^{2n+2} \\ &= \frac{\pi}{2} - \frac{\pi}{4} \sum_{n=0}^{\infty} \left[\left(\frac{1}{2}\right)_n\right]^2 \frac{(n+\frac{1}{2})}{[(n+1)!]^2} k^{2n+2} \end{aligned} \tag{6.98}$$

It is straightforward to deduce the derivative of $E(k^2)$. Defining $z \equiv k^2$ yields

$$\frac{d}{dz}E(z) = -\frac{1}{2}\int_0^{\pi/2} \frac{\sin^2\phi}{(1 - z\sin^2\phi)^{1/2}}\,d\phi = \frac{1}{2z}\int_0^{\pi/2} \frac{(1 - z\sin^2\phi - 1)}{(1 - z\sin^2\phi)^{1/2}}\,d\phi$$
$$= \frac{1}{2z}[E(z) - K(z)] \tag{6.99}$$

The derivative of $K(z)$ is a little less straightforward to obtain. Starting with

$$\frac{d}{dz}K(z) = \frac{1}{2}\int_0^{\pi/2} \frac{\sin^2\phi}{(1 - z\sin^2\phi)^{3/2}}\,d\phi = -\frac{1}{2z}\int_0^{\pi/2} \frac{(1 - z\sin^2\phi - 1)}{(1 - z\sin^2\phi)^{3/2}}\,d\phi$$
$$= \frac{1}{2z}\left[\int_0^{\pi/2} \frac{1}{(1 - z\sin^2\phi)^{3/2}}\,d\phi - K(z)\right] \tag{6.100}$$

we expand $(1 - z\sin^2\phi)^{-3/2}$ in a binomial series, using equation 6.82 or 6.84. We obtain

$$\int_0^{\pi/2} \frac{d\phi}{(1 - z\sin^2\phi)^{3/2}} = \sum_{=0}^{\infty} \frac{\Gamma(n + \frac{3}{2})}{\Gamma(\frac{3}{2})}\frac{z^n}{n!}\int_0^{\pi/2} \sin^{2n}\phi\,d\phi \tag{6.101}$$

From the properties of the β-function expressed in equation 6.94, this becomes

$$\int_0^{\pi/2} \frac{d\phi}{(1 - z\sin^2\phi)^{3/2}} = \sum_{n=0}^{\infty}\left(n + \frac{1}{2}\right)\left[\frac{\Gamma(n + \frac{1}{2})}{n!}\right]^2 z^n \tag{6.102}$$

$$\Rightarrow \quad (1 - z)\int_0^{\pi/2} \frac{d\phi}{(1 - z\sin^2\phi)^{3/2}} = \frac{\pi}{2} + \sum_{n=1}^{\infty}\left(n + \frac{1}{2}\right)\left[\frac{\Gamma(n + \frac{1}{2})}{n!}\right]^2 z^n$$
$$- \sum_{n=0}^{\infty}\left(n + \frac{1}{2}\right)\left[\frac{\Gamma(n + \frac{1}{2})}{n!}\right]^2 z^{n+1} \tag{6.103}$$

By replacing n by $n + 1$ in the first sum, and after a bit of algebraic manipulation, this becomes

$$(1 - z)\int_0^{\pi/2} \frac{d\phi}{(1 - z\sin^2\phi)^{3/2}} = \frac{\pi}{2} - \frac{1}{4}\sum_{n=0}^{\infty}\left(n + \frac{1}{2}\right)\left[\frac{\Gamma(n + \frac{1}{2})}{(n + 1)!}\right]^2 z^{n+1} = E(z) \tag{6.104}$$

$$\Rightarrow \quad \frac{d}{dz}K(z) = \frac{1}{2z}\left[\frac{E(z)}{(1 - z)} - K(z)\right] \tag{6.105}$$

The elliptic integrals cannot be evaluated in terms of elementary functions (except at $k^2 = 0$), and must be evaluated using numerical integration methods that we develop in Chapter 13, or the series representations described above. A list of selected values of these elliptic functions is presented in Table 6.1

Incomplete functions

Incomplete Γ, β, and elliptic integral functions arise occasionally in scientific problems, although their occurrence is much less frequent that that of their complete counterparts. For that reason we just list them here. For more detailed discussions of their properties,

TABLE 6.1 SELECTED LIST OF VALUES OF THE COMPLETE ELLIPTIC INTEGRALS OF THE FIRST AND SECOND KIND

k^2	$K(k^2) = \int_0^{\pi/2} \frac{d\phi}{\sqrt{1 - k^2 \sin^2 \phi}}$	$E(k^2) = \int_0^{\pi/2} \sqrt{1 - k^2 \sin^2 \phi}\, d\phi$
0.00	1.570797	1.570797
0.05	1.591004	1.550974
0.10	1.612442	1.530758
0.15	1.635257	1.510122
0.20	1.659624	1.489035
0.25	1.685751	1.467462
0.30	1.713890	1.445363
0.35	1.744351	1.422691
0.40	1.777520	1.399392
0.45	1.813884	1.375402
0.50	1.854075	1.350644
0.55	1.898952	1.325025
0.60	1.949568	1.298428
0.65	2.007599	1.270708
0.70	2.075363	1.241671
0.75	2.156516	1.211056
0.80	2.257206	1.178490
0.85	2.389018	1.143396
0.90	2.578092	1.104775
0.95	2.908338	1.060474

the reader is referred to other literature, such as the book edited by A. Erdelyi, W. Magnus, F. Oberhettinger, and F. Tricomi, *The Bateman Manuscript Project* (New York; McGraw-Hill Book Company, 1953). The book by E. T. Whittaker and G. N. Nelson, *A Course in Modern Analysis* (Cambridge: Cambridge University Press, 1927), also contains more details about these incomplete functions.

In general, an incomplete function is an integral representation in which the integrand is the same as the integrand of the complete function counterpart, and the upper limit of the integral is variable. That is, the incomplete Γ and β functions are

$$\text{(a)} \quad \Gamma_i(p, t) \equiv \int_0^t x^{p-1} e^{-x}\, dx \qquad \text{(b)} \quad \beta_i(p, q; t) \equiv \int_0^t x^{p-1}(1 - x)^{q-1}\, dx \tag{6.106}$$

The incomplete elliptic integrals are

$$\text{(c)} \quad F(k^2, \theta) = \int_0^\theta \frac{d\phi}{\sqrt{1 - k^2 \sin^2 \phi}} \qquad \text{(d)} \quad E(k^2, \theta) = \int_0^\theta \sqrt{1 - k^2 \sin^2 \phi}\, d\phi \tag{6.106}$$

Error functions

The function of the envelope of the "bell-shaped" or Gaussian curve is e^{-x^2}. The *error function* is defined as

$$\operatorname{erf}(z) \equiv \frac{2}{\sqrt{\pi}} \int_0^z e^{-x^2}\, dx \tag{6.107}$$

It is the area under the shaded portion of the Gaussian curve of Figure 6.4. The constant $2/\sqrt{\pi}$ is chosen so that the area under the entire curve from 0 to ∞ is 1.

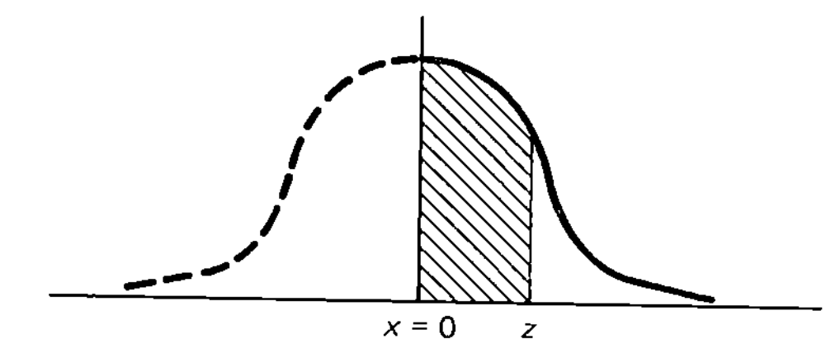

Figure 6.4 Area under a part of a Gaussian curve, the error function.

Referring to equations 6.11b and 6.107, we see that

$$\text{(a)} \quad \operatorname{erf}(\infty) = 1 \qquad \text{(b)} \quad \operatorname{erf}(0) = 0 \tag{6.108}$$

By replacing x by $-x$ in the integrand of $\operatorname{erf}(-z)$, we obtain

$$\operatorname{erf}(-z) = \frac{2}{\sqrt{\pi}} \int_0^{-z} e^{-x^2}\,dx = -\frac{2}{\sqrt{\pi}} \int_0^{z} e^{-x^2}\,dx = -\operatorname{erf}(z) \tag{6.109}$$

It is straightforward to obtain a MacLaurin series expansion for the error function. The expansion

$$e^{-x^2} = \sum_{k=0}^{\infty} (-1)^k x^{2k} \tag{6.110}$$

$$\Rightarrow \quad \operatorname{erf}(z) = \frac{2}{\sqrt{\pi}} \sum_{k=0}^{\infty} (-1)^k \frac{z^{2k+1}}{(2k+1)k!} \tag{6.111}$$

Such a series is useful for estimating the error function for small values of z.

By the Cauchy ratio test, this alternating series will converge absolutely for

$$|z|^2 \lim_{k\to\infty} \frac{1}{(k+1)(k+2)} \frac{(2k+1)}{(2k+3)} = |z|^2 \cdot 0 < 1 \tag{6.112}$$

Thus the series converges for all finite z.

However, when z is large, a large number of terms are required to obtain an acceptable estimate of $\operatorname{erf}(z)$. To obtain an accurate estimate of $\operatorname{erf}(z)$ for large z without using a large number of terms in a truncated MacLaurin series, a function is defined called the *complementary error function*,

$$\operatorname{erfc}(z) \equiv 1 - \operatorname{erf}(z) \equiv \frac{2}{\sqrt{\pi}} \int_z^{\infty} e^{-x^2}\,dx \tag{6.113}$$

$\operatorname{erfc}(z)$ is the area of the unshaded portion under the Gaussian curve in Figure 6.4. For large z, a series in $1/z$ will converge much more rapidly than the MacLaurin series of equation 6.111. Such a series is referred to as an *asymptotic series*. To obtain it, we express the complementary error function as

$$\operatorname{erfc}(z) = \frac{2}{\sqrt{\pi}} \int_z^{\infty} \frac{1}{x} x e^{-x^2}\,dx \tag{6.114}$$

and integrate by parts, with $u = 1/x$, $dv = xe^{-x^2}$. We obtain

$$\operatorname{erfc}(z) = \frac{2}{\sqrt{\pi}} \left[\frac{1}{2z} e^{-z^2} - \int_z^{\infty} \frac{1}{2x^2} e^{-x^2}\,dx \right] = \frac{2}{\sqrt{\pi}} \left[\frac{1}{2z} e^{-z^2} - \int_z^{\infty} \frac{1}{2x^3} x e^{-x^2}\,dx \right] \tag{6.115a}$$

After a second integration by parts, this becomes

$$\operatorname{erfc}(z) = \frac{2}{\sqrt{\pi}}\left[\frac{1}{2z}e^{-z^2} - \frac{1}{4z^3}e^{-z^2} + \int_z^\infty \frac{3}{4x^4}e^{-x^2}\,dx\right] \tag{6.115b}$$

Continuing this process until we obtain enough terms to achieve the required level of precision, we obtain

$$\begin{aligned}\operatorname{erfc}(z) &= \frac{2}{\sqrt{\pi}}\left[\frac{1}{2z} - \frac{1}{4z^3} + \frac{1\cdot 3}{8z^5} - \frac{1\cdot 3\cdot 5}{16z^7} + \cdots\right]e^{-z^2} \\ &= \frac{1}{z\sqrt{\pi}}\left[1 + \sum_{k=1}^{\infty}(-1)^k\frac{(2k-1)!!}{(2z^2)^k}\right]e^{-z^2}\end{aligned} \tag{6.116}$$

From this, we obtain the asymptotic series for the error function:

$$\operatorname{erf}(z) = 1 - \frac{1}{z\sqrt{\pi}}\left[1 + \sum_{k=1}^{\infty}(-1)^k\frac{(2k-1)!!}{(2z^2)^k}\right]e^{-z^2} \tag{6.117}$$

6.2 Sturm–Liouville Theory

There are many functions that occur in the solution to scientific problems that are solutions to second-order differential equations of the type studied in Chapter 5. The solutions to these differential equations are referred to as special functions or the classical functions.

Self-adjoint differential equation

A general form of a linear, homogeneous second-order differential equation is

$$[P(x)D^2 + Q(x)D + R(x)]\,y = 0 \tag{6.118a}$$

The equation is said to be *self-adjoint* if it can be written in the form

$$D[\rho(x)Dy] + R(x)y = 0 \tag{6.118b}$$

These two forms of the differential equation are the same if

$$\text{(a)}\quad P(x) = \rho(x) \quad \text{and} \quad \text{(b)}\quad Q(x) = \rho'(x) \quad \Rightarrow \quad \text{(c)}\quad Q(x) = P'(x) \tag{6.119}$$

If $Q(x) \neq P'(x)$, the differential equation in the form expressed in equation 6.118a can be made self-adjoint by multiplication by the factor

$$\mu(x) \equiv \frac{1}{P(x)}e^{\int dx Q/P} \tag{6.120}$$

$$\Rightarrow \left[e^{\int dx Q/P}D^2 + \frac{Q}{P}e^{\int dx Q/P}D + \frac{R}{P}e^{\int dx Q/P}\right]y = D\left[e^{\int dx Q/P}Dy\right] + \left[\frac{R}{P}e^{\int dx Q/P}\right]y = 0 \tag{6.121}$$

Many of the special functions are solutions to a particular form of such a self-adjoint equation called the *Sturm–Liouville equation*. The general form of the Sturm–Liouville equation is

$$\frac{d}{dx}\left(p(x)\frac{dy}{dx}\right) - q(x)y = -\lambda\rho(x)y \tag{6.122}$$

TABLE 6.2 SELECTED STURM-LIOUVILLE EQUATIONS THAT ARISE IN SCIENTIFIC PROBLEMS. Specific Sturm–Liouville functions, eigenvalues, function that makes equation self-adjoint, typical occurrences, comments.

Associated Legendre equation: $\left[(1-x^2)D^2 - 2xD - \dfrac{m^2}{(1-x^2)}\right]y = -l(l+1)y$

x	Notation	$p(x)$	$q(x)$	$\rho(x)$	λ	$\mu(x)$
$[-1, 1]$	$y \equiv P_l^m(x)$	$(1-x^2)$	$\dfrac{m^2}{(1-x^2)}$	1	$l(l+1)$	1

Arises in quantum theory of angular momentum. For spherically symmetric potential, l is orbital angular momentum quantum number, m is quantum number for z-component of orbital angular momentum. Also arises in solution to Poisson's equation for electrostatic potential created by a spherically symmetric charge distribution.

Ordinary Legendre equation: $[(1-x^2)D^2 - 2xD]y = -l(l+1)y$

x	Notation	$p(x)$	$q(x)$	$\rho(x)$	λ	$\mu(x)$
$[-1, 1]$	$y \equiv P_l(x)$	$(1-x^2)$	0	1	$l(l+1)$	1

Associated Legendre function with $m = 0$. Arises in many of the same analyses as the associated Legendre function. Basis function for numerical integration scheme known as Gauss–Legendre quadrature. If l is an integer, $P_l(x)$ is a polynomial of order l.

Associated Laguerre equation: $[xD^2 + (k - x + 1)D]y = -(n-k)y$

x	Notation	$p(x)$	$q(x)$	$\rho(x)$	λ	$\mu(x)$
$[0, \infty]$	$y \equiv L_n^k(x)$	$x^k e^{-x}$	0	$x^k e^{-x}$	$n - k$	$x^k e^{-x}$

Arises in the solution to the radial part of Schrödinger's equation in the quantum theory of the hydrogen atom. If n is an integer, $L_n^k(x)$ is a polynomial of order $n - k$.

Ordinary Laguerre equation: $[xD^2 + (1-x)D]y = -ny$

x	Notation	$p(x)$	$q(x)$	$\rho(x)$	λ	$\mu(x)$
$[0, \infty]$	$y \equiv L_n(x)$	e^{-x}	0	e^{-x}	n	e^{-x}

$L_n^k(x) = D^k L_n(x)$. If n if an integer, $L_n(x)$ is a polynomial of order n. $L_n(x)$ is a basis function for a numerical integration scheme called Gauss–Laguerre quadrature.

Hermite equation: $[D^2 - 2xD]y = -2\alpha y$

x	Notation	$p(x)$	$q(x)$	$\rho(x)$	λ	$\mu(x)$
$[-\infty, \infty]$	$y \equiv H_\alpha(x)$	e^{-x^2}	0	e^{-x^2}	2α	e^{-x^2}

Arises in quantum theory of harmonic oscillator. Is a basis function for a numerical integration scheme called Gauss–Hermite quadrature. If α is an integer, $H_\alpha(x)$ is a polynomial of order α.

Bessel equation: $[x^2D^2 + xD - \alpha^2]y = -k^2x^2y$

x	Notation	$p(x)$	$q(x)$	$\rho(x)$	λ	$\mu(x)$
$[0, \infty]$	$y \equiv J_\alpha(kx)$	x	$-\dfrac{\alpha^2}{x}$	x	k^2	$\dfrac{1}{x}$

Arises in problems in electromagnetic theory in which there is cylindrical symmetry (e.g., radiation in a cylindrical cavity, Frauenhoffer diffraction by a circular slit).

where λ is a constant, the functions $p(x)$, $q(x)$, and $\rho(x)$ are real, and $\rho(x) \geqslant 0$ for all $x \in [a, b]$. $\rho(x)$ is referred to as the *weighting function* and λ is called the *eigenvalue* of the equation.

Table 6.2 lists several of the functions that arise frequently in the study of scientific phenomena. The column labeled $\mu(x)$ lists the function defined in equation 6.120 that makes the differential equation self-adjoint.

Many of the properties of the special functions can be deduced from the properties of the solution to the general Sturm–Liouville equation. The Sturm–Liouville operator is defined as

$$L_x = \frac{d}{dx}\left(p(x)\frac{d}{dx}\right) - q(x) \tag{6.123}$$

$$\Rightarrow \quad L_x y = -\lambda\rho(x)y \tag{6.124}$$

We note that when L_x operates on y, the resulting function is the same y, multiplied by the weight function $\rho(x)$ and a constant λ. Such an equation is called an eigenvalue equation. λ is the eigenvalue, and y is the corresponding eigenfunction. For a given Sturm–Liouville equation there are, in general, an infinite number of possible eigenvalues. The corresponding eigenfunctions will differ from one another. Therefore, the eigenvalues must be distinguished by subscripts; that is, they are denoted by λ_i with corresponding eigenfunction $y \equiv \alpha_i(x)$.

Dirac bra-ket notation

Let $\alpha_i(x)$ and $\alpha_j(x)$ be two complex functions with $x \in [a, b]$ and let $\hat{\Omega}$ be an operator such as the Sturm–Liouville operator, for which operation on each $\alpha_i(x)$ has meaning. The *Dirac bra-ket* (*bracket*) of $\hat{\Omega}$ with α_i and α_j is defined by

$$\langle\alpha_i|\hat{\Omega}|\alpha_j\rangle \equiv \int_a^b \alpha_i^*(x)\hat{\Omega}\alpha_j(x)\,dx \tag{6.125}$$

For example, if $\hat{\Omega} = \mathbf{1}$, the identity operator,

$$\langle\alpha_i|\alpha_j\rangle = \int_a^b \alpha_i^*(x)\alpha_j(x)\,dx \tag{6.126}$$

and when $\hat{\Omega}$ is the Sturm–Liouville operator,

$$\langle\alpha_i|L_x|\alpha_j\rangle \equiv \int_a^b \alpha_i^*(x)L_x\alpha_j(x)\,dx \tag{6.127}$$

The exchange of α_i and α_j in the Dirac bra-ket is called transposing:

$$\langle\alpha_j|\hat{\Omega}|\alpha_i\rangle \equiv \langle\alpha_i|\hat{\Omega}|\alpha_j\rangle^T \tag{6.128}$$

The hermitean adjoint of $\hat{\Omega}$, denoted by $\hat{\Omega}^\dagger$, is defined such that the Dirac bra-ket of the operator satisfies

$$\langle\alpha_i|\hat{\Omega}^\dagger|\alpha_j\rangle = \langle\alpha_j|\hat{\Omega}|\alpha_i\rangle^* \tag{6.129}$$

The operator $\hat{\Omega}$ is said to be hermitean if

$$\langle\alpha_i|\hat{\Omega}^\dagger|\alpha_j\rangle = \langle\alpha_j|\hat{\Omega}|\alpha_i\rangle^* = \langle\alpha_i|\hat{\Omega}|\alpha_j\rangle \tag{6.130a}$$

or in terms of the integral representation of the Dirac bra-ket,

$$\int_a^b \alpha_i^*(x)\hat{\Omega}\alpha_j(x)\,dx = \int_a^b \alpha_j(x)\hat{\Omega}\alpha_i^*(x)\,dx \tag{6.130b}$$

With $\hat{\Omega} = L_x$, we consider

$$\Delta \equiv \langle \alpha_i | L_x | \alpha_j \rangle - \langle \alpha_i | L_x^\dagger | \alpha_j \rangle = \langle \alpha_i | L_x | \alpha_j \rangle - \langle \alpha_j | L_x | \alpha_i \rangle^*$$

$$= \int_a^b \left[\alpha_i^*(x) L_x \alpha_j(x) - \alpha_j(x) L_x^* \alpha_i^*(x) \right] dx \tag{6.131}$$

Since $p(x)$ and $q(x)$, and therefore L_x are real,

$$\Delta = \int_a^b \left\{ \left[\alpha_i^* \frac{d}{dx}\left(p(x) \frac{d\alpha_j}{dx} \right) - \alpha_i^* q(x) \alpha_j \right] - \left[\alpha_j \frac{d}{dx}\left(p(x) \frac{d\alpha_i^*}{dx} \right) - \alpha_j q(x) \alpha_i^* \right] \right\} dx$$

$$= \int_a^b \left[\alpha_i^* \frac{d}{dx}\left(p(x) \frac{d\alpha_j}{dx} \right) - \alpha_j \frac{d}{dx}\left(p(x) \frac{d\alpha_i^*}{dx} \right) \right] dx \tag{6.132}$$

Integrating the second integral by parts twice, we obtain

$$\int_a^b \alpha_j \frac{d}{dx}\left(p(x) \frac{d\alpha_i^*}{dx} \right) dx = \alpha_j p(x) \frac{d\alpha_i^*}{dx} \Big|_a^b - \int_a^b p(x) \frac{d\alpha_j}{dx} \frac{d\alpha_i^*}{dx} dx$$

$$= p(x) \left[\alpha_j \frac{d\alpha_i^*}{dx} - \alpha_i^* \frac{d\alpha_j}{dx} \right]_a^b + \int_a^b \alpha_i^* \frac{d}{dx}\left(p(x) \frac{d\alpha_j}{dx} \right) dx \tag{6.133}$$

$$\Rightarrow \quad \Delta = p(x) \left[\alpha_j \frac{d\alpha_i^*}{dx} - \alpha_i^* \frac{d\alpha_j}{dx} \right]_a^b \tag{6.134}$$

With the definition of Δ expressed in equation 6.131, this is known as *Green's theorem*.

The functions α_i encountered in an overwhelming number of scientific problems satisfy one of four types of boundary conditions.

$$\text{(a)} \quad \alpha_i(a) = \alpha_i(b) = 0 \quad \text{Dirichlet} \qquad \text{(b)} \quad \frac{d\alpha_i}{dx}\Big|_a = \frac{d\alpha_i}{dx}\Big|_b = 0 \quad \text{Neumann}$$

$$\text{(c)} \quad \left[\alpha_i + C \frac{d\alpha_i}{dx} \right]_a = \left[\alpha_i + C \frac{d\alpha_i}{dx} \right]_b = 0 \quad \text{Gauss or mixed} \tag{6.135}$$

It is easy to see that if α_i and α_j satisfy any one of the three conditions, then $\Delta = 0$. Therefore, in all problems in which one of the three types of boundary conditions are satisfied, L_x is hermitean.

Functions that describe periodic motion satisfy

$$\text{(d)} \quad \alpha_i(a) = \alpha_i(b) \qquad \text{and} \qquad \text{(e)} \quad \frac{d\alpha_i}{dx}\Big|_a = \frac{d\alpha_i}{dx}\Big|_b \quad \text{periodic} \tag{6.135}$$

where $(b - a)$ is the period of $\alpha_i(x)$. These are similar to the Dirichlet and Neumann conditions, but do not require the function or its derivative to be zero. Such boundary conditions also lead to $\Delta = 0$, and hermiticity for L_x.

Using the Sturm–Liouville equation, consider

$$\langle \alpha_i | L_x | \alpha_j \rangle = \int_a^b \alpha_i^* L_x \alpha_j \, dx = -\lambda_j \int_a^b \rho(x) \alpha_i^*(x) \alpha_j(x) \, dx \tag{6.136a}$$

$$\langle \alpha_j | L_x | \alpha_i \rangle = \int_a^b \alpha_j^* L_x \alpha_i \, dx = -\lambda_i \int_a^b \rho(x) \alpha_j^*(x) \alpha_i(x) \, dx \tag{6.136b}$$

Referring to equation 6.129, we have

$$\text{(a)}\quad \langle \alpha_j | L_x | \alpha_i \rangle^* = \langle \alpha_i | L_x^\dagger | \alpha_j \rangle \quad \Rightarrow \quad \text{(b)}\quad \langle \alpha_i | L_x^\dagger | \alpha_j \rangle = -\lambda_i^* \int_a^b \rho(x) \alpha_i^*(x) \alpha_j(x)\, dx \tag{6.137}$$

Since L_x is hermitean,

$$\langle \alpha_i | L_x | \alpha_j \rangle - \langle \alpha_i | L_x^\dagger | \alpha_j \rangle = 0 = (\lambda_j - \lambda_i^*) \int_a^b \rho(x) \alpha_i^*(x) \alpha_j(x)\, dx \tag{6.138}$$

When $i = j$,

$$(\lambda_i - \lambda_i^*) \int_a^b \rho(x) |\alpha_i(x)|^2\, dx = 0 \tag{6.139}$$

We have specified that $\rho(x) \geqslant 0$. Since it is the square of the magnitude of a complex function, $|\alpha_i(x)|^2$ must be $\geqslant 0$. Therefore, the integral cannot be zero. Thus the only way that equation 6.139 can be satisfied is for $\lambda_i = \lambda_i^*$. That is, every eigenvalue of L_x (or any other hermitian operator) must be real. Therefore, equation 6.138 can be written

$$(\lambda_j - \lambda_i) \int_a^b \rho(x) \alpha_i^*(x) \alpha_j(x)\, dx = 0 \tag{6.140}$$

If $i \neq j$, then either

$$\text{(a)}\quad \lambda_i = \lambda_j \qquad \text{or} \qquad \text{(b)}\quad \int_a^b \rho(x) \alpha_i^*(x) \alpha_j(x)\, dx \equiv \langle \alpha_i | \alpha_j \rangle = 0 \tag{6.141}$$

This is a slightly different notation than that expressed in equation 6.126. The integral describing $\langle \alpha_i | \alpha_j \rangle$ now includes $\rho(x)$.

If $\lambda_i \neq \lambda_j$, then α_i and α_j are said to be orthogonal. For functions $\alpha_i(x)$, the orthogonality condition expressed in equation 6.141b is analogous to the zero dot product for vectors. We see then, that if the eigenvalues for two different eigenfunctions are unequal, the eigenfunctions must be orthogonal.

However, if $\lambda_i = \lambda_j$ with $i \neq j$, the eigenvalues are said to be degenerate. In such a case, α_i is not necessarily orthogonal to α_j. If a particular equation has N eigenvalues that have the same value, there is said to be an *N-fold degeneracy*, and the N corresponding eigenfunctions are not necessarily orthogonal to one another.

Gram – Schmidt orthogonalization

When eigenfunctions are degenerate, it is possible to construct a set of orthogonal eigenfunctions as a linear combination of the degenerate eigenfunctions. The most commonly used method is the Gram–Schmidt orthogonalization technique.

Let $\alpha_1, \alpha_2, \ldots, \alpha_N$ be a set of N degenerate eigenfunctions corresponding to the eigenvalue $\lambda_1 = \lambda_2 = \cdots = \lambda_N$. Then, in general, $\alpha_1, \ldots, \alpha_N$ are not mutually orthogonal. Let $v_1, v_2, \ldots, v_N$ be a set of mutually orthogonal functions that are constructed as follows:

$$v_1 = \alpha_1 \tag{6.142}$$

Choose

$$v_2 = \alpha_2 + c v_1 \tag{6.143}$$

such that $v_1 \perp v_2$. Thus

$$\text{(a)} \quad \langle v_2|v_1\rangle = \langle \alpha_2|v_1\rangle + c\langle v_1|v_1\rangle = 0 \quad \Rightarrow \quad \text{(b)} \quad c = -\frac{\langle \alpha_2|v_1\rangle}{\langle v_1|v_1\rangle} \tag{6.144}$$

$$\Rightarrow \quad v_2 = \alpha_2 - \frac{\langle \alpha_2|v_1\rangle}{\langle v_1|v_1\rangle} v_1 \tag{6.145a}$$

Equation 6.145a provides the pattern for constructing the set of orthogonal eigenfunctions by the Gram–Schmidt method. For example,

$$v_3 = \alpha_3 - \frac{\langle \alpha_3|v_1\rangle}{\langle v_1|v_1\rangle} v_1 - \frac{\langle \alpha_3|v_2\rangle}{\langle v_2|v_2\rangle} v_2 \tag{6.145b}$$

$$\begin{aligned} \Rightarrow \quad v_k &= \alpha_k - \frac{\langle \alpha_k|v_1\rangle}{\langle v_1|v_1\rangle} v_1 - \frac{\langle \alpha_k|v_2\rangle}{\langle v_2|v_2\rangle} v_2 - \cdots - \frac{\langle \alpha_k|v_{k-1}\rangle}{\langle v_{k-1}|v_{k-1}\rangle} v_{k-1} \\ &= \alpha_k - \sum_{l=1}^{k-1} \frac{\langle \alpha_k|v_l\rangle}{\langle v_l|v_l\rangle} v_l \qquad k \geqslant 2 \end{aligned} \tag{6.146}$$

With v_3 given in equation 6.145b, consider

$$\langle v_3|v_1\rangle = \langle \alpha_3|v_1\rangle - \frac{\langle \alpha_3|v_1\rangle}{\langle v_1|v_1\rangle}\langle v_1|v_1\rangle - \frac{\langle \alpha_3|v_2\rangle}{\langle v_2|v_2\rangle}\langle v_2|v_1\rangle \tag{6.147a}$$

We have constructed v_2 so that it is orthogonal to v_1. Therefore

$$\langle v_3|v_1\rangle = \langle \alpha_3|v_1\rangle - \langle \alpha_3|v_1\rangle = 0 \tag{6.147b}$$

So $v_3 \perp v_1$. Now consider

$$\langle v_3|v_2\rangle = \langle \alpha_3|v_2\rangle - \frac{\langle \alpha_3|v_1\rangle}{\langle v_1|v_1\rangle}\langle v_1|v_2\rangle - \frac{\langle \alpha_3|v_2\rangle}{\langle v_2|v_2\rangle}\langle v_2|v_2\rangle = \langle \alpha_3|v_2\rangle - \langle \alpha_3|v_2\rangle = 0 \tag{6.148}$$

so $v_3 \perp v_2$.

Consider

$$L_x v_2 = L_x \alpha_2 - \frac{\langle \alpha_2|v_1\rangle}{\langle v_1|v_1\rangle} L_x v_1 = -\left[\lambda_2 \rho \alpha_2 - \frac{\langle \alpha_2|v_1\rangle}{\langle v_1|v_1\rangle}\lambda_1 \rho v_1\right] \tag{6.149}$$

But because of the N-fold degeneracy, $\lambda_1 = \lambda_2$ and this can be written

$$L_x v_2 = -\lambda_1 \rho\left[\alpha_2 - \frac{\langle \alpha_2|v_1\rangle}{\langle v_1|v_1\rangle} v_1\right] = -\lambda_1 \rho v_2 \tag{6.150}$$

That is, v_2 is an eigenfunction of L_x with the common eigenvalue $\lambda_1 = \lambda_2 = \cdots = \lambda_N$. From this result we see that

$$\begin{aligned} L_x v_3 &= L_x \alpha_3 - \frac{\langle \alpha_3|v_1\rangle}{\langle v_1|v_1\rangle} L_x v_1 - \frac{\langle \alpha_3|v_2\rangle}{\langle v_2|v_2\rangle} L_x v_2 \\ &= -\lambda_1 \rho\left[\alpha_3 - \frac{\langle \alpha_3|v_1\rangle}{\langle v_1|v_1\rangle} v_1 - \frac{\langle \alpha_2|v_2\rangle}{\langle v_2|v_2\rangle} v_2\right] = -\lambda_1 \rho v_3 \end{aligned} \tag{6.151}$$

Continuing this analysis, we see that each of the new functions is an eigenfunction of L_x with the common eigenvalue. Therefore, even when a degeneracy exists, one can always construct a complete set of orthogonal eigenfunctions of L_x.

Normalization

Returning to the Sturm–Liouville equation, with $y = \alpha_i$,

$$L_x \alpha_i = -\lambda_i \rho \alpha_i \tag{6.152}$$

If α_i is multiplied by an arbitrary constant C, it is trivial to see that $C\alpha_i$ also satisfies the Sturm–Liouville equation. That is, the eigenfunctions are determinable up to a multiplicative constant. To determine the constant, we impose a condition on α_i called normalization. The most general form of a normalization condition is

$$\int_a^b \rho(x)|\alpha_i(x)|^2\,dx = \langle \alpha_i|\alpha_i\rangle = N_i \tag{6.153}$$

In quantum theory, the eigenfunctions α_i (wave functions) are interpreted as probability amplitudes. That means that

$$P_i(x)\,dx \equiv |\alpha_i(x)|^2\,dx \tag{6.154}$$

is the probability that a particle in the ith quantum state is at a point x in a range dx. The probability that the particle is somewhere between the points a and b is

$$\int_a^b P_i(x)\,dx = \int_a^b |\alpha_i(x)|^2\,dx \tag{6.155}$$

If the interval $[a, b]$ is the range of points accessible to the particle, the integrated probability of equation 6.155 is the probability that the particle is somewhere, and this probability is 1. That is, in quantum theory, eigenfunctions are normalized such that

$$\int_a^b |\alpha_i(x)|^2\,dx = 1 \tag{6.156}$$

Each of the special functions listed in Table 6.2 is normalized to some constant. For some of the functions, the normalization constant is not 1.

Closure property of Sturm–Liouville functions

Let $\{\alpha_i\}$ be a complete set of Sturm–Liouville functions. We assume that a Gram–Schmidt process has been performed on those eigenfunctions for which degeneracies exist, and that the eigenfunctions have been normalized.

The basis vectors **i**, **j**, **k**, for example, form a complete set of orthonormal basis vectors such that any vector can be expanded as a linear combination of **i**, **j**, and **k**. Similarly, for $x \in [a, b]$, any function $F(x)$ can be expanded as a linear combination of the Sturm–Liouville eigenfunctions.

$$F(x) = \sum_{l=0}^{\infty} f_l \alpha_l(x) \tag{6.157}$$

This expansion is defined if we determine the coefficients f_l.

The three-dimensional vector equivalent of equation 6.157 is $\mathbf{V} = V_x\mathbf{i} + V_y\mathbf{j} + V_z\mathbf{k}$ or $\mathbf{V} = V_r\mathbf{r}_0 + V_\theta\boldsymbol{\theta}_0 + V_\phi\boldsymbol{\phi}_0$, for example. The set of basis functions $\{\alpha_l\}$ plays the role in function space that a set of basis vectors (such as **i**, **j**, **k**) play in vector space and the coefficients f_l are analogous to the vector components V_l.

Just as V_x, for example, is obtained from $\mathbf{i}\cdot\mathbf{V}$, consider

$$\langle \alpha_i|F\rangle = \sum_{l=0}^{\infty} f_l \langle \alpha_i|\alpha_l\rangle \tag{6.158a}$$

Using the orthogonality of the Sturm–Liouville eigenfunctions and the normalization as expressed in equation 6.153, we obtain an evaluation of the coefficients in the eigenfunctions expansion.

$$\langle \alpha_i | F \rangle = \sum_{l=0}^{\infty} f_l N_i \delta_{li} = f_i N_i \tag{6.158b}$$

This Dirac bra-ket is the function space equivalent of the dot product. (In fact, in analogy with the dot product for vectors, $\langle \alpha_i | F \rangle$ is referred to as the inner product of α_i with F.)

Therefore, the coefficients of the eigenfunction expansion can be calculated in terms of the Dirac bra-ket of known functions. Thus $F(x)$ can be expanded as a linear combination of the Sturm–Liouville functions.

Substituting f_i from equation 6.158b into equation 6.157 gives

$$F(x) = \sum_{l=0}^{\infty} \frac{\langle \alpha_l | F \rangle}{N_l} \alpha_l(x) \tag{6.159}$$

In terms of the integral form of the Dirac bra-ket, equation 6.159 can be written

$$\begin{aligned} F(x) &= \sum_{l=0}^{\infty} \frac{\alpha_l(x)}{N_l} \int_a^b \rho(x') \alpha_l^*(x') F(x') \, dx' \\ &= \int_a^b \rho(x') F(x') \sum_{l=0}^{\infty} \frac{1}{N_l} \alpha_l^*(x') \alpha_l(x) \, dx' \end{aligned} \tag{6.160}$$

That is, we can write any function $F(x)$ as an integral of something multiplying $F(x')$. Comparing this to the definition of the Dirac δ-symbol of equation 2.294,

$$\int_a^b F(x') \delta(x' - x) \, dx' = \begin{cases} F(x) & x \in [a, b] \\ 0 & x \notin [a, b] \end{cases} \tag{2.294}$$

we see that $\rho(x')$ multiplied by the sum in equation 6.160 satisfies the basic definition of the Dirac δ-symbol. That is,

$$\rho(x') \sum_{l=0}^{\infty} \frac{1}{N_l} \alpha_l^*(x') \alpha_l(x) = \delta(x - x') \tag{6.161}$$

This is called the closure property of the Sturm–Liouville functions. It is the generalization of equation 4.19, and the discussion that immediately follows that equation.

Generating functions

One of the ways in which the Sturm–Liouville functions can be generated is with a function known as the generating function. The generating function is denoted by $\Phi(x, z)$, and a Sturm–Liouville function is generated by $\Phi(x, z)$ as

$$\Phi(x, z) = \sum_{l=0}^{\infty} c_l \alpha_l(x) z^l \tag{6.162a}$$

Expanding the generator in a MacLaurin series in z, this can be written

$$\sum_{l=0}^{\infty} \frac{z^l}{l!} \left[\frac{\partial^l}{\partial z^l} \Phi(x, z) \right]_{z=0} = \sum_{l=0}^{\infty} c_l \alpha_l(x) z^l \tag{6.162b}$$

$$\Rightarrow \quad \alpha_l(x) = \frac{1}{l! \, c_l} \left[\frac{\partial^l}{\partial z^l} \Phi(x, z) \right]_{z=0} \tag{6.163}$$

A particular Sturm–Liouville function can usually be generated by more than one generating function. In this treatment we deal with only the most commonly used generators of the functions under consideration. For a more complete discussion of the various generators of a particular Sturm–Liouville function, the reader is referred to sources such as the text edited by A. Erdelyi, W. Magnus, F. Oberhettinger, and F. Tricomi, *The Bateman Manuscript Project* (New York: McGraw-Hill Book Company, 1953).

Some progress has been made in the development of a systematic approach to determining the different generating functions for the various Sturm–Liouville functions. However, the most reliable approach for obtaining these functions is through manipulative methods, using the properties that are unique to the particular Sturm–Liouville function being studied. The reader interested in some of the more general developments is referred to an article by L. Weisner, "Group Theoretic Origins of Certain Generating Functions," in the *Pacific Journal of Mathematics* (vol. 5, 1955, pp. 1033–1039); the book by H. M. Srivastava and H. L. Manocha, *A Treatise on Generating Functions* (New York: John Wiley & Sons, Inc., 1984); and the book by E. B. McBride, *Obtaining Generating Functions* (New York: Springer-Verlag, 1971). The discussions in these treatments are beyond the scope of this book. Therefore, we will not present any general methods for the determination of the various generating functions.

6.3 Legendre Functions

Solution to the differential equation

We first investigate the properties of the solution to the ordinary Legendre equation. As noted in Table 6.2, the ordinary Legendre equation is

$$\left[(1 - x^2)D^2 - 2xD + l(l + 1)\right]y = 0 \tag{6.164}$$

We assume a solution in series form

$$y = \sum_{m=0}^{\infty} c_m x^{m+s} \tag{6.165}$$

We note that this expansion is around $x = 0$, an ordinary point of the equation. From Fuch's theorem, we are assured of obtaining two nonsingular solutions in series. Substituting this series into equation 6.164 and performing the types of manipulations of indices discussed in Chapter 5, the series form of the ordinary Legendre equation becomes

$$c_0 s(s - 1)x^{s-2} + c_1 s(s + 1)x^{s-1} + \sum_{m=0}^{\infty} \{c_{m+2}(m + s + 2)(m + s + 1) - c_m[(m + s)(m + s + 1) - l(l + 1)]\}x^{m+s} = 0 \tag{6.166}$$

From this, we obtain the indicial equations

$$\text{(a)} \quad c_0 s(s - 1) = 0 \qquad \text{(b)} \quad c_1 s(s + 1) = 0 \tag{6.167}$$

both of which can be satisfied by choosing $s = 0$. This choice will therefore yield two independent series solutions.

From equation 6.166, the recurrence relation, with $s = 0$, is

$$c_{m+2} = \frac{[m(m + 1) - l(l + 1)]}{(m + 2)(m + 1)} c_m = -\frac{(l - m)(l + m + 1)}{(m + 1)(m + 2)} c_m \tag{6.167c}$$

From this, we note that in each independent series solution, the powers of x in two successive terms will differ by 2. Therefore, one of the independent solutions is an even series (containing only even powers of x) and the other solution will be an odd series. The absolute convergence of each series is determined by requiring the Cauchy ratio to satisfy

$$\lim_{N\to\infty} \rho_N = \lim_{N\to\infty} \left| \frac{c_{N+2}x^{N+2}}{c_N x^N} \right| = |x|^2 \lim_{N\to\infty} \frac{[N(N+1) - l(l+1)]}{(N+2)(N+1)} = |x|^2 < 1 \quad (6.168)$$

That is, the series solution converges absolutely for $-1 < x < 1$. Since the series is expanded about $x = 0$ and the equation has singularities at $x = \pm 1$, this is the expected result.

l must be real since eigenvalues of a hermitean operator are real. If l is an integer, we note from the recurrence relation of equation 6.167c when $m = l$, $c_{l+2} = 0$ for a nonzero c_l. If $c_{l+2} = 0$, then $c_{l+4} = c_{l+6} = \cdots = 0$. That is, when l is an integer, the infinite series solution for that l becomes a finite polynomial of order l. Therefore, if l is an even integer, the even series becomes the finite polynomial of even order. If l is odd, the finite polynomial is odd. In many applications, $x = \cos\theta$, where θ is the spherical polar angle. In such problems, the system can exist at $\theta = 0, \pi$. Thus one requires solutions to the Legendre differential equation that are analytic at $x = \pm 1$. These finite Legendre polynomials are those special functions that arise in the solution to many problems for which analyticity at $x = \pm 1$ is required.

Using the recurrence relations of equation 6.167c, the coefficients of the terms in the even-order Legendre polynomial are

$$c_{2m} = (-1)^m \frac{[l(l-2)\cdots(l-2m+2)][(l+1)(l+3)\cdots(l+2m+1)]}{(2m)!} c_0 \quad (6.169a)$$

with $0 \leqslant m \leqslant l/2$. If l is odd, the coefficients of the terms of the Legendre polynomial are

$$c_{2m+1} = (-1)^m \frac{[(l-1)(l-3)\cdots(l-2m+1)][(l+2)(l+4)\cdots(l+2m)]}{(2m+1)!} c_1 \quad (6.169b)$$

with $0 \leqslant m \leqslant (l-1)/2$.

For even l, using equation 6.91

$$l(l-2)(l-4)\cdots(l-2m+2) = \frac{l!!}{(l-2m)!!} = 2^m \frac{(\frac{1}{2}l)!}{(\frac{1}{2}l - m)!} \quad (6.170a)$$

$$(l+1)(l+3)(l+5)\cdots(l+2m-1) = \frac{(l+2m-1)!!}{(l-1)!!} = \frac{1}{2^m} \frac{(l+2m)!\,(\frac{1}{2}l)!}{l!\,(\frac{1}{2}l+m)!} \quad (6.170b)$$

$$\Rightarrow \quad c_{2m} = \frac{(-1)^m}{(2m)!} \frac{[(\frac{1}{2}l)!]^2}{l!\,(\frac{1}{2}l - m)!} \frac{(l+2m)!}{(\frac{1}{2}l+m)!} c_0 \quad (6.171)$$

and the even order Legendre polynomial can be written

$$P_l(x) = c_0 \sum_{m=0}^{l/2} \frac{(-1)^m}{(2m)!} \frac{[(\frac{1}{2}l)!]^2}{l!\,(\frac{1}{2}l-m)!} \frac{(l+2m)!}{(\frac{1}{2}l+m)!} x^{2m} \quad (6.172a)$$

A more commonly used form of this sum is obtained by substitution of the summation index. With $k = l/2 - m$, equation 6.172a becomes

$$P_l(x) = c_0 \sum_{k=0}^{l/2} (-1)^{l/2-k} \frac{[(\frac{1}{2}l)!]^2}{l!} \frac{(2l-2k)!}{(l-2k)!\,k!\,(l-k)!} x^{l-2k} \tag{6.172b}$$

The arbitrary constant c_0 is chosen so that

$$c_0(-1)^{l/2} \frac{[(\frac{1}{2}l)!]^2}{l!} = \frac{1}{2^l} \tag{6.173}$$

This choice is made so that, for any even l, $P_l(1) = 1$. Then equation 6.172b becomes

$$P_l(x) = \sum_{k=0}^{l/2} (-1)^k \frac{1}{2^l} \frac{(2l-2k)!}{(l-2k)!\,k!\,(l-k)!} x^{l-2k} \tag{6.174}$$

For odd l, the series coefficient of equation 6.169b can be written as

$$c_{2m+1} = \frac{(-1)^m}{(2m+1)!} \frac{(l-1)!!}{(l-2m-1)!!} \frac{(l+2m)!!}{l!!} c_1 \tag{6.175a}$$

Using equations 6.90, this can be written

$$c_{2m+1} = \frac{(-1)^m}{(2m+1)!} \frac{\left(\frac{l-1}{2}\right)!\left(\frac{l+1}{2}\right)!}{(l+1)!} \frac{(l+2m+1)!}{\left(\frac{l-2m-1}{2}\right)!\left(\frac{l+2m+1}{2}\right)!} c_1 \tag{6.175b}$$

and the odd Legendre polynomial can be written as

$$P_l(x) = c_1 \sum_{m=0}^{(l-1)/2} \frac{(-1)^m}{(2m+1)!} \frac{\left(\frac{l-1}{2}\right)!\left(\frac{l+1}{2}\right)!}{(l+1)!} \frac{(l+2m+1)!}{\left(\frac{l-2m-1}{2}\right)!\left(\frac{l+2m+1}{2}\right)!} x^{2m+1} \tag{6.176a}$$

If the summation index is replaced such that $2m + 1 = l - 2k$, this becomes

$$P_l(x) = c_1 \sum_{k=0}^{(l-1)/2} (-1)^{(l-1)/2}(-1)^k \frac{\left(\frac{l-1}{2}\right)!\left(\frac{l+1}{2}\right)!}{(l+1)!} \frac{(2l-2k)!}{(l-2k)!\,k!\,(l-k)!} x^{l-2k} \tag{6.176b}$$

By choosing

$$c_1(-1)^{(l-1)/2} \frac{\left(\frac{l-1}{2}\right)!\left(\frac{l+1}{2}\right)!}{(l+1)!} = \frac{1}{2^l} \tag{6.177}$$

[which makes $P_l(1) = 1$ for all odd l], the odd Legendre polynomial is

$$P_l(x) = \sum_{k=0}^{(l-1)/2} (-1)^k \frac{1}{2^l} \frac{(2l-2k)!}{(l-2k)!\,k!\,(l-k)!} x^{l-2k} \tag{6.178}$$

which is identical in form to the even-order Legendre polynomial, except for the upper limit of the summation index.

The symbol $[N]$ is defined as the largest integer not exceeding N. That is, it is the integer part of the number N. Then, for even l, $[\frac{1}{2}l] = \frac{1}{2}l$, and $[\frac{1}{2}l] = \frac{1}{2}(l-1)$ if l is odd. With this notation, the Legendre polynomial can be written

$$P_l(x) = \sum_{m=0}^{[l/2]} \frac{(-1)^m (2l-2m)!}{2^l m!\,(l-m)!\,(l-2m)!} x^{l-2m} \tag{6.179}$$

Since $P_l(x)$ is an odd polynomial if l is odd, and is even for even l,

$$P_l(-x) = (-1)^l P_l(x) \tag{6.180}$$

Using equation 6.179, we obtain the first few lowest-order Legendre polynomials. They are

$$\begin{array}{lll} \text{(a)}\quad P_0(x) = 1 & \text{(b)}\quad P_1(x) = x & \text{(c)}\quad P_2(x) = \frac{1}{2}(3x^2 - 1) \\ \text{(d)}\quad P_3(x) = \frac{1}{2}(5x^3 - 3x) & \text{(e)}\quad P_4(x) = \frac{1}{8}(35x^4 - 30x^2 + 3) & \end{array} \tag{6.181}$$

Generating function

The most useful and most commonly used function for generating the Legendre polynomials is

$$\Phi(x, z) = \frac{1}{\sqrt{1 - 2xz + z^2}} \tag{6.182a}$$

The Legendre polynomials are generated by expanding this function in a MacLaurin series in z. The expansion generates the Legendre polynomials as

$$\Phi(x, z) = \frac{1}{\sqrt{1 - 2xz + z^2}} = \sum_{l=0}^{\infty} z^l P_l(x) \qquad -1 < z < 1 \tag{6.182b}$$

To show the validity of equation 6.182b, we expand the generator in the binomial series, using the form expressed in equation 6.92.

$$\begin{aligned} [1 - (2xz - z^2)]^{-1/2} &= \sum_{n=0}^{\infty} \frac{(2n)!}{2^{2n}(n!)^2}(2xz - z^2)^n \\ &= \sum_{n=0}^{\infty} \frac{(2n)!}{2^{2n}(n!)^2} \sum_{k=0}^{n} \frac{n!}{k!\,(n-k)!}(-1)^k 2^{n-k} x^{n-k} z^{n+k} \end{aligned} \tag{6.183}$$

We define the coefficient of z^{n+k} as $C(n, k)$. Since k is restricted to the range $0 \leqslant k \leqslant n$, equation 6.183 has the form

$$\begin{aligned} [1 - (2xz - z^2)]^{-1/2} &= \sum_{n=0}^{\infty} \sum_{k=0}^{n} C(n,k) z^{n+k} = z^0 C(0,0) + zC(1,0) \\ &+ z^2[C(2,0) + C(1,1)] + z^3[C(3,0) + C(2,1)] + z^4[C(4,0) + C(3,1) + C(2,2)] \\ &+ z^5[C(5,0) + C(4,1) + C(3,2)] + \cdots = \sum_{l=0}^{\infty} z^l \left[\sum_{m=0}^{[l/2]} C(l-m, m) \right] \\ &= \sum_{l=0}^{\infty} z^l \left[\sum_{m=0}^{[l/2]} (-1)^m \frac{1}{2^l} \frac{(2l-2m)!}{(l-2m)!\,m!\,(l-m)!} x^{l-2m} \right] \end{aligned} \tag{6.184}$$

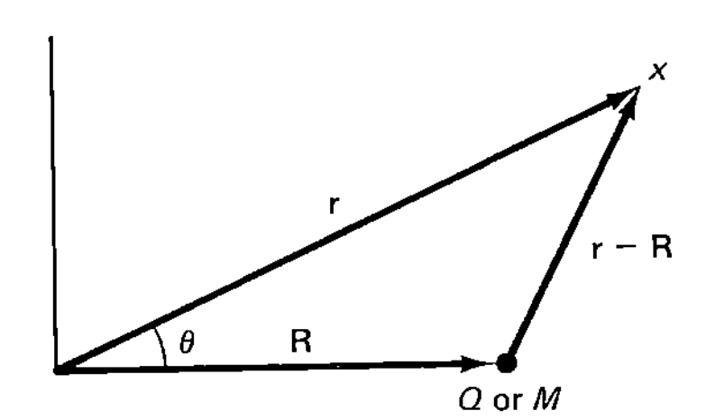

Figure 6.5 Vector configuration for determination of electrostatic or gravitational potential due to a point source.

Referring to equation 6.179, we note that the inner sum is the Legendre polynomial, $P_l(x)$, and equation 6.182b is valid.

Consider a charge Q (or a mass M) situated at a point defined by the position vector $\mathbf{R}$. The electrostatic (or gravitational) potential at a point defined by the vector $\mathbf{r}$ is the Coulomb (or Newtonian) potential

$$\text{(a)}\quad V_Q = \frac{Q}{|\mathbf{r}-\mathbf{R}|} \qquad \text{or} \qquad \text{(b)}\quad V_M = \frac{GM}{|\mathbf{r}-\mathbf{R}|} \tag{6.185}$$

In terms of the parameters specified in Figure 6.5,

$$|\mathbf{r}-\mathbf{R}| = \sqrt{R^2 + r^2 - 2rR\cos\theta} \tag{6.186}$$

$$\Rightarrow \quad |\mathbf{r}-\mathbf{R}| = R\sqrt{1 + \left(\frac{r}{R}\right)^2 - 2\left(\frac{r}{R}\right)\cos\theta} \tag{6.187a}$$

or

$$|\mathbf{r}-\mathbf{R}| = r\sqrt{1 + \left(\frac{R}{r}\right)^2 - 2\left(\frac{R}{r}\right)\cos\theta} \tag{6.187b}$$

Both forms look like the inverse of the generator of the Legendre polynomials, with $r/R = z$ in equation 6.187a and $R/r = z$ in equation 6.187b. From equation 6.182b, the MacLaurin series for the generating function converges for $|z| < 1$. Thus if we want to use the generator to expand the potential in a series involving Legendre polynomials, we must express $|\mathbf{r}-\mathbf{R}|$ in the form shown in equation 6.187a for $r < R$, and in the form shown in equation 6.187b if $r > R$. That is, the potential due to a point charge is

$$V = \frac{Q}{R\sqrt{1 + (r/R)^2 - 2(r/R)\cos\theta}} = \frac{Q}{R}\sum_{l=0}^{\infty}\left(\frac{r}{R}\right)^l P_l(\cos\theta) \qquad r < R \tag{6.188a}$$

$$V = \frac{Q}{r\sqrt{1 + (R/r)^2 - 2(R/r)\cos\theta}} = \frac{Q}{r}\sum_{l=0}^{\infty}\left(\frac{R}{r}\right)^l P_l(\cos\theta) \qquad r > R \tag{6.188b}$$

Rodriguez formula

We begin with the series description

$$P_l(x) = \sum_{m=0}^{[l/2]} \frac{(-1)^m (2l-2m)!}{2^l m!\,(l-m)!\,(l-2m)!} x^{l-2m} \tag{6.179}$$

Using

$$D^l x^{2l-2m} = \frac{(2l-2m)!}{(l-2m)!} x^{l-2m} \tag{6.189}$$

$$\Rightarrow \quad P_l(x) = \frac{1}{2^l} D^l \sum_{m=0}^{[l/2]} \frac{(-1)^m}{m!\,(l+m)!} (x^2)^{l-m} \tag{6.190}$$

Consider the quantity

$$D^l\left[\sum_{m=[l/2]+1}^{l}\frac{(-1)^m}{m!\,(l-m)!}(x^2)^{l-m}\right]=\sum_{m=[l/2]+1}^{l}\frac{(-1)^m}{m!\,(l-m)!}D^l x^{2l-2m} \tag{6.191}$$

Each power of x has the range

$$\text{(a)}\quad 0\leqslant 2l-2m\leqslant l-2\quad l\text{ even}\qquad \text{(b)}\quad 0\leqslant 2l-2m\leqslant l-1\quad l\text{ odd} \tag{6.192}$$

$$\Rightarrow\quad D^l x^{2l-2m}=0 \tag{6.193}$$

for both even and odd l. Therefore, equation 6.190 is unchanged by adding terms from $m=[\frac{1}{2}l]+1$ to $m=l$. That is, after multiplying and dividing by $l!$, the Legendre polynomial can be written

$$P_l(x)=\frac{1}{2^l l!}D^l\sum_{m=0}^{l}\frac{(-1)^m l!}{m!\,(l-m)!}(x^2)^{l-m} \tag{6.194a}$$

Replacing m by $l-k$, this becomes

$$P_l(x)=\frac{(-1)^l}{2^l l!}D^l\sum_{k=0}^{l}(-1)^k\frac{l!}{(l-k)!\,k!}(x^2)^k \tag{6.194b}$$

The sum is the binomial expansion of $(1-x^2)^l$. Therefore,

$$P_l(x)=\frac{(-1)^l}{2^l l!}D^l(1-x^2)^l \tag{6.195}$$

This is the *Rodriguez formula* for the Legendre polynomial.

Integral representations

Using Cauchy's theorem, we have shown that an analytic function such as $(1-x^2)^l$ can be written as a contour integral.

$$(1-x^2)^l=\frac{1}{2\pi i}\oint\frac{(1-z^2)^l}{(z-x)}\,dz \tag{6.196}$$

where the contour is any loop enclosing the point x. Substituting this into equation 6.195, we obtain an integral representation of the Legendre polynomial known as the *Schlafli representation*.

$$P_l(x)=\frac{(-1)^l}{2^l}\frac{1}{2\pi i}\oint\frac{(1-z^2)^l}{(z-x)^{l+1}}\,dz \tag{6.197}$$

Using this, we can develop another integral representation of $P_l(x)$ by making the substitution

$$\text{(a)}\quad z=x-i\sqrt{1-x^2}\,e^{i\phi}\quad\Rightarrow\quad\text{(b)}\quad(1-z^2)=2i\sqrt{1-x^2}\,e^{i\phi}\left(x-i\sqrt{1-x^2}\cos\phi\right) \tag{6.198}$$

$$\Rightarrow\quad P_l(x)=\frac{1}{2\pi}\int_0^{2\pi}\left(x-i\sqrt{1-x^2}\cos\phi\right)^l d\phi \tag{6.199a}$$

To make the limits symmetric, we replace ϕ by $\phi-\pi$, to obtain

$$P_l(x)=\frac{1}{2\pi}\int_{-\pi}^{\pi}\left(x+i\sqrt{1-x^2}\cos\phi\right)^l d\phi \tag{6.199b}$$

Because $\cos \phi$ is an even function, this can be written

$$P_l(x) = \frac{1}{\pi}\int_0^{\pi}\left(x + i\sqrt{1-x^2}\cos\phi\right)^l d\phi \tag{6.200a}$$

This is the *Laplace integral representation* of $P_l(x)$.

If l is replaced by $-l-1$, the Legendre differential equation is unchanged. Therefore, $P_l(x)$ and $P_{-l-1}(x)$ are not independent. Later in this chapter, we develop properties of a function called the Gauss hypergeometric function and relate the Legendre polynomial to this function. From the properties of the hypergeometric function, it will become clear that $P_{-l-1}(x) = P_l(x)$ (see equation 6.677). Therefore, replacing l by $-l-1$ in equation 6.200a, we obtain another Laplace representation for the Legendre polynomials. It is

$$P_l(x) = \frac{1}{2\pi}\int_{-\pi}^{\pi}\left(x + i\sqrt{1-x^2}\cos\phi\right)^{-l-1} d\phi = \frac{1}{\pi}\int_0^{\pi}\left(x + i\sqrt{1-x^2}\cos\phi\right)^{-l-1} d\phi \tag{6.200b}$$

Recurrence relations

In Chapter 5, we discussed recurrence relations for the coefficients of the series solution of a differential equation. In that development, we indicated that a recurrence relation relates quantities (coefficients of a power series) of higher index to quantities (coefficients) of lower index. Similarly, a recurrence relation involving Sturm-Liouville functions relates functions of higher index to ones of lower index.

For the Legendre polynomials, we start with the generating function

$$\Phi(x,z) = \frac{1}{\sqrt{1-2xz+z^2}} = \sum_{l=0}^{\infty} z^l P_l(x) \qquad -1 < z < 1 \tag{6.182b}$$

We consider

$$\frac{\partial\Phi}{\partial z} = \frac{(x-z)}{(1-2xz+z^2)^{3/2}} = \frac{(x-z)}{(1-2xz+z^2)}\frac{1}{\sqrt{1-2xz+z^2}}$$

$$= \frac{(x-z)}{(1-2xz+z^2)}\Phi(z,x) = \sum_{l=1}^{\infty} l z^{l-1} P_l(x) \tag{6.201a}$$

$$\Rightarrow \quad (1-2xz+z^2)\sum_{l=1}^{\infty} l z^{l-1} P_l(x) = (x-z)\sum_{l=0}^{\infty} z^l P_l(x) \tag{6.201b}$$

Multiplying the various factors of x and z into the sums, this becomes

$$\sum_{l=1}^{\infty} l z^{l-1} P_l(x) - 2\sum_{l=1}^{\infty} l z^l x P_l(x) + \sum_{l=1}^{\infty} l z^{l+1} P_l(x) = \sum_{l=0}^{\infty} z^l x P_l(x) - \sum_{l=0}^{\infty} z^{l+1} P_l(x) \tag{6.202a}$$

Because of the factor l multiplying z in the sums on the right-hand side, we can include the $l = 0$ term in each sum without changing equation 6.202a. Combining the sums, we obtain

$$\sum_{l=0}^{\infty} l z^{l-1} P_l(x) - \sum_{l=0}^{\infty} (2l+1) z^l x P_l(x) + \sum_{l=0}^{\infty} (l+1) z^{l+1} P_l(x) = 0 \tag{6.202b}$$

Explicitly expressing the z^0 terms, equation 6.202b can be written

$$z^0[P_1(x) + xP_0(x)] + \sum_{l=2}^{\infty} lz^{l-1}P_l(x) - \sum_{l=1}^{\infty}(2l+1)z^l xP_l(x)$$

$$+ \sum_{l=0}^{\infty}(l+1)z^{l+1}P_l(x) = 0 \qquad (6.202c)$$

We replace l by $l + 2$ in the first sum, and by $l + 1$ in the second sum. Equation 6.202c then becomes

$$z^0[P_1(x) - xP_0(x)] + \sum_{l=0}^{\infty}(l+2)z^{l+1}P_{l+2}(x) - \sum_{l=0}^{\infty}(2l+3)z^{l+1}xP_{l+1}(x)$$

$$+ \sum_{l=0}^{\infty}(l+1)z^{l+1}P_l(x) = 0 \qquad (6.202d)$$

Equating the coefficient of each power of z to zero, we obtain

$$P_1(x) = xP_0(x) \qquad l = 0 \qquad (6.203a)$$

which is valid as can be seen from equations 6.181a and 6.181b, and

$$(l+2)P_{l+2}(x) - (2l+3)xP_{l+1}(x) + (l+1)P_l(x) = 0 \qquad l \geqslant 0 \qquad (6.203b)$$

This is one of the recurrence relations that is satisfied by the Legendre polynomials.

A second recurrence relation can be deduced by considering $\partial\Phi/\partial x$.

$$\frac{\partial\Phi}{\partial x} = \frac{z}{(1-2xz+z^2)^{3/2}} = \frac{z}{(1-2zx+z^2)}\sum_{l=0}^{\infty} z^l P_l(x) = \sum_{l=0}^{\infty} z^l P_l'(x) \qquad (6.204)$$

With manipulations identical to those that led to equation 6.203b, we obtain

$$P_{l+2}'(x) - 2xP_{l+1}'(x) + P_l'(x) = P_{l+1}(x) \qquad l \geqslant 0 \qquad (6.205)$$

The Legendre polynomials satisfy other recurrence relations that are obtained using some of the properties developed in the text. The reader is referred to Problem 18.

Orthonormalization condition

It was shown that for any Sturm–Liouville function, we could obtain a complete orthogonalized set of eigenfunctions. But these eigenfunctions were determinable up to an arbitrary multiplicative constant. That constant was obtained by the normalization condition

$$\langle \alpha_l | \alpha_l \rangle = N_l \qquad (6.206)$$

For the Legendre polynomials, with the weight function $\rho(x) = 1$, the orthogonality and normalization condition (called the *orthonormality* or *orthonormalization condition*) is

$$\int_{-1}^{1} P_l(x)P_m(x)\,dx = N_l\delta_{ml} \qquad (6.207)$$

The orthogonality is assured as shown for all Sturm–Liouville functions. The normalization constant is obtained from

$$\int_{-1}^{1} [P_l(x)]^2\,dx = N_l \qquad (6.208)$$

To evaluate this integral, consider

$$\int_{-1}^{1} [\Phi(z,x)]^2\, dx = \int_{-1}^{1} \sum_{l=0}^{\infty} z^l P_l(x) \sum_{m=0}^{\infty} z^m P_m(x)\, dx = \sum_{l,m=0}^{\infty} z^{l+m} \int_{-1}^{1} P_l(x) P_m(x)\, dx$$

$$= \sum_{l,m=0}^{\infty} z^{l+m} N_l \delta_{lm} = \sum_{l=0}^{\infty} z^{2l} N_l \tag{6.209}$$

But

$$\int_{-1}^{1} [\Phi(z,x)]^2\, dx = \int_{-1}^{1} \frac{dx}{(1 - 2xz + z^2)} = \frac{1}{z} \log\left(\frac{1+z}{1-z}\right) \tag{6.210}$$

Since the series expression for the generating function converges for $-1 < z < 1$, z cannot be evaluated at a branch point of the log function. In addition, when $z \to 0$, $\log\left(\frac{1+z}{1-z}\right) \sim 2z$. Therefore, $\frac{1}{z} \log\left(\frac{1+z}{1-z}\right)$ does not have a pole at $z = 0$. That is, all z between -1 and 1 are points of analyticity of $\frac{1}{z}\log\left(\frac{1+z}{1-z}\right)$. Therefore, it can be expanded in a MacLaurin series.

$$\frac{1}{z} \log\left(\frac{1+z}{1-z}\right) = 2\sum_{l=0}^{\infty} \frac{z^{2l}}{(2l+1)} \tag{6.211}$$

$$\Rightarrow \quad \int_{-1}^{1} [\Phi(z,x)]^2\, dx = \sum_{l=0}^{\infty} z^{2l} N_l = \sum_{l=0}^{\infty} z^{2l} \frac{2}{(2l+1)} \tag{6.212}$$

Thus the orthonormalization condition is

$$\int_{-1}^{1} P_l(x) P_m(x)\, dx = \frac{2}{(2l+1)} \delta_{lm} \tag{6.213}$$

Powers of *x* in terms of Legendre polynomials

As described in equation 6.157, any function can be expanded as a linear sum of Sturm–Liouville functions. In particular, any power of x can be written as a linear combination of Legendre polynomials. For example, referring to equations 6.181, we note that

$$\begin{aligned} &\text{(a)}\quad 1 = P_0(x) \qquad \text{(b)}\quad x = P_1(x) \qquad \text{(c)}\quad x^2 = \tfrac{2}{3}P_2(x) + \tfrac{1}{3}P_0(x) \\ &\text{(d)}\quad x^3 = \tfrac{2}{5}P_3(x) + \tfrac{3}{5}P_1(x) \qquad \text{(e)}\quad x^4 = \tfrac{8}{35}P_4(x) + \tfrac{4}{7}P_2(x) + \tfrac{1}{5}P_0(x) \end{aligned} \tag{6.214}$$

To obtain a general expression for such an expansion, we again manipulate the generating function, writing

$$\sum_{l=0}^{\infty} z^l P_l(x) = (1 - 2xz + z^2)^{-1/2} = \frac{1}{(1+z^2)^{1/2}} \left[1 - \frac{2xz}{(1+z^2)}\right]^{-1/2} \tag{6.215a}$$

Using the binomial expansion as expressed in equation 6.92, this can be written

$$\sum_{l=0}^{\infty} z^l P_l(x) = \sum_{k=0}^{\infty} \frac{(2k)!}{2^{2k}(k!)^2} \frac{2^k x^k z^k}{(1+z^2)^{k+1/2}} \tag{6.215b}$$

We multiply both sides of this expression by $z^{1/2}$ and define the parameter w by

$$\text{(a)} \quad z \equiv \frac{2w}{1+\sqrt{1-4w^2}} \quad \Rightarrow \quad \text{(b)} \quad \frac{z}{1+z^2} = w \tag{6.216}$$

Then equation 6.215b becomes

$$\sum_{l=0}^{\infty} w^{l+1/2} P_l(x) \left[\frac{2}{1+\sqrt{1-4w^2}}\right]^{l+1/2} = \sum_{k=0}^{\infty} \frac{(2k)!\,x^k}{2^k (k!)^2} w^{k+1/2} \tag{6.217}$$

The expansion of $\left[\frac{2}{1+\sqrt{1-4w^2}}\right]^{l+1/2}$ in a MacLaurin series is straightforward but unwieldy. The result is

$$\left[\frac{2}{1+\sqrt{1-4w^2}}\right]^{l+1/2} = \left(l+\frac{1}{2}\right) \sum_{m=0}^{\infty} \frac{(l+2m-\frac{1}{2})!}{m!\,(l+m+\frac{1}{2})!} w^{2m} \tag{6.218}$$

Therefore, after cancellation of a factor of $w^{1/2}$, equation 6.217 becomes

$$\sum_{l=0}^{\infty} P_l(x) \left(l+\frac{1}{2}\right) \sum_{m=0}^{\infty} \frac{(l+2m-\frac{1}{2})!}{m!\,(l+m+\frac{1}{2})!} w^{l+2m} = \sum_{k=0}^{\infty} \frac{(2k)!\,x^k}{2^k (k!)^2} w^k \tag{6.219}$$

Writing the coefficients of w^{l+2m} as $C(l,m)$, the double sum on the left side of this equation is of the form

$$w^0 C(0,0) + wC(1,0) + w^2[C(2,0)+C(0,1)] + w^3[C(3,0)+C(1,1)]$$

$$+ w^4[C(4,0)+C(2,1)+C(0,2)] + \cdots = \sum_{k=0}^{\infty} w^k \sum_{m=0}^{[k/2]} C(k-2m,m) \tag{6.220}$$

Thus equation 6.219 can be written

$$\sum_{k=0}^{\infty} w^k \sum_{m=0}^{[k/2]} \left(k-2m+\frac{1}{2}\right) P_{k-2m}(x) \frac{(k-\frac{1}{2})!}{m!\,(k-m+\frac{1}{2})!} = \sum_{k=0}^{\infty} \frac{(2k)!\,x^k}{2^k (k!)^2} w^k \tag{6.221}$$

$$\Rightarrow \quad x^k = 2^k \frac{(k!)^2}{(2k)!} \left(k-\frac{1}{2}\right)! \sum_{m=0}^{[k/2]} \frac{(k-2m+\frac{1}{2})}{m!\,(k-m+\frac{1}{2})!} P_{k-2m}(x) \tag{6.222}$$

Representation of $P_l(\cos\theta)$

Since the argument of the Legendre polynomials extends over the range $[-1,1]$, we can define an angle θ by $x = \cos\theta$. Starting with the generating function, consider

$$\sum_{l=0}^{\infty} z^l P_l(x) = (1-2xz+z^2)^{-1/2} = \left[(1-xz)^2 - z^2(x^2-1)\right]^{-1/2}$$

$$= \left[1 - z\left(x-\sqrt{(x^2-1)}\right)\right]^{-1/2} \left[1 - z\left(x+\sqrt{(x^2-1)}\right)\right]^{-1/2} \tag{6.223}$$

Using the binomial expansion again (equation 6.92), we can write this as

$$\sum_{l=0}^{\infty} z^l P_l(x) = \sum_{n=0}^{\infty} \sum_{k=0}^{\infty} \frac{(2n)!}{2^{2n}(n!)^2} \frac{(2k)!}{2^{2k}(k!)^2} z^{n+k} \left[x - \sqrt{(x^2-1)}\right]^n \left[x + \sqrt{(x^2-1)}\right]^k \tag{6.224}$$

As we have done before, we denote the coefficients of the various powers of z as $C(n, k)$ and write equation 6.215 as

$$\sum_{l=0}^{\infty} z^l P_l(x) = z^0 C(0,0) + z[C(1,0) + C(0,1)]$$

$$+ z^2[C(2,0) + C(1,1) + C(0,2)] + \cdots = \sum_{l=0}^{\infty} z^l \sum_{k=0}^{l} C(l-k, k) \tag{6.225}$$

$$\Rightarrow \quad P_l(x) = \sum_{k=0}^{l} C(l-k, k)$$

$$= \sum_{k=0}^{l} \frac{(2l-2k)!\,(2k)!}{2^{2l}[(l-k)!]^2 (k!)^2} \left[x - \sqrt{(x^2-1)}\right]^{l-k} \left[x + \sqrt{(x^2-1)}\right]^k \tag{6.226}$$

With $x = \cos\theta$,

$$\text{(a)} \quad \sqrt{(x^2-1)} = i\sin\theta \quad \Rightarrow \quad \text{(b)} \quad x \pm \sqrt{(x^2-1)} = \cos\theta \pm i\sin\theta = e^{\pm i\theta} \tag{6.227}$$

and the Legendre polynomials can be written

$$P_l(\cos\theta) = \sum_{k=0}^{l} \frac{(2l-2k)!\,(2k)!}{2^{2l}[(l-k)!]^2 (k!)^2} e^{i(2k-l)\theta} \tag{6.228}$$

If l is even, we write this expression as

$$P_l(\cos\theta) = \sum_{k=0}^{l/2-1} \frac{(2l-2k)!\,(2k)!}{2^{2l}[(l-k)!]^2 (k!)^2} e^{i(2k-l)\theta} + \frac{1}{2^{2l}} \frac{(l!)^2}{[(\frac{1}{2}l)!]^4}$$

$$+ \sum_{k=l/2+1}^{l} \frac{(2l-2k)!\,(2k)!}{2^{2l}[(l-k)!]^2 (k!)^2} e^{i(2k-l)\theta} \tag{6.229}$$

Replacing k by $l-k$ in the second sum, this becomes

$$P_l(\cos\theta) = \sum_{k=0}^{l/2-1} \frac{(2l-2k)!\,(2k)!}{2^{2l}[(l-k)!]^2 (k!)^2} e^{i(2k-l)\theta} + \frac{1}{2^{2l}} \frac{(l!)^2}{[(\frac{1}{2}l)!]^4}$$

$$+ \sum_{k=0}^{l/2-1} \frac{(2l-2k)!\,(2k)!}{2^{2l}[(l-k)!]^2 (k!)^2} e^{-i(2k-l)\theta} \tag{6.230}$$

$$= \frac{1}{2^{2l}} \frac{(l!)^2}{[(\frac{1}{2})!]^4} + 2 \sum_{k=0}^{l/2-1} \frac{(2l-2k)!\,(2k)!}{2^{2l}[(l-k)!]^2 (k!)^2} \cos[(2k-l)\theta] \qquad l \text{ even}$$

If l is odd, the sum in equation 6.228 is divided as

$$P_l(\cos\theta) = \sum_{k=0}^{l/2-1} \frac{(2l-2k)!\,(2k)!}{2^{2l}[(l-k)!]^2(k!)^2} e^{i(2k-l)\theta} + \sum_{k=l/2+1}^{l} \frac{(2l-2k)!\,(2k)!}{2^{2l}[(l-k)!]^2(k!)^2} e^{i(2k-l)\theta} \tag{6.231}$$

As we did above, we replace k by $l-k$ in the second sum to obtain

$$\begin{aligned} P_l(\cos\theta) &= \sum_{k=0}^{l/2-1} \frac{(2l-2k)!\,(2k)!}{2^{2l}[(l-k)!]^2(k!)^2} e^{i(2k-l)\theta} \\ &\quad + \sum_{k=0}^{l/2-1} \frac{(2l-2k)!\,(2k)!}{2^{2l}[(l-k)!]^2(k!)^2} e^{-i(2k-l)\theta} \\ &= 2\sum_{k=0}^{l/2-1} \frac{(2l-2k)!\,(2k)!}{2^{2l}[(l-k)!]^2(k!)^2} \cos[(2k-l)\theta] \qquad l \text{ odd} \end{aligned} \tag{6.232}$$

Legendre function of the second kind

The ordinary Legendre differential equation

$$(1-x^2)D^2y - 2xDy + l(l+1)y = 0 \tag{6.164}$$

has regular singularities at $x = \pm 1$. As noted earlier, when l is an integer, one of the series solutions becomes a finite (and therefore nonsingular) polynomial. The other solution remains an infinite series which diverges at $x = \pm 1$, reflecting the singularity structure of the differential equation. Rather than investigating this second solution in the form of a singular infinite series, it is customary to develop the second solution using the method of reduction of order described in Chapter 5.

Letting

$$y_2 \equiv Q_l(x) = v(x)P_l(x) \tag{6.233}$$

$$\Rightarrow \quad (1-x^2)P_l(x)v''(x) - \left[2xP_l(x) - 2(1-x^2)P_l'(x)\right]v'(x) + \left[(1-x^2)P_l''(x) - 2xP_l'(x) + l(l+1)P_l(x)\right]v(x) = 0 \tag{6.234a}$$

Since $P_l(x)$ satisfies the Legendre differential equation, the third term is zero. Therefore, the equation for $v(x)$ reduces to

$$(1-x^2)P_l(x)v''(x) - \left[2xP_l(x) - 2(1-x^2)P_l'(x)\right]v'(x) = 0 \tag{6.234b}$$

$$\Rightarrow \quad v'(x) = \frac{C}{(1-x^2)P_l^2(x)} \tag{6.235}$$

As will be shown shortly, $Q_l(x) = v(x)P_l(x)$ has logarithmic branch points at $x = \pm 1$ arising from the singularities of $\log\left(\dfrac{x+1}{x-1}\right)$. To obtain an integral representation of $Q_l(x)$, the cut structure associated with these branch points is chosen to run from -1 to 1. That such a choice is a valid one was shown in Problem 28 of Chapter 2. When integrating equation 6.235, we can avoid adding a second constant of integration by integrating between limits over a region of analyticity of $Q_l(x)$. This region is chosen to be from ∞ to a finite point $x > 1$. That is, integrating equation 6.235,

$$\text{(a)} \quad v(x) = \int_{\infty}^{x} \frac{dx'}{(1 - x'^2) P_l^2(x')} \quad \Rightarrow \quad \text{(b)} \quad Q_l(x) = P_l(x) \int_{\infty}^{x} \frac{dx'}{(1 - x'^2) P_l^2(x')} \tag{6.236}$$

At large x, this integral behaves like

$$\text{(a)} \quad \int_{\infty}^{x} \frac{dx'}{(1 - x'^2) P_l^2(x')} \simeq \int_{\infty}^{x} \frac{dx'}{x'^{2l+2}} \rightarrow \frac{1}{x^{2l+1}}$$

$$\Rightarrow \quad \text{(b)} \quad \lim_{x \to \infty} Q_l(x) \rightarrow \lim_{x \to \infty} x^l \frac{1}{x^{2l+1}} = 0 \tag{6.237}$$

The arbitrary constant C of equation 6.235 contributes nothing to the properties of $Q_l(x)$ other than to define its magnitude. It is chosen to be 1. This second solution is called the Legendre function of the second kind.

Representations of the second Legendre function

One useful representation of $Q_l(x)$ exists in terms of a series involving the Legendre polynomials. For integer values of l, the first few lowest-index functions are

$$Q_0(x) = \int_{\infty}^{x} \frac{dx'}{(1 - x'^2)} = \frac{1}{2} \log\left(\frac{x+1}{x-1}\right) \tag{6.238a}$$

$$Q_1(x) = \int_{\infty}^{x} \frac{dx'}{(1 - x'^2) x'^2} = \frac{x}{2} \log\left(\frac{x+1}{x-1}\right) - 1 = P_1(x) Q_0(x) - P_0(x) \tag{6.238b}$$

Similarly,

$$Q_2(x) = P_2(x) Q_0(x) - \tfrac{3}{2}x = P_2(x) Q_0(x) - \tfrac{3}{2} P_1(x) \tag{6.238c}$$

In general, for integer l, defining $l1 \equiv [(l-1)/2]$

$$\begin{aligned} Q_l(x) &= P_l(x) Q_0(x) - \frac{(2l-1)}{1 \cdot l} P_{l-1}(x) - \frac{(2l-5)}{3 \cdot (l-1)} P_{l-3}(x) - \cdots \\ &= P_l(x) Q_0(x) - \sum_{k=0}^{l1} \frac{(2l-4k-1)}{(2k+1)(l-k)} P_{l-2k-1}(x) \equiv P_l(x) Q_0(x) - R_{l-1}(x) \end{aligned} \tag{6.239}$$

From equation 6.239 we see that for every l, the singularity structure of $Q_l(x)$ is the same as the singularity structure of $Q_0(x)$; that is, logarithmic branch points at $x = \pm 1$, with the cut chosen to run from -1 to 1. Using this, we can deduce an integral representation for $Q_l(x)$.

From Cauchy's theorem

$$Q_l(z) = \frac{1}{2\pi i} \oint \frac{Q_l(z')}{(z' - z)} dz' \tag{6.240}$$

where the contour encloses the region of analyticity of $Q_l(z')$. The cut structure and contour for this integral are shown in Figure 6.6. Taking the integral along the various

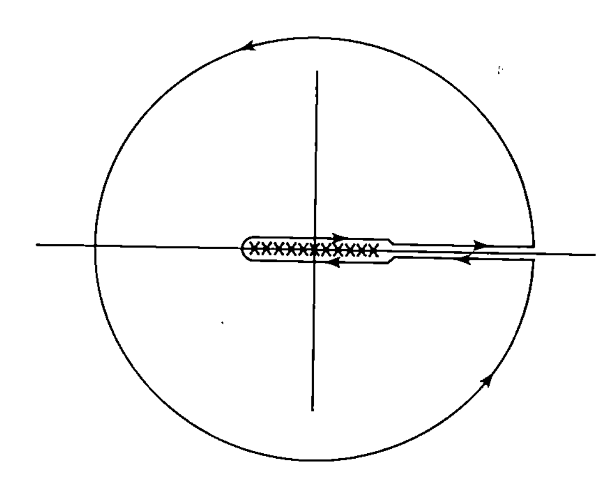

Figure 6.6 Cut structure and contour for integral representation of $Q_l(z)$.

parts of the contour, we have

$$Q_l(z) = \frac{1}{2\pi i}\left[\int_{-1}^{1} \frac{Q_l(x' + i\varepsilon)}{(x' - z)}\, dx' + \int_{1}^{\infty} \frac{Q_l(x')}{(x' - z)}\, dx' + \int_{C_\infty} \frac{Q_l(z')}{(z' - z)}\, dz' + \int_{\infty}^{1} \frac{Q_l(x')}{(x' - z)}\, dx' + \int_{1}^{-1} \frac{Q_l(x' - i\varepsilon)}{(x' - z)}\, dx'\right] \tag{6.241}$$

As shown above,

$$Q_l(z') \to \frac{1}{z'^{l+1}} \tag{6.237b}$$

for large z'. Therefore, the integral around the infinite circle is zero. Also, since the cut structure of $Q_l(x')$ has been chosen so that $Q_l(x')$ is analytic for $1 \leqslant x' \leqslant \infty$,

$$\int_{1}^{\infty} \frac{Q_l(x')}{(x' - z)}\, dx' + \int_{\infty}^{1} \frac{Q_l(x')}{(x' - z)}\, dx' = 0 \tag{6.242}$$

Therefore, equation 6.241 becomes

$$Q_l(z) = \frac{1}{2\pi i}\int_{-1}^{1} \frac{[Q_l(x' + i\varepsilon) - Q_l(x' - i\varepsilon)]}{(x' - z)}\, dx' \tag{6.243}$$

Since the polynomial $R_{l-1}(z)$ defined in equation 6.239 is analytic,

$$Q_l(x' + i\varepsilon) - Q_l(x' - i\varepsilon) = P_l(x')\,[Q_0(x' + i\varepsilon) - Q_0(x' - i\varepsilon)] \tag{6.244}$$

As shown in Chapter 2, Problem 28, the branch structure of Figure 6.6 arises from a cut running from -1 to ∞ from $\log(1 + z)$ and a cut from $+1$ to ∞ from $\log(1 - z)$. That is, the cut structure in Figure 6.6 is equivalent to the cut structure shown in Figure 6.7.

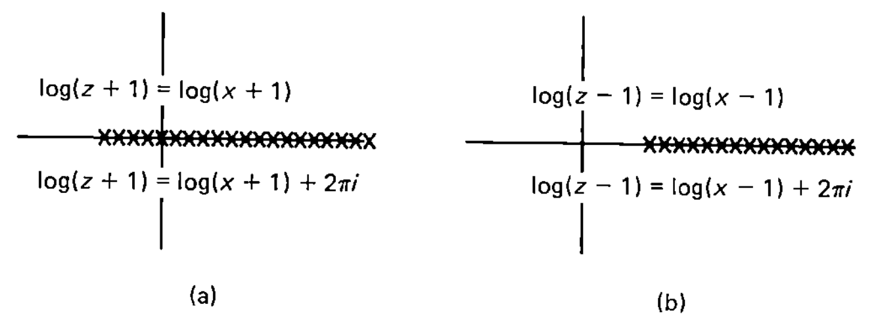

Figure 6.7 (a) Cut structure for $\log(z + 1)$; (b) cut structure of $\log(z - 1)$.

From this cut structure, we see that for $-1 \leqslant x' \leqslant 1$,

$$Q_0(x' + i\varepsilon) - Q_0(x' - i\varepsilon) \tag{6.245}$$

$$= \tfrac{1}{2}\{[\log(x' + 1) - \log(x' - 1)] - [\log(x' + 1) + 2\pi i - \log(x' - 1)]\} = -i\pi$$

$$\Rightarrow \quad Q_l(x' + i\varepsilon) - Q_l(x' - i\varepsilon) = -i\pi P_l(x') \tag{6.246}$$

$$\Rightarrow \quad Q_l(z) = \frac{1}{2}\int_{-1}^{1} \frac{P_l(x)}{(z - x)}\, dx \tag{6.247}$$

which is referred to as the *Neumann representation*.

Writing this integral as

$$\frac{1}{2}\int_{-1}^{1} \frac{P_l(z)}{(z - x)}\, dx - \frac{1}{2}\int_{-1}^{1} \frac{[P_l(z) - P_l(x)]}{(z - x)}\, dx$$
$$= \frac{1}{2}P_l(z)\log\left(\frac{z + 1}{z - 1}\right) - \frac{1}{2}\int_{-1}^{1} \frac{[P_l(z) - P_l(x)]}{(z - x)}\, dx \tag{6.248}$$

we note that the first term is $P_l(z)Q_0(z)$. Therefore, the Neumann representation of the polynomial $R_{l-1}(z)$ in equation 6.239 can be written

$$R_{l-1}(z) = \sum_{k=0}^{l1} \frac{(2l - 4k - 1)}{(2k + 1)(l - k)} P_{l-k}(z) = \frac{1}{2}\int_{-1}^{1} \frac{[P_l(z) - P_l(x')]}{(z - x')}\, dx' \tag{6.249}$$

Another integral representation involving $Q_l(z)$ can be deduced using the Neumann representation by considering

$$z^m Q_l(z) - \frac{1}{2}\int_{-1}^{1} \frac{x^m P_l(x)}{(z - x)}\, dx = \frac{1}{2}\int_{-1}^{1} \frac{(z^m - x^m)P_l(x)}{(z - x)}\, dx \tag{6.250}$$

If we write

$$(z^m - x^m) \equiv (z - x)\sum_{k=0}^{m-1} c_k z^{m-k-1}x^k = \sum_{k=0}^{m-1} c_k(z^{m-k}x^k - z^{m-k-1}x^{k+1})$$
$$= c_0(z^m - z^{m-1}x) + c_1(z^{m-1}x - z^{m-2}x^2) + c_2(z^{m-2}x^2 - z^{m-3}x^3)$$
$$+ \cdots + c_{m-1}(z^m x^{m-1} - x^m) \tag{6.251}$$

we see that this requires that $c_k = 1$ for every k. Therefore,

$$\frac{(z^m - x^m)}{(z - x)} = \sum_{k=0}^{m-1} z^{m-k-1}x^k \tag{6.252}$$

$$\Rightarrow \quad z^m Q_l(z) - \frac{1}{2}\int_{-1}^{1} \frac{x^m P_l(x)}{(z - x)}\, dx = \frac{1}{2}\sum_{k=0}^{m-1} z^{m-k-1}\int_{-1}^{1} x^k P_l(x)\, dx \tag{6.253}$$

Using equation 6.222, this becomes

$$z^m Q_l(z) - \frac{1}{2}\int_{-1}^{1} \frac{x^m P_l(x)}{(z - x)}\, dx$$
$$= \frac{1}{2}\sum_{k=0}^{m-1} z^{m-k-1}\frac{2^k(k!)^2(k - \frac{1}{2})!}{(2k)!}\sum_{r=0}^{[k/2]} \frac{(k - 2r + \frac{1}{2})}{r!\,(k - r + \frac{1}{2})!}\int_{-1}^{1} P_{k-2r}(x)P_l(x)\, dx$$
$$= \frac{1}{2}\sum_{k=0}^{m-1} z^{m-k-1}\frac{2^k(k!)^2(k - \frac{1}{2})!}{(2k)!}\sum_{r=0}^{[k/2]} \frac{(k - 2r + \frac{1}{2})}{r!\,(k - r + \frac{1}{2})!}\frac{2}{2l + 1}\delta_{r,(k-l)/2} \tag{6.254}$$

Since $r \geqslant 0$, the Kronecker δ symbol causes this double sum to be zero if $k < l$. If $m - 1 < l$, then $k < l$ for all values of k and the double sum is zero. Therefore,

$$z^m Q_l(z) = \frac{1}{2}\int_{-1}^{1} \frac{x^m P_l(x)}{(z-x)}\,dx \qquad m \leqslant l \tag{6.255a}$$

If $m - 1 = l$, the double sum will have one nonzero term, arising from $r = 0$. Therefore,

$$z^{l+1} Q_l(z) - \frac{1}{2}\int_{-1}^{1} \frac{x^{l+1} P_l(x)}{(z-x)}\,dx = \frac{2^l (l!)^2}{(2l+1)!} \tag{6.255b}$$

Let $S_N(z)$ be an Nth-order polynomial, defined by

$$S_N(z) = \sum_{m=0}^{N} c_m z^m \tag{6.256}$$

Multiplying equation 6.253 by c_m and summing, we obtain

$$S_N(z) Q_l(z) - \frac{1}{2}\int_{-1}^{1} \frac{S_N(x) P_l(x)}{(z-x)}\,dx = \frac{1}{2}\sum_{m=0}^{N} c_m \sum_{k=0}^{m-1} z^{m-k-1} \int_{-1}^{1} x^k P_l(x)\,dx \tag{6.257}$$

If $N \leqslant l$, we have just shown that $\int_{-1}^{1} x^k P_l(x)\,dx = 0$ for all k, so

$$S_N(z) Q_l(z) = \frac{1}{2}\int_{-1}^{1} \frac{S_N(x) P_l(x)}{(z-x)}\,dx \qquad N \leqslant l \tag{6.258}$$

When $N = l + 1$, the double sum on the right side of equation 6.257 reduces to a single term arising from $k = m - 1 = l$. Then equation 6.257 becomes

$$S_{l+1}(z) Q_l(z) - \frac{1}{2}\int_{-1}^{1} \frac{S_{l+1}(x) P_l(x)}{(z-x)}\,dx = c_{l+1}\frac{2^l (l!)^2}{(2l+1)!} \tag{6.259}$$

In particular, if $S_N(x)$ is the Nth Legendre polynomial, $P_N(x)$, then c_{l+1} is the coefficient of x^{l+1} in $P_{l+1}(x)$. From equation 6.179, this coefficient (for $m = 0$) is

$$c_{l+1} = \frac{(2l+2)!}{2^{l+1}[(l+1)!]^2} \tag{6.260}$$

Therefore,

$$P_N(z) Q_l(z) = \frac{1}{2}\int_{-1}^{1} \frac{P_N(x) P_l(x)}{(z-x)}\,dx \qquad N \leqslant l \tag{6.261a}$$

$$P_{l+1}(z) Q_l(z) - \frac{1}{2}\int_{-1}^{1} \frac{P_{l+1}(x) P_l(x)}{(z-x)}\,dx = \frac{1}{(l+1)} \tag{6.261b}$$

If we replace l by $l + 1$ in equation 6.261a and then set $N = l$, we obtain

$$P_l(z) Q_{l+1}(z) = \frac{1}{2}\int_{-1}^{1} \frac{P_l(x) P_{l+1}(x)}{(z-x)}\,dx \tag{6.262}$$

Combining this with equation 6.261b, we obtain

$$P_{l+1}(z) Q_l(z) - P_l(z) Q_{l+1}(z) = \frac{1}{(l+1)} \tag{6.263}$$

Generating function

$Q_l(x)$ can be obtained by expanding the generating function.

$$\Phi(z, x) = \frac{1}{\sqrt{1 - 2xz + z^2}} \log\left[\frac{x - z + \sqrt{1 - 2xz + z^2}}{x^2 - 1}\right] \tag{6.264}$$

$$= \sum_{l=0}^{\infty} z^l Q_l(x) \qquad |z| < 1, \quad |x| > 1$$

Because it is extremely unwieldy, this generator is rarely used to deduce the properties of $Q_l(z)$. Such properties are obtained by other means.

Recurrence relations

A straightforward way to obtain recurrence relations for the second Legendre function arises from the description of $Q_l(z)$ expressed in equation 6.239.

$$Q_l(z) = P_l(z)Q_0(z) - R_{l-1}(z) \tag{6.239}$$

where, with $l1 \equiv [(l - 1)/2]$,

$$R_{l-1}(z) = \sum_{k=0}^{l1} \frac{(2l - 4k - 1)}{(2k + 1)(l - k)} P_{l-k}(z) = \frac{1}{2}\int_{-1}^{1} \frac{[P_l(z) - P_l(x')]}{(z - x')}\, dx' \tag{6.249}$$

It is clear from equation 6.239 that the term $P_l(z)Q_0(z)$ must satisfy the same nonderivative recurrence relation that $P_l(z)$ satisfies. That is, from equation 6.203b,

$$[(l + 1)P_{l+1}(z) - (2l + 1)zP_l(z) + lP_{l-1}(z)]Q_0(z) = 0 \qquad l \geqslant 1 \tag{6.265}$$

Consider

$$(l + 1)R_l(z) - (2l + 1)zR_{l-1}(z) + lR_{l-2}(z) = \frac{1}{2}\Bigg[(l + 1)\int_{-1}^{1} \frac{P_{l+1}(z) - P_{l+1}(x)}{(z - x)}\, dx$$

$$-(2l + 1)\int_{-1}^{1} \frac{zP_l(z) - zP_l(x)}{(z - x)}\, dx + l\int_{-1}^{1} \frac{P_{l-1}(z) - P_{l-1}(x)}{(z - x)}\, dx\Bigg] \tag{6.266}$$

We write

$$\int_{-1}^{1} \frac{zP_l(z) - zP_l(x)}{(z - x)}\, dx = \int_{-1}^{1} \frac{zP_l(z) - (x + z - x)P_l(x)}{(z - x)}\, dx$$

$$= \int_{-1}^{1} \frac{zP_l(z) - xP_l(x)}{(z - x)}\, dx - \int_{-1}^{1} P_l(x)\, dx \tag{6.267}$$

Since $l \geqslant 1$, by the orthogonality of the Legendre polynomials,

$$\int_{-1}^{1} P_l(x)\, dx = \int_{-1}^{1} P_0(x)P_l(x)\, dx = 0 \tag{6.268}$$

Therefore, since $P_l(z)$ and $P_l(x)$ satisfy equation 6.203b, equation 6.266 becomes

$$(l+1)R_l(z) - (2l+1)zR_{l-1}(z) + lR_{l-2}(z)$$
$$= \frac{1}{2}\int_{-1}^{1} \frac{[(l+1)P_{l+1}(z) - (2l+1)zP_l(z) + lP_{l-1}(z)]}{(z-x)}\,dx$$
$$- \frac{1}{2}\int_{-1}^{1} \frac{[(l+1)P_{l+1}(x) - (2l+1)xP_l(x) + lP_{l-1}(x)]}{(z-x)}\,dx = 0 \qquad (6.269)$$

That is, the polynomial $R_{l-1}(z)$ satisfies the same nonderivative recurrence relation as $P_l(z)$. Therefore, $Q_l(z)$ must also satisfy this recurrence relation.

$$(l+1)Q_{l+1}(z) - (2l+1)zQ_l(z) + lQ_{l-1}(z) = 0 \qquad (6.270)$$

The fact that $Q_l(z)$ and $P_l(z)$ share one recurrence relation in common suggests that other recurrence relations might be satisfied by both functions. It was shown, for example, that the Legendre polynomials satisfy

$$P'_{l+2}(z) - 2zP'_{l+1}(z) + P'_l(z) = P_{l+1}(z) \qquad l \geqslant 0 \qquad (6.205)$$

Using the Neumann integral representation (equation 6.247), consider

$$Q'_{l+2}(z) - 2zQ'_{l+1}(z) + Q'_l(z) = -\frac{1}{2}\int_{-1}^{1} \frac{[P_{l+2}(x) - 2zP_{l+1}(x) + P_l(x)]}{(z-x)^2}\,dx \qquad (6.271)$$

We perform integration by parts on each integral, using

$$\int_{-1}^{1} \frac{P_n(x)}{(z-x)^2}\,dx = \left(\frac{1}{(z-1)} - \frac{(-1)^n}{(z+1)}\right) - \int_{-1}^{1} \frac{P'_n(x)}{(z-x)}\,dx \qquad (6.272)$$

$$\Rightarrow \quad Q'_{l+2}(z) - 2zQ'_{l+1}(z) + Q'_l(z)$$
$$= \left(1 + (-1)^l\right) + \frac{1}{2}\int_{-1}^{1} \frac{[P'_{l+2}(x) - 2(z-x+x)P'_{l+1}(x) + P'_l(x)]}{(z-x)}\,dx$$
$$= \left(1 + (-1)^l\right) - \int_{-1}^{1} P'_{l+1}(x)\,dx + \frac{1}{2}\int_{-1}^{1} \frac{[P'_{l+2}(x) - 2xP'_{l+1}(x) + P'_l(x)]}{(z-x)}\,dx \quad (6.273)$$

We note that

$$\int_{-1}^{1} P'_{l+1}(x)\,dx = P_{l+1}(1) - P_{l+1}(-1) = \left(1 + (-1)^l\right) \qquad (6.274a)$$

Using the recurrence relation

$$P'_{l+2}(x) - 2xP'_{l+1}(x) + P'_l(x) = P_{l+1}(x) \qquad l \geqslant 0 \qquad (6.205)$$

and the Neumann representation of $Q_l(z)$, we obtain

$$\frac{1}{2}\int_{-1}^{1} \frac{[P'_{l+2}(x) - 2xP'_{l+1}(x) + P'_l(x)]}{(z-x)}\,dx = \frac{1}{2}\int_{-1}^{1} \frac{P_{l+1}(x)}{(z-x)}\,dx = Q_{l+1}(z) \qquad (6.274b)$$

Therefore, equation 6.273 becomes

$$Q'_{l+2}(z) - 2zQ'_{l+1}(z) + Q'_l(z) = Q_{l+1}(z) \tag{6.275}$$

Thus the second Legendre function satisfies the same derivative recurrence relation satisfied by the Legendre polynomials. The reader is also referred to Problem 18.

Associated Legendre function

In Table 6.2 we have noted that the associated Legendre function satisfies

$$\left[(1-x^2)D^2 - 2xD + \left(l(l+1) - \frac{m^2}{(1-x^2)}\right)\right]y = 0 \tag{6.276}$$

and the solution is denoted by

$$y \equiv P_l^m(x) \tag{6.277}$$

By obtaining this equation from the ordinary Legendre differential equation, we will determine the relation between $P_l^m(x)$ and $P_l(x)$.

Using the Leibnitz rule for the derivative of the product of two functions

$$D^m[F(x)G(x)] = \sum_{k=0}^{m} \frac{m!}{k!(m-k)!}[D^{m-k}F(x)][D^kG(x)] \tag{6.278}$$

we take the mth derivative of the ordinary Legendre differential equation.

$$D^m[(1-x^2)D^2P_l - 2xDP_l + l(l+1)P_l]$$

$$= (1-x^2)D^{m+2}P_l - 2x(m+1)D^{m+1}P_l + [l(l+1) - m(m+1)]D^mP_l = 0 \tag{6.279}$$

Defining

$$W_m \equiv D^mP_l \tag{6.280}$$

$$\Rightarrow \quad (1-x^2)D^2W_m - 2x(m+1)DW_m + [l(l+1) - m(m+1)]W_m = 0 \tag{6.281}$$

The associated Legendre function is defined by

$$W_m(x) \equiv (1-x^2)^{-m/2}P_l^m(x) \tag{6.282}$$

$$\Rightarrow \quad P_l^m(x) = (1-x^2)^{m/2}D^mP_l(x) \tag{6.283}$$

Substituting equation 6.282 into equation 6.281, the differential equation for $P_l^m(x)$ is

$$(1-x^2)D^2P_l^m(x) - 2xDP_l^m(x) + \left(l(l+1) - \frac{m^2}{(1-x^2)}\right)P_l^m(x) = 0 \tag{6.284}$$

Since $P_l(x)$ is a polynomial of order l, it is clear that

$$P_l^m(x) = 0 \qquad m > l \tag{6.285}$$

In addition, if $m = 0$,

$$P_l^0(x) = P_l(x) \tag{6.286}$$

We also see that if m is an even integer, $P_l^m(x)$ is a polynomial of order $l - m/2$, whereas if m is an odd integer, $P_l^m(x)$ is a function with square root branch points at $x = \pm 1$.

Using equation 6.283, we explicitly express some of the lowest-index associated Legendre functions in terms of x and in terms of $x = \cos\theta$. They are

$$P_1^1(x) = (1 - x^2)^{1/2} \qquad P_1^1(\cos\theta) = \sin\theta \tag{6.287a}$$

$$P_2^1(x) = 3x(1 - x^2)^{1/2} \qquad P_2^1(\cos\theta) = 3\cos\theta\sin\theta \tag{6.287b}$$

$$P_2^2(x) = 3(1 - x^2) \qquad P_2^2(\cos\theta) = 3\sin^2\theta \tag{6.287c}$$

$$P_3^1(x) = \tfrac{3}{2}(5x^2 - 1)(1 - x^2)^{1/2} \qquad P_3^1(\cos\theta) = \tfrac{3}{2}(5\cos^2\theta - 1)\sin\theta \tag{6.287d}$$

$$P_3^2(x) = 15x(1 - x^2) \qquad P_3^2(\cos\theta) = 15\cos\theta\sin^2\theta \tag{6.287e}$$

$$P_3^3(x) = 15(1 - x^2)^{3/2} \qquad P_3^3(\cos\theta) = 15\sin^3\theta \tag{6.287f}$$

Series description of the associated Legendre function

From the series description of $P_l(x)$ (equation 6.179), it is straightforward to obtain a series representation of $P_l^m(x)$. With $lm \equiv [(l - m)/2]$, it is

$$P_l^m(x) = (1 - x^2)^{m/2} \sum_{k=0}^{lm} \frac{(-1)^k (2l - 2k)!}{2^l k!\,(l - k)!\,(l - 2k - m)!} x^{l-2k-m} \tag{6.288}$$

Generating function for $P_l^m(x)$

The generating function for $P_l^m(x)$ is obtained in a straightforward way from the generator of $P_l(x)$. Consider

$$D^m(1 - 2xz + z^2)^{-1/2} = \frac{(2m - 1)!!\,z^m}{(1 - 2xz + z^2)^{m+1/2}} = \sum_{l=0}^{\infty} z^l D^m P_l(x) \tag{6.289a}$$

Since $P_l(x)$ is a polynomial of order l, $D^m P_l(x) = 0$ for $m > l$. Thus the sum starts at $l = m$.

We multiply this expression by $(1 - x^2)^{m/2}$ and define the product as $\Psi(z, x)$. Using the expression for $(2m - 1)!!$ given in equation 6.90, we obtain one form of the generator of $P_l^m(x)$.

$$\Psi(z, x) = \frac{(2m)!}{2^m m!} \frac{z^m(1 - x^2)^{m/2}}{(1 - 2xz + z^2)^{m+1/2}} = \sum_{l=m}^{\infty} z^l P_l^m(x) \tag{6.289b}$$

Dividing by z^m and setting $l' = l - m$, this can be written

$$\frac{1}{z^m}\Psi(z, x) \equiv \Phi(z, x) = \frac{(2m)!}{2^m m!} \frac{(1 - x^2)^{m/2}}{(1 - 2xz + z^2)^{m+1/2}} = \sum_{l=0}^{\infty} z^l P_{l+m}^m(x) \tag{6.290}$$

Rodriguez formula for $P_l^m(x)$

The Rodriguez formula for $P_l^m(x)$ is a straightforward extension of the Rodriguez formula for $P_l(x)$ as expressed in equation 6.195. From the definition of $P_l^m(x)$ of equation 6.283,

$$P_l^m(x) = \frac{(-1)^l}{2^l l!}(1 - x^2)^{m/2} D^{l+m}(1 - x^2)^l \tag{6.291}$$

In our discussion up to now, we have taken m to be positive. But because it appears as m^2 in the associated differential equation, that equation is unchanged if m is negative. For negative m, equation 6.283 becomes

$$P_l^{-|m|}(x) = (1 - x^2)^{-|m|/2} D^{-|m|} P_l(x) \tag{6.292}$$

$D^{-|m|}P_l(x)$ has meaning as an $|m|$-fold integral, but $(1 - x^2)^{-|m|/2}\int dx \int dx \cdots \int P_l(x)\, dx$ does not satisfy the differential equation. Instead, $P_l^m(x)$ is defined for negative values of m by the Rodriguez formula.

$$P_l^{-|m|}(x) \equiv \frac{(-1)^l}{2^l l!}(1 - x^2)^{-|m|/2} D^{l-|m|}(1 - x^2)^l \tag{6.293}$$

Integral representations

Since the associated differential equation is unchanged when m is replaced by $-m$, as might be expected, $P_l^m(x)$ and $P_l^{-m}(x)$ are not independent. To deduce the relation between them, we develop the Schlafli and Laplace integral representations for the associated Legendre function. Taking m to be positive, the Schlafli representation is obtained straightforwardly using the same analysis that leads to the Schlafli integral for the Legendre polynomials. Using the Rodriguez formula, the representation

$$(1 - x^2)^l = \frac{1}{2\pi i}\oint \frac{(1 - z^2)^l}{(z - x)}\, dz \tag{6.196}$$

$$\Rightarrow \quad P_l^m(x) = \frac{(-1)^l}{2^l l!}\frac{(l + m)!}{2\pi i}(1 - x^2)^{m/2}\oint \frac{(1 - z^2)^l}{(z - x)^{l+m+1}}\, dz \qquad m \geqslant 0 \tag{6.294}$$

which is the Schlafli integral representation of $P_l^m(x)$.

To obtain the Laplace integral representation, we proceed as we did in developing the Laplace representation for the Legendre polynomial. We set

$$z = x - i\sqrt{1 - x^2}\, e^{i\phi} \tag{6.198a}$$

$$\Rightarrow \quad P_l^m(x) = (i)^m \frac{(l + m)!}{2\pi l!}\int_0^{2\pi}\left(x - i\sqrt{1 - x^2}\cos\phi\right)^l e^{-im\phi}\, d\phi \tag{6.295a}$$

As before, we set $\phi' = \phi - \pi$ to create symmetric limits. The result is

$$P_l^m(x) = (-i)^m \frac{(l + m)!}{2\pi l!}\int_{-\pi}^{\pi}\left(x + i\sqrt{1 - x^2}\cos\phi\right)^l e^{-im\phi}\, d\phi \tag{6.295b}$$

Writing $e^{-im\phi} = \cos(m\phi) - i\sin(m\phi)$, we note that the part of the integrand that multiplies $\cos(m\phi)$ is an even function of ϕ, and the part multiplying $\sin(m\phi)$ is an odd function. If a function $F(\phi)$ is even,

$$\text{(a)} \quad \int_{-L}^{L} F(\phi)\, d\phi = 2\int_0^L F(\phi)\, d\phi \qquad \text{and} \qquad \text{(b)} \quad \int_{-L}^{L} F(\phi)\, d\phi = 0 \tag{6.296}$$

if $F(\phi)$ is odd. Therefore, equation 6.295b can be written

$$P_l^m(x) = (-i)^m \frac{(l + m)}{2\pi l!}\int_{-\pi}^{\pi}\left(x + i\sqrt{1 - x^2}\cos\phi\right)^l \cos(m\phi)\, d\phi$$

$$= (-i)^m \frac{(l + m)!}{\pi l!}\int_0^{\pi}\left(x + i\sqrt{1 - x^2}\cos\phi\right)^l \cos(m\phi)\, d\phi \tag{6.297a}$$

This is the Laplace integral representation of $P_l^m(x)$.

Replacing m by $-m$ in equation 6.297a, we obtain

$$P_l^{-m}(x) = (-i)^{-m}\frac{(l-m)!}{\pi l!}\int_0^{\pi}\left(x + i\sqrt{1-x^2}\cos\phi\right)^l \cos(m\phi)\,d\phi \tag{6.297b}$$

Comparing this to equation 6.297a, we obtain the relation between $P_l^m(x)$ and $P_l^{-m}(x)$. It is

$$P_l^m(x) = (-1)^m\frac{(l+m)!}{(l-m)!}P_l^{-m}(x) \tag{6.298}$$

Like the ordinary Legendre differential equation, the associated differential equation is invariant in the interchange of l and $-l-1$. Thus, as with the Legendre polynomials,

$$P_{-l-1}^m(x) = P_l^m(x) \tag{6.299}$$

Replacing l by $-l-1$ in equation 6.297a yields

$$\begin{aligned} P_l^m(x) = P_{-l-1}^m(x) &= (-i)^m\frac{(-l-1+m)!}{2\pi(-l-1)!}\int_{-\pi}^{\pi}\frac{\cos(m\phi)}{\left(x+i\sqrt{1-x^2}\cos\phi\right)^{l+1}}\,d\phi \\ &= (-i)^m\frac{(-l-1+m)!}{\pi(-l-1)!}\int_0^{\pi}\frac{\cos(m\phi)}{\left(x+i\sqrt{1-x^2}\cos\phi\right)^{l+1}}\,d\phi \end{aligned} \tag{6.300}$$

Using

$$\Gamma(p)\Gamma(1-p) = \frac{\pi}{\sin \pi p} = -p\Gamma(p)\Gamma(-p) = -\Gamma(p+1)\Gamma(-p) \tag{6.301a}$$

$$\begin{aligned} \Rightarrow\quad \frac{(-l-1+m)!}{(-l-1)!} = \frac{\Gamma[-(l-m)]}{\Gamma(-l)} &= \frac{l!}{(l-m)!}\frac{\sin(\pi l)}{\sin[\pi(l-m)]} \\ &= \frac{l!}{(l-m)!}\frac{\sin(\pi l)}{\sin(\pi l)\cos(\pi m)} = (-1)^m\frac{l!}{(l-m)!} \end{aligned} \tag{6.301b}$$

Therefore, another representation of the associated Legendre function is

$$\begin{aligned} P_l^m(x) = P_{-l-1}^m(x) &= (i)^m\frac{l!}{2\pi(l-m)!}\int_{-\pi}^{\pi}\frac{\cos(m\phi)}{\left(x+i\sqrt{1-x^2}\cos\phi\right)^{l+1}}\,d\phi \\ &= (i)^m\frac{l!}{\pi(l-m)!}\int_0^{\pi}\frac{\cos(m\phi)}{\left(x+i\sqrt{1-x^2}\cos\phi\right)^{l+1}}\,d\phi \end{aligned} \tag{6.302}$$

Orthonormalization condition

As noted in Table 6.2, $P_l^m(x)$ is a Sturm–Liouville eigenfunction with eigenvalue $l(l+1)$. Since l defines the eigenvalue, from Sturm–Liouville analysis, we are assured that the set of functions $\{P_l^m(x)\}$ are mutually orthogonal in the l-index. That is,

$$\int_{-1}^{1} P_l^m(x)P_n^m(x)\,dx = N_l^m\delta_{ln} \tag{6.303a}$$

Using an analogous argument, since m does not define the eigenvalue, Sturm–Liouville analysis does not require orthogonality in the upper index. That is,

$$\int_{-1}^{1} P_l^m(x)P_l^n(x)\,dx \neq C_l^m\delta_{ln} \tag{6.303b}$$

We deduce the normalization constant of equation 6.303a, by replacing one of the associated Legendre functions for positive m with its negative m relative using equation 6.298.

$$N_l^m = \int_{-1}^{1} [P_l^m(x)]^2 \, dx = (-1)^m \frac{(l+m)!}{(l-m)!} \int_{-1}^{1} P_l^m(x) P_l^{-m}(x) \, dx \tag{6.304a}$$

Using the Rodriguez representation of $P_l^m(x)$, this becomes

$$N_l^m = (-1)^m \frac{(l+m)!}{(l-m)!} \frac{1}{2^{2l}(l!)^2} \int_{-1}^{1} \left[D^{l-m}(1-x^2)^l\right]\left[D^{l+m}(1-x^2)^l\right] dx \tag{6.304b}$$

Integration by parts m times reduces the number of derivatives from $[D^{l+m}(1-x^2)^l]$ to $[D^l(1-x^2)^l]$ and increases the number of derivatives of $[D^{l-m}(1-x^2)^l]$ to $[D^l(1-x^2)^l]$. The integrated term at each step will be zero since each integrated term will contain at least one factor of $(1-x^2)$ which is zero at ± 1. The result is

$$N_l^m = \frac{(l+m)!}{(l-m)!} \frac{1}{2^{2l}(l!)^2} \int_{-1}^{1} \left[D^l(1-x^2)^l\right]\left[D^l(1-x^2)^l\right] dx \tag{6.305a}$$

Referring to equation 6.195, we see that with the constant $\dfrac{1}{2^{2l}(l!)^2}$, this is the normalization integral for $P_l(x)$. Thus, with equation 6.213,

$$N_l^m = \frac{(l+m)!}{(l-m)!} \int_{-1}^{1} [P_l(x)]^2 \, dx = \frac{2}{2l+1} \frac{(l+m)!}{(l-m)!} \tag{6.305b}$$

$$\Rightarrow \quad \int_{-1}^{1} P_l^m(x) P_n^m(x) \, dx = \frac{2}{2l+1} \frac{(l+m)!}{(l-m)!} \delta_{ln} \tag{6.306}$$

An orthogonality relation for $P_l^m(x)$ in the upper index does exist. It is

$$\int_{-1}^{1} \frac{P_l^m(x) P_l^n(x)}{(1-x^2)} \, dx = \frac{1}{m} \frac{(l+m)!}{(l-m)!} \delta_{mn} \qquad m, n \neq 0 \tag{6.307}$$

However, it has little use in applied investigations, and so will not be discussed further.

With the orthonormalization condition of equation 6.306, it is straightforward to deduce the expansion of an arbitrary function in terms of the associated Legendre functions.

$$F_m(x) = \sum_{l=0}^{\infty} f_l^m P_l^m(x) \tag{6.308a}$$

with

$$f_l^m = \frac{(2l+1)}{2} \frac{(l+m)!}{(l-m)!} \int_{-1}^{1} F_m(x) P_l^m(x) \, dx \tag{6.308b}$$

Recurrence relations

As with the Legendre polynomials, recurrence relations can be developed for the associated Legendre functions, but since two indices define the associated functions, many more recurrence relations exist for $P_l^m(x)$ than do for $P_l(x)$. As with the Legendre polynomials, some of the recurrence relations for the associated function are developed from the generating function. Others are obtained from the recurrence relations for the Legendre polynomials, and still others are found from manipulating previously developed recurrence relations of $P_l^m(x)$.

For example, differentiation of the generator of $P_l^m(x)$ with respect to z leads to

$$(l - m + 2)P_{l+2}^m(x) - (2l + 3)xP_{l+1}^m(x) + (l + m + 1)P_l^m(x) = 0 \qquad l \geqslant 0 \tag{6.309}$$

Differentiating the generating function with respect to x yields

$$\left[[P_{l+2}^m(x)]' + \frac{mx}{(1 - x^2)}P_{l+2}^m(x)\right] - 2x\left[[P_{l+1}^m(x)]' + \frac{mx}{(1 - x^2)}P_{l+1}^m(x)\right] + \left[[P_l^m(x)]' + \frac{mx}{(1 - x^2)}P_l^m(x)\right] = (2m + 1)P_{l+1}^m(x) \qquad l \geqslant 0 \tag{6.310}$$

Another recurrence relation is obtained by writing

$$[P_l^m(x)]' = D\left[(1 - x^2)^{m/2}D^m P_l(x)\right] = \frac{1}{(1 - x^2)^{1/2}}P_l^{m+1}(x) - \frac{mx}{(1 - x^2)}P_l^m(x) \tag{6.311}$$

With this result, the recurrence relation of equation 6.310 becomes

$$P_{l+2}^{m+1}(x) - 2xP_{l+1}^{m+1}(x) + P_l^{m+1}(x) = (2m + 1)(1 - x^2)^{1/2}P_{l+1}^m(x) \tag{6.312}$$

To obtain another recurrence relation, we replace m by $m + 1$ in equation 6.309 to obtain

$$xP_{l+1}^{m+1}(x) = \frac{(l - m + 1)P_{l+2}^{m+1}(x) + (l + m + 2)P_l^{m+1}(x)}{(2l + 3)} \tag{6.313}$$

Substituting this into equation 6.312 yields

$$P_{l+2}^{m+1}(x) - (2l + 3)(1 - x^2)^{1/2}P_{l+1}^m(x) - P_l^{m+1}(x) = 0 \tag{6.314}$$

This set of recurrence relations for the associated Legendre function is by no means complete. Similar manipulations yield other such relations. For other recurrence formulas, the reader is referred to sources such as the texts by N. N. Lebedev, *Special Functions and Their Applications* (Englewood Cliffs, N.J.: Prentice-Hall, Inc., 1966), by G. Arfken, *Mathematical Methods of Physics* (New York: Academic Press, 1985), and by L. Andrews, *Special Functions for Engineers and Applied Mathematicians* (New York: Macmillan Publishing Company, 1985).

Addition theorem

Consider two position vectors $\mathbf{r}_1$ and $\mathbf{r}_2$ described in spherical coordinates (Figure 6.8). As shown in Chapter 1 (see equation 1.73a),

$$\mathbf{r}_1 = r_1(\mathbf{i}\sin\theta_1\cos\phi_1 + \mathbf{j}\sin\theta_1\sin\phi_1 + \mathbf{k}\cos\theta_1) \tag{6.315a}$$

$$\mathbf{r}_2 = r_2(\mathbf{i}\sin\theta_2\cos\phi_2 + \mathbf{j}\sin\theta_2\sin\phi_2 + \mathbf{k}\cos\phi_2) \tag{6.315b}$$

The angle between $\mathbf{r}_1$ and $\mathbf{r}_2$ is easily found from

$$\cos\Theta = \frac{\mathbf{r}_1 \cdot \mathbf{r}_2}{r_1 r_2} = \cos\theta_1\cos\theta_2 + \sin\theta_1\sin\theta_2\cos(\phi_1 - \phi_2) \tag{6.316}$$

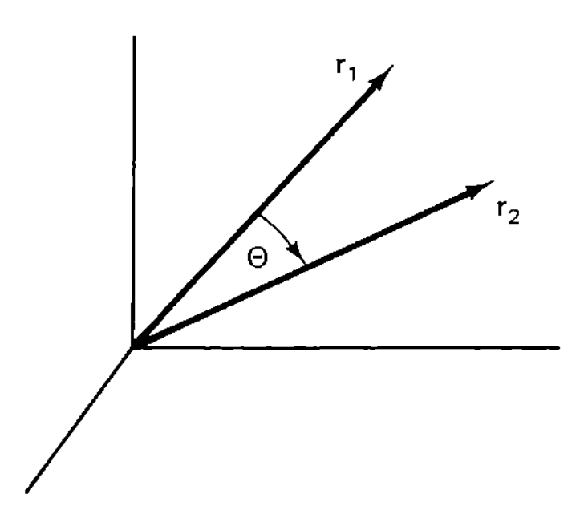

Figure 6.8 Definition of the angle between two vectors.

Expressing these functions in terms of the Legendre functions, this can be written

$$P_1(\cos\Theta) = P_1(\cos\theta_1)P_1(\cos\theta_2) + P_1^1(\cos\theta_1)P_1^1(\cos\theta_2)\cos(\phi_1 - \phi_2) \tag{6.317}$$

This is an example (for $l = 1$) of the *addition theorem for Legendre polynomials* in terms of the angles specifying two position vectors.

To derive the addition theorem for any l, we begin with the geometric series for

$$\frac{1}{M - zN} = \sum_{l=0}^{\infty} z^l \frac{N^l}{M^{l+1}} \tag{6.318}$$

We now let

$$\text{(a)} \quad M = \alpha_1 + \beta_1 \cos\psi \qquad \text{(b)} \quad N = \alpha_2 + \beta_2 \cos(\phi - \psi) \tag{6.319}$$

Integration of equation 6.318 yields

$$\int_{-\pi}^{\pi} \frac{d\psi}{[(\alpha_1 + \beta_1\cos\psi) - z(\alpha_2 + \beta_2\cos\phi\cos\psi + \beta_2\sin\phi\sin\psi)]}$$

$$= \sum_{l=0}^{\infty} z^l \int_{-\pi}^{\pi} \frac{[\alpha_2 + \beta_2\cos(\phi - \psi)]^l}{(\alpha_1 + \beta_1\cos\psi)^{l+1}}\, d\psi \tag{6.320}$$

We set $z' = e^{i\psi}$, and use Cauchy's theorem to obtain

$$\int_{-\pi}^{\pi} \frac{d\psi}{(A + B\cos\psi + C\sin\psi)} = \oint \frac{dz'}{iz'[A + B(z' + z'^{-1})/2 + C(z' - z'^{-1})/2i]}$$

$$= \frac{2\pi}{(A^2 - B^2 - C^2)^{1/2}} \tag{6.321}$$

Therefore, equation 6.320 becomes

$$\sum_{l=0}^{\infty} z^l \int_{-\pi}^{\pi} \frac{[\alpha_2 + \beta_2\cos(\phi - \psi)]^l}{(\alpha_1 + \beta_1\cos\psi)^{l+1}}\, d\psi$$

$$= \frac{2\pi}{[(\alpha_1^2 - \beta_1^2) - 2z(\alpha_1\alpha_2 - \beta_1\beta_2\cos\phi) + z^2(\alpha_2^2 - \beta_2^2)]^{1/2}} \tag{6.322}$$

Setting

$$\text{(a)}\quad \alpha_1 = \cos\theta_1 \qquad \text{(b)}\quad \alpha_2 = \cos\theta_2 \quad \text{(c)}\quad \beta_1 = i\sin\theta_1 \quad \text{(d)}\quad \beta_2 = i\sin\theta_2 \quad (6.323)$$

$$\Rightarrow \quad \sum_{l=0}^{\infty} z^l \int_{-\pi}^{\pi} \frac{[\cos\theta_2 + i\sin\theta_2\cos(\phi-\psi)]^l}{(\cos\theta_1 + i\sin\theta_1\cos\psi)^{l+1}}\,d\psi$$

$$= \frac{2\pi}{[1 - 2z(\cos\theta_1\cos\theta_2 + \sin\theta_1\sin\theta_2\cos\phi) + z^2]^{1/2}} \qquad (6.324a)$$

Referring to Figure 6.8, if $\phi = \phi_1 - \phi_2$ is the azimuthal angle between two position vectors, then from equation 6.316 and the generating function for the Legendre polynomials, equation 6.324a can be written

$$\sum_{l=0}^{\infty} z^l \int_{-\pi}^{\pi} \frac{[\cos\theta_2 + i\sin\theta_2\cos(\phi-\psi)]^l}{(\cos\theta_1 + i\sin\theta_1\cos\psi)^{l+1}}\,d\psi = \frac{2\pi}{(1 - 2z\cos\Theta + z^2)^{1/2}}$$

$$= 2\pi \sum_{l=0}^{\infty} z^l P_l(\cos\Theta) \qquad (6.324b)$$

Therefore,

$$P_l(\cos\Theta) = \frac{1}{2\pi}\int_{-\pi}^{\pi} \frac{[\cos\theta_2 + i\sin\theta_2\cos(\phi-\psi)]^l}{(\cos\theta_1 + i\sin\theta_1\cos\psi)^{l+1}}\,d\psi \qquad (6.325)$$

Since $\cos\Theta$ is an even periodic function of ϕ, with a period of 2π, as indicated in equation 6.316, $P_l(\cos\Theta)$ is also an even periodic function of ϕ, with the same period. Therefore, $P_l(\cos\Theta)$ has a Fourier cosine series in ϕ. Because $P_l(\cos\Theta)$ is a finite polynomial, its Fourier expansion will be a finite sum. That is, we can express $P_l(\cos\Theta)$ as

$$P_l(\cos\Theta) = \sum_{m=0}^{l} A_m \cos(m\phi) \qquad (6.326)$$

As developed in equations 4.16, with ϕ_0 arbitrary,

$$A_m = \frac{\varepsilon_m}{\pi}\int_{\phi_0}^{\phi_0+2\pi} P_l(\cos\Theta)\cos(m\phi)\,d\phi \qquad (6.327)$$

where

$$\varepsilon_m = \begin{cases} \frac{1}{2} & m = 0 \\ 1 & m > 0 \end{cases} \qquad (6.328)$$

Using the representation of $P_l(\cos\Theta)$ expressed in equation 6.325,

$$A_m = \frac{\varepsilon_m}{2\pi^2}\int_{-\pi}^{\pi} d\psi \int_{\phi_0}^{\phi_0+2\pi} \frac{[\cos\theta_2 + i\sin\theta_2\cos(\phi-\psi)]^l}{(\cos\theta_1 + i\sin\theta_1\cos\psi)^{l+1}}\cos(m\phi)\,d\phi \qquad (6.329)$$

To separate the integration over ϕ from the integration over ψ, we make the substitution $\phi' = \phi - \psi$, and define ϕ_0 such that $\phi_0 - \psi = -\pi$. Then

$A_m =$

$$\frac{\varepsilon_m}{2\pi_2}\int_{-\pi}^{\pi} d\psi \int_{-\pi}^{\pi} \frac{(\cos\theta_2 + i\sin\theta_2\cos\phi')^l}{(\cos\theta_1 + i\sin\theta_1\cos\psi)^{l+1}}\{\cos(m\phi')\cos(m\psi) - \sin(m\phi')\sin(m\psi)\}\,d\phi$$

$$(6.330)$$

The integrands containing $\cos(m\phi')$ and $\cos(m\psi)$ are even in their integration variables, and are therefore nonzero. The integrands containing $\sin(m\phi')$ and $\sin(m\psi)$ are odd. Thus these functions, integrated over symmetric limits, are zero. Therefore, the Fourier coefficients become

$$A_m =$$

$$\frac{\varepsilon_m}{2\pi^2}\left[\int_{-\pi}^{\pi}\frac{\cos(m\psi)}{(\cos\theta_1 + i\sin\theta_1\cos\psi)^{l+1}}\,d\psi\right]\left[\int_{-\pi}^{\pi}(\cos\theta_2 + i\sin\theta_2\cos\phi')^l\cos(m\phi')\,d\phi'\right] \tag{6.331}$$

From equations 6.297 and 6.302, we see that each of these integrals (up to a multiplicative factor) is the Laplace representation of an associated Legendre function. Therefore, the Fourier coefficient is

$$A_m = 2\varepsilon_m\frac{(l-m)!}{(l+m)!}P_l^m(\cos\theta_1)P_l^m(\cos\theta_2) \tag{6.332}$$

and the Fourier sum of equation 6.326 is

$$P_l(\cos\Theta) = 2\sum_{m=0}^{l}\varepsilon_m\frac{(l-m)!}{(l+m)!}P_l^m(\cos\theta_1)P_l^m(\cos\theta_2)\cos[m(\phi_1-\phi_2)] \tag{6.333}$$

Because ε_0 has a different value than all other ε_m, we separate the $m = 0$ term from the rest of the sum. Thus, the addition theorem for the Legendre polynomials is

$$P_l(\cos\Theta) = P_l(\cos\theta_1)P_l(\cos\theta_2) + 2\sum_{m=1}^{l}\frac{(l-m)!}{(l+m)!}P_l^m(\cos\theta_1)P_l^m(\cos\theta_2)\cos[m(\phi_1-\phi_2)] \tag{6.334}$$

Associated Legendre function of the second kind

We applied the method of reduction of order to the ordinary Legendre equation to obtain the Legendre functions of the second kind, $Q_l(x)$. A second solution to the associated Legendre differential equation, obtained by the same method, is denoted by $Q_l^m(x)$. It is defined from $Q_l(x)$ in a way very much like the definition of $P_l^m(x)$ in terms of $P_l(x)$. The definition is

$$Q_l^m(x) = (-1)^l(1-x^2)^{m/2}D^mQ_l(x) \tag{6.335}$$

Note that this definition contains a factor $(-1)^l$ not used to define $P_l^m(x)$.

Many of the properties of $Q_m^l(x)$ are closely related to the properties of the other Legendre functions. For example, the generating function of $Q_l^m(x)$ is obtained straightforwardly from the generator of $Q_l(x)$ using equation 6.335. Since $Q_l(x)$ satisfies the same recurrence relations as $P_l(x)$, one can easily deduce the fact that $Q_l^m(x)$ and $P_l^m(x)$ satisfy the same recurrence relations. In general, needed properties of $Q_l^m(x)$ can be developed using the techniques outlined here for the other Legendre functions.

From equation 6.335 it is clear that $Q_l^m(x)$ has the same logarithmic singularity structure as $Q_l(x)$. Because functions that represent physical quantities generally must be finite at points that are accessible physically (e.g., $x = \cos\theta = \pm 1$), such solutions almost always exclude $Q_l^m(x)$. For that reason its properties will not be developed here.

Spherical harmonics

A function, expressed in spherical coordinates, is said to be spherically symmetric if it is independent of θ and ϕ. For example, the potential due to a point charge or point mass has the form

$$V(\mathbf{r}) \equiv V(r, \theta, \phi) = V(r) = \frac{\text{constant}}{r} \tag{6.336}$$

This potential function is independent of the angular orientation of the field point relative to the source and so is spherically symmetric.

Many phenomena are described by partial differential equations in which functions governing the solution are spherically symmetric. Examples of this include Poisson's equation for the electrostatic potential in which the charge density does not depend on θ and ϕ, and Schrödinger's equation for a quantum system in which the potential energy function is spherically symmetric.

As discussed in Chapter 11, the solution to such partial differential equations for systems with spherical symmetry contain the associated Legendre function $P_l^m(\cos\theta)$ in combination with the exponential $e^{im\phi}$. The product of these functions, multiplied by a normalization constant, is called the *spherical harmonic* and is denoted by

$$Y_l^m(\theta, \phi) \equiv Y_l^m(\Omega) \equiv C_l^m P_l^m(\cos\theta) e^{im\phi} \tag{6.337}$$

The normalization constant is chosen such that

$$\int Y_l^{m*}(\Omega) Y_{l'}^{m'}(\Omega)\, d\Omega = C_l^{m*} C_{l'}^{m'} \int_0^{\pi} \sin\theta\, d\theta P_l^m(\cos\theta) P_{l'}^{m'}(\cos\theta) \int_0^{2\pi} e^{i(m'-m)\phi}\, d\phi$$
$$= \delta_{ll'}\delta_{mm'} \tag{6.338}$$

Orthogonality in the m index is obtained from the fact that

$$\int_0^{2\pi} e^{i(m'-m)\phi}\, d\phi = 2\pi\delta_{mm'} \tag{6.339}$$

Therefore, with $x = \cos\theta$, the orthogonormalization condition becomes

$$\int Y_l^{m*}(\Omega) Y_{l'}^{m'}(\Omega)\, d\Omega = 2\pi\delta_{mm'} C_l^{m*} C_{l'}^{m'} \int_{-1}^{1} dx\, P_l^m(x) P_{l'}^{m'}(x) \tag{6.340}$$

Using the orthonormalization condition for the associated Legendre functions (equation 6.306), this becomes

$$\int Y_l^{m*}(\Omega) Y_{l'}^{m'}(\Omega)\, d\Omega = \frac{4\pi}{2l+1} \frac{(l+m)!}{(l-m)!} |C_l^m|^2 \delta_{ll'}\delta_{mm'} \tag{6.341}$$

$$\Rightarrow \quad C_l^m = e^{i\alpha}\left[\frac{2l+1}{4\pi}\frac{(l-m)!}{(l+m)!}\right]^{1/2} \tag{6.342}$$

One commonly made choice for the undeterminable phase factor is $e^{i\alpha} = (-1)^m$. Then

$$Y_l^m(\theta, \phi) = (-1)^m \left[\frac{2l+1}{4\pi}\frac{(l-m)!}{(l+m)!}\right]^{1/2} P_l^m(\cos\theta) e^{im\phi} \tag{6.343}$$

With m positive, consider

$$Y_l^{-m}(\Omega) = (-1)^m\left[\frac{2l+1}{4\pi}\frac{(l+m)!}{(l-m)!}\right]^{1/2} P_l^{-m}(\cos\theta)e^{-im\phi} \tag{6.344}$$

Using

$$\therefore \quad P_l^m(\cos\theta) = (-1)^m\frac{(l+m)!}{(l-m)!}P_l^{-m}(\cos\theta) \tag{6.298}$$

$$\Rightarrow \quad Y_l^{-m}(\Omega) = \left[\frac{2l+1}{4\pi}\frac{(l-m)!}{(l+m)!}\right]^{1/2} P_l^m(\cos\theta)e^{-im\phi} = (-1)^m Y_l^{m*}(\Omega) \tag{6.345}$$

The addition theorem for the Legendre polynomials can be written in terms of the spherical harmonics. Starting with

$$P_l(\cos\Theta) = P_l(\cos\theta_1)P_l(\cos\theta_2) + 2\sum_{m=0}^{l}\frac{(l-m)!}{(l+m)!}P_l^m(\cos\theta_1)P_l^m(\cos\theta_2)\cos[m(\phi_1-\phi_2)] \tag{6.334}$$

we consider

$$\sum_{m=-l}^{l}\frac{(l-m)!}{(l+m)!}P_l^m(\cos\theta_1)P_l^m(\cos\theta_2)e^{im(\phi_1-\phi_2)}$$

$$= \sum_{m=-l}^{-1}\frac{(l-m)!}{(l+m)!}P_l^m(\cos\theta_1)P_l^m(\cos\theta_2)e^{im(\phi_1-\phi_2)} + P_l(\cos\theta_1)P_l(\cos\theta_2) + \sum_{m=1}^{l}\frac{(l-m)!}{(l+m)!}P_l^m(\cos\theta_1)P_l^m(\cos\theta_2)e^{im(\phi_1-\phi_2)} \tag{6.346}$$

In the first sum, we replace m by $-m$, and use equation 6.298 to obtain

$$\sum_{m=-l}^{l}\frac{(l-m)!}{(l+m)!}P_l^m(\cos\theta_1)P_l^m(\cos\theta_2)e^{im(\phi_1-\phi_2)}$$

$$= P_l(\cos\theta_1)P_l(\cos\theta_2) + \sum_{m=1}^{l}\left[\frac{(l-m)!}{(l+m)!}P_l^m(\cos\theta_1)P_l^m(\cos\theta_2)e^{im(\phi_1-\phi_2)} + \frac{(l+m)!}{(l-m)!}P_l^{-m}(\cos\theta_1)P_l^{-m}(\cos\theta_2)e^{-im(\phi_1-\phi_2)}\right]$$

$$= P_l(\cos\theta_1)P_l(\cos\theta_2) + \sum_{m=1}^{l}\frac{(l-m)!}{(l+m)!}P_l^m(\cos\theta_1)P_l^m(\cos\theta_2)[e^{im(\phi_1-\phi_2)} + e^{-im(\phi_1-\phi_2)}]$$

$$= P_l(\cos\theta_1)P_l(\cos\theta_2) + 2\sum_{m=1}^{l}\frac{(l-m)!}{(l+m)!}P_l^m(\cos\theta_1)P_l^m(\cos\theta_2)\cos[m(\phi_1-\phi_2)] \tag{6.347}$$

Referring to equation 6.334, we see that this last expression is $P_l(\cos\Theta)$. Therefore,

$$P_l(\cos\Theta) = \sum_{m=-l}^{l} \frac{(l-m)!}{(l+m)!}\left[P_l^m(\cos\theta_1)e^{im\phi_1}\right]\left[P_l^m(\cos\theta_2)e^{-im\phi_2}\right] \tag{6.348}$$

or

$$P_l(\cos\Theta) = \frac{4\pi}{2l+1}\sum_{m=-l}^{l} Y_l^m(\Omega_1)Y_l^{-m}(\Omega_2) = \frac{4\pi}{2l+1}\sum_{m=-l}^{l}(-1)^m Y_l^m(\Omega_1)Y_l^{m*}(\Omega_2) \tag{6.349}$$

6.4 Gegenbauer Functions

The *Gegenbauer functions* satisfy the Sturm–Liouville differential equation

$$\left[(1-x^2)D^2 - x(1+2\lambda)D + l(l+2\lambda)\right]y = 0 \tag{6.350}$$

and the solution is denoted by $G_l^\lambda(x)$. Referring to equation 6.164, we note the similarity between the Gegenbauer and ordinary Legendre differential equations. In fact, for $\lambda = \frac{1}{2}$, the two are identical. For this reason, the Gegenbauer functions are also referred to as the *generalized Legendre functions.*

As with the Legendre solution, when l is an integer, one of the Gegenbauer series becomes a polynomial of order l. These Gegenbauer polynomials are also called the *ultraspherical polynomials.*

Because of their somewhat limited use in scientific work, and because the methods of derivation have been presented in our detailed discussion of Legendre functions, we will simply present various properties of the Gegenbauer polynomials, with little discussion about the derivation of these properties. As such, this section will essentially be a summary of the properties of Gegenbauer polynomials. The reader interested in a more extensive discussion of the properties of Gegenbauer polynomials is referred to the text by W. Magnus, F. Oberhettinger, and R. Soni, *Formulas and Theorems for the Special Functions of Mathematical Physics*, 3rd ed. (New York: Springer-Verlag, 1966).

Series representation

With $x \in [-1, 1]$, the solution in series of the Gegenbauer differential equation leads to

$$G_l^\lambda(x) = \frac{1}{\Gamma(\lambda)}\sum_{m=0}^{[l/2]}(-1)^m\frac{\Gamma(l+\lambda-m)}{m!\,(l-2m)!}(2x)^{l-2m} \qquad \lambda > 0 \tag{6.351a}$$

$$G_l^0(x) = \sum_{m=0}^{[l/2]}(-1)^m\frac{(l-m-1)!}{m!\,(l-2m)!}(2x)^{l-2m} \qquad l > 0 \tag{6.351b}$$

$$G_0^0(x) = 1 \tag{6.351c}$$

$$\Rightarrow \quad G_l^\lambda(-x) = (-1)^l G_l^\lambda(x) \tag{6.352}$$

Using the series representation, it is also straightforward to deduce that

$$\left.\begin{aligned}&\text{(a)}\ \ G_0^\lambda(x) = 1 \qquad \text{(b)}\ \ G_1^\lambda(x) = 2\lambda x \qquad \text{(c)}\ \ G_2^\lambda(x) = 2\lambda(\lambda+1)x^2 - \lambda\\ &\text{(d)}\ \ G_3^\lambda(x) = \tfrac{4}{3}\lambda(\lambda+1)(\lambda+2)x^3 - 2\lambda(\lambda+1)x\end{aligned}\right\}\lambda > 0$$

$$\text{(e)}\ \ G_1^0(x) = 2x \qquad \text{(f)}\ \ G_2^0(x) = 2x^2 - 1 \qquad \text{(g)}\ \ G_3^0(x) = \tfrac{8}{3}x^3 - 2x \tag{6.353}$$

By algebraic manipulation, one can deduce that

$$\text{(a)}\ \ G_l^0(x) = \frac{1}{l}\left[\left(x + i\sqrt{1-x^2}\right)^l + \left(x - i\sqrt{1-x^2}\right)^l\right] \quad \Rightarrow \quad \text{(b)}\ \ G_l^0(1) = \frac{2}{l} \tag{6.354}$$

Generating function

The generator of the Gegenbauer polynomials has a form similar to that for the Legendre polynomials.

$$\Phi(z, x) \equiv (1 - 2xz + z^2)^{-\lambda} = \sum_{l=0}^{\infty} z^l G_l^\lambda(x) \qquad \lambda > 0 \qquad (6.355a)$$

and

$$\Phi(z, x) = \frac{(x - z)}{(1 - 2xz + z^2)} = \frac{1}{2} \sum_{l=0}^{\infty} (l + 1) z^l G_{l+1}^0(x) \qquad \lambda = 0 \qquad (6.355b)$$

The reader is referred to Problem 27 for two equivalent forms of equation 6.355b.

Recurrence relations

By differentiating the generating function of equation 6.355a with respect to z, one can show that for $\lambda > 0$,

$$(l + 2)G_{l+2}^\lambda(x) - 2x(l + \lambda + 1)G_{l+1}^\lambda(x) + (l + 2\lambda)G_l^\lambda(x) = 0 \qquad (6.356a)$$

Differentiation of the generator with respect to x yields

$$[G_{l+2}^\lambda(x)]' - 2x[G_{l+1}^\lambda(x)]' + [G_l^\lambda(x)]' = 2\lambda G_{l+1}^\lambda(x) \qquad (6.356b)$$

From equation 6.351a it is straightforward to see that

$$[G_l^\lambda(x)]' = \frac{d}{dx} G_l^\lambda(x) = 2\lambda G_{l-1}^{\lambda+1}(x) \qquad (6.357)$$

Therefore, equation 6.356b can also be written

$$G_{l+2}^{\lambda+1}(x) - 2xG_{l+1}^{\lambda+1}(x) + G_l^{\lambda+1}(x) = G_{l+2}^\lambda(x) \qquad (6.358a)$$

With further manipulation, we can also derive

$$(l + 2\lambda + 1)xG_{l+1}^\lambda(x) - (l + 2)G_{l+2}^\lambda(x) = 2\lambda(1 - x^2)G_l^{\lambda+1}(x) \qquad (6.358b)$$

$$(l + 2)G_{l+2}^\lambda(x) = 2\lambda[xG_{l+1}^\lambda(x) - G_l^{\lambda+1}(x)] \qquad (6.358c)$$

$$G_{l+1}^{\lambda+1}(x) = (l + 2\lambda + 1 + x)G_l^{\lambda+1}(x) \qquad (6.358d)$$

As the reader will show in Problem 27, differentiation of the generating function for $\lambda = 0$ with respect to z yields

$$(l + 2)G_{l+2}^0(x) - 2x(l + 1)G_{l+1}^0(x) + lG_l^0(x) = 0 \qquad l \geqslant 0 \qquad (6.359a)$$

and differentiation of the generator with respect to x leads to

$$(l + 2)[G_{l+2}^0(x)]' - 2x(l + 1)[G_{l+1}^0(x)]' + l[G_l^0(x)]' = 2(l + 1)G_{l+1}^0(x) \qquad (6.359b)$$

We note that differentiation of equation 6.359a yields equation 6.359b.

Rodriguez formula

The *Rodriguez formulas* for the Gegenbauer polynomials can be developed from the generating functions. They are:

$$G_l^\lambda(x) = (-1)^l \frac{\Gamma(\lambda + \frac{1}{2})\Gamma(l + 2\lambda)}{2^l l!\,\Gamma(2\lambda)\Gamma(l + \lambda + \frac{1}{2})}(1 - x^2)^{(1/2)-\lambda} D^l\left[(1 - x^2)^{l+\lambda-1/2}\right] \qquad \lambda > 0 \tag{6.360a}$$

$$G_l^0(x) = (-1)^l \frac{l\Gamma(\frac{1}{2})}{2^{l+2}\Gamma(l + \frac{1}{2})}(1 - x^2)^{1/2} D^l\left[(1 - x^2)^{l-1/2}\right] \qquad \lambda = 0 \tag{6.360b}$$

Orthonormalization condition

It is straightforward to deduce that the self-adjoint form of the Gegenbauer differential equation can be obtained by multiplying equation 6.350 by $(1 - x^2)^{\lambda - 1/2}$ to obtain

$$\left[(1 - x^2)^{\lambda+1/2} D^2 - x(1 + 2\lambda)(1 - x^2)^{\lambda-1/2} D\right] y = -l(l + 2\lambda)(1 - x^2)^{\lambda-1/2} y \tag{6.361}$$

From this, we see that the Gegenbauer weight function is

$$\rho(x) = (1 - x^2)^{\lambda - 1/2} \tag{6.362}$$

Therefore, to deduce the orthonormalization condition, we must evaluate

$$I \equiv \int_{-1}^{1} (1 - x^2)^{\lambda-1/2} [\Phi(z, x)]^2 \, dx \tag{6.363}$$

where $\Phi(z, x)$ is the Gegenbauer generating function. From this it can be shown that the orthonormalization conditions for the Gegenbauer polynomials are

$$\int_{-1}^{1} (1 - x^2)^{\lambda-1/2} G_l^\lambda(x) G_n^\lambda(x) \, dx = \frac{\pi 2^{1-2\lambda}\Gamma(l + 2\lambda)}{l!\,(l + \lambda)[\Gamma(\lambda)]^2}\delta_{ln} \qquad \lambda \neq 0 \tag{6.364a}$$

$$\int_{-1}^{1} (1 - x^2)^{-1/2} G_0^0(x) G_n^0(x) \, dx = \pi \delta_{n0} \tag{6.364b}$$

$$\int_{-1}^{1} (1 - x^2)^{-1/2} G_l^0(x) G_n^0(x) \, dx = \frac{2\pi}{l^2}\delta_{ln} \qquad l \neq 0 \tag{6.364c}$$

Integral representations

Using Cauchy's theorem and the Rodriguez formula, the *Schlafli integral* can be developed by the same method used to derive the Schlafli representation of the Legendre functions. The result is

$$G_l^\lambda(x) = (1 - x^2)^{(1/2)-\lambda} \oint \frac{(1 - z^2)^{l+\lambda-1/2}}{(z - x)^{l+1}} \, dz \tag{6.365a}$$

By manipulating the Schlafli integral, the *Laplace integral representation* is obtained. It is

$$G_l^\lambda(x) = \frac{2^{1-2\lambda}\Gamma(l + 2\lambda)}{l!\,[\Gamma(\lambda)]^2} \int_0^\pi \left(x + i\sqrt{1 - x^2}\cos\phi\right)^l \sin^{2\lambda-1}\phi \, d\phi \tag{6.365b}$$

Second solution

Like the Legendre functions of the second kind, the second Gegenbauer solution, containing the singularities of the differential equation, are obtained using the method of reduction of order discussed in Chapter 5. However, they have virtually no application to scientific problems. As such, their properties will not be discussed here and are rarely developed in other texts.

6.5 Laguerre Functions

Series representation

The series solution to the ordinary Laguerre differential equation

$$[xD^2 - (x - 1)D + n]y = 0 \tag{6.366}$$

is obtained straightforwardly using the Frobenius methods developed in Chapter 5. With

$$y = \sum_{m=0}^{\infty} c_m x^{m+s} \tag{6.367}$$

the indicial equation is

$$c_0 s^2 = 0 \tag{6.368}$$

Thus the nontrivial solution is obtained for $s = 0$. Then the recurrence relation for the coefficients is

$$c_{m+1} = \frac{(m - n)}{(m + 1)^2} c_m \tag{6.369}$$

From the Cauchy ratio

$$|x| \lim_{m \to \infty} \frac{c_{m+1}}{c_m} = |x| \lim_{m \to \infty} \frac{(m - n)}{(m + 1)^2} = 0 \cdot |x| < 1 \tag{6.370}$$

we see that absolute convergence is obtained for all finite x. We note from Table 6.2 that the Laguerre weight function is e^{-x}. Therefore, to ensure that orthonormalization integrals are finite, x must be restricted to the range $x \in [0, \infty]$.

From equation 6.369 it is clear that if n in an integer, $c_{n+1} = c_{n+2} = \cdots = 0$, and the solution is a finite polynomial. These polynomials are the *ordinary Laguerre polynomials*.

From equation 6.369 it is straightforward to develop a general expression for the Laguerre series coefficient. With that the ordinary Laguerre polynomial can be written

$$L_n(x) = \sum_{m=0}^{n} (-1)^m \frac{n!}{(n - m)!\,(m!)^2} x^m \tag{6.371}$$

where c_0 is arbitrarily taken to be 1. From this, we list the first few lowest-order polynomials

$$\text{(a)}\quad L_0(x) = 1 \qquad \text{(b)}\quad L_1(x) = 1 - x \quad \text{(c)}\quad L_2(x) = 1 - 2x + \tfrac{1}{2}x^2$$

$$\text{(d)}\quad L_3(x) = 1 - 3x + \tfrac{3}{2}x^2 - \tfrac{1}{6}x^3 \tag{6.372}$$

Unlike the Legendre and Gegenbauer polynomials, it is clear that the ordinary Laguerre polynomials do not have a specific parity. That is $L_n(-x) \neq (-1)^n L_n(x)$.

Generating function

The Laguerre polynomials can be generated from the function

$$\Phi(z, x) = \frac{e^{-xz/(1-z)}}{(1 - z)} = \sum_{n=0}^{\infty} z^n L_n(x) \tag{6.373}$$

To see that equation 6.373 is valid, we expand $\Phi(z, x)$ in a MacLaurin series. Using the binomial expansion

$$\frac{1}{(1 - z)^{r+1}} = \sum_{s=0}^{\infty} \frac{(r + s)!}{r!} \frac{z^s}{s!} \tag{6.374}$$

the generating function can be written

$$\Phi(z, x) = \sum_{r=0}^{\infty} (-1)^r \frac{x^r}{r!} \frac{z^r}{(1 - z)^{r+1}} = \sum_{r=0}^{\infty} \sum_{s=0}^{\infty} (-1)^r \frac{(r + s)!\, x^r}{(r!)^2 s!} z^{r+s} \tag{6.375}$$

Defining

$$C(r, s) \equiv (-1)^r \frac{(r + s)!\, x^r}{(r!)^2 s!} \tag{6.376}$$

we can identify the coefficients of the various powers of z by writing out equation 6.375 explicitly.

$$\begin{aligned} \Phi(z, x) &= z^0 C(0,0) + z[C(0,1) + C(1,0)] + z^2[C(0,2) + C(1,1) + C(2,0)] + \cdots \\ &= \sum_{n=0}^{\infty} z^n \sum_{k=0}^{n} C(k, n - k) \end{aligned} \tag{6.377}$$

Therefore,

$$\frac{e^{-xz/(1-z)}}{(1 - z)} = \sum_{n=0}^{\infty} z^n \sum_{k=0}^{n} (-1)^k \frac{n!}{(k!)^2 (n - k)!} x^k = \sum_{n=0}^{\infty} z^n L_n(x) \tag{6.378}$$

Recurrence relations

By differentiating the generating function with respect to z, we obtain

$$\frac{\partial \Phi}{\partial z} = \frac{(1 - z - x)}{(1 - z)^2} \Phi(z, x) \tag{6.379}$$

Writing $\Phi(z, x)$ in terms of the Laguerre polynomial sum, we have

$$(1 - z - x) \sum_{n=0}^{\infty} z^n L_n(x) = (1 - 2z + z^2) \sum_{n=1}^{\infty} n z^{n-1} L_n(x) \tag{6.380a}$$

$$\begin{aligned} \Rightarrow \quad & \sum_{n=0}^{\infty} \left[z^n L_n(x) - z^{n+1} L_n(x) - z^n x L_n(x) \right] \\ &= \sum_{n=1}^{\infty} \left[n z^{n-1} L_n(x) - 2n z^n L_n(x) + n z^{n+1} L_n(x) \right] \end{aligned} \tag{6.380b}$$

After making the appropriate changes of the summation indices so that the power of z is the same in each sum, and then equating the coefficient of like powers of z, we obtain the recurrence relation

$$(n + 2)L_{n+2}(x) - (2n + 3 - x)L_{n+1}(x) + (n + 1)L_n(x) = 0 \qquad n \geqslant 0 \tag{6.381}$$

Similarly, differentiation of the generator with respect to x leads to

$$(1 - z)\sum_{n=0}^{\infty} z^n L'_n(x) = -z\sum_{n=0}^{\infty} z^n L_n(x) \tag{6.382a}$$

$$\Rightarrow \quad \sum_{n=0}^{\infty}\left[z^{n+1}L_n(x) + z^n L'_n(x) - z^{n+1}L'_n(x)\right] = 0 \tag{6.382b}$$

By replacing n by $n + 1$ in the second summation and equating the coefficients of like powers of z to zero, we obtain the recurrence relation

$$L_n(x) + L'_{n+1}(x) - L'_n(x) = 0 \qquad n \geqslant 0 \tag{6.383}$$

Taking one derivative, the recurrence relation of equation 6.381 becomes

$$(n + 2)L'_{n+2}(x) - (2n + 3 - x)L'_{n+1}(x) + (n + 1)L'_n(x) + L_{n+1}(x) = 0 \tag{6.384}$$

Replacing n by $n + 1$ in equation 6.383 yields

$$L'_{n+2}(x) = L'_{n+1}(x) - L_{n+1}(x) \tag{6.385}$$

Replacing L'_{n+1} by $L'_n(x) - L_n(x)$ from equation 6.383, equation 6.385 can be written

$$L'_{n+2}(x) = L'_n(x) - L_n(x) - L_{n+1}(x) \tag{6.386}$$

With equation 6.383 and this result, equation 6.384 becomes

$$xL'_n(x) - (n + 1)L_{n+1}(x) + (n + 1 - x)L_n(x) = 0 \tag{6.387}$$

Rodriguez formula

Using an origin centered circular contour, consider the integral

$$\frac{1}{2\pi i}\oint \frac{\Phi(z, x)}{z^{N+1}}\,dz = \frac{1}{2\pi i}\oint \frac{e^{-xz/(1-z)}}{z^{N+1}(1 - z)}\,dz = \sum_{n=0}^{\infty} L_n(x)\frac{1}{2\pi i}\oint z^{n-N-1}\,dz \tag{6.388}$$

with $z = \rho e^{i\phi}$, we see that $\oint z^{n-N-1}\,dz = 0$ unless $n - N - 1 = -1$. Thus,

$$\frac{1}{2\pi i}\oint z^{n-N-1}\,dz = \delta_{nN} \tag{6.389}$$

Therefore, the sum in equation 6.388 reduces to a single term, and we obtain an integral representation for the Laguerre polynomial.

$$L_N(x) = \frac{1}{2\pi i}\oint \frac{e^{-xz/(1-z)}}{z^{N+1}(1 - z)}\,dz \tag{6.390}$$

We now make a substitution of variable

$$\frac{xz}{(1 - z)} = (z' - x) \tag{6.391}$$

which moves the $(N + 1)$th order pole of equation 6.390 from $z = 0$ to $z' = x$.

Dropping the prime on z', we have

$$L_N(x) = \frac{e^x}{2\pi i}\oint \frac{e^{-z}z^N}{(z-x)^{N+1}}\,dz \tag{6.392}$$

We note that the numerator is analytic everywhere inside and on the contour. Therefore, by the residue theorem, this integral becomes

$$L_N(x) = \frac{e^x}{N!}D^N(e^{-x}x^N) \tag{6.393}$$

This is the Rodriguez formula for the Laguerre polynomial.

From equation 6.392 we can easily deduce the value of $L_N(0)$. Using Cauchy's theorem,

$$L_N(0) = \frac{1}{2\pi i}\oint \frac{e^{-z}}{z}\,dz = 1 \tag{6.394}$$

Of course, this result is also obtainable from the series representation for $L_N(x)$. When $x = 0$, the only surviving term in equation 6.371 is the $m = 0$ term. That term is 1.

Orthonormalization condition

As noted in Table 6.2, it is necessary to multiply equation 6.366 by the weight function $\rho(x) = e^{-x}$ to make the ordinary Laguerre differential equation self-adjoint. Therefore, from the Sturm–Liouville analysis, the orthonormalization condition is

$$\int_0^\infty e^{-x}L_n(x)L_m(x)\,dx = N_n\delta_{nm} \tag{6.395}$$

To evaluate the normalization constant, consider

$$\begin{aligned}\int_0^\infty e^{-x}[\Phi(z,x)]^2\,dx &= \sum_{n=0}^{\infty}\sum_{m=0}^{\infty} z^{m+n}\int_0^\infty e^{-x}L_n(x)L_m(x)\,dx \\ &= \sum_{m,n=0}^{\infty} z^{m+n}N_n\delta_{mn} = \sum_{n=0}^{\infty} N_n z^{2n}\end{aligned} \tag{6.396}$$

But

$$\int_0^\infty e^{-x}[\Phi(z,x)]^2\,dx = \frac{1}{(1-z)^2}\int_0^\infty e^{-x[(1+z)/(1-z)]}\,dx = \frac{1}{(1-z^2)} = \sum_{n=0}^{\infty} z^{2n} \tag{6.397}$$

Therefore, it is easily determined that $N_n = 1$ for all n, and the orthonormalization condition for the Laguerre polynomials is

$$\int_0^\infty e^{-x}L_m(x)L_n(x)\,dx = \delta_{mn} \tag{6.398}$$

Powers of x in terms of Laguerre polynomials

From the orthonormalization condition, it is straightforward to deduce the expansion of any function $F(x)$ with $x \in [0,\infty]$ in terms of Laguerre polynomials. With

$$F(x) = \sum_{n=0}^{\infty} c_n L_n(x) \tag{6.399a}$$

we multiply $F(x)$ by $e^{-x}L_m(x)$ and integrate to obtain

$$c_m = \int_0^\infty e^{-x}L_m(x)F(x)\,dx \tag{6.399b}$$

Taking $F(x) = x^k$ for integer k, the sum of equation 6.399a will only extend to k, and with c_n given in equation 6.399b for $n \leqslant k$ ($c_n = 0$ for $n > k$),

$$x^k = \sum_{n=0}^{k} c_n L_n(x) \tag{6.400}$$

To evaluate these coefficients, consider

$$\int_0^\infty x^k e^{-x}\Phi(z,x)\,dx = \frac{1}{(1-z)}\int_0^\infty x^k e^{-x/(1-z)}\,dx = \sum_{m=0}^{k} z^m \int_0^\infty x^k e^{-x}L_m(x)\,dx \tag{6.401a}$$

But with equations 6.398 and 6.400,

$$\frac{1}{(1-z)}\int_0^\infty x^k e^{-x/(1-z)}\,dx = \sum_{n=0}^{k}\sum_{m=0}^{\infty} z^m c_n \int_0^\infty e^{-x}L_m(x)L_n(x)\,dx = \sum_{n=0}^{k} c_n z^n \tag{6.401b}$$

Substituting $y = x/(1-z)$, this becomes

$$\begin{aligned}\frac{1}{(1-z)}\int_0^\infty x^k e^{-x/(1-z)}\,dx &= (1-z)^k\int_0^\infty y^k e^{-y}\,dy = k!\,(1-z)^k \\ &= k!\sum_{n=0}^{k}(-1)^n\frac{k!}{n!\,(k-n)!}z^n\end{aligned} \tag{6.402}$$

$$\Rightarrow \quad c_n = k!\,(-1)^n\frac{k!}{n!\,(k-n)!} \tag{6.403}$$

$$\Rightarrow \quad x^k = k!\sum_{n=0}^{k}(-1)^n\frac{k!}{n!\,(k-n)!}L_n(x) \tag{6.404}$$

Since the coefficient of $L_n(x)$ is the coefficient of a binomial expansion, it is straightforward to deduce a recurrence relation for powers of x in terms of Laguerre polynomials that is similar to the formula satisfied by Bernoulli numbers (equation 3.207) or the Euler numbers (equation 3.262).

Consider the binomial expansion of $(1-L)^k$. If, after the expansion, each power of L is replaced by the corresponding Laguerre polynomial, equation 6.404 can be written

$$x^k = k!\,(1-L)^k \qquad \text{with } L^n \to L_n(x) \tag{6.405}$$

For example,

$$(1-L)^3 = L^0 - 3L^1 + 3L^2 - L^3 \tag{6.406}$$

$$\Rightarrow \quad x^3 = 3!\left[L_0(x) - 3L_1(x) + 3L_2(x) - L_3(x)\right] \tag{6.407}$$

Associated Laguerre polynomial

In the quantum theory of the hydrogen atom, the Laguerre function that is encountered is the *associated Laguerre polynomial*, defined by

$$L_n^k(x) \equiv D^k L_n(x) \tag{6.408}$$

$$\Rightarrow \quad L_n^0(x) = L_n(x) \tag{6.409}$$

[Some authors define $L_n^k(x) \equiv (-1)^k D^k L_n(x)$.] Since $L_n^k(x)$ is a polynomial of order $n - k$. Clearly,

$$L_n^k(x) = 0 \qquad k > n \tag{6.410}$$

Because the relation between the ordinary and associated Laguerre polynomials is so simple, many of the properties of $L_n^k(x)$ are easily determined from the properties of $L_n(x)$.

Differential equation

We can deduce the differential equation for $L_n^k(x)$ from the differential equation for $L_n(x)$ by taking k derivatives of equation 6.366. Consider

$$D^k[xL_n'' + (1-x)L_n' + nL_n] = 0 \tag{6.411}$$

Using the Leibnitz formula gives

$$D^k[xL_n''(x)] = xD^kL_n''(x) + kD^{k-1}L_n''(x) \tag{6.412}$$

But, for example,

$$D^kL_n''(x) = D^{k+2}L_n(x) = D^2L_n^k(x) \tag{6.413}$$

$$\Rightarrow \quad D^k[xL_n''(x)] = xD^2L_n^k(x) + kDL_n^k(x) \tag{6.414}$$

Similarly,

$$D^k[(1-x)L_n'(x)] = (1-x)DL_n^k(x) - kL_n^k(x) \tag{6.415}$$

Therefore, equation 6.411 becomes the differential equation for the associated Laguerre polynomial

$$[xD^2 + (k+1-x)D + (n-k)]L_n^k(x) = 0 \tag{6.416}$$

Series representation

By taking k derivatives of the series representation for $L_n(x)$, as expressed in equation 6.371, and using the fact that $D^kx^m = 0$ for $k > m$, the series representation for $L_n^k(x)$ is

$$L_n^k(x) = \sum_{m=k}^{n} (-1)^m \frac{n!}{(n-m)!(m-k)!m!} x^{m-k} \tag{6.417a}$$

Replacing m by $m + k$, this becomes

$$L_n^k(x) = (-1)^k \sum_{m=0}^{n-k} (-1)^m \frac{n!}{(n-m-k)!(m+k)!m!} x^m \tag{6.417b}$$

To ensure that the value of the lower index is never less than the value of the upper index, the associated Laguerre polynomial is frequently expressed as $L_{n+k}^k(x)$. By replacing n by $n + k$ in equation 6.417b, we can write the series representation as

$$L_{n+k}^k(x) = (-1)^k \sum_{m=0}^{n} (-1)^m \frac{(n+k)!}{(n-m)!(m+k)!m!} x^m \tag{6.417c}$$

From these series representations, we can determine the value of the associated Laguerre polynomial at $x = 0$. When $x = 0$, the only nonzero term in the sum is the $m = 0$ term. Therefore,

$$\text{(a)} \quad L_n^k(0) = (-1)^k \frac{n!}{(n-k)!k!} \qquad \text{or} \qquad \text{(b)} \quad L_{n+k}^k(0) = (-1)^k \frac{(n+k)!}{n!k!} \tag{6.418}$$

Generating function

The generating function of $L_n^k(x)$ is obtained by taking k derivatives of the generator of the ordinary Laguerre polynomial (equation 6.373).

$$\frac{1}{(1-z)}D^k e^{-xz/(1-z)} = \sum_{n=k}^{\infty} z^n L_n^k(x) \qquad (6.419a)$$

The first $k-1$ terms in the sum are zero due to the fact that $D^k L_n(x) = 0$ for $n < k$. Therefore, the sum starts at $n = k$. With n replaced by $n + k$, we obtain the generating function for the associated Laguerre polynomials.

$$\Phi(z, x) = (-1)^k \frac{1}{(1-z)^{k+1}} e^{-xz/(1-z)} = \sum_{n=0}^{\infty} z^n L_{n+k}^k(x) \qquad (6.419b)$$

Recurrence relations

We first note that the generating function of $L_n^k(x)$ is the product of the generator of $L_n(x)$ and $(-1)^k/(1-z)^k$. That is, equation 6.419b can be written

$$(-1)^k \sum_{n=0}^{\infty} z^n L_n(x) = (1-z)^k \sum_{n=0}^{\infty} z^n L_{n+k}^k(x) \qquad (6.420)$$

In order to indicate how we will develop a recurrence relation from this for arbitrary k, we will first present an example, considering equation 6.420 with $k = 2$.

$$\sum_{n=0}^{\infty} z^n L_n(x) = \sum_{n=0}^{\infty} z^n L_{n+2}^2(x) - 2\sum_{n=0}^{\infty} z^{n+1} L_{n+2}^2(x) + \sum_{n=0}^{\infty} z^{n+2} L_{n+2}^2(x) \qquad (6.421)$$

In the sums containing z^n, we separate the $n = 0$ and $n = 1$ terms from the rest of the sum. The same is done with the $n = 0$ term in the sum containing z^{n+1}. Then

$$z^0 L_0(x) + zL_1(x) + \sum_{n=2}^{\infty} z^n L_n(x) = z^0 L_2^2(x) + zL_3^2(x) + \sum_{n=2}^{\infty} z^n L_{n+2}^2(x)$$

$$-2zL_2^2(x) - 2\sum_{n=1}^{\infty} z^{n+1} L_{n+2}^2(x) + \sum_{n=0}^{\infty} z^{n+2} L_{n+2}^2(x) \qquad (6.422)$$

Equating the coefficients of z^0 and separately equating the coefficients of z^1, we obtain

$$\text{(a)} \quad L_0(x) = L_2^2(x) \qquad \text{(b)} \quad L_1(x) = L_3^2(x) - 2L_2^2(x) \qquad (6.423)$$

By virtue of these equalities, we can cancel the z^0 and z^1 terms from equation 6.422, leaving it in the form

$$\sum_{n=2}^{\infty} z^n L_n(x) = \sum_{n=2}^{\infty} z^n L_{n+2}^2(x) - 2\sum_{n=1}^{\infty} z^{n+1} L_{n+2}^2(x) + \sum_{n=0}^{\infty} z^{n+2} L_{n+2}^2(x) \qquad (6.424)$$

Replacing n by $n + 2$ in the sums containing z^n, and n by $n + 1$ in the sum containing z^{n+1}, this becomes

$$\sum_{n=0}^{\infty} z^{n+2} L_{n+2}(x) = \sum_{n=0}^{\infty} z^{n+2} L_{n+4}^2(x) - 2\sum_{n=0}^{\infty} z^{n+2} L_{n+3}^2(x) + \sum_{n=0}^{\infty} z^{n+2} L_{n+2}^2(x) \qquad (6.425)$$

$$\Rightarrow \quad L_{n+2}(x) = L_{n+2}^2(x) - 2L_{n+3}(x) + L_{n+2}^2(x) \qquad k = 2 \qquad (6.426)$$

We now extend this analysis to arbitrary k, by writing equation 6.420 as

$$(-1)^k \sum_{n=0}^{\infty} z^n L_n(x) = \sum_{r=0}^{k} \frac{(-1)^r k!}{r!(k-r)!} \sum_{n=0}^{\infty} z^{n+r} L_{n+k}^k(x) \tag{6.427}$$

Since the highest power of z in the double sum is z^{n+k}, as we did with $k = 2$, we separate all the powers of z lower than z^{n+k} from the rest of the sum, writing equation 6.427 as

$$\begin{aligned}(-1)^k \sum_{n=0}^{k-1} z^n L_n(x) &+ (-1)^k \sum_{n=k}^{\infty} z^n L_n(x) \\ &= \sum_{r=0}^{k} \frac{(-1)^r k!}{r!(k-r)!} \left[\sum_{n=0}^{k-r-1} z^{n+r} L_{n+k}^k(x) + \sum_{n=k-r}^{\infty} z^{n+r} L_{n+k}^k(x) \right]\end{aligned} \tag{6.428}$$

By equating the coefficients separately of each of the terms containing a power of z lower than z^{n+k}, we obtain

$$(-1)^k \sum_{n=0}^{k-1} z^n L_n(x) = \sum_{r=0}^{k} \frac{(-1)^r k!}{r!(k-r)!} \sum_{n=0}^{k-r-1} z^{n+r} L_{n+k}^k(x) \tag{6.429}$$

We now cancel these terms from equation 6.428, which then becomes

$$(-1)^k \sum_{n=k}^{\infty} z^n L_n(x) = \sum_{r=0}^{k} \frac{(-1)^r k!}{r!(k-r)!} \sum_{n=k-r}^{\infty} z^{n+r} L_{n+k}^k(x) \tag{6.430}$$

In the left-hand sum, we replace n by $n + k$, and in the sum over n on the right-hand side, n is replaced by $n + k - r$. Then equation 6.430 becomes

$$(-1)^k \sum_{n=0}^{\infty} z^{n+k} L_{n+k}(x) = \sum_{n=0}^{\infty} z^{n+k} \sum_{r=0}^{k} \frac{(-1)^r k!}{r!(k-r)!} L_{n+2k-r}^k(x) \tag{6.431}$$

$$\Rightarrow \quad L_{n+k}(x) = (-1)^k \sum_{r=0}^{k} \frac{(-1)^r k!}{r!(k-r)!} L_{n+2k-r}^k(x) \tag{6.432}$$

As we noted in developing the powers of x in terms of the Laguerre polynomials, the presence of the binomial coefficient in the sum suggests the existence of a shorthand notation for this recurrence relation something like that satisfied by the Bernoulli numbers or by the Euler numbers.

One way to express this is to consider the binomial expansion of

$$\left[1 - L_{n+2k-\mu}^k(x)\right]^k = \sum_{r=0}^{k} \frac{(-1)^r k!}{r!(k-r)!} \left[L_{n+2k-\mu}^k(x)\right]^r \tag{6.433}$$

If μ is now replaced by the power r, we obtain the right-hand sum of equation 6.432. Therefore, this recurrence relation can be written as

$$L_{n+k}(x) = (-1)^k \left[1 - L_{n+2k-\mu}^k(x)\right]^k \begin{cases} \text{with } \mu \text{ replaced by the} \\ \text{power of } L_{n+2k-\mu}^k(x) \end{cases} \tag{6.434}$$

Other recurrence relations can be obtained by differentiation of the generating function. For example, taking $\partial\Phi/\partial z$ leads to

$$[(k+1)(1-z)-x]\sum_{n=0}^{\infty} z^n L_{n+k}^k(x) = \sum_{n=0}^{\infty} n z^{n-1} L_{n+k}^k(x) - 2\sum_{n=0}^{\infty} n z^n L_{n+k}^k(x) + \sum_{n=0}^{\infty} n z^{n+1} L_{n+k}^k(x) \tag{6.435}$$

$$\Rightarrow \quad (n+2)L_{n+k+2}^k(x) - (2n+k+3-x)L_{n+k+1}^k(x) + (n+k+1)L_{n+k}^k(x) = 0 \tag{6.436}$$

Taking $\partial\Phi/\partial x$, we obtain

$$\sum_{n=0}^{\infty} z^{n+1} L_{n+k}^k(x) = \sum_{n=0}^{\infty} z^{n+1}\left[L_{n+k}^k(x)\right]' - \sum_{n=0}^{\infty} z^n \left[L_{n+k}^k(x)\right]' \tag{6.437}$$

$$\Rightarrow \quad L_{n+k}^k(x) = \left[L_{n+k}^k(x)\right]' - \left[L_{n+k+1}^k(x)\right]' \tag{6.438}$$

Using

$$\left[L_N^k(x)\right]' = L_N^{k+1}(x) \tag{6.439}$$

$$\Rightarrow \quad L_{n+k}^k(x) = L_{n+k}^{k+1}(x) - L_{n+k+1}^{k+1}(x) \tag{6.440}$$

Other recurrence relations can be obtained by differentiating the recurrence relations for the ordinary Laguerre polynomials. Taking k derivatives of equation 6.381 yields

$$(n+2)L_{n+2}^k(x) - (2n+3-x)L_{n+1}^k(x) + (n+1)L_n^k(x) + kL_{n+1}^{k-1}(x) = 0 \tag{6.441}$$

Taking k derivatives of equation 6.383 and using equation 6.439 leads to

$$\text{(a)} \quad L_n^k(x) - \left[L_{n+1}^k(x)\right]' - \left[L_n^k(x)\right]' = 0 \quad \Rightarrow \quad \text{(b)} \quad L_n^k(x) - L_{n+1}^{k+1}(x) - L_n^{k+1}(x) = 0 \tag{6.442}$$

After differentiating equation 6.386 k times, we obtain

$$\left[L_{n+2}^k(x)\right]' = \left[L_n^k(x)\right]' - L_n^k(x) - L_{n+1}^k(x) \tag{6.443a}$$

$$\Rightarrow \quad L_{n+2}^{k+1}(x) = L_n^{k+1}(x) - L_n^k(x) - L_{n+1}^k(x) \tag{6.443b}$$

Similarly, taking k derivatives of equation 6.387, we obtain

$$x\left[L_n^k(x)\right]' + k\left[L_n^{k-1}(x)\right]' - (n+1)L_{n+1}^k(x) + (n+1-x)L_n^k(x) - kL_n^{k-1}(x) = 0 \tag{6.444a}$$

$$\Rightarrow \quad xL_n^{k+1}(x) + kL_n^k(x) - (n+1)L_{n+1}^k(x) + (n+1-x)L_n^k(x) - kL_n^{k-1}(x) = 0 \tag{6.444b}$$

The differential equation can also be written as a recurrence relation for $L_n^k(x)$ in the k-index. That is,

$$\left[xD^2 + (k+1-x)D + (n-k)\right]L_n^k(x) = xL_n^{k+2}(x) + (k+1-x)L_n^{k+1}(x) + (n-k)L_n^k(x) = 0 \tag{6.445}$$

Rodriguez formula

The Rodriguez formula for the associated Laguerre polynomials can be developed quite easily from the Rodriguez formula for the ordinary Laguerre polynomials (equation 6.393). Consider

$$D^k L_n(x) = L_n^k(x) = D^k\left[\frac{e^x}{n!}D^n(e^{-x}x^n)\right] \tag{6.446}$$

As shown in Chapter 5 (equation 5.143), e^x can be moved to the left of the operator D^k by replacing D by $D + 1$. That is,

$$\text{(a)}\quad L_n^k(x) = \frac{e^x}{n!}D^n\left[(D+1)^k(e^{-x}x^n)\right] = \frac{e^x}{n!}D^n\left[e^{-x}D^k(x^n)\right] = \frac{e^x}{(n-k)!}D^n(e^{-x}x^{n-k})$$

$$\Rightarrow \quad \text{(b)}\quad L_{n+k}^k(x) = \frac{e^x}{n!}D^{n+k}(e^{-x}x^n) \tag{6.447}$$

Orthonormalization condition

To make the associated Laguerre differential equation self-adjoint, the equation must be multiplied by $x^k e^{-x}$. Thus, as is noted in Table 6.2, the weight function for $L_n^k(x)$ is $\rho(x) = x^k e^{-x}$. Therefore, the orthonormalization integral is

$$\int_0^\infty x^k e^{-x} L_{n+k}^k(x) L_{m+k}^k(x)\, dx = N_n \delta_{nm} \tag{6.448}$$

To evaluate this integral using the generating function, consider

$$\int_0^\infty x^k e^{-x}[\phi(z,x)]^2\, dx = \frac{1}{(1-z)^{2k+2}}\int_0^\infty x^k e^{-x[(1+z)/(1-z)]}\, dx$$

$$= \sum_{n=0}^\infty \sum_{m=0}^\infty z^{n+m}\int_0^\infty x^k e^{-x} L_{n+k}^k(x) L_{m+k}^k(x)\, dx = \sum_{n=0}^\infty z^{2n} N_n \tag{6.449a}$$

With the substitution $y = x[(1+z)/(1-z)]$, this becomes

$$\sum_{n=0}^\infty N_n z^{2n} = \frac{1}{(1-z)^{k+1}}\int_0^\infty y^k e^{-y}\, dy = \frac{k!}{(1-z^2)^{k+1}} = \sum_{n=0}^\infty \frac{(n+k)!}{n!} z^{2n} \tag{6.449b}$$

$$\Rightarrow \quad N_n = \frac{(n+k)!}{n!} \tag{6.450}$$

$$\Rightarrow \quad \int_0^\infty x^k e^{-x} L_{n+k}^k(x) L_{m+k}^k(x)\, dx = \frac{(n+k)!}{n!}\delta_{nm} \tag{6.451a}$$

$$\Rightarrow \quad \int_0^\infty x^k e^{-x} L_n^k(x) L_m^k(x)\, dx = \frac{n!}{(n-k)!}\delta_{nm} \tag{6.451b}$$

Powers of *x* in terms of associated Laguerre polynomials

As with any other Sturm–Liouville function, an arbitrary function that depends on $x \in [0, \infty]$ can be written as a sum over associated Laguerre polynomials. Therefore, in particular,

$$x^l = \sum_{n=0}^{l} c_n L_{n+k}^k(x) \tag{6.452}$$

$$\Rightarrow \quad c_n = \frac{n!}{(n+k)!}\int_0^\infty x^{l+k} e^{-x} L_{n+k}^k(x)\, dx \tag{6.453}$$

To evaluate these coefficients, we consider

$$\int_0^\infty x^{l+k}e^{-x}\Phi(z,x)\,dx = \frac{(-1)^k}{(1-z)^{k+1}}\int_0^\infty x^{l+k}e^{-x/(1-z)}\,dx$$

$$= \sum_{n=0}^{l} z^n \int_0^\infty x^{l+k}e^{-x}L_{n+k}^k(x)\,dx = \sum_{n=0}^{l} z^n c_n \frac{(n+k)!}{n!} \tag{6.454}$$

Substituting $y = x/(1-z)$ and expanding $(1-z)^l$ in a binomial series yields

$$\text{(a)}\quad (-1)^k(1-z)^l(l+k)! = \sum_{n=0}^{l} z^n c_n \frac{(n+k)!}{n!}$$

$$\Rightarrow \quad \text{(b)}\quad c_n = (-1)^{n+k}\frac{(l+k)!\,l!}{(n+k)!\,(l-n)!} \tag{6.455}$$

$$\Rightarrow \quad x^l = \sum_{n=0}^{l}(-1)^{n+k}\frac{(l+k)!\,l!}{(n+k)!\,(l-n)!}L_{n+k}^k(x) \tag{6.456}$$

Unlike the analogous expansion in terms of the ordinary Laguerre polynomials, the coefficients of this expansion are not the binomial coefficients. Therefore, we cannot develop a shorthand recurrence relation for the expansion of x^l in terms of $L_{n+k}^k(x)$ analogous to equation 6.405.

Second solution

Like the second solution to the other Sturm–Liouville equations, the method of reduction of order is used to obtain the second Laguerre solution, which contains the singularity of the differential equation at $x = 0$. Like the second Gegenbauer solution, the second Laguerre solution has no application to scientific problems, so we will not develop its properties.

6.6 Hermite Functions

Series representation

The Schrödinger equation with a simple harmonic oscillator potential leads to

$$[D^2 - 2xD + 2n]y = 0 \tag{6.457}$$

which is the Hermite differential equation. Assuming the series solution

$$y = \sum_{m=0}^{\infty} c_m x^{m+s} \tag{6.458}$$

$$\Rightarrow \quad \text{(a)}\quad c_0 s(s-1) = 0 \qquad \text{(b)}\quad c_1 s(s+1) = 0 \quad \text{(c)}\quad c_{m+2} = \frac{2(m+s-n)}{(m+s+2)(m+s+1)}c_m \tag{6.459}$$

Clearly, c_0 and c_1 can be chosen arbitrarily by taking $s = 0$. With this choice, then,

$$c_{m+2} = \frac{2(m-n)}{(m+2)(m+1)}c_m \tag{6.460}$$

If n is an integer, we see that $c_{n+2} = c_{n+4} = \cdots = 0$. Thus, for integer n, one of the infinite series solutions becomes a finite, nth-order polynomial. These polynomials are the Hermite polynomials.

Using the Cauchy ratio, convergence of the infinite series is determined from

$$x^2 \lim_{m\to\infty} \frac{c_{m+2}}{c_m} = x^2 \lim_{m\to\infty} \frac{2(m+s-n)}{(m+s+2)(m+s+1)} = x^2 \cdot 0 < 1 \tag{6.461}$$

Therefore, the infinite series converges for all finite values of x. Because the weight function is $\rho(x) = e^{-x^2}$, the orthogonalization integral will be finite if x is taken over the range $x \in [-\infty, \infty]$.

If n is even, the terms of the Hermite polynomial will have coefficients

$$\text{(a)}\quad c_2 = -2\frac{n}{2!}c_0 \qquad \text{(b)}\quad c_4 = 2^2\frac{n(n-2)}{4!}c_0 \qquad \text{(c)}\quad c_6 = -2^3\frac{n(n-2)(n-4)}{6!}c_0 \tag{6.462}$$

With n even, we refer to equation 6.91 to obtain the general term,

$$c_{2l} = (-1)^l \frac{2^l}{(2l)!}\frac{n!!}{(n-2l)!!}c_0 = (-1)^l \frac{2^{2l}}{(2l)!}\frac{(\frac{1}{2}n)!}{(\frac{1}{2}n-l)!}c_0 \tag{6.463}$$

The odd Hermite polynomials are determined from the coefficients

$$\text{(a)}\quad c_3 = -2\frac{(n-1)}{3!}c_1 \qquad \text{(b)}\quad c_5 = 2^2\frac{(n-1)(n-3)}{5!}c_1 \tag{6.464}$$

With n odd (n − 1 even), we again use equation 6.91 to obtain

$$c_{2l+1} = (-1)^l \frac{2^l}{(2l+1)!}\frac{(n-l)!!}{(n-2l-1)!!}c_1 = (-1)^l \frac{2^{2l}}{(2l+1)!}\frac{[\frac{1}{2}(n-1)]!}{(\frac{1}{2}n-\frac{1}{2}-l)!}c_1 \tag{6.465}$$

From these, the series representation of the Hermite polynomial is

$$H_n(x) = c_0 \sum_{l=0}^{n/2} (-1)^l \frac{(n/2)!}{(2l)!\,[\frac{1}{2}n-l]!}(2x)^{2l} \qquad n \text{ even} \tag{6.466a}$$

$$H_n(x) = \frac{c_1}{2} \sum_{l=0}^{(n-1)/2} (-1)^l \frac{[(n-1)/2]!}{(2l+1)!\,[\frac{1}{2}n-\frac{1}{2}-l]!}(2x)^{2l+1} \qquad n \text{ odd} \tag{6.466b}$$

For even n, we make the change of summation index $l \to \frac{1}{2}n - l$. Then equation 6.466a becomes

$$H_n(x) = (-1)^{n/2}\left(\frac{n}{2}\right)!c_0 \sum_{l=0}^{n/2} (-1)^l \frac{1}{(n-2l)!\,l!}(2x)^{n-2l} \tag{6.467a}$$

For odd n, the substitution $l \to \frac{1}{2}(n-1) - l$ puts equation 6.466b in the form

$$H_n(x) = (-1)^{(n-1)/2}\left(\frac{n-1}{2}\right)!\,\frac{c_1}{2} \sum_{l=0}^{(n-1)/2} (-1)^l \frac{1}{(n-2l)!\,l!}(2x)^{n-2l} \tag{6.467b}$$

Choosing c_0 and c_1 such that

$$\text{(a)}\quad (-1)^{n/2}\left(\frac{n}{2}\right)!\,c_0 = n! \qquad \text{(b)}\quad (-1)^{(n-1)/2}\left(\frac{n-1}{2}\right)!\,\frac{c_1}{2} = n! \tag{6.468}$$

$$\Rightarrow \quad H_n(x) = \sum_{l=0}^{[n/2]} (-1)^l \frac{n!}{(n-2l)!\,l!}(2x)^{n-2l} \qquad \text{all } n \tag{6.469}$$

The first few lowest-index Hermite polynomials are

$$\text{(a)}\quad H_0(x) = 1 \qquad \text{(b)}\quad H_1(x) = 2x \quad \text{(c)}\quad H_2(x) = 4x^2 - 2$$
$$\text{(d)}\quad H_3(x) = 8x^3 - 12x \quad \text{(e)}\quad H_4(x) = 16x^4 - 48x^2 + 12 \tag{6.470}$$

The Hermite polynomials have a definite parity, expressed by

$$H_n(-x) = (-1)^n H_n(x) \tag{6.471}$$

Using equation 6.469 or 6.471, we obtain

$$\text{(a)}\quad H_n(0) = 0 \qquad n \text{ odd} \qquad \text{(b)}\quad H_n(0) = (-1)^{n/2}\frac{n!}{(n/2)!} \qquad n \text{ even} \tag{6.472}$$

Generating function

To prove the validity of

$$\Phi(z, x) \equiv e^{-z^2+2xz} = \sum_{n=0}^{\infty} z^n \frac{H_n(x)}{n!} \tag{6.473}$$

we expand $\Phi(z, x)$ in a MacLaurin series.

$$e^{-z^2+2xz} = \sum_{n=0}^{\infty} \frac{(2xz - z^2)^n}{n!} = \sum_{n=0}^{\infty}\sum_{l=0}^{n} (-1)^l \frac{1}{l!\,(n-l)!}(2x)^{n-l} z^{n+l} \tag{6.474}$$

Defining

$$C(n, l) \equiv (-1)^l \frac{1}{l!\,(n-l)!}(2x)^{n-l} \tag{6.475}$$

and recognizing that we must require $l \leqslant n$, equation 6.474 can be written

$$\begin{aligned} e^{-z^2+2xz} &= z^0 C(0,0) + zC(1,0) + z^2[C(2,0) + C(1,1)] \\ &\quad + z^3[C(3,0) + C(2,1)] + z^4[C(4,0) + C(3,1) + C(2,2)] + \cdots \\ &= \sum_{n=0}^{\infty} z^n \sum_{l=0}^{[n/2]} C(n-l, l) = \sum_{n=0}^{\infty} z^n \sum_{l=0}^{[n/2]} (-1)^l \frac{1}{l!\,(n-2l)!}(2x)^{n-2l} \end{aligned} \tag{6.476}$$

Comparing this result to equation 6.469, we see that

$$e^{-z^2+2xz} = \sum_{n=0}^{\infty} z^n \frac{H_n(x)}{n!} \tag{6.473}$$

Recurrence relations

As with the other Sturm–Liouville functions we have investigated, some recurrence relations are derived by taking derivatives of the generating function.

$$\frac{\partial \Phi}{\partial z} = 2(x - z)e^{-z^2+2xz} = 2(x - z)\sum_{n=0}^{\infty} z^n \frac{H_n(x)}{n!} = \sum_{n=1}^{\infty} nz^{n-1}\frac{H_n(x)}{n!} \tag{6.477}$$

$$\Rightarrow \quad H_{n+2}(x) - 2xH_{n+1}(x) + 2(n+1)H_n(x) = 0 \qquad n \geqslant 0 \tag{6.478}$$

Taking

$$\frac{\partial \Phi}{\partial x} = 2z\Phi(z,x) = 2\sum_{n=0}^{\infty} z^{n+1}\frac{H_n(x)}{n!} = \sum_{n=0}^{\infty} z^n \frac{H_n'(x)}{n!} \tag{6.479}$$

$$\Rightarrow \quad H_{n+1}'(x) = 2(n+1)H_n(x) \qquad n \geqslant 0 \tag{6.480}$$

Substituting this into equation 6.478 and replacing n by $n-1$, we obtain

$$H_{n+1}(x) - 2xH_n(x) + H_n'(x) = 0 \qquad n \geqslant 0 \tag{6.481}$$

Differentiating equation 6.478 and using equation 6.480 to express $H_{n+2}'(x)$, we find

$$(n+1)H_{n+1}(x) - xH_{n+1}'(x) + (n+1)H_n'(x) = 0 \qquad n \geqslant 0 \tag{6.482}$$

Rodriguez formula

Consider the contour integral involving the generating function

$$\frac{1}{2\pi i}\oint \frac{\Phi(z,x)}{z^{N+1}}\,dz = \sum_{n=0}^{\infty}\frac{H_n(x)}{n!}\frac{1}{2\pi i}\oint z^{n-N-1}\,dz \tag{6.483}$$

Taking the contour to be a circle enclosing the origin, Cauchy's theorem yields

$$\frac{1}{2\pi i}\oint z^{n-N-1}\,dz = \delta_{nN} \tag{6.389}$$

Thus the sum in equation 6.483 is reduced to one term, and we obtain an integral representation of the Hermite polynomial,

$$H_n(x) = \frac{n!}{2\pi i}\oint \frac{e^{-z^2+2xz}}{z^{n+1}}\,dz \tag{6.484}$$

Writing $z^2 - 2xz = (z-x)^2 - x^2$ and replacing $(z-x)$ by z, this becomes

$$H_n(x) = e^{x^2}\frac{n!}{2\pi i}\oint \frac{e^{-z^2}}{(z+x)^{n+1}}\,dz \tag{6.485a}$$

Cauchy's theorem for an analytic function $f(z)$ yields

$$\frac{n!}{2\pi i}\oint \frac{f(z)}{(z-x)^{n+1}}\,dz = D^n f(x) \tag{2.80c}$$

Then

$$H_n(-x) = (-1)^n H_n(x) = e^{x^2}\frac{n!}{2\pi i}\oint \frac{e^{-z^2}}{(z-x)^{n+1}}\,dz = e^{x^2}D^n(e^{-x^2}) \tag{6.485b}$$

or

$$H_n(x) = (-1)^n e^{x^2} D^n(e^{-x^2}) \tag{6.486}$$

which is the Rodriguez formula for the Hermite polynomials.

Integral representation

By completing the square in the exponential of the generating function, it is straightforward to show that

$$\int_{-\infty}^{\infty} e^{-z^2+2xz}\, dz = \sqrt{\pi}\, e^{x^2} \tag{6.487}$$

$$\Rightarrow \quad \sqrt{\pi}\, D^n(e^{x^2}) = 2^n \int_{-\infty}^{\infty} z^n e^{-z^2+2xz}\, dz \tag{6.488}$$

Replacing x by ix, $D^n = d^n/dx^n$ becomes $(-i)^n D^n$ and $e^{x^2} \to e^{-x^2}$. Thus equation 6.488 becomes

$$(-i)^n \sqrt{\pi}\, D^n(e^{-x^2}) = 2^n \int_{-\infty}^{\infty} z^n e^{-z^2+2ixz}\, dz \tag{6.489}$$

Using the Rodriguez formula as expressed in equation 6.486, we obtain

$$D^n(e^{-x^2}) = (-1)^n e^{-x^2} H_n(x) \tag{6.486}$$

$$\Rightarrow \quad H_n(x) = \frac{(-2i)^n e^{x^2}}{\sqrt{\pi}} \int_{-\infty}^{\infty} z^n e^{-z^2+2ixz}\, dz \tag{6.490}$$

Orthonormalization condition

To make the Hermite differential equation self-adjoint, it must be multiplied by the weight function e^{-x^2}. Therefore, from Sturm–Liouville theory, the orthonormalization condition is

$$\int_{-\infty}^{\infty} e^{-x^2} H_n(x) H_m(x)\, dx = N_n \delta_{nm} \tag{6.491}$$

To determine the normalization constant, consider

$$\int_{-\infty}^{\infty} e^{-x^2} [\Phi(z,x)]^2\, dx = \int_{-\infty}^{\infty} e^{-x^2-2z^2+4zx}\, dx$$

$$= \sum_{n=0}^{\infty} \sum_{m=0}^{\infty} \frac{z^{n+m}}{n!\,m!} \int_{-\infty}^{\infty} e^{-x^2} H_n(x) H_m(x)\, dx = \sum_{n=0}^{\infty} \frac{z^{2n}}{(n!)^2} N_n \tag{6.492a}$$

Completing the square in the exponential, this becomes

$$e^{2z^2} \int_{-\infty}^{\infty} e^{-(x-2z)^2}\, dx = \sum_{n=0}^{\infty} \frac{z^{2n}}{(n!)^2} N_n \tag{6.492b}$$

The integral is $\sqrt{\pi}$, and after expanding e^{2z^2} in a MacLaurin series, this becomes

$$\sqrt{\pi} \sum_{n=0}^{\infty} \frac{2^n z^{2n}}{n!} = \sum_{n=0}^{\infty} \frac{z^{2n}}{(n!)^2} N_n \tag{6.493}$$

$$\Rightarrow \quad N_n = \sqrt{\pi}\, 2^n n! \tag{6.494}$$

$$\Rightarrow \quad \int_{-\infty}^{\infty} e^{-x^2} H_n(x) H_m(x)\, dx = \sqrt{\pi}\, 2^n n!\, \delta_{mn} \tag{6.495}$$

Powers of x in terms of Hermite polynomials

We write

$$x^k = \sum_{n=0}^{k} c_n H_n(x) \tag{6.496}$$

If k is even, only the even Hermite polynomials will contribute to the sum, while only the odd polynomials will be present in the sum if k is odd. Thus

$$\text{(a)} \quad x^k = \sum_{l=0}^{k/2} c_{2l} H_{2l}(x) \quad k \text{ even} \qquad \text{(b)} \quad x^k = \sum_{l=0}^{(k-1)/2} c_{2l+1} H_{2l+1}(x) \quad k \text{ odd} \tag{6.497}$$

We can determine these coefficients straightforwardly using the orthonormalization condition. The result is

$$c_N = \frac{1}{\sqrt{\pi}\, 2^N N!} \int_{-\infty}^{\infty} x^k e^{-x^2} H_N(x)\, dx \tag{6.498}$$

where $N = 2l$ if k is even and $N = 2l + 1$ for odd k.

To evaluate this integral, we consider an integral involving the generating function.

$$\int_{-\infty}^{\infty} x^k e^{-x^2} e^{-z^2+2xz}\, dx = \sqrt{\pi} \sum_{n=0}^{\infty} c_n 2^n z^n = \sum_{n=0}^{\infty} \frac{z^n}{n!} \int_{-\infty}^{\infty} x^k e^{-x^2} H_n(x)\, dx \tag{6.499}$$

$$\Rightarrow \quad \sqrt{\pi} \sum_{n=0}^{\infty} c_n 2^n z^n = \int_{-\infty}^{\infty} x^k e^{-(x-z)^2}\, dx = \int_{-\infty}^{\infty} (x+z)^k e^{-x^2}\, dx$$

$$= \sum_{n=0}^{k} \frac{k!}{n!\,(n-k)!} z^n \int_{-\infty}^{\infty} x^{k-n} e^{-x^2}\, dx \tag{6.500}$$

Because x^k can be written as a sum of Hermite polynomials, and the order of each polynomial will not exceed $k, c_n = 0,\ n > k$ in the infinite sum in equation 6.500. Therefore, this equation yields

$$c_n = \frac{1}{\sqrt{\pi}\, 2^n} \frac{k!}{n!\,(k-n)!} \int_{-\infty}^{\infty} x^{k-n} e^{-x^2}\, dx \qquad n \leqslant k \tag{6.501}$$

Since the limits of integration are symmetric about $x = 0$,

$$\int_{-\infty}^{\infty} x^{k-n} e^{-x^2}\, dx = 0 \qquad (k-n) \text{ odd} \tag{6.502}$$

To ensure that $k - n$ is even, we set $k - n \equiv 2s$. Then

$$I_{2s} \equiv \int_{-\infty}^{\infty} x^{2s} e^{-x^2}\, dx \tag{6.503}$$

is integrated by parts once, with $u = x^{2s-1}$, and $dv = xe^{-x^2}\, dx$. The result is

$$I_{2s} = \frac{(2s-1)}{2} I_{2s-2} \tag{6.504}$$

With

$$I_0 = \int_{-\infty}^{\infty} e^{-x^2}\, dx = \sqrt{\pi} \tag{6.505}$$

we have

$$\text{(a)}\quad I_2 = \frac{1}{2}\sqrt{\pi} \qquad \text{(b)}\quad I_4 = \frac{3\cdot 1}{2^2}\sqrt{\pi} \qquad \text{(c)}\quad I_6 = \frac{5\cdot 3\cdot 1}{2^3}\sqrt{\pi} \tag{6.506}$$

$$\Rightarrow \quad I_{2s} = \frac{(2s-1)!!}{2^s}\sqrt{\pi} \tag{6.507}$$

Using

$$(2s-1)!! = \frac{(2s)!}{2^s s!} \tag{6.90}$$

$$\Rightarrow \quad I_{2s} = \frac{(2s)!}{2^{2s} s!}\sqrt{\pi} \tag{6.508}$$

Therefore, with $s = (k-n)/2$,

$$\text{(a)}\quad c_n = \frac{k!}{2^k n![(k-n)/2]!} \quad k-n \text{ even} \qquad \text{(b)}\quad c_n = 0 \quad k-n \text{ odd} \tag{6.509}$$

For k even, nonzero coefficients are obtained for even n. If k is odd, the nonzero coefficients arise for odd n. With $n \equiv 2l$ or $n \equiv 2l+1$, these coefficients are

$$c_{2l} = \frac{k!}{2^k (2l)!\,((\frac{1}{2}k) - l)!} \qquad k \text{ even} \tag{6.510a}$$

$$c_{2l+1} = \frac{k!}{2^k (2l+1)!(\frac{1}{2}(k-1) - l)!} \qquad k \text{ odd} \tag{6.510b}$$

Therefore, equations 6.497 become

$$x^k = \frac{k!}{2^k} \sum_{l=0}^{k/2} \frac{1}{(2l)!\,(\frac{1}{2}k - l)!} H_{2l}(x) \qquad k \text{ even} \tag{6.511a}$$

$$x^k = \frac{k!}{2^k} \sum_{l=0}^{(k-1)/2} \frac{1}{(2l+1)!\,[\frac{1}{2}(k-1) - l]!} H_{2l+1}(x) \qquad k \text{ odd} \tag{6.511b}$$

These two expressions can be cast in a single form. In the sum for even k, we let $n = \frac{1}{2}k - l$, and in the odd k sum, we set $n = \frac{1}{2}(k-1) - l$. Recognizing that the upper limit on both sums can be written as $[\frac{1}{2}k]$, both equations can be written

$$x^k = \frac{k!}{2^k} \sum_{n=0}^{[k/2]} \frac{1}{(k-2n)!\,n!} H_{k-2n}(x) \tag{6.512}$$

Second solution and associated functions

Like the second solutions of the Gegenbauer and Laguerre equations, associated Hermite functions and a second solution to the Hermite equation have no application to scientific problems, so are not discussed in this book.

6.7 Bessel functions

Series representation

The Bessel functions are solutions to the differential equation

$$\left[x'^2D^2 + x'D + (k^2x'^2 - \alpha^2)\right]y = 0 \tag{6.513a}$$

With $x \equiv kx'$, $D = d/dx' \to k\,d/dx = kD$, and the differential equation can be expressed as

$$\left[x^2D^2 + xD + (x^2 - \alpha^2)\right]y = 0 \tag{6.513b}$$

The series solution

$$y = \sum_{m=0}^{\infty} c_m x^{m+s} \tag{6.514}$$

$$\Rightarrow \quad \text{(a)} \quad c_0(s^2 - \alpha^2) = 0 \qquad \text{(b)} \quad c_1\left[(s+1)^2 - \alpha^2\right] = 0 \tag{6.515}$$

and the recurrence relation for the coefficients

$$c_{m+2} = -\frac{c_m}{(m+s+\alpha+2)(m+s-\alpha+2)} \tag{6.516}$$

It is straightforward to demonstrate that choosing $s = \alpha$, $c_1 = 0$ yields the same solution as $s = \alpha - 1$ with $c_0 = 0$. Similarly, the solution for $s = -\alpha$, $c_1 = 0$ is the same as that for $s = -\alpha - 1$, $c_0 = 0$. Therefore, two independent solutions can be achieved in most cases by taking $s = \pm\alpha$, $c_1 = 0$. As will be seen, the exception is when α is an integer.

With $s = \alpha$, the recurrence relation for the coefficients becomes

$$c_{m+2} = -\frac{c_m}{(m+2\alpha+2)(m+2)} \tag{6.517}$$

We note that unlike the other special functions we have discussed, there is no value of α that will cause a coefficient c_{N+2} to be zero with $c_N \neq 0$. Thus the infinite Bessel series does not reduce to a finite polynomial for a specific value of α. That is, unlike the other special functions we have investigated, the Bessel functions are not polynomials.

From equation 6.517 it is straightforward to deduce that

$$c_{2m} = (-1)^m \frac{\alpha!}{2^{2m}m!\,(\alpha+m)!}c_0 \tag{6.518}$$

$$\Rightarrow \quad y = x^\alpha \alpha!\, c_0 \sum_{m=0}^{\infty} (-1)^m \frac{1}{m!\,(\alpha+m)!}\left(\frac{x}{2}\right)^{2m} \tag{6.519}$$

Setting

$$c_0 \equiv \frac{1}{2^\alpha \alpha!} \tag{6.520}$$

$$\Rightarrow \quad y(x) \equiv J_\alpha(x) = \sum_{m=0}^{\infty} (-1)^m \frac{1}{m!\,(m+\alpha)!}\left(\frac{x}{2}\right)^{2m+\alpha} \tag{6.521}$$

For $s = -\alpha$, the series is

$$J_{-\alpha}(x) = \sum_{m=0}^{\infty} (-1)^m \frac{1}{m!\,(m-\alpha)!}\left(\frac{x}{2}\right)^{2m-\alpha} \tag{6.522}$$

If α is not an integer, both Bessel functions contain fractional root branch points at the origin, reflecting the singularity structure of the differential equation. For positive α, $J_\alpha(x)$ is finite at $x = 0$, while $J_{-\alpha}(0)$ is infinite. To make $J_{-\alpha}(z)$ analytic along the positive real axis, the cut associated with the branch point at the origin is taken to run from 0 to $-\infty$. As deduced from the Cauchy ratio test, these Bessel series converge for all finite x.

Bessel functions arise most frequently in problems exhibiting cylindrical (and sometimes spherical) symmetry. In these problems, the Bessel function is a function of the radial coordinate, which extends over the range $[0, \infty]$. We therefore restrict x to this range.

If α is not an integer, as mentioned above, $J_\alpha(x)$ and $J_{-\alpha}(x)$ are independent solutions to the Bessel equation. However, if $\alpha \to n$, where n is an integer, then

$$J_{-n}(x) = \sum_{m=0}^{\infty} (-1)^m \frac{1}{m!\,(m-n)!} \left(\frac{x}{2}\right)^{2m-n} \tag{6.523}$$

But as noted in equation 6.24 or 6.26, $N! = \infty$ if N is a negative integer. Thus the terms for which $m < n$ are zero, and equation 6.523 is

$$J_{-n}(x) = \sum_{m=n}^{\infty} (-1)^m \frac{1}{m!\,(m-n)!} \left(\frac{x}{2}\right)^{2m-n} \tag{6.524}$$

With a new summation index $m' = m - n$, this becomes

$$J_{-n}(x) = (-1)^n \sum_{m=0}^{\infty} (-1)^m \frac{1}{m!\,(m+n)!} \left(\frac{x}{2}\right)^{2m+n} = (-1)^n J_n(x) \tag{6.525}$$

Thus, for integer n, $J_n(x)$ and $J_{-n}(x)$ are not independent.

From the series representation we note that

$$J_\alpha(-x) = (-1)^\alpha J_\alpha(x) \tag{6.526}$$

which means that the integer-order Bessel functions have parity described by

$$J_n(-x) = (-1)^n J_n(x) \tag{6.527}$$

We also note from the series representation that

$$J_\alpha(0) = 0 \qquad \alpha > 0 \tag{6.528}$$

$$J_0(0) = 1 \tag{6.529}$$

Generating functions for integer order

A generating function for Bessel functions of arbitrary order α cannot be found. To demonstrate that a generator of integer-order Bessel functions is

$$\Phi(z, x) = e^{(z-1/z)x/2} = \sum_{n=-\infty}^{\infty} z^n J_n(x) \tag{6.530}$$

we expand the exponential in a Laurent series.

$$e^{(z-1/z)x/2} = e^{xz/2} e^{-x/2z} = \sum_{r=0}^{\infty} \sum_{s=0}^{\infty} (-1)^s \left(\frac{x}{2}\right)^{r+s} \frac{z^{r-s}}{r!\,s!} \tag{6.531}$$

Defining

$$C(r, s) \equiv (-1)^s \frac{1}{r!\,s!} \left(\frac{x}{2}\right)^{r+s} \tag{6.532}$$

$$\Rightarrow \quad e^{(z-1/z)x/2} = z^0[C(0,0) + C(1,1) + C(2,2) + \cdots]$$
$$+z[C(1,0) + C(2,1) + C(3,2) + \cdots] + z^{-1}[C(0,1) + C(1,2) + C(2,3) + \cdots]$$
$$+z^2[C(2,0) + C(3,1) + C(4,2) + \cdots] + z^{-2}[C(0,2) + C(1,3) + C(2,4) + \cdots]$$
$$+ \cdots \tag{6.533}$$

We see that for $n \geqslant 0$, the terms involving z^n and z^{-n} are

$$z^n \sum_{m=0}^{\infty} C(n+m, m) = z^n \sum_{m=0}^{\infty} (-1)^m \frac{1}{(n+m)!\,m!} \left(\frac{x}{2}\right)^{n+2m} \tag{6.534a}$$

$$z^{-n} \sum_{m=0}^{\infty} C(m, n+m) = z^{-n} \sum_{m=0}^{\infty} (-1)^{n+m} \frac{1}{n!\,(n+m)!} \left(\frac{x}{2}\right)^{n+2m} \tag{6.534b}$$

Thus for positive n,

$$\sum_{m=0}^{\infty} (-1)^m \frac{1}{(n+m)!\,m!} \left(\frac{x}{2}\right)^{n+2m} = J_n(x) \tag{6.535a}$$

$$\sum_{m=0}^{\infty} (-1)^{m+n} \frac{1}{(n+m)!\,m!} \left(\frac{x}{2}\right)^{n+2m} = (-1)^n J_n(x) = J_{-n}(x) \tag{6.535b}$$

$$\Rightarrow \quad e^{(z-1/z)x/2} = \sum_{n=-\infty}^{\infty} z^n J_n(x) \tag{6.530}$$

Since $J_n(x)$ and $J_{-n}(x)$ are not independent, this expression can also be written

$$\begin{aligned} e^{(z-1/z)x/2} &= \sum_{n=-\infty}^{-1} z^n J_n(x) + J_0(x) + \sum_{n=1}^{\infty} z^n J_n(x) \\ &= J_0(x) + \sum_{n=1}^{\infty} \left[z^n + (-1)^n z^{-n}\right] J_n(x) \end{aligned} \tag{6.536}$$

Treating z as a complex variable, we let $z = e^{i\theta}$. Then equation 6.536 becomes

$$e^{ix\sin\theta} = J_0(x) + \sum_{n=1}^{\infty} \left[e^{in\theta} + (-1)^n e^{-in\theta}\right] J_n(x) \tag{6.537}$$

The sum is now separated into a sum over even values of n and a sum over odd n. In the sum over even indices, we replace n by $2n$ and in the sum over odd n, we replace n by $2n + 1$. Then

$$e^{ix\sin\theta} = J_0(x) + 2\sum_{n=1}^{\infty} \cos(2n\theta) J_{2n}(x) + 2i \sum_{n=0}^{\infty} \sin[(2n+1)\theta] J_{2n+1}(x) \tag{6.538}$$

Equating real parts, and separately imaginary parts, we obtain generating functions for even and odd-order Bessel functions. They are

$$\cos(x \sin\theta) = J_0(x) + 2\sum_{n=1}^{\infty} \cos(2n\theta) J_{2n}(x) \tag{6.539a}$$

$$\sin(x \sin\theta) = 2\sum_{n=0}^{\infty} \sin[(2n+1)\theta] J_{2n+1}(x) \tag{6.539b}$$

Powers of x in terms of integer-order Bessel functions

Using equation 6.539b, consider

$$\frac{\partial}{\partial\theta}\sin(x\sin\theta) = x\cos\theta[\cos(x\sin\theta)] = 2\sum_{n=0}^{\infty}(2n+1)\cos[(2n+1)\theta]J_{2n+1}(x) \quad (6.540)$$

Setting $\theta = 0$, this becomes

$$x = 2\sum_{n=0}^{\infty}(2n+1)J_{2n+1}(x) \quad (6.541)$$

From equation 6.539a,

$$\begin{aligned}\frac{\partial^2}{\partial\theta^2}\cos(x\sin\theta) &= x\sin\theta[\sin(x\sin\theta)] - x^2\cos^2\theta[\cos(x\sin\theta)] \\ &= 2\sum_{n=1}^{\infty}\left[-4n^2\cos(2n\theta)J_{2n}(x)\right]\end{aligned} \quad (6.542)$$

Again, with $\theta = 0$, we obtain

$$x^2 = 8\sum_{n=1}^{\infty}n^2 J_{2n}(x) \quad (6.543)$$

Following a process like this, one can determine any power of x in terms of sums over integer-order Bessel functions (see Problem 47).

Integral representations

Since the sine and cosine functions separately comprise complete sets of orthogonal functions, we can use equations 6.539 to develop integral representations for integer Bessel functions. It is straightforward to show that

$$\int_0^{\pi}\cos(2n\theta)\cos(2m\theta)\,d\theta = \begin{cases}\frac{\pi}{2}\delta_{nm} & m, n \neq 0 \\ \pi & m = n = 0\end{cases} \quad (6.544a)$$

$$\int_0^{\pi}\sin[(2n+1)\theta]\sin[(2m+1)\theta]\,d\theta = \frac{\pi}{2}\delta_{nm} \quad (6.544b)$$

Therefore, multiplying equation 6.539a by $\cos(2m\theta)$ and integrating yields

$$\text{(a)}\quad J_0(x) = \frac{1}{\pi}\int_0^{\pi}\cos(x\sin\theta)\,d\theta \qquad \text{(b)}\quad J_{2m}(x) = \frac{2}{\pi}\int_0^{\pi}\cos(x\sin\theta)\cos(2m\theta)\,d\theta \qquad m > 0 \quad (6.545)$$

Similarly, multiplying equation 6.539b by $\sin[(2m+1)\theta]$ and integrating leads to

$$J_{2m+1}(x) = \frac{2}{\pi}\int_0^{\pi}\sin(x\sin\theta)\sin[(2m+1)\theta]\,d\theta \qquad m > 0 \quad (6.545c)$$

To develop another integral representation of integer-order Bessel functions, consider the contour integral

$$\frac{1}{2\pi i}\oint e^{(z-1/z)x/2}\frac{dz}{z^{N+1}} = \sum_{n=0}^{\infty}J_n(x)\frac{1}{2\pi i}\oint z^{n-N-1}\,dz \quad (6.546)$$

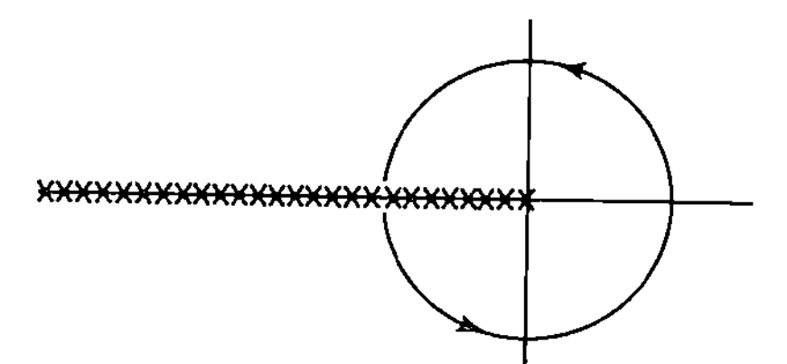

Figure 6.9 Circular contour for a noninteger Bessel function integral.

Taking the contour to be a circle centered at the origin, we have

$$\frac{1}{2\pi i}\oint z^{n-N-1}\,dz = \delta_{nN} \tag{6.389}$$

$$\Rightarrow \quad J_N(x) = \frac{1}{2\pi i}\oint e^{(z-1/z)x/2}\,\frac{dz}{z^{N+1}} \tag{6.547}$$

We let $z = e^{i\theta}$, which means that we are taking the radius of our circular contour to be 1. As mentioned above, when α is not an integer, the cut associated with the fractional root branch point is taken to run from 0 to $-\infty$. Therefore, any counterclockwise contour for noninteger α would have to be taken from $-\pi$ to π so that it will not cross the cut (see Figure 6.9). To remain consistent with noninteger α, we transverse the same contour for integer order. That is, equation 6.548 becomes

$$J_N(x) = \frac{1}{2\pi}\int_{-\pi}^{\pi} e^{i(x\sin\theta - N\theta)}\,d\theta \tag{6.548}$$

Writing the exponential in terms of its real and imaginary parts we note that since $J_N(x)$ is real,

$$J_N(x) = \frac{1}{2\pi}\int_{-\pi}^{\pi}\cos(x\sin\theta - N\theta)\,d\theta = \frac{1}{\pi}\int_{0}^{\pi}\cos(x\sin\theta - N\theta)\,d\theta \tag{6.549}$$

For noninteger order, *Poisson's integral representation* can be developed starting from the series representation. Writing $(m + \alpha)!$ in terms of the Γ-function, the Bessel function is

$$J_\alpha(x) = \sum_{m=0}^{\infty}(-1)^m \frac{1}{m!\,\Gamma(m+\alpha+1)}\left(\frac{x}{2}\right)^{2m+\alpha} \tag{6.550}$$

From the definition of the β-function of equation 6.16,

$$\frac{1}{\Gamma(m+\alpha+1)} = \frac{1}{\Gamma(m+\frac{1}{2})\Gamma(\alpha+\frac{1}{2})}\int_0^1 t^{m-1/2}(1-t)^{\alpha-1/2}\,dt \tag{6.551}$$

Substituting $t = w^2$ and using the fact that the resulting integrand is an even function, we obtain

$$\frac{1}{\Gamma(m+\alpha+1)} = \frac{2}{\Gamma(m+\frac{1}{2})\Gamma(\alpha+\frac{1}{2})}\int_0^1 w^{2m}(1-w^2)^{\alpha-1/2}\,dw$$

$$= \frac{1}{\Gamma(m+\frac{1}{2})\Gamma(\alpha+\frac{1}{2})}\int_{-1}^1 w^{2m}(1-w^2)^{\alpha-1/2}\,dw \tag{6.552}$$

Therefore,

$$J_\alpha(x) = \left(\frac{x}{2}\right)^\alpha \frac{1}{\Gamma(\alpha + \frac{1}{2})} \int_{-1}^{1} (1 - w^2)^{\alpha - 1/2} \left[\sum_{m=0}^{\infty} (-1)^m \frac{1}{m!\,\Gamma(m + \frac{1}{2})} \left(\frac{wx}{2}\right)^{2m} \right] dw \tag{6.553}$$

From equation 6.89,

$$m!\,\Gamma(m + \tfrac{1}{2}) = \frac{\sqrt{\pi}}{2^{2m}}(2m)! \tag{6.554}$$

Therefore,

$$\begin{aligned} J_\alpha(x) &= \left(\frac{x}{2}\right)^\alpha \frac{1}{\sqrt{\pi}\,\Gamma(\alpha + \frac{1}{2})} \int_{-1}^{1} (1 - w^2)^{\alpha - 1/2} \left[\sum_{m=0}^{\infty} (-1)^m \frac{(wx)^{2m}}{(2m)!} \right] dw \\ &= \left(\frac{x}{2}\right)^\alpha \frac{1}{\sqrt{\pi}\,\Gamma(\alpha + \frac{1}{2})} \int_{-1}^{1} (1 - w^2)^{\alpha - 1/2} \cos(wx)\, dw \end{aligned} \tag{6.555}$$

which is valid for $\alpha \neq -\frac{1}{2}, -\frac{3}{2}, -\frac{5}{2}, \ldots$. This can also be written

$$J_\alpha(x) = \left(\frac{x}{2}\right)^\alpha \frac{1}{\sqrt{\pi}\,\Gamma(\alpha + \frac{1}{2})} \int_{-1}^{1} (1 - w^2)^{\alpha - 1/2} e^{iwx}\, dw \tag{6.556}$$

To see this, we note that the limits of integration are symmetric about $w = 0$, and $(1 - w^2)^{\alpha - 1/2} \sin(wx)$ is an odd function. Therefore

$$\int_{-1}^{1} (1 - w^2)^{\alpha - 1/2} \sin(wx)\, dw = 0 \tag{6.557}$$

confirming the validity of equation 6.556. The result expressed in equation 6.557 can also be argued using the fact the $J_\alpha(x)$ is real.

Recurrence relations

Because there is no generating function for noninteger Bessel functions, recurrence relations between Bessel functions of general order must be deduced using other approaches. Consider

$$J_\alpha'(x) = \sum_{m=0}^{\infty} (-1)^m \frac{(2m + \alpha)}{m!\,(m + \alpha)!} \frac{x^{2m+\alpha-1}}{2^{2m+\alpha}} \tag{6.558}$$

Writing $2m + \alpha = (m + \alpha) + m$, this becomes

$$\begin{aligned} 2J_\alpha'(x) &= \sum_{m=0}^{\infty} (-1)^m \frac{1}{m!\,(m + \alpha - 1)!} \left(\frac{x}{2}\right)^{2m+\alpha-1} \\ &\quad + \sum_{m=1}^{\infty} (-1)^m \frac{1}{(m - 1)!\,(m + \alpha)!} \left(\frac{x}{2}\right)^{2m+\alpha-1} \end{aligned} \tag{6.559}$$

Because $(-1)! = \infty$, the $m = 0$ term in the second sum is zero, so the sum actually begins at $m = 1$. In the second sum, we replace m by $m + 1$ to obtain

$$\begin{aligned} 2J_\alpha'(x) &= \sum_{m=0}^{\infty} (-1)^m \frac{1}{m!\,(m + \alpha - 1)!} \left(\frac{x}{2}\right)^{2m+\alpha-1} \\ &\quad + \sum_{m=0}^{\infty} (-1)^{m+1} \frac{1}{m!\,(m + \alpha + 1)!} \left(\frac{x}{2}\right)^{2m+\alpha+1} \end{aligned} \tag{6.560a}$$

Recognizing each sum as a series description of a Bessel function, equation 6.560a is

$$2J'_\alpha(x) = J_{\alpha-1}(x) - J_{\alpha+1}(x) \tag{6.560b}$$

Consider

$$D[x^\alpha J_\alpha(x)] = x^\alpha\left(\frac{\alpha}{x}J_\alpha(x) + J'_\alpha(x)\right) \tag{6.561}$$

Using the series representation for $J_\alpha(x)$, we have

$$D[x^\alpha J_\alpha(x)] = \sum_{m=0}^{\infty}(-1)^m \frac{1}{m!\,(m+\alpha-1)!}2^\alpha\left(\frac{x}{2}\right)^{2m+2\alpha-1} = x^\alpha J_{\alpha-1}(x) \tag{6.562}$$

$$\Rightarrow \quad \frac{\alpha}{x}J_\alpha(x) + J'_\alpha(x) = J_{\alpha-1}(x) \tag{6.563}$$

Combining

$$D[x^{-\alpha}J_\alpha(x)] = -x^{-\alpha}\left(\frac{-\alpha}{x}J_\alpha(x) + J'_\alpha(x)\right) \tag{6.564}$$

with one of the results of Problem 49,

$$D[x^{-\alpha}J_\alpha(x)] = -x^{-\alpha}J_{\alpha+1}(x) \tag{6.565}$$

$$\Rightarrow \quad \frac{\alpha}{x}J_\alpha(x) - J'_\alpha(x) = J_{\alpha+1}(x) \tag{6.566}$$

Adding and subtracting equations 6.563 and 6.566 yields

$$\frac{2\alpha}{x}J_\alpha(x) = J_{\alpha-1}(x) + J_{\alpha+1}(x) \tag{6.567}$$

$$2J'_\alpha(x) = J_{\alpha-1}(x) - J_{\alpha+1}(x) \tag{6.568}$$

Half-integer Bessel functions

Bessel functions of order $\alpha = n + \frac{1}{2}$ (n an integer) can be expressed in terms of elementary functions. For example,

$$J_{1/2}(x) = \left(\frac{x}{2}\right)^{1/2}\sum_{m=0}^{\infty}(-1)^m\frac{1}{m!\,(m+\frac{1}{2})!}\left(\frac{x}{2}\right)^{2m} \tag{6.569}$$

But

$$(m+\tfrac{1}{2})! = (m+\tfrac{1}{2})\Gamma(m+\tfrac{1}{2}) \tag{6.570}$$

Using Legendre's duplication formula

$$\Gamma(m+\tfrac{1}{2}) = \frac{\sqrt{\pi}}{2^{2m}}\frac{(2m)!}{m!} \tag{6.89}$$

$$\Rightarrow \quad (m+\tfrac{1}{2})! = \frac{\sqrt{\pi}\,(2m+1)!}{2^{2m+1}m!} \tag{6.571}$$

Therefore, after multiplying one factor of x into the sum, we have

$$J_{1/2}(x) = \left(\frac{2}{\pi x}\right)^{1/2}\sum_{m=0}^{\infty}(-1)^m\frac{x^{2m+1}}{(2m+1)!} = \left(\frac{2}{\pi x}\right)^{1/2}\sin x \tag{6.572}$$

Using an identical analysis, it is straightforward to show that

$$J_{-1/2}(x) = \left(\frac{2}{\pi x}\right)^{1/2}\cos x \tag{6.573}$$

Other half-order Bessel functions can be generated from the recurrence relations of equations 6.563 and 6.566. For example, with $\alpha = \frac{1}{2}$, equation 6.563 yields

$$J_{3/2}(x) = \frac{1}{2x}J_{1/2}(x) - J'_{1/2}(x) = \left(\frac{2}{\pi x}\right)^{1/2}\left[\frac{\sin x}{x} - \cos x\right] \tag{6.574}$$

With $\alpha = -\frac{1}{2}$, equation 6.566 yields

$$J_{-3/2}(x) = -\frac{1}{2x}J_{-1/2}(x) + J'_{-1/2}(x) = -\left(\frac{2}{\pi x}\right)^{1/2}\left[\frac{\cos x}{x} + \sin x\right] \tag{6.575}$$

Orthonormalization condition

Unlike the Sturm–Liouville polynomials we have studied thus far, the Bessel functions do not satisfy an orthonormalization condition with respect to order. That is,

$$\int_0^\infty \rho(x) J_\alpha(x) J_\beta(x)\, dx \neq N_\alpha \delta_{\alpha\beta} \tag{6.576}$$

The reason that such an orthogonality condition does not exist is that the index defining the order of a Bessel function (α or β) is not the eigenvalue of the differential equation.

To understand this, we note that for any Sturm–Liouville equation, the weight function $\rho(x)$ must be finite for all allowed values of x. Dividing

$$\left[x'^2 D^2 + x'D + (k^2 x'^2 - \alpha^2)\right] y = 0 \tag{6.513a}$$

by x', we express the Bessel differential equation in the self-adjoint form

$$\left[x'D^2 + D + \left(k^2 x' - \frac{\alpha^2}{x'}\right)\right] y = \frac{d}{dx'}\left(x'\frac{dy}{dx'}\right) + \left(k^2 x' - \frac{\alpha^2}{x'}\right) y = 0 \tag{6.577}$$

If we write this in the form

$$\frac{d}{dx'}\left(x'\frac{dy}{dx'}\right) + k^2 x' y = -\left(\frac{-\alpha^2}{x'}\right) y \tag{6.578a}$$

the weight function would be identified as $\rho(x') = -1/x'$, which is negative for all x'. In addition, $1/x'$ is infinite at $x' = 0$. Since the weight function is defined to be non-negative and finite over the entire allowed range of x', $-1/x'$ cannot be $\rho(x')$. Thus, $-\alpha^2$ cannot be the eigenvalue. Writing equation 6.577 in the form

$$\frac{d}{dx'}\left(x'\frac{dy}{dx'}\right) - \frac{\alpha^2}{x'} y = -k^2 x' y \tag{6.578b}$$

we identify the Sturm–Liouville quantities for Bessel's equation as

$$\text{(a)}\quad p(x') = x' \qquad \text{(b)}\quad q(x') = \frac{\alpha^2}{x'} \qquad \text{(c)}\quad \rho(x') = x' \qquad \text{(d)}\quad \lambda = k^2 \tag{6.579}$$

Thus, as demonstrated in the general Sturm–Liouville analysis, an orthonormalization condition exists involving different k-values, not different values of α.

To develop the orthonormalization condition, we write the Bessel function as $J_\alpha(x) = J_\alpha(kx')$. With $x' = 1$, we denote the zeros of $J_\alpha(k)$ by k_m. That is, k_1 is the first zero of $J_\alpha(k)$, k_2 is the second zero of $J_\alpha(k)$, and so on, so that

$$J_\alpha(k_m) = 0 \tag{6.580}$$

(We now drop the prime on x.)

We write the Bessel differential equation in self-adjoint form, for two different eigenvalues:

$$\frac{d}{dx}\left(x\frac{d}{dx}J_\alpha(k_l x)\right) + \left(k_l^2 x - \frac{\alpha^2}{x}\right)J_\alpha(k_l x) = 0 \tag{6.581a}$$

$$\frac{d}{dx}\left(x\frac{d}{dx}J_\alpha(k_m x)\right) + \left(k_m^2 x - \frac{\alpha^2}{x}\right)J_\alpha(k_m x) = 0 \tag{6.581b}$$

Multiplying equation 6.581a by $J_\alpha(k_m x)$ and equation 6.581b by $J_\alpha(k_l x)$, subtracting, and integrating, we obtain

$$\frac{1}{(k_l^2 - k_m^2)}\int_0^1\left[J_\alpha(k_l x)\frac{d}{dx}\left(x\frac{d}{dx}J_\alpha(k_m x)\right) - J_\alpha(k_m x)\frac{d}{dx}\left(x\frac{d}{dx}J_\alpha(k_l x)\right)\right]dx$$

$$= \int_0^1 xJ_\alpha(k_l x)J_\alpha(k_m x)\,dx \tag{6.582}$$

The integral

$$\int_0^1 J_\alpha(k_m x)\frac{d}{dx}\left(x\frac{d}{dx}J_\alpha(k_l x)\right)dx$$

is integrated twice by parts. For the first integration, we take

$$\text{(a)}\quad u = J_\alpha(k_m x) \qquad \text{(b)}\quad dv = \frac{d}{dx}\left(x\frac{d}{dx}J_\alpha(k_l x)\right)dx \tag{6.583}$$

For the second, we let

$$\text{(c)}\quad u = x\frac{d}{dx}J_\alpha(k_m x) \qquad \text{(d)}\quad dv = \frac{d}{dx}J_\alpha(k_l x) \tag{6.583}$$

The result is

$$\int_0^1 J_\alpha(k_m x)\frac{d}{dx}\left(x\frac{d}{dx}J_\alpha(k_l x)\right)dx$$

$$= \int_0^1 J_\alpha(k_l x)\frac{d}{dx}\left(x\frac{d}{dx}J_\alpha(k_m x)\right)dx + x\left[J_\alpha(k_m x)\frac{d}{dx}J_\alpha(k_l x) - J_\alpha(k_l x)\frac{d}{dx}J_\alpha(k_m x)\right]_{x=0}^{x=1}$$

$$= \int_0^1 J_\alpha(k_l x)\frac{d}{dx}\left(x\frac{d}{dx}J_\alpha(k_m x)\right)dx + \left[J_\alpha(k_m x)\frac{d}{dx}J_\alpha(k_l x) - J_\alpha(k_l x)\frac{d}{dx}J_\alpha(k_m x)\right]_{x=1} \tag{6.584}$$

Therefore, equation 6.582 becomes

$$\int_0^1 xJ_\alpha(k_l x)J_\alpha(k_m x)\,dx = \frac{1}{(k_l^2 - k_m^2)}\left[J_\alpha(k_l x)\frac{d}{dx}J_\alpha(k_m x) - J_\alpha(k_m x)\frac{d}{dx}J_\alpha(k_l x)\right]_{x=1} \tag{6.585}$$

At $x = 1$, each term in the brackets is zero since $J_\alpha(k_l) = J_\alpha(k_m) = 0$. Therefore,

$$\int_0^1 xJ_\alpha(k_l x)J_\alpha(k_m x)\,dx = 0 \qquad l \neq m \tag{6.586}$$

When $k_l = k_m$, equation 6.585 is in the form $0/0$. We can therefore evaluate this limit using l'Hospital's rule. Letting $k_l = k \to k_m$, this becomes

$$\lim_{k \to k_m} \frac{1}{(k^2 - k_m^2)} \left[J_\alpha(kx) \frac{d}{dx} J_\alpha(k_m x) - J_\alpha(k_m x) \frac{d}{dx} J_\alpha(kx) \right]_{x=1}$$

$$= \lim_{k \to k_m} \frac{1}{2k} \frac{d}{dk} \left[J_\alpha(kx) \frac{d}{dx} J_\alpha(k_m x) - J_\alpha(k_m x) \frac{d}{dx} J_\alpha(kx) \right]_{x=1} \tag{6.587}$$

We can set $x = 1$ in the argument of those Bessel functions that are not being differentiated with respect to x. Thus the second term is zero, and

$$\begin{aligned} &\lim_{k \to k_m} \frac{1}{(k^2 - k_m^2)} \left[J_\alpha(kx) \frac{d}{dx} J_\alpha(k_m x) - J_\alpha(k_m x) \frac{d}{dx} J_\alpha(kx) \right]_{x=1} \\ &= \frac{J_\alpha'(k_m)}{2k_m} \frac{d}{dx} [J_\alpha(k_m x)]_{x=1} \end{aligned} \tag{6.588}$$

Writing

$$\frac{d}{dx} [J_\alpha(k_m x)]_{x=1} = k_m \frac{d}{dy} J_\alpha(y) \bigg|_{y=k_m} = k_m J_\alpha'(k_m) \tag{6.589}$$

$$\Rightarrow \quad \int_0^1 x J_\alpha(k_m x) J_\beta(k_l x)\, dx = \tfrac{1}{2} [J_\alpha'(k_m)]^2 \delta_{lm} \tag{6.590}$$

From equations 6.563 and 6.566,

$$\text{(a)} \quad J_\alpha'(k_m) = J_{\alpha-1}(k_m) \qquad \text{(b)} \quad J_\alpha'(k_m) = -J_{\alpha+1}(k_m) \tag{6.591}$$

Thus the orthonormality condition can also be expressed in three other forms.

$$\int_0^1 x J_\alpha(k_m x) J_\alpha(k_l x)\, dx = \tfrac{1}{2} [J_{\alpha-1}(k_m)]^2 \delta_{lm} \tag{6.592a}$$

$$\int_0^1 x J_\alpha(k_m x) J_\alpha(k_l x)\, dx = \tfrac{1}{2} [J_{\alpha+1}(k_m)]^2 \delta_{lm} \tag{6.592b}$$

$$\int_0^1 x J_\alpha(k_m x) J_\alpha(k_l x)\, dx = -\tfrac{1}{2} J_{\alpha-1}(k_m) J_{\alpha+1}(k_m) \delta_{lm} \tag{6.592c}$$

We note that it was not necessary to restrict x to the range [0, 1]. If we had defined the argument of the Bessel function in terms of an arbitrary x_0 as

$$J_\alpha(x) = J_\alpha\left(\frac{x'}{x_0} k_l \right) \tag{6.593}$$

our four orthonormalization conditions would have been

$$\int_0^{x_0} x J_\alpha\left(\frac{x}{x_0} k_m \right) J_\alpha\left(\frac{x}{x_0} k_l \right) dx = \tfrac{1}{2} [J_\alpha'(k_m)]^2 \delta_{lm} \tag{6.594a}$$

$$\int_0^{x_0} x J_\alpha\left(\frac{x}{x_0} k_m \right) J_\alpha\left(\frac{x}{x_0} k_l \right) dx = \tfrac{1}{2} [J_{\alpha-1}(k_m)]^2 \delta_{lm} \tag{6.594b}$$

$$\int_0^{x_0} x J_\alpha\left(\frac{x}{x_0} k_m \right) J_\alpha\left(\frac{x}{x_0} k_l \right) dx = \tfrac{1}{2} [J_{\alpha+1}(k_m)]^2 \delta_{lm} \tag{6.594c}$$

$$\int_0^{x_0} x J_\alpha\left(\frac{x}{x_0} k_m \right) J_\alpha\left(\frac{x}{x_0} k_l \right) dx = -\tfrac{1}{2} J_{\alpha-1}(k_m) J_{\alpha+1}(k_m) \delta_{lm} \tag{6.594d}$$

Neumann function

As noted earlier, when α is not an integer, $J_\alpha(x)$ and $J_{-\alpha}(x)$ are two independent solutions to the Bessel equation. However, when $\alpha = n$ is an integer,

$$J_{-n}(x) = (-1)^n J_n(x) \tag{6.525}$$

and $J_{-n}(x)$ is not the second independent solution.

Rather than resorting to the method of reduction of order, it is customary to form a second independent solution by defining a particular linear combination of $J_\alpha(x)$ and $J_{-\alpha}(x)$ and then letting $\alpha \to n$.

Clearly, for arbitrary α, $A_\alpha J_\alpha(x) + A_{-\alpha} J_{-\alpha}(x)$ will be a solution to the Bessel equation. The Neumann function is defined by

$$Y_\alpha(x) \equiv \frac{\cos(\pi\alpha) J_\alpha(x) - J_{-\alpha}(x)}{\sin(\pi\alpha)} \tag{6.595}$$

When α becomes an integer n, $Y_n(x)$ is obviously in the form $0/0$. Applying l'Hospital's rule, we obtain

$$Y_n(x) \equiv \lim_{\alpha\to n} \frac{\cos(\pi\alpha) J_\alpha(x) - J_{-\alpha}(x)}{\sin(\pi\alpha)} = \frac{(-1)^n}{\pi} \lim_{\alpha\to n} \frac{\partial}{\partial\alpha}\left[\cos(\pi\alpha) J_\alpha(x) - J_{-\alpha}(x)\right] \tag{6.596}$$

We will differentiate the Bessel function with respect to order, using its series representation. Writing $(m+\alpha)! = \Gamma(m+\alpha+1)$, and using the definition of the digamma function, we have

$$\psi(p+1) \equiv \frac{d}{dp}\log[\Gamma(p+1)] = \frac{1}{\Gamma(p+1)} \frac{d}{dp}\Gamma(p+1) \tag{6.39}$$

$$\Rightarrow \quad \text{(a)} \quad \frac{\partial}{\partial\alpha} J_\alpha(x) = \left(\frac{x}{2}\right) J_\alpha(x) - \Lambda_\alpha(x) \qquad \text{(b)} \quad \frac{\partial}{\partial\alpha} J_{-\alpha}(x) = -\left(\frac{x}{2}\right) J_{-\alpha}(x) + \Lambda_{-\alpha}(x) \tag{6.597}$$

where

$$\Lambda_{\pm\alpha}(x) \equiv \sum_{m=0}^{\infty} (-1)^m \frac{1}{m!(m\pm\alpha)!}\left(\frac{x}{2}\right)^{2m\pm\alpha} \psi(m\pm\alpha+1) \tag{6.598}$$

Therefore, with $J_n(x) = (-1)^n J_{-n}(x)$, the Neumann function of integer order is

$$Y_n(x) = \frac{1}{\pi}\left\{2\log\left(\frac{x}{2}\right) J_n(x) - \left[\Lambda_n(x) + (-1)^n \Lambda_{-n}(x)\right]\right\} \tag{6.599}$$

In Problem 6, it is shown that for $n - m \geqslant 1$,

$$\frac{\psi(m-n+1)}{\Gamma(m-n+1)} = (-1)^{n-m}(n-m-1)! \tag{6.600}$$

Therefore,

$$\Lambda_{-n}(x) = (-1)^n \sum_{m=0}^{n-1} \frac{(n-m-1)!}{m!}\left(\frac{x}{2}\right)^{2m-n} + \sum_{m=n}^{\infty} (-1)^m \frac{\psi(m-n+1)}{m!\,(m-n)!}\left(\frac{x}{2}\right)^{2m-n} \tag{6.601a}$$

In the second sum, m is replaced by $m + n$ to obtain

$$\begin{aligned}\Lambda_{-n}(x) &= (-1)^n \sum_{m=0}^{n-1} \frac{(n-m-1)!}{m!}\left(\frac{x}{2}\right)^{2m-n} \\ &\quad + (-1)^n \sum_{m=0}^{\infty} (-1)^m \frac{\psi(m+1)}{m!\,(m+n)!}\left(\frac{x}{2}\right)^{2m+n}\end{aligned} \tag{6.601b}$$

$$\begin{aligned}\Rightarrow \quad Y_n(x) &= \frac{2}{\pi}\log\left(\frac{x}{2}\right)J_n(x) - \frac{1}{\pi}\sum_{m=0}^{n-1} \frac{(n-m-1)!}{m!}\left(\frac{x}{2}\right)^{2m-n} \\ &\quad - \frac{1}{\pi}\sum_{m=0}^{\infty} (-1)^m \left[\frac{\psi(m+n+1)+\psi(m+1)}{m!\,(m+n)!}\right]\left(\frac{x}{2}\right)^{2m+n}\end{aligned} \tag{6.602}$$

Replacing n by $-n$ in equation 6.599, it is straightforward to deduce that

$$Y_{-n}(x) = (-1)^n Y_n(x) \tag{6.603}$$

Recurrence relations for the Neumann function

Since the Neumann function is a linear combination of $J_\alpha(x)$ and $J_{-\alpha}(x)$, then for noninteger α, all recurrence relations satisfied by $J_\alpha(x)$ and $J_{-\alpha}(x)$ are also satisfied by $Y_\alpha(x)$ for any order α.

Hankel functions

The Hankel functions are linear combinations of the Bessel and Neumann functions. They are defined by

$$H_\alpha^{(\pm)}(x) \equiv J_\alpha(x) \pm iY_\alpha(x) \tag{6.604}$$

They are also frequently denoted in the literature as

$$\text{(a)}\quad H_\alpha^{(1)}(x) = H_\alpha^{(+)}(x) \qquad \text{(b)}\quad H_\alpha^{(2)}(x) = H_\alpha^{(-)}(x) \tag{6.605}$$

Clearly, if x is real, the two Hankel functions are complex conjugates of one another.

Using equation 6.595 for $Y_\alpha(x)$, it is straightforward to show that

$$H_\alpha^{(\pm)}(x) = \pm\frac{i}{(\sin \pi\alpha)}\left[e^{\mp i\pi\alpha}J_\alpha(x) - J_{-\alpha}(x)\right] \tag{6.606}$$

Although the algebra and the result are quite unwieldy, it is straightforward to deduce series representations for the Hankel functions. In addition, since the Bessel and Neumann functions satisfy the same recurrence relations, the Hankel functions also satisfy the recurrence relations obeyed by the Bessel and Neumann functions.

Modified Bessel functions

Replacing x by ix in the series representation, the Bessel function becomes

$$J_\alpha(ix) = i^\alpha \sum_{m=0}^{\infty} \frac{1}{m!\,(m+\alpha)!}\left(\frac{x}{2}\right)^{2m+\alpha} \equiv i^\alpha I_\alpha(x) = e^{i\pi\alpha/2} I_\alpha(x) \tag{6.607}$$

$I_\alpha(x)$ is called *the modified Bessel function*.

When $\alpha \to -n$, with n a positive integer,

$$I_{-n}(x) = (-i)^n J_{-n}(ix) = (i)^n J_n(ix) = I_n(x) \tag{6.608}$$

By replacing x by $-ix$ in the series representation for the Bessel function, it is straightforward to show that

$$J_\alpha(-ix) = (-i)^\alpha I_\alpha(x) = e^{-i\pi\alpha/2} I_\alpha(x) \tag{6.609}$$

The modified Neumann function, denoted by $X_\alpha(x)$, is defined by

$$Y_\alpha(ix) \equiv (i)^\alpha X_\alpha(x) \tag{6.610}$$

and the modified Hankel functions are defined in a similar way. Using equation 6.606, consider

$$H_\alpha^{(+)}(ix) = \frac{i}{\sin(\pi\alpha)}\left[e^{-i\pi\alpha} J_\alpha(ix) - J_{-\alpha}(ix)\right] = (-1)^{\alpha-1}\left[\frac{I_\alpha(x) - I_{-\alpha}(x)}{\sin(\pi\alpha)}\right] \tag{6.611}$$

The modified Hankel function $K_\alpha(x)$, is defined to be real, such that

$$\text{(a)} \quad H_\alpha^{(+)}(ix) = \frac{2}{\pi}(-i)^{\alpha-1} K_\alpha(x)$$

$$\Rightarrow \quad \text{(b)} \quad K_\alpha(x) = \frac{\pi}{2}\left[\frac{I_\alpha(x) - I_{-\alpha}(x)}{\sin(\pi\alpha)}\right] \tag{6.612}$$

Consider

$$\begin{aligned} H_\alpha^{(-)}(-ix) &= \frac{-i}{\sin(\pi\alpha)}\left[e^{i\pi\alpha} J_\alpha(-ix) - J_{-\alpha}(-ix)\right] \\ &= (i)^{\alpha-1}\left[\frac{I_\alpha(x) - I_{-\alpha}(x)}{\sin(\pi\alpha)}\right] = \frac{2}{\pi}(i)^{\alpha-1} K_\alpha(x) \end{aligned} \tag{6.613}$$

Thus there is only one modified Hankel function, and

$$H_\alpha^{(\pm)}(ix) = \frac{2}{\pi}(\mp i)^{\alpha-1} K_\alpha(\pm x) \tag{6.614}$$

Spherical Bessel functions

In the quantum mechanical solution to the problem of a particle in a spherically symmetric constant potential well, which is defined by the potential energy function

$$V(\mathbf{r}) = V(r) = \begin{cases} -V_0 & r < R \\ 0 & r > R \end{cases} \tag{6.615}$$

one encounters functions related to the half-order Bessel functions.

These are designated the *spherical Bessel functions* and are defined by

$$\text{(a)} \quad j_n(x) \equiv \left(\frac{\pi}{2x}\right)^{1/2} J_{n+1/2}(x) \qquad \text{(b)} \quad \eta_n(x) \equiv (-1)^{n+1}\left(\frac{\pi}{2x}\right)^{1/2} J_{-n-1/2}(x) \tag{6.616}$$

From these definitions, if n is replaced by $-n$, it is straightforward to obtain

$$\text{(a)} \quad j_{-n}(x) = \eta_{n-1}(x) \qquad \text{(b)} \quad \eta_{-n}(x) = j_{n-1}(x) \tag{6.617}$$

Using equations 6.572, 6.573, 6.574, and 6.575, we see, for example, that

$$\begin{aligned} &\text{(a)} \quad j_0(x) = \frac{\sin x}{x} &&\text{(b)} \quad j_1(x) = \frac{\sin x}{x^2} - \frac{\cos x}{x} \\ &\text{(c)} \quad \eta_0(x) = -\frac{\cos x}{x} &&\text{(d)} \quad \eta_1(x) = -\frac{\cos x}{x^2} - \frac{\sin x}{x} \end{aligned} \tag{6.618}$$

Consistent with the definition of the spherical Bessel functions, spherical Neumann and spherical Hankel functions are also defined.

$$y_n(x) \equiv \left(\frac{\pi}{2x}\right)^{1/2} Y_{n+1/2}(x) = \left(\frac{\pi}{2x}\right)^{1/2} \frac{\cos[(n+\frac{1}{2})\pi] J_{n+1/2}(x) - J_{-n-1/2}(x)}{\sin[(n+\frac{1}{2})\pi]} \tag{6.619a}$$

But $\cos[(n+\frac{1}{2})\pi] = 0$ and $\sin[(n+\frac{1}{2})\pi] = (-1)^n$. Therefore,

$$y_n(x) = \eta_n(x) \tag{6.619b}$$

Spherical Hankel functions are defined by

$$h_n^{(\pm)}(x) \equiv j_n(x) \pm i y_n(x) = j_n(x) \pm i\eta_n(x) \tag{6.620}$$

Bessel function for small argument

It is easiest to deduce the behavior of the various Bessel functions for small x ($x \ll 1$), from the series representation. When x is small, the dominant term in the series is the one involving the smallest power of x. Thus, from the series representation (equation 6.521)

$$J_\alpha(x) \simeq \frac{1}{\alpha!}\left(\frac{x}{2}\right)^\alpha \tag{6.621}$$

$\alpha \neq$ negative integer. When α is a negative integer, $-n$,

$$J_{-n}(x) = (-1)^n J_n(x) \tag{6.525}$$

$$\Rightarrow \quad J_{-n}(x) \simeq \frac{(-1)^n}{n!}\left(\frac{x}{2}\right)^n \tag{6.622}$$

From this, the small argument Neumann function becomes

$$Y_\alpha(x) \simeq \frac{[\cos(\pi\alpha)/\alpha!](x/2)^\alpha - [1/(-\alpha)!](x/2)^{-\alpha}}{\sin \pi\alpha} \simeq -\frac{1}{(-\alpha)! \sin \pi\alpha}\left(\frac{x}{2}\right)^{-\alpha} \tag{6.623}$$

Using

$$\Gamma(\alpha)\Gamma(1-\alpha) = \Gamma(\alpha)(-\alpha)! = \frac{\pi}{\sin(\pi\alpha)} \tag{6.23b}$$

$$\Rightarrow \quad Y_\alpha(x) \simeq -\frac{\Gamma(\alpha)}{\pi}\left(\frac{x}{2}\right)^{-\alpha} \tag{6.624}$$

To avoid the infinity arising from the factorials of negative integers, this result is developed for noninteger α. When α is an integer n, we see from equation 6.602 that

$$Y_n(x) \simeq \frac{2}{\pi n!}\left(\frac{x}{2}\right)^n \log\left(\frac{x}{2}\right) - \frac{(n-1)!}{\pi}\left(\frac{x}{2}\right)^{-n} - \frac{2}{\pi n!}\left(\frac{x}{2}\right)^n[\psi(n+1)+\psi(1)] \tag{6.625a}$$

But for small x, $(x/2)^{-n}$ dominates this sum of terms. Therefore,

$$Y_n(x) \simeq -\frac{(n-1)!}{\pi}\left(\frac{x}{2}\right)^{-n} \tag{6.625b}$$

which is consistent with the result expressed in equation 6.624.

From the definition of the spherical Bessel functions expressed in equations 6.616,

$$j_n(x) \simeq \left(\frac{2}{\pi x}\right)^{1/2} \frac{1}{(n+\frac{1}{2})!}\left(\frac{x}{2}\right)^{n+1/2} \tag{6.626}$$

Using

$$(n+\tfrac{1}{2})! = \frac{\sqrt{\pi}\,(2n+1)!}{2^{2n+1}n!} \tag{6.571}$$

$$\Rightarrow \quad j_n(x) \simeq \frac{x^n}{(2n+1)!!} = \frac{n!}{(2n+1)!}(2x)^n \tag{6.627}$$

The spherical Bessel function of the second kind, for small x, is

$$\eta_n(x) \simeq (-1)^{n+1}\left(\frac{2}{\pi x}\right)^{1/2} \frac{1}{(-n-\frac{1}{2})!}\left(\frac{x}{2}\right)^{-n-1/2} \tag{6.628}$$

Using $(-n-\frac{1}{2})! = \Gamma(1-n-\frac{1}{2})$,

$$\Gamma(p)\Gamma(1-p) = \frac{\pi}{\sin \pi p} \tag{6.23b}$$

and Legendre's duplication formula,

$$\Gamma(n+\tfrac{1}{2}) = \frac{\sqrt{\pi}\,(2n)!}{2^{2n}n!} \tag{6.89}$$

$$\Rightarrow \quad (-n-\tfrac{1}{2})! = \frac{\pi}{\Gamma(n+\frac{1}{2})\sin[(n+\frac{1}{2})\pi]} = (-1)^n\frac{\sqrt{\pi}\,2^{2n}n!}{(2n)!} \tag{6.629}$$

$$\Rightarrow \quad \eta_n(x) \simeq -\frac{4}{\pi}\frac{(2n)!}{n!}\frac{1}{(2x)^{n+1}} \tag{6.630}$$

Bessel functions for large argument (asymptotic form)

We begin with the integral representation expressed in equation 6.556:

$$J_\alpha(x) = \left(\frac{x}{2}\right)^\alpha \frac{1}{\sqrt{\pi}\,\Gamma(\alpha+\frac{1}{2})}\int_{-1}^{1}(1-w^2)^{\alpha-1/2}e^{iwx}\,dw \tag{6.556}$$

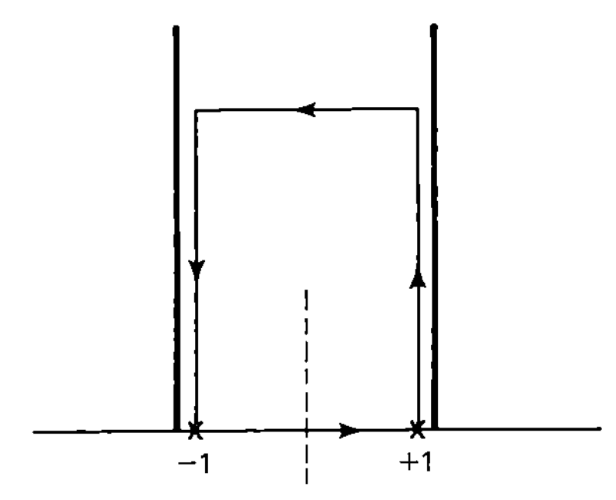

Figure 6.10 Contour and cut structure for $\oint (1 - z^2)^{\alpha - 1/2} e^{ixz}\, dz$.

Except for positive half-integer values of α, this integrand has branch points at $w = \pm 1$. Using Cauchy's theorem, we will deduce another integral description of $J_\alpha(x)$ by taking the cuts to run from the branch points to $(\pm 1 + i\infty)$. We will then consider the integral

$$I = \oint (1 - z^2)^{\alpha - 1/2} e^{ixz}\, dz \tag{6.631}$$

around the rectangular contour shown in Figure 6.10. With the complex variable $z \equiv w + iy$, the integral around the closed contour is

$$\begin{aligned} I = &\int_{-1}^{1} (1 - w^2)^{\alpha - 1/2} e^{iwx}\, dw + \int_0^\infty \left(1 - (1 + iy)^2\right)^{\alpha - 1/2} e^{ix(1+iy)} i\, dy \\ &+ \int_1^{-1} \left(1 - (w + i\infty)^2\right)^{\alpha - 1/2} e^{ix(1+i\infty)}\, dw + \int_\infty^0 \left(1 - (-1 + iy)^2\right)^{\alpha - 1/2} e^{ix(-1+iy)} i\, dy \end{aligned} \tag{6.632}$$

Because $x > 0$, the third integral, across the leg of the rectangle at $i\infty$, is zero, because of the factor $e^{-x \cdot \infty}$. Since there are no poles enclosed by this contour, $I = 0$. Therefore,

$$\begin{aligned} \int_{-1}^{1} (1 - w^2)^{\alpha - 1/2} e^{iwx}\, dw = &\ ie^{-ix} \int_0^\infty (y^2 + 2iy)^{\alpha - 1/2} e^{-xy}\, dy \\ &- ie^{ix} \int_0^\infty (y^2 - 2iy)^{\alpha - 1/2} e^{-xy}\, dy \end{aligned} \tag{6.633}$$

Defining

$$\Delta(x) \equiv ie^{-ix} \int_0^\infty (y^2 + 2iy)^{\alpha - 1/2} e^{-xy}\, dy \tag{6.634}$$

$$\Rightarrow \quad \int_{-1}^{1} (1 - w^2)^{\alpha - 1/2} e^{iwx}\, dw = \Delta(x) + \Delta^*(x) \tag{6.635}$$

After substituting $u = xy$, we obtain

$$\begin{aligned} \Delta(x) &= \frac{ie^{-ix}}{x} \int_0^\infty e^{-u} \left(\frac{2iu}{x} + \frac{u^2}{x^2} \right)^{\alpha - 1/2} du \\ &= \frac{e^{-ix}}{2} \left(\frac{2i}{x} \right)^{\alpha + 1/2} \int_0^\infty e^{-u} u^{\alpha - 1/2} \left(1 + \frac{u}{2ix}\right)^{\alpha - 1/2} du \end{aligned} \tag{6.636}$$

When x is large $(1 + u/2ix)^{\alpha - 1/2} \simeq 1$ for small values of u. The factor e^{-u} makes any contribution from large values of u negligible. Therefore, we can approximate $(1 + u/2ix)^{\alpha - 1/2} \simeq 1$ over the entire range of u. [Using the binomial expansion, it is straightforward to develop higher-order corrections to this approximation. See, for example, the text by G. Arfken, *Mathematical Methods for Physicists*, 3rd ed. (Orlando, Fla.: Academic Press, Inc., 1985), p. 616.]

Writing $i^{\alpha+1/2}$ in exponential form yields

$$\begin{aligned}\Delta(x) &\simeq \frac{1}{2}\left(\frac{2}{x}\right)^{\alpha+1/2} e^{-i[x-\alpha(\pi/2)-(\pi/4)]}\int_0^\infty u^{\alpha-1/2}e^{-u}\,du \\ &= \frac{1}{2}\left(\frac{2}{x}\right)^{\alpha+1/2}\Gamma(\alpha+\tfrac{1}{2})e^{-i[x-\alpha(\pi/2)-(\pi/4)]}\end{aligned} \tag{6.637}$$

$$\Rightarrow \int_{-1}^{1}(1-w^2)^{\alpha-1/2}e^{iwx}\,dw \simeq \left(\frac{2}{x}\right)^{\alpha+1/2}\Gamma(\alpha+\tfrac{1}{2})\cos\left(x-\alpha\frac{\pi}{2}-\frac{\pi}{4}\right) \tag{6.638}$$

$$\Rightarrow J_\alpha(x) \simeq \left(\frac{2}{\pi x}\right)^{1/2}\cos\left(x-\alpha\frac{\pi}{2}-\frac{\pi}{4}\right) \tag{6.639}$$

for large x. From equation 6.635, writing

$$J_\alpha(x) = \left(\frac{x}{2}\right)^\alpha \frac{1}{\sqrt{\pi}\,\Gamma(\alpha+\frac{1}{2})}[\Delta(x)+\Delta^*(x)] \tag{6.640}$$

and noting from equation 6.604 that

$$J_\alpha(x) = \tfrac{1}{2}\left[H_\alpha^{(+)}(x)+H_\alpha^{(-)}(x)\right] \tag{6.641}$$

we can relate $H_\alpha^{(+)}(x)$ to $\Delta^*(x)$, (so that $H_\alpha^{(+)}(x)$ is written in terms of the positive exponential) by

$$\begin{aligned}&\text{(a)}\ \ H_\alpha^{(+)}(x) = \left(\frac{x}{2}\right)^\alpha \frac{2}{\sqrt{\pi}\,\Gamma(\alpha+\frac{1}{2})}\Delta^*(x) \\ \Rightarrow\ &\text{(b)}\ \ H_\alpha^{(+)}(x) \simeq \left(\frac{2}{\pi x}\right)^{1/2} e^{i[x-\alpha(\pi/2)-(\pi/4)]}\end{aligned} \tag{6.642}$$

for large x. Similarly, $H_\alpha^{(-)}(x)$ is related to $\Delta(x)$ by

$$\begin{aligned}&\text{(a)}\ \ H_\alpha^{(-)}(x) = \left(\frac{x}{2}\right)^\alpha \frac{2}{\sqrt{\pi}\,\Gamma(\alpha+\frac{1}{2})}\Delta(x) \\ \Rightarrow\ &\text{(b)}\ \ H_\alpha^{(-)}(x) \simeq \left(\frac{2}{\pi x}\right)^{1/2} e^{i[x-\alpha(\pi/2)-(\pi/4)]}\end{aligned} \tag{6.643}$$

Therefore, referring to equation 6.604, we obtain

$$\begin{aligned}Y_\alpha(x) &= \frac{H_\alpha^{(+)}(x)-H_\alpha^{(-)}(x)}{2i} = \left(\frac{x}{2}\right)^\alpha \frac{2}{\sqrt{\pi}\,\Gamma(\alpha+\frac{1}{2})}\,\frac{\Delta^*(x)-\Delta(x)}{2i} \\ &\simeq \left(\frac{2}{\pi x}\right)^{1/2}\sin\left(x-\alpha\frac{\pi}{2}-\frac{\pi}{4}\right)\end{aligned} \tag{6.644}$$

Setting $\alpha = \pm(n+\frac{1}{2})$, we obtain asymptotic approximations for the spherical Bessel functions. They are

$$j_n(x) \simeq \frac{\cos[x-n(\pi/2)-\pi/2]}{x} = \frac{\sin[x-n(\pi/2)]}{x} \tag{6.645a}$$

$$\eta_n(x) \simeq \frac{\sin[x-n(\pi/2)-\pi/2]}{x} = -\frac{\cos[x-n(\pi/2)]}{x} \tag{6.645b}$$

6.8 Hypergeometric Functions

Gauss's hypergeometric differential equation is

$$\{x(1-x)D^2+[c-(a+b+1)x]D-ab\}y = 0 \tag{6.646a}$$

It is clear from the discussion in Chapter 5 that this equation has regular singularities at

$x = 0, 1$. Defining a new independent variable $x' \equiv 1/x$, equation 6.646a becomes

$$x'^2(x' - 1)\frac{d^2y}{dx'^2} + x'[(a + b + 1) + (2 - c)x']\frac{dy}{dx'} - aby = 0 \qquad (6.646b)$$

which has regular singularities at $x' = 0, 1$. These correspond to singularities of the original differential equation at $x = 1, \infty$. Thus equation 6.646a is singular at $x = 0$, 1, and ∞. We also note here that the differential equation and thus the solution are symmetric in the interchange of a and b.

As with the differential equations we have discussed previously, we obtain a series solution to equation 6.646a in the form

$$y = \sum_{m=0}^{\infty} \gamma_m x^{m+s} \qquad (6.647)$$

$$\Rightarrow \quad \gamma_0 s(s + c - 1) = 0 \qquad (6.648a)$$

This can be satisfied for $s = 0$ and γ_0 arbitrary. The recurrence relation for the coefficients with $s = 0$ is

$$\gamma_{m+1} = \frac{(m + a)(m + b)}{m(m + c)}\gamma_m \qquad (6.648b)$$

The general expression for the mth coefficient, obtained straightforwardly from this, is

$$\gamma_m = \frac{[(a + m - 1)(a + m - 2)\cdots(a + 1)a][(b + m - 1)(b + m - 2)\cdots(b + 1)b]}{m![(c + m - 1)(c + m - 2)\cdots(c + 1)c]}\gamma_0 \qquad (6.649)$$

It is clear that γ_m is infinite if c is an integer such that $-(m - 1) \leqslant c \leqslant 0$.

Referring to equations 6.83, we note that the product of factors in each set of brackets is a Pochammer symbol, defined by

$$\text{(a)} \quad (\lambda)_m \equiv (\lambda + m - 1)(\lambda + m - 2)\cdots(\lambda + 1)\lambda = \frac{\Gamma(\lambda + m)}{\Gamma(\lambda)}$$

$$\text{with} \quad \text{(b)} \quad (\lambda)_0 = 1 \qquad (6.83)$$

Thus, with $\gamma_0 \equiv 1$,

$$\gamma_m = \frac{(a)_m(b)_m}{(c)_m m!} \qquad (6.650)$$

The notation for the solution to the differential equation is

$$y \equiv {}_2F_1(a, b; c; x) = \sum_{m=0}^{\infty} \frac{(a)_m(b)_m}{(c)_m}\frac{x^m}{m!} \qquad (6.651)$$

The subscripts 2 and 1 indicate that there are two Pochammer symbols in the numerator and one in the denominator. Using

$$\frac{(\lambda)_{m+1}}{(\lambda)_m} = (\lambda + m) \qquad (6.652)$$

it is straightforward to determine from the Cauchy ratio test that the hypergeometric series converges absolutely for $|x| < 1$.

Generalized hypergeometric functions

The description of Gauss's hypergeometric series in equation 6.651 can be used to define a generalized hypergeometric function. With the series coefficient containing r Pochammer symbols in the numerator and s Pochammer symbols in the denominator, the general hypergeometric series is defined by

$$ {}_rF_s(\alpha_1, \alpha_2, \ldots, \alpha_r; \beta_1, \beta_2, \ldots, \beta_s; x) \equiv \sum_{m=0}^{\infty} \frac{(\alpha_1)_m(\alpha_2)_m \cdots (\alpha_r)_m}{(\beta_1)_m(\beta_2)_m \cdots (\beta_s)_m} \frac{x^m}{m!} \tag{6.653} $$

From this definition it is straightforward to see that, for example,

$$ \text{(a)} \quad {}_2F_1(a, b; a; x) = {}_1F_0(b; x) \qquad \text{(b)} \quad {}_2F_2(a, b; a, c; x) = {}_1F_1(b; c; x) $$
$$ \text{(c)} \quad {}_3F_2(a, b, d; c, d; x) = {}_2F_1(a, b; c; x) \tag{6.654} $$

and so on. That is, lower-order hypergeometric functions can be generated easily from higher-order functions.

When λ is a negative integer, the Pochammer symbol can be written

$$ \begin{aligned} (-n)_m &= (-n + m - 1)(-n + m - 2) \cdots (-n + 1)(-n) \\ &= (-1)^m n(n - 1) \cdots (n - m + 2)(n - m + 1) = (-1)^m \frac{n!}{(n - m)!} \end{aligned} \tag{6.655} $$

From this, we see that when $m \geqslant n + 1$, $(n - m)! = \infty$, and the Pochammer symbol is zero. That is,

$$ (-n)_{n+1} = (-n)_{n+2} = \cdots = 0 \tag{6.656} $$

Thus if the argument of any numerator Pochammer symbol is a negative integer, the infinite hypergeometric series becomes a finite polynomial, called the *generalized hypergeometric polynomial.*

The two hypergeometric functions that arise most frequently in scientific investigations are the *Gauss hypergeometric function*

$$ {}_2F_1(a, b; c; x) = \sum_{m=0}^{\infty} \frac{(a)_m(b)_m}{(c)_m} \frac{x^m}{m!} \tag{6.651} $$

and Kummer's *confluent hypergeometric function*

$$ {}_1F_1(a; c; x) = \sum_{m=0}^{\infty} \frac{(a)_m}{(c)_m} \frac{x^m}{m!} \tag{6.657} $$

We will restrict our discussion to these two functions and their polynomial forms,

$$ {}_2F_1(-n, b; c; x) = \sum_{m=0}^{n} \frac{(-n)_m(b)_m}{(c)_m} \frac{x^m}{m!} \tag{6.658} $$

$$ {}_1F_1(-n; c; x) = \sum_{m=0}^{n} \frac{(-n)_m}{(c)_m} \frac{x^m}{m!} \tag{6.659} $$

Confluent hypergeometric function

To see how ${}_1F_1(a; c; x)$ is related to ${}_2F_1(a, b; c; x)$, consider the term in ${}_2F_1(a, b; c; x)$:

$$ \begin{aligned} (b)_m x^m &= (b + m - 1)(b + m - 1) \cdots (b + 1)bx^m \\ &= \left(1 + \frac{m - 1}{b}\right)\left(1 + \frac{m - 2}{b}\right) \cdots \left(1 + \frac{1}{b}\right) \cdot 1(bx)^m \end{aligned} \tag{6.660} $$

We define the new independent variable $x' = bx$. This places the singularities of the hypergeometric function at $x' = 0$, b, and ∞. If we take b to be infinite, the singularities at b and ∞ "run together." This is the reason the confluent hypergeometric function is so named.

From equation 6.660,

$$\lim_{b\to\infty} (b)_m x^m = x'^m \tag{6.661}$$

$$\Rightarrow \quad \lim_{b\to\infty} {}_2F_1(a, b; c; x) = \sum_{m=0}^{\infty} \frac{(a)_m}{(c)_m}\frac{x'^m}{m!} = {}_1F_1(a; c; x') \tag{6.662}$$

To deduce Kummer's confluent hypergeometric differential equation, we make the substitution $x = x'/b$ so that $D = d/dx \to b(d/dx') = bD$. When $b \to \infty$ Gauss's differential equation becomes the Kummer differential equation

$$\lim_{b\to\infty}\left[x'\left(1 - \frac{x'}{b}\right)D^2 + \left(c - x' - \frac{a+1}{b}x'\right)D - a\right]y = \left[x'D^2 + (c - x')D - a\right]y = 0 \tag{6.663}$$

Relations between hypergeometric functions and other functions

Many of the elementary and special functions we have discussed, have representations in terms of the hypergeometric functions.

Example 6.8

For example,

$$ {}_2F_1(1, c; c; x) = \sum_{m=0}^{\infty} (1)_m \frac{x^m}{m!} \tag{6.664}$$

Using equation 6.83a, we have

$$(1)_m = \frac{\Gamma(m+1)}{\Gamma(1)} = m! \tag{6.665}$$

$$\Rightarrow \quad {}_2F_1(1, c; c; x) = \sum_{m=0}^{\infty} x^m = \frac{1}{1-x} \tag{6.666}$$

is the geometric series. □

From this, ${}_2F_1(a, b; c; x)$ can be viewed as a generalization of the geometric series. This is the source of the name "hypergeometric function." If other values are used for the parameters, other functions are obtained.

Example 6.9

Using the fact that

$$(2)_m = \frac{\Gamma(m+2)}{\Gamma(2)} = (m+1)! \tag{6.667}$$

$$\Rightarrow \quad {}_2F_1(1, 1; 2; x) = \sum_{m=0}^{\infty} \frac{x^m}{m+1} = \frac{1}{x}\sum_{m=1}^{\infty}\frac{x^m}{m} = -\frac{1}{x}\log(1-x) \quad \square \tag{6.668}$$

Example 6.10

The binomial expansion

$$\frac{1}{(1-x)^{\alpha}} = \sum_{m=0}^{\infty} \frac{\Gamma(\alpha+m)}{\Gamma(\alpha)} \frac{x^m}{m!} = \sum_{m=0}^{\infty} (\alpha)_m \frac{x^m}{m!} = {}_2F_1(\alpha, c; c; x) \tag{6.669}$$

$$\Rightarrow \quad \sin^{-1} x = \int \frac{dx}{(1-x^2)^{1/2}} = \int {}_2F_1(\tfrac{1}{2}, c; c; x^2)\, dx = \sum_{m=0}^{\infty} \frac{(\frac{1}{2})_m x^{2m+1}}{m!\,(2m+1)} \tag{6.670}$$

But

$$2m+1 = 2(m+\tfrac{1}{2}) = 2\frac{\Gamma(m+\frac{3}{2})}{\Gamma(m+\frac{1}{2})} = \frac{\Gamma(\frac{1}{2})}{\Gamma(\frac{3}{2})} \frac{\Gamma(m+\frac{3}{2})}{\Gamma(m+\frac{1}{2})} = \frac{(\frac{3}{2})_m}{(\frac{1}{2})_m} \tag{6.671}$$

Therefore,

$$\sin^{-1} x = x \sum_{m=0}^{\infty} \frac{(\frac{1}{2})_m (\frac{1}{2})_m}{(\frac{3}{2})_m} \frac{x^{2m}}{m!} = x\, {}_2F_1(\tfrac{1}{2}, \tfrac{1}{2}; \tfrac{3}{2}; x^2) \quad \square \tag{6.672}$$

Example 6.11

With $a = c$, the confluent hypergeometric series becomes

$${}_1F_1(c; c; x) = \sum_{m=0}^{\infty} \frac{x^m}{m!} = e^x \quad \square \tag{6.673}$$

Many of the Sturm–Liouville functions that we have studied are also related to the hypergeometric functions. In the following discussion, we illustrate some of the techniques for relating some of the Sturm–Liouville and hypergeometric functions.

Example 6.12

For example, the ordinary Legendre differential equation is

$$(1-x^2)\frac{d^2y}{dx^2} - 2x\frac{dy}{dx} + l(l+1)y = 0 \tag{6.163}$$

Substituting

$$x' = \frac{1-x}{2} \tag{6.674}$$

the Legendre equation becomes

$$x'(1-x')\frac{d^2y}{dx'^2} + (1-2x')\frac{dy}{dx'} + l(l+1)y = 0 \tag{6.675}$$

If we set

$$\text{(a)} \quad a+b=1 \qquad \text{(b)} \quad ab = -l(l+1) \qquad \text{(c)} \quad c=1 \tag{6.676}$$

in the Gauss hypergeometric differential equation, we obtain the ordinary Legendre differential equation in the x' variable. We can satisfy equations 6.676 by taking a and b as either $a = -l$ and $b = l+1$, or $a = l+1$ and $b = -l$. Thus

$$P_l(x) = {}_2F_1\left(-l, l+1; 1; \frac{1-x}{2}\right) = {}_2F_1\left(l+1, -l; 1; \frac{1-x}{2}\right) \tag{6.677}$$

As expected, a (or b) is a negative integer, and the resulting function is a polynomial. In addition, from the symmetry of ${}_2F_1(a, b; c; x)$ in the interchange of a and b, replacing l by $-l-1$ leaves the hypergeometric function unchanged. This confirms the fact that $P_{-l-1}(x) = P_l(x)$, as stated earlier. $\square$

Example 6.13

Consider

$$ {}_1F_1(-n; c; x^2) = \sum_{m=0}^{n} \frac{(-n)_m}{(c)_m} \frac{x^{2m}}{m!} \tag{6.678} $$

From the series representation expressed in equation 6.469, the even-order Hermite polynomial is

$$ H_{2n}(x) = \sum_{m=0}^{n} (-1)^m \frac{(2n)!}{m!(2n-2m)!} (x^2)^{n-m} \tag{6.679a} $$

or, replacing m by $n - m$, we have

$$ H_{2n}(x) = (-1)^n \sum_{m=0}^{n} (-1)^m \frac{(2n)!}{(2m)!(n-m)!} (x^2)^m \tag{6.679b} $$

In the confluent hypergeometric function, we set $c = \frac{1}{2}$ and use equation 6.89 to obtain

$$ \left(\frac{1}{2}\right)_m = \frac{\Gamma(m + \frac{1}{2})}{\Gamma(\frac{1}{2})} = \frac{(2m)!}{2^{2m} m!} \tag{6.680} $$

Referring to equation 6.655,

$$ (-n)_m = (-1)^m \frac{n!}{(n-m)!} \tag{6.655} $$

we obtain

$$ {}_1F_1(-n; \tfrac{1}{2}; x^2) = n! \sum_{m=0}^{\infty} (-1)^m \frac{(2x)^{2m}}{(2m)!(n-m)!} \tag{6.681} $$

$$ \Rightarrow \quad H_{2n}(x) = (-1)^n \frac{(2n)!}{n!} \, {}_1F_1(-n; \tfrac{1}{2}; x^2) \quad \square \tag{6.682} $$

By similar manipulations, other Sturm–Liouville functions can be related to the hypergeometric functions. We list a few additional relationships here and refer the reader to Problems 58 through 63 for still others. Many sources list such relations that are not presented here. For example, the reader is referred to the text edited by A. Erdelyi, W. Magnus, F. Oberhettinger, and F. Tricomi, *The Bateman Manuscript Project* (New York: McGraw-Hill Book Company, 1953) or the book by W. Magnus, F. Oberhettinger, and P. Soni, *Formulas and Theorems for the Special Functions of Mathematical Physics*, 3rd ed. (New York: Springer-Verlag, 1966).

Some of the Legendre functions are

$$ P_l^m(x) = \frac{(l+m)!}{(l-m)!} \frac{1}{2^m m!} (1 - x^2)^{m/2} {}_2F_1\left(m - l, m + l + 1; m + 1; \frac{1-x}{2}\right) \tag{6.683} $$

$$ Q_l(x) = \frac{\sqrt{\pi}\, l!}{\Gamma(l + \frac{3}{2})(2x)^{l+1}} \, {}_2F_1\left(\frac{l+2}{2}, \frac{l+1}{2}; l + \frac{3}{2}; \frac{1}{x^2}\right) \tag{6.684} $$

and the Gegenbauer polynomials satisfy

$$ G_{2l}^{\lambda}(x) = (-1)^l \frac{(\lambda)_l}{l!} \, {}_2F_1\left(-l, \lambda + l; \frac{1}{2}; x^2\right) \tag{6.685} $$

$$ G_{2l+1}^{\lambda}(x) = (-1)^l \frac{(\lambda)_{l+1}}{l!} \, {}_2F_1\left(-l, \lambda + l; \frac{3}{2}; x^2\right) \tag{6.686} $$

Since many of the functions we have already studied are related to hypergeometric functions, it is clear that the properties of these other functions can be deduced from the analogous properties of the appropriate hypergeometric functions. In the following discussions, we present some methods for developing some (but admittedly an incomplete list) of the properties of the Gauss and Kummer hypergeometric functions.

Integral representations

To develop integral representations of the hypergeometric functions, we write

$$\frac{(b)_m}{(c)_m} = \frac{\Gamma(b+m)}{\Gamma(c+m)}\frac{\Gamma(c)}{\Gamma(b)} = \frac{\Gamma(b+m)\Gamma(c-b)}{\Gamma(c+m)}\frac{\Gamma(c)}{\Gamma(b)\Gamma(c-b)} \tag{6.687}$$

From equations 6.16 and 6.17,

$$\frac{\Gamma(b+m)\Gamma(c-b)}{\Gamma(c+m)} = \beta(b+m, c-b) = \int_0^1 t^{b+m-1}(1-t)^{c-b-1}\,dt \tag{6.688}$$

Therefore,

$$\begin{aligned} {}_2F_1(a,b;c;x) &= \frac{\Gamma(c)}{\Gamma(b)\Gamma(c-b)}\sum_{m=0}^{\infty}\frac{(a)_m}{m!}x^m\int_0^1 t^{b+m-1}(1-t)^{c-b-1}\,dt \\ &= \frac{\Gamma(c)}{\Gamma(b)\Gamma(c-b)}\int_0^1 t^{b-1}(1-t)^{c-b-1}\sum_{m=0}^{\infty}\frac{(a)_m}{m!}(tx)^m\,dt \end{aligned} \tag{6.689}$$

Using equation 6.669 we have

$$\sum_{m=0}^{\infty}\frac{(a)_m}{m!}(tx)^m = \frac{1}{(1-xt)^a} \tag{6.690}$$

$$\Rightarrow \quad {}_2F_1(a,b;c;x) = \frac{\Gamma(c)}{\Gamma(b)\Gamma(c-b)}\int_0^1 \frac{t^{b-1}(1-t)^{c-b-1}}{(1-xt)^a}\,dt \tag{6.691a}$$

or using the symmetry of ${}_2F_1(a,b;c;x)$ in the interchange of a and b, we have

$${}_2F_1(a,b;c;x) = \frac{\Gamma(c)}{\Gamma(a)\Gamma(c-a)}\int_0^1 \frac{t^{a-1}(1-t)^{c-a-1}}{(1-xt)^b}\,dt \tag{6.691b}$$

From equations 6.654a and 6.690,

$$\frac{1}{(1-xt)^a} = {}_2F_1(a,c;c;xt) = {}_1F_0(a;xt) \tag{6.692}$$

Therefore, equations 6.691 can be written

$$\begin{aligned} {}_2F_1(a,b;c;x) &= \frac{\Gamma(c)}{\Gamma(b)\Gamma(c-b)}\int_0^1 t^{b-1}(1-t)^{c-b-1}\,{}_1F_0(a;xt) \\ &= \frac{\Gamma(c)}{\Gamma(a)\Gamma(c-a)}\int_0^1 t^{a-1}(1-t)^{c-a-1}\,{}_1F_0(b;xt) \end{aligned} \tag{6.693}$$

These are a particular expression of an integral representation for the generalized hypergeometric function. If $s \geqslant r$,

$$
{}_sF_r(\alpha_1, \alpha_2, \ldots, \alpha_s; \beta_1, \beta_2, \ldots, \beta_r; x)
$$

$$
= \frac{\Gamma(\beta_1)}{\Gamma(\alpha_1)\Gamma(\beta_1 - \alpha_1)} \int_0^1 t^{\alpha_1 - 1}(1 - t)^{\beta_1 - \alpha_1 - 1} \times {}_{s-1}F_{r-1}(\alpha_2, \alpha_3, \ldots, \alpha_s; \beta_2, \beta_3, \ldots, \beta_r; xt)\, dt \tag{6.694}
$$

From equation 6.691b, we can deduce the analogous integral representation for the confluent hypergeometric function. With $x = x'/b$,

$$
{}_2F_1\left(a, b; c; \frac{x'}{b}\right) = \frac{\Gamma(c)}{\Gamma(a)\Gamma(c - a)} \int_0^1 t^{a-1}(1 - t)^{c-a-1}\left(1 - \frac{x't}{b}\right)^{-b} dt \tag{6.695}
$$

Dropping the prime on x and taking $b \to \infty$, we use

$$
\lim_{b \to \infty}\left(1 - \frac{x't}{b}\right)^{-b} = e^{x't} \tag{6.696}
$$

$$
\Rightarrow \quad {}_1F_1(a; c; x) = \frac{\Gamma(c)}{\Gamma(a)\Gamma(c - a)} \int_0^1 t^{a-1}(1 - t)^{c-a-1} e^{xt}\, dt \tag{6.697}
$$

Generating functions

An integral representation of Gauss's hypergeometric polynomial can be obtained from the integral representation expressed in equation 6.691a. It is

$$
{}_2F_1(-n, b; c; x) = \frac{\Gamma(c)}{\Gamma(b)\Gamma(c - b)} \int_0^1 t^{b-1}(1 - t)^{c-b-1}(1 - xt)^n\, dt \tag{6.698}
$$

Consider the quantity

$$
\sum_{n=0}^{\infty} z^n \frac{(c)_n}{n!}\, {}_2F_1(-n, b; c; x) = \frac{\Gamma(c)}{\Gamma(b)\Gamma(c - b)} \int_0^1 t^{b-1}(1 - t)^{c-b-1} \sum_{n=0}^{\infty} \frac{(c)_n}{n!}[z(1 - xt)]^n\, dt \tag{6.699}
$$

Referring to equation 6.669, we have

$$
\sum_{n=0}^{\infty} \frac{(c)_n}{n!}[z(1 - xt)]^n = \frac{1}{(1 - z + xzt)^c} \tag{6.700}
$$

$$
\Rightarrow \quad \sum_{n=0}^{\infty} z^n \frac{(c)_n}{n!}\, {}_2F_1(-n, b; c; x)
$$

$$
= \frac{\Gamma(c)}{\Gamma(b)\Gamma(c - b)} \int_0^1 t^{b-1}(1 - t)^{c-b-1}(1 - z)^{-c}(1 - wt)^{-c}\, dt \tag{6.701}
$$

where

$$
w \equiv -\frac{xz}{1 - z} \tag{6.702}
$$

Again, using the integral representation in equation 6.691a, and referring again to equation 6.669, this becomes

$$
\sum_{n=0}^{\infty} z^n \frac{(c)_n}{n!}\, {}_2F_1(-n, b; c; x) = \frac{1}{(1 - z)^c} \sum_{m=0}^{\infty} \frac{(b)_m}{m!}(-w)^m = \frac{1}{(1 - z)^c} \frac{1}{(1 + w)^b} \tag{6.703}
$$

Therefore, the generating function for the Gauss hypergeometric polynomial is

$$\sum_{n=0}^{\infty} z^n \frac{(c)_n}{n!}\,{}_2F_1(-n, b; c; x) = (1-z)^{b-c}(1-z+zx)^{-b} \equiv {}_2\Phi_1(z, x) \tag{6.704}$$

To obtain the analogous generator for the confluent hypergeometric polynomials, we replace x with x'/b in equation 6.704 to obtain

$$\begin{aligned}\sum_{n=0}^{\infty} z^n \frac{(c)_n}{n!}\,{}_2F_1\left(-n, b; c; \frac{x'}{b}\right) &= (1-z)^{b-c}\left(1-z+\frac{zx'}{b}\right)^{-b} \\ &= (1-z)^{-c}\left(1+\frac{x'z}{(1-z)b}\right)^{-b}\end{aligned} \tag{6.705}$$

Using

$$\lim_{b\to\infty}\left(1+\frac{x'z}{(1-z)b}\right)^{-b} = e^{-x'z/(1-z)} \tag{6.706}$$

$$\Rightarrow \quad \sum_{n=0}^{\infty} z^n \frac{(c)_n}{n!}\,{}_1F_1(-n; c; x) = (1-z)^{-c} e^{-xz/(1-z)} \equiv {}_1\Phi_1(z, x) \tag{6.707}$$

We note that the exponential factor is the generating function for the Laguerre polynomials. Thus the Laguerre polynomials are related to the confluent hypergeometric polynomials (see Problem 61).

Recurrence relations

By differentiating the generating function for the Gauss hypergeometric polynomial with respect to z, we obtain

$$\begin{aligned}\frac{\partial}{\partial z}\,{}_2\Phi_1(z, x) &= \left[\frac{b(1-x)}{(1-z+xz)} - \frac{(b-c)}{(1-z)}\right]{}_2\Phi_1(z, x) \\ &= \sum_{n=1}^{\infty} n z^{n-1} \frac{(c)_n}{n!}\,{}_2F_1(-n, b; c; x)\end{aligned} \tag{6.708}$$

After some cumbersome algebraic manipulations, and readjustment of various summation indices, one finds that

$$\begin{aligned}&(c+n+1)(c+n)\,{}_2F_1(-n-2, b; c; x) \\ &\quad -(c+n)[(c+2n+2) - x(b+n+1)]\,{}_2F_1(-n-1, b; c; x) \\ &\quad + c(1-x)(n+1)\,{}_2F_1(-n, b; c; x) = 0\end{aligned} \tag{6.709}$$

Differentiation of ${}_2\Phi_1(z, x)$ with respect to x yields

$$\frac{\partial}{\partial x}\,{}_2\Phi_1(z, x) = \frac{-bz}{(1-z+zx)}\,{}_2\Phi_1(z, x) = \sum_{n=0}^{\infty} \frac{(c)_n}{n!} z^n\,{}_2F_1'(-n, b; c; x) \tag{6.710}$$

$$\begin{aligned}\Rightarrow \quad &(c+n)\,{}_2F_1'(-n-1, b; c; x) - (n+1)\,{}_2F_1'(-n, b; c; x) \\ &+ b(n+1)\,{}_2F_1(-n, b; c; x) = 0\end{aligned} \tag{6.711}$$

Using the results of Problem 55, part (a), we have

$${}_2F_1'(-n, b; c; x) = -\frac{nb}{c}\,{}_2F_1(-n+1, b+1; c+1; x) \tag{6.712}$$

Using this result, equation 6.711 is equivalent to

$$-(c+n)\,{}_2F_1(-n,b+1;c+1;x)+n\,{}_2F_1(-n+1,b+1;c+1;x)+c\,{}_2F_1(-n,b;c;x)=0 \tag{6.713}$$

The analogous results are obtainable for the confluent hypergeometric polynomials by using the generator of ${}_1F_1(-n;c;x)$ or by setting $x=x'/b$ in the recurrence relations for ${}_2F_1$ and taking the limit of infinite b. This second approach is quite straightforward. The results are

$$(c+n+1)(c+n)\,{}_1F_1(-n-2;c;x)-(c+n)(c+2n+2-x)\,{}_1F_1(-n-1;c;x)+c(n+1)\,{}_1F_1(-n;c;x)=0 \tag{6.714}$$

and with $d/dx=b(d/dx')$,

$$(c+n)\,{}_1F_1'(-n-1;c;x)-(n+1)\,{}_1F_1'(-n;c;x)+(n+1)\,{}_1F_1(-n;c;x)=0 \tag{6.715}$$

Using the result of Problem 55, this is easily seen to be equivalent to

$$-(c+n)\,{}_1F_1(-n;c+1;x)+n\,{}_1F_1(-n+1;c+1;x)+c\,{}_1F_1(-n;c;x)=0 \tag{6.716}$$

Other recurrence relations for these hypergeometric functions can be developed from the series representations. One such example is

$$(c-a-b)\,{}_2F_1(a,b;c;x)+a(1-x)\,{}_2F_1(a+1,b;c;x)-(c-b)\,{}_2F_1(a,b-1;c;x)=0 \tag{6.717}$$

This relation, and others like it, contain hypergeometric functions in which the values of the various parameters differ by 1. Such functions are said to be contiguous, and relations involving contiguous functions are called contiguous relations.

To verify the validity of equation 6.717, and similar contiguous relations, we first develop some properties of the Pochammer symbols that we will require. Consider

$$a(a)_m=a\frac{\Gamma(a+m)}{\Gamma(a)} \tag{6.718a}$$

$$(a)_{m-1}=\frac{\Gamma(a+m-1)}{\Gamma(a)}=\frac{1}{(a-1)}\frac{\Gamma(a+m-1)}{\Gamma(a-1)}=\frac{1}{(a-1)}(a-1)_m \tag{6.718b}$$

$$(a)_{m-1}=\frac{\Gamma(a+m-1)}{\Gamma(a)}=\frac{1}{(a+m-1)}\frac{\Gamma(a+m)}{\Gamma(a)}=\frac{1}{(a+m-1)}(a)_m \tag{6.718c}$$

$$\frac{(a)_m}{a}=\frac{\Gamma(a+m)}{a\Gamma(a)}=\frac{1}{(a+m)}\frac{\Gamma(a+m+1)}{\Gamma(a+1)} \tag{6.718d}$$

Using the series representation of ${}_2F_1(a,b;c;x)$,

$${}_2F_1(a,b;c;x)=\sum_{m=0}^{\infty}\frac{(a)_m(b)_m}{(c)_m}\frac{x^m}{m!} \tag{6.719}$$

the right-hand side of equation 6.717 becomes

$$\sum_{m=0}^{\infty}\left[(c-a-b)\frac{(a)_m(b)_m}{(c)_m}+a\frac{(a+1)_m(b)_m}{(c)_m}-(c-b)\frac{(a)_m(b-1)_m}{(c)_m}\right]\frac{x^m}{m!}$$
$$-\sum_{m=0}^{\infty}a\frac{(a+1)_m(b)_m}{(c)_m}\frac{x^{m+1}}{m!}$$
$$=\left[(c-a-b)\frac{(a)_0(b)_0}{(c)_0}+\frac{a(a+1)_0(b)_0}{(c)_0}-(c-b)\frac{(a)_0(b-1)_0}{(c)_0}\right]$$
$$+\sum_{m=1}^{\infty}\left[(c-a-b)\frac{(a)_m(b)_m}{(c)_m}+a\frac{(a+1)_m(b)_m}{(c)_m}-(c-b)\frac{(a)_m(b-1)_m}{(c)_m}\right]\frac{x^m}{m!}$$
$$-\sum_{m=0}^{\infty}a\frac{(a+1)_m(b)_m}{(c)_m}\frac{x^{m+1}}{m!} \tag{6.720a}$$

Using $(\lambda)_0 = 1$ for any λ, the term in the first set of brackets is zero. In the remaining sums, we use the properties of the Pochammer symbols expressed in equations 6.718, and also replace $m + 1$ by m in the second sum. With these manipulations, we obtain

$$(c-a-b)\,{}_2F_1(a,b;c;x)+a(1-x)\,{}_2F_1(a+1,b;c;x)-(c-b)\,{}_2F_1(a,b-1;c;x)$$
$$=\sum_{m=1}^{\infty}\left[(c-a-b)\frac{(a)_m(b)_m}{(c)_m}+(a+m)\frac{(a)_m(b)_m}{(c)_m}\right. \tag{6.720b}$$
$$\left.-(c-b)(b-1)\frac{(a)_m(b-1)_{m-1}}{(c)_m}-m(c+m-1)\frac{(a)_m(b)_{m-1}}{(c)_m}\right]\frac{x^m}{m!}$$

Using equation 6.718c to relate all Pochammer symbols involving b, this becomes

$$(c-a-b)\,{}_2F_1(a,b;c;x)+a(1-x)\,{}_2F_1(a+1,b;c;x)-(c-b)\,{}_2F_1(a,b-1;c;x)$$
$$=\sum_{m=1}^{\infty}\frac{(a)_m(b)_{m-1}}{(c)_m}[(c-a-b)(b+m-1)+(a+m)(b+m-1)$$
$$-(c-b)(b-1)-m(c+m-1)]\frac{x^m}{m!}=0 \tag{6.721}$$

Other contiguous relations can be developed using manipulations such as these. A small sample of additional contiguous relations is listed below. The reader is referred to Problems 65 and 66 for others.

$$(c-a-1)\,{}_2F_1(a,b;c;x)+a\,{}_2F_1(a+1,b;c;x)-(c-1)\,{}_2F_1(a,b;c-1;x)=0 \tag{6.722}$$

$$c(1-x)\,{}_2F_1(a,b;c;x)-c\,{}_2F_1(a-1,b;c;x)-(c-b)x\,{}_2F_1(a,b;c+1;x)=0 \tag{6.723}$$

$$c(c-1)\,{}_2F_1(a,b;c;x)+abx\,{}_2F_1(a+1,b+1;c+1;x)$$
$$-c(c-1)\,{}_2F_1(a,b;c-1;x)=0 \tag{6.724}$$

$$c(c+1)\,{}_2F_1(a,b;c;x)+a(b-c)x\,{}_2F_1(a+1,b+1;c+2;x)$$
$$-c(c+1)\,{}_2F_1(a,b+1;c+1;x)=0 \tag{6.725}$$

Second solutions

A second solution to the Gauss hypergeometric equation is achievable by the method of reduction of order. However, this approach is not the only way to develop the second independent solution.

If we define a new dependent variable, z, by

$$y \equiv x^{1-c}z \tag{6.726}$$

and substitute this into Gauss's hypergeometric equation, we obtain

$$\{x(1-x)D^2 + [(2-c) - (a+b-2c+3)x]D - (a-c+1)(b-c+1)\}z = 0 \tag{6.727}$$

This is the same form as the original differential equation, with

$$\text{(a)}\quad a' = a-c+1 \qquad \text{(b)}\quad b' = b-c+1 \qquad \text{(c)}\quad c' = 2-c \tag{6.728}$$

$$\Rightarrow \quad y_2(x) = x^{1-c}\,{}_2F_1(a-c+1, b-c+1; 2-c; x) \tag{6.729a}$$

is a second independent solution to the Gauss hypergeometric equation.

From the argument of ${}_2F_1$ in equation 6.729a, we see that $2-c$ cannot be a negative integer, so that the denominator Pochammer symbol will not be zero. That is, c cannot be $2, 3, 4, \ldots$. We also note that if $c = 1$, the second solution is ${}_2F_1(a, b; 1; x)$, and therefore not an independent solution. In fact,

$$y_2(x) = x^{1-c}\,{}_2F_1(a-c+1, b-c+1; 2-c; x) \tag{6.729a}$$

will be an independent solution only if c is noninteger. For integer values of c, one must resort to the method of reduction of order for the second solution.

From this discussion, it is indicated that there is more than one way to obtain the second solution to the Gauss equation. In fact, there are 24 separate solutions, pairs of which can equally well serve as the two independent solutions. The reader interested in these other pairs of solutions is referred to the text by H. Hochstadt, *The Functions of Mathematical Physics* (New York: John Wiley & Sons, Inc., 1971).

A similar analysis of the second solution to Kummer's confluent hypergeometric equation leads to an independent solution of the form

$$y_2(x) = x^{1-c}\,{}_1F_1(a-c+1; 2-c; x) \tag{6.729b}$$

As with Gauss's second solution, c must be restricted to be noninteger, in order that y_2 be independent of ${}_1F_1(a; c; x)$ and noninfinite.

Some of the functions we have studied are related to a function defined by a linear combination of the two independent Kummer solutions. The confluent hypergeometric solution of the second kind is defined by

$$W(a; c; x) \equiv \frac{\pi}{\sin(\pi c)}\left[\frac{{}_1F_1(a; c; x)}{\Gamma(c)\Gamma(a-c+1)} - \frac{x^{1-c}\,{}_1F_1(a-c+1; 2-c; x)}{\Gamma(a)\Gamma(2-c)}\right] \tag{6.730}$$

This is defined such that like the Neumann function, when c becomes an integer, this function is in the form $0/0$. Using l'Hospital's rule, a second independent solution can be formed for integer values of c.

The error function and the modified Hankel function are related to this second solution by

$$\operatorname{erfc}(x) \equiv 1 - \operatorname{erf}(x) = \frac{e^{-x^2}}{\sqrt{\pi}} W\left(\frac{1}{2}; \frac{1}{2}; x^2\right) \tag{6.731}$$

$$K_n(x) = \sqrt{\pi}(2x)^n e^{-x} W(n + \tfrac{1}{2}; 2n + 1; 2x) \tag{6.732}$$

6.9 Properties of the Dirac δ-symbol

Let $F(x)$ be an arbitrary function that is continuous for all $x \in [-\infty, \infty]$. We will also impose the condition, when necessary, that $F(\pm\infty) \to 0$ "fast enough." As expressed in equation 2.294, the Dirac δ-symbol is defined such that

$$\text{(a)} \quad F(x_0) = \int_{-\infty}^{\infty} F(x)\delta(x - x_0)\, dx \quad \Rightarrow \quad \text{(b)} \quad F(0) = \int_{-\infty}^{\infty} F(x)\delta(x)\, dx \tag{6.733}$$

Consider

$$I = \int_{-\infty}^{\infty} F(x)\delta(ax)\, dx \qquad a \neq 0 \tag{6.734}$$

Substituting $y = ax$, if $a > 0$, $y = |a|x$, and

$$\int_{-\infty}^{\infty} F(x)\delta(ax)\, dx = \frac{1}{|a|}\int_{-\infty}^{\infty} F\left(\frac{y}{|a|}\right)\delta(y)\, dy = \frac{1}{|a|}F(0) = \frac{1}{|a|}\int_{-\infty}^{\infty} F(x)\delta(x)\, dx \tag{6.735}$$

Since $F(x)$ is arbitrary,

$$\delta(ax) = \frac{1}{|a|}\delta(x) \tag{6.736}$$

If $a < 0$, $y = -|a|x$ and equation 6.734 becomes

$$\int_{-\infty}^{\infty} F(x)\delta(ax)\, dx = \frac{1}{|a|}\int_{-\infty}^{\infty} F\left(\frac{y}{|a|}\right)\delta(y)\, dy = \frac{1}{|a|}F(0) = \frac{1}{|a|}\int_{-\infty}^{\infty} F(x)\delta(x)\, dx \tag{6.737}$$

which is identical to equation 6.735. Thus for any nonzero value of a,

$$\delta(ax) = \frac{1}{|a|}\delta(x) \qquad a \neq 0 \tag{6.736}$$

If $a = -1$, we see from this result that

$$\delta(-x) = \delta(x) \tag{6.738a}$$

That is, the δ-symbol, viewed as a function, is even. Therefore, its first derivative is odd, its second derivative is even, and so on. Thus.

$$\text{(b)} \quad \delta'(-x) = -\delta'(x) \qquad \text{(c)} \quad \delta''(-x) = \delta''(x) \qquad \text{(d)} \quad \delta^{(n)}(-x) = (-1)^n \delta^{(n)}(x) \tag{6.738}$$

That is, all even derivatives of $\delta(x)$ are even functions, all odd derivatives of $\delta(x)$ are odd functions of x.

Consider

$$I = \int_{-\infty}^{\infty} F(x)x\delta'(x)\, dx \tag{6.739}$$

Integrating by parts, we obtain

$$\int_{-\infty}^{\infty} F(x)x\delta'(x)\,dx = xF(x)\delta(x)|_{-\infty}^{\infty} - \int_{-\infty}^{\infty} \delta(x)\frac{d}{dx}[xF(x)]\,dx \tag{6.740a}$$

Since we are assuming that $F(x) \to 0$ fast enough at $\pm\infty$ and since $\delta(x) = 0$ for $x \neq 0$, the integrated term is zero. Therefore

$$\int_{-\infty}^{\infty} F(x)x\delta'(x)\,dx = -\frac{d}{dx}[xF(x)]_{x=0} = -F(0) = -\int_{-\infty}^{\infty} F(x)\delta(x)\,dx \tag{6.740b}$$

$$\Rightarrow \quad x\delta'(x) = -\delta(x) \tag{6.741}$$

Now consider

$$I = \int_{-\infty}^{\infty} F(x)x\delta''(x)\,dx \tag{6.742}$$

Again, we integrate by parts and assume that $F(x)$ has the right behavior at $\pm\infty$ to set the integrated term to zero. The result is

$$\int_{-\infty}^{\infty} F(x)x\delta''(x)\,dx = -\int_{-\infty}^{\infty} \delta'(x)\frac{d}{dx}[xF(x)]\,dx \tag{6.743}$$

Integrating by parts a second time, again setting the integrated term to zero for appropriate $F(x)$, this becomes

$$\int_{-\infty}^{\infty} F(x)x\delta''(x)\,dx = \int_{-\infty}^{\infty} \delta(x)\frac{d^2}{dx^2}[xF(x)]\,dx = \frac{d^2}{dx^2}[xF(x)]_{x=0} = 2F'(0) \tag{6.744}$$

However,

$$F'(0) = \int_{-\infty}^{\infty} F'(x)\delta(x)\,dx = -\int_{-\infty}^{\infty} F(x)\delta'(x)\,dx \tag{6.745}$$

which is obtained by integrating by parts and setting the integrated term to zero, as before. Combining equations 6.744 and 6.745, we have

$$\text{(a)} \quad \int_{-\infty}^{\infty} F(x)x\delta''(x)\,dx = -2\int_{-\infty}^{\infty} F(x)\delta'(x)\,dx \quad \Rightarrow \quad \text{(b)} \quad x\delta''(x) = -2\delta'(x) \tag{6.746}$$

Therefore, using equation 6.741 yields

$$x^2\delta''(x) = -2x\delta'(x) = 2\delta(x) \tag{6.747}$$

In this way, it is possible to relate $x^n[d^n\delta(x)/dx^n]$ to $\delta(x)$ (see Problem 68).

Consider

$$I = \int_{-\infty}^{\infty} F(x)\delta(x^2 - a^2)\,dx \qquad a \neq 0 \tag{6.748}$$

We write this integral as

$$I = \int_{-\infty}^{0} F(x)\delta(x^2 - a^2)\,dx + \int_{0}^{\infty} F(x)\delta(x^2 - a^2)\,dx \tag{6.749}$$

and in the first integral, replace x by $-x$. The result is

$$\int_{-\infty}^{\infty} F(x)\delta(x^2 - a^2)\,dx = \int_{0}^{\infty} \delta(x^2 - a^2)[F(x) + F(-x)]\,dx \tag{6.750}$$

Defining $u = x^2 - a^2$, equation 6.750 becomes

$$\int_{-\infty}^{\infty} F(x)\delta(x^2 - a^2)\,dx = \tfrac{1}{2}\int_{-a^2}^{\infty}\left[F\left(-\sqrt{u + a^2}\right) + F\left(\sqrt{u + a^2}\right)\right]\delta(u)(u + a^2)^{-1/2}\,du$$

$$= \frac{1}{2|a|}\left[F(-|a|) + F(|a|)\right] \tag{6.751}$$

Using

$$F(\mp|a|) = \int_{-\infty}^{\infty} F(x)\delta(x \pm |a|)\,dx \tag{6.752}$$

$$\Rightarrow \quad \int_{-\infty}^{\infty} F(x)\delta(x^2 - a^2)\,dx = \frac{1}{2|a|}\int_{-\infty}^{\infty} F(x)\left[\delta(x + |a|) + \delta(x - |a|)\right]dx \tag{6.753}$$

$$\Rightarrow \quad \delta(x^2 - a^2) = \frac{1}{2|a|}\left[\delta(x + |a|) + \delta(x - |a|)\right] \tag{6.754}$$

As a final note, we demonstrate how the δ-function can be represented in terms of a limit, using the identity expressed in equations 2.299.

$$\lim_{\varepsilon\to 0}\frac{1}{(x - x_0 \pm i\varepsilon)} = \frac{1}{(x - x_0)_P} \mp i\pi\delta(x - x_0) \tag{2.299}$$

$$\Rightarrow \quad \delta(x - x_0) = \frac{1}{\pi}\,\mathrm{Im}\left[\lim_{\varepsilon\to 0}\frac{1}{(x - x_0 - i\varepsilon)}\right] = \frac{1}{\pi}\lim_{\varepsilon\to 0}\frac{\varepsilon}{(x - x_0)^2 + \varepsilon^2} \tag{6.755}$$

We note that when $x \neq x_0$, this limit is zero, and when $x = x_0$, the limit is infinite. These are the properties of the δ-symbol stated in Chapter 2, when $\delta(x - x_0)$ is treated as a function.

6.10 Summary

A summary of many of the properties of the functions developed in this chapter is presented for the convenience of the reader. This will serve as a concise reference to these properties.

Gamma and beta functions

Representations

$$\Gamma(p) \equiv \int_0^{\infty} x^{p-1}e^{-x}\,dx \tag{6.1}$$

$$\Gamma(p) = (-1)^p\int_0^1 [\log(w)]^{p-1}\,dw \tag{6.2}$$

$$\Gamma(p) = \lim_{m\to\infty} m^p\frac{m!}{p(p+1)\cdots(p+m)} \tag{6.33b}$$

$$\Gamma(p) = (p - 1)\Gamma(p - 1) = (p - 1)! \qquad \text{(6.3b) and (6.8)}$$

$$\frac{\Gamma(p)\Gamma(q)}{\Gamma(p + q)} = 2\int_0^{\pi/2}\cos^{2p-1}\phi\,\sin^{2q-1}\phi\,d\phi \equiv \beta(p, q) \tag{6.16}$$

$$\beta(p, q) = \int_0^1 t^{p-1}(1 - t)^{q-1}\,dt \tag{6.17}$$

$$\beta(p, q) = \int_0^{\infty}\frac{x^{p-1}}{(1 + x)^{p+q}}\,dx \tag{6.18}$$

Stirling approximation

$$\Gamma(n+1) = N! \simeq \sqrt{2\pi N}\,N^N e^{-N} \qquad N \text{ large} \tag{6.61}$$

Miscellaneous identities

$$\Gamma(p)\Gamma(1-p) = \frac{\pi}{\sin(\pi p)} \qquad \text{Re}(p) < 1 \tag{6.23b}$$

$$\Gamma(\tfrac{1}{2}+q)\Gamma(\tfrac{1}{2}-q) = \frac{\pi}{\cos(\pi q)} \tag{6.29}$$

$$\Gamma(p)\Gamma(-p) = -\frac{\pi}{p\sin(\pi p)} \tag{6.30}$$

$$\frac{\Gamma(p)\Gamma(p+\frac{1}{2})}{\Gamma(2p)} = \frac{\sqrt{\pi}}{2^{2p-1}} \qquad \text{(Legendre duplication formula)} \tag{6.37a}$$

$$\frac{\Gamma(q+1)\Gamma(q+\frac{1}{2})}{\Gamma(2q+1)} = \frac{\sqrt{\pi}}{2^{2q}} \qquad \text{(Legendre duplication formula)} \tag{6.37b}$$

Digamma and polygamma functions

$$\psi(p+1) \equiv \frac{d}{dp}\log[\Gamma(p+1)] = -\gamma - \frac{1}{(p+1)} + (p+1)\sum_{k=1}^{\infty}\frac{1}{k(p+1+k)}$$

(6.39) and (6.48)

$$\psi'(p+1) = \psi^{(1)}(p+1) = \sum_{k=0}^{\infty}\frac{1}{(p+1+k)^2} \tag{6.49}$$

$$\psi^{(n)}(p+1) = -\sum_{k=0}^{\infty}\frac{(-1)^n n!}{(p+1+k)^{n+1}} \tag{6.50}$$

$$\psi(p) - \psi(1-p) = \pi\cot(\pi p) \qquad \text{(Problem 6)}$$

$$\lim_{p\to N}\frac{\psi(1-p)}{\Gamma(1-p)} = (-1)^N(N-1)! \qquad N \text{ is an integer} \qquad \text{(Problem 6)}$$

Elliptic integrals

$$K(k^2) = \int_0^{\pi/2}\frac{d\phi}{\sqrt{1-k^2\sin^2\phi}} \tag{6.79}$$

$$E(k^2) \equiv \int_0^{\pi/2}\sqrt{1-k^2\sin^2\phi}\,d\phi \tag{6.80}$$

$$K(k^2) = \frac{1}{2}\sum_{n=0}^{\infty}\left[\frac{\Gamma(n+\frac{1}{2})}{n!}\right]^2 k^{2n} = \frac{\pi}{2}\sum_{n=0}^{\infty}\left[\left(\frac{1}{2}\right)_m\right]^2\frac{k^{2n}}{(n!)^2} \tag{6.95}$$

$$E(k^2) = \frac{\pi}{2} - \frac{1}{4}\sum_{n=0}^{\infty}\left[\frac{\Gamma(n+\frac{1}{2})}{(n+1)!}\right]^2(n+\tfrac{1}{2}) + k^{2n+2}$$

$$= \frac{\pi}{2} - \frac{\pi}{4}\sum_{n=0}^{\infty}\left[\left(\frac{1}{2}\right)_n\right]^2\frac{(n+\frac{1}{2})}{[(n+1)!]^2}k^{2n+2} \tag{6.98}$$

$$\frac{d}{dz}E(z) = \frac{1}{2z}[E(z)-K(z)] \tag{6.99}$$

$$\frac{d}{dz}K(z) = \frac{1}{2z}\left[\frac{E(z)}{(1-z)} - K(z)\right] \tag{6.105}$$

Pochammer symbols and double factorials

$$\text{(a)} \quad (\alpha)_m \equiv (\alpha + m - 1)(\alpha + m - 2) \cdots (\alpha + 1)\alpha = \frac{\Gamma(\alpha + m)}{\Gamma(\alpha)} \qquad \text{(b)} \quad (\alpha)_0 = 1 \tag{6.83}$$

$$(2m - 1)!! = \frac{\sqrt{\pi}}{2^m}\frac{(2m)!}{m!} \quad (6.90) \qquad (2m)!! = 2^m m! \tag{6.91}$$

Error functions

$$\text{erf}(z) \equiv \frac{2}{\sqrt{\pi}}\int_0^z e^{-x^2}\,dx \quad (6.107) \qquad \text{(a)} \quad \text{erf}(\infty) = 1 \qquad \text{(b)} \quad \text{erf}(0) = 0 \tag{6.108}$$

$$\text{erf}(-z) = -\text{erf}(z) \quad (6.109\text{b}) \qquad \text{erf}(z) = \frac{2}{\sqrt{\pi}}\sum_{k=0}^{\infty}(-1)^k\frac{z^{2k+1}}{(2k+1)k!} \tag{6.111}$$

$$\text{erfc}(z) \equiv 1 - \text{erf}(z) \equiv \frac{2}{\sqrt{\pi}}\int_z^{\infty} e^{-x^2}\,dx \tag{6.113}$$

$$\text{erfc}(z) = \frac{1}{z\sqrt{\pi}}\left[1 + \sum_{k=0}^{\infty}(-1)^k\frac{(2k-1)!!}{(2z^2)^{2k}}\right]e^{-z^2} \tag{6.116}$$

$$\text{erf}(z) = 1 - \frac{1}{z\sqrt{\pi}}\left[1 + \sum_{k=0}^{\infty}(-1)^k\frac{(2k-1)!!}{(2z^2)^{2k}}\right]e^{-z^2} \tag{6.117}$$

Legendre polynomials $P_l(x)$

Differential equation

$$\left[(1 - x^2)D^2 - 2xD + l(l+1)\right]P_l(x) = 0 \tag{6.164}$$

Series representation

$$P_l(x) = \sum_{m=0}^{[l/2]}\frac{(-1)^m(2l-2m)!}{2^l m!(l-m)!(l-2m)!}x^{l-2m} \tag{6.179}$$

Generating function

$$(1 - 2xz + z^2)^{-1/2} = \sum_{l=0}^{\infty} z^l P_l(x) \qquad |z| < 1 \tag{6.182b}$$

Rodriguez formula

$$P_l(x) = \frac{(-1)^l}{2^l l!}D^l(1 - x^2)^l \tag{6.195}$$

Integral representations

$$P_l(x) = \frac{(-1)^l}{2^l}\frac{1}{2\pi i}\oint\frac{(1-z^2)^l}{(z-x)^{l+1}}\,dz \qquad \text{(Schlafli)} \qquad (6.197)$$

$$P_l(x) = \frac{1}{2\pi}\int_{-\pi}^{\pi}\left(x + i\sqrt{1-x^2}\cos\phi\right)^l d\phi = \frac{1}{\pi}\int_0^{\pi}\left(x + i\sqrt{1-x^2}\cos\phi\right)^l d\phi \qquad \text{(Laplace)}$$

(6.199b) or (6.200a)

$$\begin{aligned} P_{-l-1}(x) = P_l(x) &= \frac{1}{2\pi}\int_{-\pi}^{\pi}\left(x + i\sqrt{1-x^2}\cos\phi\right)^{-l-1} d\phi \\ &= \frac{1}{\pi}\int_0^{\pi}\left(x + i\sqrt{1-x^2}\cos\phi\right)^{-l-1} d\phi \qquad \text{(Laplace)} \end{aligned} \qquad (6.200b)$$

Recurrence relations

$$(l+2)P_{l+2}(x) - (2l+3)P_{l+1}(x) + (l+1)P_l(x) = 0 \qquad l \geqslant 0 \qquad (6.203b)$$

$$P'_{l+2}(x) - 2xP'_{l+1}(x) + P'_l(x) = P_{l+1}(x) \qquad l \geqslant 0 \qquad (6.205)$$

$$P'_{l+2}(x) - P'_l(x) = (2l+3)P_{l+1}(x) \qquad l \geqslant 0 \qquad \text{(Problem 18)}$$

$$P'_{l+1}(x) - xP'_l(x) = (l+1)P_l(x) \qquad l \geqslant 0 \qquad \text{(Problem 18)}$$

$$xP'_{l+1}(x) - P'_l(x) = (l+1)P_{l+1}(x) \qquad l \geqslant 0 \qquad \text{(Problem 18)}$$

$$(l+1)P_l(x) - (l+1)xP_{l+1}(x) = (1-x^2)P'_{l+1}(x) \qquad l \geqslant 0 \qquad \text{(Problem 18)}$$

Orthonormalization

$$\int_{-1}^{1} P_l(x)P_n(x)\,dx = \frac{2}{2l+1}\delta_{ln} \qquad (6.213)$$

$$\int_{-1}^{1}(1-x^2)P'_l(x)P'_n(x)\,dx = \frac{2l(l+1)}{2l+1}\delta_{ln} \qquad \text{(Problem 16)}$$

Miscellaneous identities

$$x^k = \frac{2^k(k!)^2}{(2k)!}(k-\tfrac{1}{2})!\sum_{m=0}^{[k/2]}\frac{(k-2m+\frac{1}{2})}{m!(k-m+\frac{1}{2})!}P_{k-2m}(x) \qquad (6.222)$$

$$P_l(x) = P_{-l-1}(x) \qquad P_l(-x) = (-1)^l P_l(x) \qquad (6.180)$$

$$P'_l(1) = \tfrac{1}{2}l(l+1) \qquad P'_l(-1) = (-1)^{l+1}\tfrac{1}{2}l(l+1) \qquad \text{(Problem 15)}$$

For l even

$$P_l(\cos\theta) = \frac{(l!)^2}{2^{2l}[(\frac{1}{2}l)!]^4} + 2\sum_{k=0}^{l/2-1}\frac{(2l-2k)!(2k)!}{2^{2l}[(l-k)!]^2(k!)^2}\cos[(2k-l)\theta] \qquad (6.230b)$$

For l odd:

$$P_l(\cos\theta) = 2\sum_{k=0}^{(l-1)/2}\frac{(2l-2k)!(2k)!}{2^{2l}[(l-k)!]^2(k!)^2}\cos[(2k-l)\theta] \qquad (6.232b)$$

Legendre function of the second kind $Q_l(x)$

Differential equation

$$[(1-x^2)D^2 - 2xD + l(l+1)]Q_l(x) = 0 \qquad (6.164)$$

Series representation

With $l1 \equiv \left[\frac{1}{2}(l-1)\right]$

$$Q_l(x) = P_l(x)\frac{1}{2}\log\left(\frac{x+1}{x-1}\right) - \sum_{k=0}^{l1}\frac{(2l-4k-1)}{(2k+1)(l-k)}P_{l-2k-1}(x)$$

$$= P_l(x)Q_0(x) - R_{l-1}(x) \qquad (6.239)$$

Integral representations

$$Q_l(z) = \frac{1}{2}\int_{-1}^{1}\frac{P_l(x)}{(z-x)}\,dx \qquad \text{(Neumann)} \qquad (6.247)$$

$$R_{l-1}(z) = \frac{1}{2}\int_{-1}^{1}\frac{P_l(z)-P_l(x)}{(z-x)}\,dx \qquad \text{(Neumann)} \qquad (6.249)$$

$$Q_l(z) = \frac{1}{2^{l+1}}\int_{-1}^{1}\frac{(1-x^2)^l}{(z-x)^{l+1}}\,dx \qquad \text{(Schlafli)} \qquad \text{(Problem 20)}$$

$$z^m Q_l(z) - \frac{1}{2}\int_{-1}^{1}\frac{x^m P_l(x)}{(z-x)}\,dx = 0 \qquad m \leqslant l \qquad (6.255a)$$

$$z^{l+1}Q_l(z) - \frac{1}{2}\int_{-1}^{1}\frac{x^{l+1}P_l(x)}{(z-x)}\,dx = \frac{2^l(l!)^2}{(2l+1)!} \qquad (6.255b)$$

$$P_n(z)Q_l(z) - \frac{1}{2}\int_{-1}^{1}\frac{P_n(x)P_l(x)}{(z-x)}\,dx = 0 \qquad n \leqslant l \qquad (6.261a)$$

$$P_{l+1}(z)Q_l(z) - \frac{1}{2}\int_{-1}^{1}\frac{P_{l+1}(x)P_l(x)}{(z-x)}\,dx = \frac{1}{l+1} \qquad (6.261b)$$

Recurrence relations

All recurrence relations satisfied by $P_l(x)$ are also satisfied by $Q_l(x)$.

Miscellaneous identities

$$P_{l+1}(z)Q_l(z) - P_l(z)Q_{l+1}(z) = \frac{1}{l+1} \qquad (6.263)$$

Associated Legendre function $P_l^m(x)$

Differential equation

$$\left[(1-x^2)D^2 - 2xD + \left(l(l+1) - \frac{m^2}{(1-x^2)}\right)\right]P_l^m(x) = 0 \qquad (6.276)$$

Relation to Legendre polynomial $P_l(x)$

$$P_l^m(x) = (1-x^2)^{m/2}D^m P_l(x) \qquad (6.283)$$

Series representation

$$P_l^m(x) = (1-x^2)^{m/2}\sum_{k=0}^{[l/2]}\frac{(-1)^k(2l-2k)!}{2^l k!(l-k)!(l-2k-m)!}x^{l-2k-m} \qquad (6.288)$$

Generating function

$$\frac{(2m)!}{2^m m!}\frac{(1-x^2)^{m/2}}{(1-2xz+z^2)^{m+1/2}} = \sum_{l=0}^{\infty} z^l P_{l+m}^m(x) \qquad (6.290b)$$

Rodriguez formulas

$$P_l^m(x) = \frac{(-1)^l}{2^l l!}(1-x^2)^{m/2} D^{l+m}(1-x^2)^l \qquad m \geqslant 0 \tag{6.291}$$

$$P_l^{-m}(x) = \frac{(-1)^l}{2^l l!}(1-x^2)^{-m/2} D^{l-m}(1-x^2)^l \qquad m \geqslant 0 \tag{6.293}$$

Integral representations

$$P_l^m(x) = \frac{(-1)^l}{2^l l!}\frac{(l+m)!}{2\pi i}(1-x^2)^{m/2}\oint \frac{(1-z^2)^l}{(z-x)^{l+m+1}}\,dz \qquad \text{(Schlafli)} \tag{6.294}$$

$$P_l^m(x) = (-i)^m \frac{(l+m)!}{2\pi l!}\int_{-\pi}^{\pi}\left(x + i\sqrt{1-x^2}\cos\phi\right)^l \cos(m\phi)\,d\phi$$

$$= (-i)^m \frac{(l+m)!}{\pi l!}\int_{0}^{\pi}\left(x + i\sqrt{1-x^2}\cos\phi\right)^l \cos(m\phi)\,d\phi \tag{6.297a}$$

$$= (i)^m \frac{l!}{2\pi(l-m)!}\int_{-\pi}^{\pi}\left(x + i\sqrt{1-x^2}\cos\phi\right)^{-l-1} \cos(m\phi)\,d\phi$$

$$= (i)^m \frac{l!}{\pi(l-m)!}\int_{0}^{\pi}\left(x + i\sqrt{1-x^2}\cos\phi\right)^{-l-1} \cos(m\phi)\,d\phi \qquad \text{(Laplace)}$$

(6.302)

Orthonormalization

$$\int_{-1}^{1} P_l^m(x)P_n^m(x)\,dx = \frac{2}{(2l+1)}\frac{(l+m)!}{(l-m)!}\delta_{ln} \tag{6.306}$$

$$\int_{-1}^{1} \frac{P_l^m(x)P_l^n(x)}{1-x^2}\,dx = \frac{1}{m}\frac{(l+m)!}{(l-m)!}\delta_{mn} \qquad m,n \neq 0 \tag{6.307}$$

Recurrence relations

$$(l-m+2)P_{l+2}^m(x) - (2l+3)P_{l+1}^m(x) + (l+m+1)P_l^m(x) = 0 \qquad l \geqslant 0$$

(6.309)

$$\left[[P_{l+2}^m(x)]' + \frac{mx}{(1-x^2)}P_{l+2}^m(x)\right] - 2x\left[[P_{l+1}^m(x)]' + \frac{mx}{(1-x^2)}P_{l+1}^m(x)\right]$$

$$+\left[[P_l^m(x)]' + \frac{mx}{(1-x^2)}P_l^m(x)\right] = (2m+1)P_{l+1}^m(x) \qquad l \geqslant 0 \tag{6.310}$$

$$P_{l+2}^{m+1}(x) - 2xP_{l+1}^{m+1}(x) + P_l^{m+1}(x) = (2m+1)\sqrt{1-x^2}\,P_{l+1}^m(x) \qquad l \geqslant 0 \tag{6.312}$$

$$P_{l+2}^{m+1}(x) - (2l+3)\sqrt{1-x^2}\,P_{l+1}^m(x) - P_l^{m+1}(x) = 0 \qquad l \geqslant 0 \tag{6.314}$$

$$P_l^{m+2}(x) - \frac{2x}{\sqrt{1-x^2}}(m+1)P_l^{m+1}(x) + [l(l+1) - m(m+1)]P_l^m(x) = 0$$

(Problem 23)

Addition theorem

$$P_l(\cos\Theta)$$

$$= P_l(\cos\theta_1)P_l(\cos\theta_2) + 2\sum_{m=1}^{l}\frac{(l-m)!}{(l+m)!}P_l^m(\cos\theta_1)P_l^m(\cos\theta_2)\cos[m(\phi_1-\phi_2)] \tag{6.334}$$

Miscellaneous identities

$$P_l^m(x) = (-1)^m\frac{(l+m)!}{(l-m)!}P_l^{-m}(x) \qquad m \geqslant 0 \tag{6.298}$$

Associated Legendre function of the second kind $Q_l^m(x)$

Differential equation

$$\left[(1-x^2)D^2 - 2xD + \left(l(l+1) - \frac{m^2}{(1-x^2)}\right)\right]Q_l^m(x) = 0 \tag{6.276}$$

Relation to Legendre function of the second kind

$$Q_l^m(x) = (-1)^l(1-x^2)^{m/2}D^mQ_l(x) \tag{6.335}$$

Recurrence relations.

All recurrence relations satisfied by $P_l^m(x)$ are also satisfied by $Q_l^m(x)$.

Spherical harmonics $Y_l^m(\Omega)$

Definition

$$Y_l^m(\Omega) = Y_l^m(\theta,\phi) = (-1)^m\left[\frac{(2l+1)}{4\pi}\frac{(l-m)!}{(l+m)!}\right]^{1/2}P_l^m(\cos\theta)e^{im\phi} \tag{6.343}$$

Orthonormalization

$$\int_0^{2\pi}d\phi\int_0^{\pi}\sin\theta\, d\theta Y_l^{m*}(\Omega)Y_{l'}^{m'}(\Omega) = \delta_{ll'}\delta_{mm'} \tag{6.338}$$

Addition theorem

$$\begin{aligned} P_l(\cos\Theta) &= \frac{4\pi}{(2l+1)}\sum_{m=-l}^{l}Y_l^m(\Omega_1)Y_l^{-m}(\Omega_2) \\ &= \frac{4\pi}{(2l+1)}\sum_{m=-l}^{l}(-1)^mY_l^m(\Omega_1)Y_l^{m*}(\Omega_2) \end{aligned} \tag{6.349}$$

Miscellaneous identities

$$Y_l^{-m}(\Omega) = (-1)^mY_l^{m*}(\Omega) \tag{6.345}$$

Gegenbauer functions

Differential equation

$$[(1-x^2)D^2 - x(1+2\lambda)D + l(l+2\lambda)]y = 0 \tag{6.350}$$

Series representations

$$G_l^\lambda(x) = \frac{1}{\Gamma(\lambda)} \sum_{m=0}^{[l/2]} (-1)^m \frac{\Gamma(l+\lambda-m)}{m!\,(l-2m)!}(2x)^{l-2m} \qquad \lambda > 0 \tag{6.351a}$$

$$\text{(b)} \quad G_l^0(x) = \sum_{m=0}^{[l/2]} (-1)^m \frac{(l-m-1)!}{m!\,(l-2m)!}(2x)^{l-2m} \qquad l > 0 \qquad \text{(c)} \quad G_0^0(x) = 1 \tag{6.351}$$

Generating functions

$$\Phi(z,x) \equiv (1-2xz+z^2)^{-\lambda} = \sum_{l=0}^{\infty} z^l G_l^\lambda(x) \qquad \lambda > 0 \tag{6.355a}$$

$$\Phi(z,x) = \frac{(x-z)}{(1-2xz+z^2)} = \frac{1}{2} \sum_{l=0}^{\infty} (l+1) z^l G_{l+1}^0(x) \qquad \lambda = 0 \tag{6.355b}$$

Recurrence relations

$$(l+2)G_{l+2}^\lambda(x) - 2x(l+\lambda+1)G_{l+1}^\lambda(x) + (l+2\lambda)G_l^\lambda(x) = 0 \tag{6.356a}$$

$$[G_{l+2}^\lambda(x)]' - 2x[G_{l+1}^\lambda(x)]' + [G_l^\lambda(x)]' = 2\lambda G_{l+1}^\lambda(x) \tag{6.356b}$$

$$G_{l+2}^{\lambda+1}(x) - 2xG_{l+1}^{\lambda+1}(x) + G_l^{\lambda+1}(x) = G_{l+2}^\lambda(x) \tag{6.358a}$$

$$(l+2\lambda+1)xG_{l+1}^\lambda(x) - (l+2)G_{l+2}^\lambda(x) = 2\lambda(1-x^2)G_l^{\lambda+1}(x) \tag{6.358b}$$

$$(l+2)G_{l+2}^\lambda(x) = 2\lambda[xG_{l+1}^\lambda(x) - G_l^{\lambda+1}(x)] \tag{6.358c}$$

$$G_{l+1}^{\lambda+1}(x) = (l+2\lambda+1+x)G_l^{\lambda+1}(x) \tag{6.358d}$$

$$(l+2)G_{l+2}^0(x) - 2x(l+1)G_{l+1}^0(x) + lG_l^0(x) = 0 \qquad l \geqslant 0 \tag{6.359a}$$

$$(l+2)[G_{l+2}^0(x)]' - 2x(l+1)[G_{l+1}^0(x)]' + l[G_l^0(x)]' = 2(l+1)G_{l+1}^0(x) \tag{6.359b}$$

Rodriguez formulas

$$G_l^\lambda(x) = (-1)^l \frac{\Gamma(\lambda+\frac{1}{2})\Gamma(l+2\lambda)}{2^l l!\,\Gamma(2\lambda)\Gamma(l+\lambda+\frac{1}{2})}(1-x^2)^{1/2-\lambda} D^l\left[(1-x^2)^{l+\lambda-1/2}\right] \qquad \lambda > 0 \tag{6.360a}$$

$$G_l^0(x) = (-1)^l \frac{l\Gamma(\frac{1}{2})}{2^{l+2}\Gamma(l+\frac{1}{2})}(1-x^2)^{1/2} D^l\left[(1-x^2)^{l-1/2}\right] \qquad \lambda = 0 \tag{6.360b}$$

Orthonormalization

$$\int_{-1}^{1} (1-x^2)^{\lambda-1/2} G_l^\lambda(x) G_n^\lambda(x)\,dx = \frac{\pi 2^{1-2\lambda}\Gamma(l+2\lambda)}{l!\,(l+\lambda)[\Gamma(\lambda)]^2}\delta_{ln} \qquad \lambda \neq 0 \tag{6.364a}$$

$$\int_{-1}^{1} (1-x^2)^{-1/2} G_0^0(x) G_n^0(x)\,dx = \pi\delta_{n0} \tag{6.364b}$$

$$\int_{-1}^{1} (1-x^2)^{-1/2} G_l^0(x) G_n^0(x)\,dx = \frac{2\pi}{l^2}\delta_{ln} \qquad l \neq 0 \tag{6.364c}$$

Integral representations

$$G_l^\lambda(x) = (1 - x^2)^{1/2-\lambda} \oint \frac{(1 - z^2)^{l+\lambda-1/2}}{(z - x)^{l+1}} dz \qquad \text{(Schlafli)} \tag{6.365a}$$

$$G_l^\lambda(x) = \frac{2^{1-2\lambda}\Gamma(l + 2\lambda)}{l!\,[\Gamma(\lambda)]^2} \int_0^\pi \left(x + i\sqrt{1 - x^2} \cos\phi\right)^l \sin^{2\lambda-1}\phi \, d\phi \qquad \text{(Laplace)} \tag{6.365b}$$

Miscellaneous identities

$$G_l^\lambda(-x) = (-1)^l G_l^\lambda(x) \tag{6.352}$$

Ordinary Laguerre polynomials $L_n(x)$

Differential equation

$$\left[xD^2 - (x - 1)D + n\right]y = 0 \tag{6.366}$$

Series representation

$$L_n(x) = \sum_{m=0}^{n} (-1)^m \frac{n!}{(n - m)!\,(m!)^2} x^m \tag{6.371}$$

Generating function

$$\Phi(z, x) = \frac{e^{-xz/(1-z)}}{(1 - z)} = \sum_{n=0}^{\infty} z^n L_n(x) \tag{6.373}$$

Recurrence relations

$$(n + 2)L_{n+2}(x) - (2n + 3 - x)L_{n+1}(x) + (n + 1)L_n(x) = 0 \quad n \geqslant 0 \tag{6.381}$$

$$L_n(x) + L'_{n+1}(x) - L'_n(x) = 0 \qquad n \geqslant 0 \tag{6.383}$$

$$(n + 2)L'_{n+2}(x) - (2n + 3 - x)L'_{n+1}(x) + (n + 1)L'_n(x) + L_{n+1}(x) = 0 \tag{6.384}$$

$$L'_{n+2}(x) = L'_{n+1}(x) - L_{n+1}(x) \tag{6.385}$$

$$L'_{n+2}(x) = L'_n(x) - L_n(x) - L_{n+1}(x) \tag{6.386}$$

$$xL'_n(x) - (n + 1)L_{n+1}(x) + (n + 1 - x)L_n(x) = 0 \tag{6.387}$$

Addition theorem

$$L'_{n+1}(x + y) = -\sum_{m=0}^{n} L_m(x)L_{n-m}(y) \qquad \text{(Problem 34)}$$

Rodriguez formula

$$L_n(x) = \frac{e^x}{n!} D^n(e^{-x}x^n) \tag{6.393}$$

Orthonormalization

$$\int_0^\infty e^{-x} L_m(x) L_n(x) \, dx = \delta_{mn} \tag{6.398}$$

Miscellaneous identities

$$x^k = k! \sum_{n=0}^{k} (-1)^n \frac{k!}{n!\,(k - n)!} L_n(x) \tag{6.404}$$

$$x^k = k!\,(1 - L)^k \qquad \text{with } L^n \to L_n(x) \tag{6.405}$$

Associated Laguerre polynomials $L_n^k(x)$

Differential equation

$$[xD^2 + (k+1-x)D + (n-k)]L_n^k(x) = 0 \tag{6.416}$$

Series representations

$$L_n^k(x) = \sum_{m=k}^{n} (-1)^m \frac{n!}{(n-m)!(m-k)!m!} x^{m-k} \tag{6.417a}$$

$$L_n^k(x) = (-1)^k \sum_{m=0}^{n-k} (-1)^m \frac{n!}{(n-m-k)!(m+k)!m!} x^m \tag{6.417b}$$

$$L_{n+k}^k(x) = (-1)^k \sum_{m=0}^{n} (-1)^m \frac{(n+k)!}{(n-m)!(m+k)!m!} x^m \tag{6.417c}$$

Generating function

$$\Phi(z,x) = (-1)^k \frac{1}{(1-z)^{k+1}} e^{-xz/(1-z)} = \sum_{n=0}^{\infty} z^n L_{n+k}^k(x) \tag{6.419b}$$

Recurrence relations

$$(-1)^k L_{n+k}(x) = \sum_{r=0}^{k} \frac{(-1)^r k!}{r!(k-r)!} L_{n+2k-r}^k(x) \tag{6.432}$$

$$(-1)^k L_{n+k}(x) = [1 - L_{n+2k-\mu}^k(x)]^k \left\{ \begin{array}{l} \text{with } \mu \text{ replaced by the} \\ \text{power of } L_{n+2k-\mu}^k(x) \end{array} \right. \tag{6.434}$$

$$(n+2)L_{n+k+2}^k(x) - (2n+k+3-x)L_{n+k+1}^k(x) + (n+k+1)L_{n+k}^k(x) = 0 \tag{6.436}$$

$$L_{n+k}^k(x) = [L_{n+k}^k(x)]' - [L_{n+k+1}^k(x)]' \tag{6.438}$$

$$L_{n+k}^k(x) = L_{n+k}^{k+1}(x) - L_{n+k+1}^{k+1}(x) \tag{6.440}$$

$$(n+2)L_{n+2}^k(x) - (2n+3-x)L_{n+1}^k(x) + (n+1)L_n^k(x) + kL_{n+1}^{k-1}(x) = 0 \tag{6.441}$$

$$L_n^k(x) - [L_{n+1}^k(x)]' - [L_n^k(x)]' = 0 \tag{6.442a}$$

$$L_n^k(x) - L_{n+1}^{k+1}(x) - L_n^{k+1}(x) = 0 \tag{6.442b}$$

$$[L_{n+2}^k(x)]' = [L_n^k(x)]' - L_n^k(x) - L_{n+1}^k(x) \tag{6.443a}$$

$$L_{n+2}^{k+1}(x) = L_n^{k+1}(x) - L_n^k(x) - L_{n+1}^k(x) \tag{6.443b}$$

$$x[L_n^k(x)]' + k[L_n^{k-1}(x)]' - (n+1)L_{n+1}^k(x) + (n+1-x)L_n^k(x) - kL_n^{k-1}(x) = 0 \tag{6.444a}$$

$$xL_n^{k+1}(x) + kL_n^k(x) - (n+1)L_{n+1}^k(x) + (n+1-x)L_n^k(x) - kL_n^{k-1}(x) = 0 \tag{6.444b}$$

$$xL_n^{k+2}(x) + (k+1-x)L_n^{k+1}(x) + (n-k)L_n^k(x) = 0 \tag{6.445}$$

Rodriguez formulas

$$L_n^k(x) = \frac{e^x}{(n-k)!} D^n(e^{-x}x^{n-k}) \tag{6.447b}$$

$$L_{n+k}^k(x) = \frac{e^x}{n!} D^{n+k}(e^{-x}x^n) \tag{6.447c}$$

Orthonormalization

$$\int_0^\infty x^k e^{-x} L_{n+k}^k(x) L_{m+k}^k(x)\, dx = \frac{(k+n)!}{n!}\delta_{nm} \tag{6.451a}$$

$$\int_0^\infty x^k e^{-x} L_n^k(x) L_m^k(x)\, dx = \frac{n!}{(n-k)!}\delta_{nm} \tag{6.451b}$$

Miscellaneous identities

$$x^l = \sum_{n=0}^{l} (-1)^{n+k} \frac{(l+k)!\,l!}{(n+k)!\,(l-n)!} L_{n+k}^k(x) \tag{6.456}$$

Hermite polynomials $H_n(x)$

Differential equation

$$[D^2 - 2xD + 2n]H_n(x) = 0 \tag{6.457}$$

Series representation

$$H_n(x) = \sum_{l=0}^{[n/2]} (-1)^l \frac{n!}{(n-2l)!\,l!}(2x)^{n-2l} \tag{6.469}$$

Generating functions

$$\Phi(z, x) = e^{-z^2+2xz} = \sum_{n=0}^{\infty} z^n \frac{H_n(x)}{n!} \tag{6.473}$$

$$e^{z^2}\cos(2xz) = \sum_{n=0}^{\infty} (-1)^n z^{2n} \frac{H_{2n}(x)}{(2n)!} \qquad \text{(Problem 39)}$$

$$e^{z^2}\sin(2xz) = \sum_{n=0}^{\infty} (-1)^n z^{2n+1} \frac{H_{2n+1}(x)}{(2n+1)!} \qquad \text{(Problem 39)}$$

Recurrence relations

$$H_{n+2}(x) - 2xH_{n+1}(x) + 2(n+1)H_n(x) = 0 \qquad n \geqslant 0 \tag{6.478}$$

$$H'_{n+1}(x) = 2(n+1)H_n(x) \qquad n \geqslant 0 \tag{6.480}$$

$$H_{n+1}(x) - 2xH_n(x) + H'_n(x) = 0 \qquad n \geqslant 0 \tag{6.481}$$

$$(n+1)H_{n+1}(x) - xH'_{n+1}(x) + (n+1)H'_n(x) = 0 \qquad n \geqslant 0 \tag{6.482}$$

Rodriguez formula

$$H_n(x) = (-1)^n e^{x^2} D^n(e^{-x^2}) \tag{6.486}$$

Integral representation

$$H_n(x) = \frac{(-2i)^n}{\sqrt{\pi}} e^{x^2} \int_{-\infty}^{\infty} z^n e^{-z^2+2ixz}\, dz \tag{6.490}$$

Orthonormalization

$$\int_{-\infty}^{\infty} e^{-x^2} H_n(x) H_m(x)\, dx = \sqrt{\pi}\, 2^n n!\, \delta_{mn} \tag{6.495}$$

Addition theorem

$$H_n(x+y) = 2^{-n/2} \sum_{m=0}^{n} \frac{n!}{m!\,(n-m)!} H_m(\sqrt{2}x) H_{n-m}(\sqrt{2}y) \qquad \text{(Problem 40)}$$

Miscellaneous identities

$$H_n(-x) = (-1)^n H_n(x) \tag{6.471}$$

$$x^k = \frac{k!}{2^k} \sum_{n=0}^{[k/2]} \frac{1}{(k-2n)!\,n!} H_{k-2n}(x) \tag{6.512}$$

$$P_n(x) = \frac{1}{\sqrt{\pi}\, n!} \int_{-\infty}^{\infty} e^{-t^2} t^n H_n(xt)\, dt \qquad \text{(Problem 42)}$$

Ordinary Bessel functions $J_\alpha(x)$

Differential equations

$$\text{(a)} \quad [x'^2 D^2 + x'D + (k^2 x'^2 - \alpha^2)]y = 0 \qquad \text{(b)} \quad [x^2 D^2 + xD + (x^2 - \alpha^2)]y = 0 \tag{6.513}$$

Series representation

$$J_\alpha(x) = \sum_{m=0}^{\infty} (-1)^m \frac{1}{m!\,(m+\alpha)!} \left(\frac{x}{2}\right)^{2m+\alpha} \tag{6.521}$$

Generating functions for integer order

$$\Phi(z,x) = e^{(z-1/z)x/2} = \sum_{n=-\infty}^{\infty} z^n J_n(x) \tag{6.530}$$

$$\Phi(z,x) = J_0(x) + \sum_{n=1}^{\infty} [z^n + (-1)^n z^{-n}] J_n(x) \tag{6.536}$$

$$\cos(x \sin\theta) = J_0(x) + 2\sum_{n=0}^{\infty} \cos(2n\theta) J_{2n}(x) \tag{6.539a}$$

$$\sin(x \sin\theta) = 2\sum_{n=0}^{\infty} \sin[(2n+1)\theta] J_{2n+1}(x) \tag{6.539b}$$

$$\cosh(x \sinh w) = J_0(x) + 2\sum_{n=0}^{\infty} \cosh(2nw) J_{2n}(x) \qquad \text{(Problem 46)}$$

$$\sinh(x \sinh w) = 2\sum_{n=0}^{\infty} \sinh[(2n+1)w] J_{2n+1}(x) \qquad \text{(Problem 46)}$$

Addition theorem

$$J_n(x+y) = \sum_{l=-\infty}^{\infty} J_l(x) J_{n-l}(y) \qquad \text{(Problem 45)}$$

Integral representations

$$\text{(a)} \quad J_0(x) = \frac{1}{\pi}\int_0^{\pi} \cos(x \sin\theta)\, d\theta \qquad \text{(b)} \quad J_{2m}(x) = \frac{2}{\pi}\int_0^{\pi} \cos(x\sin\theta)\cos(2m\theta)\, d\theta$$

$$\text{(c)} \quad J_{2m+1}(x) = \frac{2}{\pi}\int_0^{\pi} \sin(x\sin\theta)\sin[(2m+1)\theta]\, d\theta \tag{6.545}$$

$$J_N(x) = \frac{1}{2\pi i}\oint e^{(z-1/z)x/2}\frac{dz}{z^{N+1}} \tag{6.547}$$

$$J_N(x) = \frac{1}{2\pi}\int_{-\pi}^{\pi} e^{i(x\sin\theta - N\theta)}\, d\theta \tag{6.548}$$

$$J_N(x) = \frac{1}{\pi}\int_0^{\pi} \cos(x\sin\theta - N\theta)\, d\theta \tag{6.549}$$

$$J_\alpha(x) = \left(\frac{x}{2}\right)^\alpha \frac{1}{\sqrt{\pi}\,\Gamma(\alpha+\frac{1}{2})}\int_{-1}^{1}(1-w^2)^{\alpha-1/2}\cos(wx)\, dw \tag{6.555}$$

$$J_\alpha(x) = \left(\frac{x}{2}\right)^\alpha \frac{1}{\sqrt{\pi}\,\Gamma(\alpha+\frac{1}{2})}\int_{-1}^{1}(1-w^2)^{\alpha-1/2}e^{iwx}\, dw \tag{6.556}$$

Recurrence relations

$$2J'_\alpha(x) = J_{\alpha-1}(x) - J_{\alpha+1}(x) \tag{6.560b}$$

$$\frac{\alpha}{x}J_\alpha(x) + J'_\alpha(x) = J_{\alpha-1}(x) \tag{6.563}$$

$$\frac{\alpha}{x}J_\alpha(x) - J'_\alpha(x) = J_{\alpha+1}(x) \tag{6.566}$$

$$\frac{2\alpha}{x}J_\alpha(x) = J_{\alpha-1}(x) + J_{\alpha+1}(x) \tag{6.567}$$

$$2J'_\alpha(x) = J_{\alpha-1}(x) - J_{\alpha+1}(x) \tag{6.568}$$

Differentiation formulas

$$D[x^{-\alpha}J_\alpha(x)] = -x^{-\alpha}J_{\alpha+1}(x) \qquad D[x^{\alpha}J_{-\alpha}(x)] = -x^{\alpha}J_{-\alpha+1}(x) \quad \text{(Problem 49)}$$

$$D[x^{-\alpha}J_{-\alpha}(x)] = x^{-\alpha}J_{-\alpha-1}(x) \quad \text{(Problem 49)} \qquad D[x^{\alpha}J_\alpha(x)] = x^{\alpha}J_{\alpha-1}(x) \tag{6.562}$$

Bessel functions of half-integer order

$$J_{1/2}(x) = \left(\frac{2}{\pi x}\right)^{1/2}\sin x \tag{6.572}$$

$$J_{-1/2}(x) = \left(\frac{2}{\pi x}\right)^{1/2}\cos x \tag{6.573}$$

$$J_{3/2}(x) = \left(\frac{2}{\pi x}\right)^{1/2}\left[\frac{\sin x}{x} - \cos x\right] \tag{6.574}$$

$$J_{-3/2}(x) = -\left(\frac{2}{\pi x}\right)^{1/2}\left[\frac{\cos x}{x} + \sin x\right] \tag{6.575}$$

Orthonormalization

With k_m defined so that $J_\alpha(k_m) = 0$

$$\int_0^1 x J_\alpha(k_m x) J_\alpha(k_l x)\, dx = \tfrac{1}{2}[J'_\alpha(k_m)]^2 \delta_{lm} \tag{6.590}$$

$$\int_0^1 x J_\alpha(k_m x) J_\alpha(k_l x)\, dx = \tfrac{1}{2}[J_{\alpha-1}(k_m)]^2 \delta_{lm} \tag{6.592a}$$

$$\int_0^1 x J_\alpha(k_m x) J_\alpha(k_l x)\, dx = \tfrac{1}{2}[J_{\alpha+1}(k_m)]^2 \delta_{lm} \tag{6.592b}$$

$$\int_0^1 x J_\alpha(k_m x) J_\alpha(k_l x)\, dx = -\tfrac{1}{2} J_{\alpha-1}(k_m) J_{\alpha+1}(k_m) \delta_{lm} \tag{6.592c}$$

Small x behavior

$$J_\alpha(x) \simeq \frac{1}{\alpha!}\left(\frac{x}{2}\right)^\alpha \qquad \alpha \neq \text{negative integer} \tag{6.621}$$

$$J_{-n}(x) \simeq \frac{(-1)^n}{n!}\left(\frac{x}{2}\right)^n \qquad n = \text{positive integer} \tag{6.622}$$

Large x behavior

$$J_\alpha(x) \simeq \left(\frac{2}{\pi x}\right)^{1/2} \cos\left(x - \alpha\frac{\pi}{2} - \frac{\pi}{4}\right) \tag{6.639}$$

Miscellaneous identities

$$J_{-n}(x) = (-1)^n J_n(x) \qquad n = \text{integer} \tag{6.525}$$

$$J_\alpha(-x) = (-1)^\alpha J_\alpha(x) \tag{6.526}$$

$$x = 2\sum_{n=0}^{\infty} (2n+1) J_{2n+1}(x) \tag{6.541}$$

$$x^2 = 8\sum_{n=1}^{\infty} n^2 J_{2n}(x) \tag{6.543}$$

Neumann function $Y_\alpha(x)$

Definition

$$Y_\alpha(x) \equiv \frac{\cos(\pi\alpha) J_\alpha(x) - J_{-\alpha}(x)}{\sin(\pi\alpha)} \qquad \alpha \neq \text{integer} \tag{6.595}$$

Series representation

$$Y_n(x) = \frac{2}{\pi}\log\left(\frac{x}{2}\right) J_n(x) - \frac{1}{\pi}\sum_{m=0}^{n-1} \frac{(n-m-1)!}{m!}\left(\frac{x}{2}\right)^{2m-n} \tag{6.602}$$

Small x behavior

$$Y_\alpha(x) \simeq -\frac{\Gamma(\alpha)}{\pi}\left(\frac{x}{2}\right)^{-\alpha} \tag{6.624}$$

Large x behavior

$$Y_\alpha(x) \simeq \left(\frac{2}{\pi x}\right)^{1/2} \sin\left(x - \alpha\frac{\pi}{2} - \frac{\pi}{4}\right) \tag{6.644}$$

Miscellaneous identities

$$Y_{-n}(x) = (-1)^n Y_n(x) \tag{6.603}$$

Hankel functions $H_\alpha^\pm(x)$

Definition

$$H_\alpha^{(\pm)}(x) \equiv J_\alpha(x) \pm iY_\alpha(x) \tag{6.604}$$

$$H_\alpha^{(\pm)}(x) = \pm\frac{1}{\sin(\pi\alpha)}\left[e^{\mp i\pi\alpha}J_\alpha(x) - J_{-\alpha}(x)\right] \tag{6.606}$$

Small x behavior

$$H_\alpha^{(\pm)}(x) \simeq Y_\alpha(x) \simeq -\frac{\Gamma(\alpha)}{\pi}\left(\frac{x}{2}\right)^{-\alpha}$$

Large x behavior

$$H_\alpha^{(\pm)}(x) \simeq \left(\frac{2}{\pi x}\right)^{1/2} e^{\pm i[x-\alpha(\pi/2)-(\pi/4)]} \qquad \text{(6.642b) and (6.643b)}$$

Miscellaneous identities

$$H_{-n}(x) = (-1)^n H_n(x)$$

Modified Ordinary Bessel functions $I_\alpha(x)$

Definition

$$I_\alpha(x) \equiv e^{-i\pi\alpha/2}J_\alpha(x) = e^{i\pi\alpha/2}J_\alpha(-ix) \qquad \text{(6.607) and (6.609)}$$

Generating function

$$e^{(z+1/z)x/2} = \sum_{n=-\infty}^{\infty} z^n I_n(x) \qquad \text{(Problem 50)}$$

Recurrence relations

$$\frac{2\alpha}{x}I_\alpha(x) = I_{\alpha-1}(x) - I_{\alpha+1}(x) \qquad 2I_\alpha'(x) = I_{\alpha-1}(x) + I_{\alpha+1}(x) \quad \text{(Problem 51)}$$

Miscellaneous identities

$$I_{-n}(x) = I_n(x) \tag{6.608}$$

Modified Neumann functions $X_\alpha(x)$

Definition

$$X_\alpha(x) \equiv (-i)^\alpha Y_\alpha(x) \tag{6.610}$$

Modified Hankel functions $K_\alpha(x)$

Definition

$$K_\alpha(x) \equiv \frac{\pi}{2}\left[\frac{I_\alpha(x) - I_{-\alpha}(x)}{\sin(\pi\alpha)}\right] \tag{6.612b}$$

Miscellaneous identities

$$K_\alpha(\pm x) = \frac{\pi}{2}(\pm i)^{\alpha-1} H_\alpha^{(\pm)}(x) \tag{6.614}$$

Spherical Bessel functions $j_n(x)$, $\eta_n(x)$

Definition

$$\text{(a)} \quad j_n(x) \equiv \left(\frac{\pi}{2x}\right)^{1/2} J_{n+1/2}(x) \qquad \text{(b)} \quad \eta_n(x) \equiv (-1)^{n+1}\left(\frac{\pi}{2x}\right)^{1/2} J_{-n-1/2}(x) \tag{6.616}$$

Integral representation

$$j_n(x) = \frac{1}{n!}\left(\frac{x}{2}\right)^n \int_0^{\pi/2} \cos(x \sin\phi)\cos^{2n+1}\phi \, d\phi \qquad \text{(Problem 52)}$$

Small x behavior

$$j_n(x) \simeq \frac{n!}{(2n+1)!}(2x)^n \quad (6.627) \qquad \eta_n(x) \simeq -\frac{4}{\pi}\frac{(2n)!}{n!}\frac{1}{(2x)^{n+1}} \quad (6.630)$$

Large x behavior

$$j_n(x) \simeq \frac{\cos[x - n(\pi/2) - \pi/2]}{x} = \frac{\sin[x - n(\pi/2)]}{x} \tag{6.645a}$$

$$\eta_n(x) \simeq \frac{\sin[x - n(\pi/2) - \pi/2]}{x} = -\frac{\cos[x - n(\pi/2)]}{x} \tag{6.645b}$$

Miscellaneous identities

$$\text{(a)} \quad j_{-n}(x) = \eta_{n-1}(x) \qquad \text{(b)} \quad \eta_{-n}(x) = j_{n-1}(x) \tag{6.617}$$

Spherical Neumann function $y_n(x)$

Relation to other spherical functions

$$y_n(x) = \eta_n(x) \tag{6.619b}$$

Spherical Hankel functions $h_n^{\pm}(x)$

Relation to other spherical functions

$$h_n^{(\pm)}(x) \equiv j_n(x) \pm iy_n(x) = j_n(x) \pm i\eta_n(x) \tag{6.620}$$

Gauss's Hypergeometric functions ${}_2F_1(a, b; c; x)$

Differential equations

$$\{x(1-x)D^2 + [c - (a+b+1)x]D - ab\}y = 0 \tag{6.646a}$$

$$x'^2(x'-1)\frac{d^2y}{dx'^2} + x'[(a+b+1) + (2-c)x']\frac{dy}{dx'} - aby = 0 \tag{6.646b}$$

Series solution

$${}_2F_1(a, b; c; x) = \sum_{m=0}^{\infty} \frac{(a)_m(b)_m}{(c)_m}\frac{x^m}{m!} \tag{6.651}$$

Relation to other functions

$$\frac{1}{(1-x)} = {}_2F_1(1, c; c; x) \quad (6.666) \qquad \frac{1}{x}\log(1-x) = {}_2F_1(1, 1; 2; x) \quad (6.668)$$

$$\frac{1}{(1-x)^{\alpha}} = {}_2F_1(\alpha, c; c; x) \quad (6.669) \qquad \sin^{-1} x = x\,{}_2F_1(\tfrac{1}{2}, \tfrac{1}{2}; \tfrac{3}{2}; x^2) \quad (6.672)$$

$$P_l(x) = {}_2F_1\left(-l, l+1; 1; \frac{1-x}{2}\right) \tag{6.677}$$

$$P_l^m(x) = \frac{(l+m)!}{(l-m)!}\frac{1}{2^m m!}(1-x^2)^{m/2}{}_2F_1\left(m-l, m+l+1; m+1; \frac{1-x}{2}\right) \tag{6.683}$$

$$P_{2l}(x) = (-1)^l\frac{(2l)!}{2^{2l}(l!)^2}\,{}_2F_1(-l, l+\tfrac{1}{2}; \tfrac{1}{2}; x^2) \qquad \text{(Problem 61)}$$

$$P_{2l+1}(x) = (-1)^l\frac{(2l+1)!}{2^{2l}(l!)^2}\,{}_2F_1(-l, l+\tfrac{3}{2}; \tfrac{3}{2}; x^2) \qquad \text{(Problem 61)}$$

$$Q_l(x) = \frac{\sqrt{\pi}\,l!}{\Gamma(l+\frac{3}{2})(2x)^{l+1}}\,{}_2F_1\left(\frac{l+2}{2}, \frac{l+1}{2}; l+\tfrac{3}{2}; \frac{1}{x^2}\right) \tag{6.684}$$

$$G_{2l}^{\lambda}(x) = (-1)^l\frac{(\lambda)_l}{l!}\,{}_2F_1(-l, \lambda+l; \tfrac{1}{2}; x^2) \tag{6.685}$$

$$G_{2l+1}^{\lambda}(x) = (-1)^l\frac{(\lambda)_{l+1}}{l!}\,{}_2F_1(-l, \lambda+l; \tfrac{3}{2}; x^2) \tag{6.686}$$

$$K(k^2) = \frac{\pi}{2}\,{}_2F_1(\tfrac{1}{2}, \tfrac{1}{2}; 1; k^2) \qquad E(k^2) = \frac{\pi}{2}\,{}_2F_1(\tfrac{1}{2}, -\tfrac{1}{2}; 1; k^2) \quad \text{(Problem 59)}$$

Generating function for hypergeometric polynomials

$${}_2\Phi_1(z, x) \equiv (1-z)^{b-c}(1-z+zx)^{-b} = \sum_{n=0}^{\infty} z^n\frac{(c)_n}{n!}\,{}_2F_1(-n, b; c; x) \tag{6.704}$$

Recurrence and contiguous relations

$$\begin{aligned}&(c+n+1)(c+n)\,{}_2F_1(-n-2,b;c;x)\\&-(c+n)[(c+2n+2)-x(b+n+1)]\,{}_2F_1(-n-1,b;c;x)\\&+c(1-x)(n+1)\,{}_2F_1(-n,b;c;x)=0\end{aligned} \tag{6.709}$$

$$(c+n)\,{}_2F_1'(-n-1,b;c;x)-(n+1)\,{}_2F_1'(-n,b;c;x)+b(n+1)\,{}_2F_1(-n,b;c;x)=0 \tag{6.711}$$

$$\begin{aligned}&-(c+n)\,{}_2F_1(-n,b+1;c+1;x)+n\,{}_2F_1(-n+1,b+1;c+1;x)\\&+c\,{}_2F_1(-n,b;c;x)=0\end{aligned} \tag{6.713}$$

$$(c-a-b)\,{}_2F_1(a,b;c;x)+a(1-x)\,{}_2F_1(a+1,b;c;x)-(c-b)\,{}_2F_1(a,b-1;c;x)=0 \tag{6.717}$$

$$(c-a-1)\,{}_2F_1(a,b;c;x)+a\,{}_2F_1(a+1,b;c;x)-(c-1)\,{}_2F_1(a,b;c-1;x)=0 \tag{6.722}$$

$$c(1-x)\,{}_2F_1(a,b;c;x)-c\,{}_2F_1(a-1,b;c;x)-(c-b)x\,{}_2F_1(a,b;c+1;x)=0 \tag{6.723}$$

$$\begin{aligned}&c(c-1)\,{}_2F_1(a,b;c;x)+abx\,{}_2F_1(a+1,b+1;c+1;x)\\&-c(c-1)\,{}_2F_1(a,b;c-1;x)=0\end{aligned} \tag{6.724}$$

$$\begin{aligned}&c(c+1)\,{}_2F_1(a,b;c;x)+a(b-c)x\,{}_2F_1(a+1,b+1;c+2;x)\\&-c(c+1)\,{}_2F_1(a,b+1;c+1;x)=0\end{aligned} \tag{6.725}$$

$$c\,{}_2F_1(a,b;c;x)+ax\,{}_2F_1(a+1,b+1;c+1;x)-c\,{}_2F_1(a,b+1;c;x)=0 \tag{Problem 65}$$

$$(b-a)\,{}_2F_1(a,b;c;x)+a\,{}_2F_1(a+1,b;c;x)-b\,{}_2F_1(a,b+1;c;x)=0 \tag{Problem 66}$$

Integral representations

$${}_2F_1(a,b;c;x)=\frac{\Gamma(c)}{\Gamma(b)\Gamma(c-b)}\int_0^1\frac{t^{b-1}(1-t)^{c-b-1}}{(1-xt)^a}\,dt \tag{6.691a}$$

$${}_2F_1(a,b;c;x)=\frac{\Gamma(c)}{\Gamma(a)\Gamma(c-a)}\int_0^1\frac{t^{a-1}(1-t)^{c-a-1}}{(1-xt)^b}\,dt \tag{6.691b}$$

$$\begin{aligned}{}_2F_1(a,b;c;x)&=\frac{\Gamma(c)}{\Gamma(b)\Gamma(c-b)}\int_0^1 t^{b-1}(1-t)^{c-b-1}\,{}_1F_0(a;xt)\\&=\frac{\Gamma(c)}{\Gamma(a)\Gamma(c-a)}\int_0^1 t^{a-1}(1-t)^{c-a-1}\,{}_1F_0(b;xt)\end{aligned} \tag{6.693}$$

Differential formula

$$D[{}_2F_1(a,b;c;x)]=\frac{ab}{c}\,{}_2F_1(a+1,b+1;c+1;x) \tag{Problem 55}$$

Second solution

$$y_2(x) = x^{1-c}\,{}_2F_1(a-c+1, b-c+1; 2-c; x) \tag{6.729a}$$

Kummer's hypergeometric function ${}_1F_1(a;c;x)$

Differential equation

$$[x'D^2 + (c-x')D - a]y = 0 \tag{6.663}$$

Series solution

$$ {}_1F_1(a;c;x) = \sum_{m=0}^{\infty} \frac{(a)_m}{(c)_m}\frac{x^m}{m!} \tag{6.657}$$

Relation to Gauss's hypergeometric function

$$ {}_1F_1(a;c;x) = \lim_{b\to\infty} {}_2F_1(a,b;c;x/b) \tag{6.662}$$

Kummer function of the second kind

$$W(a;c;x) \equiv \frac{\pi}{\sin(\pi c)}\left[\frac{{}_1F_1(a;c;x)}{\Gamma(c)\Gamma(a-c+1)} - \frac{x^{1-c}\,{}_1F_1(a-c+1;2-c;x)}{\Gamma(a)\Gamma(2-c)}\right] \tag{6.730}$$

Relation to other functions

$$e^x = {}_1F_1(c;c;x) \quad (6.673) \qquad H_{2n}(x) = (-1)^n\frac{(2n)!}{n!}\,{}_1F_1(-n;\tfrac{1}{2};x^2) \quad (6.682)$$

$$H_{2n+1}(x) = \frac{(-1)^n(2n+1)!}{n!}2x\,{}_1F_1(-n;\tfrac{3}{2};x^2) \qquad \text{(Problem 63)}$$

$$(1-x)e^{-x} = {}_1F_1(2;1;x) \qquad \frac{1-e^{-x}}{x} = {}_1F_1(1;2;-x) \qquad \text{(Problem 60)}$$

$$L_n(x) = {}_1F_1(-n;1;x) \qquad \text{(Problem 62)}$$

$$\operatorname{erfc}(x) \equiv 1 - \operatorname{erf}(x) = \frac{e^{-x^2}}{\sqrt{\pi}}W(\tfrac{1}{2};\tfrac{1}{2};x^2) \tag{6.731}$$

$$K_n(x) = \sqrt{\pi}(2x)^n e^{-x}W(n+\tfrac{1}{2};2n+1;2x) \tag{6.732}$$

$$J_\alpha(x) = \frac{(2x)^\alpha e^{-ix}}{\sqrt{\pi}}\frac{\Gamma(\alpha+\frac{1}{2})}{\Gamma(2\alpha+1)}\,{}_1F_1(\alpha+\tfrac{1}{2};2\alpha+1;2ix) \qquad \text{(Problem 64)}$$

$$e^x\sinh x = x\,{}_1F_1(1;2;2x) \qquad \text{(Problem 64)}$$

Generating function

$$(1-z)^{-c}e^{-xz/(1-z)} = \sum_{n=0}^{\infty} z^n\frac{(c)_n}{n!}\,{}_1F_1(-n;c;x) \tag{6.707}$$

Recurrence and contiguous relations

$$(c+n+1)(c+n)\,{}_1F_1(-n-2;c;x) - (c+n)(c+2n+2-x)\,{}_1F_1(-n-1;c;x)$$
$$+c(n+1)\,{}_1F_1(-n;c;x) = 0 \tag{6.714}$$

$$(c+n)\,{}_1F_1'(-n-1;c;x) - (n+1)\,{}_1F_1'(-n;c;x) + (n+1)\,{}_1F_1(-n;c;x) = 0 \tag{6.715}$$

$$-(c+n)\,{}_1F_1(-n;c+1;x) + n\,{}_1F_1(-n+1;c+1;x) + c\,{}_1F_1(-n;c;x) = 0 \tag{6.716}$$

Integral representation

$$ {}_1F_1(a;c;x) = \frac{\Gamma(c)}{\Gamma(a)\Gamma(c-a)} \int_0^1 t^{a-1}(1-t)^{c-a-1} e^{xt}\, dt \tag{6.697} $$

Differentiation formula

$$ D[{}_1F_1(a;c;x)] = \frac{a}{c}\,{}_1F_1(a+1;c+1;x) \qquad \text{(Problem 55)} $$

Second solution

$$ y_2(x) = x^{1-c}\,{}_1F_1(a-c+1;2-c;x) \tag{6.729b} $$

Generalized hypergeometric function $_rF_s(\alpha_1, \alpha_2, \ldots, \alpha_r; \beta_1, \beta_2, \ldots, \beta_s; x)$

Series representation

$$ {}_rF_s(\alpha_1,\alpha_2,\ldots,\alpha_r;\beta_1,\beta_2,\ldots,\beta_s;x) = \sum_{m=0}^{\infty} \frac{(\alpha_1)_m(\alpha_2)_m \cdots (\alpha_r)_m}{(\beta_1)_m(\beta_2)_m \cdots (\beta_s)_m} \frac{x^m}{m!} \tag{6.653} $$

Relationship between hypergeometric functions

$$ \text{(a)}\quad {}_2F_1(a,b;a;x) = {}_1F_0(b;x) \qquad \text{(b)}\quad {}_2F_2(a,b;a,c;x) = {}_1F_1(b;c;x) $$

$$ \text{(c)}\quad {}_3F_2(a,b,d;c,d;x) = {}_2F_1(a,b;c;x) \tag{6.654} $$

Integral representation

$$ {}_sF_r(\alpha_1,\alpha_2,\ldots,\alpha_s;\beta_1,\beta_2,\ldots,\beta_r;x) $$

$$ = \frac{\Gamma(\beta_1)}{\Gamma(\alpha_1)\Gamma(\beta_1-\alpha_1)} \int_0^1 t^{\alpha_1-1}(1-t)^{\beta_1-\alpha_1-1}\,{}_{s-1}F_{r-1}(\alpha_2,\alpha_3,\ldots,\alpha_s;\beta_2,\beta_3,\ldots,\beta_r;xt)\, dt \tag{6.694} $$

Dirac δ-function

Definition

$$ F(x_0) = \int_{-\infty}^{\infty} F(x)\delta(x-x_0)\, dx \tag{6.733a} $$

Miscellaneous identities

$$ \delta(ax) = \frac{1}{|a|}\delta(x) \qquad a \neq 0 \tag{6.736} $$

$$ \text{(a)}\quad \delta(-x) = \delta(x) \qquad \text{(b)}\quad \delta'(-x) = -\delta'(x) \qquad \text{(c)}\quad \delta''(-x) = \delta''(x) \tag{6.738} $$

$$ x\delta'(x) = -\delta(x) \quad (6.741) \qquad x^2\delta''(x) = 2\delta(x) \tag{6.747} $$

$$ x^3\delta'''(x) = -6\delta(x) \qquad \text{(Problem 68)} $$

$$ \delta(x^2 - a^2) = \frac{1}{2|a|}[\delta(x+|a|) + \delta(x-|a|)] \tag{6.754} $$

$$ \delta(x-x_0) = \frac{1}{\pi} \lim_{\varepsilon \to 0} \frac{\varepsilon}{(x-x_0)^2 + \varepsilon^2} \tag{6.755} $$

PROBLEMS

1. Stirling's approximation of $N!$ is accurate to at least two decimal places for $N \geqslant 9$. Estimate the value of each of the following numbers with at least two-decimal-place accuracy using Stirling's approximation.
   (a) $\Gamma(12.3)$
   (b) $(\pi^2)!$
   (c) $\frac{1}{2}!$ (Compare your result to the exact value.)
2. (a) Deduce a Stirling approximation of $\beta(N+1, M+1)$ for large M and N.
   (b) Estimate the value of $\beta(12.3, 15.1)$.
3. (a) Evaluate

$$\int_0^1 \left(\frac{1-t}{t}\right)^{9/2} dt$$

   exactly.
   (b) Using the same analysis as that employed in part (a), deduce an expression for

$$\int_0^1 \left(\frac{1-t}{t}\right)^{N+1/2} dt$$

   for integer N.
4. Let $p + q = N + 1$, with N an integer $\geqslant 0$. Show that

$$\int_0^{\pi/2} \sin^{2p-1}\phi \cos^{2N+2p+1}\phi \, d\phi = \beta(p, p-N-1)$$

$$= \frac{(N-p)(N-p-1)(N-p-2)\cdots(1-p)}{N!} \times \frac{\pi}{\sin(\pi p)}$$

   Thus, an integral of the above form can be evaluated exactly.
5. Using the fact that the Stirling approximation of $\Gamma(N)$ is accurate to two decimal places for $N \geqslant 10$, estimate the value of
   (a) $\displaystyle\int_0^1 \frac{(1-t)^{1/4}}{t^{1/3}} dt$
   (b) $\displaystyle\int_0^{\pi/2} \tan^{7/3}\phi \, d\phi$
   to two decimal places.
6. (a) Show that the digamma function satisfies

$$\psi(p) - \psi(1-p) = \pi \cot \pi p$$

   (b) Use the result of part (a) to show that if N is an integer

$$\lim_{p \to N} \frac{\psi(1-p)}{\Gamma(1-p)} = (-1)^N (N-1)!$$

7. Determine the [1, 1] Padé approximant to:
   (a) $\displaystyle \operatorname{erf}(x) \equiv \frac{2}{\sqrt{\pi}} \int_0^x e^{-y^2} dy$
   (b) $\displaystyle K(k^2) \equiv \int_0^{\pi/2} \frac{d\phi}{\sqrt{1 - k^2 \sin^2\phi}}$
   (c) $\displaystyle E(k^2) \equiv \int_0^{\pi/2} \sqrt{1 - k^2 \sin^2\phi} \, d\phi$
8. Prove the following.
   (a) $\displaystyle \frac{2}{\sqrt{\pi}} \int_{x_1}^{x_2} e^{-y^2} dy = \operatorname{erf}(x_2) - \operatorname{erf}(x_1)$
   (b) $\displaystyle \frac{1}{\sqrt{\pi}} \int_{-x}^{x} e^{-y^2} dy = \operatorname{erf}(x)$
   (c) The Laplace transform of $\operatorname{erf}(x)$ is

$$\text{L.T.}[\operatorname{erf}(x)] = \frac{1}{s} e^{s^2/4} \left[1 - \operatorname{erf}\left(\frac{s}{2}\right)\right]$$

9. Show that

$$\int_0^{\pi} \frac{\cos\theta}{(\alpha + \beta\cos\theta)^{1/2}} d\theta$$

$$= k\left(\frac{2}{\beta}\right)^{1/2} \left[\left(1 - \frac{2}{k^2}\right) K(k^2) + \frac{2}{k^2} E(k^2)\right]$$

   where $K(k^2)$ and $E(k^2)$ are the complete elliptic functions and $k^2 \equiv 2\beta/(\alpha + \beta)$.
10. (a) Using the substitution $x = \sqrt{1 - \sin\phi}$ in the integral $\int_0^{\pi/2} d\phi / \sqrt{\sin\phi}$, show that the elliptic integral of the first kind satisfies

$$K\left(k^2 = \frac{1}{2}\right) = \frac{[\Gamma(\frac{1}{4})]^2}{4\sqrt{\pi}}$$

   (b) Use a similar substitution in the integral $\int_0^{\pi/2} \sqrt{\cos\phi} \, d\phi$ to evaluate the elliptic integral of the second kind $E(k^2 = \frac{1}{2})$.
11. Let the set $(1, x, x^2)$ be the bases for a function space defined by $x \in [0, 1]$. Use these to determine a set of orthogonal functions that are normalized to 1, with a weight function $\rho(x) = 1$.
12. The Humbert polynomials $h_l^{\lambda}(x)$ can be generated by the function

$$\Phi(z, x) \equiv (1 - 2xz + z^3)^{-\lambda} = \sum_{l=0}^{\infty} z^l h_l^{\lambda}(x)$$

   Prove that

$$\left[h_{l+3}^{\lambda}(x)\right]' - 2x\left[h_{l+2}^{\lambda}(x)\right]' + \left[h_l^{\lambda}(x)\right]' = 2\lambda h_{l+2}^{\lambda}(x)$$

and

$$(l+3)h_{l+3}^{\lambda}(x) - 2x(l+\lambda+2)h_{l+2}^{\lambda}(x) + (l+3\lambda)h_l^{\lambda}(x) = 0$$

where $[h_l^{\lambda}(x)]' = D[h_l^{\lambda}(x)]$.

13. The generating function of the Legendre polynomials is

$$\Phi(z,x) = (1-2xz+z^2)^{-1/2} = \sum_{l=0}^{\infty} z^l P_l(x)$$

and that of the Hermite polynomials is

$$\Phi(z,x) = e^{-z^2+2xz} = \sum_{l=0}^{\infty} z^l \frac{H_l(x)}{l!}$$

Both generating functions depend on z and x in the combination $\Phi(z,x) = \Phi(z^2 - 2xz)$. If a polynomial $w_l(x)$ is generated by such a function as

$$\Phi(z^2-2xz) = \sum_{l=0}^{\infty} z^l w_l(x)$$

(a) Show that $w_l(x)$ has the following properties

$$w_0(x) = \text{constant}$$

$$xw'_{l+1}(x) - (l+1)w_{l+1}(x) - w'_l(x) = 0$$

(b) Show that $P_l(x)$ and $H_l(x)/l!$ satisfy these equations.

14. Use the Leibnitz formula for the derivative of a product of two functions to show that a second series representation of the Legendre polynomials is

$$P_l(x) = (-1)^l \sum_{m=0}^{l} \left[\frac{l!}{m!(l-m)!}\right]^2 \times \left(\frac{1+x}{2}\right)^m \left(\frac{1-x}{2}\right)^{l-m}$$

15. (a) Use the generating function for the Legendre polynomials to prove that

$$P_l'(1) = \tfrac{1}{2}l(l+1)$$

(b) From the result of part (a), show that

$$P_l'(-1) = (-1)^{l+1}\tfrac{1}{2}l(l+1)$$

16. Use the Legendre differential equation to show that the Legendre polynomials satisfy the orthonormality condition

$$\int_{-1}^{1}(1-x^2)P_l'(x)P_n'(x)\,dx = \frac{2l(l+1)}{(2l+1)}\delta_{ln}$$

17. Evaluate the integral

$$I_{ln} \equiv \int_{-1}^{1} xP_l(x)P_n(x)\,dx$$

(a) Using a recurrence relation and the orthonormalization condition satisfied by the Legendre polynomials

(b) Using the generating function of the Legendre polynomials and any properties of this integral developed in part (a)

18. It was shown in the text that the Legendre polynomials satisfy the recurrence relations

$$(l+2)P_{l+2}(x) - (2l+3)P_{l+1}(x) + (l+1)P_l(x) = 0 \qquad l \geqslant 0 \quad (6.203\text{b})$$

and

$$P'_{l+2}(x) - 2xP'_{l+1}(x) + P'_l(x) = P_{l+1}(x) \quad (6.205)$$

(a) By differentiating the first recurrence relation and combining the result with the second, show that

$$P'_{l+2}(x) - P'_l(x) = (2l+3)P_{l+1}(x) \qquad l \geqslant 0$$

(b) By combining the three recurrence relations listed above, show that
(1) $P'_{l+1}(x) - xP'_l(x) = (l+1)P_l(x), \qquad l \geqslant 0$
(2) $xP'_{l+1}(x) - P'_l(x) = (l+1)P_{l+1}(x), \qquad l \geqslant 0$
(3) $(1-x^2)P'_{l+1}(x) = (l+1)[P_l(x) - xP_{l+1}(x)], \qquad l \geqslant 0$
Note that in none of these manipulations will it be specified that the Legendre function used is the Legendre polynomial. Therefore, all manipulations will be valid for second Legendre functions as well. Because equations 6.203b and 6.205 are valid when $P_l(x)$ is replaced by $Q_l(x)$, the other four recurrence relations are also satisfied by $Q_l(x)$.

19. By analyzing a product of Legendre generating functions, show that

$$P_l(\tfrac{1}{2}) = \sum_{k=0}^{2l}(-1)^{l+k}P_{2l-k}(\tfrac{1}{2})P_k(\tfrac{1}{2})$$

20. Use the Rodriguez formula for the Legendre polynomials to show that

$$Q_l(z) = \frac{1}{2^{l+1}}\int_{-1}^{1}\frac{(1-x^2)^l}{(z-x)^{l+1}}\,dx$$

21. Prove that

$$\frac{1}{(z-x)} = \sum_{l=0}^{\infty}(2l+1)Q_l(z)P_l(x)$$

22. Find:

(1) $P_l(0)$

(2) $P_l^m(0)$

(a) From the series representation of each Legendre function

(b) From the generator of each Legendre function

23. Show that the associated Legendre differential equation for $P_l^m(x)$ is equivalent to the recurrence relation

$$P_l^{m+2}(x) - 2x\frac{(m+1)}{\sqrt{1-x^2}}P_l^{m+1}(x) + [l(l+1) - m(m+1)]P_l^m(x) = 0$$

24. Prove that

$$P_l^l(\cos\theta) = C_l \sin^l\theta$$

and determine the constant C_l.

25. Use the Neumann representation of the second Legendre function to derive

$$Q_{l+2}'(x) - Q_l'(x) = (2l+3)Q_{l+1}(x)$$

26. Show that the function $R_{l-1}(x)$ that relates the second Legendre function to the Legendre polynomials by

$$Q_l(x) = P_l(x)Q_0(x) - R_{l-1}(x) \quad (6.239)$$

satisfies the differential equation

$$(1-x^2)R_{l-1}''(x) - 2xR_{l-1}'(x) + l(l+1)R_{l-1}(x) = 2P_l'(x)$$

27. (a) Show that the generator of the Gegenbauer polynomial $G_l^0(x)$,

$$\Phi(z,x) = \frac{(x-z)}{(1-2xz+z^2)} = \frac{1}{2}\sum_{l=0}^{\infty}(l+1)z^l G_{l+1}^0(x) \quad (6.355b)$$

is equivalent to

(1) $\dfrac{(1-z^2)}{(1-2xz+z^2)} = 1 + \sum_{l=1}^{\infty} l z^l G_l^0(x)$

and is also equivalent to

(2) $\dfrac{(1-xz)}{(1-2xz+z^2)} = \dfrac{1}{2}\sum_{l=0}^{\infty} l z^l G_l^0(x)$

(b) Show that by differentiation of one of these forms with respect to z, one obtains

$$(l+2)G_{l+2}^0(x) - 2x(l+1)G_{l+1}^0(x) + lG_l^0(x) = 0$$

[Hint: Using one of these three forms, you may find the algebraic identity $2x^2 + z^2 - 2xz - 1 = 2x(x-z) - (1-z^2)$ useful.]

(c) Use one of the generators to prove that

$$(l+2)\left[G_{l+2}^0(x)\right]' - 2x(l+1)\left[G_{l+1}^0(x)\right]' + l\left[G_l^0(x)\right]' = 2(l+1)G_{l+1}^0(x)$$

[*Note:* This recurrence relation can also be obtained by differentiation of the recurrence relation of part (b).]

28. The Tschebyschev (Chebyscheff) differential equations are:

$$\text{Type I: } \left[(1-x^2)D^2 - xD + n^2\right]y = 0 \qquad y \equiv T_n(x)$$

$$\text{Type II: } \left[(1-x^2)D^2 - 3xD + n(n+2)\right]y = 0 \qquad y \equiv U_n(x)$$

(a) Find the function $\mu(x)$ that makes each of these equations self-adjoint.

(b) Identify the Sturm–Liouville functions $p(x)$, $q(x)$, $\rho(x)$ and the eigenvalue λ for each of these equations.

(c) What are the relations between the Gegenbauer polynomial and each of the Chebyscheff polynomials using the fact that $T_n(1) = 1$ and $U_n(1) = n + 1$?

(d) Deduce the series representation for each of the Chebyscheff polynomials.

(e) What is the Rodriguez formula for each Chebyscheff polynomial?

(f) Determine the orthonormalization condition for each Chebyscheff polynomial.

(g) By setting $x = \cos\theta$ in each equation, deduce the form of each Chebyscheff equation in terms of θ. Show that $T_n(\cos\theta) = \cos(n\theta)$ satisfies the type I equation and that $U_n(\cos\theta) = \sin[(n+1)\theta]/\sin\theta$ satisfies the type II equation.

(h) Use the generating function of each of the Chebyscheff polynomials to evaluate

(1) $\int_{-1}^{1} T_n(x)\,dx$

(2) $\int_{-1}^{1} U_n(x)\,dx$

29. Use the generating function of the Laguerre polynomials to prove that $L_n(0) = 1$.

30. Find the expansion of $e^{-\alpha x}$ in terms of the Laguerre polynomials for $\text{Re}(\alpha) > 0$.

31. For the ordinary Laguerre polynomials, derive the fact that

(a) $L_n'(0) = -n, \quad n \geqslant 1$

(b) $L_n''(0) = \frac{1}{2}n(n-1), \quad n \geqslant 2$

(c) Deduce an expression for $D^m L_n(x)|_{x=0}$, $n \geqslant m$.

32. Prove that the Laplace transform of the ordinary Laguerre polynomial is

$$\int_0^\infty e^{-sx} L_n(x)\, dx = \frac{(s-1)^n}{s^{n+1}}$$

33. Use the generating function for the ordinary Laguerre polynomials to show that

$$x \sum_{n=0}^{\infty} \frac{z^n}{(z+x)^{n+1}} L_n(x) = e^{-z} \qquad \text{for all } x > 0$$

[Hint: Use the substitution $w = t/(1-t)$ in the generator $\phi(t, x)$.]

34. Using the generator of the ordinary Laguerre polynomials, prove that

$$L'_{n+1}(x+y) = -\sum_{m=0}^{n} L_m(x) L_{n-m}(y)$$

35. Prove that

$$L^k_{n+k}(0) = (-1)^k \frac{(n+k)!}{n!k!}$$

36. Show that

$$\int_0^\infty e^{-\alpha x} x^{k+1} L^k_{n+k}(x)\, dx$$

$$= (-1)^k \frac{(n+k)!}{n!} \frac{(\alpha-1)^n}{\alpha^{n+k+1}}$$

for $\mathrm{Re}(\alpha) > 0$.

37. Using the recurrence relations and orthonormalization conditions for the associated Laguerre polynomials, derive the following:

(a) $\int_0^\infty e^{-x} x^{k+2} L^k_n(x) L^k_{n+2}(x)\, dx = \frac{(n+2)!}{(n-k)!}$

(b) $\int_0^\infty e^{-x} x^{k+2} L^k_n(x) L^k_{n+1}(x)\, dx =$

$-2(2n-k+2)\frac{(n+1)!}{(n-k)!}$

(c) $\int_0^\infty e^{-x} x^{k+2} [L^k_n(x)]^2\, dx =$

$(6n^2 + k^2 - 6nk + 6n - 3k + 2)\frac{n!}{(n-k)!}$

38. The Schrödinger equation for the one-dimensional quantum oscillator is

$$-\frac{\hbar^2}{2m} D^2 \psi(x) + \frac{1}{2} m\omega^2 x^2 \psi(x) = E\psi(x)$$

(a) The wave function for the nth quantum state of this system is

$$\psi_n(x) = c_n e^{-u^2/2} H_n(u)$$

where $u \equiv (m\omega/\hbar)^{1/2} x$. From this, determine the energy of the oscillator in the nth quantum state.

(b) Determine the normalization constant c_n such that

$$\int_{-\infty}^{\infty} \psi_n(x) \psi_l(x)\, dx = \delta_{nl}$$

(c) Derive the generating function $\Phi(z, x)$ of $\psi_n(x)$ so that

$$\Phi(z, x) = \sum_{n=0}^{\infty} z^n \left[\frac{2^n}{n!}\right]^{1/2} \psi_n(x)$$

(d) Use the generator derived in part (c) to determine the Rodriguez formula for $\psi_n(x)$.

(e) Show that the Fourier transform of $\psi_n(x)$ is

$$\text{F.T.}[\psi_n(x)] \equiv \frac{1}{\sqrt{2\pi}} \int_{-\infty}^{\infty} e^{ikx} \psi_n(x)\, dx$$

$$= \left(\frac{\hbar}{m\omega}\right)^{1/2} n! e^{-\Lambda^2/2} \sum_{m=0}^{[n/2]} \frac{(2i\Lambda)^{n-2m}}{m!(n-2m)!}$$

where $\Lambda \equiv k(\hbar/m\omega)^{1/2}$.

39. (a) Starting with the generator of the Hermite polynomials, show that the even-order polynomials can be generated by

$$e^{z^2} \cos(2xz) = \sum_{n=0}^{\infty} (-1)^n z^{2n} \frac{H_{2n}(x)}{(2n)!}$$

and the odd-order Hermite polynomials can be obtained from

$$e^{z^2} \sin(2xz) = \sum_{n=0}^{\infty} (-1)^n z^{2n+1} \frac{H_{2n+1}(x)}{(2n+1)!}$$

(b) From these generating functions, deduce the values of $H_{2n}(0)$ and $H_{2n+1}(0)$.

40. Use the generating function of the Hermite polynomials to show that the polynomials satisfy the addition theorem

$$H_n(x+y) = 2^{-n/2} \sum_{m=0}^{n} \frac{n!}{m!(n-m)!} H_m(\sqrt{2}\, x) H_{n-m}(\sqrt{2}\, y)$$

41. Using the recurrence relations and orthonormalizations condition for the Hermite polynomials established in the text, evaluate

$$\int_{-\infty}^{\infty} x^k e^{-x^2} H_n(x) H_m(x)\, dx$$

for $k = 1, 2,$ and 3.

42. Start with the series representation of the Hermite polynomials to prove that the Legendre polynomial can be written

$$P_n(x) = \frac{1}{n!\sqrt{\pi}} \int_{-\infty}^{\infty} e^{-t^2} t^n H_n(xt)\, dx$$

43. (a) Show that

$$x^2 D^2 y + (1 - 2a) x D y + \left[(bcx^c)^2 + (a^2 - c^2\alpha^2) \right] y = 0$$

has a solution in terms of the ordinary Bessel function $y = x^a J_\alpha(bx^c)$.

(b) Determine one of the solutions to

$$xD^2 y + 2Dy + 4y = 0$$

in terms of ordinary Bessel functions.

44. Deduce the values of the Bessel functions:

(a) $J_0(0)$

(b) $J_0'(0)$

(c) $J_1(0)$

(d) $J_1'(0)$

45. (a) Use the generating function for the integer-order Bessel functions to derive the addition theorem

$$J_n(x + y) = \sum_{l=-\infty}^{\infty} J_l(x) J_{n-l}(y)$$

(b) Using the results of part (a) and of Problem 44, show that

(1) $J_0(2x) = [J_0(x)]^2 + 2 \sum_{n=1}^{\infty} (-1)^n [J_n(x)]^2$

(2) $[J_0(x)]^2 + 2 \sum_{n=1}^{\infty} [J_n(x)]^2 = 1$

46. Prove that

(a) $\cosh(x \sinh w) = J_0(x) + 2 \sum_{n=1}^{\infty} \cosh(2nw) J_{2n}(x)$

(b) $\sinh(x \sinh w) = 2 \sum_{n=0}^{\infty} \sinh[(2n + 1)w] J_{2n+1}(x)$

47. Find x^3 and x^4 in terms of sums over integer-order Bessel functions.

48. Show that the Laplace transform of $x^\alpha J_\alpha(kx)$ is

$$\text{L.T.}[x^\alpha J_\alpha(kx)] = \int_0^{\infty} e^{-sx} x^\alpha J_\alpha(kx) = \frac{(2k)^\alpha \Gamma(\alpha + \frac{1}{2})}{\sqrt{\pi}\,(k^2 + s^2)^{\alpha + 1/2}}$$

49. Derive the following:

(a) $D[x^{-\alpha} J_\alpha(x)] = -x^{-\alpha} J_{\alpha+1}(x)$

(b) $D[x^{\alpha} J_{-\alpha}(x)] = -x^{\alpha} J_{-\alpha+1}(x)$

(c) $D[x^{-\alpha} J_{-\alpha}(x)] = x^{-\alpha} J_{-\alpha-1}(x)$

(d) Show that $D[x^\alpha Y_\alpha(x)] = x^\alpha Y_{\alpha-1}(x)$ and that each of the conditions above is also satisfied by the Neumann function $Y_\alpha(x)$.

Note that from these results it is clear that both $J_\alpha(x)$ and $Y_\alpha(x)$ satisfy the recurrence relations expressed in equations 6.560b, 6.563, 6.566, 6.567, and 6.568.

50. Prove that the integer-order modified Bessel functions are generated by

$$e^{(z+1/z)x/2} = \sum_{n=-\infty}^{\infty} z^n I_n(x)$$

51. Show that the modified Bessel functions satisfy:

(a) $\frac{2\alpha}{x} I_\alpha(x) = I_{\alpha-1}(x) - I_{\alpha+1}(x)$

(b) $2I_\alpha'(x) = I_{\alpha-1}(x) + I_{\alpha+1}(x)$

52. (a) Use the Poisson integral representation for the Bessel function

$$J_\alpha(x) = \left(\frac{x}{2}\right)^\alpha \frac{1}{\sqrt{\pi}\,\Gamma(\alpha + \frac{1}{2})} \int_{-1}^{1} e^{iwx} (1 - w^2)^{\alpha - 1/2}\, dw \tag{6.555}$$

to show that the spherical Bessel function can be represented as

$$j_l(x) = \frac{1}{l!} \left(\frac{x}{2}\right)^l \int_0^{\pi/2} \cos(x \sin\phi) \cos^{2l+1}\phi\, d\phi$$

(b) From the result of part (a) evaluate the following integrals in terms of elementary functions.

(1) $\int_0^{\pi/2} \cos(x \sin\phi) \cos\phi\, d\phi$

(2) $\int_0^{\pi/2} \cos(x \sin\phi) \cos^3\phi\, d\phi$

(c) From the results of part (a) and properties of the spherical Bessel functions developed in the text, show that

$$\int_0^{\pi/2} \cos^{2l+1}\phi\, d\phi = \frac{2^{2l}(l!)^2}{(2l + 1)!}$$

53. By substituting $y = u/\sqrt{x}$ into Bessel's differential equation, show that for large x, the solution must be of the form

$$y = A \frac{\sin x}{\sqrt{x}} + B \frac{\cos x}{\sqrt{x}}$$

54. Deduce the functional forms of the half-order Bessel functions listed below in terms of elementary functions.
(a) $I_{1/2}(x)$
(b) $I_{-3/2}(x)$
(c) $Y_{1/2}(x)$
(d) $H^{(+)}_{1/2}(x)$
(e) $H^{(-)}_{-1/2}(x)$

55. (a) Prove that

$$D[{}_2F_1(a, b; c; x)] = \frac{ab}{c}\,{}_2F_1(a+1, b+1; c+1; x)$$

(b) Show that

$$D[{}_1F_1(a; c; x)] = \frac{a}{c}\,{}_1F_1(a+1; c+1; x)$$

(c) Continue this analysis to deduce an expression for
(1) $D^k[{}_2F_1(a, b; c; x)]$
(2) $D^k[{}_1F_1(a; c; x)]$

56. Find the elementary function that has the hypergeometric functional representation ${}_2F_1(a, 1; 2; x)$ for
(a) $a = 0$
(b) $a = 1$
(c) $a \neq 0, 1$

57. Show that

$${}_2F_1(a, b; c; 1) = \frac{\Gamma(c)\Gamma(c-a-b)}{\Gamma(c-a)\Gamma(c-b)}$$

58. Prove that ${}_2F_1(a, b; a+b+1-c; x)$ is a solution to Gauss's hypergeometric differential equation.

59. Show that the complete elliptic functions are related to the hypergeometric function by:
(a) $K(k^2) = \frac{\pi}{2}\,{}_2F_1(\frac{1}{2}, \frac{1}{2}; 1; k^2)$
(b) $E(k^2) = \frac{\pi}{2}\,{}_2F_1(\frac{1}{2}, -\frac{1}{2}; 1; k^2)$

60. Show that:
(a) ${}_1F_1(2; 1; -x) = (1-x)e^{-x}$
(b) ${}_1F_1(1; 2; -x) = \dfrac{1-e^{-x}}{x}$

61. Prove that the Legendre polynomials can be written as:
(a) $P_{2l}(x) = (-1)^l \dfrac{(2l)!}{2^{2l}(l!)^2}\,{}_2F_1(-l, l+\frac{1}{2}; \frac{1}{2}; x^2)$

(b) $P_{2l+1}(x) = (-1)^l \dfrac{(2l+1)!}{2^{2l}(l!)^2}\,{}_2F_1(-l, l+\frac{3}{2}; \frac{3}{2}; x^2)$

62. Show that the Laguerre polynomials can be written

$$L_n(x) = {}_1F_1(-n; 1; x)$$

63. Prove that the odd-order Hermite polynomials can be expressed as

$$H_{2n+1}(x) = \frac{(-1)^n(2n+1)!}{n!}2x\,{}_1F_1(-n; \tfrac{3}{2}; x^2)$$

64. (a) Use the Poisson integral representation for the Bessel function

$$J_\alpha(x) = \left(\frac{x}{2}\right)^\alpha \frac{1}{\sqrt{\pi}\,\Gamma(\alpha+\frac{1}{2})}\int_{-1}^{1} e^{iwx}(1-w^2)^{\alpha-1/2}\,dw \tag{6.555}$$

and make the substitution $t = (1+w)/2$ to show that

$$J_\alpha(x) = \frac{(2x)^\alpha e^{-ix}}{\sqrt{\pi}}\frac{\Gamma(\alpha+\frac{1}{2})}{\Gamma(2\alpha+1)}\,{}_1F_1(\alpha+\tfrac{1}{2}; 2\alpha+1; 2ix)$$

(b) Use the result of part (a) to prove that

$$x\,{}_1F_1(1; 2; 2x) = e^x \sinh x$$

65. Prove the validity of the contiguous relation

$$c\,{}_2F_1(a, b; c; x) + ax\,{}_2F_1(a+1, b+1; c+1; x) - c\,{}_2F_1(a, b+1; c; x) = 0$$

66. Using the contiguous relation of equation 6.722 and the symmetry of Gauss's hypergeometric function in the interchange of a and b, derive

$$(b-a){}_2F_1(a, b; c; x) + a\,{}_2F_1(a+1, b; c; x) - b\,{}_2F_1(a, b+1; c; x) = 0$$

67. Show that the Laplace transform of ${}_1F_1(a; c; x)$ is

$$\text{L.T.}[{}_1F_1(a; c; x)] = \int_0^\infty e^{-sx}\,{}_1F_1(a; c; x)\,dx = \frac{1}{s}\,{}_2F_1\left(a; 1; c; \frac{1}{s}\right)$$

68. (a) Start with the integral

$$\int_{-\infty}^{\infty} x^2 D^3\delta(x)\,dx$$

to show that $x^3D^3\delta(x) = -6\delta(x)$.
(b) From this analysis and similar ones developed in the text, deduce a general expression for $x^nD^n\delta(x)$.

69. Evaluate
(a) $\displaystyle\int_{-\infty}^{\infty} \frac{e^{3ix}}{(x^2+4)}\delta(x^2-1)\,dx$
(b) $\displaystyle\int_{0}^{\infty} \frac{e^{3ix}}{(x^2+4)}\delta(x^2-1)\,dx$

CHAPTER 7

CALCULUS OF VARIATIONS

7.1 Statement of the Problem and Formulation of the Solution

Consider the problem of an object being moved over a rough surface from point p_0 to point p_1. The work done against the frictional force $\mathbf{f}$ is

$$W = \int_{p_0}^{p_1} \mathbf{f} \cdot d\mathbf{s} \tag{7.1}$$

Clearly, the amount of work done depends on the path along which the object is moved. The minimum work is done by taking the shortest path between the points. If the object moves over a flat surface, this minimal path is a straight line joining the points. Over the surface of a sphere, the shortest path is along the arc of a circle. This minimal curve is called a *geodesic*.

In contrast to finding the minimum of a function with respect to an independent variable using rules of differential calculus, the *calculus of variations* involves determining the path of integration such that an integral is a minimum with respect to the path or contour. The actual problems involve determination of the path that will make the integral an extremum, but that extremum is very often a minimum.

If the work integral can be constructed in terms of a parameter ε such that different values of ε characterize different paths taken, the extremum of W with respect to path could be determined from $dW/d\varepsilon = 0$. This is a simple example of the type of problem one solves by methods of variational calculus.

Example 7.1

Consider the problem of finding the shortest distance between two points in a flat plane. Clearly, there are an infinite number of different paths that can be taken between the two points. A representative sample is shown in Figure 7.1.

Let ds be the scalar line element along a path. The distance between p_0 and p_1 is

$$S = \int_{p_0}^{p_1} ds = \int_{(x_0, y_0)}^{(x_1, y_1)} \sqrt{dx^2 + dy^2} = \int_{x_0}^{x_1} \sqrt{1 + y'^2}\, dx \tag{7.2}$$

where $y' = dy/dx$.

A very general way to parametrize the paths is to let $\mu(x)$ be an arbitrary function and take $y(x)$ along a path to be

$$y(x, \varepsilon) = y(x) + f(\varepsilon)\mu(x) \tag{7.3}$$

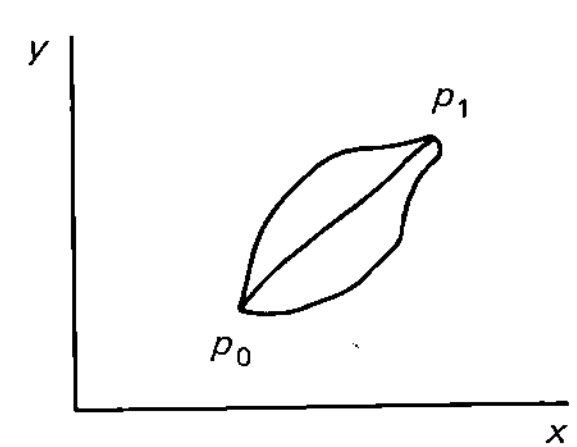

Figure 7.1 Various paths between p_0 and p_1.

where, by definition, $\varepsilon = 0$ defines the optimal path. That is, $y(x,0) \equiv y(x)$ along this optimal path. Therefore, we must require

$$f(0) = 0 \tag{7.4}$$

Referring to Figure 7.1, we also note that all paths converge at p_0 and p_1. Therefore, for any ε,

$$\text{(a)} \quad y(x_0,\varepsilon) = y(x_0) = y(x_0,0) \qquad \text{(b)} \quad y(x_1,\varepsilon) = y(x_1) = y(x_1,0) \tag{7.5}$$

$$\Rightarrow \quad \mu(x_0) = \mu(x_1) = 0 \tag{7.6}$$

Consider the variation of S with path. We denote this by

$$\delta S = \frac{\partial S}{\partial \varepsilon}\,\delta\varepsilon = \delta\varepsilon\,\frac{\partial}{\partial \varepsilon}\int_{x_0}^{x_1}\sqrt{1 + y'^2}\,dx \tag{7.7}$$

Since the endpoints do not vary with path, the only path dependence arises from y' in the integrand. From equation 7.3

$$y'(x,\varepsilon) = y'(x) + f(\varepsilon)\mu'(x) \tag{7.8}$$

$$\Rightarrow \quad \frac{\delta S}{\delta \varepsilon} = \int_{x_0}^{x_1}\frac{\partial}{\partial \varepsilon}\sqrt{1 + y'^2}\,dx = \frac{df}{d\varepsilon}\int_{x_0}^{x_1}\left[\frac{y'}{\sqrt{1 + y'^2}}\right]\mu'(x)\,dx \tag{7.9}$$

Integrating by parts, with

$$\text{(a)} \quad u = \frac{y'}{\sqrt{1 + y'^2}} \qquad \text{(b)} \quad dv = \mu'(x)\,dx \tag{7.10}$$

$$\Rightarrow \quad \frac{\delta S}{\delta \varepsilon} = \frac{df}{d\varepsilon}\left\{\mu(x)\frac{y'}{\sqrt{1 + y'^2}}\Bigg|_{x_0}^{x_1} - \int_{x_0}^{x_1}\mu(x)\frac{d}{dx}\left[\frac{y'}{\sqrt{1 + y'^2}}\right]dx\right\} \tag{7.11}$$

Since $\mu(x)$ is zero at the endpoints, the integrated term is zero, and

$$\frac{\delta S}{\delta \varepsilon} = -\frac{df}{d\varepsilon}\int_{x_0}^{x_1}\mu(x)\frac{d}{dx}\left[\frac{y'}{\sqrt{1 + y'^2}}\right]dx \tag{7.12}$$

The path will be an extremum when $\delta S = 0$. For any path of finite length, one can always conceive of a somewhat longer path. Thus the extremum is a minimum.

δS will be zero if $df/d\varepsilon = 0$, or $f =$ constant. However, since $f = 0$ when $\varepsilon = 0$, f would have to be zero for all ε, and y would not vary with path. Thus $df/d\varepsilon \neq 0$, $f(\varepsilon)$ is not restricted any further. As such, there is no loss of generality in taking $df/d\varepsilon = 1$, or $f = \varepsilon$. We will do so from this point on.

From equation 7.12, the minimum path arises when

$$\int_{x_0}^{x_1}\mu(x)\frac{d}{dx}\left[\frac{y'}{\sqrt{1 + y'^2}}\right]dx = 0 \tag{7.13}$$

Since $\mu(x)$ is arbitrary, this integral will be zero for

$$\text{(a)} \quad \frac{d}{dx}\left[\frac{y'}{\sqrt{1 + y'^2}}\right] = 0 \quad \Rightarrow \quad \text{(b)} \quad \frac{y'}{\sqrt{1 + y'^2}} = \text{constant} \tag{7.14}$$

This path of minimum distance arises for

$$\text{(a)} \quad y' = \alpha = \text{constant} \qquad \Rightarrow \qquad \text{(b)} \quad y = \alpha x + \beta \tag{7.15}$$

a straight line. □

As in this simple example, the solution to the minimum path problem is a differential equation, the solution to which defines the path that minimizes the integral.

Example 7.2

As a second example, consider the shortest path between points defined by (θ_0, ϕ_0) and (θ_1, ϕ_1) on the surface of a sphere of radius r. The line element for any path is given by

$$ds^2 = r^2\,d\theta^2 + r^2 \sin^2\theta\,d\phi^2 \tag{7.16}$$

A path on the surface of a sphere is defined by specifying how θ varies with ϕ, or how ϕ varies with θ. That is, we can either parametrize θ or ϕ as

$$\text{(a)} \quad \theta(\phi, \varepsilon) = \theta(\phi) + \varepsilon\mu(\phi) \qquad \text{(b)} \quad \phi(\theta, \varepsilon) = \phi(\theta) + \varepsilon\mu(\theta) \tag{7.17}$$

In both parametrizations, μ is zero at the endpoints.

Choosing the first parametrization, the distance is written

$$S = r\int_{\phi_0}^{\phi_1} \sqrt{\sin^2\theta + \theta'^2}\,d\phi \tag{7.18a}$$

where $\theta' = d\theta/d\phi$. Since θ is the dependent variable in this parametrization, the path variable ε occurs in both $\sin^2\theta$ and in θ'^2. If the parametrization is made in terms of ϕ, the distance integral is

$$S = r\int_{\theta_0}^{\theta_1} \sqrt{1 + \sin^2\theta\phi'^2}\,d\theta \tag{7.18b}$$

and the path variable occurs in the factor $\phi' = d\phi/d\theta$ only. The results will be the same using either parametrization, but the analysis will be less cumbersome if we express the distance between the points as

$$S = r\int_{\theta_0}^{\theta_1} \sqrt{1 + \sin^2\theta\phi'^2}\,d\theta \tag{7.18b}$$

The minimum distance is found from

$$\frac{\delta S}{\delta\varepsilon} = r\int_{\theta_0}^{\theta_1} \frac{\partial}{\partial\varepsilon}\sqrt{1 + \sin^2\theta\phi'^2}\,d\theta = r\int_{\theta_0}^{\theta_1} \frac{\sin^2\theta\phi'}{\sqrt{1 + \sin^2\theta\,\phi'^2}}\mu'(\theta)\,d\theta = 0 \tag{7.19}$$

As in Example 7.1, we integrate once by parts and set the integrated term to zero by using the fact that $\mu(\theta_0) = \mu(\theta_1) = 0$. We obtain

$$-r\int_{\theta_0}^{\theta_1} \mu(\theta)\frac{d}{d\theta}\left[\frac{\sin^2\theta\,\phi'}{\sqrt{1 + \sin^2\theta\,\phi'^2}}\right] d\theta = 0 \tag{7.20}$$

and since $\mu(\theta)$ is arbitrary,

$$\text{(a)} \quad \frac{d}{d\theta}\left[\frac{\sin^2\theta\,\phi'}{\sqrt{1 + \sin^2\theta\,\phi'^2}}\right] = 0 \qquad \Rightarrow \qquad \text{(b)} \quad \frac{\sin^2\theta\,\phi'}{\sqrt{1 + \sin^2\theta\,\phi'^2}} = \text{constant} \equiv \beta \tag{7.21}$$

Therefore, the minimum path is characterized by

$$\phi(\theta) = \beta\int \frac{d\theta}{\sin\theta\sqrt{\sin^2\theta - \beta^2}} \tag{7.22}$$

Taking $\theta_0 < \theta < \theta_1$, we must ensure that the square root is real by requiring that

$$\beta^2 < \sin^2\theta_0 \tag{7.23}$$

This integral can be evaluated by a series of substitutions. Setting

$$\text{(a)}\quad \sin\theta = \beta\sec\alpha \qquad \Rightarrow \qquad \text{(b)}\quad \phi(\theta) = \int \frac{\cos\alpha}{\sqrt{(1-\beta^2)-\sin^2\alpha}}\,d\alpha \tag{7.24}$$

We then let

$$\text{(c)}\quad \sin\alpha = \sqrt{1-\beta^2}\,\sin\rho \qquad \Rightarrow \qquad \text{(d)}\quad \phi(\theta) = \rho \tag{7.24}$$

The resulting path is described in terms of $\phi(\theta)$ or in terms of $\theta(\phi)$ as

$$\text{(a)}\quad \sin[\phi(\theta)] = \frac{\sqrt{\sin^2\theta - \beta^2}}{\sqrt{1-\beta^2}\,\sin\theta} \qquad \text{(b)}\quad \sin[\theta(\phi)] = \frac{\beta}{\sqrt{\cos^2\phi + \beta^2\sin^2\phi}} \qquad \square \tag{7.25}$$

Fermat's principle and the laws of reflection and refraction

In some situations, the extremization with respect to path can be accomplished by first evaluating the integrals involved, and then extremizing the result with respect to path parameters. In such problems, extremizing is achieved by the standard method of setting the derivatives with respect to the parameters that define the path to zero.

Example 7.3

As an example, Fermat's principle for the traversal of a light ray is a minimization principle. Its application leads to the laws of reflection and refraction of a light ray. The principle states that a light ray will follow a path from a point p_0 to p_1 in such a way that the time of traversal is a minimum.

Consider a light ray reflected from a surface as shown in Figure 7.2 The path of the light ray can be changed by varying x. That is, x is the variable that parametrizes the path. Since the speed of light is $v = c/n = ds/dt$, the time to travel from p_0 to p_1 is

$$t = \frac{1}{c}\int_{p_0}^{p_1} n\,ds \tag{7.26}$$

where n is the index of refraction of the medium in which the light is traveling. For reflection, the index of refraction of the medium before encountering the surface is the same as the index after the reflection. Thus n is constant, and

$$t = \frac{n}{c}(s_0 + s_1) = \frac{n}{c}\left(\sqrt{h_0^2 + x^2} + \sqrt{h_1^2 + (L-x)^2}\right) \tag{7.27}$$

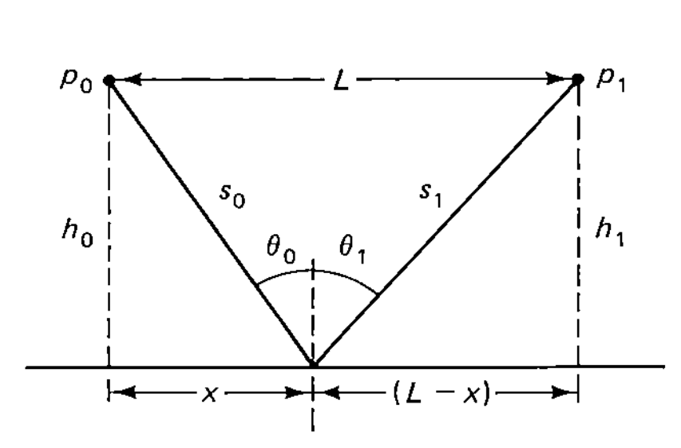

Figure 7.2 Reflection of a light ray.

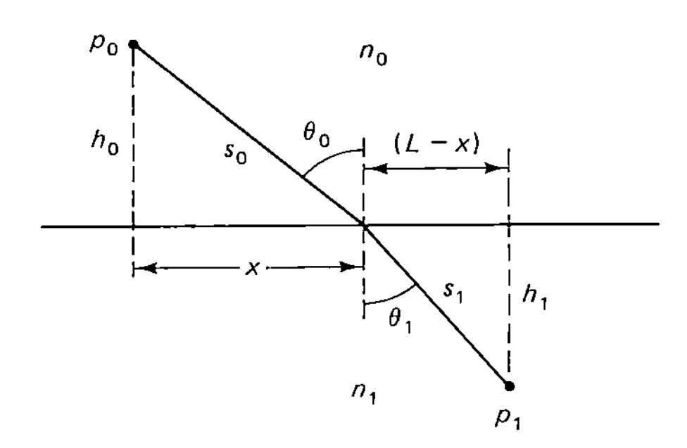

Figure 7.3 Refraction of a light ray.

The minimization process, then, is the standard method of setting the derivative of the function to zero.

$$\text{(a)}\quad \frac{dt}{dx} = 0 = \frac{n}{c}\left(\frac{x}{s_0} - \frac{(L-x)}{s_1}\right) = \frac{n}{c}(\sin\theta_0 - \sin\theta_1) \quad \Rightarrow \quad \text{(b)}\quad \theta_0 = \theta_1 \tag{7.28}$$

That is, the law of reflection is that the angle of incidence and angle of reflection are equal. By taking the second derivative and setting $\theta_0 = \theta_1$, it is easily verified that the second derivative is positive. Therefore, the extremum of t is a minimum.

If the light ray refracts when it encounters the surface, then, referring to Figure 7.3, we see that the time of traversal from p_0 and p_1 is

$$t = \frac{1}{c}(n_0 s_0 + n_1 s_1) = \frac{1}{c}\left(n_0\sqrt{h_0^2 + x^2} + n_1\sqrt{h_1^2 + (L-x)^2}\right) \tag{7.29}$$

Minimizing t with respect to x leads to

$$n_0\frac{x}{s_0} - n_1\frac{(L-x)}{s_1} = n_0 \sin\theta_0 - n_1 \sin\theta_1 = 0 \tag{7.30}$$

This is *Snell's law of refraction*. As with reflection, the second derivative is positive, indicating that the extremum of t is a minimum. □

Euler's equation

Using these examples as guides, the general problem addressed by the method of variational calculus can be stated as the problem of extremizing (minimizing) an integral of the form

$$S \equiv \int_{p_0}^{p_1} L(x, y, y')\,dx \tag{7.31}$$

with respect to the path taken from p_0 to p_1. This minimum integral will be obtained by defining $y(x)$ in the terms of a parameter that characterizes the path. The path is parametrized as $y(x, \varepsilon)$, and we set the variation of the integral with respect to ε to zero. Since the endpoints are fixed, the variation of S with path is

$$\delta S = \int_{p_0}^{p_1} \delta L(x, y, y')\,dx \tag{7.32a}$$

The independent variable x describes the position of any point along a particular path; it is therefore not dependent on the path chosen. Therefore, $L(x, y, y')$ varies with path only because y and y' vary with path. Thus equation 7.32a becomes

$$\delta S = \int_{p_0}^{p_1} \left[\frac{\partial L}{\partial y}\delta y + \frac{\partial L}{\partial y'}\delta y'\right] dx \tag{7.32b}$$

With

$$\delta y' = \delta\left(\frac{dy}{dx}\right) = \frac{d}{dx}\delta y \tag{7.33}$$

the second term in equation 7.32b is integrated by parts. Thus

$$\int_{p_0}^{p_1} \frac{\partial L}{\partial y'}\delta y'\,dx = \int_{p_0}^{p_1} \frac{\partial L}{\partial y'}\frac{d}{dx}\delta y\,dx = \frac{\partial L}{\partial y'}\delta y\Big|_{p_0}^{p_1} - \int_{p_0}^{p_1} \frac{d}{dx}\left(\frac{\partial L}{\partial y'}\right)\delta y\,dx \tag{7.34}$$

Since y does not vary with path at the endpoints, the integrated term is zero and

$$\delta S = \int_{p_0}^{p_1} \left[\frac{\partial L}{\partial y} - \frac{d}{dx}\left(\frac{\partial L}{\partial y'} \right) \right] \delta y \, dx \tag{7.35}$$

The minimum path is obtained by setting $\delta S = 0$. Since δy is an arbitrary, nonzero variation of y with path [δy is essentially the function $\mu(x)$ in the previous examples], δS can be zero only if

$$\frac{\partial L}{\partial y} - \frac{d}{dx}\left(\frac{\partial L}{\partial y'} \right) = 0 \tag{7.36}$$

This is the *one-dimensional Euler equation*. It will result in a differential equation which has a solution $y(x)$ that specifies the minimum path from p_0 to p_1.

Ignorable coordinates

We now return to the problem of determining the equation of minimum path between two points in a flat plane in terms of Euler's equation. Referring to equation 7.2, we identify the function

$$L(x, y, y') = \sqrt{1 + y'^2} \tag{7.37}$$

$$\Rightarrow \quad \frac{\partial L}{\partial y} - \frac{d}{dx}\left(\frac{\partial L}{\partial y'} \right) = -\frac{d}{dx}\left[\frac{y'}{\sqrt{1 + y'^2}} \right] = 0 \tag{7.38}$$

which is identical to equation 7.14a and therefore results in a straight-line minimum path.

We note that in this example, the function L does not depend on y. In general, if $L(x, y, y') = L(x, y')$, then Euler's equation becomes

$$\text{(a)} \quad -\frac{d}{dx}\left(\frac{\partial L}{\partial y'} \right) = 0 \quad \Rightarrow \quad \text{(b)} \quad \left(\frac{\partial L}{\partial y'} \right) = \text{constant} \tag{7.39}$$

When the integrand does not depend on y, then y is said to be an ignorable coordinate.

Brachistochrone problem

Example 7.4

Historically, the problem that gave birth to variational calculus is that of determining the path of a particle, acted on by a constant gravitational force mg, such that the time of travel between two points is minimized. For example, if a bead is to slide down a frictionless wire that is extended between two points, what should be the shape of the wire if the bead is to reach the second point in minimum time? This problem was first investigated by John Bernoulli and is known as the *brachistochrone problem* (from the Greek *brachistos*, meaning minimum, and *chronos*, which means time).

Let ds be an infinitesimal distance along the path taken by the particle. Then $v = ds/dt$, and the time to travel from p_0 to p_1 is

$$t = \int_{p_0}^{p_1} \frac{ds}{v} = \int_{x_0}^{x_1} \frac{1}{v} \sqrt{1 + y'^2} \, dx \tag{7.40}$$

Since the total energy of the particle E is a constant, the velocity at any point on the path is

$$v = \sqrt{\frac{2E}{m} - 2gy} \tag{7.41}$$

$$\Rightarrow \quad t = \int_{x_0}^{x_1} \frac{\sqrt{1 + y'^2}}{\sqrt{\frac{2E}{m} - 2gy}}\, dx \tag{7.42}$$

Identifying

$$L(x, y, y') = \frac{\sqrt{1 + y'^2}}{\sqrt{\frac{2E}{m} - 2gy}} \tag{7.43}$$

$$\Rightarrow \quad \frac{y''}{(1 + y'^2)} - \frac{g}{[(2E/m) - 2gy]} = 0 \quad \square \tag{7.44}$$

This is quite a difficult differential equation to solve. A first integral of a differential equation is said to be achieved if one can reduce the order of the equation by one.

For example, Newton's second law is a second-order differential equation of the form

$$m\frac{d^2y}{dt^2} = F\left(y, \frac{dy}{dt}, t\right) \tag{7.45}$$

If the total energy is constant, and if the potential energy depends on derivatives of y no higher than the first, then

$$E = \text{constant} = \frac{1}{2}m\left(\frac{dy}{dt}\right)^2 + V\left(y, \frac{dy}{dt}\right) \tag{7.46}$$

which is a first-order differential equation. Thus, by describing the motion of a particle in terms of its constant total energy instead of in terms of Newton's law, one achieves a first integral of equation 7.45.

Since y and y' depend on x, the function

$$F(x, y, y') \equiv L(x, y, y') - y'\frac{\partial L}{\partial y'} \tag{7.47}$$

can be expressed as $F(x)$. Then

$$\frac{dF}{dx} = \frac{\partial L}{\partial x} + \frac{\partial L}{\partial y}y' + \frac{\partial L}{\partial y'}y'' - y''\frac{\partial L}{\partial y'} - y'\frac{d}{dx}\left(\frac{\partial L}{\partial y'}\right) \tag{7.48}$$

Using Euler's equation, the second and last terms on the right side of the equation add to zero and this becomes

$$\frac{dF}{dx} = \frac{\partial L}{\partial x} \tag{7.49}$$

Therefore, if L is explicitly independent of x, $dF/dx = 0$, or F is constant.

We note from equation 7.43 that L does not contain any explicit x-dependence. Therefore,

$$F = L - y'\frac{\partial L}{\partial y'} = \frac{1}{\sqrt{1 + y'^2}\sqrt{\frac{2E}{m} - 2gy}} = \text{constant} \tag{7.50}$$

Since F is constant, this is a first-order differential equation, and a first integral has been obtained. Although this too, looks formidable, the substitution

$$y' = \tan\left(\frac{\alpha}{2}\right) \tag{7.51}$$

allows us to manipulate equation 7.50 into the form

$$y = \frac{1}{2g}\left[\frac{2E}{m} - \frac{1}{F^2}\cos^2\left(\frac{\alpha}{2}\right)\right] \tag{7.52}$$

Differentiating this expression, we have

$$y' = \frac{1}{2gF^2}\cos\left(\frac{\alpha}{2}\right)\sin\left(\frac{\alpha}{2}\right)\frac{d\alpha}{dx} \tag{7.53a}$$

Combining this with equation 7.51, we obtain

$$\frac{1}{2gF^2}\cos\left(\frac{\alpha}{2}\right)\sin\left(\frac{\alpha}{2}\right)\frac{d\alpha}{dx} = \tan\left(\frac{\alpha}{2}\right) \tag{7.53b}$$

$$\Rightarrow \quad \cos^2\left(\frac{\alpha}{2}\right)\frac{d\alpha}{dx} = 2gF^2 \tag{7.54}$$

$$\Rightarrow \quad x = x_0 + \frac{1}{4gF^2}(\alpha + \sin\alpha) \tag{7.55}$$

The equation of the path can be obtained by solving equation 7.52 for α and substituting the result into equation 7.55 to obtain the equation of path in the form $x = x(y)$. It is less cumbersome to leave the solution in the parametrized form given in equations 7.52 and 7.55, expressing the solution as $y = y[\alpha(x)]$.

7.2 Isoperimetric Constraints and Lagrange Multipliers

Example 7.5

Consider the problem of a uniform massive rope fixed at two points (Figure 7.4). The constant linear mass density (mass/unit length) will be denoted by σ. Then the mass of a small segment of the rope of length ds is $dm = \sigma\, ds$. Let this mass element be a distance y above the position defined as the zero level of potential energy. The potential energy of this segment is

$$\text{(a)} \quad dV = gy\,dm = \sigma gy\,ds \quad \Rightarrow \quad \text{(b)} \quad V = \sigma g\int_{p_0}^{p_1} y\,ds = \sigma g\int_{p_0}^{p_1} y\sqrt{1 + y'^2}\,dx \tag{7.56}$$

The shape that the rope will assume is the one that will minimize the potential energy. In this problem the "path" is the shape of the rope.

However, the solution to the problem is not found simply by minimizing the potential energy with respect to shape. Since the rope cannot stretch, its length is fixed. Therefore, there is a constraint on the system in the form

$$l = \int_{p_0}^{p_1} ds = \int_{p_0}^{p_1}\sqrt{1 + y'^2}\,dx = \text{constant} \quad \square \tag{7.57}$$

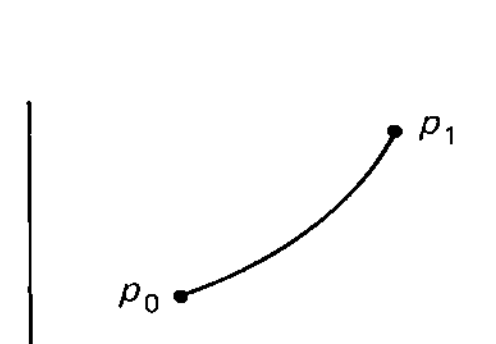

Figure 7.4 Rope with ends fixed acted on by gravitional force.

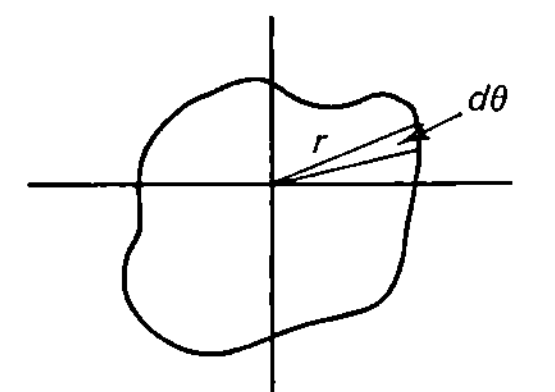

Figure 7.5 Arbitrary shape of a loop enclosing area A.

Example 7.6

As a second example of a problem in which a constraint is imposed on the system, consider a string of fixed length that has its ends tied together to form a closed loop. If it is placed on a flat plane, what shape should the loop take to enclose that largest possible area?

As shown in Figure 7.5 the arbitrarily shaped loop is segmented into strips of radius r, and of area

$$\text{(a)}\quad dA = \tfrac{1}{2}r^2\,d\theta \qquad \Rightarrow \qquad \text{(b)}\quad A = \tfrac{1}{2}\int_0^{2\pi} r^2(\theta)\,d\theta \tag{7.58}$$

This is a problem in maximizing the area with respect to shape. Like the problem of the rope, this area is constrained by the fixed length of the string, which is expressed as

$$l = \oint ds = \int_0^{2\pi} r(\theta)\,d\theta = \text{constant} \qquad \square \tag{7.59}$$

It is clear that the mathematical structure of this problem is identical to that of the rope. In both cases we will have to find the extremum of an integral-defined function with respect to shape subject to a constraint in the form of a second integral being constant. The term *isoperimetric* means constant perimeter.

The general form of the isoperimetric problem can be stated as one in which the integral

$$S \equiv \int_{p_0}^{p_1} L(y, y', x)\,dx \tag{7.60a}$$

must be extremized, subject to the constraint

$$l = \int_{p_0}^{p_1} M(y, y', x)\,dx = \text{constant} \tag{7.60b}$$

We cannot parametrize the path or shape of these functions with a single shape parameter ε. If we did, then in particular, since $l = l(\varepsilon)$, the length of the system would not be constant for different values of ε (different shapes). Therefore, we must parametrize S and l with two parameters ε_1 and ε_2. Therefore,

$$S = S(\varepsilon_1, \varepsilon_2) \tag{7.61a}$$

is subject to the constraint

$$l = l(\varepsilon_1, \varepsilon_2) = \text{constant} \tag{7.61b}$$

Because of this constraint, ε_1 and ε_2 are not independent, so there is still only one independent shape (path) parameter.

With two shape parameters,

$$y(x, \varepsilon_1, \varepsilon_2) = y(x) + \varepsilon_1\mu_1(x) + \varepsilon_2\mu_2(x) \tag{7.62}$$

with

$$\mu_1(x_0) = \mu_1(x_1) = \mu_2(x_0) = \mu_2(x_1) = 0 \tag{7.63}$$

The variation of S and the constancy of l result in

$$\text{(a)}\quad \delta S = \frac{\partial S}{\partial \varepsilon_1}\delta\varepsilon_1 + \frac{\partial S}{\partial \varepsilon_2}\delta\varepsilon_2 \quad \text{and} \quad \text{(b)}\quad \delta l = \frac{\partial l}{\partial \varepsilon_1}\delta\varepsilon_1 + \frac{\partial l}{\partial \varepsilon_2}\delta\varepsilon_2 = 0 \tag{7.64}$$

At the extremum, $\delta S = 0$. Thus we have two quantities with zero variation. One way to solve this is to eliminate one ε from the equation for the constant l and substitute into the expression for S. Then the variation of S will be a variation with respect to one shape parameter.

A more straightforward approach is to define a new function of ε_1 and ε_2 that is a linear combination of S and l.

$$\overline{S} \equiv S + \lambda l \tag{7.65}$$

λ is called the *Lagrange multiplier*. Consider

$$\delta \overline{S} = \delta S + \lambda \, \delta l \tag{7.66}$$

But since $\delta l = 0$, $\delta \overline{S} = \delta S$. Thus, making $\overline{S}$ an extremum is equivalent to making S an extremum.

In some cases the constraint can be written in the form

$$l = \int_{p_0}^{p_1} M(x, y, y') \, dx \tag{7.67}$$

Then we define

$$\overline{S} = \int_{p_0}^{p_1} (L + \lambda M) \, dx \tag{7.68}$$

Defining the function

$$\overline{L} = L + \lambda M \tag{7.69}$$

the problem is reduced to the original one of extremizing

$$\overline{S} = \int_{p_0}^{p_1} \overline{L}(y, y', x) \, dx \tag{7.70}$$

As developed above, the extremum of $\overline{S}$ leads to the Euler equation for $\overline{L}$

$$\begin{aligned} &\text{(a)} \quad \frac{\partial \overline{L}}{\partial y} - \frac{d}{dx}\left(\frac{\partial \overline{L}}{\partial y'}\right) = 0 \\ \Rightarrow \quad &\text{(b)} \quad \left[\frac{\partial L}{\partial y} - \frac{d}{dx}\left(\frac{\partial L}{\partial y'}\right)\right] + \lambda\left[\frac{\partial M}{\partial y} - \frac{d}{dx}\left(\frac{\partial M}{\partial y'}\right)\right] = 0 \end{aligned} \tag{7.71}$$

For the problem of the shape of the rope fixed at two points,

$$\overline{V} = \sigma g \int_{x_0}^{x_1} y \sqrt{1 + y'^2} \, dx + \lambda \int_{x_0}^{x_1} \sqrt{1 + y'^2} \, dx \tag{7.72}$$

$$\Rightarrow \quad \overline{L} = (\lambda + \sigma g) \sqrt{1 + y'^2} \tag{7.73}$$

We note that $\overline{L}$ has no explicit x dependence. Therefore, a first integral can be obtained in terms of the constant

$$F = \overline{L} - y' \frac{\partial \overline{L}}{\partial y'} = \frac{(\lambda + \sigma g y)}{\sqrt{1 + y'^2}} \tag{7.74}$$

$$\Rightarrow \quad y' = \left[\frac{(\lambda + \sigma g y)^2}{F^2} - 1\right]^{1/2} \tag{7.75}$$

The substitution

$$\frac{(\lambda + \sigma g y)}{F} = \cosh u \tag{7.76}$$

casts the first order differential equation into the form

$$\frac{F}{\sigma g}\frac{du}{dx} = 1 \tag{7.77}$$

$$\Rightarrow \quad \text{(a)} \quad u = \frac{\sigma g}{F}x + \alpha \qquad \text{(b)} \quad \frac{(\lambda + \sigma g y)}{F} = \cosh\left(\frac{\sigma g}{F}x + \alpha\right) \tag{7.78}$$

We impose the conditions that the values of y at the endpoints are fixed. Therefore,

$$\text{(a)} \quad y(x_0) = y_0 \qquad \text{(b)} \quad y(x_1) = y_1 \tag{7.79}$$

In addition, the length of the rope is specified. Thus

$$l = \int_{x_0}^{x_1} \sqrt{1 + y'^2}\, dx = \int_{x_0}^{x_1} \cosh\left(\frac{\sigma g}{F}x + \alpha\right) dx = \frac{F}{\sigma g}\sinh\left(\frac{\sigma g}{F}x + \alpha\right) \tag{7.79c}$$

The three independent conditions of equations 7.79 will yield values for λ, F, and α.

We return to the problem of finding the shape of a loop of fixed perimeter that will enclose the largest area. Referring to equations 7.58b and 7.59, we define

$$\overline{A} = \tfrac{1}{2}\int_0^{2\pi} r^2\, d\theta + \lambda \int_0^{2\pi} r\, d\theta \tag{7.80}$$

$$\Rightarrow \quad \overline{L} = \tfrac{1}{2}r^2 + \lambda r \tag{7.81}$$

Since $\overline{L}$ does not depend on $r' = dr/d\theta$, the Euler equation becomes

$$\text{(a)} \quad \frac{\partial \overline{L}}{\partial r} = r + \lambda = 0 \qquad \Rightarrow \qquad \text{(b)} \quad r = -\lambda = \text{constant} \tag{7.82}$$

The equation $r =$ constant defines a circle. The reader is also referred to problem 5.

7.3 Several Dependent Variables, the Principle of Least Action, and Lagrange's Equations

The problem of finding the extremum of an integral in which the integrand L depends on several dependent variables is a straightforward extension of the single-variable problem. One very important example of this multivariable problem arises in the *principle of least action* or *Hamilton's principle*. This will be used to illustrate the method of extremizing an integral when the integrand depends on more than one dependent variable.

Beginning with Newton's second law in the form

$$\sum_{i=1}^{N} m_i \frac{d^2\mathbf{r}_i}{dt^2} - \mathbf{F}_{\text{net}} = 0 \tag{7.83}$$

we denote the difference in the position vector of the ith particle along two different paths as $\delta\mathbf{r}_i$. With the endpoints fixed,

$$\delta\mathbf{r}_i|_{t_0} = \delta\mathbf{r}_i|_{t_1} = 0 \tag{7.84}$$

Then

$$\sum_{i=1}^{N} \int_{t_0}^{t_1} \left(m_i \frac{d^2\mathbf{r}_i}{dt^2} - \mathbf{F}_{\text{net}}\right) \cdot \delta\mathbf{r}_i\, dt = 0 \tag{7.85}$$

The first term is integrated by parts once to obtain

$$\int_{t_0}^{t_1} m_i \frac{d^2\mathbf{r}_i}{dt^2} \cdot \delta\mathbf{r}_i \, dt = m_i \frac{d\mathbf{r}_i}{dt} \cdot \delta\mathbf{r}_i \Big|_{t_0}^{t_1} - \int_{t_0}^{t_1} m_i \frac{d\mathbf{r}_i}{d} \cdot \frac{d}{dt}\delta\mathbf{r}_i \, dt$$

$$= -\int_{t_0}^{t_1} m_i \frac{d\mathbf{r}_i}{dt} \cdot \delta\left(\frac{d\mathbf{r}_i}{dt}\right) dt \tag{7.86}$$

Consider the variation in the square of the velocity:

$$\delta\left(v_i^2\right) = \delta\left(\frac{d\mathbf{r}_i}{dt} \cdot \frac{d\mathbf{r}_i}{dt}\right) = 2\frac{d\mathbf{r}_i}{dt} \cdot \delta\left(\frac{d\mathbf{r}_i}{dt}\right) \tag{7.87}$$

which is part of the integrand of equation 7.86. Therefore, equation 7.85 can be written

$$\sum_{i=1}^{N} \int_{t_0}^{t_1} \left[\delta\left(\tfrac{1}{2}m_i v_i^2\right) + \mathbf{F}_{\text{net}} \cdot \delta\mathbf{r}_i\right] dt = 0 \tag{7.88}$$

$T \equiv \sum_{n=1}^{N} \frac{1}{2}m_i v_i^2$ is recognized as the total kinetic energy of the system. The quantity $\sum_{n=1}^{N} \mathbf{F}_{\text{net}} \cdot \delta\mathbf{r}_i \equiv -\delta W$ is the negative variation of the work done by the force $\mathbf{F}_{\text{net}}$. If that net force is conservative (and can therefore be written as the gradient of a potential energy), then

$$\sum_{i=1}^{N} \mathbf{F}_{\text{net}} \cdot \delta\mathbf{r}_i = -\delta V \tag{7.89}$$

$$\Rightarrow \quad \delta \int_{t_0}^{t_1} (T - V)\, dt = 0 \tag{7.90}$$

The Lagrangian

The quantity $T - V$ is called the *Lagrangian function* of the N particle system. Such a system has $3N$ independent position components which we designate as y_i and $3N$ independent velocity components which we denote by $v_i = dy_i/dt \equiv \dot{y}_i$ $(1 \leqslant i \leqslant 3N)$. With this notation, the Lagrangian is

$$L(y_1, y_2, \ldots, y_{3N}, \dot{y}_1, \dot{y}_2, \ldots, \dot{y}_{3N}, t) \equiv T - V \tag{7.91}$$

The action is defined by the integral of equation 7.90.

$$S \equiv \int_{t_0}^{t_1} L\, dt \tag{7.92}$$

Starting at a point $\mathbf{r}_0$, at time t_0, a system can move in an infinite number of different ways so that it will arrive at a point $\mathbf{r}_1$ at a time t_1. Equation 7.90, which is the *principle of least action*, states that the motion of a system of objects will evolve in time such that S will be a minimum.

For example, a particle is dropped from some point y_0 above the ground at time t_0. At time t_1 it reaches the ground. Two of an infinite number of possible evolutions of the motion are:

1. The particle accelerates uniformly at 9.8 meters/sec^2 and reaches the ground at time t_1.
2. The particle accelerates at a rate of 2 meters/sec^2 for a short time. Then its acceleration increases to 12 meters/sec^2 for another period of time, and it hits the ground at time t_1.

The evolutionary "path" that the system will take is the one that minimizes the action integral (path 1 in the example above).

In these examples, we note that the particle's position and velocity at a specified instant will be different for different evolutionary paths. That is, the action varies with the evolutionary path because the dynamic variables vary with the path. The time variable describes the motion along one of the paths, and so is path independent. In addition, the positions of the particle at t_0 and t_1 are path independent. The particle must be at the release point y_0 at t_0 and must be on the ground at y_1 at time t_1. Therefore,

$$\delta S = \int_{t_0}^{t_1} \delta L(y_1, y_2, \ldots, y_{3N}, \dot{y}_1, \dot{y}_2, \ldots, \dot{y}_{3N}, t)\, dt$$

$$= \sum_{i=1}^{3N} \int_{t_0}^{t_1} \left(\frac{\partial L}{\partial y_i} \delta y_i + \frac{\partial L}{\partial \dot{y}_i} \delta \dot{y}_i \right) dt \tag{7.93}$$

The analysis of each term in this expression is identical to that for the one-dimensional problem. Thus

$$\int_{t_0}^{t_1} \frac{\partial L}{\partial \dot{y}_i} \delta \dot{y}_i \, dt = \int_{t_0}^{t_1} \frac{\partial L}{\partial \dot{y}_i} \delta \left(\frac{dy_i}{dt} \right) dt = \int_{t_0}^{t_1} \frac{\partial L}{\partial \dot{y}_i} \frac{d\, \delta y_i}{dt}\, dt \tag{7.94a}$$

When integrated by parts once, this becomes

$$\int_{t_0}^{t_1} \frac{\partial L}{\partial \dot{y}_i} \delta \dot{y}_i \, dt = \frac{\partial L}{\partial \dot{y}_i} \delta y_i \Big|_{t_0}^{t_1} - \int_{t_0}^{t_1} \frac{d}{dt} \left(\frac{\partial L}{\partial \dot{y}_i} \right) \delta y_i \, dt \tag{7.94b}$$

The integrated term is zero, since y_i does not vary with path at the end points t_0 and t_1. Therefore, equation 7.93 becomes

$$\delta S = \sum_{i=1}^{3N} \int_{t_0}^{t_1} \delta y_i \left[\frac{\partial L}{\partial y_i} - \frac{d}{dt} \left(\frac{\partial L}{\partial \dot{y}_i} \right) \right] dt \tag{7.95}$$

To minimize this action integral, we set $\delta S = 0$. If there are no constraints imposed on the system, each δy_i is independent and arbitrary. Therefore, the action will be a minimum only when each term in the brackets is zero. That is, S is minimized by

$$\frac{\partial L}{\partial y_i} - \frac{d}{dt} \left(\frac{\partial L}{\partial \dot{y}_i} \right) = 0 \tag{7.96}$$

There are $3N$ such equations, one for each coordinate y_i. Another way to state this is to say that there are $3N$ degrees of freedom. These are the *Euler or Euler–Lagrange equations* (also called the *Lagrange equations*) of motion.

Example 7.7

For the problem of the falling object, if y is the height of the object above the ground at some instant, then

$$L = \tfrac{1}{2}m\dot{y}^2 - mgy \tag{7.97}$$

and the Lagrange equation of motion,

$$\ddot{y} = -g = \text{constant} \tag{7.98}$$

This verifies that the motion described above in the first example is the one that minimizes the action. □

The Hamiltonian

Consider

$$\begin{aligned} \frac{d}{dt}\left[\sum_{i=1}^{3N} \dot{y}_i \frac{\partial L}{\partial \dot{y}_i} - L\right] &= \sum_{i=1}^{3N}\left[\ddot{y}_i \frac{\partial L}{\partial \dot{y}_i} + \dot{y}_i \frac{d}{dt}\left(\frac{\partial L}{\partial \dot{y}_i}\right) - \dot{y}_i \frac{\partial L}{\partial y_i} - \ddot{y}_i \frac{\partial L}{\partial \dot{y}_i}\right] - \frac{\partial L}{\partial t} \\ &= \sum_{i=1}^{3N} \dot{y}_i\left[\frac{d}{dt}\left(\frac{\partial L}{\partial \dot{y}_i}\right) - \frac{\partial L}{\partial y_i}\right] - \frac{\partial L}{\partial t} \end{aligned} \tag{7.99a}$$

Since L satisfies the Euler–Lagrange equation for each i, this becomes

$$\frac{d}{dt}\left[\sum_{i=1}^{3N} \dot{y}_i \frac{\partial L}{\partial \dot{y}_i} - L\right] = -\frac{\partial L}{\partial t} \tag{7.99b}$$

Therefore, if L does not contain any explicit time dependence,

$$H \equiv \sum_{i=1}^{3N} \dot{y}_i \frac{\partial L}{\partial \dot{y}_i} - L = \text{constant} \tag{7.100}$$

H is called the *Hamiltonian function* of the system.

Constraints and Lagrange multipliers

If constraints are imposed on a system, the number of degrees of freedom is reduced. Using the isoperimetric examples discussed above as guides, the equations that describe the constraints can be expressed in the form

$$\phi_k = \int_{t_0}^{t_1} M_k(y_1, y_2, \ldots, \dot{y}_1, \dot{y}_2, \ldots, t)\, dt = \text{constant} \tag{7.101}$$

If there are n such constraints, there will be $3N - n$ remaining degrees of freedom. That is, out of the $3N$ position variables y_i and velocities $\dot{y}_i, n$ of them will be dependent.

Extremizing S requires that

$$\delta S = \sum_{i=1}^{3N} \int_{t_0}^{t_1} \left[\frac{\partial L}{\partial y_i} - \frac{d}{dt}\left(\frac{\partial L}{\partial \dot{y}_i}\right)\right] \delta y_i\, dt = 0 \tag{7.102}$$

Because of the constraint equations, not all y_i are independent. Therefore, not all δy_i are independent. Thus the term in brackets cannot be set to zero for each index i.

As in the isoperimetric problem, we can include the constraints by noting that the variation of the constant ϕ_k is zero. This leads to

$$\delta\phi_k = 0 = \sum_{i=1}^{n} \int_{t_0}^{t_1} \left[\frac{\partial M_k}{\partial y_i} - \frac{d}{dt}\left(\frac{\partial M_k}{\partial \dot{y}_i}\right)\right] \delta y_i\, dt = 0 \qquad 1 \leqslant k \leqslant n \tag{7.103}$$

We now define the modified Lagrangian,

$$\bar{L}(\mathbf{y},\dot{\mathbf{y}},t) = L(\mathbf{y},\dot{\mathbf{y}},t) + \sum_{k=1}^{n} \lambda_k M_k(\mathbf{y},\dot{\mathbf{y}},t) \tag{7.104}$$

and from this, the modified action integral,

$$\bar{S} = \int_{t_0}^{t_1} \bar{L}(\mathbf{y},\dot{\mathbf{y}},t)\, dt \tag{7.105}$$

Clearly, with equation 7.103, we have

$$\delta\bar{S} = \delta S \tag{7.106}$$

Thus an extremum of $\bar{S}$ is an extremum of S.

Example 7.8

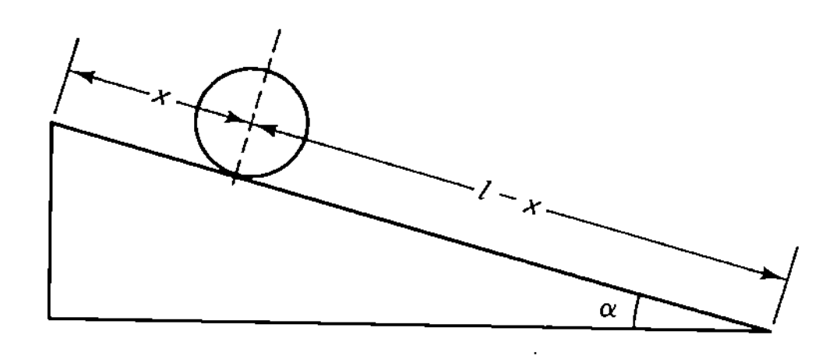

Figure 7.6 Object rolling down an inclined ramp.

As an example, consider a round object (sphere, cylinder, etc.) of mass M, radius R, rolling down a ramp inclined at an angle α as shown in Figure 7.6. We assume that the rolling occurs without slipping. Therefore, the distance traveled by the center of mass in time t is related to the angle through which the object has rotated about its center of mass by

$$x = R\theta \tag{7.107a}$$

or since $x - R\theta = 0$, the constraint can also be written in the form

$$\int_{t_0}^{t_1} (x - R\theta)\, dt = \text{constant} \tag{7.107b}$$

Therefore, with one constraint,

$$\bar{L} = \tfrac{1}{2} M\dot{x}^2 + \tfrac{1}{2} I\dot{\theta}^2 - Mg(l - x)\sin\alpha + \lambda(x - R\theta) \tag{7.108}$$

and the Euler–Lagrange equations for this system become

$$\text{(a)} \quad M\ddot{x} - Mg\sin\alpha - \lambda = 0 \qquad \text{(b)} \quad I\ddot{\theta} + \lambda R = 0 \tag{7.109}$$

These are differential equations involving constant accelerations and their solutions are straightforward. For simplicity, it will be assumed that the object starts at rest at the top of the ramp. Thus $x_0 = l$, $v_0 = \omega_0 = 0$, and we will define $\theta_0 \equiv 0$. Then

$$\text{(a)} \quad x = l + \frac{1}{2}\left(\frac{\lambda}{M} + g\sin\alpha\right)t^2 \qquad \text{(b)} \quad \theta = -\frac{1}{2}\frac{\lambda R}{I}t^2 \tag{7.110}$$

Eliminating λ between the two equations yields

$$x = l + \frac{1}{2} g \sin\alpha t^2 - \frac{I}{MR}\theta \tag{7.111}$$

Setting

$$\theta = \frac{x}{R} \tag{7.107a}$$

$$\Rightarrow \quad x(t) = R\theta(t) = \frac{1}{1 + I/MR^2}\left[l + \frac{1}{2} g \sin\alpha t^2\right] \quad \square \tag{7.112}$$

7.4 Several Independent Variables

For the sake of brevity, we consider the problem of determining the condition for extremizing an integral, the integrand of which contains two independent variables. The extension to more than two variables is straightforward.

For a system with N degrees of freedom, we consider

$$S \equiv \int_{t_0}^{t_1}\int_{u_0}^{u_1} L\left(\mathbf{y}, \frac{\partial \mathbf{y}}{\partial t}, \frac{\partial \mathbf{y}}{\partial u}, t, u\right) dt\, du \tag{7.113}$$

This is the type of integral that one would extremize in a field theory. The dependent variables y_i are fields that depend on independent space and time variables represented by t and u.

As with the single-independent-variable problem, we can parametrize the path by

$$y_i(t, u, \varepsilon) = y_i(t, u) + \varepsilon\mu_i(t, u) \tag{7.114}$$

with

$$\mu_i(t_0, u_0) = \mu_i(t_1, u_0) = \mu_i(t_0, u_1) = \mu_i(t_1, u_1) = 0 \tag{7.115}$$

That is, $\delta y_i = 0$ at the end points of each integral. Therefore,

$$\delta S = \sum_{i=1}^{N}\int_{t_0}^{t_1}\int_{u_0}^{u_1}\left[\frac{\partial L}{\partial y_i}\delta y_i + \frac{\partial L}{\partial(\partial y_i/\partial t)}\delta\left(\frac{\partial y_i}{\partial t}\right) + \frac{\partial L}{\partial(\partial y_i/\partial u)}\delta\left(\frac{\partial y_i}{\partial u}\right)\right] dt\, du \tag{7.116}$$

Using $\delta(\partial y_i/\partial t) = \partial(\delta y_i)/\partial t$ and $\delta y_i(t_0, u) = \delta y_i(t_1, u) = 0$, the second term on the right hand side of equation 7.116 can be written

$$\int_{t_0}^{t_1}\int_{u_0}^{u_1}\left[\frac{\partial L}{\partial(\partial y_i/\partial t)}\delta\left(\frac{\partial y_i}{\partial t}\right)\right] dt\, du = \int_{u_0}^{u_1}\left[\frac{\partial L}{\partial(\partial y_i/\partial t)}\delta y_i\bigg|_{t_0}^{t_1} - \int_{t_0}^{t_1}\delta y_i\frac{\partial}{\partial t}\left(\frac{\partial L}{\partial(\partial y_i/\partial t)}\right) dt\right] du$$

$$= -\int_{u_0}^{u_1}\int_{t_0}^{t_1}\delta y_i\frac{\partial}{\partial t}\left(\frac{\partial L}{\partial(\partial y_i/\partial t)}\right) dt\, du \tag{7.117a}$$

In an identical way, writing the third term as

$$\int_{t_0}^{t_1}\int_{u_0}^{u_1}\left[\frac{\partial L}{\partial(\partial y_i/\partial u)}\delta\left(\frac{\partial y_i}{\partial u}\right)\right] du\, dt = -\int_{t_0}^{t_1}\int_{u_0}^{u_1}\delta y_i\frac{\partial}{\partial u}\left(\frac{\partial L}{\partial(\partial y_i/\partial u)}\right) du\, dt \tag{7.117b}$$

$$\Rightarrow \quad \delta S = \sum_{i=1}^{N}\int_{t_0}^{t_1}\int_{u_0}^{u_1}\left\{\frac{\partial L}{\partial y_i} - \frac{\partial}{\partial t}\left[\frac{\partial L}{\partial(\partial y_i/\partial t)}\right] - \frac{\partial}{\partial u}\left[\frac{\partial L}{\partial(\partial y_i/\partial u)}\right]\right\}\delta y_i\, dt\, du \tag{7.118}$$

Without constraints, the independence of each δy_i requires that for δS to be zero,

$$\frac{\partial L}{\partial y_i} - \frac{\partial}{\partial t}\left[\frac{\partial L}{\partial(\partial y_i/\partial t)}\right] - \frac{\partial}{\partial u}\left[\frac{\partial L}{\partial(\partial y_i/\partial u)}\right] = 0 \tag{7.119}$$

If n constraints are imposed on the system in the form

$$\phi_k = \int_{t_0}^{t_1}\int_{u_0}^{u_1} M_k\left(\mathbf{y}, \frac{\partial \mathbf{y}}{\partial t}, \frac{\partial \mathbf{y}}{\partial u}, t, u\right) dt\, du = \text{constant} \tag{7.120}$$

they will be included by extremizing the modified integral

$$\bar{S} \equiv \int_{t_0}^{t_1} \int_{u_0}^{u_1} \bar{L}\left(\mathbf{y}, \frac{\partial \mathbf{y}}{\partial t}, \frac{\partial \mathbf{y}}{\partial u}, t, u\right) dt\, du \tag{7.121}$$

where

$$\bar{L} \equiv L + \sum_{k=1}^{n} \lambda_k M_k \tag{7.122}$$

This, of course, is a straightforward extension of our analysis involving a single independent variable.

7.5 Rayleigh – Ritz Variational Method

The *Rayleigh–Ritz method* is a variational technique for obtaining an estimate of an eigenvalue. Consider an eigenvalue equation in the form

$$\hat{\Omega}_0 y_\alpha(x) = \lambda_\alpha \rho(x) y_\alpha(x) \tag{7.123}$$

(the Sturm–Liouville equation, for example), where the eigenfunctions $y_\alpha(x)$ satisfy specified boundary conditions. Consider an eigenvalue equation that is of the same form as equation 7.123,

$$\hat{\Omega} Y(x) = \Lambda \rho(x) Y(x) \tag{7.124}$$

If $\hat{\Omega}$ is not $\hat{\Omega}_0$, then Λ will not be one of the λ_α and $Y(x)$ will not be one of the $y_\alpha(x)$.

The eigenvalue is obtained straightforwardly as

$$\Lambda = \frac{\displaystyle\int_{x_0}^{x_1} Y^*(x) \hat{\Omega} Y(x)\, dx}{\displaystyle\int_{x_0}^{x_1} Y^*(x) \rho(x) Y(x)\, dx} = \frac{\langle Y | \hat{\Omega} | Y \rangle}{\langle Y | Y \rangle} \tag{7.125}$$

If $\hat{\Omega}$ is not very different from $\hat{\Omega}_0$, then Λ and λ_α should almost be equal. The best estimate of Λ will be obtained if $(\Lambda - \lambda_\alpha)^2$ is minimized. That is, we require that

$$\delta\left(\frac{\langle Y|\hat{\Omega}|Y\rangle}{\langle Y|Y\rangle} - \lambda_\alpha\right)^2 = 2\left(\frac{\langle Y|\hat{\Omega}|Y\rangle}{\langle Y|Y\rangle} - \lambda_\alpha\right)\delta\left(\frac{\langle Y|\hat{\Omega}|Y\rangle}{\langle Y|Y\rangle} - \lambda_\alpha\right) = 0 \tag{7.126}$$

The method of minimizing the bracket of equation 7.126b involves judiciously choosing a trial function that will satisfy equation 7.126b. Since Λ will be different than λ_α, the first bracket of equation 7.126b will not be zero. Because λ_α does not vary, the best estimate of Λ is obtained by choosing a trial eigenfunction with variable parameters and requiring that

$$\delta\left(\frac{\langle Y|\hat{\Omega}|Y\rangle}{\langle Y|Y\rangle} - \lambda_\alpha\right) = \delta\frac{\langle Y|\hat{\Omega}|Y\rangle}{\langle Y|Y\rangle} = 0 \tag{7.127}$$

Since $\langle Y|\hat{\Omega}|Y\rangle / \langle Y|Y\rangle$ will be a function of variable parameters, the minimization of this function will be the customary method of setting derivatives with respect to the variational parameters to zero. Once the values of those parameters are determined, equation 7.125 will then yield an estimate of Λ.

One standard approach is to choose a trial function in the form

$$Y(x) = y_0(x) + \sum_{i=1}^{N} c_i y_i(x) \tag{7.128}$$

such that at the end points,

$$\begin{aligned} &\text{(a)} \quad y_0(x_0) = Y(x_0) \qquad \text{(b)} \quad y_0(x_1) = Y(x_1) \\ &\Rightarrow \quad \text{(c)} \quad y_i(x_0) = y_i(x_i) = 0 \qquad i \geqslant 1 \end{aligned} \tag{7.129}$$

Almost a Legendre operator

Example 7.9

As an example, let us find the lowest eigenvalue of the differential equation

$$\hat{\Omega}Y = \left[(1-x^2)D^2 - 2xD + \alpha x^2\right]Y = -\Lambda Y \tag{7.130}$$

with $Y(\pm 1) = 1$. We note that if $\alpha = 0$, $\hat{\Omega}$ is the Legendre operator corresponding to $l = 1$. We designate this Legendre operator as $\mathscr{L}$. That is,

$$\hat{\Omega} = \mathscr{L} + \alpha x^2 \tag{7.131}$$

To find the lowest eigenvalue of $\hat{\Omega}$, we assume a trial function with one parameter that satisfies $Y(\pm 1) = 1$.

$$Y(x) = P_0(x) + a[P_2(x) - P_0(x)] = 1 + \tfrac{3}{2}a(x^2 - 1) \tag{7.132a}$$

One could also consider other one parameter trial functions, such as

$$Y(x) = P_0(x) + a[P_4(x) - 2P_2(x) + P_0(x)] \tag{7.132b}$$

Such a function also satisfies $Y(\pm 1) = 1$, but requires more computation than the choice expressed in equation 7.132a. Thus, as a first estimate, we use the less involved trial function.

Recall that the Legendre polynomials satisfy

$$\left[(1-x^2)D^2 - 2xD\right]P_l(x) = -l(l+1)P_l(x) \tag{6.163}$$

with

$$\int_{-1}^{1} P_l(x)P_m(x)\,dx = \frac{2}{2l+1}\delta_{lm} \tag{6.213}$$

Then, with $\rho(x) = 1$, and with the eigenvalue defined by equation 7.130 as $-\Lambda$,

$$-\Lambda = \frac{\langle Y|\hat{\Omega}|Y\rangle}{\langle Y|Y\rangle} = \frac{\langle P_0 + a(P_2 - P_0)|\mathscr{L} + \alpha x^2|P_0 + a(P_2 - P_0)\rangle}{\langle P_0 + a(P_2 - P_0)|P_0 + a(P_2 - P_0)\rangle} \tag{7.133}$$

Using equations 6.163 and 6.213, it is straightforward to show that

$$\langle P_0 + a(P_2 - P_0)|\mathscr{L}|P_0 + a(P_2 - P_0)\rangle = -\tfrac{12}{5}a^2 \tag{7.134a}$$

$$\langle P_0 + a(P_2 - P_0)|P_0 + a(P_2 - P_0)\rangle = \tfrac{12}{5}a^2 - 4a + 2 \tag{7.134b}$$

and using equation 7.132a, we have

$$\alpha\langle P_0 + a(P_2 - P_0)|x^2|P_0 + a(P_2 - P_0)\rangle = \alpha\left(\tfrac{12}{35}a^2 - \tfrac{4}{5}a + \tfrac{2}{3}\right) \tag{7.134c}$$

Therefore,

$$-\Lambda = \frac{\left(\frac{12}{35}\alpha - \frac{12}{5}\right)a^2 - \frac{4}{5}\alpha a + \frac{2}{3}\alpha}{\frac{12}{5}a^2 - 4a + 2} \tag{7.135}$$

Setting $\partial\Lambda/\partial a = 0$ leads to

$$\frac{6}{5}\left(\frac{2}{35}\alpha + 1\right)a^2 + \frac{6}{5}\left(\frac{\alpha}{7} - \frac{4}{3}\right)a + \left(\frac{1}{3} - \frac{\alpha}{5}\right) = 0 \tag{7.136}$$

For this illustration, we take $\alpha = 0.01$ so that $\hat{\Omega}$ is only slightly different from $\mathscr{L}$. The two values that satisfy equation 7.136 are $a = 1.0743$ and $a = 0.2569$. For $a = 1.0743$ and $a = 0.2569$ respectively, we obtain

$$\text{(a)} \quad \Lambda = 5.8557 \qquad \text{(b)} \quad \Lambda = 0.1358 \tag{7.137}$$

The approximation to the eigenfunction corresponding to $\Lambda = 0.1358$ is

$$Y(x) \simeq 1 + \tfrac{3}{2}(0.2569)(x^2 - 1) = 0.3854x^2 + 0.6147 \quad \square \tag{7.138}$$

To improve the estimate of the eigenvalue and eigenfunction, one might consider a trial function containing two variational parameters. For example, one possible choice that satisfies the boundary conditions is

$$Y(x) = P_0(x) + a[P_2(x) - P_0(x)] + b[P_2(x) - P_0(x)]^2 \tag{7.139}$$

There are, of course, many other two parameter choices.

Helium atom

Example 7.10

The energy of the helium atom in its lowest energy or ground state can be estimated by a Raleigh–Ritz variational method. This example illustrates a somewhat different approach to the choice of trial eigenfunction.

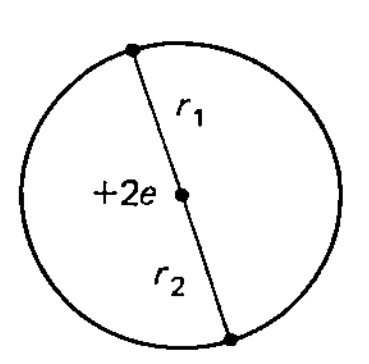

Figure 7.7 Configuration of the helium atom.

The helium atom is a charge $+2e$ nucleus with two electrons, each with charge $-e$ in the lowest allowed orbit around the nucleus (see Figure 7.7). We designate the coordinates of the two electrons by (r_1, θ_1, ϕ_1) and (r_2, θ_2, ϕ_2). The Schrödinger equation for the wave function of the system is

$$\left[-\frac{\hbar^2}{2m}\nabla_1^2\Psi(\mathbf{r}_1, \mathbf{r}_2) - \frac{2e^2}{r_1}\Psi(\mathbf{r}_1, \mathbf{r}_2)\right] + \left[-\frac{\hbar^2}{2m}\nabla_2^2\Psi(\mathbf{r}_1, \mathbf{r}_2) - \frac{2e^2}{r_2}\Psi(\mathbf{r}_1, \mathbf{r}_2)\right]$$
$$+ \frac{e^2}{|\mathbf{r}_1 - \mathbf{r}_2|}\Psi(\mathbf{r}_1, \mathbf{r}_2) = E\Psi(\mathbf{r}_1, \mathbf{r}_2) \tag{7.140}$$

If there were only one electron in the atom (the helium ion), the system would be like a hydrogen atom with a nuclear charge of $2e$. The Schrödinger equation for such a system would be

$$-\frac{\hbar^2}{2m}\nabla^2\Psi(\mathbf{r}) - \frac{2e^2}{r}\Psi(\mathbf{r}) = E\Psi(\mathbf{r}) \tag{7.141}$$

We see, then, that the Schrödinger equation for the helium atom is a sum of two hydrogen-like terms and an interaction term. If we make the assumption that the interaction energy is significantly smaller than the energy of each electron in its interaction with the nucleus, we can treat the helium atom as two hydrogen-like atoms with a small correction. As such, we construct a trial wave function as the product of two hydrogen atom wave functions interacting with a nucleus with charge number Z.

$$\Psi(\mathbf{r}_1, \mathbf{r}_2) = U_0(\mathbf{r}_1)U_0(\mathbf{r}_2) \tag{7.142}$$

The ground-state wave function for such a hydrogen-like atom is

$$U_0(\mathbf{r}) \propto L_1^1\left(2Z\frac{r}{r_0}\right)e^{-Zr/r_0}Y_{00}(\Omega) \tag{7.143}$$

where $L_1^1(x)$ is the (1, 1) associated Laguerre polynomial and

$$r_0 = \frac{\hbar^2}{me^2} \tag{7.144}$$

is the radius of the electron in hydrogen in its ground state.

Using equations 6.372b and 6.408, $L_1^1(x) = -1$, and referring to equation 6.343, $Y_{00}(\Omega) = 1/\sqrt{4\pi}$. Combining these as a single constant, the ground-state hydrogen wave function is

$$\text{(a)}\quad U_0(\mathbf{r}) = Ce^{-Zr/r_0} \quad \text{with} \quad \text{(b)}\quad \int U_0^*(\mathbf{r})U_0(\mathbf{r})r^2\,dr\,d\Omega = 4\pi C^2\int_0^\infty r^2 e^{-2Zr/r_0}\,dr = 1 \tag{7.145}$$

This leads to

$$U_0(\mathbf{r}) = \left(\frac{Z^3}{\pi r_0^3}\right)^{1/2} e^{-Zr/r_0} \tag{7.146}$$

$$\Rightarrow \quad \Psi(\mathbf{r}_1, \mathbf{r}_2) = U_0(\mathbf{r}_1)U_0(\mathbf{r}_2) = \frac{Z^3}{\pi r_0^3}e^{-Z(r_1-r_2)/r_0} \tag{7.147}$$

A standard choice for the variational parameter is Z. The argument advanced is that the presence of another electron shields each electron somewhat from the nuclear charge of $+2e$. That is, each electron is influenced by a nuclear charge $Ze \neq 2e$. It is noted that Z is made variable in the wave function, but the nuclear charge number is fixed as 2 in the Schrödinger operator.

Writing the term in the Schrödinger equation for each electron as $\hat{H}_i\Psi(\mathbf{r}_1, \mathbf{r}_2)$, and using the fact that when normalized

$$\langle\Psi(\mathbf{r}_1, \mathbf{r}_2)|\Psi(\mathbf{r}_1, \mathbf{r}_2)\rangle = \langle U_0(\mathbf{r}_1)|U_0(\mathbf{r}_1)\rangle\langle U_0(\mathbf{r}_2)|U_0(\mathbf{r}_2)\rangle = 1 \tag{7.148}$$

the energy eigenvalue is

$$E = \int\Bigg[U_0(\mathbf{r}_1)U_0(\mathbf{r}_2)\hat{H}_1U_0(\mathbf{r}_1)U_0(\mathbf{r}_2) + U_0(\mathbf{r}_1)U_0(\mathbf{r}_2)\hat{H}_2U_0(\mathbf{r}_1)U_0(\mathbf{r}_2)$$
$$+U_0(\mathbf{r}_1)U_0(\mathbf{r}_2)\frac{e^2}{|\mathbf{r}_1 - \mathbf{r}_2|}U_0(\mathbf{r}_1)U_0(\mathbf{r}_2)\Bigg]r_1^2\,dr_1 r_2^2\,dr_2\,d\Omega_1\,d\Omega_2 \tag{7.149}$$

Since $\mathbf{r}_1$ and $\mathbf{r}_2$ are independent

$$\int U_0(\mathbf{r}_1)U_0(\mathbf{r}_2)\hat{H}_1U_0(\mathbf{r}_1)U_0(\mathbf{r}_2)r_1^2\,dr_1\,d\Omega_1\,r_2^2\,dr_2\,d\Omega_2$$
$$= \int|U_0(\mathbf{r}_2)|^2 r_2^2\,dr_2\,d\Omega_2 \int U_0(\mathbf{r}_1)\hat{H}_1U_0(\mathbf{r}_1)r_1^2\,dr_1\,d\Omega_1 = \int U_0(\mathbf{r}_1)\hat{H}_1U_0(\mathbf{r}_1)r_1^2\,dr_1\,d\Omega_1 \tag{7.150}$$

An identical expression is obtained for the term containing $\hat{H}_2$. Thus

$$E = 2\int U_0(\mathbf{r})\hat{H}U_0(\mathbf{r})r^2\,dr\,d\Omega + \int U_0^2(\mathbf{r}_1)U_0^2(\mathbf{r}_2)\frac{e^2}{|\mathbf{r}_1 - \mathbf{r}_2|}r_1^2\,dr_1\,d\Omega_1\,r_2^2\,dr_2\,d\Omega_2 \tag{7.151}$$

The ground-state wave function contains no angular dependence. Therefore, using equation 1.201c, we have

$$\nabla^2 U_0(\mathbf{r}) = \frac{1}{r}\frac{d^2}{dr^2}[rU_0(r)] \tag{7.152}$$

Referring to equation 7.146, after a straightforward computation, we obtain

$$2\int U_0(\mathbf{r})\hat{H}U_0(\mathbf{r})r^2\,dr\,d\Omega = \frac{2Z}{r_0}\left[\frac{\hbar^2}{2m}\frac{Z}{r_0} - 2e^2\right] \tag{7.153a}$$

Since r_0 is not variable (see Problem 14), we set

$$r_0 = \frac{\hbar^2}{me^2} \tag{7.144}$$

$$\Rightarrow \quad 2\int U_0(\mathbf{r})\hat{H}U_0(\mathbf{r})r^2\,dr\,d\Omega = \frac{e^2}{r_0}Z(Z-4) \tag{7.153b}$$

To evaluate the electron–electron interaction term, we write

$$\frac{1}{|\mathbf{r}_1 - \mathbf{r}_2|} = \frac{1}{\sqrt{r_1^2 + r_2^2 - 2r_1r_2\cos\Theta}} \tag{7.154}$$

where Θ is the angle between $\mathbf{r}_1$ and $\mathbf{r}_2$. Referring to equations 6.188, the interaction potential energy can be expressed as

$$\frac{e^2}{|\mathbf{r}_1 - \mathbf{r}_2|} = e^2\sum_{l=0}^{\infty}\frac{r_2^l}{r_1^{l+1}}P_l(\cos\Theta) \qquad r_2 < r_1 \tag{7.155a}$$

$$\frac{e^2}{|\mathbf{r}_1 - \mathbf{r}_2|} = e^2\sum_{l=0}^{\infty}\frac{r_1^l}{r_2^{l+1}}p_l(\cos\Theta) \qquad r_1 < r_2 \tag{7.155b}$$

From the addition theorem for the Legendre polynomials, as expressed in equation 6.334,

$$P_l(\cos\Theta) \tag{6.334}$$

$$= P_l(\cos\theta_1)P_l(\cos\theta_2) + 2\sum_{m=1}^{l}\frac{(l-m)!}{(l+m)!}P_l^m(\cos\theta_1)P_l^m(\cos\theta_2)\cos[m(\phi_1 - \phi_2)]$$

Integrating over the ϕ variables, we have

$$\int_0^{2\pi}\int_0^{2\pi}d\phi_1\,d\phi_2\cos[m(\phi_1 - \phi_2)] = 4\pi^2\delta_{m0} \tag{7.156a}$$

Thus there is no contribution to the interaction energy from the summation over m in equation 6.334. Therefore, the integrals over θ_1 and θ_2 are

$$\int_0^{\pi}\int_0^{\pi}\sin\theta_1\,d\theta_1\sin\theta_2\,d\theta_2\,P_l(\cos\theta_1)P_l(\cos\theta_2)$$

$$= \int_{-1}^{1}d(\cos\theta_1)P_l(\cos\theta_1)P_0(\cos\theta_1)\int_{-1}^{1}d(\cos\theta_2)P_l(\cos\theta_2)P_0(\cos\theta_2) = 4\delta_{l0} \tag{7.156b}$$

Using equations 7.155, the interaction energy can be evaluated as

$$\int U_0^2(\mathbf{r}_1)U_0^2(\mathbf{r}_2)\frac{e^2}{|\mathbf{r}_1 - \mathbf{r}_2|}r_1^2\,dr_1\,d\Omega_1 r_2^2\,dr_2\,d\Omega_2$$

$$= 16\pi^2e^2\int_0^{\infty}e^{-2Zr_1/r_0}r_1^2\,dr_1\left[\frac{1}{r_1}\int_0^{r_1}e^{-2Zr_2/r_0}r_2^2\,dr_2 + \int_0^{r_1}\frac{1}{r_2}e^{-2Zr_2/r_0}r_2^2\,dr_2\right] = \frac{5}{8}\frac{Ze^2}{r_0} \tag{7.157}$$

Thus the energy of the ground state of helium is found by minimizing

$$E = \frac{e^2}{r_0}\left[Z(Z-4) + \tfrac{5}{8}Z\right] \tag{7.158}$$

with respect to Z. Setting $dE/dz = 0$, we find that the Z value for minimum E is $Z = \frac{27}{16}$

(which is slightly less than 2). This yields a ground-state energy

$$E = -\left(\frac{27}{16}\right)^2 \frac{e^2}{r_0} = -2.848\frac{e^2}{r_0} \tag{7.159a}$$

The measured energy of the electrons in the ground state of helium is

$$E = -2.904\,\frac{e^2}{r_0} \quad \square \tag{7.159b}$$

7.6 Estimating the Solution to a Differential Equation

An estimate to the solution of a certain type of second-order differential equations can also be obtained using a variational method. Consider a differential equation of the form

$$P(x)y'' + Q(x)y' + E(x,y) = 0 \tag{7.160}$$

As discussed in Chapter 6, this can be cast into self-adjoint form

$$\frac{d}{dx}[p(x)y'] + F(x,y) = 0 \tag{7.161}$$

by multiplying the equation by the factor $e^{-\int(Q/P)dx}$.

We have seen that we obtain a second-order differential equation by extremizing the integral $\int_{x_0}^{x_1} L(x, y, y')\,dx$. In order to employ variational methods to find a solution to the differential equation, we attempt to determine the simplest function $L(x, y, y')$ such that by setting

$$\delta \int_{x_0}^{x_1} L(x,y,y')\,dx = 0 \tag{7.162}$$

we obtain the differential equation we are trying to solve.

We have seen that if equation 7.162 is satisfied, $L(x, y, y')$ is a solution to the Euler–Lagrange equation

$$\frac{d}{dx}\left(\frac{\partial L}{\partial y'}\right) - \frac{\partial L}{\partial y} = 0 \tag{7.36}$$

Comparing this to equation 7.161, we can equate

$$\frac{\partial L}{\partial y'} = p(x)y' \tag{7.163}$$

$$\Rightarrow \quad L = \tfrac{1}{2}p(x)y'^2 + \Phi(x,y) \tag{7.164}$$

We then equate

$$\frac{\partial L}{\partial y} = \frac{\partial \Phi}{\partial y} = -F(x,y) \tag{7.165}$$

$$\Rightarrow \quad \Phi(x,y) = -\int F(x,y)\partial y + G(x) \tag{7.166}$$

Thus with

$$L(x,y,y') = \tfrac{1}{2}p(x)y'^2 - \int F(x,y)\partial y + G(x) \tag{7.167}$$

the Euler–Lagrange equation becomes the differential equation being investigated.

It is clear from Euler's equation that $G(x)$ is not determinable and is therefore arbitrary. As such, it is chosen to be zero.

The solution to the differential equation is estimated by selecting a trial function that satisfies the boundary conditions and also extremizes

$$S = \int_{x_0}^{x_1} L(x, y, y')\, dx = \int_{x_0}^{x_1} \left[\frac{1}{2} p(x) y'^2 - \int F(x, y) \partial y \right] dx \tag{7.168}$$

Example 7.11

As an example, consider the equation

$$\text{(a)} \quad y'' + y + x = 0 \quad \text{with} \quad \text{(b)} \quad y(0) = y(1) = 0 \quad x \in [0, 1] \tag{7.169}$$

Using standard techniques developed in Chapter 5, the solution is obtained straightforwardly to be

$$y(x) = -x + \frac{\sin(x)}{\sin(1)} \tag{7.170}$$

Comparing this differential equation to equation 7.161, we see that

$$\text{(a)} \quad p(x) = 1 \qquad \text{(b)} \quad F(x, y) = x + y \tag{7.171}$$

$$\Rightarrow \quad L(x, y, y') = \tfrac{1}{2}\left(y'^2 - y^2 - 2xy\right) \tag{7.172}$$

A simple trial function that satisfies the required boundary conditions is

$$y(x) = ax(1 - x) \tag{7.173}$$

where a is the variable parameter. Substituting this into equation 7.172 yields

$$S = \tfrac{1}{2} \int_0^1 \left[a^2(1 - 2x)^2 - a^2x^2(1 - x)^2 - ax^2(1 - x) \right] dx = \tfrac{3}{20}a^2 - \tfrac{1}{12}a \tag{7.174}$$

Thus S is extremized in the usual way by setting

$$\text{(a)} \quad \frac{dS}{da} = \tfrac{3}{10}a - \tfrac{1}{12} = 0 \quad \Rightarrow \quad \text{(b)} \quad a = \tfrac{5}{18} \tag{7.175}$$

$$\Rightarrow \quad y(x) = \tfrac{5}{18}x(1 - x) \tag{7.176}$$

In Table 7.1 we present the values of the exact and estimated solutions at selected values of x between 0 and 1. As can be seen, the agreement is of the same order of magnitude but is not satisfactory at most of the points.

An improved solution is obtained by considering the two-parameter trial function

$$y(x) = ax(1 - x) + bx^2(1 - x) \tag{7.177}$$

$$\Rightarrow \quad S = \tfrac{9}{30}a^2 + \tfrac{13}{105}b^2 + \tfrac{9}{30}ab - \tfrac{1}{6}a - \tfrac{1}{10}b \tag{7.178}$$

TABLE 7.1 COMPARISON OF EXACT AND ONE-PARAMETER TRIAL SOLUTIONS

x	y_{exact}	y_{trial}	Percent difference
0.1	0.0186	0.0250	34.1
0.2	0.0361	0.0444	23.1
0.3	0.0512	0.0583	13.9
0.4	0.0628	0.0667	6.2
0.5	0.0697	0.0694	0.4
0.6	0.0710	0.0667	6.1
0.7	0.0656	0.0583	11.1
0.8	0.0525	0.0444	15.3
0.9	0.0309	0.0250	19.1

Setting $\partial S/\partial a = 0$ and separately $\partial S/\partial b = 0$, we obtain

$$\text{(a)}\quad 18a - 9b = 5 \qquad \text{(b)}\quad 63a + 52b = 21 \tag{7.179}$$

from which we find $a = 0.1924$ and $b = 0.1707$. That is, the improved trial solution is

$$y(x) = 0.1924x(1-x) + 0.1707x^2(1-x) \tag{7.180}$$

Table 7.2 lists the values of the exact solution and this two-parameter trial solution at selected values of x. As can be seen, this is a much better estimate of the solution than the one-parameter trial function. Another indication that this two-parameter trial function approximates the solution well is that the trial function has a maximum at $x = 0.5720$, whereas the maximum of the exact solution occurs at $x = 0.5708$.

TABLE 7.2 COMPARISON OF EXACT AND TWO-PARAMETER TRIAL SOLUTIONS

x	y_{exact}	y_{trial}	Percent difference
0.1	0.0186	0.0189	1.1
0.2	0.0361	0.0362	0.4
0.3	0.0512	0.0512	0.1
0.4	0.0628	0.0626	0.3
0.5	0.0697	0.0694	0.4
0.6	0.0710	0.0708	0.4
0.7	0.0656	0.0655	0.1
0.8	0.0525	0.0526	0.2
0.9	0.0309	0.0311	0.8

Of course, the trial solution of equation 7.177 is not the only two-parameter function one can use. With an identical analysis one can show that

$$y = ax(1-x) + bx^2(1-x)^2 \tag{7.181}$$

leads to $a = 0.2731$ and $b = 0.0234$. The resulting values of this trial function are essentially the same poor ones as those obtained with the one-parameter fit displayed in Table 7.1. Without the exact solution for comparison, one must resort to other methods to deduce whether one particular trial solution is better than another. In this example we obtained the same values for a one-parameter and a two-parameter solution. This illustrates that even though the two-trial functions yield the same solution, indicating stability of the method, that does not mean that we have found the correct solution.

One approach to ascertaining that our solution is the correct one is a substitution of the trial solution back into the differential equation. In Table 7.3 we present the values of

TABLE 7.3 COMPARISON OF DIFFERENTIAL EQUATION FOR TWO DIFFERENT TWO-PARAMETER TRIAL SOLUTIONS

	$y'' + y + x$ with:	
x	$y = 0.1924x(1-x) + 0.1707x^2(1-x)$	$y = 0.2731x(1-x) + 0.0234x^2(1-x)^2$
0.1	−0.0270	−0.4467
0.2	−0.0120	−0.3468
0.3	0.0005	−0.2468
0.4	0.0095	−0.1467
0.5	0.0139	−0.0467
0.6	0.0128	0.0533
0.7	0.0052	0.1532
0.8	−0.0101	0.2532
0.9	−0.0340	0.3533

the differential equation of our example in the form

$$y'' + y + x = 0 \tag{7.169a}$$

for the two two-parameter trial solutions. As can be seen, the values of $y'' + y + x$ are closer to zero using the trial solution of equation 7.180 than those obtained with the trial function of equation 7.181. This indicates that for this differential equation, one can obtain improvements to the solution by considering a trial function of the form

$$y(x) = x(1-x) \sum_{n=0}^{N} c_n x^n \tag{7.182}$$

rather than using a trial function of the form

$$y(x) = \sum_{n=1}^{N} c_n x^n (1-x)^n \quad \square \tag{7.183}$$

Sturm – Liouville equation

As a final example, we consider the Sturm–Liouville equation in the form

$$\frac{d}{dx}[p(x)y'] + [\lambda\rho(x) - q(x)]y = 0 \tag{7.184}$$

Comparing the first term with $\frac{d}{dx}(\partial L/\partial y')$ and the second with $-\partial L/\partial y$, we obtain

$$L(x, y, y') = \tfrac{1}{2}\{p(x)y'^2 - [\lambda\rho(x) - q(x)]y^2\} \tag{7.185}$$

$$\Rightarrow \quad S = \tfrac{1}{2}\int_{x_0}^{x_1}\{p(x)y'^2 - [\lambda\rho(x) - q(x)]y^2\}\,dx \tag{7.186}$$

The variable parameters in the trial solution are then adjusted to extremize S.

For the Legendre differential equation, $p(x) = (1 - x^2)$, $q(x) = 0$, $\lambda = l(l+1)$, and $\rho(x) = 1$. Therefore, the estimate of the Legendre polynominal is obtained by extremizing

$$S = \tfrac{1}{2}\int_{-1}^{1}\left[(1-x^2)y'^2 - l(l+1)y^2\right]dx \tag{7.187}$$

For example, to estimate $P_2(x)$, we set $l = 2$. Using the knowledge that the second Legendre polynominal contains only even powers of x, we estimate the solution as

$$y_{\text{trial}} = a + bx^2 \tag{7.188}$$

The boundary conditions $y(\pm 1) = 1$ requires $a + b = 1$, so there is one independent variable parameter.

With $l = 2$, this trial function yields

$$S = -\tfrac{4}{3}b^2 + 4b - 3 \tag{7.189}$$

Thus, setting $dS/db = 0$, we find that

$$\text{(a)} \quad b = \tfrac{3}{2} \quad \Rightarrow \quad \text{(b)} \quad a = -\tfrac{1}{2} \tag{7.190}$$

$$\Rightarrow \quad y_{\text{trial}} = \tfrac{1}{2}(3x^2 - 1) \tag{7.191}$$

which is exactly $P_2(x)$.

PROBLEMS

1. Show that the shortest distance between two points in three dimensions is a straight line.
2. Find the equation $z = z(\phi)$ for the shortest distance between two points on the curved surface of a cylinder of radius ρ.
3. An optical medium has a continuously increasing index of refraction given by

$$n = \frac{c}{v}e^{\lambda y}$$

where y is the vertical depth of a point in the medium. $\lambda = 0$ for a vacuum and $\lambda > 0$ for an optically dense medium. Referring to Figure 7.37 if a light ray from a source at point p_0 in a vacuum enters such a medium at a point (x_a, y_a), arriving at a detector at point $p_1 = (x_b, y_b)$, what is the equation of the path of the light ray in the optically dense medium?

4. (a) Prove that if the slope of the curve $y = y(x)$ does not vary with path at the endpoints p_0 and p_1, the condition for the minimum of the integral

$$S = \int_{p_0}^{p_1} L(x, y, y', y'')dx$$

is

$$\frac{d^2}{dx^2}\left(\frac{\partial L}{\partial y''}\right) - \frac{d}{dx}\left(\frac{\partial L}{\partial y'}\right) + \frac{\partial L}{\partial y} = 0$$

(b) From the results of part (a), deduce the Euler equation for the minimum of

$$S = \int_{p_0}^{p_1} L(x, y, y', y'', \ldots, y^{(n)})\, dx$$

subject to the conditions that at the endpoints p_0 and p_1, $y, y', y'', \ldots, y^{(n-1)}$, are independent of the path taken, where

$$\left(y^{(k)} \equiv \frac{d^k y}{dx^k}\right)''$$

(c) Show that if $L(x, y, y', y'')$ does not depend on y, then

$$\frac{\partial L}{\partial y'} - \frac{d}{dx}\left(\frac{\partial L}{\partial y''}\right) = \text{constant}$$

and thus yields a first integral of the differential equation.

(d) Show that if $L(x, y, y', y'')$ does not depend on x, then

$$L - y'\left[\frac{\partial L}{\partial y'} - \frac{d}{dx}\left(\frac{\partial L}{\partial y''}\right)\right] = \text{constant}$$

and thus yields a first integral of the differential equation.

5. (a) Using Stokes's theorem, show that the area enclosed by a loop can be written

$$A = \tfrac{1}{2}\oint(x\,dy - y\,dx)$$

(b) From the result of part (a), show that the shape of the loop of fixed length that encloses the largest area is a circle.

6. Find the function $y(x)$ that extremizes the potential energy integral without constraints.

$$V = \int_{x_0}^{x_1} y\sqrt{1 + y'^2}\,dx$$

Note that this is the same integral that was considered for the rope strung between two fixed points. However, without constraints, there is no constant length. This, then, would describe the shape of a soap film or other extendable elastic medium.

7. Determine the functions $y_1(x)$ and $y_2(x)$ that extremize

$$S = \int_{x_0}^{x_1}\left[2y_1y_2 - 2y_1^2 + (y_1')^2 - (y_2')^2\right]dx$$

8. A string of fixed length l is to be formed in the shape of a rectangle of sides x and y. Using the method of Lagrange multipliers, determine the values of x and y such that the area of the rectangle is a maximum.
9. An ellipse has a semimajor axis of length a oriented along the x-axis. Its semiminor axis, of length b, is oriented along the y-axis. The points on the ellipse are thus constrained by the equation

$$\frac{x^2}{a^2} + \frac{y^2}{b^2} = 1$$

What are the lengths of the sides of the largest rectangle that can be inscribed inside this ellipse?

10. An Atwood machine is constructed by connecting masses m_1 and m_2 together by a wire of fixed length l that passes over an ideal pulley as shown in Figure 7.8.

(a) Write the constraint (that the wire has a fixed length l) as an integral

$$\int_{t_0}^{t_1} M(y_1, y_2, t)dt$$

where $y_1(t)$ and $y_2(t)$ describe the vertical positions of the two masses. Construct a modified

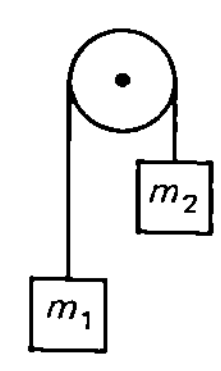

Figure 7.8 Atwood machine.

Lagrangian involving a Lagrange multiplier and obtain the equations of motion for the two masses.

(b) Obtain solutions to the equations of motion. Eliminate the Lagrange multiplier from the solutions to obtain a single equation relating $y_1(t)$ and $y_2(t)$.

11. Using Lagrange's equations of motion, show that a particle moving in a two-dimensional force free space moves in a straight line.

12. For the eigenvalue problem

$$\left[(1-x^2)D^2 - 2xD - 0.01x^2\right]Y = -\Lambda Y$$

(a) Show that the trial function

$$Y(x) = P_0(x) + a[P_3(x) - P_1(x)]$$

cannot be used to obtain an estimate of the lowest eigenvalue.

(b) Obtain an estimate of the eigenvalue $\Lambda \simeq l(l+1)$ that is closest to 2 ($l = 1$) using a trial function

$$Y(x) = P_1(x) + a[P_3(x) - P_1(x)]$$

13. Find an estimate of the lowest eigenvalue of the operator

$$\hat{\Omega} = (1-x^2)D^2 - 3xD$$
$$= (1-x^2)D^2 - (2x+x)D$$

subject to $y(\pm 1) = \pm 1$. Note that $\hat{\Omega}$ is almost a Legendre operator.

14. Consider a treatment of the helium atom in which the argument is made that the presence of another electron causes the ground-state radius of each electron's orbit to be different from the Bohr radius $\hbar^2/me^2 \equiv R_0$. Thus, by letting r_0 be variable but keeping the proton number $Z = z$ constant, find an estimate of the ground-state energy of helium.

15. Estimate the energy of the one-dimensional quantum harmonic oscillator that satisfies the Schrödinger equation

$$-\frac{\hbar^2}{2m}\frac{d^2\Psi}{dx^2} + \frac{1}{2}m\omega^2x^2\Psi = E\Psi$$

using the trial wave function

$$\Psi = \frac{1}{1+\beta x^2}$$

Let β be the variable parameter. The exact value for the energy of the ground state is $E_0 = \frac{1}{2}\hbar\omega$.

16. Estimate the solution to

$$y'' - 3y' + 2y = x$$

subject to $y(0) = y'(0) = 0$, and $y \in [0, 1]$. Take the one-parameter function $y = ax^2$ as a trial solution. Compare your results to the exact solution $y(x) = \frac{1}{4}e^{2x} - e^x + \frac{1}{2}x + \frac{3}{4}$ at $x = 0.25$, 0.50, and 0.75.

17. The associated Legendre function $P_2^2(x) = 3(1 - x^2)$. Show that if one chooses a trial function that is exact [i.e., $y = A(1-x^2)$, where A is a multiplicative normalizing constant], the integral

$$S = \frac{1}{2}\int_{x_0}^{x_1}\{p(x)y'^2 - [\lambda\rho(x) - q(x)]y^2\}\,dx = 0$$

for the $l = m = 2$ associated Legendre equation. (*Note*: This is a general result. If the exact solution to the Euler–Lagrange equation is chosen as a trial function, then, in general, $S = \int_{x_0}^{x_1} L\,dx = 0$.)

18. The associated Legendre function $P_1^1(x) = (1 - x^2)^{1/2}$. The trial function

$$Y(x) = (1-x^2)(1+bx^2)$$

satisfies the same boundary conditions as $P_1^1(x)$. That is, $Y(\pm 1) = P_1^1(\pm 1) = 0$, and $Y(0) = P_1^1(0) = 1$. Use this trial function to obtain an estimate of $P_1^1(x)$. Compare the estimate to $P_1^1(x)$ at $x = 0.25$, 0.50, and 0.75.

CHAPTER 8

DETERMINANTS AND MATRICES

8.1 Determinants

The evaluation of a 3×3 determinant was introduced in Chapter 1 as a method of expressing cross products. Jacobian determinants of rank 2 and of rank 3 were multipliers of the differential area and volume elements in a transformation from one coordinate system to another. For both the 2×2 and 3×3 determinants, the value can be obtained by multiplication of elements along diagonals with an appropriate sign assignment. For a determinant of rank higher than 3, however, such a multiplication scheme yields incorrect results.

Evaluation of a determinant

The value of a determinant of rank $N \times N$ is described in terms of a Levi–Civita ε-symbol with N indices, $\varepsilon_{ijk\ldots lmn}$. The properties of the N-index symbol are those specified in Chapter 1 for the three-index symbol. That is

1. $\varepsilon_{ijk\ldots lmn} = 0$ if any two indices have the same value. For example, $\varepsilon_{12342756\ldots} = 0$ since the second and fifth indices are both 2.
2. The interchange of any two indices negates the ε-symbol. For example, $\varepsilon_{12497835} = -\varepsilon_{12897435}$ since the third and sixth indices have been interchanged.
3. $\varepsilon_{123456789\ldots N} \equiv +1$.

With this definition of the N-index Levi–Civita symbol, the value of a determinant of rank N is

$$|A| = \begin{vmatrix} a_{11} & a_{12} & a_{13} & \cdots & a_{1N} \\ a_{21} & a_{22} & a_{23} & \cdots & a_{2N} \\ \vdots & & & & \\ a_{N1} & a_{N2} & a_{N3} & \cdots & a_{NN} \end{vmatrix} = \begin{cases} \displaystyle\sum_{i,j,\ldots,m,n} \varepsilon_{ijk\cdots n} a_{1i} a_{2j} a_{3k} \cdots a_{Nn} \\ \displaystyle\sum_{i,j,\ldots,m,n} \varepsilon_{ijk\cdots n} a_{i1} a_{j2} a_{k3} \cdots a_{nN} \end{cases} \tag{8.1}$$

That is, when multiplying elements and summing, one can specify a particular row index for each element and sum over every column index for that element, or one can specify a particular column index for each element and sum over all row indices.

From equation 8.1 it is clear that if every element in the pth row was zero, then $a_{pj} = 0$ for every value of j, and the determinant would be zero. Similarly, if every element in the qth column were zero, the determinant would be zero because $a_{iq} = 0$ for every value of i.

As an example, we will evaluate a 3×3 determinant using equation 8.1. The reader will easily see that this expansion yields the correct result expressed in Chapter 1, equation 1.12b. Equation 8.1 for a 3×3 determinant becomes

$$|A| = \begin{vmatrix} a_{11} & a_{12} & a_{13} \\ a_{21} & a_{22} & a_{23} \\ a_{31} & a_{32} & a_{33} \end{vmatrix} = \sum_{i,j,k} \varepsilon_{ijk} a_{1i} a_{2j} a_{3k} = \varepsilon_{123} a_{11} a_{22} a_{33} + \varepsilon_{132} a_{11} a_{23} a_{32}$$

$$+ \varepsilon_{213} a_{12} a_{21} a_{33} + \varepsilon_{231} a_{12} a_{23} a_{31} + \varepsilon_{312} a_{13} a_{21} a_{32} + \varepsilon_{321} a_{13} a_{22} a_{31} \quad (8.2)$$

where we have omitted those terms containing ε-symbols that are zero because two indices have the same value. With $\varepsilon_{123} = +1$, permuting indices leads to $\varepsilon_{132} = -1$, $\varepsilon_{213} = -1$, $\varepsilon_{231} = +1$, $\varepsilon_{312} = +1$, and $\varepsilon_{321} = -1$. Therefore, as expressed in equation 1.12b,

$$|A| = a_{11}a_{22}a_{33} - a_{11}a_{23}a_{32} - a_{12}a_{21}a_{33} + a_{12}a_{23}a_{31} + a_{13}a_{21}a_{32} - a_{13}a_{22}a_{31} \quad (8.3)$$

Laplace expansion

It is clear that the method described above becomes extremely cumbersome for large determinants. An alternative to the basic expansion of equation 8.1 is the *Laplace expansion*. To describe this approach for a determinant of rank N, we define a determinant of rank $N - 1$ called the *minor* of the element a_{ij}.

The minor of a_{ij} is obtained by removing the ith row and the jth column from the original determinant. That is, the minor M_{ij} is defined by drawing lines through the ith row and the jth column of $|A|$, and omitting those elements that lie on those lines.

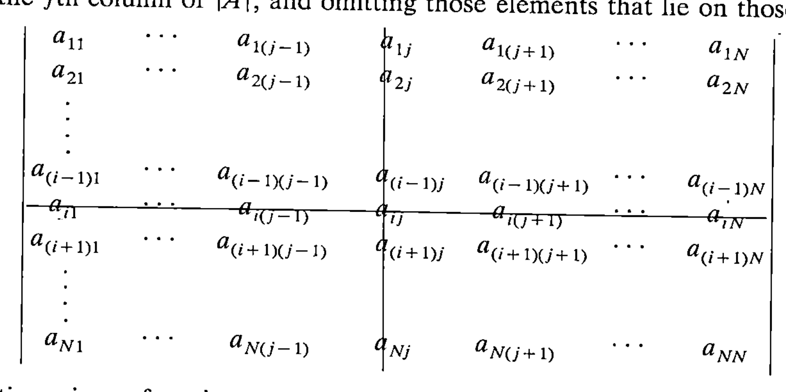

The resulting minor of a_{ij} is

$$M_{ij} = \begin{vmatrix} a_{11} & a_{12} & \cdots & a_{1(j-1)} & a_{1(j+1)} & \cdots & a_{1N} \\ a_{21} & a_{22} & \cdots & a_{2(j-1)} & a_{2(j+1)} & \cdots & a_{2N} \\ \vdots & & & & & & \\ a_{(i-1)1} & a_{(i-1)2} & \cdots & a_{(i-1)(j-1)} & a_{(i-1)(j+1)} & \cdots & a_{(i-1)N} \\ a_{(i+1)1} & a_{(i+1)2} & \cdots & a_{(i+1)(j-1)} & a_{(i+1)(j+1)} & \cdots & a_{(i+1)N} \\ \vdots & & & & & & \\ a_{N1} & a_{N2} & \cdots & a_{N(j-1)} & a_{N(j+1)} & \cdots & a_{NN} \end{vmatrix} \quad (8.4)$$

The *cofactor* of the element a_{ij} is also called the *signed minor* of a_{ij} and is defined by

$$\text{cof}(a_{ij}) = (-1)^{i+j} M_{ij} \tag{8.5}$$

The Laplace expansion of $|A|$, then, is

$$|A| = \sum_{j=1}^{N} a_{ij}\,\text{cof}(a_{ij}) \qquad \text{for any row index } i \tag{8.6a}$$

or

$$|A| = \sum_{i=1}^{N} a_{ij}\,\text{cof}(a_{ij}) \qquad \text{for any column index } j \tag{8.6b}$$

The expansion expressed in equation 8.6a is called expansion about the ith row. Equation 8.6b describes the Laplace expansion about the jth column. For determinants of rank 4 or higher, Laplace expansion is essential for evaluating the determinant.

Example 8.1

As an example, consider

$$|A| = \begin{vmatrix} 5 & 0 & 1 & 3 \\ 2 & 3 & 7 & 4 \\ 4 & 1 & 9 & 3 \\ 2 & 2 & 5 & 7 \end{vmatrix} \tag{8.7}$$

expanded about the third column. The expansion is

$$\begin{aligned} |A| &= 1 \cdot (-1)^{1+3} \begin{vmatrix} 2 & 3 & 4 \\ 4 & 1 & 3 \\ 2 & 2 & 7 \end{vmatrix} + 7 \cdot (-1)^{2+3} \begin{vmatrix} 5 & 0 & 3 \\ 4 & 1 & 3 \\ 2 & 2 & 7 \end{vmatrix} \\ &\quad + 9 \cdot (-1)^{3+3} \begin{vmatrix} 5 & 0 & 3 \\ 2 & 3 & 4 \\ 2 & 2 & 7 \end{vmatrix} + 5 \cdot (-1)^{4+3} \begin{vmatrix} 5 & 0 & 3 \\ 2 & 3 & 4 \\ 4 & 1 & 3 \end{vmatrix} \end{aligned} \tag{8.8}$$

The resulting 3×3 determinants are evaluated by multiplying elements along diagonals, yielding the result $|A| = 355$. □

To indicate the validity of the Laplace expansion for a determinant of rank N, we will consider an arbitrary 3×3 determinant as a specific example and generalize these results.

From equation 8.1, the 3×3 determinant can be expressed as

$$|A| = \sum_{ijk=1}^{3} \varepsilon_{ijk} a_{1i} a_{2j} a_{3k} \tag{8.9}$$

There is no loss of generality in making a Laplace expansion about the first row. Thus it is claimed that equation 8.9 is identical to

$$|A| = \sum_{j=1}^{3} a_{1j}\,\text{cof}(a_{1j}) \tag{8.10}$$

However,

$$\operatorname{cof}(a_{11}) = (-1)^{1+1}\begin{vmatrix} a_{22} & a_{23} \\ a_{32} & a_{33} \end{vmatrix} = +1 \sum_{\alpha\beta \neq 1} \varepsilon_{\alpha\beta} a_{2\alpha} a_{3\beta} \tag{8.11a}$$

$$\operatorname{cof}(a_{12}) = (-1)^{1+2}\begin{vmatrix} a_{21} & a_{23} \\ a_{31} & a_{33} \end{vmatrix} = -1 \sum_{\alpha\beta \neq 2} \varepsilon_{\alpha\beta} a_{2\alpha} a_{3\beta} \tag{8.11b}$$

$$\operatorname{cof}(a_{13}) = (-1)^{1+3}\begin{vmatrix} a_{21} & a_{22} \\ a_{31} & a_{32} \end{vmatrix} = +1 \sum_{\alpha\beta \neq 3} \varepsilon_{\alpha\beta} a_{2\alpha} a_{3\beta} \tag{8.11c}$$

with $\varepsilon_{12} = \varepsilon_{23} = \varepsilon_{13} = +1$. Extending this analysis to a determinant of rank N,

$$\operatorname{cof}(a_{1j}) = (-1)^{1+j} \sum_{\alpha\beta \cdots \gamma \neq j} \varepsilon_{\alpha\beta \cdots \gamma} a_{2\alpha} a_{3\beta} \cdots a_{N\gamma} \tag{8.12}$$

Consider the terms in the sum of equations 8.11:

$$(-1)^{1+1}\varepsilon_{\alpha\beta} = \begin{cases} +\varepsilon_{23} = +1 = \varepsilon_{123} \\ +\varepsilon_{32} = -1 = \varepsilon_{132} \end{cases} \qquad \alpha, \beta \neq 1 \tag{8.13a}$$

$$(-1)^{1+2}\varepsilon_{\alpha\beta} = \begin{cases} -\varepsilon_{13} = -1 = \varepsilon_{213} \\ -\varepsilon_{31} = +1 = \varepsilon_{231} \end{cases} \qquad \alpha, \beta \neq 2 \tag{8.13b}$$

$$(-1)^{1+3}\varepsilon_{\alpha\beta} = \begin{cases} +\varepsilon_{12} = +1 = \varepsilon_{312} \\ +\varepsilon_{21} = -1 = \varepsilon_{321} \end{cases} \qquad \alpha, \beta \neq 3 \tag{8.13c}$$

Therefore, in general,

$$(-1)^{1+j}\varepsilon_{\alpha\beta \cdots \gamma} = \varepsilon_{j\alpha\beta \cdots \gamma} \qquad \alpha, \beta, \ldots, \gamma \neq j \tag{8.13d}$$

$$\Rightarrow \quad \operatorname{cof}(a_{1j}) = \sum_{\alpha\beta \cdots \gamma \neq j} \varepsilon_{j\alpha\beta \cdots \gamma} a_{2\alpha} a_{3\beta} \cdots a_{N\gamma} \tag{8.14}$$

Since the ε-symbol is zero, any of the Greek indices has the same value as j, the restriction indicated under the sum can be dropped. Therefore, with equation 8.14, the Laplace expansion of the generalization of equation 8.10 becomes

$$\sum_j a_{1j} \operatorname{cof}(a_{1j}) = \sum_{j\alpha\beta \cdots \gamma} \varepsilon_{j\alpha\beta \cdots \gamma} a_{1j} a_{2\alpha} a_{3\beta} \cdots a_{N\gamma} = |A| \tag{8.15}$$

As is evident from the discussion above, the Laplace expansion of a rank N determinant results in N determinants each of rank $N - 1$. It would be a much simpler process to evaluate the rank N determinant if there was only one resulting determinant of rank $N - 1$. This would occur if all but one of the elements of one row or of one column was zero. Then, expansion about that row or column would result in one nonzero term in the Laplace sum, leaving a single $(N - 1) \times (N - 1)$ determinant to evaluate. This can be accomplished by performing allowed operations on the rows and columns of the determinant. We develop some properties of determinants from which we will be able to ascertain these permitted operations.

1. The interchange of any two rows (or the interchange of any two columns) yields a determinant that is the negative of the original determinant. From the definition of equation 8.1,

$$|A| = \sum \varepsilon_{ijk \cdots n} a_{1i} a_{2j} a_{3k} \cdots a_{Nn} \tag{8.1}$$

Interchanging rows 1 and 2 results in the determinant

$$|B| = \sum \varepsilon_{ijk \cdots n} a_{2i} a_{1j} a_{3k} \cdots a_{Nn} \tag{8.16a}$$

Since multiplication of the elements is ordinary algebraic multiplication, the order in which the elements is written is immaterial. Thus the determinant of equation 8.16a can be written

$$|B| = \sum \varepsilon_{ijk \cdots n} a_{1j} a_{2i} a_{3k} \cdots a_{Nn} \tag{8.16b}$$

Since i and j are dummy indices, we can rename them. We let $j = i'$ and $i = j'$. Then, dropping the primes, this becomes

$$|B| = \sum \varepsilon_{jik \cdots n} a_{1i} a_{2j} a_{3k} \cdots a_{Nn} \tag{8.17a}$$

Interchanging i and j on the ε-symbol, we obtain

$$|B| = -\sum \varepsilon_{ijk \cdots n} a_{1i} a_{2j} a_{3k} \cdots a_{Nn} = -|A| \tag{8.17b}$$

2. From property 1 if any two rows are the same (or if any two columns are the same), the determinant is zero. If row_i is the same as row_j, an interchange of row_i and row_j leaves the determinant unchanged. But that interchange also negates the determinant. Thus if $\text{row}_i = \text{row}_j$, $|A| = -|A|$, or $|A| = 0$.

3. If a determinant is multiplied by a constant, it is the same as multiplying all the elements in any one row (or all the elements in any one column) by the constant. Using equation 8.1, we obtain

$$k|A| = k\sum \varepsilon_{ijk \cdots n} a_{1i} a_{2j} a_{3k} \cdots a_{Nn} = \sum \varepsilon_{ijk \cdots n} a_{1i}(ka_{2j}) a_{3k} \cdots a_{Nn} \tag{8.18}$$

The second sum is the defining sum for a determinant like $|A|$, but with each element of row_2 of $|A|$ multiplied by k.

A corollary to this property is that if all elements in any row (or all elements in any column) contain a common factor, that multiplier can be factored from the row (or column).

Example 8.2

For example, since all elements in the second column have a common factor 2,

$$|A| = \begin{vmatrix} 5 & 2 & 1 \\ 4 & 6 & 0 \\ 3 & 18 & 6 \end{vmatrix} = 2\begin{vmatrix} 5 & 1 & 1 \\ 4 & 3 & 0 \\ 3 & 9 & 6 \end{vmatrix} \tag{8.19a}$$

The factor 2 could then be multiplied into another row or column (row_1 for example). That is, $|A|$ is the same as

$$|A| = \begin{vmatrix} 10 & 2 & 2 \\ 4 & 3 & 0 \\ 3 & 9 & 6 \end{vmatrix} \tag{8.19b}$$

We note that every element of row_3 is a multiple of 3 which can be factored out of row_3 and, if necessary, multiplied into another row or column without changing the value of $|A|$. □

4. Let two $N \times N$ determinants have $N - 1$ rows (or columns) that are identical, and let one of the rows (columns) differ. Then the sum of the two determinants is a determinant, the rows (columns) of which are the rows (columns) that are common to the two constituent determinants. The row (column) of the combined determinant corresponding to the row (column) that is different in the constituents, is the sum of the elements in the differing rows (columns). For example, if the two determinants are of rank 3, then

$$|C| = \begin{vmatrix} a_{11} & a_{12} & a_{13} \\ a_{21} & a_{22} & a_{23} \\ a_{31} & a_{32} & a_{33} \end{vmatrix} + \begin{vmatrix} a_{11} & b_{12} & a_{13} \\ a_{21} & b_{22} & a_{23} \\ a_{31} & b_{32} & a_{33} \end{vmatrix} = \begin{vmatrix} a_{11} & (a_{12}+b_{12}) & a_{13} \\ a_{21} & (a_{22}+b_{22}) & a_{23} \\ a_{31} & (a_{32}+b_{32}) & a_{33} \end{vmatrix} \tag{8.20}$$

For two determinants of rank N, using equation 8.1,

$$\text{(a)} \quad |A| = \sum \varepsilon_{ij\cdots k} a_{1i} a_{2j} a_{3k} \cdots a_{Nk} \qquad \text{(b)} \quad |B| = \sum \varepsilon_{ij\cdots k} a_{1i} b_{2j} a_{3k} \cdots a_{Nl} \tag{8.21}$$

$$\Rightarrow \quad |A| + |B| = \sum \varepsilon_{ij\cdots k} a_{1i}(a_{2j} + b_{2j}) a_{3k} \cdots a_{Nl} \tag{8.21c}$$

which is a determinant with the elements of row_2 of $|A|$ added to the elements of row_2 of $|B|$, and all elements that $|A|$ and $|B|$ have in common are left unaffected.

5. A determinant is unchanged if a constant multiple of one row (column) is added to another row (column). (A multiple of a row *cannot* be added to a column.) Consider

$$|B| = \begin{vmatrix} a_{11} & a_{12} & a_{13} & \cdots & a_{1N} \\ a_{11} & a_{12} & a_{13} & \cdots & a_{1N} \\ a_{31} & a_{32} & a_{33} & \cdots & a_{3N} \\ \vdots & & & & \\ a_{i1} & a_{i2} & a_{i3} & \cdots & a_{iN} \\ \vdots & & & & \\ a_{N1} & a_{N2} & a_{N3} & \cdots & a_{NN} \end{vmatrix} = 0 \tag{8.22}$$

Therefore,

$$|A| = \begin{vmatrix} a_{11} & a_{12} & a_{13} & \cdots & a_{1N} \\ a_{21} & a_{22} & a_{23} & \cdots & a_{2N} \\ a_{31} & a_{32} & a_{33} & \cdots & a_{3N} \\ \vdots & & & & \\ a_{i1} & a_{i2} & a_{i3} & \cdots & a_{iN} \\ \vdots & & & & \\ a_{N1} & a_{N2} & a_{N3} & \cdots & a_{NN} \end{vmatrix} + K \begin{vmatrix} a_{11} & a_{12} & a_{13} & \cdots & a_{1N} \\ a_{11} & a_{12} & a_{13} & \cdots & a_{1N} \\ a_{31} & a_{32} & a_{33} & \cdots & a_{3N} \\ \vdots & & & & \\ a_{i1} & a_{i2} & a_{i3} & \cdots & a_{iN} \\ \vdots & & & & \\ a_{N1} & a_{N2} & a_{N3} & \cdots & a_{NN} \end{vmatrix} \tag{8.23a}$$

We multiply K into row_2 of the second determinant and note that only the second rows of these two determinants are different. Therefore, using property 4, we obtain

$$|A| = \begin{vmatrix} a_{11} & a_{12} & \cdots & a_{1N} \\ (a_{12} + Ka_{11}) & (a_{22} + Ka_{12}) & \cdots & (a_{2N} + Ka_{1N}) \\ a_{31} & a_{32} & \cdots & a_{3N} \\ a_{i1} & a_{i2} & \cdots & a_{iN} \\ a_{N1} & a_{N2} & \cdots & a_{NN} \end{vmatrix} \tag{8.23b}$$

That is, $|A|$ is unchanged when a constant multiple of row_1 is added to row_2. To denote such an operation we use the notation $\text{row}_2 \rightarrow \text{row}_2 + K\,\text{row}_1$.

Manipulation of elements

It is property 5 that allows us to manipulate determinants such that all but one specified element in a row or column is zero. The Laplace expansion of a $N \times N$ determinant around that row or column results in a single $(N-1) \times (N-1)$ determinant. We will illustrate the method by an example, with which we will indicate a systematic approach to the manipulations required.

Example 8.3

Using the determinant of Example 8.1,

$$|A| = \begin{vmatrix} 5 & 0 & 1 & 3 \\ 2 & 3 & 7 & 4 \\ 4 & 1 & 9 & 3 \\ 2 & 2 & 5 & 7 \end{vmatrix} \tag{8.7a}$$

we will use property 5 to manipulate this determinant until all elements of col_3 are zero except the $(4,3)$ element. Since we want to make column elements zero, we will be adding multiples of rows to other rows. Because we want the third column elements of rows 1, 2 and 3 to be zero, the systematic approach is to add multiples of row_4 to each of the other rows. That is, row_4 will become what we call the working row. The manipulations that yield all elements of col_3 zero except the $(4,3)$ element are

$$\text{(a)}\quad \text{row}_1 \to \text{row}_1 - \tfrac{1}{5}\,\text{row}_4 \qquad \text{(b)}\quad \text{row}_2 \to \text{row}_2 - \tfrac{7}{5}\,\text{row}_4 \qquad \text{(c)}\quad \text{row}_3 \to \text{row}_3 - \tfrac{9}{5}\,\text{row}_4 \tag{8.24}$$

$$\Rightarrow \quad |A| = \begin{vmatrix} \frac{23}{5} & -\frac{2}{5} & 0 & \frac{8}{5} \\ -\frac{4}{5} & \frac{1}{5} & 0 & -\frac{29}{5} \\ \frac{2}{5} & -\frac{13}{5} & 0 & -\frac{48}{5} \\ 2 & 2 & 5 & 7 \end{vmatrix} = \frac{1}{5^3} \begin{vmatrix} 23 & -2 & 0 & 8 \\ -4 & 1 & 0 & -29 \\ 2 & -13 & 0 & -48 \\ 2 & 2 & 5 & 7 \end{vmatrix} \tag{8.25}$$

This determinant is now expanded around column 3. We obtain

$$|A| = (-1)^{4+3}\frac{5}{5^3}\begin{vmatrix} 23 & -2 & 8 \\ -4 & 1 & -29 \\ 2 & -13 & -48 \end{vmatrix} \tag{8.26}$$

This 3×3 determinant is evaluated by multiplying elements along diagonals. We find $|A| = -\frac{1}{25}(-8875) = 355$. □

As is noted above, there is a considerable amount of work involved in reducing a determinant of rank N to a single $(N-1) \times (N-1)$ determinant by Laplace expansion. Once this is accomplished, of course, the resulting $(N-1) \times (N-1)$ determinant must be similarly manipulated to obtain a single $(N-2) \times (N-2)$ determinant. Clearly, for determinants of any appreciable size, (6×6 for example) such continued manipulations become prohibitive for human beings and computers are needed. However, it is possible to evaluate reasonably large determinants by hand by either of two methods; one is called triangularization, and the other is known as pivotal condensation.

Triangularization

Using property 5 it is possible to manipulate a determinant until it is in a *triangular form*. Using row_1 as the working row, the determinant can be cast into the form

$$|A| = \begin{vmatrix} a_{11} & a_{12} & a_{13} & \cdots & & a_{1N} \\ 0 & a_{22} & a_{23} & \cdots & & a_{2N} \\ 0 & 0 & a_{33} & \cdots & & a_{3N} \\ \vdots & & & & & \\ 0 & 0 & 0 & \cdots & 0 & a_{NN} \end{vmatrix} \tag{8.27a}$$

Alternatively, with row_N as the working row, the determinant can be cast into the form

$$|A| = \begin{vmatrix} a_{11} & 0 & 0 & \cdots & 0 \\ a_{21} & a_{22} & 0 & \cdots & 0 \\ a_{31} & a_{32} & a_{33} & \cdots & 0 \\ \vdots & & & & \\ a_{N1} & a_{N2} & a_{N3} & \cdots & a_{NN} \end{vmatrix} \tag{8.27b}$$

That is, the determinant can be manipulated until all elements either below or above the main diagonal are zero. A Laplace expansion of the *upper triangularized* determinant of equation 8.27a around the first column results in the single determinant

$$|A| = a_{11} \begin{vmatrix} a_{22} & a_{23} & a_{24} & \cdots & & a_{2N} \\ 0 & a_{33} & a_{34} & \cdots & & a_{3N} \\ 0 & 0 & a_{44} & \cdots & & a_{4N} \\ \vdots & & & & & \\ 0 & 0 & 0 & \cdots & 0 & a_{NN} \end{vmatrix} \tag{8.28a}$$

Expanding this determinant about the first column yields

$$|A| = a_{11}a_{22} \begin{vmatrix} a_{33} & a_{34} & a_{35} & \cdots & & a_{3N} \\ 0 & a_{44} & a_{45} & \cdots & & a_{4N} \\ 0 & 0 & a_{55} & \cdots & & a_{5N} \\ \vdots & & & & & \\ 0 & 0 & 0 & \cdots & 0 & a_{NN} \end{vmatrix} \tag{8.28b}$$

Continuing this process, we eventually obtain

$$|A| = a_{11}a_{22}a_{33} \cdots a_{NN} = \prod_{i=1}^{N} a_{ii} \tag{8.29}$$

Expansions of the *lower triangularized* form about the last column at each step of the process yields the same result. Thus, when cast in triangular form, the determinant is the product of its diagonal elements.

Example 8.4

To illustrate, we again use the determinant

$$|A| = \begin{vmatrix} 5 & 0 & 1 & 3 \\ 2 & 3 & 7 & 4 \\ 4 & 1 & 9 & 3 \\ 2 & 2 & 5 & 7 \end{vmatrix} \tag{8.7}$$

The elements of col_1 can be manipulated to zero using row_1 as a working row. The manipulations that achieve this are

(a) $\text{row}_2 \rightarrow \text{row}_2 - \frac{2}{5}\,\text{row}_1$ (b) $\text{row}_3 \rightarrow \text{row}_3 - \frac{4}{5}\,\text{row}_1$ (c) $\text{row}_4 \rightarrow \text{row}_4 - \frac{2}{5}\,\text{row}_1$ (8.30)

$$\Rightarrow \quad |A| = \begin{vmatrix} 5 & 0 & 1 & 3 \\ 0 & 3 & \frac{33}{5} & \frac{14}{5} \\ 0 & 1 & \frac{41}{5} & \frac{3}{5} \\ 0 & 2 & \frac{23}{5} & \frac{29}{5} \end{vmatrix} \tag{8.31}$$

Using row_2 as the working row, the operations

$$\text{(a)}\quad \text{row}_3 \to \text{row}_3 - \tfrac{1}{3}\,\text{row}_2 \qquad \text{(b)}\quad \text{row}_4 \to \text{row}_4 - \tfrac{2}{3}\,\text{row}_2 \tag{8.32}$$

$$\Rightarrow \quad |A| = \begin{vmatrix} 5 & 0 & 1 & 3 \\ 0 & 3 & \frac{33}{5} & \frac{14}{5} \\ 0 & 0 & 6 & -\frac{1}{3} \\ 0 & 0 & \frac{1}{5} & \frac{59}{15} \end{vmatrix} \tag{8.33}$$

With row_3 as the working row,

$$\text{row}_4 \to \text{row}_4 - \tfrac{1}{30}\,\text{row}_3 \tag{8.34}$$

$$\Rightarrow \quad |A| = \begin{vmatrix} 5 & 0 & 1 & 3 \\ 0 & 3 & \frac{33}{5} & \frac{14}{5} \\ 0 & 0 & 6 & -\frac{1}{3} \\ 0 & 0 & 0 & \frac{71}{18} \end{vmatrix} \tag{8.35}$$

Therefore, $|A| = 5 \cdot 3 \cdot 6 \cdot \frac{71}{18} = 355.$ □

Pivotal condensation

The method of *pivotal condensation* reduces an $N \times N$ determinant to a single $(N-1) \times (N-1)$ determinant without requiring any operations on rows and columns to make elements zero. Although any element will serve as the pivotal element (see Problem 5), the simplest way to apply the method is to use a_{11} as the pivotal element. As such, a_{11} must be nonzero. If a_{11} is zero, one must interchange the first row or column with another row or column so that the resulting $a_{11} \neq 0$.

With $a_{11} \neq 0$, consider

$$a_{11}^{N-1}|A| = a_{11}^{N-1} \begin{vmatrix} a_{11} & a_{12} & a_{13} & a_{14} & \cdots & a_{1N} \\ a_{21} & a_{22} & a_{23} & a_{24} & \cdots & a_{2N} \\ a_{31} & a_{32} & a_{33} & a_{34} & \cdots & a_{3N} \\ a_{41} & a_{42} & a_{43} & a_{44} & \cdots & a_{4N} \\ \vdots & & & & & \\ a_{N1} & a_{N2} & a_{N3} & a_{N4} & \cdots & a_{NN} \end{vmatrix} \tag{8.36a}$$

Multiplying one factor of a_{11} into each row of the determinant except the first row, we obtain

$$a_{11}^{N-1}|A| = \begin{vmatrix} a_{11} & a_{12} & a_{13} & a_{14} & \cdots & a_{1N} \\ a_{11}a_{21} & a_{11}a_{22} & a_{11}a_{23} & a_{11}a_{24} & \cdots & a_{11}a_{2N} \\ a_{11}a_{31} & a_{11}a_{32} & a_{11}a_{33} & a_{11}a_{34} & \cdots & a_{11}a_{3N} \\ a_{11}a_{41} & a_{11}a_{42} & a_{11}a_{43} & a_{11}a_{44} & \cdots & a_{11}a_{4N} \\ \vdots & & & & & \\ a_{11}a_{N1} & a_{11}a_{N2} & a_{11}a_{N3} & a_{11}a_{N4} & \cdots & a_{11}a_{NN} \end{vmatrix} \tag{8.36b}$$

Performing the operations

$$\text{(a)}\quad \text{row}_2 \to \text{row}_2 - a_{21}\text{row}_1 \qquad \text{(b)}\quad \text{row}_3 \to \text{row}_3 - a_{31}\text{row}_1$$

$$\text{(c)}\quad \text{row}_4 \to \text{row}_4 - a_{41}\text{row}_1 \cdots \text{(d)}\quad \text{row}_N \to \text{row}_N - a_{N1}\text{row}_1 \tag{8.37}$$

this becomes

$$
a_{11}^{N-1}|A| = \begin{vmatrix} a_{11} & a_{12} & a_{13} & \cdots & a_{1N} \\ 0 & (a_{11}a_{22} - a_{21}a_{12}) & (a_{11}a_{23} - a_{21}a_{13}) & \cdots & (a_{11}a_{2N} - a_{21}a_{1N}) \\ 0 & (a_{11}a_{32} - a_{31}a_{12}) & (a_{11}a_{33} - a_{31}a_{13}) & \cdots & (a_{11}a_{3N} - a_{31}a_{1N}) \\ 0 & (a_{11}a_{42} - a_{41}a_{12}) & (a_{11}a_{43} - a_{41}a_{13}) & \cdots & (a_{11}a_{4N} - a_{41}a_{1N}) \\ \vdots & & & & \\ 0 & (a_{11}a_{N2} - a_{N1}a_{12}) & (a_{11}a_{N3} - a_{N1}a_{13}) & \cdots & (a_{11}a_{NN} - a_{N1}a_{1N}) \end{vmatrix} \tag{8.38}
$$

We note that the elements in rows 2 through N and columns 2 through N can be written as 2×2 determinants. Therefore, expanding the determinant around the first column, we obtain

$$
a_{11}^{N-1}|A| = a_{11} \begin{vmatrix} \begin{vmatrix} a_{11} & a_{12} \\ a_{21} & a_{22} \end{vmatrix} & \begin{vmatrix} a_{11} & a_{13} \\ a_{21} & a_{23} \end{vmatrix} & \begin{vmatrix} a_{11} & a_{14} \\ a_{21} & a_{24} \end{vmatrix} & \cdots & \begin{vmatrix} a_{11} & a_{1N} \\ a_{21} & a_{2N} \end{vmatrix} \\ \begin{vmatrix} a_{11} & a_{12} \\ a_{31} & a_{32} \end{vmatrix} & \begin{vmatrix} a_{11} & a_{13} \\ a_{31} & a_{33} \end{vmatrix} & \begin{vmatrix} a_{11} & a_{14} \\ a_{31} & a_{34} \end{vmatrix} & \cdots & \begin{vmatrix} a_{11} & a_{1N} \\ a_{31} & a_{3N} \end{vmatrix} \\ \begin{vmatrix} a_{11} & a_{12} \\ a_{41} & a_{42} \end{vmatrix} & \begin{vmatrix} a_{11} & a_{13} \\ a_{41} & a_{43} \end{vmatrix} & \begin{vmatrix} a_{11} & a_{14} \\ a_{41} & a_{44} \end{vmatrix} & \cdots & \begin{vmatrix} a_{11} & a_{1N} \\ a_{41} & a_{4N} \end{vmatrix} \\ \vdots & & & & \\ \begin{vmatrix} a_{11} & a_{12} \\ a_{N1} & a_{N2} \end{vmatrix} & \begin{vmatrix} a_{11} & a_{13} \\ a_{N1} & a_{N3} \end{vmatrix} & \begin{vmatrix} a_{11} & a_{14} \\ a_{N1} & a_{N4} \end{vmatrix} & \cdots & \begin{vmatrix} a_{11} & a_{1N} \\ a_{N1} & a_{NN} \end{vmatrix} \end{vmatrix} \tag{8.39a}
$$

or

$$
|A| = \frac{1}{a_{11}^{N-2}} \begin{vmatrix} \begin{vmatrix} a_{11} & a_{12} \\ a_{21} & a_{22} \end{vmatrix} & \begin{vmatrix} a_{11} & a_{13} \\ a_{21} & a_{23} \end{vmatrix} & \begin{vmatrix} a_{11} & a_{14} \\ a_{21} & a_{24} \end{vmatrix} & \cdots & \begin{vmatrix} a_{11} & a_{1N} \\ a_{21} & a_{2N} \end{vmatrix} \\ \begin{vmatrix} a_{11} & a_{12} \\ a_{31} & a_{32} \end{vmatrix} & \begin{vmatrix} a_{11} & a_{13} \\ a_{31} & a_{33} \end{vmatrix} & \begin{vmatrix} a_{11} & a_{14} \\ a_{31} & a_{34} \end{vmatrix} & \cdots & \begin{vmatrix} a_{11} & a_{1N} \\ a_{31} & a_{3N} \end{vmatrix} \\ \begin{vmatrix} a_{11} & a_{12} \\ a_{41} & a_{42} \end{vmatrix} & \begin{vmatrix} a_{11} & a_{13} \\ a_{41} & a_{43} \end{vmatrix} & \begin{vmatrix} a_{11} & a_{14} \\ a_{41} & a_{44} \end{vmatrix} & \cdots & \begin{vmatrix} a_{11} & a_{1N} \\ a_{41} & a_{4N} \end{vmatrix} \\ \vdots & & & & \\ \begin{vmatrix} a_{11} & a_{12} \\ a_{N1} & a_{N2} \end{vmatrix} & \begin{vmatrix} a_{11} & a_{13} \\ a_{N1} & a_{N3} \end{vmatrix} & \begin{vmatrix} a_{11} & a_{14} \\ a_{N1} & a_{N4} \end{vmatrix} & \cdots & \begin{vmatrix} a_{11} & a_{1N} \\ a_{N1} & a_{NN} \end{vmatrix} \end{vmatrix} \tag{8.39b}
$$

That is, determinant of rank N can be reduced to a single $(N-1) \times (N-1)$ determinant without performing any manipulations (other than assuring that $a_{11} \neq 0$) by generating elements of the reduced determinant in the form of 2×2 determinants.

To illustrate the pattern one must follow to form these 2×2 determinants, we write $|A|$ with the elements used to generate the elements of the reduced determinant surrounded by a box. For example, the $(1, 1)$ element of the reduced determinant is obtained by the 2×2 determinant containing a_{11}, a_{12}, a_{21}, and a_{22}, in the pattern shown in Figure 8.1.

For the $(1, 2)$ element of the reduced determinant, we take the elements in the pattern shown in Figure 8.2. The $(2, 1)$ element of the reduced determinant is obtained using the pattern shown in Figure 8.3 and the $(3, 2)$ element of the reduced determinant is found using the pattern shown in Figure 8.4.

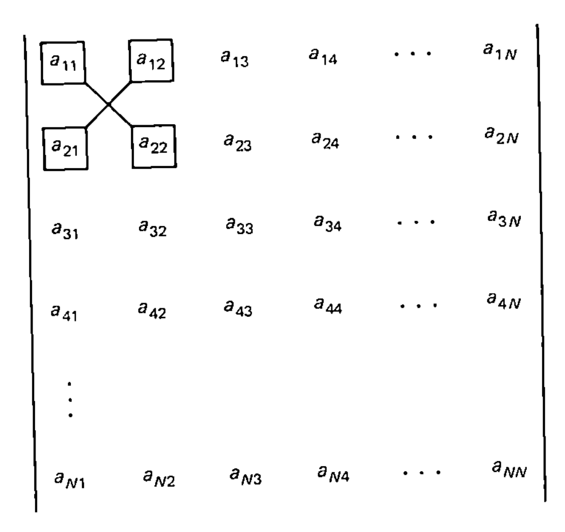

Figure 8.1 Elements and pattern of 2×2 determinant for (1, 1) element of reduced determinant.

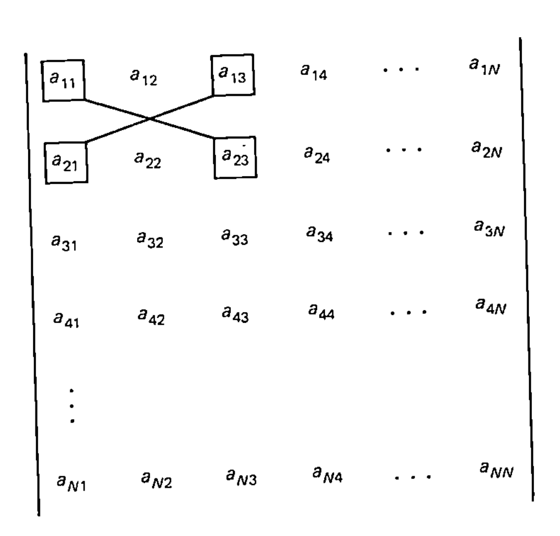

Figure 8.2 Elements and pattern of 2×2 determinant for (1, 2) element of reduced determinant.

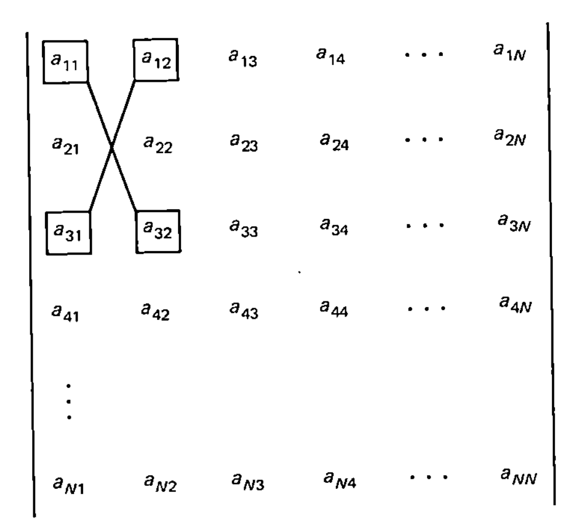

Figure 8.3 Elements and pattern of 2×2 determinant for (2, 1) element of reduced determinant.

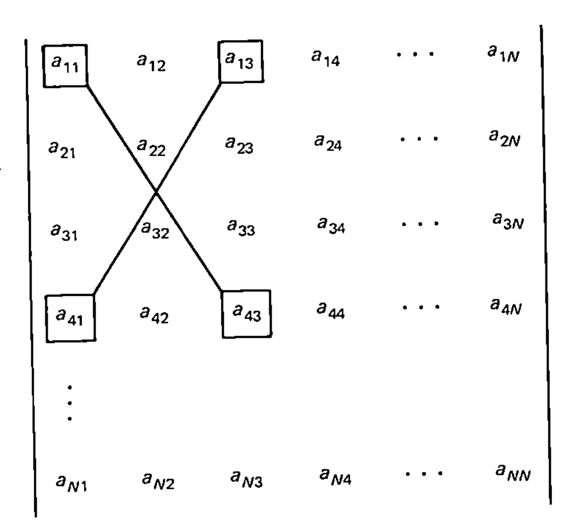

Figure 8.4 Elements and pattern of 2×2 determinant for (3, 2) element of reduced determinant.

The key to the ease of the method of pivotal condensation depends on the user becoming familiar with the pattern for generating the elements of the reduced determinant.

Example 8.5

As an example, we will again consider the determinant

$$|A| = \begin{vmatrix} 5 & 0 & 1 & 3 \\ 2 & 3 & 7 & 4 \\ 4 & 1 & 9 & 3 \\ 2 & 2 & 5 & 7 \end{vmatrix} \tag{8.7a}$$

We note that the pivotal element, $a_{11} = 5$, is not zero. Using pivotal condensation, $|A|$ can be reduced to the 3×3 determinant

$$|A| = \frac{1}{5^{4-2}} \begin{vmatrix} \begin{vmatrix} 5 & 0 \\ 2 & 3 \end{vmatrix} & \begin{vmatrix} 5 & 1 \\ 2 & 7 \end{vmatrix} & \begin{vmatrix} 5 & 3 \\ 2 & 4 \end{vmatrix} \\ \begin{vmatrix} 5 & 0 \\ 4 & 1 \end{vmatrix} & \begin{vmatrix} 5 & 1 \\ 4 & 9 \end{vmatrix} & \begin{vmatrix} 5 & 3 \\ 4 & 3 \end{vmatrix} \\ \begin{vmatrix} 5 & 0 \\ 2 & 2 \end{vmatrix} & \begin{vmatrix} 5 & 1 \\ 2 & 5 \end{vmatrix} & \begin{vmatrix} 5 & 3 \\ 2 & 7 \end{vmatrix} \end{vmatrix} = \frac{1}{25} \begin{vmatrix} 15 & 33 & 14 \\ 5 & 41 & 3 \\ 10 & 23 & 29 \end{vmatrix} \tag{8.40a}$$

The resulting 3×3 determinant can, of course, be evaluated by multiplication of elements along diagonals. However, as a further illustration of pivotal condensation, we will reduce it to a 2×2 determinant before obtaining the value.

$$|A| = \frac{1}{25} \begin{vmatrix} 15 & 33 & 14 \\ 5 & 41 & 3 \\ 10 & 23 & 29 \end{vmatrix} = \frac{1}{25} \frac{1}{15^{3-2}} \begin{vmatrix} \begin{vmatrix} 15 & 33 \\ 5 & 41 \end{vmatrix} & \begin{vmatrix} 15 & 14 \\ 5 & 3 \end{vmatrix} \\ \begin{vmatrix} 15 & 33 \\ 10 & 23 \end{vmatrix} & \begin{vmatrix} 15 & 14 \\ 10 & 29 \end{vmatrix} \end{vmatrix} = \frac{1}{375} \begin{vmatrix} 450 & -25 \\ 15 & 295 \end{vmatrix} = 355 \quad \square \tag{8.40b}$$

Cramer's rule

A set of N linear equations in N unknowns can be solved by *Cramer's rule*, which relates the solutions to ratios of determinants.

Consider the set of N simultaneous, inhomogeneous linear equations

$$a_{11}x_1 + a_{12}x_2 + \cdots + a_{1N}x_N = c_1 \tag{8.41a}$$

$$a_{21}x_1 + a_{22}x_2 + \cdots + a_{2N}x_N = c_2 \tag{8.41b}$$

$$\vdots$$

$$a_{N1}x_1 + a_{N2}x_2 + \cdots + a_{NN}x_N = c_N \tag{8.41c}$$

where $\{x_i\}$ is the set of unknowns. The coefficients a_{ij} and the inhomogeneous terms c_i are all known.

Let $|A|$ be the determinant of the coefficients. Then

$$x_1|A| = x_1 \begin{vmatrix} a_{11} & a_{12} & a_{13} & \cdots & a_{1N} \\ a_{21} & a_{22} & a_{23} & \cdots & a_{2N} \\ a_{31} & a_{32} & a_{33} & \cdots & a_{3N} \\ \vdots \\ a_{N1} & a_{N2} & a_{N3} & \cdots & a_{NN} \end{vmatrix} = \begin{vmatrix} a_{11}x_1 & a_{12} & a_{13} & \cdots & a_{1N} \\ a_{21}x_1 & a_{22} & a_{23} & \cdots & a_{2N} \\ a_{31}x_1 & a_{32} & a_{33} & \cdots & a_{3N} \\ \vdots \\ a_{N1}x_1 & a_{N2} & a_{N3} & \cdots & a_{NN} \end{vmatrix} \tag{8.42}$$

Recall that if any two columns of a determinant are identical, the determinant is zero. Using this, we add $N - 1$ zeros to $x_1|A|$ in the form

$$0 = \sum_{j=2}^{N} x_j \begin{vmatrix} a_{1j} & a_{12} & a_{13} & \cdots & a_{1N} \\ a_{2j} & a_{22} & a_{23} & \cdots & a_{2N} \\ a_{3j} & a_{32} & a_{33} & \cdots & a_{3N} \\ \vdots \\ a_{Nj} & a_{N2} & a_{N3} & \cdots & a_{NN} \end{vmatrix} \tag{8.43}$$

Since $j = 2, 3, \ldots,$ the first column of the jth determinant is identical to one of the other columns. Thus, each determinant is zero. We also note that columns 2 through N of each determinant are identical to columns 2 through N of $x_1|A|$. We multiply each x_j into the first column of the jth determinant and add this to $x_1|A|$. Using property 4 as illustrated in equation 8.20, we obtain

$$x_1|A| = \begin{vmatrix} \sum_{j=1}^{N} a_{1j}x_j & a_{12} & a_{13} & \cdots & a_{1N} \\ \sum_{j=1}^{N} a_{2j}x_j & a_{22} & a_{23} & \cdots & a_{2N} \\ \vdots & & & & \\ \sum_{j=1}^{N} a_{Nj}x_j & a_{N2} & a_{N3} & \cdots & a_{NN} \end{vmatrix} \tag{8.44}$$

Referring to equations 8.41, each sum in the first column is

$$\sum_{j=1}^{N} a_{ij}x_j = c_i \tag{8.45}$$

$$\Rightarrow \quad \text{(a)} \quad x_1|A| = \begin{vmatrix} c_1 & a_{12} & a_{13} & \cdots & a_{1N} \\ c_2 & a_{22} & a_{23} & \cdots & a_{2N} \\ \vdots & & & & \\ c_N & a_{N2} & a_{N3} & \cdots & a_{NN} \end{vmatrix} \equiv |A_1| \quad \Rightarrow \quad \text{(b)} \quad x_1 = \frac{|A_1|}{|A|} \tag{8.46}$$

This is easily generalized to any unknown. In the solution for x_i, the ith column of the coefficient determinant is replaced by the column of inhomogeneous terms to form the determinant $|A_i|$. Then Cramer's rule is

$$x_i = \frac{|A_i|}{|A|} \tag{8.47}$$

Example 8.6

As an example, we apply Cramer's rule to the set of equations

$$\text{(a)} \quad 5x_1 + 2x_2 + x_3 = 2 \qquad \text{(b)} \quad 2x_1 - 3x_2 + 3x_3 = 1 \qquad \text{(c)} \quad 7x_1 + 4x_2 - 2x_3 = -5 \tag{8.48}$$

The coefficient determinant is

$$|A| = \begin{vmatrix} 5 & 2 & 1 \\ 2 & -3 & 3 \\ 7 & 4 & -2 \end{vmatrix} = 49 \tag{8.49a}$$

and, by replacing column 1 by the column of inhomogeneous values, we obtain

$$|A_1| = \begin{vmatrix} 2 & 2 & 1 \\ 1 & -3 & 3 \\ -5 & 4 & -2 \end{vmatrix} = -49 \tag{8.49b}$$

Therefore, $x_1 = -1$. By identical analysis,

$$\text{(c)} \quad |A_2| = \begin{vmatrix} 5 & 2 & 1 \\ 2 & 1 & 3 \\ 7 & -5 & -2 \end{vmatrix} = 98 \qquad \text{(d)} \quad |A_3| = \begin{vmatrix} 5 & 2 & 2 \\ 2 & -3 & 1 \\ 7 & 4 & -5 \end{vmatrix} = 147 \tag{8.49}$$

so $x_2 = 2$ and $x_3 = 3$. □

If the set of equations is homogeneous (all $c_i = 0$), every $|A_i| = 0$. If the coefficient determinant is not zero, the only solution is the trivial one; that is, all $x_i = 0$. Therefore, the only way that a nontrivial solution to a set of homogeneous equations can exist is if the coefficient determinant is zero.

If the coefficient determinant is zero for a set of N homogeneous equations, the equations are not independent. That means that one of the equations can be obtained by combining two (or more) of the other equations, or one of the equations is a multiple of another equation. In that case, if a subset of $N - 1$ equations is independent (the coefficient determinant of the subset of equations is not zero), one can find a solution to the subset by Cramer's rule.

Example 8.7

As an example, consider

$$\text{(a)} \quad -5x_1 + 2x_2 - x_3 = 0 \qquad \text{(b)} \quad 3x_1 + x_2 + 5x_3 = 0 \qquad \text{(c)} \quad 3x_1 - x_2 + x_3 = 0 \tag{8.50}$$

Equation 8.50a is obtained by taking $\frac{1}{6}$ of equation 8.50b $-$ $\frac{11}{6}$ of equation 8.50c. Thus these equations are not independent. This is also evidenced by the fact that the coefficient determinant

$$|A| = \begin{vmatrix} -5 & 2 & -1 \\ 3 & 1 & 5 \\ 3 & -1 & 1 \end{vmatrix} = 0 \tag{8.51}$$

Taking equations 8.50b and 8.50c as the independent equations, they can be written in the form

$$\text{(a)} \quad \frac{x_2}{x_1} + 5\frac{x_3}{x_1} = -3 \qquad \text{(b)} \quad \frac{x_2}{x_1} - \frac{x_3}{x_1} = 3 \tag{8.52}$$

The new unknown quantities are $y_2 \equiv x_2/x_1$ and $y_3 \equiv x_3/x_1$, and the solution to equations 8.52 by Cramer's rule is

$$\text{(a)} \quad y_2 = \frac{\begin{vmatrix} -3 & 5 \\ 3 & -1 \end{vmatrix}}{\begin{vmatrix} 1 & 5 \\ 1 & -1 \end{vmatrix}} = 2 \quad \text{and} \quad \text{(b)} \quad y_3 = \frac{\begin{vmatrix} 1 & -3 \\ 1 & 3 \end{vmatrix}}{\begin{vmatrix} 1 & 5 \\ 1 & -1 \end{vmatrix}} = -1 \quad \square \tag{8.53}$$

8.2 Arithmetic of Matrices

Unlike a determinant, which is an array that represents a single value, or scalar quantity, a matrix is a two-dimensional array that represents an operator. It is not a measurable or evaluatable entity but is designed to "do something" to another entity.

It is written very much like the array that denotes a determinant. Whereas the determinant array is surrounded by "absolute value bars," the matrix array is surrounded by symbols that represent parentheses. Unlike a determinant, however, the number of rows of a matrix does not have to be the same as the number of columns. A general $M \times N$ matrix will be denoted with a boldface character and is of the form

$$\mathbf{A} = \begin{bmatrix} a_{11} & a_{12} & a_{13} & \cdots & a_{1N} \\ a_{21} & a_{22} & a_{23} & \cdots & a_{2N} \\ \vdots & & & & \\ a_{M1} & a_{M2} & a_{M3} & \cdots & a_{MN} \end{bmatrix} \tag{8.54}$$

Addition and subtraction

Let a matrix $\mathbf{C}$ be defined as the sum or difference of matrices $\mathbf{A}$ and $\mathbf{B}$. Then

$$\text{(a)} \quad \mathbf{C} = \mathbf{A} \pm \mathbf{B} \qquad \Rightarrow \qquad \text{(b)} \quad c_{ij} = a_{ij} \pm b_{ij} \tag{8.55}$$

Clearly, if the addition or subtraction of $\mathbf{A}$ and $\mathbf{B}$ is to have meaning, the number of rows of $\mathbf{A}$ must be the same as the number of rows of $\mathbf{B}$, and the number of columns of the two matrices must also be the same. For example,

$$\begin{bmatrix} 3 & 1 & 0 & 2 \\ 5 & 2 & 4 & 1 \\ 2 & -1 & 1 & 3 \end{bmatrix} + \begin{bmatrix} 2 & 3 & 1 & -1 \\ 0 & 1 & -2 & 2 \\ 4 & 3 & -1 & -2 \end{bmatrix} = \begin{bmatrix} 5 & 4 & 1 & 1 \\ 5 & 3 & 2 & 3 \\ 6 & 2 & 0 & 1 \end{bmatrix} \tag{8.56a}$$

whereas

$$\begin{bmatrix} 3 & 1 & 0 & 2 \\ 5 & 2 & 4 & 1 \\ 2 & -1 & 1 & 3 \end{bmatrix} + \begin{bmatrix} 2 & 3 & 1 \\ 0 & 1 & -2 \\ 4 & 3 & -1 \end{bmatrix} \tag{8.56b}$$

is not defined.

Multiplication by a scalar

Let K be a number and $\mathbf{A}$ be a matrix with elements denoted by a_{ij}. The meaning of $K\mathbf{A}$ is a matrix $\mathbf{B}$ with elements

$$b_{ij} = Ka_{ij} \tag{8.57}$$

That is, multiplication of a matrix by a scalar means multiplying every element of the matrix by the scalar.

Multiplication of matrices

Let the matrix $\mathbf{C}$ be defined as the product of $\mathbf{A}$ and $\mathbf{B}$ such that

$$\mathbf{C} = \mathbf{AB} \tag{8.58}$$

$$\Rightarrow \quad c_{ij} = \sum_{k=1}^{N} a_{ik} b_{kj} \tag{8.59}$$

Because k denotes the column index of the elements of $\mathbf{A}$ and the row index of the elements of $\mathbf{B}$, it is clear that if this product is to be defined, the number of columns of $\mathbf{A}$ and the number of rows of $\mathbf{B}$ must be the same. But, we note, there is no such restriction on the number of rows of $\mathbf{A}$ or the number of columns of $\mathbf{B}$.

The range of values of i is the number of rows of $\mathbf{A}$ and thus the number of rows of $\mathbf{C}$. Similarly, the range of j values designates the number of columns of $\mathbf{B}$ and therefore the number of columns of $\mathbf{C}$.

From this analysis, we can deduce the following: Let $\mathbf{A}$ be a $(M \times N)$ matrix and $\mathbf{B}$ be a $(P \times Q)$ matrix. The product $\mathbf{AB}$ is a matrix $\mathbf{C}$ that is $(M \times N) \cdot (P \times Q)$. This product is only defined if $N = P$. If $N = P$, then $\mathbf{C}$ is a $(M \times Q)$ matrix.

We also deduce from this that if $N = P$ but $M \neq Q$, the product **AB** is defined while the product **BA** is not. Taking $M = Q$, let

$$\text{(a)}\quad \mathbf{D} \equiv \mathbf{BA} \quad\Rightarrow\quad \text{(b)}\quad d_{ij} = \sum_{k=1}^{Q} b_{ik}a_{kj} \tag{8.60}$$

Since Q may be different from N, it is clear that $\mathbf{C} = \mathbf{AB}$ and $\mathbf{D} = \mathbf{BA}$ are not equal.

If the number of rows of a matrix is the same as the number of columns, the matrix is called a *square matrix*. In general, matrix multiplication is not commutative, even if the matrices being multiplied are square matrices (and, of course, are of the same order).

Example 8.8

As an example, consider the product of

$$\text{(a)}\quad \mathbf{A} = \begin{bmatrix} 3 & 0 & 1 \\ 7 & 2 & 0 \\ 1 & 1 & 2 \end{bmatrix} \quad \text{and} \quad \text{(b)}\quad \mathbf{B} = \begin{bmatrix} 5 & 3 & 2 \\ 0 & 3 & 7 \\ 1 & 0 & 1 \end{bmatrix} \tag{8.61}$$

We describe a practical method for finding the elements of the product matrix. The ijth element of the product **AB** can be obtained by laying the ith row of **A** next to the jth column of **B**, multiplying the adjacent elements, and adding the products. For example, the (1, 1) element of **AB** is obtained by placing the first row of **A** next to the first column of **B**. The resulting (1, 1) element is

$$\begin{array}{c} 3 \cdot 5 \\ + \\ 0 \cdot 0 \\ + \\ 1 \cdot 1 \\ \hline 16 \end{array}$$

The (2, 3) element of **AB** is obtained by placing column 2 of **A** next to column 3 of **B**, multiplying adjacent elements, and adding. The result is

$$\begin{array}{c} 7 \cdot 2 \\ + \\ 2 \cdot 7 \\ + \\ 0 \cdot 1 \\ \hline 28 \end{array}$$

Using this prescription,

$$\mathbf{AB} = \begin{bmatrix} 16 & 9 & 7 \\ 35 & 27 & 28 \\ 7 & 6 & 11 \end{bmatrix} \tag{8.62a}$$

Since **A** and **B** are square matrices, both **AB** and **BA** are defined. It is straightforward to show that

$$\mathbf{BA} = \begin{bmatrix} 39 & 8 & 9 \\ 28 & 13 & 14 \\ 4 & 1 & 3 \end{bmatrix} \neq \mathbf{AB} \quad \square \tag{8.62b}$$

Identity matrix

The $N \times N$ *identity matrix*, denoted by **1**, is defined with a 1 for each element on the diagonal, and 0 for each off-diagonal element. That is,

$$\text{(a)} \quad \mathbf{1} \equiv \begin{bmatrix} 1 & 0 & 0 & 0 & 0 & \cdots & 0 \\ 0 & 1 & 0 & 0 & 0 & \cdots & 0 \\ 0 & 0 & 1 & 0 & 0 & \cdots & 0 \\ \vdots & & & & & & \\ 0 & 0 & 0 & 0 & 0 & \cdots & 1 \end{bmatrix} \quad \Rightarrow \quad \text{(b)} \quad (\mathbf{1})_{ij} = \delta_{ij} \tag{8.63}$$

where δ_{ij} is the Kronecker δ-symbol. Consider

$$(\mathbf{1} \cdot \mathbf{A})_{ij} = \sum_{k=1}^{N} (\mathbf{1})_{ik} a_{kj} = \sum_{k=1}^{N} \delta_{ik} a_{kj} = a_{ij} \tag{8.64a}$$

That is, the product of the identity matrix with any square matrix **A** of the same rank results in the same matrix **A**. Postmultiplying **A** by **1** leads to the same result. Thus,

$$\mathbf{1} \cdot \mathbf{A} = \mathbf{A} \cdot \mathbf{1} = \mathbf{A} \tag{8.64b}$$

Arithmetic of partitioned matrices

It is sometimes convenient to express matrices in terms of elements that are themselves matrices. The rules of addition, subtraction, and multiplication of such partitioned matrices are the same as those for matrices in which the elements are algebraic quantities (numbers).

Let the matrices **A** and **B** be partitioned in the form

$$\text{(a)} \quad \mathbf{A} = \begin{bmatrix} \mathbf{A}_{11} & \mathbf{A}_{12} \\ \mathbf{A}_{21} & \mathbf{A}_{22} \end{bmatrix} \qquad \text{(b)} \quad \mathbf{B} = \begin{bmatrix} \mathbf{B}_{11} & \mathbf{B}_{12} \\ \mathbf{B}_{21} & \mathbf{B}_{22} \end{bmatrix} \tag{8.65}$$

where $\mathbf{A}_{ij}$ and $\mathbf{B}_{ij}$ are matrices. Then

$$\mathbf{A} \pm \mathbf{B} = \begin{bmatrix} \mathbf{A}_{11} \pm \mathbf{B}_{11} & \mathbf{A}_{12} \pm \mathbf{B}_{12} \\ \mathbf{A}_{21} \pm \mathbf{B}_{21} & \mathbf{A}_{22} \pm \mathbf{B}_{22} \end{bmatrix} \tag{8.66a}$$

Of course, the number of rows and separately the number of columns of the submatrices $\mathbf{A}_{ij}$ and $\mathbf{B}_{ij}$ must be the same so that the addition and subtraction are defined.

Multiplication of **A** and **B** becomes

$$\mathbf{AB} = \begin{bmatrix} \mathbf{A}_{11}\mathbf{B}_{11} + \mathbf{A}_{12}\mathbf{B}_{21} & \mathbf{A}_{11}\mathbf{B}_{12} + \mathbf{A}_{12}\mathbf{B}_{22} \\ \mathbf{A}_{21}\mathbf{B}_{11} + \mathbf{A}_{22}\mathbf{B}_{21} & \mathbf{A}_{21}\mathbf{B}_{12} + \mathbf{A}_{22}\mathbf{B}_{22} \end{bmatrix} \tag{8.66b}$$

Again, **A** and **B** must be partitioned so that the products of all submatrices are defined.

As an example, the Pauli spin matrices are the operators that describe the spin angular momentum of a nonrelativistic quantum particle with spin angular momentum $\frac{1}{2}\hbar$. These matrices are

$$\text{(a)} \quad \boldsymbol{\sigma}_x \equiv \boldsymbol{\sigma}_1 = \begin{bmatrix} 0 & 1 \\ 1 & 0 \end{bmatrix} \qquad \text{(b)} \quad \boldsymbol{\sigma}_y \equiv \boldsymbol{\sigma}_2 = \begin{bmatrix} 0 & -i \\ i & 0 \end{bmatrix} \qquad \text{(c)} \quad \boldsymbol{\sigma}_z \equiv \boldsymbol{\sigma}_3 = \begin{bmatrix} 1 & 0 \\ 0 & -1 \end{bmatrix} \tag{8.67}$$

In the relativistic quantum theory of such spin-$\frac{1}{2}$ particles, the matrix operators that arise in the Dirac equation can be written in partitioned form as

$$\text{(a)}\quad \boldsymbol{\alpha}_r = \begin{bmatrix} \mathbf{0} & \boldsymbol{\sigma}_r \\ \boldsymbol{\sigma}_r & \mathbf{0} \end{bmatrix} \quad \text{and} \quad \text{(b)}\quad \boldsymbol{\alpha}_4 = \begin{bmatrix} \mathbf{1} & \mathbf{0} \\ \mathbf{0} & -\mathbf{1} \end{bmatrix} \tag{8.68}$$

where $\mathbf{1}$, $\mathbf{0}$, and $\boldsymbol{\sigma}_r$ are the 2×2 unit matrix, null matrix, and rth Pauli matrix, respectively.

It is straightforward to show that

$$\boldsymbol{\sigma}_1\boldsymbol{\sigma}_2 = -\boldsymbol{\sigma}_2\boldsymbol{\sigma}_1 = i\boldsymbol{\sigma}_3, \qquad \boldsymbol{\sigma}_2\boldsymbol{\sigma}_3 = -\boldsymbol{\sigma}_3\boldsymbol{\sigma}_2 = i\boldsymbol{\sigma}_1, \qquad \boldsymbol{\sigma}_3\boldsymbol{\sigma}_1 = -\boldsymbol{\sigma}_1\boldsymbol{\sigma}_3 = i\boldsymbol{\sigma}_2 \tag{8.69a}$$

$$\boldsymbol{\sigma}_1^2 = \boldsymbol{\sigma}_2^2 = \boldsymbol{\sigma}_3^2 = \mathbf{1} \tag{8.69b}$$

The notation

$$\mathbf{AB} + \mathbf{BA} \equiv [\mathbf{A}, \mathbf{B}] \tag{8.70}$$

is called the *anticommutator* of $\mathbf{A}$ with $\mathbf{B}$. From equation 8.69a, we see that

$$[\boldsymbol{\sigma}_r, \boldsymbol{\sigma}_s] = 0 \qquad r \neq s \tag{8.71}$$

That is, the Pauli matrices anticommute. In addition, equation 8.69a can be written

$$\boldsymbol{\sigma}_r\boldsymbol{\sigma}_s = i \sum_{t=1}^{3} \varepsilon_{rst}\boldsymbol{\sigma}_t \tag{8.72}$$

Using equations 8.66b and 8.71, the reader will show in Problem 12 that the Dirac matrices satisfy

$$[\boldsymbol{\alpha}_r, \boldsymbol{\alpha}_s] = 0 \quad r \neq s \qquad \text{(b)}\quad [\boldsymbol{\alpha}_r, \boldsymbol{\alpha}_4] = 0 \qquad \text{(c)}\quad \boldsymbol{\alpha}_1^2 = \boldsymbol{\alpha}_2^2 = \boldsymbol{\alpha}_3^2 = \mathbf{1}$$

$$\text{(d)}\quad \boldsymbol{\alpha}_r\boldsymbol{\alpha}_s \neq i \sum_{t=1}^{3} \varepsilon_{rst}\boldsymbol{\alpha}_t \qquad r, s, t \leqslant 3 \tag{8.73}$$

Determinant of a product of matrices

Consider

$$|\mathbf{C}| \equiv |\mathbf{AB}| = \sum_{ij\cdots n} \varepsilon_{ij\cdots n} c_{1i} c_{2j} \cdots c_{Nn} = \sum_{ij\cdots n} \varepsilon_{ij\cdots n} \sum_{\alpha\beta\cdots\nu} a_{1\alpha} b_{\alpha i} a_{2\beta} b_{\beta j} \cdots a_{N\nu} b_{\nu n} \tag{8.74}$$

If the sequence $\alpha, \beta, \ldots, \nu$ is an even number of permutations of $1, 2, \ldots, N$, then, from equation 8.1,

$$\sum_{ij\cdots n} \varepsilon_{ij\cdots n} b_{\alpha i} b_{\beta j} \cdots b_{\nu n} = |\mathbf{B}| \tag{8.75a}$$

and if $\alpha, \beta, \ldots, \nu$ is an odd permutation of the sequence $1, 2, \ldots, N$, then

$$\sum_{ij\cdots n} \varepsilon_{ij\cdots n} b_{\alpha i} b_{\beta j} \cdots b_{\nu n} = -|\mathbf{B}| \tag{8.75b}$$

That is,

$$\sum_{ij\cdots n} \varepsilon_{ij\cdots n} b_{\alpha i} b_{\beta j} \cdots b_{\nu n} = \varepsilon_{\alpha\beta\cdots\nu}|\mathbf{B}| \tag{8.76}$$

$$\Rightarrow \quad |\mathbf{AB}| = |\mathbf{B}| \sum_{\alpha\beta\cdots\nu} \varepsilon_{\alpha\beta\cdots\nu} a_{1\alpha} a_{2\beta} \cdots a_{N\nu} = |\mathbf{B}||\mathbf{A}| \tag{8.77}$$

That is, the determinant of a product of matrices is the product of the individual determinants.

Trace or spur of a matrix

The *trace* or *spur* of a matrix is defined only if the matrix is a square array. For such a matrix, the trace is defined as the sum of the diagonal elements.

$$\mathrm{Tr}(\mathbf{A}) \equiv \sum_{i=1}^{N} a_{ii} \tag{8.78}$$

Consider the trace of a product of two matrices:

$$\mathrm{Tr}(\mathbf{C}) \equiv \mathrm{Tr}(\mathbf{AB}) = \sum_{i=1}^{N} c_{ii} = \sum_{i=1}^{N}\sum_{k=1}^{N} a_{ik}b_{ki} \tag{8.79a}$$

Both the sums, and order of multiplication of the elements of **A** and **B** can be interchanged. Therefore,

$$\mathrm{Tr}(\mathbf{AB}) = \sum_{k=1}^{N}\sum_{i=1}^{N} b_{ki}a_{ik} = \sum_{k=1}^{N} (\mathbf{BA})_{kk} = \mathrm{Tr}(\mathbf{BA}) \tag{8.79b}$$

That is, the trace of a product of two matrices is invariant to the order of multiplication.

Consider the trace of a product of three matrices.

$$\mathrm{Tr}(\mathbf{D}) \equiv \mathrm{Tr}(\mathbf{ABC}) = \sum_{i=1}^{N} d_{ii} = \sum_{i=1}^{N}\sum_{j=1}^{N}\sum_{k=1}^{N} a_{ij}b_{jk}c_{ki} \tag{8.80}$$

We note that an interchange of the order of multiplication of b_{jk} and c_{ki}, for example, results in the triple sum $\sum_{k=1}^{N}\sum_{i=1}^{N}\sum_{j=1}^{N} a_{ij}c_{ki}b_{jk}$. But there is no way to sum over the indices so that $c_{ki}b_{jk}$ will satisfy the definition for the elements of a product of two matrices. However, if c_{ki} is placed in front of a_{ij}, the sums over i and j satisfy the definition of the (k, k) element of the product **CAB**. That is,

$$\mathrm{Tr}(\mathbf{ABC}) = \mathrm{Tr}(\mathbf{CAB}) = \mathrm{Tr}(\mathbf{BCA}) \tag{8.81}$$

Thus we see that the trace of a product is invariant to the cyclic interchange of the order of multiplication of the matrices.

Transposing a matrix

The *transpose* of a matrix is the simple operation of interchanging the rows and columns of the matrix. It is denoted by $\mathbf{A}^T$ or $\tilde{\mathbf{A}}$. That is

$$\text{(a)}\quad \mathbf{A} = \begin{bmatrix} a_{11} & a_{12} & a_{13} & \cdots & a_{1N} \\ a_{21} & a_{22} & a_{23} & \cdots & a_{2N} \\ a_{31} & a_{32} & a_{33} & \cdots & a_{3N} \\ \vdots & & & & \\ a_{N1} & a_{N2} & a_{N3} & \cdots & a_{NN} \end{bmatrix} \Rightarrow \text{(b)}\quad \tilde{\mathbf{A}} = \begin{bmatrix} a_{11} & a_{21} & a_{31} & \cdots & a_{N1} \\ a_{12} & a_{22} & a_{32} & \cdots & a_{N2} \\ a_{13} & a_{23} & a_{33} & \cdots & a_{N3} \\ \vdots & & & & \\ a_{1N} & a_{2N} & a_{3N} & \cdots & a_{NN} \end{bmatrix} \tag{8.82}$$

Another way of describing transposition is to view **A** as "flipped" around the main diagonal.

Consider the determinant of the transpose of a matrix.

$$|\tilde{\mathbf{A}}| = \sum \varepsilon_{ijk \cdots n} \tilde{a}_{1i} \tilde{a}_{2j} \tilde{a}_{3k} \cdots \tilde{a}_{Nn} = \sum \varepsilon_{ijk \cdots n} a_{i1} a_{j2} a_{k3} \cdots a_{nN} = |\mathbf{A}| \quad (8.83)$$

That is, the determinant of a matrix is invariant under transposition.

Let $\mathbf{C} = \mathbf{AB}$. Then the (i, j) element of $\tilde{\mathbf{C}}$ is

$$(\tilde{\mathbf{C}})_{ij} = c_{ji} = \sum_k a_{jk} b_{ki} = \sum_k (\tilde{\mathbf{B}})_{ik} (\tilde{\mathbf{A}})_{kj} = (\tilde{\mathbf{B}} \tilde{\mathbf{A}})_{ij} \quad (8.84)$$

That is, the transpose of a product of matrices is the product of the transposed individual matrices multiplied in the reverse order. Thus

$$\text{(a)} \quad \mathbf{D} \equiv \mathbf{ABC} \quad \Rightarrow \quad \text{(b)} \quad \tilde{\mathbf{D}} = \tilde{\mathbf{C}} \tilde{\mathbf{B}} \tilde{\mathbf{A}} \quad (8.85)$$

Matrix inversion

Since a matrix is an operator, inversion is the matrix equivalent of algebraic division. This is analogous to the inverse of the Heaviside D operator, for example,

$$D^{-1} \equiv \int dx \quad (5.103)$$

Inversion of a matrix has meaning only for square arrays.

The inverse of the matrix $\mathbf{A}$ is defined such that

$$\mathbf{AA}^{-1} = \mathbf{A}^{-1}\mathbf{A} = \mathbf{1} \quad (8.86)$$

The Laplace expansion of the determinant of $\mathbf{A}$ around the ith row is

$$|\mathbf{A}| = \sum_{k=1}^{N} a_{ik} \operatorname{cof}(a_{ik}) \quad (8.6a)$$

We have established the fact that a determinant is zero if the ith and jth rows are identical. Consider the Laplace expansion of a determinant for which $a_{ik} = a_{jk}$ for every k. Then

$$|\mathbf{A}| = \begin{vmatrix} a_{11} & a_{12} & a_{13} & \cdots & a_{1N} \\ a_{21} & a_{22} & a_{23} & \cdots & a_{2N} \\ \vdots & & & & \\ a_{i1} & a_{i2} & a_{i3} & \cdots & a_{iN} \\ \vdots & & & & \\ a_{j1} & a_{j2} & a_{j3} & \cdots & a_{jN} \\ \vdots & & & & \\ a_{N1} & a_{N2} & a_{N3} & \cdots & a_{NN} \end{vmatrix} = 0$$

The Laplace expansion of this determinant about the ith row is

$$|\mathbf{A}| = \sum_{k=1}^{N} a_{ik} \operatorname{cof}(a_{ik}) = 0 \quad (8.87a)$$

Since the ith row is identical to the jth row, this can be written

$$|\mathbf{A}| = \sum_{k=1}^{N} a_{jk} \operatorname{cof}(a_{ik}) = 0 \qquad j \neq i \quad (8.87b)$$

$$\Rightarrow \quad \sum_{k=1}^{N} a_{ik} \operatorname{cof}(a_{jk}) = |\mathbf{A}| \delta_{ij} \quad (8.88)$$

If we view $\text{cof}(a_{jk}) \equiv c_{jk}$ as the elements of a matrix called the cofactor matrix $\mathbf{C}$, then c_{jk} is the (k, j) element of the transpose of $\mathbf{C}$. That is, equation 8.88 can be written

$$\text{(a)} \quad \frac{1}{|\mathbf{A}|} \sum_{k=1}^{N} a_{ik}\tilde{c}_{kj} = \delta_{ij} \qquad \text{or} \qquad \text{(b)} \quad \frac{1}{|\mathbf{A}|}\mathbf{A}\tilde{\mathbf{C}} = \mathbf{1} \tag{8.89}$$

Comparing this to equation 8.86, we see that

$$\mathbf{A}^{-1} = \frac{1}{|\mathbf{A}|}\tilde{\mathbf{C}} \tag{8.90}$$

One fact becomes clear immediately; if $|\mathbf{A}| = 0$, $\mathbf{A}^{-1}$ does not exist. Such a matrix is said to be singular.

To summarize, the rules for determining $\mathbf{A}^{-1}$ are:

1. Determine if $\mathbf{A}^{-1}$ exists by determining if $|\mathbf{A}| \neq 0$.

If $\mathbf{A}$ is not singular, then

2. Form the cofactor matrix, the elements of which are the cofactors of the corresponding elements of $\mathbf{A}$.
3. Transpose the cofactor matrix.
4. Divide the transposed cofactor matrix by the determinant of $\mathbf{A}$.

Example 8.9

As an example, we will determine the inverse of

$$\mathbf{A} = \begin{bmatrix} 3 & 5 & 1 \\ 2 & 0 & 3 \\ 7 & 1 & 4 \end{bmatrix} \tag{8.91}$$

Since $|\mathbf{A}| = 58$, $\mathbf{A}$ is not singular. It is straightforward to determine the cofactor matrix

$$\mathbf{C} = \begin{bmatrix} -3 & 13 & 2 \\ -19 & 5 & 32 \\ 15 & -7 & -10 \end{bmatrix} \tag{8.92}$$

$$\Rightarrow \quad \mathbf{A}^{-1} = \frac{1}{|\mathbf{A}|}\tilde{\mathbf{C}} = \frac{1}{58}\begin{bmatrix} -3 & -19 & 15 \\ 13 & 5 & -7 \\ 2 & 32 & -10 \end{bmatrix} \quad \square \tag{8.93}$$

It is important to check if the matrix found is the correct inverse. This is done straightforwardly by multiplying the original matrix by the purported inverse to determine if the product is the identity matrix. In example 8.9, the product of $\mathbf{A}$ of equation 8.91 with $\mathbf{A}^{-1}$ of equation 8.93 does yield the unit matrix.

It has been assumed that there is only one inverse. However, because matrix multiplication is not commutative, a premultiplicative inverse might not be the same as a postmultiplicative inverse. Let us assume that there are two inverses $\mathbf{A}_L^{-1}$ and $\mathbf{A}_R^{-1}$ such that

$$\mathbf{A}_L^{-1}\mathbf{A} = \mathbf{A}\mathbf{A}_R^{-1} = \mathbf{1} \tag{8.94}$$

If we premultiply $\mathbf{A}\mathbf{A}_R^{-1}$ by $\mathbf{A}_L^{-1}$ and postmultiply $\mathbf{A}_L^{-1}\mathbf{A}$ by $\mathbf{A}_R^{-1}$, we obtain

$$\text{(a)} \quad \mathbf{A}_L^{-1}\mathbf{A}\mathbf{A}_R^{-1} = \mathbf{A}_L^{-1}\mathbf{1} = \mathbf{A}_L^{-1} \qquad \text{(b)} \quad \mathbf{A}_L^{-1}\mathbf{A}\mathbf{A}_R^{-1} = \mathbf{1}\cdot\mathbf{A}_R^{-1} = \mathbf{A}_R^{-1} \tag{8.95}$$

$$\Rightarrow \quad \mathbf{A}_L^{-1} = \mathbf{A}_R^{-1} \tag{8.96}$$

and the matrix has only one inverse.

Inverse of a product of matrices

Since the inverse of a matrix is formed essentially from the transpose of the cofactor matrix, the inverse of a product of matrices will involve the transpose of a product of cofactor matrices. Since transposing a product of matrices results in the product of transposed matrices in reverse order, the inverse of a product of matrices will be the product of the inverses of the individual matrices, multiplied in reverse order. That is, if

$$\text{(a)} \quad \mathbf{D} \equiv \mathbf{ABC} \quad \Rightarrow \quad \text{(b)} \quad \mathbf{D}^{-1} = \mathbf{C}^{-1}\mathbf{B}^{-1}\mathbf{A}^{-1} \tag{8.97}$$

It is clear that this is the only combination of inverses of **A**, **B**, and **C** that will result in $\mathbf{DD}^{-1} = \mathbf{1}$. That is, only multiplication in the order

$$\mathbf{ABCC}^{-1}\mathbf{B}^{-1}\mathbf{A}^{-1} = \mathbf{1} \tag{8.98}$$

results in the identity matrix.

Solution of simultaneous equations

Using Cramer's rule, the solution to a set of N simultaneous equations is found one unknown at a time. By matrix inversion, the entire set of unknowns can be found in a single operation. The set of equations

$$a_{11}x_1 + a_{12}x_2 + \cdots + a_{1N}x_N = c_1 \tag{8.41a}$$

$$a_{21}x_1 + a_{22}x_2 + \cdots + a_{2N}x_N = c_2 \tag{8.41b}$$

$$\vdots$$

$$a_{N1}x_1 + a_{N2}x_2 + \cdots + a_{NN}x_N = c_N \tag{8.41c}$$

can be written in matrix form. We define two $N \times 1$ matrices, the elements of which are the unknowns and the inhomogeneous terms.

$$\mathbf{X} \equiv \begin{bmatrix} x_1 \\ x_2 \\ \vdots \\ x_N \end{bmatrix} \quad \text{and} \quad \mathbf{C} \equiv \begin{bmatrix} c_1 \\ c_2 \\ \vdots \\ c_N \end{bmatrix} \tag{8.99}$$

Then the set of simultaneous equations can be written as the matrix equation

$$\mathbf{AX} = \mathbf{C} \tag{8.100}$$

The solution, obtained by premultiplying by $\mathbf{A}^{-1}$, is

$$\mathbf{X} = \mathbf{A}^{-1}\mathbf{C} \tag{8.101}$$

Example 8.10

As an example, consider

$$\text{(a)} \quad 2x + y - z = -1 \qquad \text{(b)} \quad 3x - 2y = -7 \qquad \text{(c)} \quad x - 2y + 2z = -3 \tag{8.102}$$

The coefficient matrix

$$\mathbf{A} = \begin{bmatrix} 2 & 1 & -1 \\ 3 & -2 & 0 \\ 1 & -2 & 2 \end{bmatrix} \tag{8.103}$$

has a nonzero determinant $|\mathbf{A}| = -10$. The inverse of the coefficient matrix is

$$\mathbf{A}^{-1} = \frac{1}{10}\begin{bmatrix} 4 & 0 & 2 \\ 6 & -5 & 3 \\ 4 & -5 & 7 \end{bmatrix} \tag{8.104}$$

$$\Rightarrow \quad \mathbf{X} = \begin{bmatrix} x \\ y \\ z \end{bmatrix} = \frac{1}{10}\begin{bmatrix} 4 & 0 & 2 \\ 6 & -5 & 3 \\ 4 & -5 & 7 \end{bmatrix}\begin{bmatrix} -1 \\ -7 \\ -3 \end{bmatrix} = \begin{bmatrix} -1 \\ 2 \\ 1 \end{bmatrix} \quad \square \tag{8.105}$$

Gauss – Jordan elimination method for A^{-1}

The calculation of the cofactor matrix, and from it, the inverse, is a cumbersome process for a matrix of any appreciable size. The *Gauss–Jordan method* is an arithmetically simpler technique for finding $\mathbf{A}^{-1}$. This is achieved by manipulating the matrix $\mathbf{A}$ with specific allowed operations, and performing the same operations on the identity matrix. The goal of the manipulations is to change $\mathbf{A}$ into the identity matrix. When this has been accomplished, $\mathbf{1}$ will have been converted into $\mathbf{A}^{-1}$.

All operations are performed only on the rows of $\mathbf{A}$, or only on the columns of $\mathbf{A}$. One cannot perform some of the operations on rows, and others on columns. The allowed operations are:

1. Multiplication of all the elements in a row (column) by a constant
2. Addition of a multiple of any one row (column) to any other row (column)
3. Interchange of any two rows (columns)

Let $\mathbf{M}_1, \mathbf{M}_2, \ldots, \mathbf{M}_n$ be matrices such that premultiplication of $\mathbf{A}$ by one of them performs one of the three allowed operations on the rows of $\mathbf{A}$. For example, let $\mathbf{M}*\mathbf{A}$ be a matrix identical to $\mathbf{A}$ except that row_2 of $\mathbf{A}$ is multiplied by a constant K. That is,

$$(\mathbf{M}*\mathbf{A})_{ij} = \begin{cases} a_{ij} & i \neq 2 \\ Ka_{ij} & i = 2 \end{cases} \tag{8.106}$$

The matrix $\mathbf{M}$ that accomplishes this has elements

$$\text{(a)} \quad m_{ij} = \begin{cases} \delta_{ij} & i \neq 2 \\ K\delta_{ij} & i = 2 \end{cases} \quad \Rightarrow \quad \text{(b)} \quad \mathbf{M} = \begin{bmatrix} 1 & 0 & 0 & 0 & \cdots & 0 \\ 0 & K & 0 & 0 & \cdots & 0 \\ 0 & 0 & 1 & 0 & \cdots & 0 \\ \vdots & & & & & \\ 0 & 0 & 0 & 0 & \cdots & 0 \end{bmatrix} \tag{8.107}$$

performs the desired manipulation of $\mathbf{A}$.

Similarly,

$$\mathbf{M} = \begin{bmatrix} 0 & 1 & 0 & 0 & \cdots & 0 \\ 1 & 0 & 0 & 0 & \cdots & 0 \\ 0 & 0 & 1 & 0 & \cdots & 0 \\ \vdots & & & & & \\ 0 & 0 & 0 & 0 & \cdots & 0 \end{bmatrix} \tag{8.108}$$

will interchange row_1 with row_2, and

$$\mathbf{M} = \begin{bmatrix} 1 & K & 0 & 0 & \cdots & 0 \\ 0 & 1 & 0 & 0 & \cdots & 0 \\ 0 & 0 & 1 & 0 & \cdots & 0 \\ \vdots & & & & & \\ 0 & 0 & 0 & 0 & \cdots & 0 \end{bmatrix} \tag{8.109}$$

will replace row_1 with $row_1 + K\ row_2$.

If

$$(\mathbf{M}_1 \cdot \mathbf{M}_2 \cdot \mathbf{M}_3 \cdots \mathbf{M}_n)\mathbf{A} = \mathbf{1} \tag{8.110}$$

then clearly, the product of **M** matrices multiplying **A** is equivalent to operating on the rows of **A** to convert **A** into **1**. But equation 8.110 can also be written

$$(\mathbf{M} \cdot \mathbf{M}_2 \cdot \mathbf{M}_3 \cdots \mathbf{M}_n \cdot \mathbf{1})\mathbf{A} = \mathbf{1} \tag{8.111}$$

In this form, the product of **M** matrices performs the same operations on the rows of **1**. Since $(\mathbf{M}_1 \cdot \mathbf{M}_2 \cdots \mathbf{M}_n \cdot \mathbf{1})$, when multiplied by **A**, yields the identity matrix,

$$\mathbf{M}_1 \cdot \mathbf{M}_2 \cdot \mathbf{M}_3 \cdots \mathbf{M}_n \cdot \mathbf{1} = \mathbf{A}^{-1} \tag{8.112}$$

Postmultiplication by matrices **M** operates on the columns of **A** and **1**. Thus, column manipulation can be performed to achieve the same result. That is, if

$$\mathbf{A} \cdot \mathbf{M}_1 \cdot \mathbf{M}_2 \cdot \mathbf{M}_3 \cdots \mathbf{M}_n = (\mathbf{1} \cdot \mathbf{M}_1 \cdot \mathbf{M}_2 \cdot \mathbf{M}_3 \cdots \mathbf{M}_n) = \mathbf{1} \tag{8.113a}$$

$$\Rightarrow \quad \mathbf{1} \cdot \mathbf{M}_1 \cdot \mathbf{M}_2 \cdot \mathbf{M}_3 \cdots \mathbf{M}_n = \mathbf{A}^{-1} \tag{8.113b}$$

It is not possible to derive this result with both premultiplied and postmultiplied matrices **M**. Operations on both rows and columns are achieved by multiplications in the form $(\mathbf{M}_1 \cdot \mathbf{M}_2 \cdots) \cdot \mathbf{A} \cdot (\mathbf{M}_3 \cdot \mathbf{M}_4 \cdots)$. But this product cannot be written in the form $(\mathbf{M}_1 \cdot \mathbf{M}_2 \cdot \mathbf{M}_3 \cdots \mathbf{1}) \cdot \mathbf{A}$. Thus manipulation only on rows, or only on columns, is permitted, but not on both.

Example 8.11

Consider

$$\mathbf{A} = \begin{bmatrix} 3 & 2 & 1 \\ 2 & 2 & 1 \\ 1 & 1 & 4 \end{bmatrix} \tag{8.114}$$

To ascertain that **A** is not singular, we note that $|A| = 7$.

It is easier to keep track of the operations performed on the two matrices by writing **A** and **1** next to one another.

$$\mathbf{A} = \begin{bmatrix} 3 & 2 & 1 \\ 2 & 2 & 1 \\ 1 & 1 & 4 \end{bmatrix} \qquad \mathbf{1} = \begin{bmatrix} 1 & 0 & 0 \\ 0 & 1 & 0 \\ 0 & 0 & 1 \end{bmatrix} \tag{8.115}$$

As with the allowed operations on determinants, we will illustrate a systematic approach to applying the Gauss–Jordan method. The first steps are to manipulate the **A** matrix to get zeros for the off-diagonal elements. Starting with the first column, we can convert the (2, 1) and (3, 1) elements of the **A** matrix to zero, using row_1 as the working row. The manipulations that accomplish this are

$$\text{(a)} \quad \text{row}_2 \to \text{row}_2 - \tfrac{2}{3}\,\text{row}_1 \qquad \text{(b)} \quad \text{row}_3 \to \text{row}_3 - \tfrac{1}{3}\,\text{row}_1 \tag{8.116}$$

$$\Rightarrow \quad \mathbf{A} \to \begin{bmatrix} 3 & 2 & 1 \\ 0 & \frac{2}{3} & \frac{1}{3} \\ 0 & \frac{1}{3} & \frac{11}{3} \end{bmatrix} \qquad \mathbf{1} \to \begin{bmatrix} 1 & 0 & 0 \\ -\frac{2}{3} & 1 & 0 \\ -\frac{1}{3} & 0 & 1 \end{bmatrix} \tag{8.117}$$

To obtain zeros for the (1, 2) and (3, 2) elements in the second column, we use row_2 as the working row. The required operations are

$$\text{(a)} \quad \text{row}_1 \to \text{row}_1 - 3\,\text{row}_2 \qquad \text{(b)} \quad \text{row}_3 \to \text{row}_3 - \tfrac{1}{2}\,\text{row}_2 \tag{8.118}$$

$$\Rightarrow \quad \mathbf{A} \to \begin{bmatrix} 3 & 0 & 0 \\ 0 & \frac{2}{3} & \frac{1}{3} \\ 0 & 0 & \frac{7}{2} \end{bmatrix} \qquad \mathbf{1} \to \begin{bmatrix} 3 & -3 & 0 \\ -\frac{2}{3} & 1 & 0 \\ 0 & -\frac{1}{2} & 1 \end{bmatrix} \tag{8.119}$$

To replace the (2, 3) element in the matrix that was **A**, row_3 is used as the working row and we set

$$\text{row}_2 \to \text{row}_2 - \tfrac{2}{21}\,\text{row}_3 \tag{8.120}$$

$$\Rightarrow \quad \mathbf{A} \to \begin{bmatrix} 3 & 0 & 0 \\ 0 & \frac{2}{3} & 0 \\ 0 & 0 & \frac{7}{2} \end{bmatrix} \qquad \mathbf{1} \to \begin{bmatrix} 3 & -3 & 0 \\ -\frac{2}{3} & \frac{22}{21} & -\frac{2}{21} \\ 0 & -\frac{1}{2} & 1 \end{bmatrix} \tag{8.121}$$

The following operations convert the matrix that began as **A** into the identity:

$$\text{(a)} \quad \text{row}_1 \to \tfrac{1}{3}\,\text{row}_1 \qquad \text{(b)} \quad \text{row}_2 \to \tfrac{3}{2}\,\text{row}_2 \qquad \text{(c)} \quad \text{row}_3 \to \tfrac{2}{7}\,\text{row}_3 \tag{8.122}$$

$$\Rightarrow \quad \mathbf{A} \to \begin{bmatrix} 1 & 0 & 0 \\ 0 & 1 & 0 \\ 0 & 0 & 1 \end{bmatrix} = \mathbf{1} \qquad \mathbf{1} \to \begin{bmatrix} 1 & -1 & 0 \\ -1 & \frac{11}{7} & -\frac{1}{7} \\ 0 & -\frac{1}{7} & \frac{2}{7} \end{bmatrix} = \mathbf{A}^{-1} \tag{8.123}$$

By multiplying the resulting $\mathbf{A}^{-1}$ by **A**, it is straightforward to confirm that **1** has been converted into $\mathbf{A}^{-1}$. □

Choleski–Turing method

The allowed row manipulations used in the Gauss–Jordan method to convert $\mathbf{A} \to \mathbf{1}$ and $\mathbf{1} \to \mathbf{A}^{-1}$, also provide a method for solving simultaneous equations. Let a sequence of row operations generated by the premultiplication of **A** by the matrices $\mathbf{M}_1\mathbf{M}_2 \cdots \mathbf{M}_n$, cast **A** into a triangular form. That is,

$$\mathbf{M}_1\mathbf{M}_2 \cdots \mathbf{M}_n\mathbf{A} = \begin{bmatrix} \alpha_{11} & \alpha_{12} & \alpha_{13} & \cdots & & \alpha_{1N} \\ 0 & \alpha_{22} & \alpha_{23} & \cdots & & \alpha_{2N} \\ 0 & 0 & \alpha_{33} & \cdots & & \alpha_{3N} \\ \vdots & & & & & \\ 0 & 0 & 0 & \cdots & 0 & \alpha_{NN} \end{bmatrix} \tag{8.124a}$$

or

$$\mathbf{M}_1\mathbf{M}_2 \cdots \mathbf{M}_n\mathbf{A} = \begin{bmatrix} \alpha_{11} & 0 & 0 & \cdots & 0 \\ \alpha_{21} & \alpha_{22} & 0 & \cdots & 0 \\ \alpha_{31} & \alpha_{32} & \alpha_{33} & \cdots & 0 \\ \vdots & & & & \\ \alpha_{N1} & \alpha_{N2} & \alpha_{N3} & \cdots & \alpha_{NN} \end{bmatrix} \tag{8.124b}$$

where, in general, $\alpha_{ij} \neq 0$. With

$$\mathbf{X} \equiv \begin{bmatrix} x_1 \\ x_2 \\ \vdots \\ x_N \end{bmatrix} \quad \text{and} \quad \mathbf{C} \equiv \begin{bmatrix} c_1 \\ c_2 \\ \vdots \\ c_N \end{bmatrix} \tag{8.99}$$

the operations

$$\mathbf{M}_1 \cdot \mathbf{M}_2 \cdot \cdot \mathbf{M}_n \cdot \mathbf{A} \cdot \mathbf{X} = \mathbf{M}_1 \cdot \mathbf{M}_2 \cdot \cdot \mathbf{M}_n \cdot \mathbf{C} \tag{8.125}$$

cast the set of simultaneous equations, in matrix representation, in a triangular form. Either

$$\begin{bmatrix} \alpha_{11} & \alpha_{12} & \alpha_{13} & \cdots & & \alpha_{1N} \\ 0 & \alpha_{22} & \alpha_{23} & \cdots & & \alpha_{2N} \\ 0 & 0 & \alpha_{33} & \cdots & & \alpha_{3N} \\ \vdots & & & & & \\ 0 & 0 & 0 & \cdots & 0 & \alpha_{NN} \end{bmatrix} \begin{bmatrix} x_1 \\ x_2 \\ x_3 \\ \vdots \\ x_N \end{bmatrix} = \begin{bmatrix} \gamma_1 \\ \gamma_2 \\ \gamma_3 \\ \vdots \\ \gamma_N \end{bmatrix} \tag{8.126a}$$

or

$$\begin{bmatrix} \alpha_{11} & 0 & 0 & \cdots & 0 \\ \alpha_{21} & \alpha_{22} & 0 & \cdots & 0 \\ \alpha_{31} & \alpha_{32} & \alpha_{33} & \cdots & 0 \\ \vdots & & & & \\ \alpha_{N1} & \alpha_{N2} & \alpha_{N3} & \cdots & \alpha_{NN} \end{bmatrix} \begin{bmatrix} x_1 \\ x_2 \\ x_3 \\ \vdots \\ x_N \end{bmatrix} = \begin{bmatrix} \gamma_1 \\ \gamma_2 \\ \gamma_3 \\ \vdots \\ \gamma_N \end{bmatrix} \tag{8.126b}$$

With the matrix in upper triangular form as in equation 8.126a, for example, if we begin the solution with multiplication of the Nth row, we obtain

$$\alpha_{NN} x_N = \gamma_N \tag{8.127a}$$

and we have an immediate solution for x_N. The solution obtained from multiplication of the $(N - 1)$th row is

$$\alpha_{(N-1)(N-1)} x_{N-1} + \alpha_{(N-1)N} x_N = \gamma_{N-1} \tag{8.127b}$$

Since the value of x_N is known from equation 8.127a, the value of x_{N-1} can easily be determined. When the matrix is in lower triangular form, we start with multiplication of the first row of equation 8.126b. This leads to analogous simple solutions for x_1, then x_2, and so on. Thus after row manipulation of the matrix into triangular form, and performing the same manipulations on the column of inhomogeneous terms, one can determine the solutions to a set of simultaneous equations with very simple arithmetic. This approach is known as the *Choleski–Turing method* of solution.

Example 8.12

To illustrate, consider the set of simultaneous equations of example 8.10

$$\text{(a)}\quad 2x + y - z = -1 \qquad \text{(b)}\quad 3x - 2y = -7 \qquad \text{(c)}\quad x - 2y + 2z = -3 \tag{8.102}$$

We place the coefficient matrix and inhomogeneous column side by side,

$$\mathbf{A} = \begin{bmatrix} 2 & 1 & -1 \\ 3 & -2 & 0 \\ 1 & -2 & 2 \end{bmatrix} \qquad \mathbf{C} = \begin{bmatrix} -1 \\ -7 \\ -3 \end{bmatrix} \tag{8.128}$$

and, using row_1 as the working row, perform the manipulations

$$\text{(a)} \quad row_2 \to row_2 - \tfrac{3}{2}\,row_1 \qquad \text{(b)} \quad row_3 \to row_3 - \tfrac{1}{2}\,row_1 \tag{8.129}$$

$$\Rightarrow \quad \mathbf{A} \to \begin{bmatrix} 2 & 1 & -1 \\ 0 & -\frac{7}{2} & \frac{3}{2} \\ 0 & -\frac{5}{2} & \frac{5}{2} \end{bmatrix} \qquad \mathbf{C} \to \begin{bmatrix} -1 \\ -\frac{11}{2} \\ -\frac{5}{2} \end{bmatrix} \tag{8.130}$$

Using row_2 as the working row, the manipulation

$$row_3 \to row_3 - \tfrac{5}{7}\,row_2 \tag{8.131}$$

$$\Rightarrow \quad \mathbf{A} \to \begin{bmatrix} 2 & 1 & -1 \\ 0 & -\frac{7}{2} & \frac{3}{2} \\ 0 & 0 & \frac{10}{7} \end{bmatrix} \qquad \mathbf{C} \to \begin{bmatrix} -1 \\ -\frac{11}{2} \\ \frac{10}{7} \end{bmatrix} \tag{8.132}$$

Then from multiplication of the third row, we obtain

$$\tfrac{10}{7}z = \tfrac{10}{7} \qquad \Rightarrow \qquad z = 1 \tag{8.133a}$$

Then from multiplication of the second row, we find

$$-\tfrac{7}{2}y + \tfrac{3}{2}z = -\tfrac{11}{2} \qquad \Rightarrow \qquad y = 2 \tag{8.133b}$$

Multiplication of the first row, then, yields

$$2x + y - z = -1 \qquad \Rightarrow \qquad x = -1 \quad \square \tag{8.133c}$$

Instead of leaving the coefficient matrix in triangular form, let us continue to perform row operations to convert it into a diagonal matrix (or if we then divide the diagonal elements by constants, the unit matrix). Of course, we simultaneously perform the same row manipulations on the column of inhomogeneous terms. That is,

$$\mathbf{M}_1 \cdot \mathbf{M}_2 \cdots \mathbf{M}_n \cdot \mathbf{AX} = \mathbf{M}_1 \cdot \mathbf{M}_2 \cdots \mathbf{M}_n\mathbf{C} \tag{8.134}$$

casts the matrix form of simultaneous equations into

$$\begin{bmatrix} \alpha_{11} & 0 & 0 & \cdots & & 0 \\ 0 & \alpha_{22} & 0 & \cdots & & 0 \\ 0 & 0 & \alpha_{33} & \cdots & & 0 \\ \vdots & & & & & \\ 0 & 0 & 0 & \cdots & 0 & \alpha_{NN} \end{bmatrix} \begin{bmatrix} x_1 \\ x_2 \\ x_3 \\ \vdots \\ x_N \end{bmatrix} = \begin{bmatrix} \gamma_1 \\ \gamma_2 \\ \gamma_3 \\ \vdots \\ \gamma_N \end{bmatrix} \tag{8.135a}$$

or

$$\begin{bmatrix} 1 & 0 & 0 & \cdots & & 0 \\ 0 & 1 & 0 & \cdots & & 0 \\ 0 & 0 & 1 & \cdots & & 0 \\ \vdots & & & & & \\ 0 & 0 & 0 & \cdots & 0 & 1 \end{bmatrix} \begin{bmatrix} x_1 \\ x_2 \\ x_3 \\ \vdots \\ x_N \end{bmatrix} = \begin{bmatrix} \delta_1 \\ \delta_2 \\ \delta_3 \\ \vdots \\ \delta_N \end{bmatrix} \tag{8.135b}$$

In these forms, the solution for each unknown is obtained independently, without a need for prior solution of other unknowns. In the form expressed in equation 8.135a, $x_i = \gamma_i/\alpha_{ii}$, whereas if the coefficient matrix is converted into the identity matrix as in equation 8.135b, the column of inhomogeneous terms has been converted into the column of solutions.

For the equations of example 8.12, some of the work required to diagonalize the coefficient matrix has already been completed in that example. Starting with the triangularized form in equation 8.132,

$$\mathbf{A} \to \begin{bmatrix} 2 & 1 & -1 \\ 0 & -\frac{7}{2} & \frac{3}{2} \\ 0 & 0 & \frac{10}{7} \end{bmatrix} \qquad \mathbf{C} \to \begin{bmatrix} -1 \\ -\frac{11}{2} \\ \frac{10}{7} \end{bmatrix} \tag{8.132}$$

the manipulations

$$\text{(a)} \quad \text{row}_1 \to \text{row}_1 + \tfrac{7}{10}\,\text{row}_3 \qquad \text{(b)} \quad \text{row}_2 \to \text{row}_2 - \tfrac{21}{20}\,\text{row}_3 \tag{8.136}$$

$$\Rightarrow \quad \mathbf{A} \to \begin{bmatrix} 2 & 1 & 0 \\ 0 & -\frac{7}{2} & 0 \\ 0 & 0 & \frac{10}{7} \end{bmatrix} \qquad \mathbf{C} \to \begin{bmatrix} 0 \\ -7 \\ \frac{10}{7} \end{bmatrix} \tag{8.137}$$

Then

$$\text{row}_1 \to \text{row}_1 + \tfrac{2}{7}\,\text{row}_2 \tag{8.138}$$

$$\Rightarrow \quad \mathbf{A} \to \begin{bmatrix} 2 & 0 & 0 \\ 0 & -\frac{7}{2} & 0 \\ 0 & 0 & \frac{10}{7} \end{bmatrix} \qquad \mathbf{C} \to \begin{bmatrix} -2 \\ -7 \\ \frac{10}{7} \end{bmatrix} \tag{8.139}$$

from which we obtain $[x, y, z] = [(-2/2), (-7/-\frac{7}{2}), (\frac{10}{7}/\frac{10}{7})] = [-1, 2, 1]$.

Of course, if we perform one or more steps to convert $\mathbf{A}$ to the identity matrix, by

$$\text{(a)} \quad \text{row}_1 \to \tfrac{1}{2}\,\text{row}_1 \qquad \text{(b)} \quad \text{row}_2 \to -\tfrac{2}{7}\,\text{row}_2 \qquad \text{(c)} \quad \text{row}_3 \to \tfrac{7}{10}\,\text{row}_3 \tag{8.140}$$

$$\mathbf{C} \to \begin{bmatrix} -1 \\ 2 \\ 1 \end{bmatrix} = \begin{bmatrix} x \\ y \\ z \end{bmatrix} \tag{8.141}$$

8.3 Functions of Matrices

A function of a matrix is defined through its Laurent series. Let $F(z)$ be a function of a complex variable that has a Laurent series expansion about the point z_0.

$$F(z) = \sum_{n=-\infty}^{\infty} a_n(z_0)(z - z_0)^n \tag{2.128}$$

The matrix $F(\mathbf{A})$ is defined by

$$F(\mathbf{A}) = \sum_{n=-\infty}^{\infty} a_n(z_0)(\mathbf{A} - z_0\mathbf{1})^n \tag{8.142}$$

The meaning of a matrix to a negative power is

$$\mathbf{M}^{-n} \equiv (\mathbf{M}^{-1})^n \tag{8.143}$$

In many situations, the function under consideration is analytic at $z = 0$ and therefore has a MacLaurin series representation. For example,

$$e^z = \sum_{n=0}^{\infty} \frac{z^n}{n!} \tag{3.1}$$

$$\Rightarrow \quad e^{\mathbf{A}} \equiv \sum_{n=0}^{\infty} \frac{1}{n!}\mathbf{A}^n \tag{8.144}$$

Similarly, using equation 3.6a, we obtained

$$\frac{1}{(1-z)} = \sum_{n=0}^{\infty} z^n \tag{3.84}$$

$$\Rightarrow \quad (\mathbf{1} - \mathbf{A})^{-1} = \sum_{n=0}^{\infty} \mathbf{A}^n \tag{8.145}$$

Example 8.13

Consider the matrix

$$\text{(a)} \quad \mathbf{A} = \begin{bmatrix} 1 & -1 \\ 1 & -1 \end{bmatrix} \quad \Rightarrow \quad \text{(b)} \quad \mathbf{A}^2 = \begin{bmatrix} 0 & 0 \\ 0 & 0 \end{bmatrix} \tag{8.146}$$

Such a matrix is called a nilpotent matrix of order 2. Therefore,

$$\text{(a)} \quad e^{\mathbf{A}} = \mathbf{1} + \mathbf{A} = \begin{bmatrix} 2 & -1 \\ 1 & 0 \end{bmatrix} \quad \text{and} \quad \text{(b)} \quad (\mathbf{1} - \mathbf{A})^{-1} = \mathbf{1} + \mathbf{A} = \begin{bmatrix} 2 & -1 \\ 1 & 0 \end{bmatrix} \quad \square \tag{8.147}$$

8.4 Matrices with Special Properties

Orthogonal matrices

The rotation of coordinate axes through an angle θ in a two-dimensional space can be achieved by a matrix operation. Referring to Figure 8.5 we can write

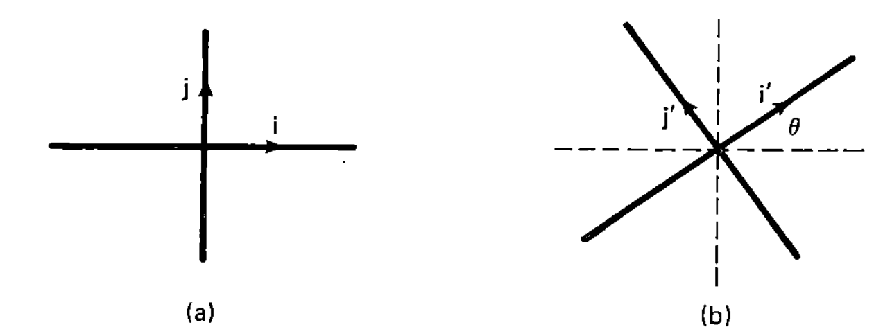

Figure 8.5 (a) Two-dimensional coordinate system; (b) rotated two-dimensional coordinate system.

$$\text{(a)} \quad \mathbf{i}' = \mathbf{i}\cos\theta + \mathbf{j}\sin\theta \qquad \text{(b)} \quad \mathbf{j}' = -\mathbf{i}\sin\theta + \mathbf{j}\cos\theta \tag{8.148}$$

$$\Rightarrow \quad \begin{bmatrix} \mathbf{i}' \\ \mathbf{j}' \end{bmatrix} = \begin{bmatrix} \cos\theta & \sin\theta \\ -\sin\theta & \cos\theta \end{bmatrix} \begin{bmatrix} \mathbf{i} \\ \mathbf{j} \end{bmatrix} \tag{8.149}$$

We can describe an arbitrary vector $\mathbf{V}$ in the nonrotated frame of Figure 8.5a, or in the rotated frame of Figure 8.5b, as

$$\text{(a)} \quad \mathbf{V} = V_x\mathbf{i} + V_y\mathbf{j} \qquad \text{or} \qquad \text{(b)} \quad \mathbf{V}' = V_x'\mathbf{i}' + V_y'\mathbf{j}' \tag{8.150}$$

Clearly, $\mathbf{V}$ and $\mathbf{V}'$ are the same vector described in terms of different basis vectors. Thus, with $\mathbf{V}' = \mathbf{V}$, using equations 8.148, it is straightforward to deduce that

$$\text{(a)} \quad V_x = V_x'\cos\theta - V_y'\sin\theta \qquad \text{(b)} \quad V_y = V_y'\sin\theta + V_x'\cos\theta \tag{8.151}$$

$$\Rightarrow \quad \begin{bmatrix} V_x \\ V_y \end{bmatrix} = \begin{bmatrix} \cos\theta & \sin\theta \\ \sin\theta & \cos\theta \end{bmatrix} \begin{bmatrix} V_x' \\ V_y' \end{bmatrix} \tag{8.152}$$

From this description of a vector in terms of a column matrix, such $N \times 1$ matrices are also referred to as *column vectors*.

To obtain $\begin{bmatrix} V_x' \\ V_y' \end{bmatrix}$ in terms of $\begin{bmatrix} V_x \\ V_y \end{bmatrix}$ the matrix of equation 8.152 must be inverted. It is straightforward to ascertain that

$$\mathbf{S} \equiv \begin{bmatrix} \cos\theta & -\sin\theta \\ \sin\theta & \cos\theta \end{bmatrix} \tag{8.153}$$

$$\Rightarrow \quad \mathbf{S}^{-1} \equiv \begin{bmatrix} \cos\theta & \sin\theta \\ -\sin\theta & \cos\theta \end{bmatrix} = \tilde{\mathbf{S}} \tag{8.154}$$

We also note that

$$|\mathbf{S}| = 1 \tag{8.155}$$

The dot product can be represented in terms of the column vectors as

$$\mathbf{V} \cdot \mathbf{V} = \begin{bmatrix} V_x & V_y \end{bmatrix} \begin{bmatrix} V_x \\ V_y \end{bmatrix} = \tilde{\mathbf{V}}\mathbf{V} \tag{8.156}$$

Using equation 8.152, it is straightforward to determine that

$$\tilde{\mathbf{V}}'\mathbf{V}' = \tilde{\mathbf{V}}\mathbf{V} \tag{8.157}$$

These properties characterize an *orthogonal matrix*. They are:

1. $\tilde{\mathbf{S}} = \mathbf{S}^{-1}$ or, equivalently, $\mathbf{S}\tilde{\mathbf{S}} = \mathbf{1}$.
2. $|\mathbf{S}| = 1$, which follows from property 1.
3. The lengths of two vectors connected by an orthogonal transformation are equal. Stated another way, the length of a vector is invariant under an orthogonal transformation.

Rotations in three dimensions

As noted in Figures 8.5, when a coordinate system is rotated in two dimensions, it is rotated about a point, and one angle is required to define an arbitrary rotation. A rotation in three dimensions occurs about an axis. In three dimensions, three variables are required to define the position of an object. Therefore, three angular coordinates are necessary to describe an arbitrary rotation of an axis system.

In the dynamics of solid objects (rigid bodies), the customary way of describing the orientation of the body with respect to an external coordinate frame is to fix a second set of axes on the object that initially coincides with the space axes. An arbitrary reorientation of this body-centered axis system with respect to the fixed laboratory or space coordinates is then generated by a set of three rotations about body-centered axes. The three angles that specify these rotations are called the *Euler angles*. There are many possible sets of Euler rotations that will generate a general reorientation of the body axes. A commonly used set of these rotations is defined below. [The reader is also referred to the book by T. Bradbury, *Theoretical Mechanics* (New York: John Wiley & Sons, Inc., 1968) and by H. Goldstein, *Classical Mechanics*, 2nd ed. (Reading, Mass.: Addison-Wesley Publishing Company, 1980).]

1. A rotation of the body centered (x, y, z) axes about the z-axis through an angle ϕ, to an axis system defined by $(x', y', z' = z)$ (see Figure 8.6).
2. A rotation of the body centered (x', y', z') axes about the x'-axis through an angle θ to an axis system defined by $(x'' = x', y'', z'')$ (Figure 8.7).
3. A rotation of the body centered (x'', y'', z'') axes about the z''-axis through an angle ψ to an axis system defined by $(X, Y, Z = z'')$ (Figure 8.8).

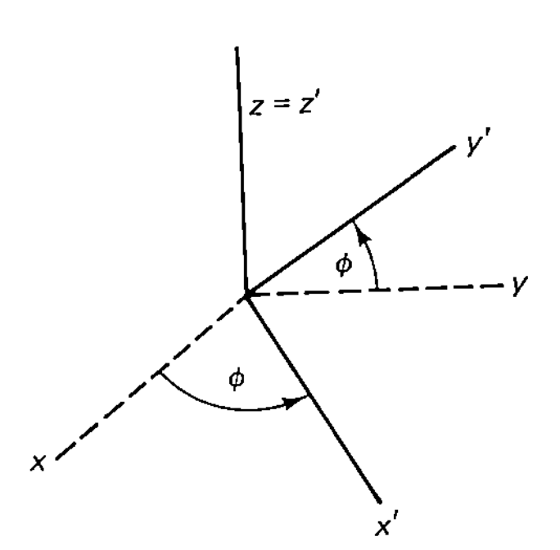

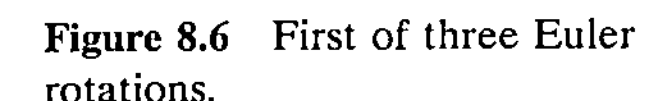
Figure 8.6 First of three Euler rotations.

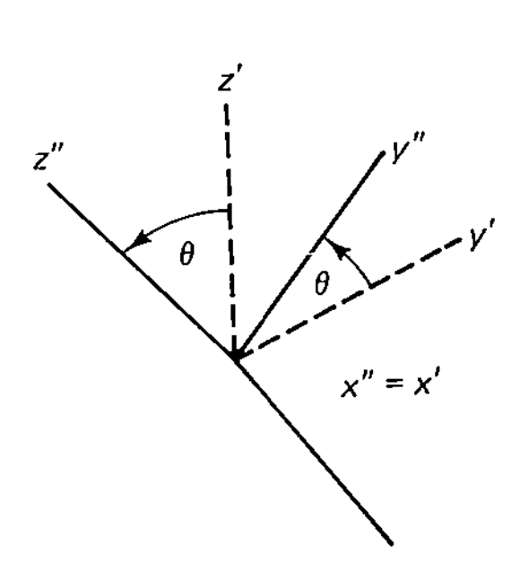

Figure 8.7 Second of three Euler rotations.

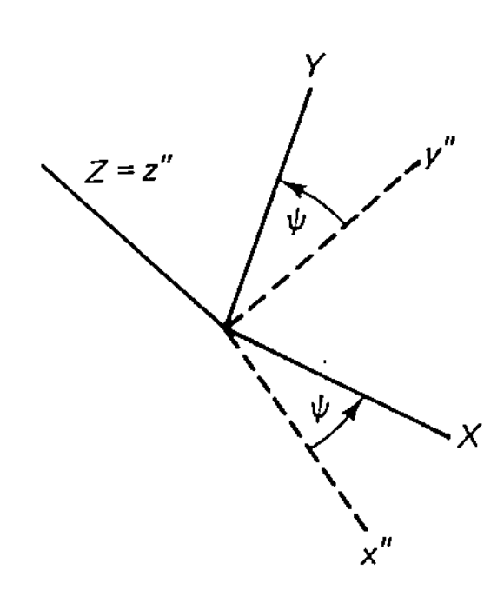

Figure 8.8 Third of three Euler rotations.

Referring to Figures 8.6, 8.7, and 8.8 and equations 8.152 and 8.154, the matrix equation that describes this rotation is

$$\text{(a)}\quad \begin{bmatrix} x' \\ y' \\ z' \end{bmatrix} = \begin{bmatrix} \cos\phi & \sin\phi & 0 \\ -\sin\phi & \cos\phi & 0 \\ 0 & 0 & 1 \end{bmatrix} \begin{bmatrix} x \\ y \\ z \end{bmatrix} \qquad \text{(b)}\quad \begin{bmatrix} x'' \\ y'' \\ z'' \end{bmatrix} = \begin{bmatrix} 1 & 0 & 0 \\ 0 & \cos\theta & \sin\theta \\ 0 & -\sin\theta & \cos\theta \end{bmatrix} \begin{bmatrix} x' \\ y' \\ z' \end{bmatrix}$$

$$\text{(c)}\quad \begin{bmatrix} X \\ Y \\ Z \end{bmatrix} = \begin{bmatrix} \cos\psi & \sin\psi & 0 \\ -\sin\psi & \cos\psi & 0 \\ 0 & 0 & 1 \end{bmatrix} \begin{bmatrix} x'' \\ y'' \\ z'' \end{bmatrix} \tag{8.158}$$

It is customary to designate each rotation matrix with a subscript corresponding to the axis about which the rotation is made, and to describe the angle as a functional argument. The matrix equations that describe the Euler rotations are

$$\text{(a)}\quad \mathbf{r}' = \mathbf{D}_z(\phi)\mathbf{r} \qquad \text{(b)}\quad \mathbf{r}'' = \mathbf{D}_x(\theta)\mathbf{r}' \qquad \text{(c)}\quad \mathbf{R} = \mathbf{D}_z(\psi)\mathbf{r}'' \tag{8.159}$$

where boldface characters are being used to denote column vectors. Combining these three equations, we can write the reorientation of the body-centered axes in one matrix equation:

$$\mathbf{R} = \mathbf{D}_z(\psi)\mathbf{D}_x(\theta)\mathbf{D}_z(\phi)\mathbf{r} \equiv \mathscr{D}(\phi, \theta, \psi)\mathbf{r} \tag{8.160}$$

Symmetric and antisymmetric matrices

A matrix is *symmetric* if the matrix and its transpose are equal. That is

$$\text{(a)}\quad \tilde{\mathbf{A}} = \mathbf{A} \qquad \Rightarrow \qquad \text{(b)}\quad a_{ij} = a_{ji} \tag{8.161}$$

An *antisymmetric* matrix is one for which

$$\text{(a)}\quad \tilde{\mathbf{A}} = -\mathbf{A} \qquad \Rightarrow \qquad \text{(b)}\quad a_{ij} = -a_{ji} \tag{8.162}$$

Setting $i = j$ for an antisymmetric matrix, we obtain the property that

$$\text{(a)}\quad a_{ii} = -a_{ii} \qquad \Rightarrow \qquad \text{(b)}\quad a_{ii} = 0 \tag{8.163}$$

Let $\mathbf{B}$ be a general matrix. $\mathbf{B}$ can be written

$$\mathbf{B} = \tfrac{1}{2}(\mathbf{B} + \tilde{\mathbf{B}}) + \tfrac{1}{2}(\mathbf{B} - \tilde{\mathbf{B}}) \tag{8.164}$$

We note that the matrix in the first parentheses is symmetric and the matrix in the second parentheses is antisymmetric. Thus any matrix can be written as a sum of symmetric and antisymmetric parts.

Inversion of a symmetric matrix

A method for inverting a symmetric matrix is presented which may prove arithmetically easier than the approach presented earlier for a general matrix.

Using the property that the transpose of a product of two matrices is the same as the product of the individual transposed matrices multiplied in reverse order (equation 8.84), it is clear that the matrix

$$\mathbf{A} \equiv \mathbf{B}\tilde{\mathbf{B}} \tag{8.165a}$$

is symmetric. Writing this in terms of the elements, we have

$$a_{ij} = \sum_{k=1}^{N} b_{ik}\tilde{b}_{kj} = \sum_{k=1}^{N} b_{ik}b_{jk} \tag{8.165b}$$

However, $\mathbf{B}$ is not unique. For example,

$$\text{(a)} \quad \mathbf{B} = \begin{bmatrix} 1 & 0 \\ 2 & i \end{bmatrix} \quad \text{and} \quad \text{(b)} \quad \mathbf{B} = \begin{bmatrix} 0 & 1 \\ i & 2 \end{bmatrix} \tag{8.166}$$

are two of an infinite number of matrices that satisfy

$$\mathbf{B}\tilde{\mathbf{B}} = \begin{bmatrix} 1 & 2 \\ 2 & 3 \end{bmatrix} \tag{8.167}$$

If a matrix $\mathbf{B}$ can be found that can easily be inverted, the inversion of $\mathbf{A}$ is simplified, using

$$\mathbf{A}^{-1} = \tilde{\mathbf{B}}^{-1}\mathbf{B}^{-1} \tag{8.168}$$

Let $\mathbf{B}$ be a lower triangular matrix (the development using an upper triangular matrix proceeds in the same way):

$$\mathbf{B} = \begin{bmatrix} b_{11} & 0 & 0 & \cdots & 0 \\ b_{21} & b_{22} & 0 & \cdots & 0 \\ b_{31} & b_{32} & b_{33} & \cdots & 0 \\ \vdots & & & & \\ b_{N1} & b_{N2} & b_{N3} & \cdots & b_{NN} \end{bmatrix} \tag{8.169}$$

$$\Rightarrow \quad b_{ik} = 0 \qquad k > i \tag{8.170}$$

For $i = 1$, equation 8.165b becomes

$$\text{(a)} \quad a_{1j} = a_{j1} = \sum_{k=1}^{N} b_{1k}b_{jk} = b_{11}b_{j1} \qquad \Rightarrow \qquad \text{(b)} \quad b_{j1} = \frac{a_{j1}}{b_{11}} \tag{8.171}$$

Setting $j = 1$, we obtain

$$b_{11}^2 = a_{11} \tag{8.172}$$

$$\Rightarrow \quad b_{j1} = \frac{a_{j1}}{\sqrt{a_{11}}} \tag{8.173}$$

With $i = 2$, equation 8.171a is

$$a_{2j} = a_{j2} = \sum_{k=1}^{N} b_{2k}b_{jk} = b_{21}b_{j1} + b_{22}b_{j2} \tag{8.174}$$

Using equation 8.173, this becomes

$$a_{2j} = a_{j2} = \frac{a_{21}a_{j1}}{a_{11}} + b_{22}b_{j2} \tag{8.175}$$

Setting $j = 2$, we obtain

$$b_{22} = \left[a_{22} - \frac{a_{21}^2}{a_{11}}\right]^{1/2} \tag{8.176}$$

$$\Rightarrow \quad b_{j2} = \frac{a_{j2}}{b_{22}} - \frac{a_{21}a_{j1}}{b_{22}a_{11}} \tag{8.177}$$

Using similar analysis, we obtain

$$a_{3j} = a_{j3} = b_{31}b_{j1} + b_{32}b_{j2} + b_{33}b_{j3} \tag{8.178}$$

Setting $j = 3$, we find

$$b_{33} = \left[a_{33}^2 - b_{31}^2 - b_{32}^2\right]^{1/2} \tag{8.179}$$

where b_{31} is found from equation 8.178 and b_{32} is obtained from equations 8.175 and 8.176. Once the value of b_{33} is found, each b_{j3} can be determined.

Proceeding in this way, all the elements of a triangular **B** can be determined. Clearly, the elements of **B** can be complex even though **A** is real. This is shown in the example of equations 8.166.

Example 8.14

As an example, we will use this approach to invert the symmetric matrix

$$\mathbf{A} = \begin{bmatrix} 4 & 1 & 2 \\ 1 & 2 & 1 \\ 2 & 1 & 1 \end{bmatrix} \tag{8.180}$$

by determining the elements of

$$\mathbf{B} = \begin{bmatrix} b_{11} & 0 & 0 \\ b_{21} & b_{22} & 0 \\ b_{31} & b_{32} & b_{33} \end{bmatrix} \tag{8.181}$$

From equation 8.172, we obtain

$$b_{11} = \sqrt{a_{11}} = 2 \tag{8.182a}$$

Then, using equations 8.173, 8.176, 8.177, and 8.179 we find

$$\text{(b)} \quad b_{21} = \tfrac{1}{2} \qquad \text{(c)} \quad b_{31} = 1 \qquad \text{(d)} \quad b_{22} = \frac{\sqrt{7}}{2} \qquad \text{(e)} \quad b_{32} = \frac{1}{\sqrt{7}} \qquad \text{(f)} \quad b_{33} = \frac{i}{\sqrt{7}} \tag{8.182}$$

$$\Rightarrow \quad \mathbf{B} = \begin{bmatrix} 2 & 0 & 0 \\ \frac{1}{2} & \frac{\sqrt{7}}{2} & 0 \\ 1 & \frac{1}{\sqrt{7}} & \frac{i}{\sqrt{7}} \end{bmatrix} \tag{8.183}$$

Using the Gauss–Jordan method to invert a matrix, for example, we can see that there is less work involved in inverting a triangular matrix than a general matrix, since for the triangular matrix, half the off-diagonal elements are already zero. Inverting **B** of equation 8.183, we obtain

$$\mathbf{B}^{-1} = \begin{bmatrix} \frac{1}{2} & 0 & 0 \\ \frac{-1}{2\sqrt{7}} & \frac{2}{\sqrt{7}} & 0 \\ \frac{3i}{\sqrt{7}} & \frac{2i}{\sqrt{7}} & -i\sqrt{7} \end{bmatrix} \tag{8.184}$$

We note that $\mathbf{B}^{-1}$ is in the same triangular form as $\mathbf{B}$, and $b_{ii}^{-1} = 1/b_{ii}$. This is a general result, as the reader will show in Problem 26. Thus

$$\mathbf{A}^{-1} = \tilde{\mathbf{B}}^{-1}\mathbf{B}^{-1} = \begin{bmatrix} \frac{1}{2} & \frac{-1}{2\sqrt{7}} & \frac{3i}{\sqrt{7}} \\ 0 & \frac{2}{\sqrt{7}} & \frac{2i}{\sqrt{7}} \\ 0 & 0 & -i\sqrt{7} \end{bmatrix} \begin{bmatrix} \frac{1}{2} & 0 & 0 \\ \frac{-1}{2\sqrt{7}} & \frac{2}{\sqrt{7}} & 0 \\ \frac{3i}{\sqrt{7}} & \frac{2i}{\sqrt{7}} & -i\sqrt{7} \end{bmatrix}$$

$$= \begin{bmatrix} -1 & -1 & 3 \\ -1 & 0 & 2 \\ 3 & 2 & -7 \end{bmatrix} \quad \square \tag{8.185}$$

Involutory, periodic, indempotent, and nilpotent matrices

If the square of a matrix is the identity matrix

$$\mathbf{A}^2 = \mathbf{1} \tag{8.186}$$

then the matrix is said to be *involutory*. The three Pauli matrices given in equations 8.67 are examples of involutory matrices.

A matrix $\mathbf{A}$ is called *periodic* with period k if

$$\mathbf{A}^{k+1} = \mathbf{A} \tag{8.187}$$

A periodic matrix is *idempotent* if

$$\mathbf{A}^2 = \mathbf{A} \tag{8.188}$$

If

$$\mathbf{A}^p = \mathbf{0} \tag{8.189}$$

$\mathbf{A}$ is said to be *nilpotent* of order p.

Hermitian and antihermitian matrices

The concept of a hermitian operator was introduced in Chapter 6. In terms of the Dirac bra-ket, an operator is hermitian if

$$\langle \alpha_i | \hat{\Omega}^\dagger | \alpha_j \rangle = \langle \alpha_j | \hat{\Omega} | \alpha_i \rangle^* = \langle \alpha_i | \hat{\Omega} | \alpha_j \rangle \tag{6.130a}$$

We can view the Dirac bra-ket as a matrix element, and it is frequently referred to as such. It is then very straightforward to extend the idea of hermiticity to matrices. The hermitian adjoint of a matrix is defined by

$$\mathbf{A}^\dagger \equiv \tilde{\mathbf{A}}^* \tag{8.190}$$

The matrix is *hermitian* if

$$\text{(a)}\quad \mathbf{A}^\dagger = \mathbf{A} \qquad \Rightarrow \qquad \text{(b)}\quad a_{ji}^* = a_{ij} \tag{8.191}$$

From this, we immediately ascertain that a real symmetric matrix is also hermitian. If all the matrix elements are imaginary, the hermitian matrix is antisymmetric.

A matrix is said to be *antihermitian* if

$$\text{(a)}\quad \mathbf{A}^\dagger = -\mathbf{A} \qquad \Rightarrow \qquad \text{(b)}\quad a_{ji}^* = -a_{ij} \tag{8.192}$$

Clearly, if all elements are real, the matrix is also antisymmetric, and if all the elements are imaginary, the matrix is symmetric.

Any matrix can be written as a sum of hermitian and anithermitian parts. That is, $\mathbf{A}$ can be written

$$\mathbf{A} = \tfrac{1}{2}(\mathbf{A} + \mathbf{A}^\dagger) + \tfrac{1}{2}(\mathbf{A} - \mathbf{A}^\dagger) \tag{8.193}$$

Since the hermitian adjoint of a matrix involves transposition, and because transposing a product of matrices is the same as the product of individual transposed matrices taken in reverse order, we can write a hermitian matrix as the product

$$\mathbf{A} = \mathbf{B}\mathbf{B}^\dagger \tag{8.194a}$$

which is analogous to equation 8.165a for a symmetric matrix. In terms of the elements of $\mathbf{A}$ and $\mathbf{B}$, this equation is

$$a_{ij} = \sum_{k=1}^{N} b_{ik}\tilde{b}_{kj}^* = \sum_{k=1}^{N} b_{ik}b_{jk}^* \tag{8.194b}$$

As argued for the symmetric matrices, there are many $\mathbf{B}$ matrices that satisfy equation 8.194a. However, if we require $\mathbf{B}$ to be triangular, a unique $\mathbf{B}$ matrix can be found. Since $\mathbf{B}^\dagger = \tilde{\mathbf{B}}^*$, the approach to finding the elements of $\mathbf{B}$ that satisfy equation 8.194a is the same as that for the determination of the $\mathbf{B}$ matrix of equation 8.165a. The reader is referred to Problems 27 for an example of that analysis.

Unitary and antiunitary matrices

A *unitary matrix* is one for which the hermitian adjoint is its inverse. That is,

$$\text{(a)}\quad \mathbf{A}^\dagger = \mathbf{A}^{-1} \qquad \Rightarrow \qquad \text{(b)}\quad \mathbf{A}\mathbf{A}^\dagger = \mathbf{1} \tag{8.195}$$

An *antiunitary matrix* satisfies

$$\text{(a)}\quad \mathbf{A}^\dagger = -\mathbf{A}^{-1} \qquad \Rightarrow \qquad \text{(b)}\quad \mathbf{A}\mathbf{A}^\dagger = -1 \tag{8.196}$$

8.5 Eigenvalue Equations and Similarity Transformations

When an $N \times N$ matrix $\mathbf{A}$ multiplies an $N \times 1$ column vector $\mathbf{X}$, the result is another $N \times 1$ column vector $\mathbf{Y}$.

$$\mathbf{A}\mathbf{X} = \mathbf{Y} \tag{8.197}$$

where both $\mathbf{X}$ and $\mathbf{Y}$ are described in the same coordinate frame.

Let $\mathbf{A}$ represent some meaningful operator. For example, in the Heisenberg formulation of quantum theory, each dynamical quantity such as momentum, energy, and coordinates, is represented by a matrix. Let $\mathbf{D}$ be a transformation matrix that takes the vectors $\mathbf{X}$ and $\mathbf{Y}$ into vectors $\mathbf{X}'$ and $\mathbf{Y}'$ described in a new coordinate system (for example, a rotated coordinate system like that described in equations 8.160). We have shown that the vectors transform as

$$\text{(a)} \quad \mathbf{X}' = \mathbf{DX} \qquad \text{(b)} \quad \mathbf{Y}' = \mathbf{DY} \tag{8.198}$$

We now investigate the transformation of the matrix $\mathbf{A}$ into the new coordinate frame.

The matrix $\mathbf{A}'$ in the new coordinate system is defined by

$$\mathbf{Y}' = \mathbf{A}'\mathbf{X}' \tag{8.199}$$

$$\Rightarrow \quad \text{(a)} \quad \mathbf{DY} = \mathbf{A}'\mathbf{DX} \quad \Rightarrow \quad \text{(b)} \quad \mathbf{Y} = \mathbf{D}^{-1}\mathbf{A}'\mathbf{DX} \tag{8.200}$$

Comparing this to equation 8.197, we obtain

$$\text{(a)} \quad \mathbf{A} = \mathbf{D}^{-1}\mathbf{A}'\mathbf{D} \quad \Rightarrow \quad \text{(b)} \quad \mathbf{A}' = \mathbf{DAD}^{-1} \tag{8.201}$$

Two matrices that are related in this way are said to be *similar*, and the transformation is called a *similarity transformation*.

Relations between similar matrices

1. Two matrices that are connected by a similarity transformation have the same determinant. The proof is very straightforward. Using equation 8.77, we have

$$|\mathbf{A}'| = |\mathbf{DAD}^{-1}| = |\mathbf{D}||\mathbf{A}||\mathbf{D}^{-1}| = |\mathbf{A}||\mathbf{D}||\mathbf{D}^{-1}| = |\mathbf{A}||\mathbf{DD}^{-1}| = |\mathbf{A}| \tag{8.202}$$

2. The trace of a matrix is invariant under a similarity transformation. From equation 8.81,

$$\mathrm{Tr}(\mathbf{A}') = \mathrm{Tr}(\mathbf{DAD}^{-1}) = \mathrm{Tr}(\mathbf{D}^{-1}\mathbf{DA}) = \mathrm{Tr}(\mathbf{A}) \tag{8.203}$$

3. If $\mathbf{A}'$ is similar to $\mathbf{A}$, then $\tilde{\mathbf{A}}'$ is similar to $\tilde{\mathbf{A}}$. Consider

$$\mathbf{A}' = \mathbf{DAD}^{-1} \tag{8.201b}$$

$$\Rightarrow \quad \tilde{\mathbf{A}}' = \tilde{\mathbf{D}}^{-1}\tilde{\mathbf{A}}\tilde{\mathbf{D}} \equiv \mathbf{E}\tilde{\mathbf{A}}\mathbf{E}^{-1} \tag{8.204}$$

If $\mathbf{D}$ is an orthogonal transformation, then $\tilde{\mathbf{D}}^{-1} = \mathbf{E} = \mathbf{D}$, and

$$\tilde{\mathbf{A}}' = \mathbf{D}\tilde{\mathbf{A}}\mathbf{D}^{-1} \tag{8.205}$$

Such a similarity transformation is referred to as an *orthogonal transformation*.

4. If $\mathbf{A}'^{\dagger}$ is similar to $\mathbf{A}^{\dagger}$ if $\mathbf{A}'$ is similar to $\mathbf{A}$. Let

$$\text{(a)} \quad \mathbf{A}' = \mathbf{DAD}^{-1} \quad \Rightarrow \quad \text{(b)} \quad \mathbf{A}'^{\dagger} = (\mathbf{DAD}^{-1})^{\dagger} = (\mathbf{D}^{-1})^{\dagger}\mathbf{A}^{\dagger}\mathbf{D}^{\dagger} \equiv \mathbf{E}\mathbf{A}^{\dagger}\mathbf{E}^{-1} \tag{8.206}$$

If $\mathbf{D}$ is unitary, then $\mathbf{D}^{-1} = \mathbf{D}^{\dagger}$ and this becomes

$$\mathbf{A}'^{\dagger} = \mathbf{D}\mathbf{A}^{\dagger}\mathbf{D}^{-1} \tag{8.207}$$

Then the similarity transformation is called a *unitary transformation*.

5. If $\mathbf{A}'$ is similar to $\mathbf{A}$, then $\mathbf{A}'^{-1}$ is similar to $\mathbf{A}^{-1}$.

$$\mathbf{A}'^{-1} = (\mathbf{DAD}^{-1})^{-1} = \mathbf{DA}^{-1}\mathbf{D}^{-1} \tag{8.208}$$

6. If two or more matrices transform from one coordinate frame to another via the same similarity transformation, the sums and products of these matrices transform via the same similarity transformation. For example,

$$\text{(a)} \quad \mathbf{A}' = \mathbf{DAD}^{-1} \qquad \text{(b)} \quad \mathbf{B}' = \mathbf{DBD}^{-1} \tag{8.209}$$

$$\Rightarrow \quad \text{(a)} \quad (\mathbf{A}' + \mathbf{B}') = \mathbf{D}(\mathbf{A} + \mathbf{B})\mathbf{D}^{-1} \qquad \text{(b)} \quad \mathbf{A}'\mathbf{B}' = (\mathbf{DAD}^{-1})(\mathbf{DBD}^{-1}) = \mathbf{D}(\mathbf{AB})\mathbf{D}^{-1} \tag{8.210}$$

Since matrix equations involve the equality of sums and products of matrices, matrix equations have the same form when the matrices undergo similarity transformations. When equations retain their form under transformations, the equations are said to be *covariant*.

Eigenvalues and eigenvectors of a matrix

A matrix is called a *diagonal matrix* if all off-diagonal elements are zero. That is, the elements are of the form

$$a_{ij} = a_{ii}\delta_{ij} \tag{8.211}$$

One very useful feature of a similarity transformation is that it can be used to transform a matrix into diagonal form.

To see how this is done, consider

$$\mathbf{Y} = \mathbf{AX} \tag{8.197}$$

In many problems of scientific interest, $\mathbf{Y}$ is a multiple of $\mathbf{X}$. That is,

$$\mathbf{AX} = \lambda\mathbf{X} \tag{8.212}$$

This equation has the same form as the Sturm–Liouville eigenvalue equation given in equation 6.124. $\mathbf{A}$ is the matrix equivalent of the differential operator L_x, $\mathbf{X}$ is the matrix equivalent of the eigenfunction $\alpha_i(x)$, the weight function is 1, and λ is the eigenvalue.

Example 8.15

Referring to equation 8.67c, consider multiplying (operating on) the vectors

$$\text{(a)}\quad \mathbf{X}_+ \equiv \begin{bmatrix} 1 \\ 0 \end{bmatrix} \qquad \text{(b)}\quad \mathbf{X}_- \equiv \begin{bmatrix} 0 \\ 1 \end{bmatrix} \tag{8.213}$$

by the Pauli matrix $\boldsymbol{\sigma}_z$. It is straightforward to show that

$$\text{(a)}\quad \boldsymbol{\sigma}_z\mathbf{X}_+ = +\mathbf{X}_+ \qquad \text{(b)}\quad \boldsymbol{\sigma}_z\mathbf{X}_- = -\mathbf{X}_- \tag{8.214}$$

Thus $\mathbf{X}_+$ and $\mathbf{X}_-$ are eigenvectors of $\boldsymbol{\sigma}_z$ with the eigenvalues ± 1, respectively. We also notice from equation 8.67c that $\boldsymbol{\sigma}_z$ is a diagonal matrix with diagonal elements ± 1. □

To show that, in general, the diagonal elements of a diagonal matrix are its eigenvalues, we write the eigenvalue equation of equation 8.212 in the form

$$(\mathbf{A} - \lambda\mathbf{1})\mathbf{X} = \begin{bmatrix} (a_{11} - \lambda) & a_{12} & a_{13} & \cdots & a_{1N} \\ a_{21} & (a_{22} - \lambda) & a_{23} & \cdots & a_{2N} \\ a_{31} & a_{32} & (a_{33} - \lambda) & \cdots & a_{3N} \\ \vdots & & & & \\ a_{N1} & a_{N2} & a_{N3} & \cdots & (a_{NN} - \lambda) \end{bmatrix} \begin{bmatrix} x_1 \\ x_2 \\ x_3 \\ \\ x_N \end{bmatrix} = 0 \tag{8.215}$$

This is the matrix expression of a set of N homogeneous equations. In order that the solution be nontrivial, the coefficient determinant must be zero. That is,

$$|\mathbf{A} - \lambda\mathbf{1}| = 0 \tag{8.216}$$

This is called the *secular equation* and $|\mathbf{A} - \lambda\mathbf{1}|$ is the *secular determinant*. The values of λ that satisfy this equation are the *eigenvalues* of $\mathbf{A}$. The vector $\mathbf{X}$ that corresponds to a particular eigenvalue is the *eigenvector*.

When the secular determinant is expanded, the result is a polynomial in λ of order N. Therefore, in general, there will be N eigenvalues, although they are not all necessarily distinct. That is, just as with the Sturm–Liouville operator, degeneracies may exist among the eigenvalues of the matrix operator.

If $\mathbf{A}$ is in diagonal form, the secular equation is

$$\begin{vmatrix} (a_{11}-\lambda) & 0 & 0 & \cdots & 0 \\ 0 & (a_{22}-\lambda) & 0 & \cdots & 0 \\ 0 & 0 & (a_{33}-\lambda) & \cdots & 0 \\ \vdots & & & & \\ 0 & 0 & 0 & \cdots & (a_{NN}-\lambda) \end{vmatrix} \tag{8.217}$$

$$= (a_{11}-\lambda)(a_{22}-\lambda)\cdots(a_{NN}-\lambda) = 0$$

Therefore, it is easy to see that in diagonal form, the diagonal elements of a matrix are its eigenvalues. That is, in diagonal form

$$\mathbf{A} = \begin{bmatrix} \lambda_1 & 0 & 0 & \cdots & 0 \\ 0 & \lambda_2 & 0 & \cdots & 0 \\ 0 & 0 & \lambda_3 & \cdots & 0 \\ \vdots & & & & \\ 0 & 0 & 0 & \cdots & \lambda_N \end{bmatrix} \tag{8.218}$$

When $\mathbf{A}$ is in diagonal form, the eigenvalue equation is

$$\begin{bmatrix} (\lambda_1-\lambda) & 0 & 0 & \cdots & 0 \\ 0 & (\lambda_2-\lambda) & 0 & \cdots & 0 \\ 0 & 0 & (\lambda_3-\lambda) & \cdots & 0 \\ \vdots & & & & \\ 0 & 0 & 0 & \cdots & (\lambda_N-\lambda) \end{bmatrix} \begin{bmatrix} x_1 \\ x_2 \\ x_3 \\ \vdots \\ x_N \end{bmatrix} = \begin{bmatrix} (\lambda_1-\lambda)x_1 \\ (\lambda_2-\lambda)x_2 \\ (\lambda_3-\lambda)x_3 \\ \vdots \\ (\lambda_N-\lambda)x_N \end{bmatrix} = 0 \tag{8.219}$$

We first consider the case in which there are no degeneracies (all λ_i are distinct). For example, if $\lambda = \lambda_1 \neq \lambda_2 \neq \lambda_3 \neq \cdots \neq \lambda_N$, equation 8.219b can only be satisfied by $x_2 = x_3 = \cdots = x_N = 0$, and x_1 arbitrary. Thus $\mathbf{X}_1$ is a vector that has x_1 as the first element and zeros for all other elements. With identical arguments for $\lambda = \lambda_2$, $\lambda = \lambda_3, \ldots, \lambda = \lambda_N$, we deduce that in the space in which $\mathbf{A}$ is diagonal, and when there is no degeneracy, the eigenvectors are of the form

$$\mathbf{X}_i = x_i \begin{bmatrix} 0 \\ \vdots \\ 0 \\ 1 \\ 0 \\ \vdots \\ 0 \end{bmatrix} \tag{8.220}$$

a column vector with 1 in the ith row, and with all other elements 0.

As is done for the Sturm–Liouville eigenfunctions, the multiplicative constant x_i is determined by a normalization condition,

$$\mathbf{X}_i^\dagger \mathbf{X}_i = \tilde{\mathbf{X}}_i^* \mathbf{X}_i = N_i \tag{8.221}$$

For almost all applications, including the diagonalization of a matrix, the normalization factor is chosen to be $N_i = 1$. Referring to equation 8.220, this yields $x_i = e^{i\phi}$. The phase angle ϕ is arbitrary and is customarily taken to be 0. Thus $x_i = 1$.

If there is a degeneracy, there will be more than one non-zero x_j for that eigenvalue. For example, if $\lambda_1 = \lambda_2$, we see from equation 8.219b that when $\lambda = \lambda_1 = \lambda_2$, both x_1 and x_2 are arbitrary. Therefore, both $\mathbf{X}_1$ and $\mathbf{X}_2$ will be in the form

$$\mathbf{X}_1, \mathbf{X}_2 = \begin{bmatrix} x_1 \\ x_2 \\ 0 \\ \vdots \\ 0 \end{bmatrix} \tag{8.222}$$

while $\mathbf{X}_3$ through $\mathbf{X}_N$ will be in the form expressed in equation 8.220 with $x_i = 1$.

Just as we did with degenerate Sturm–Liouville eigenfunctions, we can orthogonalize degenerate eigenvectors by the Gram–Schmidt orthogonalization method. Defining

$$\langle X_i | X_j \rangle \equiv \mathbf{X}_i^\dagger \mathbf{X}_j \neq \delta_{ij} \tag{8.223}$$

we can obtain a set of orthonormalized eigenvectors $\mathbf{Y}_i$ from

$$\text{(a)} \quad \mathbf{Y}_1 = \mathbf{X}_1 \qquad \text{and} \qquad \text{(b)} \quad \mathbf{Y}_k = \mathbf{X}_k - \sum_{j=1}^{k-1} \frac{\langle X_k | X_j \rangle}{\langle X_j | X_j \rangle} \mathbf{X}_j \qquad k \geqslant 2 \tag{8.224}$$

Applying this to the degenerate eigenvectors of equation 8.222, it is straightforward to show that $\mathbf{Y}_1$ and $\mathbf{Y}_2$ can also be expressed in the same form as the nondegenerate eigenvectors of equation 8.220, and are then both normalized to 1. Therefore, whether there is a degeneracy or not, when $\mathbf{A}$ is in diagonal form, its eigenvectors will be in the form of equation 8.220 with $x_i = 1$.

We have determined that eigenvectors will transform according to

$$\mathbf{X}' = \mathbf{D}\mathbf{X} \tag{8.198a}$$

and matrices will transform according to the similarity transformation

$$\mathbf{A}' = \mathbf{D}\mathbf{A}\mathbf{D}^{-1} \tag{8.201b}$$

To see how such a transformation affects eigenvalues, consider the eigenvalue equation in the primed coordinate frame. As discussed above, matrix equations are covariant (they have the same form in both the primed and unprimed frames). Therefore, in the primed frame, the eigenvalue equation has the form

$$(\mathbf{A}' - \lambda' \mathbf{1})\mathbf{X}' = 0 \tag{8.225}$$

Using equations 8.198a and 8.210b, this becomes

$$\text{(a)} \quad (\mathbf{D}\mathbf{A}\mathbf{D}^{-1} - \lambda'\mathbf{1})\mathbf{D}\mathbf{X} = \mathbf{D}(\mathbf{A} - \lambda'\mathbf{1})\mathbf{X} = 0 \qquad \text{or} \qquad \text{(b)} \quad (\mathbf{A} - \lambda'\mathbf{1})\mathbf{X} = 0 \tag{8.226}$$

Comparing this to the eigenvalue equation in the unprimed frame, we see that $\lambda = \lambda'$; that is, eigenvalues are unchanged by a similarity transformation.

Diagonalization of a matrix

We are now prepared to construct the matrix that will diagonalize $\mathbf{A}$ by a similarity transformation. With the eigenvectors orthonormalized, consider a matrix $\mathbf{S}$, the columns of which are the column eigenvectors of the matrix $\mathbf{A}$. That is, in a frame in which $\mathbf{A}$ is not diagonal, we denote the ith eigenvector by

$$\mathbf{X}_i = \begin{bmatrix} x_{i1} \\ x_{i2} \\ \vdots \\ x_{iN} \end{bmatrix} \tag{8.227}$$

Then

$$\mathbf{S} = \begin{bmatrix} x_{11} & x_{21} & x_{31} & \cdots & x_{N1} \\ x_{12} & x_{22} & x_{32} & & x_{N2} \\ x_{13} & x_{23} & x_{33} & & x_{N3} \\ \vdots & & & & \\ x_{1N} & x_{2N} & x_{3N} & & x_{NN} \end{bmatrix} \equiv \Big[[\mathbf{X}_1] \quad [\mathbf{X}_2] \quad [\mathbf{X}_3] \quad \cdots \quad [\mathbf{X}_N] \Big] \tag{8.228}$$

Consider

$$\mathbf{S}^\dagger \mathbf{S} = \begin{bmatrix} [\mathbf{X}_1^*] \\ [\mathbf{X}_2^*] \\ \vdots \\ [\mathbf{X}_N^*] \end{bmatrix} \Big[[\mathbf{X}_1] \quad [\mathbf{X}_2] \quad \cdots \quad [\mathbf{X}_N] \Big] \tag{8.229}$$

where $[\mathbf{X}_i^*]$ is a row of $\mathbf{S}^\dagger$, the elements of which are the complex conjugated elements of $\mathbf{X}_i$. Then

$$\mathbf{S}^\dagger \mathbf{S} = \begin{bmatrix} [\mathbf{X}_1^*][\mathbf{X}_1] & [\mathbf{X}_1^*][\mathbf{X}_2] & \cdots & [\mathbf{X}_1^*][\mathbf{X}_N] \\ [\mathbf{X}_2^*][\mathbf{X}_1] & [\mathbf{X}_2^*][\mathbf{X}_2] & \cdots & [\mathbf{X}_2^*][\mathbf{X}_N] \\ \vdots & & & \\ [\mathbf{X}_N^*][\mathbf{X}_1] & [\mathbf{X}_N^*][\mathbf{X}_2] & \cdots & [\mathbf{X}_N^*][\mathbf{X}_N] \end{bmatrix} \tag{8.230}$$

The orthonormalization condition

$$\mathbf{X}_i^\dagger \mathbf{X}_j = \delta_{ij} \tag{8.223}$$

$$\Rightarrow \quad \mathbf{S}^\dagger \mathbf{S} = \begin{bmatrix} 1 & 0 & 0 & \cdots & 0 \\ 0 & 1 & 0 & \cdots & 0 \\ 0 & 0 & 1 & \cdots & 0 \\ \vdots & & & & \\ 0 & 0 & 0 & \cdots & 1 \end{bmatrix} = \mathbf{1} \tag{8.231}$$

That is, $\mathbf{S}$ is a unitary matrix.

Consider

$$\mathbf{AS} = \mathbf{A}\Big[[\mathbf{X}_1] \; [\mathbf{X}_2] \; [\mathbf{X}_3] \; \cdots \; [\mathbf{X}_N] \Big] \tag{8.232}$$

The multiplication of $\mathbf{A}$ into the first column of $\mathbf{S}$ is the product $\mathbf{AX}_1 = \lambda_1 \mathbf{X}_1$. Similarly, multiplying $\mathbf{A}$ into the jth column of $\mathbf{S}$ is $\mathbf{AX}_j = \lambda_j \mathbf{X}_j$. Therefore,

$$\mathbf{AS} = \Big[[\lambda_1 \mathbf{X}_1] \; [\lambda_2 \mathbf{X}_2] \; [\lambda_3 \mathbf{X}_3] \; \cdots \; [\lambda_N \mathbf{X}_N] \Big] \tag{8.233}$$

from which we obtain

$$\mathbf{S}^\dagger\mathbf{A}\mathbf{S} = \begin{bmatrix} [\mathbf{X}_1^*][\lambda_1\mathbf{X}_1] & [\mathbf{X}_1^*][\lambda_2\mathbf{X}_2] & \cdots & [\mathbf{X}_1^*][\lambda_N\mathbf{X}_N] \\ [\mathbf{X}_2^*][\lambda_1\mathbf{X}_1] & [\mathbf{X}_2^*][\lambda_2\mathbf{X}_2] & \cdots & [\mathbf{X}_2^*][\lambda_N\mathbf{X}_N] \\ \vdots & & & \\ [\mathbf{X}_N^*][\lambda_1\mathbf{X}_1] & [\mathbf{X}_N^*][\lambda_2\mathbf{X}_2] & \cdots & [\mathbf{X}_N^*][\lambda_N\mathbf{X}_N] \end{bmatrix} \quad (8.234)$$

Again, from the orthonormalization condition, we have

$$[\mathbf{X}_i^*]\Big[\lambda_j\mathbf{X}_1\Big] = \lambda_j\mathbf{X}_i^\dagger\mathbf{X}_j = \lambda_j\delta_{ij} \quad (8.235a)$$

$$\Rightarrow \quad \mathbf{S}^+\mathbf{A}\mathbf{S} = \begin{bmatrix} \lambda_1 & 0 & 0 & \cdots & 0 \\ 0 & \lambda_2 & 0 & \cdots & 0 \\ \vdots & & & & \\ 0 & 0 & 0 & \cdots & \lambda_N \end{bmatrix} \quad (8.235b)$$

Thus $\mathbf{S}^\dagger\mathbf{A}\mathbf{S}$ yields $\mathbf{A}$ in diagonalized form. Since $\mathbf{S}$ is unitary,

$$\mathbf{S}^\dagger\mathbf{A}\mathbf{S} = (\mathbf{S}^\dagger)\mathbf{A}(\mathbf{S}^\dagger)^{-1} \quad (8.236a)$$

which is a similarity transformation. That is, $\mathbf{S}^\dagger$ is the transformation matrix defined as $\mathbf{D}$ in equation 8.201b. Therefore, vectors transform under this similarity transformation as

$$\mathbf{X}' = \mathbf{S}^\dagger\mathbf{X} \quad (8.236b)$$

In order to diagonalize $\mathbf{A}$, we see from the development above that we must first solve the secular equation for the eigenvalues and from those, obtain the eigenvectors. If $\mathbf{A}$ were the only matrix involved in a particular problem, and the eigenvectors the only vectors, going through the procedure above would be a waste of time. Once the eigenvalues are known, $\mathbf{A}$ in diagonal form is known to have the eigenvalues as the diagonal elements, and each eigenvector is known to contain 1 in the appropriate row and 0 everywhere else. However, there are often other significant matrices and vectors involved in a problem, and a knowledge of their forms in the frame in which $\mathbf{A}$ is diagonal is critical to the problem. In the Heisenberg formulation of quantum mechanics, for example, one usually solves a problem in a frame in which the energy matrix is diagonal, and therefore, its diagonal elements are the energies of the quantum states of the system. It is essential to also know the forms of the momentum and coordinate matrices in this space.

If the motion of a rotating rigid body were investigated in the coordinate system fixed on the body, vectors fixed on the body (angular momentum, for example) would be easily described. For example, if the body-centered z-axis were defined as the axis of rotation, the angular momentum vector in the body coordinates would be

$$\mathbf{L} = \begin{bmatrix} 0 \\ 0 \\ I\omega \end{bmatrix} \quad (8.237)$$

To describe those vectors in a laboratory-based coordinate system, one would have to know the orthogonal transformation between the body-centered and the laboratory-centered coordinate frames.

Example 8.16

As an example, let us diagonalize the matrix

$$\mathbf{A} = \begin{bmatrix} 1 & 1 & 0 \\ 1 & 0 & 1 \\ 0 & 1 & 1 \end{bmatrix} \quad (8.238)$$

by a similarity transformation. The secular equation is

$$\begin{vmatrix} (1-\lambda) & 1 & 0 \\ 1 & -\lambda & 1 \\ 0 & 1 & (1-\lambda) \end{vmatrix} = (1-\lambda)(\lambda^2 - \lambda - 2) = 0 \tag{8.239}$$

$$\begin{bmatrix} 1 & 1 & 0 \\ 1 & 0 & 1 \\ 0 & 1 & 1 \end{bmatrix}\begin{bmatrix} x_1 \\ x_2 \\ x_3 \end{bmatrix} = 2\begin{bmatrix} x_1 \\ x_2 \\ x_3 \end{bmatrix} \tag{8.240}$$

$$\Rightarrow \quad \text{(a)} \quad x_1 + x_2 = 2x_1 \qquad \text{(b)} \quad x_1 + x_3 = 2x_2 \qquad \text{(c)} \quad x_2 + x_3 = 2x_3 \tag{8.241}$$

This set of equations has one solution

$$x_1 = x_2 = x_3 \tag{8.242}$$

$$\Rightarrow \quad \mathbf{X}_2 = x_1\begin{bmatrix} 1 \\ 1 \\ 1 \end{bmatrix} \tag{8.243a}$$

which, when normalized, becomes

$$\mathbf{X}_2 = \frac{1}{\sqrt{3}}\begin{bmatrix} 1 \\ 1 \\ 1 \end{bmatrix} \tag{8.243b}$$

By identical analysis,

$$\text{(a)} \quad \mathbf{X}_1 = \frac{1}{\sqrt{2}}\begin{bmatrix} 1 \\ 0 \\ -1 \end{bmatrix} \qquad \text{(b)} \quad \mathbf{X}_{-1} = \frac{1}{\sqrt{6}}\begin{bmatrix} 1 \\ -2 \\ 1 \end{bmatrix} \tag{8.244}$$

Thus the unitary matrix that diagonalizes $\mathbf{A}$ by the transformation $\mathbf{S}^\dagger\mathbf{A}\mathbf{S}$ is

$$\mathbf{S} = \begin{bmatrix} \frac{1}{\sqrt{2}} & \frac{1}{\sqrt{6}} & \frac{1}{\sqrt{3}} \\ 0 & \frac{-2}{\sqrt{6}} & \frac{1}{\sqrt{3}} \\ \frac{-1}{\sqrt{2}} & \frac{1}{\sqrt{6}} & \frac{1}{\sqrt{3}} \end{bmatrix} \tag{8.245}$$

with

$$\mathbf{S}^\dagger\mathbf{A}\mathbf{S} = \begin{bmatrix} 1 & 0 & 0 \\ 0 & -1 & 0 \\ 0 & 0 & 2 \end{bmatrix} \tag{8.246}$$

and

$$\text{(a)} \quad \mathbf{S}^\dagger\mathbf{X}_1 = \begin{bmatrix} 1 \\ 0 \\ 0 \end{bmatrix} \qquad \text{(b)} \quad \mathbf{S}^\dagger\mathbf{X}_{-1} = \begin{bmatrix} 0 \\ 1 \\ 0 \end{bmatrix} \qquad \text{(c)} \quad \mathbf{S}^\dagger\mathbf{X}_2 = \begin{bmatrix} 0 \\ 0 \\ 1 \end{bmatrix} \quad \square \tag{8.247}$$

Example 8.17

There is no degeneracy for the matrix in Example 8.16. For the matrix

$$\mathbf{A} = \begin{bmatrix} 0 & 1 & 1 \\ 1 & 0 & 1 \\ 1 & 1 & 0 \end{bmatrix} \tag{8.248}$$

the secular equation is

$$-\lambda^3 + 3\lambda + 2 = 0 \tag{8.249}$$

which has solution $\lambda = 2, -1, -1$. Thus there is a two-fold degeneracy. The eigenvalue equation and normalization yield a unique eigenvector for $\lambda = 2$. It is

$$\mathbf{X}_2 = \frac{1}{\sqrt{3}}\begin{bmatrix} 1 \\ 1 \\ 1 \end{bmatrix} \tag{8.250}$$

For $\lambda = -1$, the eigenvalue equation yields

$$x_1 + x_2 + x_3 = 0 \tag{8.251}$$

so the two eigenvectors for $\lambda = -1$ are of the form

$$\mathbf{X}_{-1} = \begin{bmatrix} x_1 \\ x_2 \\ -(x_1 + x_2) \end{bmatrix} \tag{8.252}$$

Referring to equation 8.250, $\mathbf{X}_{-1}$ will be orthogonal to $\mathbf{X}_2$ for any choice of x_1, x_2.

As can be seen, there are an infinite number of choices that can be made to represent $\mathbf{X}_{-1}$. For example, one reasonable choice is to take $x_1 + x_2 = 0$. Then, after normalizing,

$$\mathbf{X}_{-1} = \frac{1}{\sqrt{2}}\begin{bmatrix} 1 \\ -1 \\ 0 \end{bmatrix} \tag{8.253}$$

Any other independent choice of x_1 and x_2 will yield a second eigenvector corresponding to $\lambda = -1$. However, after that second eigenvector is determined, it will be necessary to employ the Gram–Schmidt process to determine a second vector orthogonal to the one expressed in equation 8.253. The orthogonalization process can be avoided by selecting values of x_1 and x_2 that make the second eigenvector orthogonal to the $\mathbf{X}_{-1}$ of equation 8.253. Thus we set

$$\begin{bmatrix} 1 & -1 & 0 \end{bmatrix}\begin{bmatrix} x_1 \\ x_2 \\ -(x_1 + x_2) \end{bmatrix} = 0 \tag{8.254}$$

from which we obtain $x_1 = x_2$. The second eigenvector corresponding to $\lambda = -1$, which is orthogonal to $\mathbf{X}_{-1}$ given in equation 8.253, is

$$\mathbf{X}'_{-1} = \frac{1}{\sqrt{2}}\begin{bmatrix} 1 \\ 1 \\ -2 \end{bmatrix} \quad \square \tag{8.255}$$

Coupled differential equations: small oscillations

The problem of several particles that interact with one another is often described by a set of coupled differential equations. In many cases, these can be cast into matrix form, and by diagonalizing the matrices involved, the differential equations can be decoupled.

One such example is that of the problem of a multiparticle system in oscillation. Let the positions of the particles in such a system be described by coordinates $x_1, x_2, \ldots, x_N$. The kinetic energy of the aggregate of particles is

$$T = \tfrac{1}{2}M_1\dot{x}_1^2 + \tfrac{1}{2}M_2\dot{x}_2^2 + \cdots + \tfrac{1}{2}M_N\dot{x}_N^2 \tag{8.256}$$

We define the equilibrium position of each particle by $x_i = 0$. When each particle is at its equilibrium position, the potential energy will be a minimum. We define this minimum to be the zero of potential energy. In addition, the potential energy is a minimum at the origin. Thus

$$\text{(a)}\quad V(\mathbf{0}) \equiv 0 \qquad \text{and} \qquad \text{(b)}\quad \left.\frac{\partial V}{\partial x_i}\right|_{x_i=0} = 0 \tag{8.257}$$

Therefore, a MacLaurin expansion of $V(\mathbf{x})$ about $\mathbf{x} = 0$ yields

$$\begin{aligned} V(\mathbf{x}) &= V(\mathbf{0}) + \sum_{i=1}^{N} x_i \frac{\partial V}{\partial x_i}\bigg|_{x_i=0} + \sum_{i,j=1}^{N} x_i x_j \frac{1}{2!}\frac{\partial^2 V}{\partial x_i \partial x_j}\bigg|_{x_i=x_j=0} + \cdots \\ &= \frac{1}{2}\sum_{i,j=1}^{N} x_i x_j \frac{\partial^2 V}{\partial x_i \partial x_j}\bigg|_{x_i=x_j=0} + \cdots \end{aligned} \tag{8.258}$$

If the amplitude of oscillation of each particle is small, we can ignore terms of higher order than the leading nonzero quadratic term. If the particles interact via Hooke's law potentials ($V = \frac{1}{2}kx^2$), the quadratic term is the only nonzero term in the expansion for any amplitude.

We define the expansion of the potential energy in terms of

$$K_{ij} \equiv \frac{\partial^2 V}{\partial x_i \partial x_j}\bigg|_{x_i=x_j=0.} \tag{8.259}$$

$$\Rightarrow \quad V(\mathbf{x}) \simeq \frac{1}{2}\sum_{i,j=1}^{N} K_{ij}x_i x_j = \frac{1}{2}\left[K_{11}x_1^2 + K_{12}x_1x_2 + K_{21}x_2x_1 + K_{22}x_2^2 + \cdots\right] \tag{8.260}$$

Therefore, for this system, the Lagrangian introduced in Chapter 7 becomes

$$L = \frac{1}{2}\left[M_1\dot{x}_1^2 + M_2\dot{x}_2^2 + \cdots + M_N\dot{x}_N^2\right] - \frac{1}{2}\left[K_{11}x_1^2 + K_{12}x_1x_2 + K_{21}x_2x_1 + \cdots\right] \tag{8.261}$$

Therefore, the Lagrange equations of motion (see Chapter 7) are

$$M_i\ddot{x}_i + K_{i1}x_1 + K_{i2}x_2 + \cdots = M_i\ddot{x}_i + \sum_{j=1}^{N} K_{ij}x_j = M_i\ddot{x}_i + \sum_{j=1}^{N} \frac{K_{ij}}{M_j}(M_jx_j) = 0 \tag{8.262}$$

Defining a column vector $\mathbf{Z}$ with elements M_ix_i and a matrix $\mathbf{\Lambda}$ with elements $\frac{K_{ij}}{M_j}$, equation 8.262 can be written in matrix form as

$$\begin{bmatrix} M_1\ddot{x}_1 \\ \vdots \\ M_N\ddot{x}_N \end{bmatrix} + \begin{bmatrix} \frac{K_{11}}{M_1} & \frac{K_{12}}{M_2} & \cdots & \frac{K_{1N}}{M_N} \\ \vdots & & & \\ \frac{K_{N1}}{M_1} & \frac{K_{N2}}{M_2} & \cdots & \frac{K_{NN}}{M_N} \end{bmatrix} \begin{bmatrix} M_1x_1 \\ \vdots \\ M_Nx_N \end{bmatrix} \equiv \ddot{\mathbf{Z}} + \mathbf{\Lambda Z} = 0 \tag{8.263}$$

For simplicity, let the masses be the same. Then this set of equations becomes

$$M\begin{bmatrix} \ddot{x}_1 \\ \ddot{x}_2 \\ \vdots \\ \ddot{x}_N \end{bmatrix} + \begin{bmatrix} K_{11} & K_{12} & \cdots & K_{1N} \\ K_{21} & K_{22} & \cdots & K_{2N} \\ \vdots & & & \\ K_{N1} & K_{N2} & \cdots & K_{NN} \end{bmatrix}\begin{bmatrix} x_1 \\ x_2 \\ \vdots \\ x_N \end{bmatrix} \equiv M\ddot{\mathbf{X}} + \mathbf{KX} = 0 \tag{8.264}$$

Let $\mathbf{S}$ be the unitary matrix that diagonalizes $\mathbf{K}$. Premultiplying equation 8.264b by $\mathbf{S}^\dagger$, we obtain

$$M\mathbf{S}^\dagger\ddot{\mathbf{X}} + \mathbf{S}^\dagger\mathbf{KX} = M\mathbf{S}^\dagger\ddot{\mathbf{X}} + \mathbf{S}^\dagger\mathbf{KSS}^\dagger\mathbf{X} = 0 \tag{8.265}$$

where we have inserted the identity matrix in the form $\mathbf{SS}^\dagger$ between $\mathbf{K}$ and $\mathbf{X}$.

Writing

$$\text{(a)}\quad \mathbf{K}' = \mathbf{S}^\dagger\mathbf{K}\mathbf{S} = \begin{bmatrix} K'_{11} & 0 & \cdots & 0 \\ 0 & K'_{22} & \cdots & 0 \\ \vdots & & & \\ 0 & 0 & \cdots & K'_{NN} \end{bmatrix} \quad \text{and} \quad \text{(b)}\quad \mathbf{Y} = \mathbf{S}^\dagger\mathbf{X} = \begin{bmatrix} y_1 \\ y_2 \\ \vdots \\ y_N \end{bmatrix} \tag{8.266}$$

equation 8.265 can now be expressed in component form as

$$M\ddot{y}_j + K'_{jj}y_j = 0 \tag{8.267}$$

That is, by diagonalizing the matrix **K**, the differential equations have been decoupled.

The solution to equation 8.267 is

$$y_j = \alpha_j e^{i\omega_j t} + \beta_j e^{-i\omega_j t} \tag{8.268}$$

where

$$\omega_j = \left[\frac{K'_{jj}}{M}\right]^{1/2} \tag{8.269}$$

In diagonal form, we know that K'_{jj} is the jth eigenvalue of **K**. Thus this eigenvalue analysis also yields the natural frequencies of the oscillating system.

Example 8.18

As an example, we consider a "tube" of length L, with a small enough diameter to constrain protons placed in it to move in one dimension. If two protons are placed in the tube, and the ends of the tube are charged to $\frac{4}{3}e$ (e is the charge of a proton), the equilibrium position of each proton will be $\frac{1}{3}L$ from one end of the tube as shown in Figure 8.9.

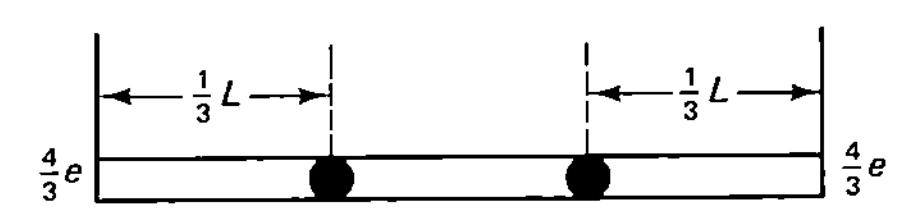

Figure 8.9 Two protons confined to move in one dimension.

When the protons are displaced small distances x_1 and x_2 from their equilibrium positions, the potential energy of the system is

$$V = \frac{\frac{4}{3}e^2}{(\frac{1}{3}L + x_1)} + \frac{\frac{4}{3}e^2}{(\frac{2}{3}L - x_1)} + \frac{\frac{4}{3}e^2}{(\frac{2}{3}L + x_2)} + \frac{\frac{4}{3}e^2}{(\frac{1}{3}L - x_2)} + \frac{e^2}{(\frac{1}{3}L - x_1 + x_2)} - 15\frac{e^2}{L} \tag{8.270}$$

where the constant $-15(e^2/L)$ ensures that $V(x_1 = x_2 = 0) = 0$. It is easy to verify that at equilibrium

$$\left.\frac{\partial V}{\partial x_1}\right|_0 = \left.\frac{\partial V}{\partial x_2}\right|_0 = 0 \tag{8.271}$$

Therefore, with x_1 and x_2 small,

$$\begin{aligned} V &\simeq \frac{1}{2}\left.\frac{\partial^2 V}{\partial x_1^2}\right|_0 x_1^2 + \frac{1}{2}\left\{\left.\frac{\partial^2 V}{\partial x_1\,\partial x_2}\right|_0 x_1x_2 + \left.\frac{\partial^2 V}{\partial x_2\,\partial x_1}\right|_0 x_2x_1\right\} + \frac{1}{2}\left.\frac{\partial^2 V}{\partial x_2^2}\right|_0 x_2^2 \\ &= \frac{1}{2}\left(135\frac{e^2}{L^3}\right)x_1^2 - \frac{1}{2}\left(27\frac{e^2}{L^3}\right)x_1x_2 - \frac{1}{2}\left(27\frac{e^2}{L^3}\right)x_2x_1 + \frac{1}{2}\left(135\frac{e^2}{L^3}\right)x_2^2 \end{aligned} \tag{8.272}$$

We note that we do not combine the term $\left.\frac{\partial^2 V}{\partial x_1 \partial x_2}\right|_0 x_1 x_2$ with $\left.\frac{\partial^2 V}{\partial x_2 \partial x_1}\right|_0 x_2 x_1$ even though these terms are equal. They are separate terms in the expansion of the potential. Thus the coupled equations of motion, in matrix form, are

$$M\begin{bmatrix} \ddot{x}_1 \\ \ddot{x}_2 \end{bmatrix} + \frac{e^2}{L^3}\begin{bmatrix} 135 & -27 \\ -27 & 135 \end{bmatrix}\begin{bmatrix} x_1 \\ x_2 \end{bmatrix} = 0 \tag{8.273}$$

From the secular equation

$$\begin{vmatrix} 135-\mu & -27 \\ -27 & 135-\mu \end{vmatrix} = (135-\mu)^2 - 27^2 = 0 \tag{8.274}$$

the eigenvalues of the $\mathbf{K}$ matrix, $\lambda = \mu e^2/L^3$, are

$$\text{(a)}\quad \lambda = (135+27)\frac{e^2}{L^3} = 162\frac{e^2}{L^3} \qquad \text{(b)}\quad \lambda = (135-27)\frac{e^2}{L^3} = 108\frac{e^2}{L^3} \tag{8.275}$$

Therefore, the protons will oscillate with angular frequencies

$$\text{(a)}\quad \omega_1 = \left[\frac{162e^2}{ML^3}\right]^{1/2} \qquad \text{(b)}\quad \omega_2 = \left[\frac{108e^2}{ML^3}\right]^{1/2} \tag{8.276}$$

The eigenvectors are determined by

$$\text{(a)}\quad \begin{bmatrix} 135 & -27 \\ -27 & 135 \end{bmatrix}\begin{bmatrix} \alpha \\ \beta \end{bmatrix} = 162\begin{bmatrix} \alpha \\ \beta \end{bmatrix} \qquad \text{(b)}\quad \begin{bmatrix} 135 & -27 \\ -27 & 135 \end{bmatrix}\begin{bmatrix} \alpha \\ \beta \end{bmatrix} = 108\begin{bmatrix} \alpha \\ \beta \end{bmatrix} \tag{8.277}$$

$$\Rightarrow \quad \text{(a)}\quad \mathbf{X}_{162} = \frac{1}{\sqrt{2}}\begin{bmatrix} 1 \\ -1 \end{bmatrix} \qquad \text{(b)}\quad \mathbf{X}_{108} = \frac{1}{\sqrt{2}}\begin{bmatrix} 1 \\ 1 \end{bmatrix} \tag{8.278}$$

From this, we obtain the unitary matrix that diagonalizes $\mathbf{K}$. It is

$$\text{(a)}\quad \mathbf{S} = \frac{1}{\sqrt{2}}\begin{bmatrix} 1 & 1 \\ -1 & 1 \end{bmatrix} \quad \Rightarrow \quad \text{(b)}\quad \mathbf{S}^\dagger = \frac{1}{\sqrt{2}}\begin{bmatrix} 1 & -1 \\ 1 & 1 \end{bmatrix} \tag{8.279}$$

The vector $\mathbf{X}$, in the space in which $\mathbf{K}$ is diagonal, is

$$\begin{bmatrix} y_1 \\ y_2 \end{bmatrix} = \mathbf{S}^\dagger\begin{bmatrix} x_1 \\ x_2 \end{bmatrix} = \frac{1}{\sqrt{2}}\begin{bmatrix} x_1 - x_2 \\ x_1 + x_2 \end{bmatrix} \tag{8.280}$$

and the decoupled equations of motion become

$$\text{(a)}\quad \ddot{y}_1 + 162\frac{e^2}{ML^3}y_1 = 0 \qquad \text{(b)}\quad \ddot{y}_2 + 108\frac{e^2}{ML^3}y_2 = 0 \tag{8.281}$$

These are easily solved using methods developed in Chapter 5. The solutions are

$$\text{(a)}\quad y_1 = 2a_1e^{i\omega_1 t} + 2b_1e^{-i\omega_1 t} \qquad \text{(b)}\quad y_2 = 2a_2e^{i\omega_2 t} + 2b_2e^{-i\omega_2 t} \tag{8.282}$$

Using equation 8.280, the solution for the physical position variables becomes

$$x_1 = \tfrac{1}{2}(y_2 + y_1) = a_2e^{i\omega_2 t} + b_2e^{-i\omega_2 t} + a_1e^{i\omega_1 t} + b_1e^{-i\omega_1 t} \tag{8.283a}$$

$$x_2 = \tfrac{1}{2}(y_2 - y_1) = a_2e^{i\omega_2 t} + b_2e^{-i\omega_2 t} - a_1e^{i\omega_1 t} - b_1e^{-i\omega_1 t} \tag{8.283b}$$

The coefficients are determined by initial conditions, of course. □

Simultaneous diagonalization of commuting matrices

To say that $\mathbf{A}$ and $\mathbf{B}$ are commuting matrices means that

$$\mathbf{AB} - \mathbf{BA} = \mathbf{0} \tag{8.284}$$

This is often denoted by

$$(\mathbf{A}, \mathbf{B}) = \mathbf{0} \tag{8.285}$$

$(\mathbf{A}, \mathbf{B})$ is called the *commutator* of $\mathbf{A}$ with $\mathbf{B}$. (Recall the definition of the anticommutator given in equation 8.70.)

Let λ_i be a nondegenerate eigenvalue of $\mathbf{A}$ with corresponding eigenvector $\mathbf{X}_i$. That is,

$$\mathbf{AX}_i = \lambda_i \mathbf{X}_i \tag{8.212}$$

Premultiplication by $\mathbf{B}$ yields

$$\mathbf{BAX}_i = \lambda_i \mathbf{BX}_i \tag{8.286}$$

Using the fact that $\mathbf{A}$ and $\mathbf{B}$ commute, this can be written

$$\mathbf{ABX}_i = \lambda_i \mathbf{BX}_i \tag{8.287}$$

$\mathbf{BX}_i$ is a column vector which we define as $\mathbf{Y}_i$. Therefore, equation 8.287 is

$$\mathbf{AY}_i = \lambda_i \mathbf{Y}_i \tag{8.288}$$

Since λ_i is a nondegenerate eigenvalue, the eigenvector corresponding to it is unique, up to a multiplicative constant. That is,

$$\mathbf{Y}_i = \mu_i \mathbf{X}_i = \mathbf{BX}_i \tag{8.289}$$

Therefore, $\mathbf{X}_i$ is also an eigenvector of $\mathbf{B}$ with eigenvalue μ_i. Thus, if two matrices commute, the eigenvector of every nondegenerate eigenvalue of one matrix is also an eigenvector of the other.

Now consider an N-fold degeneracy for $\lambda = \lambda_1$. Then

$$\mathbf{AX}_i = \lambda_1 \mathbf{X}_i \qquad 1 \leqslant i \leqslant N \tag{8.290}$$

Using the commutativity of $\mathbf{A}$ and $\mathbf{B}$, we have

$$\mathbf{BAX}_i = \lambda_1 \mathbf{BX}_i = \mathbf{ABX}_i \tag{8.291}$$

That is, $\mathbf{BX}_i$ is an eigenvector of $\mathbf{A}$ with eigenvalue λ_1. Because of the N-fold degeneracy, there are N possible vectors that correspond to the eigenvalue λ_1. Therefore,

$$\mathbf{BX}_i = \sum_{k=1}^{N} m_{ik} \mathbf{X}_k \qquad 1 \leqslant i \leqslant N \tag{8.292}$$

Viewing m_{ik} as an element of a matrix $\mathbf{M}$, let the matrix $\mathbf{D}$ be a transformation matrix that diagonalizes $\mathbf{M}$. That is, we take

$$\mathbf{DMD}^{-1} = \begin{bmatrix} \mu_1 & 0 & 0 & \cdots & 0 \\ 0 & \mu_2 & 0 & \cdots & 0 \\ 0 & 0 & \mu_3 & \cdots & 0 \\ \vdots & & & & \\ 0 & 0 & 0 & \cdots & \mu_N \end{bmatrix} \equiv \mathbf{M}' \tag{8.293}$$

$$\Rightarrow \quad \sum_{k=1}^{N} d_{ik} m_{kj} = \sum_{k=1}^{N} m'_{ik} d_{kj} = \mu_i d_{ij} \tag{8.294}$$

We now define a set of vectors

$$\mathbf{Y}_i \equiv \sum_{j=1}^{N} d_{ij}\mathbf{X}_j \tag{8.295}$$

Consider

$$\mathbf{Y}_l^{\dagger}\mathbf{Y}_m = \sum_{j=1}^{N}\sum_{k=1}^{N} \mathbf{X}_k^{\dagger} d_{lk}^{*} d_{mj}\mathbf{X}_j \tag{8.296}$$

With $\mathbf{X}_k^{\dagger}\mathbf{X}_j = \delta_{kj}$, this becomes

$$\mathbf{Y}_l^{\dagger}\mathbf{Y}_m = \sum_{k=1}^{N} d_{mk}\tilde{d}_{kl}^{*} = (\mathbf{D}\mathbf{D}^{\dagger})_{ml} \tag{8.297}$$

Thus if $\mathbf{D}$ is unitary, the new vectors are naturally mutually orthogonal to one another. If $\mathbf{D}$ is not unitary, the Gram–Schmidt process must be employed to orthogonalize them. In either case, a set of orthonormalized vectors $\mathbf{Y}_i$ can be generated.

Consider

$$\mathbf{B}\mathbf{Y}_i = \sum_{j=1}^{N} d_{ij}\mathbf{B}\mathbf{X}_j = \sum_{j=1}^{N} d_{ij}\sum_{k=1}^{N} m_{jk}\mathbf{X}_k \tag{8.298}$$

But from equation 8.294,

$$\sum_{j=1}^{N} d_{ij}m_{jk} = \mu_i d_{ik} \tag{8.294}$$

$$\Rightarrow \quad \mathbf{B}\mathbf{Y}_i = \mu_i \sum_{k=1}^{N} d_{ik}\mathbf{X}_k = \mu_i\mathbf{Y}_i \tag{8.299}$$

That is, $\mathbf{Y}_i$ is an eigenvector of $\mathbf{B}$ with eigenvalue μ_i.

Now consider

$$\mathbf{A}\mathbf{Y}_i = \sum_{k=1}^{N} d_{ik}\mathbf{A}\mathbf{X}_k = \lambda_1 \sum_{k=1}^{N} d_{ik}\mathbf{X}_k = \lambda_1\mathbf{Y}_i \tag{8.300}$$

Thus for the N-fold degenerate subset of eigenvalues $\lambda_1 = \lambda_2 = \cdots = \lambda_N$, we can determine an orthonormal set of vectors $\mathbf{Y}_i$ that are eigenvectors of the commuting matrices $\mathbf{A}$ and $\mathbf{B}$. Therefore, with or without a degeneracy, we see that commuting matrices have common eigenvectors. Thus the unitary transformation matrix with these eigenvectors as columns, is the same for both matrices. So if $\mathbf{A}$ and $\mathbf{B}$ commute, they can be diagonalized by the same unitary transformation (which is also termed simultaneous diagonalization of the commuting matrices $\mathbf{A}$ and $\mathbf{B}$).

Eigenvalues and eigenvectors of hermitian matrices

The hermitian matrix is of particular interest in the investigation of physical phenomena. In Chapter 6 we studied hermitian operators and showed that their eigenvalues were real, and the eigenfunctions of the nondegenerate eigenvalues were naturally orthogonal. Because the eigenvalues are real, they are suitable for representing physical quantities.

To demonstrate that hermitian matrices have these same properties, let $\mathbf{A}^{\dagger} = \mathbf{A}$ and consider the eigenvalue equation

$$\mathbf{A}\mathbf{X}_i = \lambda_i\mathbf{X}_i \tag{8.301}$$

The analysis that follows for the hermitian matrix is in direct one-to-one correspondence with the analysis of the hermitian operator discussed in Chapter 6.

The hermitian adjoint of equation 8.301 is

$$\mathbf{X}_i^\dagger \mathbf{A}^\dagger = \lambda_i^* \mathbf{X}_i^\dagger \tag{8.302a}$$

We use the fact that $\mathbf{A}$ is hermitian and change the index to j in writing this as

$$\mathbf{X}_j^\dagger \mathbf{A} = \lambda_j^* \mathbf{X}_j^\dagger \tag{8.302b}$$

$\mathbf{X}_j^\dagger$ is a row vector, the elements of which are the complex conjugates of the elements of $\mathbf{X}_j$.

We premultiply equation 8.301 by $\mathbf{X}_j^\dagger$ and postmultiply equation 8.302b by $\mathbf{X}_i$ to obtain

$$\text{(a)} \quad \mathbf{X}_j^\dagger \mathbf{A}\mathbf{X}_i = \lambda_i \mathbf{X}_j^\dagger \mathbf{X}_i \qquad \text{and} \qquad \text{(b)} \quad \mathbf{X}_j^\dagger \mathbf{A}\mathbf{X}_i = \lambda_j^* \mathbf{X}_j^\dagger \mathbf{X}_i \tag{8.303}$$

$$\Rightarrow \quad (\lambda_i - \lambda_j^*)\mathbf{X}_j^\dagger \mathbf{X}_i = 0 \tag{8.304}$$

For $j = i$, this becomes

$$(\lambda_i - \lambda_i^*)\mathbf{X}_i^\dagger \mathbf{X}_i = 0 \tag{8.305}$$

Since $\mathbf{X}_i^\dagger \mathbf{X}_j$ is the square of the length of the vector $\mathbf{X}_i$ (which we will eventually normalize to 1), $\mathbf{X}_i^\dagger \mathbf{X}_i \neq 0$. Therefore,

$$\lambda_i = \lambda_i^* \tag{8.306}$$

That is, every eigenvalue is real. Thus equation 8.304 becomes

$$(\lambda_i - \lambda_j)\mathbf{X}_j^\dagger \mathbf{X}_i = 0 \tag{8.307}$$

For $j \neq i$, either $\lambda_i = \lambda_j$ (a degeneracy), or $\mathbf{X}_j^\dagger \mathbf{X}_i = 0$. That is, for nondegenerate eigenvalues, the eigenvectors are naturally orthogonal. If a degeneracy does exist, however, we can generate a complete set of orthogonal eigenvectors with the Gram–Schmidt process, or an equivalent method such as was described in Example 8.17.

Sylvester's theorem and an iterative method for estimating the eigenvalues and eigenvectors of a hermitian matrix

As a final note, consider a matrix that has real, nondegenerate eigenvalues. To ensure that such eigenvalues are real, we will take the matrix $\mathbf{A}$ to be hermitian. Thus, since we are assuming a nondegenerate system, all eigenvectors are orthogonal without resorting to an orthogonalization technique.

Let $Q(x)$ be an Nth-order polynomial, the roots of which are the eigenvalues of $\mathbf{A}$. That is,

$$Q(x) = (x - \lambda_1)(x - \lambda_2) \cdots (x - \lambda_N) = \prod_{i=1}^{N} (x - x_i) \tag{8.308}$$

If $P(x)$ is some polynomial, we can write the ratio $P(x)/Q(x)$ in terms of partial fractions.

$$\frac{P(x)}{Q(x)} = \frac{P(x)}{\prod_{i=1}^{N}(x - \lambda_i)} \equiv \sum_{j=1}^{N} \frac{\alpha_j}{(x - \lambda_j)} \tag{8.309}$$

$$\Rightarrow \quad P(x) = \sum_{j=1}^{N} \alpha_j \frac{\prod_{i=1}^{N}(x - \lambda_i)}{(x - \lambda_j)} = \sum_{j=1}^{N} \alpha_j \prod_{\substack{i=1 \\ i \neq j}}^{N} (x - \lambda_i) \tag{8.310}$$

To determine the constants α_j, using equation 8.309, we consider

$$\lim_{x \to \lambda_k} \frac{(x - \lambda_k)P(x)}{\prod_{i=1}^{N}(x - \lambda_i)} = \frac{P(\lambda_k)}{\prod_{\substack{i=1 \\ i \neq k}}^{N}(\lambda_k - \lambda_i)} = \lim_{x \to \lambda_k} \sum_{j=1}^{N} \alpha_j \frac{(x - \lambda_k)}{(x - \lambda_j)} = \alpha_k \tag{8.311}$$

$$\Rightarrow \quad P(x) = \sum_{j=1}^{N} P(\lambda_j) \prod_{\substack{i=1 \\ i \neq j}}^{N} \frac{(x - \lambda_i)}{(\lambda_j - \lambda_i)} \tag{8.312}$$

$$\Rightarrow \quad P(\mathbf{A}) = \sum_{j=1}^{N} P(\lambda_j) \prod_{\substack{i=1 \\ i \neq j}}^{N} \frac{(\mathbf{A} - \lambda_i \mathbf{1})}{(\lambda_j - \lambda_i)} \tag{8.313}$$

This is *Sylvester's theorem.*

In particular, if $P(\mathbf{A})$ is some positive power of $\mathbf{A}$, Sylvester's equation becomes

$$\mathbf{A}^k = \sum_{j=1}^{N} \lambda_j^k \prod_{\substack{i=1 \\ i \neq j}}^{N} \frac{(\mathbf{A} - \lambda_i \mathbf{1})}{(\lambda_j - \lambda_i)} \tag{8.314}$$

If λ_1 is the eigenvalue with the largest magnitude (λ_i can be positive or negative), then when k is large enough, the $j = 1$ term will dominate the sum of equation 8.314. That is, for large enough k,

$$\mathbf{A}^k \simeq \lambda_1^k \prod_{i=2}^{N} \frac{(\mathbf{A} - \lambda_i \mathbf{1})}{(\lambda_1 - \lambda_i)} \equiv \lambda_1^k \mathbf{B} \tag{8.315}$$

where $\mathbf{B}$ is a matrix whose elements are unimportant for this development.

If $\mathbf{X}_0$ is an arbitrary vector, then

$$\mathbf{A}^k \mathbf{X}_0 = \lambda_1^k \mathbf{B} \mathbf{X}_0 \tag{8.316a}$$

Replacing k by $k' + 1$, then dropping the prime, this becomes

$$\mathbf{A}^{k+1} \mathbf{X}_0 = \lambda_1^{k+1} \mathbf{B} \mathbf{X}_0 \tag{8.316b}$$

Let

$$\text{(a)} \quad \mathbf{A}^k\mathbf{X}_0 \equiv \begin{bmatrix} x_{k,1} \\ x_{k,2} \\ \vdots \\ x_{k,N} \end{bmatrix} \quad \text{and} \quad \text{(b)} \quad \mathbf{A}^{k+1}\mathbf{X}_0 \equiv \begin{bmatrix} x_{k+1,1} \\ x_{k+1,2} \\ \vdots \\ x_{k+1,N} \end{bmatrix} \tag{8.317}$$

We also define the vector

$$\mathbf{B}\mathbf{X}_0 \equiv \begin{bmatrix} \beta_1 \\ \beta_2 \\ \vdots \\ \beta_N \end{bmatrix} \tag{8.318}$$

Therefore, in component form, equations 8.316 become

$$\text{(a)} \quad x_{k,i} = \lambda_1^k \beta_i \qquad \text{(b)} \quad x_{k+1,i} = \lambda_1^{k+1}\beta_i \tag{8.319}$$

Thus, for nonzero elements $x_{k,i}$, $x_{k+1,i}$, and β_i, we can divide these two equations to obtain

$$\lambda_1 = \frac{x_{k+1,i}}{x_{k,i}} \tag{8.320}$$

for every value of i.

We designate the eigenvalue with the second largest magnitude as λ_2. The relative size of λ_2/λ_1 determines the level of precision obtained by approximating the sum of equation 8.314 by the $j = 1$ term, and therefore, the number of iterations needed to deduce a specified level of precision for λ_1.

In addition to an estimate of the eigenvalue, we also obtain an estimate of the corresponding eigenvector. To see this, let

$$\mathbf{A}^k\mathbf{X}_0 \equiv \mathbf{X}_1 \tag{8.321}$$

Then equation 8.316a becomes

$$\mathbf{X}_1 = \lambda_1^k \mathbf{B}\mathbf{X}_0 \tag{8.322}$$

Therefore, equation 8.316b can be expressed as

$$\mathbf{A}\mathbf{X}_1 = \lambda_1 \mathbf{X}_1 \tag{8.323}$$

That is, $\mathbf{X}_1 = \mathbf{A}^k\mathbf{X}_0$ is the eigenvector corresponding to λ_1.

Example 8.19

As an example, consider the hermitian matrix

$$\mathbf{A} = \begin{bmatrix} 6 & 0 & 4 \\ 0 & 1 & 0 \\ 4 & 0 & 6 \end{bmatrix} \tag{8.324}$$

which has eigenvalues 10, 2, and 1. Taking

$$\mathbf{X}_0 = \begin{bmatrix} 1 \\ 0 \\ 0 \end{bmatrix} \tag{8.325}$$

$$\Rightarrow \quad \text{(a)} \quad \mathbf{AX}_0 = \begin{bmatrix} 6 \\ 0 \\ 4 \end{bmatrix} \qquad \text{(b)} \quad \mathbf{A}^2\mathbf{X}_0 = \begin{bmatrix} 52 \\ 0 \\ 48 \end{bmatrix} \qquad \text{(c)} \quad \mathbf{A}^3\mathbf{X}_0 = \begin{bmatrix} 504 \\ 0 \\ 496 \end{bmatrix} \qquad \text{(d)} \quad \mathbf{A}^4\mathbf{X}_0 = \begin{bmatrix} 5008 \\ 0 \\ 4992 \end{bmatrix} \tag{8.326}$$

Dividing the nonzero elements of $\mathbf{A}^2\mathbf{X}_0$ by the corresponding elements of $\mathbf{AX}_0$, we obtain

$$\text{(a)} \quad \lambda_1 \simeq \frac{52}{6} = 8.67 \qquad \text{and} \qquad \text{(b)} \quad \lambda_1 \simeq \frac{48}{4} = 12 \tag{8.327}$$

$$\Rightarrow \quad \bar{\lambda}_1 = 10.33 \tag{8.328}$$

The ratios of the nonzero elements of $\mathbf{A}^3\mathbf{X}_0$ to those of $\mathbf{A}^2\mathbf{X}_0$ are

$$\text{(a)} \quad \lambda_1 \simeq \frac{504}{52} = 9.69 \qquad \text{and} \qquad \text{(b)} \quad \lambda_1 \simeq \frac{496}{48} = 10.33 \tag{8.329}$$

$$\Rightarrow \quad \bar{\lambda}_1 = 10.01 \tag{8.330}$$

The corresponding ratios of $\mathbf{A}^4\mathbf{X}_0$ to $\mathbf{A}^3\mathbf{X}_0$ elements yield

$$\text{(a)} \quad \lambda_1 \simeq \frac{5008}{504} = 9.94 \qquad \text{and} \qquad \text{(b)} \quad \lambda_1 \simeq \frac{4992}{496} = 10.06 \tag{8.331}$$

$$\Rightarrow \quad \bar{\lambda}_1 = 10.00 \tag{8.332}$$

As can be seen, the largest eigenvalue is converging to 10, and the eigenvector (when normalized) is converging to

$$\mathbf{X}_1 = N_1 \begin{bmatrix} 1 \\ 0 \\ 1 \end{bmatrix} = \frac{1}{\sqrt{2}} \begin{bmatrix} 1 \\ 0 \\ 1 \end{bmatrix} \tag{8.333}$$

To find the eigenvalue with the second largest magnitude, consider the matrix

$$\mathbf{A}_1 \equiv \mathbf{A} - \lambda_1 \mathbf{X}_1 \mathbf{X}_1^\dagger \tag{8.334}$$

Then

$$\mathbf{A}_1\mathbf{X}_1 = \mathbf{AX}_1 - \lambda_1 \mathbf{X}_1 \mathbf{X}_1^\dagger \mathbf{X}_1 = \mathbf{AX}_1 - \lambda_1 \mathbf{X}_1 = 0\mathbf{X}_1 \tag{8.335}$$

Thus one of the eigenvalues of $\mathbf{A}_1$ is 0, with corresponding eigenvector $\mathbf{X}_1$. With $\mathbf{X}_i \perp \mathbf{X}_1$ for $i \neq 1$, consider

$$\mathbf{A}_1\mathbf{X}_i = \mathbf{AX}_i - \lambda_1 \mathbf{X}_1 \mathbf{X}_1^\dagger \mathbf{X}_i = \mathbf{AX}_i = \lambda_i \mathbf{X}_i \qquad i \neq 1 \tag{8.336}$$

Therefore, all the eigenvalues of $\mathbf{A}$ are the eigenvalues of $\mathbf{A}_1$ except the largest eigenvalue λ_1. λ_1 for $\mathbf{A}$ has been replaced by 0 for $\mathbf{A}_1$. Thus the largest eigenvalue of $\mathbf{A}_1$ is λ_2, and we can perform the same analysis with $\mathbf{A}_1$ to obtain an estimate of λ_2.

We apply this to our example. With

$$\mathbf{X}_1 = \frac{1}{\sqrt{2}} \begin{bmatrix} 1 \\ 0 \\ 1 \end{bmatrix} \tag{8.333}$$

$$\Rightarrow \quad \mathbf{A}_1 = \mathbf{A} - 10\mathbf{X}_1\mathbf{X}_1^\dagger = \begin{bmatrix} 6 & 0 & 4 \\ 0 & 1 & 0 \\ 4 & 0 & 6 \end{bmatrix} - 5 \begin{bmatrix} 1 \\ 0 \\ 1 \end{bmatrix} \begin{bmatrix} 1 & 0 & 1 \end{bmatrix} = \begin{bmatrix} 1 & 0 & -1 \\ 0 & 1 & 0 \\ -1 & 0 & 1 \end{bmatrix} \tag{8.337}$$

Solving the secular equation for $\mathbf{A}_1$ verifies that its eigenvalues are 2, 1, and 0.

Starting with the same initial vector

$$\mathbf{X}_0 = \begin{bmatrix} 1 \\ 0 \\ 0 \end{bmatrix} \tag{8.325}$$

$$\Rightarrow \quad \text{(a)} \quad \mathbf{A}_1\mathbf{X}_0 = \begin{bmatrix} 1 \\ 0 \\ -1 \end{bmatrix} \quad \text{(b)} \quad \mathbf{A}_1^2\mathbf{X}_0 = \begin{bmatrix} 2 \\ 0 \\ -2 \end{bmatrix} \quad \text{(c)} \quad \mathbf{A}_1^3\mathbf{X}_0 = \begin{bmatrix} 4 \\ 0 \\ -4 \end{bmatrix} \tag{8.338}$$

Clearly, the ratios of the corresponding elements of $\mathbf{A}_1^2\mathbf{X}_0$ and $\mathbf{A}_1\mathbf{X}_0$ are the same as the ratios of the elements of $\mathbf{A}_1^3\mathbf{X}_0$ and $\mathbf{A}_1^2\mathbf{X}_0$. For all ratios, we obtain $\lambda_2 = 2$. It is also evident that the normalized eigenvector for $\lambda_2 = 2$ is

$$\mathbf{X}_2 = \frac{1}{\sqrt{2}}\begin{bmatrix} 1 \\ 0 \\ -1 \end{bmatrix} \tag{8.339}$$

Applying the same approach to find the third eigenvalue and eigenvector, we define

$$\mathbf{A}_2 \equiv \mathbf{A}_1 - \lambda_2\mathbf{X}_2\mathbf{X}_2^\dagger \tag{8.340}$$

which has eigenvalues 1, 0, and 0. Using equations 8.337 and 8.339, we obtain

$$\mathbf{A}_2 = \begin{bmatrix} 0 & 0 & 0 \\ 0 & 1 & 0 \\ 0 & 0 & 0 \end{bmatrix} \tag{8.341}$$

Since $\mathbf{A}_2$ is in diagonal form, we see immediately that its eigenvalues are $1, 0, 0$.

It is clear that with

$$\mathbf{X}_0 = \begin{bmatrix} 1 \\ 0 \\ 0 \end{bmatrix} \tag{8.325}$$

$$\Rightarrow \quad \mathbf{A}_2\mathbf{X}_0 = \begin{bmatrix} 0 \\ 0 \\ 0 \end{bmatrix} \tag{8.342}$$

Therefore, we must begin with a vector that does not yield the null vector when multiplied by any power of the matrix. If, for example, we take

$$\mathbf{X}_0 = \begin{bmatrix} 1 \\ 1 \\ 1 \end{bmatrix} \tag{8.343}$$

$$\Rightarrow \quad \text{(a)} \quad \mathbf{A}_2\mathbf{X}_0 = \begin{bmatrix} 0 \\ 1 \\ 0 \end{bmatrix} \quad \text{(b)} \quad \mathbf{A}_2^2\mathbf{X}_0 = \begin{bmatrix} 0 \\ 1 \\ 0 \end{bmatrix} \tag{8.344}$$

which yields $\lambda_3 = 1$, and the normalized vector

$$\mathbf{X}_3 = \begin{bmatrix} 0 \\ 1 \\ 0 \end{bmatrix} \quad \square \tag{8.345}$$

PROBLEMS

1. (a) Expand the 4×4 determinant

$$|A| = \begin{vmatrix} a_{11} & a_{12} & 0 & 0 \\ a_{21} & a_{22} & a_{23} & 0 \\ 0 & a_{32} & a_{33} & a_{34} \\ 0 & 0 & a_{43} & a_{44} \end{vmatrix}$$

in terms of the Levi–Civita four-index symbol, and by evaluating these symbols, express $|A|$ in terms of its elements, in an expression analogous to that for a general 3×3 determinant given in equation 8.3.

(b) Use the result of part (a) to evaluate

$$|A| = \begin{vmatrix} 1 & -1 & 0 & 0 \\ 2 & 1 & 4 & 0 \\ 0 & 2 & -2 & 3 \\ 0 & 0 & -3 & 2 \end{vmatrix}$$

2. Expand the determinant

$$|A| = \begin{vmatrix} 1 & 0 & 3 & 2 \\ 2 & 1 & 2 & 2 \\ 0 & 1 & 4 & 3 \\ 4 & 2 & 0 & 1 \end{vmatrix}$$

around the row or column specified. Evaluate all resulting 3×3 determinants by multiplying elements along diagonals, and obtain a value for $|A|$.

(a) Expand about column 2.

(b) Expand about row 3.

3. Manipulate the 4×4 determinant of Problem 2 as specified:

(a) Manipulate the original 4×4 determinant until all the elements in column 2 are zero except the $(3, 2)$ element.

(b) Manipulate the original 4×4 determinant until all the elements in row 4 are zero except the $(4, 2)$ element.

For each part, expand that resulting determinant about that specified row or column. In each case, further reduce the resulting 3×3 determinant to a single 2×2 determinant by manipulating the 3×3 determinant until all elements in its second column are zero except the $(3, 2)$ element. Then expand the 3×3 determinant about the second column, and evaluate the resulting 2×2 determinant.

4. Consider the set of linear equations

$$\begin{aligned} 2x + 5y - z + 4w &= 13 \\ -3x - 2z + 6w &= 12 \\ 2y + 3z + 10w &= 0 \\ 5x + 3y + 2z &= -5 \end{aligned}$$

(a) Evaluate the determinant of the coefficient matrix by manipulating it into an upper triangular form (with all elements below the main diagonal being zero).

(b) Solve for w by Cramer's rule. Evaluate $|A_w|$ by manipulating it into a lower triangular form.

(c) Solve for x and y by Cramer's rule. Evaluate $|A_x|$ and $|A_y|$ by reducing each determinant to a single 3×3 determinant using the method of pivotal condensation. Evaluate the resulting 3×3 determinant in each case by multiplication of the elements along diagonals.

(d) Solve for z using one of the equations above and the results of parts (a), (b), and (c).

5. For a general 4×4 determinant, derive the method of pivotal condensation using the nonzero $(3, 4)$ element as the pivotal element.

6. Evaluate the determinant

$$|A| = \begin{vmatrix} 0 & 2 & 1 & 1 & 2 \\ 4 & 1 & 2 & 0 & 1 \\ 3 & 1 & 2 & 1 & 0 \\ 2 & 5 & 3 & 1 & 4 \\ 1 & 2 & 1 & 2 & 3 \end{vmatrix}$$

Use the method of pivotal condensation for all reductions to ultimately obtain a single 3×3 determinant. Evaluate the resulting 3×3 determinant by multiplying its elements along diagonals. Complete the evaluation in 3 minutes or less.

7. Reduce

$$|A| = \begin{vmatrix} 5 & 4 & 3 & 2 & 1 \\ 4 & 3 & 2 & 1 & 0 \\ 3 & 2 & 1 & 0 & -1 \\ 2 & 1 & 0 & -1 & -2 \\ 1 & 0 & -1 & -2 & -3 \end{vmatrix}$$

to a 4×4 determinant using the method of pivotal condensation. Evaluate the resulting 4×4 determinant by any method you choose.

8. (a) Evaluate

$$|A| = \begin{vmatrix} 1 & 4 & 2 & 4 \\ 2 & 3 & 1 & 3 \\ 3 & 2 & 3 & 2 \\ 4 & 1 & 4 & 1 \end{vmatrix} + \begin{vmatrix} 1 & 4 & 2 & 4 \\ 2 & 3 & 1 & 3 \\ 2 & 3 & 2 & 3 \\ 4 & 1 & 4 & 1 \end{vmatrix}$$

by the method of pivotal condensation.

(b) Prove that

$$|A| = \begin{vmatrix} 1 & 4 & 2 & 4 \\ 2 & 3 & 1 & 3 \\ 3 & 2 & 3 & 3 \\ 4 & 1 & 4 & 1 \end{vmatrix} - \begin{vmatrix} 1 & 3 & 2 & 4 \\ 2 & 1 & 1 & 3 \\ 3 & -1 & 2 & 3 \\ 4 & -3 & 4 & 1 \end{vmatrix} = 0$$

without evaluating the determinants.

9. Prove that the equations

$$5x - 2y + z = 4$$
$$x + 2z = -7$$
$$3x - 2y - 3z = 18$$

are not linearly independent.

10. Find the value(s) of α for which a nontrivial solution to the following set of equations exists.

$$\alpha x + 2y - z = 0$$
$$3x + y - \alpha z = 0$$
$$3x - y + z = 0$$

11. Without performing matrix multiplication, use the properties of the Pauli spin matrices to prove that

(a) $\boldsymbol{\sigma}_1\boldsymbol{\sigma}_2\boldsymbol{\sigma}_1\boldsymbol{\sigma}_3 = -i\boldsymbol{\sigma}_1$

(b) $\boldsymbol{\sigma}_1\boldsymbol{\sigma}_2\boldsymbol{\sigma}_1\boldsymbol{\sigma}_3\boldsymbol{\sigma}_2\boldsymbol{\sigma}_3 = \mathbf{1}$

12. The Dirac matrices that arise in the equation for the relativistic spin-$\frac{1}{2}$ particle are defined by

$$\boldsymbol{\alpha}_r = \begin{bmatrix} \mathbf{0} & \boldsymbol{\sigma}_r \\ \boldsymbol{\sigma}_r & \mathbf{0} \end{bmatrix} \quad \text{and} \quad \boldsymbol{\alpha}_4 = \begin{bmatrix} \mathbf{1} & \mathbf{0} \\ \mathbf{0} & -\mathbf{1} \end{bmatrix} \quad (8.68)$$

(a) Use the properties of the Pauli matrices as expressed in equations 8.69 and the multiplication rules for partitioned matrices to prove that

(1) $[\boldsymbol{\alpha}_r, \boldsymbol{\alpha}_s] = \boldsymbol{\alpha}_r\boldsymbol{\alpha}_s + \boldsymbol{\alpha}_s\boldsymbol{\alpha}_r = \mathbf{0}, \quad r \neq s,$ $r, s = 1, 2, 3$

(2) $[\boldsymbol{\alpha}_r, \boldsymbol{\alpha}_4] = \mathbf{0}$

(3) $\boldsymbol{\alpha}_r^2 = \boldsymbol{\alpha}_4^2 = \mathbf{1}$

(4) $\boldsymbol{\alpha}_r\boldsymbol{\alpha}_s \neq i\sum_{t=1}^{3} \varepsilon_{rst}\boldsymbol{\alpha}_t, \quad r, s, t = 1, 2, 3$

(b) Use the rules for the multiplication of partitioned matrices to determine the five Dirac $\boldsymbol{\gamma}$ matrices defined by

$$\boldsymbol{\gamma}_r \equiv \boldsymbol{\alpha}_4\boldsymbol{\alpha}_r \quad (r = 1, 2, 3), \quad \boldsymbol{\gamma}_4 = \boldsymbol{\alpha}_4, \quad \boldsymbol{\gamma}_5 = \boldsymbol{\gamma}_1\boldsymbol{\gamma}_2\boldsymbol{\gamma}_3\boldsymbol{\gamma}_4$$

(c) Using the properties of the Dirac $\boldsymbol{\alpha}$ matrices, show that the anticommutator of the $\boldsymbol{\gamma}$ matrices obey

$$[\boldsymbol{\gamma}_i, \boldsymbol{\gamma}_j] = 2\delta_{ij}\mathbf{1} \qquad i, j = 1, 2, 3, 4$$

(d) From the results of part (c), show that

(1) $\boldsymbol{\gamma}_i\boldsymbol{\gamma}_j\boldsymbol{\gamma}_i\boldsymbol{\gamma}_k = \boldsymbol{\gamma}_j\boldsymbol{\gamma}_k(2\delta_{ij} - 1),$

$$i, j, k = 1, 2, 3, 4$$

(2) $(\boldsymbol{\gamma}_i\boldsymbol{\gamma}_j\boldsymbol{\gamma}_k)^2 = (2\delta_{ik} + 2\delta_{jk} + 2\delta_{ij} - 1)\mathbf{1},$

$$i, j, k = 1, 2, 3, 4$$

13. For a general $N \times N$ matrix **A**, find the matrix **M** by which **A** must be multiplied, that performs the following operation on **A**:

(a) Replaces $\text{col}_2 \to \text{col}_2 - K \cdot \text{col}_4$

(b) Interchanges $\text{col}_2 \leftrightarrow \text{col}_4$

(c) Replaces $\text{col}_3 \to K \cdot \text{col}_3$

Indicate whether **M** premultiplies or postmultiplies **A**.

14. Given the set of equations

$$8x + y + 8z = 3$$
$$7x + y + 4z = 0$$
$$2x + z = -1$$

(a) Determine the inverse of the coefficient matrix by determination of the cofactor matrix.

(b) Determine the inverse of the coefficient matrix by the Gauss–Jordan method indicating each row or column operation made.

(c) Find the values of x, y, and z that satisfy the set of equations using the result of part (a) or (b).

(d) Find the solution to this set of equations using the Choleski–Turing triangularization method. Manipulate the matrix until all elements above the main diagonal are zero.

(e) Starting with the triangular form obtained in part (d), further manipulate the coefficient matrix and column of inhomogeneous terms until the matrix is diagonal, and determine the solution from the result.

15. **A** is an indempotent matrix and $f(z)$ is a function that is analytic at the origin. Prove that

$$f(\mathbf{A}) = \mathbf{A}f(1) + (\mathbf{A} - \mathbf{1})f(0)$$

16. (a) Show that the Pauli spin matrices satisfy

$$\boldsymbol{\sigma}_k^2 = \mathbf{1} \qquad k = 1, 2, 3$$

(b) Use this property to prove that

$$e^{i\theta\boldsymbol{\sigma}_k} = \mathbf{1}\cos\theta + i\boldsymbol{\sigma}_k\sin\theta \qquad k = 1, 2, 3$$

17. For the matrix

$$\mathbf{A} = \begin{bmatrix} \frac{1}{2} & 0 & \frac{1}{2} \\ 0 & 1 & 0 \\ \frac{1}{2} & 0 & \frac{1}{2} \end{bmatrix}$$

find the matrix:

(a) $\log(\mathbf{1} + \mathbf{A})$

(b) $\sin(\mathbf{A})$

(c) $e^{\mathbf{A}}$

(d) From the results of part (c), show that $e^{-\mathbf{A}}$ is the inverse of $e^{\mathbf{A}}$.

18. If $\mathbf{A}$ is a hermitian matrix, prove that $f(\mathbf{A}) = e^{iK\mathbf{A}}$ is a unitary matrix for any real K.

19. $\mathbf{A}$ and $\mathbf{B}$ are involutory, anticommuting matrices (such as the Pauli matrices).

(a) Show that

$$e^{i\mathbf{A}\theta} = \mathbf{1}\cos\theta + i\mathbf{A}\sin\theta$$

(b) Prove that

$$\tfrac{1}{2}(\mathbf{A} + \mathbf{B})^2 = \mathbf{1}$$

(c) Show that

$$e^{i\mathbf{A}\theta}e^{i\mathbf{B}\theta} \neq e^{i(\mathbf{A}+\mathbf{B})\theta}$$

This result illustrates the general property that for matrices $\mathbf{M}$ and $\mathbf{N}$,

$$e^{\mathbf{M}}e^{\mathbf{N}} \neq e^{\mathbf{M}+\mathbf{N}}$$

Therefore

$$e^{\mathbf{M}}e^{\mathbf{N}} \neq e^{\mathbf{N}}e^{\mathbf{M}}$$

20. **(a)** Deduce the coefficients c_0, c_1, c_2, c_3 so that the 2×2 hermitian matrix $\mathbf{A}$ can be written

$$\mathbf{A} = \begin{bmatrix} 3 & 2+i \\ 2-i & 4 \end{bmatrix} = c_0\mathbf{1} + \sum_{i=1}^{3} c_i\boldsymbol{\sigma}_i \equiv c_0\mathbf{1} + \mathbf{c}\cdot\boldsymbol{\sigma}$$

Prove that for any 2×2 matrix, written as

$$\mathbf{A} = c_0\mathbf{1} + \mathbf{c}\cdot\boldsymbol{\sigma}$$

its square is

$$\mathbf{A}^2 = \left(c_0^2 + \mathbf{c}\cdot\mathbf{c}\right)\mathbf{1} + 2c_0(\mathbf{c}\cdot\boldsymbol{\sigma})$$

21. Show that the square root of

$$\mathbf{A} = \begin{bmatrix} 13 & 0 & 12 \\ 0 & 4 & 0 \\ 12 & 0 & 13 \end{bmatrix}$$

is

$$\mathbf{B} = \begin{bmatrix} 3 & 0 & 2 \\ 0 & 2 & 0 \\ 2 & 0 & 3 \end{bmatrix} = \sqrt{\mathbf{A}}$$

22. For the matrix

$$\mathbf{A} = \begin{bmatrix} 3 & 0 & 2i \\ 0 & 4 & 1 \\ 1+i & 3 & 1 \end{bmatrix}$$

(a) Find the symmetric and antisymmetric matrices that add to $\mathbf{A}$.

(b) Find the hermitian and antihermitian matrices that add to $\mathbf{A}$.

23. **(a)** If $\mathbf{V}$ is a column vector with complex elements, prove that the matrix formed by the outer product

$$\mathbf{A} \equiv \mathbf{V}\mathbf{V}^\dagger$$

is hermitian.

(b) If $\mathbf{V}$ is normalized to 1, show that $\mathbf{A}$ is indempotent.

(c) Find $\mathbf{A}$ for

$$\mathbf{V} = \begin{bmatrix} -2 \\ -i \\ 2i \end{bmatrix}$$

24. Fill in the following table for the Pauli $\boldsymbol{\sigma}$ matrices given in equations 8.67, and the Dirac $\boldsymbol{\alpha}$ matrices given in equations 8.68 by placing a Y (Yes) or N (No) in the appropriate box to indicate if the specified matrix has the corresponding property.

	σ_1	σ_2	σ_3	α_1	α_2	α_3	α_4
Orthogonal							
Symmetric							
Antisymmetric							
Involutory							
Indempotent							
Nilpotent							
Hermitian							
Antihermitian							
Unitary							
Antiunitary							

25. **(a)** For the symmetric matrix

$$\mathbf{A} = \begin{bmatrix} 4 & 1 & 2 \\ 1 & 2 & 1 \\ 2 & 1 & 1 \end{bmatrix}$$

find the upper triangular matrix $\mathbf{B}$ (that has

zeros below the main diagonal), such that $\mathbf{A}$ can be written

$$\mathbf{A} = \mathbf{B}\tilde{\mathbf{B}}$$

(b) Determine $\mathbf{B}^{-1}$ and from it find $\mathbf{A}^{-1}$.

26. (a) Prove that a lower triangular matrix (with zeros above the main diagonal) has an inverse that is lower triangular.
(b) Prove that the diagonal elements of the inverse of a triangular matrix $\mathbf{B}$ satisfy

$$(b^{-1})_{ii} = \frac{1}{b_{ii}}$$

27. The hermitian matrix $\mathbf{A}$ is defined as

$$\mathbf{A} = \begin{bmatrix} 4 & e^{i\theta} & e^{-i\theta} \\ e^{-i\theta} & 1 & 0 \\ e^{i\theta} & 0 & 1 \end{bmatrix}$$

(a) Find the triangular matrix $\mathbf{B}$ that has zeros below the main diagonal such that

$$\mathbf{A} = \mathbf{B}\mathbf{B}^{\dagger}$$

(b) Find $\mathbf{B}^{-1}$ and from it determine $\mathbf{A}^{-1}$.

28. Prove that if a matrix is symmetric in one representation, it is symmetric in all representations connected by an orthogonal transformation. (Such a proof is identical for a matrix that is antisymmetric.)

29. Prove that a matrix that is hermitian in one representation is hermitian in all representations connected by a unitary transformation.

30. (a) Prove that if $\mathbf{A}'$ is similar to $\mathbf{A}$ in the form

$$\mathbf{A}' = \mathbf{D}\mathbf{A}\mathbf{D}^{-1}$$

then $(\mathbf{A}')^n$ is similar to $(\mathbf{A})^n$ via the same similarity transformation for all positive n.

(b) Prove that if $\mathbf{A}'$ is similar to $\mathbf{A}$ in the form

$$\mathbf{A}' = \mathbf{D}\mathbf{A}\mathbf{D}^{-1}$$

then $(\mathbf{A}')^{-n}$ is similar to $(\mathbf{A})^{-n}$ via the same similarity transformation for all positive n.

(c) Using the results of parts (a) and (b), prove that if

$$\mathbf{A}' = \mathbf{D}\mathbf{A}\mathbf{D}^{-1}$$

then any function that can be expressed in a Laurent series

$$f(z) = \sum_{n=-\infty}^{\infty} a_n(z_0)(z - z_0)^n$$

satisfies

$$f(\mathbf{A}') = \mathbf{D}f(\mathbf{A})\mathbf{D}^{-1}$$

31. For the matrix

$$\mathbf{A} = \begin{bmatrix} 3 & 2 \\ 2 & 3 \end{bmatrix}$$

(a) Find the eigenvalues and corresponding eigenvectors of $\mathbf{A}$.
(b) Determine the unitary matrix that diagonalizes $\mathbf{A}$ via a unitary transformation.
(c) In the space in which $\mathbf{A}$ is diagonal, what are $\boldsymbol{\sigma}_1$, $\boldsymbol{\sigma}_2$, and $\boldsymbol{\sigma}_3$?
(d) In the text it was shown that

$$\mathbf{X}_+ = \begin{bmatrix} 1 \\ 0 \end{bmatrix} \qquad (8.213a)$$

and

$$\mathbf{X}_- = \begin{bmatrix} 0 \\ 1 \end{bmatrix} \qquad (8.213b)$$

are eigenvectors of $\boldsymbol{\sigma}_3$. In the representation in which $\mathbf{A}$ is diagonal, what are $\mathbf{X}_+$ and $\mathbf{X}_-$?

32. Find the eigenvalues and corresponding eigenvectors of

(a) $\begin{bmatrix} 2 & i & 0 \\ -i & 0 & 1+i \\ 0 & 1-i & 2 \end{bmatrix}$

(b) $\begin{bmatrix} 5 & 0 & 2 \\ 0 & 1 & 0 \\ 2 & 0 & 2 \end{bmatrix}$

(c) $\begin{bmatrix} 4 & 0 & 3i \\ 0 & 5 & 0 \\ -3i & 0 & -4 \end{bmatrix}$

33. For the matrix

$$\mathbf{A} = \begin{bmatrix} 1 & 2 & 0 & 2 \\ 2 & 0 & -1 & -1 \\ 0 & -1 & 2 & 2 \\ 2 & -1 & 1 & 0 \end{bmatrix}$$

find the eigenvalues corresponding to the orthonormal eigenvectors

$$\mathbf{X}_1 = \frac{1}{\sqrt{3}}\begin{bmatrix} 1 \\ 1 \\ -1 \\ 0 \end{bmatrix} \qquad \mathbf{X}_2 = \frac{1}{\sqrt{3}}\begin{bmatrix} 1 \\ 0 \\ 1 \\ 1 \end{bmatrix}$$

$$\mathbf{X}_3 = \frac{1}{\sqrt{3}}\begin{bmatrix} -1 \\ 1 \\ 0 \\ 1 \end{bmatrix} \qquad \mathbf{X}_4 = \frac{1}{\sqrt{3}}\begin{bmatrix} 0 \\ 1 \\ 1 \\ -1 \end{bmatrix}$$

34. Construct an orthonormal set of eigenvectors from

$$\mathbf{X}_1 = \begin{bmatrix} 0 \\ 1 \\ 2 \end{bmatrix} \qquad \mathbf{X}_2 = \begin{bmatrix} 2 \\ 0 \\ 1 \end{bmatrix} \qquad \mathbf{X}_3 = \begin{bmatrix} 1 \\ 2 \\ 0 \end{bmatrix}$$

35. A matrix $\mathbf{A}$ has eigenvalues -1, -2, and -3. Its corresponding eigenvectors are

$$\mathbf{X}_{-1} = \begin{bmatrix} \frac{\sqrt{3}}{2} \\ \frac{1}{2} \\ 0 \end{bmatrix} \qquad \mathbf{X}_{-2} = \begin{bmatrix} -\frac{1}{2} \\ \frac{\sqrt{3}}{2} \\ 0 \end{bmatrix} \qquad \mathbf{X}_{-3} = \begin{bmatrix} 0 \\ 0 \\ 1 \end{bmatrix}$$

(a) What is the determinant of **A**?

(b) What is the matrix **A** in nondiagonal form?

(c) What is the matrix $\mathbf{A}^n$ in nondiagonal form for any positive integer n?

36. (a) A matrix has eigenvalues λ_1, λ_2, and λ_3. The corresponding orthonormal eigenvectors can be written in the form

$$\mathbf{X}_1 = \begin{bmatrix} 1 \\ 0 \\ 0 \end{bmatrix} \qquad \mathbf{X}_2 = \begin{bmatrix} 0 \\ \cos\theta \\ \sin\theta \end{bmatrix} \qquad \mathbf{X}_3 = \begin{bmatrix} 0 \\ -\sin\theta \\ \cos\theta \end{bmatrix}$$

What is **A** in nondiagonal form in terms of θ for any $\lambda_1, \lambda_2, \lambda_3$?

(b) What is the form of **A** if $\lambda_2 = \lambda_3$?

37. (a) Prove that the diagonal elements of a triangular matrix are its eigenvalues.

(b) Show that the transformation matrix

$$\mathbf{D} = \begin{bmatrix} \frac{1}{4} & 0 & 0 \\ 0 & \frac{1}{8} & 0 \\ \frac{1}{4} & -\frac{11}{8} & 1 \end{bmatrix}$$

the inverse of which is

$$\mathbf{D}^{-1} = \begin{bmatrix} 4 & 0 & 0 \\ 0 & 8 & 0 \\ -1 & 11 & 1 \end{bmatrix}$$

triangularizes the matrix

$$\mathbf{A} = \begin{bmatrix} 0 & -15 & 20 \\ 4 & -12 & 16 \\ 3 & 1 & 7 \end{bmatrix}$$

via the similarity transformation

$$\mathbf{A}' = \mathbf{DAD}^{-1}$$

From this, determine the eigenvalues of **A**.

38. Let

$$\mathbf{AX}_i = \lambda_i \mathbf{X}_i$$

(a) Prove that the eigenvalues of $\mathbf{A}^{-1}$ are the inverses of the eigenvalues of **A** and that the corresponding eigenvectors of $\mathbf{A}^{-1}$ are the same as the corresponding eigenvectors of **A**. That is,

$$\mathbf{A}^{-1}\mathbf{X}_i = \lambda_i^{-1}\mathbf{X}_i$$

where $\mathbf{X}_i$ is the eigenvector of **A** corresponding to the eigenvalue λ_i.

(b) Prove that any functions $F(z)$ that has a Laurent expansion

$$F(z) = \sum_{n=-\infty}^{\infty} a_n(0) z^n$$

satisfies

$$F(\mathbf{A})\mathbf{X}_i = F(\lambda_i)\mathbf{X}_i$$

(c) If **A** is a diagonal matrix, prove that

$$F(\mathbf{A}) = \begin{bmatrix} F(\lambda_1) & 0 & 0 & 0 & \cdots & 0 \\ 0 & F(\lambda_2) & 0 & 0 & \cdots & 0 \\ 0 & 0 & F(\lambda_3) & 0 & \cdots & 0 \\ \vdots & & & & & \\ 0 & 0 & 0 & 0 & \cdots & F(\lambda_N) \end{bmatrix}$$

39. Show that the diagonal elements of a hermitian matrix are real and the diagonal elements of antihermitian matrix are imaginary.

40. (a) Prove that an antihermitian matrix has imaginary eigenvalues and the corresponding nondegenerate eigenvectors are naturally orthogonal.

(b) Prove that if a matrix is unitary, its eigenvalues can be written in the form $\lambda_k = e^{i\phi_k}$ (complex numbers of unit length), and the eigenvectors of nondegenerate eigenvalues are naturally orthogonal.

41. Two identical masses hang from a ceiling, connected by identical springs as shown in Figure 8.10. Find the frequencies of oscillation and the set of decoupled equations of motion of the system.

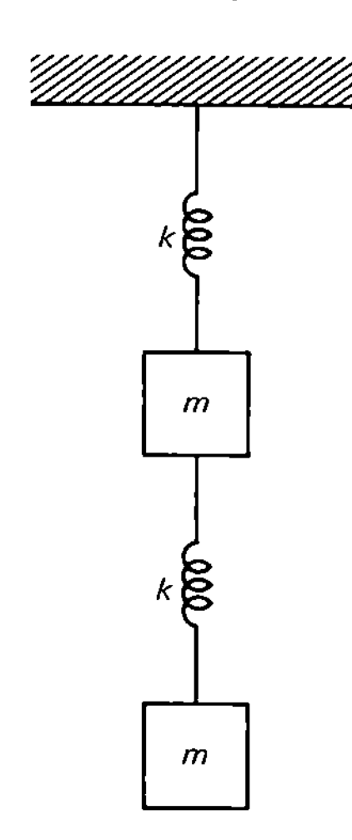

Figure 8.10 Coupled oscillators.

42. Estimate the three real, distinct eigenvalues and the three corresponding orthonormalized eigenvectors of

$$\mathbf{A} = \begin{bmatrix} 3 & 2 & 2 \\ 2 & 2 & 0 \\ 2 & 0 & 4 \end{bmatrix}$$

using Sylvester's theorem and the method of iteration described in the text.

CHAPTER 9

TENSOR ANALYSIS

Tensor analysis is an extension of the analysis of vectors developed in Chapter 1. Tensor analysis has been applied successfully to many physical systems. For example, tensors are used in the three-dimensional descriptions of the theory of elasticity of deformable media, the dynamics of particles and solid bodies, and electromagnetic phenomena. The reader interested in the examples of such applications is referred to the book by A. J. McConnell, *Applications of Tensor Analysis* (New York: Dover Publications, Inc., 1957). Other texts also present examples in some of these areas. The reader is referred to the book by G. Arfken, *Mathematical Methods for Physicists*, 3rd ed. (Orlando Fla.: Academic Press, Inc. 1985) or to the book by L. A. Pipes and L. R. Harvill, *Applied Mathematics for Engineers and Physicists*, 3rd ed. (New York: McGraw-Hill. Book Company, 1970).

Tensor analysis has made its greatest impact in the description of general relativity in four-dimensional curved space-time, which are the three spatial dimensions and one dimension representing time. By a *curved space*, one means that the geometry that describes the space is not flat. Thus the shortest distance between two points is not a straight line. (For two-dimensional motion along the surface of a sphere, for example, the shortest distance between two points is the arc of a circle.) The geometry of curved space is called *Riemannian geometry*, and the curve of minimal distance is called a *geodesic*.

9.1 Definitions and Tensor Algebra

In Chapter 8 we discussed the transformation from one coordinate system to another in matrix form. For example, rotations in three dimensions are describable in terms of matrices, the elements of which are functions of the Euler angles. Such coordinate transformations in a space of N dimensions are of the form

$$x_i' = \sum_{j=1}^{N} d_{ij} x_j \tag{9.1}$$

Such a transformation is a linear one. That is, both the x and x' coordinates appear in the transformation equation to the first power. However, a more general set of transformation equations would be of the form

$$x_i' = x_i'(x_1, x_2, \ldots, x_N) \tag{9.2}$$

For example, the equations that describe the transformation from Cartesian to spherical coordinates are nonlinear

$$\text{(a)}\quad r = \sqrt{x^2 + y^2 + z^2} \qquad \text{(b)}\quad \theta = \cos^{-1}\left(\frac{z}{\sqrt{x^2 + y^2 + z^2}}\right) \qquad \text{(c)}\quad \phi = \tan^{-1}\left(\frac{y}{x}\right) \tag{9.3}$$

Covariance and contravariance

For a general coordinate transformation, consider a small displacement of a component of the position vector x_i'. It can be related to a small displacement of x_j by

$$dx_i' = \sum_{j=1}^{N} \frac{\partial x_i'}{\partial x_j} dx_j \tag{9.4}$$

The gradient of a scalar function was another vector defined in Chapter 1. A scalar function is one that has the same value at corresponding points in two coordinate frames. That is, if $\mathbf{r}'$ and $\mathbf{r}$ are two points related by a transformation, then the function $\Phi(\mathbf{r})$ is a scalar if

$$\text{(a)}\quad \Phi(\mathbf{r}') = \Phi(\mathbf{r}) \qquad \text{or} \qquad \text{(b)}\quad \Phi' = \Phi \tag{9.5}$$

Such a function is also referred to as an *invariant*.

The transformation of a component of the gradient of a scalar function is described by

$$\frac{\partial \Phi}{\partial x_i'} = \sum_{j=1}^{N} \frac{\partial \Phi}{\partial x_j} \frac{\partial x_j}{\partial x_i'} \tag{9.6}$$

$\partial x_i'/\partial x_j$, the coefficient of dx_j in equation 9.4 and $\partial x_j/\partial x_i'$, the coefficient of $\partial\Phi/\partial x_j$ in equation 9.6, are called *transformation coefficients*.

We note that the coefficient for the transformation of a component of the displacement vector dx_i differs in form from the coefficient for the transformation of a component of the gradient vector. That is, the coefficient for the transformation from dx to dx' is of the form $\partial x'/\partial x$. The coefficient for the transformation of a component of the gradient is of the form $\partial x/\partial x'$. Thus dx_i and $\partial \phi/\partial x_i$ are components of different types of vectors.

A vector that transforms like the displacement, with coefficients of the form $\partial x'/\partial x$, is called a *contravariant vector*. A *covariant vector* is one that transforms the way the gradient transforms, with coefficients $\partial x/\partial x'$.

To distinguish between the two, the components of a covariant vector will be denoted with a subscripted index (which is the notation that has been used thus far for all vector components). The components of a contravariant vector will be denoted by a superscript. Therefore, since the components of the position vector transform like a contravariant vector, they must be written as x^i. To distinguish between the ith component of a contravariant vector and a quantity raised to the ith power, the standard notation is that p^i is the ith component of a contravariant vector and $(p)^i$ (with parentheses around it) is the ith power of the quantity p.

There is no confusion in referring to x^i as a contravariant vector, even though we understand that it is really the ith component of the contravariant vector $\mathbf{x}$. As such, we will henceforth refer to the component of a tensor, as the tensor. We will not distinguish between a tensor and its component unless such a reference causes confusion.

Since position vectors are contravariant, equations 9.4 and 9.6 must be modified slightly. They must be rewritten

$$\text{(a)} \quad dx'^i = \sum_{j=1}^{N} \frac{\partial x'^i}{\partial x^j} dx^j \quad \text{and} \quad \text{(b)} \quad \frac{\partial \Phi}{\partial x'^i} = \sum_{j=1}^{N} \frac{\partial x^j}{\partial x'^i} \frac{\partial \Phi}{\partial x^j} \tag{9.7}$$

With this notation, then, a contravariant vector p^i and a covariant vector q_i transform according to

$$\text{(a)} \quad p'^i = \sum_{j=1}^{N} \frac{\partial x'^i}{\partial x^j} p^j \quad \text{and} \quad \text{(b)} \quad q'_i = \sum_{j=1}^{N} \frac{\partial x^j}{\partial x'^i} q_j \tag{9.8}$$

Since all coordinates in a given space are independent,

$$\text{(a)} \quad \frac{\partial x'^i}{\partial x'^j} = \delta^i_j \qquad \text{(b)} \quad \frac{\partial x^i}{\partial x^j} = \delta^i_j \tag{9.9}$$

But since $x'^i = x'^i(x^1, x^2, \ldots, x^N)$ and $x^i = x^i(x'^1, x'^2, \ldots, x'^N)$, we can write equations 9.9 as

$$\text{(a)} \quad \delta^i_j = \sum_{k=1}^{N} \frac{\delta x'^i}{\partial x^k} \frac{\partial x^k}{\partial x'^j} \qquad \text{(b)} \quad \delta^i_j = \sum_{k=1}^{N} \frac{\partial x^i}{\partial x'^k} \frac{\partial x'^k}{\partial x^j} \tag{9.10}$$

Thus if we view the coefficients $\partial x'^i/\partial x^k$ as the (i, k) element of a transformation matrix **T**, then $\partial x^i/\partial x'^k$ is the (i, k) element of T^{-1}, since their product is the unit matrix. Therefore, in matrix form, the transformation equations can be written

$$\text{(a)} \quad \mathbf{X}'^{\text{contravariant}} = \mathbf{T}\mathbf{X}^{\text{contravariant}} \qquad \text{(b)} \quad \mathbf{X}'_{\text{covariant}} = \mathbf{T}^{-1}\mathbf{X}_{\text{covariant}} \tag{9.11}$$

Thus far we have discussed the transformation properties of two types of vectors and a scalar, the value of which does not change under a transformation. The scalar is written without indices on it and each vector has one index on it. The *rank* of a tensor is the number of indices required to define it. Therefore, the scalar or invariant is a tensor of rank 0, and a vector is a rank 1 tensor. It is straightforward to extend the definitions above to quantities that have several (two or more) indices.

In general, some of the indices of a multiple index quantity can be covariant, others can be contravariant. The quantity will be a tensor only if each contravariant index introduces the transformation coefficient of the form in the sum of equation 9.8a, and each covariant index results in a transformation coefficient like that expressed in the sum of equation 9.8b. For example, the two-index quantity T^{ij} is a doubly contravariant tensor only if

$$T'^{ij} = \sum_{k,l=1}^{N} \frac{\partial x'^i}{\partial x^k} \frac{\partial x'^j}{\partial x^l} T^{kl} \tag{9.12a}$$

Similarly, T_{ij} is a doubly covariant tensor only if it transforms as

$$T'_{ij} = \sum_{k,l=1}^{N} \frac{\partial x^k}{\partial x'^i} \frac{\partial x^l}{\partial x'^j} T_{kl} \tag{9.12b}$$

and T^i_j is a tensor of mixed variance only if

$$T'^i_j = \sum_{k,l=1}^{N} \frac{\partial x'^i}{\partial x^k} \frac{\partial x^l}{\partial x'^i} T^k_l \tag{9.12c}$$

The reader is cautioned that not all quantities described by indices are tensors. As an example, if A_i is a covariant vector, the two-index quantity $C_{ij} \equiv \partial A_i/\partial x^j$ is not a second-rank doubly covariant tensor. To see this, we note that since A_i is a covariant vector,

$$A'_i = \sum_{k=1}^{N} \frac{\partial x^k}{\partial x'^i} A_k \tag{9.13a}$$

$$\Rightarrow \quad \frac{\partial A'_i}{\partial x'^j} = \sum_{k=1}^{N} \frac{\partial}{\partial x'^j}\left[\frac{\partial x^k}{\partial x'^i} A_k\right] = \sum_{\substack{k=1\\ \ell=1}}^{N} \left[\frac{\partial x^k}{\partial x'^i}\frac{\partial x^m}{\partial x'^j}\frac{\partial A_k}{\partial x^m} + A_k \frac{\partial^2 x^k}{\partial x'^i \partial x'^j}\right] \tag{9.13b}$$

Because of the presence of the second term, equation 9.13b is not the correct transformation rule for a rank 2 doubly covariant tensor. However, if the transformation from $\mathbf{x}$ to $\mathbf{x}'$ is a linear transformation, $\partial^2 x^k/\partial x'^i \partial x'^j = 0$. Thus $\partial A_i/\partial x^j$ is a second-rank doubly covariant tensor only if the transformation is linear.

Summation convention

Referring to equations 9.12 as examples, we note that the summations are performed over indices that appear on two different quantities. For example, in equation 9.12a, the index k appears as a contravariant index in the denominator of $\partial x'^i/\partial x^k$ and as a contravariant index on the quantity T^{kl}. Let us treat a contravariant index in the denominator as if it were a covariant index. Then the sum over k in equation 9.12a occurs over a product of two quantities. On one of the quantities, the repeated index k is the covariant index, and on the other, k is contravariant. In equation 9.12b, k is a contravariant index on $\partial x^k/\partial x'^i$. This multiplies the quantity T_{kl} on which k is a covariant index.

This recognition allows us to define a summation convention. If two tensors are multiplied, and a covariant index on one of the factors is repeated as a contravariant index on the other factor, a summation is implied. This allows us to dispense with the summation sign and equations 9.12 can be written

$$\text{(a)} \quad T'^{ij} = \frac{\partial x'^i}{\partial x^k}\frac{\partial x'^j}{\partial x^l} T^{kl} \qquad \text{(b)} \quad T'_{ij} = \frac{\partial x^k}{\partial x'^i}\frac{\partial x^l}{\partial x'^j} T_{kl} \qquad \text{(c)} \quad T_j'^i = \frac{\partial x'^i}{\partial x^k}\frac{\partial x^l}{\partial x'^j} T_l^k \tag{9.14}$$

We note that the same guidelines are satisfied for l to be a summation index.

As an addendum to the summation convention, if two indices are of the same variance and a sum is to be performed, it must be specifically stated. If two indices are of opposite variance and are repeated, but no sum is to be performed, that too must be specifically stated. For example, of A_i and B_i are vectors, then $A_i B_i$ is to be interpreted as the (i, i) element of a two-index symbol formed by the outer product of two covariant vectors. It does not mean $\sum_{i=1}^{N} A_i B_i$. Similarly, if the index i with the correct variances in the product $A^i B_i$ is not to be summed over, this must be written $A^i B_i$ (*no sum*).

It is not necessary that the repeated index with the correct variance appear on two different factors in a product. The summation convention also applies to the quantity T_i^i, for example. That is, T_i^i requires that we sum over i unless otherwise noted. Thus if T_j^i is viewed as a matrix element, T_i^i is the trace of the matrix.

A *free index* is an index that is not summed over. That is, such an index is free to be assigned any value. Returning to the product $A^i B_i$, we note that this is the tensor equivalent of the dot product of two vectors. Since i is summed over, there are no free indices. Any quantity without free indices is an invariant. Let us consider the transformation of such a quantity. Since A^i is a contravariant vector and B_i is a covariant vector, their transformation equations are

$$\text{(a)} \quad A'^i = \frac{\partial x'^i}{\partial x^j} A^j \quad \text{and} \quad \text{(b)} \quad B'_i = \frac{\partial x^k}{\partial x'^i} B_k \tag{9.15}$$

$$\Rightarrow \quad A'^i B'_i = \frac{\partial x'^i}{\partial x^j} \frac{\partial x^k}{\partial x'^i} A^j B_k \tag{9.16}$$

Since the sum over i is implied, equation 9.10b yields

$$\frac{\partial x'^i}{\partial x^j} \frac{\partial x^k}{\partial x'^i} = \frac{\partial x^k}{\partial x^j} = \delta_j^k \tag{9.17}$$

$$\Rightarrow \quad A'^i B'_i = \delta_j^k A^j B_k = A^j B_j \tag{9.18a}$$

In Chapter 3 we defined any index that is summed over as a dummy index. Such an index can be renamed as long as the new symbol is not the same as an index used elsewhere on the tensor or product of tensors. This is known as the *dummy index rule*. j is such a dummy index in the product $A^j B_j$. We can therefore replace it with i, since i is not used elsewhere in the product. Therefore,

$$A'^i B'_i = A^i B_i \tag{9.18b}$$

and we see that this inner product is invariant under the transformation.

Arithmetic of tensors

1. *Addition, subtraction, and equality.* Two tensors can be added, subtracted, or equated only if the individual tensors have the same number of indices, and the indices are of the same variance. The same symbol must be used to denote corresponding indices. For example,

$$T^{ij} + U^{ij} = V^{ij} \tag{9.19}$$

has meaning as an arithmetic equation involving tensors. The following examples are combinations that are not defined as tensors or tensor equalities:

$$\text{(a)} \quad T^{ij} + U^{kl} \qquad \text{(b)} \quad T^{ij} - U^i_j \qquad \text{(c)} \quad T^{ij} = U^{ij}_k \tag{9.20}$$

2. *Multiplication of tensors: outer and inner products.* The outer product of T^{ij}_k and U^l_m is denoted by

$$T^{ij}_k U^l_m = V^{ijl}_{km} \tag{9.21}$$

This product is a tensor of rank 5 of mixed variance. To see this, we must determine how V^{ijl}_{km} transforms, given that T^{ij}_k and U^l_m transform like the indicated tensors. That is,

$$\text{(a)} \quad T'^{ij}_k = \frac{\partial x'^i}{\partial x^p} \frac{\partial x'^j}{\partial x^q} \frac{\partial x^r}{\partial x'^k} T^{pq}_r \qquad \text{(b)} \quad U'^l_m = \frac{\partial x'^l}{\partial x^s} \frac{\partial x^t}{\partial x'^m} U^s_t \tag{9.22}$$

$$\Rightarrow \quad T'^{ij}_k U'^l_m = V'^{ijl}_{km} = \frac{\partial x'^i}{\partial x^p} \frac{\partial x'^j}{\partial x^q} \frac{\partial x^r}{\partial x'^k} \frac{\partial x'^l}{\partial x^s} \frac{\partial x^t}{\partial x'^m} T^{pq}_r U^s_t = \frac{\partial x'^i}{\partial x^p} \frac{\partial x'^j}{\partial x^q} \frac{\partial x^r}{\partial x'^k} \frac{\partial x'^l}{\partial x^s} \frac{\partial x^t}{\partial x'^m} V^{pqs}_{rt} \tag{9.23}$$

We note that there are five transformation coefficients of the required forms. Three of them have x' in the numerator and two have x' in the denominator. This is the transformation rule for a rank 5 tensor with three contravariant indices and two covariant indices: the indicated variance of V^{ijl}_{km}. Thus V^{ijl}_{km} is such a tensor.

The inner product of two tensors is defined like the outer product except that a covariant index on the tensor is a repeat of a contravariant index on the other and is therefore summed over. For example,

$$T^{ij}_{k}U^{k}_{m} = V^{ij}_{m} \tag{9.24}$$

Since k is summed over, there are three free indices. So the result is a three-index quantity. To show that V^{ij}_{m} is a third-rank tensor of the indicated variance, we determine how it transforms using the transformation rules for T^{ij}_{k} and U^{k}_{m}.

$$\text{(a)} \quad T'^{ij}_{k} = \frac{\partial x'^i}{\partial x^p}\frac{\partial x'^j}{\partial x^q}\frac{\partial x^r}{\partial x'^k}T^{pq}_{r} \qquad \text{(b)} \quad U'^{l}_{m} = \frac{\partial x'^l}{\partial x^s}\frac{\partial x^t}{\partial x'^m}U^{s}_{t} \tag{9.22}$$

$$\Rightarrow \quad T'^{ij}_{k}U'^{k}_{m} = V'^{ij}_{k} = \frac{\partial x'^i}{\partial x^p}\frac{\partial x'^j}{\partial x^q}\frac{\partial x^r}{\partial x'^k}\frac{\partial x'^k}{\partial x^s}\frac{\partial x^t}{\partial x'^m}T^{pq}_{r}U^{s}_{t} \tag{9.25}$$

But

$$\frac{\partial x^r}{\partial x'^k}\frac{\partial x'^k}{\partial x^s} = \frac{\partial x^r}{\partial x^s} = \delta^r_s \tag{9.17b}$$

$$\Rightarrow \quad V'^{ij}_{k} = \frac{\partial x'^i}{\partial x^p}\frac{\partial x'^j}{\partial x^q}\delta^r_s\frac{\partial x^t}{\partial x'^m}T^{pq}_{r}U^{s}_{t} = \frac{\partial x'^i}{\partial x^p}\frac{\partial x'^j}{\partial x^q}\frac{\partial x^t}{\partial x'^m}T^{pq}_{r}U^{r}_{t} = \frac{\partial x'^i}{\partial x^p}\frac{\partial x'^j}{\partial x^q}\frac{\partial x^t}{\partial x'^m}V^{pq}_{t} \tag{9.26}$$

This is the transformation rule for a third-rank tensor of the indicated mixed variance.

3. *Contraction*. If a multi-indexed tensor of mixed variance has one of its covariant indices set equal to one of its contravariant indices and a sum over that index is performed, the operation is called *contraction* on the specified index. The result is a tensor, the rank of which is smaller by two that the rank of the original tensor.

For example, a possible contraction of T^{ijk}_{lm} is obtained by setting $j = l$ and summing. It is denoted by $V^{ik}_{m} = T^{ilk}_{lm}$. Since T^{ilk}_{lm} is a tensor, it transforms according to

$$T'^{ilk}_{lm} = V'^{ik}_{m} = \frac{\partial x'^i}{\partial x^p}\frac{\partial x'^l}{\partial x^q}\frac{\partial x'^k}{\partial x^r}\frac{\partial x^s}{\partial x'^l}\frac{\partial x^t}{\partial x'^m}T^{pqr}_{st} \tag{9.27}$$

Again, from

$$\frac{\partial x'^l}{\partial x^q}\frac{\partial x^s}{\partial x'^l} = \frac{\partial x^s}{\partial x^q} = \delta^s_q \tag{9.17b}$$

$$\Rightarrow \quad V'^{ik}_{m} = \frac{\partial x'^i}{\partial x^p}\frac{\partial x'^k}{\partial x^r}\delta^s_q\frac{\partial x^t}{\partial x'^m}T^{pqr}_{st} = \frac{\partial x'^i}{\partial x^p}\frac{\partial x'^k}{\partial x^r}\frac{\partial x^t}{\partial x'^m}T^{psr}_{st} = \frac{\partial x'^i}{\partial x^p}\frac{\partial x'^k}{\partial x^r}\frac{\partial x^t}{\partial x'^m}V^{pr}_{t} \tag{9.28}$$

This is the transformation rule for a third-rank tensor of the indicated variance.

4. *Quotient rule*. If the (inner or outer) product of two factors is a tensor, and if one of the factors is a tensor, the other factor must also be a tensor with the variance required. For example, let $C_{ij} \equiv A^{kl}_{ij}B_{kl}$. If C_{ij} is a doubly covariant tensor and A^{kl}_{ij} is a fourth-rank tensor of mixed variance, then B_{kl} must be a doubly covariant tensor. Similarly, if $C^{kl}_{ij} \equiv A^{k}_{i}B^{l}_{j}$ is a rank 4 tensor, and if A^{k}_{i} is a second-rank tensor of mixed variance, then B^{l}_{j} must also be a second-rank tensor of mixed variance.

As a corollary to the quotient law, if the (inner or outer) product of factors is a tensor, and if one of the factors is not a tensor, the other factor must not be a tensor. In each of the examples above, if the quantity denoted by C is a tensor, and the quantity denoted by A is not a tensor, then B cannot be a tensor.

5. *Dummy index rule*. As described above, the symbol used for any repeated index in the inner product of two tensors, or in the contraction of a multi-index tensor, can be replaced by any other symbol, as long as that symbol does not occur elsewhere on one of the tensors in the product, or on the contracted tensor. As examples, the index j can be replaced by n in equation 9.29a, and t can replace k in equation 9.29b so that

$$\text{(a)}\quad A^{kl}_{ij}B^{j}_{m} = A^{kl}_{in}B^{n}_{m} \qquad \text{(b)}\quad C^{ijk}_{klm} = C^{ijt}_{tlm} \tag{9.29}$$

6. *Symmetrization and antisymmetrization of indices*. Any tensor with two or more indices (rank 2 or higher) can be written as a sum of two tensors, one of which is symmetric in the interchange of any pair of indices, and the other that is antisymmetric in the interchange of that pair of indices. As an example,

$$A^{ijk}_{lm} = \tfrac{1}{2}\left(A^{ijk}_{lm} + A^{ikj}_{lm}\right) + \tfrac{1}{2}\left(A^{ijk}_{lm} - A^{ikj}_{lm}\right) \tag{9.30a}$$

is a sum of tensors that are respectively symmetric and antisymmetric in the interchange of the indices j and k. Similarly,

$$A^{ijk}_{lm} = \tfrac{1}{2}\left(A^{ijk}_{lm} + A^{ijl}_{km}\right) + \tfrac{1}{2}\left(A^{ijk}_{lm} - A^{ijl}_{km}\right) \tag{9.30b}$$

is a sum of tensors, one of which is symmetric in k and l, the other is antisymmetric in k and l.

9.2 Riemannian Geometry and the Equation of a Geodesic

Line elements and the metric tensor

As described above, an invariant or scalar quantity is one that is unaffected by a transformation. That is, it has the same value in both coordinate frames. One of the important invariants of a space is the differential line element, the distance between two neighboring points.

In Cartesian coordinates, for example,

$$ds^2 = dx^2 + dy^2 + dz^2 = (dx^1)^2 + (dx^2)^2 + (dx^3)^2 \tag{9.31}$$

With $(r, \theta, \phi) \equiv (x^1, x^2, x^3)$, the square of the line element in spherical coordinates is

$$ds^2 = dr^2 + r^2\, d\theta^2 + r^2 \sin^2\theta\, d\phi^2 = (dx^1)^2 + (x^1)^2(dx^2)^2 + (x^1)^2 \sin^2(x^2)(dx^3)^2 \tag{9.32}$$

That is, in general, the coefficients of the squares of the differential coordinates can be functions of the coordinates.

We have shown that the differential coordinate dx^i is a contravariant vector. This differential line element can be put into the form of an invariant by defining a quantity g_{ij} such that

$$ds^2 = g_{ij}\, dx^i\, dx^j \tag{9.33a}$$

By the quotient law, if ds^2 is an invariant and dx^i and dx^j are contravariant vectors, g_{ij} must be second-rank covariant tensor. g_{ij} is called the *metric tensor* of the space.

Since the order of multiplication of dx^i and dx^j is immaterial, equation 9.33a can be written

$$ds^2 = g_{ij}\, dx^j\, dx^i \tag{9.33b}$$

Recognizing i and j as dummy indices, these symbols can be interchanges and ds^2 can be written

$$ds^2 = g_{ji}\, dx^i\, dx^j \tag{9.34}$$

Comparing this to equation 9.33a, it is clear that g_{ij} is symmetric in the interchange of its indices.

In Cartesian and spherical coordinates, the metric tensors, in matrix form, are

$$\text{(a)} \quad \mathbf{g} = \begin{bmatrix} 1 & 0 & 0 \\ 0 & 1 & 0 \\ 0 & 0 & 1 \end{bmatrix} = \mathbf{1} \qquad \text{(b)} \quad \mathbf{g} = \begin{bmatrix} 1 & 0 & 0 \\ 0 & (x^1)^2 & 0 \\ 0 & 0 & (x^1)^2 \sin^2(x^2) \end{bmatrix} \tag{9.35}$$

An invariant can also be obtained by the inner product of a covariant and contravariant vector. If there were such a quantity as a covariant differential coordinate, dx_i, then

$$ds^2 = dx^i \, dx_i \tag{9.36}$$

would be an invariant. Comparing equation 9.36 to equation 9.33a, we see that we can define the covariant form of the differential coordinate as

$$dx_i = g_{ij} \, dx^j \tag{9.37}$$

The covariant form of the contravariant vector dx^i, obtained this way, is called the *dual* of dx^i.

It is straightforward to extend this definition to a tensor of any rank. The covariant dual of the contravariant index j for a tensor of any rank is defined by

$$T^{\cdots\cdots ab \cdots}_{\cdots hik \cdots} = g_{ij} T^{\cdots\cdots ajb \cdots}_{\cdots hk \cdots} \tag{9.38}$$

That is, the inner product of g_{ij} with $T^{\cdots\cdots ajb \cdots}_{\cdots hk \cdots}$ has removed the index j from the string of contravariant indices and added the index i to the string of covariant indices. (We will discuss the position in which i is to be placed later.) Thus the number of contravariant indices and the number of covariant indices on tensors that are dual to one another differ by 1. It is clear from equation 9.35a that in Cartesian space, covariant and contravariant forms of a tensor are the same. That is, in Cartesian space, tensors are self-dual.

Another designation for the metric tensor g_{ij}, then, is that it is an *index-lowering operator*. From this description, it is possible to define two other forms of the metric tensor. The form that is of mixed variance is defined such that

$$g_{ij} \equiv g_{ik} g_j^k \tag{9.39a}$$

But equation 9.39a can also be written in terms of the Kronecker δ-symbol as

$$g_{ij} = g_{ik} \delta_j^k \tag{9.39b}$$

$$\Rightarrow \quad g_j^k = \begin{cases} 1 & \text{for } j = k \\ 0 & \text{for } j \neq k \end{cases} = \delta_j^k \tag{9.40}$$

Using the quotient rule, we see that $g_j^k = \delta_j^k$ is a second-rank tensor of mixed variance. The contravariant form of the metric tensor, then, is dual to g_j^k. Therefore,

$$g_j^k = g_{jl} g^{lk} = \delta_j^k \tag{9.41}$$

If g_{jl} is viewed as the (j, l) element of a matrix, equation 9.41 is the equation for the (j, k) element of a product of two matrices. This product, then, yields the unit matrix. Therefore, the matrix with elements g^{ij} is the inverse of the matrix with elements g_{ij}.

Physical components

When vector components are added or subtracted, the combination has meaning as a tensor sum or difference only if both vectors are covariant or if both vectors are contravariant. However, one must be careful that the vectors that are added or subtracted also represent the same type of physical quantities. For example, in Cartesian coordinates, defining two contravariant vectors u^i and v^i, with $(u^1, u^2, u^3) \equiv (x, y, z)$ and $(v^1, v^2, v^3) \equiv (x', y', z')$ both vectors represent distances so that both the a and b components are described in units of length (centimeters). In spherical coordinates, the vector components $(w^1, w^2, w^3) \equiv (r, \theta, \phi)$ represent different types of physical quantities. Thus the tensor sum $u^i + v^i$ has meaning both as a tensor sum and as a sum of physical quantities. However, even though the sum $v^i + w^i$ has meaning as a tensor sum, it is invalid as a sum of physical quantities. (For example, $y' + \theta$ is not defined physically, although it is defined as a tensor sum.)

In spherical coordinates, with $(x^1, x^2, x^3) \equiv (r, \theta, \phi)$, the square of the line element is

$$ds^2 = g_{ij}\, dx^i\, dx^j = (dx^1)^2 + (x^1)^2(dx^2)^2 + (x^1)^2 \sin^2(x^2)(dx^3)^2 \qquad (9.42a)$$

Since the metric tensor is diagonal, this can be written

$$ds^2 = \sum_{i=1}^{3} g_{ii}(dx^i)^2 = \sum_{i=1}^{3} \left(\sqrt{g_{ii}}\, dx^i\right)^2 \qquad (9.42b)$$

The quantity $\sqrt{g_{ii}}\, dx^i$ has the same physical units (centimeters) for all three values of i. From this analysis we generalize the definition of physical components for any vector. If v^i is the tensor component of a vector, the physical component corresponding to v^i is

$$\nu^i \equiv \sqrt{g_{ii}}\, v^i \qquad (9.43)$$

Geometry of covariant and contravariant components

We are now in a position to describe the difference between covariant and contravariant vector components geometrically. For simplicity, we consider a two-dimensional space. We first consider a space in which the axes are perpendicular to one another, and label these axes X and Y, with corresponding basis vectors (Figure 9.1). We designate these basis vectors as **i** and **j** (or $\boldsymbol{\rho}_0$ and $\boldsymbol{\phi}_0$, for example).

Any vector in the space can be written as a sum of its components $\mathbf{v}_X$ and $\mathbf{v}_Y$. One can view the construction of these components in two ways:

1. By constructing perpendiculars to the axes from the tip of **v**, the components are vectors directed along the axes from the origin to the point of intersection of the perpendiculars.
2. By constructing a parallelogram formed by the two axes and lines drawn through the tip of **v** parallel to the two axes, the components are vectors directed along the axes from the origin to the intersection of the sides of the parallelogram.

For the axis system in which the axes are orthogonal, the two views yield the same components of **v**.

In a two-dimensional axis system in which the axes are not orthogonal, these two different constructions lead to different components of **v**, as shown in Figure 9.2. Clearly, $\mathbf{v}_{a_i} \neq \mathbf{v}_{b_i}$.

To associate these components with covariant and contravariant components, we must first identify the types of basis vectors that define this space. We will use the

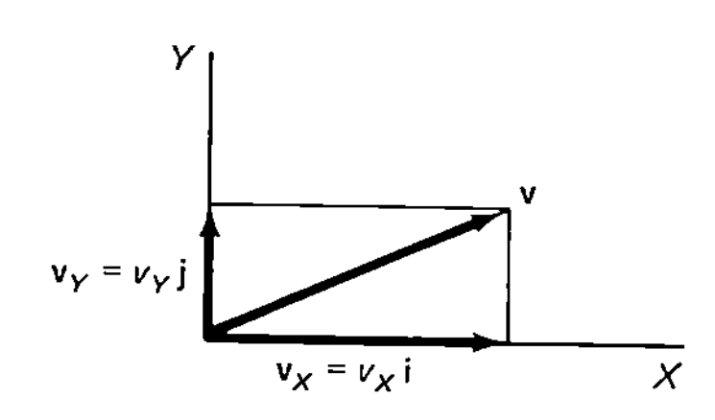

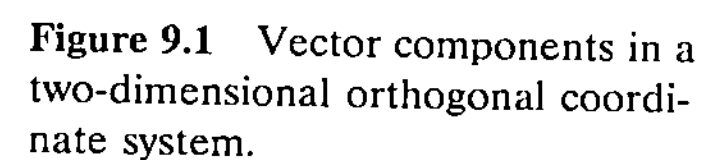

Figure 9.1 Vector components in a two-dimensional orthogonal coordinate system.

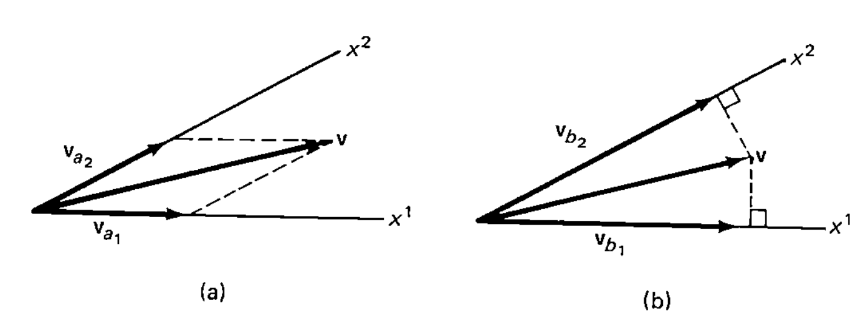

Figure 9.2 (a) Components found by constructing a parallelogram; (b) components found by constructing perpendiculars to the axes.

symbol $\mathbf{e}$ to denote these basis vectors. We have shown that the components of the differential of the position vector transform like contravariant vectors. Therefore, the summation convention requires that the basis vectors be designated with a covariant index so that the vector line element $d\mathbf{r}$ can be written

$$d\mathbf{r} = dx^i \, \mathbf{e}_i \tag{9.44}$$

But the square of the line element is

$$ds^2 = d\mathbf{r} \cdot d\mathbf{r} = g_{ij} \, dx^i \, dx^j = \mathbf{e}_i \cdot \mathbf{e}_j \, dx^i \, dx^j \tag{9.45}$$

$$\Rightarrow \quad g_{ij} = \mathbf{e}_i \cdot \mathbf{e}_j \tag{9.46}$$

We can also define the contravariant dual of $\mathbf{e}_i$ by

$$\mathbf{e}^i = g^{ij}\mathbf{e}_j \tag{9.47}$$

$$\Rightarrow \quad \mathbf{e}^i \cdot \mathbf{e}^j = g^{ik}\mathbf{e}_k \cdot g^{jl}\mathbf{e}_l = g^{ik}g^{jl}g_{kl} = g^{ik}\delta^j_k = g^{ij} \tag{9.48}$$

and

$$\mathbf{e}^i \cdot \mathbf{e}_j = g^{ik}\mathbf{e}_k \cdot \mathbf{e}_j = g^{ik}g_{kj} = \delta^i_j \tag{9.49}$$

Thus we see that orthonormality of basis vectors exists only for the dot product of a covariant basis vector with a contravariant basis vector.

We also note that

$$\text{(a)} \quad |\mathbf{e}_i| = \sqrt{\mathbf{e}_i \cdot \mathbf{e}_i} = \sqrt{g_{ii}} \quad (\textit{no sum}) \qquad \text{(b)} \quad |\mathbf{e}^i| = \sqrt{\mathbf{e}^i \cdot \mathbf{e}^i} = \sqrt{g^{ii}} \quad (\textit{no sum}) \tag{9.50}$$

Since g_{ii} and g^{ii} are not 1, the $\mathbf{e}^i$ and $\mathbf{e}_i$ are not unit basis vectors. The unit basis vectors formed from them are

$$\text{(a)} \quad \mathbf{u}_i = \frac{\mathbf{e}_i}{|\mathbf{e}_i|} \qquad \text{(b)} \quad \mathbf{u}^i = \frac{\mathbf{e}^i}{|\mathbf{e}^i|} \tag{9.51}$$

The scalar products of these unit basis vectors are

$$\text{(a)} \quad \mathbf{u}_i \cdot \mathbf{u}_j = \frac{\mathbf{e}_i \cdot \mathbf{e}_j}{\sqrt{g_{ii}g_{jj}}} = \frac{g_{ij}}{\sqrt{g_{ii}g_{jj}}} \qquad \text{(b)} \quad \mathbf{u}^i \cdot \mathbf{u}^j = \frac{\mathbf{e}^i \cdot \mathbf{e}^j}{\sqrt{g^{ii}g^{jj}}} = \frac{g^{ij}}{\sqrt{g^{ii}g^{jj}}}$$

$$\text{(c)} \quad \mathbf{u}_i \cdot \mathbf{u}^j = \frac{\mathbf{e}_i \cdot \mathbf{e}^j}{\sqrt{g_{ii}g^{jj}}} = \frac{1}{\sqrt{g_{ii}g^{jj}}}\delta^j_i \tag{9.52}$$

Since the inner product of equation 9.52c is zero for $i \neq j$, and since in matrix form, $g^{ij} = (g_{ij})^{-1}$, then $g_{ii}g^{ii} = 1$. Therefore,

$$\mathbf{u}_i \cdot \mathbf{u}^j = \delta_i^j \tag{9.53}$$

Since

$$d\mathbf{r} = \mathbf{e}_i \, dx^i \tag{9.44}$$

$\mathbf{e}_i$ and not $\mathbf{e}^i$ is the basis vector of a space. Thus an arbitrary vector is written

$$\mathbf{v} = v^i \mathbf{e}_i = v^i \sqrt{g_{ii}}\, \mathbf{u}_i \tag{9.54}$$

But from equation 9.43, $v^i\sqrt{g_{ii}} = \nu^i$ is the physical component of $\mathbf{v}$, so

$$\mathbf{v} = \nu^i \mathbf{u}_i \tag{9.55}$$

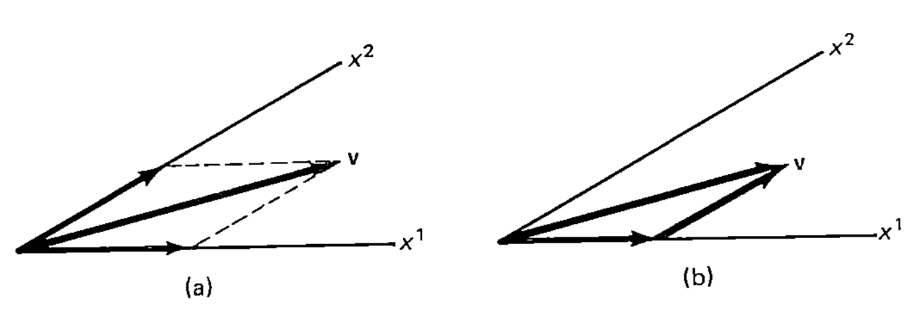

Figure 9.3 (a) Parallelogram method for adding components; (b) polygon method for adding components.

Consider vector addition by the parallelogram or equivalent polygon method illustrated in Figures 9.3. These figures represent the vector sum

$$\mathbf{v} = \nu^1 \mathbf{u}_1 + \nu^2 \mathbf{u}_2 = v^1 \mathbf{e}_1 + v^2 \mathbf{e}_2 \tag{9.56}$$

Therefore, the components shown in Figure 9.3a, which are the sides of a parallelogram, are the contravariant components of a vector.

To see that the components of Figure 9.3b are the covariant components, we write the vector $\mathbf{v}$ in terms of the contravariant bases

$$\mathbf{v} = v_i \mathbf{e}^i = \nu_i \mathbf{u}^i \tag{9.57}$$

The ith component of $\mathbf{v}$ is obtained from

$$\mathbf{v} \cdot \mathbf{u}_i = \nu_j \mathbf{u}^j \cdot \mathbf{u}_i = \nu_j \delta_i^j = \nu_i = \sqrt{g_{ii}}\, v_i \tag{9.58}$$

But

$$\mathbf{v} \cdot \mathbf{u}_i = |\mathbf{v}|\,|\mathbf{u}_i| \cos\theta_i = |\mathbf{v}| \cos\theta_i \tag{9.59}$$

Referring to Figure 9.2b, we note that $|\mathbf{v}| \cos\theta_i$ is the length of the component along the ith axis obtained by constructing the perpendicular to that axis. Since

$$\nu_i = |\mathbf{v}| \cos\theta_i \tag{9.60}$$

the components depicted in Figure 9.2b are the covariant components.

Minimum distance between two points and the equation of the geodesic

By specifying the line element $ds = \sqrt{ds^2}$, we can determine the distance between two points p_0 and p_1. This distance is

$$S = \int_{p_0}^{p_1} ds = \int_{p_0}^{p_1} \left(g_{ij}\, dx^i\, dx^j\right)^{1/2} \tag{9.61a}$$

Of course, the distance traveled between p_0 and p_1 depends on the path taken. We will determine the shortest distance between points p_0 and p_1 by minimizing S with respect to path. The path corresponding to this minimum distance is called the *geodesic*. As mentioned earlier, the geodesic is not a straight line in every space.

Let λ be a variable that parametrizes the path taken. (For example, the time of travel could parametrize the path and we would take $\lambda = t$.) Equation 9.61a can be written

$$S = \int_{\lambda_0}^{\lambda_1} \left(g_{ij} \frac{dx^i}{d\lambda} \frac{dx^j}{d\lambda} \right)^{1/2} d\lambda \tag{9.61b}$$

Although it may not be a true velocity, we define

$$\text{(a)} \quad \frac{dx^i}{d\lambda} \equiv \dot{x}^i \quad \text{and} \quad \text{(b)} \quad \left(g_{ij} \frac{dx^i}{d\lambda} \frac{dx^j}{d\lambda} \right)^{1/2} \equiv L = \frac{ds}{d\lambda} \tag{9.62}$$

Since the analysis that follows corresponds closely to the determination of the Lagrange equations of motion that we discussed in Chapter 7, the similarity of notation could be helpful in understanding the derivation of the equation of the geodesic.

As with the variation of the action, the variation of distance S with path is

$$\delta S = \int_{p_0}^{p_1} \delta L \, d\lambda = \int_{p_0}^{p_1} \left[\frac{\partial L}{\partial x^i} \delta x^i + \frac{\partial L}{\partial \dot{x}^i} \delta \dot{x}^i \right] d\lambda \tag{9.63}$$

where

$$\delta \dot{x}^i = \delta \left(\frac{dx^i}{d\lambda} \right) = \frac{d}{d\lambda} (\delta x^i) \tag{9.64}$$

Integrating the second term by parts once, and using equation 9.64, we obtain

$$\int_{p_0}^{p_1} \frac{\partial L}{\partial \dot{x}^i} \frac{d}{d\lambda} (\delta x^i) \, d\lambda = \frac{\partial L}{\partial \dot{x}^i} \delta x^i \bigg|_{p_0}^{p_1} - \int_{p_0}^{p_1} \frac{d}{d\lambda} \left(\frac{\partial L}{\partial \dot{x}^i} \right) \delta x^i \, d\lambda \tag{9.65}$$

Since every path starts at p_0 and terminates at p_1, there is no variation of the coordinates x^i at the endpoints. Thus the integrated term is zero, and equation 9.63 becomes

$$\delta S = \int_{p_0}^{p_1} \left[\frac{\partial L}{\partial x^i} - \frac{d}{d\lambda} \left(\frac{\partial L}{\partial \dot{x}^i} \right) \right] \delta x^i \, d\lambda \tag{9.66}$$

The distance will be a minimum when $\delta S = 0$. Since δx^i is not zero for different paths, the equation for the geodesic is

$$\frac{\partial L}{\partial x^i} - \frac{d}{d\lambda} \left(\frac{\partial L}{\partial \dot{x}^i} \right) = 0 \tag{9.67}$$

where L is given by equation 9.62b.

In realistic situations, $g_{ij} \neq g_{ij}(\dot{x}^k)$. Therefore, using $\partial \dot{x}^m / \partial \dot{x}^n = \delta_n^m$,

$$\frac{\partial L}{\partial \dot{x}^i} = \frac{1}{2L} g_{jk} \frac{\partial}{\partial \dot{x}^i} (\dot{x}^j \dot{x}^k) = \frac{1}{2L} g_{jk} (\delta_i^j \dot{x}^k + \delta_i^k \dot{x}^j) = \frac{1}{2L} (g_{ik} \dot{x}^k + g_{ji} \dot{x}^j) \tag{9.68a}$$

Using the fact that k, in the first term, and j in the second term, are dummy indices, and the fact that the metric tensor is symmetric, this becomes

$$\frac{\partial L}{\partial \dot{x}^i} = \frac{1}{L} g_{ij} \dot{x}^j = \frac{1}{L} g_{ij} \frac{dx^j}{d\lambda} \tag{9.68b}$$

Since $L = ds/d\lambda$, this can be written

$$\frac{\partial L}{\partial \dot{x}^i} = g_{ij}\frac{dx^j}{ds} \tag{9.68c}$$

Therefore,

$$\frac{d}{d\lambda}\left(\frac{\partial L}{\partial \dot{x}^i}\right) = \frac{d}{ds}\left(g_{ij}\frac{dx^j}{ds}\right)\frac{ds}{d\lambda} \tag{9.69}$$

With $\dot{x}^j = dx^j/d\lambda$ and $L = ds/d\lambda$, we have

$$\frac{\partial L}{\partial x^i} = \frac{1}{2L}\frac{\partial g_{jk}}{\partial x^i}\dot{x}^j\dot{x}^k = \frac{1}{2}\frac{\partial g_{jk}}{\partial x^i}\frac{dx^j}{d\lambda}\frac{dx^k}{d\lambda}\frac{d\lambda}{ds} = \frac{1}{2}\frac{\partial g_{jk}}{\partial x^i}\frac{dx^j}{ds}\frac{dx^k}{ds}\frac{ds}{d\lambda} \tag{9.70}$$

Combining equations 9.69 and 9.70, the equation of the geodesic is

$$\frac{1}{2}\frac{\partial g_{jk}}{\partial x^i}\frac{dx^j}{ds}\frac{dx^k}{ds} - \frac{d}{ds}\left(g_{ij}\frac{dx^j}{ds}\right) = 0 \tag{9.71}$$

The second term of equation 9.71 can be expanded as

$$\frac{d}{ds}\left(g_{ij}\frac{dx^j}{ds}\right) = \frac{\partial g_{ij}}{\partial x^m}\frac{dx^m}{ds}\frac{dx^j}{ds} + g_{ij}\frac{d^2x^j}{ds^2} \tag{9.72}$$

$$\Rightarrow \quad \frac{1}{2}\frac{\partial g_{jk}}{\partial x^i}\frac{dx^j}{ds}\frac{dx^k}{ds} - \frac{\partial g_{ij}}{\partial x^m}\frac{dx^m}{ds}\frac{dx^j}{ds} - g_{ij}\frac{d^2x^j}{ds^2} = 0 \tag{9.73}$$

Christoffel symbols

We note that in the second term of equation 9.73, that m and j are dummy indices. Therefore, we can write this term as

$$\frac{\partial g_{ij}}{\partial x^m}\frac{dx^m}{ds}\frac{dx^j}{ds} = \frac{1}{2}\left(\frac{\partial g_{ij}}{\partial x^m}\frac{dx^m}{ds}\frac{dx^j}{ds} + \frac{\partial g_{im}}{\partial x^j}\frac{dx^j}{ds}\frac{dx^m}{ds}\right) = \frac{1}{2}\left(\frac{\partial g_{ij}}{\partial x^m} + \frac{\partial g_{im}}{\partial x^j}\right)\frac{dx^j}{ds}\frac{dx^m}{ds} \tag{9.74}$$

Similarly, the index k in the first term of equation 9.73 is a dummy index, and can be replaced by m. Then the equation of the geodesic becomes

$$\frac{1}{2}\left(\frac{\partial g_{ij}}{\partial x^m} + \frac{\partial g_{im}}{\partial x^j} - \frac{\partial g_{mj}}{\partial x^i}\right)\frac{dx^j}{ds}\frac{dx^m}{ds} + g_{ij}\frac{d^2x^j}{ds^2} = 0 \tag{9.75}$$

The factor

$$\frac{1}{2}\left(\frac{\partial g_{ij}}{\partial x^m} + \frac{\partial g_{im}}{\partial x^j} - \frac{\partial g_{mj}}{\partial x^i}\right) \equiv [j, m; i] \tag{9.76}$$

is called the *Christoffel symbol of the first kind*. In terms of this symbol, the equation of the geodesic is

$$[j, m; i]\frac{dx^j}{ds}\frac{dx^m}{ds} + g_{ij}\frac{d^2x^j}{ds^2} = 0 \tag{9.77}$$

From its definition, we note that the Christoffel symbol is symmetric in the interchange of the two indices preceding the semicolon.

If equation 9.77 is multiplied by the contravariant metric tensor and we use the fact that $g^{ki}g_{ij} = \delta^k_j$, we obtain

$$\frac{d^2x^k}{ds^2} + g^{ki}[j, m; i]\frac{dx^j}{ds}\frac{dx^m}{ds} = 0 \tag{9.78}$$

The product

$$g^{ki}[j,m;i] \equiv \left\{ {k \atop j\;\;m} \right\} \tag{9.79}$$

is called the *Christoffel symbol of the second kind*. Since the first Christoffel symbol is symmetric in the interchange of the two indices that precede the semicolon, the second Christoffel symbol is symmetric in the interchange of its lower indices. With this definition, then, the equation for the shortest distance between points can be written

$$\frac{d^2x^k}{ds^2} + \left\{ {k \atop j\;\;m} \right\} \frac{dx^j}{ds}\frac{dx^m}{ds} = 0 \tag{9.80}$$

Example 9.1

To illustrate the computation of these Christoffel symbols, we consider a three-dimensional space described in cylindrical coordinates by the square of the line element

$$ds^2 = d\rho^2 + \rho^2\, d\phi^2 + dz^2 \tag{9.81}$$

Thus with $x^1 = \rho$, $x^2 = \phi$, and $x^3 = z$, the covariant metric tensor, in matrix form, is

$$\text{(a)}\quad \mathbf{g}_{\text{covariant}} = \begin{bmatrix} 1 & 0 & 0 \\ 0 & \rho^2 & 0 \\ 0 & 0 & 1 \end{bmatrix} \quad \Rightarrow \quad \text{(b)}\quad \mathbf{g}^{\text{contravariant}} = \begin{bmatrix} 1 & 0 & 0 \\ 0 & 1/\rho^2 & 0 \\ 0 & 0 & 1 \end{bmatrix} \tag{9.82}$$

since the contravariant metric, in matrix form, is the inverse of the covariant metric.

The nonzero elements of the first Christoffel symbol arise for terms that satisfy

$$\frac{\partial g_{mn}}{\partial x^p} \neq 0 \tag{9.83}$$

From equation 9.82a, we see that this will occur for $m = n = 2$, $p = 1$. Then

$$\frac{\partial g_{22}}{\partial x^1} = 2\rho \tag{9.84}$$

$$\Rightarrow \quad [i,j;k] = \frac{1}{2}\left(\frac{\partial g_{jk}}{\partial x^i} + \frac{\partial g_{ik}}{\partial x^j} - \frac{\partial g_{ij}}{\partial x^k} \right) = \rho\left(\delta_{j2}\delta_{k2}\delta_{i1} + \delta_{i2}\delta_{k2}\delta_{j1} - \delta_{i2}\delta_{j2}\delta_{k1}\right) \tag{9.85}$$

Therefore, there are only three nonzero components of the first Christoffel symbol.

$$\text{(a)}\quad [2,1;2] = [1,2;2] = \rho \qquad \text{(b)}\quad [2,2;1] = -\rho \tag{9.86}$$

From

$$\left\{ {k \atop j\;\;m} \right\} = g^{ki}[j,m;i] \tag{9.79}$$

we see that the three nonzero components of the second Christoffel symbol are

$$\text{(a)}\quad \left\{ {2 \atop 1\;\;2} \right\} = \left\{ {2 \atop 2\;\;1} \right\} = \frac{1}{\rho} \qquad \text{(b)}\quad \left\{ {1 \atop 1\;\;1} \right\} = -\rho \tag{9.87}$$

The three equations for the geodesic curves are

$$\frac{d^2x^1}{ds^2} + \left\{ {1 \atop 1\;\;1} \right\} \frac{dx^1}{ds}\frac{dx^1}{ds} = \frac{d^2\rho}{ds^2} - \rho\left(\frac{d\rho}{ds}\right)^2 = 0 \tag{9.88a}$$

$$\frac{d^2x^2}{ds^2} + \left\{ {2 \atop 1\;\;2} \right\} \frac{dx^1}{ds}\frac{dx^2}{ds} + \left\{ {2 \atop 2\;\;1} \right\} \frac{dx^2}{ds}\frac{dx^1}{ds} = \frac{d^2\phi}{ds^2} + \frac{2}{\rho}\frac{d\rho}{ds}\frac{d\phi}{ds} = 0 \tag{9.88b}$$

$$\frac{d^2x^3}{ds^2} = \frac{d^2z}{ds^2} = 0 \quad \square \tag{9.88c}$$

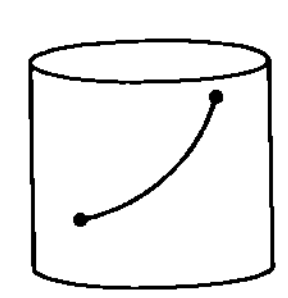

Figure 9.4 Geodesic on the curved surface of a cylinder.

Example 9.2

As a simple example, on the curved surface of a cylinder, ρ = constant. Therefore, equation 9.88a is satisfied trivially, and equation 9.88b becomes

$$\frac{d^2\phi}{ds^2} = 0 \tag{9.89}$$

Thus along with equation 9.88c, the parametrized equations of the geodesic are

$$\text{(a)} \quad \phi = \alpha_1 s + \beta_1 \qquad \text{(b)} \quad z = \alpha_2 s + \beta_2 \tag{9.90}$$

Eliminating s, the equations for the shortest curve between two points on the curved surface of a cylinder can be written

$$\text{(a)} \quad \rho = \text{constant} \qquad \text{(b)} \quad z = A\phi + B \tag{9.91}$$

Such a curve is depicted in Figure 9.4. □

As the reader will show in Problem 4, the Christoffel symbol of the first kind transforms like

$$[j, m; i]' = \frac{\partial x^k}{\partial x'^j}\frac{\partial x^l}{\partial x'^m}\frac{\partial x^n}{\partial x'^i}[k, l; n] + g_{kl}\frac{\partial x^k}{\partial x'^i}\frac{\partial^2 x^l}{\partial x'^j \partial x'^m} \tag{9.92}$$

Therefore, for a general transformation, the first Christoffel symbol is not a tensor. However, if the transformation is linear, then

$$\frac{\partial^2 x^l}{\partial x'^j \partial x'^m} = 0 \tag{9.93}$$

$$\Rightarrow \quad [j, m; i]' = \frac{\partial x^k}{\partial x'^j}\frac{\partial x^l}{\partial x'^m}\frac{\partial x^n}{\partial x'^i}[k, l; n] \tag{9.94}$$

This is the transformation rule for a rank 3, covariant tensor.

As the reader will also show in Problem 4, the second Christoffel symbol transforms as

$$\begin{Bmatrix} k \\ j \quad m \end{Bmatrix}' = \frac{\partial x'^k}{\partial x^l}\frac{\partial x^r}{\partial x'^j}\frac{\partial x^s}{\partial x'^m}\begin{Bmatrix} l \\ r \quad s \end{Bmatrix} + \frac{\partial x'^k}{\partial x^l}\frac{\partial^2 x^l}{\partial x'^j \partial x'^m} \tag{9.95}$$

As with the first Christoffel symbol, this is not the transformation of a tensor. If the transformation is linear, the second term is zero and

$$\begin{Bmatrix} k \\ j \quad m \end{Bmatrix}' = \frac{\partial x'^k}{\partial x^l}\frac{\partial x^r}{\partial x'^j}\frac{\partial x^s}{\partial x'^m}\begin{Bmatrix} l \\ r \quad s \end{Bmatrix} \tag{9.96}$$

That is, for a linear transformation, the second Christoffel symbol transforms like a rank 3 tensor that is singly contravariant and doubly covariant.

From equation 9.79 we see that the second Christoffel symbol is the dual of the first Christoffel symbol, such that the index after the semicolon of the symbol of the first kind has been raised by g^{ij} to obtain the symbol of the second kind. Given the transformation properties of the first Christoffel symbol, the transformation of the Christoffel symbol of the second is the expected one.

Covariant derivative

Writing a vector $\mathbf{A}$ as

$$\mathbf{A} = A^i \mathbf{e}_i \tag{9.54a}$$

let us now investigate how the vector $\mathbf{A}$ changes when we move from one point to another. This change is

$$d\mathbf{A} = dA^i \mathbf{e}_i + A^i d\mathbf{e}_i \tag{9.97}$$

Since the covariant basis vectors form a complete covariant set, the vector $d\mathbf{e}_i$ can be expanded in that basis. Since $d\mathbf{e}_i$ is an infinitesimal, the expansion coefficient must be infinitesimal. That is,

$$d\mathbf{e}_i = \left(dG_i^j\right)\mathbf{e}_j \tag{9.98}$$

From equation 9.97, the size of $d\mathbf{e}_i$ depends on the size of $d\mathbf{A}$. But $d\mathbf{A}$ depends on how far apart the two points are. That is, $d\mathbf{A}$ depends on $d\mathbf{r}$, the vector displacement between the two points. The only way this can be introduced is to let dG_i^j be proportional to coordinate displacements. Therefore, we must have

$$dG_i^j = \Gamma_{ik}^j \, dx^k \tag{9.99}$$

$$\Rightarrow \quad d\mathbf{e}_i = \Gamma_{ik}^j \, dx^k \mathbf{e}_j \tag{9.100}$$

For example, in spherical coordinates, it was shown that

$$\text{(a)} \quad d\mathbf{r}_0 = d\theta \, \boldsymbol{\theta}_0 + \sin\theta \, d\phi \, \boldsymbol{\phi}_0 \qquad \text{(b)} \quad d\boldsymbol{\theta}_0 = -d\theta \mathbf{r}_0 + \cos\theta \, d\phi \, \boldsymbol{\phi}_0$$
$$\text{(c)} \quad d\boldsymbol{\phi}_0 = -\sin\theta \, d\phi \, \mathbf{r}_0 - \cos\theta \, d\phi \, \boldsymbol{\theta}_0 \tag{1.77}$$

If (r, θ, ϕ) are designated as (x^1, x^2, x^3), then

$$\text{(a)} \quad \Gamma_{12}^1 = 1 \qquad \text{(b)} \quad \Gamma_{13}^3 = \sin(x^2) \qquad \text{(c)} \quad \Gamma_{22}^1 = -1$$
$$\text{(d)} \quad \Gamma_{23}^3 = \cos(x^2) \qquad \text{(e)} \quad \Gamma_{33}^1 = -\sin(x^2) \qquad \text{(f)} \quad \Gamma_{33}^2 = -\cos(x^2) \tag{9.101}$$

All other Γ symbols are zero.

Substituting equation 9.100 into equation 9.97, the change in the vector $\mathbf{A}$ becomes

$$d\mathbf{A} = dA^i \mathbf{e}_i + A^i \Gamma_{ik}^j \, dx^k \mathbf{e}_j \tag{9.102}$$

Writing

$$dA^i \mathbf{e}_i = dA^j \mathbf{e}_j = \frac{\partial A^j}{\partial x^k} \, dx^k \mathbf{e}_j \tag{9.103}$$

$$\Rightarrow \quad d\mathbf{A} = \left(\frac{\partial A^j}{\partial x^k} + A^i \Gamma_{ik}^j \right) dx^k \mathbf{e}_j \tag{9.104}$$

In the Cartesian coordinate system, the displacement of a vector is given by

$$d\mathbf{A} = \frac{\partial A^j}{\partial x^k} \, dx^k \mathbf{e}_j \tag{9.105}$$

We see that in the more general Riemann geometry, there is an additional term in the derivative that arises because the space is curved. Comparing equations 9.104 and 9.105, we see that a noncurved or flat space is one in which $\Gamma^j_{ik} = 0$.

Since $d\mathbf{A}$ is a vector in the space, it can be expanded in terms of the basis vectors. We define a contravariant component of $d\mathbf{A}$ by

$$d\mathbf{A} \equiv \delta A^j \mathbf{e}_j \tag{9.106}$$

Referring to equation 9.104, we see that this contravariant vector is

$$\delta A^j = \left(\frac{\partial A^j}{\partial x^k} + A^i \Gamma^j_{ik} \right) dx^k \tag{9.107}$$

The combination of terms in equation 9.107 is called the *covariant derivative* of A^j and is denoted by

$$\left(\frac{\partial A^j}{\partial x^k} + A^i \Gamma^j_{ik} \right) \equiv A^j{}_{,k} \tag{9.108a}$$

A second, less often used notation for the covariant derivative is

$$A^j{}_{,k} \equiv (A^j)_k \tag{9.108b}$$

The covariant derivative of a vector transforms like a tensor, whereas the partial derivative of that vector with respect to a coordinate does not (see Problem 6). Covariant differentiation is, then, the tensor generalization of partial differentiation.

Equation 9.106 can be written in terms of the covariant derivative, as

$$d\mathbf{A} = A^j{}_{,k}\, dx^k \mathbf{e}_j \tag{9.109}$$

It was shown earlier that

$$g_{ij} = \mathbf{e}_i \cdot \mathbf{e}_j \tag{9.46}$$

$$\Rightarrow \quad \frac{\partial g_{ij}}{\partial x^k} = \frac{\partial \mathbf{e}_i}{\partial x^k} \cdot \mathbf{e}_j + \mathbf{e}_i \cdot \frac{\partial \mathbf{e}_j}{\partial x^k} \tag{9.110}$$

But

$$d\mathbf{e}_i = \Gamma^l_{ik}\, dx^k \mathbf{e}_l \tag{9.100}$$

$$\Rightarrow \quad \frac{\partial \mathbf{e}_i}{\partial x^k} = \Gamma^l_{ik} \mathbf{e}_l \tag{9.111}$$

$$\Rightarrow \quad \frac{\partial g_{ij}}{\partial x^k} = \Gamma^l_{ik} \mathbf{e}_l \cdot \mathbf{e}_j + \Gamma^l_{jk} \mathbf{e}_i \cdot \mathbf{e}_l = \Gamma^l_{ik} g_{lj} + \Gamma^l_{jk} g_{li} \tag{9.112a}$$

By interchanging the positions of these indices, we also have

$$\text{(b)} \quad \frac{\partial g_{jk}}{\partial x^i} = \Gamma^l_{ji} g_{lk} + \Gamma^l_{ki} g_{lj} \qquad \text{(c)} \quad \frac{\partial g_{ki}}{\partial x^j} = \Gamma^l_{kj} g_{li} + \Gamma^l_{ij} g_{lk} \tag{9.112}$$

Combining these terms with the appropriate signs, the Christoffel symbol of the first kind can then be expressed in terms of the Γ symbols

$$\begin{aligned} [i, j; k] &= \tfrac{1}{2}\left(\frac{\partial g_{jk}}{\partial x^i} + \frac{\partial g_{ki}}{\partial x^j} - \frac{\partial g_{ij}}{\partial x^k} \right) \\ &= \tfrac{1}{2}\left[\left(\Gamma^l_{ji} g_{lk} + \Gamma^l_{ki} g_{lj} \right) + \left(\Gamma^l_{kj} g_{li} + \Gamma^l_{ij} g_{lk} \right) - \left(\Gamma^l_{ik} g_{lj} + \Gamma^l_{jk} g_{li} \right) \right] \\ &= \tfrac{1}{2}\left[g_{il}\left(\Gamma^l_{kj} - \Gamma^l_{jk} \right) + g_{jl}\left(\Gamma^l_{ki} - \Gamma^l_{ik} \right) + g_{kl}\left(\Gamma^l_{ji} + \Gamma^l_{ij} \right) \right] \end{aligned} \tag{9.113}$$

Consider the vector line element

$$d\mathbf{s} = dx^i \mathbf{e}_i \tag{9.114}$$

$$\Rightarrow \quad d^2\mathbf{s} = d^2x^i\mathbf{e}_i + dx^i\, d\mathbf{e}_i = d^2x^i\mathbf{e}_i + \frac{\partial \mathbf{e}_i}{\partial x^j}\, dx^i\, dx^j \tag{9.115}$$

Recognizing that i and j are dummy indices, we see that

$$\frac{\partial \mathbf{e}_i}{\partial x^j}\, dx^i\, dx^j = \frac{\partial \mathbf{e}_j}{\partial x^i}\, dx^i\, dx^j \tag{9.116}$$

$$\Rightarrow \quad \frac{\partial \mathbf{e}_i}{\partial x^j} = \frac{\partial \mathbf{e}_j}{\partial x^i} \tag{9.117}$$

Using equation 9.111, then,

$$\Gamma_{ij}^k \mathbf{e}_k = \Gamma_{ji}^k \mathbf{e}_k \tag{9.118}$$

That is, Γ_{ij}^k is symmetric in the interchange of its covariant indices. Therefore, the second differential of the line element can be written

$$d^2\mathbf{s} = \left(d^2x^k + \Gamma_{ij}^k\, dx^i\, dx^j\right)\mathbf{e}_k \tag{9.119}$$

Using the symmetry of the Γ symbol, equation 9.113 becomes

$$[i, j; k] = g_{kl}\Gamma_{ij}^l \tag{9.120}$$

Using the fact that $g^{mk}g_{kl} = \delta_l^m$, we premultiply this equation by g^{mk} to obtain

$$\Gamma_{ij}^m = g^{mk}[i, j; k] \tag{9.121}$$

Comparing this to equation 9.79, we see that the Γ symbol is the same as the Christoffel symbol of the second kind

$$\Gamma_{ij}^k = \left\{ {k \atop i\ \ j} \right\} \tag{9.122}$$

That is, the covariant derivative of equation 9.108a can be written

$$A^i_{\ ,j} = \frac{\partial A^i}{\partial x^j} + A^k \left\{ {i \atop k\ \ j} \right\} \tag{9.123}$$

It was indicated earlier, and will be shown by the reader in Problem 4, that the second Christoffel symbol does not transform like a tensor, unless the transformation is linear. In a similar way, $\partial A^i/\partial x^j$ does not transform like a tensor except under a linear transformation. However, the covariant derivative, $A^i_{\ ,j}$ transforms as a second-rank tensor of mixed variance. The reader will demonstrate this in Problem 6.

Since the covariant derivative of A^i is a second-rank mixed variance tensor, a doubly covariant tensor can be formed from it by multiplication by g_{ik}. That is,

$$A_{i,j} = g_{il}A^l_{\ ,j} \tag{9.124}$$

which is the covariant derivative of a covariant vector. From equations 9.108a, or 9.123,

$$A_{i,j} = g_{il}\left(\frac{\partial A^l}{\partial x^j} + A^m\Gamma_{mj}^l\right) \tag{9.125}$$

Consider

$$g_{il}\frac{\partial A^l}{\partial x^j} = \frac{\partial\left(g_{il}A^l\right)}{\partial x^j} - A^l\frac{\partial g_{il}}{\partial x^j} = \frac{\partial A_i}{\partial x^j} - A^l\frac{\partial g_{il}}{\partial x^j} \tag{9.126}$$

Then

$$A_{i,j} = \frac{\partial A_i}{\partial x^j} - A^l \frac{\partial g_{il}}{\partial x^j} + g_{il} A^m \Gamma^l_{mj} \tag{9.127a}$$

In the third term, since both l, and m are dummy indices, we can rename l to be m, and m to be l. Then equation 9.127a becomes

$$A_{i,j} = \frac{\partial A_i}{\partial x^j} - \left[\frac{\partial g_{il}}{\partial x^j} + g_{im}\Gamma^m_{lj}\right] A^l = \frac{\partial A_i}{\partial x^j} - g^{lk}\left[\frac{\partial g_{il}}{\partial x^j} + g_{im}\Gamma^m_{lj}\right] A_k \tag{9.127b}$$

In Problem 5 the reader will demonstrate that

$$g^{lk}\left[\frac{\partial g_{il}}{\partial x^j} - g_{im}\Gamma^m_{lj}\right] = \Gamma^k_{ij} \tag{9.128}$$

$$\Rightarrow \quad A_{i,j} = \frac{\partial A_i}{\partial x^j} - \Gamma^k_{ij} A_k \tag{9.129}$$

Covariant differentiation of covariant and contravariant vectors can be extended to tensors of higher rank. Referring to equations 9.108a (or 9.123) and 9.129, we note that the covariant derivative of a contravariant vector results in the addition of the partial derivative of the vector and the term containing the second Christoffel symbol. The covariant derivative of a covariant vector contains the difference of such terms. Therefore, it is clear that the covariant derivative of tensors of higher rank contain the ordinary derivative of the tensor, plus the addition of terms containing the second Christoffel symbol for each contravariant index and the subtraction of terms containing Γ symbols for each covariant index. For example,

$$\text{(a)} \quad A^{ij}{}_{,k} = \frac{\partial A^{ij}}{\partial x^k} + \Gamma^i_{mk} A^{mj} + \Gamma^j_{kn} A^{in} \qquad \text{(b)} \quad A^i_{j,k} = \frac{\partial A^i_j}{\partial x^k} + \Gamma^i_{mk} A^m_j - \Gamma^n_{jk} A^i_n$$

$$\text{(c)} \quad A^{ij}_{kl,m} = \frac{\partial A^{ij}_{kl}}{\partial x^m} + \Gamma^i_{rm} A^{rj}_{kl} + \Gamma^j_{rm} A^{ir}_{kl} - \Gamma^r_{km} A^{ij}_{rl} - \Gamma^r_{lm} A^{ij}_{kr} \tag{9.130}$$

Let A^i and B_j be two vectors. Recognizing that their outer product is a second-rank tensor of mixed variance (see Problem 2), we refer to equation 9.130b, and consider

$$\left(A^i B_j\right)_{,k} = \frac{\partial\left(A^i B_j\right)}{\partial x^k} + \Gamma^i_{rk} A^r B_j - \Gamma^r_{jk} A^i B_r = A^i \frac{\partial B_j}{\partial x^k} + B_j \frac{\partial A^i}{\partial x^k} + \Gamma^i_{rk} A^r B_j - \Gamma^r_{jk} A^i B_r$$

$$= A^i\left[\frac{\partial B_j}{\partial x^k} - \Gamma^r_{jk} B_r\right] + \left[\frac{\partial A^i}{\partial x^k} + \Gamma^i_{rk} A^r\right] B_j = A^i B_{j,k} + A^i{}_{,k} B_j \tag{9.131}$$

That is, the covariant derivative of a product of tensors obeys the same distribution rule that the ordinary derivative of a product of functions obeys.

Generalization of the divergence and curl

Writing the vector $\mathbf{A}$ as $A^i \mathbf{e}_i$, the generalized divergence of $\mathbf{A}$, is

$$\Box \cdot \mathbf{A} \equiv \mathbf{e}^k \frac{\partial}{\partial x^k} \cdot \left(A^i \mathbf{e}_i\right) = \mathbf{e}^k \cdot \left(\frac{\partial A^i}{\partial x^k}\mathbf{e}_i + A^i \frac{\partial \mathbf{e}_i}{\partial x^k}\right) \tag{9.132a}$$

We point out that because the component of $\mathbf{A}$ is written as a contravariant vector, the basis vector associated with it must be in covariant form. Since $\partial/\partial x^k$ transforms like a covariant vector, the basis vector associated with it must be the contravariant form, $\mathbf{e}^k$.

With

$$\frac{\partial \mathbf{e}_i}{\partial x^k} = \Gamma^r_{ik}\mathbf{e}_r \tag{9.111}$$

$$\Rightarrow \quad \square \cdot \mathbf{A} = \frac{\partial A^i}{\partial x^k}\mathbf{e}^k \cdot \mathbf{e}_i + A^i\Gamma^r_{ik}\mathbf{e}^k \cdot \mathbf{e}_r \tag{9.132b}$$

But

$$\mathbf{e}^m \cdot \mathbf{e}_l = \delta_l^m \tag{9.49}$$

$$\Rightarrow \quad \square \cdot \mathbf{A} = \frac{\partial A^k}{\partial x^k} + A^i\Gamma^k_{ik} = A^k{}_{,k} \tag{9.133}$$

That is, the contracted covariant derivative of a contravariant vector is a generalization of our concept of the divergence of a vector. As such, $A^k{}_{,k}$ is also referred to as the *covariant divergence* of A^k.

In Problem 5 the reader will show that

$$\Gamma^k_{ik} = \frac{1}{2}g^{jk}\frac{\partial g_{jk}}{\partial x^i} \tag{9.134}$$

Thus an alternative way of expressing the generalized divergence is

$$\square \cdot \mathbf{A} = A^k{}_{,k} = \frac{\partial A^k}{\partial x^k} + \frac{1}{2}g^{jk}\frac{\partial g_{jk}}{\partial x^i}A^i \tag{9.135}$$

$$\Rightarrow \quad \square^2 F \equiv \square \cdot \square F = \mathbf{e}^l\frac{\partial}{\partial x^l} \cdot \left(\frac{\partial F}{\partial x^k}\mathbf{e}^k\right) = g^{kl}\frac{\partial^2 F}{\partial x^k \partial x^l} - \Gamma^k_{lm}g^{lm}\frac{\partial F}{\partial x^k} \tag{9.136}$$

where F is an invariant function (see Problem 7).

Using the fact that the second Christoffel symbols are symmetric in the covariant indices,

$$A_{i,j} - A_{j,i} = \frac{\partial A_i}{\partial x^j} - \frac{\partial A_j}{\partial x^i} - \Gamma^k_{ij}A_k + \Gamma^k_{ji}A_k = \frac{\partial A_i}{\partial x^j} - \frac{\partial A_j}{\partial x^i} \tag{9.137}$$

which would be the kth component of $\nabla \times \mathbf{A}$ in three-dimensional Cartesian space. We point out that this *generalized curl* is a second-rank covariant tensor. The reader will show in Problem 8 that $(A^i{}_{,j} - A^j{}_{,i})$ is not in the same form as equation 9.137b, and therefore is not a generalization of the curl.

Riemann – Christoffel curvature tensor and the Riemann curvature invariant

As noted earlier, the partial derivative of a vector is generalized to the covariant derivative of the vector. In Euclidean geometry, we know that $\partial^2 A_i/\partial x^j \partial x^k = \partial^2 A_i/\partial x^k \partial x^j$. We now investigate whether the second covariant derivative is unchanged when the differentiation indices are interchanged. As such, we consider

$$D_{ijk} \equiv (A_{i,j})_{,k} - (A_{i,k})_{,j} \tag{9.138}$$

Defining

$$C_{mn} \equiv A_{m,n} \tag{9.139}$$

$$\Rightarrow \quad C_{mn,r} = \frac{\partial C_{mn}}{\partial x^r} - \Gamma^l_{nr} C_{ml} - \Gamma^l_{mr} C_{ln} \tag{9.140a}$$

$$\Rightarrow \quad (A_{m,n})_{,r} = \frac{\partial}{\partial x^r} A_{m,n} - \Gamma^l_{nr} A_{m,l} - \Gamma^l_{mr} A_{l,n}$$

$$= \frac{\partial}{\partial x^r}\left[\frac{\partial A_m}{\partial x^n} - \Gamma^l_{mn} A_l\right] - \Gamma^l_{nr}\left[\frac{\partial A_m}{\partial x^l} - \Gamma^s_{ml} A_s\right] - \Gamma^l_{mr}\left[\frac{\partial A_l}{\partial x^n} - \Gamma^s_{nl} A_s\right] \tag{9.140b}$$

Using this expression in equation 9.138, the symmetry of the second Christoffel symbol in its lower indices results in the cancellation of many terms. After some manipulation, equation 9.138 becomes

$$(A_{i,j})_{,k} - (A_{i,k})_{,j} = \left[\frac{\partial \Gamma^l_{ik}}{\partial x^j} - \frac{\partial \Gamma^l_{ij}}{\partial x^k} + \Gamma^l_{mj}\Gamma^m_{ik} - \Gamma^l_{mk}\Gamma^m_{ij}\right] A_l \equiv R_{ijk\cdot}{}^{l} A_l \tag{9.141}$$

The term in brackets, which is denoted by $R_{ijk\cdot}{}^{l}$ is called the *Riemann–Christoffel curvature tensor*. The dot behind the index k indicates where a lowered index would be placed if $R_{ijk\cdot}{}^{l}$ were multiplied by g_{lm} (with a sum over l implied). That is, for example,

$$\text{(a)} \quad g_{lm} R_{ijk\cdot}{}^{l} = R_{ijkm} \qquad \text{(b)} \quad g_{lm} R_{ijk\cdot}{}^{l} = R_{ijmk} \tag{9.142}$$

Using

$$[i, j; m] = g_{ml}\Gamma^l_{ij} \tag{9.120}$$

$$g_{ml}\frac{\partial \Gamma^l_{ik}}{\partial x^j} = \frac{\partial\left(g_{ml}\Gamma^l_{ik}\right)}{\partial x^j} - \Gamma^l_{ik}\frac{\partial g_{ml}}{\partial x^j} \tag{9.143}$$

it is straightforward to show that

$$R_{ijkm} = \frac{1}{2}\left(\frac{\partial^2 g_{im}}{\partial x^j \partial x^k} + \frac{\partial^2 g_{jk}}{\partial x^i \partial x^m} - \frac{\partial^2 g_{ik}}{\partial x^j \partial x^m} - \frac{\partial^2 g_{jm}}{\partial x^i \partial x^k}\right) + \left([i, m; t]\Gamma^t_{jk} - [i, k; t]\Gamma^t_{jm}\right) \tag{9.144}$$

$$\Rightarrow \quad \text{(a)} \quad R_{ijkm} = -R_{jikm} \qquad \text{(b)} \quad R_{ijkm} = -R_{ijmk} \qquad \text{(b)} \quad R_{ijkm} = +R_{kmij} \tag{9.145}$$

$$R_{ijkm} + R_{ikmj} + R_{imjk} = 0 \tag{9.146}$$

After rather unwieldy but straightforward manipulation, we also find that

$$R_{ijk\cdot}{}^{l}{}_{,n} + R_{ikn\cdot}{}^{l}{}_{,j} + R_{inj\cdot}{}^{l}{}_{,k} = 0 \tag{9.147}$$

This is known as *Bianchi's identity*. It contains the covariant derivative of $R_{ijk\cdot}{}^{l}$ with respect to n, and two other terms obtained by cyclic permutations of j, k, and n.

Let $\mathbf{v}$ and $\mathbf{w}$ be any two of the basis vectors $\mathbf{e}$. Then, from equations 9.46, 9.48, and 9.49,

$$\text{(a)} \quad v^i w_j = \delta^i_j \qquad \Rightarrow \qquad \text{(b)} \quad v^i w^j = g^{ij} \qquad \text{(c)} \quad v_i w_j = g_{ij} \tag{9.148}$$

We define vectors $\mathbf{x}$ and $\mathbf{y}$ that are linear combinations of $\mathbf{v}$ and $\mathbf{w}$:

$$\text{(a)}\quad x^i = av^i + bw^i \qquad \text{(b)}\quad y^i = cv^i + dw^i \tag{9.149}$$

with a, b, c, and d being invariants.

From the orthonormality of $\mathbf{v}$ and $\mathbf{w}$ expressed in equation 9.148a, it is straightforward to show that

$$(g_{ik}g_{jl} - g_{il}g_{jk})x^i y^j x^k y^l = (\mathbf{x}\cdot\mathbf{x})(\mathbf{y}\cdot\mathbf{y}) - (\mathbf{x}\cdot\mathbf{y})^2 = a^2d^2 + b^2c^2 - 2abcd \tag{9.150}$$

which is a tensor invariant referred to as the *Gaussian curvature.*

Consider the tensor invariant

$$R_{ijkl}x^i y^j x^k y^l = R_{ijkl}(av^i + bw^i)(cv^j + dw^j)(av^k + bw^k)(cv^l + dw^l) \tag{9.151}$$

Using

$$\text{(a)}\quad R_{ijkl} = -R_{jikl} \qquad \text{(b)}\quad R_{ijkl} = -R_{ijkl} \tag{9.145}$$

it is clear that any term containing $R_{ijkl}v^i v^j$ or $R_{ijkl}v^k v^l$ (or an identical term with v replaced by w) is zero. Therefore,

$$R_{ijkl}x^i y^j x^k y^l = R_{ijkl}[(adv^i w^j + bcw^i v^j)(adv^k w^l + bcw^k v^l)] \tag{9.152}$$
$$= a^2d^2R_{ijkl}v^i w^j v^k w^l + b^2c^2R_{ijkl}w^i v^j w^k v^l + abcdR_{ijkl}w^i v^j v^k w^l + abcdR_{ijkl}v^i w^j w^k v^l$$

Using the dummy index rule and equation 9.146, the second, third, and fourth terms can be written

$$b^2c^2R_{ijkl}w^i v^j w^k v^l = b^2c^2R_{jilk}v^i w^j v^k w^l = b^2c^2R_{ijkl}v^i w^j v^k w^l \tag{9.153a}$$

$$abcdR_{ijkl}w^i v^j v^k w^l = abcdR_{jikl}v^i w^j v^k w^l = -abcdR_{ijkl}v^i w^j v^k w^l \tag{9.153b}$$

$$abcdR_{ijkl}v^i w^j w^k v^l = abcdR_{ijlk}v^i w^j v^k w^l = -abcdR_{ijlk}v^i w^j v^k w^l \tag{9.153c}$$

$$\Rightarrow \quad R_{ijkl}x^i y^j x^k y^l = (a^2d^2 + b^2c^2 - 2abcd)\,R_{ijkl}v^i w^j v^k w^l \tag{9.154}$$

Referring to equation 9.150, we note that

$$K \equiv \frac{R_{ijkl}x^i y^j x^k y^l}{(g_{ik}g_{jl} - g_{il}g_{jk})x^i y^j x^k y^l} \tag{9.155}$$

is not only a tensor invariant, but is also invariant as to the choice of the vectors $\mathbf{x}$ and $\mathbf{y}$. Thus we could choose $\mathbf{x} = \mathbf{v}$ and $\mathbf{y} = \mathbf{w}$ so that $a = 1$, $b = c = 0$, and $d = 1$ without changing the value of K. Then the Gaussian curvature becomes

$$(g_{ik}g_{jl} - g_{il}g_{jk})x^i y^j x^k y^l = (\mathbf{x}\cdot\mathbf{x})(\mathbf{y}\cdot\mathbf{y}) - (\mathbf{x}\cdot\mathbf{y})^2 = 1 \tag{9.156}$$

$$\Rightarrow \quad K = R_{ijkl}v^i w^j v^k w^l \tag{9.157}$$

K is called the *Riemann curvature of the space.* A space is said to be flat if $K = 0$.

If $\mathbf{v} = \mathbf{x}$ and $\mathbf{w} = \mathbf{y}$ are unit basis vectors as defined in equations 9.51, then

$$(g_{ik}g_{jl} - g_{il}g_{jl})v^i w^j v^k w^l \neq 1 \tag{9.158}$$

In particular, in a two-dimensional space, since there are only two unit basis vectors to select from, we can only take

$$\text{(a)} \quad \mathbf{v} = (1,0) \qquad \text{(b)} \quad \mathbf{w} = (0,1) \tag{9.159}$$

$$\Rightarrow \quad (g_{ik}g_{jl} - g_{il}g_{jk})v^i w^j v^k w^l = (g_{ik}g_{jl} - g_{il}g_{jk})\delta_1^i\delta_2^j\delta_1^k\delta_2^l = (g_{11}g_{22} - g_{12}g_{21}) = \det(\mathbf{g}) \tag{9.160}$$

As the reader will ascertain when solving Problem 11, if the metric is diagonal, there is only one independent component of the Riemann–Christoffel curvature tensor. All nonzero components can be related to R_{1212} using equation 9.146. Therefore, in two dimensions,

$$K = \frac{R_{1212}}{\det(\mathbf{g})} \tag{9.161}$$

Example 9.3

To give the reader an indication of the geometric meaning of the Riemann curvature K, consider the problem of a two-dimensional space defined by the surface of a sphere of radius r. The invariant line element of such a space is

$$ds^2 = r^2\,d\theta^2 + r^2\sin^2\theta\,d\phi^2 \tag{9.162}$$

$$\Rightarrow \quad \text{(a)} \quad \mathbf{g}_{\text{covariant}} = \begin{bmatrix} r^2 & 0 \\ 0 & r^2\sin^2\theta \end{bmatrix} \qquad \text{(b)} \quad \mathbf{g}_{\text{contravariant}} = \begin{bmatrix} \dfrac{1}{r^2} & 0 \\ 0 & \dfrac{1}{r^2\sin^2\theta} \end{bmatrix} \tag{9.163}$$

With $x^1 = \theta$ and $x^2 = \phi$, it is straightforward to ascertain that the three nonzero Christoffel symbols of the first kind are

$$[1,2;2] = [2,1;2] = \frac{1}{2}\frac{\partial g_{22}}{\partial\theta} = r^2\sin\theta\cos\theta \tag{9.164a}$$

$$[2,2;1] = -\frac{1}{2}\frac{\partial g_{22}}{\partial\theta} = -r^2\sin\theta\cos\theta \tag{9.164b}$$

From these results, the three nonzero second Christoffel symbols are found to be

$$\text{(a)} \quad \Gamma_{12}^2 = \Gamma_{21}^2 = g^{22}[1,2;2] = \cot\theta \qquad \text{(b)} \quad \Gamma_{22}^1 = g^{11}[2,2;1] = -\sin\theta\cos\theta \tag{9.165}$$

Therefore, the one independent Riemann–Christoffel curvature tensor component is

$$R_{1212} = -\frac{1}{2}\frac{\partial^2 g_{22}}{\partial\theta^2} + [1,2;2]\Gamma_{12}^2 = r^2\sin^2\theta \tag{9.166}$$

Thus the Riemann curvature in the two-dimensional world on the surface of a sphere is

$$K = \frac{R_{1212}}{|g|} = \frac{1}{r^2} \tag{9.167}$$

That is, the Riemann curvature is a (reciprocal) measure of the radius of curvature of the space. □

If we consider the same curved surface in higher dimensions, however, the space is not necessarily curved; that is, K can be zero. The reader will show this in Problem 13 for the space described by spherical coordinates.

Example 9.4

We consider the example of a three-dimensional space described by cylindrical coordinates. With $x^1 = \rho$, $x^2 = \phi$, and $x^3 = z$, cylindrical space is spanned by the three orthogonal unit vectors $\boldsymbol{\rho}_0 = (1,0,0)$, $\boldsymbol{\phi}_0 = (0,1,0)$, and $\mathbf{k} = (0,0,1)$. If we take $\mathbf{v}$ and $\mathbf{w}$ to be any two of these, ρ_0 and ϕ_0 for example, then with $ds^2 = d\rho^2 + \rho^2 d\phi^2 + dz^2$,

$$K = \frac{R_{1212}}{g_{11}g_{22}} \tag{9.168}$$

From equation 9.144,

$$R_{1212} = \frac{1}{2}\left[\frac{\partial^2 g_{12}}{\partial x^1 \partial x^2} + \frac{\partial^2 g_{21}}{\partial x^2 \partial x^1} - \frac{\partial^2 g_{11}}{(\partial x^2)^2} - \frac{\partial^2 g_{22}}{(\partial x^1)^2}\right] + [1,2;m]\Gamma_{21}^m - [1,1;m]\Gamma_{22}^m \tag{9.169}$$

Referring to equations 9.86 and 9.87, there are three nonzero Christoffel symbols in a space described by cylindrical coordinates. From these equations we see that

$$\text{(a)} \quad [1,2;m]\Gamma_{21}^m = [1,2;2]\Gamma_{12}^2 = 1 \qquad \text{(b)} \quad [1,1;m]\Gamma_{22}^m = 0 \tag{9.170}$$

We also point out that the only nonzero second derivative of the metric tensor components in equation 9.169 is

$$\frac{\partial^2 g_{22}}{(\partial x^1)^2} = \frac{\partial^2 \rho^2}{\partial \rho^2} = 2 \tag{9.171}$$

$$\Rightarrow \quad K = \frac{1}{\rho^2} R_{1212} = \frac{1}{\rho^2}\left(\frac{1}{2}(-2) + 1\right) = 0 \quad \square \tag{9.172}$$

Thus we see that even though we describe the higher-dimensional space in terms of coordinates in which natural surfaces are curved (the surface defined by ρ = constant in this example), the space is a flat space. In general, any space that can be reached by a transformation from a flat space is also flat in the sense that the Riemann curvature is zero. Such a space is also referred to as a *Euclidean space.*

We note that in the two-dimensional space defined by the surface of the sphere, the Riemann curvature is the inverse product of two radii of curvature (which are the same for a sphere). In general, this is the meaning of the Riemann curvature—that at each point, K is the inverse of a product of two radii of curvature. These can change from point to point, such as on the surface of an ellipsoid of revolution. It is also possible for such radii to be negative. For example, if a space is defined by the surfaces of a saddle, when one radius of curvature of the saddle is defined to be positive, the other must be negative. In that case, K will be negative.

Ricci tensor and Einstein's field equations of general relativity

The *Ricci tensor* is defined from the Riemann–Christoffel curvature tensor by the contraction

$$R_{ij} \equiv R_{ijk.}{}^{k} \tag{9.173}$$

Contraction on l and k in the Bianchi identity given in equation 9.147 yields

$$R_{ijk.,n}{}^{k} + R_{ikn.,j}{}^{k} + R_{inj.,k}{}^{k} = R_{ij,n} + R_{ikn.,j}{}^{k} + R_{inj.,k}{}^{k} = 0 \tag{9.174}$$

Using equations 9.146 we obtain

$$R_{ikn.}{}^{k} = -R_{ink.}{}^{k} = -R_{in} \tag{9.175}$$

$$\Rightarrow \quad R_{ij,n} - R_{in,j} + R_{inj.,k}{}^{k} = 0 \tag{9.176}$$

In Problem 9 the reader will show that

$$g^{mn}{}_{,r} = 0 \tag{9.177}$$

Premultiplying equation 9.151 by g^{mi} and using the fact that covariant differentiation of a product is the product of covariant derivatives (equation 9.131) yields

$$\text{(a)}\quad \left(g^{mi}R_{ij}\right)_{,n} - \left(g^{mi}R_{in}\right)_{,j} + \left(g^{mi}R_{inj.}^{\ \ \ \ k}\right)_{,k} = 0 \qquad \text{(b)}\quad R^{m}_{.j,n} - R^{m}_{.n,j} + R^{m\ \ k}_{.nj.,k} = 0 \tag{9.178}$$

The fully contracted Ricci tensor, called the *Ricci invariant*, is defined by

$$R \equiv R^{m}_{.m} \tag{9.179}$$

Contracting n with m in equation 9.178b results in

$$R^{n}_{.j,n} - R_{,j} + R^{n\ \ k}_{.nj.,k} = 0 \tag{9.180}$$

From the identities given in equations 9.145, this can be written

$$R^{n\ \ k}_{.nj.} = g^{ni}g^{kl}R_{injl} = -g^{ni}g^{kl}R_{lnji} = +g^{ni}g^{kl}R_{ljni} = g^{kl}R_{ljn.}^{\ \ \ \ n} = g^{kl}R_{lj} = R^{k}_{.j} \tag{9.181}$$

$$\Rightarrow \quad R^{n}_{.j,n} - R_{,j} + R^{k}_{.j,k} = 0 \tag{9.182a}$$

Since k is a dummy index in the third term, it can be replaced by n, and we see that the first and third terms are the same. Thus

$$R^{n}_{.j,n} - \tfrac{1}{2}R_{,j} = 0 \tag{9.182b}$$

Using equation 9.177, we multiply this by g^{mj} to obtain

$$\left(g^{mj}R^{\ n}_{.j}\right)_{,n} - \tfrac{1}{2}\left(g^{mj}R\right)_{,j} = R^{nm}{}_{,n} - \tfrac{1}{2}\left(g^{mn}R\right)_{,n} = \left(R^{nm} - \tfrac{1}{2}g^{nm}R\right)_{,n} = 0 \tag{9.183}$$

From this description of the covariant derivative as the generalization of the divergence, equation 9.183b can be interpreted as the statement that $R^{nm} - \frac{1}{2}g^{nm}R$ is a divergenceless tensor.

Referring to Maxwell's equation

$$\nabla \cdot \mathbf{E} = \frac{\rho}{\varepsilon_0} \tag{1.153a}$$

we recall that this is Poisson's equation in terms of the electric field. If the charge density is zero, **E** is a divergenceless field. Therefore, equation 9.183b can be viewed as Poisson's equation for the tensor $R^{nm} - \frac{1}{2}g^{nm}R$ when there is no source (the generalization of Laplace's equation).

It was pointed out in Chapter 1 that since the curl of the gradient of a scalar function is zero, the magnetic potential could be modified by a gauge transformation in the form $\mathbf{A}' = \mathbf{A} + \nabla\chi$ and the magnetic field $\mathbf{B} \equiv \nabla \times \mathbf{A}$ is invariant under such a gauge transformation. Since

$$g^{nm}{}_{,r} = 0 \tag{9.177}$$

we see that any constant multiple of g^{nm} can be added to $R^{nm} - \frac{1}{2}g^{nm}R$ without affecting the solution to equation 9.183b. That is, using equation 9.177,

$$\left(R^{nm} - \tfrac{1}{2}g^{nm}R - \Lambda g^{nm}\right)_{,n} = 0 \tag{9.184}$$

has the same solution as equation 9.183b. Λ is referred to as the *cosmological constant* and is often set to zero.

The solution to equation 9.184 is a second-rank contravariant tensor, the covariant derivative of which is zero. That is, the solution to equation 9.184 is

$$R^{nm} - \tfrac{1}{2}g^{nm}R - \Lambda g^{nm} = KT^{nm} \tag{9.185}$$

with

$$T^{nm}{}_{,n} = 0 \tag{9.186}$$

T^{nm} is called the *gravitational stress-energy tensor*. It embodies the description of the space-time distribution of the masses that are the source of the gravitational field. Equation 9.185 is the tensor representation of Einstein's field equations for general relativity.

If we multiply equation 9.185 by g_{ln} and then contract l with m, we obtain

$$(1 - \tfrac{1}{2}N)R - \Lambda N = KT_m^m \equiv KT \tag{9.187}$$

where $\delta_m^m = \text{Tr}(\mathbf{1}) = N$ is the dimension of the space. We note that in a two-dimensional space, the contracted stress-energy tensor must be the constant $-2\Lambda/K$. We also note that for a space of any dimension, if Λ is taken to be zero, and if there are no sources (so that $T = 0$), then $R = 0$. That is, in a space in which there are no sources, and if the cosmological constant is taken to be zero, Einstein's field equations become

$$R^{nm} = 0 \tag{9.188}$$

9.3 Applications to Relativity

The special theory of relativity involves spaces in which there are no gravitational sources and objects are not accelerated. The general theory of relativity is a theory of gravitation, involving gravitational sources and accelerating objects.

Special theory of relativity

The special theory of relativity is essentially a theory of observations made by two observers that are moving with respect to one another. The relative velocity is constant and is comparable to the velocity of light. Such a nonaccelerating coordinate system is called an *inertial frame*.

Special relativity is based on two postulates. From the results of the Michelson–Morley experiments, Einstein postulated the following:

1. **A measurement of the speed of electromagnetic radiation will yield the same results to observers in all inertial reference frames regardless of their relative motion.**

For example, if one inertial frame is approaching a source of radiation and another is receding from that source, observers in both frames will measure the same value for the speed of the radiation. This value is designated by $c \simeq 3 \times 10^8$ meters/sec.

2. **Observers in inertial coordinate systems that are in motion relative to one another will describe a phenomenon with equations that have the same form in both systems.**

Referring to Chapter 8, in which the covariance of matrix equations under similarity transformations was discussed, this principle can also be stated as:

2. **The equations that describe a phenomenon are covariant under a transformation between inertial coordinate systems that are in motion relative to one another.**

For example, Maxwell's equations have the same form in two inertial coordinate systems that are in relative motion.

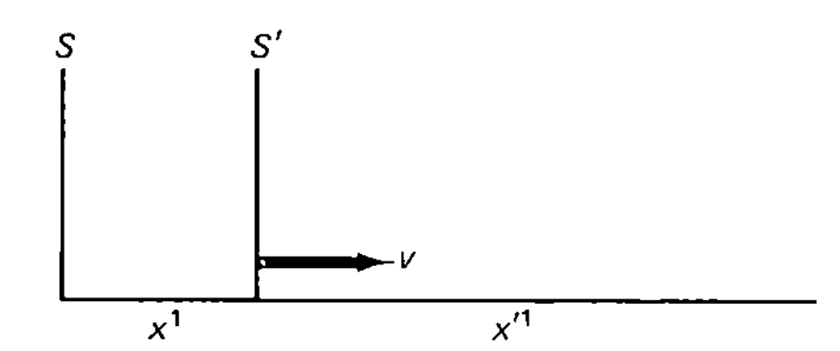

Figure 9.5 Coordinate systems in relative motion.

To deduce the transformation equations from one inertial frame to a second inertial frame that is in motion relative to the first, we begin by defining the coordinates of a four-dimensional space-time as

$$\text{(a)} \quad x^1 = x \qquad \text{(b)} \quad x^2 = y \qquad \text{(c)} \quad x^3 = z \qquad \text{(d)} \quad x^4 = ct \tag{9.189}$$

Let the two frames, which we label S and S', have a relative velocity v along their x and x' (or x^1 and x'^1) axes as shown in Figure 9.5. Based on the two postulates of special relativity, Einstein deduced that the transformation equations between the coordinates of S and S' were a set of equations developed by Lorentz. These involve the parameters

$$\text{(a)} \quad \beta \equiv \frac{v}{c} \qquad \text{(b)} \quad \gamma \equiv \left(1 - \beta^2\right)^{-1/2} \tag{9.190}$$

The Lorentz transformation equations for relative motion along the x^1, x'^1 axes are

$$\text{(a)} \quad x'^1 = \gamma\left(x^1 - \beta x^4\right) \qquad \text{(b)} \quad x'^2 = x^2 \qquad \text{(c)} \quad x'^3 = x^3 \qquad \text{(d)} \quad x'^4 = \gamma\left(x^4 - \beta x^1\right) \tag{9.191}$$

Since the relative velocity of the two frames is a constant, β and γ are constants. Thus the Lorentz transformation is a linear one.

If an event occurs at a space-time point (x^1, x^2, x^3, x^4) in the S frame, these equations allow us to determine the space-time point (x'^1, x'^2, x'^3, x'^4) in S' at which the event occurs. This set of equations can be expressed in matrix form:

$$\text{(a)} \quad \begin{bmatrix} x'^1 \\ x'^2 \\ x'^3 \\ x'^4 \end{bmatrix} = \begin{bmatrix} \gamma & 0 & 0 & -\beta\gamma \\ 0 & 1 & 0 & 0 \\ 0 & 0 & 1 & 0 \\ -\beta\gamma & 0 & 0 & \gamma \end{bmatrix} \begin{bmatrix} x^1 \\ x^2 \\ x^3 \\ x^4 \end{bmatrix} \quad \text{or} \quad \text{(b)} \quad \mathbf{X}' = \mathbf{L}\mathbf{X} \tag{9.192}$$

Knowing the Lorentz transformation also allows us to determine the transformation coefficients $\partial x'^i / \partial x^j$.

Because one of the coordinates, x^4, has a distinctly different physical character from the other three, it is sometimes necessary to discuss aspects of a problem in terms of just the three spatial coordinates. In those instances, indices will take on the values 1, 2, and 3. In other aspects of a problem, in which all four coordinates are involved, the indices will range from 1 to 4. A convention has been adopted by many authors that if an index is a Greek letter, that index ranges from 1 to 4, while a Latin index takes on the values 1, 2, or 3. For example, the Lorentz transformations involve all four coordinates. Thus these transformation equations, written in component form, are

$$x'^\mu = L^\mu_\nu x^\nu \tag{9.193a}$$

where v is summed from 1 to 4 and μ is a free index that can have any value from 1 to 4. The square of the length of a vector $\mathbf{V}$ is the sum of the squares of its three spatial components. This would be written

$$\mathbf{V} \cdot \mathbf{V} = V^i V_i \tag{9.193b}$$

with i summed from 1 to 3.

In addition to the transformation equations, it is necessary to determine the metric tensor in order to specify a system. This is defined from the invariant square of the line element. In a four-dimensional space this is written

$$ds^2 = g_{\mu\nu}\, dx^\mu\, dx^\nu \tag{9.194}$$

In a Cartesian space in which there are no sources,

$$ds^2 = (dx^1)^2 + (dx^2)^2 + (dx^3)^2 - c^2\, dt^2 = (dx^1)^2 + (dx^2)^2 + (dx^3)^2 - (dx^4)^2 \tag{9.195}$$

Therefore, the elements of the metric tensor are

$$\text{(a)}\quad g_{ij} = \delta_{ij} \qquad \text{(b)}\quad g_{i4} = 0 \qquad \text{(c)}\quad g_{44} = -1 \tag{9.196}$$

In matrix form

$$\mathbf{g} = \begin{bmatrix} 1 & 0 & 0 & 0 \\ 0 & 1 & 0 & 0 \\ 0 & 0 & 1 & 0 \\ 0 & 0 & 0 & -1 \end{bmatrix} \tag{9.197}$$

It is straightforward to show that $\mathbf{g}$ is involutory. That is, $\mathbf{g}^2 = \mathbf{1}$, or $\mathbf{g}^{-1} = \mathbf{g}$. Thus

$$g^{\nu\mu} = g_{\nu\mu} \tag{9.198}$$

Some authors define $x^4 = ict$, in which case

$$ds^2 = (dx^1)^2 + (dx^2)^2 + (dx^3)^2 + (dx^4)^2 \tag{9.199}$$

$$\Rightarrow \quad \mathbf{g} = \mathbf{1} \tag{9.200}$$

This is called the *Minkowski metric*. We will use the metric of equation 9.197.

Clearly, for this metric for all μ, ν, γ,

$$\frac{\partial g_{\mu\nu}}{\partial x^\gamma} = 0 \tag{9.201}$$

$$\Rightarrow \quad \text{(a)}\quad [\mu, \nu; \lambda] = 0 \quad \Rightarrow \quad \text{(b)}\quad \Gamma^\lambda_{\mu\nu} = 0 \tag{9.202}$$

Therefore, in this space, the equation of a geodesic is

$$\frac{d^2 x^\mu}{ds^2} = 0 \tag{9.203a}$$

If the origin of the coordinate system can be defined as the point at which the event occurs, then $x^i = 0 \Rightarrow dx^i = 0 \Rightarrow ds^2 = -c^2\, dt^2$. Then equation 9.203a can be written

$$\frac{d^2 x^\mu}{dt^2} = 0 \tag{9.203b}$$

which is the statement that there is no acceleration.

Using equations 9.201 and 9.202, the Riemann–Christoffel curvature tensor is also zero.

$$\text{(a)}\quad R_{\mu\nu\lambda\sigma} = 0 \quad \Rightarrow \quad \text{(b)}\quad R^{\mu}_{\cdot\nu} = 0 \quad \Rightarrow \quad \text{(c)}\quad R = 0 \tag{9.204}$$

Covariance of Maxwell's equations

In a space in which there is no material, Maxwell's equations are

$$\text{(a)}\quad \nabla\cdot\mathbf{E}=\frac{\rho}{\varepsilon_0}\qquad \text{(b)}\quad \nabla\cdot\mathbf{B}=0$$
$$\text{(c)}\quad \nabla\times\mathbf{E}=-\frac{\partial\mathbf{B}}{\partial t}\qquad \text{(d)}\quad \nabla\times\mathbf{B}=\mu_0\mathbf{j}+\mu_0\varepsilon_0\frac{\partial\mathbf{E}}{\partial t}\tag{1.153}$$

The magnetic vector potential is defined by

$$\mathbf{B}=\nabla\times\mathbf{A}\tag{1.165}$$

Therefore, equation 1.153c becomes

$$\text{(a)}\quad \nabla\times\left(\mathbf{E}+\frac{\partial\mathbf{A}}{\partial t}\right)=0\quad\Rightarrow\quad\text{(b)}\quad \mathbf{E}=-\nabla\phi-\frac{\partial\mathbf{A}}{\partial t}\tag{9.205}$$

If we define two four-dimensional vectors called the *4-vector potential*, and the *4-vector current* as

$$\text{(a)}\quad A_\mu\equiv\left(A_1,A_2,A_3,-\frac{1}{c}\phi\right)\qquad\text{(b)}\quad j_\mu\equiv(j_1,j_2,j_3,c\rho)\tag{9.206}$$

then Maxwell's equations can be written in tensor form by defining a second-rank tensor called the *electromagnetic field tensor*. Its elements are

$$F_{\mu\nu}\equiv\frac{\partial}{\partial x^\mu}A_\nu-\frac{\partial}{\partial x^\nu}A_\mu\tag{9.207a}$$

Since the Christoffel symbols are zero for a metric with constant elements, the covariant derivative and the partial derivative of a tensor are the same. Thus equation 9.207a can also be written

$$F_{\mu\nu}=A_{\nu,\mu}-A_{\mu,\nu}\tag{9.207b}$$

Clearly,

$$\text{(a)}\quad F_{\mu\mu}=0\quad(\textit{no sum})\qquad\text{and}\qquad\text{(b)}\quad F_{\mu\nu}=-F_{\nu\mu}\tag{9.208}$$

That is, in matrix form, $F_{\mu\nu}$ is antisymmetric.

Consider

$$F_{12}=\frac{\partial A_2}{\partial x^1}-\frac{\partial A_1}{\partial x^2}=(\nabla\times\mathbf{A})_3=B_3\tag{9.209a}$$

and, referring to equation 9.205b,

$$F_{14}=\frac{\partial A_4}{\partial x^1}-\frac{\partial A_1}{\partial x^4}=-\frac{1}{c}\frac{\partial\phi}{\partial x^1}-\frac{1}{c}\frac{\partial A_1}{\partial t}=\frac{E_1}{c}\tag{9.209b}$$

With these as examples, one can generate the elements of $F_{\mu\nu}$. In matrix form, the covariant form of F is

$$\mathbf{F}_{\text{covariant}} = \begin{bmatrix} 0 & B_3 & -B_2 & \dfrac{E_1}{c} \\ -B_3 & 0 & B_1 & \dfrac{E_2}{c} \\ B_2 & -B_1 & 0 & \dfrac{E_3}{c} \\ -\dfrac{E_1}{c} & -\dfrac{E_2}{c} & -\dfrac{E_3}{c} & 0 \end{bmatrix} \tag{9.210a}$$

In Problem 17 the reader will show that the contravariant matrix form of this tensor is

$$\mathbf{F}^{\text{contravariant}} = \begin{bmatrix} 0 & B_3 & -B_2 & -\dfrac{E_1}{c} \\ -B_3 & 0 & B_1 & -\dfrac{E_2}{c} \\ B_2 & -B_1 & 0 & -\dfrac{E_3}{c} \\ \dfrac{E_1}{c} & \dfrac{E_2}{c} & \dfrac{E_3}{c} & 0 \end{bmatrix} \tag{9.210b}$$

Maxwell's four equations can be combined into two tensor equations. Consider

$$\text{(a)} \quad \frac{\partial F_{\mu\nu}}{\partial x^{\lambda}} + \frac{\partial F_{\nu\lambda}}{\partial x^{\mu}} + \frac{\partial F_{\lambda\mu}}{\partial x^{\nu}} = 0 \quad \text{and} \quad \text{(b)} \quad \frac{\partial F^{\mu\nu}}{\partial x^{\nu}} = \mu_0 j^{\mu} \tag{9.211}$$

where

$$j^{\mu} = g^{\mu\nu} j_{\nu} = (j_1, j_2, j_3, -c\rho) \tag{9.212}$$

All of the indices in equation 9.211a are free indices. As the reader will show in Problem 16, if any two have the same value, equation 9.211a is identically zero. Thus, to obtain nontrivial results, μ, ν, and λ must have different values.

If we choose $\mu = 1$, $\nu = 2$, $\lambda = 3$, we obtain

$$\frac{\partial F_{12}}{\partial x^3} + \frac{\partial F_{23}}{\partial x^1} + \frac{\partial F_{31}}{\partial x^2} = \frac{\partial B_3}{\partial x^3} + \frac{\partial B_1}{\partial x^1} + \frac{\partial B_2}{\partial x^2} = \nabla \cdot \mathbf{B} = 0 \tag{9.213a}$$

This is the Maxwell equation given in equation 1.153b. The same result is obtained if some other permutation of 1, 2, and 3 is chosen for μ, ν, and λ.

If we take $\mu = 1$, $\nu = 2$, and $\lambda = 4$, and use the fact that the diagonal elements of F are zero, equation 9.211a becomes

$$\frac{\partial F_{12}}{\partial x^4} + \frac{\partial F_{24}}{\partial x^1} + \frac{\partial F_{41}}{\partial x^2} = \frac{1}{c}\frac{\partial B_3}{\partial t} + \frac{1}{c}\frac{\partial E_2}{\partial x^1} - \frac{\partial E_1}{\partial x^2} = \frac{1}{c}\left[\frac{\partial B_3}{\partial t} + (\nabla \times \mathbf{E})_3\right] = 0 \tag{9.213b}$$

which is a component of the Maxwell equation expressed in equation 1.153c.

There is one free index in equation 9.211b. If we choose a spatial index for μ ($\mu = 1$ for example), with equations 9.210b and 9.212 we obtain

$$\frac{\partial F^{11}}{\partial x^1} + \frac{\partial F^{21}}{\partial x^2} + \frac{\partial F^{31}}{\partial x^3} + \frac{\partial F^{41}}{\partial x^4} = \mu_0 j^1 \tag{9.214}$$

$$\Rightarrow \quad \frac{\partial B_3}{\partial x^2} - \frac{\partial B_2}{\partial x^3} - \frac{1}{c^2}\frac{\partial E_1}{\partial t} = \left(\nabla \times \mathbf{B} - \frac{1}{c^2}\frac{\partial \mathbf{E}}{\partial t}\right)_1 = \mu_0 j_1 \tag{9.215}$$

We see that this is a component of equation 1.153d if

$$\mu_0 \varepsilon_0 = \frac{1}{c^2} \tag{9.216}$$

This relation between electric and magnetic constants and the speed of light is evidence of the electromagnetic nature of light.

If $\mu = 4$, equation 9.211b becomes

$$\text{(a)} \quad \frac{\partial F^{14}}{\partial x^1} + \frac{\partial F^{24}}{\partial x^2} + \frac{\partial F^{34}}{\partial x^3} + \frac{\partial F^{44}}{\partial x^4} = \mu_0 j^4 \quad \Rightarrow \quad \text{(b)} \quad -\frac{1}{c}\left[\frac{\partial E_1}{\partial x^1} + \frac{\partial E_2}{\partial x^2} + \frac{\partial E_3}{\partial x^3}\right] = -\mu_0 c\rho \tag{9.217}$$

Using equation 9.216, this becomes

$$\nabla \cdot \mathbf{E} = \frac{\rho}{\varepsilon_0} \tag{9.218}$$

which is identical to equation 1.153a.

Therefore, Maxwell's four equations can be expressed as two tensor equations. The fact that they are tensor equations means that they are covariant under Lorentz transformations.

General relativity

When gravitational sources are present, the metric tensor is not a clear-cut quantity as it is in special relativity. Einstein postulated that all gravitational effects are accounted for by the metric tensor. To construct a general relativistic model, one constructs a source or stress-energy tensor that will describe the properties of the matter distribution in space-time. Of course, referring to equation 9.186, the covariant derivative of this tensor must be zero. One then solves the Einstein field equations for the elements of the metric tensor.

One phenomenon well described by general relativity is an effect known as the advance of the perihelion of the planet Mercury. It has been known since Kepler's work that all planets follow elliptic orbits about the sun. Perihelion is the point in the planet's orbit that is closest to the sun. This is the point P in Figure 9.6. It is observed that the major axis of the orbit slowly rotates about the sun in the direction of the planet's orbital motion. This is also viewed as the advance of the perihelion point, as illustrated in Figure 9.7.

Most of this advance can be accounted for by the gravitational forces exterted on Mercury by the other planets and asteroids in the solar system. However, a small part of that advance, 43 seconds of arc per century, cannot be accounted for by these gravitational effects, but is well described by general relativity

It has been observed during a solar eclipse that starlight passing near the sun is bent, or deflected from its straight path, as depicted in Figure 9.8. The amount of

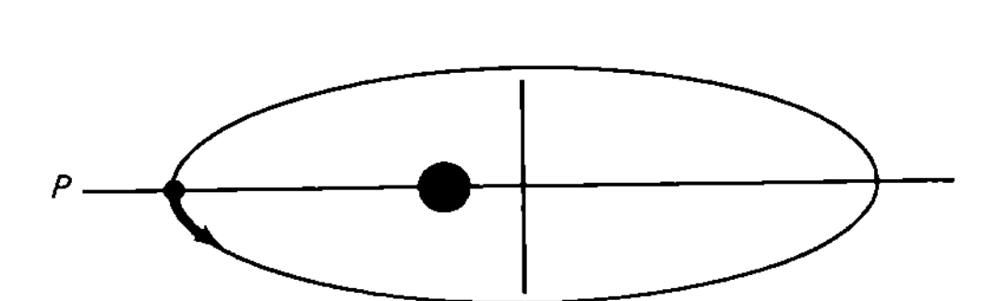

Figure 9.6 Elliptic orbit of Mercury.

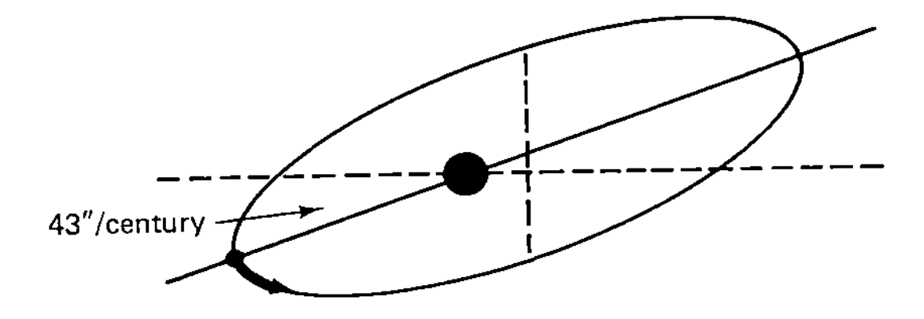

Figure 9.7 Advance of Mercury's perhelion.

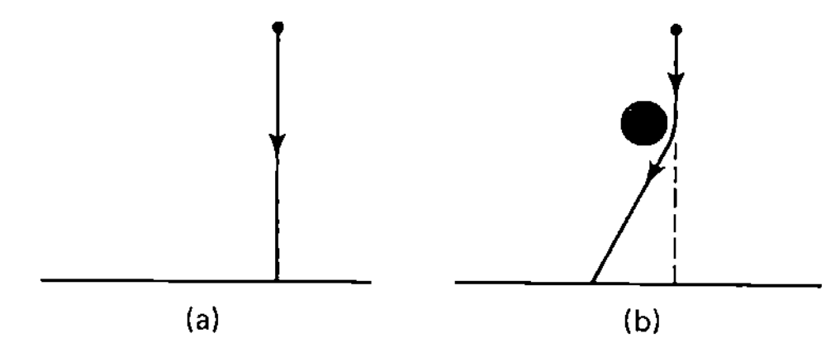

Figure 9.8 (a) Starlight observed without intervening gravitational source; (b) starlight deflected by the gravitational field of the sun.

deflection depends on how close the beam of light gets to the sun. For grazing incidence, as illustrated in Figure 9.8b, the angle of deflection predicted by general relativity is $\simeq 1.75''$ of arc, in good agreement with measured values.

Both of these phenomena are well described by general relativity using metric tensors that are spherically symmetric. The square of the four-dimensional line element in spherical coordinates in a space in which there are no sources is

$$ds^2 = dr^2 + r^2\,d\theta^2 + r^2\sin^2\theta\,d\phi^2 - c^2\,dt^2 \qquad (9.219)$$

The successful models have been ones that are modifications of this line element.

It is not the intent in this book to solve the field equations of general relativity but to give the reader a sense of the types of models that are considered. As such, we will simply list some of the most successful.

The Schwarzchild metric is developed to describe the geometry of space-time in a region around a single spherically symmetric nonrotating mass. It is designed to describe the fundamental properties of objects such as stars and black holes. The components of the Schwarzchild metric are specified by the square of the line element

$$ds^2 = \frac{1}{(1 - 2m/rc^2)}dr^2 + r^2\,d\theta^2 + r^2\sin^2\theta\,d\phi^2 - \left(1 - \frac{2m}{rc^2}\right)c^2\,dt^2 \qquad (9.220)$$

Thus the Schwarzchild metric tensor elements are

$$\text{(a)}\quad g_{\mu\nu} = 0 \quad \mu \neq \nu \qquad \text{(b)}\quad g_{11} = \frac{1}{\left(1 - \dfrac{2m}{rc^2}\right)} \qquad \text{(c)}\quad g_{22} = r^2$$
$$\text{(d)}\quad g_{33} = r^2\sin^2\theta \qquad \text{(e)}\quad g_{44} = -\left(1 - \frac{2m}{rc^2}\right) \qquad (9.221)$$

The Einstein and deSitter models are developed in an attempt to describe the relativistic effects of an entire universe uniformly filled with nonrotating matter (something like a universe filled with a uniform fluid). These models are applicable to global or universal problems. That is, they are cosmological models.

The Einstein metric is defined from the line element

$$ds^2 = \frac{1}{(1 - r^2/R^2)}dr^2 + r^2\,d\theta^2 + r^2\sin^2\theta\,d\phi^2 - c^2\,dt^2 \qquad (9.222)$$

and the deSitter model is described by the line element

$$ds^2 = \frac{1}{(1 - r^2/R^2)}dr^2 + r^2\,d\theta^2 + r^2\sin^2\theta\,d\phi^2 - \left(1 - \frac{r^2}{R^2}\right)c^2\,dt^2 \qquad (9.223)$$

Finally, for completeness, we present the Kerr metric, which is designed to describe the problem of the geometry of space-time in the vicinity of a rotating nonsymmetric object. The Kerr metric is defined in terms of two parameters. α is a multiple of the mass of the object and β^2 is related to the object's rotational angular momentum. Defining

$$\begin{aligned} &(\text{a}) \quad \mu \equiv r^2 - 2\alpha r + \beta^2 && (\text{b}) \quad \nu \equiv r^2 + \beta^2 \cos^2\theta \\ &(\text{c}) \quad \lambda \equiv \mu - \beta^2 \sin^2\theta && (\text{d}) \quad \sigma \equiv \left(r^2 + \beta^2\right)^2 - \mu\beta^2 \sin^2\theta \end{aligned} \tag{9.224}$$

the Kerr metric is obtainable from

$$ds^2 = \frac{\nu}{\mu} dr^2 + \nu\, d\theta^2 + \frac{\sigma}{\nu} \sin^2\theta\, d\phi^2 - \frac{\lambda}{\nu} c^2\, dt^2 - 4\frac{\alpha\beta r \sin^2\theta}{\nu} d\phi\, c\, dt \tag{9.225}$$

9.4 MAXWELL'S EQUATIONS AND QUATERNIONS

Among the other esthetically pleasing aspects of the tensor treatment of electrodynamics, we have seen that by expressing the current and charge density and the electric and magnetic potentials as elements of four-dimensional vectors, and the electric and magnetic field components as elements of a second-rank tensor, Maxwell's four vector equations reduce to two tensor equations.

Quaternions are not an alternative form of tensors. We include a brief discussion of them at this point to demonstrate that when the electric and magnetic fields are expressed in complex quaternion notation, Maxwell's four vector equations can be written as one quaternion equation.

Quaternions are a mathematical structure that is an extension of the mathematics of complex variables. The bases of complex variables are 1 and i, with $i^2 = -1$. This means that any complex quantity z can be written as a linear combination of 1 and i as $z = 1 \cdot x + i \cdot y$.

In quaternion structure, there are three distinct bases, the squares of which are all -1. We will designate these by q_1, q_2, and q_3. The bases of the quaternion system are 1, q_1, q_2, and q_3 and any quaternion can be written as

$$Q = 1 \cdot w + x \cdot q_1 + y \cdot q_2 + z \cdot q_3 \tag{9.226}$$

The properties of these bases are

$$(\text{a}) \quad q_1^2 = q_2^2 = q_3^2 = -1 \qquad (\text{b}) \quad q_r q_s = \sum_{t=1}^{3} \varepsilon_{rst} q_t \tag{9.227}$$

and they anticommute, so

$$[q_r, q_s] = 0 \qquad r \neq s \tag{9.227c}$$

We point out that this algebraic structure is similar to the equivalent properties of the Pauli spin matrices, the slight differences being that $\boldsymbol{\sigma}_r^2 = \mathbf{1}$ instead of $-\mathbf{1}$, and $\boldsymbol{\sigma}_r \boldsymbol{\sigma}_s = i \sum_{t=1}^{3} \varepsilon_{rst} \boldsymbol{\sigma}_t$, whereas the analogous algebraic structure of quaternions does not contain the complex factor i. It is straightforward to show that $(-1)^{r+1} i\boldsymbol{\sigma}_r$ and q_r have the same algebra.

With $i = \sqrt{-1}$ one of the bases of the complex variables, we define a complex quaternion field, a complex quaternion differential operator, and a complex quaternion current density by

$$\text{(a)}\quad F \equiv \sum_{r=1}^{3} q_r\left[B_r + i\frac{E_r}{c}\right] \qquad \text{(b)}\quad \Box \equiv \frac{i}{c}\frac{\partial}{\partial t} + \sum_{r=1}^{3} q_r \frac{\partial}{\partial x_r} \qquad \text{(c)}\quad J \equiv -ic\rho + \sum_{r=1}^{3} q_r j_r \tag{9.228}$$

Then the four Maxwell equations expressed in equations 1.153 are embodied in the one complex quaternion equation,

$$\Box F = \mu_0 J \tag{9.229}$$

This equation represents

$$\sum_{r=1}^{3} \frac{q_r}{c}\frac{\partial}{\partial t}\left[iB_r - \frac{E_r}{c}\right] + \sum_{r=1}^{3}\sum_{s=1}^{3} q_r q_s \frac{\partial}{\partial x_r}\left[B_s + i\frac{E_s}{c}\right] = -i\mu_0 c\rho + \mu_0 \sum_{r=1}^{3} q_r j_r \tag{9.230}$$

The double sum can be separated into a sum of terms for which r and s are equal, plus a sum of terms for which $r \neq s$.

$$\sum_{r=1}^{3}\sum_{s=1}^{3} q_r q_s \frac{\partial}{\partial x_r}\left[B_s + i\frac{E_s}{c}\right] = \sum_{r=1}^{3} q_r^2 \frac{\partial}{\partial x_r}\left[B_r + i\frac{E_r}{c}\right] + \sum_{\substack{r=1 \\ r\neq s}}^{3}\sum_{s=1}^{3} q_r q_s \frac{\partial}{\partial x_r}\left[B_s + i\frac{E_s}{c}\right] \tag{9.231}$$

Since $q_r^2 = -1$, the first term on the right-hand side becomes

$$\sum_{r=1}^{3} q_r^2 \frac{\partial}{\partial x_r}\left[B_r + i\frac{E_r}{c}\right] = -\nabla\cdot\mathbf{B} - \frac{i}{c}\nabla\cdot\mathbf{E} \tag{9.232}$$

Using equation 9.227b, the second term can be written

$$\sum_{\substack{r=1 \\ r\neq s}}^{3}\sum_{s=1}^{3} q_r q_s \frac{\partial}{\partial x_r}\left[B_s + i\frac{E_s}{c}\right] = \sum_{r=1}^{3}\sum_{s=1}^{3}\sum_{v=1}^{3} \varepsilon_{rsv} q_v \frac{\partial}{\partial x_r}\left[B_s + i\frac{E_s}{c}\right] \tag{9.233}$$

where ε_{rsv} contains the restriction $r \neq s$. Therefore, equation 9.229 becomes

$$\sum_{r=1}^{3} \frac{q_r}{c}\frac{\partial}{\partial t}\left[iB_r - \frac{E_r}{c}\right] - \nabla\cdot\mathbf{B} - \frac{i}{c}\nabla\cdot\mathbf{E} + \sum_{r=1}^{3}\sum_{s=1}^{3}\sum_{v=1}^{3} \varepsilon_{rsv} q_v \frac{\partial}{\partial x_r}\left[B_s + i\frac{E_s}{c}\right]$$

$$= -i\mu_0 c\rho + \mu_0 \sum_{v=1}^{3} q_v j_v \tag{9.234}$$

$$= -\nabla\cdot\mathbf{B} - \frac{i}{c}\nabla\cdot\mathbf{E} + \sum_{v=1}^{3} q_v\left\{\frac{1}{c}\frac{\partial}{\partial t}\left[iB_v - \frac{E_v}{c}\right] + \sum_{r=1}^{3}\sum_{s=1}^{3} \varepsilon_{rsv}\frac{\partial}{\partial x_r}\left[B_s + i\frac{E_s}{c}\right]\right\}$$

Equating the coefficients of the quaternion bases yields

$$-\nabla\cdot\mathbf{B} - \frac{i}{c}\nabla\cdot\mathbf{E} = -i\mu_0 c\rho \tag{9.235a}$$

$$\frac{1}{c}\frac{\partial}{\partial t}\left[iB_v - \frac{E_v}{c}\right] + \sum_{r=1}^{3}\sum_{s=1}^{3} \varepsilon_{rsv}\frac{\partial}{\partial x_r}\left[B_s + i\frac{E_s}{c}\right] = \frac{1}{c}\frac{\partial}{\partial t}\left[iB_v - \frac{E_v}{c}\right]$$

$$+ \sum_{r=1}^{3}\sum_{s=1}^{3} \varepsilon_{vrs}\frac{\partial}{\partial x_r}\left[B_s + i\frac{E_s}{c}\right] = \frac{1}{c}\frac{\partial}{\partial t}\left[iB_v - \frac{E_v}{c}\right] + \left\{\nabla\times\left[\mathbf{B} + i\frac{\mathbf{E}}{c}\right]\right\}_v = \mu_0 j_v \tag{9.235b}$$

By equating the real parts and separately the imaginary parts of equation 9.235a, we obtain

$$\text{(a)} \quad \nabla \cdot \mathbf{B} = 0 \qquad \text{(b)} \quad \nabla \cdot \mathbf{E} = \mu_0 c^2 \rho = \frac{\rho}{\varepsilon_0} \tag{9.236}$$

where we have used the identity

$$c^2 = \frac{1}{\mu_0 \varepsilon_0} \tag{9.216}$$

Equating the real parts and separately the imaginary parts of equation 9.235b, we obtain

$$\text{(a)} \quad (\nabla \times \mathbf{B})_v - \frac{1}{c^2}\frac{\partial E_v}{\partial t} = \mu_0 j_v \qquad \text{(b)} \quad \frac{\partial B_v}{\partial t} + (\nabla \times \mathbf{E})_v = 0 \tag{9.237}$$

which are the two remaining Maxwell equations in component form.

Thus, as stated, with the quaternion algebra defined in equations 9.228, Maxwell's four vector equations can be expressed as a single quaternion equation,

$$\Box F = \mu_0 J \tag{9.229}$$

PROBLEMS

1. T_{ij} is a second-rank, antisymmetric tensor, and v^i is a contravariant vector. Prove that $T_{ij}v^i v^j$ is zero in all coordinate frames.

2. **(a)** In the products listed below, the A and B quantities are tensors of the indicated rank and variance. Prove that the quantities C transform like tensors of the indicated rank and variance.

(1) $C^i_j = A^i B_j$
(2) $C^{ij} = A^i B^j$
(3) $C_{ij} = A_i B_j$
(4) $C^{ik}_j = A^i_j B^k$
(5) $C^i_j = A^k_j B^i_k$
(6) $C = A^i_j B^j_i$
(7) $C^{ik}_{jl} = A^{ik}_j B_l$

(b) Prove that with A and B vectors of the indicated variance,

(1) $\sum_{i=1}^{N} A_i B_i$
(2) $\sum_{i=1}^{N} A^i B^i$

are not tensor invariants.

(c) Prove that the Kronecker δ symbol δ^i_j, which is a second-rank tensor of mixed variance, is the same in all frames.

3. Determine the nonzero components of the Christoffel symbols of the first and second kind in a three-dimensional space described by the spherical coordinates $(x^1, x^2, x^3) = (r, \theta, \phi)$.

4. Prove that the Christoffel symbols transform according to

$$[i, j; k]' = \frac{\partial x^l}{\partial x'^i}\frac{\partial x^m}{\partial x'^j}\frac{\partial x^n}{\partial x'^k}[l, m; n] + g_{lm}\frac{\partial x^l}{\partial x'^k}\frac{\partial^2 x^m}{\partial x'^i \partial x'^j}$$

and

$$\Gamma'^k_{ij} = \frac{\partial x'^k}{\partial x^l}\frac{\partial x^m}{\partial x'^i}\frac{\partial x^n}{\partial x'^j}\Gamma^l_{mn} + \frac{\partial x'^k}{\partial x^l}\frac{\partial^2 x^l}{\partial x'^i \partial x'^j}$$

5. **(a)** Prove that the second Christoffel symbol can be written

$$\Gamma^k_{ij} = g^{kl}\left(\frac{\partial g_{il}}{\partial x^j} - g_{im}\Gamma^m_{lj}\right)$$

(b) Use the results of part (a) to show that the contracted second Christoffel symbol is

$$\Gamma^k_{jk} = \Gamma^k_{kj} = \frac{1}{2}g^{kl}\frac{\partial g_{lk}}{\partial x^j}$$

(c) Use the results of Problem 4 to show that Γ^k_{ki} is a covariant vector.

6. Using the results of Problem 4, show that the covariant derivative of a contravariant vector $A^i{}_{,j}$ is a second-rank tensor of mixed variance.
7. For an invariant function F,
   (a) Prove that the generalization of $\nabla^2 F$, which is $\Box^2 F \equiv \Box \cdot \Box F$, is

$$\Box^2 F = g^{kl}\frac{\partial^2 F}{\partial x^k \partial x^l} - g^{lm}\Gamma^k_{lm}\frac{\partial F}{\partial x^k}$$

   (b) Using the metric for a three-dimensional space described by spherical coordinates (equation 9.35b), find $\Box^2 F$ in spherical coordinates. Compare your results with $\nabla^2 F$ given in spherical coordinates in equation 1.201c.
8. (a) Determine the form of

$$A^i{}_{,j} - A^j{}_{,i}$$

   and compare it to the form of $A_{i,j} - A_{j,i}$ in equation 9.137b. Is $A^i{}_{,j} - A^j{}_{,i}$ a component of $\nabla \times \mathbf{A}$ in three dimensions?
   (b) Deduce the covariant derivative of a scaler function F and show that $F_{,i}$ is a covariant vector.
   (c) Prove that

$$(F_{,i})_{,j} - (F_{,j})_{,i} = 0$$

   Interpret this result in terms of operations with ∇ on F.
9. Prove that $g^{ij}{}_{,k} = 0$.
10. (a) Prove that the Ricci tensor satisfies $R_{ijkk} = 0$ (*no sum*).
    (b) Use the results of part (a) to show that if the metric tensor is diagonal, the Ricci tensor $R_{ij} = 0$.
11. (a) Show that in a two-dimensional space defined by

$$ds^2 = (dx)^2 - F(x)c^2\,dt^2$$

    there is only one independent curvature tensor component. That is, all nonzero components of R_{ijkm} are related to R_{1212}. Find R_{1212}.
    (b) What is the Riemann curvature of this space in terms of $F(x)$?
    (c) Using the results of Problem 10, determine the components of the Ricci tensor.
12. Find the Riemann curvature of a two-dimensional space confined to the surface of a cylinder of radius ρ.
13. The square of the line element in three-dimensional space in spherical coordinates is

$$ds^2 = dr^2 + r^2\,d\theta^2 + r^2\sin^2\theta\,d\phi^2$$

    with $(x^1, x^2, x^3) = (r, \theta, \phi)$. Prove that this space is flat.
14. A space is defined by the square of the line element

$$ds^2 = dx^2 - \left(1 + \frac{\alpha x}{c^2}\right)c^2\,dt^2$$

    with $(x^1, x^2) = (x, ct)$. c is the speed of light, and α is an invariant.
    (a) Find ds^2 (an invariant) in a coordinate system S' that is related to S by the transformation

$$x' = x\cosh\left(\frac{\alpha t}{c}\right) + \frac{c^2}{\alpha}\left[\cosh\left(\frac{\alpha t}{c}\right) - 1\right]$$

$$t' = \left[\frac{c}{\alpha} + \frac{x}{c}\right]\sinh\left(\frac{\alpha t}{c}\right)$$

    What can you conclude from the result?
    (b) Find all the nonzero components of the second Christoffel symbol in the S frame.
15. The unit vector $\mathbf{t}$ is defined by $\mathbf{t} \equiv \mathbf{ds}/ds = (dx^i/ds)\mathbf{e}_i$. By considering $\mathbf{dt}/ds$, prove that along a geodesic, $\mathbf{t}$ is a constant. What is the interpretation of this result?
16. (a) Evaluate the tensor form of the source-free Maxwell equation (equation 9.211a) for $\mu = 2$, $\nu = 3$, and $\lambda = 1$.
    (b) For $\mu = \nu \neq \lambda$, show that equation 9.211a is an identity.
17. In a flat space-time $(ds^2 = dx^2 + dy^2 + dz^2 - c^2\,dt^2)$:
    (a) What is the contravariant form of the 4-vector potential the covariant form of which is $A_\mu = (A_1, A_2, A_3, -\phi/c)$?
    (b) Show that the fully contravariant form of the electromagnetic field tensor is that expressed in equation 9.210b.
    (c) Prove that the electromagnetic field tensor of mixed variance is

$$F_{\mu\cdot}{}^{\nu} = \frac{\partial A^\nu}{\partial x^\mu} - g^{\nu\lambda}\frac{\partial A_\mu}{\partial x^\lambda}$$

    (d) Show that in matrix form, $F_{\mu\cdot}{}^{\nu} = \tilde{F}^{\nu}{}_{\cdot\mu}$
18. (a) Using equations 9.210, prove that $\mathbf{E}\cdot\mathbf{E}/c^2 - \mathbf{B}\cdot\mathbf{B}$ is a Lorentz invariant.
    (b) Show that $\mathbf{A}\cdot\mathbf{A} - \phi^2/c^2$, $\mathbf{j}\cdot\mathbf{j} - \rho^2/c^2$ and $\mathbf{A}\cdot\mathbf{j} + \rho\phi/c^2$ are Lorentz invariants.
19. The electric and magnetic fields are measured to be $\mathbf{E}$ and $\mathbf{B}$ in a frame labeled S. The sources of these fields in S are $\mathbf{j}$ and ρ.
    (a) What are the fields $\mathbf{E}'$ and $\mathbf{B}'$ in a frame S' that is moving with a velocity $\mathbf{v} = \beta c\mathbf{i}$ as seen by an S-based observer?
    (b) What are the potentials $\mathbf{A}'$ and ϕ' and the sources $\mathbf{j}'$ and ρ' as determined by an observer in S'?

(c) Using the results of part (a), find the force on a charge Q as measured by an observer in S if the charge is at rest in S' and thus experiences the electrostatic force $\mathbf{F}' = Q\mathbf{E}'$. (Charge is a Lorentz invariant.)

20. A second-rank tensor of mixed variance is defined by

$$G^{\mu}_{\cdot\nu} \equiv \left(F^{\mu\lambda}F_{\lambda\nu} - \tfrac{1}{4}\delta^{\mu}_{\nu}F^{\lambda\sigma}F_{\lambda\sigma}\right)$$

where $F_{\mu\nu}$ is the electromagnetic field tensor.

(a) Show that

$$G^{i}_{\cdot 4} = \frac{1}{c}(\mathbf{E} \times \mathbf{B})_i = \frac{\mu_0}{c}S^i$$

where S^i is the ith component of the Poynting vector.

(b) Prove that

$$G^{4}_{\cdot 4} = -\frac{1}{2}\left(\frac{1}{c^2}E^2 + B^2\right) = -\frac{1}{\mu_0}U$$

where U is the electromagnetic energy density.

21. The Minkowski metric in matrix form is the identity matrix.

(a) In Minkowski space find the form of the 4-vector potential so that

$$\mathbf{F} = \begin{bmatrix} 0 & B_3 & -B_2 & -i\dfrac{E_1}{c} \\ -B_3 & 0 & B_1 & -i\dfrac{E_2}{c} \\ B_2 & -B_1 & 0 & -i\dfrac{E_3}{c} \\ i\dfrac{E_1}{c} & i\dfrac{E_2}{c} & i\dfrac{E_3}{c} & 0 \end{bmatrix}$$

(b) From the results of part (a), determine the 4-vector current so that Maxwell's equations with sources can be written as

$$\frac{\partial F_{\lambda\nu}}{\partial x^{\nu}} = \mu_0 j_\lambda$$

CHAPTER 10

INTRODUCTION TO GROUP THEORY

There are many situations in which certain properties of a physical system are unaffected by changes in the system. For example, as discussed in Chapter 8, the length of a vector is unaffected by rotations of the vector or the coordinate system. The Coulomb potential due to a point charge depends only on the length of the position vector from the point charge to the field point, not on the angular orientation of the field point. Therefore, the potential is unchanged or invariant under a rotation of the position vector from one field point to a second that is the same distance from the charge as the first field point. The collection of these points forms an equipotential surface.

In a simple solid, the potential that an electron experiences is periodic, arising from the spatially periodic positions of identical lattice ions. Therefore, the potential energy of an electron in a lattice potential of periodicity λ is unaffected by a translation from a point x to a point $x + N\lambda$, where N is an integer.

There are operations, such as rotations, which leave geometric shapes unchanged. For example, the rotation of a rectangle by 180° does not affect the orientation of the rectangle. If the operations involve transformations about a point, they are referred to as *point transformations*.

Such invariants are described as symmetries of the system under specified transformations. Often properties can be deduced from these symmetries. In some cases these transformations form a mathematical group, and the mathematical properties of that group can be used to deduce the physical properties of the system.

The groups formed by point transformations are called *point groups*. An important application of point groups arises in the study of the structure of crystals, in which the lattices have a high degree of geometric symmetry. The groups of transformations about a crystal lattice point which leave the lattice orientation invariant are called *crystal point groups*.

10.1 Abstract Theory of Finite Groups

Definitions and properties of groups

A *set* is defined as a collection of entities called *elements*. A mathematical group is a set of elements along with a rule for combining the elements. We denote the abstract operation for combining the elements by the symbol "$*$". For example, if x and y are two elements of the group, and if the rule for combining elements is addition, then $x * y = x + y$.

To define a group, the set of elements and the operation for combining them must satisfy what are called the four *group axioms*.

1. *Closure.* If x and y are elements of the group (x and $y \in G$), then the combination $x * y$ is also an element of the group.
2. *Associativity.* The combination of three elements is associative. That is if x, y, and $z \in G$, then $(x * y) * z = x * (y * z)$.
3. *Identity element.* There is an identity element $e \in G$ such that for every $x \in G$, $e * x = x * e = x$.
4. *Inverse element.* For every element $x \in G$ there exists an element $x^{-1} \in G$ called the *inverse* of x such that $x * x^{-1} = x^{-1} * x = e$.

If every element of G satisfies these axioms, G is said to be a *group* under the specified operation.

Examples of sets of elements under specified operations that are not groups include:

Example 10.1

The set of real numbers $x \in [1, 2]$ under multiplication:

(a) Is not closed. For example, $1.5 \times 1.6 = 2.40 \notin [1, 2]$.
(b) Is associative.
(c) Contains an identity element, $e = 1$.
(d) Does not contain an inverse for every element. For example, $(1.5)^{-1} = \frac{2}{3} \notin [1, 2]$. □

Example 10.2

The set of real numbers $x \in [-\infty, \infty]$ under the operation $x * y = x^2 + y$:

(a) Is closed.
(b) Is not associative. $(x * y) * z = (x^2 + y) * z = (x^2 + y)^2 + z$, whereas $x * (y * z) = x * (y^2 + z) = x^2 + y^2 + z$.
(c) Contains no identity. From $e * x = e^2 + x = x$, we could identify $e = 0$. However, $x * e = x^2 + e \neq x$ for any e. Thus there is no identity.
(d) Does not contain an inverse, since such an element has no meaning without an identity element. □

Example 10.3

As an example of a set of elements that does comprise a group, consider the set of real numbers under addition.

(a) If x and y are real numbers, then $x + y$ is also a real number and therefore is an element of G.
(b) If x, y, and z are real numbers, then $(x + y) + z = x + (y + z)$.
(c) Since 0 is a real number, for any real number x, $x + 0 = 0 + x = x$. Thus, under addition, 0 is the identity element for the real numbers.
(d) If x is a real number, then $-x$ is a real number, and $x + (-x) = (-x) + x = 0$. Therefore, $-x$ is the inverse of x under addition.

Since the four group axioms are satisfied for the real numbers under addition, this is a group. □

Example 10.4

A second example of a group is the set consisting of 1 and -1 under ordinary multiplication. Clearly

(a) $1 \times 1 = 1 \in G$, $1 \times (-1) = -1 \in G$, and $(-1) \times (-1) = 1 \in G$, so closure is satisfied.

(b) $(1 \times 1) \times 1 = 1 \times (1 \times 1)$, $(-1 \times 1) \times (-1) = -1 \times (1 \times (-1))$, and so on, so the combination of elements is associative.

(c) 1 is the identity element since $1 \times 1 = 1$ and $1 \times (-1) = -1$.

(d) Since 1 is the identity element, each element is its own inverse. That is, $1 \times 1 = 1$, or $1^{-1} = 1$. Similarly, $(-1) \times (-1) = 1$, so $(-1)^{-1} = -1$. □

One noticeable difference between the groups of these two examples is that the first group contains an infinite number of elements while the second group has two. The number of elements in the group is called the *order* of the group. Thus the group in Example 10.3 is of infinite order, and the group of Example 10.4 is a finite group of order 2.

As will be described later, certain sets of matrices under standard matrix multiplication form a group. We have seen that matrix multiplication is not commutative. That is, in general for matrices **A** and **B**, $\mathbf{AB} \neq \mathbf{BA}$. Similarly, the combination of two abstract elements of a group is generally not commutative. If $x * y = y * x$ for all $x, y \in G$, the group is said to be *Abelian*.

Subsets and subgroups

If S is a set of elements and T is a set of elements such that all the elements of T are also elements of S, T is called a *subset* of S. From this definition, S can be considered a subset of itself. For that reason, T is defined as a *proper subset* of S if there is at least one element of S that is not a member of T.

The definition of a *proper subgroup* is the expected extension of the definition of the proper subset. If the elements of H form a proper subset of G and if H and G are both groups under the same rule for combining elements, then H is a proper subgroup of G.

Group multiplication table

If the group is of finite order, one common way to describe the combination of elements is by the group multiplication table. The multiplication table is a two-dimensional table that describes all combinations of every pair of elements of the group. Each row of the table is headed by the first element, and each column is headed by the second element of the combination. The product is the entry in the table at the intersection of the specified row and column.

Example 10.5

To illustrate, consider the group of Example 10.4, consisting of 1 and -1 under ordinary multiplication. This group is called C_2. The multiplication table for this group is shown in Figure 10.1 on page 518. □

This group can be described in more abstract form. Instead of referring to the elements as ± 1, let us call them e and x, where e is the identity element. Then these group elements have the multiplication table shown in Figure 10.2. If e and x were matrices and x were involutory, then referring to equation 8.186, it is clear that matrix multiplication would produce the table of Figure 10.2. Thus C_2 can also be described as

	1	−1
1	1	−1
−1	−1	1

Figure 10.1 Group multiplication table for the group C_2.

	e	x
e	e	x
x	x	e

Figure 10.2 Generalization of group multiplication table for the group C_2.

the group consisting of an involutory matrix and the unit matrix. Note that C_2 is Abelian.

We now demonstrate that C_2 as defined by the multiplication table of Figure 10.2 is the only group of order 2. Instead of taking $x * x = e$, the only other possible assignment of the product is $x * x = x$. That is, if x were a matrix, it would be an idempotent. With this assignment, the closure property required of a group is still satisfied. However, with such an assignment, there is no element in this group that serves as the inverse of x. Thus the set consisting of e and x such that $x * x = x$ does not form a group. Therefore, C_2 is the only finite group of order 2.

If we can deduce general properties of C_2 in a rather simple way (using, for example, the representation of C_2 in terms of $1, -1$ under ordinary multiplication), these properties will be satisfied by other representations of C_2 (in terms of matrices, for example). This is one compelling motivation for the study of abstract group theory.

Consider the group C_3 which has three elements. We denote them by e, x, and y. To construct the multiplication table, we must deduce the products $x * x$, $y * y$, $x * y$, and $y * x$. If $x * x = x$ and $y * y = y$, then x and y do not have inverses. Therefore, either $x * x = y * y = e$ (so x and y are self-inverses) or $x * y = e$.

We can rule out the choice that x and y are self-inverses. To see this, let $x * x = e$. Then the possible assignments for $x * y$ are $x * y = e$, $x * y = x$ or $x * y = y$. But if $x * x = e$ and $x * y = e$ then premultiplying by x we obtain $x * x * x = x * x * y$ or $x = y$. But x and y are distinct elements. Therefore, we cannot have an assignment of products that leads to the equality of distinct elements. Thus, if $x * x = e$, it is not possible to have $x * y = e$.

If we consider $x * x = e$ and take $x * y = x$, then $x * x * y = x * x$ or $y = e$. Similarly, taking $x * x = e$ and $x * y = y$ leads to $x = e$. Therefore, with $x * x = e$, we have ruled out all possible combinations for $x * y$. So we conclude that we cannot assign $x * x = e$. By identical reasoning, $y * y \neq e$. Therefore, we must require $x * y = y * x = e$.

Since $x * x \neq x$, and $x * x \neq e$, we must assign $x * x = y$, and by an identical argument, $y * y = x$. Therefore, the only possible table for C_3 is that shown in Figure 10.3. Because there is only one possible multiplication table for three elements, there exists only one finite group of order 3. We also note that the table for C_3 is symmetric about its main diagonal. Therefore, the multiplication of the elements is commutative. That is, like the group C_2, C_3 is Abelian.

Referring to their multiplication tables, we note that the tables are constructed so that the element that designates a specified row is also the element heading that specific

	e	x	y
e	e	x	y
x	x	y	e
y	y	e	x

Figure 10.3 Group multiplication table for the group C_3.

column. For example, x denotes the second row and also the second column of the table for C_3. With such a construction, we see that the tables for C_2 and C_3 are symmetric about the diagonal that runs from the upper left corner of the table to the lower right corner (the main diagonal when referring to matrices). The table of any Abelian group will have this symmetry.

We point out that every element in the group appears once in every row and once in every column of the table. This will be true for any finite group. For an arbitrary finite group, let the row of a table be designated by w. If the element x appeared in both the r column and the s column, that would mean that $w * r = x$ and $w * s = x$, or $w * r = w * s$. Premultiplication by w^{-1} yields $r = s$. This is a contradiction since the elements of the group must be distinct. Therefore, in the multiplication table, every row must contain one and only one entry of each element of the group. An identical argument leads to the same conclusion about column entries.

Isomorphism and homomorphism

As the reader will show in Problem 6, the three cube roots of $+1$, ($e^{2\pi ik/3}$ $k = 0, 1, 2$) under ordinary complex number multiplication, form a group with the same multiplication table as C_3. Similarly, in Problem 7 the reader will show that there are sets of matrices under matrix multiplication that generate the table of C_3. When two groups containing different types of elements and different rules of combination produce the same group multiplication table, the groups are said to be *isomorphic* to one another. The terminology means that they have the same form. From an abstract point of view, if two groups are isomorphic, they are actually the same group. As such, each element of one group will correspond to one and only one element of the second group. For example, each of the three cube roots of $+1$ in Problem 6 correspond to a different element of C_3. That is, there is a one-to-one correspondence between the elements of the two groups. As such, the group of cube roots of $+1$ under ordinary complex multiplication is said to be a different representation of the group C_3.

If more than one element of one group corresponds to each element of a second group, the first group is said to be *homomorphic* to the second, and the correspondence is said to be a many (two or more)-to-one correspondence.

Example 10.6

As an example, the set of all real numbers excluding zero form a group under multiplication, as do the set of positive real numbers under multiplication. By squaring the elements of the group of all real numbers, we obtain the elements of the group of positive real numbers. However, two elements of the group of all real numbers correspond to each element of the group of positive real numbers. The terminology of this correspondence is that the group of all real numbers is homomorphic to (or more precisely a two-to-one homomorphism of) the group of positive real numbers under the mapping of squaring. □

Cyclic nature of elements

As the order of the finite group increases, it quickly becomes cumbersome to express the structure in terms of a large multiplication table. Thus it is important to develop other, more concise methods for specifying the structure of the group.

To this end we consider the result of repeated combinations of an element with itself. We designate k combinations of the element x by

$$\underbrace{x * x * \cdots * x}_{k \text{ times}} \equiv x^k \tag{10.1}$$

Because the group is finite, for each element x, there must be some integer n such that the sequence

$$x^0 (= e), x, x^2, \ldots, x^n (= e) \tag{10.2a}$$

begins to repeat itself. That is, there are only n distinct powers of x, so that

$$x^{n+1} = x \tag{10.2b}$$

Thus the sequence expressed in equation 10.2a is cyclic for every $x \in G$. The power n is called the *period* of the element. For example, for the group C_3, e is cyclic of period 1, while x and y are each cyclic of period 3. That is, for C_3

$$\text{(a)} \quad x * x * x = y * x = e \qquad \text{(b)} \quad y * y * y = x * y = e \tag{10.3}$$

Clearly, the identity element of any group has a period 1. As the reader will show in Problem 10, if an element x of any finite group has period n, then both x^{-1} and yxy^{-1} have period n, where y is any element of the group distinct from x. The element yxy^{-1} is called the *conjugate* of x with respect to y.

It is intuitive that if x is an element of a group G, and x has a period n, then the elements $e, x, x^2, x^3, \ldots, x^{n-1}$ form a subgroup of G. It is possible the elements $e, x, x^2, x^3, \ldots, x^{n-1}$ comprise the entire group, as in C_3. A group or subgroup whose elements can be generated by repeated combinations of a single element x, is called a *cyclic group* or *subgroup*, and the element x is called the *generator* of the group or subgroup. Clearly, a cyclic group or subgroup is Abelian, and the period of the generator is the order of the group or subgroup. A cyclic group of order N is denoted by C_N.

Permutation group and Cayley's theorem

Consider an operation that interchanges two items. Such an operation is called a *permutation*.

Example 10.7

As an example, let us label three items by g_1 g_2, and g_3. There are $3! = 6$ possible permutations of the ordering of these items. They are:

e:	$g_1, g_2, g_3 \rightarrow g_1, g_2, g_3$	(no change in the ordering)	(10.4a)
p:	$g_1, g_2, g_3 \rightarrow g_2, g_3, g_1$	(moves the first item to the end, moves all other items forward one position)	(10.4b)
q:	$g_1, g_2, g_3 \rightarrow g_3, g_1, g_2$	(moves the last item to the front, moves all other items back one position)	(10.4c)
r:	$g_1, g_2, g_3 \rightarrow g_1, g_3, g_2$	(leaves the first item alone, interchanges second and third items)	(10.4d)
s:	$g_1, g_2, g_3 \rightarrow g_3, g_2, g_1$	(leaves the second item alone, interchanges first and third items)	(10.4e)
t:	$g_1, g_2, g_3 \rightarrow g_2, g_1, g_3$	(leaves the last item alone, interchanges first and second items)	(10.4f)

Such a group of permutations of N items is called the *permutation group* or the *symmetric group*. It contains $N!$ elements and is designated by S_N.

A standard way of representing the permutations is by a symbol that contains the original ordering of the items in one row, and the reordered arrangement in a row below it. For example, the six elements of equations 10.4 are written

$$\text{(a)} \quad e \equiv \begin{pmatrix} g_1 & g_2 & g_3 \\ g_1 & g_2 & g_3 \end{pmatrix} \qquad \text{(b)} \quad p \equiv \begin{pmatrix} g_1 & g_2 & g_3 \\ g_2 & g_3 & g_1 \end{pmatrix} \qquad \text{(c)} \quad q \equiv \begin{pmatrix} g_1 & g_2 & g_3 \\ g_3 & g_1 & g_2 \end{pmatrix}$$

$$\text{(d)} \quad r \equiv \begin{pmatrix} g_1 & g_2 & g_3 \\ g_1 & g_3 & g_2 \end{pmatrix} \qquad \text{(e)} \quad s \equiv \begin{pmatrix} g_1 & g_2 & g_3 \\ g_3 & g_2 & g_1 \end{pmatrix} \qquad \text{(f)} \quad t \equiv \begin{pmatrix} g_1 & g_2 & g_3 \\ g_2 & g_1 & g_3 \end{pmatrix} \tag{10.5}$$

We note that the ordering of the items in the first row is unimportant, as long as the type of permutations described in equations 10.4 are preserved. For example, the elements

$$\text{(a)} \quad r = \begin{pmatrix} g_1 & g_2 & g_3 \\ g_1 & g_3 & g_2 \end{pmatrix} \quad \text{and} \quad \text{(b)} \quad r = \begin{pmatrix} g_2 & g_1 & g_3 \\ g_2 & g_3 & g_1 \end{pmatrix} \tag{10.6}$$

are identical since both indicate that the first item is left alone, and the second and third items are interchanged, as indicated in equation 10.4d.

Although the elements are written in the form of matrices with two rows, they are not matrices and do not combine with other permutations by standard matrix multiplication. For example, consider the permutation of g_1, g_2, g_3 defined by p, followed by the permutation of the result defined by r. p takes the ordering $g_1, g_2, g_3 \rightarrow g_2, g_3, g_1$. r then takes $g_2, g_3, g_1 \rightarrow g_2, g_1, g_3$. Thus $p * r$ takes $g_1, g_2, g_3 \rightarrow g_2, g_1, g_3$, which is the permutation denoted by t. In the notation of equations 10.5, this is written

$$p * r = \begin{pmatrix} g_1 & g_2 & g_3 \\ g_2 & g_3 & g_1 \end{pmatrix}\begin{pmatrix} g_2 & g_3 & g_1 \\ g_2 & g_1 & g_3 \end{pmatrix} = \begin{pmatrix} g_1 & g_2 & g_3 \\ g_2 & g_1 & g_3 \end{pmatrix} = t \tag{10.7}$$

From this example, we see that the six elements of this group of permutations are not independent. As the reader will demonstrate in Problem 8 in constructing the multiplication table for S_3, $p^2 = q$, $p^2 * r = s$, and, as just demonstrated, $p * r = t$. Therefore, the group has only three independent elements, e, p, and r, for example. It is straightforward to see from this discussion that these three independent elements do not form a subgroup since closure is not satisfied. However, as the reader can deduce from the table constructed in Problem 8 $\{e, p, q = p^2\}$ does constitute a subgroup of S_3. □

The importance of the permutation group arises from the very powerful theorem known as *Cayley's theorem*:

Every finite group of order N is a subgroup of S_N.

Its proof is fairly straightforward. Let G be a finite group of order N. We will order the elements of G as

$$G = \{g_0, g_1, \ldots, g_N\} \tag{10.8}$$

Let γ be one of the elements of G. Because G is a group, the combination of γ with each of the g_l will be an element of G. But unless $\gamma = e$, $\gamma * g_l \neq g_l$.

We denote $\gamma * G$ by the ordered sequence

$$\gamma * G = \{\gamma * g_0, \gamma * g_1, \ldots, \gamma * g_N\} \tag{10.9}$$

This is the set of elements of G in a different order than the one specified in equation 10.8. Therefore, we can describe the effect of γ on the elements of G by a permutation. That is,

$$\gamma * G = \begin{pmatrix} g_0 & g_1 & g_2 & \cdots & g_N \\ \gamma * g_0 & \gamma * g_1 & \gamma * g_2 & \cdots & \gamma * g_N \end{pmatrix} \tag{10.10}$$

Thus we can associate an element of the permutation group with each element of $\gamma \in G$. As argued above, if α and β are two distinct elements of G, then $\alpha * g_l \neq \beta * g_l$. Therefore, α and β cannot create the same reordering of the elements of G. That is, there is a one-to-one correspondence between the elements of G and a subset of S_N. Therefore, the correspondence is an isomorphism. Since the ordering of the elements in the first row of a permutation is immaterial, the initial ordering of the elements of G does not affect the correspondence between the elements of S_N and those of G.

Example 10.8

As an example, consider C_3. If we order the elements of C_3 as $C_3 = \{e, x, y\}$, then, referring to the table in Figure 10.3,

$$\text{(a)}\quad e * C_3 = \{e, x, y\} \qquad \text{or} \qquad \text{(b)}\quad e * C_3 = \begin{pmatrix} e & x & y \\ e & x & y \end{pmatrix} \tag{10.11}$$

which is the identity element of S_3. In addition,

$$\text{(a)}\quad x * C_3 = \{x, y, e\} \qquad \text{or} \qquad \text{(b)}\quad x * C_3 = \begin{pmatrix} e & x & y \\ x & y & e \end{pmatrix} \tag{10.12}$$

Referring to equation 10.4b or 10.5b, we see that this is the element p of S_3. Similarly,

$$\text{(c)}\quad y * C_3 = \{y, e, x\} \qquad \text{or} \qquad \text{(d)}\quad y * C_3 = \begin{pmatrix} e & x & y \\ y & e & x \end{pmatrix} \tag{10.12}$$

which is the element q of S_3. Therefore, C_3 is isomorphic to the subgroup $\{e, p, q\} = \{e, p, p^2\}$ of S_3. □

Cosets and Lagrange's theorem

Thus far we have discussed abstract groups containing one, two, and three elements. We have seen that there is only one group of each of these orders. The lowest order for which there is more than one group is order 4. One group is the cyclic group C_4 with elements $\{e, x, x^2, x^3\}$ with $x^4 = e$. The other group, distinct from C_4, is called the *dihedral group* D_2 with elements $\{e, x, y, z\}$. This group reflects the operations on a rectangle that leave its orientation unchanged.

Consider the rectangle shown in Figure 10.4. We have labeled the corners to distinguish them. However, the rectangle can undergo various rotations without changing the orientation of Figure 10.4. Because of its geometric symmetry, the rectangle looks the same as the orientation of Figure 10.4 under the following operations:

- e. Nothing is done to the rectangle.
- x. The rectangle is rotated by 180° out of the plane of the page about the x_1 axis. The corners undergo the following interchanges: 2 ↔ 3 and 1 ↔ 4.
- y. The rectangle is rotated by 180° out of the plane of the page about the x_2 axis. The corners undergo the following interchanges: 1 ↔ 2 and 3 ↔ 4.
- z. The rectangle is rotated in the plane of the page by 180° about its center. The corners undergo the following interchanges: 1 ↔ 3 and 2 ↔ 4.

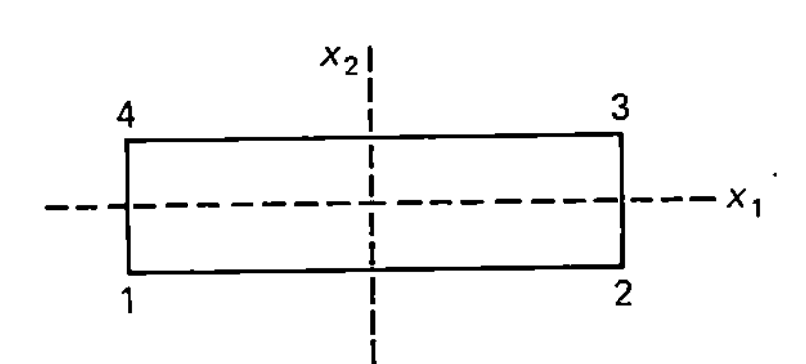

Figure 10.4 Rectangle orientation with axes.

	e	x	y	z
e	e	x	y	z
x	x	e	z	y
y	y	z	e	x
z	z	y	x	e

Figure 10.5 Multiplication table for D_2.

We can construct the multiplication table for the group from these transformations. For example, any of these rotations, performed twice, returns the rectangle to the labeling of Figure 10.4. Thus $x * x = y * y = z * z = e$. The successive rotations specified by $x * y$ has the overall effect of interchanging corners 1 and 3, and interchanging corners 2 and 4. That is the effect of the operation specified by z. Thus $x * y = z$. Using this type of analysis, the multiplication table for D_2 can be determined to be as shown in Figure 10.5.

Since $x * x = e$, $\{e, x\}$ satisfies the four group axioms, and so is a proper subgroup of D_2. We note that the order of this subgroup is an integer divisor of the order of D_2. Similarly, it is clear that $\{e, y\}$ and $\{e, z\}$ are also proper subgroups of D_2 and their orders are also divisors of the order of D_2. This is an example of a general property of subgroups known as *Lagrange's theorem*.

The order of a subgroup is an integer divisor of the order of the group.

To prove this theorem, we let G be a group of order N and H be a subgroup of G of order n. If $n = N$, then $H = G$ and the theorem is trivially valid. If there is one element in G that is not an element of H, then $n < N$ and H is a proper subgroup of G.

Let H be a proper subgroup of G with elements

$$H = \{h_0(\equiv e), h_1, h_2, \ldots, h_{n-1}\} \tag{10.13}$$

Let $\alpha \in G$, with $\alpha \notin H$. Consider the sets of elements $\alpha * h_i$, $0 \leqslant i \leqslant n - 1$. Since $\alpha \notin H$, the elements $\alpha * h_i \notin H$. That is, if $\alpha * h_i = h_j \in H$, then $\alpha = h_j * h_i^{-1} \in H$, which contradicts our requirement that $\alpha \notin H$.

The set of elements $\{\alpha * h_i\}$ with $0 \leqslant i \leqslant n - 1$ is called the *left coset* of H and is denoted by $\alpha * H$. Clearly, $\alpha * H$ is also of order n. Since the elements of $\alpha * H \notin H$, one possibility is that G is comprised of two cosets; H and $\alpha * H$. This is denoted by

$$G = H \oplus \alpha * H \tag{10.14a}$$

Since the order of each of these cosets is n, the order of G would be $N = 2n$.

We point out that since H is a proper subgroup of G, it contains the identity element. Therefore, any coset of H cannot contain the identity, and thus a coset does not form a group.

If H and $\alpha * H$ do not span the entire group G, there must be at least one more element $\beta \in G$ with $\beta \notin H$ and $\beta \notin \alpha * H$ such that

$$G = H \oplus \alpha * H \oplus \beta * H \tag{10.14b}$$

Then since each coset is of order $n, N = 3n$. This process is continued until all the distinct left cosets of G are determined. Then

$$G = H \oplus \alpha * H \oplus \beta * H \oplus \cdots \oplus \mu * H \tag{10.14c}$$

Since H and each of its cosets is of order n, the order of G is

$$N = (1 + \text{number of cosets of } H)n \tag{10.15}$$

and the order of the subgroup H is an integer divisor of the order of G.

From this proof we see that the integer divisor tells us how many independent cosets of H are contained in G. The divisor $N/n \equiv \nu$ is called the index of H in G. Clearly, this analysis could have been carried out using right cosets in the form $H * \alpha$ with identical results. The reader should be aware that, in general, $\alpha * H \neq H * \alpha$.

From this analysis we can conclude that if the order of G is a prime number, the order of its subgroups can only be 1 and N. That is, the only proper subgroup of G is of order 1. Such a subgroup can only contain the identity element. As such, each element of the group other than the identity element, when combined with itself a specified number of times, will generate every other element of the group. That is, all groups of prime order are cyclic (and therefore, Abelian) groups. For example, there is only one group of order 17: the cyclic group C_{17}.

Correspondence between the elements of a group and the elements of a permutation group

Recall that Cayley's theorem states that a group of order N is isomorphic to a subgroup of the permutation group S_N. We use the multiplication table for D_2 to illustrate the method of identifying the elements of a group with the elements of the permutation group. Since the order of D_2 is four, the permutations we determine will be the subgroup of S_4 to which D_2 is isomorphic.

Example 10.9

From the multiplication table of D_2, we can write the elements of the row labeled by e as the first row of a permutation. If we write the row for any element as the second row of the permutation, we have constructed the permutation equivalent for that element of D_2. In this way we obtain

$$x \to \begin{pmatrix} e & x & y & z \\ x & e & z & y \end{pmatrix} = \begin{pmatrix} g_1 & g_2 & g_3 & g_4 \\ g_2 & g_1 & g_4 & g_3 \end{pmatrix} \tag{10.16a}$$

$$y \to \begin{pmatrix} e & x & y & z \\ x & e & z & y \end{pmatrix} = \begin{pmatrix} g_1 & g_2 & g_3 & g_4 \\ g_3 & g_4 & g_1 & g_2 \end{pmatrix} \tag{10.16b}$$

$$z \to \begin{pmatrix} e & x & y & z \\ x & e & z & y \end{pmatrix} = \begin{pmatrix} g_1 & g_2 & g_3 & g_4 \\ g_4 & g_3 & g_2 & g_1 \end{pmatrix} \tag{10.16c}$$

$$e \to \begin{pmatrix} e & x & y & z \\ e & x & y & z \end{pmatrix} = \begin{pmatrix} g_1 & g_2 & g_3 & g_4 \\ g_1 & g_2 & g_3 & g_4 \end{pmatrix} \tag{10.16d}$$

We have constructed the subgroup of S_4 to which D_2 is isomorphic. □

Conjugate classes

Earlier, we defined the conjugate of the element x via p as $p^{-1} * x * p$. Conjugacy for an element is a generalization of the similarity transformation of a matrix.

If $y = p^{-1} * x * p$, then y is said to be *conjugate* to x. Let the elements of a group G be $\{\dots, p, q, \dots, x, y, \dots\}$. Then the set of elements $\{\dots, p^{-1} * x * p, q^{-1} * x * q, \dots, x^{-1} * x * x (= x), y^{-1} * x * y, \dots\}$ is the set of all elements conjugate to x. This set of elements is called the *conjugate class* or the *equivalence class* of x. Conjugacy satisfies the following three conditions:

1. *Reflexivity.* x is self-conjugate. Clearly, $x^{-1} * x * x = x$.
2. *Symmetry.* If x is conjugate to y, then y is conjugate to x. Let $y = p^{-1} * x * p$, and let $q = p^{-1} \in G$. Then $x = q^{-1} * y * q$, and y is conjugate to x.
3. *Transitivity.* If x is conjugate to y and y is conjugate to z, then x is conjugate to z. Let $y = p^{-1} * x * p$ and $z = q^{-1} * y * q$. Then $z = (q^{-1} * p^{-1}) * x * (p * q)$. As with matrices, if $w = p * q$, then $w * w^{-1} = e$ requires that $w^{-1} = (p * q)^{-1} = q^{-1} * p^{-1}$. Therefore, $z = (p * q)^{-1} * x * (p * q)$.

We note here that for every group, the identity element satisfies $p^{-1} * e * p = e$. Therefore, the identity element forms an equivalence class containing only itself. Also, by definition, elements that belong to different equivalence classes cannot be conjugate to each other. Therefore, except for the identity class, equivalence classes do not contain the identity element, and therefore, are not subgroups of G.

Example 10.10

As an illustration, let us determine the equivalence classes of the elements of D_2. From the table of Figure 10.5, we deduce that

$$\text{(a)}\quad e^{-1} * x * e = x \qquad \text{(b)}\quad x^{-1} * x * x = x \qquad \text{(c)}\quad y^{-1} * x * y = y * x * y = z * y = x$$

$$\text{(d)}\quad z^{-1} * x * z = z * x * z = y * z = x \tag{10.17}$$

That is, x forms an equivalence class of D_2 by itself. It is straightforward to show that the same is true of y and z. Thus D_2 has four equivalence classes, each containing one element.

If a group is Abelian, then for all elements g_i and g_j,

$$g_j^{-1} * g_i * g_j = g_j^{-1} * g_j * g_i = g_i \tag{10.18}$$

Therefore, every element of the group forms an equivalence class containing only itself. Conversely, if all equivalence classes contain one element, the group is Abelian. □

The reader will also show in Problem 10 that if x has a period n, then every element of the equivalence class to which x belongs also has period n.

Invariant subgroups and factor groups

Let H be a proper subgroup of G with elements $\{h_i\}$, $0 \leqslant i \leqslant n-1$, with $h_0 \equiv e$. For an element $\alpha \in G$, consider the set of elements

$$\{\alpha^{-1} * h_i * \alpha\} \equiv \alpha^{-1} * H * \alpha \tag{10.19}$$

for each $h_i \in H$. This set forms a subgroup of G that is isomorphic to H. It is called a *conjugate subgroup* of H. Different subgroups are formed for different α.

To prove that $\alpha^{-1} * H * \alpha$ is a subgroup of G, we show that the elements satisfy the four group axioms.

1. *Closure.* Since H is a subgroup, $h_i * h_j \equiv h_k \in H$. Then

$$(\alpha^{-1} * h_i * \alpha) * (\alpha^{-1} * h_j * \alpha) = \alpha^{-1} * h_i * h_j * \alpha = \alpha^{-1} * h_k * \alpha \in \alpha^{-1} * H * \alpha \tag{10.20}$$

2. *Associativity.* Since H is a subgroup, $(h_i * h_j) * h_k = h_i * (h_j * h_k)$. Therefore,

$$(\alpha^{-1} * h_i * \alpha * \alpha^{-1} * h_j * \alpha) * \alpha^{-1} * h_k * \alpha = \alpha^{-1} * (h_i * h_j) * h_k * \alpha$$

$$= \alpha^{-1} * h_i * (h_j * h_k) * \alpha = \alpha^{-1} * h_i * \alpha * (\alpha^{-1} * h_k * \alpha * \alpha^{-1} * h_k * \alpha) \tag{10.21}$$

3. *Identity element.* Since H is a subgroup, $e \in H$. Thus,

$$\alpha^{-1} * e * \alpha = e \in \alpha^{-1} * H * \alpha \tag{10.22}$$

4. *Inverse element.* Since H is a subgroup, $h_i^{-1} \in H$. As shown above, $(p * q)^{-1} = q^{-1} * p^{-1}$. Thus

$$(\alpha^{-1} * h_i * \alpha)^{-1} = \alpha^{-1} * h_i^{-1} * \alpha \tag{10.23}$$

If H is a subgroup such that for every element $\alpha \in G$, $\alpha^{-1} * H * \alpha = H$, then H is called an *invariant* or *self-conjugate subgroup*.

Example 10.11

As an example, we know that $C_3 = \{e, p, p^2\}$ is a subgroup of the permutation group S_3. Consider the effect of $\alpha^{-1} * C_3 * \alpha$ for each $\alpha \in S_3$. Referring to equations 10.5 and Problem 8, the elements of S_3 can be written $\{e, p, p^2, r, p^2 * r, p * r\}$. Clearly,

$$e^{-1} * C_3 * e = \{e^{-1} * e * e, e^{-1} * p * e, e^{-1} * p^2 * e\} = C_3 \tag{10.24a}$$

$$p^{-1} * C_3 * p = \{p^{-1} * e * p, p^{-1} * p * p, p^{-1} * p^2 * p\} = C_3 \tag{10.24b}$$

$$(p^2)^{-1} * C_3 * p^2 = \{p^{-2} * e * p^2, p^{-2} * p * p^2, p^{-2} * p^2 * p^2\} = C_3 \tag{10.24c}$$

Using the fact that the elements r, s, and t of S_3 are their own inverses, we obtain

$$r^{-1} * C_3 * r = \{e, r^{-1} * p * r, r^{-1} * p^2 * r\} = \{e, p^2, p\} = C_3 \tag{10.24d}$$

$$s^{-1} * C_3 * s = \{e, s^{-1} * p * s, s^{-1} * p^2 * s\} = \{e, p^2, p\} = C_3 \tag{10.24e}$$

$$t^{-1} * C_3 * t = \{e, t^{-1} * p * t, t^{-1} * p^2 * t\} = \{e, p^2, p\} = C_3 \tag{10.24f}$$

Thus C_3 is an invariant subgroup of S_3. □

Recall that the index of a subgroup is the ratio of the order of the group to the order of the subgroup. If H is a subgroup of G of index 2, it is an invariant subgroup. To see this, let α be an element of G. If $\alpha \in H$, then $\alpha^{-1} * H * \alpha = H$ and the statement is trivially valid. Since the index of H is 2, if $\alpha \notin H$, G can be written either as

$$\text{(a)} \quad G = H \oplus \alpha * H \qquad \text{or} \qquad \text{(b)} \quad G = H \oplus H * \alpha \tag{10.25}$$

Since $\alpha \notin H$, and H is a proper subgroup of G, then $\alpha * H \neq H$ and $H * \alpha \neq H$. Therefore,

$$\text{(a)} \quad \alpha * H = H * \alpha \qquad \text{or} \qquad \text{(b)} \quad H = \alpha^{-1} * H * \alpha \tag{10.26}$$

and H is an invariant subgroup.

Factor groups

Let $\{g_l * H\}$ and $\{g_m * H\}$ be two distinct left cosets of H, where H is an invariant subgroup of the group G. The product of two such cosets is

$$\{g_l * H\} * \{g_m * H\} = \{g_l * H * g_m * H\} = \{g_l * g_m * H * H\} \tag{10.27a}$$

Since H is a subgroup, the product of its elements, represented by $H * H$, are the elements of H. That is $H * H = H$. Since $g_l * g_m \equiv g_k \in G$, the product of two cosets of H is a coset of H.

$$\{g_l * H\} * \{g_m * H\} = \{g_k * H\} \tag{10.27b}$$

Since the elements of G satisfy the associativity axiom of a group, the product of cosets is associative. That is,

$$\{g_l * H\} * [\{g_m * H\} * \{g_k * H\}] = [\{g_l * H\} * \{g_m * H\}] * \{g_k * H\} \tag{10.27c}$$

Referring to equation 10.27b, if $g_l = e$, then $\{e * H\} = H$ and $g_k = g_m$. Therefore, equation 10.27b becomes

$$\{H * g_m * H\} = \{g_m * H\} \tag{10.28}$$

Thus, under coset multiplication, H is the identity element.

If in equations 10.27, $g_l = g_m^{-1}$, so that $g_k = e$, then

$$\{g_m^{-1} * H\} * \{g_m * H\} = H \tag{10.29}$$

That is, since H is the identity element under coset multiplication, $\{g_m^{-1} * H\}$ is the inverse of $\{g_m * H\}$.

If there are n distinct left (and therefore right) cosets of an invariant subgroup, these cosets form a group of order n under coset multiplication. The group is called the *factor group* of G by H and is often denoted by G/H.

Example 10.12

As an example, $C_3 = \{e, p, p^2\}$ is an invariant subgroup of the permutation group S_3. In fact, it is straightforward to show that

$$S_3 = C_3 \oplus r * C_3 = \{e, p, p^2\} \oplus \{r, r * p, r * p^2\} = \{e, p, p^2\} \oplus \{r, t, s\} \tag{10.30}$$

Thus C_3 and $r * C_3$ are elements of the factor group S_3/C_3. It is a group of order 2 under coset multiplication. Since $r * r = e$,

$$\{r * C_3\} * \{r * C_3\} = C_3 \tag{10.31}$$

That is, C_3 is the identity element of this group.

Thus, referring to the multiplication table of C_2 in Figure 10.1b, we see that the factor group S_3/C_3 is isomorphic to $C_2 = \{e, x\}$. The correspondence is $C_3 \leftrightarrow e$ and $r * C_3 \leftrightarrow x$. □

Direct products

A group G is said to be the *direct product* of two distinct subgroups H and H' if:

1. Every $h_i \in H$ commutes with $h'_j \in H'$.
2. Every element $g_k \in G$ can be obtained from a product of $h_i \in H$ with $h'_j \in H'$. That is $g_k = h_i * h'_j$, so a direct product group is composed of elements that are ordered pairs $h_i * h'_j$, such that $h_i \in H$ and $h'_j \in H'$.

If these conditions are satisfied, the direct product is denoted as $G = H \otimes H'$.

Example 10.13

For example, the cyclic group C_6 is $C_6 = \{e, w, w^2, w^3, w^4, w^5\}$ with $w^6 = e$. If $w^3 \equiv x$, we see that $\{e, w^3\} = \{e, x\} = C_2$. Similarly, with $w^2 \equiv x$, $\{e, w^2, w^4\} = \{e, x, x^2\} = C_3$. Consider the direct product of $C_2 = \{e, w^3\}$ with $C_3 = \{e, w^2, w^4\}$. With $w^7 = w^6 * w = e * w = w$,

$$C_2 \otimes C_3 = \{e * e, e * w^2, e * w^4, w^3 * e, w^3 * w^2, w^3 * w^4\} = \{e, w^2, w^4, w^3, w^5, w\} = C_6 \quad \square \tag{10.32}$$

We note that even though both $C_2 \otimes C_3$ and S_3 have six elements, $S_3 \neq C_2 \otimes C_3$. One indication of this is that the elements of S_3 do not commute, whereas $C_2 \otimes C_3$ is Abelian.

Matrix representation of a group

As stated in Cayley's theorem, every finite group is isomorphic to a subgroup of the permutation group. Thus by discussing matrix representations of the elements of the permutation group, we will be investigating matrix representations of finite groups in general. As such, we will consider as an example the matrix representation of S_3.

Example 10.14

Consider a column vector containing the elements g_1, g_2, and g_3.

$$\mathbf{V} \equiv \begin{bmatrix} g_1 \\ g_2 \\ g_3 \end{bmatrix} \tag{10.33}$$

It is straightforward to show that when each of the matrices

$$\text{(a)}\quad \mathbf{M}_1(e) \equiv \begin{bmatrix} 1 & 0 & 0 \\ 0 & 1 & 0 \\ 0 & 0 & 1 \end{bmatrix} = \mathbf{1} \qquad \text{(b)}\quad \mathbf{M}_1(p) \equiv \begin{bmatrix} 0 & 1 & 0 \\ 0 & 0 & 1 \\ 1 & 0 & 0 \end{bmatrix} \qquad \text{(c)}\quad \mathbf{M}_1(q) \equiv \begin{bmatrix} 0 & 0 & 1 \\ 1 & 0 & 0 \\ 0 & 1 & 0 \end{bmatrix} \tag{10.34}$$

$$\text{(d)}\quad \mathbf{M}_1(r) \equiv \begin{bmatrix} 1 & 0 & 0 \\ 0 & 0 & 1 \\ 0 & 1 & 0 \end{bmatrix} \qquad \text{(e)}\quad \mathbf{M}_1(s) \equiv \begin{bmatrix} 0 & 0 & 1 \\ 0 & 1 & 0 \\ 1 & 0 & 0 \end{bmatrix} \qquad \text{(f)}\quad \mathbf{M}_1(t) \equiv \begin{bmatrix} 0 & 1 & 0 \\ 1 & 0 & 0 \\ 0 & 0 & 1 \end{bmatrix}$$

multiplies the column vector $\mathbf{V}$, it rearranges the elements of $\mathbf{V}$ in the same way as its abstract counterpart does, as described in equations 10.4 or 10.5. Therefore, these matrices, under ordinary matrix multiplication, form a group that is isomorphic to S_3. This group is known as a *matrix representation* of S_3. □

Care must be taken about the ordering of matrix multiplication. For example, for S_3 as shown above, $p * r = t$ means that the permutation specified by p is performed first, followed by the r permutation. In matrix form, the elements of the vector $\mathbf{V}$ of equation 10.33 are affected by $\mathbf{M}(p)$ and then by $\mathbf{M}(r)$ in the form $\mathbf{M}(r)\mathbf{M}(p)\mathbf{V}$. From this example we see that

$$\mathbf{M}(g_k * g_l) = \mathbf{M}(g_l)\mathbf{M}(g_k) \tag{10.35}$$

In general, a set of matrices under matrix multiplication that is isomorphic or homomorphic to a group G is a matrix representation of G denoted by Γ. If the matrices of the representation are $N \times N$, the matrix representation is said to be *N-dimensional*.

There are many possible matrix representations of a group G. For example, from a representation Γ, it is possible to generate a matrix representation Γ' by transforming each matrix by a similarity transformation. That is, for some similarity transformation matrix $\mathbf{U}$,

$$\mathbf{M}'(g_l) \equiv \mathbf{U}^{-1}\mathbf{M}(g_l)\mathbf{U} \tag{10.36}$$

for every $g_l \in G$, forms a second matrix representation Γ' that is clearly the same dimension as Γ. The group properties of the two sets of matrices are identical. As such, they are said to be *equivalent representations*.

In general, there is more than one matrix representation of a group. The group of matrices defined in equations 10.34 will be designated by the representation Γ_1. As an example, the reader will show in Problem 14 that the matrices

$$\text{(a)}\quad \mathbf{M}_2(e) = \begin{bmatrix} 1 & 0 \\ 0 & 1 \end{bmatrix} \qquad \text{(b)}\quad \mathbf{M}_2(p) = \begin{bmatrix} 0 & 1 \\ -1 & -1 \end{bmatrix} \qquad \text{(c)}\quad \mathbf{M}_2(q) = \begin{bmatrix} -1 & -1 \\ 1 & 0 \end{bmatrix} \tag{10.37}$$

$$\text{(d)}\quad \mathbf{M}_2(r) = \begin{bmatrix} 0 & 1 \\ 1 & 0 \end{bmatrix} \qquad \text{(e)}\quad \mathbf{M}_2(s) = \begin{bmatrix} 1 & 0 \\ -1 & -1 \end{bmatrix} \qquad \text{(f)}\quad \mathbf{M}_2(t) = \begin{bmatrix} -1 & -1 \\ 0 & 1 \end{bmatrix}$$

produce the same multiplication table as the abstract elements of S_3. Thus they comprise another matrix representation of S_3, which we designate Γ_2. The reader will also show in Problem 14 that Γ_2' and Γ_2'', which are the sets of matrices

$$\text{(a)}\quad \mathbf{M}_2'(e) = \begin{bmatrix} 1 & 0 \\ 0 & 1 \end{bmatrix} \qquad \text{(b)}\quad \mathbf{M}_2'(p) = \frac{1}{2}\begin{bmatrix} -1 & -\sqrt{3} \\ \sqrt{3} & -1 \end{bmatrix}$$

$$\text{(c)}\quad \mathbf{M}_2'(q) = \frac{1}{2}\begin{bmatrix} -1 & \sqrt{3} \\ -\sqrt{3} & -1 \end{bmatrix} \qquad \text{(d)}\quad \mathbf{M}_2'(r) = \begin{bmatrix} 1 & 0 \\ 0 & -1 \end{bmatrix}$$

$$\text{(e)}\quad \mathbf{M}_2'(s) = \frac{1}{2}\begin{bmatrix} -1 & \sqrt{3} \\ \sqrt{3} & 1 \end{bmatrix} \qquad \text{(f)}\quad \mathbf{M}_2'(t) = \frac{1}{2}\begin{bmatrix} -1 & -\sqrt{3} \\ -\sqrt{3} & 1 \end{bmatrix} \tag{10.38}$$

$$\text{(a)}\quad \mathbf{M}_2''(e) = \begin{bmatrix} 1 & 0 \\ 0 & 1 \end{bmatrix} \qquad \text{(b)}\quad \mathbf{M}_2''(p) = \begin{bmatrix} \omega & 0 \\ 0 & \omega^2 \end{bmatrix} \qquad \text{(c)}\quad \mathbf{M}_2''(q) = \begin{bmatrix} \omega^2 & 0 \\ 0 & \omega \end{bmatrix}$$

$$\text{(d)}\quad \mathbf{M}_2''(r) = \begin{bmatrix} 0 & 1 \\ 1 & 0 \end{bmatrix} \qquad \text{(e)}\quad \mathbf{M}_2''(s) = \begin{bmatrix} 0 & \omega \\ \omega^2 & 0 \end{bmatrix} \qquad \text{(f)}\quad \mathbf{M}_2''(t) = \begin{bmatrix} 0 & \omega^2 \\ \omega & 0 \end{bmatrix} \tag{10.39}$$

are also matrix representations of S_3. The quantity ω in the matrices of Γ_2'' is the cube root of 1. That is, $\omega = e^{2\pi i/3} = \frac{1}{2}(-1 + i\sqrt{3})$. Thus $\omega^* = \omega^2$ and $\omega^3 = +1$. As will be seen later, S_3 has only one representation of dimension 2. It will then become clear that Γ_2, Γ_2', and Γ_2'' must be equivalent to one another, related by a similarity transformation.

Reducible representations

Let $\mathbf{M}_1(g_l)$ and $\mathbf{M}_2(g_l)$ be elements of two different matrix representations Γ_1 and Γ_2 of the same group. We can form another representation by making $\mathbf{M}_1$ and $\mathbf{M}_2$ the partitions of a new matrix.

$$\mathbf{M}(g_l) \equiv \begin{bmatrix} \mathbf{M}_1(g_l) & \mathbf{0} \\ \mathbf{0} & \mathbf{M}_2(g_l) \end{bmatrix} \tag{10.40a}$$

The arrangement of the partitions along the main diagonal of the matrices is immaterial.

$$\mathbf{M}'(g_l) \equiv \begin{bmatrix} \mathbf{M}_2(g_l) & \mathbf{0} \\ \mathbf{0} & \mathbf{M}_1(g_l) \end{bmatrix} \tag{10.40b}$$

is an element of the same representation as the matrix $\mathbf{M}(g_l)$ of equation 10.40a. If all the matrices of the representation Γ can be written in this same partitioned form, Γ is said to be *reducible* as the direct sum of Γ_1 and Γ_2. This is denoted by

$$\Gamma = \Gamma_1 \oplus \Gamma_2 \tag{10.41}$$

We stress that it must be possible to partition every matrix of the representation in the same form for it to be reducible. If there is at least one matrix of the representation cannot be cast in the same partitioned form as the others, the representation is *irreducible*.

If Γ and Γ' are representations related by a similarity transformation, and if Γ is reducible as the direct sum of Γ_1 and Γ_2, then Γ' is also reducible as the direct sum of Γ_1' and Γ_2' such that Γ_1 is equivalent to Γ_1' and Γ_2 is equivalent to Γ_2'.

The reduction or resolution of a general representation into irreducible representations is unique. That is, there is only one set of matrices $\mathbf{M}_i$ that will appear along the main diagonal of

$$\mathbf{M}(g_l) \equiv \begin{bmatrix} \mathbf{M}_1(g_l) & \mathbf{0} & \cdots & \mathbf{0} \\ \mathbf{0} & \mathbf{M}_2(g_l) & & \vdots \\ \vdots & & \ddots & \vdots \\ \mathbf{0} & \cdots & & \mathbf{M}_n(g_l) \end{bmatrix} \tag{10.42}$$

Except for unimportant reorderings of the matrices $\mathbf{M}_k$, this reduction is unique.

Schur's lemma

The fundamental question of the theory of representations is how to resolve a general representation into a direct sum of irreducible representations. The approach requires the statement of the theorem

> **Let $\Gamma = \{\mathbf{M}(g_l)\}$ and $\Gamma' = \{\mathbf{M}'(g_l)\}$ be two representations of the group G, with dimensions N and N' respectively. Let A be a $N \times N'$ matrix independent of g_l such that $\mathbf{A}\mathbf{M}(g_l) = \mathbf{M}'(g_l)\mathbf{A}$ for all $g_i \in G$. There are three possible situations.**
>
> **(a) $N \neq N'$ and $\mathbf{A} = \mathbf{0}$. Then Γ and Γ' are inequivalent representations.**
>
> **(b) $N = N'$ and A is singular. Then $\mathbf{A} = \mathbf{0}$. No information about the relation between Γ and Γ' can be deduced from this condition.**
>
> **(c) $N = N'$ and A is nonsingular so that $\mathbf{A}^{-1}$ exists. Then $\mathbf{M}'(g_l) = \mathbf{A}\mathbf{M}(g_l)\mathbf{A}^{-1}$. That is, Γ and Γ' are equivalent representations.**

This is known as *Schur's lemma*. Its somewhat involved proof will not be presented here. There are many good presentations of the proof, and the interested reader is referred to sources such as the texts by E. Wigner, *Group Theory and Its Application to the Quantum Mechanics of Atomic Spectra* (New York, Academic Press, Inc., 1959, p. 73), M. Tinkham, *Group Theory and Quantum Mechanics* (New York: McGraw-Hill Book Company, 1964, p. 21), or J. Lomont, *Applications of Finite Groups* (New York: Academic Press, 1959, p. 47).

Let $\Gamma = \Gamma' = \{\mathbf{M}(g_l)\}$ be an irreducible representation of a group G. That is, $\mathbf{M}(g_l) = \mathbf{M}'(g_l)$. Let $\mathbf{A}$ be a matrix that commutes with every $\mathbf{M}(g_l)$ in Γ. That is, $\mathbf{A}\mathbf{M}(g_l) = \mathbf{M}(g_l)\mathbf{A}$ for all $g_l \in G$. Then $\mathbf{A}$ must be a square matrix. Let λ be one of the eigenvalues of $\mathbf{A}$, which means that $\det(\mathbf{A} - \lambda\mathbf{1}) = 0$. Therefore, $\mathbf{A} - \lambda\mathbf{1}$ is a singular matrix. Since $\mathbf{A}$ commutes with every $\mathbf{M} \in \Gamma$, then clearly $\mathbf{A} - \lambda\mathbf{1}$ also commutes with every $\mathbf{M} \in \Gamma$. From condition (b) of Schur's lemma, then $\mathbf{A} - \lambda\mathbf{1} = \mathbf{0}$, or $\mathbf{A} = \lambda\mathbf{1}$. From this, we deduce the corollary to Schur's lemma:

> **A representation of a group is irreducible if and only if a multiple of the unit matrix is the only matrix that will commute with all the matrices in the representation.**

That is, if $\Gamma = \{\mathbf{M}(g_l)\}$ is a matrix representation of the group G, and if $\mathbf{AM} = \mathbf{MA}$ for every g_l, then $\mathbf{A}$ must be the identity matrix if Γ is irreducible. Conversely, for some number λ, if $\mathbf{A} = \lambda\mathbf{1}$ is the only matrix that will commute with each $\mathbf{M}(g_l)$, then Γ is irreducible.

If the representation is reducible, it is straightforward to find a matrix that is not the unit matrix which commutes with all matrices of the representation. As an example, for the reducible representation given in equation 10.42

$$\mathbf{M}(g_l) \equiv \begin{bmatrix} \mathbf{M}_1(g_l) & \mathbf{0} & \cdots & \mathbf{0} \\ \mathbf{0} & \mathbf{M}_2(g_l) & & \vdots \\ \vdots & \vdots & & \vdots \\ \mathbf{0} & \cdots & & \mathbf{M}_n(g_l) \end{bmatrix} \tag{10.42}$$

the matrix

$$\mathbf{I} \equiv \begin{bmatrix} a_1\mathbf{1}_1 & \mathbf{0} & \cdots & \mathbf{0} \\ \mathbf{0} & a_2\mathbf{1}_2 & & \vdots \\ \vdots & \vdots & & \vdots \\ \mathbf{0} & \cdots & & a_n\mathbf{1}_n \end{bmatrix} \tag{10.43}$$

commutes with $\mathbf{M}(g_l)$ for every i. In this matrix $\mathbf{I}$, a_k is a complex number and $\mathbf{1}_k$ is the identity matrix of dimension k. Since the a_k are in general different for different indices k, $\mathbf{I}$ is not the unit matrix.

From this analysis it is straightforward to show that

All irreducible representations of an Abelian group are one-dimensional.

Let Γ be an N-dimensional representation of an Abelian group. Then since G is Abelian,

$$\mathbf{M}(g_1)\mathbf{M}(g_k) = \mathbf{M}(g_k)\mathbf{M}(g_1) \qquad \text{for all } g_k \in G \tag{10.44}$$

From the corollary to Schur's lemma above,

$$\mathbf{M}(g_1) = \lambda_1\mathbf{1} \tag{10.45}$$

Thus since g_1 represents any element of the group, the matrix representation of each element of G is of the form

$$\mathbf{M}(g_1) = \begin{bmatrix} \lambda_1 & 0 & 0 & \cdots & 0 \\ 0 & \lambda_1 & 0 & \cdots & 0 \\ \vdots & & & & \\ 0 & 0 & 0 & \cdots & \lambda_1 \end{bmatrix} \tag{10.46}$$

$\mathbf{M}(g_1)$ is in the partitioned form of a representation of N one-dimensional irreducible representations.

Regular representation

A second method of labeling a multiplication table of a group is to label the rows of the multiplication table by the elements of the group, and the columns by that element's inverse. For example, the table for C_3 defined by this arrangement is that shown in Figure 10.6. With such a designation of rows and columns, the identity element will always lie along the main diagonal. We note that because C_3 is Abelian, this description of its table is symmetric about the diagonal that runs from the upper right corner to the lower left corner. This form of the multiplication table for any Abelian group will have this property.

The regular matrix representation for a given element is obtained from this form of the multiplication table. The table is treated as a matrix. For a given element, each

	e	$x^{-1} = y$	$y^{-1} = x$
e	e	y	x
x	x	e	y
y	y	x	e

Figure 10.6 Alternative form of group multiplication table for the group C_3.

entry of that element in the table is replaced by a 1, and the entries of all other elements are replaced by zeros.

Example 10.15

For example, the matrix of the regular representation of the element x of C_3 is obtained by treating the table of Figure 10.6 as a 3×3 matrix, placing 1 in the location of each x, and 0 in all locations in which x does not appear. In this way we see that the matrices of the regular representation of C_3 are

$$\text{(a)}\quad \mathbf{M}(e) = \begin{bmatrix} 1 & 0 & 0 \\ 0 & 1 & 0 \\ 0 & 0 & 1 \end{bmatrix} \qquad \text{(b)}\quad \mathbf{M}(x) = \begin{bmatrix} 0 & 0 & 1 \\ 1 & 0 & 0 \\ 0 & 1 & 0 \end{bmatrix} \qquad \text{(c)}\quad \mathbf{M}(y) = \begin{bmatrix} 0 & 1 & 0 \\ 0 & 0 & 1 \\ 1 & 0 & 0 \end{bmatrix} \quad \square$$

(10.47)

From this construction it is obvious that the dimension of the regular representation is the same as the order of the group.

Orthogonality theorem

Let G be a group of order μ and let Γ_1 and Γ_2 be inequivalent representations of dimensions d_1 and d_2. The orthogonality theorem states that

$$\text{(a)}\quad \sum_{l=1}^{\mu} [\mathbf{M}_1(g_l^{-1})]_{\alpha\beta}[\mathbf{M}_2(g_l)]_{\lambda\sigma} = 0 \qquad \text{(b)}\quad \sum_{l=1}^{\mu} [\mathbf{M}_1(g_l^{-1})]_{\alpha\beta}[\mathbf{M}_1(g_l)]_{\lambda\sigma} = \frac{\mu}{d_1}\delta_{\alpha\lambda}\delta_{\beta\sigma} \tag{10.48}$$

for any $\alpha, \beta, \lambda, \sigma$. $[\mathbf{M}_r(g_l)]_{\alpha\beta}$ is the $\alpha\beta$ element of the matrix $\mathbf{M}_r(g_l)$. Combining these results, we can write

$$\sum_{l=1}^{\mu} [\mathbf{M}_r(g_l^{-1})]_{\alpha\beta}[\mathbf{M}_s(g_l)]_{\lambda\sigma} = \frac{\mu}{d_r}\delta_{\alpha\lambda}\delta_{\beta\sigma}\delta_{rs} \tag{10.49}$$

for any $\alpha, \beta, \lambda, \sigma$. Again the proof will be omitted, but the reader is referred to page 23 of the text by Tinkham, or to Wigner's book, page 79, for its details.

As shown in Problem 15, the matrix representation of any group element satisfies

$$\mathbf{M}(g_l^{-1}) = \mathbf{M}^{-1}(g_l) \tag{10.50}$$

Thus if the representation is unitary, $\mathbf{M}(g_l^{-1}) = \mathbf{M}^{\dagger}(g_l)$ and equations 10.49 can be written

$$\sum_{l=1}^{\mu} [\mathbf{M}_r^*(g_l)]_{\beta\alpha}[\mathbf{M}_s(g_l)]_{\lambda\sigma} = \frac{\mu}{d_r}\delta_{\alpha\lambda}\delta_{\beta\sigma}\delta_{rs} \tag{10.51}$$

for any $\alpha, \beta, \lambda, \sigma$.

It can be shown that any representation in terms of nonsingular matrices is equivalent to a unitary representation (see Tinkham's book, page 20, for example). This is known as *Maschke's theorem*.

Character of a representation

A function $F(g_l)$ on a group G of order μ is a specified relation that assigns a complex number to each group element g_l. For any specific element $g_1 \in G$,

$$\sum_{l=1}^{\mu} F(g_1 * g_l) = \sum_{l=1}^{\mu} F(g_l * g_1) = \sum_{l=1}^{\mu} F(g_l) \tag{10.52}$$

Since $g_1 * g_l = g_k \in G$, the elements in the summation occur in a different order in $\sum_{l=1}^{\mu} F(g_1 * g_l)$ then they do in $\sum_{l=1}^{\mu} F(g_l)$, but the summation is still taken over all the elements of the group. This property is called the invariance of the left (or right) summation over the group elements. It is valid for any function on the group.

One important function is the character of the representation. If $\Gamma = \{\mathbf{M}(g_l)\}$ is a matrix representation of a group G, the character of that representation is defined as the set of numbers that are the traces of the matrices of the representation. The traces are denoted by

$$\chi(g_l) \equiv \mathrm{Tr}[\mathbf{M}(g_l)] \tag{10.53a}$$

and the set of traces of a representation Γ is denoted by

$$\chi(\Gamma) \equiv \{\chi(g_l)\} \tag{10.53b}$$

It is easy to see that all equivalent representations have the same character. If two representations are equivalent, they are related by a similarity transformation. That is,

$$\mathbf{M}'(g_l) = \mathbf{U}^{-1}\mathbf{M}(g_l)\mathbf{U} \tag{10.36}$$

But as was shown in Chapter 8, the trace of a matrix is invariant under a similarity transformation. Therefore, if Γ and Γ' are equivalent representations, $\chi(\Gamma) = \chi(\Gamma')$.

As described earlier, the equivalence class of an element g_1 of a group G is the set of all elements g_l formed by the combination

$$g_l = g_k^{-1} * g_1 * g_k \qquad \text{for all } g_k \in G \tag{10.54a}$$

Therefore, using equation 10.50, the matrix representations of the elements in a given equivalence class are related by the similarity transformation

$$\mathbf{M}(g_l) = \mathbf{M}^{-1}(g_k)\mathbf{M}(g_1)\mathbf{M}(g_k) \tag{10.54b}$$

and since the trace is invariant under similarity transformations, every element in a given equivalence class has the same character. For this reason, the character is often referred to as the character of an equivalence class instead of the character of an element.

Let Γ_r and Γ_s be irreducible, inequivalent unitary representations of a group G of order μ. Referring to equation 10.51, we can form the traces of the matrices by setting $\alpha = \beta$ and summing, and doing the same with λ and σ. Then equation 10.51 can be cast in terms of the characters of the representations as

$$\chi^{\dagger}(\Gamma_r)\chi(\Gamma_s) \equiv \sum_{l=1}^{\mu} \chi_r^*(g_l)\chi_s(g_l) = \frac{\mu}{\nu_r}\delta_{rs} \sum_{\alpha=1}^{\nu_r} \sum_{\lambda=1}^{\nu_s} \delta_{\alpha\lambda}\delta_{\alpha\lambda} = \frac{\mu}{\nu_r}\delta_{rs} \sum_{\alpha=1}^{\nu_r} \delta_{\alpha\alpha} = \mu\delta_{rs} \tag{10.55}$$

Let us now construct a *character vector*, the elements of which are the characters of the elements of the representation.

$$\boldsymbol{\chi}(\Gamma) \equiv \begin{bmatrix} \chi(e) \\ \chi(g_1) \\ \chi(g_2) \\ \vdots \\ \chi(g_\mu) \end{bmatrix} \tag{10.56}$$

Then the orthogonality theorem in terms of the character vectors of the unitary representations can be written

$$\boldsymbol{\chi}^\dagger(\Gamma_r)\boldsymbol{\chi}(\Gamma_s) = \mu\delta_{rs} \tag{10.57}$$

That is, the character vectors of inequivalent representations are mutually orthogonal and are normalized to $\sqrt{\mu}$.

Setting $r = s$, we have

$$\boldsymbol{\chi}^\dagger(\Gamma_r)\boldsymbol{\chi}(\Gamma_r) = \sum_{l=1}^{\mu} |\chi(g_l)|^2 = \mu \tag{10.58}$$

Therefore, to determine if a particular representation is irreducible, one can determine the character for each element in the representation and form the sum of the squares of the characters. If that sum adds to the order of the group, the representation is irreducible.

Since every character of a given equivalence class has the same value, we can form a vector that is smaller (unless the group is Abelian), but contains the same information as the character vector of equation 10.56. Let $\chi(\mathscr{C}_l)$ be the character of the equivalence class $\mathscr{C}_l$, which has ν_l elements. Let t be the number of distinct equivalence classes of G. Then for the representation Γ, we form the vector

$$\mathbf{V}(\Gamma) \equiv \begin{bmatrix} \eta_1\chi(C_1) \\ \eta_2\chi(C_2) \\ \eta_3\chi(C_3) \\ \vdots \\ \eta_t\chi(C_t) \end{bmatrix} \tag{10.59}$$

where

$$\eta_l^2 \equiv \frac{\nu_l}{\mu} \tag{10.60}$$

Let Γ_r and Γ_s be two inequivalent irreducible representations, and consider

$$\mathbf{V}^\dagger(\Gamma_r)\mathbf{V}(\Gamma_s) = \sum_{l=1}^{t} \eta_l^2\chi_r(\mathscr{C}_l)\chi_s(\mathscr{C}_l) = \frac{1}{\mu}\sum_{l=1}^{t} \nu_l\chi_r(\mathscr{C}_l)\chi_s(\mathscr{C}_l) \tag{10.61}$$

But ν_l is the number of elements in the equivalence class $\mathscr{C}_l$, and since all elements in $\mathscr{C}_l$ have the same character,

$$\nu_l\chi_r(\mathscr{C}_l)\chi_s(\mathscr{C}_l) = \sum_{k=1}^{\nu_l} \chi_r(g_k)\chi_s(g_k) \qquad \text{for all } g_k \in \mathscr{C}_l \tag{10.62}$$

Therefore,

$$\mathbf{V}^{\dagger}(\Gamma_r)\mathbf{V}(\Gamma_s) = \frac{1}{\mu}\sum_{l=1}^{\mu}\chi_r(g_l)\chi_s(g_l) = \frac{1}{\mu}\sum_{l=1}^{t}\nu_l\chi_r(\mathscr{C}_l)\chi_s(\mathscr{C}_l) = \delta_{rs} \tag{10.63}$$

That is, the vectors formed from the characters of the different equivalence classes are orthonormal with respect to irreducible representation. Since $\mathbf{V}(\Gamma)$ is a t-dimensional vector, there can be no more than t such mutually orthogonal vectors. Therefore, the number of inequivalent irreducible representations is t, the number of different equivalence classes.

Let a group G have N irreducible representations, $\Gamma_1, \Gamma_2, \ldots, \Gamma_N$. Since the trace of a partitioned matrix is the sum of the traces of those partitions that lie along the main diagonal, the trace of a reducible matrix is the sum of the traces of each of the partitioned irreducible matrices. Let the matrices of an arbitrary representation Γ be partitioned such that the matrices of Γ_1 appear along the main diagonal n_1 times, Γ_2 appears n_2 times, and so on. That means that Γ is reducible into Γ_1 n_1 times, Γ_2 n_2 times, and so on. That is, the character of any element g_l must be of the form

$$\chi(g_l) = n_1\chi_1(g_l) + n_2\chi_2(g_l) + \cdots + n_N\chi_N(g_l) \tag{10.64a}$$

or, written in terms of the characters of the representations,

$$\chi(\Gamma) = \sum_{l=1}^{N} n_l\chi(\Gamma_l) \tag{10.64b}$$

Therefore, looking at $\chi(\Gamma)$ as the column vector of equation 10.56, and using equation 10.57, we have

$$\text{(a)}\quad \boldsymbol{\chi}^{\dagger}(\Gamma_s)\boldsymbol{\chi}(\Gamma) = \sum_{l=1}^{\mu}\chi_s^*(g_l)\chi(g_l) = \mu n_s \quad \text{or} \quad \text{(b)}\quad \frac{1}{\mu}\boldsymbol{\chi}^{\dagger}(\Gamma_s)\boldsymbol{\chi}(\Gamma) = n_s \tag{10.65}$$

That is, the product described in equation 10.65b gives us the number of times the irreducible representation Γ_s is contained in the reducible representation Γ. In particular, if, in equation 10.65b, $\Gamma = \Gamma_s$, then Γ_s appears once in Γ_s and equation 10.65b becomes

$$|\chi(\Gamma_s)|^2 = \sum_{l=1}^{\mu}|\chi(g_l)|^2 = \mu \tag{10.66}$$

which is the result obtained in equation 10.58.

Referring to the method of construction of the regular representation from the alternative form of the multiplication table, we see that $\mathbf{M}(e) = \mathbf{1}$ has nonzero diagonal elements, and $\mathbf{M}(g_l)$ must have zeros along the main diagonal for all $g_l \neq e$. Thus $\chi_{\text{reg}}(e) = \mu$, the order of the group, and $\chi_{\text{reg}}(g_l \neq e) = 0$. Applying equation 10.65b to the regular representation, we obtain

$$n_s = \frac{1}{\mu}\chi^{\dagger}(\Gamma_s)\chi(\Gamma_{\text{reg}}) = \frac{1}{\mu}\sum_{l=1}^{\mu}\chi_s^*(g_l)\chi_{\text{reg}}(g_l) = \frac{1}{\mu}\chi_s^*(e)\chi_{\text{reg}}(e) = d_s \tag{10.67}$$

where, as before, d_s is the dimension of the sth representation, Γ_s. That is, the number of times Γ_s appears in the regular representation is the same as the dimension of that irreducible representation.

Combining this result with equation 10.64a, we write

$$\chi_{\text{reg}}(g_l) = \sum_{s=1}^{N} d_s \chi_s(g_l) \tag{10.68}$$

Since $\chi_{\text{reg}}(e) = \mu$, we set $g_l = e$. Then

$$\mu = \sum_{s=1}^{N} d_s \chi_s(e) = \sum_{s=1}^{N} d_s^2 \tag{10.69}$$

Thus the sum of the squares of the dimensionalities of the irreducible representations is the order of the group.

We now apply equation 10.66 to the regular representation. Since $\chi(g_l \neq e) = 0$,

$$\sum_{l=1}^{\mu} |\chi_{\text{reg}}(g_l)|^2 = |\chi_{\text{reg}}(e)|^2 \tag{10.70a}$$

But the dimension of the regular representation is the order of the group,

$$|\chi_{\text{reg}}(e)|^2 = \mu^2 \tag{10.70b}$$

Since a representation is irreducible only if

$$\sum_{l=1}^{\mu} |\chi(g_l)|^2 = \mu \tag{10.66}$$

we see that for all groups of order $\mu > 1$, the regular representation is reducible.

Example 10.16

As an example, we will find the irreducible representations of the sixth-order group S_3. In Problem 9 the reader will show that the equivalence classes of S_3 are

$$\text{(a)}\quad \mathscr{C}_1 = \{e\} \qquad \text{(b)}\quad \mathscr{C}_2 = \{p, q\} \qquad \text{(c)}\quad \mathscr{C}_3 = \{r, s, t\} \tag{10.71}$$

Therefore, S_3 has three irreducible representations.

For S_3 it is fairly straightforward to deduce the dimensionalities of these representations from the dimensionality identity of equation 10.69. After very few attempts, one can easily deduce that the only triplet of integers that satisfy

$$\text{(a)}\quad d_1^2 + d_2^2 + d_3^2 = 6 \qquad \text{is} \qquad \text{(b)}\quad (d_1, d_2, d_3) = (2, 1, 1) \tag{10.72}$$

That is, the irreducible representations of S_3 are of dimensions 2, 1, and 1.

This straightforward determination of the dimensions of the irreducible representations of S_3 is possible because S_3 has a small number of irreducible representations, and the sums of the squares of their dimensions is a relatively small value (6). Such a straightforward determination is not possible for groups of large order with a large number of irreducible representations.

In equations 10.34, 10.37, 10.38, and 10.39, we presented four matrix representations of S_3 (see Problem 14). Clearly, they cannot all be irreducible representations, since S_3 has only three. In addition, Γ_1 of equations 10.34 is three-dimensional. Since the irreducible representations of S_3 are either one- or two-dimensional, Γ_1 cannot be irreducible. This can also be seen by applying the normalization condition of equation 10.58 or equation 10.66.

$$\sum_{l=1}^{6} |\chi_1(g_l)|^2 = |\chi_1(e)|^2 + |\chi_1(p)|^2 + |\chi_1(q)|^2 + |\chi_1(r)|^2 + |\chi_1(s)|^2 + |\chi_1(t)|^2$$

$$= 9 + 1 + 0 + 1 + 4 + 1 \neq 6 \tag{10.73}$$

Applying this sum to Γ_2, we obtain

$$\sum_{l=1}^{6} |\chi_2(g_l)|^2 = 6 \tag{10.74}$$

so Γ_2 is an irreducible representation of S_3.

Clearly, the other two-dimensional representations labeled Γ_2' and Γ_2'' cannot be other irreducible representations that are not equivalent to Γ_2. We have determined that the other two irreducible representations of S_3 must both be one dimensional. We designate these one-dimensional representations Γ_3 and Γ_4, and will determine them by using the concept of a character table. □

Character table

A convenient way to represent the characters of a group is in the form of a table. The character table is constructed by designating the rows in the table by the irreducible representations of the group, and the columns by the distinct equivalence classes. Often, the number of elements in the equivalence class that heads the column is included in that column heading. The ordering of the rows representing the irreducible representations or the columns representing the equivalence classes is immaterial.

As noted, S_3 has three distinct equivalence classes, $\mathscr{C}_1 = \{e\}$, $\mathscr{C}_2 = \{p, q\}$, and $\mathscr{C}_3 = \{r, s, t\}$. Therefore, taking the traces of the matrices of equations 10.36, we obtain the row entries of the character table for the irreducible representation Γ_2 (Figure 10.7).

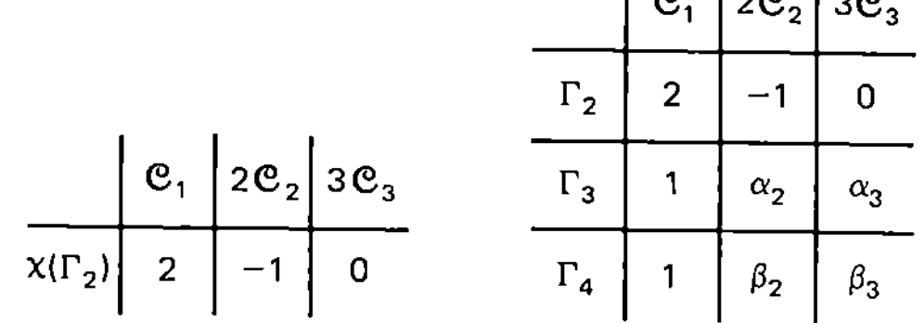

	$\mathcal{C}_1$	$2\mathcal{C}_2$	$3\mathcal{C}_3$
$\chi(\Gamma_2)$	2	−1	0

Figure 10.7 Partial character table for S_3.

	$\mathcal{C}_1$	$2\mathcal{C}_2$	$3\mathcal{C}_3$
Γ_2	2	−1	0
Γ_3	1	α_2	α_3
Γ_4	1	β_2	β_3

Figure 10.8 Incomplete character table for S_3.

In addition, we point out that in any marix representation, the identity element e will always be represented by the unit matrix. Therefore, in any representation, the character for the class containing e will be the dimension of the representation. We see that for Γ_2, which is two-dimensional, $\chi_2(\mathscr{C}_1) = 2$. Since the other two irreducible representations are one-dimensional, $\chi_3(\mathscr{C}_1) = \chi_4(\mathscr{C}_1) = 1$. Thus the full character table of S_3 is of the form shown in Figure 10.8.

The other two irreducible representations, being one-dimensional, are six numbers, one for each element. The sums of their squares must add to 6, and none of numbers can be zero (so that an inverse under multiplication will exist). The characters of Γ_3 and Γ_4 must satisfy the orthonormality condition as expressed in equation 10.63,

$$\sum_{l=1}^{3} \nu_l \chi_r(\mathscr{C}_l)\chi_s(\mathscr{C}_l) = \chi_r(\mathscr{C}_1)\chi_s(\mathscr{C}_1) + 2\chi_r(\mathscr{C}_2)\chi_s(\mathscr{C}_2) + 3\chi_r(\mathscr{C}_3)\chi_s(\mathscr{C}_3) = 6\delta_{rs} \tag{10.75}$$

We note that for any irreducible representation, the number of elements in the equivalence classes add to the order of the group.

$$\sum_{l=1}^{\mu} \nu_l = \mu \tag{10.76}$$

Thus one set of characters can be chosen to be the number 1 for each class. If the representation is one-dimensional, the matrices of that representation are numbers. Therefore, by choosing the characters to be one, each of the matrices of that representation is also the number 1. That is, we can choose one of the one-dimensional representations of S_3 to be

$$\mathbf{M}_3(e) = \mathbf{M}_3(p) = \mathbf{M}_3(q) = \mathbf{M}_3(r) = \mathbf{M}_3(s) = \mathbf{M}_3(t) = 1 \tag{10.77}$$

Therefore, $\alpha_2 = \alpha_3 = 1$ in Figure 10.8.

To determine the characters of Γ_4, we apply the orthogonality theorem as expressed in equation 10.75. We have designated the irreducible representations Γ_2, Γ_3, and Γ_4. Thus r and s will take on the values 2, 3, and 4. Referring to the table in Figure 10.8, with $s = 4$, we set $r = 2$ to obtain

$$2 + 2(-1)\beta_2 + 0\beta_3 = 0 \tag{10.78a}$$

so $\beta_2 = 1$. From $s = 4$, $r = 3$, with $\beta_2 = \alpha_2 = \alpha_3 = 1$, we obtain

$$1 + 2 + 3\beta_3 = 0 \tag{10.78b}$$

so $\beta_3 = -1$. We note that the normalization, obtained from equation 10.75 with $r = s = 4$, is also satisfied with these assignments. Therefore, the complete character table for S_3 is as shown in Figure 10.9.

	$\mathcal{C}_1$	$2\mathcal{C}_2$	$3\mathcal{C}_3$
Γ_2	2	−1	0
Γ_3	1	1	1
Γ_4	1	1	−1

Figure 10.9 Complete character table for S_3.

Like Γ_3, since Γ_4 is one-dimensional, the matrices of this representation will be the same as the characters. Therefore,

$$\text{(a)}\quad \mathbf{M}_4(e) = \mathbf{M}_4(p) = \mathbf{M}_4(q) = 1 \qquad \text{(b)}\quad \mathbf{M}_4(r) = \mathbf{M}_4(s) = \mathbf{M}_4(t) = -1 \tag{10.79}$$

Representations generated by basis vectors

The concept of the matrix representation of a group element was introduced by example for S_3. The elements of the vector $\mathbf{V}$ of equation 10.33 were rearranged by the matrices of equations 10.34 in the same way the ordering of $\{g_1, g_2, g_3\}$ was permuted by the abstract elements of S_3. From this example we see that the abstract operations that make up a group, when they are represented by matrices, must operate on column vectors. Such vectors are represented in the Dirac notation by a ket. That is, in a space spanned by the column vectors $\mathbf{V}_1, \mathbf{V}_2, \ldots, \mathbf{V}_N$

$$\mathbf{V}_k \equiv |k\rangle \tag{10.80}$$

In applying group theory to problems of scientific interest, one is interested in determining the effect an operator has on column vectors that describe the physical system. Let Ω_i be one of the operators belonging to a group G of order μ. When Ω_i operates on a specified basis vector the result is another vector. That is,

$$\Omega_i|l\rangle = |F\rangle \tag{10.81}$$

where $|F\rangle$ is a vector in the space that can be written as a linear combination of the basis vectors of the space. That is, in a vector space spanned by N vectors,

$$\Omega_i|l\rangle = \sum_{k=1}^{N} M_{kl}(\Omega_i)|k\rangle \tag{10.82}$$

The numbers $M_{kl}(\Omega_i)$ are the coefficients in the expansion of $\Omega_i|l\rangle$ in terms of the basis kets. Clearly, these coefficients can be identified as the elements of a matrix $\mathbf{M}(\Omega_i)$ that forms a matrix representation of $\Omega_i \in G$. In fact, using the orthonormality of the Dirac bra-ket, $\langle n|k\rangle = \delta_{nk}$, premultiplication of equation 10.82 by $\langle n|$, yields

$$\langle n|\Omega_i|l\rangle = \sum_{k=1}^{N} M_{kl}(\Omega_i)\langle n|k\rangle = M_{nl}(\Omega_i) \tag{10.83}$$

which is a previous description of a matrix element of an operator in terms of the Dirac bra-ket.

Example 10.17

As an example of deducing a matrix representation for the operators of a group in terms of basis vectors, we consider the dihedral group D_2 of operations that leave the orientation of the rectangle of Figure 10.4 invariant. The vector space that describes this orientation is the two-dimensional space spanned by the unit vectors

$$\text{(a)}\quad \mathbf{V}_1 = \begin{bmatrix} 1 \\ 0 \end{bmatrix} = \mathbf{i} \equiv |1\rangle \qquad \text{(b)}\quad \mathbf{V}_2 = \begin{bmatrix} 0 \\ 1 \end{bmatrix} = \mathbf{j} \equiv |2\rangle \tag{10.84}$$

As noted earlier, the four elements of D_2 are denoted by

e (leaves the rectangle unchanged)
x (rotates the rectangle around $\mathbf{V}_1 = \mathbf{i}$ by 180°)
y (rotates the rectangle around $\mathbf{V}_2 = \mathbf{j}$ by 180°)
z (rotates the rectangle around its center by 180°)

Therefore, for $k = 1, 2$,

$$\text{(a)}\quad e|k\rangle = |k\rangle \qquad \Rightarrow \qquad \text{(b)}\quad M_{nk}(e) = \langle n|e|k\rangle = \langle n|k\rangle = \delta_{nk} \tag{10.85}$$

That is, as expected, $\mathbf{M}(e) = \mathbf{1}$.

When the rectangle is rotated around $\mathbf{i}$, the result is $\mathbf{i} \to \mathbf{i}$, $\mathbf{j} \to -\mathbf{j}$. This is the transformation denoted by x. Thus

$$\text{(a)}\quad x|1\rangle = |1\rangle \qquad \text{(b)}\quad x|2\rangle = -|2\rangle \tag{10.86}$$

$$\Rightarrow \quad \mathbf{M}(x) = \begin{bmatrix} 1 & 0 \\ 0 & -1 \end{bmatrix} \tag{10.87}$$

By an identical analysis, we find that

$$\text{(a)}\quad y|1\rangle = -|1\rangle \qquad \text{(b)}\quad y|2\rangle = |2\rangle \tag{10.88}$$

$$\Rightarrow \quad \mathbf{M}(y) = \begin{bmatrix} -1 & 0 \\ 0 & 1 \end{bmatrix} \tag{10.89}$$

$$\text{(a)}\quad z|1\rangle = -|1\rangle \qquad \text{(b)}\quad z|2\rangle = -|2\rangle \tag{10.90}$$

$$\Rightarrow \quad \mathbf{M}(z) = \begin{bmatrix} -1 & 0 \\ 0 & -1 \end{bmatrix} \quad \square \tag{10.91}$$

Representations generated by basis functions

In Chapter 6 we saw that matrix representations of operators could also be defined in terms of integrals involving the operators operating on functions. That is, the Dirac bra-ket

$$\langle \alpha_i | \Omega | \alpha_j \rangle \equiv \int_a^b \alpha_i^*(x) \Omega \alpha_j(x)\, dx \tag{6.125}$$

can be viewed as the (i, j) element of the matrix for Ω.

As with vectors, basis functions can also be represented in the Dirac notation by a ket. For example, in the functions space spanned by Legendre polynomials, we denote

$$P_l(x) \equiv |l\rangle \tag{10.92}$$

Often, in scientific problems, transformations such as rotations and translations are made on the coordinate vector $\mathbf{r} = (x, y, z)$. For example, a rotation of the form

$$\mathbf{r}' = R_z(\phi)\mathbf{r} \tag{10.93}$$

$$\Rightarrow \quad \text{(a)} \quad x' = x \cos \phi + y \sin \phi \qquad \text{(b)} \quad y' = -x \sin \phi + y \cos \phi \qquad \text{(c)} \quad z' = z \tag{10.94}$$

while a translation along the x-axis by a fixed interval λ of the form

$$\mathbf{r}' = T_\lambda \mathbf{r} \tag{10.95}$$

$$\Rightarrow \quad \text{(a)} \quad x' = x + \lambda \qquad \text{(b)} \quad y' = y \qquad \text{(c)} \quad z' = z \tag{10.96}$$

In many situations, physically significant functions are invariant under such a transformation. Let $|l\rangle \equiv f_l(\mathbf{r})$ be such a function, and let

$$\mathbf{r}' = \Omega_i \mathbf{r} \tag{10.97}$$

be a transformation such that

$$f_l(\Omega_i \mathbf{r}) = g_l(\mathbf{r}') \tag{10.98}$$

That is, by transforming the coordinates, a different functional dependence on $\mathbf{r}'$ is obtained. For example, let

$$\Rightarrow \quad \text{(a)} \quad f(\mathbf{r}) = x^2 + y^2 + z^2 \qquad \text{(b)} \quad f(\mathbf{r}') = x'^2 + y'^2 + z'^2 \tag{10.99}$$

If Ω_i is the translation operator of equation 10.95 resulting in the coordinates of equations 10.96, then

$$f_l(T_\lambda \mathbf{r}) = g_l(\mathbf{r}') = (x' + \lambda)^2 + y'^2 + z'^2 = r'^2 + \lambda^2 + 2\lambda x' = f(\mathbf{r}') + \lambda^2 + 2\lambda x' \tag{10.99c}$$

That is, under translations, the functional form of the basis functions is changed.

If the physical functions are invariant under the transformation Ω_i, then

$$g_l(\mathbf{r}') = f_l(\Omega_i \mathbf{r}) = f_l(\mathbf{r}) \tag{10.100}$$

As will be seen later, there are different ways to describe an operator. For example, in a space spanned by column vectors, the operator can be written as a matrix that generates the desired transformations of the basis column vectors. In a function space, a particular combination of derivatives will generate the same transformation among the basis functions. In such a case, the matrix and the combination of derivatives will be different descriptions of the same abstract operator.

Let P_{Ω_i} be a description of an operator that operates on the functions $f_l(\mathbf{r})$ such that

$$g_l(\mathbf{r}') = P_{\Omega_i} f_l(\mathbf{r}') \tag{10.101}$$

Referring to equation 10.100, invariance under the transformation Ω_i requires that

$$P_{\Omega_i} f_l(\Omega_i \mathbf{r}) = f_l(\mathbf{r}) \tag{10.102a}$$

We now make the substitution $\mathbf{r}' = \Omega_i \mathbf{r}$, or $\mathbf{r} = \Omega_i^{-1}\mathbf{r}'$. That is, to obtain invariance of the basis functions, the operator P_{Ω_i} operating on the basis function $f_l(\mathbf{r})$ satisfies

$$P_{\Omega_i} f_l(\mathbf{r}') = f_l(\Omega_i^{-1}\mathbf{r}') \tag{10.102b}$$

This condition of invariance holds for any $\mathbf{r}'$, so we can drop the prime on the designation of the position vector.

Example 10.18

To illustrate the straightforward nature of the identity of equations 10.102, consider a function of one variable, $F(x)$. Let $P_{S(\phi)}$ be an operation that rotates $F(x)$ about the origin by an angle ϕ, as shown in Figure 10.10. This operation changes the orientation of $F(x)$ with respect to the axes, but does not change the function. That is, $F(x)$ is invariant under such a transformation.

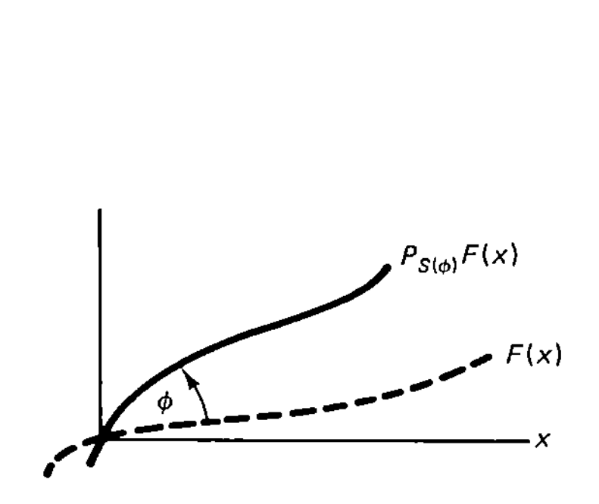

Figure 10.10 Rotation of $F(x)$.

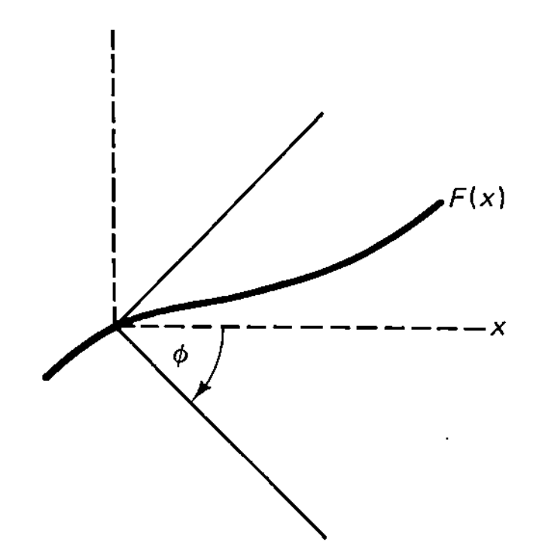

Figure Figure 10.11 Rotation of the axes.

This same change in orientation can be achieved by a rotation of the axes through an angle $-\phi$ as shown in Figure 10.11. As discussed in Chapter 8, the operator (matrix) for a rotation through ϕ is

$$\mathbf{S}(\phi) = \begin{bmatrix} \cos\phi & \sin\phi \\ -\sin\phi & \cos\phi \end{bmatrix} \tag{8.154}$$

and $\mathbf{S}(\phi)$ is an orthogonal transformation. Thus

$$\mathbf{S}(-\phi) = \begin{bmatrix} \cos\phi & -\sin\phi \\ \sin\phi & \cos\phi \end{bmatrix} = \tilde{\mathbf{S}}(\phi) = \mathbf{S}^{-1}(\phi) \tag{10.103}$$

$$\Rightarrow \quad P_{S(\phi)}F(x) = F[S(-\phi)x] = F[S^{-1}(\phi)x] \quad \square \tag{10.104}$$

Example 10.19

As a second example, a translation of $F(x)$ by a distance λ changes the orientation of the function with respect to the axes as shown in Figure 10.12, but does not change the function. The same reorientation can be achieved by translating the axes a distance $-\lambda$ as shown in Figure 10.13. Therefore,

$$P_{T(\lambda)}F(x) = F[T(-\lambda)x] \tag{10.105}$$

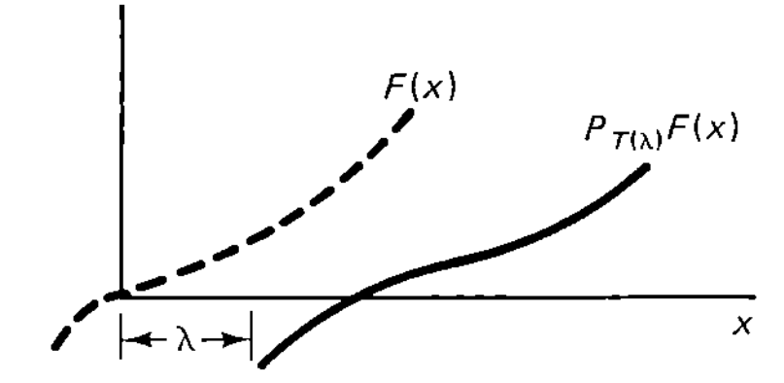

Figure 10.12 Translation of $F(x)$.

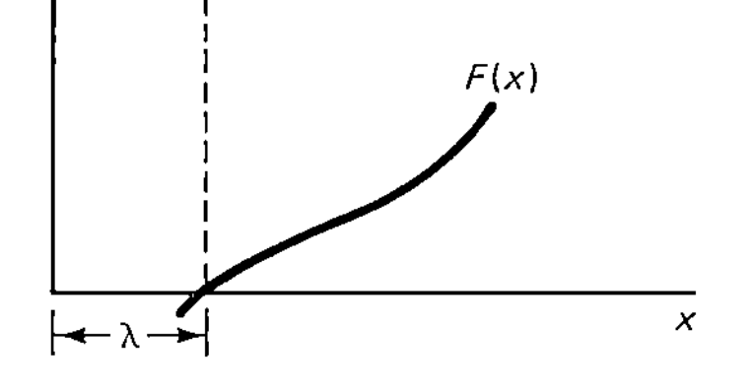

Figure Figure 10.13 Translation of the axes.

If the axes are translated by a distance λ by $T(\lambda)$, and then returned to their original configuration by a translation of $-\lambda$ by $T(-\lambda)$, clearly,

$$\text{(a)}\quad \mathbf{T}(\lambda)\mathbf{T}(-\lambda) = \mathbf{1} \quad \Rightarrow \quad \text{(b)}\quad \mathbf{T}(-\lambda) = \mathbf{T}^{-1}(\lambda) \tag{10.106}$$

$$\Rightarrow \quad P_{T(\lambda)}F(x) = F\left[T^{-1}(\lambda)x\right] \quad \square \tag{10.107}$$

To obtain the matrix representation of an operator that operates on the basis functions, we expand $f_l(\Omega_i^{-1}\mathbf{r})$ in terms of the basis functions $f_l(\mathbf{r})$. The coefficients of the expansion are the elements of the matrix representation of P_{Ω_i}. If N basis functions $|l\rangle = f_l(\mathbf{r})$ span the function space, then

$$P_{\Omega_i}f_l(\mathbf{r}) = f_l\left(\Omega_i^{-1}\mathbf{r}\right) = \sum_{k=1}^{N} M_{lk}(\Omega_i)f_k(\mathbf{r}) \tag{10.108}$$

Example 10.20

As an example, consider the six-dimensional space spaced by the functions

$$\begin{array}{llll} \text{(a)}\ f_1(\mathbf{r}) = x^2 & \text{(b)}\ f_2(\mathbf{r}) = y^2 & \text{(c)}\ f_3(\mathbf{r}) = z^2 \\ \text{(d)}\ f_4(\mathbf{r}) = xy & \text{(e)}\ f_5(\mathbf{r}) = yz & \text{(f)}\ f_6(\mathbf{r}) = xz \end{array} \tag{10.109}$$

For a rotation about the z-axis by 45°, we will determine the matrix representation of the operator that leaves the basis functions invariant. For simplicity, we will abbreviate the designation of the rotation matrix $\mathbf{R}_z(45°)$ by $\mathbf{R}$. The rotation matrix for this transformation, which is orthogonal, is

$$\text{(a)}\quad \mathbf{R} = \begin{bmatrix} \frac{1}{\sqrt{2}} & \frac{1}{\sqrt{2}} & 0 \\ -\frac{1}{\sqrt{2}} & \frac{1}{\sqrt{2}} & 0 \\ 0 & 0 & 1 \end{bmatrix} \quad \Rightarrow \quad \text{(b)}\quad \mathbf{R}^{-1} = \tilde{\mathbf{R}} = \begin{bmatrix} \frac{1}{\sqrt{2}} & -\frac{1}{\sqrt{2}} & 0 \\ \frac{1}{\sqrt{2}} & \frac{1}{\sqrt{2}} & 0 \\ 0 & 0 & 1 \end{bmatrix} \tag{10.110}$$

$$\Rightarrow \quad P_R f_1(\mathbf{r}) = f_1(\mathbf{R}^{-1}\mathbf{r}) \tag{10.111}$$

where

$$\mathbf{R}^{-1}\mathbf{r} = \begin{bmatrix} \frac{1}{\sqrt{2}}x - \frac{1}{\sqrt{2}}y \\ \frac{1}{\sqrt{2}}x + \frac{1}{\sqrt{2}}y \\ z \end{bmatrix} \tag{10.112}$$

Therefore, referring to equations 10.109,

$$P_R f_1(\mathbf{r}) = f_1(\mathbf{R}^{-1}\mathbf{r}) = \left(\frac{1}{\sqrt{2}}x - \frac{1}{\sqrt{2}}y\right)^2 = \frac{1}{2}f_1(\mathbf{r}) + \frac{1}{2}f_2(\mathbf{r}) - f_4(\mathbf{r}) \tag{10.113a}$$

$$P_R f_2(\mathbf{r}) = f_2(\mathbf{R}^{-1}\mathbf{r}) = \left(\frac{1}{\sqrt{2}}x + \frac{1}{\sqrt{2}}y\right)^2 = \frac{1}{2}f_1(\mathbf{r}) + \frac{1}{2}f_2(\mathbf{r}) + f_4(\mathbf{r}) \tag{10.113b}$$

$$P_R f_3(\mathbf{r}) = f_3(\mathbf{R}^{-1}\mathbf{r}) = z^2 = f_3(\mathbf{r}) \tag{10.113c}$$

This is the expected result since the rotation is about the z-axis.

$$P_R f_4(\mathbf{r}) = f_4(\mathbf{R}^{-1}\mathbf{r}) = \left(\frac{1}{\sqrt{2}}x + \frac{1}{\sqrt{2}}y\right)\left(\frac{1}{\sqrt{2}}x - \frac{1}{\sqrt{2}}y\right) = \frac{1}{2}f_1(\mathbf{r}) - \frac{1}{2}f_2(\mathbf{r}) \tag{10.113d}$$

$$P_R f_5(\mathbf{r}) = f_5(\mathbf{R}^{-1}\mathbf{r}) = \frac{1}{\sqrt{2}}xz + \frac{1}{\sqrt{2}}yz = \frac{1}{\sqrt{2}}f_6(\mathbf{r}) + \frac{1}{\sqrt{2}}f_5(\mathbf{r}) \tag{10.113e}$$

$$P_R f_6(\mathbf{r}) = f_6(\mathbf{R}^{-1}\mathbf{r}) = \frac{1}{\sqrt{2}}xz - \frac{1}{\sqrt{2}}yz = \frac{1}{\sqrt{2}}f_6(\mathbf{r}) - \frac{1}{\sqrt{2}}f_5(\mathbf{r}) \tag{10.113f}$$

Therefore, the marix representation of $\mathbf{R}_z(45°)$ in this function space is

$$\mathbf{M}[R_z(45°)] = \begin{bmatrix} -\frac{1}{2} & \frac{1}{2} & 0 & -1 & 0 & 0 \\ \frac{1}{2} & \frac{1}{2} & 0 & 1 & 0 & 0 \\ 0 & 0 & 1 & 0 & 0 & 0 \\ \frac{1}{2} & -\frac{1}{2} & 0 & 0 & 0 & 0 \\ 0 & 0 & 0 & 0 & \frac{1}{\sqrt{2}} & \frac{1}{\sqrt{2}} \\ 0 & 0 & 0 & 0 & -\frac{1}{\sqrt{2}} & \frac{1}{\sqrt{2}} \end{bmatrix} \quad \square \tag{10.114}$$

10.2 Theory and Application of Continuous Groups

A group containing elements that are generated by continuous parameters are groups of infinite order. As an example, rotations in two dimensions are parametrized by an angle θ, such that, in matrix form, a group element is represented by

$$\mathbf{R}(\theta) = \begin{bmatrix} \cos\theta & \sin\theta \\ -\sin\theta & \cos\theta \end{bmatrix} \tag{8.154}$$

Clearly, $\mathbf{R}(\theta)$ and $\mathbf{R}(\theta + d\theta)$ are two different elements of the group of rotations, and for $0 \leqslant \theta \leqslant 2\pi$, there are an infinite number of such elements. The allowed range of values of the parameters that define the group elements is called the *group space*.

Not all infinite groups are continuous groups. For example, the set of all integers under addition is an infinite group, but there are an infinite number of real numbers between two successive integers that are not elements of the group. That is, there is a discontinuity between the elements of the group. Since two elements of this group are not connected by the variation of a continuous parameter, the group of integers under addition is an infinite, but not a continuous group.

In a continuous group, elements such as $\mathbf{R}(\theta)$ and $\mathbf{R}(\theta + d\theta)$ that differ infinitesimally are called *adjacent elements*. In order that the elements form a group, there must always be one value of the parameters that defines the identity of the group. For example, referring to equation 8.154, $\mathbf{R}(0) = \mathbf{1}$ is the identity element for the group of two-dimensional rotations. Therefore, in a continuous group, there will always be a collection of elements adjacent to the identity. This group is known as the *infinitesimal group*. For the rotations, the infinitesimal group is $\mathbf{R}(d\theta)$.

In general terms, a continuous (and thus infinite) group whose elements are generated by analytic functions of a set of parameters over the group space is known as a *Lie* (pronounced "Lee") *group* (for Sophus Lie, the mathematician who made extensive contributions to the study of such groups). The group of rotations in two dimensions is such a group since its elements are defined in matrix form by analytic functions $\sin\theta$ and $\cos\theta$ over the group space $\theta \in [0, 2\pi]$.

Let the set of parameters $\lambda_1, \lambda_2, \ldots, \lambda_N \equiv \boldsymbol{\lambda}$ be the parameters that generate the elements of a Lie group. The elements of the group are denoted by $g(\boldsymbol{\lambda})$. For example, for the group of two-dimensional rotations, $g(\boldsymbol{\lambda}) = \mathbf{R}(\theta)$.

Let $\boldsymbol{\lambda}_0$ be the set of parameters that define the identity element as $e = g(\boldsymbol{\lambda}_0)$. The elements adjacent to the identity are of the form $g(\boldsymbol{\lambda}_0 + d\boldsymbol{\lambda})$. Since the group is a Lie group, the functional dependence of the elements on the parameters $\boldsymbol{\lambda}$ is analytic. Thus a Taylor expansion can be made, and we can write

$$g(\boldsymbol{\lambda}_0 + d\boldsymbol{\lambda}) = e + \boldsymbol{\gamma} \cdot d\boldsymbol{\lambda} \tag{10.115}$$

The quantities $\boldsymbol{\gamma}$ are called the generators of the infinitesmal group.

Algebra of a group

Many of the properties of infinite, continuous groups can be determined by studying the properties of the generators of the group. Since the order of such a group is infinite, while the number of generators is finite, the study of the properties of the generators becomes essential.

For example, $g(\boldsymbol{\lambda}) = \mathbf{R}(\theta)$ describes rotations in two dimensions through an angle θ. As described above, if a system is invariant under such rotations, then

$$P_R\psi_\alpha(\mathbf{r}) = \psi_\alpha(\mathbf{R}^{-1}\mathbf{r}) \tag{10.116}$$

Let H be a significant operator that describes the state of a system. For example, in quantum theory, dynamical quantities such as momentum, coordinates, energy, and so on, are described in terms of operators. In particular, the energy operator is called the *Hamiltonian*. The non-quantum-mechanical Hamiltonian function was introduced in Chapter 7 (equation 7.100). Let the eigenvalue equation be

$$H\psi_\alpha(\mathbf{r}) = \Lambda_\alpha\psi_\alpha(\mathbf{r}) \tag{10.117}$$

Operating on $H\psi_\alpha(\mathbf{r})$ with P_R, we obtain

$$P_R H\psi_\alpha(\mathbf{r}) = P_R\Lambda_\alpha\psi_\alpha(\mathbf{r}) = \Lambda_\alpha\psi_\alpha(\mathbf{R}^{-1}\mathbf{r}) \tag{10.118a}$$

But from equation 10.116,

$$\Lambda_\alpha\psi_\alpha(\mathbf{R}^{-1}\mathbf{r}) = H\psi_\alpha(\mathbf{R}^{-1}\mathbf{r}) = HP_R\psi_\alpha(\mathbf{r}) \tag{10.118b}$$

Therefore,

$$(P_R H - HP_R)\psi_\alpha(\mathbf{r}) = 0 \tag{10.119}$$

The quantity

$$(P_R H - HP_R) \equiv (P_R, H) \tag{10.120}$$

is the commutator of the two operators, and if the system is invariant under transformations generated by $\mathbf{R}$, that commutator is zero.

Since $g(\boldsymbol{\lambda})$ can be written in terms of its generator for any $\boldsymbol{\lambda}$, it is clear that if g commutes with H, then the generators also commute with H. Thus, rather than dealing with the commutators of an infinite number of elements $g(\boldsymbol{\lambda})$, it is sufficient to work with the commutators of a finite number of generators.

To define an algebra, we begin with a group element that is generated by a single generator γ_α. That is, g_α is obtained by taking all but the αth parameter to be constant so that

$$\text{(a)} \quad d\boldsymbol{\lambda} = (0, 0, \ldots, d\lambda_\alpha, 0, \ldots, 0) \quad \Rightarrow \quad \text{(b)} \quad g_\alpha \equiv e + \boldsymbol{\gamma} \cdot d\boldsymbol{\lambda} = e + \gamma_\alpha \, d\lambda_\alpha \tag{10.121}$$

For example, one possible rotation generated by the Euler angles ϕ, θ, and ψ is the one for which θ and ψ are constant, so that $d\theta = d\psi = 0$. Referring to equation 10.121b, consider

$$(e + \gamma_\alpha \, d\lambda_\alpha)(e - \gamma_\alpha \, d\lambda_\alpha) = e - \gamma_\alpha^2 (d\lambda_\alpha)^2 \tag{10.122}$$

Taking the square of the infinitesimal $(d\lambda_\alpha)^2 = 0$, we see that

$$\text{(a)} \quad (e + \gamma_\alpha \, d\lambda_\alpha)(e - \gamma_\alpha \, d\lambda_\alpha) = g_\alpha(e - \gamma_\alpha \, d\lambda_\alpha) = e \quad \Rightarrow \quad \text{(b)} \quad g_\alpha^{-1} = e - \gamma_\alpha \, d\lambda_\alpha \tag{10.123}$$

Let **U** be an operator that is significant in that its operation on a basis state (such as an eigenfunction of the Hamiltonian, for example) has a meaning in a problem. It is possible that **U** is not one of the group generators. However, one of the properties that characterizes the algebra of a group is that any such operator can be written as a linear combination of the generators. That is, the generators act as "basis" operators of an algebra in the same way that orthogonal vectors (such as $\mathbf{i}, \mathbf{j}, \mathbf{k}$ or $\mathbf{r}_0, \boldsymbol{\theta}_0, \boldsymbol{\phi}_0$) are the bases of a vector space, or orthogonal functions [$P_l(x)$, for example] serve as the bases of a function space. This means that any operator **U** that can operate on the bases can be written

$$\mathbf{U} = \sum_{\mu=1}^{M} a_\mu \gamma_\mu \tag{10.124}$$

Thus if the null operator is written in terms of the generators, the equation

$$\mathbf{0} = \sum_{\mu=1}^{M} a_\mu \gamma_\mu \tag{10.125}$$

$$\Rightarrow \quad a_\mu = 0 \tag{10.126}$$

for every value of μ.

Structure constants

One extremely important theorem about the structure of Lie groups is that the commutator of any two generators of a Lie group can be written as a linear combination of the group generators. That is,

$$(\gamma_\alpha, \gamma_\beta) = \sum_{\mu=1}^{M} C_{\alpha\beta}^{\mu} \gamma_\mu \tag{10.127}$$

where the $C_{\alpha\beta}^{\mu}$ are complex numbers called the *structure constants* of the Lie algebra.

We note from their definition that these structure constants are antisymmetric in the interchange of their lower indices.

$$C^{\mu}_{\alpha\beta} = -C^{\mu}_{\beta\alpha} \tag{10.128}$$

Any set of generators γ_μ that satisfy equations 10.124 and 10.127 form the basis operators for the algebra of a group.

The generators also satisfy the Jacobi identity

$$\big((\gamma_\alpha, \gamma_\beta), \gamma_\varepsilon\big) + \big((\gamma_\beta, \gamma_\varepsilon), \gamma_\alpha\big) + \big((\gamma_\varepsilon, \gamma_\alpha), \gamma_\beta\big) = 0 \tag{10.129a}$$

(See Problem 8 of Chapter 1 for the analogous identity involving vectors.) Written in terms of the structure constants, this becomes

$$\sum_\mu C^{\mu}_{\alpha\beta}(\gamma_\mu, \gamma_\varepsilon) + \sum_\mu C^{\mu}_{\beta\varepsilon}(\gamma_\mu, \gamma_\alpha) + \sum_\mu C^{\mu}_{\varepsilon\alpha}(\gamma_\mu, \gamma_\beta)$$

$$= \sum_{\mu,\rho} (C^{\mu}_{\alpha\beta}C^{\rho}_{\mu\varepsilon} + C^{\mu}_{\beta\varepsilon}C^{\rho}_{\mu\alpha} + C^{\mu}_{\varepsilon\alpha}C^{\rho}_{\mu\beta})\gamma_\rho = 0 \tag{10.129b}$$

Thus the Jacobi identity in terms of the structure constants is

$$\sum_\mu (C^{\mu}_{\alpha\beta}C^{\rho}_{\mu\varepsilon} + C^{\mu}_{\beta\varepsilon}C^{\rho}_{\mu\alpha} + C^{\mu}_{\varepsilon\alpha}C^{\rho}_{\mu\beta}) = 0 \tag{10.130}$$

The structure constants can be viewed as matrix elements in the form

$$C^{\mu}_{\alpha\beta} \equiv (C_\alpha)^{\mu}_{\beta} \tag{10.131}$$

That is, $C^{\mu}_{\alpha\beta}$ is the (μ, β) element of the matrix C_α. Then writing

$$\sum_\mu C^{\mu}_{\alpha\beta}C^{\rho}_{\mu\varepsilon} = \sum_\mu (C_\alpha)^{\mu}_{\beta}(-C_\varepsilon)^{\rho}_{\mu} = -[(-C_\alpha)(-C_\varepsilon)]^{\rho}_{\beta} \tag{10.132a}$$

$$\sum_\mu C^{\mu}_{\beta\varepsilon}C^{\rho}_{\mu\beta} = -\sum_\mu (-C_\varepsilon)^{\mu}_{\beta}(-C_\alpha)^{\rho}_{\mu} = -[(-C_\varepsilon)(-C_\alpha)]^{\rho}_{\beta} \tag{10.132b}$$

$$\sum_\mu C^{\mu}_{\varepsilon\alpha}C^{\rho}_{\mu\beta} = -\sum_\mu C^{\mu}_{\alpha\varepsilon}(-C_\mu)^{\rho}_{\beta} \tag{10.132c}$$

equation 10.130 becomes the matrix equation

$$(-C_\alpha, -C_\varepsilon)^{\rho}_{\beta} = -\sum_\mu C^{\mu}_{\alpha\varepsilon}(-C_\mu)^{\rho}_{\beta} \tag{10.133}$$

Viewed in this way, we see that the matrix $-C_\alpha$ has the same algebraic structure as the abstract generator γ_α. As such, the structure constants can be used as the elements of a matrix representation of the generators. For this reason, if two algebras have the same structure constants, they are said to be *equivalent representations* of the same algebra.

In general, the type of Lie group that has application in scientific problems is one that contains no invariant subgroups (except for the subgroup of the identity element). Such a group is called a *simple group*. Our discussion from this point on will focus on the properties of simple Lie groups.

In many applications, physically meaningful operators are identified with the generators of a group. Some of these quantities may commute with one another. Let $\{\mathbf{D}_a\}$ denote the set of generators of an algebra that commute with one another, and $\{\mathbf{N}_\alpha\}$ designate the remaining noncommuting generators. That is, the set of generators $\{\gamma_A\}$ is comprised of the sets $\{\mathbf{D}_a\}$ and $\{\mathbf{N}_\alpha\}$. These generators satisfy

$$\text{(a)} \quad (\mathbf{D}_a, \mathbf{D}_b) = \sum_k C_{ab}^k \mathbf{D}_k + \sum_\mu C_{ab}^\mu \mathbf{N}_\mu = 0 \qquad \text{(b)} \quad (\mathbf{D}_a, \mathbf{N}_\alpha) = \sum_k C_{a\alpha}^k \mathbf{D}_k + \sum_\mu C_{a\alpha}^\mu \mathbf{N}_\mu \neq 0$$

$$\text{(c)} \quad (\mathbf{N}_\alpha, \mathbf{N}_\beta) = \sum_k C_{\alpha\beta}^k \mathbf{D}_k + \sum_\mu C_{\alpha\beta}^\mu \mathbf{N}_\mu \neq 0 \tag{10.134}$$

Note that we are denoting the indices of the commuting generators by lowercase Latin letters, and those of the noncommuting generators by Greek letters. Capital Latin letters will be used if the index can be either lowercase Latin or Greek. We have intentionally not specified the range of values of the Latin and Greek indices. This will be discussed later.

Referring to equations 10.125 and 10.126, equation 10.134a yields

$$C_{ab}^k = C_{ab}^\mu = 0 \tag{10.135}$$

As shown in Chapter 8, matrices that commute can be simultaneously diagonalized (diagonalized by a single similarity transformation). As such, we can choose the matrix representations of each of the $\mathbf{D}_a$ to be diagonal. For that reason, the generators $\mathbf{D}$ are also referred to as the *diagonal generators*. The matrices representing the $\mathbf{N}_\alpha$ will be *nondiagonal*. (This is the reason for the designations $\mathbf{D}$ and $\mathbf{N}$.)

Since the structure constants can serve as the matrices representing these generators, the structure constants corresponding to the $\mathbf{D}_a$ will be diagonal. That is,

$$(C_a)_\alpha^\mu = v_a(\alpha)\delta_\alpha^\mu \tag{10.136a}$$

Similarly, since the Greek indices take on different values than the Latin indices, and because the matrix C_a must be diagonal,

$$(C_a)_\alpha^b = 0 \tag{10.136b}$$

$$\Rightarrow \quad (\mathbf{D}_a, \mathbf{N}_\alpha) = v_a(\alpha)\mathbf{N}_\alpha \tag{10.137}$$

We define a vector $\mathbf{v}(\alpha)$, called the *root vector*, the components of which are $v_a(\alpha)$.

The commutator $(\mathbf{N}_\alpha, \mathbf{N}_\beta)$ can be deduced from the Jacobi identity

$$\big((\mathbf{N}_\alpha, \mathbf{N}_\beta), \mathbf{D}_a\big) + \big((\mathbf{N}_\beta, \mathbf{D}_a), \mathbf{N}_\alpha\big) + \big((\mathbf{D}_a, \mathbf{N}_\alpha), \mathbf{N}_\beta\big) = 0 \tag{10.138}$$

Using equations 10.134 and 10.137, this becomes

$$\sum_\mu C_{\alpha\beta}^\mu(\mathbf{N}_\mu, \mathbf{D}_a) - v_a(\beta)(\mathbf{N}_\beta, \mathbf{N}_\alpha) + v_a(\alpha)(\mathbf{N}_\alpha, \mathbf{N}_\beta)$$

$$= \sum_\mu C_{\alpha\beta}^\mu v_a(\mu)\mathbf{N}_\mu + [v_a(\alpha) + v_a(\beta)]\left\{\sum_k C_{\alpha\beta}^k \mathbf{D}_k + \sum_\mu C_{\alpha\beta}^\mu \mathbf{N}_\mu\right\} = 0 \tag{10.139}$$

As noted in equations 10.125 and 10.126, since the sum over the generators is zero, the coefficient of each generator is zero. Thus

$$\text{(a)} \quad C_{\alpha\beta}^k[v_a(\alpha) + v_a(\beta)] = 0 \qquad \text{(b)} \quad C_{\alpha\beta}^\mu[v_a(\alpha) + v_a(\beta) + v_a(\mu)] = 0 \tag{10.140}$$

From equation 10.140a, we deduce that either

$$\text{(a)} \quad v_a(\beta) = -v_a(\alpha) \qquad \text{or} \qquad \text{(b)} \quad C_{\alpha\beta}^k = 0 \tag{10.141}$$

That is, the nonzero structure constants with two Greek subscripts and a Latin superscript are of the form

$$C^k_{\alpha,-\alpha} \equiv w^k(\alpha) \tag{10.142}$$

Similarly, using equation 10.140b, we see that either

$$\text{(a)} \quad v_a(\mu) = -[v_a(\alpha) + v_a(\beta)] \quad \text{or} \quad \text{(b)} \quad C^{\mu}_{\alpha\beta} = 0 \tag{10.143}$$

That is, the nonzero structure constants from this set are of the form

$$C^{-(\alpha+\beta)}_{\alpha\beta} \equiv A_{\alpha\beta} \tag{10.144}$$

By a somewhat involved analysis, it can be shown that the root vectors $\mathbf{v}(\alpha)$ and $\mathbf{w}(\alpha)$ can be taken to be the same. The reader interested in the details of the process is referred to C. Fronsdal's article "Group Theory," in *Elementary Particle Physics and Field Theory*, (New York: W.A. Benjamin, Inc., 1962). This equality requires that the roots $\mathbf{v}(\alpha)$ be normalized as

$$\sum_{\alpha} v_a(\alpha) v_b(\alpha) = \delta_{ab} \tag{10.145}$$

From that analysis, one can develop the following fundamental theorem about the root vectors:

1. If $\mathbf{v}(\boldsymbol{\alpha})$ and $\mathbf{v}(\boldsymbol{\beta})$ are two root vectors, the quantity

$$2\frac{\mathbf{v}(\boldsymbol{\alpha}) \cdot \mathbf{v}(\boldsymbol{\beta})}{\mathbf{v}(\boldsymbol{\alpha}) \cdot \mathbf{v}(\boldsymbol{\alpha})} \equiv \mathbf{N}_1 = \textbf{integer} \tag{10.146a}$$

If $\boldsymbol{\alpha}$ and $\boldsymbol{\beta}$ are interchanged, the theorem also states that

$$2\frac{\mathbf{v}(\boldsymbol{\alpha}) \cdot \mathbf{v}(\boldsymbol{\beta})}{\mathbf{v}(\boldsymbol{\beta}) \cdot \mathbf{v}(\boldsymbol{\beta})} \equiv \mathbf{N}_2 = \textbf{integer} \tag{10.146b}$$

$\mathbf{N}_2$ can be different from N_1.

2. If $\mathbf{v}(\boldsymbol{\alpha})$ and $\mathbf{v}(\boldsymbol{\beta})$ are two root vectors, the vector

$$\mathbf{v}(\boldsymbol{\alpha}) - \left[2\frac{\mathbf{v}(\boldsymbol{\alpha}) \cdot \mathbf{v}(\boldsymbol{\beta})}{\mathbf{v}(\boldsymbol{\alpha}) \cdot \mathbf{v}(\boldsymbol{\alpha})}\right]\mathbf{v}(\boldsymbol{\beta}) = \mathbf{v}(\boldsymbol{\alpha}) - \mathbf{N}_1\mathbf{v}(\boldsymbol{\beta}) \tag{10.146c}$$

is also a root.

As a corollary to part 1 of this theorem, let $\mathbf{v}(\alpha)$ and $c\mathbf{v}(\alpha)$ be roots. Thus both $2\frac{\mathbf{v}(\alpha) \cdot \mathbf{v}(\beta)}{\mathbf{v}(\beta) \cdot \mathbf{v}(\beta)}$ and $c\left[2\frac{\mathbf{v}(\alpha) \cdot \mathbf{v}(\beta)}{\mathbf{v}(\beta) \cdot \mathbf{v}(\beta)}\right]$ must be integers, which requires c to be an integer. Similarly, interchanging α and β, $2\frac{\mathbf{v}(\alpha) \cdot \mathbf{v}(\beta)}{\mathbf{v}(\alpha) \cdot \mathbf{v}(\alpha)}$ and $\frac{c}{c^2}\left[2\frac{\mathbf{v}(\alpha) \cdot \mathbf{v}(\beta)}{\mathbf{v}(\alpha) \cdot \mathbf{v}(\alpha)}\right]$ must be integers, which requires $1/c$ also to be an integer. $c = \pm 1$ are the only integers with inverses that are also integers. Thus if $\mathbf{v}(\alpha)$ is a root, $-\mathbf{v}(\alpha)$ is also a root and no other integer multiple of $\mathbf{v}(\alpha)$ can be a root. Thus we can identify

$$\mathbf{v}(-\alpha) = -\mathbf{v}(\alpha) \tag{10.147}$$

Therefore, for a rank 1 algebra, there are only two roots, $\mathbf{v}(\alpha) = \pm v(\alpha)$.

In summary, then, with $\mathbf{w}(\alpha) = \mathbf{v}(\alpha)$, the generators satisfy

$$\begin{gathered} \text{(a)} \quad (\mathbf{D}_a, \mathbf{D}_b) = 0 \qquad \text{(b)} \quad (\mathbf{D}_a, \mathbf{N}_\alpha) = v_a(\alpha)\mathbf{N}_\alpha \\ \text{(c)} \quad (\mathbf{N}_\alpha, \mathbf{N}_{-\alpha}) = \sum_{k=1}^{l} v_k(\alpha)\mathbf{D}_k = \mathbf{v}(\alpha) \cdot \mathbf{D} \qquad \text{(d)} \quad (\mathbf{N}_\alpha, \mathbf{N}_\beta) = A_{\alpha\beta}\mathbf{N}_{\alpha+\beta} \end{gathered} \tag{10.148}$$

Therefore, the labeling for the $\mathbf{N}_\alpha$ will be $\mathbf{N}_{\pm 1}, \mathbf{N}_{\pm 2}, \ldots$. The diagonal generators will be labeled $\mathbf{D}_1, \mathbf{D}_2, \ldots$.

From equation 10.148b we see that the number of distinct roots is the number of nondiagonal generators, and the rank of the group is the number of diagonal generators. Therefore, the total number of generators of the group is the number of roots plus the rank of the group.

Root diagrams

If we represent the root vectors on a vector diagram (called the *root diagram*), the rank of the group is the dimensionality of the root diagram. For an algebra of rank 1, the root vector has one component. Thus a vector diagram of such a root will be one-dimensional, with roots $\pm v$. For an algebra of rank 2 or higher the roots can be found using the fundamental theorem for the roots expressed in equations 10.146. For such an algebra, let the angle between $\mathbf{v}(\alpha)$ and $\mathbf{v}(\beta)$ be θ. Then, with N_1, N_2 integers

$$\text{(a)}\quad 2\frac{\mathbf{v}(\alpha)\cdot\mathbf{v}(\beta)}{\mathbf{v}(\beta)\cdot\mathbf{v}(\beta)} = 2\frac{|\mathbf{v}(\alpha)|}{|\mathbf{v}(\beta)|}\cos\theta = N_1 \qquad \text{(b)}\quad 2\frac{\mathbf{v}(\alpha)\cdot\mathbf{v}(\beta)}{\mathbf{v}(\alpha)\cdot\mathbf{v}(\alpha)} = 2\frac{|\mathbf{v}(\beta)|}{|\mathbf{v}(\alpha)|}\cos\theta = N_2 \tag{10.149}$$

Therefore, from the product of these we obtain

$$4\cos^2\theta = N_1 N_2 = \text{integer} \tag{10.150}$$

If we define $|\mathbf{v}(\alpha)| \geqslant |\mathbf{v}(\beta)|$, then from equations 10.149, $N_1 \geqslant N_2$. Since $\cos^2\theta \leqslant 1$, N_1N_2 cannot exceed 4, only a small set of angles and integers satisfy this equation. For example, if $N_1 = N_2 = 1$, $\theta = \pi/3$. Therefore, from equation 10.149a,

$$\frac{|\mathbf{v}(\alpha)|}{|\mathbf{v}(\beta)|} = 1 \tag{10.151}$$

In this way the entries in Table 10.1 can be determined for all possible solutions to equations 10.149 and 10.150.

TABLE 10.1 ANGLE AND INTEGER SOLUTIONS FOR ROOTS OF AN ALGEBRA

θ	(N_1, N_2)	$\frac{\lvert\mathbf{v}(\alpha)\rvert}{\lvert\mathbf{v}(\beta)\rvert}$	Name of Rank 2 Group	Simple
$\frac{\pi}{2}$	$(N_1, 0)$	—	O_4	No
$\frac{\pi}{3}$	(1, 1)	1	SU(3)	Yes
$\frac{\pi}{4}$	(2, 1)	$\sqrt{2}$	C_2	Yes
$\frac{\pi}{6}$	(3, 1)	$\sqrt{3}$	G_2	Yes
π	(4, 1)	2	$\theta = \pi$; These are 1 dimensional	No
π	(2, 2)	1		No

Weights of representations of Lie algebras

Let $|\mathbf{m},\mathbf{j}\rangle$ denote the basis states of a representation of an algebra. In this notation, $\mathbf{m}$ is the set of eigenvalues of the diagonal generators, and $\mathbf{j}$ is the set of all other labels. That is,

$$\mathbf{D}_a|\mathbf{m},\mathbf{j}\rangle = m_a|\mathbf{m},\mathbf{j}\rangle \tag{10.152}$$

The vector $\mathbf{m}$, with components m_a, is called the *weight* of the representation. Clearly, the number of components of the weight is the same as the number of components of the root vectors. The distinction between the roots and the weights is that the roots characterize a particular algebra, and the weights characterize a specific representation of an algebra.

Using the commutation relations expressed in equation 10.148b, consider

$$\mathbf{D}_a\mathbf{N}_\alpha|\mathbf{m},\mathbf{j}\rangle = \mathbf{N}_\alpha\mathbf{D}_a|\mathbf{m},\mathbf{j}\rangle + v_a(\alpha)\mathbf{N}_\alpha|\mathbf{m},\mathbf{j}\rangle = [m_a + v_a(\alpha)]\mathbf{N}_\alpha|\mathbf{m},\mathbf{j}\rangle \tag{10.153}$$

That is, if $|\mathbf{m},\mathbf{j}\rangle$ is a basis state of the representation with weight vector $\mathbf{m}$, then either $\mathbf{N}_\alpha|\mathbf{m},\mathbf{j}\rangle$ is also a basis state of the representation with weight $[\mathbf{m} + \mathbf{v}(\alpha)]$ or $\mathbf{N}_\alpha|\mathbf{m},\mathbf{j}\rangle = 0$. In this way, basis states of a given representation can be generated from one another, and if we know one weight of the representation, the other weights can be generated from the roots.

As shown in equation 10.153, $\mathbf{N}_\alpha$ creates a state with a higher weight, and by an identical analysis, $\mathbf{N}_{-\alpha}$ creates a lower-weight state. From this, the operators are often referred to as *raising and lowering operators*, and also as *creation and annihilation operators*.

The weights satisfy a theorem that is similar to the fundamental theorem for the roots.

If m is a weight and v(α) is a root, then

$$\text{(a)}\quad 2\left[\frac{\mathbf{m}\cdot\mathbf{v}(\boldsymbol{\alpha})}{\mathbf{v}(\boldsymbol{\alpha})\cdot\mathbf{v}(\boldsymbol{\alpha})}\right] = N = \textbf{integer}$$

$$\text{(b)}\quad \mathbf{m}' = \mathbf{m} - 2\left[\frac{\mathbf{m}\cdot\mathbf{v}(\boldsymbol{\alpha})}{\mathbf{v}(\boldsymbol{\alpha})\cdot\mathbf{v}(\boldsymbol{\alpha})}\right]\mathbf{v}(\boldsymbol{\alpha}) = \mathbf{m} - N\mathbf{v}(\boldsymbol{\alpha}) \textbf{ is a weight.} \tag{10.154}$$

For the details of the proof, the interested reader is again referred to C. Fronsdal's article "Group Theory," in *Elementary Particle Physics and Field Theory*, Brandeis Lecture Series (New York: W. A. Benjamin, Inc., 1962).

Clearly, if $\mathbf{m} = 0$, $\mathbf{m}' = 0$. Thus nonzero weights cannot be generated from $\mathbf{m} = 0$. Therefore, any weights not obtainable from equations 10.154 must be zero, and only the nonzero weights of a representation can be generated using this theorem. Referring to equation 10.154b, we see that a weight generated from $\mathbf{v}(\alpha)$ is the same as the weight obtained from $\mathbf{v}(-\alpha) = -\mathbf{v}(\alpha)$.

If $\mathbf{m}$ and $\mathbf{m}'$ are two weights of a representation, and if the first nonvanishing component of $\mathbf{m} - \mathbf{m}'$ is positive, $\mathbf{m}$ is said to be a higher weight than $\mathbf{m}$. There is a *highest* or *dominant* weight $\mathbf{M}$, that is higher than any other weight in the representation. If two irreducible representations have the same highest weight, they are equivalent representations (related by a similarity transformation).

It was shown by Cartan [see E. Cartan, *Complete Works* (Paris: Société Gauthier-Villars, 1952)] that for a simple group of rank l, there are l dominant weights $\mathbf{M}_1, \mathbf{M}_2, \ldots, \mathbf{M}_l$, which are called the *fundamental dominant weights*, such that the dominant weight $\mathbf{M}$ of any irreducible representation of the group can be written as a linear combination of these fundamental dominant weights. That is,

$$\mathbf{M} = \sum_{i=1}^{l} \lambda_i \mathbf{M}_i \tag{10.155}$$

The numbers λ_i are nonnegative integers. The l irreducible representations for which the $\mathbf{M}_i$ are the dominant weights are called the *fundamental irreducible representations* of the group. For example,

$$\text{(a)} \quad \lambda_1 = 1, \lambda_2 = \lambda_3 = \cdots = \lambda_l = 0 \quad \Rightarrow \quad \text{(b)} \quad \mathbf{M} = \mathbf{M}_1 \tag{10.156}$$

Thus a fundamental dominant weight is obtained by setting one of the λ_i to 1, and all others to 0.

Recall that $\mathbf{N}_\alpha$ is a raising operator, such that $\mathbf{N}_\alpha|\mathbf{m},\mathbf{j}\rangle$ has weight $\mathbf{m} + \mathbf{v}(\alpha)$. Since $\mathbf{M}$ is the highest weight of a representation, then $\mathbf{N}_\alpha|\mathbf{M},\mathbf{j}\rangle = 0$. Similarly, there will be a lowest weight $\mathbf{M}'$ such that $\mathbf{N}_{-\alpha}|\mathbf{M}',\mathbf{j}\rangle = 0$.

Let $|\mathbf{M},\mathbf{j}\rangle$ be the state with the highest weight. We define

$$\text{(a)} \quad \mathbf{N}_{-\alpha}|\mathbf{M},\mathbf{j}\rangle \equiv \Gamma|\mathbf{M} - \mathbf{v}(\alpha),\mathbf{j}\rangle \qquad \text{(b)} \quad \langle\mathbf{M},\mathbf{j}|\mathbf{N}_\alpha = \Gamma^*\langle\mathbf{M} - \mathbf{v}(\alpha),\mathbf{j}| \tag{10.157}$$

Consider

$$\langle\mathbf{M},\mathbf{j}|N_\alpha\mathbf{N}_{-\alpha} - \mathbf{N}_{-\alpha}\mathbf{N}_\alpha|\mathbf{M},\mathbf{j}\rangle = \langle\mathbf{M},\mathbf{j}|\mathbf{v}(\alpha)\cdot\mathbf{D}|\mathbf{M},\mathbf{j}\rangle = \mathbf{v}(\alpha)\cdot\mathbf{M} \tag{10.158}$$

But since $\mathbf{M}$ is the dominant weight,

$$\mathbf{N}_\alpha|\mathbf{M},\mathbf{j}\rangle = 0 \tag{10.159}$$

$$\Rightarrow \quad \langle\mathbf{M},\mathbf{j}|\mathbf{N}_\alpha\mathbf{N}_{-\alpha}|\mathbf{M},\mathbf{j}\rangle = |\Gamma|^2 = \mathbf{v}(\alpha)\cdot\mathbf{M} \tag{10.160}$$

$$\Rightarrow \quad \mathbf{N}_{-\alpha}|\mathbf{M},\mathbf{j}\rangle \equiv \pm[\mathbf{v}(\alpha)\cdot\mathbf{M}]^{1/2}|\mathbf{M} - \mathbf{v}(\alpha),\mathbf{j}\rangle \tag{10.161}$$

Similarly, as the reader will show in Problem 22, if $\mathbf{M}'$ is the lowest weight of the representation,

$$\mathbf{N}_\alpha|\mathbf{M}',\mathbf{j}\rangle \equiv \pm[-\mathbf{v}(\alpha)\cdot\mathbf{M}']^{1/2}|\mathbf{M}' + \mathbf{v}(\alpha),\mathbf{j}\rangle \tag{10.162}$$

Casimir operators

We see from this analysis that $\mathbf{N}_\alpha\mathbf{N}_{-\alpha}$ is a diagonal operator, although it is not a generator. In general, $\mathbf{N}_\alpha\mathbf{N}_{-\alpha}$ does not commute with all the generators of the group. It can be shown that the quantity

$$\mathbf{Z} \equiv \sum_A \gamma_A\gamma_A = \sum_k \mathbf{D}_k\mathbf{D}_k + \sum_\mu \mathbf{N}_\mu\mathbf{N}_\mu \tag{10.163}$$

commutes with each γ_A. The reader interested in the details of this proof is referred to the text by M. Hamermesh, *Group Theory and Its Applications to Physical Problems* (Reading, Mass.: Addison-Wesley Publishing Company, 1962, p. 317). An operator such as $\mathbf{Z}$ that commutes with all the generators of an algebra is called a *Casimir operator*.

SO(2)

A group that is generated by a single generator (a one-parameter group) has a rank 0 algebra. One description of this group is in terms of the rotations in two dimensions about a point, as expressed in equation 8.154. We note that for any value of θ, the group element satisfies the property that its determinant is 1. That is, the element is said to be unimodular, which means that it has a modulus 1. The Lie groups with unimodular elements are called *special* and the name of the group contains the letter S to denote this.

It is also noted that the matrix of equation 8.154 is orthogonal, describing a transformation (rotation) in two dimensions. With these properties, the group of elements as defined by equation 8.153 is called the *special orthogonal group* in two dimensions, SO(2). It is straightforward to show that these matrices form a group.

1. *Closure.* From the matrix of equation 8.154 , it is easy to see that

$$\mathbf{R}(\theta_2)\mathbf{R}(\theta_1) = \mathbf{R}(\theta_1 + \theta_2) \tag{10.164}$$

It is possible that $\theta_1 + \theta_2$ might be outside the parameter space $0 \leqslant \theta_1 + \theta_2 \leqslant 2\pi$. Thus we define $\mathbf{R}(\theta \pm 2\pi) = \mathbf{R}(\theta)$.

2. *Associativity.* Clearly, from the closure property,

$$[\mathbf{R}(\theta_1)\mathbf{R}(\theta_2)]\mathbf{R}(\theta_3) = \mathbf{R}(\theta_1)[\mathbf{R}(\theta_2)\mathbf{R}(\theta_3)] = \mathbf{R}(\theta_1 + \theta_2 + \theta_3) \tag{10.165}$$

3. *Existence of an identity element.* Letting $\theta_2 = 0$ in equation 10.164, we have

$$\mathbf{R}(0)\mathbf{R}(\theta_1) = \mathbf{R}(\theta_1) \tag{10.166}$$

$$\Rightarrow \quad \mathbf{R}(0) = \mathbf{1} \tag{10.167}$$

4. *Existence of an inverse element.* If we set $\theta_2 = -\theta_1$ in equation 10.164, we see that

$$\mathbf{R}(-\theta_1)\mathbf{R}(\theta_1) = \mathbf{R}(0) = \mathbf{1} \tag{10.168}$$

$$\Rightarrow \quad \mathbf{R}^{-1}(\theta) = \mathbf{R}(-\theta) \tag{10.103}$$

To obtain the generator of SO(2), we look for the element that is adjacent to the identity element. This is obtained by taking the angle to be infinitesimally different from 0. That is,

$$\mathbf{R}(d\theta) = \mathbf{1} + \boldsymbol{\gamma}\, d\theta \tag{10.169}$$

Because SO(2) is a Lie group, $\mathbf{R}(\theta + d\theta)$ can be expanded in a Taylor series. That is,

$$\mathbf{R}(\theta + d\theta) = \mathbf{R}(\theta) + \frac{d\mathbf{R}}{d\theta}\, d\theta \tag{10.170a}$$

Using equations 10.164 and 10.169, we can also write

$$\mathbf{R}(\theta + d\theta) = \mathbf{R}(\theta)\mathbf{R}(d\theta) = \mathbf{R}(\theta) + \mathbf{R}(\theta)\boldsymbol{\gamma}\, d\theta \tag{10.170b}$$

$$\Rightarrow \quad \frac{d\mathbf{R}}{d\theta} = \mathbf{R}(\theta)\boldsymbol{\gamma}\, d\theta \tag{10.171}$$

With $\mathbf{R}(0) = \mathbf{1}$, this has the straightforward solution

$$\mathbf{R}(\theta) = \mathbf{R}(0)e^{\boldsymbol{\gamma}\theta} = e^{\boldsymbol{\gamma}\theta} \tag{10.172}$$

This development does not depend on a particular representation of the rotation operator. Therefore, all representations of the group of rotations in two dimensions can be described by the exponential of equation 10.172b.

It is customary to define the generator in the form

$$\text{(a)} \quad \mathbf{L} \equiv i\boldsymbol{\gamma} \qquad \Rightarrow \qquad \text{(b)} \quad \mathbf{R}(\theta) = e^{-i\mathbf{L}\theta} \tag{10.173}$$

To determine the two-dimensional representation of $\mathbf{L}$, we return to equation 10.169. With $\cos(d\theta) = 1$, and $\sin(d\theta) = d\theta$, we have

$$\mathbf{R}(d\theta) = \begin{bmatrix} 1 & d\theta \\ -d\theta & 1 \end{bmatrix} = \mathbf{1} + \begin{bmatrix} 0 & 1 \\ -1 & 0 \end{bmatrix} d\theta \tag{10.174}$$

$$\Rightarrow \quad \text{(a)} \quad \boldsymbol{\gamma} = \begin{bmatrix} 0 & 1 \\ -1 & 0 \end{bmatrix} \quad \text{or} \quad \text{(b)} \quad \mathbf{L} = \begin{bmatrix} 0 & -i \\ i & 0 \end{bmatrix} = \boldsymbol{\sigma}_y \tag{10.175}$$

where $\boldsymbol{\sigma}_y$ is the y component of the Pauli spin matrices introduced in Chapter 8 (equations 8.67).

It is straightforward to show that this representation of $\mathbf{L}$ satisfies

$$\text{(a)} \quad \mathbf{L}^{2k} = (-i)^k \mathbf{1} \qquad \text{(b)} \quad \Rightarrow \quad \mathbf{L}^{2k+1} = (-i)^k \mathbf{L} \tag{10.176}$$

Therefore, in this representation,

$$\mathbf{R}(\theta) = e^{-i\mathbf{L}\theta} = \mathbf{1} - i\mathbf{L}\theta - \mathbf{1}\frac{\theta^2}{2!} + i\mathbf{L}\frac{\theta^3}{3!} + \mathbf{1}\frac{\theta^4}{4!} + \cdots = \mathbf{1}\cos\theta - i\mathbf{L}\sin\theta \tag{10.177}$$

Referring to equation 10.164, it is clear that the elements of SO(2) commute with one another. That is, SO(2) is an Abelian group. Therefore, all the irreducible representations of SO(2) are one-dimensional.

To deduce a general representation of $\mathbf{L}$, we define an eigenstate of $\mathbf{L}$ to be $|l\rangle$ such that

$$\mathbf{L}|l\rangle = \lambda_l |l\rangle \tag{10.178}$$

Using equation 10.175b, it is straightforward to determine from the secular equation that in the two-dimensional representation, the eigenvalues of $\mathbf{L}$ are $\lambda_l = \pm 1$. Then in this representation,

$$\mathbf{R}(\theta)|l\rangle = \cos\theta \mathbf{1}|l\rangle - i\sin\theta \mathbf{L}|l\rangle = (\cos\theta \mp i\sin\theta)|l\rangle = e^{-i\lambda_l\theta}|l\rangle \tag{10.179}$$

Thus the representation of equation 8.154 can be written

$$\mathbf{R}(\theta) = \begin{bmatrix} e^{-i\theta} & 0 \\ 0 & e^{i\theta} \end{bmatrix} \tag{10.180}$$

In other representations, the eigenvalues of $\mathbf{L}$ are not ± 1. However, in any representation,

$$\mathbf{R}(\theta)|l\rangle = e^{-i\mathbf{L}\theta}|l\rangle = e^{-i\lambda_l\theta}|l\rangle \tag{10.181}$$

Using the property that rotations must be single valued,

$$\mathbf{R}(\theta + 2\pi)|l\rangle = e^{-i\lambda_l(\theta+2\pi)}|l\rangle = e^{-i\lambda_l\theta}|l\rangle \tag{10.182}$$

which requires that $e^{-i\lambda_l 2\pi} = 1$, or that $\lambda_l = l$ is an integer. Thus

$$\mathbf{L}|l\rangle = l|l\rangle \tag{10.183}$$

Comparing equation 10.179 with $\lambda_l = l$ to equation 10.180, we see that for the two-dimensional representation, $l = \pm 1$.

SO(3)

The extension of rotations into three dimensions was discussed in Chapter 8 in terms of the Euler angles. The matrices of this group, being unimodular and orthogonal, define the group SO(3). The three Euler angles are the group parameters. A rotation about one of the axes in three-dimensional space is exactly what SO(2) describes. Thus SO(2) is a subgroup of SO(3).

For example, in three dimensions, the rotation matrix that describes a rotation about the z-axis by an angle ϕ is

$$\mathbf{R}_z(\phi) = \begin{bmatrix} \cos\phi & \sin\phi & 0 \\ -\sin\phi & \cos\phi & 0 \\ 0 & 0 & 1 \end{bmatrix} \tag{10.184}$$

Therefore, the three-dimensional representation of the generator of this transformation is found from

$$\mathbf{R}_z(d\phi) \equiv \mathbf{1} + \mathbf{B}_z\, d\phi \tag{10.185}$$

Letting the angle in equation 10.184 be $d\phi$, and using the small-angle values of $\sin\phi$ and $\cos\phi$, we obtain

$$\mathbf{B}_z = \begin{bmatrix} 0 & 1 & 0 \\ -1 & 0 & 0 \\ 0 & 0 & 0 \end{bmatrix} \tag{10.186}$$

or defining $\mathbf{J}_z \equiv i\mathbf{B}_z$, we have

$$\mathbf{J}_z = \begin{bmatrix} 0 & i & 0 \\ -i & 0 & 0 \\ 0 & 0 & 0 \end{bmatrix} \tag{10.187}$$

Using the same analysis that leads to equation 10.172b, it is straightforward to show that

$$\mathbf{R}_z(\phi) = e^{i\mathbf{J}_z\phi} \tag{10.188}$$

Using analogous definitions of $\mathbf{J}_x$ and $\mathbf{J}_y$ in terms of $\mathbf{R}_x(\phi)$ and $\mathbf{R}_y(\phi)$ (see equation 8.158b, for example), the generators of rotations about the x and y axes are

$$\text{(a)}\quad \mathbf{J}_x = \begin{bmatrix} 0 & 0 & 0 \\ 0 & 0 & i \\ 0 & -i & 0 \end{bmatrix} \qquad \text{(b)}\quad \mathbf{J}_y = \begin{bmatrix} 0 & 0 & i \\ 0 & 0 & 0 \\ -i & 0 & 0 \end{bmatrix} \tag{10.189}$$

Using the set of Euler angles specified in Chapter 8, Figures 8.6, 8.7, and 8.8 and equations 8.158 as the group parameters, the elements of SO(3) are

$$\mathbf{R}(\phi,\theta,\psi) = \mathbf{R}_z(\psi)\mathbf{R}_x(\theta)\mathbf{R}_z(\phi) = e^{i\mathbf{J}_z\psi}e^{-i\mathbf{J}_x\theta}e^{-i\mathbf{J}_z\phi} \tag{10.190}$$

Relabeling the generators $\mathbf{J}_1$, $\mathbf{J}_2$, $\mathbf{J}_3$, it is straightforward to show that they satisfy

$$(\mathbf{J}_r, \mathbf{J}_s) = i\sum_{t=1}^{3} \varepsilon_{rst}\mathbf{J}_t \tag{10.191}$$

ε_{rst} is the Levi–Civita symbol. With

$$\mathbf{J}^2 = \mathbf{J}_1^2 + \mathbf{J}_2^2 + \mathbf{J}_3^2 \tag{10.192}$$

the generators also satisfy

$$(\mathbf{J}^2, \mathbf{J}_r) = 0 \qquad \text{for } r = 1,2,3 \tag{10.193}$$

Since $\mathbf{J}^2$ commutes with all generators of SO(3) (and therefore all the elements of the group), it is a Casimir operator of the group.

As illustrated by example in Chapter 8, Problem 19, if $\mathbf{A}$ and $\mathbf{B}$ are noncommuting operators under multiplication, then

$$e^{\mathbf{A}}e^{\mathbf{B}} \neq e^{\mathbf{A}+\mathbf{B}} \tag{10.194}$$

$$\Rightarrow \quad \mathbf{R}(\phi, \theta, \psi) = e^{-i\mathbf{J}_3\psi}e^{i\mathbf{J}_1\theta}e^{-i\mathbf{J}_3\phi} \neq e^{-i(\mathbf{J}_3\psi + i\mathbf{J}_1\theta + i\mathbf{J}_3\phi)} \tag{10.195}$$

If two operators $\mathbf{A}$ and $\mathbf{B}$ commute, then

$$e^{\mathbf{A}}e^{\mathbf{B}} = e^{\mathbf{B}}e^{\mathbf{A}} = e^{(\mathbf{A}+\mathbf{B})} \tag{10.196}$$

Since $\mathbf{J}_r$ commutes with itself,

$$e^{i\mathbf{J}_r\alpha}e^{-i\mathbf{J}_r\alpha} = e^{(i\mathbf{J}_r\alpha - i\mathbf{J}_r\alpha)} = \mathbf{1} \tag{10.197}$$

$$\Rightarrow \quad \left(e^{-i\mathbf{J}_r\alpha}\right)^{-1} = e^{i\mathbf{J}_r\alpha} \tag{10.198}$$

for each $\mathbf{J}_r$. Therefore, it is straightforward to see that

$$\mathbf{R}^{-1}(\phi, \theta, \psi) = e^{i\mathbf{J}_3\psi}e^{i\mathbf{J}_1\theta}e^{i\mathbf{J}_3\phi} \tag{10.199}$$

As described in Chapter 8, if two matrices commute, they have the same eigenvectors and both can be diagonalized by the same similarity transformation. Thus in SO(3), $\mathbf{J}^2$ and one other generator $\mathbf{J}_r$ can be diagonalized simultaneously. By convention, $\mathbf{J}_3$ is chosen as the diagonal generator. Therefore, the bases of the space are the eigenvectors of $\mathbf{J}_3$, which are also eigenvectors of $\mathbf{J}^2$. Of course, $\mathbf{J}^2$ is not a generator of SO(3). Thus $\mathbf{J}_3$ is the single diagonal generator, which means that SO(3) is a group of rank 1.

In addition to describing the nondiagonal generators by $\mathbf{J}_1$ and $\mathbf{J}_2$, alternative forms of the generators are defined from these by

$$\mathbf{J}_\pm \equiv \mathbf{J}_1 \pm i\mathbf{J}_2 \tag{10.200}$$

These are the operators specific to SO(3) that were designated $\mathbf{N}_{\pm\alpha} = \mathbf{N}_{\pm 1}$ in general terms in equation 10.148c.

Using equations 10.189, we note that

$$\mathbf{J}_- = \mathbf{J}_+^\dagger \tag{10.201}$$

As the reader will show in Problem 24, the generators satisfy the following commutation relations:

$$\text{(a)} \quad (\mathbf{J}_3, \mathbf{J}_\pm) = \pm\mathbf{J}_\pm \qquad \text{(b)} \quad (\mathbf{J}_+, \mathbf{J}_-) = 2\mathbf{J}_3 \tag{10.202}$$

Using

$$\text{(a)} \quad \mathbf{J}_1 = \tfrac{1}{2}(\mathbf{J}_+ + \mathbf{J}_-) \qquad \text{(b)} \quad \mathbf{J}_2 = -\tfrac{1}{2}i(\mathbf{J}_+ - \mathbf{J}_-) \tag{10.203}$$

and equation 10.202b, it is straightforward to show that

$$\mathbf{J}^2 = \mathbf{J}_3^2 - \mathbf{J}_3 + \mathbf{J}_+\mathbf{J}_- = \mathbf{J}_3^2 + \mathbf{J}_3 + \mathbf{J}_-\mathbf{J}_+ \tag{10.204}$$

We define the eigenstates in any representation from the diagonal generators. Thus, as we did in SO(2), we set

$$\mathbf{J}_3|m\rangle = \lambda_m|m\rangle \tag{10.205a}$$

For now, we designate the states by one index m. Since both $\mathbf{J}_3$ and $\mathbf{J}^2$ are diagonal operators, we expect a second index, corresponding to eigenvalues of $\mathbf{J}^2$ to also specify the basis states.

Using equation 10.188 we obtain

$$\mathbf{R}_z(\phi)|m\rangle = e^{-i\mathbf{J}_3\phi}|m\rangle = e^{-i\lambda_m\phi}|m\rangle \tag{10.205b}$$

Since $\mathbf{R}_z(\phi)$ is single-valued, $\mathbf{R}_z(\phi + 2\pi) = \mathbf{R}_z(\phi)$ requires that $\lambda_m = m$ be an integer. Thus

$$\mathbf{J}_3|m\rangle = m|m\rangle \tag{10.206}$$

To see that $\mathbf{J}_+$ is the raising or creation operator, we use equation 10.202a and consider

$$\mathbf{J}_3\mathbf{J}_+ = (\mathbf{J}_3, \mathbf{J}_+) + \mathbf{J}_+\mathbf{J}_3 = \mathbf{J}_+ + \mathbf{J}_+\mathbf{J}_3 \tag{10.207}$$

$$\Rightarrow \quad \mathbf{J}_3\mathbf{J}_+|m\rangle = \mathbf{J}_+|m\rangle + \mathbf{J}_+\mathbf{J}_3|m\rangle = (m+1)\mathbf{J}_+|m\rangle \tag{10.208}$$

We see that either $\mathbf{J}_+|m\rangle$ is an eigenvector of $\mathbf{J}_3$ with eigenvalue $(m+1)$ or $\mathbf{J}_+|m\rangle = 0$. Since

$$\mathbf{J}_3|m+1\rangle = (m+1)|m+1\rangle \tag{10.209}$$

we see that if $\mathbf{J}_+|m\rangle \neq 0$,

$$\mathbf{J}_+|m\rangle = C_m^+|m+1\rangle \tag{10.210a}$$

As the reader will show in Problem 25, $\mathbf{J}_-$ is a lowering operator. That is, either $\mathbf{J}_-|m\rangle = 0$ or

$$\mathbf{J}_-|m\rangle = C_m^-|m-1\rangle \tag{10.210b}$$

The constants $C_m^{\pm}$ are found using equation 10.202b. Consider

$$(\mathbf{J}_+\mathbf{J}_- - \mathbf{J}_-\mathbf{J}_+)|m\rangle = 2\mathbf{J}_3|m\rangle \tag{10.211}$$

$$\Rightarrow \quad C_{m-1}^+C_m^- - C_{m+1}^-C_m^+ = 2m \tag{10.212}$$

Since $\mathbf{J}_- = \mathbf{J}_+^\dagger$, the coefficients satisfy

$$C_m^- = (C_{m-1}^+)^* \tag{10.213}$$

Therefore, equation 10.212 becomes

$$|C_{m-1}^+|^2 - |C_m^+|^2 = 2m \tag{10.214}$$

The simplest (polynomial) solution to this difference equation is

$$|C_m^+|^2 = \alpha - m(m+1) \tag{10.215}$$

The constant α cannot be determined from the difference equation since it cancels when two adjacent coefficients are subtracted.

As noted above, there will be "highest" and "lowest" weight eigenstates, defined by

$$\text{(a)} \quad \mathbf{J}_+|\text{highest}\rangle = 0 \qquad \text{and} \qquad \text{(b)} \quad \mathbf{J}_-|\text{lowest}\rangle = 0 \tag{10.216}$$

Let $|j\rangle$ designate the highest state. Then

$$\mathbf{J}_+|j\rangle = C_j^+|j\rangle = 0 \tag{10.217}$$

which requires that $|C_j^+|^2 = \alpha - j(j+1) = 0$. Thus

$$\alpha = j(j+1) \tag{10.218}$$

$$\Rightarrow \quad |C_m^+|^2 = j(j+1) - m(m+1) \tag{10.219a}$$

and from equation 10.213,

$$|C_m^-|^2 = j(j+1) - m(m-1) \tag{10.219b}$$

Referring to equation 10.204, for the state $m = j$, we have

$$\mathbf{J}^2|j\rangle = (\mathbf{J}_3^2 + \mathbf{J}_3 + \mathbf{J}_-\mathbf{J}_+)|j\rangle = (j^2 + j + 0)|j\rangle = j(j+1)|j\rangle \tag{10.220}$$

To see what the label j signifies, we take $m < j$ and consider

$$\mathbf{J}^2|m\rangle = (\mathbf{J}_3^2 + \mathbf{J}_3 + \mathbf{J}_-\mathbf{J}_+)|m\rangle = (m^2 + m + |C_m^+|^2)|m\rangle = j(j+1)|m\rangle \tag{10.221}$$

That is, j designates the eigenvalue of $\mathbf{J}^2$. That eigenvalue is $j(j+1)$ for all states of the representation.

Let $|\eta\rangle$ designate the lowest-index eigenstate, in that $\mathbf{J}_-|\eta\rangle = 0$. Then, from equation 10.204,

$$\mathbf{J}_+\mathbf{J}_-|\eta\rangle = (\mathbf{J}^2 - \mathbf{J}_3^2 + \mathbf{J}_3)|\eta\rangle = [j(j+1) - \eta^2 + \eta]|\eta\rangle = 0 \tag{10.222a}$$

$$\Rightarrow \quad \eta^2 - \eta = j^2 + j \tag{10.222b}$$

$$\Rightarrow \quad \text{(a)} \quad \eta = -j \qquad \text{or} \qquad \text{(b)} \quad \eta = j + 1 \tag{10.223}$$

Clearly, $\eta = j + 1$ has no meaning since the lowest-index eigenstate cannot have a larger j-value than the highest-index state. Therefore, the lowest eigenstate corresponds to $\eta = -j$. From this we see that the states of the jth representation are labeled by values of m that range from $-j$ to j. Therefore, the representation characterized by j is $(2j + 1)$-dimensional.

Because the states are eigenstates of both $\mathbf{J}^2$ and $\mathbf{J}_3$, they must be designated by both eigenvalues j and m. Thus the notation for a state in a particular representation will be $|m, j\rangle$. We will sometimes omit one of the pair when it is convenient to do so and does not cause confusion.

The allowed values of j are obtained from the fact that the dimension of the representation designated by j must be an integer > 0. That is, $2j + 1 = 1, 2, 3, 4, \ldots$ or $j = 0, \frac{1}{2}, 1, \frac{3}{2}, \ldots$.

For $j = 0$, there is only one basis state, $|m = 0\rangle$, such that

$$\text{(a)} \quad \mathbf{J}_\pm|0\rangle = 0|0\rangle \qquad \text{and} \qquad \text{(b)} \quad \mathbf{J}_3|0\rangle = 0|0\rangle \tag{10.224}$$

Thus in this representation, the generators are represented by one-dimensional null matrices.

The two-dimensional representation is characterized by $2j + 1 = 2$, or $j = \frac{1}{2}$. Thus the bases of this representation will be $|m = \frac{1}{2}\rangle$ and $|m = -\frac{1}{2}\rangle$. With m and m' taking on these values, the matrices of the generators of this representation are

$$\text{(a)} \quad \langle m'|\mathbf{J}^2|m\rangle = \tfrac{1}{2}(\tfrac{1}{2} + 1)\delta_{mm'} \qquad \text{or} \qquad \text{(b)} \quad \mathbf{J}^2 = \frac{3}{4}\begin{bmatrix} 1 & 0 \\ 0 & 1 \end{bmatrix} \tag{10.225}$$

The identity

$$\text{(a)}\quad \langle m'|\mathbf{J}_3|m\rangle = m\delta_{mm'} \qquad \Rightarrow \qquad \text{(b)}\quad \mathbf{J}_3 = \frac{1}{2}\begin{bmatrix}1 & 0\\ 0 & -1\end{bmatrix} = \tfrac{1}{2}\boldsymbol{\sigma}_z \tag{10.226}$$

Since $|\tfrac{1}{2}\rangle$ is the highest index state of the representation, $\mathbf{J}_+|\tfrac{1}{2}\rangle = 0$. Therefore, the matrix elements of the raising operator are

$$\langle \pm\tfrac{1}{2}|\mathbf{J}_+|\tfrac{1}{2}\rangle = 0 \tag{10.227a}$$

Because $\mathbf{J}_+|-\tfrac{1}{2}\rangle \sim |\tfrac{1}{2}\rangle$ and $\langle -\tfrac{1}{2}|\tfrac{1}{2}\rangle = 0$,

$$\langle -\tfrac{1}{2}|\mathbf{J}_+|-\tfrac{1}{2}\rangle \sim \langle -\tfrac{1}{2}|\tfrac{1}{2}\rangle = 0 \tag{10.227b}$$

Using equations 10.219, the only nonzero element is

$$\langle\tfrac{1}{2}|\mathbf{J}_+|-\tfrac{1}{2}\rangle = \sqrt{\tfrac{1}{2}(\tfrac{1}{2}+1) - (\tfrac{1}{2})(-\tfrac{1}{2}+1)} = 1 \tag{10.227c}$$

so labeling the rows of $\mathbf{J}_\pm$ by m and the columns by m' we obtain

$$\begin{matrix} m' = & \tfrac{1}{2} & -\tfrac{1}{2} \end{matrix}$$
$$\mathbf{J}_+ = \begin{bmatrix}0 & 1\\ 0 & 0\end{bmatrix}\begin{matrix} m=\tfrac{1}{2}\\ m=-\tfrac{1}{2}\end{matrix} \tag{10.228}$$

$$\begin{matrix} \tfrac{1}{2} & -\tfrac{1}{2} \end{matrix}$$
$$\Rightarrow \quad \mathbf{J}_- = \mathbf{J}_+^\dagger = \begin{bmatrix}0 & 0\\ 1 & 0\end{bmatrix}\begin{matrix} \tfrac{1}{2}\\ -\tfrac{1}{2}\end{matrix} \tag{10.229}$$

Therefore, referring to equations 10.203, we have

$$\text{(a)}\quad \mathbf{J}_1 = \frac{1}{2}\begin{bmatrix}0 & 1\\ 1 & 0\end{bmatrix} = \tfrac{1}{2}\boldsymbol{\sigma}_x \qquad \text{(b)}\quad \mathbf{J}_2 = \frac{1}{2}\begin{bmatrix}0 & -i\\ i & 0\end{bmatrix} = \tfrac{1}{2}\boldsymbol{\sigma}_y \tag{10.230}$$

The matrix representation of $\mathbf{R}(\phi, \theta, \psi)$ in the representation with bases $|m, j\rangle$ is found from

$$\mathbf{R}|m, j\rangle = \sum_{m'=-j}^{m} [\mathscr{D}^j(\phi, \theta, \psi)]_{mm'}|m', j\rangle \tag{10.231a}$$

$$\Rightarrow \quad [\mathscr{D}^j(\phi, \theta, \psi)]_{mm'} = \langle m', j|\mathbf{R}|m, j\rangle \tag{10.231b}$$

The symbol $\mathscr{D}$ is used by many authors to describe the matrix representation of the rotation operator. This arises from the German word *darstellung*, one translation of which is *representation*. Using equation 10.190b, this representation is

$$[\mathscr{D}^j(\phi, \theta, \psi)]_{mm'} = \langle m', j|e^{-i\mathbf{J}_3\psi}e^{-i\mathbf{J}_1\theta}e^{-i\mathbf{J}_3\phi}|m, j\rangle \tag{10.232}$$

Because $\mathbf{J}_3$ is diagonal,

$$e^{-i\mathbf{J}_3\phi}|m, j\rangle = e^{-im\phi}|m, j\rangle \tag{10.233}$$

$$\Rightarrow \quad \langle m, j|e^{-i\mathbf{J}_3\phi} = \left[e^{i\mathbf{J}_3^\dagger\phi}|m, j\rangle\right]^\dagger \tag{10.234a}$$

Since $\mathbf{J}_3$ is diagonal, with real (integer) eigenvalues, $\mathbf{J}_3^\dagger = \mathbf{J}_3$. Therefore,

$$\langle m, j|e^{-i\mathbf{J}_3\phi} = \left[e^{im\phi}|m, j\rangle\right]^\dagger = m\langle m, j|e^{-im\phi} \tag{10.234b}$$

$$\Rightarrow \quad \left[\mathscr{D}^j(\phi, \theta, \psi)\right]_{mm'} = \langle m', j|e^{-im'\psi}e^{-i\mathbf{J}_1\theta}e^{-im\phi}|m, j\rangle$$

$$= e^{-im'\psi}\langle m', j|e^{-i\mathbf{J}_1\theta}|m, j\rangle e^{-im\phi} \equiv e^{-im'\psi}d^j_{mm'}(\theta)e^{-im\phi} \tag{10.235}$$

In the two-dimensional representation ($j = \frac{1}{2}$), $\mathbf{J}_1 = \frac{1}{2}\boldsymbol{\sigma}_x$, and since $\boldsymbol{\sigma}_x^2 = \mathbf{1}$, it is straightforward to show that

$$e^{-i\mathbf{J}_1\theta} = e^{-i\sigma_x\theta/2} = \mathbf{1}\cos(\tfrac{1}{2}\theta) - i\boldsymbol{\sigma}_x\sin(\tfrac{1}{2}\theta) \tag{10.236}$$

(see Problem 16 of Chapter 8). Therefore,

$$\mathbf{d}^{1/2}(\theta) = \begin{bmatrix} \cos(\frac{1}{2}\theta) & -i\sin(\frac{1}{2}\theta) \\ -i\sin(\frac{1}{2}\theta) & \cos(\frac{1}{2}\theta) \end{bmatrix} \tag{10.237}$$

Therefore, labeling the row and columns of $\mathscr{D}^j$ in the same way the rows and columns of $\mathbf{J}_\pm$ are labeled in equation 10.228,

$$\mathscr{D}^{1/2}(\phi, \theta, \psi) = \begin{bmatrix} e^{-i\psi/2}\cos(\frac{1}{2}\theta)e^{-i\phi/2} & -ie^{i\psi/2}\sin(\frac{1}{2}\theta)e^{-i\phi/2} \\ -ie^{-i\psi/2}\sin(\frac{1}{2}\theta)e^{i\phi/2} & e^{i\psi/2}\cos(\frac{1}{2}\theta)e^{i\phi/2} \end{bmatrix} \tag{10.238}$$

Relation to spherical harmonics and characters of representations

As the reader will recall from our discussion in Chapter 6, the spherical harmonics $Y_{lm}(\Omega)$ formed the bases for the description of rotations in three dimensions. That is, the angles θ and ϕ described the angular orientation of any point in three dimensions, and any function of θ and ϕ could be expanded as

$$F(\theta, \phi) = \sum_{l=0}^{\infty}\sum_{m=-l}^{l} C_{lm}Y_{lm}(\theta, \phi) \tag{10.239}$$

If a reorientation of the axes is made through Euler angles α, β, γ, a point in space described originally by spherical angles θ and ϕ will now be described by new angles θ' and ϕ'. Therefore, the spherical harmonics $Y_{lm}(\theta, \phi) \to Y_{lm}(\theta', \phi')$, which can be expanded in terms of the spherical harmonics $Y_{lm}(\theta, \phi)$ as

$$Y_{lm}(\theta', \phi') = \sum_{l', m'} C_{l'm'}Y_{l'm'}(\theta, \phi) \tag{10.240a}$$

A rotation of axes cannot change the number of independent basis functions. Thus if there are $2l + 1$ bases in the unrotated space, there must be that number of basis functions in the rotated space. That is, l and l' are the same, and the sum over l' collapses to one term, $l' = l$. Then

$$Y_{lm}(\theta', \phi') = \mathbf{R}(\alpha, \beta, \gamma)Y_{lm}(\theta, \phi) = \sum_{m'=-l}^{l} C_{lm'}Y_{lm'}(\theta, \phi) \tag{10.240b}$$

Comparing this to the more abstract form of the representation of $\mathbf{R}(\alpha, \beta, \gamma)$ as expressed in equation 10.231a, we see that the spherical harmonics form a set of bases for the representation designated by j and m. That is,

$$|m, j\rangle = Y_{lm}(\theta, \phi) \tag{10.241}$$

We note that since l must be integer in this treatment, the spherical harmonics serve as a set of bases only for integer value of $j = l$.

We also note from equation 10.231a that in such a representation

$$\mathbf{R}(\alpha,\beta,\gamma)Y_{lm}(\theta,\phi) = \sum_{m'=-l}^{l} [\mathscr{D}^l(\alpha,\beta,\gamma)]_{mm'} Y_{lm'}(\theta,\phi)$$

$$= \sum_{m'=-l}^{l} e^{-im'\gamma} d^l_{mm'}(\beta) e^{-im\alpha} Y_{lm'}(\theta,\phi) \qquad (10.242)$$

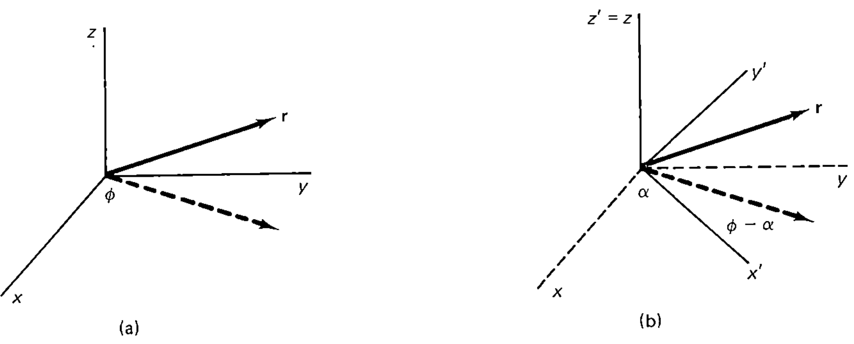

Figure 10.14 (a) Angle ϕ defined by the projection of the position vector; (b) azimuthal angle in a rotated coordinate system.

The angle ϕ is defined as the angle between the x-axis and the projection of the position vector $\mathbf{r}$ into the x-y plane, as shown in Figure 10.14a. If a rotation is made about the z-axis through an angle α, the x'-axis will make an angle $\phi - \alpha$ with this projection. Thus

$$\mathbf{R}(\alpha,0,0)Y_{lm}(\theta,\phi) = Y_{lm}(\theta,\phi-\alpha) = e^{-im\alpha}Y_{lm}(\theta,\phi) \qquad (10.243)$$

This is an eigenvalue equation for the operator $\mathbf{R}(\alpha,0,0)$. The eigenfunctions are $Y_{lm}(\theta,\phi)$ with eigenvalues $e^{-im\alpha}$. Therefore, the $(2l+1)$-dimensional matrix representation of this operator is

$$\mathscr{D}^l(\alpha,0,0) = \begin{bmatrix} e^{-il\alpha} & 0 & 0 & \cdots & 0 & 0 \\ 0 & e^{-i(l-1)\alpha} & & \cdots & 0 & 0 \\ \vdots & & & & & \\ 0 & 0 & \cdots & \cdots & e^{i(l-1)\alpha} & 0 \\ 0 & 0 & \cdots & \cdots & 0 & e^{il\alpha} \end{bmatrix} \qquad (10.244)$$

Thus, in this representation, the character is given by

$$\chi_l(\alpha) = \sum_{m=-l}^{l} e^{-im\alpha} = e^{il\alpha} \sum_{m=-l}^{l} e^{-i(m+l)\alpha} \qquad (10.245a)$$

Replacing $m + l$ with m' (dropping the prime), this becomes

$$\chi_l(\alpha) = e^{il\alpha} \sum_{m=0}^{2l} e^{-im\alpha} \tag{10.245b}$$

We recognize this sum as the finite geometric sum. With $z = e^{-i\alpha}$

$$S \equiv \sum_{m=0}^{2l} e^{-im\alpha} = \sum_{m=0}^{2l} z^m \tag{10.246}$$

Referring to equation 3.7,

$$S = \frac{1 - z^{2l+1}}{1 - z} = \frac{1 - e^{-(2l+1)i\alpha}}{1 - e^{-i\alpha}} = e^{-il\alpha} \frac{(e^{(l+1/2)i\alpha} - e^{-(l+1/2)i\alpha})}{(e^{1/2i\alpha} - e^{-1/2i\alpha})}$$

$$= e^{-il\alpha} \frac{\sin[(l + \frac{1}{2})\alpha]}{\sin(\frac{1}{2}\alpha)} \tag{10.247}$$

Therefore, from equation 10.245b, the character of the lth representation is

$$\chi_l(\alpha) = \frac{\sin[(l + \frac{1}{2})\alpha]}{\sin(\frac{1}{2}\alpha)} \tag{10.248}$$

Referring to equation 10.244, we note that $\mathscr{D}^l(0,0,0) = \mathbf{1}$. Thus $\mathrm{Tr}[\mathscr{D}^l(0,0,0)] = \lim_{\alpha \to 0} \chi_l(\alpha)$ is the dimension of the lth representation. From equation 10.248 we obtain

$$\lim_{\alpha \to 0} \chi_l(\alpha) = \lim_{\alpha \to 0} \frac{\sin[(l + \frac{1}{2})\alpha]}{\sin(\frac{1}{2}\alpha)} = 2l + 1 \tag{10.249}$$

Therefore, the $l = 0$ representation is one-dimensional, and every element of the group (every rotation) is represented by the number 1. The $l = 1$ representation is the three-dimensional representation described by the Euler rotations expressed in equations 10.186 and 10.187.

Reduction of symmetry

When an electron is in the spherically symmetric field of a nucleus, it has rotational symmetry, and its properties are therefore invariant under the transformations of SO(3). Under such a symmetry, the bases of a representation, defined by l, can be taken to be the spherical harmonics $Y_{lm}(\theta, \phi)$ as expressed in equation 10.240a. If the electron is placed in a lattice of ions, as in a crystal, this spherical symmetry is destroyed.

In general, the symmetry of the lattice will be a finite symmetry, as we discuss in more detail in the example that follows. As such, the invariance of the electron's properties under an infinite number of SO(3) transformations is reduced to invariance under the transformation of a finite group.

Example 10.21

As an example, consider placing an electron in the center of a lattice in which the ions are positioned at the corners of a cube. In such a lattice, the electron's properties are invariant under all rotations of the cube that leave it in its original orientation. This is a straightforward extension of the example of the transformations that leave the orientation of the rectangle of Figure 10.4 invariant.

The orientation of the cube will be invariant under the following transformations:

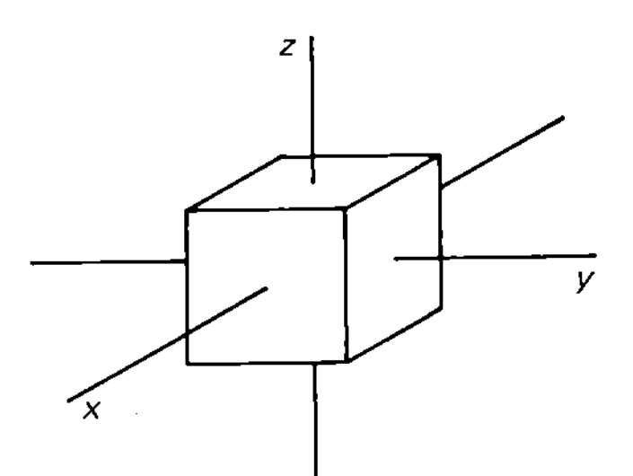

Figure 10.15 x, y, and z symmetry axes of a cube.

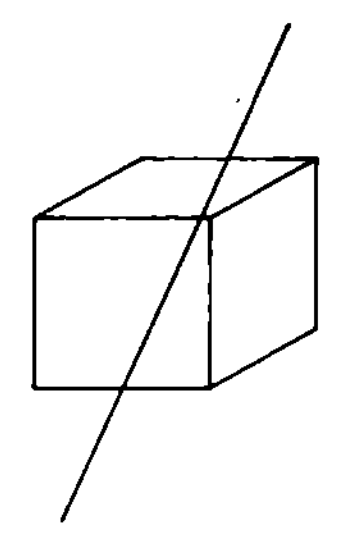

Figure 10.16 Symmetry axis of a cube passing between the midpoints of diagonally opposite sides.

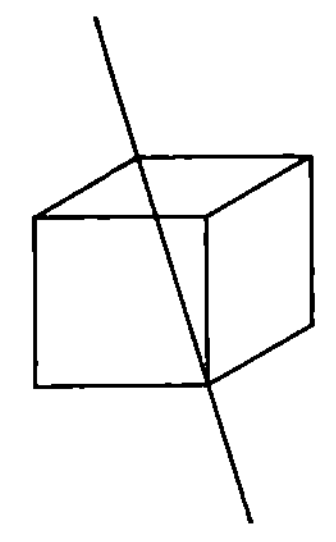

Figure 10.17 Symmetry axis of a cube passing from diagonally opposite corners.

1. No rotations. This is the identity element of the group.
2. Rotations about the axes passing through the center of the cube perpendicular to the faces, as shown in Figure 10.15.
   a) Rotations about the x, y, and z axes by $\pm\pi/2$. These are denoted by $R_x(\pm\pi/2)$, $R_y(\pm\pi/2)$, and $R_z(\pm\pi/2)$. Clearly, $R_i^{-1}(\pi/2) = R_i(-\pi/2)$. There are six elements of this type.
   b) Rotations about the x, y, and z axes by π, denoted by $R_x(\pi)$, $R_y(\pi)$, and $R_z(\pi)$. Clearly, $R_i^{-1}(\pi) = R_i(\pi)$. There are three elements of this type.
3. Rotations about axes passing through the center of the cube between the midpoints of diagonally opposite sides. An example of one such axis is shown in Figure 10.16. There are six elements of this type, two such axes for each pair of parallel faces and three pairs of parallel faces. Rotations about these axes by π will leave the orientation of the cube invariant. These rotations will be designated $S_i(\pi)$.
4. Rotations about axes passing through the center of the cube between the diagonally opposite corners, such as is shown in Figure 10.17. There are four such pairs of corners, and thus four axes of this type. Rotations about these axes through angles of $\pm 2\pi/3$ leave the orientation of the cube invariant. Thus there are eight elements of this type. These rotations will be designated by $T_i(\pm 2\pi/3)$. Clearly, $T_i^{-1}(2\pi/3) = T_i(-2\pi/3)$.

From this analysis we note that the order of the group is 24. We designate this group $\mathcal{G}$.

By rotating a cube through specified transformations, it is a straightforward process to see that the transformations $R_i(\pm\pi/2)$, $R_i(\pi)$, $S_i(\pi)$, and $T_i(\pm 2\pi/3)$ are all mutually commuting transformations. For example, consider a cube that has its faces and/or corners marked for identification (like the rectangle of Figure 10.4). We first rotate the cube by $\pi/2$ about its x-axis, then rotate it through an angle of $2\pi/3$ about the ith diagonal axis such as is shown in Figure 10.17. We obtain the same reorientation of the cube after these rotations as we would by first rotating the cube about the ith diagonal axis, followed by a rotation about the x-axis through $\pi/2$. That is,

$$R_x\left(\frac{\pi}{2}\right)T_i\left(\frac{2\pi}{3}\right) = T_i\left(\frac{2\pi}{3}\right)R_x\left(\frac{\pi}{2}\right) \tag{10.250a}$$

By this type of analysis, it is straightforward to show that all the commutators

$$\left(R_i\left(\pm\frac{\pi}{2}\right), R_j(\pi)\right) = \left(R_i\left(\pm\frac{\pi}{2}\right), S_j(\pi)\right) = \left(R_i\left(\pm\frac{\pi}{2}\right), T_j\left(\pm\frac{2\pi}{3}\right)\right)$$
$$= \left(R_i(\pi), S_j(\pi)\right) = \left(R_i(\pi), T_j\left(\pm\frac{2\pi}{3}\right)\right) = \left(S_i(\pi), T_j\left(\pm\frac{2\pi}{3}\right)\right) = 0 \tag{10.250b}$$

By similar modeling, one can also see that

$$\text{(c)}\quad \left(R_i\left(\frac{\pi}{2}\right), R_j\left(-\frac{\pi}{2}\right)\right) \neq 0 \qquad \text{(d)}\quad \left(T_i\left(\frac{2\pi}{3}\right), T_j\left(-\frac{2\pi}{3}\right)\right) \neq 0 \tag{10.250}$$

Therefore, for example,

$$\text{(a)}\quad R_j^{-1}(\pi)R_i\left(\pm\frac{\pi}{2}\right)R_j(\pi) = R_i\left(\pm\frac{\pi}{2}\right) \qquad \text{(b)}\quad S_j^{-1}(\pi)R_i\left(\pm\frac{\pi}{2}\right)S_j(\pi) = R_i\left(\pm\frac{\pi}{2}\right)$$

$$\text{(c)}\quad T_j^{-1}\left(\pm\frac{2\pi}{3}\right)R_i\left(\pm\frac{\pi}{2}\right)T_j\left(\pm\frac{2\pi}{3}\right) = R_i\left(\pm\frac{\pi}{2}\right) \tag{10.251}$$

Thus the elements $R_i(\pm\pi/2)$ comprise a single equivalence class. By identical analysis, $R_i(\pi)$, $S_i(\pi)$, and $T_j(\pm 2\pi/3)$ each comprise separate equivalence classes. Therefore, the group of rotations that leaves the orientation of a cube invariant has five equivalence classes:

$$\{e\} \equiv \mathscr{C}_1 \text{ has one element} \qquad \left\{R_i\left(\pm\frac{\pi}{2}\right)\right\} \equiv \mathscr{C}_2 \text{ has six elements}$$

$$\{R_i(\pi)\} \equiv \mathscr{C}_3 \text{ has three elements} \qquad \{S_i(\pi)\} \equiv \mathscr{C}_4 \text{ has six elements}$$

$$\left\{T_j\left(\pm\frac{2\pi}{3}\right)\right\} \equiv \mathscr{C}_5 \text{ has eight elements}$$

Therefore, $\mathscr{G}$ has five irreducible representations, the dimensions of which must satisfy

$$\sum_{i=1}^{5} d_i^2 = 24 \tag{10.252a}$$

The only set of integers that satisfies this condition is

$$d_i = \{1, 1, 2, 3, 3\} \tag{10.252b}$$

Referring to equation 10.63, the characters of $\mathscr{G}$ satisfy

$$\sum_{l=1}^{5} \nu_l \chi_r(\mathscr{C}_l)\chi_s(\mathscr{C}_l) = \chi_r(\mathscr{C}_1)\chi_s(\mathscr{C}_1) + 6\chi_r(\mathscr{C}_2)\chi_s(\mathscr{C}_2) + 3\chi_r(\mathscr{C}_3)\chi_s(\mathscr{C}_3)$$

$$+6\chi_r(\mathscr{C}_4)\chi_s(\mathscr{C}_4) + 8\chi_r(\mathscr{C}_5)\chi_s(\mathscr{C}_5) = 24\delta_{rs} \tag{10.253}$$

Using the fact that the character of the identity in each irreducible representation is the dimension of the representation, the character table for $\mathscr{G}$ is of the form shown in Figure 10.18. As argued earlier, the matrices and therefore the characters of one of the one-dimensional irreducible representations can be assigned to be 1. As such, we take $\alpha_i = 1$, $2 \leqslant i \leqslant 5$. Then the orthonormality condition expressed in equation 10.253 becomes

$$1 + 6\beta_2 + 3\beta_3 + 6\beta_4 + 8\beta_5 = 0 \qquad r = 1, s = 2 \tag{10.254a}$$

$$1 + 6\beta_2^2 + 3\beta_3^2 + 6\beta_4^2 + 8\beta_5^2 = 24 \qquad r = s = 2 \tag{10.254b}$$

These equations are satisfied by $\beta_2 = \beta_4 = -1$, $\beta_3 = \beta_5 = 1$. Therefore, the character table of $\mathscr{G}$ becomes that shown in Figure 10.19. Now, with $r = 3$, we obtain three equations for the γ_i from equation 10.253; one for each of the values $s = 1$, 2, and 3. The equations are

$$\text{(a)}\quad 2 + 6\gamma_2 + 3\gamma_3 + 6\gamma_4 + 8\gamma_5 = 0 \qquad \text{(b)}\quad 2 - 6\gamma_2 + 3\gamma_3 - 6\gamma_4 + 8\gamma_5 = 0$$

$$\text{(c)}\quad 4 + 6\gamma_2^2 + 3\gamma_3^2 + 6\gamma_4^2 + 8\gamma_5^2 = 24 \tag{10.255}$$

which are satisfied by $\gamma_2 = \gamma_4 = 0, \gamma_3 = 2$, and $\gamma_5 = -1$.

	$1\mathcal{C}_1$	$6\mathcal{C}_2$	$3\mathcal{C}_3$	$6\mathcal{C}_4$	$8\mathcal{C}_5$
$\chi(\mathcal{C}_1)$	1	α_2	α_3	α_4	α_5
$\chi(\mathcal{C}_2)$	1	β_2	β_3	β_4	β_5
$\chi(\mathcal{C}_3)$	2	γ_2	γ_3	γ_4	γ_5
$\chi(\mathcal{C}_4)$	3	δ_2	δ_3	δ_4	δ_5
$\chi(\mathcal{C}_5)$	3	ϵ_2	ϵ_3	ϵ_4	ϵ_5

Figure 10.18 Character table for the group of symmetry of the cube.

	$1\mathcal{C}_1$	$6\mathcal{C}_2$	$3\mathcal{C}_3$	$6\mathcal{C}_4$	$8\mathcal{C}_5$
$\chi(\mathcal{C}_1)$	1	1	1	1	1
$\chi(\mathcal{C}_2)$	1	-1	1	-1	1
$\chi(\mathcal{C}_3)$	2	γ_2	γ_3	γ_4	γ_5
$\chi(\mathcal{C}_4)$	3	δ_2	δ_3	δ_4	δ_5
$\chi(\mathcal{C}_5)$	3	ϵ_2	ϵ_3	ϵ_4	ϵ_5

Figure 10.19 Character table for the group of symmetry of the cube.

	$1\mathcal{C}_1$	$6\mathcal{C}_2$	$3\mathcal{C}_3$	$6\mathcal{C}_4$	$8\mathcal{C}_5$
$\chi(\mathcal{C}_1)$	1	1	1	1	1
$\chi(\mathcal{C}_2)$	1	-1	1	-1	1
$\chi(\mathcal{C}_3)$	2	0	2	0	-1
$\chi(\mathcal{C}_4)$	3	1	-1	-1	0
$\chi(\mathcal{C}_5)$	3	-1	-1	1	0

Figure 10.20 Character table for the group of symmetry of the cube.

These calculations are again applied to obtain values for the δ_i and ε_i. The resulting character table for $\mathscr{G}$ is shown in Figure 10.20. To determine the representations of SO(3) that retain the symmetry of $\mathscr{G}$, we find the characters of SO(3) that correspond to the equivalence classes of $\mathscr{G}$. Since these classes are specified by angles, we can obtain the corresponding characters of SO(3) from equations 10.248 for those angles. That is, $\mathscr{C}_1 \Rightarrow \alpha = 0$, so

$$\chi_l^{SO(3)}(\mathscr{C}_1) = \lim_{\alpha\to 0} \frac{\sin[(l+\frac{1}{2})\alpha]}{\sin(\frac{1}{2}\alpha)} = 2l+1 \tag{10.256a}$$

Since $\mathscr{C}_2 \Rightarrow \alpha = \pm\dfrac{\pi}{2}$

$$\chi_l^{SO(3)}(\mathscr{C}_2) = \frac{\sin[(l+\frac{1}{2})(\pi/2)]}{\sin(\pi/4)} = \sin\left(\frac{l\pi}{2}\right) + \cos\left(\frac{l\pi}{2}\right) \tag{10.256b}$$

Similarly, with $\mathscr{C}_3 \Rightarrow \pi$, $\mathscr{C}_4 \Rightarrow \pi$, and $\mathscr{C}_5 \Rightarrow \pm 2\pi/3$,

$$\chi_l^{SO(3)}(\mathscr{C}_3) = \chi_l^{SO(3)}(\mathscr{C}_4) = \cos(l\pi) = (-1)^l \tag{10.256c}$$

$$\chi_l^{SO(3)}(\mathscr{C}_5) = \frac{1}{\sqrt{3}}\sin\left(\frac{2\pi l}{3}\right) + \cos\left(\frac{2\pi l}{3}\right) \tag{10.256d}$$

In the quantum theory of an electron in a spherically symmetric potential, one finds that for a specific l, the electron has the same energy for all values of m. That is, due to the invariance under SO(3) transformations, there is a $(2l-1)$-fold degeneracy. In terms of representation theory, the energy of an electron in one eigenstate will differ from its energy in another eigenstate only if those eigenstates are states of different irreducible representations. The states will be degenerate (have the same energy) if they are states of the same irreducible representation.

We now investigate what happens to that $(2l-1)$-fold degeneracy when the electron is placed in a field that has cubic symmetry of the group $\mathscr{G}$. To determine this, we deduce how many times a given representation of the SO(3) occurs in a particular irreducible representation of $\mathscr{G}$.

Using equations 10.256 for $l = 1$, the characters of SO(3) are $\chi(\mathscr{C}_1) = 3$, $\chi(\mathscr{C}_2) = 1$, $\chi(\mathscr{C}_3) = \chi(\mathscr{C}_4) = -1$, and $\chi(\mathscr{C}_5) = 0$. Referring to the character table of $\mathscr{G}$, we see that the character of SO(3) for $l = 1$ is identical to χ_4 of $\mathscr{G}$. That is, $\chi_{l=1}^{SO(3)}$ occurs once in the fourth irreducible representation of $\mathscr{G}$ and therefore in none of the others. Thus the $l = 1$

representation of SO(3) is an irreducible representation of $\mathscr{G}$, and the degeneracy created by SO(3) symmetry is not affected by the reduced symmetry of $\mathscr{G}$.

For $l = 2$, the characters of SO(3) are $\chi(\mathscr{C}_1) = 5$, $\chi(\mathscr{C}_2) = -1$, $\chi(\mathscr{C}_3) = \chi(\mathscr{C}_4) = 1$, and $\chi(\mathscr{C}_5) = -1$. This does not correspond to a single irreducible representation of $\mathscr{G}$. Thus the $l = 2$ representation of SO(3) is reducible. To determine the number of times this representation of SO(3) occurs in the irreducible representations of $\mathscr{G}$, we apply equation 10.65b.

$$n_r = \frac{1}{\mu}\sum_{i=1}^{5} \nu_i \chi_r^{\mathscr{G}}(\mathscr{C}_i)\chi^{\mathrm{SO(3)}}(\mathscr{C}_i) = \frac{1}{24}\Big[\chi_r^{\mathscr{G}}(\mathscr{C}_1)\chi^{\mathrm{SO(3)}}(\mathscr{C}_1) + 6\chi_r^{\mathscr{G}}(\mathscr{C}_2)\chi^{\mathrm{SO(3)}}(\mathscr{C}_2)$$

$$+3\chi_r^{\mathscr{G}}(\mathscr{C}_3)\chi^{\mathrm{SO(3)}}(\mathscr{C}_3) + 6\chi_r^{\mathscr{G}}(\mathscr{C}_4)\chi^{\mathrm{SO(3)}}(\mathscr{C}_4) + 8\chi_r^{\mathscr{G}}(\mathscr{C}_5)\chi^{\mathrm{SO(3)}}(\mathscr{C}_5)\Big] \tag{10.257}$$

$$\Rightarrow \quad \text{(a)} \quad n_1 = \tfrac{1}{24}[5 - 6 + 3 + 6 - 8] = 0 \qquad \text{(b)} \quad n_2 = \tfrac{1}{24}[5 + 6 + 3 - 6 - 8] = 0$$

$$\text{(c)} \quad n_3 = \tfrac{1}{24}[10 - 0 + 6 + 0 + 8] = 1 \tag{10.258}$$

$$\text{(d)} \quad n_4 = \tfrac{1}{24}[15 - 6 - 3 - 6 - 0] = 0 \qquad \text{(e)} \quad n_5 = \tfrac{1}{24}[15 + 6 - 3 + 6 - 0] = 1$$

Thus the $l = 2$ representation of SO(3) is contained once in the third irreducible representation of $\mathscr{G}$ (which is two-dimensional and thus contains two states), and once in the fifth irreducible representation of $\mathscr{G}$ (which is three-dimensional). So by placing an electron in the $l = 2$ state into a lattice with cubic symmetry, the fivefold degenerate energy of SO(3) is split into two energies. These two energy levels are each still degenerate, but under the symmetry of $\mathscr{G}$, one level has a twofold degeneracy, the other a threefold degeneracy. □

Other descriptions of the generators of SO(3)

The operators and bases of SO(3) can be described in forms other than those discussed above. For example, consider the basis functions

$$|m, j\rangle \equiv \alpha_{j,m} x^{j+m} y^{j-m} \tag{10.259}$$

and consider the operation

$$x\frac{\partial}{\partial y}|m, j\rangle = \alpha_{j,m}(j - m)x^{j+m+1}y^{j-(m+1)} = (j - m)\frac{\alpha_{j,m}}{\alpha_{j,m+1}}|m + 1, j\rangle \tag{10.260}$$

Referring to equations 10.210a and 10.219, we can identify

$$\mathbf{J}_+ \equiv x\frac{\partial}{\partial y} \tag{10.261}$$

if the coefficient of $|m + 1, j\rangle$ of equation 10.260 can be written as

$$(j - m)\frac{\alpha_{j,m}}{\alpha_{j,m+1}} = \sqrt{j(j + 1) - m(m + 1)} \tag{10.262}$$

A little algebraic manipulation shows that this is satisfied by

$$\alpha_{j,m} = \frac{1}{\sqrt{(j + m)!(j - m)!}} \tag{10.263}$$

Thus in a function space spanned by

$$|m, j\rangle = \frac{x^{j+m}y^{j-m}}{\sqrt{(j + m)!(j - m)!}} \tag{10.264}$$

the operators $\mathbf{J}_{\pm}$, $\mathbf{J}_3$, and $\mathbf{J}^2$ can be represented by the derivatives

$$\text{(a)}\quad \mathbf{J}_+ = x\frac{\partial}{\partial y} \qquad \text{(b)}\quad \mathbf{J}_- = y\frac{\partial}{\partial x} \qquad \text{(c)}\quad \mathbf{J}_3 = \frac{1}{2}\left(x\frac{\partial}{\partial x} - y\frac{\partial}{\partial y}\right) \tag{10.265}$$

$$\Rightarrow \mathbf{J}^2 = \frac{3}{4}\left(x\frac{\partial}{\partial x} + y\frac{\partial}{\partial y}\right) + \frac{1}{4}\left(x^2\frac{\partial^2}{\partial x^2} + y^2\frac{\partial^2}{\partial y^2}\right) + \frac{1}{2}xy\frac{\partial^2}{\partial x\,\partial y} \tag{10.266}$$

The reader will demonstrate in Problem 27 that in addition to

$$\mathbf{J}_+|m, j\rangle = \sqrt{j(j+1) - m(m+1)}\;|m+1, j\rangle \tag{10.267a}$$

the operators specified in equations 10.265b, 10.265c, and 10.266 satisfy

$$\begin{gathered}\text{(b)}\quad \mathbf{J}_-|m, j\rangle = \sqrt{j(j+1) - m(m-1)}\;|m-1, j\rangle \\ \text{(c)}\quad \mathbf{J}_3|m, j\rangle = m|m, j\rangle \qquad \text{(d)}\quad \mathbf{J}^2|m, j\rangle = j(j+1)\,|m, j\rangle\end{gathered} \tag{10.267}$$

In Problem 28, the reader will also show in that for integer values of $j = l$, one can represent the operators $\mathbf{J}_1$, $\mathbf{J}_2$, and $\mathbf{J}_3$ by

$$\mathbf{J}_k = -i(\mathbf{r} \times \nabla)_k \tag{10.268}$$

in a space in which the bases are the spherical harmonics $Y_{lm}(\theta, \phi)$.

SU(2)

The set of all unitary matrices under matrix multiplication form a group called $U(n)$. When the additional constraint that the determinant of each matrix is 1, (the matrix is unimodular) the group is called the *special unitary group*, SU(n).

For example, SU(2) is the group of all unitary unimodular 2×2 matrices. The most general form of such a matrix is

$$\text{(a)}\quad \mathbf{A} = \begin{bmatrix} a & b \\ -b^* & a^* \end{bmatrix} \qquad \text{with} \qquad \text{(b)}\quad |a|^2 + |b|^2 = 1 \tag{10.269}$$

Since the matrices contain two arbitrary complex (and therefore, four real) quantities subject to one constant, there are three independent parameters that define SU(2).

We can also write the general SU(2) matrix as

$$\mathbf{A} = \begin{bmatrix} x_0 + ix_1 & x_2 + ix_3 \\ -x_2 + ix_3 & x_0 - ix_1 \end{bmatrix} \tag{10.270a}$$

The constraint of equation 10.269b becomes

$$x_0^2 + x_1^2 + x_2^2 + x_3^2 = 1 \tag{10.270b}$$

Taking the real quantities x_i to be Cartesian coordinates of a four-dimensional Euclidean space, these four parameters of SU(2), subject to the constraint of equation 10.270b, describe the surface of a sphere of radius 1.

We can also write the elements a and b of equations 10.269 in terms of three real parameters α, β, and γ, as

$$\text{(a)} \quad a = \cos\gamma e^{i\alpha} \qquad \text{and} \qquad \text{(b)} \quad b = \sin\gamma e^{i\beta} \tag{10.271}$$

Then the general form of the matrix of SU(2) is

$$\mathbf{A} = \begin{bmatrix} \cos\gamma e^{i\alpha} & \sin\gamma e^{i\beta} \\ -\sin\gamma e^{i\beta} & \cos\gamma e^{-i\alpha} \end{bmatrix} \tag{10.272}$$

If we set

$$\text{(a)} \quad \alpha = -\tfrac{1}{2}(\psi + \phi) \qquad \text{(b)} \quad \beta = \tfrac{1}{2}(\psi - \phi - \pi) \qquad \text{(c)} \quad \gamma = \tfrac{1}{2}\theta \tag{10.273}$$

we see from equation 10.238 that $\mathbf{A}$ is the same as the matrix $\mathscr{D}^{1/2}(\phi, \theta, \psi)$ of SO(3).

We note here that both SU(2) and SO(3) are described by three parameters, and therefore are generated by three generators. (Both groups have two roots, $\pm v$, and are of rank 1.) This, plus the equality of the matrices of their two-dimensional representations, strongly suggest that there is a general correspondence (homomorphism or isomorphism) between the two groups.

To see what this correspondence is, consider a rotation about the z-axis through an angle ϕ. This rotation matrix, as described in terms of an element of SO(3), is

$$\mathbf{R}_z(\phi) = \begin{bmatrix} \cos\phi & \sin\phi & 0 \\ -\sin\phi & \cos\phi & 0 \\ 0 & 0 & 1 \end{bmatrix} \tag{10.184}$$

From equations 10.272 and 10.273, with $\theta = \psi = 0$, this rotation is described in SU(2) by

$$\mathbf{A}(\phi) = \begin{bmatrix} e^{-i\phi/2} & 0 \\ 0 & e^{i\phi/2} \end{bmatrix} \tag{10.274a}$$

But a rotation through an angle ϕ is identical to a rotation through $\phi + 2\pi$. Therefore, in SU(2), the rotation is also described by

$$\mathbf{A}(\phi + 2\pi) = \begin{bmatrix} -e^{-i\phi/2} & 0 \\ 0 & -e^{i\phi/2} \end{bmatrix} = -\mathbf{A}(\phi) \tag{10.274b}$$

Thus, in general, two elements of SU(2) correspond to one element of SO(3), which means that the correspondence between the groups is a 2-to-1 homomorphism of SU(2) onto SO(3).

Because of this homomorphism, one description of the basis functions expressed in equation 10.264 for SO(3),

$$|m, j\rangle \equiv f_m(x, y) \equiv \frac{x^{j+m}y^{j-m}}{\sqrt{(j+m)!(j-m)!}} \tag{10.264}$$

can be taken as the basis functions of SU(2). That is, for an operator $\Omega \in \text{SU(2)}$,

$$P_\Omega f_m(x, y) = f_m(x', y') = \sum_{m'=-j}^{j} \mathscr{D}^j_{mm'}(\Omega) f_{m'}(x, y) \tag{10.275}$$

Referring to equation 10.269a, any element of SU(2) generates the transformation

$$\begin{bmatrix} x' \\ y' \end{bmatrix} = \begin{bmatrix} a & b \\ -b^* & a^* \end{bmatrix}\begin{bmatrix} x \\ y \end{bmatrix} = \begin{bmatrix} ax + by \\ -b^*x + a^*y \end{bmatrix} \tag{10.276}$$

$$\Rightarrow P_\Omega f_m(x, y) = \left[\frac{(ax + by)^{j+m}(-b^*x + a^*y)^{j-m}}{\sqrt{(j+m)!(j-m)!}}\right]$$

$$= \frac{1}{\sqrt{(j+m)!(j-m)!}} \sum_{k=0}^{j+m} \frac{(j+m)!}{k!(j+m-k)!} a^{j+m-k} x^{j+m-k} b^k y^k$$

$$\times \sum_{l=0}^{j-m} \frac{(j-m)!}{l!(j-m-l)!} (-b^*)^{j-m-l} x^{j-m-l} (a^*)^l y^l$$

$$= \sum_{k=0}^{j+m} \sum_{l=0}^{j-m} (-1)^{j-m-l} \frac{\sqrt{(j+m)!(j-m)!}}{k!l!(j+m-k)!(j-m-l)!}$$

$$\times a^{j+m-k}(a^*)^l b^k (b^*)^{j-m-l} x^{2j-k-l} y^{k+l} \tag{10.277}$$

Let $l = j - k + m'$, so that

$$x^{2j-k-l} y^{k+l} = x^{j+m'} y^{j-m'} = \sqrt{(j+m')!(j-m')!}\, f_{m'}(x, y) \tag{10.278}$$

We note that m' has a maximum value when $k = l = 0$. The minimum value of m' is obtained for $k = j + m$ and $l = j - m$. Thus $-j \leqslant m' \leqslant j$. Therefore, the elements of a matrix representation of an operator $\Omega \in \text{SU}(2)$ are given by

$$P_\Omega f_m(x, y) = \sum_{m'=-j}^{j} \mathscr{D}^j_{mm'}(\Omega) f_{m'}(x, y) \tag{10.279}$$

where

$$\mathscr{D}^j_{mm'} = \sum_{k=0}^{j+m} (-1)^{m'-m+k} \frac{\sqrt{(j+m)!(j-m)!(j+m')!(j-m')!}}{k!(j-m'-k)!(j+m-k)!(m'-m+k)!}$$

$$\times a^{j+m-k}(a^*)^{j-m'-k} b^k (b^*)^{m'-m+k} \tag{10.280}$$

For any representation, there are $2j + 1$ possible values of m and m'. That is, the representation characterized by j is $(2j + 1)$-dimensional.

Example 10.22

As an example, the matrix for $j = \frac{1}{2}$ representation is found as follows:

$$\mathscr{D}^{1/2}_{1/2,1/2} = \sum_{k=0}^{1} (-1)^k \frac{\sqrt{1!0!1!0!}}{k!(-k)!(1-k)!k!} a^{1-k}(a^*)^{-k} b^k (b^*)^k \tag{10.281}$$

Since $(-k)! = 1$ for $k = 0$, and $(-k)! = \infty$ for $k = 1$, only the $k = 0$ term in the sum is nonzero. Thus

$$\mathscr{D}^{1/2}_{1/2,1/2} = a \tag{10.282a}$$

Similarly, it is straightforward to obtain

$$\text{(b)}\quad \mathscr{D}^{1/2}_{1/2,-1/2} = b \qquad \text{(c)}\quad \mathscr{D}^{1/2}_{-1/2,1/2} = -b^* \qquad \text{(d)}\quad \mathscr{D}^{1/2}_{-1/2,-1/2} = a^* \tag{10.282}$$

$$\Rightarrow \mathscr{D}^{1/2} = \begin{bmatrix} a & b^* \\ -b^* & a^* \end{bmatrix} \quad \square \tag{10.283}$$

By identical analysis, the reader will show in Problem 29 that for $j = 1$,

$$\mathscr{D}^1 = \begin{bmatrix} a^2 & \sqrt{2}\,ab & b^2 \\ -\sqrt{2}\,ab^* & |a|^2 - |b|^2 & \sqrt{2}\,a^*b \\ b^{*2} & -\sqrt{2}\,a^*b^* & a^{*2} \end{bmatrix} \tag{10.284}$$

which is unitary, of course, since a and b are constrained as expressed in equation 10.269b.

Because of the homomorphism of SU(2) onto SO(3), the three generators of SU(2) can be chosen as a linear combination of the generators of SO(3). Denoting the generators of SU(2) by $\boldsymbol{\tau}_k$, we can write

$$\boldsymbol{\tau}_k = \sum_{l=1}^{3} t_{kl}\mathbf{J}_l \tag{10.285}$$

There is no loss of generality in taking the two sets of generators to be the same (i.e., $t_{kl} = \delta_{kl}$). Thus the one diagonal generator of SU(2) is $\boldsymbol{\tau}_3$. The label j of SO(3) is designated τ for SU(2).

Referring to equation 10.155, let λ be the single integer multiplier of the dominant weight of the fundamental irreducible representation that yields the dominant weight of all higher-dimensional irreducible representations. In an article by R. Behrends, J. Dreitlein, C. Fronsdal, and W. Lee, "Simple Groups and Strong Interaction Symmetries," *Reviews of Modern Physics* (vol. 34, 1962, p. 8), it is shown that the dimensions of the irreducible representations of SU(2) are given by

$$\eta = \lambda + 1 \tag{10.286}$$

Since the dimension of an irreducible representation of SO(3), and thus SU(2), is $2\tau + 1$, we see that $\lambda = 2\tau$. SU(2) has one fundamental irreducible representation which is two-dimensional, defined by $\lambda = 1$ or $\tau = \frac{1}{2}$.

From the orthogonality conditions for the roots expressed in equation 10.145,

$$\text{(a)}\quad [v(1)]^2 + [-v(1)]^2 = 1 \quad \Rightarrow \quad \text{(b)}\quad |v(1)| = |v(-1)| = \frac{1}{\sqrt{2}} \tag{10.287}$$

Therefore, the one-dimensional root diagram of SU(2) is that shown in Figure 10.21.

The states of the fundamental representation are characterized by $m = \pm\frac{1}{2}$. That is, the weight vectors of the $\tau = \frac{1}{2}$ representation are one-component vectors, shown in the weight diagram of Figure 10.22.

$$-v(1) = -\frac{1}{\sqrt{2}} \longleftrightarrow v(1) = \frac{1}{\sqrt{2}}$$

Figure 10.21 Root diagram of SU(2).

$$-\tfrac{1}{2} \longleftrightarrow +\tfrac{1}{2}$$

Figure 10.22 Weight diagram of $\tau = \frac{1}{2}$ representation of SU(2).

Isotopic spin, nucleons, and pions

Many phenomena involving nuclei (protons and neutrons) are independent of whether the particle involved is a proton or a neutron. It is also noted that the proton and neutron have almost the same mass ($m_n/m_p - 1 = 1.4 \times 10^{-3}$). It is therefore conjectured that the proton and neutron are different states or isotopes of a single particle called the *nucleon*.

Designating the states of the fundamental irreducible representation by the particle names, we assign

$$\text{(a)} \quad |p\rangle = |+\tfrac{1}{2}\rangle \qquad \text{(b)} \quad |n\rangle = |-\tfrac{1}{2}\rangle \tag{10.288}$$

$$\Rightarrow \quad \text{(a)} \quad \boldsymbol{\tau}_3|p\rangle = \tfrac{1}{2}|p\rangle \qquad \text{(b)} \quad \boldsymbol{\tau}_3|n\rangle = -\tfrac{1}{2}|n\rangle \tag{10.289}$$

With this assignment, the $\tau = \frac{1}{2}$ weight diagram is relabeled as shown in Figure 10.23.

The charge of the proton is $+e$ (one electronic charge) and the neutron's charge is zero. A charge operator is defined by

$$\mathbf{Q} = e(\boldsymbol{\tau}_3 + \tfrac{1}{2}) \tag{10.290}$$

$$\Rightarrow \quad \text{(a)} \quad \mathbf{Q}|p\rangle = e|p\rangle \qquad \text{(b)} \quad \mathbf{Q}|n\rangle = 0|n\rangle \tag{10.291}$$

The three-dimensional irreducible representation corresponds to $\lambda = 2$ or $\tau = 1$ and thus has three weights, $m_1 = 1$, $m_2 = 0$, $m_3 = -1$. The weight diagram for this representation is shown in Figure 10.24.

We note that the charge operator $\mathbf{Q} = e(\boldsymbol{\tau}_3 + \frac{1}{2})$ yields charges for the three states of this representation of

$$\text{(a)} \quad q_1 = \tfrac{3}{2}e \qquad \text{(b)} \quad q_2 = \tfrac{1}{2}e \qquad \text{(c)} \quad q_3 = -\tfrac{1}{2}e \tag{10.292}$$

n ⟵⟶ p

Figure 10.23 Weight diagram of $\tau = \frac{1}{2}$ representation of SU(2).

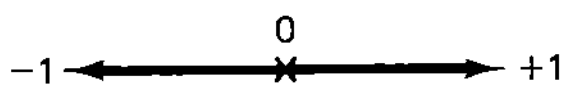

−1 ⟵ 0 ⟶ +1

Figure 10.24 Weight diagram of $\tau = 1$ representation of SU(2).

π^- ⟵ π^0 ⟶ π^+

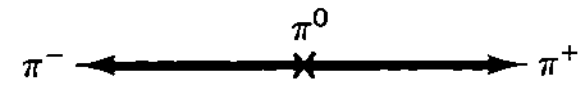

Figure 10.25 Weight diagram of $\tau = 1$ representation of SU(2).

These values do not correspond to the charges of any known particles. However, if the charge formula is modified to be

$$\mathbf{Q} = e(\boldsymbol{\tau}_3 + \tfrac{1}{2}B) \tag{10.293}$$

such that $B = 1$ for $\tau = \frac{1}{2}$ (nucleon) representation, and $B = 0$ for $\tau = 1$ (i.e., $B = 2 - 2\tau$) the charge operator yields correct values for the charges of three particles of very similar masses that have been designated π^+, π°, π^-. Collectively, they are viewed as the three isotropic states of a single entity called the *pion*. The weight diagram, in terms of the pion states, is shown in Figure 10.25. The quantity B is called the *baryon number*. From this designation, nucleons are baryons, pions are not. They are called *mesons*.

As noted earlier, many experimental results are independent of whether the experiment is performed on protons or on neutrons. The same statement can be made about some phenomena involving pions. The fact that these particles can be classified in irreducible representations of SU(2) means that the forces involved in these phenomena affect all states of a given representation of SU(2) in the same way. As such, these forces are invariant under SU(2) transformations.

SU(3)

Other baryons have been discovered with properties like the nucleon. They have the same intrinsic angular momentum $\frac{1}{2}\hbar$, and their quantum wave functions have the same behavior under the reflection $\mathbf{r} \to -\mathbf{r}$. Their masses are also similar to the nucleon masses. They are designated Σ^+, Σ°, Σ^-, Λ°, Ξ°, Ξ^-. Similarly, mesons exist which, like pions, have intrinsic angular momentum 0, and whose wave functions behave like pion wave functions under reflection. Their masses are similar to the masses of the pions. They are called K^+, K°, $\bar{K}^\circ$, K^-, η°.

It appears as though we can classify these particles as charge states of SU(2) the way the nucleons and pions are. However, their charges are not correctly given by the charge operator of equations 10.293. For example, if we try to classify the Σ triplet in the three-dimensional representation, with $\tau_3 = +1, 0, -1$, and $B = 1$, the charges of the three states would be $Q(\Sigma^+) = \frac{3}{2}e$, $Q(\Sigma^\circ) = \frac{1}{2}e$, and $Q(\Sigma^-) = -\frac{1}{2}e$ which are not the observed charges. Similarly, it is not possible to classify the three Σs and Λ° in a four-dimensional representation. Attempts to classify the four K particles and the η° in various SU(2) representations also fail (see Problem 31).

Referring to Table 10.1, we see that for an algebra of rank 2, the number of vectors in the root diagram is $2\pi/\theta$. For example, the group SU(3), characterized by $\theta = \pi/3$, has six root vectors. (It is also occasionally referred to as A_2.) Since SU(3) has six roots and is a group of rank 2, it is generated by eight generators.

Let $\mathbf{v}(1)$ be the largest root in that $|\mathbf{v}(\beta)| \leqslant |\mathbf{v}(1)|$ for all β. Let $\mathbf{v}(2)$ be the root adjacent to (and 60° rotated from) $\mathbf{v}(1)$. Referring to Table 10.1, we see that $\mathbf{v}(2)$ is the same length as $\mathbf{v}(1)$. A third root is generated from these using the fundamental theorem for roots. From Table 10.1, for SU(3), $N_1 = N_2 = 1$. Therefore,

$$\mathbf{v}(3) = \mathbf{v}(1) - N_1\mathbf{v}(2) = \mathbf{v}(1) - \mathbf{v}(2) \tag{10.294}$$

The remaining three roots are $-\mathbf{v}(1)$, $-\mathbf{v}(2)$, and $-\mathbf{v}(3)$. From equations 10.294

$$|\mathbf{v}(3)| = \sqrt{|\mathbf{v}(1)|^2 + |\mathbf{v}(2)|^2 - 2|\mathbf{v}(1)||\mathbf{v}(2)|\cos 60^\circ} = |\mathbf{v}(1)| \tag{10.295}$$

That is, all six root vectors of SU(3) have the same length. This length can be determined from the normalization conditions expressed in equation 10.145, which for SU(3) is

$$\text{(a)} \quad \sum_{a=1}^{2} \sum_{\alpha=\pm 1}^{\pm 3} v_a(\alpha) v_a(\alpha) = \sum_{a=1}^{2} \delta_{aa} = 2 \quad \Rightarrow \quad \text{(b)} \quad 6|\mathbf{v}(1)|^2 = 2 \tag{10.296}$$

In this way we can construct the root vector diagram of SU(3). If we designate $\mathbf{v}(1)$ to lie along the "x-axis" of our root diagram, then

$$\text{(a)} \quad \mathbf{v}(1) = \left(\frac{1}{\sqrt{3}}, 0\right) \quad \text{(b)} \quad \mathbf{v}(2) = \left(\frac{1}{2\sqrt{3}}, \frac{1}{2}\right) \quad \text{(c)} \quad \mathbf{v}(3) = \left(-\frac{1}{2\sqrt{3}}, \frac{1}{2}\right) \tag{10.297}$$

The root diagram of SU(3) is shown in Figure 10.26.

The root diagrams for the other algebras are constructed in an identical way. (The reader is referred to the *Reviews of Modern Physics* article by Behrends et al. referred to earlier.)

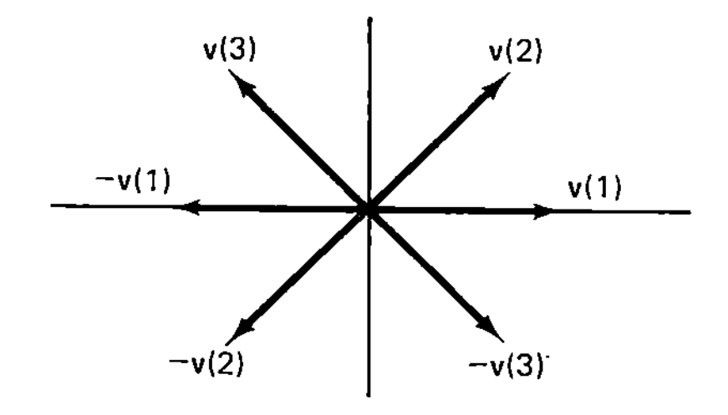

Figure 10.26 Root diagram for SU(3).

Let $\mathbf{M} \equiv (M_1, M_2)$ be the dominant weight of an irreducible representation of SU(3). As specified in equation 10.154a,

$$2\frac{\mathbf{M} \cdot \mathbf{v}(\alpha)}{\mathbf{v}(\alpha) \cdot \mathbf{v}(\alpha)} = \text{integer} \tag{10.154a}$$

Therefore, referring to equations 10.297, using the roots $\mathbf{v}(2)$ and $\mathbf{v}(3)$,

$$\text{(a)} \quad 2\frac{(M_1/2\sqrt{3} + M_2/2)}{\frac{1}{3}} = \lambda_1 \qquad \text{and} \qquad \text{(b)} \quad 2\frac{(-M_1/2\sqrt{3} + M_2/2)}{\frac{1}{3}} = \lambda_2 \tag{10.298}$$

where λ_1 and λ_2 are integers. From this we obtain the dominant weight of any irreducible representation of SU(3) to be

$$\mathbf{M} = \lambda_1\left(\frac{1}{2\sqrt{3}}, \frac{1}{6}\right) + \lambda_2\left(\frac{1}{2\sqrt{3}}, -\frac{1}{6}\right) \tag{10.299}$$

Thus, as expected, there are two fundamental irreducible representations with fundamental dominant weights

$$\text{(a)} \quad \mathbf{m}_1 = \left(\frac{1}{2\sqrt{3}}, \frac{1}{6}\right) \qquad \text{(b)} \quad \mathbf{m}_2 = \left(\frac{1}{2\sqrt{3}}, -\frac{1}{6}\right) \tag{10.300}$$

The dimensionality of any irreducible representation of SU(3) is calculated in the *Reviews of Modern Physics* article by Behrends et al. The result is

$$D(\lambda_1, \lambda_2) = \left[1 + \tfrac{1}{2}(\lambda_1 + \lambda_2)\right]\left[(1 + \lambda_1)(1 + \lambda_2)\right] \tag{10.301}$$

We note that if λ_1 and λ_2 are interchanged, the dimensionality of the representations is unchanged. Thus there are two irreducible representations of a given dimensionality. The dimensions of the two inequivalent fundamental irreducible representations are found by setting $\lambda_1 = 1$, $\lambda_2 = 0$ and $\lambda_1 = 0$, $\lambda_2 = 1$. We see that they are three-dimensional representations. A standard notation for designating a representation is to write the dimension of the representation underlined. In such a notation, the two inequivalent three-dimensional representations of SU(3) are written as $\underline{3}$ $\underline{3}^*$.

In the *Reviews of Modern Physics* article, it is shown that two irreducible representations of SU(3) are equivalent only when $\lambda_1 = \lambda_2$. Therefore, the two fundamental irreducible representations are inequivalent. The smallest dimension (except for the one-dimensional representations) for which the two representations are equivalent is the eight-dimensional representation obtained by setting $\lambda_1 = \lambda_2 = 1$. In general, setting $\lambda_1 = \lambda_2 \equiv \lambda$ in equation 10.301, we note that the dimensionality of any two equivalent representations must be the cube of an integer.

In order to classify the known particles in SU(3), we first note that SU(3) has two diagonal generators. One of these can be identified with τ_3. The second describes a new property of the particles. Because we cannot fit the Σ, Λ°, Ξ and the K, η into representations of SU(2), Gell-Mann postulated that particles had another property which he called *strangeness* S. Those particles that can be classified in SU(2) subject to the charge formula, have $S = 0$ (nucleons and pions). Those that cannot are strange particles in that they have $S \neq 0$. The charge formula is then modified to be

$$Q = e\left[\tau_3 + \tfrac{1}{2}(B + S)\right] \tag{10.302}$$

which does not affect nucleons and pions. The values of S are assigned to the strange

TABLE 10.2

(a) τ_3, S assignments for baryons

$B = 1$	p	n	Σ^+	Σ°	Σ^-	Λ°	Ξ°	Ξ^-
τ_3	$\frac{1}{2}$	$-\frac{1}{2}$	1	0	-1	0	$\frac{1}{2}$	$-\frac{1}{2}$
S	0	0	-1	-1	-1	-1	-2	-2

(b) τ_3, S assignments for mesons

$B = 0$	K^+	K°	π^+	π°	π^-	η°	$\bar{K}^\circ$	K^-
τ_3	$\frac{1}{2}$	$-\frac{1}{2}$	1	0	-1	0	$\frac{1}{2}$	$-\frac{1}{2}$
S	1	1	0	0	0	0	-1	-1

particles to give the correct charge values using equation 10.302. The assignment of τ_3 and S for each of the particles is given in Table 10.2.

Since each aggregate contains eight particles, it is natural to try to classify them as the states of an eight-dimensional representation, which is one of the irreducible representations of SU(3), obtained for $\lambda_1 = \lambda_2 = 1$. Referring to equation 10.299, the dominant weight of $\underset{\sim}{8}$ is

$$\mathbf{M}\left(\frac{1}{\sqrt{3}}, 0\right) \tag{10.303}$$

We can generate the other weights of the octet from the dominant weight. Applying equation 10.154a to the dominant weight, we obtain

$$\mathbf{M} - 2\frac{\mathbf{M} \cdot \mathbf{v}(1)}{\mathbf{v}(1) \cdot \mathbf{v}(1)}\mathbf{v}(1) = -\left(\frac{1}{\sqrt{3}}, 0\right) = -\mathbf{M} \tag{10.304a}$$

$$\mathbf{M} - 2\frac{\mathbf{M} \cdot \mathbf{v}(2)}{\mathbf{v}(2) \cdot \mathbf{v}(2)}\mathbf{v}(2) = \left(\frac{1}{2\sqrt{3}}, -\frac{1}{2}\right) \equiv \mathbf{m}_a \tag{10.304b}$$

$$\mathbf{M} - 2\frac{\mathbf{M} \cdot \mathbf{v}(3)}{\mathbf{v}(3) \cdot \mathbf{v}(3)}\mathbf{v}(3) = \left(\frac{1}{2\sqrt{3}}, \frac{1}{2}\right) \equiv \mathbf{m}_b \tag{10.304c}$$

Since $\mathbf{M}$ is the highest weight of the representation, $-\mathbf{M}$ must be lower than any other weight.

Applying equation 10.154b to $\mathbf{m}_a$ and $\mathbf{m}_b$ we obtain other weights,

$$\text{(d)}\quad \left(-\frac{1}{2\sqrt{3}}, \frac{1}{2}\right) = -\mathbf{m}_a \qquad \text{(e)}\quad -\left(\frac{1}{2\sqrt{3}}, \frac{1}{2}\right) = -\mathbf{m}_b \qquad \text{(f)}\quad \mathbf{m}_c = (0, 0) \tag{10.304}$$

Using the theorem for weights, we obtain seven of the eight required weights. Since no additional weights can be obtained using equations 10.154b, the remaining weight $\mathbf{m}_c$ must be zero. The weights are shown on the weight diagram in Figure 10.27 (page 574).

To describe these states in terms of the eight baryons or mesons, we define a diagonal operator called the *hypercharge* as

$$\text{(a)}\quad Y \equiv B + S \qquad \text{so that} \qquad \text{(b)}\quad Q = e\left(\tau_3 + \tfrac{1}{2}Y\right) \tag{10.305}$$

We identify the first component of each weight with τ_3 and the second component with

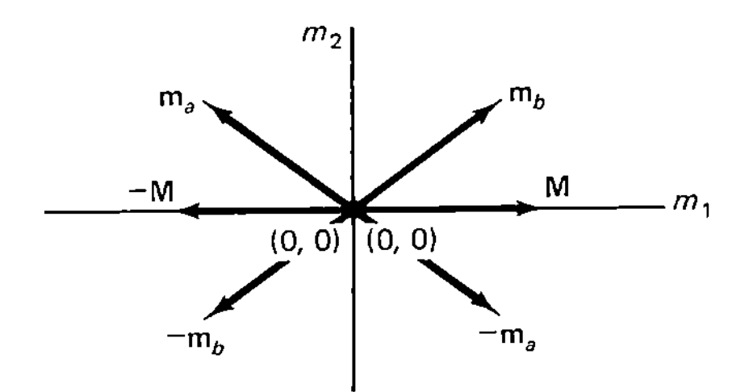

Figure 10.27 Weight diagram for the eight-dimensional representation of SU(3).

Y such that

$$\text{(a)} \quad \tau_3 \equiv \sqrt{3}\, m_1 \qquad \text{(b)} \quad Y = 2m_2 \tag{10.306}$$

The weight diagram can be labeled with the names of the particles, as shown in Figure 10.28.

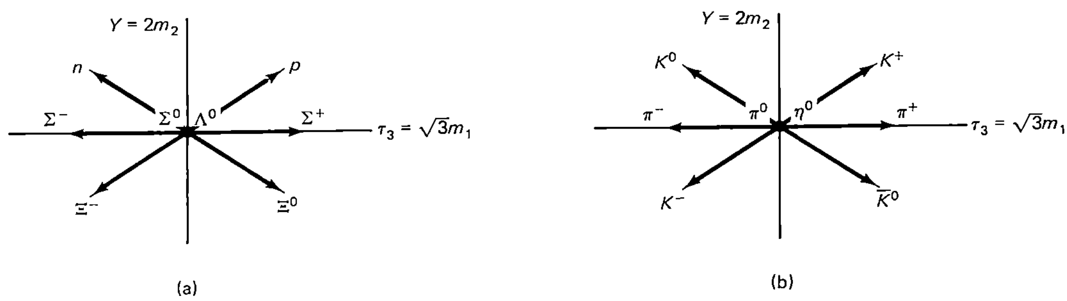

Figure 10.28 Classification of (a) baryons and (b) mesons in the octet of SU(3).

Quarks

We now return to $\underset{\sim}{3}$ and $\underset{\sim}{3}^*$. For $\underset{\sim}{3}$ ($\lambda_1 = 1, \lambda_2 = 0$), the dominant weight is

$$\mathbf{M}_1 \equiv \mathbf{m}_u = \tfrac{1}{6}(\sqrt{3}\,, 1) \tag{10.307a}$$

The weights generated from $\mathbf{m}_u$ using equation 10.154b are

$$\text{(b)} \quad \mathbf{M}_2 \equiv \mathbf{m}_d = \tfrac{1}{6}(-\sqrt{3}\,, 1) \qquad \text{(c)} \quad \mathbf{M}_3 \equiv \mathbf{m}_s = (0, -\tfrac{1}{3}) \tag{10.307}$$

An identical analysis for the weights of $\underset{\sim}{3}^*$ yields the following weights:

$$\text{(a)} \quad \mathbf{M}_1^* \equiv \mathbf{m}_{\bar{u}} = \tfrac{1}{6}(\sqrt{3}\,, -1) \qquad \text{(b)} \quad \mathbf{M}_2^* \equiv \mathbf{m}_{\bar{d}} = \tfrac{1}{6}(-\sqrt{3}\,, -1) \qquad \text{(c)} \quad \mathbf{M}_3^* \equiv \mathbf{m}_{\bar{s}} = (0, \tfrac{1}{3}) \tag{10.308}$$

The reason for designating the weights $\bar{u}$, $\bar{d}$, and $\bar{s}$ will be discussed shortly.

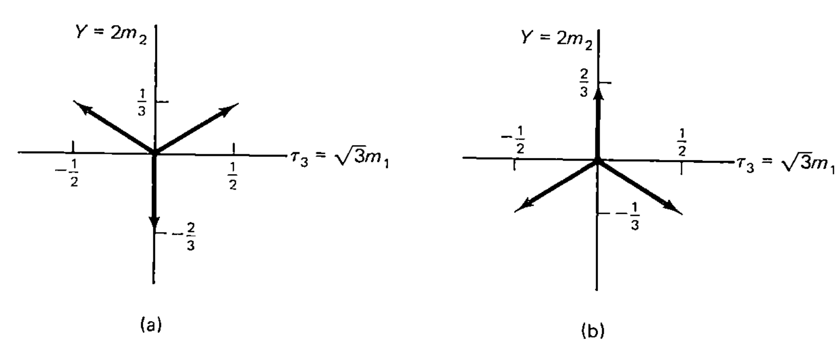

Figure 10.29 Weight diagram for (a) $\underset{\sim}{3}$ and (b) $\underset{\sim}{3}^*$.

The weight diagrams, in terms of τ_3 and Y, are shown in Figure 10.29. The three particle states corresponding to the weights of Figure 10.29a have been named *quarks*, and separately, are designated *up* (u), *down* (d), and *strange* (s) quarks. Those of Figure 10.29b are called *antiquarks*, $\bar{u}$, $\bar{d}$, and $\bar{s}$. The bar over the symbol indicates an antiparticle. Along with related properties, the antiparticle is an object that has the same mass as its particle counterpart, and opposite charge, strangeness, and baryon number.

Applying the charge formula expressed in equation 10.305b, we see that these states have fractional charges. Referring to Figures 10.29, we have

$$\text{(a)}\quad Q_u = e\left(\tfrac{1}{2} + \tfrac{1}{6}\right) = \tfrac{2}{3}e \qquad \text{(b)}\quad Q_d = e\left(-\tfrac{1}{2} + \tfrac{1}{6}\right) = -\tfrac{1}{3}e \qquad \text{(c)}\quad Q_s = e\left(0 - \tfrac{1}{3}\right) = -\tfrac{1}{3}e$$

$$\text{(d)}\quad Q_{\bar{u}} = \tfrac{1}{3}e \qquad \text{(e)}\quad Q_{\bar{d}} = -\tfrac{2}{3}e \qquad \text{(f)}\quad Q_{\bar{s}} = \tfrac{1}{3}e \tag{10.309}$$

No particle has ever been detected that carries a charge that is a fraction of the electron charge. Therefore, if quarks actually exist (and there is strong evidence that they do), it is postulated that they are permanently confined inside the observed particles (baryons and mesons).

Both the quarks and antiquarks are conjectured to have intrinsic spin angular momentum $\frac{1}{2}\hbar$. Three quarks are needed to construct a baryon. Thus, the quarks have baryon number $\frac{1}{3}$, and therefore, integral strangeness. The antiquarks have baryon number $-\frac{1}{3}$, and the strangeness of each antiquark is the negative of the corresponding quark.

The mesons are made by confining a quark and an antiquark, that is, a particle from $\underset{\sim}{3}$ and one from $\underset{\sim}{3}^*$. In terms of the representations of SU(3), the octet of mesons must arise from the direct product

$$\underset{\sim}{3} \otimes \underset{\sim}{3}^* = \underset{\sim}{8} \oplus \underset{\sim}{1} \tag{10.310}$$

In quantum theory, when two spin-$\frac{1}{2}\hbar$ quarks combine, their spins can align themselves antiparallel ($\uparrow\downarrow$), resulting in particles with $0\hbar$ spin. These are the mesons shown in Figure 10.28b. The quarks can also align themselves with their spins parallel ($\uparrow\uparrow$), resulting in particles of spin $1\hbar$. Therefore, one should also expect to find an octet of spin 1 mesons arising from $\underset{\sim}{3} \otimes \underset{\sim}{3}^*$. Such an octet has been well established. The weight diagram for this aggregate of particles is identical to that of Figure 10.28b if we

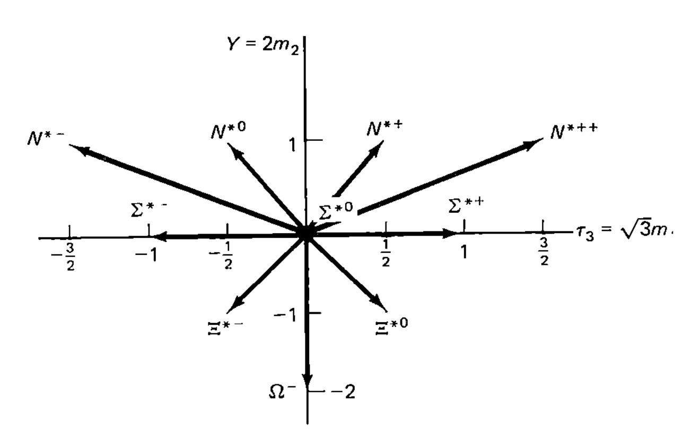

Figure 10.30 Decuplet of baryon states.

replace

$$\text{(a)}\quad (\pi^+, \pi^\circ, \pi^-) \rightarrow (\rho^+, \rho^\circ, \rho^-) \qquad \text{(b)}\quad (K^+, K^\circ, \bar{K}^\circ, K^-) \rightarrow (K^{*+}, K^{*\circ}, \bar{K}^{*\circ}, K^{*-}) \tag{10.311}$$

and the η° state is replaced by a linear combination of two observed particles

$$\eta^\circ \rightarrow \alpha\omega^\circ + \beta\phi^\circ \tag{10.311c}$$

Since three quarks are required to build a baryon, the octet of baryons arises from the direct product

$$\underset{\sim}{3} \otimes \underset{\sim}{3} \otimes \underset{\sim}{3} = \underset{\sim}{10} \oplus \underset{\sim}{8} \oplus \underset{\sim}{8} \oplus \underset{\sim}{1} \tag{10.312}$$

In addition to the octet of baryons, the states of $\underset{\sim}{10}$ also have charges that are integral multiples of the electronic charge. A group of particles has been detected that fits the characteristics of this 10-dimensional representation. The particle names are N^* or Δ, Σ^*, Ξ^*, and Ω^-. The weight diagram of $\underset{\sim}{10}$ is shown in Figure 10.30 with the states labeled by the particle names.

In addition to the success of these particle classifications in irreducible representations of SU(3), relations between the masses of these particles have been derived assuming that the forces involved in the interactions of these particles are approximately SU(3) invariant. These successes lend great credence to the idea that an SU(3) symmetry of baryon and meson interactions is a valid model of these forces. The reader interested in more detail about these ideas and calculations is referred to several excellent sources on the subject. Among these are the books by D. Lichtenberg, *Unitary Symmetry and Elementary Particles*, 2nd ed. (New York: Academic Press, 1978) and by N. Swamy and M. Samuel, *Group Theory Made Easy for Scientists and Engineers* (New York: John Wiley & Sons, Inc., 1979).

Matrix representations

With the weights and roots established for a given representation, it is straightforward to determine the matrix representations of the generators. For example, if we label the states of $\underset{\sim}{3}$ of SU(3) by $|\underset{\sim}{3}, \mathbf{m}, \mathbf{J}\rangle \equiv |\underset{\sim}{3}, A\rangle$, then A will have the values 1, 2, 3, or u, d, s. For diagonal generators,

$$\mathbf{D}_i |\underset{\sim}{3}, A\rangle = m_i(A) |\underset{\sim}{3}, A\rangle \tag{10.313}$$

$$\Rightarrow \quad (\mathbf{D}_i)_{AA} = m_i(A) \tag{10.314}$$

Thus, from equations 10.307,

$$\text{(a)}\quad (\mathbf{D}_1)_{11} \equiv (\mathbf{D}_1)_{uu} = (\mathbf{m}_u)_1 = \frac{\sqrt{3}}{6} \qquad \text{(b)}\quad (\mathbf{D}_1)_{22} \equiv (\mathbf{D}_1)_{dd} = (\mathbf{m}_d)_1 = -\frac{\sqrt{3}}{6}$$

$$\text{(c)}\quad (\mathbf{D}_1)_{33} \equiv (\mathbf{D}_1)_{ss} = (\mathbf{m}_s)_1 = 0$$

$$(10.315)$$

$$\text{(d)}\quad (\mathbf{D}_2)_{11} \equiv (\mathbf{D}_2)_{uu} = (\mathbf{m}_u)_2 = \tfrac{1}{6} \qquad \text{(e)}\quad (\mathbf{D}_2)_{22} \equiv (\mathbf{D}_2)_{dd} = (\mathbf{m}_d)_2 = \tfrac{1}{6}$$

$$\text{(f)}\quad (\mathbf{D}_2)_{33} \equiv (\mathbf{D}_2)_{ss} = (\mathbf{m}_s)_2 = -\tfrac{1}{3}$$

$$\Rightarrow \quad \text{(a)}\quad \mathbf{D}_1 = \frac{\sqrt{3}}{6}\begin{bmatrix} 1 & 0 & 0 \\ 0 & -1 & 0 \\ 0 & 0 & 0 \end{bmatrix} \qquad \text{(b)}\quad \mathbf{D}_2 = \frac{1}{6}\begin{bmatrix} 1 & 0 & 0 \\ 0 & 1 & 0 \\ 0 & 0 & -2 \end{bmatrix} \qquad (10.316)$$

To obtain the matrix representation of the nondiagonal generators, we recall (see equation 10.153) that $\mathbf{N}_\alpha|\mathbf{m}, \mathbf{J}\rangle = 0$ if $\mathbf{m} + \mathbf{v}(\alpha)$ is not a weight of the representation. Starting with the highest state of the representation, either $\mathbf{m}_u - \mathbf{v}(1)$ is a weight or $\mathbf{N}_{-1}|\underset{\sim}{3}, u\rangle = 0$. Since

$$\mathbf{m}_u - \mathbf{v}(1) = \tfrac{1}{6}(-\sqrt{3}, 1) = \mathbf{m}_d \qquad (10.317\text{a})$$

then, referring to equation 10.161,

$$\mathbf{N}_{-1}|\underset{\sim}{3}, u\rangle = [\mathbf{m}_u \cdot \mathbf{v}(1)]^{1/2}|\underset{\sim}{3}, d\rangle = \frac{1}{\sqrt{6}}|\underset{\sim}{3}, d\rangle \qquad (10.317\text{b})$$

Since

$$\mathbf{m}_d - \mathbf{v}(1) = \tfrac{1}{6}(-3\sqrt{3}, 1) \qquad (10.318\text{a})$$

is not a weight

$$\mathbf{N}_{-1}|\underset{\sim}{3}, d\rangle = 0 \qquad (10.318\text{b})$$

By identical analysis

$$\mathbf{N}_{-1}|\underset{\sim}{3}, s\rangle = 0 \qquad (10.318\text{c})$$

Defining the basis states in column vector form as

$$\text{(a)}\quad |\underset{\sim}{3}, u\rangle = \begin{bmatrix} 1 \\ 0 \\ 0 \end{bmatrix} \qquad \text{(b)}\quad |\underset{\sim}{3}, d\rangle = \begin{bmatrix} 0 \\ 1 \\ 0 \end{bmatrix} \qquad \text{(c)}\quad |\underset{\sim}{3}, s\rangle = \begin{bmatrix} 0 \\ 0 \\ 1 \end{bmatrix} \qquad (10.319)$$

the matrix representation of $\mathbf{N}_{-1}$ in $\underset{\sim}{3}$ is

$$\mathbf{N}_{-1} = \frac{1}{\sqrt{6}}\begin{bmatrix} 0 & 0 & 0 \\ 1 & 0 & 0 \\ 0 & 0 & 0 \end{bmatrix} \qquad (10.320\text{a})$$

The reader will show in Problem 32 that

$$\text{(b)}\quad \mathbf{N}_{-2} = \frac{1}{\sqrt{6}}\begin{bmatrix} 0 & 0 & 0 \\ 0 & 0 & 0 \\ 1 & 0 & 0 \end{bmatrix} \qquad \text{(c)}\quad \mathbf{N}_{-3} = \frac{1}{\sqrt{6}}\begin{bmatrix} 0 & 0 & 0 \\ 0 & 0 & 0 \\ 0 & 1 & 0 \end{bmatrix} \qquad (10.320)$$

Using $\mathbf{N}_\alpha = \mathbf{N}_{-\alpha}^\dagger$ and the fact that the elements of $\mathbf{N}_{-\alpha}$ are real, the matrix representations of $\mathbf{N}_1$, $\mathbf{N}_2$, and $\mathbf{N}_3$ are simply the transpose of the matrices of equations 10.320.

Referring to equations 10.317b, we note that $\mathbf{N}_{-1}$ can be expressed in terms of the bra and ket descriptions of the bases. Writing

$$\text{(a)} \quad \mathbf{N}_{-1} = \frac{1}{\sqrt{6}} |\underset{\sim}{3}, d\rangle\langle \underset{\sim}{3}, u| \quad \Rightarrow \quad \text{(b)} \quad \mathbf{N}_{-1}|\underset{\sim}{3}, u\rangle = \frac{1}{\sqrt{6}} |\underset{\sim}{3}, d\rangle\langle \underset{\sim}{3}, u|\underset{\sim}{3}, u\rangle = \frac{1}{\sqrt{6}} |\underset{\sim}{3}, d\rangle \tag{10.321}$$

Thus equation 10.321a is an alternative way to express a generator. In such a notation, for example,

$$\mathbf{D}_1 = \frac{\sqrt{3}}{6} |\underset{\sim}{3}, u\rangle\langle \underset{\sim}{3}, u| - \frac{\sqrt{3}}{6} |\underset{\sim}{3}, d\rangle\langle \underset{\sim}{3}, d| + 0 |\underset{\sim}{3}, s\rangle\langle \underset{\sim}{3}, s| \tag{10.322}$$

Lorentz group

We end this chapter with an introductory discussion of the group of Lorentz transformations. As we have noted, SO(3) is the group of rotations defined by the Euler angles. Such rotations leave the magnitude of the position vector, $r^2 = x^2 + y^2 + z^2$ invariant. The Lorentz group is the group of all transformations that leave the square of the four-component line element $s^2 = x^2 + y^2 + z^2 - c^2t^2$ invariant. Of course, any rotation that leaves $x^2 + y^2 + z^2$ invariant will also leave $x^2 + y^2 + z^2 - c^2t^2$ invariant. Thus SO(3) is a subgroup of the Lorentz group.

In Chapter 9, we discussed a set of Lorentz transformations for the relative motion of two coordinate systems along the x-axis of one of them. This transformation also leaves the space-time element invariant. In matrix form, the Lorentz transformation that connects points in these two frames can be written

$$\begin{bmatrix} X^1 \\ X^2 \\ X^3 \\ X^4 \end{bmatrix} = \begin{bmatrix} \gamma & 0 & 0 & -\beta\gamma \\ 0 & 1 & 0 & 0 \\ 0 & 0 & 1 & 0 \\ -\beta\gamma & 0 & 0 & \gamma \end{bmatrix} \begin{bmatrix} x^1 \\ x^2 \\ x^3 \\ x^4 \end{bmatrix} \tag{9.192a}$$

where $x^4 \equiv ct$. There is, of course, a transformation matrix for relative motion along the y-axis and another for motion along the z-axis. Thus the group of Lorentz transformations consist of these three space-time transformations plus the rotations of SO(3). That is, there are six generators of the Lorentz group.

The Lorentz transformations involving one space and the time coordinate can be cast in a form similar to the rotation matrix. We note that $\beta = (v/c) \in [0, 1]$ and $\gamma = (1 - \beta^2)^{-1/2} \in [1, \infty]$. Thus, using equation 9.192a, we can parametrize this type of transformation by

$$\text{(a)} \quad \gamma = \cosh\phi \qquad \text{(b)} \quad \beta\gamma = \sinh\phi \quad \Rightarrow \quad \text{(c)} \quad \beta = \tanh\phi \tag{10.323}$$

Then the Lorentz transformation matrix of equation 9.192a becomes

$$L_{xt}(\phi) = \begin{bmatrix} \cosh\phi & 0 & 0 & -\sinh\phi \\ 0 & 1 & 0 & 0 \\ 0 & 0 & 1 & 0 \\ -\sinh\phi & 0 & 0 & \cosh\phi \end{bmatrix} \tag{10.324}$$

We will refer to this as a rotation in the x-t plane of four-dimensional space-time. We note that like the three space rotations of SO(3), these Lorentz rotations satisfy

$$L_{xt}(\phi_1)L_{xt}(\phi_2) = L_{xt}(\phi_1 + \phi_2) \tag{10.325}$$

The generators associated with the x-t rotation are defined from

$$\text{(a)} \quad L_{xt}(d\phi) \equiv \mathbf{1} + i\mathbf{\Lambda}_{xt}\, d\phi \quad \Rightarrow \quad \text{(b)} \quad \mathbf{\Lambda}_{xt} = \begin{bmatrix} 0 & 0 & 0 & i \\ 0 & 0 & 0 & 0 \\ 0 & 0 & 0 & 0 \\ i & 0 & 0 & 0 \end{bmatrix} \tag{10.326}$$

Following the analysis used for SO(2) and SO(3), with the use of equations 10.325,

$$\text{(a)} \quad L_{xt}(\phi)L_{xt}(d\phi) = L_{xt}(\phi + d\phi) \quad \Rightarrow \quad \text{(b)} \quad \frac{1}{L_{xt}}\frac{dL_{xt}}{d\phi} = i\mathbf{\Lambda}_{xt} \tag{10.327}$$

so like the spatial rotations of SO(3), the x-t "rotations" can be written as

$$L_{xt}(\phi) = e^{i\Lambda_{xt}\phi} \tag{10.328}$$

We see that the Lorentz group has three generators associated with a rotation in a space-time plane. We have denoted these by $\mathbf{\Lambda}_{xt}$, $\mathbf{\Lambda}_{yt}$, and $\mathbf{\Lambda}_{zt}$.

The generators of SO(3), $\mathbf{J}_x$, $\mathbf{J}_y$, and $\mathbf{J}_z$ are also generators of the Lorentz group. $\mathbf{J}_x$, for example, generates a rotation about the x-axis which can also be described as a rotation in the y-z plane. As such, the generators of the spatial rotations will be denoted by $\mathbf{\Lambda}_{xy}$, $\mathbf{\Lambda}_{yx}$, and $\mathbf{\Lambda}_{zx}$, with

$$(\mathbf{J}_x, \mathbf{J}_y, \mathbf{J}_z) \rightarrow (\mathbf{\Lambda}_{yz}, \mathbf{\Lambda}_{zx}, \mathbf{\Lambda}_{xy}) \tag{10.329}$$

Since the three spatial rotations do not affect x^4, the four-dimensional representations of these generators are constructed by augmenting each of the 3×3 matrices of $\mathbf{J}_x$, $\mathbf{J}_y$, and $\mathbf{J}_z$ with a fourth row and column containing zeros. For example, the three-dimensional representation of $\mathbf{J}_x$ is

$$\mathbf{J}_x = \begin{bmatrix} 0 & 0 & 0 \\ 0 & 0 & i \\ 0 & -i & 0 \end{bmatrix} \tag{10:189a}$$

$$\Rightarrow \quad \mathbf{\Lambda}_{yz} = \begin{bmatrix} 0 & 0 & 0 & 0 \\ 0 & 0 & i & 0 \\ 0 & -i & 0 & 0 \\ 0 & 0 & 0 & 0 \end{bmatrix} \tag{10.330}$$

The commutation relations among the six generators is straightforward to deduce. If the indices are relabeled, with Greek indices taking on the values 1, 2, 3, 4, then

$$\text{(a)} \quad \mathbf{\Lambda}_{\alpha\beta}\mathbf{\Lambda}_{\beta\gamma} - \mathbf{\Lambda}_{\beta\gamma}\mathbf{\Lambda}_{\alpha\beta} = i\mathbf{\Lambda}_{\alpha\gamma} \quad (\textit{no sum on } \beta) \qquad \text{(b)} \quad \mathbf{\Lambda}_{\alpha\beta}\mathbf{\Lambda}_{\gamma\delta} - \mathbf{\Lambda}_{\gamma\delta}\mathbf{\Lambda}_{\alpha\beta} = 0 \tag{10.331}$$

where α, β, γ, δ have different values. In this description

$$\mathbf{\Lambda}_{\alpha\beta} = -\mathbf{\Lambda}_{\beta\alpha} \tag{10.332}$$

Representation of the Lorentz group

Because we can treat transformations in a space-time plane as rotations, there are algebraic similarities between the six generators of the Lorentz group and the generators of SO(3). As such, we will deduce a representation of the Lorentz group in a development similar to that of SO(3). We define basis functions as

$$\begin{aligned} &\text{(a)} \quad U_{++} \equiv z + ct = x^3 + x^4 && \text{(b)} \quad U_{+-} \equiv x + iy = x^1 + ix^2 \\ &\text{(c)} \quad U_{-+} \equiv x - iy = x^1 - ix^2 && \text{(d)} \quad U_{--} = -z + ct = -x^3 + x^4 \end{aligned} \tag{10.333}$$

In analogy with equations 10.265, we define the operators in the form

$$\mathbf{\Lambda}_{12} \equiv -i\left(x\frac{\partial}{\partial y} - y\frac{\partial}{\partial x}\right) \equiv -i\left(x^1\partial_2 - x^2\partial_1\right) \tag{10.334a}$$

for a spatial rotation in the x-y plane. We have introduced the notation $\partial_\alpha \equiv \partial/\partial x^\alpha$. For a rotation in the z-t plane, the operator is defined as

$$\mathbf{\Lambda}_{34} \equiv -\left(\frac{z}{c}\frac{\partial}{\partial t} + ct\frac{\partial}{\partial z}\right) = -\left(x^3\partial_4 + x^4\partial_3\right) \tag{10.334b}$$

From the $\mathbf{\Lambda}$ operators, we define

$$\text{(a)}\quad \mathbf{A}_i \equiv \tfrac{1}{2}(\mathbf{\Lambda}_{jk} + \mathbf{\Lambda}_{i4}) \qquad \text{(b)}\quad \mathbf{B}_i \equiv \tfrac{1}{2}(\mathbf{\Lambda}_{jk} - \mathbf{\Lambda}_{i4}) \tag{10.335}$$

with $i \neq j \neq k$, and i, j, k cyclic permutations of 1, 2, 3. For example,

$$\mathbf{A}_2 \equiv \tfrac{1}{2}(\mathbf{\Lambda}_{31} - \mathbf{\Lambda}_{24}) \tag{10.335c}$$

The raising and lowering operators of the Lorentz group are defined by

$$\begin{aligned} &\text{(a)}\quad \mathbf{E}_+ \equiv -\mathbf{A}_1 + i\mathbf{A}_2 \qquad \text{(b)}\quad \mathbf{E}_- \equiv -\mathbf{A}_1 - i\mathbf{A}_2 \qquad \text{(c)}\quad \mathbf{E}_3 \equiv -\mathbf{A}_3 \\ &\text{(d)}\quad \mathbf{F}_+ \equiv \mathbf{B}_1 + i\mathbf{B}_2 \qquad \text{(e)}\quad \mathbf{F}_- \equiv \mathbf{B}_1 - i\mathbf{B}_2 \qquad \text{(f)}\quad \mathbf{F}_3 \equiv \mathbf{B}_3 \end{aligned} \tag{10.336}$$

As the reader will indicate in Problem 34, these operators satisfy

$$\begin{aligned} &\text{(a)}\quad (\mathbf{E}_+, \mathbf{E}_3) = \mathbf{E}_+ \qquad \text{(b)}\quad (\mathbf{E}_-, \mathbf{E}_3) = -\mathbf{E}_- \qquad \text{(c)}\quad (\mathbf{E}_+, \mathbf{E}_-) = 2\mathbf{E}_3 \\ &\text{(d)}\quad (\mathbf{F}_+, \mathbf{F}_3) = \mathbf{F}_+ \qquad \text{(e)}\quad (\mathbf{F}_-, \mathbf{F}_3) = -\mathbf{F}_- \qquad \text{(f)}\quad (\mathbf{F}_+, \mathbf{F}_-) = -2\mathbf{F}_3 \qquad \text{(g)}\quad (\mathbf{E}_i, \mathbf{F}_j) = 0 \end{aligned} \tag{10.337}$$

In Problem 35, the reader will also indicate that, for example,

$$\begin{aligned} \mathbf{E}_- U_{++} &= \left[-\tfrac{1}{2}(\mathbf{\Lambda}_{23} + \mathbf{\Lambda}_{14}) - i\tfrac{1}{2}(\mathbf{\Lambda}_{31} + \mathbf{\Lambda}_{24})\right](x^3 + x^4) \\ &= \tfrac{1}{2}\left[i(x^2\partial_3 - x^3\partial_2) + (x^1\partial_4 + x^4\partial_1) - (x^3\partial_1 - x^1\partial_3) + i(x^2\partial_4 + x^4\partial_2)\right](x^3 + x^4) = U_{+-} \end{aligned} \tag{10.338a}$$

$$\mathbf{F}_3 U_{+-} = \tfrac{1}{2}\left[-i(x^1\partial_2 - x^2\partial_1) + (x^3\partial_4 + x^4\partial_3)\right](x^1 + ix^2) = \tfrac{1}{2}U_{+-} \tag{10.338b}$$

The entries of Table 10.3 are generated in this way.

We note that the $\mathbf{E}$ operators affect only the second index of the eigenfunction. That is, with * representing either + or −,

$$\text{(a)}\quad \mathbf{E}_+ U_{*+} = 0 \qquad \text{(b)}\quad \mathbf{E}_+ U_{*-} = U_{*+} \qquad \text{(c)}\quad \mathbf{E}_- U_{*+} = U_{*-} \qquad \text{(d)}\quad \mathbf{E}_- U_{*-} = 0 \tag{10.339}$$

TABLE 10.3 OPERATORS AND EIGENFUNCTIONS OF THE LORENTZ GROUP

	E_+	E_-	E_3	F_+	F_-	F_3
U_{++}	0	U_{+-}	$\frac{1}{2}U_{++}$	0	U_{-+}	$\frac{1}{2}U_{++}$
U_{+-}	U_{++}	0	$-\frac{1}{2}U_{+-}$	0	U_{--}	$\frac{1}{2}U_{+-}$
U_{-+}	0	U_{--}	$\frac{1}{2}U_{-+}$	U_{++}	0	$-\frac{1}{2}U_{-+}$
U_{--}	U_{-+}	0	$-\frac{1}{2}U_{--}$	U_{+-}	0	$-\frac{1}{2}U_{--}$

The $\mathbf{F}$ operators similarly affect only the first index. In addition,

$$\text{(e)}\quad \mathbf{E}_3 U_{*\pm} = \pm\tfrac{1}{2} U_{*\pm} \qquad \text{(f)}\quad \mathbf{F}_3 U_{\pm *} = \pm\tfrac{1}{2} U_{\pm *} \tag{10.339}$$

We see that $\mathbf{E}_\pm$ and $\mathbf{F}_\pm$ are raising and lowering operators, and that $\mathbf{E}_3$ and $\mathbf{F}_3$ are diagonal operators with eigenvalues $\pm\frac{1}{2}$. In the relativistic quantum theory of spin-$\frac{1}{2}$ particles, $\mathbf{E}_3$ and $\mathbf{F}_3$ play a role similar to the $j = \frac{1}{2}$ representation of $\mathbf{J}_3$ of SO(3), and $\mathbf{E}_\pm$ and $\mathbf{F}_\pm$ are related to $\mathbf{J}_\pm$.

As final note, we can easily deduce a matrix representation of these generators and states. We define

$$\text{(a)}\quad U_{++} = \begin{bmatrix} 1 \\ 0 \\ 0 \\ 0 \end{bmatrix} \quad \text{(b)}\quad U_{+-} = \begin{bmatrix} 0 \\ 1 \\ 0 \\ 0 \end{bmatrix} \quad \text{(c)}\quad U_{-+} = \begin{bmatrix} 0 \\ 0 \\ 1 \\ 0 \end{bmatrix} \quad \text{(d)}\quad U_{--} = \begin{bmatrix} 0 \\ 0 \\ 0 \\ 1 \end{bmatrix} \tag{10.340}$$

Then the matrix forms of the operators

$$\text{(a)}\quad \mathbf{E}_+ = \begin{bmatrix} 0 & 1 & 0 & 0 \\ 0 & 0 & 0 & 0 \\ 0 & 0 & 0 & 1 \\ 0 & 0 & 0 & 0 \end{bmatrix} \quad \text{(b)}\quad \mathbf{E}_- = \begin{bmatrix} 0 & 0 & 0 & 0 \\ 1 & 0 & 0 & 0 \\ 0 & 0 & 0 & 0 \\ 0 & 0 & 1 & 0 \end{bmatrix} \quad \text{(c)}\quad \mathbf{E}_3 = \frac{1}{2}\begin{bmatrix} 1 & 0 & 0 & 0 \\ 0 & -1 & 0 & 0 \\ 0 & 0 & 1 & 0 \\ 0 & 0 & 0 & -1 \end{bmatrix}$$

$$\text{(d)}\quad \mathbf{F}_+ = \begin{bmatrix} 0 & 0 & 1 & 0 \\ 0 & 0 & 0 & 1 \\ 0 & 0 & 0 & 0 \\ 0 & 0 & 0 & 0 \end{bmatrix} \quad \text{(e)}\quad \mathbf{F}_- = \begin{bmatrix} 0 & 0 & 0 & 0 \\ 0 & 0 & 0 & 0 \\ 1 & 0 & 0 & 0 \\ 0 & 1 & 0 & 0 \end{bmatrix} \quad \text{(f)}\quad \mathbf{F}_3 = \frac{1}{2}\begin{bmatrix} 1 & 0 & 0 & 0 \\ 0 & 1 & 0 & 0 \\ 0 & 0 & -1 & 0 \\ 0 & 0 & 0 & -1 \end{bmatrix} \tag{10.341}$$

produce Table 10.3. In this form we see that $\mathbf{E}_\pm$ and $\mathbf{E}_3$ generate transformations among the upper two entries in the $\mathbf{U}$ columns and separately among the lower two entries. $\mathbf{F}_\pm$ and $\mathbf{F}_3$ generate transformations between the upper and lower pair of entries in the $\mathbf{U}$ columns.

PROBLEMS

1. Show that the following are not groups. Specify which group axioms are not satisfied.
   - **(a)** The set of integers under subtraction.
   - **(b)** The set of real numbers under multiplication.
   - **(c)** The set of real numbers excluding zero under division.
   - **(d)** The set of complex numbers under the operation

$$z_1 * z_2 \equiv z_1 z_2 + z_1 + z_2$$

2. Show that the following are groups. Identify the identity and inverse elements of each group.
   - **(a)** The set of real numbers > 0 under multiplication.
   - **(b)** The set of complex numbers of magnitude 1 under multiplication.
   - **(c)** The infinite set of matrices

$$\mathbf{R}(\theta) = \begin{bmatrix} \cos\theta & \sin\theta \\ -\sin\theta & \cos\theta \end{bmatrix}$$

   for all $\theta \in [0, 2\pi]$ under ordinary matrix multiplication with the restriction that $\mathbf{R}(\theta + 2\pi) = \mathbf{R}(\theta)$. Two elements of this group are $\mathbf{R}(\theta_1)$ and $\mathbf{R}(\theta_2)$.

3. **(a)** We define left and right identity elements for a group such that $e_L * x = x$ and $x * e_R = x$. Prove that $e_L = e_R$ and thus there is only one identity element.
   - **(b)** We define left and right inverses such that $x_L^{-1} * x = e$ and $x * x_R^{-1} = e$. Use the fact that there is only one identity element to prove that $x_L^{-1} = x_R^{-1}$.

4. Prove that if $z \equiv x * y$, then $z^{-1} = y^{-1} * x^{-1}$.

5. **(a)** Show that the set consisting of the Pauli matrices and the identity matrix

$$\boldsymbol{\sigma}_1 = \begin{bmatrix} 0 & 1 \\ 1 & 0 \end{bmatrix} \quad \boldsymbol{\sigma}_2 = \begin{bmatrix} 0 & -i \\ i & 0 \end{bmatrix}$$

$$\boldsymbol{\sigma}_3 = \begin{bmatrix} 1 & 0 \\ 0 & -1 \end{bmatrix} \quad \mathbf{1} = \begin{bmatrix} 1 & 0 \\ 0 & 1 \end{bmatrix}$$

do not form a group under matrix multiplication. Identify which of the group axioms do not hold.

(b) Show that the group of the quaternions, defined by $\pm i\sigma_1$, $\pm i\sigma_2$, $\pm i\sigma_3$, $\pm \mathbf{1}$ form a group, and construct its multiplication table. Is the group Abelian?

6. **(a)** The cube roots of $+1$ are

$$\omega_1 = e^{2\pi i/3}$$

$$\omega_2 = e^{4\pi i/2}$$

$$\omega_3 = e^{6\pi i/3} = +1$$

Show that these numbers under complex number multiplication form a group isomorphic to the cyclic group C_3.

(b) Is the set $\{\omega_1, \omega_2, \omega_3\}$ a group under addition? If not, which group axioms do not hold?

(c) Does the set of the three cube roots of -1,

$$\Omega_1 = e^{i\pi/3}$$

$$\Omega_2 = e^{3i\pi/3} = -1$$

$$\Omega_3 = e^{5i\pi/3}$$

form a group under complex number multiplication? If so, construct the group multiplication table; if not, specify which group axioms do not hold.

7. Consider the two sets of matrices

$$\mathbf{A}_2 = \begin{bmatrix} 0 & 1 \\ -1 & -1 \end{bmatrix} \qquad \mathbf{B}_2 = \begin{bmatrix} -1 & -1 \\ 1 & 0 \end{bmatrix}$$

$$\mathbf{1} = \begin{bmatrix} 1 & 0 \\ 0 & 1 \end{bmatrix}$$

and

$$\mathbf{A}_3 = \begin{bmatrix} 0 & 1 & 0 \\ 0 & 0 & 1 \\ 1 & 0 & 0 \end{bmatrix} \qquad \mathbf{B}_3 = \begin{bmatrix} 0 & 0 & 1 \\ 1 & 0 & 0 \\ 0 & 1 & 0 \end{bmatrix}$$

$$\mathbf{1} = \begin{bmatrix} 1 & 0 & 0 \\ 0 & 1 & 0 \\ 0 & 0 & 1 \end{bmatrix}$$

Prove that each set of matrices under matrix multiplication forms a group that is isomorphic to C_3.

8. The elements of S_3 are written as

$$q = \begin{pmatrix} g_1 & g_2 & g_3 \\ g_3 & g_1 & g_2 \end{pmatrix} \qquad r = \begin{pmatrix} g_1 & g_2 & g_3 \\ g_1 & g_3 & g_2 \end{pmatrix}$$

$$e = \begin{pmatrix} g_1 & g_2 & g_3 \\ g_1 & g_2 & g_3 \end{pmatrix} \qquad p = \begin{pmatrix} g_1 & g_2 & g_3 \\ g_2 & g_3 & g_1 \end{pmatrix}$$

$$s = \begin{pmatrix} g_1 & g_2 & g_3 \\ g_3 & g_2 & g_1 \end{pmatrix} \qquad t = \begin{pmatrix} g_1 & g_2 & g_3 \\ g_2 & g_1 & g_3 \end{pmatrix}$$

(a) Prove that $q = p^2$.
(b) Find p^{-1} and r^{-1}.
(c) Construct the multiplication table for S_3.
(d) Find the matrices for the elements p and s in the regular representation, denoted by Γ_R.

9. Show that S_3 has the following equivalence classes:

$$\mathscr{C}_1 \equiv \{e\}$$

$$\mathscr{C}_2 \equiv \{p, q\}$$

$$\mathscr{C}_3 \equiv \{r, s, t\}$$

10. A group element x is cyclic with period n.
(a) Prove that x^{-1} is cyclic with period n.
(b) y is any element of the group except x and e. z is conjugate to x through y; that is, $z = y^{-1}xy$. Prove that $z = y^{-1}xy$ and $z^{-1} = y^{-1}x^{-1}y$ are cyclic with period n. (Thus any two elements in the same equivalence class have the same period.)

11. There are two distinct groups of order 4. One is the dihedral group D_2 defined by the multiplication table of Figure 10.5. The other is the cyclic group $C_4 \equiv \{e = x^4, x, x^2, x^3\}$.
(a) Construct the multiplication table for C_4.
(b) C_4 has one proper subgroup. Identify its elements.
(c) Find the subgroup of the permutation group S_4 to which C_4 is isomorphic.

12. Referring to Figure 10.5, find the left and right cosets of the subgroup $\{e, x\}$ of D_2.

13. The lowest order non-Abelian group is the sixth-order group of rotations of an equilateral triangle that leave the orientation of the triangle unchanged. Referring to Figure 10.31, the six transformations are:

e. Nothing is done to the triangle.
p. The triangle is rotated about its center by $2\pi/3$ in the plane of the page. The corners undergo the following interchanges: $1 \to 2$, $2 \to 3$, $3 \to 1$.
q. The triangle is rotated about its center by $-2\pi/3$ in the plane of the page. The corners undergo the following interchanges: $1 \to 3$, $3 \to 2$, $2 \to 1$.

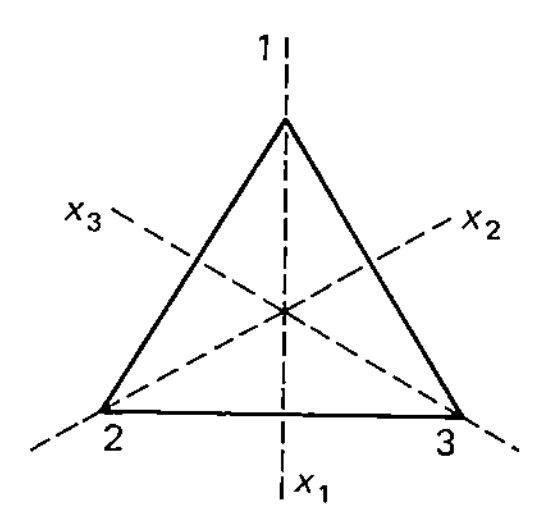

Figure 10.31 Symmetry axes of equilateral triangle.

r. The triangle is rotated about the x_1-axis by π out of the plane of the page. The corners undergo the following interchanges: $1 \to 1$, $2 \to 3$, $3 \to 2$.

s. The triangle is rotated about the x_2-axis by π out of the plane of the page. The corners undergo the following interchanges: $1 \to 3$, $2 \to 2$, $3 \to 1$.

t. The triangle is rotated about the x_3-axis by π out of the plane of the page. The corners undergo the following interchanges: $1 \to 2$, $2 \to 1$, $3 \to 3$.

Express these elements in terms of permutations of the three corners. Identify the group to which this group is isomorphic.

14. Indicate that the matrices of Γ_2 of equations 10.37 Γ_2' of equations 10.38 and Γ_2'' of equations 10.39 are matrix representations of S_3 by demonstrating that the products $p * q = e$, $q * s = t$, and $s * t = p$ are satisfied by the appropriate matrix products in each of these representations.

15. Prove that for any matrix representation of a group $G \equiv \{g_i\}$, $\mathbf{M}(g_i^{-1}) = \mathbf{M}^{-1}(g_i)$. [Therefore, for a unitary representation, $\mathbf{M}(g_i^{-1}) = \mathbf{M}^{\dagger}(g_i)$.]

16. (a) Show that the cyclic group C_{10} satisfies $C_{10} = C_2 \otimes C_5$.
(b) Consider the cyclic group $C_N = \{e = x^N, x, x^2, \ldots, x^{N-1}\}$. For N even, $\frac{1}{2}N$ odd, show that $C_N = C_2 \otimes C_{N/2}$.
(c) For the cyclic group C_N, indicate that if $\frac{1}{2}N$ is even, then $C_N \neq C_2 \otimes C_{N/2}$ by showing that $C_8 \neq C_2 \otimes C_4$.

17. Find the six-dimensional representation of the rotation denoted by $\mathbf{R}_y(60°)$ for which the six basis functions expressed in equations 10.109 are invariant.

18. (a) Prove that the six functions given in equations 10.109 are not basis functions for the operator that generates translations in the x-direction given by $T(\lambda)\mathbf{r} = (x + \lambda, y, z)$.
(b) Show that the set of functions $g_i(\mathbf{r}) = \{1, x\}$ are the bases for the two-dimensional representation of $T(\lambda)$. Determine the two-dimensional representation of $T(\lambda)$ so that the basis functions are invariant under these translations.

19. The continuous Lie group of translations in one dimension are characterized by the operations $\mathbf{T}(\lambda)\psi_\alpha(x, y, z) = \psi_\alpha(x + \lambda, y, z)$. Show that
(a) $\mathbf{T}(\lambda_1)\mathbf{T}(\lambda_2) = \mathbf{T}(\lambda_1 + \lambda_2)$
(b) $\mathbf{T}(0) = \mathbf{1}$
(c) $\mathbf{T}^{-1}(\lambda) = \mathbf{T}(-\lambda)$
(d) The generator of this group, $\mathbf{K}$, is defined by

$$\mathbf{T}(d\lambda) = \mathbf{1} + \mathbf{K}\, d\lambda$$

Show that the translation operator can be written

$$\mathbf{T}(\lambda) = e^{\mathbf{K}\lambda}$$

(e) Using the basis functions of the two-dimensional representation deduced in Problem 18, part (b), find the two-dimensional representation of $\mathbf{K}$.

20. γ_A, γ_B, and γ_C are three generators of a Lie group. Prove that

$$(\gamma_A\gamma_B, \gamma_C) = \gamma_A(\gamma_B, \gamma_C) + (\gamma_A, \gamma_C)\gamma_B$$

21. Referring to Table 10.1:
(a) How many generators does the rank 2 group C_2 have? (This is not the finite cyclic group C_2.)
(b) How many generators does the rank 3 group have if its root vectors are mutually perpendicular?
(c) How many generators does a rank 4 group with mutually orthogonal root vectors have?

22. Show that if $\mathbf{M}'$ is the lowest weight of a representation, then

$$\mathbf{N}_\alpha|\mathbf{M}', \mathbf{j}\rangle = \pm[-\mathbf{v}(\alpha) \cdot \mathbf{M}']^{1/2}|\mathbf{M}' + \mathbf{v}(\alpha), \mathbf{j}\rangle$$

23. Let $\mathbf{r}$ and $\mathbf{s}$ be three-component vectors. Show that when the generators of SO(3) $\mathbf{J}_1$, $\mathbf{J}_2$, and $\mathbf{J}_3$ are viewed as components of an operator vector (like Pauli matrices), the sum of commutators given below satisfies

$$\sum_{\substack{k=1 \\ l=1}}^{3} (r_k\mathbf{J}_k, s_l\mathbf{J}_l) = i\mathbf{r} \times \mathbf{s} \cdot \mathbf{J}$$

24. Show that in any representation, the generators of SO(3) satisfy

$$(\mathbf{J}_3, \mathbf{J}_\pm) = \pm\mathbf{J}_\pm$$

$$(\mathbf{J}_+, \mathbf{J}_-) = 2\mathbf{J}_3$$

25. Show that in any representation, the $\mathbf{J}_-$ of SO(3) satisfies

$$\mathbf{J}_-|m, j\rangle \propto |m - 1, j\rangle$$

26. (a) Using the result of $\mathbf{J}_\pm$ on the basis states $|m, j\rangle$, show that one representation of the generators of SO(3) is

$$\mathbf{J}_1 = \frac{1}{\sqrt{2}}\begin{bmatrix} 0 & 1 & 0 \\ 1 & 0 & 1 \\ 0 & 1 & 0 \end{bmatrix}$$

$$\mathbf{J}_2 = \frac{1}{\sqrt{2}}\begin{bmatrix} 0 & -i & 0 \\ i & 0 & -i \\ 0 & i & 0 \end{bmatrix}$$

$$\mathbf{J}_3 = \begin{bmatrix} 1 & 0 & 0 \\ 0 & 0 & 0 \\ 0 & 0 & -1 \end{bmatrix}$$

$$\mathbf{J}^2 = 2 * 1$$

(b) Prove that in this representation,

$$e^{i\mathbf{J}_1\theta} = \mathbf{1} + (\cos\theta - 1)\mathbf{J}_1^2 - i\sin\theta\mathbf{J}_1$$

(c) From the results of part (b), determine the matrix $d^{j=1}(\theta)$ defined in equation 10.235.

27. Consider the state

$$|m, j\rangle \equiv \frac{x^{j+m}y^{j-m}}{\sqrt{(j+m)!(j-m)!}}$$

(a) From the definition $\mathbf{J}_- \equiv y(\partial/\partial x)$, show that

$$\mathbf{J}_-|m, j\rangle = \sqrt{j(j+1) - m(m-1)}\,|m-1, j\rangle$$
$$= C_m^-|m-1, j\rangle$$

(b) Show that the definition $\mathbf{J}_3 \equiv \frac{1}{2}[x(\partial/\partial x) - y(\partial/\partial y)]$ results in

$$\mathbf{J}_3|m, j\rangle = m|m, j\rangle$$

(c) Prove that the definitions $\mathbf{J}_+ \equiv x(\partial/\partial y)$, $\mathbf{J}_- \equiv y(\partial/\partial x)$, and $\mathbf{J}_3 \equiv \frac{1}{2}[x(\partial/\partial x) - y(\partial/\partial y)]$ result in

$$\mathbf{J}^2 = \frac{1}{4}\left(x^2\frac{\partial^2}{\partial x^2} + y^2\frac{\partial^2}{\partial y^2}\right) + \frac{3}{4}\left(x\frac{\partial}{\partial x} + y\frac{\partial}{\partial y}\right)$$
$$+ \frac{1}{2}xy\frac{\partial^2}{\partial x\,\partial y}$$

and that $\mathbf{J}^2|m, j\rangle = j(j+1)|m, j\rangle$.

28. For integer values of $j = l$, the bases of SO(3) can be taken to be the spherical harmonics $|m, l\rangle = Y_{lm}(\theta, \phi)$. Let

$$\mathbf{J}_1 = -i(\mathbf{r} \times \nabla)_x = -i\left(y\frac{\partial}{\partial z} - z\frac{\partial}{\partial y}\right)$$
$$= i\left(\sin\phi\frac{\partial}{\partial\theta} + \cot\theta\cos\phi\frac{\partial}{\partial\phi}\right)$$

$$\mathbf{J}_2 = -i(\mathbf{r} \times \nabla)_y = -i\left(z\frac{\partial}{\partial x} - x\frac{\partial}{\partial z}\right)$$
$$= i\left(-\cos\phi\frac{\partial}{\partial\theta} + \cot\theta\sin\phi\frac{\partial}{\partial\phi}\right)$$

$$\mathbf{J}_3 = -i(\mathbf{r} \times \nabla)_z = -i\left(x\frac{\partial}{\partial y} - y\frac{\partial}{\partial x}\right)$$
$$= -i\frac{\partial}{\partial\phi}$$

(a) Prove that $\mathbf{J}_3 Y_{lm}(\theta, \phi) = mY_{lm}(\theta, \phi)$

(b) Deduce the form of $\mathbf{J}^2$ in terms of spherical coordinates and show that

$$\mathbf{J}^2 Y_{lm}(\theta, \phi) = l(l+1)Y_{lm}(\theta, \phi)$$

[*Hint*: Express the differential equation for $P_l^m(x)$ in terms of $x = \cos\theta$.]

29. Show that the three-dimensional representation of the SU(2) matrix is

$$\mathfrak{D} = \begin{bmatrix} a^2 & \sqrt{2}ab & b^2 \\ -\sqrt{2}ab^* & |a|^2 - |b|^2 & \sqrt{2}a^*b \\ b^{*2} & -\sqrt{2}a^*b^* & a^{*2} \end{bmatrix}$$

Show that $\mathfrak{D}$ is unitary.

30. (a) Deduce the group of operations that leaves the orientation of the rectangular box of Figure 10.32 invariant. The lengths a_x, a_y, and a_z of the three sides are different.

(b) Determine the character table of this group.

(c) Determine what happens to the degeneracy of the $l = 1$ and $l = 2$ state under SO(3) rotations when an electron is placed in a lattice constructed of positive ions placed at the corners of the box of Figure 10.32.

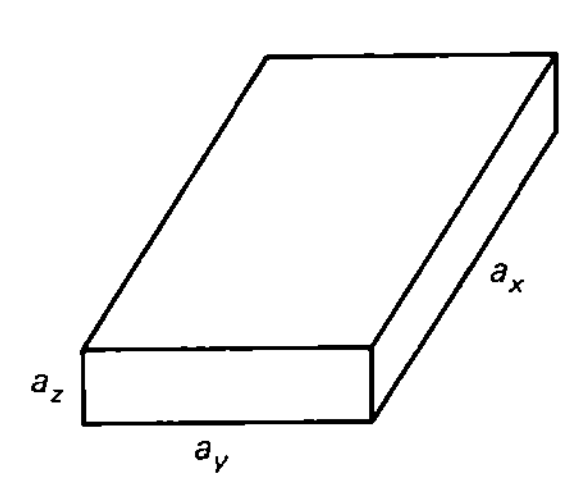

Figure 10.32 Rectangular solid.

31. (a) Prove that the charge values of $\Sigma^+, \Sigma^\circ, \Sigma^-, \Lambda^\circ$ are not correctly given by the four-dimensional representation of SU(2).

(b) Show that $K^+, K^\circ, \bar{K}^\circ, K^-, \eta^\circ$ cannot be classified in the five-dimensional representation of SU(2).

32. Show that in $\underset{\sim}{3}$ of SU(3), the nondiagonal generators can be written

$$\mathbf{N}_{-2} = \frac{1}{\sqrt{6}}\begin{bmatrix} 0 & 0 & 0 \\ 0 & 0 & 0 \\ 1 & 0 & 0 \end{bmatrix}$$

$$\mathbf{N}_{-3} = \frac{1}{\sqrt{6}}\begin{bmatrix} 0 & 0 & 0 \\ 0 & 0 & 0 \\ 0 & 1 & 0 \end{bmatrix}$$

Express $\mathbf{N}_{-2}$, $\mathbf{N}_{-3}$, $\mathbf{N}_2$, $\mathbf{N}_3$, and $\mathbf{D}_2$ in the ket-bra notation in terms of the up, down, and strange states of $\underset{\sim}{3}$ of SU(3).

33. Indicate that the generators of the Lorentz group satisfy equations 10.331 by demonstrating that

$$\mathbf{\Lambda}_{xy}\mathbf{\Lambda}_{yt} - \mathbf{\Lambda}_{yt}\mathbf{\Lambda}_{xy} = i\mathbf{\Lambda}_{xt}$$

$$\mathbf{\Lambda}_{xy}\mathbf{\Lambda}_{yz} - \mathbf{\Lambda}_{yz}\mathbf{\Lambda}_{xy} = i\mathbf{\Lambda}_{xz}$$

$$\mathbf{\Lambda}_{xy}\mathbf{\Lambda}_{zt} - \mathbf{\Lambda}_{zt}\mathbf{\Lambda}_{xy} = \mathbf{0}$$

34. Show that the generators of the Lorentz group satisfy:

(a) $(\mathbf{E}_+, \mathbf{E}_3) = \mathbf{E}_+$

(b) $(\mathbf{F}_+, \mathbf{F}_-) = -2\mathbf{F}_3$

(c) $(\mathbf{E}_+, \mathbf{F}_-) = \mathbf{0}$

35. Prove that following entries in Table 10.3 for the generators of the Lorentz group are correct.

(a) $\mathbf{E}_+ U_{++} = 0$

(b) $\mathbf{E}_- U_{-+} = U_{--}$

(c) $\mathbf{E}_3 U_{--} = -\frac{1}{2}U_{--}$

CHAPTER 11

PARTIAL DIFFERENTIAL EQUATIONS

Many physical systems are described by differential equations in which the unknown function depends on several independent variables. Therefore, the derivatives of the unknown function are partial derivatives. A large majority of these systems are described by equations that contain either first- or second-order partial derivatives. As such, we restrict our discussion primarily to first- and second-order partial differential equations.

One type of equation that arises in scientific problems is the first-order, nonlinear partial differential equation, the general form of which is

$$F\left(\Psi, \left(\frac{\partial \Psi}{\partial x_i}\right)^k, \left(\frac{\partial \Psi}{\partial t}\right)^l\right) = f(\mathbf{r}, t) \tag{11.1a}$$

The Hamilton–Jacobi equation is one important example of such an equation. It describes the dynamics of a system of N particles in terms of $3N$ position vector components $\{x_1, y_1, z_1, x_2, \ldots, z_N\}$, which we designate by the $3N$ component vector $\mathbf{X}$, and by $3N$ momentum components, which we denote by $\mathbf{P}$. (In Hamilton–Jacobi theory, the dynamics are described in terms of momenta rather than in terms of velocities.)

A second type of equation that one often encounters in scientific investigation is the first- or second-order linear equation, which is of the form

$$F\left(\Psi, \nabla^2\Psi, \frac{\partial^n \Psi}{\partial t^n}\right) = f(\mathbf{r}, t) \tag{11.1b}$$

where, most often, $n = 0$, 1, or 2. Examples of this type of equation include:

1. *Poisson's equation* for the electrostatic potential:

$$\nabla^2 V(\mathbf{r}) = -\frac{\rho(\mathbf{r})}{\varepsilon_0} \tag{11.2}$$

2. The *classical wave equation* for a wave with velocity c:

$$\nabla^2 \Psi(\mathbf{r}, t) - \frac{1}{c^2}\frac{\partial^2 \Psi(\mathbf{r}, t)}{\partial t^2} = f(\mathbf{r}, t) \tag{11.3}$$

where $f(\mathbf{r}, t)$ describes the source of the wave. For example, applying the procedures to Maxwell's equations with nonzero sources that were used in Problem 34 of Chapter 1 it is straightforward to deduce that if equation 11.3 is the wave equation for the electric field, the source function is

$$f(\mathbf{r}, t) = \frac{1}{\varepsilon_0}\nabla_\rho + \mu_0 \frac{\partial \mathbf{j}}{\partial t} \tag{11.4a}$$

For the wave equation for the magnetic field, the source term is

$$f(\mathbf{r}, t) = -\mu_0 \nabla \times \mathbf{j} \tag{11.4b}$$

3. The *diffusion equation*:

$$\Lambda \nabla^2 T(\mathbf{r}, t) - \frac{\partial T}{\partial t} = f(\mathbf{r}, t) \tag{11.5}$$

where Λ is the diffusion coefficient. If $T(\mathbf{r}, t)$ is the temperature of the medium, this equation describes thermal diffusion, or heat conduction. As with the wave equation, $f(\mathbf{r}, t)$ represents a source of the quantity that is diffusing. For example, if equation 11.5 describes heat conduction,

$$f(\mathbf{r}, t) = -\frac{Q(\mathbf{r}, t)}{\rho C} \tag{11.6}$$

where $Q(\mathbf{r}, t)$ is the amount of heat generated at the point $\mathbf{r}$ at time t per unit volume, per second. ρ is the density of the diffusing medium and C is its specific heat.

4. The *Schrödinger equation* for a quantum wave function of a particle in the potential energy field $V(\mathbf{r}, t)$:

$$-\frac{\hbar^2}{2m}\nabla^2 \Psi(\mathbf{r}, t) + V(\mathbf{r}, t)\Psi(\mathbf{r}, t) = i\hbar \frac{\partial \Psi}{\partial t} \tag{11.7}$$

11.1 Separation of Variables

The method of separation of variables is applicable to certain forms of the first- and second-order equations described above. The technique succeeds because of the property that each of the variables that the unknown function depends on is an independent variable. The solution is obtained by expressing the unknown function in terms of separate one-variable functions, each of which is dependent on a single independent variable.

For the linear, homogeneous first- and second-order equations of the form expressed in equation 11.1b, this separation is written as a product of functions of the independent variables. Using $\mathbf{Q}$ to represent all the independent variables (coordinates, momenta, etc.) except t, this separation is in the form

$$\Psi(\mathbf{Q}, t) = U_1(q_1)U_2(q_2) \cdots U_n(q_n)\tau(t) \tag{11.8a}$$

Many nonlinear equations such as those in the form expressed in equation 11.1a are separated by writing $\Psi(\mathbf{Q}, t)$ as a sum of one-variable functions:

$$\Psi(\mathbf{Q}, t) = U_1(q_1) + U_2(q_2) + \cdots + U_n(q_n) + \tau(t) \tag{11.8b}$$

In the following discussions, we illustrate the methods of separation using examples that arise in the study of problems of scientific interest.

Hamilton – Jacobi equation

For a system of N particles, the positions of which are defined by $3N$ components of position vectors, defined by $\{x_1, x_2, \ldots, x_{3N}\}$ the Hamilton–Jacobi equation is

$$H\left(x_1, x_2, \ldots, x_{3N}, \frac{\partial S}{\partial x_1}, \frac{\partial S}{\partial x_2}, \ldots, \frac{\partial S}{\partial x_{3N}}\right) + \frac{\partial S}{\partial t} = 0 \tag{11.9}$$

where H is the Hamiltonian function introduced in Chapter 7, equation 7.100. $S(s_1, x_2, \ldots, x_{3N}, t)$ is called *Hamilton's principal function*. It is defined such that the momentum component p_i is given by

$$p_i \equiv \frac{\partial S}{\partial x_i} \tag{11.10}$$

Most often, the Hamiltonian is the total energy of the system of particles, expressed in terms of coordinates and momenta. Since the kinetic energy of a particle is expressed in terms of the square of the momenta, the Hamilton–Jacobi equation depends on $(\partial S/\partial x_i)^2$ and is therefore a nonlinear first-order partial differential equation for S. As noted above, this type of equation is separated by expressing S as a sum of functions of the independent variables. For example, denoting the $3N$-component position vector by $\mathbf{X}$, we write

$$S(\mathbf{X}, t) = S_1(\mathbf{X}) + \tau(t) \tag{11.11}$$

$S_1(\mathbf{X})$ is called *Hamilton's characteristic function*. Designating a $3N$-component gradient operator by ∇', such that $(\nabla')_i = \partial/\partial x_i$, the Hamilton–Jacobi equation can be written

$$H(\mathbf{X}, \nabla' S_1) = -\frac{d\tau}{dt} \tag{11.12}$$

We see that the left-hand side of equation 11.12 contains only $\mathbf{X}$-dependence and the right-hand side depends only on t. Since $\mathbf{X}$ and t are independent variables, changing t, for example, cannot affect anything that depends only on $\mathbf{X}$. If $\mathbf{X}$ is held fixed, the left-hand side of equation 11.12 is constant. Therefore, even if t is varied, the right-hand side must also be constant. (By an identical argument, for a fixed value of t, the right-hand side of equation 11.12 is constant. Therefore, even if $\mathbf{X}$ is varied, the left-hand side must remain constant.) That is, equation 11.12 can be separated into two equations:

$$\text{(a)}\quad H(\mathbf{X}, \nabla' S_1) = E = \text{constant} \qquad \text{(b)}\quad \frac{d\tau}{dt} = -E = \text{constant} \tag{11.13}$$

$$\Rightarrow \quad \tau(t) = -E(t - t_0) \tag{11.14}$$

$$\Rightarrow \quad S(\mathbf{X}, t) = S_1(\mathbf{X}) - E(t - t_0) \tag{11.15}$$

From this analysis, it is understandable why E is referred to as a *separation constant*.

To determine $S_1(\mathbf{X})$, one must know the specific physical system being investigated.

Example 11.1

For example, the atomic electron that is influenced by a nuclear charge Ze has potential energy $V(\rho) = -Ze^2/\rho$. We define the plane of motion of the electron to be the x-y plane and describe the motion of the electron in terms of circular coordinates (ρ, ϕ) (cylindrical coordinates with $z = 0$). Then the Hamiltonian (energy) is

$$H = \frac{1}{2m}\left[p_\rho^2 + \frac{1}{\rho^2}p_\phi^2\right] - \frac{Ze^2}{\rho} \tag{11.16}$$

Referring to equations 11.10 and 11.11, the separated form of the Hamilton–Jacobi equation, in terms of the characteristic function, is

$$\frac{1}{2m}\left[\left(\frac{\partial S_1}{\partial \rho}\right)^2 + \frac{1}{\rho^2}\left(\frac{\partial S_1}{\partial \phi}\right)^2\right] - \frac{Ze^2}{\rho} = E \tag{11.17}$$

where $S_1 = S_1(\rho, \phi)$. This nonlinear equation can be further separated by expressing S_1 as the sum

$$S_1(\rho, \phi) = S_\rho(\rho) + S_\phi(\phi) \tag{11.18}$$

$$\Rightarrow \quad \frac{1}{2m}\left[\left(\frac{dS_\rho}{d\rho}\right)^2 + \frac{1}{\rho^2}\left(\frac{dS_\phi}{d\phi}\right)^2\right] - \frac{Ze^2}{\rho} = E \tag{11.19a}$$

$$\Rightarrow \quad \left(\frac{dS_\phi}{d\phi}\right)^2 = \rho^2\left[2m\left(E + \frac{Ze^2}{\rho}\right) - \left(\frac{dS_\rho}{d\rho}\right)^2\right] \tag{11.19b}$$

Since the left-hand side of equation 11.19b contains all the ϕ-dependence while all the ρ-dependence is on the right side, the same argument used to separate the space- and time-dependent parts of the principal function can be used to separate the ϕ-dependence from the ρ-dependence. That is,

$$\text{(a)} \quad \frac{dS_\phi}{d\phi} = \alpha = \text{constant} \quad \Rightarrow \quad \text{(b)} \quad S_\phi = \alpha\phi \tag{11.20}$$

α is another separation constant.

From equation 11.20a we can now write equation 11.19a as

$$\frac{1}{2m}\left[\left(\frac{dS_\rho}{d\rho}\right)^2 + \frac{\alpha^2}{\rho^2}\right] - \frac{Ze^2}{\rho} = E \tag{11.21}$$

from which S_ρ can be determined. Thus, in terms of the integral expression for S_ρ, Hamilton's principal function is

$$S(\mathbf{r}, t) = \int\left[2mE + \frac{2mZe^2}{\rho} - \frac{\alpha^2}{\rho^2}\right]^{1/2} d\rho + \alpha\phi - E(t - t_0) \quad \square \tag{11.22}$$

Hamilton – Jacobi theory

To complete this example, we note some of the features of Hamilton–Jacobi theory. For a more complete description, the reader is referred to any advanced text on classical mechanics. One excellent source is the text by H. Goldstein, *Classical Mechanics*, 2nd ed. (Reading, Mass.: Addison-Wesley Publishing Company, 1980), Chapter 9.

In Hamilton–Jacobi theory, one finds that Hamilton's characteristic function satisfies

$$\frac{\partial S_1}{\partial E} = t - t_0 \tag{11.23a}$$

where t_0 is an arbitrary initial time. The theory also specifies that partial derivatives of $S_1(\mathbf{X})$ with respect to the other separation constants will be constant. That is,

$$\frac{\partial S_1}{\partial \alpha_i} = a_i = \text{constant} \tag{11.23b}$$

Referring to the equation 11.22, with $S_1 = S_\rho + S_\phi$, we apply equation 11.23a to

$$S_1(\rho, \phi) = \int \left[2mE + \frac{2mZe^2}{\rho} - \frac{\alpha^2}{\rho^2}\right]^{1/2} d\rho + \alpha\phi \tag{11.24}$$

$$\Rightarrow \quad t - t_0 = m \int \left[2mE + \frac{2mZe^2}{\rho} - \frac{\alpha^2}{\rho^2}\right]^{-1/2} d\rho \tag{11.25a}$$

When this integral is evaluated, one can determine the time dependence of the radial position of the electron. Applying equation 11.23b to $S_1(\rho, \phi)$ yields

$$\frac{\partial S_1}{\partial \alpha} = a = \phi - \alpha \int \left[2mE + \frac{2mZe^2}{\rho} - \frac{\alpha^2}{\rho^2}\right]^{-1/2} \frac{d\rho}{\rho^2} \tag{11.25b}$$

Evaluation of this integral yields the description of the orbital path of the electron in the form $\rho(\phi)$.

First- and second-order linear equations

The most commonly used method for separating homogeneous ($f(\mathbf{r}, t) = 0$) first- and second-order linear partial differential equations is in the form of a product of functions of the independent variables.

Example 11.2

For example, a first-order linear equation of the form

$$A_1(x)A_2(y)\frac{\partial \Psi}{\partial x} + B_1(x)B_2(y)\frac{\partial \Psi}{\partial y} = 0 \tag{11.26a}$$

can be cast into the form

$$\alpha(x)\frac{\partial \Psi}{\partial x} + \beta(y)\frac{\partial \Psi}{\partial y} = 0 \tag{11.26b}$$

by dividing by $A_2(y)B_1(x)$. This equation is separable by writing

$$\Psi(x, y) = U_1(x)U_2(y) \tag{11.27}$$

$$\Rightarrow \quad \alpha(x)U_2(y)\frac{dU_1}{dx} + \beta(y)U_1(x)\frac{dU_2}{dy} = 0 \tag{11.28a}$$

Dividing the equation by $U_1(x)U_2(y)$, this becomes

$$\alpha(x)\frac{1}{U_1(x)}\frac{dU_1}{dx} + \beta(y)\frac{1}{U_2(y)}\frac{dU_2}{dy} = 0 \tag{11.28b}$$

Because x and y are independent variables

$$\alpha(x)\frac{1}{U_1(x)}\frac{dU_1}{dx} = -\beta(y)\frac{1}{U_2(y)}\frac{dU_2}{dy} = \gamma = \text{constant} \tag{11.29}$$

and equation 11.26b is separated. □

The procedure for a linear homogeneous second-order equation is identical.

Example 11.3

As an illustrative example, we consider the time-dependent Schrödinger equation (see equation 11.7). To separate the $\mathbf{r}$-dependence from time dependence, $\Psi(\mathbf{r}, t)$ is written as the product

$$\Psi(\mathbf{r}, t) \equiv U(\mathbf{r})\tau(t) \tag{11.30}$$

Since ∇^2 only affects functions of $\mathbf{r}$ and $\partial/\partial t$ only operates on time-dependent functions, substituting equation 11.30 into equation 11.7 yields

$$\left[-\frac{\hbar^2}{2m}\nabla^2 U(\mathbf{r}) + \nabla(\mathbf{r}, t)U(\mathbf{r})\right]\tau(t) = i\hbar U(\mathbf{r})\frac{d\tau(t)}{dt} \qquad (11.31a)$$

Dividing by $\Psi = U(\mathbf{r})\tau(t)$, this becomes

$$\left[-\frac{\hbar^2}{2m}\frac{1}{U(\mathbf{r})}\nabla^2 U(\mathbf{r}) + V(\mathbf{r}, t)\right] = i\hbar\frac{1}{\tau(t)}\frac{d\tau(t)}{dt} \qquad (11.31b)$$

We note that if the potential is time-independent, then

$$\left[-\frac{\hbar^2}{2m}\frac{1}{U(\mathbf{r})}\nabla^2 U(\mathbf{r}) + V(\mathbf{r})\right] = i\hbar\frac{1}{\tau(t)}\frac{d\tau(t)}{dt} \qquad (11.32)$$

and the left-hand side of this equation contains only $\mathbf{r}$-dependence, and all the t-dependence is on the right-hand side. Since varying one of these independent variables cannot affect anything dependent entirely on the other variable, both sides of equation 11.32 must be constant. That is, with E = constant,

$$\text{(a)} \quad i\hbar\frac{1}{\tau(t)}\frac{d\tau(t)}{dt} = E \quad \text{and} \quad \text{(b)} \quad -\frac{\hbar^2}{2m}\frac{1}{U(\mathbf{r})}\nabla^2 U(\mathbf{r}) + V(\mathbf{r}) = E \qquad (11.33)$$

It is clear from the discussion above that the separation is possible only if the potential is independent of time. For such a potential, the time-dependent part of $\Psi(\mathbf{r}, t)$ is the solution to equation 11.33a:

$$\tau(t) = \tau(0)e^{-i(E/\hbar)t} \qquad (11.34)$$

Equation 11.33b can be written

$$-\frac{\hbar^2}{2m}\nabla^2 U(\mathbf{r}) + V(\mathbf{r})U(\mathbf{r}) = EU(\mathbf{r}) \qquad (11.35)$$

This eigenvalue equation is the *time-independent Schrödinger equation.*

The Schrödinger operator $-\hbar^2/2m\nabla^2 + V(\mathbf{r})$ is also called the *quantum Hamiltonian operator*. It is denoted by $\hat{H}$, and the time-independent Schrödinger equation is often written in the compact form

$$\hat{H}U(\mathbf{r}) = EU(\mathbf{r}) \qquad (11.36)$$

The quantum mechanical *momentum operator* is

$$\hat{p} = -i\hbar\nabla \qquad (11.37)$$

Therefore, the quantum Hamiltonian can be written as

$$\hat{H} = -\frac{\hbar^2}{2m}\nabla^2 + V(\mathbf{r}) = \frac{\hat{p}^2}{2m} + V(\mathbf{r}) \qquad (11.38)$$

which classically would be the total energy of the particle. As such, the separation constant E becomes the total energy of the particle.

Since the time-independent Schrödinger equation is a partial differential equation for $U(\mathbf{r})$, it can sometimes be separated into three ordinary differential equations in the three spatial variables by the same approach used above to separate the $\mathbf{r}$-dependence from the t-dependence. The types of potentials for which this method of separation is applicable are those that depend on only one of the three independent spatial variables, or possibly a sum of potentials, each depending on the independent variable in a specified way (see Problem 8). One important class of problems involving such a one-variable potential is the system exhibiting spherical symmetry. That is, $V(\mathbf{r}) = V(r)$, which is independent of θ and ϕ. For such a problem it is most natural to express the Schrödinger equation in spherical coordinates:

$$-\frac{\hbar^2}{2m}\left[\frac{1}{r}\frac{\partial^2(rU)}{\partial r^2} + \frac{1}{r^2\sin\theta}\frac{\partial}{\partial\theta}\left(\sin\theta\frac{\partial U}{\partial\theta}\right) + \frac{1}{r^2\sin^2\theta}\frac{\partial^2 U}{\partial\phi^2}\right] + V(r)U = EU \tag{11.39}$$

The separation is achieved by writing

$$U(\mathbf{r}) = R(r)P(\theta)\Phi(\phi) \tag{11.40}$$

Along with a little algebraic manipulation, this substitution casts equation 11.39 into the form

$$P(\theta)\Phi(\phi)\frac{1}{r}\frac{d^2(rR)}{dr^2} + R(r)\Phi(\phi)\frac{1}{r^2\sin\theta}\frac{d}{d\theta}\left(\sin\theta\frac{dP}{d\theta}\right) + R(r)P(\theta)\frac{1}{r^2\sin^2\theta}\frac{d^2\Phi}{d\phi^2}$$
$$+ \frac{2m}{\hbar^2}[E - V(r)]R(r)P(\theta)\Phi(\phi) = 0 \tag{11.41a}$$

or if we divide by $R(r)P(\theta)\Phi(\phi)$, and multiply by $r^2\sin^2\theta$, this becomes

$$r^2\sin^2\theta\left[\frac{1}{rR(r)}\frac{d^2(rR)}{dr^2} + \frac{2m}{\hbar^2}[E - V(r)]\right] + \frac{\sin\theta}{P(\theta)}\frac{d}{d\theta}\left(\sin\theta\frac{dP}{d\theta}\right) + \frac{1}{\Phi(\phi)}\frac{d^2\Phi}{d\phi^2} = 0 \tag{11.41b}$$

We note that because the potential is independent of ϕ, all the ϕ-dependence is contained in the last term of equation 11.41b. Since r, θ, and ϕ are independent, varying ϕ cannot affect the terms dependent on r and θ. Therefore, the ϕ-dependent term must be constant. For pedagogic reasons, we express this constant as

$$\text{(a)}\quad \frac{1}{\Phi(\phi)}\frac{d^2\Phi}{d\phi^2} = -m^2 \qquad \text{or} \qquad \text{(b)}\quad (D^2 + m^2)\Phi = 0 \tag{11.42}$$

$$\Rightarrow \quad \Phi(\phi) = \alpha_m e^{im\phi} + \beta_m e^{-im\phi} \tag{11.43}$$

From physical considerations, $\Phi(\phi)$ is expected to be single-valued. That is, $\Phi(\phi + 2\pi) = \Phi(\phi)$. This requires $e^{\pm i2\pi m} = 1$ which restricts m to be an integer. It is for this reason that the constant of equation 11.42a was chosen in the form $-m^2$. Had this constant been expressed in the form $\Phi''/\Phi = c$, we would have concluded that $\sqrt{-c}$ had to be an integer.

Referring to equation 11.43, if m is allowed to take on both positive and negative values, $e^{im\phi}$ and $e^{-im\phi}$ are duplicate terms. That is,

$$\Phi(\phi) = \alpha_m e^{i|m|\phi} + \beta_m e^{-i|m|\phi} \qquad m > 0 \tag{11.44a}$$

$$\Phi(\phi) = \alpha_{-|m|} e^{-i|m|\phi} + \beta_{-|m|} e^{i|m|\phi} \qquad m < 0 \tag{11.44b}$$

Except for the names of the coefficients of the exponentials, the two solutions are identical. Therefore, to avoid this duplication, we can either retain both terms in the solution and restrict $m \geqslant 0$, or allow m to be both positive and negative and express the solution in terms of one exponential. The standard choice is

$$\Phi(\phi) = \alpha_m e^{im\phi} \tag{11.45}$$

with m taking on positive and negative integer values.

With $\Phi''/\Phi = -m^2$, we divide equation 11.41b by $\sin^2\theta$ to obtain

$$r^2\left[\frac{1}{rR(r)}\frac{d^2(rR)}{dr^2} + \frac{2m}{\hbar^2}(E - V(r))\right] + \frac{1}{\sin\theta P(\theta)}\frac{d}{d\theta}\left(\sin\theta\frac{dP}{d\theta}\right) - \frac{m^2}{\sin^2\theta} = 0 \tag{11.46}$$

Since the potential is independent of θ, we see that all the θ-dependence is contained in the last two terms, and all the r-dependence is in the first terms. Thus the sum of r-dependent terms must be constant, and the sum of terms containing θ must also be constant. Therefore,

$$\frac{1}{\sin\theta P(\theta)}\frac{d}{d\theta}\left(\sin\theta\frac{dP}{d\theta}\right)-\frac{m^2}{\sin^2\theta}=\lambda \tag{11.47}$$

With $x = \cos\theta$, and $\lambda = -l(l+1)$, where l is an integer, it is straightforward to demonstrate that this is the differential equation for the associated Legendre functions. Thus the solution is a linear combination of $P_l^m(\cos\theta)$ and $Q_l^m(\cos\theta)$. The solution $Q_l^m(\cos\theta)$ is usually omitted since this function is logarithmically infinite at $\theta = 0$, π $(x = \pm 1)$. If the physical system can exist at $\theta = 0$ and π, the solution must be finite at these angles.

With $\lambda = -l(l+1)$, after division by r^2, equation 11.46 becomes

$$\frac{1}{rR(r)}\frac{d^2(rR)}{dr^2}+\frac{2m}{\hbar^2}[E-V(r)]+\frac{l(l+1)}{r^2}=0 \tag{11.48}$$

Clearly, the solution to this radial equation depends on the potential $V(r)$ that defines the particular physical system. However, for a single-variable potential, the partial differential equation has been separated into ordinary differential equations. When the solution to the radial equation is obtained, the complete solution to the Schrödinger equation with a spherically symmetric potential is

$$\Psi(\mathbf{r},t)=C_{lm}R(r)P_l^m(\cos\theta)e^{im\phi}e^{-iEt/\hbar}=N_{lm}R(r)Y_{lm}(\theta,\phi)e^{-iEt/\hbar} \tag{11.49}$$

where the function $Y_{lm}(\theta,\phi)$ is the spherical harmonic function expressed in equation 6.343. □

Inhomogeneous linear second-order equations

In some applications a system will be described by an inhomogeneous second-order linear partial differential equation, in which the inhomogeneous term depends on just the position vector, or is only time-dependent. For example, a vibrating medium satisfies the wave equation expressed in equation 11.3. If the inhomogeneous term is independent of t, or is independent of $\mathbf{r}$, such an equation can be separated into two equations.

$f(\mathbf{r},t) = F(\mathbf{r})$

Using the driven wave equation as an example, if $f(\mathbf{r},t) = F(\mathbf{r})$, then

$$\nabla^2\Psi(\mathbf{r},t)-\frac{1}{c^2}\frac{\partial^2\Psi}{\partial t^2}=F(\mathbf{r}) \tag{11.50}$$

can be separated into a homogeneous wave equation and an inhomogeneous equation that depends on $\mathbf{r}$. To do so, we express

$$\Psi(\mathbf{r},t)\equiv\chi(\mathbf{r},t)+U(\mathbf{r}) \tag{11.51}$$

$$\Rightarrow\quad \nabla^2\chi(\mathbf{r},t)-\frac{1}{c^2}\frac{\partial^2\chi}{\partial t^2}+\nabla^2U(\mathbf{r})=F(\mathbf{r}) \tag{11.52}$$

Since $\chi(\mathbf{r},t)$ and $U(\mathbf{r})$ are as yet unspecified, we choose $U(\mathbf{r})$ to satisfy

$$\nabla^2U(\mathbf{r})=F(\mathbf{r}) \tag{11.53a}$$

which is basically Poisson's equation. Then $\chi(\mathbf{r},t)$ satisfies the homogeneous wave equation

$$\nabla^2\chi(\mathbf{r},t)-\frac{1}{c^2}\frac{\partial^2\chi}{\partial t^2}=0 \tag{11.53b}$$

which can be separated into ordinary differential equations by writing $\chi(\mathbf{r}, t)$ as a product of one-variable functions.

$f(\mathbf{r}, t) = F(t)$

If $f(\mathbf{r}, t) = F(t)$, we take

$$\Psi(\mathbf{r}, t) \equiv \chi(\mathbf{r}, t) + U(t) \tag{11.54}$$

$$\Rightarrow \quad \nabla^2\chi(\mathbf{r}, t) - \frac{1}{c^2}\frac{\partial^2\chi}{\partial t^2} - \frac{1}{c^2}\frac{d^2U}{dt^2} = F(t) \tag{11.55}$$

By choosing $U(t)$ to satisfy

$$\frac{1}{c^2}\frac{d^2U}{dt^2} = -F(t) \tag{11.56}$$

$\chi(\mathbf{r}, t)$ again satisfies the homogeneous wave equation.

We note that $\Psi(\mathbf{r}, t)$, as expressed in either equation 11.51 or 11.54, is the sum of the solutions to the homogeneous and inhomogeneous differential equations. Therefore, $\Psi(\mathbf{r}, t)$ is the complete solution to the partial differential equation.

Example 11.4

As an example, consider the equation for a vibrating string of length L (a one-spatial-dimension medium), forced by an external driving term, which we take to be $kx(L - x)$. The amplitude satisfies the inhomogeneous wave equation

$$\frac{\partial^2\Psi}{\partial x^2} - \frac{1}{c^2}\frac{\partial^2\Psi}{\partial t^2} = kx(L - x) \tag{11.57}$$

For this example we will impose the following boundary and initial conditions:

$$\text{(a)}\quad \Psi(0, t) = 0 \qquad \text{(b)}\quad \Psi(L, t) = 0 \qquad \text{(c)}\quad \left.\frac{\partial\Psi}{\partial t}\right|_{t=0} = 0 \qquad \text{(d)}\quad \left.\frac{\partial\Psi}{\partial x}\right|_{x=\frac{1}{2}L} = \cos\left(\frac{4\pi c}{L}t\right) \tag{11.58}$$

Because the inhomogeneous term depends on x, we take

$$\Psi(x, t) = \chi(x, t) + U(x) \tag{11.59}$$

$$\Rightarrow \quad \frac{\partial^2\chi}{\partial x^2} - \frac{1}{c^2}\frac{\partial^2\chi}{\partial t^2} + \frac{d^2U}{dx^2} = kx(L - x) \tag{11.60}$$

By setting

$$\text{(a)}\quad \frac{d^2U}{dx^2} = kx(L - x) \qquad \Rightarrow \qquad \text{(b)}\quad \frac{\partial^2\chi}{\partial x^2} - \frac{1}{c^2}\frac{\partial^2\chi}{\partial t^2} = 0 \tag{11.61}$$

The solution for $U(x)$ is found straightforwardly to be

$$U(x) = U_0 + U_1x + \tfrac{1}{6}kx^3(L - \tfrac{1}{2}x) \tag{11.62}$$

The solution for $\chi(x, t)$ is obtained by taking

$$\chi(x, t) = X(x)\tau(t) \tag{11.63a}$$

Substituting this into equation 11.61b and dividing by χ, we obtain

$$\frac{1}{X}\frac{d^2X}{dt^2} - \frac{1}{c^2}\frac{1}{\tau}\frac{d^2\tau}{dt^2} = 0 \tag{11.63b}$$

Since x and t are independent variables, each term must be a constant. For arithmetic simplicity, because we know that the solution will be oscillatory (periodic), we choose the separation constant to be $-m^2$. Then

$$\text{(a)}\quad \frac{1}{X}\frac{d^2X}{dt^2} = -m^2 \qquad \text{and} \qquad \text{(b)}\quad \frac{1}{c^2}\frac{1}{\tau}\frac{d^2\tau}{dt^2} = -m^2 \tag{11.64}$$

$$\Rightarrow \quad \text{(a)}\quad X(x) = A_x \sin(mx) + B_x \cos(mx) \qquad \text{(b)}\quad \tau(t) = A_t \sin(mct) + B_t \cos(mct) \tag{11.65}$$

$$\Rightarrow \quad \Psi(x,t) = [A_x \sin(mx) + B_x \cos(mx)][A_t \sin(mct) + B_t \cos(mct)] + U_0 + U_1 x + \tfrac{1}{6}kx^3(L - \tfrac{1}{2}x) \tag{11.66}$$

From the boundary condition expressed in equation 11.58a, we obtain

$$\Psi(0,t) = B_x[A_t \sin(mct) + B_t \cos(mct)] + U_0 = 0 \tag{11.67}$$

This is satisfied for all values of t by taking $B_x = U_0 = 0$. Thus

$$\Psi(x,t) = A_x \sin(mx)[A_t \sin(mct) + B_t \cos(mct)] + U_1 x + \tfrac{1}{6}kx^3(L - \tfrac{1}{2}x) \tag{11.68}$$

The boundary condition expressed in equation 11.58b requires that

$$\Psi(L,t) = A_x \sin(mL)[A_t \sin(mct) + B_t \cos(mct)] + U_1 L + \tfrac{1}{12}kL^4 \tag{11.69}$$

If we take $A_x = 0$, Ψ would be time independent. To avoid this unrealistic result, we instead require

$$\text{(a)}\quad \sin(mL) = 0 \quad \Rightarrow \quad \text{(b)}\quad m = \frac{n\pi}{L} \qquad n = 0, \pm 1, \pm 2, \ldots \tag{11.70}$$

The boundary condition expressed in equation 11.69 also yields

$$U_1 = -\tfrac{1}{12}kL^3 \tag{11.70c}$$

We note that the term $\sin[(n\pi/L)x] = 0$ for $n = 0$. In addition, the solution for negative values of n is not independent of the positive n solution. As such, we can restrict the values of n to be $n \geqslant 1$.

Since equation 11.69 is a solution for each value of n, the complete solution is a sum of all allowed terms. Thus combining A_x with A_t and B_t, the solution is

$$\Psi(x,t) = \sum_{n=1}^{\infty} \sin\left(\frac{n\pi}{L}x\right)\left[A_n \sin\left(\frac{n\pi c}{L}t\right) + B_n \cos\left(\frac{n\pi c}{L}t\right)\right] + \frac{1}{12}kx(2Lx^2 - x^3 - L^3) \tag{11.71}$$

Applying the initial condition of equation 11.58c yields

$$\left.\frac{\partial \Psi}{\partial t}\right|_{t=0} = 0 = \sum_{n=1}^{\infty} A_n \frac{n\pi c}{L} \sin\left(\frac{n\pi}{L}x\right) \tag{11.72}$$

This is in the form of a Fourier sine series for the function $f(x) = 0$. The only way such a series can be zero is if each coefficient in the series is zero. Therefore, with $A_n = 0$, and

$$\Psi(x,t) = \sum_{n=1}^{\infty} B_n \sin\left(\frac{n\pi}{L}x\right)\cos\left(\frac{n\pi c}{L}t\right) + \frac{1}{12}kx(2Lx^2 - x^3 - L^3) \tag{11.73}$$

from equation 11.58d we obtain

$$\left.\frac{\partial \Psi}{\partial x}\right|_{x=L/2} = \sum_{n=1}^{\infty} \frac{n\pi}{L} B_n \cos\left(\frac{n\pi}{2}\right)\cos\left(\frac{n\pi c}{L}t\right) + \frac{k}{12}\left(\frac{3}{2}L^3 - \frac{1}{2}L^3 - L^3\right)$$

$$= \sum_{n=1}^{\infty} \frac{n\pi}{L} B_n \cos\left(\frac{n\pi}{2}\right)\cos\left(\frac{n\pi c}{L}t\right) = \cos\left(\frac{4\pi c}{L}t\right) \tag{11.74}$$

This, too, is a Fourier cosine series for the function $\cos[(4\pi c/L)t]$. From the orthogonality of the cosine functions, the only nonzero coefficient of this series is

$$B_4 = \frac{L}{4\pi} \tag{11.75}$$

Therefore, the complete solution to the partial differential equation with the specified boundary and initial conditions is

$$\Psi(x,t) = \frac{L}{4\pi}\sin\left(\frac{4\pi}{L}x\right)\cos\left(\frac{4\pi c}{L}t\right) + \frac{1}{12}kx(2Lx^2 - x^3 - L^3) \quad \square \tag{11.76}$$

To generalize this analysis, we write the inhomogeneous, second-order (in the spatial variables) linear equation in the three-dimensional analog of the Sturm–Liouville equation:

$$\nabla\cdot[p(\mathbf{r})\nabla\Psi(\mathbf{r},t)] + W(\mathbf{r})\Psi(\mathbf{r},t) + \alpha\frac{\partial^n\Psi(\mathbf{r},t)}{\partial t^n} = f(\mathbf{r},t) \tag{11.77}$$

As indicated in equations 11.2, 11.3, 11.5, and 11.7, in the majority of problems of scientific interest, the order of differentiation with respect to time is $n = 0$, $n = 1$, or $n = 2$.

If $f(\mathbf{r}, t) = F(\mathbf{r})$, it is straightforward to see that by writing

$$\Psi(\mathbf{r},t) = \chi(\mathbf{r},t) + U(\mathbf{r}) \tag{11.78}$$

equation 11.77 is separable into an inhomogeneous equation satisfied by $U(\mathbf{r})$,

$$\nabla\cdot[p(\mathbf{r})\nabla U(\mathbf{r})] + W(\mathbf{r})U(\mathbf{r}) = F(\mathbf{r}) \tag{11.79a}$$

and the homogeneous equation satisfied by $\chi(\mathbf{r}, t)$,

$$\nabla\cdot[p(\mathbf{r})\nabla\chi(\mathbf{r},t)] + W(\mathbf{r})\chi(\mathbf{r},t) + \alpha\frac{\partial^n\chi(\mathbf{r},t)}{\partial t^n} = 0 \tag{11.79b}$$

If $f(\mathbf{r}, t) = F(t)$, then writing

$$\Psi(\mathbf{r},t) = \chi(\mathbf{r},t) + U(t) \tag{11.80}$$

$$\Rightarrow \quad \nabla\cdot[p(\mathbf{r})\nabla\chi(\mathbf{r},t)] + W(r)\chi(\mathbf{r},t) + \alpha\frac{\partial^n\chi(\mathbf{r},t)}{\partial t^n} = 0 \tag{11.79b}$$

and

$$\text{(a)}\quad \alpha\frac{d^nU}{dt^n} = F(t) \quad \Rightarrow \quad \text{(b)}\quad U(t) = U_0(t) + \frac{1}{\alpha}D^{-n}F(t) \tag{11.81}$$

The solution to the homogeneous equation is obtained by the method of separation of variables outlined above. If $f(\mathbf{r}, t) = F(\mathbf{r})$, the type of inhomogeneous equation for $U(\mathbf{r})$ expressed in equation 11.79a is solved by two widely used techniques. Under certain conditions, integral transforms can be used. A second method of solution is the Green's function approach.

11.2 Integral Transform Methods

Just as integral transforms can be used to solve a wide range of ordinary differential equations, the solution to many partial differential equations can also be found using integral transform methods. As we do for ordinary differential equations, we restrict our presentation to methods of solution based on Fourier and Laplace transforms. We illustrate these methods by examples.

Fourier transform methods

Since Fourier transforms can be written in terms of several variables, Fourier methods are suited to solving the space-dependent part of certain partial differential equations of the type expressed in equation 11.79a.

Referring to

$$\nabla \cdot [p(\mathbf{r}) \nabla U(\mathbf{r})] + W(\mathbf{r})U(\mathbf{r}) = F(\mathbf{r}) \tag{11.79a}$$

we point out that Fourier transform methods are applicable to this type of equation only if $p(\mathbf{r})$ and $W(\mathbf{r})$ are constants.

Example 11.5

As an example, consider the wave equation for a wave impressed on an elastic medium driven by an external source. This system satisfies

$$\nabla^2\Psi(\mathbf{r}, t) - \frac{1}{c^2}\frac{\partial^2\Psi(\mathbf{r}, t)}{\partial t^2} = f(\mathbf{r}, t) \tag{11.3}$$

The *Helmholtz equation* is obtained from this by taking $f(\mathbf{r}, t)$ to be oscillatory in the form

$$f(\mathbf{r}, t) = F(\mathbf{r})e^{i\omega t} \tag{11.82}$$

This suggests that the time dependence of the solution will be of the form $e^{i\omega t}$. Therefore, we take

$$\Psi(\mathbf{r}, t) = U(\mathbf{r})e^{i\omega t} \tag{11.83}$$

$$\Rightarrow \quad (\nabla^2 + \Lambda^2)U(\mathbf{r}) = F(\mathbf{r}) \tag{11.84}$$

$\Lambda \equiv \omega/c$ is called the *wave number*.

Equation 11.84 is the Helmholtz equation. It is soluble by Fourier transform methods because $p(\mathbf{r}) = 1$ and $W(\mathbf{r}) = \Lambda^2$ are constants for the Helmholtz equation (see equation 11.79a). To obtain the solution, we define the Fourier transform of $U(\mathbf{r})$ by

$$\text{(a)} \quad \Upsilon(\mathbf{k}) \equiv \int U(\mathbf{r})e^{-i\mathbf{k}\cdot\mathbf{r}}\frac{d^3k}{(2\pi)^{3/2}} \quad \Rightarrow \quad \text{(b)} \quad U(\mathbf{r}) = \int \Upsilon(\mathbf{k})e^{i\mathbf{k}\cdot\mathbf{r}}\frac{d^3k}{(2\pi)^{3/2}} \tag{11.85}$$

The Fourier transform of $F(\mathbf{r})$ is defined as

$$\text{(a)} \quad \Phi(\mathbf{k}) \equiv \int F(\mathbf{r})e^{-i\mathbf{k}\cdot\mathbf{r}}\frac{d^3k}{(2\pi)^{3/2}} \quad \Rightarrow \quad \text{(b)} \quad F(\mathbf{r}) = \int \Phi(\mathbf{k})e^{i\mathbf{k}\cdot\mathbf{r}}\frac{d^3k}{(2\pi)^{3/2}} \tag{11.86}$$

Writing $U(\mathbf{r})$ in terms of its Fourier integral, we perform operations like

$$\frac{\partial^2}{\partial x^2}\int \Upsilon(\mathbf{k})e^{i(xk_x + yk_y + zk_z)}\frac{d^3k}{(2\pi)^{3/2}} = -\int \Upsilon(\mathbf{k})(k_x^2)e^{i\mathbf{k}\cdot\mathbf{r}}\frac{d^3k}{(2\pi)^{3/2}} \tag{11.87}$$

to express the Helmholtz equation as

$$(\nabla^2 + \lambda^2)U(\mathbf{r}) = \int \Upsilon(\mathbf{k})(-k^2 + \Lambda^2)e^{i\mathbf{k}\cdot\mathbf{r}}\frac{d^3k}{(2\pi)^{3/2}} = \int \Phi(\mathbf{k})e^{i\mathbf{k}\cdot\mathbf{r}}\frac{d^3k}{(2\pi)^{3/2}} \tag{11.88}$$

where $k^2 = k_x^2 + k_y^2 + k_z^2$. As discussed in Chapter 4, since Fourier transforms are unique, we can equate the integrands of the two transforms to obtain

$$\Upsilon(\mathbf{k}) = \frac{\Phi(\mathbf{k})}{\Lambda^2 - k^2} \tag{11.89}$$

$$\Rightarrow \quad U(\mathbf{r}) = \int e^{i\mathbf{k}\cdot\mathbf{r}}\frac{\Phi(\mathbf{k})}{\Lambda^2 - k^2}\frac{d^3k}{(2\pi)^{3/2}} \tag{11.90}$$

So like its one-dimensional counterpart, the solution to this partial differential equation is reduced to the problem of evaluating a Fourier integral. □

It is straightforward to see from this example that if either $p(\mathbf{r})$ or $W(\mathbf{r})$ is not constant, the Fourier transform of the differential equation for $U(\mathbf{r})$ does not result in an algebraic equation for $\Upsilon(\mathbf{k})$.

Example 11.6

To illustrate the method of solution by integral transform methods when the equation contains both space and time derivatives, we consider the time-dependent Schrödinger equation. Referring to equation 11.79a,

$$\text{(a)}\quad p(\mathbf{r}) = -\frac{\hbar^2}{2m} \qquad \text{and} \qquad \text{(b)}\quad W(\mathbf{r}) = V(\mathbf{r}) \tag{11.91}$$

Therefore, to be amenable to solution by integral transform, the potential must be constant. The equation for such a system is

$$-\frac{\hbar^2}{2m}\nabla^2\Psi(\mathbf{r},t) + V_0\Psi(\mathbf{r},t) = i\hbar\frac{\partial\Psi}{\partial t} \tag{11.92}$$

Clearly, we can choose the zero of potential energy such that $V_0 = 0$.

Defining the spatial Fourier inverse of $\Psi(\mathbf{r}, t)$ by

$$\Psi(\mathbf{r},t) \equiv \int Z(\mathbf{k},t)e^{i\mathbf{k}\cdot\mathbf{r}}\frac{d^3k}{(2\pi)^{3/2}} \tag{11.93}$$

$$\Rightarrow \quad \int \frac{\hbar^2k^2}{2m}Z(\mathbf{k},t)e^{i\mathbf{k}\cdot\mathbf{r}}\frac{d^3k}{(2\pi)^{3/2}} = \int i\hbar\frac{\partial Z}{\partial t}e^{i\mathbf{k}\cdot\mathbf{r}}\frac{d^3k}{(2\pi)^{3/2}} \tag{11.94}$$

which reduces the partial differential equation in both space and time variables to an equation for $Z(\mathbf{k}, t)$ containing only time derivatives. It is

$$\frac{\hbar^2k^2}{2m}Z(\mathbf{k},t) = i\hbar\frac{\partial Z(\mathbf{k},t)}{\partial t} \tag{11.95}$$

This differential equation cannot be further reduced by defining a Fourier time inverse of $Z(\mathbf{k}, t)$. To illustrate, let

$$Z(\mathbf{k},t) \equiv \int \zeta(\mathbf{k},\omega)e^{i\omega t}\frac{d\omega}{\sqrt{2\pi}} \tag{11.96}$$

Then equation 11.95 would become

$$\int \frac{\hbar^2k^2}{2m}\zeta(\mathbf{k},\omega)e^{i\omega t}\frac{d\omega}{\sqrt{2\pi}} = -\int \hbar\omega\zeta(\mathbf{k},\omega)e^{i\omega t}\frac{d\omega}{\sqrt{2\pi}} \tag{11.97}$$

$$\Rightarrow \quad \left(\frac{\hbar^2k^2}{2m} + \hbar\omega\right)\zeta(\mathbf{k},\omega) = 0 \tag{11.98}$$

One possible choice is to take

$$\frac{\hbar^2k^2}{2m} + \hbar\omega = 0 \tag{11.99a}$$

In addition to leaving $\zeta(\mathbf{k}, t)$ arbitrary, this choice also requires that ω be a function of $\mathbf{k}$. But since $\mathbf{r}$ and t are independent variables, ω and $\mathbf{k}$ must also be independent. Therefore, equation 11.99a is not acceptable.

If we take

$$\zeta(\mathbf{k},t) = 0 \tag{11.99b}$$

this yields $Z(\mathbf{k}, t) = 0$ from which $\Psi(\mathbf{r}, t) = 0$, the trivial solution. □

Thus, as long as an equation contains derivatives with respect to both space and time without initial conditions, a complete reduction of the differential equation to algebraic form is not possible by Fourier methods.

Laplace transform methods

The failure of Fourier transforms to obtain a complete reduction of the differential equation to an integral arises from the absence of initial conditions on $\Psi(\mathbf{r}, t)$. If the system is specified such that

$$\text{(a)}\quad \Psi(\mathbf{r}, t) = 0 \qquad t < 0 \qquad \text{(b)}\quad \Psi(\mathbf{r}, 0) \equiv g(\mathbf{r}) \tag{11.100}$$

the solution can be expressed as an integral using a combination of Fourier and Laplace transforms.

As seen by the developments that lead to equation 11.95, the Fourier transform can be used to reduce the spatial dependence of $\Psi(\mathbf{r}, t)$ to algebraic dependence of $Z(\mathbf{k}, t)$ on $\mathbf{k}$. Since $\Psi(\mathbf{r}, t) = 0$ for negative t, $Z(\mathbf{k}, t) = 0$ for $t < 0$. Because $\Psi(\mathbf{r}, 0) \equiv g(\mathbf{r})$ is a specified function, $Z(\mathbf{k}, 0) \equiv \Phi(\mathbf{k})$, the Fourier inverse of $g(\mathbf{r})$, which, in principle, is known. Therefore, the use of Fourier transform methods to reduce the original differential equation to the partial differential equation

$$\frac{\hbar^2 k^2}{2m} Z(\mathbf{k}, t) = i\hbar \frac{\partial Z(\mathbf{k}, t)}{\partial t} \tag{11.95}$$

is valid.

Further reduction to algebraic form is accomplished using Laplace transform methods for the time variable. Referring to equation 11.100a, we note that $\Psi(\mathbf{r}, t)$, and therefore $Z(\mathbf{k}, t)$, has the correct time constraint that permits application of Laplace transforms. As such, we define the Laplace transform of $Z(\mathbf{k}, t)$ to be

$$\zeta(\mathbf{k}, s) \equiv \int_0^\infty Z(\mathbf{k}, t) e^{-st}\, dt \tag{11.101}$$

Integrating by parts, we obtain

$$\int_0^\infty \frac{\partial Z}{\partial t} e^{-st}\, dt = e^{-st} Z(\mathbf{k}, t)\Big|_0^\infty + s\zeta(\mathbf{k}, s) \tag{11.102a}$$

We take $e^{-st} Z(\mathbf{k}, t)$ to be zero at $t = \infty$. Then, with $Z(\mathbf{k}, 0) \equiv \Phi(\mathbf{k})$, this becomes

$$\int_0^\infty \frac{\partial Z}{\partial t} e^{-st}\, dt = -\Phi(\mathbf{k}) + s\zeta(\mathbf{k}, s) \tag{11.102b}$$

Therefore, the Laplace transform of equation 11.95 is

$$\text{(a)}\quad \frac{\hbar^2 k^2}{2m} \zeta(\mathbf{k}, s) = i\hbar[s\zeta(\mathbf{k}, s) - \Phi(\mathbf{k})]$$

$$\Rightarrow \quad \text{(b)}\quad \zeta(\mathbf{k}, s) = \frac{\Phi(\mathbf{k})}{[s + i(\hbar k^2/2m) + i(V_0/\hbar)]} \tag{11.103}$$

The inverse Laplace transform, obtained by evaluating the Bromwich integral (or using Laplace transform tables) is

$$Z(\mathbf{k}, t) = \frac{1}{2\pi i} \int_{c-i\infty}^{c+i\infty} \frac{\Phi(\mathbf{k})}{[s + i(\hbar k^2/2m)]} e^{st}\, ds \tag{11.104}$$

The integrand has a simple pole on the negative imaginary axis, and is therefore enclosed by the Bromwich contour. Therefore,

$$Z(\mathbf{k}, t) = \Phi(\mathbf{k}) e^{-i\hbar k^2 t/2m} \tag{11.105}$$

$$\Rightarrow \quad \Psi(\mathbf{r}, t) = \int \frac{d^3k}{(2\pi)^{3/2}} \Phi(\mathbf{k}) e^{-i\hbar k^2 t/2m} e^{i\mathbf{k}\cdot\mathbf{r}} \tag{11.106}$$

Therefore, as with the time-independent Helmholtz equation, the solution is reduced to evaluating an integral.

11.3 Green's Function Methods

The Green's function methods are applicable to partial differential equations of the form expressed in equation 11.79a.

$$\nabla \cdot [p(\mathbf{r}) \nabla U(\mathbf{r})] + W(\mathbf{r}) U(\mathbf{r}) = F(\mathbf{r}) \tag{11.79a}$$

In Chapter 6 we introduced the four types of boundary conditions satisfied by Sturm–Liouville functions. These boundary conditions are:

1. *Dirichlet conditions:* $y(x) = 0$ at the boundaries (limits of x)
2. *Neumann conditions:* $dy/dx = 0$ at the boundaries
3. *Mixed or Gauss conditions:* $\alpha y(x) + \beta(dy/dx) = 0$ at the boundaries
4. *Periodic conditions:* $y\big|_{\substack{\text{lower}\\\text{boundary}}} = y\big|_{\substack{\text{upper}\\\text{boundary}}}$ and $y'\big|_{\substack{\text{lower}\\\text{boundary}}} = y'\big|_{\substack{\text{upper}\\\text{boundary}}}$

In three dimensions these become:

1. *Dirichlet conditions:* $U(\mathbf{r}) = 0$ at the boundaries (limits of x, y, and z or r, θ, and ϕ, etc.)
2. *Neumann conditions:* $\nabla U(\mathbf{r}) = 0$ at the boundaries
3. *Mixed or Gauss conditions:* $\alpha U(\mathbf{r}) + \boldsymbol{\beta} \cdot \nabla U = 0$ at the boundaries
4. *Periodic conditions:* $U\big|_{\substack{\text{lower}\\\text{boundary}}} = U\big|_{\substack{\text{upper}\\\text{boundary}}}$ and $\nabla U\big|_{\substack{\text{lower}\\\text{boundary}}} = \nabla U\big|_{\substack{\text{upper}\\\text{boundary}}}$

The Green's function is defined by the operator and the boundary conditions of the particular inhomogeneous partial differential equation. Henceforth, when we refer to the operator, this will include the boundary conditions satisfied by the function being operated on by the differential operator.

The Green's function for the operator of equation 11.79a is defined so that

$$U(\mathbf{r}) = \int G(\mathbf{r}, \mathbf{r}') F(\mathbf{r}')\, d^3r' \tag{11.107}$$

To obtain the differential equation satisfied by $G(\mathbf{r}, \mathbf{r}')$, we operate on $U(\mathbf{r})$ in equation 11.107 with the operator of equation 11.79a. Then

$$\int \Big\{ \nabla_r \cdot [p(\mathbf{r})\, \nabla_r G(\mathbf{r}, \mathbf{r}')] + W(\mathbf{r}) G(\mathbf{r}, \mathbf{r}') \Big\} F(\mathbf{r}')\, d^3r' = F(\mathbf{r}) \tag{11.108}$$

where ∇_r represents partial differentiation with respect to the $\mathbf{r}$ coordinates rather than the $\mathbf{r}'$ coordinates. Since the Dirac δ-function satisfies

$$\int \delta(\mathbf{r} - \mathbf{r}') F(\mathbf{r}')\, d^3r' = F(\mathbf{r}) \tag{11.109}$$

the Green's function must satisfy

$$\nabla_r \cdot [p(\mathbf{r}) \nabla_r G(\mathbf{r},\mathbf{r}')] + W(\mathbf{r})G(\mathbf{r},\mathbf{r}') = \delta(\mathbf{r}-\mathbf{r}') + H(\mathbf{r},\mathbf{r}') \quad (11.110)$$

with

$$\int H(\mathbf{r},\mathbf{r}')F(\mathbf{r}')\, d^3r' = 0 \quad (11.111)$$

In many (but not all) examples, $H(\mathbf{r},\mathbf{r}') = 0$.

The Green's function of equation 11.110 satisfies the same boundary conditions that $U(\mathbf{r})$ satisfies. For example,

$$\text{(a)}\quad U(x=a, y=b, z=c) = 0 \quad \Rightarrow \quad \text{(b)}\quad G(x=a, y=b, z=c, \mathbf{r}') = 0 \quad (11.112)$$

To understand the meaning of $\delta(\mathbf{r}-\mathbf{r}')$, consider the identity

$$\int \delta(\mathbf{r}-\mathbf{r}')\, d^3r' = 1 \quad (11.113)$$

in Cartesian coordinates. The one-dimensional δ-function satisfies $\int \delta(x-x')\,dx' = 1$ and $d^3r' = dx'\,dy'\,dz'$. Therefore, equation 11.113 will be satisfied if

$$\delta(\mathbf{r}-\mathbf{r}') = \delta(x-x')\delta(y-y')\delta(z-z') \quad (11.114)$$

In spherical coordinates, $d^3r' = r'^2 \sin\theta'\, dr'\, d\theta'\, d\phi'$. Since $\int \delta(r-r')\,dr' = 1$ and $\int \delta(\theta - \theta')\,d\theta' = 1$, equation 11.113 will be valid if

$$\delta(\mathbf{r}-\mathbf{r}') = \frac{1}{r'^2 \sin\theta'} \delta(r-r')\delta(\theta-\theta')\delta(\phi-\phi') \quad (11.115a)$$

so that in spherical coordinates, equation 11.113 becomes

$$\begin{aligned}\int \delta(\mathbf{r}-\mathbf{r}')\, d^3r' &= \int r'^2 \sin\theta\, dr'\, d\theta'\, d\phi' \frac{1}{r'^2 \sin\theta'} \delta(r-r')\delta(\theta-\theta')\delta(\phi-\phi') \\ &= \int_0^\infty \delta(r-r')\, dr' \int_0^\pi \delta(\theta-\theta')\, d\theta' \int_0^{2\pi} \delta(\phi-\phi')\, d\phi' = 1\end{aligned} \quad (11.115b)$$

It can be shown that the Green's function also satisfies

$$\nabla_{r'} \cdot [p(\mathbf{r}') \nabla_{r'} G(\mathbf{r},\mathbf{r}')] + W(\mathbf{r}')G(\mathbf{r},\mathbf{r}') = \delta(\mathbf{r}-\mathbf{r}') + H(\mathbf{r},\mathbf{r}') \quad (11.116)$$

by developing the three-dimensional Green's theorem. This is the extension of the one-dimensional Green's theorem developed for the Sturm–Liouville equation as expressed in equations 6.131 and 6.134. If the Green's function satisfies equation 11.116, then $G(\mathbf{r},\mathbf{r}')$ satisfies the boundary conditions of $U(\mathbf{r})$ in the $\mathbf{r}'$ variables. For example,

$$\text{(a)}\quad U(x=a, y=b, z=c) = 0 \quad \Rightarrow \quad \text{(b)}\quad G(\mathbf{r}, x'=a, y'=b, z'=c) = 0 \quad (11.117)$$

Consider

$$\begin{aligned}\Delta \equiv \int \Big\{ & U(\mathbf{r}')\big[\nabla_{r'} \cdot [p(\mathbf{r}') \nabla_{r'} G(\mathbf{r},\mathbf{r}')] + W(\mathbf{r}')G(\mathbf{r},\mathbf{r}')\big] \\ & - G(\mathbf{r},\mathbf{r}')\big[\nabla_{r'} \cdot [p(\mathbf{r}') \nabla_{r'} U(\mathbf{r}')] + W(\mathbf{r}')U(\mathbf{r}')\big]\Big\}\, d^3r'\end{aligned} \quad (11.118)$$

Using equations 11.79a and 11.116 yields

$$\Delta = \int U(\mathbf{r}')\delta(\mathbf{r}-\mathbf{r}')\, d^3r' + \int H(\mathbf{r},\mathbf{r}')U(\mathbf{r}')\, d^3r' - \int G(\mathbf{r},\mathbf{r}')F(\mathbf{r}')\, d^3r'$$

$$= U(\mathbf{r}) + \int H(\mathbf{r},\mathbf{r}')U(\mathbf{r}')\, d^3r' - \int G(\mathbf{r},\mathbf{r}')F(\mathbf{r}')\, d^3r' \qquad (11.119)$$

As will be shown later, the requirement that $H(\mathbf{r},\mathbf{r}')$ be orthogonal to $F(\mathbf{r}')$ (equation 11.111) also requires $H(\mathbf{r},\mathbf{r}')$ be orthogonal to $U(\mathbf{r}')$. That is,

$$\int H(\mathbf{r},\mathbf{r}')U(\mathbf{r}')\, d^3r' = 0 \qquad (11.120)$$

$$\Rightarrow \quad \Delta = U(\mathbf{r}) - \int G(\mathbf{r},\mathbf{r}')F(\mathbf{r}')\, d^3r' \qquad (11.121)$$

Returning to equation 11.118, after canceling the terms containing $W(\mathbf{r}')$, Δ can be written

$$\Delta \equiv \int \Big\{U(\mathbf{r}')\, \nabla_{r'} \cdot [p(\mathbf{r}')\, \nabla_{r'} G(\mathbf{r},\mathbf{r}')] - G(\mathbf{r},\mathbf{r}')\, \nabla_{r'} \cdot [p(\mathbf{r}')\, \nabla_{r'} U(\mathbf{r}')]\Big\}\, d^3r' \qquad (11.122)$$

By subtracting the identities

$$U(\mathbf{r}')\, \nabla_{r'} \cdot [p(\mathbf{r}')\, \nabla_{r'} G(\mathbf{r},\mathbf{r}')] = \nabla_{r'} \cdot [U(\mathbf{r}')p(\mathbf{r}')\, \nabla_{r'} G(\mathbf{r},\mathbf{r}')] - p(\mathbf{r}')\, \nabla_{r'} U(\mathbf{r}') \cdot \nabla_{r'} G(\mathbf{r},\mathbf{r}') \qquad (11.123a)$$

$$G(\mathbf{r},\mathbf{r}')\, \nabla_{r'} \cdot [p(\mathbf{r}')\, \nabla_{r'} U(\mathbf{r}')] = \nabla_{r'} \cdot [G(\mathbf{r},\mathbf{r}')p(\mathbf{r}')\, \nabla_{r'} U(\mathbf{r}')] - p(\mathbf{r}')\, \nabla_{r'} G(\mathbf{r},\mathbf{r}') \cdot \nabla_{r'} U(\mathbf{r}') \qquad (11.123b)$$

and integrating, we obtain

$$\Delta \equiv \int \Big\{U(\mathbf{r}')\, \nabla_{r'} \cdot [p(\mathbf{r}')\, \nabla_{r'} G(\mathbf{r},\mathbf{r}')] - G(\mathbf{r},\mathbf{r}')\, \nabla_{r'} \cdot [p(\mathbf{r}')\, \nabla_{r'} U(\mathbf{r}')]\Big\}\, d^3r'$$

$$= \int \nabla_{r'} \cdot \Big\{[U(\mathbf{r}')p(\mathbf{r}')\, \nabla_{r'} G(\mathbf{r},\mathbf{r}')] - [G(\mathbf{r},\mathbf{r}')p(\mathbf{r}')\, \nabla_{r'} U(\mathbf{r}')]\Big\}\, d^3r' \qquad (11.124)$$

From Gauss's theorem (equation 1.128), this integral can be expressed as an integral over a surface at the boundary. That is

$$\Delta = -\int_{\text{boundary}} p(\mathbf{r}')[G(\mathbf{r},\mathbf{r}')\, \nabla_{r'} U(\mathbf{r}') - U(\mathbf{r}')\, \nabla_{r'} G(\mathbf{r},\mathbf{r}')] \cdot \mathbf{n}\, dS' \qquad (11.125)$$

From equations 11.118 and 11.125, we obtain

$$\int \Big[U(\mathbf{r}')\Big\{\nabla_{r'} \cdot [p(\mathbf{r}')\, \nabla_{r'} G(\mathbf{r},\mathbf{r}')] + W(\mathbf{r}')G(\mathbf{r},\mathbf{r}')\Big\}$$

$$-G(\mathbf{r},\mathbf{r}')\Big\{\nabla_{r'} \cdot [p(\mathbf{r}')\, \nabla_{r'} U(\mathbf{r}')] + W(\mathbf{r}')U(\mathbf{r}')\Big\}\Big]\, d^3r'$$

$$= \int_{\text{boundary}} p(\mathbf{r}')[G(\mathbf{r},\mathbf{r}')\, \nabla_{r'} U(\mathbf{r}') - U(\mathbf{r}')\, \nabla_{r'} G(\mathbf{r},\mathbf{r}')] \cdot \mathbf{n}\, dS' \qquad (11.126)$$

This is the three-dimensional expression of Green's theorem.

In analogy with the one-dimensional derivation, since both $U(\mathbf{r}')$ and $G(\mathbf{r}, \mathbf{r}')$ satisfy one of the four conditions at the boundaries, $\Delta = 0$. Therefore, from equation 11.121, we obtain

$$U(\mathbf{r}) = \int G(\mathbf{r}, \mathbf{r}') F(\mathbf{r}')\, d^3r' \tag{11.107}$$

Therefore, if we can determine the Green's function for an operator, the problem of solving a partial differential equation is reduced to the task of evaluating an integral.

Eigenfunction expansion of Green's function

The Green's function can be expressed in terms of a sum involving the solutions to the eigenvalue equation of the operator. We define the eigenvalue equation corresponding to the operator of equation 11.79a by

$$\nabla \cdot [p(\mathbf{r}) \nabla u_n(\mathbf{r})] + W(\mathbf{r}) u_n(\mathbf{r}) = \lambda_n u_n(\mathbf{r}) \tag{11.127}$$

From this eigenvalue equation, we see that if $\lambda_n = 0$ is an eigenvalue of the operator, the homogeneous form of equation 11.79a will have a nontrivial solution. If $\lambda_n = 0$ is not an eigenvalue, the only solution to the homogeneous equation is the trivial one, $u_n(\mathbf{r}) = 0$.

As we saw in the study of the various Sturm–Liouville equations in Chapter 6, the solutions of an eigenvalue equation form a mutually orthogonal (or Gram–Schmidt orthogonalizable) set of functions. Therefore, any function of $\mathbf{r}$ can be expanded in a sum over the $u_n(\mathbf{r})$. As such, we can express the solution and the source or driving function as

$$\text{(a)} \quad U(\mathbf{r}) = \sum_n \alpha_n u_n(\mathbf{r}) \qquad \text{(b)} \quad F(\mathbf{r}) = \sum_n \beta_n u_n(\mathbf{r}) \tag{11.128}$$

By taking the eigenfunctions to be normalized to 1, the coefficients are given by

$$\text{(a)} \quad \alpha_n = \langle u_n | U \rangle = \int u_n^*(\mathbf{r}) U(\mathbf{r})\, d^3r \qquad \text{(b)} \quad \beta_n = \langle u_n | F \rangle = \int u_n^*(\mathbf{r}) F(\mathbf{r})\, d^3r \tag{11.129}$$

Since $F(\mathbf{r})$ is a known function, in principle, β_n is determinable.

Substituting equations 11.128 into

$$\nabla \cdot [p(\mathbf{r}) \nabla U(\mathbf{r})] + W(\mathbf{r}) U(\mathbf{r}) = F(\mathbf{r}) \tag{11.79a}$$

and using the eigenvalue equation, we obtain

$$\sum_n \alpha_n \lambda_n u_n(\mathbf{r}) = \sum_n \beta_n u_n(\mathbf{r}) \tag{11.130}$$

Thus, from the orthogonality of the eigenfunctions,

$$\alpha_n = \frac{\beta_n}{\lambda_n} = \frac{1}{\lambda_n} \int u_n^*(\mathbf{r}') F(\mathbf{r}')\, d^3r' \tag{11.131}$$

$$\Rightarrow \quad U(\mathbf{r}) = \sum_n \frac{u_n(\mathbf{r})}{\lambda_n} \int u_n^*(\mathbf{r}') F(\mathbf{r}')\, d^3r' \tag{11.132a}$$

or interchanging the order of summation and integration,

$$U(\mathbf{r}) = \int \left[\sum_n \frac{u_n(\mathbf{r}) u_n^*(\mathbf{r}')}{\lambda_n} \right] F(\mathbf{r}')\, d^3r' \tag{11.132b}$$

Comparing equation 11.132b to

$$U(\mathbf{r}) = \int G(\mathbf{r}, \mathbf{r}')F(\mathbf{r}')\, d^3\mathbf{r}' \tag{11.107}$$

$$\Rightarrow \quad G(\mathbf{r}, \mathbf{r}') = \sum_n \frac{u_n(\mathbf{r})u_n^*(\mathbf{r}')}{\lambda_n} \tag{11.133}$$

This is the *eigenfunction expansion* of the Green's function.

If one of the eigenvalues is zero (e.g., $\lambda_k = 0$), either the Green's function is infinite (due to the $n = k$ term) and therefore no solution exists, or the $n = k$ term is not included in the sum (in which case the Green's function is finite and a solution exists). From equations 11.129 we see that the $n = k$ term will be absent if the kth eigenfunction is orthogonal to $F(\mathbf{r})$. That is,

$$\int u_k^*(\mathbf{r}')F(\mathbf{r}')\, d^3r' = 0 \tag{11.134}$$

which requires that the coefficient $\beta_k = 0$.

We first determine the Green's function equation for the case in which zero is not an eigenvalue. Then the expansion of $F(\mathbf{r})$ in equation 11.129a contains all the eigenfunctions. Since these eigenfunctions form a complete orthonormal set, like the Green's function, the function $H(\mathbf{r}, \mathbf{r}')$ introduced in equation 11.110 can be expanded in an eigenfunction series which we express as

$$H(\mathbf{r}, \mathbf{r}') = \sum_n \gamma_n u_n^*(\mathbf{r})u_n^*(\mathbf{r}') \tag{11.135}$$

(If $\{u_n(\mathbf{r})\}$ form a complete set of basis functions, then $\{u_n^*(r)\}$ also form a complete set of basis functions.) Then equation 11.111 becomes

$$\int H(\mathbf{r}, \mathbf{r}')F(\mathbf{r}')\, d^3r' = 0 = \sum_n \gamma_n u_n^*(\mathbf{r}) \int u_n^*(\mathbf{r}')F(\mathbf{r}')\, d^3r' \tag{11.136}$$

Since the expansion of $F(\mathbf{r}')$ contains every eigenfunction,

$$\int u_n^*(\mathbf{r}')F(\mathbf{r}')\, d^3r' \neq 0 \tag{11.137}$$

for every n. Therefore, the integral of equation 11.136 can be zero only if every γ_n is zero. Thus if zero is not an eigenvalue of the operator, $H(\mathbf{r}, \mathbf{r}') = 0$, and either in terms of $\mathbf{r}$ or $\mathbf{r}'$, the Green's function equation is

$$\nabla \cdot [p\nabla G(\mathbf{r}, \mathbf{r}')] + WG(\mathbf{r}, \mathbf{r}') = \delta(\mathbf{r} - \mathbf{r}') \tag{11.138a}$$

$$\Rightarrow \quad \nabla \cdot [p\nabla G(\mathbf{r}, \mathbf{r}')] + WG(\mathbf{r}, \mathbf{r}') = 0 \qquad \mathbf{r} \neq \mathbf{r}' \tag{11.138b}$$

where the functions p and W depend on $\mathbf{r}$ if $\nabla = \nabla_r$, and p and W are functions of $\mathbf{r}'$ if $\nabla = \nabla_{r'}$.

If zero is an eigenvalue, then in order for a solution to exist, the eigenfunction corresponding to $\lambda_k = 0$ must be absent from the eigenfunction series for $F(\mathbf{r}')$. Then

$$\int u_k^*(\mathbf{r}')F(\mathbf{r}')\, d^3r' = 0 \tag{11.134}$$

This requires that $\gamma_n = 0$, $n \neq k$, but γ_k does not have to be zero. That is, if $\lambda_k = 0$ is an eigenvalue,

$$H(\mathbf{r},\mathbf{r}') = \gamma_k u_k^*(\mathbf{r}) u_k^*(\mathbf{r}') \tag{11.139}$$

$$\Rightarrow \quad \nabla \cdot [p\nabla G(\mathbf{r},\mathbf{r}')] + WG(\mathbf{r},\mathbf{r}') = \delta(\mathbf{r}-\mathbf{r}') + \gamma_k u_k^*(\mathbf{r}) u_k^*(\mathbf{r}') \tag{11.140a}$$

$$\Rightarrow \quad \nabla \cdot [p\nabla G(\mathbf{r},\mathbf{r}')] + WG(\mathbf{r},\mathbf{r}') = \gamma_k u_k^*(\mathbf{r}) u_k^*(\mathbf{r}') \qquad \mathbf{r} \neq \mathbf{r}' \tag{11.140b}$$

In summary, the steps involved in applying Green's function methods to solve

$$\nabla \cdot [p\nabla U(\mathbf{r})] + WU(\mathbf{r}) = F(\mathbf{r}) \tag{11.79a}$$

are:

1. Solve the eigenvalue equation

$$\nabla \cdot [p\nabla u_n(\mathbf{r})] + Wu_n(\mathbf{r}) = \lambda_n u_n(\mathbf{r}) \tag{11.127}$$

2. If zero is not an eigenvalue, or if $\lambda_k = 0$ is an eigenvalue and

$$\int u_k^*(\mathbf{r}) F(\mathbf{r})\, d^3r = 0$$

then a solution for $U(r)$ exists. If $\lambda_k = 0$ is an eigenvalue and

$$\int u_k^*(\mathbf{r}) F(\mathbf{r})\, d^3r \neq 0$$

equation 11.79a does not have a solution.

3. If $\lambda_k = 0$ is an eigenvalue, then

$$H(\mathbf{r},\mathbf{r}') = \gamma_k u_k^*(\mathbf{r}) u_k^*(\mathbf{r}') \tag{11.139}$$

If zero is not an eigenvalue, then $H(\mathbf{r},\mathbf{r}') = 0$.

4. If a solution exists, solve the Green's function equation

$$\nabla_{r'} \cdot [p(\mathbf{r}')\, \nabla_{r'} G(\mathbf{r},\mathbf{r}')] + W(\mathbf{r}')G(\mathbf{r},\mathbf{r}') = \delta(\mathbf{r}-\mathbf{r}') + \gamma_k u_k^*(\mathbf{r}) u_k^*(\mathbf{r}') \tag{11.140a}$$

to determine $G(\mathbf{r},\mathbf{r}')$, with $\gamma_k = 0$ if zero is not an eigenvalue.

5. Evaluate

$$U(\mathbf{r}) = \int G(\mathbf{r},\mathbf{r}') F(\mathbf{r}')\, d^3r' \tag{11.107}$$

As a final note, we have shown that if $\lambda_k = 0$, then for a solution to exist, the $n = k$ term must be absent from the Green's function series. That is,

$$G(\mathbf{r},\mathbf{r}') = \sum_{n \neq k} \frac{u_n(\mathbf{r}) u_n^*(\mathbf{r}')}{\lambda_n} \tag{11.141}$$

$$\Rightarrow \quad U(\mathbf{r}) = \sum_{n \neq k} \frac{u_n(\mathbf{r})}{\lambda_n} \int u_n^*(\mathbf{r}') F(\mathbf{r}')\, d^3r' = \sum_{n \neq k} \alpha_n u_n(\mathbf{r}) \tag{11.142}$$

Thus the eigenfunction series for $U(\mathbf{r})$ must not contain a term in $u_k(\mathbf{r})$. Therefore, when $\lambda_k = 0$,

$$\int H(\mathbf{r},\mathbf{r}') U(\mathbf{r}')\, d^3r' = \gamma_k u_k^*(\mathbf{r}) \sum_{n \neq k} \alpha_n \int u_k^*(\mathbf{r}') u_n(\mathbf{r}')\, d^3r' = 0 \tag{11.143}$$

Of course, when zero is not an eigenvalue, $H(\mathbf{r}, \mathbf{r}') = 0$. Thus, if a solution exists, whether zero is an eigenvalue or not,

$$\int H(\mathbf{r}, \mathbf{r}')U(\mathbf{r}')\, d^3r' = 0 \tag{11.120}$$

as stated earlier.

Example 11.7

As an example, consider the static (time-independent) equation for the amplitude of a rectangular drum head driven by a forcing function. The amplitude satisfies the time-independent wave equation

$$\frac{\partial^2 U}{\partial x^2} + \frac{\partial^2 U}{\partial y^2} = f(x, y) \tag{11.144a}$$

If the drum head is fastened at the edges, the boundary conditions are

$$U(0, y) = U(a, y) = U(x, 0) = U(x, b) = 0 \tag{11.144b}$$

The eigenvalue equation for this operator is

$$\frac{\partial^2 u_n}{\partial x^2} + \frac{\partial^2 u_n}{\partial y^2} = \lambda_n u_n \tag{11.145}$$

Since there are two variables, two indices will be required to specify the eigenfunctions. As such, n represents two indices. The solution to the eigenvalue equation is achieved by writing

$$u_n(x, y) \equiv v_l(x) w_m(y) \tag{11.146}$$

$$\Rightarrow \quad \frac{1}{v_l(x)} \frac{d^2 v_l}{dx^2} + \frac{1}{w_m(y)} \frac{d^2 w_m}{dy^2} = \lambda_{lm} \tag{11.147}$$

Referring to the ideas described in Section 11.1, each term must be a constant. Thus

$$\text{(a)} \quad \frac{1}{v_l(x)} \frac{d^2 v_l}{dx^2} \equiv -k_x^2 = \text{constant} \qquad \text{(b)} \quad \frac{1}{w_m(y)} \frac{d^2 w_m}{dy^2} \equiv -k_y^2 = \text{constant} \tag{11.148}$$

$$\Rightarrow \quad \lambda_{lm} = -\left(k_x^2 + k_y^2\right) \tag{11.149}$$

The solutions to equations 11.148 are

$$v_l(x) = A_l \sin(k_x x) + B_l \cos(k_x x) \tag{11.150a}$$

$$w_m(y) = C_m \sin(k_y y) + D_m \cos(k_y y) \tag{11.150b}$$

Applying the boundary conditions $v_l(0) = v_l(a) = 0$, we obtain $B_l = 0$ and

$$k_x = \frac{l\pi}{a} \qquad l = 0, \pm 1, \pm 2, \ldots \tag{11.151a}$$

The boundary conditions $w_m(0) = w_m(b) = 0$ yield $D_m = 0$ and

$$k_y = \frac{m\pi}{b} \qquad m = 0, \pm 1, \pm 2, \ldots \tag{11.151b}$$

Therefore, the normalized functions are

$$\text{(a)} \quad v_l(x) = \left(\frac{2}{a}\right)^{1/2} \sin\left(\frac{l\pi}{a}x\right) \qquad \text{(b)} \quad w_m(y) = \left(\frac{2}{a}\right)^{1/2} \sin\left(\frac{m\pi}{b}y\right) \tag{11.152}$$

$$\Rightarrow \quad u_{lm}(x, y) = \frac{2}{ab} \sin\left(\frac{l\pi}{a}x\right) \sin\left(\frac{m\pi}{b}y\right) \tag{11.153}$$

The associated eigenvalues are

$$\lambda_{lm} = -\pi^2\left(\frac{l^2}{a^2} + \frac{m^2}{b^2}\right) \tag{11.154}$$

We note that for $l = m = 0$, $\lambda_{00} = 0$ is an eigenvalue. However, the associated eigenfunction is $u_{00}(x, y) = 0$. Therefore, the Green's function for this operator is

$$G(x, x'; y, y') = \sum_{\substack{l=1\\m=1}} \frac{u^*_{lm}(x, y)u_{lm}(x', y')}{\lambda_{lm}}$$

$$= -\frac{4}{ab\pi^2}\sum_{\substack{l=1\\m=1}} \frac{\sin\left(\frac{l\pi}{a}x\right)\sin\left(\frac{l\pi}{a}x'\right)\sin\left(\frac{m\pi}{b}y\right)\sin\left(\frac{m\pi}{b}y'\right)}{(l^2/a^2 + m^2/b^2)} \tag{11.155}$$

$$\Rightarrow \quad U(x, y) = \int_0^a\int_0^b G(x, x'; y, y')F(x', y')\,dx'\,dy' \tag{11.156}$$

$$= -\frac{4}{ab\pi^2}\sum_{\substack{l=1\\m=1}} \frac{\sin\left(\frac{l\pi}{a}x\right)\sin\left(\frac{m\pi}{b}y\right)}{(l^2/a^2 + m^2/b^2)}\int_0^a\int_0^b \sin\left(\frac{l\pi}{a}x'\right)\sin\left(\frac{m\pi}{b}y'\right)f(x', y')\,dx'\,dy' \quad \square$$

Green's function in one dimension

The Green's function method can also be used to solve ordinary inhomogeneous differential equations. The Green's function will satisfy a differential equation that we write in the self-adjoint form of a Sturm–Liouville equation. That is, each one-dimensional Green's function satisfies an equation of the form

$$\frac{d}{dx'}\left[p(x')\frac{d}{dx'}G(x, x')\right] - q(x')G(x, x') = \delta(x - x') + H(x, x') \tag{11.157}$$

When viewed as a function, it was noted that the δ-function was infinite when its argument was zero. That is, $\delta(x - x')$ is singular at $x = x'$. (Such a singularity is referred to as a δ-function singularity.) As we see from equation 11.157, three terms involving G, dG/dx' and d^2G/dx'^2 add to a function containing a δ-function singularity at $x' = x$. Since $p(x')$ and $q(x')$ do not depend on x, they cannot contain this singularity. Therefore, the singularity resides in either G, dG/dx' or d^2G/dx'^2.

If a function is singular, its derivative has a higher singularity. This is most easily seen by considering a function containing a pole of order N. Such a function, near the pole, is of the form

$$F(z) \cong \frac{R(z_0)}{(z - z_0)^N} \tag{11.158}$$

From this, we note that dF/dz has a pole of order $(N + 1)$.

If we assigned the δ-function singularity to $G(x, x')$, then dG/dx' and d^2G/dx'^2 would have singularities of higher order. Since the singularity of equation 11.157 is a δ-function singularity, $G(x, x')$ cannot contain this singularity. By identical reasoning, we also conclude that dG/dx' cannot contain the δ-function singularity. Therefore, this singularity must arise from the second derivative. That is,

$$\frac{d^2G}{dx'^2} \sim \delta(x - x') \tag{11.159}$$

As shown in equation 4.162, the derivative of the Heaviside step function is

$$\frac{d}{dx'}\Theta(x - x') = \delta(x - x') \tag{4.162}$$

But $\Theta(x - x')$ is representative of a discontinuous function (see equation 4.143). That is, a function that is discontinuous at x_0 is of the form

$$F(x) = A(x) + B(x)\Theta(x - x_0) \tag{11.160}$$

Thus, from equation 11.159, we conclude that dG/dx' is discontinuous at $x = x'$. Therefore, G is continuous at $x = x'$ (with a discontinuous slope). A representation of a typical Green's function is pictured in Figure 11.1. The derivative of such a function would look like that shown in Figure 11.2.

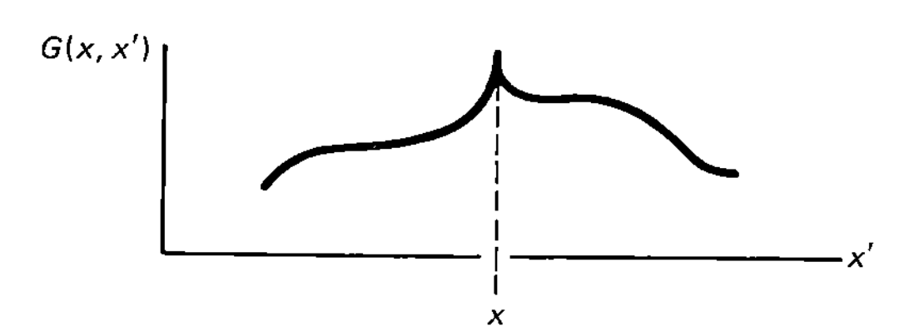

Figure 11.1 Typical Green's function with discontinuous slope.

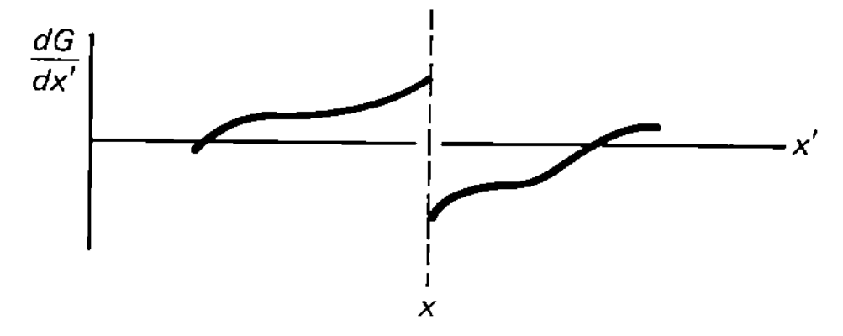

Figure 11.2 Derivative of typical Green's function.

Because of the discontinuity in the slope, we define two Green's functions.

$$G(x, x') \equiv \begin{cases} G_S(x, x') & x' < x \\ G_L(x, x') & x' > x \end{cases} \tag{11.161}$$

G_L and G_S are referred to as the *large* and *small segments* of the Green's function, or simply the *large* and *small Green's functions* (Figure 11.3).

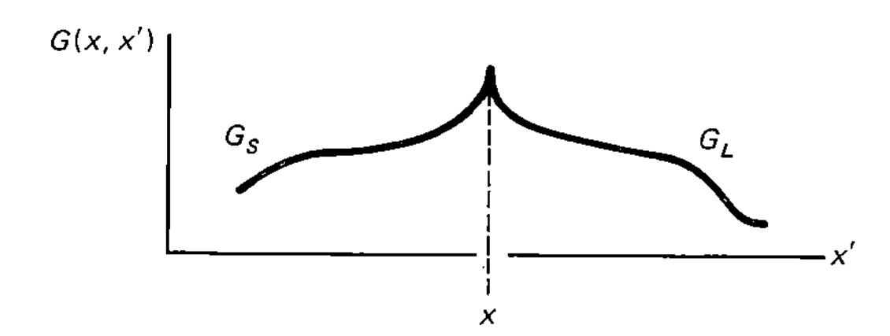

Figure 11.3 Large and small Green's functions.

As discussed above, the Green's function is continuous at $x = x'$. That is,

$$G_L(x, x) = G_S(x, x) \tag{11.162}$$

Since the Green's function has a discontinuous derivative at $x' = x$, we must evaluate $(dG_L/dx' - dG_S/dx')_{x'=x}$. Returning to the differential equation

$$\frac{d}{dx'}\left[p(x')\frac{dG(x, x')}{dx'}\right] - q(x')G(x, x') = \delta(x - x') + H(x, x') \tag{11.157}$$

we consider

$$\lim_{\varepsilon\to 0}\int_{x-\varepsilon}^{x+\varepsilon}\left[\frac{d}{dx'}\left[p(x')\frac{dG(x, x')}{dx'}\right] - q(x')G(x, x')\right]dx'$$

$$= \lim_{\varepsilon\to 0}\int_{x-\varepsilon}^{x+\varepsilon}\delta(x - x')\,dx' + \lim_{\varepsilon\to 0}\int_{x-\varepsilon}^{x+\varepsilon}H(x, x')\,dx' \tag{11.163}$$

Because both $G(x, x')$ and $H(x, x')$ are continuous at $x' = x$,

$$\lim_{\varepsilon \to 0} \int_{x-\varepsilon}^{x+\varepsilon} q(x')G(x, x')\, dx' = \lim_{\varepsilon \to 0} \int_{x-\varepsilon}^{x+\varepsilon} H(x, x')\, dx' = 0 \tag{11.164}$$

$$\Rightarrow \quad \lim_{\varepsilon \to 0} \int_{x-\varepsilon}^{x+\varepsilon} \frac{d}{dx'}\left[p(x') \frac{dG(x, x')}{dx'} \right] dx' = p(x') \lim_{\varepsilon \to 0} \left. \frac{dG(x, x')}{dx'} \right|_{x-\varepsilon}^{x+\varepsilon}$$

$$= \lim_{\varepsilon \to 0} \int_{x-\varepsilon}^{x+\varepsilon} \delta(x - x')\, dx' = 1 \tag{11.165}$$

From the definitions of the large and small parts of the Green's function, $G = G_L$ at $x' = x + \varepsilon$, and at $x' = x - \varepsilon$, $G = G_S$. Therefore, after taking $\varepsilon \to 0$, equation 11.165 becomes

$$\left(\frac{dG_L}{dx'} - \frac{dG_S}{dx'} \right)_{x'=x} = \frac{1}{p(x)} \tag{11.166}$$

The solution to the one-dimensional inhomogeneous Sturm–Liouville equation by Green's function methods, then, requires that we solve

$$\frac{d}{dx'}\left[p(x') \frac{d}{dx'} G(x, x') \right] - q(x')G(x, x') = H(x, x') \qquad x \neq x' \tag{11.167}$$

for the large and small segments of G, then apply the appropriate boundary conditions and the continuity and discontinuity conditions of equations 11.162 and 11.166. The solution $U(x)$ is then obtained by evaluating

$$U(x) = \int_a^b G(x, x')F(x')\, dx' \tag{11.168}$$

$H(x, x') = 0$

We now develop the condition for the existence of a solution when $H(x, x') = 0$. We begin with the continuity and discontinuity conditions

$$G_L(x, x) = G_S(x, x) \tag{11.162}$$

and

$$\left(\frac{dG_L}{dx'} - \frac{dG_S}{dx'} \right)_{x'=x} = \frac{1}{p(x)} \tag{11.166}$$

Since $H(x, x') = 0$, each segment of the Green's function satisfies

$$\frac{d}{dx'}\left[p(x') \frac{d}{dx'} G(x, x') \right] - q(x')G(x, x') = 0 \qquad x \neq x' \tag{11.169}$$

with $x \in [a, b]$. Let $\Gamma(x')$ be the solution to the homogeneous differential equation

$$\frac{d}{dx'}\left[p(x') \frac{d}{dx'} \Gamma(x') \right] - q(x')\Gamma(x') = 0 \tag{11.170}$$

$\Gamma_L(x')$ is defined by imposing the boundary condition at $x' = b$, and $\Gamma_S(x')$ is the same solution subject to the boundary condition at $x' = a$. Then the Green's function solution to equation 11.169 is

$$\text{(a)} \quad G_L(x, x') = A_L(x)\Gamma_L(x') \qquad \text{and} \qquad \text{(b)} \quad G_S(x, x') = A_S(x)\Gamma_S(x') \tag{11.171}$$

The continuity and discontinuity conditions require that

$$\text{(a)}\quad A_L(x)\Gamma_L(x) - A_S(x)\Gamma_S(x) = 0 \qquad \text{(b)}\quad A_L(x)\Gamma_L'(x) - A_S(x)\Gamma_S'(x) = \frac{1}{p(x)} \tag{11.172}$$

This is a set of inhomogeneous equations [assuming that $1/p(x) \neq 0$] in two unknowns, $A_L(x)$ and $A_S(x)$. A solution will exist if the determinant of the coefficient matrix is nonzero. That is,

$$W(x) \equiv \begin{vmatrix} \Gamma_L(x) & \Gamma_S(x) \\ \Gamma_L'(x) & \Gamma_S'(x) \end{vmatrix} \neq 0 \tag{11.173}$$

Referring to Chapter 5 (equation 5.409), this is the Wronskian determinant of the two functions $\Gamma_L(x)$ and $\Gamma_S(x)$. Following the analysis presented in Chapter 5, we see that the condition for a solution to exist when $H(x, x') = 0$, is that the Wronskian of the large and small segments of the Green's function be nonzero. If the Wronskian is nonzero, then

$$\text{(a)}\quad A_L(x) = -\frac{\Gamma_S(x)}{p(x)W(x)} \qquad \text{(b)}\quad A_S(x) = -\frac{\Gamma_L(x)}{p(x)W(x)} \tag{11.174}$$

We return to the homogeneous differential equation satisfied by $\Gamma_L(x)$ and $\Gamma_S(x)$.

$$\text{(a)}\quad \frac{d}{dx}\left[p(x)\frac{d}{dx}\Gamma_L(x)\right] - q(x)\Gamma_L(x) = 0 \qquad \text{(b)}\quad \frac{d}{dx}\left[p(x)\frac{d}{dx}\Gamma_S(x)\right] - q(x)\Gamma_S(x) = 0 \tag{11.175}$$

Multiplying equation 11.175a by $\Gamma_S(x)$ and equation 11.175b by $\Gamma_L(x)$ and subtracting, we obtain

$$\Gamma_S(x)\frac{d}{dx}\left[p(x)\frac{d}{dx}\Gamma_L(x)\right] - \Gamma_L(x)\frac{d}{dx}\left[p(x)\frac{d}{dx}\Gamma_S(x)\right]$$

$$= \frac{d}{dx}\left[p(x)\left(\Gamma_S(x)\frac{d}{dx}\Gamma_L(x) - \Gamma_L(x)\frac{d}{dx}\Gamma_S(x)\right)\right] = 0 \tag{11.176}$$

$$\Rightarrow \quad p(x)[\Gamma_S(x)\Gamma_L'(x) - \Gamma_L(x)\Gamma_S'(x)] = \text{constant} \equiv \frac{1}{\alpha} \tag{11.177a}$$

As noted above, the quantity in brackets is the Wronskian determinant, $W(x)$. Therefore,

$$p(x)W(x) = \frac{1}{\alpha} \tag{11.177b}$$

This is known as *Abel's formula*. With it, equations 11.174 become

$$\text{(a)}\quad A_L(x) = \alpha\Gamma_L(x) \qquad \text{(b)}\quad A_S(x) = \alpha\Gamma_S(x) \tag{11.178}$$

$$\Rightarrow \quad G(x, x') = \begin{cases} \alpha\Gamma_S(x)\Gamma_L(x') & x' > x \\ \alpha\Gamma_L(x)\Gamma_S(x') & x' < x \end{cases} \tag{11.179}$$

Example 11.8

As an example, consider the differential equation

$$\text{(a)}\quad \frac{d^2U}{dx^2} + U(x) = 1 \quad \text{with} \quad \text{(b)}\quad U(0) = U\left(\frac{\pi}{2}\right) = 0 \tag{11.180}$$

The Sturm–Liouville functions and source term for this differential equation are

$$\text{(a)} \quad p(x) = 1 \qquad \text{(b)} \quad q(x) = -1 \qquad \text{and} \qquad \text{(c)} \quad F(x) = 1 \tag{11.181}$$

Therefore, the solution to equations 11.180 is

$$U(x) = \int_0^{\pi/2} G(x,x') \cdot 1\, dx' = \int_0^x G_S(x,x')\, dx' + \int_x^{\pi/2} G_L(x,x')\, dx' \tag{11.182}$$

To determine the eigenvalues of the operator of equations 11.180, we solve the eigenvalue equation

$$\text{(a)} \quad \frac{d^2 u_n}{dx^2} + u_n = \lambda_n u_n \qquad \text{or} \qquad \text{(b)} \quad \frac{d^2 u_n}{dx^2} + (1 - \lambda_n) u_n = 0 \tag{11.183}$$

$$\Rightarrow \quad u_n(x) = A\cos\left(\sqrt{1-\lambda_n}\, x\right) + B\sin\left(\sqrt{1-\lambda_n}\, x\right) \tag{11.184}$$

Applying the boundary conditions

$$\text{(a)} \quad u_n(0) = 0 \quad \Rightarrow \quad \text{(b)} \quad A = 0 \quad \text{and} \quad \text{(c)} \quad u_n\left(\frac{\pi}{2}\right) = 0$$

$$\Rightarrow \quad \text{(d)} \quad \sin\left(\sqrt{1-\lambda_n}\, \frac{\pi}{2}\right) = 0 \tag{11.185}$$

Thus

$$\text{(a)} \quad \sqrt{1-\lambda_n} = 2n \quad \Rightarrow \quad \text{(b)} \quad \lambda_n = 1 - 4n^2 \qquad n = 0, \pm 1, \pm 2, \ldots \tag{11.186}$$

Therefore, the corresponding eigenfunctions are

$$u_n(x) = B_n \sin(2nx) \tag{11.187}$$

Since the sine functions are naturally orthogonal over $x \in [0, \pi/2]$, it is not necessary to Gram–Schmidt orthogonalize them. B_n is determined from the normalization integral

$$\int_0^{\pi/2} |u_n(x)|^2\, dx = B_n^2 \int_0^{\pi/2} \sin^2(2nx)\, dx = 1 \tag{11.188}$$

$$\Rightarrow \quad u_n(x) = \left[\frac{4}{\pi}\right]^{1/2} \sin(2nx) \tag{11.189}$$

Since there is no value of n for which $\lambda_n = 0$, zero is not an eigenvalue of the operator. Therefore,

$$H(x,x') = 0 \tag{11.190}$$

$$\Rightarrow \quad \frac{d^2 G_i}{dx'^2} + G_i(x,x') = 0 \tag{11.191}$$

where i is L or S. The solution to this equation is straightforward. We express the solution for the two segments of the Green's function as

$$G_S(x,x') = A_S(x)\sin x' + B_S(x)\cos x' \tag{11.192a}$$

$$G_L(x,x') = A_L(x)\sin x' + B_L(x)\cos x' \tag{11.192b}$$

A_S, B_S, A_L, and B_L are determined from the boundary and continuity conditions. Since $x' = 0 \leqslant x$, the boundary condition at $x' = 0$ is applied to G_S. Similarly, $x' = \pi/2 \geqslant x$, so the boundary condition at $x' = \pi/2$ is applied to G_L. Therefore,

$$\text{(a)} \quad G_S(x,0) = 0 = B_S(x) \qquad \text{(b)} \quad G_L\left(x, \frac{\pi}{2}\right) = 0 = A_L(x) \tag{11.193}$$

$$\Rightarrow \quad \text{(a)} \quad G_S(x,x') = A_S(x)\sin x' \qquad \text{(b)} \quad G_L(x,x') = B_L(x)\cos x' \tag{11.194}$$

From the continuity of the Green's function and the discontinuity of the derivative at $x' = x$, with $p(x) = 1$, we obtain

$$\text{(a)} \quad B_L(x) = A_S(x)\tan x \text{(b)} \quad B_L(x)\sin x + A_S(x)\cos x = -1 \tag{11.195}$$

$$\Rightarrow \quad \text{(a)} \quad A_S(x) = -\cos x \text{(b)} \quad B_L(x) = -\sin x \tag{11.196}$$

$$\Rightarrow \quad \text{(a)} \quad G_S(x, x') = -\cos x \sin x' \qquad \text{(b)} \quad G_L(x, x') = -\sin x \cos x' \tag{11.197}$$

Since $H(x, x') = 0$ in this example, the fact that the Green's function exists means that the Wronskian determinant must be nonzero. From equations 11.197 we see that

$$\text{(a)} \quad \Gamma_L(x') = \cos x' \qquad \text{(b)} \quad \Gamma_S(x') = \sin x' \tag{11.198}$$

and the multiplicative constant of equation 11.179 is $\alpha = -1$. Therefore, the Wronskian determinant is

$$W(x') = \begin{vmatrix} \Gamma_L(x') & \Gamma_S(x') \\ \Gamma_L'(x') & \Gamma_S'(x') \end{vmatrix} = \begin{vmatrix} \cos x' & \sin x' \\ -\sin x' & \cos x' \end{vmatrix} \neq 0 \tag{11.199}$$

Referring to equations 11.197, the solution to the original differential equation is

$$U(x) = -\cos x \int_0^x \sin x' \, dx' - \sin x \int_x^{\pi/2} \cos x' \, dx' = 1 - \sin x - \cos x \tag{11.200}$$

The eigenfunction series for the Green's function is expressed in terms of the functions

$$u_n(x) = \left[\frac{4}{\pi}\right]^{1/2} \sin(2nx) \tag{11.189}$$

$n \geqslant 1$. Therefore, with $\lambda_n = 1 - 4n^2$,

$$G(x, x') = \frac{4}{\pi} \sum_{n=1}^{\infty} \frac{\sin(2nx)\sin(2nx')}{1 - 4n^2} \tag{11.201}$$

from which the solution to the inhomogeneous equation with the source term $F(x) = 1$ is

$$U(x) = \frac{4}{\pi} \sum_{n=1}^{\infty} \frac{\sin(2nx)}{1 - 4n^2} \int_0^{\pi/2} \sin(2nx') \, dx' = \frac{4}{\pi} \sum_{n=1}^{\infty} \frac{\sin(2nx)}{1 - 4n^2} \frac{\left[(-1)^n - 1\right]}{2n} \tag{11.202a}$$

Since all terms with even n will be zero, this can be written

$$U(x) = -\frac{4}{\pi} \sum_{k=0}^{\infty} \frac{\sin[2(2k+1)x]}{(2k+1)(4k+1)(4k+3)} \tag{11.202b}$$

where n has been replaced by $2k + 1$.

Comparing the form of the Green's function expressed in equation 11.201 to that of equations 11.197, we obtain

$$\frac{4}{\pi} \sum_{n=1}^{\infty} \frac{\sin(2nx)\sin(2nx')}{1 - 4n^2} = \begin{cases} -\cos x \sin x' & x' \leqslant x \\ -\sin x \cos x' & x' \geqslant x \end{cases} \tag{11.203}$$

Similarly, a comparison of equations 11.200 and 11.202b yields

$$\frac{4}{\pi} \sum_{k=0}^{\infty} \frac{\sin[2(2k+1)x]}{(2k+1)(4k+1)(4k+3)} = \sin x + \cos x - 1 \tag{11.204}$$

We can use results such as this to deduce numerical identities. For example, if we set $x = x' = \pi/4$ in equation 11.203, we obtain

$$\frac{4}{\pi} \sum_{n=1}^{\infty} \frac{\sin^2(n\pi/2)}{1 - 4n^2} = -\frac{1}{2} \tag{11.205}$$

With $\sin(n\pi/2) = 0$ if n is even, and $\sin(n\pi/2) = (-1)^{n+1}$ for odd n, equation 11.205 becomes

$$\text{(a)}\quad \frac{4}{\pi}\sum_{\substack{n=1\\ n\ \text{odd}}}^{\infty}\frac{1}{4n^2-1} = \frac{1}{2} \quad\Rightarrow\quad \text{(b)}\quad \pi = 8\sum_{l=0}^{\infty}\frac{1}{(4l+1)(4l+3)} \tag{11.206}$$

where we have set $n = 2l + 1$.

A similar numerical identity can be obtained by setting $x = \pi/4$ in equation 11.204. We obtain

$$\frac{4}{\pi}\sum_{k=0}^{\infty}\frac{\sin[(2k+1)\pi/2]}{(2k+1)(4k+1)(4k+3)} = \sqrt{2} - 1 \tag{11.207a}$$

With $\sin[(2k+1)\pi/2] = (-1)^k$, this becomes

$$\pi = \frac{4}{(\sqrt{2}-1)}\sum_{k=0}^{\infty}\frac{(-1)^k}{(2k+1)(4k+1)(4k+3)} \qquad \square \tag{11.207b}$$

Example 11.9

As an second example, we again consider the differential equation

$$\text{(a)}\quad \frac{d^2U}{dx^2} + U(x) = 1 \qquad \text{with} \qquad \text{(b)}\quad U(0) = U(\pi) = 0 \tag{11.208}$$

Again, the eigenvalue equation is

$$\begin{aligned} &\text{(a)}\quad \frac{d^2u_n}{dx^2} + (1-\lambda_n)u_n = 0 \\ \Rightarrow\quad &\text{(b)}\quad u_n(x) = A\cos\left(\sqrt{1-\lambda_n}\,x\right) + B\sin\left(\sqrt{1-\lambda_n}\,x\right) \end{aligned} \tag{11.209}$$

The boundary conditions

$$\text{(a)}\quad u_n(0) = 0 \quad\Rightarrow\quad \text{(b)}\quad A = 0 \quad \text{and} \quad \text{(c)}\quad u_n(\pi) = 0 \quad\Rightarrow\quad \text{(d)}\quad \sin\left(\sqrt{1-\lambda_n}\,\pi\right) = 0 \tag{11.210}$$

From equation 11.210d, the eigenvalues are

$$\lambda_n = 1 - n^2 \qquad n = 0, \pm 1, \pm 2, \ldots \tag{11.211a}$$

with corresponding eigenfunctions

$$u_n(x) = B_n\sin(nx) \tag{11.211b}$$

As in Example 11.8, the $n = 0$ eigenfunction is zero, and the negative and positive n eigenfunctions are not independent. Thus we can restrict n to be $n \geqslant 1$.

We note that for $n = 1$, $\lambda_1 = 0$ is an eigenvalue. Therefore, a solution will exist only if $u_1(x)$ is orthogonal to $F(x)$. But

$$\int_0^{\pi}u_1(x)F(x)\,dx = B_1\int_0^{\pi}\sin(x)\cdot 1\,dx = 2 \neq 0 \tag{11.212}$$

Therefore, there is no solution to equation 11.208a subject to the boundary conditions expressed in equation 11.208b.

To see that no solution exists, it is straightforward to solve the differential equation by standard methods. The solution to the homogeneous form of equation 11.208a is

$$U_0(x) = A\sin x + B\cos x \tag{11.213a}$$

and the particular solution is

$$U_p(x) = (D^2+1)^{-1} \cdot 1 = [1 - O(D^2)] \cdot 1 = 1 \tag{11.213b}$$

$$\Rightarrow \quad U(x) = 1 + A \sin x + B \cos x \tag{11.214}$$

The boundary conditions $U(0) = U(\pi) = 0$ require

$$\text{(a)} \quad B = -1 \qquad \text{and} \qquad \text{(b)} \quad B = +1 \tag{11.215}$$

From this inconsistency we see that there is no solution to equations 11.208. □

Example 11.10

The equation

$$\frac{d^2U}{dx^2} + U = \cos x \tag{11.216a}$$

subject to the periodic boundary condition

$$U(0) = U(\pi) \tag{11.216b}$$

has a solution. As seen above, the eigenvalues are

$$\lambda_n = 1 - n^2 \qquad n \geqslant 1 \tag{11.211a}$$

so that $\lambda_1 = 0$. The associated eigenfunction is $u_1(x) = A \sin x$. For this example, then

$$\int_0^\pi u_1^*(x) F(x)\, dx = A^* \int_0^\pi \sin x \cos x \, dx = 0 \tag{11.217}$$

Thus the differential equation has a solution.

The Green's function equation is determined by noting that

$$H(x, x') = \gamma u_1^*(x) u_1^*(x') = \gamma \sin x \sin x' \tag{11.218}$$

$$\Rightarrow \quad \frac{d^2G}{dx'^2} + G = \gamma \sin x \sin x' \tag{11.219}$$

The homogeneous solution to this equation is

$$G_0(x, x') = A(x) \sin x' + B(x) \cos x' \tag{11.220}$$

and the particular solution is

$$\begin{aligned} G_p(x, x') &= \gamma \sin x (D_{x'}^2 + 1)^{-1} \sin x' \\ &= \gamma \sin x \, \mathrm{Im}[(D_{x'} - i)^{-1}(D_{x'} + i)^{-1} e^{ix'}] = \gamma \sin x \, \mathrm{Im}\left(\frac{x' e^{ix'}}{2i}\right) = -\tfrac{1}{2}\gamma x' \sin x \cos x' \end{aligned} \tag{11.221}$$

$$\Rightarrow \quad G_L(x, x') = -\tfrac{1}{2}\gamma x' \sin x \cos x' + A_L(x) \sin x' + B_L(x) \cos x' \tag{11.222a}$$

$$G_S(x, x') = -\tfrac{1}{2}\gamma x' \sin x \cos x' + A_S(x) \sin x' + B_S(x) \cos x' \tag{11.222b}$$

The boundary conditions yield

$$\text{(a)} \quad B_L(x) = -\tfrac{1}{2}\gamma\pi \sin x \qquad \text{(b)} \quad B_S(x) = 0 \tag{11.223}$$

$$\Rightarrow \quad G_L(x, x') = -\tfrac{1}{2}\gamma(x' - \pi) \sin x \cos x' + A_L(x) \sin x' \tag{11.224a}$$

$$G_S(x, x') = -\tfrac{1}{2}\gamma x' \sin x \cos x' + A_S(x) \sin x' \tag{11.224b}$$

The continuity and discontinuity conditions yield

$$\text{(a)}\quad A_S(x) = A_L(x) + \tfrac{1}{2}\gamma\pi\cos x \qquad \text{(b)}\quad [A_L(x) - A_S(x)]\cos x - \tfrac{1}{2}\gamma\pi\sin^2 x = 1 \tag{11.225}$$

Combining equations 11.225 leads to $\gamma = -2/\pi$, so the Green's function becomes

$$G_L(x, x') = \frac{1}{\pi}(x' - \pi)\sin x \cos x' + A_L(x)\sin x' \tag{11.226a}$$

$$G_S(x, x') = \frac{1}{\pi}x' \sin x \cos x' + A_L(x)\sin x' - \cos x \sin x' \tag{11.226b}$$

$$\Rightarrow \quad U(x) = \int_0^x G_S(x, x')\cos x'\, dx' + \int_x^\pi G_L(x, x')\cos x'\, dx'$$

$$= \frac{1}{\pi}\sin x \int_0^x x' \cos^2 x'\, dx' + A_L(x)\int_0^x \sin x' \cos x'\, dx' - \cos x \int_0^x \sin x' \cos x'\, dx'$$

$$+ \frac{1}{\pi}\sin x \int_x^\pi (x' - \pi)\cos^2 x'\, dx' + A_L(x)\int_x^\pi \sin x' \cos x'\, dx' \tag{11.227a}$$

We note that the integrals involving $x' \cos^2 x'$ and those multiplying $A_L(x)$ can be combined to a single integral over the range $[0, \pi]$. Thus

$$U(x) = \frac{1}{\pi}\sin x \int_0^\pi x' \cos^2 x'\, dx' + A_L(x)\int_0^\pi \sin x' \cos x'\, dx'$$

$$- \cos x \int_0^x \sin x' \cos x'\, dx' - \sin x \int_x^\pi \cos^2 x'\, dx' \tag{11.227b}$$

Because $\sin x'$ is orthogonal to $\cos x'$ over $[0, \pi]$, the integral multiplying the indeterminate function $A_L(x)$ is zero. Thus $A_L(x)$ does not affect the solution, and could have been set to zero. The solution is

$$U(x) = \left(\frac{x}{2} - \frac{\pi}{4}\right)\sin x \quad \square \tag{11.228}$$

Green's function for the electrostatic potential

The Green's function for Poisson's equation can be determined from a knowledge of the potential for a point charge; that is, the Coulomb potential. Poisson's equation in the MKS system for the electrostatic potential due to any charge distribution with density $\rho(\mathbf{r})$ is

$$\nabla^2 V(\mathbf{r}) = -\frac{\rho(\mathbf{r})}{\varepsilon_0} \tag{11.2}$$

with $V(\infty) = 0$. The potential obtained from the Green's function approach is

$$V(\mathbf{r}) = -\frac{1}{\varepsilon_0}\int G(r, r')\rho(r')\, d^3r' \tag{11.229}$$

If zero is not an eigenvalue of the operator ∇^2 with the specified boundary conditions, then $H(\mathbf{r}, \mathbf{r}') = 0$ and the Green's function satisfies either

$$\text{(a)}\quad \nabla_r^2 G(\mathbf{r}, \mathbf{r}') = \delta(\mathbf{r} - \mathbf{r}') \qquad \text{or} \qquad \text{(b)}\quad \nabla_{r'}^2 G(\mathbf{r}, \mathbf{r}') = \delta(\mathbf{r} - \mathbf{r}') \tag{11.230}$$

To determine this Green's function, consider the Poisson's equation for a point charge. Referring to Figure 11.4, the point charge occupies a volume of zero size at the

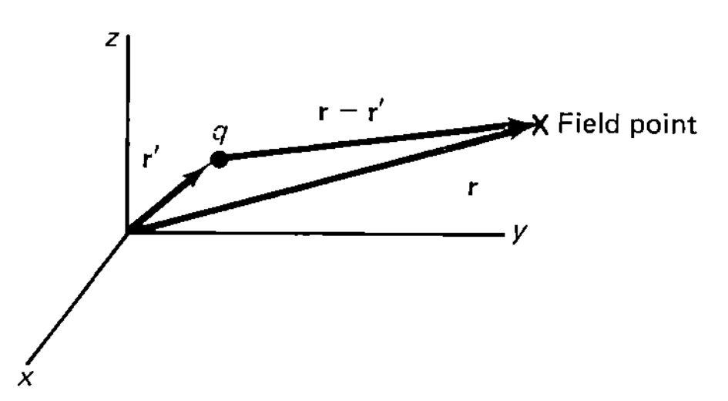

Figure 11.4 Charge q in the vicinity of a field point.

point r'. Therefore, the charge density is given by

$$\rho(\mathbf{r}) = \begin{cases} 0 & \mathbf{r} \neq \mathbf{r}' \\ \infty & \mathbf{r} = \mathbf{r}' \end{cases} \tag{11.231}$$

such that

$$\int \rho(\mathbf{r}')\, d^3r' = q \tag{11.232}$$

$$\Rightarrow \quad \rho(\mathbf{r}) = q\delta(\mathbf{r} - \mathbf{r}') \tag{11.233}$$

Since $\int \delta(\mathbf{r} - \mathbf{r}')\, d^3r' = 1$, $\delta(\mathbf{r} - \mathbf{r}')$ has units of inverse volume. Thus $q\delta(\mathbf{r} - \mathbf{r}')$ has the correct units of charge density.

With equation 11.233, Poisson's equation for a point charge is

$$\nabla^2 V(\mathbf{r}) = -\frac{q}{\varepsilon_0}\delta(\mathbf{r} - \mathbf{r}') \tag{11.234}$$

Comparing this to equations 11.230 for the Green's function, we see that

$$G(\mathbf{r}, \mathbf{r}') = -\frac{\varepsilon_0}{q} V_{\text{point charge}} \tag{11.235}$$

But the potential for a point charge is the well-known Coulomb potential

$$V_{\text{point charge}} = \frac{1}{4\pi\varepsilon_0} \frac{q}{|\mathbf{r} - \mathbf{r}'|} \tag{11.236}$$

$$\Rightarrow \quad G(\mathbf{r}, \mathbf{r}') = -\frac{1}{4\pi} \frac{1}{|\mathbf{r} - \mathbf{r}'|} \tag{11.237}$$

is the Green's function for the Poisson operator. Substituting this into equation 11.229, the potential due to any charge distribution is given by

$$V(\mathbf{r}) = \frac{1}{4\pi\varepsilon_0} \int \frac{\rho(\mathbf{r}')}{|\mathbf{r} - \mathbf{r}'|}\, d^3r' \tag{11.238}$$

subject to the potential being zero at infinite distances from the distribution.

Example 11.11

As an example, we consider a spherically symmetric charge density. That is, $\rho(r, \theta, \phi) = \rho(r)$. Writing

$$\frac{1}{|\mathbf{r} - \mathbf{r}'|} = \frac{1}{\sqrt{r^2 + r'^2 - 2rr'\cos\theta}} \tag{11.239}$$

we note that this function can be manipulated into the form of the generating function for the Legendre polynomials. Referring to equations 6.188, we express this as

$$\frac{1}{\sqrt{r^2 + r'^2 - 2rr'\cos\theta}} = \sum_{l=0}^{\infty} \frac{r'^l}{r^{l+1}} P_l(\cos\theta) \qquad r' < r \qquad (11.240a)$$

$$\frac{1}{\sqrt{r^2 + r'^2 - 2rr'\cos\theta}} = \sum_{l=0}^{\infty} \frac{r^l}{r'^{l+1}} P_l(\cos\theta) \qquad r' > r \qquad (11.240b)$$

In the integral of equation 11.238, we use equation 11.240a for that part of the range of integration for which $r' < r$. Over the interval for which $r' > r$, equation 11.240b is used. Therefore,

$$V(\mathbf{r}) = \frac{1}{4\pi\varepsilon_0} \int_0^{2\pi} d\phi \sum_{l=0}^{\infty} \left[\int_0^{r} \rho(r') \frac{r'^l}{r^{l+1}} r'^2 \, dr' + \int_r^{\infty} \rho(r') \frac{r^l}{r'^{l+1}} r'^2 \, dr' \right] \int_0^{\pi} P_l(\cos\theta) \sin\theta \, d\theta$$

$$(11.241)$$

From the orthogonality of the Legendre polynomials, with $x = \cos\theta$ and $P_0(x) = 1$,

$$\int_0^{\pi} P_l(\cos\theta) \sin\theta \, d\theta = \int_{-1}^{1} P_l(x) P_0(x) \, dx = 2\delta_{l,0} \qquad (11.242)$$

That is, only the $l = 0$ term in the sum is nonzero, and equation 11.241 becomes

$$V(r) = \frac{1}{\varepsilon_0} \left[\frac{1}{r} \int_0^{r} \rho(r') r'^{r} \, dr' + \int_r^{\infty} \rho(r') r' \, dr' \right] \qquad (11.243)$$

Thus we see that because the charge density is spherically symmetric, the potential is also spherically symmetric. □

Green's functions by integral transform methods

We recall that the Dirac δ-function can be expressed in terms of an integral obtained from integral transforms. That is, as deduced in equation 4.106a, the Fourier representation of the δ function is

$$\delta(x - x') = \frac{1}{2\pi} \int_{-\infty}^{\infty} e^{ik(x-x')} \, dk \qquad (4.106a)$$

or, referring to Table 4.2 of Laplace transforms,

$$\delta(t - t') = \frac{1}{2\pi i} \int_{c-i\infty}^{c+i\infty} e^{s(t-t')} \, ds \qquad (11.244)$$

Example 11.12

To illustrate the method of finding the Green's function using integral transform methods, we consider the Helmholtz equation as expressed in equation 11.84 as an example. The Helmholtz equation is

$$(\nabla^2 + \Lambda^2) U(\mathbf{r}) = F(\mathbf{r}) \qquad (11.84)$$

Therefore, if the boundary conditions are such that $H(\mathbf{r}, \mathbf{r}') = 0$, the Green's function equation is

$$(\nabla_{r'}^2 + \Lambda^2) G(\mathbf{r}, \mathbf{r}') = \delta(\mathbf{r} - \mathbf{r}') \qquad (11.245)$$

Since $\delta(\mathbf{r} - \mathbf{r}')$ was shown to be a product of three one-dimensional δ functions (see equations 11.114 and 11.115a), it is straightforward to deduce that the Fourier representation of $\delta(\mathbf{r}-\mathbf{r}')$ is a product of three one-dimensional Fourier integrals. Thus

$$\delta(\mathbf{r}-\mathbf{r}') = \frac{1}{(2\pi)^3}\int d^3k' e^{i\mathbf{k}'\cdot(\mathbf{r}-\mathbf{r}')} \tag{11.246}$$

Defining the Fourier inverse of the Green's function by

$$G(\mathbf{r},\mathbf{r}') \equiv \int \frac{d^3k'}{(2\pi)^{3/2}}\Phi(\mathbf{r},\mathbf{k}')e^{-i\mathbf{k}'\cdot\mathbf{r}'} \tag{11.247}$$

equation 11.245 becomes

$$\int \frac{d^3k'}{(2\pi)^{3/2}}(-k'^2+\Lambda^2)\Phi(\mathbf{r},\mathbf{k}')e^{-i\mathbf{k}'\cdot\mathbf{r}'} = \frac{1}{(2\pi)^3}\int d^3k' e^{i\mathbf{k}'\cdot\mathbf{r}}e^{-i\mathbf{k}'\cdot\mathbf{r}'} \tag{11.248}$$

$$\Rightarrow \quad \Phi(\mathbf{r},\mathbf{k}') = -\frac{1}{(2\pi)^{3/2}}\frac{e^{i\mathbf{k}'\cdot\mathbf{r}}}{(k'^2-\Lambda^2)} \tag{11.249}$$

$$\Rightarrow \quad G(\mathbf{r},\mathbf{r}') = -\frac{1}{(2\pi)^3}\int d^3k' \frac{e^{i\mathbf{k}'\cdot(\mathbf{r}-\mathbf{r}')}}{(k'^2-\Lambda^2)} \tag{11.250}$$

This integral can be evaluated by defining the k_z' axis such that

$$\mathbf{k}'\cdot(\mathbf{r}-\mathbf{r}') = k'|\mathbf{r}-\mathbf{r}'|\cos\theta \tag{11.251}$$

$$G(\mathbf{r},\mathbf{r}') = -\frac{1}{(2\pi)^3}\int_0^{2\pi} d\phi \int_0^{\pi} \sin\theta\, d\theta \int_0^{\infty} k'^2\, dk' \frac{e^{ik'|\mathbf{r}-\mathbf{r}'|\cos\theta}}{(k'^2-\Lambda^2)}$$

$$= -\frac{1}{(2\pi)^3}2\pi\int_0^{\infty}\frac{k'^2\, dk'}{(k'^2-\Lambda^2)}\int_{-1}^{1} dx\, e^{ik'|\mathbf{r}-\mathbf{r}'|x}$$

$$= -\frac{1}{(2\pi)^2}\frac{1}{i|\mathbf{r}-\mathbf{r}'|}\int_0^{\infty}\frac{k'\, dk'}{(k'^2-\Lambda^2)}\left\{e^{ik'|\mathbf{r}-\mathbf{r}'|} - e^{-ik'|\mathbf{r}-\mathbf{r}'|}\right\} \tag{11.252}$$

In the integral involving the negative exponential, we replace k' by $-k'$ to obtain

$$G(\mathbf{r},\mathbf{r}') = -\frac{1}{(2\pi)^2}\frac{1}{i|\mathbf{r}-\mathbf{r}'|}\int_{-\infty}^{\infty}\frac{k'\, dk'}{(k'^2-\Lambda^2)}e^{ik'|\mathbf{r}-\mathbf{r}'|} \tag{11.253}$$

Since $|\mathbf{r}-\mathbf{r}'| > 0$, this integral can be evaluated using Cauchy's theorem by closing the contour in the upper-half complex k' plane. Giving Λ a small positive imaginary part, the poles are displaced from the contour as shown in Figure 11.5. Therefore, the Green's

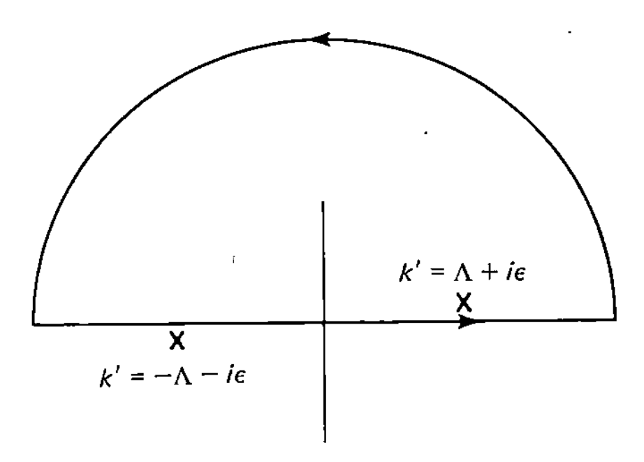

Figure 11.5 Poles and contour for the Green's function for the Helmholtz equation.

function for the Helmholtz operator is

$$G(\mathbf{r},\mathbf{r}') = -\frac{1}{(2\pi)^2}\frac{1}{i|\mathbf{r}-\mathbf{r}'|}2\pi iR(\Lambda + i\varepsilon) = -\frac{e^{i\Lambda|\mathbf{r}-\mathbf{r}'|}}{4\pi|\mathbf{r}-\mathbf{r}'|} \tag{11.254}$$

Thus for a general driving function $F(\mathbf{r})$, the solution to the Helmholtz equation [with boundary conditions such that $H(\mathbf{r},\mathbf{r}') = 0$] is

$$U(\mathbf{r}) = -\int d^3r'\frac{e^{i\Lambda|\mathbf{r}-\mathbf{r}'|}}{4\pi|\mathbf{r}-\mathbf{r}'|}F(\mathbf{r}') \quad \square \tag{11.255}$$

Symmetry of the Green's function

As the reader will note, in those examples in which $H(\mathbf{r},\mathbf{r}') = 0$, the Green's function is symmetric in the interchange of $\mathbf{r}$ and $\mathbf{r}'$. For example, the reader will note the symmetry of the one-dimensional Green's function expressed in equation 11.179. If $H(\mathbf{r},\mathbf{r}') \neq 0$, the Green's function does not possess this symmetry, as evidenced in the example expressed in equations 11.226.

To demonstrate the existence of this symmetry property for the Sturm–Liouville operator, consider the one-dimensional Green's function equation with $H(x, x') = 0$. Then

$$\frac{d}{dx'}\left(p(x')\frac{dG(x,x')}{dx'}\right) - q(x')G(x,x') = \delta(x-x') \tag{11.256a}$$

Replacing x by y, this can also be written

$$\frac{d}{dx'}\left(p(x')\frac{dG(y,x')}{dx'}\right) - q(x')G(y,x') = \delta(y-x') \tag{11.256b}$$

We multiply equation 11.256a by $G(y, x')$, multiply equation 11.256b by $G(x, x')$, and subtract. The terms involving $q(x')$ cancel, and we obtain

$$\begin{aligned}
G(y,x')\frac{d}{dx'}\left(p(x')\frac{dG(x,x')}{dx'}\right) &- G(x,x')\frac{d}{dx'}\left(p(x')\frac{dG(y,x')}{dx'}\right) \\
&= \frac{d}{dx'}\left[G(y,x')p(x')\frac{dG(x,x')}{dx'} - G(x,x')p(x')\frac{dG(y,x')}{dx'}\right] \\
&= G(y,x')\delta(x-x') - G(x,x')\delta(y-x')
\end{aligned} \tag{11.257}$$

This is now integrated over the range of x'. Because of the discontinuity in the derivative of the Green's function, the range of integration must be divided into two segments. That is, since the first term on the left-hand side of equation 11.257 contains the derivative of $G(x, x')$, we must write

$$\begin{aligned}
&\int_{x_0}^{x_f}\frac{d}{dx'}\left[G(y,x')p(x')\frac{dG(x,x')}{dx'}\right]dx' \\
&= \int_{x_0}^{x}\frac{d}{dx'}\left[G_S(y,x')p(x')\frac{dG_S(x,x')}{dx'}\right]dx' + \int_{x}^{x_f}\frac{d}{dx'}\left[G_L(y,x')p(x')\frac{dG_L(x,x')}{dx'}\right]dx' \\
&= p(x)G_S(y,x)\left.\frac{dG_S(x,x')}{dx'}\right|_{x'=x} - p(x_0)G_S(y,x_0)\left.\frac{dG_S(x,x')}{dx'}\right|_{x'=x_0} \\
&+ p(x_f)G_L(y,x_f)\left.\frac{dG_L(x,x')}{dx'}\right|_{x'=x_f} - p(x)G_L(y,x)\left.\frac{dG_L(x,x')}{dx'}\right|_{x'=x}
\end{aligned} \tag{11.258}$$

If the Green's function satisfies the Dirichlet or Neumann boundary conditions, the terms evaluated at x_0 and x_f are separately zero. If the boundary conditions are Gauss or periodic conditions, these terms combine to zero. For all four boundary conditions, then, the first integral of equation 11.257 becomes

$$\int_{x_0}^{x_f} \frac{d}{dx'}\left[G(y,x')p(x')\frac{dG(x,x')}{dx'}\right]dx'$$

$$= p(x)G(y,x)\left[\frac{dG_S(x,x')}{dx'} - \frac{dG_L(x,x')}{dx'}\right]_{x'=x} \tag{11.259}$$

But from the discontinuity condition for the derivative of the Green's function, (equation 11.166)

$$p(x)\left[\frac{dG_S(x,x')}{dx'} - \frac{dG_L(x,x')}{dx'}\right]_{x'=x} = -1 \tag{11.260}$$

$$\Rightarrow \quad \int_{x_0}^{x_f} \frac{d}{dx'}\left[G(y,x')p(x')\frac{dG(x,x')}{dx'}\right]dx' = -G(y,x) \tag{11.261a}$$

Similarly, replacing y by x, the integral of the second term in equation 11.257 becomes

$$\int_{x_0}^{x_f} \frac{d}{dx'}\left[G(x,x')p(x')\frac{dG(y,x')}{dx'}\right]dx' = -G(x,y) \tag{11.261b}$$

and the integral of equation 11.257 becomes

$$-G(y,x) + G(x,y) = \int_{x_0}^{x_f}[G(y,x')\delta(x-x') - G(x,x')\delta(y-x')]\,dx'$$

$$= G(y,x) - G(x,y) \tag{11.262}$$

$$\Rightarrow \quad G(x,y) = G(y,x) \tag{11.263}$$

In one dimension, if not expressed as an eigenfunction series, the Green's function is defined in segments. For example, $G(x, y) = G_S(x, y)$ for $y < x$. That is, G is the small segment if the second variable is less than the first, and G is the large segment if the second variable is larger than the first. Therefore, if $y < x$, $G(x, y) = G_S(x, y)$ and $G(y, x) = G_L(y, x)$. Similarly, $G(x, y) = G_L(x, y)$ and $G(y, x) = G_S(y, x)$ if $y > x$. Thus the symmetry of the Green's function in terms of the large and small pieces is

$$\text{(a)}\quad G_S(x,y) = G_L(y,x) \quad x > y \qquad \text{(b)}\quad G_L(x,y) = G_S(y,x) \quad x < y \tag{11.264}$$

PROBLEMS

1. Find the complete solution to

$$\left(\frac{\partial\Psi}{\partial x}\right)^2 + x^2\left(\frac{\partial\Psi}{\partial y}\right)^2 = x^2y^2$$

2. The classical Hamiltonian for a cylindrically symmetric harmonic oscillator in the x-y plane is

$$H = \frac{1}{2m}\left[p_\rho^2 + \frac{1}{\rho^2}p_\phi^2\right] + \frac{1}{2}k\rho^2$$

Find Hamilton's principal function for this system.

3. Find Hamilton's characteristic function $S_1(x, y)$ for the Hamiltonian

$$H = \frac{1}{2m}\left(p_x^2 + p_y^2\right) + k\left(x^2 + 2y^2\right)$$

4. Solve the equation

$$\frac{y}{x}\frac{\partial \Psi}{\partial x} + e^{x+y}\frac{\partial \Psi}{\partial y} = 0$$

by separating $\Psi(x, y)$ as a product of functions of the independent variables.

5. A quantum particle is influenced by a cylindrically symmetric potential $V(\rho, \phi, z) = V(\rho)$.
   (a) Separate the time-independent Schrödinger equation into ordinary differential equations.
   (b) Find the quantum wave function for the potential $V(\rho) = \alpha^2/\rho^2$ subject to $\Psi(\rho, \phi, z, t) = 0$ for $|z| \geqslant L$, $\Psi(\rho, \phi, z, t) = \Psi(\rho, \phi, -z, t)$, $\Psi(R, \phi, z, t) = 0$, and $\Psi(0, \phi, z, t)$ is finite and single-valued.

6. Separate Laplace's equation in spherical coordinates.

7. Charge is confined to a three-dimensional box of sides l_1, l_2, and l_3. It is distributed inside the box in such a way that the volume charge density is

$$\rho(x, y, z) = \begin{cases} \lambda(2l_1 x + 3y^2 + 2l_3 z) & \text{inside the box} \\ 0 & \text{outside the box} \end{cases}$$

   (a) Find the total charge inside the box (in terms of λ, l_1, l_2, l_3).
   (b) Separate Poisson's equation for this charge density at all points inside the box. (*Hint:* Not all linear equations are separable in terms of a *product* of one-variable functions.)
   (c) Separate Poisson's equation for points outside the box.

8. If the proton had an electric dipole moment p, the electron in the hydrogen atom would have a potential energy

$$V(r, \theta, \phi) = -\frac{e^2}{r} + \frac{ep\cos\theta}{r^2}$$

Separate the Schrödinger equation in spherical coordinates for this potential into three ordinary differential equations.

9. Find the solution to the thermal diffusion equation to determine the temperature distribution of a one-dimensional bar. The differential equation for $T(x, t)$ is

$$K\frac{\partial^2 T}{\partial x^2} = \frac{\partial T}{\partial t}$$

Use the initial and boundary conditions

$$T(0, 0) = T_0$$

$$T(\tfrac{1}{2}L, 0) = \sqrt{3}\,T_0$$

$$T(L, 0) = 2T_0$$

$$T(x, \infty) = 0$$

10. A thermally conducting one-dimensional bar, driven by a heat source, satisfies

$$K\frac{\partial^2 T}{\partial x^2} - \frac{\partial T}{\partial t} = F(x, t)$$

Find the temperature at all $x \in [0, L]$ for all $t \geqslant 0$ if:
   (a) $F(x, t) = F(x) = F_0 e^{2x/L}$ subject to

$$T(0, t) = 0$$

$$\left.\frac{\partial T}{\partial x}\right|_{x=L} = 0$$

$$T(x, \infty) \neq \infty$$

$$\left.\frac{\partial T}{\partial t}\right|_{t=0} = -\sin\left(\frac{5\pi x}{2L}\right)$$

   (b) $F(x, t) = F(t) = F_0 e^{-\omega t}$ (ω real) subject to

$$\left.\frac{\partial T}{\partial x}\right|_{\substack{x=0\\t=0}} = 0$$

$$\left.\frac{\partial T}{\partial x}\right|_{x=L} = 0 \qquad \text{for all } t$$

$$T(x, \infty) = 0$$

$$\left.\frac{\partial T}{\partial x}\right|_{x=L/2} = e^{-9\pi^2 K t/L^2}$$

11. Show that

$$\frac{\partial^2 U}{\partial x^2} - \frac{\partial^2 U}{\partial y^2} = F(x, y)$$

subject to $U(0, y) = U(a, y) = U(x, 0) = U(x, b) = 0$ will not have a solution if $a/b = N$ or $a/b = 1/N$, where N is any integer > 0.

12. The amplitude (displacement from equilibrium) of a cubic gel that is acted on by an external force satisfies the static inhomogeneous wave equation

$$\frac{\partial^2 U}{\partial x^2} + \frac{\partial^2 U}{\partial y^2} + \frac{\partial^2 U}{\partial z^2} = xyz$$

subject to

$U(x, y, z) = 0$ for $x = 0$ and $x = a$, for all y and z

$U(x, y, z) = 0$ for $y = 0$ and $y = b$, for all x and z

$U(x, y, z) = 0$ for $z = 0$ and $z = a$, for all x and y

Find the eigenfunction expansion of the Green's function for this system, and from that find $U(x, y, z)$.

13. Determine $\Psi(x, t)$ in terms of an integral by integral transform methods, if

$$\frac{\partial^2 \Psi}{\partial x^2} = \frac{\partial \Psi}{\partial t}$$

subject to

$$\Psi(x,\infty) = 0$$

$$\Psi(x,t) = 0 \qquad \text{for } t < 0$$

$$\Psi(x,0) = e^{-|x|}$$

14. Find the solution to the wave equation

$$\frac{\partial^2 \Psi}{\partial x^2} - \frac{1}{c^2}\frac{\partial^2 \Psi}{\partial t^2} = e^{-\beta t} \qquad \beta > 0$$

subject to

$$\Psi(x,\infty) = \left.\frac{\partial \Psi}{\partial t}\right|_{t=\infty} = 0$$

$$\Psi(x,t) = 0 \qquad \text{for } t < 0$$

$$\left.\frac{\partial \Psi}{\partial t}\right|_{t=0} = 0$$

$$\Psi(x,0) = 1$$

You may leave the solution in the form of an integral.

15. Find the solution to

$$x\frac{d^2U}{dx^2} + \frac{dU}{dx} = x$$

subject to $U(1) = 0$, $U(0)$ finite, and $x \in [0,1]$, by Green's function methods. Find the Green's function by direct solution of the Green's function equation. [*Hint:* The eigenvalue equation

$$x\frac{d^2u_n}{dx^2} + \frac{du_n}{dx} = \lambda_n u_n$$

is not quite the Bessel equation for $J_0(x)$ (see Table 6.2). Assume a series solution to the eigenvalue equation to demonstrate that $\lambda = 0$ is not an eigenvalue.]

16. Solve the differential equation

$$\frac{d^2y}{dx^2} - 2\frac{dy}{dx} + y = e^x$$

subject to $y(0) = y(1) = 0$, and $x \in [0,1]$. Use Green's function methods to obtain the solution. Find the Green's function by solving the Green's function equation directly. (*Caution:* The differential equation is not self-adjoint.)

17. Find the solution to

$$\frac{d^2U}{dx^2} = 2x^2 - 1$$

subject to $U(0) = U(1) = 0$, $x \in [0,1]$, by Green's function methods.

(a) Find the Green's function by direct solution of the Green's function equation.

(b) Find the Green's function in terms of an eigenfunction expansion.

(c) Combine the results of parts (a) and (b) to show that

$$\frac{2}{\pi^2}\sum_{k=1}^{\infty}\frac{\sin(k\pi x)\sin(k\pi x')}{k^2} = \begin{cases} x'(1-x) & x' \leqslant x \\ x(1-x') & x' \geqslant x \end{cases}$$

(d) From the results of part (c), show that

$$\pi^2 = 8\sum_{l=0}^{\infty}\frac{1}{(2l+1)^2}$$

(e) From the results of parts (a) and (c), show that

$$\frac{x^3}{3} - \frac{x^2}{2} + \frac{x}{6} = \frac{4}{\pi^2}\sum_{l=1}^{\infty}\frac{\sin(2l\pi x)}{l^2} \qquad x \in [0,1]$$

(f) Use the results of part (e) to prove that

$$\pi^2 = 64\sum_{k=0}^{\infty}\frac{(-1)^k}{(2k+1)^2}$$

18. The amplitude of a static wave on a string satisfies the equation

$$\frac{d^2U}{dx^2} + k^2U = F(x)$$

subject to $U(0) = U(L) = 0$

(a) Find the eigenfunction representation of the Green's function for this operator.

(b) Use the result of part (a) to determine the solution to this equation for $F(x) = F_0|L - 2x|$.

19. Find the solution to

$$\frac{d^2U}{dx^2} = 2x^2 - 1$$

subject to $U(0) = U(1) - U'(1) = 0$ by Green's function methods. (*Caution:* Zero is an eigenvalue of this operator.)

20. $\Psi(x)$ satisfies the differential equation

$$\frac{d}{dx}\left(p(x)\frac{d\Psi}{dx}\right) = F(x)$$

subject to boundary conditions such that the eigenvalue equation

$$\frac{d}{dx}\left(p(x)\frac{du_n}{dx}\right) - \lambda_n u_n = 0$$

has only a trivial solution for $\lambda_n = 0$ (i.e., zero is not an eigenvalue of the operator).

(a) Find the Green's function for this operator in terms of an integral involving $p(x)$.

(b) Use the result of part (a) to prove that a Green's function does not exist for the Legendre operator

$$\frac{d}{dx}\left((1-x^2)\frac{d}{dx}\right)$$

with $\Psi(x)$ subject to $\Psi(\pm 1)$ finite.

21. Consider the differential equation

$$x^2\frac{d^2U}{dx^2} + x\frac{dU}{dx} - U = x$$

subject to $U(0) = U(1) = 0$, $x \in [0, 1]$.

(a) Prove that the Wronskian of this operator is nonzero.

(b) Show that the Green's function for this operator is

$$G(x, x') = \begin{cases} c_L \dfrac{x}{x'}(1 - x'^2) & x' > x \\ c_S \dfrac{x'}{x}(1 - x^2) & x' < x \end{cases}$$

22. Using Green's function methods, find the solution to

$$\frac{d^2U}{dx^2} - \Lambda^2 U = e^{-x^2}$$

subject to $U(\pm\infty)$ finite, $x \in [-\infty, \infty]$. Take $H(x, x')$ to be zero, and determine the Green's function by Fourier transform methods. [*Hint:* Express your result in terms of the complementary error function erfc(z) expressed in equation 6.113.]

23. Using Green's function methods, find the solution to

$$\frac{d^2U}{dx^2} + \Lambda^2 U = e^{-x}$$

subject to $U(x) = 0$ for $x \leqslant 0$, $U'(0) = 1$, and $U(\infty)$ finite, $x \in [0, \infty]$. Determine the Green's function by Laplace transform methods. Take $H(x, x') = 0$.

24. Find the solution to the conduction equation

$$\Lambda\nabla^2 T(\mathbf{r}, t) = \frac{\partial T}{\partial t}$$

subject to

$$T(\mathbf{r}, 0) = T_0$$

$$T(\mathbf{r}, t) = 0 \qquad \text{for } t < 0$$

$$T(\mathbf{r}, \infty) = 0$$

as an integral over the Green's function. Find the Green's function by integral transform methods.

25. Charge is distributed symmetrically throughout a sphere of radius R, such that $\rho(r)$ increases radially as

$$\rho(\mathbf{r}) = \rho(r) = \frac{Q}{\frac{4}{3}\pi R^{N+3}} r^N \qquad 0 \leqslant r \leqslant R$$

$$\rho(\mathbf{r}) = 0 \qquad r > R$$

Find the electrostatic potential at all points in space (for $r \leqslant R$ and for $r > R$). Deduce the potential at any point if $\rho(r)$ is a constant (i.e. $N = 0$). Show that if $\rho(r) =$ constant, the potential for points outside the sphere is the expected Coulomb potential for a point charge.

26. A dipole is defined as two point charges of equal magnitude q and opposite sign separated by a small distance L (see Figure 11.6). The dipole moment is defined as $p \equiv qL$. $\mathbf{k}$ is the unit vector along the z-axis.

(a) Deduce the charge density for this distribution.

(b) Find the potential of a dipole at all points in space from the Green's function. Using $|\mathbf{r}| \gg L$, show that the lowest-order term in the potential is the expected dipole potential,

$$V(\mathbf{r}) \simeq \frac{p\cos\theta}{4\pi\varepsilon_0 r^2}$$

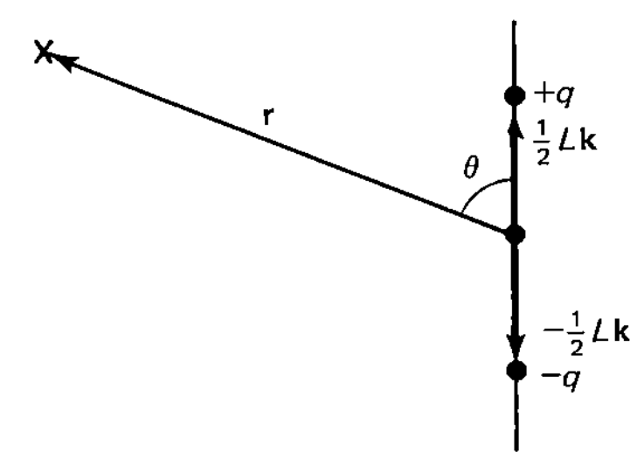

Figure 11.6 Electric dipole aligned along the z-axis.

CHAPTER 12

INTEGRAL EQUATIONS

12.1 Fredholm and Volterra Equations and Differential Equations

In Chapter 5 we discussed differential equations, which involve an unknown function operated on by differential operators. An *integral equation* is one in which the function to be determined appears under an integral. Many physical systems can be described by integral equations called Fredholm and Volterra equations.

The most general form of the *Fredholm equation* is

$$A(x)\Psi(x) = g(x) + \lambda \int_a^b K(x,t)\Psi(t)\,dt \tag{12.1a}$$

where λ, a and b are constants, independent of x. $A(x)$, $g(x)$, and $K(x,t)$ are known and $\Psi(x)$ is the unknown function. $g(x)$ is the inhomogeneous term, and $K(x,t)$ is called the *kernel* of the integral equation.

If $A(x) = 0$, then

$$g(x) = \lambda \int_a^b K(x,t)\Psi(t)\,dt \tag{12.1b}$$

is called the *Fredholm equation of the first kind*. If $A(x) \neq 0$ for all $x \in [a,b]$, we can divide equation 12.1a by $A(x)$. By defining a new inhomogeneous function and a new kernel (but retaining the same symbols to represent them), we obtain the *Fredholm equation of the second kind*, which is of the form

$$\Psi(x) = g(x) + \lambda \int_a^b K(x,t)\Psi(t)\,dt \tag{12.1c}$$

This form can also be obtained by setting $A(x) = 1$ in equation 12.1a.

If $A(x) = 0$ for at least one value of x, we cannot divide the integral equation by $A(x)$. In that case, equation 12.1a cannot be modified further. Such an equation is called the *Fredholm equation of the third kind*.

The *Volterra equation*

$$A(x)\Psi(x) = g(x) + \lambda \int_a^x K(x,t)\Psi(t)\,dt \tag{12.2a}$$

is similar in form to the Fredholm equation, the difference between the two being that the upper limit on the integral in the Volterra equation is the variable x rather than a constant. Like the Fredholm equations, the *Volterra equation of the first kind*

$$g(x) = \lambda \int_a^x K(x,t)\Psi(t)\,dt \tag{12.2b}$$

is obtained if $A(x) = 0$. If $A(x) \neq 0$ for all values of x, we can set $A(x) = 1$ to obtain the *Volterra equation of the second kind*

$$\Psi(x) = g(x) + \lambda \int_a^x K(x,t)\Psi(t)\,dt \tag{12.2c}$$

If $A(x) = 0$ for at least one value of x, equation 12.2a is referred to as the *Volterra equation of the third kind.*

Many differential equations that describe scientific phenomena can be cast into the form of one of these integral equations. In some cases, the integral equation has a more straightforward solution than the corresponding differential equation.

Example 12.1

For example, consider the second-order linear differential equation

$$\frac{d^2y}{dx^2} + P(x)\frac{dy}{dx} + Q(x)y = 0 \tag{12.3}$$

subject to initial conditions, $y(x_0) \equiv y_0$ and $y'(x_0) \equiv y_0'$, which are known constants. Letting

$$y(x) \equiv U(x)\Psi(x) \tag{12.4}$$

$$\Rightarrow \quad U(x)\frac{d^2\Psi}{dx^2} + \left(P(x)U(x) + 2\frac{dU}{dx}\right)\frac{d\Psi}{dx} + \left(\frac{d^2U}{dx^2} + P(x)\frac{dU}{dx} + Q(x)U(x)\right)\Psi = 0 \tag{12.5}$$

We choose $U(x)$ such that

$$\text{(a)}\quad P(x)U(x) + 2\frac{dU}{dx} = 0 \quad \Rightarrow \quad \text{(b)}\quad U(x) = U_0 e^{-\frac{1}{2}\int P(x)\,dx} \tag{12.6}$$

With this form of $U(x)$, the combination of terms multiplying $\Psi(x)$ in equation 12.5 is

$$\frac{d^2U}{dx^2} + P(x)\frac{dU}{dx} + Q(x)U(x) = \left(Q - \frac{1}{2}\frac{dP}{dx} + \frac{1}{4}P^2\right)U(x) \equiv R(x)U(x) \tag{12.7}$$

Therefore, the general second-order homogeneous linear differential equation can be cast into the simpler form

$$\frac{d^2\Psi}{dx^2} + R(x)\Psi(x) = 0 \tag{12.8}$$

The initial conditions become

$$\text{(a)}\quad \Psi(x_0) = \frac{y_0}{U(x_0)} \equiv \Psi_0 \qquad \text{(b)}\quad \Psi'(x_0) = \frac{y_0' - \frac{1}{2}P(x_0)y_0}{U(x_0)} \equiv \Psi_0' \tag{12.9}$$

To express equation 12.8 in the form of an integral equation, consider

$$\int_{x_0}^{s} \frac{d^2\Psi}{dt^2}\, dt = \frac{d\Psi}{ds} - \Psi_0' = -\int_{x_0}^{s} R(t)\Psi(t)\, dt \tag{12.10}$$

$$\Rightarrow \quad \int_{x_0}^{x} \frac{d\Psi}{ds}\, ds = \Psi(x) - \Psi(x_0) = \Psi_0' \int_{x_0}^{x} ds - \int_{x_0}^{x} ds \int_{x_0}^{s} R(t)\Psi(t)\, dt$$

$$= (x - x_0)\Psi_0' - \int_{x_0}^{x} \Phi(s)\, ds \tag{12.11}$$

where

$$\Phi(s) \equiv \int_{x_0}^{s} R(t)\Psi(t)\, dt \tag{12.12}$$

Integrating by parts, with $\Phi(s) = u$ and $ds = dv$, we obtain

$$\int_{x_0}^{x} \Phi(s)\, ds = s\Phi(s)\Big|_{x_0}^{x} - \int_{x_0}^{x} s\frac{d\Phi}{ds}\, ds \tag{12.13}$$

Since

$$\Phi(x_0) = \int_{x_0}^{x_0} R(t)\Psi(t)\, dt = 0 \tag{12.14}$$

the integrated term of equation 12.13 is $x\Phi(x)$. By defining the indefinite integral

$$G(t) \equiv \int R(t)\Psi(t)\, dt \tag{12.15}$$

$\Phi(s)$ can be written

$$\Phi(s) = G(s) - G(x_0) \tag{12.16}$$

$$\Rightarrow \quad \frac{d\Phi}{ds} = \frac{dG}{ds} = R(s)\Psi(s) \tag{12.17}$$

Therefore, using the definition of $\Phi(x)$ from equation 12.12, equation 12.13 becomes

$$\int_{x_0}^{x} \Phi(s)\, ds = x\Phi(x) - \int_{x_0}^{x} sR(s)\Psi(s)\, ds = \int_{x_0}^{x} (x - s)R(s)\Psi(s)\, ds \tag{12.18}$$

Therefore, the differential equation in the form of equation 12.8 can be expressed as the Volterra integral equation

$$\Psi(x) = \Psi_0 + (x - x_0)\Psi_0' - \int_{x_0}^{x} (x - s)R(s)\Psi(s)\, ds \tag{12.19a}$$

Then, using equations 12.4 and 12.9, we obtain

$$y(x) = \frac{U(x)}{U(x_0)} y_0 + (x - x_0)\frac{U(x)}{U(x_0)}\left[y_0' + \frac{1}{2}P(x_0)y_0\right] - \int_{x_0}^{x} (x - s)\frac{U(x)}{U(s)} R(s)y(s)\, ds \tag{12.19b}$$

Since the function U appears only in a ratio in the equation for $y(x)$, the indeterminate constant U_0 cancels. As such, there is no loss of generality in setting $U_0 = 1$. □

Example 12.2

As a second example, consider the inhomogeneous form of the second-order linear differential equation expressed in equation 12.3 subject to boundary conditions $y(a) = y_a$ and $y(b) = y_b$, where y_a and y_b are known constants.

As shown above, by setting

$$y(x) \equiv U(x)\Psi(x) \tag{12.4}$$

and taking $U(x)$ to be

$$U(x) = U_0 e^{-\frac{1}{2}\int P(x)\,dx} \tag{12.20}$$

$$\Rightarrow \quad \frac{d^2\Psi}{dx^2} = \frac{f(x)}{U(x)} - R(x)\Psi(x) \tag{12.21}$$

The boundary conditions then become

$$\text{(a)} \quad \Psi(a) = \frac{y(a)}{U(a)} \equiv \Psi_a \qquad \text{(b)} \quad \Psi(b) = \frac{y(b)}{U(b)} \equiv \Psi_b \tag{12.22}$$

Consider solving equation 12.21 by Green's function methods using $f(x)/U(x) - R(x)y(x)$ as the source term, and d^2/dx^2 with the boundary conditions of equations 12.22 as the operator. As discussed in Chapter 11, the Green's function is the solution to

$$\frac{d^2G}{dx'^2} = \delta(x - x') + H(x, x') \tag{12.23}$$

and the solution to the differential equation is

$$\begin{aligned} \Psi(x) &= \int_a^b G(x, x')\frac{f(x')}{U(x')}\,dx' - \int_a^b G(x, x')R(x')\Psi(x')\,dx' \\ &\equiv g(x) - \int_a^b G(x, x')R(x')\Psi(x')\,dx' \end{aligned} \tag{12.24a}$$

$$\Rightarrow \quad y(x) = U(x)g(x) - U(x)\int_a^b G(x, x')\frac{R(x')}{U(x')}y(x')\,dx' \tag{12.24b}$$

This is an inhomogeneous Fredholm integral equation of the second kind.

Using the techniques developed in Chapter 11, it is straightforward to determine that if zero is not an eigenvalue of the operator, and therefore $H(x, x') = 0$, the Green's function is

$$G(x, x') = \begin{cases} \dfrac{(x' - b)(x - a) + (x' - a)\Psi(b) - (x' - b)\Psi(a)}{(b - a)} & x' > x \\[2ex] \dfrac{(x' - a)(x - b) + (x' - a)\Psi(b) - (x' - b)\Psi(a)}{(b - a)} & x' < x \end{cases} \tag{12.25}$$

Noting that $g(x)$ involves an integral containing $f(x')/U(x')$, it is clear that equation 12.24b is independent of the indeterminate constant U_0. Therefore, as in the analysis of the Volterra equation, U_0 can be set to 1 without loss of generality. □

We will now develop methods of solution for the Fredholm and Volterra equations expressed in equations 12.1 and 12.2. In the development of these techniques, we will take $K(x, y)$ to be analytic over the range of x and y.

12.2 Methods of Solution for Fredholm Equations

Separable or degenerate kernel

If $K(x, y)$ can be written as a finite sum of products of one-variable functions

$$K(x, y) = \sum_{l=1}^{N} A_l(x)B_l(y) \tag{12.26}$$

K is said to be a *separable* or *degenerate kernel*. An exact solution to Fredholm equations can be obtained for such a kernel.

The Fredholm equation of the second kind for a separable kernel becomes

$$\Psi(x) = g(x) + \lambda \sum_{n=1}^{N} A_n(x) \int_a^b B_n(y)\Psi(y)\, dy \tag{12.27}$$

We note that $\int_a^b B_n(y)\Psi(y)\, dy$ contains no x-dependence, and is therefore a constant. Defining

$$\beta_n \equiv \int_a^b B_n(y)\Psi(y)\, dy = \text{constant} \tag{12.28}$$

$$\Rightarrow \quad \Psi(x) = g(x) + \lambda \sum_{n=1}^{N} A_n(x)\beta_n \tag{12.29}$$

with $\psi(y)$ given in equation 12.29,

$$\beta_l = \int_a^b B_l(y)\Psi(y)\, dy = \int_a^b B_l(y) g(y)\, dy + \lambda \sum_{n=1}^{N} \beta_n \int_a^b B_l(y) A_n(y)\, dy \tag{12.30}$$

Defining the constants

$$\text{(a)} \quad \alpha_l \equiv \int_a^b B_l(y) g(y)\, dy \qquad \text{and} \qquad \text{(b)} \quad M_{ln} \equiv \int_a^b B_l(y) A_n(y)\, dy \tag{12.31}$$

(which can, in principle, be determined), equation 12.30 becomes

$$\beta_l = \alpha_l + \lambda \sum_{n=1}^{N} M_{ln}\beta_n \tag{12.32}$$

This is a set of N simultaneous equations in the N unknowns β_l which can be solved using matrix inversion as described in Chapter 8. That is, we define column vectors $\boldsymbol{\alpha}$ and $\boldsymbol{\beta}$, and determine the unknown β_l from

$$\boldsymbol{\beta} = (1 - \lambda \mathbf{M})^{-1}\boldsymbol{\alpha} \tag{12.33}$$

Once the β_l are determined, $\Psi(x)$ is found from equation 12.29.

Example 12.3

As an example, consider

$$\Psi(x) = x + 2\int_0^1 (xy^2 + x^2 y)\Psi(y)\, dy = x + 2\left[x\int_0^1 y^2\Psi(y)\, dy + x^2\int_0^1 y\Psi(y)\, dy\right] \tag{12.34a}$$

Since each integral is a constant, this can be written

$$\Psi(x) = x + 2(x\beta_1 + x^2\beta_2) \tag{12.34b}$$

where

$$\beta_1 = \int_0^1 y^2\Psi(y)\, dy = \int_0^1 y^2[y + 2(y\beta_1 + y^2\beta_2)]\, dy = \tfrac{1}{4} + \tfrac{1}{2}\beta_1 + \tfrac{2}{5}\beta_2 \tag{12.35a}$$

$$\beta_2 = \int_0^1 y\Psi(y)\, dy = \int_0^1 y[y + 2(y\beta_1 + y^2\beta_2)]\, dy = \tfrac{1}{3} + \tfrac{2}{3}\beta_1 + \tfrac{1}{2}\beta_2 \tag{12.35b}$$

Therefore, $\beta_1 = -\frac{31}{2}$ and $\beta_2 = -20$. With these values, equation 12.34b yields

$$\Psi(x) = -30x - 40x^2 \quad \square \tag{12.36}$$

If $g(x)$ in equation 12.1c is zero, the equation is the homogeneous Fredholm equation of the second kind. It only has nontrivial solutions for certain values of λ. These are called the *eigenvalues* of the kernel.

If the kernel is degenerate and $g(x) = 0$, then

$$\beta_l = \alpha_l + \lambda \sum_{n=1}^{N} M_{ln}\beta_n \tag{12.32}$$

becomes

$$\text{(a)} \quad \beta_l = \lambda \sum_{n=1}^{N} M_{ln}\beta_n$$

$$\Rightarrow \quad \text{(b)} \quad \sum_{n=1}^{N} (\delta_{ln} - \lambda M_{ln})\beta_n = \sum_{n=1}^{N} (\mathbf{1} - \lambda \mathbf{M})_{ln}\beta_n = 0 \tag{12.37}$$

This system of equations has solution only if

$$\det(\mathbf{1} - \lambda \mathbf{M}) = 0 \tag{12.38}$$

This is the secular equation for determination of the eigenvalues of the matrix $\mathbf{M}$.

We note from this analysis that for a degenerate kernel, there will be a finite number of eigenvalues, that number being the number of terms in the sum of equation 12.26.

Neumann series for a nondegenerate kernel

If the kernel is not separable, clearly the method described above cannot be applied to the Fredholm equation. One approach to the solution for a nondegenerate kernel involves an iterative process that leads to an infinite series in λ. Referring to equation 12.1c, we rename the external variable x as y_0. Then

$$\Psi(y_0) = g(y_0) + \lambda \int_a^b K(y_0, y_1)\Psi(y_1)\, dy_1 \tag{12.39}$$

Expressing Ψ as a function of y_1, we can substitute the entire right-hand side of the equation for $\Psi(y_1)$ in the integral. That is,

$$\begin{aligned} \Psi(y_0) &= g(y_0) + \lambda \int_a^b K(y_0, y_1)\left[g(y_1) + \lambda \int_a^b K(y_1, y_2)\Psi(y_2)\, dy_2\right] dy_1 \\ &= g(y_0) + \lambda \int_a^b K(y_0, y_1) g(y_1)\, dy_1 + \lambda^2 \int_a^b \int_a^b K(y_0, y_1) K(y_1, y_2)\Psi(y_2)\, dy_1\, dy_2 \end{aligned} \tag{12.40}$$

A substitution of equation 12.39 for $\Psi(y_2)$ under the double integral yields

$$\begin{aligned} \Psi(y_0) &= g(y_0) + \lambda \int_a^b K(y_0, y_1) g(y_1)\, dy_1 + \lambda^2 \int_a^b \int_a^b K(y_0, y_1) K(y_1, y_2) g(y_2)\, dy_1\, dy_2 \\ &+ \lambda^3 \int_a^b \int_a^b \int_a^b K(y_0, y_1) K(y_1, y_2) K(y_2, y_3)\Psi(y_3)\, dy_1\, dy_2\, dy_3 \end{aligned} \tag{12.41}$$

It is straightforward to see that a continuation of this process leads to the infinite series

$$\Psi(x) = g(x) + \sum_{n=1}^{\infty} \lambda^n I_n(x) \tag{12.42}$$

where I_n is the n-fold integral

$$I_n \equiv \int_a^b \left[\prod_{m=0}^{n} K(y_m, y_{m+1}) \right] g(y_{m+1})\, d^n y \qquad n \geqslant 1 \tag{12.43a}$$

The series of equation 12.42 is called the *Neumann series* for $\Psi(x)$. (Although we have written only one integral sign in equation 12.43a, it is understood that an n-fold multiple integration is to be performed.) With $I_0(x) \equiv g(x)$,

$$I_{n+1}(x) = \int_a^b K(x, y) I_n(y)\, dy \qquad n \geqslant 0 \tag{12.43b}$$

Example 12.4

As an example, consider

$$\Psi(x) = x + \lambda \int_0^1 xy\Psi(y)\, dy \tag{12.44}$$

Clearly, since the kernel is degenerate, the equation can be solved exactly. The solution is

$$\Psi(x) = \frac{3x}{(3 - \lambda)} \tag{12.45}$$

To develop the Neumann series, we note that with $K(y_m, y_{m+1}) = y_m y_{m+1}$ and with $g(y_m) = y_m$,

$$I_n = \int_0^1 \underbrace{(xy_1)(y_1 y_2) \cdots (y_{n-1} y_n)}_{n \text{ pairs}}\, y_n\, dy_1\, dy_2 \cdots dy_n = x\left[\int_0^1 y^2\, dy\right]^n = \frac{x}{3^n} \tag{12.46}$$

Therefore, the Neumann series solution is

$$\Psi(x) = x + x \sum_{n=1}^{\infty} \frac{\lambda^n}{3^n} \tag{12.47}$$

which sums to the exact solution given in equation 12.45. □

Solution by reciprocal kernel method

We define the sequence of functions

$$\text{(a)} \quad K_1(x, y) \equiv K(x, y) \qquad \text{(b)} \quad K_2(x, y) = \int_a^b K(x, z) K_1(z, y)\, dz$$

$$\text{(c)} \quad K_3(x, y) = \int_a^b K(x, z) K_2(z, y)\, dz \cdots \qquad \text{(d)} \quad K_n(x, y) = \int_a^b K(x, z) K_{n-1}(z, y)\, dz \tag{12.48}$$

where the constant λ has been absorbed into the kernel. That is, in this treatment, $\lambda K(x, y) \to K(x, y)$.

We define M to be the maximum of $K(x, y)$ over the range $x, y \in [a, b]$. That is,

$$|K(x, y)| \leqslant M \tag{12.49}$$

The function $\chi(x, y)$, defined by the series

$$-\chi(x, y) \equiv \sum_{n=1}^{\infty} K_n(x, y) \tag{12.50}$$

is referred to as the kernel *reciprocal* to $K(x, y)$.

Such a series will not converge for any arbitrary kernel. As discussed in Chapter 3, a necessary and sufficient condition for convergence of a series is that the Cauchy ratio be less than 1. Thus, for sufficiently large index, the reciprocal kernel converges if

$$|K_n(x, y)| < |K_{n-1}(x, y)| \tag{12.51a}$$

To deduce the restrictions this places on the kernel, we express this inequality as

$$\left|\int_a^b K(x, z)K_{n-1}(z, y)\, dz\right| < |K_{n-1}(x, y)| \tag{12.51b}$$

But since M is the bound or maximum of $K(x, y)$, this can be written

$$\left|\int_a^b K(x, z)K_{n-1}(z, y)\, dz\right| < M|K_{n-1}(x, y)|(b - a) \tag{12.52}$$

Therefore, the inequality of equation 12.51b is assured and the reciprocal kernel series converges if

$$\text{(a)}\quad M|K_{n-1}(x, y)|(b - a) < |K_{n-1}(x, y)| \quad \Rightarrow \quad \text{(b)}\quad M < \frac{1}{(b - a)} \tag{12.53}$$

To develop the solution in terms of the reciprocal kernel, we note that $K_1(x, y) = K(x, y)$ and write

$$\chi(x, y) + K(x, y) = -[K_2(x, y) + K_3(x, y) + \cdots] \tag{12.54}$$

Using equations 12.50 and 12.51, this can be written

$$\chi(x, y) + K(x, y) = -\int_a^b [K(x, z)K_1(z, y) + K(x, z)K_2(z, y) + \cdots]\, dz$$

$$= -\int_a^b K(x, z)[K_1(z, y) + K_2(z, y) + \cdots]\, dz = \int_a^b K(x, z)\chi(z, y)\, dz \tag{12.55a}$$

We note that this integral equation for $\chi(x, y)$ is symmetric in the interchange of the functions χ and K. That is, equation 12.55a can be written.

$$\chi(x, y) + K(x, y) = \int_a^b \chi(x, z)K(z, y)\, dz \tag{12.55b}$$

It is for this reason that they are referred to as *complementary* or *reciprocal* functions.

We return to the Fredholm integral equation of the second kind,

$$\Psi(z) = g(z) + \int_a^b K(z, y)\Psi(y)\, dy \tag{12.56}$$

where, as noted above, λ has been combined with the kernel. We multiply the Fredholm equation by the reciprocal kernel and integrate to obtain

$$\int_a^b \chi(x,z)\Psi(z)\,dz = \int_a^b \chi(x,z)g(z)\,dz + \int_a^b\int_a^b \chi(x,z)K(z,y)\Psi(y)\,dz\,dy \tag{12.57}$$

With equation 12.55b, this becomes

$$\int_a^b \chi(x,z)\Psi(z)\,dz = \int_a^b \chi(x,z)g(z)\,dz + \int_a^b [\chi(x,y) + K(x,y)]\Psi(y)\,dy \tag{12.58a}$$

$$\Rightarrow \quad \int_a^b K(x,y)\Psi(y)\,dy = -\int_a^b \chi(x,z)g(z)\,dz \tag{12.58b}$$

from which the inhomogeneous Fredholm integral equation becomes

$$\Psi(x) = g(x) + \int_a^b K(x,y)\Psi(y)\,dy = g(x) - \int_a^b \chi(x,y)g(y)\,dy \tag{12.59}$$

Thus if the reciprocal kernel can be determined by summing the series of equation 12.50, the problem of solving the integral equation is reduced to the task of evaluating an integral.

Example 12.5

To illustrate, we consider

$$\Psi(x) = \tfrac{5}{6}x + \int_0^1 \left(\frac{1}{2}xy\right)\Psi(y)\,dy \tag{12.60}$$

which can be solved exactly using the fact that the kernel is degenerate.

To sum the reciprocal kernel series of equation 12.50, we identify

$$\text{(a)} \quad K_1(x,y) = \left(\frac{1}{2}xy\right) \quad \Rightarrow \quad \text{(b)} \quad K_2(x,y) = \int_0^1 \left(\frac{1}{2}xz\right)\left(\frac{1}{2}zy\right)dz = \frac{1}{2^2}\frac{1}{3}xy \tag{12.61}$$

$$\Rightarrow \quad \text{(c)} \quad K_3(x,y) = \int_0^1 \left(\frac{1}{2}xz\right)\left(\frac{1}{2^2}\frac{1}{3}zy\right)dz = \frac{1}{2^3}\frac{1}{3^2}xy$$

$$\Rightarrow \quad K_n(x,y) = \frac{1}{2^n}\frac{1}{3^{n-1}}xy \tag{12.62}$$

Therefore, the reciprocal kernel is

$$\chi(x,y) = -\sum_{n=1}^{\infty}\frac{1}{2^n}\frac{1}{3^{n-1}}xy = -3xy\sum_{n=1}^{\infty}\frac{1}{6^n} = -3xy\left[\frac{1}{1-\frac{1}{6}} - 1\right] = -\frac{3}{5}xy \tag{12.63}$$

$$\Rightarrow \quad \Psi(x) = \tfrac{5}{6}x - \int_0^1 (-\tfrac{3}{5}xy)(\tfrac{5}{6}y)\,dy = x \quad \square \tag{12.64}$$

Solution by the Fredholm resolvent

Fredholm's approach to solving the Fredholm equation of the second kind begins by writing the integral as a limit of sum. That is,

$$\int_a^b K(x,y)\Psi(y)\,dy \simeq \lim_{\substack{\Delta y \to 0 \\ N \to \infty}} \sum_{n=1}^{N} K(x,y_n)\Psi(y_n)\,\Delta y \tag{12.65}$$

Therefore, before the limit is approached,

$$\Psi(x) \simeq g(x) + \lambda \sum_{n=1}^{N} K(x, y_n)\Psi(y_n)\,\Delta y \tag{12.66}$$

In sequence, we now set $x = y_1, x = y_2, \ldots, x = y_N$ to obtain a set of N simultaneous equations in the form

$$\Psi(y_m) \simeq g(y_m) + \lambda \sum_{n=1}^{N} K(y_m, y_n)\Psi(y_n)\,\Delta y \tag{12.67a}$$

or with $K_{mn} \equiv K(y_m, y_n)$,

$$\sum_{n=1}^{N} [\delta_{mn} - \lambda K_{mn}\,\Delta y]\Psi(y_n) = g(y_m) \tag{12.67b}$$

Using Cramer's rule, we can find each $\Psi(y_l)$ by

$$\Psi(y_l) = \frac{\begin{vmatrix} (1-\lambda K_{11}\,\Delta y) & -\lambda K_{12}\,\Delta y & \cdots & g(y_1) & \cdots & -\lambda K_{1N}\,\Delta y \\ -\lambda K_{21}\,\Delta y & (1-\lambda K_{22}\,\Delta y) & \cdots & g(y_2) & & -\lambda K_{2N}\,\Delta y \\ \vdots & & & & & \\ -\lambda K_{N1}\,\Delta y & -\lambda K_{N2}\,\Delta y & \cdots & g(y_N) & & (1-\lambda K_{NN}\,\Delta y) \end{vmatrix}}{|\mathbf{1} - \lambda \mathbf{K} \Delta y|} \tag{12.68}$$

where, of course, the column containing $g(y_1), g(y_2), \ldots$ is the lth column.

The denominator determinant is called the *Fredholm determinant*. Such a determinant can be expanded in powers of λ, as

$$D(\lambda) \equiv |\mathbf{1} - \lambda \mathbf{K}\,\Delta y| = 1 - \lambda \sum_{n=1}^{N} K_{nn}\,\Delta y + \frac{\lambda^2}{2!} \sum_{m,n=1}^{N} \begin{vmatrix} K_{mm} & K_{mn} \\ K_{nm} & K_{nn} \end{vmatrix} \Delta y_m\,\Delta y_n$$

$$- \frac{\lambda^3}{3!} \sum_{lmn=1}^{N} \begin{vmatrix} K_{ll} & K_{lm} & K_{ln} \\ K_{ml} & K_{mm} & K_{mn} \\ K_{nl} & K_{nm} & K_{nn} \end{vmatrix} \Delta y_l\,\Delta y_m\,\Delta y_n + \cdots \tag{12.69}$$

The reader interested in the details of this expansion is referred to the text by H. Hochstadt, *Integral Equations* (New York: John Wiley & Sons, Inc., 1973), page 236, and to the example illustrated in Problem 10.

Since $\Delta y = \Delta y_1 = \Delta y_2 = \cdots = \Delta y_N$, we have, for example, written

$$\Delta y^3 = \Delta y_l\,\Delta y_m\,\Delta y_n \tag{12.70}$$

In the limit as each $\Delta y_i \to 0$ and $N \to \infty$, the sums become integrals and the Fredholm determinant becomes

$$D(\lambda) = 1 - \lambda \int_a^b K(x,x)\,dx + \frac{\lambda^2}{2!} \int_a^b \int_a^b \begin{vmatrix} K(x,x) & K(x,y) \\ K(y,x) & K(y,y) \end{vmatrix} dx\,dy$$

$$- \frac{\lambda^3}{3!} \int_a^b \int_a^b \int_a^b \begin{vmatrix} K(x,x) & K(x,y) & K(x,z) \\ K(y,x) & K(y,y) & K(y,z) \\ K(z,x) & K(z,y) & K(z,z) \end{vmatrix} dx\,dy\,dz + \cdots \tag{12.71}$$

It can be shown that the series expansion of the Fredholm determinant is absolutely convergent. The reader interested in a proof of this convergence is referred to the book by W. Lovitt, *Linear Integral Equations* (New York: Dover Publications, Inc., 1950), pages 32–34, or M. Bocher's text *An Introduction to the Study of Integral Equations* (New York: Hafner Press, 1960), pages 30–32.

The first term $\int_a^b K(x, x)\, dx$ is called the *trace term*. Clearly, it becomes cumbersome to evaluate terms in the Fredholm determinant much beyond the trace term. Thus if the convergence is rapid enough, the Fredholm determinant may be well approximated by the trace term, or it might be possible to estimate

$$D(\lambda) \simeq 1 - \lambda \int_a^b K(x, x)\, dx + \frac{\lambda^2}{2!} \int_a^b \int_a^b \begin{vmatrix} K(x, x) & K(x, y) \\ K(y, x) & K(y, y) \end{vmatrix} dx\, dy \tag{12.72}$$

Returning to equation 12.68, we expand the numerator determinant around the lth column containing $g(y_1), \ldots, g(y_N)$ to obtain

$$\begin{vmatrix} (1 - \lambda K_{11}\,\Delta y) & -\lambda K_{12}\,\Delta y & \cdots & g(y_1) & \cdots & -\lambda K_{1N}\,\Delta y \\ -\lambda K_{21}\,\Delta y & (1 - \lambda K_{22}\,\Delta y) & \cdots & g(y_2) & \cdots & -\lambda K_{2N}\,\Delta y \\ \vdots & & & & & \\ -\lambda K_{N1}\,\Delta y & -\lambda K_{N2}\,\Delta y & \cdots & g(y_N) & \cdots & (1 - \lambda K_{NN}\,\Delta y) \end{vmatrix}$$

$$= \sum_{m=1}^{N} C_{lm}\, g(y_m) \tag{12.73}$$

where C_{lm} is the cofactor of the element $g(y_m)$.

If $l = m$, the expansion of C_{ll} is similar to that of $D(\lambda)$, the Fredholm determinant. That is,

$$\begin{vmatrix} (1 - \lambda K_{11}\,\Delta y) & \cdots & -\lambda K_{1(l-1)}\,\Delta y & -\lambda K_{1(l+1)}\,\Delta y & \cdots & -\lambda K_{1N}\,\Delta y \\ \vdots & & & & & \\ -\lambda K_{(l-1)1}\,\Delta y & \cdots & (1 - \lambda K_{(l-1)(l-1)}\,\Delta y) & -\lambda K_{(l-1)(l+1)}\,\Delta y & \cdots & -\lambda K_{(l-1)N}\,\Delta y \\ -\lambda K_{(l+1)1}\,\Delta y & \cdots & -\lambda K_{(l+1)(l-1)}\,\Delta y & (1 - \lambda K_{(l+1)(l+1)}\,\Delta y) & \cdots & -\lambda K_{(l+1)N}\,\Delta y \\ \vdots & & & & & \\ -\lambda K_{N1}\,\Delta y & \cdots & -\lambda K_{N(l-1)}\,\Delta y & -\lambda K_{N(l+1)}\,\Delta y & \cdots & (1 - \lambda K_{NN}\,\Delta y) \end{vmatrix}$$

$$= 1 - \lambda \sum_{\substack{r \neq l \\ r=1}}^{N} K_{rr}\,\Delta y + \frac{\lambda^2}{2!} \sum_{\substack{rs \neq l \\ rs=1}}^{N} \begin{vmatrix} K_{rr} & K_{rs} \\ K_{sr} & K_{ss} \end{vmatrix} (\Delta y)^2 - \frac{\lambda^3}{3!} \sum_{\substack{rts \neq l \\ rst=1}}^{N} \begin{vmatrix} K_{rr} & K_{rs} & K_{rt} \\ K_{sr} & K_{ss} & K_{st} \\ K_{tr} & K_{ts} & K_{tt} \end{vmatrix} (\Delta y)^3 + \cdots \tag{12.74}$$

If $l \neq m$, the expansion of the cofactor is

$$C_{lm} = \lambda K_{lm}\,\Delta y - \lambda^2 \sum_{r=1}^{N} \begin{vmatrix} K_{lm} & K_{lr} \\ K_{rm} & K_{rr} \end{vmatrix} (\Delta y)^2 + \frac{\lambda^3}{2!} \sum_{rs=1}^{N} \begin{vmatrix} K_{lm} & K_{lr} & K_{ls} \\ K_{rm} & K_{rr} & K_{rs} \\ K_{sm} & K_{sr} & K_{ss} \end{vmatrix} (\Delta y)^3 - \cdots \tag{12.75}$$

When we approach the limits $N \to \infty$, each $\Delta y \to 0$, the cofactor of equation 12.74 becomes identical to the Fredholm determinant $D(\lambda)$. [In this limit, we can lift the restrictions on the sums. By doing so, we are including a finite number of terms that become zero in an infinite sum. Thus in the limit, $C_{ll} \to D(\lambda)$.]

For $l \neq m$, the limiting process applied to C_{lm} becomes

$$C_{lm} \to \lambda K(x,y) - \lambda^2 \int_a^b \begin{vmatrix} K(x,y) & K(x,z_1) \\ K(z_1,y) & K(z_1,z_1) \end{vmatrix} dz_1$$

$$+ \frac{\lambda^3}{2!} \int_a^b \int_a^b \begin{vmatrix} K(x,y) & K(x,z_1) & K(x,z_2) \\ K(z_1,y) & K(z_1,z_1) & K(z_1,z_2) \\ K(z_2,y) & K(z_2,z_1) & K(z_2,z_2) \end{vmatrix} dz_1\, dz_2 - \cdots \equiv \mathfrak{D}(x,y;\lambda) \tag{12.76}$$

Therefore, from equation 12.68,

$$\Psi(x) \to \frac{g(x)D(\lambda) + \int_a^b \mathfrak{D}(x,y;\lambda)g(y)\,dy}{D(\lambda)} = g(x) + \frac{1}{D(\lambda)} \int_a^b \mathfrak{D}(x,y;\lambda)g(y)\,dy \tag{12.77}$$

The ratio $\mathfrak{D}(x,y;\lambda)/D(\lambda)$ is called the *Fredholm resolvent* and will be denoted by $\mathfrak{R}(x,y;\lambda)$.

Referring to equation 12.67b, we note that if $g(x) = 0$, the eigenvalues are found, as expected, by setting the secular determinant to zero. In the limit $N \to \infty$, this will yield an infinite number of solutions to the secular equation. Therefore, from the analyses above, we see that a separable kernel of the Fredholm equation has a finite number of eigenvalues, while a nonseparable kernel has an infinite number of eigenvalues.

Example 12.6

To illustrate the method of determining the Fredholm resolvent, we again consider the integral equation of Example 12.4:

$$\Psi(x) = x + \lambda \int_0^1 xy\Psi(y)\,dy \tag{12.44}$$

which has the solution

$$\Psi(x) = \frac{3x}{3-\lambda} \tag{12.45}$$

as obtained earlier. We will develop the solution by determining the Fredholm resolvent. For the Fredholm determinant, the trace term and the integrals multiplying λ^2 and λ^3 respectively are

$$\text{(a)}\quad T_1 = \int_0^1 K(y,y)\,dy = \int_0^1 y^2\,dy = \tfrac{1}{3} \qquad \text{(b)}\quad T_2 = \int_0^1 \int_0^1 \begin{vmatrix} x^2 & xy \\ yx & y^2 \end{vmatrix} dx\,dy = 0$$

$$\text{(c)}\quad T_3 = \int_0^1 \int_0^1 \int_0^1 \begin{vmatrix} x^2 & xy & xz \\ yx & y^2 & yz \\ zx & zy & z^2 \end{vmatrix} dx\,dy\,dz = 0 \tag{12.78}$$

All terms beyond the trace term are zero, and therefore the Fredholm determinant is

$$D(\lambda) = 1 - \frac{\lambda}{3} \tag{12.79}$$

The Fredholm numerator $\mathfrak{D}(x, y; \lambda)$ series is

$$\mathfrak{D}(x, y; \lambda) = \lambda xy - \lambda^2 \int_0^1 \begin{vmatrix} xy & xz \\ zy & z^2 \end{vmatrix} dz + \frac{\lambda^3}{2!} \int_0^1 \int_0^1 \begin{vmatrix} xy & xz & xw \\ zy & z^2 & zw \\ wy & wz & w^2 \end{vmatrix} dz\, dw - \cdots \tag{12.80a}$$

It can be shown straightforwardly that all the integrals in this series are zero. Thus

$$\mathfrak{D}(x, y; \lambda) = \lambda xy \tag{12.80b}$$

$$\Rightarrow \quad \mathfrak{R}(x, y; \lambda) = \frac{\mathfrak{D}(x, y; \lambda)}{D(\lambda)} = \frac{\lambda xy}{(1 - \lambda/3)} = \frac{3\lambda xy}{3 - \lambda} \tag{12.81}$$

$$\Rightarrow \quad \Psi(x) = g(x) + \int_a^b \frac{\mathfrak{D}(x, y; \lambda)}{D(\lambda)} g(y)\, dy = x + \frac{3\lambda}{3 - \lambda} \int_0^1 (xy) y\, dy = \frac{3x}{3 - \lambda} \quad \square \tag{12.82}$$

Eigenvalues of the Fredholm kernel

In most cases the value of the constant λ determines the existence of the solution to the homogeneous and nonhomogeneous Fredholm equations. The possible alternatives are:

1. If λ is an eigenvalue of the kernel of a Fredholm equation, (a) the homogeneous equation has a nontrivial solution which is the eigenfunction corresponding to λ, and (b) there is no solution to the inhomogeneous Fredholm equation for that value of λ.
2. If λ is not an eigenvalue of the kernel of a Fredholm equation, (a) the only solution to the homogeneous equation is the trivial one, $\psi_n(x) = 0$, and (b) a solution to the inhomogeneous equation exists.

These options collectively are known as the *Fredholm alternatives*.

In some cases it is possible for the inhomogeneous solution to exist when λ is an eigenvalue. Let $\psi_n(x)$ be an eigensolution to the homogeneous Fredholm equation,

$$\psi_n(x) = \lambda_n \int_a^b K(x, y) \psi_n(y)\, dy \tag{12.83}$$

and $\Psi(x)$ satisfy the inhomogeneous Fredholm equation,

$$\Psi(x) = g(x) + \lambda \int_a^b K(x, y) \Psi(y)\, dy \tag{12.84}$$

The inhomogeneous solution will exist for $\lambda = \lambda_n$ if and only if

$$\int_a^b \psi_n(x) g(x)\, dx = 0 \tag{12.85}$$

To see this, we take the eigensolutions to the homogeneous equation to form a complete, orthonormal set of functions (using a Gram–Schmidt orthogonalization, if necessary). Since the basis functions of a function space must be infinite in number, the set of eigenfunctions, and therefore the number of eigenvalues must be infinite. As such, the kernel cannot be degenerate.

We expand the solution to the inhomogeneous equation in terms of the eigenfunctions,

$$\Psi(x) = \sum_{n=1}^{\infty} \alpha_n \psi_n(x) \tag{12.86}$$

and take the eigenfunctions to be normalized so that

$$\int_a^b \psi_l^*(x)\psi_n(x)\,dx = \delta_{ln} \tag{12.87}$$

$$\Rightarrow \quad \alpha_n = \int_a^b \psi_n^*(x)\Psi(x)\,dx \tag{12.88}$$

If the inhomogeneous solution is to exist, at least one of these coefficients, α_k, must be nonzero. Using equation 12.86, the inhomogeneous equation can be written

$$\sum_{n=1}^{\infty} \alpha_n \psi_n(x) = g(x) + \lambda \sum_{n=1}^{\infty} \alpha_n \int_a^b K(x, y)\psi_n(y)\,dy \tag{12.89}$$

But from equation 12.83,

$$\int_a^b K(x, y)\psi_n(y)\,dy = \frac{\psi_n(x)}{\lambda_n} \tag{12.90}$$

Thus equation 12.89 becomes

$$\text{(a)} \quad \sum_{n=1}^{\infty} \alpha_n \psi_n(x) = g(x) + \sum_{n=1}^{\infty} \alpha_n \frac{\lambda}{\lambda_n} \psi_n(x) \qquad \text{or} \qquad \text{(b)} \quad \sum_{n=1}^{\infty} \alpha_n \left(1 - \frac{\lambda}{\lambda_n}\right)\psi_n(x) = g(x) \tag{12.91}$$

Using the orthonormalization condition, we obtain

$$\sum_{n=1}^{\infty} \alpha_n \left(1 - \frac{\lambda}{\lambda_n}\right) \int_a^b \psi_k^*(x)\psi_n(x)\,dx = \alpha_k\left(1 - \frac{\lambda}{\lambda_k}\right) = \int_a^b \psi_k^*(x)g(x)\,dx \tag{12.92}$$

Therefore, if a solution is to exist for $\lambda = \lambda_k$ (and thus $\alpha_k \neq 0$), we must require that

$$\int_a^b \psi_k^*(x)g(x)\,dx = 0 \tag{12.93}$$

Another property of Fredholm equations is that if λ is an eigenvalue of the kernel $K(x, y)$, it is also an eigenvalue of the kernel $K(y, x)$ even if $K(x, y)$ is not symmetric in the interchange of the arguments.

Let

$$\psi_k(x) = \lambda_k \int_a^b K(x, y)\psi_k(y)\,dy \tag{12.94}$$

The hermitian adjoint of $K(x, y)$ is defined by interchanging the arguments and taking the complex conjugate of $K(x, y)$. If $K(x, y)$ were viewed as the x, y element of a matrix $\mathbf{K}$, this definition of hermiticity is identical to that for matrices. The eigenvalue equation for $K^*(y, x)$ is

$$\phi_l(x) = \Lambda_l \int_a^b K^*(y, x)\phi_l(y)\,dy \tag{12.95}$$

If $K(x, y) \neq K^*(y, x)$, it is expected that Λ_l cannot be one of the eigenvalues λ_k and $\phi_l(x)$ cannot be any one of the $\psi_k(x)$. But since the $\psi_k(x)$ form a complete orthonormal set of functions, we can expand $\phi_l(x)$ as

$$\text{(a)} \quad \phi_l(x) = \sum_k \beta_{lk}\psi_k(x) \qquad \text{with} \qquad \text{(b)} \quad \beta_{lk} = \int_a^b \psi_k^*(x)\phi_l(x)\,dx \tag{12.96}$$

Substituting $\phi_l(x)$ as given in equation 12.95b, this becomes

$$\beta_{lk} = \Lambda_l \int_a^b \psi_k^*(x)\left[\int_a^b K^*(y, x)\phi_l(y)\,dy\right] dx \tag{12.97}$$

Interchanging the order of integration, we note from equation 12.95 that

$$\int_a^b K^*(y, x)\psi_k^*(x)\,dx = \frac{\psi_k^*(y)}{\lambda_k^*} \tag{12.98}$$

$$\Rightarrow \quad \beta_{lk} = \frac{\Lambda_l}{\lambda_k^*}\int_a^b \phi_l(y)\psi_k^*(y)\,dy = \frac{\Lambda_l}{\lambda_k^*}\beta_{lk} \tag{12.99}$$

If $\beta_{lk} \neq 0$, $\Lambda_l = \lambda_k^*$. But since Λ_l is an eigenvalue of $K^*(y, x)$, $\Lambda_l^* = \lambda_k$ is an eigenvalue of $K(y, x)$. Therefore, as stated above, if λ is an eigenvalue of $K(x, y)$, it is also an eigenvalue of $K(y, x)$.

The converse of this, then, is that if λ is not an eigenvalue of $K(x, y)$, it is also not an eigenvalue of $K(y, x)$. Therefore, by the Fredholm alternative [assuming that $g(x)$ is not orthogonal to any eigenfunction], if $\Psi_1(x)$ and $\Psi_2(x)$ are the solutions to

$$\text{(a)} \quad \Psi(x) = g(x) + \lambda\int_a^b K(x, y)\Psi(y)\,dy \qquad \text{(b)} \quad \Psi(x) = g(x) + \lambda\int_a^b K(y, x)\Psi(y)\,dy \tag{12.100}$$

respectively, we can state that if $K(x, y) \neq K(y, x)$, then $\Psi_1(x) \neq \Psi_2(x)$.

It is possible for a Fredholm kernel to have no eigenvalues and eigenfunctions. The reader is referred to Problems 13 and 14 for examples of such cases.

Hilbert–Schmidt theory for a hermitian kernel

As expected from the definition above, a hermitian kernel is defined by the property

$$K(x, y) = K^*(y, x) \tag{12.101}$$

Therefore, a real symmetric kernel is hermitian. Like its matrix counterpart, a hermitian kernel has real eigenvalues and all eigenfunctions corresponding to nondegenerate eigenvalues are mutually orthogonal (or they are Gram–Schmidt orthogonalizable if there are degeneracies).

The proof of this property parallels the proof of the reality of eigenvalues and natural orthogonality of nondegenerate eigenvectors of a hermitian matrix. Starting from the eigenvalue equation and using the hermiticity of the kernel

$$\psi_n(x) = \lambda_n\int_a^b K(x, y)\psi_n(y)\,dy \tag{12.102a}$$

$$\Rightarrow \quad \psi_l^*(x) = \lambda_l^*\int_a^b K^*(x, y)\psi_l^*(y)\,dy = \lambda_l^*\int_a^b K(y, x)\psi_l^*(y)\,dy \tag{12.102b}$$

Multiplying equation 12.102a by $\psi_l^*(x)$, equation 12.102b by $\psi_n(x)$, integrating over x, and subtracting, we obtain

$$\lambda_n \int_a^b \psi_l^*(x) \int_a^b K(x, y)\psi_n(y)\, dy\, dx - \lambda_l^* \int_a^b \psi_n(x) \int_a^b K(y, x)\psi_l^*(y)\, dy\, dx = 0 \tag{12.103a}$$

By interchanging the order of x and y integration, and then interchanging the names of the variables ($x \leftrightarrow y$), this becomes

$$(\lambda_n - \lambda_l^*) \int_a^b \psi_l^*(x) \int_a^b K(x, y)\psi_n(y)\, dy = 0 \tag{12.103b}$$

But from the eigenvalue equation,

$$\int_a^b K(x, y)\psi_n(y)\, dy = \frac{\psi_n(x)}{\lambda_n} \tag{12.90}$$

$$\Rightarrow \quad \frac{(\lambda_n - \lambda_l^*)}{\lambda_n} \int_a^b \psi_n^*(x)\psi_l(x)\, dx = 0 \tag{12.104}$$

If it is finite, the eigenvalue λ_n cannot cause this expression to be zero and can therefore be removed from the denominator. Thus, for $l = n$,

$$(\lambda_n - \lambda_n^*) \int_a^b |\psi_n(x)|^2\, dx = 0 \tag{12.105}$$

As argued for eigenfunctions of hermitian operators in Chapter 6 and for the eigenvectors of hermitian matrices in Chapter 8, unless $\psi_n(x) = 0$,

$$\int_a^b |\psi_n(x)|^2\, dx \neq 0 \tag{12.106}$$

Therefore, $\lambda_n = \lambda_n^*$, which verifies the reality of the eigenvalues.

If $n \neq l$, then

$$(\lambda_n - \lambda_l) \int_a^b \psi_n^*(x)\psi_l(x)\, dx = 0 \tag{12.107}$$

For nondegenerate eigenvalues, $\lambda_n \neq \lambda_l$. Thus equation 12.107 can only be satisfied if

$$\int_a^b \psi_n^*(x)\psi_l(x)\, dx = 0 \tag{12.108}$$

After normalizing the integral expressed in equation 12.106 to 1, we have

$$\int_a^b \psi_n^*(x)\psi_l(x)\, dx = \delta_{nl} \tag{12.109}$$

which proves the statement of the natural orthogonality of the nondegenerate Fredholm eigenfunctions of the hermitian kernel.

For those Fredholm equations for which eigenvalues and eigenfunctions exist, the eigenfunctions form a complete, orthonormal set. Therefore, any kernel can be expanded in an eigenfunction series. We write the kernel as

$$\text{(a)} \quad K(x, y) = \sum_n \psi_n(x) A_n(y) \quad \text{where} \quad \text{(b)} \quad A_n(y) = \int_a^b K(x, y)\psi_n^*(x)\, dx \tag{12.110}$$

Using the hermiticity of the kernel, this can be written

$$\text{(a)} \quad A_n(y) = \int_a^b K^*(y, x)\psi_n^*(x)\,dx \quad \Rightarrow \quad \text{(b)} \quad A_n^*(x) = \int_a^b K(x, y)\psi_n(y)\,dy \tag{12.111}$$

where x and y have been renamed y and x. From the eigenvalue equation and the property that λ_n is real, we obtain

$$\text{(a)} \quad A_n^*(x) = \int_a^b K(x, y)\psi_n(y)\,dy = \frac{\psi_n(x)}{\lambda_n} \quad \Rightarrow \quad \text{(b)} \quad A_n(x) = \frac{\psi_n^*(x)}{\lambda_n} \tag{12.112}$$

$$\Rightarrow \quad K(x, y) = \sum_n \frac{\psi_n(x)\psi_n^*(y)}{\lambda_n} \tag{12.113}$$

$$\Rightarrow \quad \Psi(x) = g(x) + \lambda\int_a^b K(x, y)\Psi(y)\,dy = g(x) + \sum_n \frac{\lambda}{\lambda_n}\psi_n(x)\int_a^b \psi_n^*(y)\Psi(y)\,dy \tag{12.114}$$

We define the constant α_n as

$$\alpha_n \equiv \int_a^b \psi_n^*(y)\Psi(y)\,dy \tag{12.115}$$

$$\Rightarrow \quad \Psi(x) = g(x) + \sum_l \alpha_l \frac{\lambda}{\lambda_l}\psi_l(x) \tag{12.116}$$

Substituting equation 12.116 into equation 12.115, we obtain

$$\alpha_n = \int_a^b \psi_n^*(y)g(y)\,dy + \sum_l \alpha_l \frac{\lambda}{\lambda_l}\int_a^b \psi_n^*(y)\psi_l(y)\,dy \tag{12.117}$$

We use the fact that the eigenfunctions are orthonormalized, and define the constant β_n as

$$\beta_n \equiv \int_a^b \psi_n^*(y)g(y)\,dy \tag{12.118}$$

Then equation 12.117 yields

$$\alpha_n = \frac{\lambda_n}{\lambda_n - \lambda}\beta_n = \frac{\lambda_n}{\lambda_n - \lambda}\int_a^b \psi_n^*(y)g(y)\,dy \tag{12.119}$$

Therefore, the inhomogeneous Fredholm equation becomes

$$\begin{aligned}\Psi(x) &= g(x) + \sum_l \frac{\lambda}{\lambda_l}\psi_l(x)\frac{\lambda_l}{\lambda_l - \lambda}\int_a^b \psi_l^*(y)g(y)\,dy \\ &= g(x) + \lambda\int_a^b\left[\sum_l \frac{\psi_l(x)\psi_l^*(y)}{(\lambda_l - \lambda)}\right]g(y)\,dy\end{aligned} \tag{12.120}$$

Comparing this to the Fredholm solution of equation 12.77b, we see that the term in brackets is related to the Fredholm resolvent by

$$\Re(x, y; \lambda) = \lambda \sum_l \frac{\psi_l(x)\psi_l^*(y)}{(\lambda_l - \lambda)} \tag{12.121}$$

Example 12.7

As an example, consider

$$\Psi(x) = 1 + \lambda \int_0^{\pi} \sin(x + y)\Psi(y)\, dy \tag{12.122}$$

Since the kernel is degenerate, it is straightforward to determine that the exact solution to this equation is

$$\Psi(x) = 1 + \frac{1}{\pi\left[(4/\pi^2) - \lambda^2\right]}\left[4\lambda \sin x + \frac{8}{\pi} \cos x\right] \tag{12.123}$$

Noting that the kernel is real and symmetric, we can apply the Hilbert–Schmidt approach. The first step is to determine the eigenvalues of the kernel

$$K(x, y) = \sin(x + y) \tag{12.124}$$

by solving

$$\begin{aligned} \psi_n(x) = \lambda_n \int_0^{\pi} \sin(x + y)\psi_n(y)\, dy &= \lambda_n\left[\sin x \int_0^{\pi} \cos y\psi_n(y)\, dy + \cos x \int_0^{\pi} \sin y\psi_n(y)\, dy\right] \\ &\equiv \lambda_n[\alpha_n \sin x + \beta_n \cos x] \end{aligned} \tag{12.125}$$

Using equation 12.125, the constants α_n and β_n satisfy

$$\alpha_n = \int_0^{\pi} \cos y\psi_n(y)\, dy = \lambda_n \int_0^{\pi} \cos y[\alpha_n \sin y + \beta_n \cos y]\, dy = \frac{\pi}{2}\lambda_n\beta_n \tag{12.126a}$$

$$\beta_n = \int_0^{\pi} \sin y\psi_n(y)\, dy = \lambda_n \int_0^{\pi} \sin y[\alpha_n \sin y + \beta_n \cos y]\, dy = \frac{\pi}{2}\lambda_n\alpha_n \tag{12.126b}$$

Expressing these equations in matrix form, we have

$$\text{(a)} \quad \begin{bmatrix} 1 & -\lambda_n\frac{\pi}{2} \\ -\lambda_n\frac{\pi}{2} & 1 \end{bmatrix}\begin{bmatrix} \alpha_n \\ \beta_n \end{bmatrix} = 0 \quad \Rightarrow \quad \text{(b)} \quad \lambda_n \equiv \lambda_{\pm} = \pm\frac{2}{\pi} \tag{12.127}$$

Substituting these values into either of equations 12.126, we obtain an expression for $\beta_{\pm}$ in terms of $\alpha_{\pm}$. From this we find that

$$\psi_{\pm}(x) = \pm\frac{2}{\pi}\alpha_{\pm}(\sin x \pm \cos x) \equiv c_{\pm}(\sin x \pm \cos x) \tag{12.128}$$

The constants $c_{\pm}$ are determined from the normalization condition

$$\int_0^{\pi} |\psi_{\pm}(x)|^2\, dx = |c_{\pm}|^2 \int_0^{\pi} (\sin x \pm \cos x)^2\, dx = \pi|c_{\pm}|^2 = 1 \tag{12.129}$$

$$\Rightarrow \quad \psi_{\pm}(x) = \frac{1}{\sqrt{\pi}}(\sin x \pm \cos x) \tag{12.130}$$

The Fredholm resolvent, expressed in terms of these eigenfunctions, is

$$\Re(x, y; \lambda) = \frac{\lambda}{\pi}\left[\frac{(\sin x + \cos x)(\sin y - \cos y)}{[(2/\pi) - \lambda]} - \frac{(\sin x - \cos x)(\sin y - \cos y)}{[(2/\pi) + \lambda]}\right]$$

$$= \frac{\lambda}{\pi}\left[\frac{2\lambda \cos(x - y) + (4/\pi)\sin(x + y)}{[(4/\pi^2) - \lambda^2]}\right] \qquad (12.131)$$

$$\Rightarrow \quad \Psi(x) = 1 + \frac{\lambda}{\pi}\int_0^{\pi}\left[\frac{2\lambda \cos(x - y) + (4/\pi)\sin(x + y)}{[(4/\pi^2) - \lambda^2]}\right] dy$$

$$= 1 + \frac{1}{\pi[(4/\pi^2) - \lambda^2]}\left[4\lambda \sin x + \frac{8}{\pi}\cos x\right] \quad \square \qquad (12.132)$$

Green's function solution

A Green's function method can also be developed to solve the inhomogeneous Fredholm equation

$$\Psi(x) = g(x) + \lambda\int_a^b K(x, y)\Psi(y)\, dy \qquad (12.1c)$$

The Green's function is defined as the solution to

$$G(x, x') = \delta(x - x') + \lambda\int_a^b G(x, y)K(y, x')\, dy \qquad (12.133)$$

We multiply this equation by $\Psi(x')$ and integrate to obtain

$$\int_a^b G(x, x')\Psi(x')\, dx' = \Psi(x) + \lambda\int_a^b G(x, y)\int_a^b K(y, x')\Psi(x')\, dx'\, dy \qquad (12.134)$$

But

$$\lambda\int_a^b K(y, x')\Psi(x')\, dx' = \Psi(y) - g(y) \qquad (12.135)$$

so that equation 12.134 becomes

$$\int_a^b G(x, x')\Psi(x')\, dx' = \Psi(x) + \int_a^b G(x, y)\Psi(y)\, dy - \int_a^b G(x, y)g(y)\, dy \qquad (12.136a)$$

$$\Rightarrow \quad \Psi(x) = \int_a^b G(x, y)g(y)\, dy \qquad (12.136b)$$

Using an identical analysis, it is straightforward to demonstrate that for the Fredholm equation

$$\Psi(x) = g(x) + \lambda\int_a^b \Psi(y)K(y, x)\, dy \qquad (12.137)$$

the Green's function, defined by

$$G(x, x') = \delta(x - x') + \lambda\int_a^b K(x, y)G(y, x')\, dy \qquad (12.138)$$

$$\Rightarrow \quad \Psi(x) = \int_a^b g(y)G(y, x)\, dy \qquad (12.139)$$

We recall that for second-order differential equations, we have to associate the δ-function singularity with $d^2G(x, y)/dy^2$. Thus the Green's function for the differential operator is continuous. Referring to equation 12.133, we note that for the Fredholm integral equation, the Green's function itself contains the δ-function singularity. As such, we write this Green's function as

$$G(x, y) = \delta(x - y) + H(x, y) \tag{12.140}$$

and the Green's function solution of equation 12.136b becomes

$$\Psi(x) = \int_a^b \delta(x - y)g(y)\,dy + \int_a^b H(x, y)g(y)\,dy = g(x) + \int_a^b H(x, y)g(y)\,dy \tag{12.141}$$

Comparing this to the Fredholm solution of equation 12.77b, we see that $H(x, x')$ is the Fredholm resolvent,

$$\Re(x, y; \lambda) = H(x, y) \tag{12.142}$$

Therefore, if the kernel is hermitian, for example,

$$H(x, x') = \lambda \sum_n \frac{\psi_l(x)\psi_l^*(x')}{(\lambda_l - \lambda)} \tag{12.143}$$

Substituting

$$G(x, x') = \delta(x - x') + H(x, x') \tag{12.140}$$

into

$$G(x, x') = \delta(x - x') + \lambda \int_a^b G(x, y)K(y, x')\,dy \tag{12.133}$$

$$\Rightarrow \quad H(x, x') = \lambda \int_a^b [\delta(x - y) + H(x, y)]K(y, x')\,dy$$

$$= \lambda K(x, x') + \lambda \int_a^b H(x, y)K(y, x')\,dy \tag{12.144}$$

Displacement kernel and solution by Fourier transform methods

A displacement or difference kernel is of the form

$$K(x, y) = K(x - y) \tag{12.145}$$

If $x, y \in [-\infty, \infty]$, a Fredholm equation with such a kernel is soluble by Fourier transform methods.

With

$$\Psi(x) = g(x) + \lambda \int_{-\infty}^{\infty} K(x - y)\Psi(y)\,dy \tag{12.146}$$

we define the Fourier inverses of $\Psi(x)$ and $g(x)$ as

$$\text{(a)} \quad F(k) \equiv \int_{-\infty}^{\infty} \Psi(x)e^{-ikx}\frac{dx}{\sqrt{2\pi}} \quad \text{and} \quad \text{(b)} \quad \gamma(k) \equiv \int_{-\infty}^{\infty} g(x)e^{-ikx}\frac{dx}{\sqrt{2\pi}} \tag{12.147}$$

$$\Rightarrow \quad \text{(a)} \quad \Psi(x) = \int_{-\infty}^{\infty} F(k)e^{ikx}\frac{dk}{\sqrt{2\pi}} \quad \text{and} \quad \text{(b)} \quad g(x) = \int_{-\infty}^{\infty} \gamma(k)e^{ikx}\frac{dk}{\sqrt{2\pi}} \tag{12.148}$$

Using equations 12.148, equation 12.146 can be written as

$$\int_{-\infty}^{\infty} F(k)e^{ikx}\frac{dk}{\sqrt{2\pi}} = \int_{-\infty}^{\infty} \gamma(k)e^{ikx}\frac{dx}{\sqrt{2\pi}} + \lambda\int_{-\infty}^{\infty} K(x-y)\int_{-\infty}^{\infty} F(k)e^{iky}\frac{dk}{\sqrt{2\pi}}\,dy \tag{12.149a}$$

Interchanging the order of integration in the double integral, we have

$$\int_{-\infty}^{\infty} F(k)e^{ikx}\frac{dk}{\sqrt{2\pi}} = \int_{-\infty}^{\infty} \gamma(k)e^{ikx}\frac{dk}{\sqrt{2\pi}} + \lambda\int_{-\infty}^{\infty} F(k)\frac{dk}{\sqrt{2\pi}}\int_{-\infty}^{\infty} K(x-y)e^{iky}\,dy \tag{12.149b}$$

We now set $z = x - y$. Then

$$\int_{-\infty}^{\infty} K(x-y)e^{iky}\,dy = e^{ikx}\int_{-\infty}^{\infty} K(z)e^{-ikz}\,dz \tag{12.150}$$

Referring to equations 12.147, we see that up to a factor of $\sqrt{2\pi}$, the integral over z is the Fourier transform of $K(z)$. Defining

$$\chi(k) \equiv \int_{-\infty}^{\infty} K(z)e^{-ikz}\frac{dz}{\sqrt{2\pi}} \tag{12.151}$$

$$\Rightarrow \quad \int_{-\infty}^{\infty} F(k)e^{ikx}\frac{dk}{\sqrt{2\pi}} = \int_{-\infty}^{\infty} \gamma(k)e^{ikx}\frac{dk}{\sqrt{2\pi}} + \lambda\int_{-\infty}^{\infty} F(k)\sqrt{2\pi}\,\chi(k)e^{ikx}\frac{dk}{\sqrt{2\pi}} \tag{12.152}$$

Since Fourier transforms are unique, we can equate the integrands to obtain

$$F(k) = \frac{\gamma(k)}{\left[1 - \lambda\sqrt{2\pi}\,\chi(k)\right]} \tag{12.153}$$

$$\Rightarrow \quad \Psi(x) = \int_{-\infty}^{\infty} \frac{\gamma(k)}{\left[1 - \lambda\sqrt{2\pi}\,\chi(k)\right]}e^{ikx}\frac{dk}{\sqrt{2\pi}} \tag{12.154}$$

Example 12.8

As an example, we consider the equation

$$\Psi(x) = 1 + \tfrac{1}{2}\int_{-\infty}^{\infty} e^{-(x-y)^2}\Psi(y)\,dy \tag{12.155}$$

The required Fourier inverses are

$$\gamma(k) = \int_{-\infty}^{\infty} 1 \cdot e^{-ikx}\frac{dx}{\sqrt{2\pi}} = \sqrt{2\pi}\,\delta(k) \tag{12.156a}$$

$$\chi(k) = \int_{-\infty}^{\infty} e^{-z^2}e^{-ikz}\frac{dz}{\sqrt{2\pi}} = e^{-k^2/4}\int_{-\infty}^{\infty} e^{-[z+ik/2]^2}\frac{dz}{\sqrt{2\pi}} \tag{12.156b}$$

$$\equiv e^{-k^2/4}\int_{-\infty}^{\infty} e^{-t^2}\frac{dt}{\sqrt{2\pi}} = \frac{1}{\sqrt{2}}e^{-k^2/4}$$

$$\Rightarrow \quad F(k) = \sqrt{2\pi}\,\delta(k)\left[1 - \tfrac{1}{2}\sqrt{\pi}\,e^{-k^2/4}\right]^{-1} \tag{12.157a}$$

$$\Rightarrow \quad \Psi(x) = \int_{-\infty}^{\infty} e^{ikx}\sqrt{2\pi}\,\delta(k)\left[1 - \tfrac{1}{2}\sqrt{\pi}\,e^{-k^2/4}\right]^{-1}\frac{dk}{\sqrt{2\pi}} = \frac{1}{\left(1 - \tfrac{1}{2}\sqrt{\pi}\right)} \tag{12.157b}$$

That is, $\Psi(x)$ is a constant. □

Padé approximant

In Chapter 3 (equation 3.178) we introduced the Padé approximant as an estimate of a power series. That is, the series $S(x) = \sum_{n=0}^{\infty} s_n x^n$ is sometimes well approximated by the $[M, N]$ Padé approximant

$$S^{[M,N]}(x) = \frac{P_M(x)}{Q_N(x)} = \frac{p_0 + p_1 x + \cdots + p_M x^M}{1 + q_1 x + \cdots + q_N x^N} \tag{3.179b}$$

The Neumann series for the Fredholm equation as expressed in equation 12.42 is such a power series in λ. As such, the Padé approximant, expressed as a ratio of polynomials in λ, can be used to estimate $\Psi(x)$.

Writing the Neumann series for $\Psi(x)$ as

$$\Psi(x) = g(x) + \sum_{n=1}^{\infty} \lambda^n I_n(x) \tag{12.42}$$

the Padé approximant is of the form

$$\Psi^{[M,N]}(x) = \frac{A_0(x) + \lambda A_1(x) + \lambda^2 A_2(x) + \cdots + \lambda^M A_M(x)}{1 + \lambda B_1(x) + \lambda^2 B_2(x) + \cdots + \lambda^N B_N(x)} \tag{12.158}$$

The functions $A_l(x)$ and $B_l(x)$ are defined such that

$$\sum_{m=0}^{M} \lambda^m A_m(x) - \left[1 + \sum_{n=1}^{N} \lambda^n B_n(x)\right]\left[\sum_{l=0}^{\infty} \lambda^l I_l(x)\right] = O(\lambda^{M+N+1}) \tag{12.159}$$

For example, the [1, 1] approximant is

$$\Psi^{[1,1]}(x) = \frac{A_0(x) + \lambda A_1(x)}{1 + \lambda B_1(x)} \tag{12.160}$$

such that

$$\begin{aligned}
&[A_0(x) + \lambda A_1(x)] - [1 + \lambda B_1(x)][g(x) + \lambda I_1(x) + \lambda^2 I_2(x) + O(\lambda^3)] \\
&= [A_0(x) - g(x)] + \lambda[A_1(x) - B_1(x)g(x) - I_1(x)] \\
&+ \lambda^2[-B_1(x)I_1(x) - I_2(x)] + O(\lambda^3) = O(\lambda^3)
\end{aligned} \tag{12.161}$$

Equating the coefficients of corresponding powers of λ, we obtain

$$\text{(a)}\quad A_0(x) = g(x) \qquad \text{(b)}\quad A_1(x) = B_1(x)g(x) + I_1(x) \qquad \text{(c)}\quad B_1(x) = -\frac{I_2(x)}{I_1(x)} \equiv \rho(x) \tag{12.162}$$

$$\Rightarrow \quad \Psi^{[1,1]}(x) = \frac{g(x) + \lambda[I_1(x) - g(x)\rho(x)]}{[1 - \lambda\rho(x)]} \tag{12.163}$$

Example 12.9

As an example, we will determine the [1, 1] Padé approximant for

$$\Psi(x) = x + \tfrac{1}{2}\int_0^1 (x + y)\Psi(y)\,dy \tag{12.164}$$

which has the exact solution

$$\Psi_e(x) = \tfrac{4}{23}(9x + 2) \tag{12.165}$$

The Neumann series in λ for this equation is

$$\Psi_N(x) = x + \lambda\int_0^1 (x + y)y\,dy + \lambda^2\int_0^1 (x + y)\int_0^1 (y + z)z\,dz\,dy + \cdots \tag{12.166}$$

$$\Rightarrow \text{(a)} \quad I_1(x) = \int_0^1 (x + y)y\,dy = \tfrac{1}{6}(3x + 2)$$

$$\text{(b)} \quad I_2(x) = \int_0^1 (x + y)\int_0^1 (y + z)z\,dz\,dy = \tfrac{1}{12}(7x + 4) \quad \Rightarrow \quad \text{(c)} \quad \rho(x) = \frac{1}{2}\frac{(7x + 4)}{(3x + 2)} \tag{12.167}$$

$$\Rightarrow \quad \Psi^{[1,1]}(x) = \frac{x + \lambda\left[\dfrac{1}{6}(3x + 2) - \dfrac{x}{2}\dfrac{(7x + 4)}{(3x + 2)}\right]}{\left[1 - \dfrac{1}{2}\lambda\dfrac{(7x + 4)}{(3x + 2)}\right]} \tag{12.168a}$$

In the current example, $\lambda = \frac{1}{2}$. Thus

$$\Psi^{[1,1]}(x) = 2\left[\frac{4x^2 + 4x + \frac{2}{3}}{5x + 4}\right] \tag{12.168b}$$

With $\lambda = \frac{1}{2}$, the Neumann series, truncated after the λ^2 term becomes

$$\Psi_2(x) = x + \lambda I_1(x) + \lambda^2 I_2(x) = \tfrac{67}{48}x + \tfrac{1}{4} \tag{12.169}$$

Table 12.1 presents a comparison of these truncated Neumann and Padé estimates to the exact solution to equation 12.164 at selected values of x. As can be seen, the [1, 1] Padé approximant is significantly more accurate than the truncated Neumann series. □

TABLE 12.1 COMPARISON OF APPROXIMATED AND EXACT VALUES OF A SOLUTION TO A FREDHOLM EQUATION

x	$\Psi^{[1,1]}$	Ψ_2	Ψ_e
0.00	0.333	0.250	0.348
0.25	0.730	0.599	0.739
0.50	1.128	0.948	1.130
0.75	1.527	1.297	1.522
1.00	1.926	1.646	1.913

Picard method

The Picard method of approximating a solution is an iterative approach. To illustrate the technique, we consider the inhomogeneous Fredholm equation

$$\Psi(x) = g(x) + \lambda\int_a^b K(x,y)\Psi(y)\,dy \tag{12.1c}$$

The iterations are begun by making an initial guess at the unknown function and substituting that guess (zeroth iteration) for $\Psi(y)$ in the integral to generate the first approximation to $\Psi(x)$. This first iteration is then substituted for $\Psi(y)$ in the integral and a second estimate of $\Psi(x)$ is obtained from the integral equation. The process is continued until the maximum difference between two successive approximations is less than some specified value. That is,

$$\text{Max}\left|\Psi^{(N)}(x) - \Psi^{(N-1)}(x)\right| < \varepsilon \tag{12.170}$$

where ε is some small specified value.

For example, one could choose the zeroth iterate to be

$$\Psi^{(0)}(y) = g(y) \tag{12.171}$$

Then the first iteration becomes

$$\Psi^{(1)}(x) = g(x) + \lambda\int_a^b K(x,y)\Psi^{(0)}(y)\,dy = g(x) + \lambda\int_a^b K(x,y)g(y)\,dy \tag{12.172}$$

Using this for $\Psi(y)$ under the integral, the second estimate of $\Psi(x)$ becomes

$$\Psi^{(2)}(x) = g(x) + \lambda\int_a^b K(x,y)g(y)\,dy + \lambda^2\int_a^b K(x,y)\int_a^b K(y,z)g(z)\,dz\,dy \tag{12.173}$$

and so on. As can be seen, choosing $g(x)$ as the initial estimate of $\Psi(x)$ results in a truncated Neumann sum.

However, one can make choices for the initial trial function other than $g(x)$. For example, if the [1, 1] Padé approximant is a better estimate of $\Psi(x)$ than $g(x)$, then choosing

$$\Psi^{(0)}(y) = \Psi^{[1,1]}(y) \tag{12.174}$$

will lead to a much more rapid convergence to an accurate estimate of $\Psi(x)$.

Example 12.10

To illustrate this, we return to the Fredholm equation of Example 12.9:

$$\Psi(x) = x + \tfrac{1}{2}\int_0^1 (x+y)\Psi(y)\,dy \tag{12.164}$$

Referring to equation 12.168b, we take the initial guess to be

$$\Psi^{(0)}(y) = \Psi^{[1,1]}(y) = 2\left[\frac{4y^2 + 4y + \frac{2}{3}}{5y+4}\right] \tag{12.175}$$

Then the first iteration becomes

$$\Psi^{(1)}(x) = x + \int_0^1 (x+y)\left[\frac{4y^2 + 4y + \frac{2}{3}}{5y+4}\right] dy \equiv Ax + B \tag{12.176}$$

where

$$\text{(a)}\quad A = 1 + \int_0^1 \left[\frac{4y^2 + 4y + \frac{2}{3}}{5y+4}\right] dy = 1.5643 \qquad \text{(b)}\quad B = \int_0^1 y\left[\frac{4y^2 + 4y + \frac{2}{3}}{5y+4}\right] dy = 0.3485 \tag{12.177}$$

$$\Rightarrow \quad \Psi^{(1)}(x) = 1.5643x + 0.3485 \tag{12.178}$$

Using this for $\Psi(y)$ in the integral, the second iteration becomes

$$\Psi^{(2)}(x) = x + \tfrac{1}{2}\int_0^1 (x+y)(1.5643y + 0.3485)\,dy = 1.5653x + 0.3478 \tag{12.179}$$

We note that this is not significantly different from the first iterate, and both compare very well with the exact solution given in equation 12.165:

$$\Psi(x) = \tfrac{4}{23}(9x + 2) = 1.5652x + 0.3478 \quad \square \tag{12.165}$$

Expansion of the kernel

If $K(x, y)$ can be well approximated by a truncated Taylor series in $y \in [a, b]$, an estimate of the solution can be obtained by taking

$$\text{(a)}\quad K(x, y) \simeq \sum_{n=0}^{N} \frac{1}{n!} K^{(n)}(x, y_0)(y - y_0)^n \quad \text{where} \quad \text{(b)}\quad K^{(n)}(x, y_0) \equiv \left.\frac{\partial^n K}{\partial y^n}\right|_{y=y_0} \tag{12.180}$$

Since the derivatives depend on x and not on y, this approximated kernel is degenerate, and the method of solution proceeds as described above for such kernels.

Example 12.11

To illustrate this approach, we note that for $x, y \in [0, 1]$, $\sin(xy)$ is reasonably well approximated by the first two terms in the MacLaurin series. That is, we take

$$\sin(xy) \simeq xy - \frac{x^3 y^3}{3!} \tag{12.181}$$

over $[0, 1]$. We note that this approximation is more accurate for small values of x, y. Thus we can estimate that the largest error occurs when x, y are as large as possible. In this example, that would be for $x = y = 1$. This largest error is

$$E = \sin(1) - \left(1 - \frac{1}{3!}\right) = 8.14 \times 10^{-3} \tag{12.182}$$

This method of estimating errors is not foolproof for all kernels and the maximum error should be determined by graphical or well-founded analytic methods.

With such a small error in this example, we expect

$$\Psi(x) = x + \tfrac{1}{2}\int_0^1 \sin(xy)\Psi(y)\, dy \tag{12.183}$$

to be well approximated by

$$\Psi(x) \simeq x + \frac{1}{2}\int_0^1 \left(xy - \frac{x^3 y^3}{3!}\right)\Psi(y)\, dy = x + \frac{1}{2}\left[Ax - \frac{x^3}{6}B\right] \tag{12.184}$$

where

$$A = \int_0^1 y\Psi(y)\, dy = \int_0^1 \left[y^2 + \tfrac{1}{2}y^2 A - \tfrac{1}{12}y^4 B\right] dy \tag{12.185a}$$

$$B = \int_0^1 y^3\Psi(y)\, dy = \int_0^1 \left[y^4 + \tfrac{1}{2}y^4 A - \tfrac{1}{12}y^6 B\right] dy \tag{12.185b}$$

Evaluating these integrals and solving the resulting equations, we obtain

$$\text{(a)}\quad A = 0.3953 \qquad \text{(b)}\quad B = 0.2367 \tag{12.186}$$

$$\Rightarrow \quad \Psi(x) \simeq 1.1977x - 0.0197x^3 \quad \square \tag{12.187}$$

12.3 Volterra Equations

Some of the techniques that are applicable to Fredholm equations can also be used to find or estimate the solution to Volterra equations.

Neumann series

Just as was done for the inhomogeneous Fredholm equation of the second kind, an estimate of the solution to the inhomogeneous Volterra equation of the second kind,

$$\Psi(x) = g(x) + \lambda \int_a^x K(x, y)\Psi(y)\, dy \tag{12.2c}$$

can be obtained by developing a Neumann series in λ. Substituting the entire expression on the right-hand side of equation 12.2c for $\Psi(y)$ under the integral, we obtain

$$\Psi(x) = g(x) + \lambda \int_a^x K(x, y_1)\left[g(y_1) + \lambda \int_a^{y_1} K(y_1, y_2)\Psi(y_2)\, dy_2\right] dy_1 \tag{12.188a}$$

A second substitution of the right-hand side of equation 12.2c for $\Psi(y_2)$ yields

$$\Psi(x) = g(x) + \lambda \int_a^x K(x, y_1) g(y_1)\, dy_1 + \lambda^2 \int_a^x \int_a^{y_1} K(y_1, y_2) K(y_1, y_2) g(y_2)\, dy_2\, dy_1$$

$$+ \lambda^2 \int_a^x \int_a^{y_1} \int_a^{y_2} K(y_1, y_2) K(y_1, y_2) K(y_2, y_3) \Psi(y_3)\, dy_3\, dy_2\, dy_1 \tag{12.188b}$$

Continuing this process yields the Neumann series solution,

$$\Psi(x) = \sum_{n=0}^{\infty} \lambda_n I_n(x) \tag{12.189}$$

where

$$I_n(x) = \int_a^x K(x, y_1) \int_a^{y_1} K(y_1, y_2) \cdots \int_a^{y_{n-1}} K(y_{n-1}, y_n) g(y_n)\, dy_n\, dy_{n-1} \cdots dy_1 \tag{12.190a}$$

$$\Rightarrow \quad I_{n+1}(x) = \int_a^x K(x, y) I_n(y)\, dy \tag{12.190b}$$

If $K(x, y)$ and $g(x)$ are continuous over the interval $x, y \in [a, b]$, the Neumann series given in equation 12.189 is uniformly convergent over this interval. The reader interested in a proof of this convergence is referred to the text by J. Cushing, *Applied Analytical Mathematics for Physical Scientists* (New York: John Wiley & Sons, Inc., 1975), page 184.

Separable kernel

If the kernel is separable or degenerate, the Volterra equation is not directly soluble in the way that the Fredholm equation is. However, although more work may be required, a solution is obtainable.

For example, the simplest degenerate kernel is

$$K(x, y) = A(x)B(y) \tag{12.191}$$

Then equation 12.2c becomes

$$\Psi(x) = g(x) + \lambda A(x) \int_a^x B(y)\Psi(y)\, dy \equiv g(x) + \lambda A(x) C(x) \tag{12.192}$$

Substituting $\Psi(x)$ into the definition of $C(x)$, we have

$$C(x) \equiv \int_a^x B(y)\Psi(y)\, dy = \int_a^x B(y) g(y)\, dy + \lambda \int_a^x B(y) A(y) C(y)\, dy \tag{12.193}$$

We note that $C(x)$ also satisfies a Volterra equation. However, the kernel of this Volterra equation, $B(y)A(y)$, contains no x-dependence. Therefore, if we differentiate equation 12.193, we obtain

$$\frac{dC}{dx} = B(x) g(x) + \lambda B(x) A(x) C(x) \tag{12.194}$$

Comparing this to equation 5.88, we note that this is a Bernoulli differential equation with index 0. Referring to equation 5.93, the solution for $C(x)$ is

$$C(x) e^{-\lambda \int A(x) B(x)\, dx} - \int B(x) g(x) e^{-\lambda \int A(x) B(x)\, dx}\, dx = \alpha \tag{12.195}$$

We can determine the constant of integration α by noting from equation 12.193 that $C(a) = 0$.

Example 12.12

As an example, we consider

$$\Psi(x) = \sqrt{x} + \lambda \int_0^x \sqrt{xy}\, \Psi(y)\, dy \tag{12.196}$$

The solution is of the form

$$\text{(a)}\quad \Psi(x) = \sqrt{x}\,[1 + \lambda C(x)] \quad \Rightarrow \quad \text{(b)}\quad C(x) = \int_0^x \sqrt{y}\, \Psi(y)\, dy = \int_0^x [y + \lambda y C(y)]\, dy \tag{12.197}$$

$$\Rightarrow \quad \frac{dC}{dx} = x + \lambda x C(x) \tag{12.198}$$

Then with $A(x) = B(x) = g(x) = \sqrt{x}$, the solution for $C(x)$ is

$$\text{(a)}\quad C(x) e^{-\lambda \int x\, dx} - \int x e^{-\lambda \int x\, dx}\, dx = \alpha \quad \Rightarrow \quad \text{(b)}\quad C(x) = \alpha e^{\lambda x^2/2} - \frac{1}{\lambda} \tag{12.199}$$

$\alpha = 1/\lambda$ is obtained by setting $C(0) = 0$. Therefore,

$$\Psi(x) = \sqrt{x}\, e^{\lambda x^2/2} \quad \square \tag{12.200}$$

If the separable kernel contains more than one term,

$$K(x, y) = \sum_{n=1}^{N} A_n(x) B_n(y) \tag{12.201}$$

one obtains sets of coupled first-order differential equations for the functions $C_n(x)$. In general, the equations are rather unwieldy to solve if there is more than one term in the separated kernel (see Problem 24).

Reciprocal kernel

As was done for the inhomogeneous Fredholm equation, a reciprocal kernel solution can be developed for inhomogeneous Volterra equations. Let

$$-\chi(x,y) \equiv K(x,y) + \int_y^x K(x,y_1)K(y_1,y)\,dy_1$$

$$+ \int_y^x K(x,y_1)\,dy_1 \int_{y_1}^x K(y_1,y_2)K(y_2,y)\,dy_2$$

$$+ \int_y^x K(x,y_1)\,dy_1 \int_{y_1}^x K(y_1,y_2)\,dy_2 \int_{y_2}^x K(y_2,y_3)K(y_3,y)\,dy_3 + \cdots \tag{12.202}$$

or with $K_1(x,y) \equiv K(x,y)$ we can define the $(n+1)$th iteration as

$$K_{n+1}(x,y) = \int_y^x K(x,y_1)K_n(y_1,y)\,dy_1 = \int_y^x K_n(x,y_1)K(y_1,y)\,dy_1 \tag{12.203}$$

As with the reciprocal kernel for the Fredholm equation, we have absorbed λ in the kernel $[\lambda K(x,y) \to K(x,y)]$.

We note that the integrals involved in the definition of the reciprocal kernel for the Fredholm equation are evaluated over the interval $[a,b]$, which are the limits of the integral in the Fredholm equation. For the Volterra equation, the lower limit on the integral is a constant, whereas the lower limits on the integrals defining the Volterra reciprocal kernel are the y-variables.

Like the Fredholm kernel, the Volterra reciprocal kernel satisfies an integral equation. Writing

$$-\chi(x,y) = \sum_{n=1}^{\infty} K_n(x,y) = K(x,y) + \sum_{n=2}^{\infty} K_n(x,y)$$

$$= K(x,y) + \int_y^x K(x,y_1) \sum_{n=2}^{\infty} K_{n-1}(y_1,y)\,dy_1 \tag{12.204}$$

By substituting $n-1 \to n$, we obtain

$$\sum_{n=2}^{\infty} K_{n-1}(y_1,y) = \sum_{n=1}^{\infty} K_n(y_1,y) = -\chi(y_1,y) \tag{12.205}$$

$$\Rightarrow \quad -\chi(x,y) = K(x,y) - \int_y^x K(x,y_1)\chi(y_1,y)\,dy_1 \tag{12.206a}$$

Similarly, we can write equation 12.204 as

$$-\chi(x,y) = K(x,y) + \int_y^x \left(\sum_{n=2}^{\infty} K_{n-1}(x,y_1) \right) K(y_1,y)\,dy_1$$

$$= K(x,y) - \int_y^x \chi(x,y_1)K(y_1,y)\,dy_1 \tag{12.206b}$$

Using this second form of the equation for $\chi(x, y)$, consider

$$-\int_a^x \chi(x, y)\Psi(y)\,dy = \int_a^x K(x, y)\Psi(y)\,dy - \int_a^x \left[\int_y^x \chi(x, y_1)K(y_1, y)\,dy_1\right]\Psi(y)\,dy \tag{12.207}$$

We will use the Volterra equation to express

$$\int_a^x K(x, y)\Psi(y)\,dy = \Psi(x) - g(x) \tag{12.208}$$

In order to obtain an integral of this form in the double integral, the order of integration must be interchanged. Because the lower limit of the inner integral is an integration variable, this cannot be done straightforwardly.

To see how to interchange the orders of integration, consider

$$I \equiv \int_a^x A(x, y_1)\left[\int_a^{y_1} B(y_1, y)\,dy\right] dy_1 \tag{12.209}$$

Integration over y implies that $a \leqslant y \leqslant y_1$. This restriction can be imposed in terms of the Heaviside Θ-function defined in equation 4.143. That is,

$$\int_a^x A(x, y_1)\left[\int_a^{y_1} B(y_1, y)\,dy\right] dy_1 = \int_a^x A(x, y_1)\int_a^x B(y_1, y)\Theta(y_1 - y)\,dy\,dy_1 \tag{12.210a}$$

Now, since both sets of limits are independent of the integration variables, the order of integration can be interchanged, and

$$\int_a^x A(x, y_1)\left[\int_a^{y_1} B(y_1, y)\,dy\right] dy_1 = \int_a^x \int_a^x A(x, y_1)B(y_1, y)\Theta(y_1 - y)\,dy_1\,dy \tag{12.210b}$$

Since the Θ-function restricts $y_1 \geqslant y$, this becomes

$$\int_a^x A(x, y_1)\left[\int_a^{y_1} B(y_1, y)\,dy\right] dy_1 = \int_a^x \left[\int_y^x A(x, y_1)B(y_1, y)\,dy_1\right] dy \tag{12.211}$$

$$\Rightarrow \quad \int_a^x \left[\int_y^x \chi(x, y_1)K(y_1, y)\,dy_1\right]\Psi(y)\,dy$$

$$= \int_a^x \chi(x, y_1)\,dy_1 \int_a^{y_1} K(y_1, y)\Psi(y)\,dy_1 = \int_a^x \chi(x, y_1)[\Psi(y_1) - g(y_1)]\,dy_1 \tag{12.212}$$

Therefore, equation 12.207 becomes

$$-\int_a^x \chi(x, y)\Psi(y)\,dy = \Psi(x) - g(x) - \int_a^x \chi(x, y_1)\Psi(y_1)\,dy_1 + \int_a^x \chi(x, y_1)g(y_1)\,dy_1 \tag{12.213a}$$

$$\Rightarrow \quad \Psi(x) = g(x) - \int_a^x \chi(x, y)g(y)\,dy \tag{12.213b}$$

where we have dropped the subscript on the integration variable.

Example 12.13

Referring to the equation of Example 12.12 with $\lambda = 1$,

$$\Psi(x) = \sqrt{x} + \int_0^x \sqrt{xy}\,\Psi(y)\,dy \tag{12.214}$$

we have

$$K_1(x,y) = K(x,y) = \sqrt{xy} \tag{12.215a}$$

$$K_2(x,y) = \int_y^x K_1(x,y_1)K(y_1,y)\,dy_1 = \int_y^x \sqrt{xy_1}\sqrt{y_1 y}\,dy_1 = \tfrac{1}{2}(x^2-y^2)\sqrt{xy} \tag{12.215b}$$

$$K_3(x,y) = \int_y^x K_2(x,y_1)K(y_1,y)\,dy_1 = \int_y^x \sqrt{xy_1}\,\tfrac{1}{2}(x^2-y_1^2)\sqrt{y_1 y}\,dy_1$$

$$= \frac{1}{2}\sqrt{xy}\int_y^x (x^2-y_1^2)\frac{1}{2}\,dy_1^2 = \frac{1}{2\cdot 4}\sqrt{xy}\,(x^2-y^2)^2 = \frac{\sqrt{xy}}{2!}\left[\frac{(x^2-y^2)}{2}\right]^2 \tag{12.215c}$$

$$K_4(x,y) = \frac{1}{2\cdot 4}\sqrt{xy}\int_y^x (x^2-y_1^2)^2\frac{1}{2}\,dy_1^2 = \frac{1}{2\cdot 4\cdot 6}\sqrt{xy}\,(x^2-y^2)^3 = \frac{\sqrt{xy}}{3!}\left[\frac{(x^2-y^2)}{2}\right]^3 \tag{12.215d}$$

$$\Rightarrow \quad K_n(x,y) = \frac{\sqrt{xy}}{(n-1)!}\left[\frac{1}{2}(x^2-y^2)\right]^{n-1} \tag{12.216}$$

Therefore,

$$\chi(x,y) = \sqrt{xy}\sum_{n=1}^{\infty}\frac{1}{(n-1)!}\left[\frac{1}{2}(x^2-y^2)\right]^{n-1} = \sqrt{xy}\sum_{n=0}^{\infty}\frac{1}{n!}\left[\frac{1}{2}(x^2-y^2)\right]^{n}$$
$$= \sqrt{xy}\,e^{(x^2-y^2)/2} \tag{12.217}$$

$$\Rightarrow \quad \Psi(x) = \sqrt{x} + \int_0^x \sqrt{xy}\,e^{(x^2-y^2)/2}\sqrt{y}\,dy = \sqrt{x}\,e^{x^2/2} \tag{12.218}$$

Padé approximant

Since the development of the Padé approximant from the Neumann series is independent of the specific forms of the integrals involved, the Padé approximant for the solution to a Volterra equation is identical in form to that for a Fredholm solution. For example,

$$\Psi^{[1,1]}(x) = \frac{g(x) + \lambda[I_1(x) - g(x)\rho(x)]}{[1-\lambda\rho(x)]} \tag{12.163}$$

where

$$\rho(x) \equiv -\frac{I_2(x)}{I_1(x)} \tag{12.162c}$$

For the Volterra equations, the integrals, defined in the Neumann series, are

$$\text{(a)}\quad I_1(x) = \int_a^x K(x,y)g(y)\,dy \qquad \text{(b)}\quad I_2(x) = \int_a^x K(x,y_1)\left[\int_a^{y_1} K(y_1,y)g(y)\,dy\right]dy_1 \tag{12.219}$$

Example 12.14

Using the equation of Example 12.12,

$$\Psi(x) = \sqrt{x} + \lambda \int_0^x \sqrt{xy}\, \Psi(y)\, dy \tag{12.196}$$

the Neumann series, up to λ^2, is

$$\Psi_2(x) = \sqrt{x} + \lambda \int_0^x \sqrt{xy}\sqrt{y}\, dy + \lambda^2 \int_0^x \sqrt{xy_1}\left[\int_0^{y_1} \sqrt{y_1 y}\sqrt{y}\, dy\right] dy_1 \tag{12.220}$$

$$\Rightarrow \quad \text{(a)} \quad I_1(x) = \sqrt{x}\int_0^x y\, dy = \tfrac{1}{2}x^{5/2} \qquad \text{(b)} \quad I_2(x) = \sqrt{x}\int_0^x y_1\left[\int_0^{y_1} y\, dy\right] dy_1 = \tfrac{1}{8}x^{9/2} \tag{12.221}$$

$$\Rightarrow \quad \Psi^{[1,1]}(x) = \sqrt{x}\left(\frac{1 + \frac{1}{4}\lambda x^2}{1 - \frac{1}{4}\lambda x^2}\right) \tag{12.222}$$

In Table 12.2 we present the values of the truncated Neumann and [1, 1] Padé estimates of the solution to this equation, along with the exact values at selected values of x. As in Table 12.1, we take $\lambda = 1$. □

TABLE 12.2 COMPARISON OF APPROXIMATED AND EXACT VALUES OF A SOLUTION TO A VOLTERRA EQUATION

x	$\Psi^{[1,1]}$	Ψ_2	Ψ_e
0.00	0.0000	0.0000	0.0000
0.25	0.5159	0.5159	0.5159
0.50	0.8014	0.8010	0.8013
0.75	1.1495	1.1438	1.1473
1.00	1.6667	1.6250	1.6487

Picard method

As discussed for the Fredholm equation, this iterative approach involves substituting a good initial guess for $\Psi(y) \equiv \Psi^{(0)}(y)$ under the integral of the Volterra equation. In this way, a first estimate, $\Psi^{(1)}(x)$ is generated from the integral equation. $\Psi^{(1)}(x)$ is then substituted for $\Psi(y)$ in the integral to generate $\Psi^{(2)}(x)$, and so on.

Like the Picard method for the Fredholm equation, choosing $\Psi^{(0)}(y) = g(y)$ produces the Neumann series, and thus is not necessarily the best guess for $\Psi^{(0)}(y)$. A better estimate of $\Psi^{(0)}(y)$ (the [1, 1] Padé approximant, for example) will yield a better, more rapidly convergent estimate of $\Psi(x)$.

Example 12.15

For example, for the equation

$$\Psi(x) = \sqrt{x} + \lambda \int_0^x \sqrt{xy}\, \Psi(y)\, dy \tag{12.196}$$

if we choose

$$\Psi^{(0)}(y) = \Psi^{[1,1]}(y) = \sqrt{y}\left(\frac{1 + \frac{1}{4}\lambda y^2}{1 - \frac{1}{4}\lambda y^2}\right) \tag{12.223}$$

then, with $\lambda = 1$,

$$\Psi^{(1)}(x) = \sqrt{x} + \int_0^x \sqrt{xy}\,\sqrt{y}\left(\frac{1 + \frac{1}{4}\lambda y^2}{1 - \frac{1}{4}\lambda y^2}\right) dy = \sqrt{x}\left[1 - \tfrac{1}{2}x^2 - 4\log(1 - \tfrac{1}{4}x^2)\right] \tag{12.224a}$$

$$\Rightarrow \quad \Psi^{(2)}(x) = \sqrt{x} + \int_0^x \sqrt{xy}\,\sqrt{y}\left[1 - \tfrac{1}{2}y^2 - 4\log\left(1 - \tfrac{1}{4}y^2\right)\right] dy$$

$$= \sqrt{x}\left[1 + \tfrac{1}{2}x^2 - \tfrac{1}{8}x^4 + 8\left\{\left(1 - \tfrac{1}{4}x^2\right)\log\left(1 - \tfrac{1}{4}x^2\right) - \left(1 - \tfrac{1}{4}x^2\right) + 1\right\}\right] \tag{12.224b}$$

Table 12.3 lists a comparison of these first and second iterations to the exact solution at selected values of x. As can be seen, the agreement is quite good. □

TABLE 12.3 COMPARISON OF APPROXIMATED AND EXACT VALUES OF A SOLUTION TO A VOLTERRA EQUATION

x	$\Psi^{(1)}$	$\Psi^{(2)}$	Ψ_e
0.00	0.0000	0.0000	0.0000
0.25	0.5159	0.5159	0.5159
0.50	0.8013	0.8013	0.8013
0.75	1.1474	1.1473	1.1473
1.00	1.6507	1.6489	1.6487

Difference kernel and the Laplace transform method

As shown in Problem 21 of Chapter 4, the convolution theorem for the Laplace transform is

$$\text{L.T.}\left[\int_0^x Q(x-y)R(y)\,dy\right] = \text{L.T.}[Q(x)]\,\text{L.T.}[R(x)] \tag{12.225}$$

A Volterra equation with a displacement or difference kernel

$$\Psi(x) = g(x) + \lambda\int_0^x K(x-y)\Psi(y)\,dy \tag{12.226}$$

has a straightforward solution using Laplace transforms. We define

$$\text{(a)}\quad P(s) \equiv \text{L.T.}[\Psi(x)] = \int_0^\infty e^{-sx}\Psi(x)\,dx \qquad \text{(b)}\quad \gamma(s) \equiv \text{L.T.}[g(x)] = \int_0^\infty e^{-sx}g(x)\,dx$$

$$\text{(c)}\quad \Phi(s) \equiv \text{L.T.}[K(x)] = \int_0^\infty e^{-sx}K(x)\,dx \tag{12.227}$$

Using equation 12.225, the Laplace transform of the Volterra equation becomes

$$P(s) = \gamma(s) + \lambda\,\text{L.T.}\left[\int_0^x K(x-y)\Psi(y)\,dy\right] = \gamma(s) + \lambda\Phi(s)P(s) \tag{12.228}$$

$$\Rightarrow \quad P(s) = \frac{\gamma(s)}{1 - \lambda\Phi(s)} \tag{12.229}$$

Thus, solving the Volterra equation with a difference kernel is reduced to the task of evaluating the Bromwich integral:

$$\Psi(x) = \frac{1}{2\pi i}\int_{c-i\infty}^{c+i\infty} \frac{\gamma(s)}{1-\lambda\Phi(s)} e^{sx}\, ds = \frac{1}{2\pi i}\oint \frac{\gamma(s)}{1-\lambda\Phi(s)} e^{sx}\, ds \tag{12.230}$$

(The reader is referred to Figure 4.10 for the Bromwich contour.)

Example 12.16

As an example, we will solve the Volterra equation

$$\Psi(x) = x + \int_0^x \sin(x-y)\Psi(y)\, dy \tag{12.231}$$

Since the kernel is separable, this equation is soluble using techniques discussed earlier (see Problem 24.)

The required Laplace transforms are

$$\text{(a)}\quad \gamma(s) = \int_0^\infty x e^{-sx}\, dx = \frac{1}{s^2} \qquad \text{(b)}\quad \Phi(s) = \int_0^\infty \sin(x) e^{-sx}\, dx = \frac{1}{s^2+1} \tag{12.232}$$

$$\Rightarrow \quad \text{(c)}\quad P(s) = \frac{s^2+1}{s^4} = \frac{1}{s^4} + \frac{1}{s^2}$$

$$\Rightarrow \quad \Psi(x) = \frac{1}{2\pi i}\oint \left(\frac{1}{s^4} + \frac{1}{s^2}\right) e^{sx}\, ds \tag{12.233}$$

Referring to Figure 4.10, we note that the second- and fourth-order poles at $s = 0$ are inside the Bromwich contour. Therefore, by Cauchy's theorem,

$$\Psi(x) = R_2(0) + R_4(0) = \frac{d}{ds}(e^{sx})_{s=0} + \frac{1}{3!}\frac{d^3}{ds^3}(e^{sx})_{s=0} = x + \tfrac{1}{6}x^3 \qquad \square \tag{12.234}$$

12.4 Singular Integral Equations for Scientific Problems

Abel's equation

Many physical systems are described by integral equations that have singular kernels. The first such equation to appear in print was Abel's equation, in 1811.

Consider a particle of mass m, starting from rest a height x above some reference point (see Figure 12.1). It slides down a frictionless wire under the influence of a constant gravitational acceleration a. When it is a distance y above the reference point, its speed is given by

$$\frac{1}{2}m\left(\frac{ds}{dt}\right)^2 = ma(x-y) \tag{12.235}$$

where ds is an incremental distance along the wire. At time $t = 0$, $y = x$, so $s = 0$.

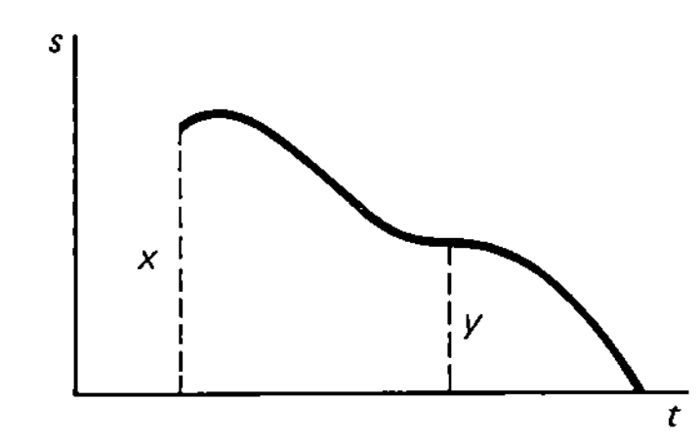

Figure 12.1 Path of a particle moving under constant acceleration.

Therefore, the time it takes the particle to reach the reference point, $y = 0$, is given by

$$T(x) = \int_0^S \frac{ds}{\sqrt{2a}\,(x-y)^{1/2}} = \int_x^0 \frac{-1}{\sqrt{2a}}\left(\frac{ds}{dy}\right)\frac{dy}{(x-y)^{1/2}} \tag{12.236}$$

where S is the total length of the particle's path. The negative sign is introduced because, as defined, s increases as y decreases. Thus $ds/dy < 0$.

Defining

$$\Psi(y) \equiv \frac{1}{\sqrt{2a}}\left(\frac{ds}{dy}\right) \tag{12.237}$$

$$\Rightarrow \quad T(x) = \int_0^x \frac{\Psi(y)}{\sqrt{(x-y)}}\,dy \tag{12.238}$$

This is *Abel's equation.*

If one knows the shape of the path taken by the particle (the shape of the wire), equation 12.238 gives the time it takes to travel that path. However, Abel's problem is to determine the path such that the particle reaches the reference point in a specified time $T(x)$. For this problem, $\Psi(y)$ is the unknown function, and equation 12.238 becomes a Volterra equation of the first kind.

The solution to this equation can be obtained by considering

$$\int_0^z \frac{T(x)}{\sqrt{(z-x)}}\,dx = \int_0^z \frac{dx}{\sqrt{(z-x)}} \int_0^x \frac{\Psi(y)}{\sqrt{(x-y)}}\,dy \tag{12.239}$$

Using equation 12.211, we can interchange the order of integration so that

$$\int_0^z \frac{T(x)}{\sqrt{(z-x)}}\,dx = \int_0^z \Psi(y)\,dy \int_y^z \frac{dx}{\sqrt{(z-x)}\,\sqrt{(x-y)}} \tag{12.240}$$

Making the substitution

$$x = y + (z-y)t \tag{12.241}$$

$$\Rightarrow \quad \int_y^z \frac{dx}{\sqrt{(z-x)}\,\sqrt{(x-y)}} = \int_0^1 t^{-1/2}(1-t)^{-1/2}\,dt = \beta\left(\tfrac{1}{2},\tfrac{1}{2}\right) = \frac{\left[\Gamma\left(\frac{1}{2}\right)\right]^2}{\Gamma(1)} = \pi \tag{12.242}$$

$$\Rightarrow \quad \text{(a)} \quad \int_0^z \Psi(y)\,dy = \frac{1}{\pi}\int_0^z \frac{T(x)}{\sqrt{(z-x)}}\,dx \quad \Rightarrow \quad \text{(b)} \quad \Psi(z) = \frac{1}{\pi}\frac{d}{dz}\int_0^z \frac{T(x)}{\sqrt{(z-x)}}\,dx \tag{12.243}$$

Since $(x-y)^{-1/2}$ is a difference kernel, Abel's equation can also be solved using Laplace transform methods (see Problem 29).

The solution to the generalized Abel equation

$$T(x) = \int_0^x \frac{\Psi(y)}{(x-y)^\alpha}\,dy \tag{12.244}$$

can be obtained using the approach above. As the reader will demonstrate in Problem 29, multiplying the equation by $1/(z-x)^{1-\alpha}$ and proceeding as above yields

$$\Psi(y) = \frac{\sin(\pi\alpha)}{\pi}\frac{d}{dz}\int_0^z \frac{T(x)}{(z-x)^{1-\alpha}}\,dx \tag{12.245}$$

Lippmann – Schwinger equation

As will be noted, the kernel of the Abel equation approaches infinity like $1/\varepsilon^{\alpha}$ as $\varepsilon \to 0$, with $\alpha < 1$. Singularities of this form are referred to as *weak singularities*. Some of the integral equations describing physical systems have kernels that contain pole singularities.

The Lippmann–Schwinger equation describes the scattering of a quantum mechanical particle moving in a potential field. One form of the equation for the scattering wave function of a particle in a particular angular momentum state l is

$$\Psi_l(p^2, p'^2) = V(p^2, p'^2) - \frac{2}{\pi}\int_0^{\infty} \frac{V(p^2, k^2)}{(k^2 - k_0^2 - i\varepsilon)}\Psi_l(k^2, p'^2)k^2\,dk \tag{12.246}$$

where the momenta of the incoming (beam) particle and the outgoing (scattered) particle are defined as $\hbar\mathbf{p}$ and $\hbar\mathbf{p}'$, respectively. $V(p^2, p'^2)$ is related to the potential function that causes the particle to scatter, and k_0 is related to the energy of the scattered particle by

$$\hbar^2 k_0^2 = 2mE \tag{12.247}$$

The measurable quantity in the quantum scattering process is a quantity called the *phase shift*. It depends on the energy of the particle and is denoted $\delta(k_0^2)$. In the representation expressed in equation 12.246, the phase shift is related to the wave function by

$$\Psi_l(k_0^2, k_0^2) = -\frac{\sin\delta(k_0^2)}{k_0}e^{i\delta(k_0^2)} \tag{12.248}$$

If the potential is zero, the energy of the particle is its kinetic energy. Thus (infinitely) far from the potential source,

$$E = \frac{\hbar^2 p^2}{2m} = \frac{\hbar^2 p'^2}{2m} \tag{12.249a}$$

But from equation 12.247,

$$E = \frac{\hbar^2 k_0^2}{2m} \tag{12.249b}$$

That is, far from the scattering center, $p = p' = k_0$. Since the Lippmann–Schwinger amplitude depends on momenta $p \neq k_0$ and $p' \neq k_0$, $\Psi_l(p^2, p'^2)$ is referred to as the *off-energy-shell amplitude*. When $p = p' = k_0$, the amplitude is called the *on-shell amplitude*. Since the phase shift is determined from the on-shell amplitude, it is the on-shell amplitude that is measurable.

An analytic form of the solution to equation 12.246 cannot be developed for a general potential $V(p^2, p'^2)$. Thus an approximation, such as a truncated Neumann series or a Padé approximant, is the only analytic expression one can obtain for $\Psi_l(p^2, p'^2)$. The reader is referred to Problem 31 for examples of these approximations to the wave function. Most often, a solution is obtained numerically. Some of the numerical techniques used to achieve this are discussed in Chapter 13.

However, if $V(p^2, p'^2)$ is separable, it is possible to obtain an analytic solution for $\Psi_l(p^2, p'^2)$. The reader is referred to Problem 30 for an example. Let

$$V(p^2, p'^2) \equiv \sum_{r=1}^{N} A_r(p^2) B_r(p'^2) \tag{12.250}$$

Then equation 12.246 becomes

$$\Psi_l(p^2, p'^2) = \sum_{r=1}^{N} A_r(p^2) B_r(p'^2) - \frac{2}{\pi} \sum_{r=1}^{N} A_r(p^2) \int_0^\infty \frac{B_r(k^2)}{(k^2 - k_0^2 - i\varepsilon)} \Psi_l(k^2, p'^2) k^2 \, dk \tag{12.251}$$

Since the integral contains no p^2-dependence, it can be written

$$I_r(p'^2) \equiv \int_0^\infty \frac{B_r(k^2)}{(k^2 - k_0^2 - i\varepsilon)} \Psi_l(k^2, p'^2) k^2 \, dk \tag{12.252}$$

and the Lippmann–Schwinger equation becomes

$$\Psi_l(p^2, p'^2) = \sum_{r=1}^{N} A_r(p^2) B_r(p'^2) - \frac{2}{\pi} \sum_{r=1}^{N} A_r(p^2) I_r(p'^2) \tag{12.253}$$

$$\Rightarrow \quad I_r(p'^2) = \int_0^\infty \frac{B_r(k^2)}{(k^2 - k_0^2 - i\varepsilon)} \left[\sum_{s=1}^{N} A_s(k^2) B_s(p'^2) - \frac{2}{\pi} \sum_{s=1}^{N} A_s(k^2) I_s(p'^2) \right] k^2 \, dk \tag{12.254}$$

Defining

$$\text{(a)} \quad M_{rs} \equiv \int_0^\infty \frac{B_r(k^2) A_s(k^2)}{(k^2 - k_0^2 - i\varepsilon)} k^2 \, dk \qquad \text{(b)} \quad Q_r(p'^2) \equiv \sum_{s=1}^{N} M_{rs} B_s(p'^2) \tag{12.255}$$

$$\Rightarrow \quad I_r(p'^2) = Q_r(p'^2) - \frac{2}{\pi} \sum_{s=1}^{N} M_{rs} I_s(p'^2) \tag{12.256}$$

For a given value of p'^2, we define N-component column vectors $\mathbf{I}(p'^2)$ and $\mathbf{Q}(p'^2)$ and an $N \times N$ matrix $\mathbf{M}$, with components I_r, Q_r, and M_{rs}, respectively. Then the solution to equation 12.256 is

$$\mathbf{I}(p'^2) = \left(\mathbf{1} + \frac{2}{\pi}\mathbf{M}\right)^{-1} \mathbf{Q}(p'^2) \tag{12.257}$$

Substituting the components of $\mathbf{I}(p'^2)$ into equation 12.253 yields the Lippmann–Schwinger amplitude for a separable potential.

It is possible to manipulate the Lippmann–Schwinger equation into a form in which the pole singularity is removed. This can be particularly important in obtaining numerical solutions.

We recognize that because the phase shifts are measurable, the on-shell amplitude $\Psi_l(k_0^2, k_0^2)$ is the only amplitude that can be determined experimentally. We therefore begin by setting $p' = k_0$. From equation 12.246, this half-on-shell amplitude satisfies

$$\Psi_l(p^2, k_0^2) = V(p^2, k_0^2) - \frac{2}{\pi}\int_0^\infty \frac{V(p^2, k^2)}{(k^2 - k_0^2 - i\varepsilon)} \Psi_l(k^2, k_0^2) k^2\, dk \tag{12.258}$$

A function $f(p^2)$ is defined in terms of the fully on-shell amplitude by

$$\Psi_l(p^2, k_0^2) \equiv f(p^2)\Psi_l(k_0^2, k_0^2) \tag{12.259}$$

$$\Rightarrow \quad f(k_0^2) = 1 \tag{12.260}$$

Then equation 12.258 becomes

$$f(p^2)\Psi_l(k_0^2, k_0^2) = V(p^2, k_0^2) - \frac{2}{\pi}\Psi_l(k_0^2, k_0^2)\int_0^\infty \frac{V(p^2, k^2)}{(k^2 - k_0^2 - i\varepsilon)} f(k^2) k^2\, dk \tag{12.261}$$

By setting $p^2 = k_0^2$, the fully on-shell amplitude is

$$\Psi_l(k_0^2, k_0^2) = \frac{V(k_0^2, k_0^2)}{\left[1 + \frac{2}{\pi}\int_0^\infty \frac{V(p^2, k^2)}{(k^2 - k_0^2 - i\varepsilon)} f(k^2) k^2\, dk\right]} \tag{12.262}$$

Therefore, the on-shell amplitude is determined in terms of an integral over the function $f(k^2)$. Dividing equation 12.261 by $\Psi_l(k_0^2, k_0^2)$ and using equation 12.262, we obtain

$$f(p^2) = \frac{V(p^2, k_0^2)}{\Psi_l(k_0^2, k_0^2)} - \frac{2}{\pi}\int_0^\infty \frac{V(p^2, k^2)}{(k^2 - k_0^2 - i\varepsilon)} f(k^2) k^2\, dk$$

$$= \frac{V(p^2, k_0^2)}{V(k_0^2, k_0^2)} - \frac{2}{\pi}\int_0^\infty \left[\frac{V(p^2, k_0^2)V(k_0^2, k^2)}{V(k_0^2, k_0^2)} - V(p^2, k^2)\right] \frac{f(k^2)}{(k^2 - k_0^2 - i\varepsilon)} k^2\, dk \tag{12.263}$$

As can be seen, when $k^2 \to k_0^2$, the terms in brackets become zero. Therefore,

$$\lim_{k^2 \to k_0^2} \left[\frac{V(p^2, k_0^2)V(k_0^2, k^2)}{V(k_0^2, k_0^2)} - V(p^2, k^2)\right] \frac{1}{(k^2 - k_0^2 - i\varepsilon)}$$

$$= \frac{d}{d(k^2)}\left[\frac{V(p^2, k_0^2)V(k_0^2, k^2)}{V(k_0^2, k_0^2)} - V(p^2, k^2)\right]_{k^2 \to k_0^2} \tag{12.264}$$

is finite, and the kernel of the equation for $f(p^2)$ does not contain a pole singularity.

Bethe – Salpeter equation

The Bethe–Salpeter equation describes the scattering of two particles via forces that the particles exert on one another (whereas the Lippmann–Schwinger equation describes the scattering of a single particle due to its presence in a potential field). Many of the techniques that have been presented for the Lippmann–Schwinger equation are applicable to the Bethe–Salpeter equation.

There have also been several approximation methods proposed specifically for the Bethe–Salpeter equation. Because of the complexity of the equation arising in part from the fact that it is a two-body equation, we have chosen not to present any details of these methods. However, the reader interested in such approaches for the simplest forms of the Bethe–Salpeter equation is referred to two articles and the references therein. One by the author and A. Pagnamenta, "Various Approximations to the Bethe–Salpeter Amplitude," appears in *Physical Review* (vol. 181, 1969, p. 2098). A second, by the author, "Improvement of the Blankenbecler–Sugar Approximation," appears in *Physical Review* (ser. D, vol. 2, 1970, p. 1738).

Muskhelishvili – Omnes equation

Referring to equation 12.246, we note that the pole in the Lippmann–Schwinger kernel is at a fixed value of k: namely, $k = k_0 - i\varepsilon$. That is, the position of the pole does not vary with the external momentum variable p^2.

In Abel's equation the kernel is of the form $(x - y)^{-\alpha}$. Thus the weak singularity of this kernel depends on the external variable x. There are physical systems described by integral equations in which the singularity of the kernel is a pole, and the position of the pole depends linearly on the external variable.

A large number of integral equations of this type arise from dispersion relations which were introduced in Chapter 2. Such dispersion integral equations are commonly of the form

$$\operatorname{Re}\Psi(x) = g(x) + \lambda P\int_a^b \frac{K(x, y)}{(y - x)}\operatorname{Im}\Psi(y)\,dy \tag{12.265}$$

This type of pole singularity is called a *Cauchy singularity*. Under certain physical constraints, the dispersion integral equation can be cast into a form for which a method of solution can be developed.

The Muskhelishvili–Omnes equation (which we will refer to as the MO equation) is such an equation. The simplest form of the MO equation is

$$\Psi(x) = g(x) + \frac{1}{\pi}\int_C \frac{h(y)}{(y - x - i\varepsilon)}\Psi(y)\,dy \tag{12.266}$$

where C is some contour along the real axis. $h(y)$ is the nonsingular part of the MO kernel, and we note here that it does not depend on the external variable x. The reader interested in a derivation and more complete discussion of the physics of the MO equation is referred to the book by G. Barton, *Introduction to Dispersion Techniques in Field Theory* (New York: W. A. Benjamin, Inc., 1965), pages 143–155. The reader should also consult the article by R. Omnes, "On the Solution of Certain Integral Equations of Quantum Field Theory," in *Il Nuovo Cimento* (vol. 8, 1958, pp. 316–326), and the book by N. Muskhelishvili, *Singular Integral Equations* (Leyden, The Netherlands: Nordhoff International Publishing, 1977), particularly Chapter 6.

To obtain a solution to the MO equation, let z be a complex variable and define

$$F(z) \equiv \frac{1}{2\pi i}\int_C \frac{h(y)}{(y - z)}\Psi(y)\,dy \tag{12.267}$$

$$\Rightarrow \quad F(x + i\varepsilon) \equiv \frac{1}{2\pi i}\int_C \frac{h(y)}{(y - x - i\varepsilon)}\Psi(y)\,dy \tag{12.268}$$

Once $F(z)$ is determined, the solution to the MO equation is

$$\Psi(x) = g(x) + 2iF(x + i\varepsilon) \tag{12.269}$$

Using the identity developed in equations 2.299,

$$\frac{1}{(y - x \pm i\varepsilon)} = P\frac{1}{(y - x)} \mp i\pi\delta(y - x) \tag{12.270}$$

in the defining equation for $F(z)$, we obtain

$$F(x + i\varepsilon) - F(x - i\varepsilon) = h(x)\Psi(x) \tag{12.271}$$

Since $F(z)$ is discontinuous for all x (for all real z), it has a cut along the real axis. $F(x + i\varepsilon)$ is the value of $F(z)$ just above the cut and $F(x - i\varepsilon)$ is its value below the cut. The discontinuity across that cut is $h(x)\Psi(x)$. Substituting for $\Psi(x)$ from equation 12.269, this becomes

$$[1 - 2ih(x)]F(x + i\varepsilon) - F(x - i\varepsilon) = g(x)h(x) \tag{12.272}$$

We let $F_0(z)$ be the solution to the homogeneous form of this equation ($g(x) = 0$) and define

$$1 - 2ih(x) \equiv A(x)e^{-2i\alpha(x)} \tag{12.273}$$

$$\Rightarrow \quad A(x)e^{-2i\alpha(x)}F_0(x + i\varepsilon) - F_0(x - i\varepsilon) = 0 \tag{12.274a}$$

$$\Rightarrow \quad \log[F_0(x + i\varepsilon)] - \log[F_0(x - i\varepsilon)] = 2i\alpha(x) - \log[A(x)] \tag{12.274b}$$

Thus, like $F(z)$, $\log[F_0(z)]$ has a cut along the real axis, the discontinuity across that cut being $2i\alpha(x) - \log[A(x)]$. Using equation 12.270, equation 12.274b can be obtained from

$$\log[F_0(z)] = \frac{1}{2\pi i}\int_C \frac{\log[F_0(y)]}{(y - z)}\,dz$$

$$= \frac{1}{\pi}\int_C \frac{\alpha(y)}{(y - z)}\,dy - \frac{1}{2\pi i}\int_C \frac{\log[A(y)]}{(y - z)}\,dy + R(z) \tag{12.275}$$

where $R(z)$ is analytic on the real axis. That is,

$$R(x + i\varepsilon) - R(x - i\varepsilon) = 0 \tag{12.276}$$

Therefore,

$$F_0(z) \sim e^{R(z)} \tag{12.277}$$

Since $F_0(z)$ satisfies the homogeneous equation, and $R(z)$ satisfies equation 12.276, this exponential will be an overall multiplicative factor in equation 12.274a and will cancel. As such, $R(z)$ can be taken to be zero, so that

$$\log[F_0(z)] = \frac{1}{\pi}\int_C \frac{\alpha(y)}{(y - z)}\,dy - \frac{1}{2\pi i}\int_C \frac{\log[A(y)]}{(y - z)}\,dy \tag{12.278}$$

The function $\Phi(z)$ is defined such that

$$F(z) \equiv \Phi(z)F_0(z) \tag{12.279}$$

$$\Rightarrow \quad \Phi(x+i\varepsilon) - \Phi(x-i\varepsilon) = \frac{F(x+i\varepsilon)}{F_0(x+i\varepsilon)} - \frac{F(x-i\varepsilon)}{F_0(x-i\varepsilon)} \tag{12.280}$$

Using equation 12.274a to express equation $F_0(x - i\varepsilon)$ in terms of $F_0(x + i\varepsilon)$, this becomes

$$\Phi(x+i\varepsilon) - \Phi(x-i\varepsilon) = \frac{1}{A(x)e^{-2i\alpha(x)}F_0(x+i\varepsilon)}\left[A(x)e^{-2i\alpha(x)}F(x+i\varepsilon) - F(x-i\varepsilon)\right] \tag{12.281}$$

We note that the combination in brackets is given in equation 12.272 to be $g(x)h(x)$. Therefore,

$$\Phi(x+i\varepsilon) - \Phi(x-i\varepsilon) = \frac{g(x)h(x)e^{2i\alpha(x)}}{A(x)F_0(x+i\varepsilon)} \equiv \Gamma(x) \tag{12.282}$$

That is, $\Gamma(x)$ is the discontinuity across the cut on the real axis of the function $\Phi(z)$. Therefore, again using equation 12.270, $\Phi(z)$ can be written

$$\Phi(z) = \frac{1}{2\pi i}\int_C \frac{\Gamma(y)}{(y-z)}\,dy + P(z) \tag{12.283}$$

where $P(z)$ is an arbitrary function that is analytic on the real axis, so that

$$P(x+i\varepsilon) - P(x-i\varepsilon) = 0 \tag{12.284}$$

Unlike $R(z)$ in equation 12.275, $P(z)$ cannot be taken to be zero since $\Phi(z)$ does not satisfy a homogeneous equation.

Since $F_0(x + i\varepsilon)$ is known in terms of $A(x)$ and $\alpha(x)$ from equation 12.278, and thus $\Gamma(x)$ is known from equation 12.282, $\Phi(z)$ is known up to an arbitrary function that is analytic on the real axis. Therefore, $F(z)$ is known from equation 12.279. Thus, up to an arbitrary analytic function, $\Psi(x)$ is obtained from equation 12.269. The result is

$$\Psi(x) = g(x) + \frac{F_0(x+i\varepsilon)}{\pi}\int_C \frac{\Gamma(y)}{(y-x-i\varepsilon)}\,dy + 2iF_0(x+i\varepsilon)P(x) \tag{12.285}$$

Under certain constraints, $P(x)$ will be zero, and the solution will be completely determined. As noted in the article by R. Omnes in *Il Nuovo Cimento* cited above, if the MO equation is applied to quantum processes in which pi mesons occur in either the initial or final state, the non-singular part of the kernel is of the form

$$h(x) = \sin\delta(x)e^{-i\delta(x)} \tag{12.286}$$

where $\delta(x)$ is a phase shift parameter used to describe the process. As noted by Omnes, with $x \in [1, \infty]$, if $\delta(x)$ is zero at $x = 1$, then $P(x) = 0$ and the solution to the MO equation is uniquely determined. The reader is referred to Problem 32 for an example of such a solution.

PROBLEMS

1. (a) A damped harmonic oscillator satisfies the differential equation of motion

$$m\frac{d^2y}{dt^2} - 2\beta m\frac{dy}{dt} + ky = 0$$

where $2\beta m$ is the damping coefficient and $\omega^2 = k/m$ is the square of the natural frequency of the oscillator. Determine the Volterra integral equation satisfied by $y(t)$ if

(1) $y(0) = A, \left.\frac{dy}{dt}\right|_{t=0} = 0$

(2) $y(0) = 0, \left.\frac{dy}{dt}\right|_{t=0} = \omega A$

(b) Deduce the Volterra equation for an undamped oscillator ($\beta \to 0$) for each set of initial conditions. Show that $y(t) = A\cos(\omega t)$ is the solution to the resulting Volterra equation for the initial conditions expressed in (1) and the Volterra equation obtained using the conditions expressed in (2) has the solution $y(t) = A\sin(\omega t)$.

2. (a) Determine the Volterra integral equation satisfied by $\Psi(x)$ equivalent to

$$\frac{d^2\Psi}{dx^2} + R(x)\Psi = g(x)$$

subject to $\Psi(x_0) = \Psi_0$ and $\Psi'(x_0) = \Psi_0'$.

(b) From the result of part (a), show that

$$\frac{d^2\Psi}{dx^2} + x\Psi = x^2$$

with $\Psi_0 = \Psi_0' = 0$ satisfies

$$\Psi(x) = \tfrac{1}{6}x^3 - \int_0^x (x-s)s\Psi(s)\,ds$$

3. Find the homogeneous differential equation and initial conditions for $\Psi(x)$, if $\Psi(x)$ satisfies

$$\Psi(x) = 1 + \tfrac{1}{2}x - \int_0^x (x-y)e^{-y}\Psi(y)\,dy$$

4. Deduce the Fredholm integral equation satisfied by $y(x) = P_1(x) = x$, the first Legendre polynomial. $y(x)$ satisfies

$$(1-x^2)\frac{d^2y}{dx^2} - 2x\frac{dy}{dx} + 2y = 0$$

subject to $y(1) = 1$, $y(-1) = -1$.

5. Show that if $\Psi(x)$ is the solution to

$$\frac{d^2\Psi}{dx^2} + x\Psi = 1 \qquad \text{with } \Psi(0) = \Psi(1) = 0$$

then $\Psi(x)$ satisfies the Fredholm integral equation

$$\Psi(x) = -\tfrac{1}{2}x(1-x) - \int_0^1 G(x,x')x'\Psi(x')\,dx'$$

with

$$G(x,x') = \begin{cases} x(1-x') & x' > x \\ x'(1-x) & x' < x \end{cases}$$

6. Find the solution to each of the following inhomogeneous Fredholm equations.

(a) $\Psi(x) = 1 + \int_0^{\pi/4} \cos(x-y)\Psi(y)\,dy$

(b) $\Psi(x) = x - \tfrac{1}{2}\int_0^{\infty} e^{-(x+y)}\Psi(y)\,dy$

(c) $\Psi(x) = x + 2\int_0^1 (x^2y - y^2)\Psi(y)\,dy$

7. Determine the solution to each of the following inhomogeneous Fredholm equations by summing the Neumann series.

(a) $\Psi(x) = 1 + \tfrac{1}{2}\int_0^{\pi/2} \cos(x-y)\Psi(y)\,dy$

(b) $\Psi(x) = x - \int_0^{\infty} e^{-(x+y)}\Psi(y)\,dy$

8. (a) Find the first three terms in the Neumann series for

$$\Psi(x) = \sin x + \int_0^{\pi/2} \cos(x-y)\Psi(y)\,dy$$

(b) Find the first three terms in the Neumann series for

$$\Psi(x) = x + \tfrac{1}{2}\int_0^1 (x^2+y^2)\Psi(y)\,dy$$

(c) Find the first two terms in the Neumann series for

$$\Psi(x) = 1 - \tfrac{1}{2}\int_0^{\pi/2} \cos(xy)\Psi(y)\,dy$$

(d) Find the first two terms in the Neumann series for

$$\Psi(x) = x + 2\int_0^{\infty} e^{-xy}\Psi(y)\,dy$$

9. Determine the reciprocal kernel, and the solution from that reciprocal kernel, for each of the Fredholm equations below.

(a) $\Psi(x) = x^2 - \tfrac{1}{4}\int_1^2 \frac{x}{y}\Psi(y)\,dy$

(b) $\Psi(x) = x + \tfrac{1}{3}\int_0^{\infty} e^{-(x+y)}\Psi(y)\,dy$

10. Prove that the indicated expansion of the determinant below is valid.

$$\begin{vmatrix} (1-\lambda K_{11}\Delta y) & -\lambda K_{12}\Delta y & -\lambda K_{13}\Delta y \\ -\lambda K_{21}\Delta y & (1-\lambda K_{22}\Delta y) & -\lambda K_{23}\Delta y \\ -\lambda K_{31}\Delta y & -\lambda K_{32}\Delta y & (1-\lambda K_{33}\Delta y) \end{vmatrix}$$

$$= 1 - \lambda \sum_{n=1}^{3} K_{nn}\,\Delta y$$

$$+ \frac{\lambda^2}{2!}\sum_{m,n=1}^{3}\begin{vmatrix} K_{mm} & K_{mn} \\ K_{nm} & K_{nn} \end{vmatrix}\Delta y^2 + O(\lambda^3)$$

11. By determining the numerator and denominator of the Fredholm resolvent up to the power of λ specified, use Fredholm's solution to estimate $\Psi(x)$ for each of the Fredholm equations below.

(a) $\Psi(x) = x + \lambda\int_0^1 (x-y)\Psi(y)\,dy$ up to λ^3

(b) $\Psi(x) = 1 - \lambda\int_0^\pi \sin(x-y)\Psi(y)\,dy$ up to λ^2

12. Find the eigenvalues and eigenfunctions for each of the Fredholm kernels listed below.

(a) $K(x, y) = \cos(x+y), \quad x, y \in [-\pi, \pi]$

(b) $K(x, y) = \cos(x+y), \quad x, y \in [0, \pi/4]$

(c) $K(x, y) = e^{-(x+y)}, \quad x, y \in [0, \infty]$

13. (a) $K(x, y) = A(x)B(y)$ is a degenerate Fredholm kernel. Prove that if $A(x)$ and $B(x)$ are orthogonal over the interval $[a, b]$, which means $\int_a^b A(x)B(x)\,dx = 0$, then the only solution to the eigenvalue equation

$$\psi_n(x) = \lambda_n\int_a^b K(x, y)\Psi_n(y)\,dy$$

is the trivial one. Thus such a kernel does not have any eigenvalues or eigenfunctions.

(b) Determine the solution to the inhomogeneous Fredholm equation

$$\Psi(x) = g(x) + \lambda\int_a^b A(x)B(y)\Psi(y)\,dy$$

if $A(x)$ and $B(x)$ are orthogonal over $[a, b]$.

(c) Use the results of part (b) to determine the solution to

$$\Psi(x) = x + \tfrac{1}{2}\int_0^\pi \sin(x)\cos(y)\Psi(y)\,dy$$

14. (a) Let

$$K(x, y) = A_1(x)B_1(y) + A_2(x)B_2(y)$$

and let $A_1(x)$ and $A_2(x)$ be orthogonal to both $B_1(x)$ and $B_2(x)$ over the interval $[a, b]$. That is,

$$\int_a^b A_i(x)B_j(x)\,dx = 0$$

for all combinations of $i, j = 1, 2$. Show that the only solution to the Fredholm eigenvalue equation

$$\psi_n(x) = \lambda_n\int_a^b K(x, y)\psi_n(y)\,dy$$

is the trivial one. Thus for such a kernel, there are no eigenvalues and eigenfunctions.

(b) Determine the solution to the inhomogeneous Fredholm equation

$$\Psi(x) = g(x) + \lambda\int_a^b [A_1(x)B_1(y) + A_2(x)B_2(y)]\Psi(y)\,dy$$

if each $A_i(x)$ is orthogonal to each $B_j(x)$.

(c) What is the solution to

$$\Psi(x) = x + \tfrac{1}{2}\int_{-1}^{1} [x^2y^3 + x^4y^5]\Psi(y)\,dy$$

15. Prove that an antihermitian Fredholm kernel, defined by

$$K^*(x, y) = -K(y, x)$$

has imaginary eigenvalues, and eigenfunctions that are naturally orthogonal in the absence of degeneracies.

16. Determine the Fredholm resolvent in terms of an eigenfunction expansion for

$$\Psi(x) = x - 2\int_0^\infty (x+y)e^{-(x+y)}\Psi(y)\,dy$$

17. Use Fourier transform methods to determine the solution to

$$\Psi(x) = e^{-x^2} + \lambda\int_{-\infty}^{\infty}(x-y)e^{-(x-y)^2}\Psi(y)\,dy$$

You may leave your result in the form of a Fourier integral. By careful analysis, the resulting Fourier integral can be evaluated. The result is

$$\Psi(x) = \frac{2\sqrt{\pi}}{\lambda}\frac{e^{-\alpha_0 x}}{(1+\frac{1}{2}\alpha_0^2)} \qquad x > 0$$

$$\Psi(x) = 0 \qquad x < 0$$

where α_0 is the solution to the transcendental equation

$$\alpha_0 e^{\alpha_0^2/4} = \frac{1}{\lambda}\left(\frac{2}{\pi}\right)^{1/2}$$

18. **(a)** Denote the Neumann series for the Fredholm equation

$$\Psi(x) = g(x) + \lambda \int_a^b K(x, y)\Psi(y)\, dy$$

as

$$\Psi(x) = \sum_{n=0}^{\infty} \lambda^n I_n(x)$$

where

$$I_0(x) \equiv g(x)$$

$$I_{n+1}(x) = \int_a^b K(x, y) I_n(y)\, dy$$

Develop expressions for the numerator and denominator functions of the [2, 2] Padé approximant for $\Psi(x)$.

(b) Prepare a table comparing $\Psi^{[2,2]}(x)$ to the exact solution of

$$\Psi(x) = x + \tfrac{1}{2}\int_0^1 (x + y)\Psi(y)\, dy$$

at those values of x listed in Table 12.1.

19. Prove that the [2, 2] Padé approximant to

$$\Psi(x) = g(x) + \lambda \int_a^b U(x)V(y)\Psi(y)\, dy$$

does not exist. Show that the [1, 1] Padé approximant does exist. [*Hint:* Try to determine the functions in the denominator of $\Psi^{[2,2]}(x)$.]

20. If $\Psi(x)$ satisfies

$$\Psi(x) = x - \tfrac{1}{2}\int_0^1 e^{-xy}\Psi(y)\, dy$$

estimate $\Psi(x)$ by approximating the exponential kernel by the first two terms in its MacLaurin expansion. Estimate the error incurred in making this approximation.

21. **(a)** It is straightforward to determine that the solution to

$$\Psi(x) = x + \tfrac{1}{2}\int_0^\infty e^{-(x+y)}\Psi(y)\, dy$$

is

$$\Psi(x) = x + \tfrac{2}{3}e^{-x}$$

As an initial guess, let $\Psi^{(0)}(x) = x + e^{-x}$ and deduce the third Picard iteration for $\Psi(x)$. Would an initial guess $\Psi^{(0)}(x) = x + \frac{1}{2}e^{-x}$ improve, worsen, or not affect the convergence of the Picard iterations?

(b) $\Psi(x)$ is the solution to

$$\Psi(x) = 1 + \tfrac{1}{2}\int_0^\pi \sin x \sin y\, \Psi(y)\, dy$$

Make a reasonable initial guess for $\Psi^{(0)}(x)$ based on the obvious form of the solution. Use this to generate the third Picard iterate $\Psi^{(3)}(x)$. Compare $\Psi^{(3)}(x)$ to the exact solution

$$\Psi(x) = 1 + \frac{1}{(1 - \frac{1}{4}\pi)} \sin x = 1 + 4.6598 \sin x$$

22. In Example 12.12 it was shown that the solution to

$$\Psi(x) = \sqrt{x} + \lambda \int_0^x \sqrt{xy}\, \Psi(y)\, dy \quad (12.196)$$

is

$$\Psi(x) = \sqrt{x}\, e^{\lambda x^2/2} \quad (12.200)$$

Develop the Neumann series for this Volterra equation and sum the series to obtain this solution.

23. Use the fact that the kernel of each of the Volterra equations listed below is separable to determine the solution to:

(a) $\Psi(x) = x - \frac{1}{2}\int_0^x x^{1/2}y^{-3/2}\Psi(y)\, dy$

(b) $\Psi(x) = x^2 - \int_0^x \left(\frac{x}{y}\right)\Psi(y)\, dy$

24. Determine the set of coupled first-order differential equations satisfied by $C_1(x)$ and $C_2(x)$ defined by

$$C_i(x) \equiv \int_0^x B_i(y)\Psi(y)\, dy$$

where $\Psi(x)$ is the solution to the Volterra equation

$$\Psi(x) = g(x) + \lambda \int_0^x [A_1(x)B_1(y) + A_2(x)B_2(y)]\Psi(y)\, dy$$

Show that C_1 and C_2 satisfy

$$\frac{dC_1}{dx}B_2 = \frac{dC_2}{dx}B_1$$

25. Find the kernel reciprocal to $e^{-(x-y)}$ and use that reciprocal kernel to solve

$$\Psi(x) = x + \int_0^x e^{-(x-y)}\Psi(y)\, dy$$

26. **(a)** Find the [1, 1] Padé approximant for $\Psi(x)$, where $\Psi(x)$ is the solution to

$$\Psi(x) = x - \tfrac{1}{2}\int_0^x e^{-(x-y)}\Psi(y)\, dy$$

Solve this Volterra equation exactly and compare the exact solution to $\Psi^{[1,1]}(x)$ at $x = 0.0$, 0.5, and 1.0.

(b) Noting that the exact solution to the equation of part (a) is

$$\Psi(x) = x - \tfrac{1}{2}C(x)e^{-x}$$

take $\Psi^{(0)}(x) = x - \frac{1}{2}e^{-x}$ as an initial guess and determine the second Picard iterate and compare $\Psi^{(2)}(x)$ to the exact values of $\Psi(x)$ at $x = 0.0, 0.5$, and 1.0.

27. Using Laplace transform methods, determine the solution to:

(a) $\Psi(x) = \sin x - 2\int_0^x \cos(x-y)\Psi(y)\,dy$

(b) $\Psi(x) = x - \int_0^x e^{-(x-y)}\Psi(y)\,dy$

28. A bead slides along a frictionless wire, starting from rest at a height x above the ground. Use Abel's equation to find the equation for the shape of the wire such that the bead will reach the ground in a time given by $T(x) = \alpha\sqrt{x}$. [*Hint:* You might find it useful to use a substitution $y = x(1 - \sin\theta)/2$ in evaluating an integral.]

29. Solve the generalized Abel equation

$$T(x) = \int_0^x \frac{\Psi(y)}{(x-y)^{\alpha}}\,dy \qquad 0 < \alpha < 1$$

(a) By considering

$$\int_0^z \frac{T(x)}{(z-x)^{1-\alpha}}\,dx$$

(b) By recognizing that the kernel $1/(x-y)^{\alpha}$ is a difference kernel and applying Laplace transform methods for $T(x) = Ax$. (*Hint:* Refer to Example 4.7 and the Bromwich contour of Figure 4.11.)

30. (a) Find the analytic solution to the Lippmann–Schwinger equation for the Yamaguchi potential

$$V(p^2, p'^2) \equiv \frac{\alpha}{(p^2+\beta^2)(p'^2+\beta^2)}$$

(b) Find the function $f(p^2)$ defined for the Lippmann–Schwinger amplitude by

$$f(p^2)\Psi(k_0^2, k_0^2) = \Psi(p^2, k_0^2)$$

for the Yamaguchi potential.

31. (a) Using the Yamaguchi potential

$$V(p^2, p'^2) \equiv \frac{\alpha}{(p^2+\beta^2)(p'^2+\beta^2)}$$

determine the Neumann series for the Lippmann–Schwinger amplitude truncated at the λ^2 term where $\lambda = -2/\pi$ for the Lippmann–Schwinger equation.

(b) From the results of part (a), determine the [1, 1] Padé approximant to the Lippmann–Schwinger amplitude for the Yamaguchi potential.

32. Find the solution to the Muskhelishvili–Omnes equation for

$$h(x) = \sin\delta(x)e^{-i\delta(x)} \qquad x \in [1,\infty]$$

$$g(x) = 1$$

$$\delta(x) = \frac{(x-1)}{x^2}$$

You may leave your result in terms of integrals involving elementary functions. As noted in the Omnes article in *Il Nuovo Cimento* cited in the text, if $\delta(1) = 0$, the additive arbitrary analytic function is zero and the solution is uniquely determined.

CHAPTER 13

NUMERICAL METHODS

Many problems encountered in scientific work are not soluble analytically. With the advent of high-speed computers, numerically approximate results can often be quickly obtained. The Euler–MacLaurin integration formula, introduced in Chapter 3, and the use of Sylvester's theorem for determining the eigenvalues and eigenvectors of a matrix, as described in Chapter 8, are examples of methods for estimating results numerically. In this chapter we present numerical techniques for solving other problems commonly encountered in scientific investigations.

13.1 Interpolation

Polynomial interpolation

Often, one is presented with the description of a function in tabulated form. That is, discrete values of the function are listed at specified points. The results of an experiment are customarily represented in this way. Using interpolation methods, one can estimate values of the function at points not listed in the table.

Example 13.1

For example, the values of a function are listed in Table 13.1 at five different values of its argument. Values of x outside the range listed in the table are called *extrapolated values*, while the values of x within the range of tabulated values are called *interpolated values* of x. Interpolation by these methods often yields accurate approximations of $f(x)$. Interpolation can also be used to estimate $f(x)$ at extrapolated values of x. However, extrapolation is often very unreliable.

TABLE 13.1 SAMPLE OF TABULATED DATA

x	$f(x)$
0.00	0.52
0.10	0.61
0.15	0.62
0.23	0.63
0.30	0.65

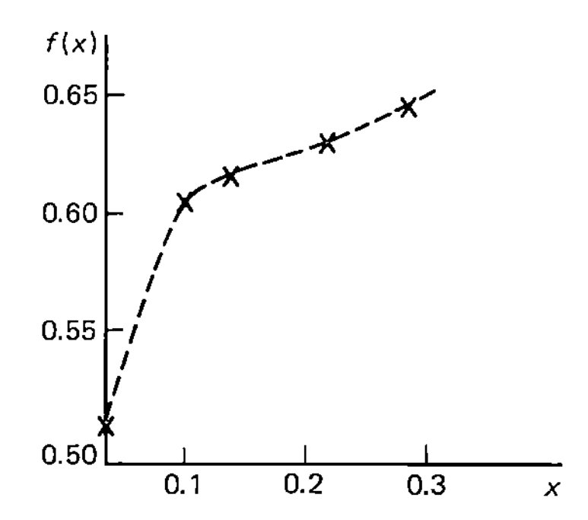

Figure 13.1 Graphical representation of data of Table 13.1.

The graphical method of interpolation involves drawing an approximate graph of the function, and reading an approximate value of $f(x)$ from the graph. For example, the graph of Figure 13.1 is a representation of the data in Table 13.1. From it, we estimate that $f(0.27) \simeq 0.64$.

One of the most widely used analytic methods of interpolation is called *polynomial interpolation*. This technique involves approximating the function by a polynomial over the range of values given.

In general, one can generate a polynomial of order $N - 1$ for a data set containing N points. For example, if five data points are specified, such as the data in Table 13.1, we write

$$f(x) = Ax^4 + Bx^3 + Cx^2 + Dx + E \tag{13.1}$$

The coefficients are determined by substituting each of the tabulated values for x, and $f(x)$. This assures us that $f(x)$ will be exact at the tabulated values of x. Since five values of $f(x)$ are known, we can determine five coefficients. That is why a fourth-order polynomial is used to approximate the function of Table 13.1. The method then results in five simultaneous equations for the coefficients. □

It is possible to obtain an approximation by a polynomial of order $N - 1$ without resorting to solving simultaneous equations. The method is known as *Lagrange interpolation*. Consider the $(N - 1)$th-order polynomial term

$$\mu_l(x) = \frac{(x - x_1)(x - x_2) \cdots (x - x_{l-1})(x - x_{l+1}) \cdots (x - x_N)}{(x_l - x_1)(x_l - x_2) \cdots (x_l - x_{l-1})(x_l - x_{l+1}) \cdots (x_l - x_N)} \tag{13.2}$$

where $x_1, x_2, \ldots, x_N$ are the x-values at which the function is known (e.g., the five x-values listed in Table 13.1).

We note that $\mu_l(x)$ has the property that

$$\begin{aligned} &\text{(a)} \quad \mu_l(x_k) = \begin{cases} 1 & k = l \\ 0 & k \neq l \end{cases} = \delta_{kl} \\ \Rightarrow \quad &\text{(b)} \quad \mu_l(x_k)f(x_k) = \begin{cases} f(x_l) & k = l \\ 0 & k \neq l \end{cases} = f(x_k)\delta_{kl} \end{aligned} \tag{13.3}$$

The Lagrange interpolation polynomial is a sum of such terms, one for each entry in the table. That is,

$$f(x) = \sum_{l=1}^{N} \mu_l(x) f(x_l) \tag{13.4}$$

From equation 13.3b we see that $f(x)$ will have the exact value listed in the table, at the corresponding tabulated values of x. If there are N table entries, each $\mu_l(x)$, and thus the interpolating polynomial, will be a polynomial of order $N - 1$. Unless the function varies wildly in between the tabulated values, the interpolation of equation 13.4 should yield a reasonably accurate approximation to the function at all interpolated values of x.

We note that for a five-point data set, the Lagrange interpolating polynomial will be fourth order. Therefore, the fourth-order Lagrange interpolation polynomial is equivalent to the polynomial of equation 13.1. It is obtained without the need to solve five simultaneous equations for coefficients.

Example 13.2

As an example, consider the subset of Table 13.1 shown in Table 13.2. The Lagrange interpolation of this table is the second-order polynomial

$$f(x) = \frac{(x - 0.10)(x - 0.15)}{(0.00 - 0.10)(0.00 - 0.15)} f(0.00) + \frac{(x - 0.00)(x - 0.15)}{(0.10 - 0.00)(0.10 - 0.15)} f(0.10) + \frac{(x - 0.00)(x - 0.10)}{(0.15 - 0.00)(0.15 - 0.10)} f(0.15) \quad \square \tag{13.5}$$

TABLE 13.2 SET OF THREE DATA POINTS FOR INTERPOLATION

x	$f(x)$
0.00	0.52
0.10	0.61
0.15	0.62

A polynomial interpolation can be achieved using a polynomial other than the one described in equation 13.2; a Sturm–Liouville polynomial, for example. In certain problems, a particular polynomial will be naturally convenient for approximating a function. Such a situation will arise when we discuss the numerical evaluation of certain types of integrals.

Let $Z_N(x)$ be a polynomial that has zeros $x_1, x_2, \ldots, x_N$. We will assume that $Z_N(x)$ has no repeated roots. We define the function

$$\mu_l(x) = \frac{Z_N(x)}{(x - x_l)Z_N'(x_l)} \tag{13.6}$$

where x_l is one of the zeros of $Z_N(x)$. Consider

$$\mu_l(x_k) = \frac{Z_N(x_k)}{(x_k - x_l)Z_N'(x_l)} \tag{13.7}$$

Because x_k is a root of $Z_N(x)$, $Z_N(x_k) = 0$. Since $Z_N(x)$ has no repeated roots, $(x_k - x_l) \neq 0$ for $l \neq k$ and thus $\mu_l(x_k) = 0$. If $l = k$, then $Z_N(x)/(x - x_l)$ is in the form 0/0. Therefore,

$$\lim_{x \to x_l} \frac{Z_N(x)}{(x - x_l)} = Z_N'(x_l) \tag{13.8}$$

so $\mu_l(x_l) = 1$. So, like the polynomials of equation 13.2,

$$\mu_l(x_k) = \delta_{kl} \tag{13.9}$$

Thus the polynomials $\mu_l(x)$ defined in equation 13.6 are equivalent to those defined in equation 13.2, and a polynomial interpolation over the polynomial $Z_N(x)$,

$$f(x) = \sum_{l=1}^{N} \mu_l(x) f(x_l) \tag{13.4}$$

is equivalent to the Lagrange interpolation. Like the interpolation over the polynomials defined in equation 13.2, this approximation guarantees that $f(x)$ will have the tabulated value $f(x_l)$ at $x = x_l$.

For some functions, Lagrange-like interpolation using a polynomial in x can be very inefficient for describing the function. For example, consider a polynomial approximation of a function that decreases rapidly at large values of x. If x is larger than the largest x_k used in the definition of $\mu_l(x)$ in equation 13.2, then $\mu_l(x)$ will increase as x increases. Similarly, if x is larger than the largest root of $Z_N(x)$, then $\mu_l(x)$, as defined in equation 13.6 will increase with increasing x. Multiplying large numbers arising from $\mu_l(x)$ by small values of $f(x)$ and adding these products may result in significant inaccuracies. In such cases, one might achieve a more reasonable fit by using a polynomial in $1/x$. If $f(x)$ decays exponentially at large x, a polynomial in e^{-x} or e^{-x^2} could lead to more accurate representation of $f(x)$.

Example 13.3

For example, consider a function that is to be interpolated over a region of large x, and in that region, the function decreases like $e^{-\alpha x}$. In analogy with equation 13.2, consider a term

$$\mu_l(x) = \frac{(e^{-x} - e^{-x_1})(e^{-x} - e^{-x_2}) \cdots (e^{-x} - e^{-x_{l-1}})(e^{-x} - e^{-x_{l+1}}) \cdots (e^{-x} - e^{-x_N})}{(e^{-x_l} - e^{-x_1})(e^{-x_l} - e^{-x_2}) \cdots (e^{-x_l} - e^{-x_{l-1}})(e^{-x_l} - e^{-x_{l+1}}) \cdots (e^{-x_l} - e^{-x_N})} \tag{13.10a}$$

where, as before, the points $x_1, \ldots, x_N$ are the x-values at which $f(x)$ is known. Like its polynomial counterpart, this term satisfies

$$\mu_l(x_k) = \delta_{lk} \tag{13.9}$$

Therefore, an exponentially decreasing function might be more accurately interpolated by

$$f(x) = \sum_{l=1}^{N} \mu_l(x) f(x_l) \tag{13.10b}$$

using the function $\mu_l(x)$ given in equation 13.10a rather than by one of the polynomials expressed in either equation 13.2 or 13.6. □

It is straightforward to generalize this argument to any form of interpolative function. Let $q(x)$ be a function that is appropriate for interpolating a function $G(x)$. Then, with

$$\mu_l(x) = \frac{[q(x) - q(x_1)][q(x) - q(x_2)] \cdots [q(x) - q(x_{l-1})][q(x) - q(x_{l+1})] \cdots [q(x) - q(x_N)]}{[q(x_l) - q(x_1)] \cdots [q(x_l) - q(x_{l-1})][q(x_l) - q(x_{l+1})] \cdots [q(x_l) - q(x_N)]} \tag{13.11a}$$

we would interpolate $G(x)$ as

$$G(x) = \sum_{l=1}^{N} \mu_l(x) G(x_l) \tag{13.11b}$$

Example 13.4

As an example, Table 13.3 on page 672 displays the results of interpolating the function $e^{-\alpha x}$ in $x \in [-1, 1]$, for specified values of $\alpha \in [-1, 1]$ using a 10-term polynomial interpolation. The points $x_1, x_2, \ldots, x_{10}$ are chosen to be equally spaced between -1 and 1.

As can be seen in Table 13.3a, the polynomial interpolation yields reasonable results. However, there are small but noticable inaccuracies. In Table 13.3b we list results obtained using a 10-term interpolation over e^{-x}. That is, $q(x) = e^{-x}$ in equation 13.11a. We see that the results in Table 13.3b are noticeably better than those obtained with the polynomial interpolation. □

TABLE 13.3

(a) Polynomial interpolation of $e^{-\alpha x}$

α		$x = 0.7$	$x = 0.1$	$x = 0$	$x = -0.3$
0.9	Interpolated	0.532605	0.914012	1.000037	1.309915
	Exact	0.532592	0.913931	1.000000	1.309964
0.3	Interpolated	0.810603	0.970533	1.000039	1.094132
	Exact	0.810584	0.970446	1.000000	1.094174
−0.1	Interpolated	1.072533	1.010141	1.000039	0.970407
	Exact	1.072508	1.010050	1.000000	0.970446
−0.5	Interpolated	1.419101	1.051365	1.000038	0.860675
	Exact	1.419068	1.051271	1.000000	0.860708

(b) Interpolation of $e^{-\alpha x}$ using $q(x) = e^{-x}$

α		$x = 0.7$	$x = 0.1$	$x = 0$	$x = -0.3$
0.9	Interpolated	0.532592	0.913931	1.000000	1.309964
	Exact	0.532592	0.913931	1.000000	1.309964
0.3	Interpolated	0.810584	0.970446	1.000000	1.094174
	Exact	0.810584	0.970446	1.000000	1.094174
−0.1	Interpolated	1.072508	1.010050	1.000001	0.970446
	Exact	1.072508	1.010050	1.000000	0.970446
−0.5	Interpolated	1.419068	1.051271	1.000017	0.860708
	Exact	1.419068	1.051271	1.000000	0.860708

Spline interpolation

A spline is a draftsman's device that can be bent so that more than two points can be connected even if the points do not lie on a line. A spline interpolation is an interpolation, in which one fits a function to a small subset of points in a data set by a polynomial.

Consider a graphical representation of a discrete set of data as shown in Figure 13.2. The simplest spline interpolation is to approximate $f(x)$ as a constant within a range around each point. For example, we can define

$$f(x) \equiv \tfrac{1}{2}[f(x_2) + f(x_1)] \qquad x_1 \leqslant x \leqslant x_2 \tag{13.12a}$$

$$f(x) = \tfrac{1}{2}[f(x_3) + f(x_2)] \qquad x_2 \leqslant x \leqslant x_3 \tag{13.12b}$$

$$\Rightarrow \quad f(x) = \tfrac{1}{2}[f(x_{l+1}) + f(x_l)] \qquad x_l \leqslant x \leqslant x_{l+1} \tag{13.13}$$

This least accurate spline interpolation is called the *cardinal spline*. It is described graphically in Figure 13.3.

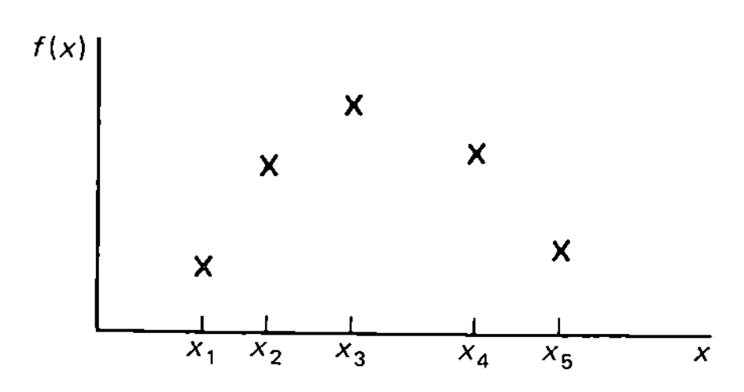

Figure 13.2 Set of discrete data points.

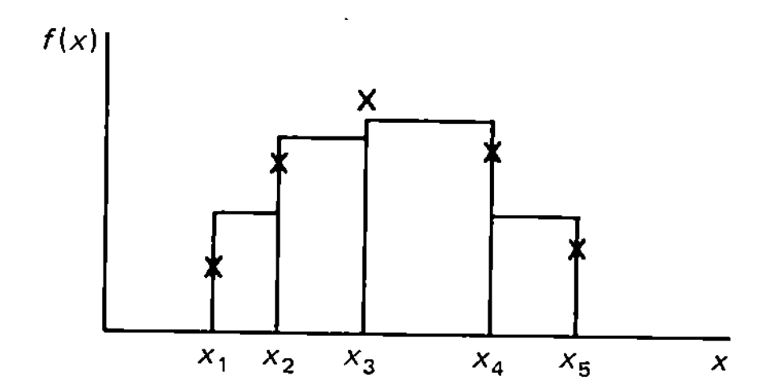

Figure 13.3 Cardinal spline interpolation of a discrete set of data points.

One glaring defect of the cardinal spline is that it makes the function discontinuous at the given data points. Thus, as noted in Chapter 4 (see equation 4.162), the first derivative of $f(x)$ has a δ-function singularity at each point.

The next-higher-order spline is the *linear spline*. For this interpolation, we approximate $f(x)$ by a straight line between two adjacent points.

$$f(x) = a_l x + b_l \qquad x_l \leqslant x \leqslant x_{l+1} \tag{13.14}$$

$$\Rightarrow \quad \text{(a)} \quad f(x_l) = a_l x_l + b_l \qquad \text{(b)} \quad f(x_{l+1}) = a_l x_{l+1} + b_l \tag{13.15}$$

for each interval. The graphical representation of this spline is shown in Figure 13.4. As can be seen, this interpolation yields a continuous $f(x)$. However, the first derivative of $f(x)$ is discontinuous at each x_l, and thus the second derivative has a δ-function singularity at each x_l.

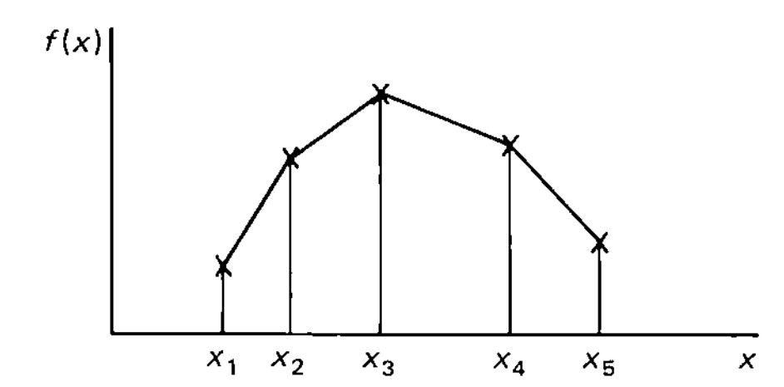

Figure 13.4 Linear spline interpolation of a discrete set of data points.

For large data sets, greater accuracy is obtained by using higher-order splines. The quadratic spline fits the function to a quadratic polynomial using three successive values of x. That is,

$$f(x) = a_l x^2 + b_l x + c_l \qquad x_l \leqslant x \leqslant x_{l+2} \tag{13.16}$$

$$\Rightarrow \quad \text{(a)} \quad f(x_l) = a_l x_l^2 + b_l x_l + c_l \qquad \text{(b)} \quad f(x_{l+1}) = a_l x_{l+1}^2 + b_l x_{l+1} + c_l$$

$$\text{(c)} \quad f(x_{l+2}) = a_l x_{l+2}^2 + b_l x_{l+2} + c_l \tag{13.17}$$

Example 13.5

As an example, consider the five data points of Table 13.1. By approximating $f(x)$ as a linear function

$$f(x) = ax + b \tag{13.18}$$

over each interval, we obtain

$$f(x) = 0.9000x + 0.5200 \qquad 0.00 \leqslant x \leqslant 0.10 \tag{13.19a}$$

$$f(x) = 0.2000x + 0.5900 \qquad 0.10 \leqslant x \leqslant 0.15 \tag{13.19b}$$

$$f(x) = 0.1250x + 0.6013 \qquad 0.15 \leqslant x \leqslant 0.23 \tag{13.19c}$$

$$f(x) = 0.2857x + 0.5673 \qquad 0.23 \leqslant x \leqslant 0.30 \tag{13.19d}$$

The quadratic spline description of the function is determined by solving for the coefficients a_l, b_l, and c_l in each of the two intervals. The result is

$$f(x) = -4.667x^2 + 0.433x + 0.520 \qquad 0 \leqslant x \leqslant 0.15 \tag{13.20a}$$

$$f(x) = 1.071x^2 - 0.282x + 0.638 \qquad 0.15 \leqslant x \leqslant 0.30 \quad \square \tag{13.20b}$$

An easier way to generate a higher-order spline interpolation is to use a Lagrange interpolation over the limited range specified by the spline interpolation. For example, the quadratic spline of equation 13.16 can be written as

$$f(x) = \frac{(x - x_{l+1})(x - x_{l+2})}{(x_l - x_{l+1})(x_l - x_{l+2})} f(x_l) + \frac{(x - x_l)(x - x_{l+2})}{(x_{l+1} - x_l)(x_{l+1} - x_{l+2})} f(x_{l+1})$$
$$+ \frac{(x - x_l)(x - x_{l+1})}{(x_{l+2} - x_l)(x_{l+2} - x_{l+1})} f(x_{l+2}) \qquad x_l \leqslant x \leqslant x_{l+2} \tag{13.21}$$

In this way, the algebra required to find the coefficients in equation 13.16 is avoided.

The cubic spline is one of the more frequently used interpolation methods. A cubic spline interpolation approximates $f(x)$ by a third-order polynomial over four successive data points. For most data sets, the cubic spline is sufficient to describe the function accurately. If one uses a very high-order spline interpolation, one is approaching the Lagrange approximation of interpolating $f(x)$ over the entire data set with one polynomial.

Like the Lagrange approach, the spline approximation usually gives a reasonably good estimate of $f(x)$ at interpolated values of x. However, as with Lagrange interpolation, extrapolation beyond the data set is unreliable.

Just as we could generate a Lagrange-like interpolation in terms of a general function $q(x)$ as in equations 13.11, we can generate a spline-like interpolation over a limited range of x using a general function $q(x)$. For example, a spline-like interpolation of $G(x)$ using the function $q(x)$ over pairs of successive data points is of the form

$$G(x) = \frac{[q(x) - q(x_{l+1})]}{[q(x_l) - q(x_{l+1})]} \frac{[q(x) - q(x_{l+2})]}{[q(x_l) - q(x_{l+1})]} G(x_l)$$
$$+ \frac{[q(x) - q(x_l)]}{[q(x_{l+1}) - q(x_l)]} \frac{[q(x) - q(x_{l+2})]}{[q(x_{l+1}) - q(x_{l+2})]} G(x_{l+1})$$
$$+ \frac{[q(x) - q(x_l)]}{[q(x_{l+2}) - q(x_l)]} \frac{[q(x) - q(x_{l+1})]}{[q(x_{l+2}) - q(x_{l+1})]} G(x_{l+2}) \tag{13.22}$$

for $x_l \leqslant x \leqslant x_{l+2}$.

Operator interpolation

An operator interpolation of $f(x)$ can be developed if the data points are equally spaced. Let

$$x_{n+1} - x_n \equiv h \tag{13.23}$$

Since the points are equally spaced, h is independent of the index n. We define an operator E, called the *raising operator*, such that

$$Ef(x_n) = f(x_{n+1}) \tag{13.24}$$

We also define a *difference operator*, Δ, such that

$$\text{(a)} \quad \Delta f(x_n) = f(x_{n+1}) - f(x_n) \quad \Rightarrow \quad \text{(b)} \quad \Delta f(x_n) = Ef(x_n) - f(x_n) = (E - 1)f(x_n) \tag{13.25}$$

$$\Rightarrow \quad \Delta = (E - 1) \tag{13.26}$$

Consider successive applications of E on $f(x_1) \equiv f_1$.

$$\text{(a)} \quad Ef_1 = f_2 = f(x_1 + h) \qquad \text{(b)} \quad E^2 f_1 = Ef_2 = f(x_1 + 2h) \cdots \qquad \text{(c)} \quad E^n f_1 = f(x_1 + nh) \tag{13.27}$$

For n = integer, $x_1 + nh$ is one of the data points.

To interpolate to a point not in the data set, we let $n \to \alpha$, $\alpha \neq$ integer. That is, with $x = x_1 + \alpha h$ and referring to equation 13.26, equation 13.27c is replaced by

$$f(x) = E^\alpha f_1 = (1 + \Delta)^\alpha f_1 \tag{13.28}$$

We then expand the operator $(1 + \Delta)^\alpha$ in a binomial series to obtain

$$f(x) = \left(1 + \alpha\Delta + \frac{\alpha(\alpha - 1)}{2!}\Delta^2 + \frac{\alpha(\alpha - 1)(\alpha - 2)}{3!}\Delta^3 + \cdots\right)f_1 \tag{13.29}$$

To evaluate the various powers of Δ operating on f_1, consider

$$\text{(a)}\quad \Delta f_1 = f_2 - f_1 \qquad \text{(b)}\quad \Delta^2 f_1 = f_3 - f_2 - f_2 + f_1 = f_3 - 2f_2 + f_1$$
$$\text{(c)}\quad \Delta^3 f_1 = f_4 - 3f_3 + 3f_2 - f_1 \tag{13.30}$$

etc.

If the differences $f_{l+1} - f_l$ are small, this operator method will yield accurate results.

As can be seen from the pattern for $\Delta^k f_1$, the coefficients of the various tabulated f-values are the coefficients of a binomial expansion. This allows us to deduce a recurrence relation similar to the one developed for the Bernoulli and Euler numbers. If we expand $(f - 1)^k$ in a binomial series, then replace f^n in the expansion by f_{n+1}, the recurrence relation becomes

$$\Delta^k f_1 = (f - 1)^k \tag{13.31}$$

For example

$$\Delta^3 f_1 = (f - 1)^3 = f^3 - 3f^2 + 3f^1 - f^0 \to f_4 - 3f_3 + 3f_2 - f_1 \tag{13.32}$$

which agrees with $\Delta^3 f_1$ as expressed in equation 13.30c.

From this recurrence relation, we see that if the data set contains N points, f_N will arise from $\Delta^{N-1}f_1$. The binomial series of equation 13.29 is an infinite series for $\alpha \neq$ integer, but because there are a finite number of data points, and since f_N is the last entry in the table, the series of equation 13.29 must be terminated at the Δ^{N-1} term.

Example 13.6

As an example, consider the tabulated form of e^x shown in Table 13.4. To find $e^{0.33}$, we note that

$$\text{(a)}\quad x_1 = 0 \qquad \text{(b)}\quad h = 0.1 \quad \Rightarrow \quad \text{(c)}\quad x_1 + \alpha h = 0.33 \quad \Rightarrow \quad \text{(d)}\quad \alpha = 3.3 \tag{13.33}$$

Since there are five entries in Table 13.4, the expansion of $(1 + \Delta)^\alpha$ must be terminated at Δ^4. Therefore,

$$e^{0.33} \simeq \left(1 + 3.3\Delta + \frac{(3.3)(2.3)}{2!}\Delta^2 + \frac{(3.3)(2.3)(1.3)}{3!}\Delta^3 + \frac{(3.3)(2.3)(1.3)(0.3)}{3!}\Delta^4\right)f_1 \tag{13.34}$$

TABLE 13.4 REPRESENTATIVE VALUES OF e^x

x	e^x
0	1.000000
0.1	1.105171
0.2	1.221403
0.3	1.349859
0.4	1.491825

Referring to Table 13.4, and using the recurrence relation expressed in equation 13.31,

$$\text{(a)} \quad f_1 = 1.000000 \qquad \text{(b)} \quad \Delta f_1 = f_2 - f_1 = 0.105171 \qquad \text{(c)} \quad \Delta^2 f_1 = f_3 - 2f_2 + f_1 = 0.011061$$

$$\text{(d)} \quad \Delta^3 f_1 = f_4 - 3f_3 + 3f_2 - f_1 = 0.001163 \tag{13.35}$$

$$\text{(e)} \quad \Delta^4 f_1 = f_5 - 4f_4 + 6f_3 - 4f_2 + f_1 = 0.000122$$

$$\Rightarrow \quad e^{0.33} \simeq 1.390968 \tag{13.36}$$

which agrees with the exact value to six decimal places. □

Taylor series by operator method

A Taylor series estimate of $f(x)$ can be developed using the E and Δ operators. Let us expand $f(x_1 + \alpha h)$ in a Taylor series around x_1. For a data set containing N points, we will truncate the series at the $(N - 1)$th derivative. As will be seen later, the derivative will be approximated by powers of Δ. Since we can evaluate powers of the difference operator no higher than $\Delta^{N-1} f_1$, we will not be able to estimate derivatives higher than $D^{N-1} f(x_1)$.

The Taylor expansion of $f(x_1 + \alpha h)$ is

$$f(x_1 + \alpha h) = f(x_1) + \alpha h f'(x_1) + \frac{\alpha^2 h^2}{2!} f''(x_1) + \cdots = \sum_{n=0}^{\infty} \frac{\alpha^n h^n}{n!} D^n f(x_1) \tag{13.37}$$

To approximate $D^n f(x_1)$, consider

$$f_{k+1} = E f_k = f(x_k + h) \tag{13.38}$$

Expanding this in a Taylor series around x_k, we obtain

$$E f_k = f(x_k) + hDf(x_k) + \frac{h^2}{2!} D^2 f(x_k) + \cdots = \sum_{l=0}^{\infty} \frac{(hD)^l}{l!} f(x_k) \tag{13.39a}$$

$$\Rightarrow \quad E = \sum_{l=0}^{\infty} \frac{(hD)^l}{l!} = e^{hD} \tag{13.39b}$$

Like the function of a matrix operator, the function of any operator is simply a shorthand notation for a series involving the operator. Thus, from equation 13.39b, we have

$$D = \frac{1}{h} \log E = \frac{1}{h} \log(1 + \Delta) = \frac{1}{h} \left(\Delta - \frac{\Delta^2}{2} + \frac{\Delta^3}{3} - \cdots \right) \tag{13.40}$$

Since we can only evaluate powers of Δ operating on $f(x_1)$ up to Δ^{N-1}, the series of equation 13.40 is truncated at Δ^{N-1} for a data set of N points.

Example 13.7

To illustrate, we again consider the approximation to $e^{0.33}$ using the data of Table 13.4. For this data set, we will truncate all expansions at fourth order. Therefore, equation 13.37 becomes

$$e^x = f(x) \simeq f(x_1) + \alpha h D f(x_1) + \frac{\alpha^2 h^2}{2!} D^2 f(x_1) + \frac{\alpha^3 h^3}{3!} D^3 f(x_1) + \frac{\alpha^4 h^4}{4!} D^4 f(x_1) \tag{13.41}$$

where, from equation 13.40, omitting all powers of Δ higher than Δ^4,

$$hDf(x_1) \approx \left(\Delta - \frac{\Delta^2}{2} + \frac{\Delta^3}{3} - \frac{\Delta^4}{4}\right)f(x_1) \tag{13.42a}$$

$$h^2D^2f(x_1) \approx \left(\Delta - \frac{\Delta^2}{2} + \frac{\Delta^3}{3} - \frac{\Delta^4}{4}\right)^2 f(x_1) \approx \left(\Delta^2 - \Delta^3 + \frac{11}{12}\Delta^4\right)f(x_1) \tag{13.42b}$$

$$h^3D^3f(x_1) \approx \left(\Delta - \frac{\Delta^2}{2} + \frac{\Delta^3}{3} - \frac{\Delta^4}{4}\right)^3 f(x_1) \approx \left(\Delta^3 - \frac{1}{2}\Delta^4\right)f(x_1) \tag{13.42c}$$

$$h^4D^4f(x_1) \approx \left(\Delta - \frac{\Delta^2}{2} + \frac{\Delta^3}{3} - \frac{\Delta^4}{4}\right)^4 f(x_1) \approx \Delta^4 f(x_1) \tag{13.42d}$$

With these approximations, and the values of e^x from Table 13.4, equation 13.41 becomes

$$e^{0.33} \approx f_1 + \alpha\left(\Delta f_1 - \frac{\Delta^2}{2}f_1 + \frac{\Delta^3}{3}f_1 - \frac{\Delta^4}{4}f_1\right) + \frac{\alpha^2}{2}\left(\Delta^2 f_1 - \Delta^3 f_1 + \frac{2}{3}\Delta^4 f_1\right)$$
$$+ \frac{\alpha^3}{6}\left(\Delta^3 f_1 - \frac{3}{2}\Delta^4 f_1\right) + \frac{\alpha^4}{24}\Delta^4 f_1 = 1.390802 \tag{13.43}$$

which is not quite as accurate as the estimate obtained using the expansion of equation 13.29. □

13.2 Least Squares Curve Fitting

In many cases, theoretical analysis leads one to expect tabulated data from an experiment, when represented graphically, to describe a particular type of function. However, because of inherent uncertainties in measurement, the data points will not fall exactly on the expected curve.

As a simple example, adding a mass M to a spring will stretch the spring a distance x from equilibrium, the weight Mg being the deforming force causing the displacement. Hooke's law for this system, expressed in terms of the stiffness (spring) constant k, is

$$Mg = kx \tag{13.44}$$

Therefore, a graph of mass versus displacement is expected to be a straight line, and the spring constant can be determined from the slope of the linear graph.

Because experimental fluctuations will cause the points to deviate somewhat from a straight line, if $M(x)$ is described by a polynomial, the graph of M versus x will not be a straight line, but would typically look like the dashed curve of Figure 13.5. Therefore, interpolation is not a valid way to describe the data. But if one constructs a

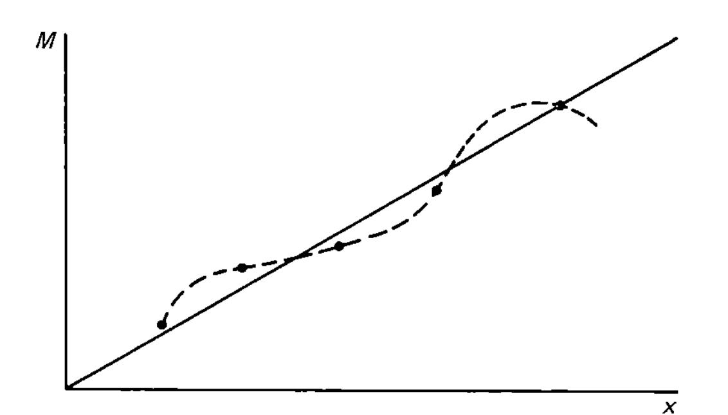

Figure 13.5 Typical graphical description of mass versus displacement. Dashed curve is interpolated polynomial; solid line is linear fit to data.

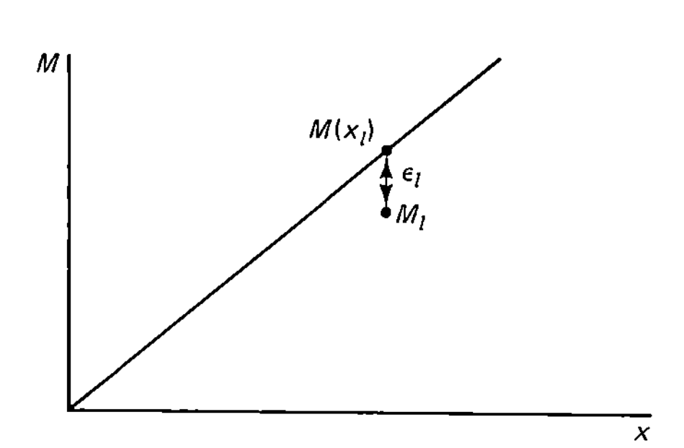

Figure 13.6 Typical deviation of measured value from expected functional value.

best fit to the expected straight line, one has a better graphical representation of the data, and the spring constant can be found from the slope of that best straight line. Such a best straight line is the solid curve shown in Figure 13.5.

Least squares criterion

The criterion for what is meant by the best fit to a curve is called the *principle of least squares*. Let F_l be the measured value of a function at a point x_l. If the N data points are expected to lie on a curve $F(x)$, then, due to experimental deviations of measured data, in general,

$$F(x_l) \neq F_l \tag{13.45}$$

The deviation (error) in the lth value of F is

$$\varepsilon_l \equiv F(x_l) - F_l \tag{13.46}$$

A typical error is shown in Figure 13.6 for the mass–spring system. The least squares criterion states that the best curve to fit the data is that which makes the function

$$E \equiv \sum_{l=1}^{N} \varepsilon_l^2 = \sum_{l=1}^{N} [F(x_l) - F_l]^2 \tag{13.47}$$

a minimum. That is achieved by defining $F(x)$ in terms of adjustable parameters, and minimizing E with respect to those parameters.

Therefore, one major difference between describing a function by interpolation, and describing that function by a least squares optimum curve is that the interpolated function

$$F(x) = \sum_{l=1}^{N} \mu_l(x) F_l \tag{13.4}$$

requires (guarantees) that $F(x)$ will have the exact value of the measured F_l at $x = x_l$. A least squares curve does not require that $F(x_l)$ match the measured F_l at any point.

Returning to the example of the mass–spring system, the best fit curve $F(x) = M(x)$ is a straight line, which we parametrize by

$$M(x) = \alpha x + \beta \tag{13.48}$$

$$\Rightarrow \quad E = \sum_{l=1}^{N} (\alpha x_l + \beta - M_l)^2 \tag{13.49}$$

be a minimum. For such a linear curve, there are two undetermined parameters, α and β. To minimize E, we set

$$\text{(a)} \quad \frac{\partial E}{\partial \alpha} = 2\sum_{l=1}^{N} x_l(\alpha x_l + \beta - M_l) = 0 \qquad \text{(b)} \quad \frac{\partial E}{\partial \beta} = 2\sum_{l=1}^{N} (\alpha x_l + \beta - M_l) = 0 \tag{13.50}$$

from which the values of α and β are determined.

Average value and justification of the criterion

To justify that the principle of least squares is the correct method for finding the best curve, consider the simple problem of determining the average value $\bar{x}$ for a set of N numbers $\{x_l\}$. The expected value of any measurement is $\bar{x}$. That is, the expected curve is

$$F(x) = \bar{x} = \text{constant} \tag{13.51}$$

Thus $\bar{x}$ becomes the only adjustable parameter in the equation for the expected curve. But in general, the lth data point will differ from $\bar{x}$ by $(x_l - \bar{x})$. By definition, the average value is that value of x such that there is as much data that is greater than $\bar{x}$ as there is data less then $\bar{x}$. Therefore, if we add the errors (deviations from $\bar{x}$), we obtain

$$\sum_{l=1}^{N} (x_l - \bar{x}) = 0 \tag{13.52}$$

The reason this sum is zero is because there are negative errors that cancel positive errors. Thus, to get a sense of the size of the errors, the negative signs must be eliminated. One way to achieve this is to square the individual errors. Therefore, for the average value,

$$E = \sum_{l=1}^{N} (x_l - \bar{x})^2 \tag{13.53}$$

Since $\bar{x}$ is the only adjustable parameter, we set

$$\text{(a)} \quad \frac{\partial E}{\partial \bar{x}} = -2\sum_{l=1}^{N} (x_l - \bar{x}) = 0 \quad \Rightarrow \quad \text{(b)} \quad \sum_{l=1}^{N} x_l = \bar{x}\sum_{l=1}^{N} 1 = N\bar{x} \tag{13.54}$$

Thus with this description of E, we obtain the expected result,

$$\bar{x} = \frac{1}{N}\sum_{l=1}^{N} x_l \tag{13.55}$$

From equation 13.54a, we obtain

$$\frac{\partial^2 E}{\partial \bar{x}^2} = 2\sum_{l=1}^{N} 1 = 2N > 0 \tag{13.56}$$

we see that E is a minimum for this definition.

It is also possible to eliminate the negative signs on the errors by other means. For example, we could define

$$E' \equiv \sum_{l=1}^{N} |\varepsilon_l| = \sum_{l=1}^{N} |F(x_l) - F_l| \tag{13.57}$$

Applying this to the example of the average value, this becomes

$$E' = \sum_{l=1}^{N} |x_l - \bar{x}| = \sum_{l=1}^{N} \left[(x_l - \bar{x})^2\right]^{1/2} \tag{13.58}$$

$$\Rightarrow \quad \frac{\partial E'}{\partial \bar{x}} = -\sum_{l=1}^{N'} \frac{(x_l - \bar{x})}{|x_l - \bar{x}|} = -\sum_{l=1}^{N'} \text{sign}(x_l - \bar{x}) = 0 \tag{13.59}$$

where the meaning of N' will be discussed shortly.

Equation 13.59 requires that $\bar{x}$ be that point such that there are as many data points larger than $\bar{x}$ as there are data points smaller than $\bar{x}$ without regard to the sizes of the differences. This is the median of the data, which is not the same as the average.

Example 13.8

For example, for the data set

$$x = \{1.0, 1.1, 7.0, 9.5, 10.3\}\bar{x} = 7.0. \tag{13.60}$$

which has an odd number of points, $\bar{x} = 7.0$. This is the only choice that yields two data points with $(x_l - \bar{x}) > 0$ and two points with $(x_l - \bar{x}) < 0$. Clearly, by this criterion, $\bar{x}$

would be the same for the data set

$$x = \{0.3, 5.7, 7.0, 7.1, 33.8\} \tag{13.60b}$$

This is not what is meant by an average value.

From this example we note that for a data set with an odd number of points, $\bar{x}$ has one value, the central point of the data set. Since the difference $(x_{\text{center}} - \bar{x})$ is zero, the central point is omitted from the sum for E' so that $\partial E'/\partial \bar{x}$ does not have a term of the form $0/0$. Thus for an odd number of points, $N' = N - 1$.

Now consider the set of four points

$$x = \{1.0, 1.1, 7.0, 9.5\} \tag{13.61}$$

In order that there be an equal number of positive errors ε_l as there are negative errors, $\bar{x}$ can have any value in the range

$$1.1 < \bar{x} < 7.0 \tag{13.62}$$

which is clearly not the meaning of an average value.

From this example it is straightforward to deduce that for a data set with an even number of points, E' can be minimized for a range of $\bar{x}$-values. Since $\bar{x}$ cannot take on the value of any point in the data set, none of the terms are omitted from equation 13.59 and thus $N' = N$. □

The negative sign on the errors can also be eliminated by defining

$$E'' \equiv \sum_{l=1}^{N} (F(x_l) - F_l)^4 \tag{13.63}$$

Using the previous examples as guides, it is straightforward to show that when applied to the average value, one finds that minimizing E'' leads to a value of $\bar{x}$ that is the solution to

$$\sum_{l=1}^{N} x_l^3 - 3\bar{x} \sum_{l=1}^{N} x_l^2 + 3\bar{x}^2 \sum_{l=1}^{N} x_l - \bar{x}^3 = 0 \tag{13.64}$$

Again, this is not how an average value is defined.

From these examples, we see that the principle of least squares (minimizing the error function of equation 13.47) is the only criterion that yields the correct expression for the average value.

Example 13.9

To illustrate the principle of least squares, let the data of Table 13.5 represent data for the displacement of a spring for various masses. We define the straight line fit to these data by

$$M(x) = \alpha x + \beta \tag{13.48}$$

$$\Rightarrow \quad E = \sum_{l=1}^{4} (\alpha x_l + \beta - M_l)^2 \tag{13.65}$$

TABLE 13.5 HYPOTHETICAL DATA FOR THE STRETCHING OF A SPRING BY VARIOUS MASSES

x (cm)	M (grams)
3.0	12.3
6.0	30.5
9.0	60.0
12.0	77.5

The values of α and β are determined from

$$\frac{\partial E}{\partial \alpha} = 2\left[\alpha \sum_{l=1}^{4} x_l^2 + \beta \sum_{l=1}^{4} x_l - \sum_{l=1}^{4} x_l M_l\right] = 0 \tag{13.66a}$$

$$\frac{\partial E}{\alpha \beta} = 2\left[\alpha \sum_{l=1}^{4} x_l + \beta \sum_{l=1}^{4} 1 - \sum_{l=1}^{4} M_l\right] = 0 \tag{13.66b}$$

For the data in Table 13.5, these equations become

$$\text{(a)} \quad 270\alpha + 30\beta = 1689.9 \qquad \text{(b)} \quad 30\alpha + 4\beta = 180.3 \tag{13.67}$$

Equations 13.66, or equivalently equations 13.67, are referred to as the normal equations for the linear fit. From them we obtain $\alpha = 7.5$ and $\beta = -11.2$. Therefore, the best linear fit to the data of Table 13.5 is

$$M(x) = 7.5x - 11.2 \tag{13.68}$$

The slope of the mass versus displacement curve is k/g, so that for these data, the best estimate of the spring constant is

$$k = \alpha g = 7350.0 \text{ dynes/cm} \tag{13.69}$$

and the best estimate of the vertical intercept, which should be zero, is -11.2. Figure 13.7 displays the data of Table 13.5 and the best straight-line fit to that data.

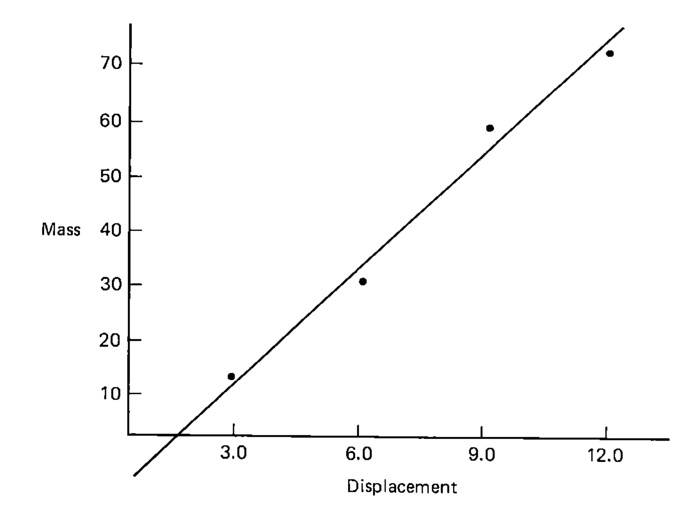

Figure 13.7 Best linear fit to Hooke's law data.

The minimum error, called the *root-mean square* (rms) *error*, is defined by

$$\varepsilon_{\text{rms}} = \left(\frac{E}{N}\right)^{1/2} = \left[\frac{1}{N}\sum_{l=1}^{N} (F(x_l) - F_l)^2\right]^{1/2} \tag{13.70}$$

For the data of Table 13.5,

$$\varepsilon_{\text{rms}} = \left[\tfrac{1}{4}\sum_{l=1}^{4} (7.5x_l - 11.2 - M_l)^2\right]^{1/2} = 2.6 \quad \square \tag{13.71}$$

It is straightforward to see how the least squares process is extended to the determination of the best fit of data to a higher-order polynomial.

Example 13.10

For example, we will fit the data shown in Table 13.6 to a parabola by defining

$$F(x) = \alpha_0 + \alpha_1 x + \alpha_2 x^2 \tag{13.72}$$

TABLE 13.6 DATA TO BE FIT TO A PARABOLA

x	F
0.2	0.275
0.4	1.405
0.6	3.411
0.8	5.019
1.0	8.888

This is accomplished by determining the values of α_0, α_1, and α_2 that minimize

$$E = \sum_{l=1}^{5} \left(\alpha_0 + \alpha_1 x_l + \alpha_2 x_l^2 - F_l\right)^2 \tag{13.73}$$

Setting

$$\frac{\partial E}{\partial \alpha_0} = 2\sum_{l=1}^{5} (\alpha_0 + \alpha_1 x_l + \alpha_2 x_l^2 - F_l) = 2\left[5\alpha_0 + \alpha_1 \sum_{l=1}^{5} x_l + \alpha_2 \sum_{l=1}^{5} x_l^2 - \sum_{l=1}^{5} F_l\right] = 0 \tag{13.74a}$$

$$\frac{\partial E}{\partial \alpha_1} = 2\left[\alpha_0 \sum_{l=1}^{5} x_l + \alpha_1 \sum_{l=1}^{5} x_l^2 + \alpha_2 \sum_{l=1}^{5} x_l^3 - \sum_{l=1}^{5} x_l F_l\right] = 0 \tag{13.74b}$$

$$\frac{\partial E}{\partial \alpha_2} = 2\left[\alpha_0 \sum_{l=1}^{5} x_l^2 + \alpha_1 \sum_{l=1}^{5} x_l^3 + \alpha_2 \sum_{l=1}^{5} x_l^4 - \sum_{l=1}^{5} x_l^2 F_l\right] = 0 \tag{13.74c}$$

results in the set of coupled equations

$$\text{(a)}\quad 5\alpha_0 + 3\alpha_1 + 2.2\alpha_2 = 18.988 \qquad \text{(b)}\quad 3\alpha_0 + 2.2\alpha_1 + 1.8\alpha_2 = 15.567$$

$$\text{(c)}\quad 2.2\alpha_0 + 1.8\alpha_1 + 1.57\alpha_2 = 13.564 \tag{13.75}$$

From the solution for the α-parameters, we find the best parabolic fit to be

$$F(x) = 0.0876 - 0.4657x + 9.0714x^2 \tag{13.76}$$

The data of Table 13.6 and the best-fit parabola are displayed in Figure 13.8. □

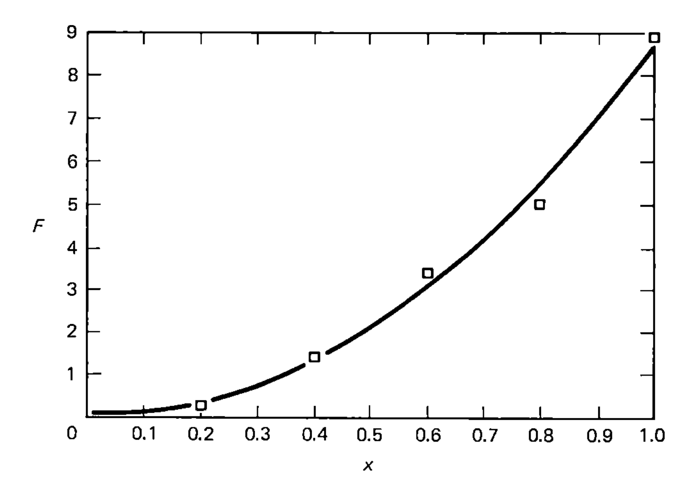

Figure 13.8 Best parabolic fit to the data of Table 13.6.

From these discussions it is easy to see that a set of data points $\{(x_l, F_l), 1 \leqslant l \leqslant N\}$ can be fit to a polynomial of order M,

$$F(x) = \alpha_0 + \alpha_1 x + \alpha_2 x^2 + \cdots + \alpha_M x^M \tag{13.77}$$

by minimizing

$$E = \sum_{l=1}^{N} \left(\alpha_0 + \alpha_1 x_l + \alpha_2 x_l^2 + \cdots + \alpha_M x_l^M - F_l\right)^2 \tag{13.78}$$

with respect to the α-parameters. This procedure results in a set of M simultaneous equations for these parameters, which are then solved by matrix inversion (or other techniques) as described in Chapter 8.

Clearly, the number of points in the data set is independent of the order of the polynomial to which the data are to be fit. Thus the order of the polynomial can be larger than the size of the data set. This is another difference between a least squares polynomial and one obtained by interpolation. The order of an interpolation polynomial to describe a set of N data points can be no larger than $N - 1$.

It is not necessary to restrict the functional dependence of the best curve to a polynomial. The least squares principle can be applied to fitting data to a more general function

$$F(x) = \sum_{k=1}^{M} \alpha_k q_k(x) \tag{13.79}$$

where each $q_k(x)$ is a specified function.

Example 13.11

TABLE 13.7 DATA TO BE FIT TO A POLYNOMIAL IN e^x

x	$F \simeq e^{3x/2}$
0.2	1.351
0.4	1.820
0.6	2.457
0.8	3.322
1.0	4.482

For example, the data shown in Table 13.7 might be better described by taking

$$\text{(a)}\quad q_k(x) = e^{kx} \quad \text{rather than} \quad \text{(b)}\quad q_k(x) = x^k \tag{13.80}$$

Taking

$$F(x) = \alpha_0 + \alpha_1 e^x + \alpha_2 e^{2x} \tag{13.81}$$

$$\Rightarrow \quad E = \sum_{l=1}^{5} \left[\alpha_0 + \alpha_1 e^{x_l} + \alpha_2 e^{2x_l} - F_l\right]^2 \tag{13.82}$$

Minimizing E, leads to

$$F(x) = -0.3045 + 1.0196 e^x + 0.2730 e^{2x} \tag{13.83}$$

with an rms error of 0.0036. An analogous fit of these data to a parabola, by minimizing

$$E = \sum_{l=1}^{5} \left(\alpha_0 + \alpha_1 x_l + \alpha_2 x_l^2 - F_l\right)^2 \tag{13.84}$$

$$\Rightarrow \quad F(x) = 1.1622 + 0.4320x + 2.8750x^2 \tag{13.85}$$

with an rms error of 0.0180, which is four times larger than the rms error incurred using the sum of exponentials in equation 13.83. □

If data in an N-point data set are odd and periodic (of period $\lambda = 2\pi/k$), it might be best to fit it to

$$F(x) = \sum_{m=1}^{M} \alpha_m \sin(mkx) \tag{13.86}$$

by minimizing

$$E = \sum_{l=1}^{N} \left[\sum_{m=1}^{M} \alpha_m \sin(mkx_l) - F_l \right]^2 \tag{13.87}$$

One might try to fit a set of data that decreases with increasing x to the function

$$F(x) = \sum_{m=0}^{M} \alpha_m \frac{1}{x^m} \tag{13.88a}$$

by minimizing

$$E = \sum_{l=1}^{N} \left[\sum_{m=0}^{M} \alpha_m \frac{1}{x_l^m} - F_l \right]^2 \tag{13.88b}$$

It might be more appropriate to fit such decreasing data to

$$F(x) = \sum_{m=0}^{M} \alpha_m e^{-mx} \tag{13.89a}$$

by making

$$E = \sum_{l=1}^{N} \left[\sum_{m=0}^{M} \alpha_m e^{-mx_l} - F_l \right]^2 \tag{13.89b}$$

a minimum (see Problem 6).

In some instances a set of data may contain points that are not as reliable as other data in the set, but ones that the experimenter cannot discard. As such, these less reliable points may be weighted less than their more reliable counterparts by assigning a weight to each data point and defining a weighted error function

$$E_w \equiv \sum_{l=1}^{N} w_l (F(x_l) - F_l)^2 \tag{13.90}$$

The weights are determined by the experimenter based on an estimate of the reliability of each data point. They are not additional adjustable parameters to be used to minimize E_w. As with unweighted data, all adjustable parameters are contained in the function $F(x)$ that defines the curve.

The least squares criterion can also be used to approximate a continuous function $G(x)$ by a polynomial. Several examples of Sturm–Liouville functions were presented in Chapter 6, many of which were polynomials. One of the properties of the sets of Sturm–Liouville polynomials $\{U_l(x)\}$ is that they are real and mutually orthogonal.

Therefore, as expressed in equations 6.141b and 6.153, they satisfy

$$\langle U_l | U_m \rangle \equiv \int_a^b \rho(x) U_l(x) U_m(x)\, dx = N_l \delta_{lm} \tag{13.91}$$

If we wish to approximate a real function $G(x)$ by a polynomial of order M, we can write the trial function as

$$G(x) \simeq \sum_{m=1}^{M} \alpha_m U_m(x) \tag{13.92}$$

where the range of x for $U_m(x)$ is that for $G(x)$. For example if $G(x)$ is defined over the range $x \in [-1, 1]$, one would take $U_l(x)$ to be the Legendre polynomial $P_l(x)$, or perhaps the Gegenbauer or Chebyscheff polynomial (see Problem of Chapter 6), which are defined over the range $x \in [-1, 1]$ rather than a Laguerre polynomial for which $x \in [0, \infty]$.

The error function for continuous data, $G(x)$, is defined analogous to that for discrete data, as

$$E \equiv \int_a^b \rho(x) \left[\sum_{m=1}^{M} \alpha_m U_m(x) - G(x) \right]^2 dx \tag{13.93}$$

This error is minimized with respect to the α-parameters to obtain the normal equations. That is

$$\frac{\partial E}{\partial \alpha_k} = 2 \int_a^b \rho(x) U_k(x) \left[\sum_{m=1}^{M} \alpha_m U_m(x) - G(x) \right] dx = 0 \tag{13.94}$$

Using the orthogonality condition of equation 13.91, this becomes

$$\sum_{m=1}^{M} \alpha_m \int_a^b \rho(x) U_k(x) U_m(x)\, dx = N_k \alpha_k = \int_a^b \rho(x) U_k(x) G(x)\, dx \tag{13.95a}$$

$$\Rightarrow \quad \alpha_k = \frac{1}{N_k} \int_a^b \rho(x) U_k(x) G(x)\, dx \tag{13.95b}$$

Thus by representing a continuous function by a sum of orthogonal polynomials, each coefficient is determined independently, rather than as one element of set of solutions to M simultaneous equations.

Example 13.12

As an example, we will fit

$$G(x) = \cos\left(\frac{\pi}{2} x\right) \tag{13.96}$$

by a parabola over the interval $x \in [-1, 1]$. For this range of x, we ue one of the Sturm–Liouville polynomials defined over this range, such as the Legendre polynomials. As such, we express a second-order polynomial in the form

$$G(x) \simeq \alpha_0 + \alpha_1 P_1(x) + \alpha_2 P_2(x) \tag{13.97}$$

Then, with $\rho(x) = 1$,

$$E = \int_{-1}^{1} \left[\alpha_0 + \alpha_1 P_1(x) + \alpha_2 P_2(x) - \cos\left(\frac{\pi}{2} x\right) \right]^2 dx \tag{13.98}$$

Minimizing E with respect to the α-parameter leads to the normal equations

$$\alpha_k = \frac{2k+1}{2} \int_{-1}^{1} P_k(x) \cos\left(\frac{\pi}{2}x\right) dx \tag{13.99}$$

$$\Rightarrow \quad \text{(a)} \quad \alpha_0 = \frac{1}{2} \int_{-1}^{1} \cos\left(\frac{\pi}{2}x\right) dx = \frac{2}{\pi} = 0.6366 \qquad \text{(b)} \quad \alpha_1 = \frac{3}{2} \int_{-1}^{1} x \cos\left(\frac{\pi}{2}x\right) dx = 0$$

(13.100)

$$\text{(c)} \quad \alpha_2 = \frac{5}{2} \int_{-1}^{1} \tfrac{1}{2}(3x^2 - 1) \cos\left(\frac{\pi}{2}x\right) dx = \frac{10}{\pi}\left(1 - \frac{12}{\pi^2}\right) = -0.6871$$

$$\Rightarrow \quad \cos\left(\frac{\pi}{2}x\right) \simeq 0.6366 P_0(x) - 0.6871 P_2(x) = 0.9802 - 1.0307x^2 \tag{13.101}$$

Figure 13.9 shows the agreement between $\cos(\frac{1}{2}\pi x)$ and this polynomial approximation over the interval $x \in [-1, 1]$. The solid curve is the cosine function and the discrete points represent the parabola. As can be seen, the parabolic approximation is quite good. □

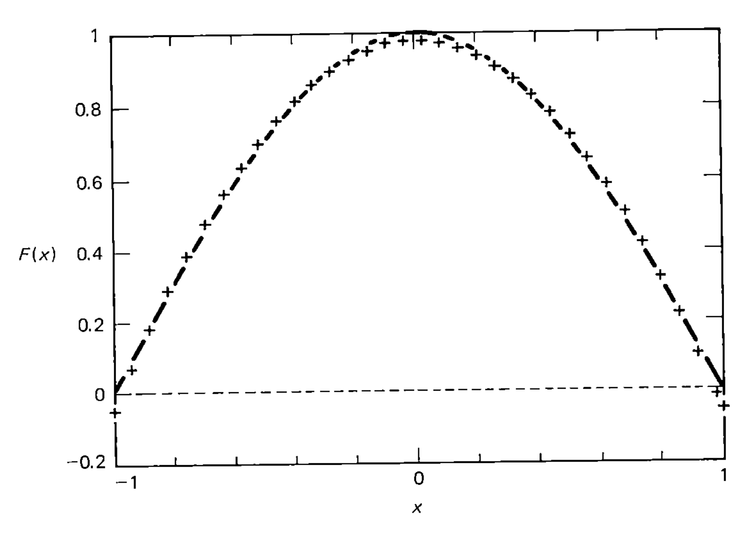

Figure 13.9 Comparison between $\cos(\frac{1}{2}\pi x)$ and parabolic approximation.

13.3 Numerical Integration

Often, an integral of the form $\int_a^b f(x)\,dx$ cannot be evaluated analytically. In addition, either the interval $[a, b]$ or the function $f(x)$ makes it inconvenient or perhaps impossible to expand $f(x)$ in a series that could then be integrated term by term. In this section we develop methods whereby the integral is approximated by

$$\int_a^b f(x)\,dx \simeq \sum_{l=1}^{N} w_l f(x_l) \tag{13.102}$$

The sum on the right-hand side is called the *quadrature sum*. The term quadrature comes from the fact that two parameters are required to define the sum. The numbers w_l are called the *weights* and the x_l are called the *abscissae* of the quadrature.

Unlike the Euler–MacLaurin approach, or expansion of the function in a series, quadrature methods do not require evaluation of the derivatives of $f(x)$. That is particularly important if $f(x)$ is known in tabulated form, at discrete values of x.

Newton–Coates quadratures

All the Newton–Coates quadratures are developed by dividing $[a, b]$ into a large number of equal-sized segments. That is, for two successive points

$$x_{l+1} - x_l \equiv \Delta x \tag{13.103}$$

with Δx independent of l.

The simplest quadrature is obtained by approximating $f(x)$ by a constant in each segment. That is, the function is approximated by a cardinal spline. The evaluation of the integral is based on the fact that a definite integral is the area under the $f(x)$ curve between a and b. For the simplest Newton–Coates scheme, this area is approximated by the sum of the areas of the rectangles shown in Figure 13.10. For this reason, the quadrature is called the *rectangular rule*.

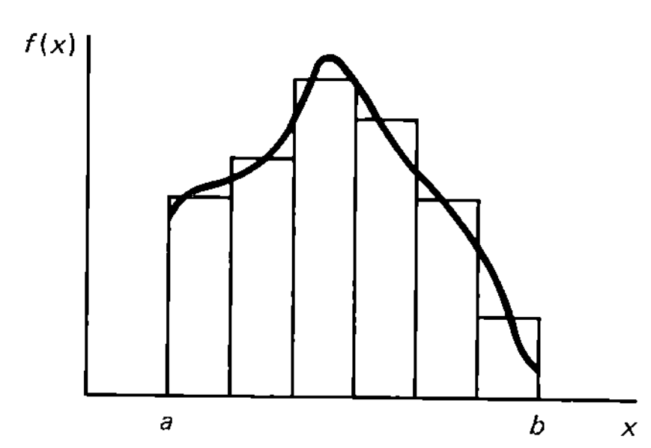

Figure 13.10 Cardinal spline interpolation for simplest Newton–Coates integration.

If x_l is the midpoint of the lth interval and $f(x_l)$ is the value of $f(x)$ at that midpoint, then

$$\text{(a)} \quad \int_a^b f(x)\,dx \simeq \sum_{l=1}^{N} f(x_l)\,\Delta x \qquad \text{with} \qquad \text{(b)} \quad \Delta x = \frac{(b-a)}{N} \tag{13.104}$$

Note that all the weights of the quadrature have the same value, Δx.

Example 13.13

As an example, consider

$$\int_0^1 \log(1+x)\,dx \simeq \sum_{l=1}^{N} \log(1+x_l)\,\Delta x \tag{13.105}$$

If the interval $[0, 1]$ is divided into five segments, $\Delta x = \frac{1}{5}$, so $x_1 = 0 + \frac{1}{2}\Delta x = \frac{1}{10}$, $x_2 = x_1 + \Delta x = \frac{3}{10}$, $x_3 = \frac{5}{10}$, $x_4 = \frac{7}{10}$, and $x_5 = \frac{9}{10}$. Therefore

$$\int_0^1 \log(1+x)\,dx \simeq \tfrac{1}{5}\left[\log(1+\tfrac{1}{10}) + \log(1+\tfrac{3}{10}) + \cdots + \log(1+\tfrac{9}{10})\right] = 0.3871 \tag{13.106a}$$

which compares reasonably well with the exact result:

$$\int_0^1 \log(1+x)\,dx = (1+x)[\log(1+x) - 1]_0^1 = 0.3863 \quad \square \tag{13.106b}$$

Caution should be exercised in using such a crude approximation. In the example above, the logarithm function is a very smoothly varying function. For functions that vary more wildly, this simplest Newton–Coates technique will not yield reasonable results.

The accuracy of the quadrature is improved by approximating the function with a higher-order spline interpolation within selected segments. For example, to improve on the rectangular rule, one can use the linear spline to approximate $f(x)$ by a straight line between two successive points. That is,

$$f(x) \simeq ax + b \qquad x_l \leq x \leq x_{l+1} \tag{13.107}$$

where, as shown above, a and b are evaluated in terms of $f(x_l)$ and $f(x_{l+1})$. As mentioned earlier, for higher-order splines it is easier to achieve such a fit by a Lagrange interpolation over the limited range of x. Thus the linear interpolation between x_l and x_{l+1} can be written as

$$f(x) = \frac{(x - x_{l+1})}{(x_l - x_{l+1})} f(x_l) + \frac{(x - x_l)}{(x_{l+1} - x_l)} f(x_{l+1}) \tag{13.108a}$$

Writing

$$x_{l+1} - x_l = \Delta x \tag{13.103}$$

$$\Rightarrow \quad f(x) = \frac{1}{\Delta x}[-(x - x_{l+1})f(x_l) + (x - x_l)f(x_{l+1})] \tag{13.108b}$$

The interpolation of $f(x)$ between $f(x_l)$ and $f(x_{l+1})$ forms a trapezoid as shown in Figure 13.11. Integrating the function as expressed in equation 13.108b, the area of the trapezoid is

$$A_l = \int_{x_l}^{x_{l+1}} f(x)\, dx = \int_{x_l}^{x_l + \Delta x} f(x)\, dx = \frac{\Delta x}{2}[f(x_l) + f(x_{l+1})] \tag{13.109}$$

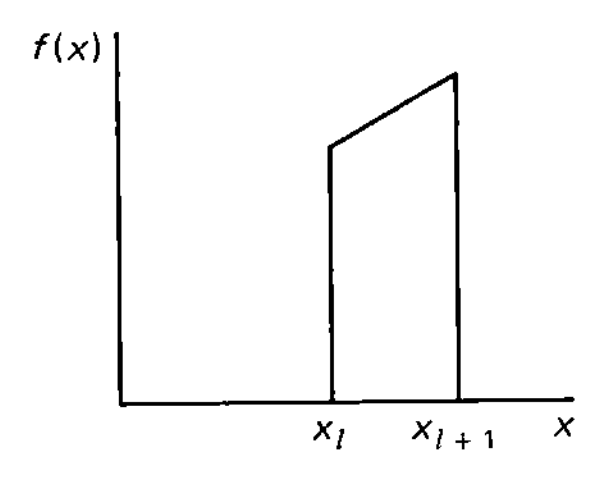

Figure 13.11 Trapezoid formed by linear spline interpolation.

Therefore, approximating the integral by the sum of the areas of all such trapezoids in the interval $[a, b]$ yields

$$\begin{aligned}\int_a^b d(x)\, dx &\simeq \frac{\Delta x}{2}[f(x_1) + f(x_2)] + \frac{\Delta x}{2}[f(x_2) + f(x_3)] \\ &\quad + \frac{\Delta x}{2}[f(x_3) + f(x_4)] + \cdots + \frac{\Delta x}{2}[f(x_{N-1}) + f(x_N)] \\ &= \Delta x[\tfrac{1}{2}f(x_1) + f(x_2) + f(x_3) + \cdots + f(x_{N-1}) + \tfrac{1}{2}f(x_N)] \end{aligned} \tag{13.110}$$

This Newton–Coates quadrature scheme is called the *trapezoidal rule*.

If we approximate $f(x)$ by a quadratic spline between x_l and x_{l+2}, then

$$\begin{aligned} f(x) &= -\frac{(x - x_{l+1})(x - x_{l+2})}{2(\Delta x)^2} f(x_l) - \frac{(x - x_l)(x - x_{l+2})}{(\Delta x)^2} f(x_{l+1}) \\ &\quad + \frac{(x - x_l)(x - x_{l+1})}{2(\Delta x)^2} f(x_{l+2}) \end{aligned} \tag{13.111}$$

where we have used

$$\text{(a)}\quad \Delta x = x_{l+1} - x_l = x_{l+2} - x_{l+1} \qquad \text{(b)}\quad 2\,\Delta x = x_{l+2} - x_l \tag{13.112}$$

Therefore, the area under the segments of the parabola formed by the quadratic spline approximation of $f(x)$ is

$$A_l = \int_{x_l}^{x_{l+2}} f(x)\,dx = \int_{x_l}^{x_l+2\,\Delta x} f(x)\,dx = \frac{\Delta x}{3}[f(x_l) + 4f(x_{l+1}) + f(x_{l+2})] \tag{13.113}$$

Adding the areas under all such parabolic segments yields

$$\int_a^b f(x)\,dx \simeq \frac{\Delta x}{3}\{[f(x_1) + 4f(x_2) + f(x_3)] + [f(x_3) + 4f(x_4) + f(x_5)d]$$
$$+ \cdots + [f(x_{N-2}) + 4f(x_{N-1}) + f(x_N)]\}$$
$$= \frac{\Delta x}{3}[f(x_1) + 4f(x_2) + 2f(x_3) + 4f(x_4) + \cdots + 2f(x_{N-2}) + 4f(x_{N-1}) + f(x_N)] \tag{13.114}$$

This quadrature is called the *Simpson* $\frac{1}{3}$ *rule.*

Using a cubic spline interpolation of $f(x)$ for $x_l \leqslant x \leqslant x_{l+3}$, one obtains the Newton–Coates quadrature rule called *Simpson's* $\frac{3}{8}$ *rule*, which approximates the integral by

$$\int_a^b f(x)\,dx \simeq \tfrac{3}{8}\Delta x\{[f(x_1) + 3f(x_2) + 3f(x_3) + f(x_4)]$$
$$+ [f(x_4) + 3f(x_5) + 3f(x_6) + f(x_7)] + \cdots$$
$$+ [f(x_{N-3}) + 3f(x_{N-2}) + 3f(x_{N-1}) + f(x_N)]\}$$
$$= \tfrac{3}{8}\Delta x[f(x_1) + 3f(x_2) + 3f(x_3) + 2f(x_4) + 3f(x_5) + 3f(x_6) + 2f(x_7) + \cdots$$
$$+ 2f(x_{N-3}) + 3f(x_{N-2}) + 3f(x_{N-1}) + f(x_N)] \tag{13.115}$$

With Simpson's $\frac{1}{3}$ rule, the intervals must be taken two at a time. Therefore, we must divide the range of integration into an even number of intervals. In the $\frac{3}{8}$ rule, the range of integration must be divided into a number of intervals that is divisible by 3. In most applications, the $\frac{1}{3}$ rule is sufficiently accurate, and the $\frac{3}{8}$ rule does not yield significant improvement over the $\frac{1}{3}$ rule. As such, the $\frac{3}{8}$ rule is rarely used.

Example 13.14

As examples, we estimate $\int_0^1 \log(1 + x)\,dx$ by the trapezoidal and Simpson's $\frac{1}{3}$ rule. For the trapezoidal rule, we can use the same five intervals used when applying the rectangular rule. Taking $\Delta x = 0.2$, and therefore

$$\begin{array}{llll} \text{(a)}\ x_1 = 0.0 & \text{(b)}\ x_2 = 0.2 & \text{(c)}\ x_3 = 0.4 \\ \text{(d)}\ x_4 = 0.6 & \text{(e)}\ x_5 = 0.8 & \text{(f)}\ x_6 = 1.0 \end{array} \tag{13.116}$$

$$\Rightarrow \int_0^1 \log(1 + x)\,dx \simeq 0.2[\tfrac{1}{2}\log(1) + \log(1.2) + \cdots \log(1.8) + \tfrac{1}{2}\log(2.0)] = 0.3846 \tag{13.117}$$

Simpson's $\frac{1}{3}$ rule requires that the range of integration be divided into an even number of intervals. If we use six intervals, $\Delta x = \frac{1}{6}$ and

$$\text{(a)}\quad x_1 = 0 \qquad \text{(b)}\quad x_2 = \tfrac{1}{6} \qquad \text{(c)}\quad x_3 = \tfrac{1}{3} \qquad \text{(d)}\quad x_4 = \tfrac{1}{2}$$

$$\text{(e)}\quad x_5 = \tfrac{2}{3} \qquad \text{(f)}\quad x_6 = \tfrac{5}{6} \qquad \text{(g)}\quad x_7 = 1 \tag{13.118}$$

$$\Rightarrow \quad \int_0^1 \log(1+x)\,dx \simeq \tag{13.119}$$

$$\tfrac{1}{18}[\log(1) + 4\log(\tfrac{7}{6}) + 2\log(\tfrac{4}{3}) + 4\log(\tfrac{3}{2}) + 2\log(\tfrac{5}{3}) + 4\log(\tfrac{11}{6}) + \log(2)] = 0.3863$$

which is identical to the exact result to four decimal places. □

Gaussian quadratures

All Newton–Coates quadrature rules are based on interpolating the integrand over a number of segments of equal width, and approximating the integral by a sum of the areas of those segments. By requiring the points to be equally spaced, the weights of the quadrature rule are fixed by the spacing and the interpolation used.

If one were to make a full Lagrange interpolation of $f(x)$ using the $f(x_l)$ of an N-point Newton–Coates formula, one could fit $f(x)$ to a polynomial of order $N-1$. That is, with $\Delta x \equiv x_{l+1} - x_l$,

$$f(x) \simeq \frac{(x-x_2)(x-x_3)\cdots(x-x_N)}{(x_1-x_2)(x_1-x_3)\cdots(x_1-x_N)} f(x_1) + \cdots$$

$$= \frac{(x-x_2)(x-x_3)\cdots(x-x_N)}{(-\Delta x)(-2\,\Delta x)\cdots[-(N-1)\Delta x]} f(x_1) + \cdots \tag{13.120}$$

Thus if $f(x)$ were a polynomial of order $N-1$ or smaller, such an interpolation would be an exact description of $f(x)$, and a Newton–Coates formula using a Lagrange polynomial interpolation would be exact. All Newton–Coates quadrature rules using smaller spline interpolations are of this level of accuracy.

Newton–Coates quadrature rules are determined by using the property that a definite integral is the area under the functional curve. Gaussian quadrature rules are obtained by a different approach. The weights and abscissae are determined by imposing the constraint that if $f(x)$ is a polynomial of specified order or lower, the quadrature sum must be the exact value of the integral. That is, for an N-point Gaussian quadrature rule

$$\int_a^b \rho(x) f(x)\,dx = \sum_{l=1}^{N} w_l f(x_l) \tag{13.121}$$

must be exact for $f(x)$ a polynomial of specified order. $\rho(x)$ is a weighting function that is nonnegative in the interval $[a, b]$. The particular form of $\rho(x)$ depends on the quadrature being used and will be discussed shortly.

The simplest form of a polynomial of order k is x^k. Thus the Gaussian quadrature rule requires that

$$I_k \equiv \int_a^b \rho(x) x^k\,dx = \sum_{l=1}^{N} w_l x_l^k \tag{13.122}$$

be exact for certain values of k. Since $2N$ parameters are to be determined (N w_l's and N x_l's), $2N$ equations are needed to solve for them. Thus k can range over the lowest

$2N$ values, $0 \leqslant k \leqslant 2N - 1$. Therefore, for an N-point Gaussian quadrature scheme, the quadrature sum will be the exact value of the integral if $f(x)$ is a polynomial of order $2N - 1$ or lower. As such, for an N-point rule, with $N > 1$, Gaussian quadrature rules are significantly more accurate than Newton–Coates schemes.

Obviously, the $2N - 1$ equations expressed in equation 13.122 are highly nonlinear and very difficult to solve. To make the solution easier, we develop a method for determining abscissae separately from the weights.

Let $Q(x)$ be any polynomial of order $2N - 1$. Then

$$\int_a^b \rho(x) Q(x)\, dx = \sum_{l=1}^{N} w_l Q(x_l) \tag{13.123}$$

must be exact. Let $\Pi(x)$ be a polynomial of order N, the roots of which are the abscissae of the quadrature rule. Thus we can write

$$\Pi(x) = (x - x_1)(x - x_2) \cdots (x - x_N) \tag{13.124}$$

If $B_{N-1}(x)$ is an arbitrary polynomial of order $N - 1$, then $B_{N-1}(x)\Pi(x)$ is a polynomial of order $2N - 1$. Therefore

$$\int_a^b \rho(x) B_{N-1}(x) \Pi(x)\, dx = \sum_{l=1}^{N} w_l B_{N-1}(x_l) \Pi(x_l) \tag{13.125}$$

is exact. Because the points x_l are the roots of $\Pi(x)$, $\Pi(x_l) = 0$ and therefore

$$\int_a^b \rho(x) B_{N-1}(x) \Pi(x)\, dx = 0 \tag{13.126}$$

is an exact result.

Let $\{Z_l(x)\}$ be a set of mutually orthogonal polynomials. We studied several examples of such Sturm–Liouville polynomials in Chapter 6. For example, the Legendre polynomials are orthogonal over the interval $[-1, 1]$ with a weighting function $\rho(x) = 1$. That means that

$$\int_{-1}^{1} P_l(x) P_m(x)\, dx = \frac{2}{(2l + 1)} \delta_{lm} \tag{6.213}$$

Similarly, the Hermite polynomials are orthogonal over the interval $[-\infty, \infty]$ with a weighting function $\rho(x) = e^{-x^2}$.

$$\int_{-\infty}^{\infty} e^{-x^2} H_l(x) H_m(x)\, dx = 2^m m! \sqrt{\pi}\, \delta_{lm} \tag{6.495}$$

We take $Z_l(x)$ to satisfy

$$\int_a^b \rho(x) Z_l(x) Z_m(x)\, dx = C_m \delta_{ml} \tag{13.127}$$

As discussed in Chapter 6, any function whose argument $x \in [a, b]$ can be expanded in an infinite series in $Z_l(x)$. Since $B_{N-1}(x)$ is a polynomial of order $N - 1$, that infinite series becomes a finite sum, and we can write

$$B_{N-1}(x) = \sum_{l=0}^{N-1} A_l Z_l(x) \tag{13.128}$$

Then equation 13.126 becomes

$$\sum_{l=0}^{N-1} A_l \int_a^b \rho(x) \Pi(x) Z_l(x)\, dx = 0 \tag{13.129}$$

Since $B_{N-1}(x)$ is arbitrary, the A_l are arbitrary. As such, we cannot take A_l to be zero. Thus since every $Z_l(x)$ in equation 13.129 is of order $\leqslant N-1$, the only way that we can ensure that the integral will be exactly zero is to make $\Pi(x)$ orthogonal to each $Z_l(x)$ in the sum of equation 13.128. Because $\Pi(x)$ is a Nth-order polynomial, this requires that

$$\Pi(x) = \alpha Z_N(x) \tag{13.130}$$

where α is an arbitrary constant. Thus the abscissae of the quadrature are the roots of the Nth-order polynomial $Z_N(x)$. For example, for an integral over $[-1, 1]$, one commonly used quadrature rule uses abscissae that are the roots of the Nth Legendre polynomial.

To find the weights, let $q(x)$ and $r(x)$ be polynomials of order $N-1$. Then

$$Q(x) = q(x)Z_N(x) + r(x) \tag{13.131}$$

is a polynomial of order $2N-1$. Since $Z_N(x_l) = 0$,

$$Q(x_l) = r(x_l) \tag{13.132}$$

Referring to equation 13.6, consider the polynomial

$$\mu_l(x) = \frac{Z_N(x)}{(x-x_l)Z_N'(x_l)} \tag{13.6}$$

where $Z_N(x)$ is the Nth-order member of the set of orthogonal polynomials $\{Z_l(x)\}$. As shown earlier,

$$\mu_l(x_k) = \delta_{kl} \tag{13.9}$$

Since $\mu_l(x)$ is a polynomial of order $N-1$, we can write $r(x)$ as a sum of these polynomials

$$r(x) = \sum_{l=1}^{N} c_l \mu_l(x) \tag{13.133}$$

where, using equation 13.9, we have

$$c_l = r(x_l) \tag{13.134}$$

Thus

$$r(x) = \sum_{l=1}^{N} r(x_l)\mu_l(x) \tag{13.135}$$

$$\Rightarrow \quad Q(x) = q(x)Z_N(x) + \sum_{l=1}^{N} r(x_l)\mu_l(x) \tag{13.136}$$

Since $Q(x)$ is a polynomial of order $2N-1$,

$$\int_a^b \rho(x)Q(x)\,dx = \sum_{l=1}^{N} w_l Q(x_l) \tag{13.137}$$

is exact. Referring to equation 13.132, this becomes

$$\int_a^b \rho(x)Q(x)\,dx = \sum_{l=1}^{N} w_l r(x_l) \tag{13.138a}$$

or using equation 13.136,

$$\int_a^b \rho(x)Q(x)\,dx = \int_a^b \rho(x)q(x)Z_N(x)\,dx + \sum_{l=1}^{N} r(x_l)\int_a^b \rho(x)\mu_l(x)\,dx \tag{13.138b}$$

Because $q(x)Z_N(x)$ is of order $2N - 1$. and the x_l are the roots of $Z_N(x)$,

$$\int_a^b \rho(x) q(x) Z_N(x)\, dx = \sum_{l=1}^{N} w_l q(x_l) Z_N(x_l) = 0 \tag{13.139a}$$

is exact. Thus

$$\sum_{l=1}^{N} w_l r(x_l) = \sum_{l=1}^{N} r(x_l) \int_a^b \rho(x) \mu_l(x)\, dx \tag{13.139b}$$

$$\Rightarrow \quad w_l = \int_a^b \rho(x) \mu_l(x)\, dx = \frac{1}{Z_N'(x_l)} \int_a^b \rho(x) \frac{Z_N(x)}{(x - x_l)}\, dx \tag{13.140}$$

Approximating the integrand by polynomial interpolation,

$$f(x) = \sum_{l=1}^{N} f(x_l) \mu_l(x) \tag{13.4}$$

$$\Rightarrow \quad \int_a^b \rho(x) f(x)\, dx = \sum_{l=1}^{N} f(x_l) \int_a^b \rho(x) \mu_l(x)\, dx \tag{13.141}$$

where, as before, $\mu_l(x)$ is given by

$$\mu_l(x) = \frac{Z_N(x)}{(x - x_l) Z_N'(x_l)} \tag{13.6}$$

From equation 13.140

$$\int_a^b \rho(x) \mu_l(x)\, dx = w_l \tag{13.142}$$

$$\Rightarrow \quad \int_a^b \rho(x) f(x)\, dx = \sum_{l=1}^{N} w_l f(x_l) \tag{13.143}$$

Thus the use of an N-point Gaussian quadrature rule is identical to an N-point interpolation of the integrand. Therefore, a quadrature rule is as accurate as the accuracy of interpolating the integrand $f(x)$ over the $(N-1)$th-order polynomial involving $Z_N(x)$ as in equation 13.4. We can easily determine how accurately the values of $f(x)$ are approximated by the interpolation. Such a determination is critical, since, in practice, a quadrature rule is used only when the integral cannot be evaluated analytically. Therefore, without a comparative value of the integral, this determination of the accuracy of the result is the only way to tell if the answer obtained by the quadrature rule is correct.

We use the terminology that a particular Gaussian quadrature is a quadrature over a set of specified mutually orthogonal polynomials. The abscissae are the roots of that polynomial, and the weights are found from an integral involving that polynomial.

Legendre quadratures

As an example, we consider the Gauss–Legendre quadrature. The Legendre polynomials satisfy the orthogonality condition of equation 6.213 with a weighting function $\rho(x) = 1$. The range of integration must be the same as the range of x-values appropriate to the polynomial. For the Legendre polynomials, $x \in [-1, 1]$. Thus the Legendre quadrature rule is

$$\int_{-1}^{1} f(x)\, dx \simeq \sum_{l=1}^{N} w_l f(x_l) \tag{13.144}$$

where the x_l are the roots of the Nth Legendre polynomial, and the weights are

$$w_l = \frac{1}{P_N'(x_l)} \int_{-1}^{1} \frac{P_N(x)}{(x - x_l)}\, dx \tag{13.145}$$

Using the Neumann integral representation of the second Legendre function,

$$Q_N(z) = \frac{1}{2} \int_{-1}^{1} \frac{P_N(x)}{(z - x)}\, dx \tag{6.247}$$

the weights of the N-point Legendre quadrature can also be written as

$$w_l = -2\frac{Q_N(x_l)}{P_N'(x_l)} \tag{13.146a}$$

Since

$$Q_N(x) = P_N(x)Q_0(x) - R_{N-1}(x) \tag{6.239}$$

and $P_N(x_l) = 0$, the weights can be written

$$w_l = 2\frac{R_{N-1}(x_l)}{P_N'(x_l)} \tag{13.146b}$$

If the range of integration is not $[-1, 1]$, it is straightforward to transform to $[-1, 1]$. Examples of such transformations were described in Chapter 3 in connection with the Euler–MacLaurin integration scheme.

If a and b are finite, the substitution

$$\text{(a)} \quad x = -1 + 2\frac{(y - a)}{(b - a)} \qquad \text{with} \qquad \text{(b)} \quad dy = \tfrac{1}{2}(b - a)\, dx \tag{13.147}$$

takes $y \in [a, b]$ to $x \in [-1, 1]$. Similarly, if a is finite and nonzero and $b = \infty$,

$$\text{(a)} \quad x = -1 + 2\frac{a}{y} \qquad \text{with} \qquad \text{(b)} \quad dy = -2\frac{a}{(1 + x)^2}\, dx \tag{13.148}$$

will transform an integral over $y \in [a, \infty]$ to an integral over $x \in [-1, 1]$. Substitutions such as

$$\text{(a)} \quad y = \tan\left(\frac{\pi(1 + x)}{4}\right) \qquad \text{with} \qquad \text{(b)} \quad dy = \frac{\pi}{4}\sec^2\left(\frac{\pi(1 + x)}{4}\right) dx$$

or (13.149)

$$\text{(c)} \quad y = \frac{(1 + x)}{(1 - x)} \qquad \text{with} \qquad \text{(d)} \quad dy = 2\frac{dx}{(1 - x)^2}$$

are examples of transformations that take $y \in [0, \infty]$ to $x \in [-1, 1]$. If $y \in [-\infty, \infty]$, then the substitution

$$\text{(a)} \quad y = -\cot\left(\frac{\pi(1 + x)}{2}\right) \qquad \text{with} \qquad \text{(b)} \quad dy = \frac{\pi}{2}\csc^2\left(\frac{\pi(1 + x)}{4}\right) dx \tag{13.150}$$

takes x to $[-1, 1]$. Therefore, any integral can be transformed into an integral over $[-1, 1]$, which can then be evaluated by Gauss–Legendre quadrature.

Table 13.8 lists the weights and abscissae for some of the smaller Legendre quadrature sets. For more complete listings of Legendre and other quadrature sets, the

TABLE 13.8 SELECTED SETS OF GAUSS – LEGENDRE QUADRATURE POINTS

x_l		w_l
	$N = 2$	
± 0.57735		1.00000
	$N = 3$	
0.00000		0.88889
± 0.77460		0.55556
	$N = 4$	
± 0.33998		0.65215
± 0.86114		0.34785
	$N = 5$	
0.00000		0.56889
± 0.53847		0.47863
± 0.90618		0.23693
	$N = 10$	
± 0.14887		0.29552
± 0.43340		0.26927
± 0.67941		0.21909
± 0.86506		0.14945
± 0.97391		0.06667

reader is referred to the text by A. Stroud and D. Secrest, *Gaussian Quadrature Formulas* (Englewood Cliffs, N.J.: Prentice-Hall, Inc., 1966), or the book edited by M. Abramowitz and I. A. Stegun, *Handbook of Mathematical Functions* (Washington, D.C.: National Bureau of Standards, 1964).

Example 13.15

As an example we evaluate

$$\int_1^2 \cos^2 y\, dy = \tfrac{1}{4}[2 + \sin(4) - \sin(2)] = 0.08348 \tag{13.151}$$

using a three-point Legendre quadrature. Using equations 13.147, we first convert [1, 2] to the interval [−1, 1]. Then, referring to Table 13.8, we have

$$\int_1^2 \cos^2 y\, dy = \int_{-1}^1 \cos^2\left[\tfrac{1}{2}(x+3)\right] dx = \Big\{0.88889 \cos^2\left[\tfrac{1}{2}(3 + 0.00000)\right]$$

$$+ 0.55556 \cos^2\left[\tfrac{1}{2}(3 + 0.77460)\right] + 0.55556 \cos^2\left[\tfrac{1}{2}(3 - 0.77460)\right]\Big\} = 0.08346 \tag{13.152}$$

Laguerre quadratures

The Laguerre polynomials satisfy the orthogonality condition

$$\int_0^\infty e^{-x} L_m(x) L_n(x)\, dx = \delta_{mn} \tag{6.398}$$

Therefore, the Gauss–Laguerre quadratures are suitable for approximating integrals of the form

$$\int_0^\infty e^{-x} f(x)\, dx \simeq \sum_{l=0}^{N} w_l f(x_l) \tag{13.153}$$

TABLE 13.9 SELECTED SETS OF GAUSS – LAGUERRE QUADRATURE POINTS

x_l	w_l	$w_l e^{x_l}$
	$N = 2$	
0.58579	0.85355E00	1.53333
3.41421	0.14645E00	4.45096
	$N = 3$	
0.41577	0.71109E00	1.07769
2.29428	0.27852E00	2.76214
6.28995	0.10389E-01	5.60109
	$N = 4$	
0.32255	0.60315E00	0.83274
1.74576	0.35742E00	2.04810
4.53662	0.38888E-01	3.63115
9.39507	0.53929E-03	6.48715
	$N = 5$	
0.26356	0.52176E00	0.67909
1.41340	0.39867E00	1.63849
3.59643	0.75942E-01	2.76944
7.08581	0.36118E-02	4.31566
12.64080	0.23370E-04	7.21919
	$N = 10$	
0.13779	0.30844E00	0.35401
0.72945	0.40112E00	0.83190
1.80834	0.21807E00	1.33029
3.40143	0.62087E-01	1.86306
5.55250	0.95015E-02	2.45026
8.33015	0.75301E-03	3.12276
11.84379	0.28259E-04	3.93415
16.27926	0.42493E-06	4.99241
21.99659	0.18396E-08	6.57220
29.92070	0.99118E-12	9.78470

Some of the smaller Gauss–Laguerre quadrature sets are listed in Table 13.9. We are using the notation used for computer output to represent powers of 10. If the integrand explicitly contains the factor e^{-x}, accurate results will be obtained using the Laguerre quadratures.

Example 13.16

As an example, we use a Laguerre quadrature rule to evaluate

$$\int_0^\infty \frac{e^{-x}}{(e^{-x}+1)^3}\,dx = \frac{1}{2}\left[1 - \left(\frac{1}{2}\right)^2\right] = 0.37500 \tag{13.154}$$

Using the five-point Laguerre rule, we obtain

$$\int_0^\infty \frac{e^{-x}}{(e^{-x}+1)^3}\,dx \simeq \sum_{l=1}^{5} w_l \frac{1}{(e^{-x_l}+1)^3} = 0.37544 \tag{13.155a}$$

A 10-point Laguerre rule yields

$$\int_0^\infty \frac{e^{-x}}{(e^{-x}+1)^3}\,dx \simeq \sum_{l=1}^{10} w_l \frac{1}{(e^{-x_l}+1)^3} = 0.37500 \tag{13.155b}$$

As we see in this example, the accuracy of the quadrature rule increases with increasing size of the quadrature set. □

If $f(x)$ does not contain the factor e^{-x} explicitly, it is still possible to evaluate $\int_0^\infty f(x)\,dx$ using Laguerre quadratures. Note that Table 13.9 for Gauss–Laguerre quadratures has three columns, the last containing the product $w_l e^{x_l}$. This is because we can write

$$\int_0^\infty f(x)\,dx = \int_0^\infty e^{-x}[e^x f(x)]\,dx \simeq \sum_{l=1}^{N} w_l e^{x_l} f(x_l) \tag{13.156}$$

However, unless

$$\lim_{x\to\infty} e^x f(x) = 0 \tag{13.157}$$

using larger Laguerre quadrature sets will not result in more accurate values for the approximated integral.

The author's experience has been that in most cases, if the integrand does not explicitly contain the factor e^{-x}, Laguerre rules are not well suited to evaluating the integral. When e^{-x} does not appear explicitly in the integrand, a better estimate of the integral is obtained by transforming the range of integration to $[-1,1]$ and using a Legendre quadrature. We illustrate this fact with two examples.

Example 13.17

Consider

$$\int_0^\infty \frac{x}{(1+x^2)^2}\,dx = \frac{1}{2} \tag{13.158}$$

Since

$$\lim_{x\to\infty} e^x \frac{x}{(1+x^2)^2} = \infty \tag{13.159}$$

this integral, evaluated by Laguerre quadrature rule, will not converge to the correct result by increasing the size of the Laguerre quadrature sets. To see this, we write

$$\int_0^\infty \frac{x}{(1+x^2)^2}\,dx = \int_0^\infty e^{-x}\frac{xe^x}{(1+x^2)^2}\,dx \simeq \sum_{l=1}^{N} w_l e^{x_l}\frac{x_l}{(1+x_l^2)^2} \tag{13.160}$$

Referring to Table 13.9, the three-point Laguerre quadrature yields

$$\int_0^\infty \frac{x}{(1+x^2)^2}\,dx \simeq 1.07769\frac{0.41577}{(1+0.41577^2)^2} + 2.76214\frac{2.76214}{(1+2.76214^2)^2}$$

$$+\,5.60109\frac{6.28995}{(1+6.28995^2)^2} = 0.50866 \tag{13.161}$$

which is an inaccuracy of 0.00866. The numbers in the quadrature set are accurate to five decimal places. Therefore, the inaccuracy of this approximation is significantly larger than the inaccuracy of the quadrature set and therefore cannot be due to round-off errors. Using a six-point quadrature yields a value of 0.48705, which is a larger inaccuracy of 0.01295.

This is an example of the general result that if the integrand does not satisfy the limiting condition expressed in equation 13.157, increasing the size of the Laguerre quadrature set increases the inaccuracy of the result. By using larger quadrature sets, the values of the largest abscissae increase. For example, the largest x-value in the four-point

Laguerre set is 9.39507, while the largest x in the 10-point set is 29.92070. Thus as the larger x-values increase with the use of larger quadrature sets, the product $f(x_l)e^{x_l}$ makes larger contributions to the quadrature sum at those x-values where the contributions should be getting smaller. Thus if equation 13.157 is not satisfied, the Laguerre quadrature rule is unsuited for evaluating the integral under investigation.

Returning to our example, if we transform the range of integration to $[-1, 1]$ and use the Legendre quadrature, the results converge to the correct result as the size of the quadrature set increases. Setting

$$x = \tan\left(\frac{\pi}{4}(1 + z)\right) \tag{13.162}$$

$$\Rightarrow \quad \int_0^\infty \frac{x}{(1 + x^2)^2}\, dx = \frac{\pi}{4}\int_{-1}^{1} dz \frac{\tan\left[\frac{\pi}{4}(1 + z)\right]}{\left\{1 + \tan^2\left[\frac{\pi}{4}(1 + z)\right]\right\}^2}$$

$$= \frac{\pi}{4}\int_{-1}^{1} \sin\left(\frac{\pi}{4}(1 + z)\right)\cos^3\left(\frac{\pi}{4}(1 + z)\right) dz \simeq \frac{\pi}{4}\sum_{l=1}^{N} w_l \sin\left(\frac{\pi}{4}(1 + z_l)\right)\cos^3\left(\frac{\pi}{4}(1 + z_l)\right) \tag{13.163}$$

The result obtained for a three-point Legendre quadrature is 0.50035. With a six-point rule, we obtain 0.50003. Thus since $e^x f(x)$ does not converge as stated in equation 13.157, transformation to $[-1, 1]$ and the use of the Legendre quadrature generally yields more accurate results. □

Example 13.18

As a second set of examples, we consider the integral

$$\int_0^\infty x^{k-1} e^{-x^k}\, dx = \frac{1}{k} \tag{13.164}$$

Clearly, for $k > 1$,

$$\lim_{x\to\infty} e^x x^{k-1} e^{-x^k} = 0 \tag{13.165}$$

Thus using the Laguerre quadrature rule,

$$\int_0^\infty x^{k-1} e^{-x^k}\, dx \simeq \sum_{l=1}^{N} (w_l e^{x_l}) x_l^{k-1} e^{-x_l^k} \tag{13.166}$$

we expect accuracy to increase with increasing size of the quadrature set. In Table 13.10, we present the results of using the Laguerre rule for four- and 10-point quadrature sets, with $k = 2$, 3, and 4. We also transform the range of integration to $[-1, 1]$ using equation 13.162 and use four- and 10-point Legendre schemes. These results are also presented in Table 13.10. As can be seen, the results using Laguerre quadratures are converging to the correct values, but for the same-size quadrature set, the Legendre results are more accurate in all cases. This is an illustration of the statement made above that unless the integrand contains the factor e^{-x} explicitly, one will obtain more accurate results transforming the range of integration to $[-1, 1]$ and using a Legendre quadrature scheme. □

TABLE 13.10 RESULTS OF EXAMPLES USING LAGUERRE AND LEGENDRE QUADRATIC RULES

	Laguerre			Legendre		
	$k = 2$	$k = 3$	$k = 4$	$k = 2$	$k = 3$	$k = 4$
4-Point	0.41178	0.11430	0.02865	0.47880	0.21608	0.11459
10-Point	0.49578	0.31872	0.24438	0.49931	0.33763	0.24965
Exact	$\frac{1}{2}$	$\frac{1}{3}$	$\frac{1}{4}$	$\frac{1}{2}$	$\frac{1}{3}$	$\frac{1}{4}$

TABLE 13.11 SELECTED DATA FOR GAUSS–HERMITE QUADRATURES

x_l	w_l	$w_l e^{x_l^2}$
	$N = 2$	
±0.70711	0.88623E00	1.46114
	$N = 3$	
0.00000	0.11816E01	1.18164
±1.22474	0.29541E00	1.32393
	$N = 4$	
±0.52465	0.80491E00	1.05996
±1.65068	0.81313E-01	1.24023
	$N = 5$	
0.00000	0.94531E00	0.94531
±0.95857	0.39362E00	0.98658
±2.02018	0.19953E-01	1.18149
	$N = 10$	
±0.34290	0.61086E00	0.68708
±1.03661	0.24014E00	0.70330
±1.75668	0.33874E-01	0.74144
±2.53273	0.13436E-02	0.82067
±3.43616	0.76404E-05	1.02545

Hermite quadratures

The Gauss–Hermite quadrature rule is used to approximate integrals containing an exponential of x^2. The quadrature rule is

$$\int_{-\infty}^{\infty} e^{-x^2} f(x)\, dx \simeq \sum_{l=1}^{N} w_l f(x_l) \tag{13.167}$$

where the abscissae are the roots of the Nth Hermite polynomial. Table 13.11 lists a few of the smaller Hermite quadrature sets.

Like the Laguerre quadrature, integrals over $[-\infty, \infty]$ that do not contain e^{-x^2} can be evaluated with the Hermite quadrature by writing

$$\int_{-\infty}^{\infty} f(x)\, dx = \int_{-\infty}^{\infty} e^{-x^2}\left(e^{x^2} f(x)\right) dx \simeq \sum_{l=1}^{N} w_l e^{x_l^2} f(x_l) \tag{13.168}$$

As with the Laguerre quadrature rule, if the integrand of equation 13.168 satisfies

$$\lim_{x \to \infty} e^{x^2} f(x) = 0 \tag{13.169}$$

accuracy of the Hermite quadrature rule will increase with increasing size of the quadrature set. However, in general, unless the integrand contains the explicit factor e^{-x^2}, using a Hermite quadrature rule of a given size will be less accurate than transforming $[-\infty, \infty]$ to $[-1, 1]$ and using a Legendre quadrature set of the same size.

As discussed for the Laguerre quadrature, if the integrand does not satisfy the convergence condition expressed in equation 13.169, the accuracy of the result will decrease with increasing size of the quadrature set. As with the Laguerre quadrature sets, this occurs because as the size of the quadrature set increases, the contribution of $f(x_l)e^{x_l^2}$ to the quadrature sum increases at the large x-values. These contributions should be decreasing because the integrand decreases as x increases. In this case it is essential that the range of integration be transformed to $[-1,1]$ and a Legendre quadrature rule be used.

Principal value integrals

One commonly used method for evaluating a principal value integral involves writing the integral in the form

$$P\int_a^b \frac{f(x)}{(x-x_0)}\,dx = \int_a^b \frac{f(x)-f(x_0)}{(x-x_0)}\,dx + f(x_0)P\int_a^b \frac{1}{(x-x_0)}\,dx \tag{13.170}$$

where $a < x_0 < b$. Since

$$P\frac{1}{(x-x_0)} = \mathrm{Re}\left(\frac{1}{(x-x_0 \pm i\varepsilon)}\right) \tag{13.171}$$

the second integral becomes

$$P\int_a^b \frac{1}{(x-x_0)}\,dx = \left(\log|x-x_0|\right)_a^b = \log\left(\frac{b-x_0}{x_0-a}\right) \tag{13.172}$$

The principal value notation for the first integral is not applicable since the integrand no longer contains a pole. Therefore, it is not necessary to avoid the point $x = x_0$, and the first integral can be evaluated by a quadrature sum

$$\int_a^b \frac{f(x)-f(x_0)}{(x-x_0)}\,dx \simeq \sum_{l=1}^{N} w_l \frac{f(x_l)-f(x_0)}{(x_l-x_0)} \tag{13.173}$$

If x_0 has the value of one of the quadrature points, x_k, the $l = k$ term in the sum is replaced by the derivative $f'(x_k)$. That is,

$$\lim_{x_0 \to x_k} \frac{f(x_k)-f(x_0)}{(x_k-x_0)} = f'(x_k) \tag{13.174}$$

A second approach, which does not use quadrature rules, is based on the integral representation for the Legendre function of the second kind,

$$Q_N(z) = \frac{1}{2}\int_{-1}^{1} \frac{P_N(x)}{(z-x)}\,dx \tag{6.247}$$

As discussed above, any range of integration can be transformed to $[-1,1]$. As such, it is quite general to consider

$$I(x_0) \equiv P\int_{-1}^{1} \frac{f(x)}{(x-x_0)}\,dx \tag{13.175}$$

where $f(x)$ is assumed to be nonsingular at x_0.

Since it is nonsingular, $f(x)$ is interpolated over the Legendre polynomials as

$$f(x) \simeq \sum_{l=1}^{N} \mu_l(x) f(x_l) \tag{13.4}$$

with

$$\mu_l(x) \equiv \frac{P_N(x)}{(x-x_l)P_N'(x_l)} \tag{13.176}$$

Then

$$I(x_0) = \sum_{l=1}^{N} \frac{f(x_l)}{P_N'(x_l)} \int_{-1}^{1} \frac{P_N(x)}{(x-x_0)(x-x_l)}\, dx$$

$$= \sum_{l=1}^{N} \frac{f(x_l)}{(x_l - x_0)P_N'(x_l)} \int_{-1}^{1} P_N(x)\left[\frac{1}{(x-x_l)} - \frac{1}{(x-x_0)}\right] dx \qquad (13.177)$$

Using the integral representation for $Q_N(z)$ above, this becomes

$$I(x_0) = -2\sum_{l=1}^{N} \frac{f(x_l)}{P_N'(x_l)}\left[\frac{Q_N(x_l) - Q_N(x_0)}{(x_l - x_0)}\right] \qquad (13.178)$$

where for the $l = k$ term, the quantity in brackets is replaced by the derivative $Q_N'(x_k)$ if $x_0 = x_k$. Using recurrence relations satisfied by $P_N(x)$ and $Q_N(x)$, one can straightforwardly generate the needed values of $P_N'(x)$, $Q_N(x)$, and $Q_N'(x)$. The reader is referred to examples in Section 13.6.

Example 13.19

In Chapter 2, methods of complex analysis were used to show that

$$P\int_{-\infty}^{\infty} \frac{dx}{(x^2 + x_1^2)(x - x_0)} = -\frac{\pi x_0}{x_1(x_0^2 + x_1^2)} \qquad (2.303)$$

$$\Rightarrow \quad P\int_{-\infty}^{\infty} \frac{dx}{(x^2+4)(x-3)} = -0.36249 \qquad (13.179)$$

To illustrate the technique described above, we choose a finite interval symmetric around $x = 3$ and write

$$P\int_{-\infty}^{\infty} \frac{dx}{(x^2+4)(x-3)} = \int_{-\infty}^{0} \frac{dx}{(x^2+4)(x-3)} + P\int_{0}^{6} \frac{dx}{(x^2+4)(x-3)}$$

$$+ \int_{6}^{\infty} \frac{dx}{(x^2+4)(x-3)} \qquad (13.180)$$

Since the integrands over the semi-infinite ranges are not singular we evaluate them by transforming the ranges to $[-1, 1]$ and using a Gauss–Legendre quadrature rule. With

$$\text{(a)} \quad z = -1 + \frac{4}{\pi}\left[\frac{\pi}{2} + \tan^{-1} x\right] \quad \text{or} \quad \text{(b)} \quad x = -\cot\left[\frac{\pi}{4}(1+z)\right] \qquad (13.181)$$

$$\Rightarrow \quad \int_{-\infty}^{0} \frac{dx}{(x^2+4)(x-3)}$$

$$= -\frac{\pi}{4}\int_{-1}^{1} \frac{\csc^2[\pi(1+z)/4]}{\{\cot^2[\pi(1+z)/4] + 4\}\{\cot[\pi(1+z)/4] + 3\}}\, dz$$

$$\simeq \sum_{l=1}^{N} w_l \frac{\csc^2[\pi(1+z_l)/4]}{\{\cot^2[\pi(1+z_l)/4] + 4\}\{\cot[\pi(1+z_l)/4] + 3\}} \qquad (13.182)$$

Similarly,

$$z = -1 + \frac{12}{x} \tag{13.183}$$

$$\Rightarrow \int_6^\infty \frac{dx}{(x^2+4)(x-3)} = 12\int_{-1}^1 \frac{dz}{(148+8z+4z^2)(9+6z-3z^2)}$$

$$\simeq 12\sum_{l=1}^N w_l \frac{1}{(148+8z_l+4z_l^2)(9+6z_l-3z_l^2)} \tag{13.184}$$

Setting

$$z = -1 + \frac{x}{3} \tag{13.185}$$

$$\Rightarrow \quad P\int_0^6 \frac{dx}{(x^2+4)(x-3)} = P\int_{-1}^1 \frac{dz}{z(9z^2+18z+13)} \tag{13.186}$$

The pole at $x = 3$ has been transformed to a pole at $z = 0$. We then interpolate the part of the integrand that is analytic at $z = 0$ as

$$\frac{1}{(9z^2+18z+13)} \simeq \sum_{l=1}^N \mu_l(z)\frac{1}{(9x_l^2+18x_l+13)} \tag{13.187}$$

$$\Rightarrow \quad P\int_0^6 \frac{dx}{(x^2+4)(x-3)} \simeq -2\sum_{l=1}^N \frac{1}{P_N'(x_l)}\frac{1}{(9x_l^2+18x_l+13)}\,\mathrm{Re}\left[\frac{Q_N(x_l)-Q_N(0)}{x_l}\right] \tag{13.188}$$

If we use an odd number of Legendre quadrature points, one of the x_l in the set is 0. For $x_l = 0$, the term in brackets becomes $Q_N'(0)$ as described earlier. Since zero is not a value of a quadrature set with an even number of points, using such a set eliminates the need to calculate $Q_N'(0)$.

When each of the three integrals is approximated by a five-point Legendre quadrature rule, we obtain

$$P\int_{-\infty}^\infty \frac{dx}{(x^2+4)(x-3)} \simeq -0.37027 \tag{13.189a}$$

This is a reasonable but not very accurate result. However, with a 10-point quadrature set, we obtain

$$P\int_{-\infty}^\infty \frac{dx}{(x^2+4)(x-3)} \simeq -0.36249 \tag{13.189b}$$

which is exact to five decimal places. □

Referring to the discussions of the Legendre, Laguerre, and Hermite quadrature rules, we note that these sets of commonly used Sturm–Liouville polynomials, the roots of which are the quadrature abscissae, have the property that no root is also the end point of the appropriate range of x.

For example, for an integral over $x \in [-1, 1]$ a Legendre quadrature would be used, and for every Legendre polynomial, $P_N(\pm 1) \neq 0$. Similarly, Laguerre polynomials are not zero at $x = 0$ and $x = \infty$, and the Hermite polynomials satisfy $H_N(\pm\infty) \neq 0$. This property of the orthogonal polynomials used for Gaussian quadrature rules lends itself to the development of a third method for evaluating Cauchy principal value integrals.

To illustrate the method, we consider $P\int_a^b [f(x)/(x - x_0)]\, dx$. As discussed in Chapter 2, the principal value implies integration over all points except x_0. When $x < x_0$, the quantity $1/(x - x_0)$ is negative. $1/(x - x_0)$ is positive for $x > x_0$.

To concretize the discussion, we will take $f(x)$ to be positive in the region around x_0. The illustration of Figure 13.12 is a typical representation of the integrand in the region around x_0. Thus a numerical scheme that approximates the principal value integral must contain points that get close to x_0 but never be equal to it. In addition, the scheme must evaluate the integrand at equal distances from x_0 on the two sides of the division. This is necessary so that the large positive values of the integrand in the region just above x_0 are canceled by equally large negative values of the integrand just below x_0. Because the abscissae of the Gaussian quadratures never take on the values of the endpoint of the integral, the Gauss quadrature rules accomplish this in a natural way.

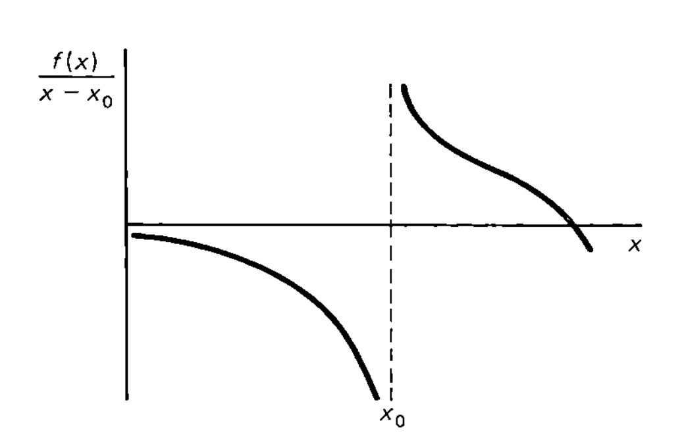

Figure 13.12 Representation of $f(x)/(x - x_0)$ around x_0.

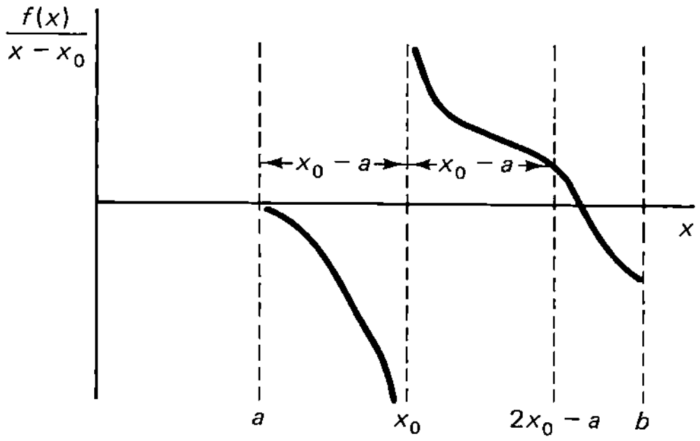

Figure 13.13 Division of interval $[a, b]$ for evaluating principal value integral.

For the sake of demonstration, we will take x_0 to be closer to the lower limit than to the upper limit of the integral. The first step is to divide the range of integration into three regions (see Figure 13.13). The first range is from the lower limit a to x_0, the second from x_0 to $2x_0 - a$, and the third from $2x_0 - a$ to b. This division makes the widths of the intervals around x_0 equal. The principal value integral is then written

$$P\int_a^b \frac{f(x)}{(x - x_0)}\, dx = \int_a^{x_0} \frac{f(x)}{(x - x_0)}\, dx + \int_{x_0}^{2x_0 - a} \frac{f(x)}{(x - x_0)}\, dx + \int_{2x_0 - a}^{b} \frac{f(x)}{(x - x_0)}\, dx \tag{13.190}$$

We now transform each of these integrals to the interval $[-1, 1]$. That is, for the three integrals, we substitute

$$\text{(a)}\quad z = -1 + 2\frac{(x - x_0)}{(a - x_0)} \qquad \text{(b)}\quad z = -1 + 2\frac{(x - x_0)}{(x_0 - a)} \qquad \text{(c)}\quad z = -1 + 2\frac{(x - 2x_0 + a)}{(b - 2x_0 + a)} \tag{13.191}$$

respectively. Then the first and second integrals become

$$\int_a^{x_0} \frac{f(x)}{(x - x_0)}\, dx = \int_{-1}^{1} \frac{f\left[a + \frac{1}{2}(x_0 - a)(1 + z)\right]}{(z - 1)}\, dz$$

$$\simeq \sum_{l=1}^{N} w_l \frac{f\left[a + \frac{1}{2}(x_0 - a)(1 + z_l)\right]}{(z_l - 1)} \tag{13.192a}$$

$$\int_{x_0}^{2x_0 - a} \frac{f(x)}{(x - x_0)}\, dx = \int_{-1}^{1} \frac{f\left[x_0 + \frac{1}{2}(x_0 - a)(1 + z)\right]}{(z + 1)}\, dz$$

$$\simeq \sum_{l=1}^{N} w_l \frac{f\left[x_0 + \frac{1}{2}(x_0 - a)(1 + z_l)\right]}{(z_l + 1)} \tag{13.192b}$$

As can be seen, the singularity at $x = x_0$ in the first integral has been transformed to the singularity at $z = 1$, and the singularity at $x = x_0$ in the second integral is now at $z = -1$. We note that the Legendre quadrature is symmetric about zero. That is, $z_1 = -z_N$ where z_1 is the quadrature point closest to -1 and z_N is the point closest to 1. Therefore

$$\frac{1}{(z_N - 1)} = -\frac{1}{(z_1 + 1)} \tag{13.193}$$

If z_N is close to 1, and therefore z_1 is close to -1, $f(z_N)$ in the quadrature sum of equation 13.192a has essentially the same value as $f(z_1)$ in the quadrature sum of equation 13.192b. Thus when the quadrature sums are added, the $l = N$ term of the first sum and the $l = 1$ term of the second sum will essentially add to zero, even though each of these terms is large individually. That is the reason it is important that the two intervals around the singularity be the same width and that the same quadrature set be used to approximate the integrals over these intervals.

Because the third integral is over a range of x-values far from the singularity, this integral is evaluated by a quadrature sum with no additional manipulations.

Example 13.20

As an example, we will evaluate

$$P\int_0^3 \frac{dx}{(x + 1)(x - \frac{1}{2})} = \frac{2}{3}\,\mathrm{Re}\int_0^3 \left[\frac{dx}{(x - \frac{1}{2} + i\varepsilon)} - \frac{dx}{(x + 1)}\right] = \frac{2}{3}\log(\tfrac{5}{4}) = 0.148762 \tag{13.194}$$

Noting that $\frac{1}{2}$ is closer to 0 than to 3, we first divide the interval from 0 to 3 as

$$P\int_0^3 \frac{dx}{(x + 1)(x - \frac{1}{2})} = \int_0^{1/2} \frac{dx}{(x + 1)(x - \frac{1}{2})}$$

$$+ \int_{1/2}^{1} \frac{dx}{(x + 1)(x - \frac{1}{2})} + \int_1^3 \frac{dx}{(x + 1)(x - \frac{1}{2})} \tag{13.195a}$$

Transforming each interval to $[-1, 1]$, we obtain

$$P\int_0^3 \frac{dx}{(x+1)(x-\frac{1}{2})} = 4\int_{-1}^1 \frac{dz}{(z+5)(z-1)} + 4\int_{-1}^1 \frac{dz}{(z+7)(z+1)} + \int_{-1}^1 \frac{dz}{(z+3)(z+\frac{3}{2})}$$

$$\simeq 4\sum_{l=1}^N w_l \frac{1}{(z_l+5)(z_l-1)} + 4\sum_{l=1}^N w_l \frac{1}{(z_l+7)(z_l+1)} + \sum_{l=1}^N w_l \frac{1}{(z_l+3)(z_l+\frac{3}{2})} \quad (13.195b)$$

Each of these sums is evaluated using a five-point Legendre quadrature. We obtain

$$P\int_0^3 \frac{dx}{(x+1)(x-\frac{1}{2})} \simeq 0.148663 \quad (13.196)$$

so the approximation is quite reasonable. With a 10-point Legendre rule, the result is identical to the exact value to six decimal places. □

When the range of integration is $[-\infty, \infty]$, there are two approaches that one can employ to evaluate principal value integrals. These will be illustrated by example, again evaluating the integral

$$P\int_{-\infty}^{\infty} \frac{dx}{(x^2+x_1^2)(x-x_0)} = -\frac{\pi x_0}{x_1(x_0^2+x_1^2)} \quad (2.303)$$

numerically.

One approach begins by arbitrarily choosing some point $x = a$ below the pole, and writing the integral as follows:

$$P\int_{-\infty}^{\infty} \frac{dx}{(x^2+x_1^2)(x-x_0)} = \int_{-\infty}^{a} \frac{dx}{(x^2+x_1^2)(x-x_0)} + \int_{a}^{x_0} \frac{dx}{(x^2+x_1^2)(x-x_0)}$$

$$+ \int_{x_0}^{2x_0-a} \frac{dx}{(x^2+x_1^2)(x-x_0)} + \int_{2x_0-a}^{\infty} \frac{dx}{(x^2+x_1^2)(x-x_0)} \quad (13.197)$$

Transformations are now performed to transform each range of integration to $[-1, 1]$. Each of these integrals is then evaluated as described above.

Example 13.21

To illustrate, we again consider

$$P\int_{-\infty}^{\infty} \frac{dx}{(x^2+4)(x-3)} = -0.36249 \quad (13.179)$$

Referring to equation 13.197, we take $a = 0$. Then

$$P\int_{-\infty}^{\infty} \frac{dx}{(x^2+4)(x-3)} = \int_{-\infty}^{0} \frac{dx}{(x^2+4)(x-3)} + \int_{0}^{3} \frac{dx}{(x^2+4)(x-3)}$$

$$+ \int_{3}^{6} \frac{dx}{(x^2+4)(x-3)} + \int_{6}^{\infty} \frac{dx}{(x^2+4)(x-3)} \quad (13.198)$$

Transforming each interval to $[-1, 1]$ and evaluating the integrals by a five-point Legendre quadrature yields

$$P\int_{-\infty}^{\infty} \frac{dx}{(x^2+4)(x-3)} \simeq -0.36249 \quad (13.199)$$

which is identical to the exact result to five decimal places. □

Another approach, using the same basic idea is to transform the range of integration to a finite range, then evaluate the resulting integral by the outlined prescription for finite integrals.

Example 13.22

To illustrate, we transform to finite limits by setting $x = 2\tan y$. Then

$$P\int_{-\infty}^{\infty} \frac{dx}{(x^2+4)(x-3)} = P\int_{-\pi/2}^{\pi/2} \frac{dy}{(2\tan y - 3)} \tag{13.200}$$

which has transformed the pole to $y = \tan^{-1}(\frac{3}{2}) = 0.98279$. Since the pole is closer to $\pi/2$ than to $-\pi/2$, the integral is written as

$$P\int_{-\pi/2}^{\pi/2} \frac{dy}{(2\tan y - 3)} = \int_{-\pi/2}^{2\tan^{-1}(3/2)-\pi/2} \frac{dy}{(2\tan y - 3)} + \int_{2\tan^{-1}(3/2)-\pi/2}^{\tan^{-1}(3/2)} \frac{dy}{(2\tan y - 3)}$$
$$+ \int_{\tan^{-1}(3/2)}^{\pi/2} \frac{dy}{(2\tan y - 3)} \tag{13.201}$$

As before, each of these ranges is then transformed to $[-1, 1]$ and then evaluated by a Legendre quadrature. Using a five-point quadrature rule for each integral, the result is

$$P\int_{-\infty}^{\infty} \frac{dx}{(x^2+4)(x-3)} \simeq -0.36244 \tag{13.202}$$

which is fairly accurate and comparable to the result obtained by other methods. □

Monte Carlo integration

The Monte Carlo method of evaluating an integral requires computer software that has a random number generator. The scheme involves generating a large number of random numbers.

The method is based on the fact that the integral $\int_a^b F(x)\,dx$ is the area under the curve $F(x)$ versus x. A representative curve is shown in Figure 13.14. Note that we have also indicated a value F_{max} which is no smaller than the largest value of $F(x)$ in the interval $[a, b]$. Since $F_{max} \geqslant F(x)$ for all $x \in [a, b]$, the area under the curve described by $F(x)$ is less than the area of the rectangle.

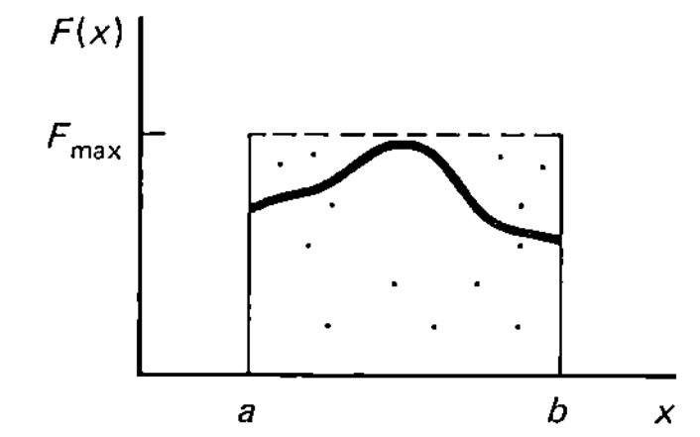

Figure 13.14 Representative function $F(x)$ with indication of maximum value of function.

The process requires generating several pairs of random numbers. For each pair of generated numbers, we take one number to represent a value of x and the other to represent a value of $y = F$. Many generators produce random numbers between 0 and

1. For random numbers $N_1, N_2 \in [0, 1]$ values of $x \in [a, b]$ and $y \in [0, F_{max}]$ are obtained from

$$\text{(a)} \quad x = (b - a)N_1 + a \qquad \text{(b)} \quad y = N_2 F_{max} \tag{13.203}$$

The next step is to determine the value of the function $F(x)$ at the randomly generated x, and compare it to the randomly generated value of y. If the randomly generated y-value is less than or equal to the value of $F(x)$ at the randomly generated value of x, we call this a *hit*. If the random y is greater than $F(x)$ at the random x, we call it a *miss*. The points under the $F(x)$ curve in Figure 13.14 represent hits and those above the $F(x)$ curve are misses. The probability that a pair of random numbers creates a hit is the ratio of the area under the $F(x)$ curve to the area of the rectangle. Thus, after generating a large number of pairs,

$$\frac{\text{area under } F(x) \text{ curve}}{\text{area of rectangle}} \simeq \frac{\text{number of hits}}{\text{total number of pairs generated}} \tag{13.204a}$$

Since the area of the rectangle is $(b - a)F_{max}$ and the area under the $F(x)$ curve is the integral being evaluated, this becomes

$$\text{area under } F(x) = \int_a^b F(x)\,dx \simeq \frac{\text{number of hits}}{\text{total number of pairs generated}}(b - a)F_{max} \tag{13.204b}$$

We point out that the height of the rectangle can be greater than F_{max} but it cannot be less than F_{max}. But as the height of the rectangle increases, the more misses there will be. Then one must generate a larger number of pairs of random numbers to achieve reliable statistics (a representative fraction of hits) to obtain an accurate result. As such, for the sake of efficiency, F_{max} should be as close to (but never less than) the largest value of $F(x)$ in $[a, b]$.

Example 13.23

To illustrate the process, we present a Fortran program to evaluate the integral

$$I_1 \equiv \int_0^2 (1 + x^3)\,dx = 6 \tag{13.205}$$

In this example, the maximum value of the function is easily obtained from evaluating the integrand at the upper limit. That is, $F_{max} = 9$. Thus, for this integral, equation 13.204b becomes

$$I_1 \simeq 18\frac{\text{number of hits}}{\text{total number}} \tag{13.206}$$

We assume that the Fortran package contains a random number generator such that calling the program RND(1) generates a random number between 0 and 1. The flowchart for the code to evaluate this integral is given in Figure 13.15. The Fortran program to perform this task is presented in Figure 13.16 on page 708.

Using this code, the author has evaluated the integral of Example 13.23. For each choice of number of pairs generated, two runs were executed. Typical results are presented in Table 13.12. □

TABLE 13.12 TYPICAL MONTE CARLO VALUES FOR $\int_0^2 (1 + x^3)\,dx$

	500	Percent Error	750	Percent Error	1000	Percent Error	5000	Percent Error
First run	6.4440	7.4	6.0720	1.2	5.5800	7.0	5.9652	0.6
Second run	6.1200	2.0	6.1440	2.4	5.7960	3.4	5.9904	0.2

Since the Monte Carlo technique is a statistical one, it is not possible to obtain consistent results from one execution of the program to the next. The accuracy of the result of any one run depends on how good the package's random number generator is.

Multiple integrals can also be evaluated using Monte Carlo methods by a straightforward extension of the analysis above. For example, the double integral

$$I_2 \equiv \int_{x=a}^{x=b} \int_{y=c}^{y=d} F(x, y)\, dx\, dy \tag{13.207}$$

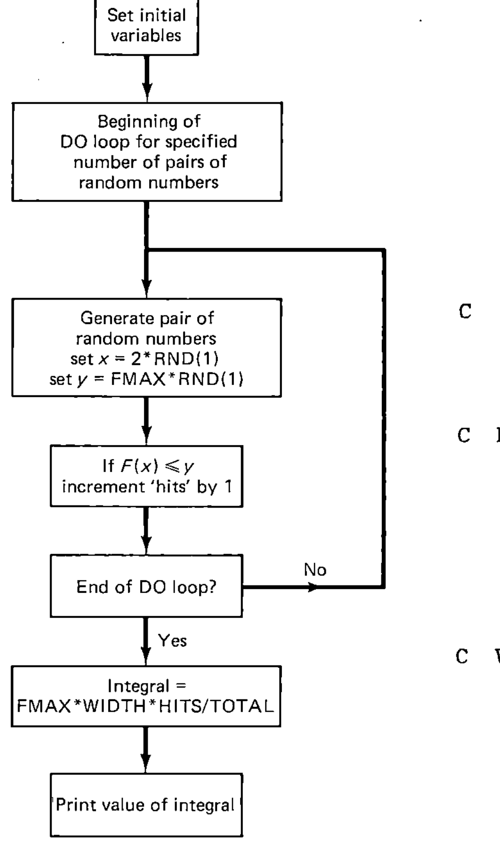

Figure 13.15 Flowchart for Monte Carlo evaluation of an integral.

```
C  INITIALIZE PARAMETERS
     FMAX = 9.
     WIDTH = 2.
     NPTS = 5000
     HITS = 0.
C  DO LOOP
     DO 100 I = 1, NPTS
        X = 2.*RND(1)
        Y = FMAX*RND(1)
        FN = F(X)
        IF Y <= FN THEN
           HITS = HITS + 1
        ENDIF
     100 CONTINUE
C  VALUE OF INTEGRAL
     FINT = FMAX*WIDTH*HITS/NPTS
     PRINT (6,*)"INTEGRAL = ",FINT
     STOP
     FUNCTION F(X)
     F = 1. + X**3
     RETURN
     END
```

Figure 13.16 Fortran code for Monte Carlo evaluation of integral of $F(x)$.

Multiple integrals can also be evaluated using Monte Carlo methods by a straightforward extension of the analysis above. For example, the double integral

$$I_2 \equiv \int_{x=a}^{x=b} \int_{y=c}^{y=d} F(x, y)\, dx\, dy \tag{13.207}$$

is evaluated by generating a triplet of random numbers, one representing $x \in [a, b]$, one for $y \in [c, d]$, and the third for $z \in [0, F_{max}]$. If the randomly generated z is less than or equal to the integrand at the random values of x and y, the number of hits is incremented by 1. In this way

$$I_2 \simeq (b-a)(d-c)F_{max} \frac{\text{number of hits}}{\text{total number of triplets}} \tag{13.208}$$

Example 13.24

Using the random number generator RND(1) to obtain random numbers between 0 and 1, a Fortran program that will evaluate

$$I_2 = \int_{x=0}^{x=2} \int_{y=0}^{y=3} (x+y)^{1/2}\, dx\, dy = \tfrac{4}{15}(5^{5/2} - 2^{5/2} - 3^{5/2}) = 9.2417 \tag{13.209}$$

is shown in Figure 13.17. Results obtained by the author using this code are presented in Table 13.13. □

```
C  INITIALIZE PARAMETERS
     FMAX = SQRT(5.)
     XWIDTH = 2.
     YWIDTH = 3.
     NPTS = 5000
     HITS = 0.
C  DO LOOP
     DO 100 I = 1, NPTS
        X = 2.*RND(1)
        Y = 3.*RND(1)
        Z = FMAX*RND(1)
        FN = F(X,Y)
        IF Z <= FN THEN
           HITS = HITS + 1
        ENDIF
     100 CONTINUE
C  VALUE OF INTEGRAL
     FINT = FMAX*XWIDTH*YWIDTH*HITS/NPTS
     PRINT (6,*)"INTEGRAL = ",FINT
     STOP
     FUNCTION F(X)
     F = SQRT(X+Y)
     RETURN
     END
```

Figure 13.17 Fortran code for Monte Carlo evaluation of integral of $F(x, y)$.

TABLE 13.13 TYPICAL MONTE CARLO VALUES FOR $\int_0^2 dx \int_0^3 dy (x+y)^{1/2}$

	500	Percent Error	750	Percent Error	1000	Percent Error	5000	Percent Error
First run	9.1232	1.3	9.5525	3.4	9.2573	0.2	9.1763	0.7
Second run	8.9353	3.3	8.8012	3.3	9.3110	0.7	9.3056	0.7

13.4 Numerical Solutions to Differential Equations

This section is devoted to the development of methods for numerically estimating the solution to first- and second-order differential equations with specified initial or boundary conditions. As discussed in Chapter 5, if the differential equation is inhomogeneous, the inclusion of the initial or boundary conditions yields the complete solution. If we do not impose initial or boundary conditions on an inhomogeneous equation, the resulting numerical values will represent only the particular solution.

The general form of a first-order linear differential equation is

$$\frac{dy}{dx} = f(x, y) \tag{13.210a}$$

which has one boundary condition expressed in the form $y(x_0) = y_0$. The second-order differential equation we will investigate is of the form

$$y''(x) + P(x)y'(x) + Q(x)y(x) = f(x, y, y') \tag{13.210b}$$

subject to initial conditions $y(x_0) = y_0$ and $y'(x_0) = y_0'$, or boundary conditions $y(a) = y_a$ and $y(b) = y_b$.

Taylor series approximation for first-order equations

The Taylor series method assumes a solution for $y(x)$ in the form of a Taylor series around x_0.

$$y(x) = y(x_0) + y'(x_0)(x - x_0) + \frac{1}{2!}y''(x_0)(x - x_0)^2 + \cdots \tag{13.211}$$

One aspect of this approach requires a word of caution (which applies to the Taylor series approximation for inhomogeneous equations of all orders). Since the solution is not known, we do not know the radius of convergence of the Taylor series for $y(x)$. For a second-order equation, some information about the singularity structure of $y(x)$ can be deduced from the functions $P(x)$ and $Q(x)$ defined in equation 13.210b, and the explicit x-dependence of the inhomogeneous term, f. However, for an inhomogeneous equation of any order, f also contains implicit x-dependence through its dependence on y and its derivatives. Therefore, without knowing $y(x)$, the complete singularity structure of f is not known, and therefore we cannot be assured a priori that the Taylor series converges at the value of x at which the solution is being estimated.

By specifying the accuracy to which we want to evaluate y at a given value of x, we terminate the series by dropping all terms that are smaller than the desired accuracy.

The only task is to determine the various derivatives of y at x_0. For the first-order differential equation, we have

$$y'(x_0) = f(x_0, y_0) \tag{13.212a}$$

The second derivative can be found by differentiating the differential equation once and evaluating x at x_0. That is,

$$y''(x_0) = \left.\frac{df}{dx}\right|_{x=x_0} = \left[\frac{\partial f}{\partial x} + \frac{\partial f}{\partial y}\frac{dy}{dx}\right]_{x=x_0} = \left[\frac{\partial f}{\partial x} + f\frac{\partial f}{\partial y}\right]_{x=x_0} \tag{13.212b}$$

Similarly,

$$y'''(x_0) = \left.\frac{d^2 f}{dx^2}\right|_{x=x_0} = \frac{\partial}{\partial x}\left[\frac{\partial f}{\partial x} + \frac{\partial f}{\partial y}\frac{dy}{dx}\right]_{x=x_0} + f\frac{\partial}{\partial y}\left[\frac{\partial f}{\partial x} + \frac{\partial f}{\partial y}\frac{dy}{dx}\right]_{x=x_0} \tag{13.212c}$$

This process is continued for all required derivatives of y at x_0.

Example 13.25

As an example, consider

$$\text{(a)}\quad \frac{dy}{dx} = x^2y^2 \qquad \text{with} \qquad \text{(b)}\quad y(0) = 3 \tag{13.213}$$

Clearly, this equation is separable, and an analytic solution can be obtained. It is

$$\text{(a)}\quad y(x) = \frac{3}{(1 - x^3)} \qquad \Rightarrow \qquad \text{(b)}\quad y(0.5) = 3.4286 \tag{13.214}$$

Stating the result to four decimal places implies that we require four-decimal-place accuracy in the numerical solution.

From the differential equation, we obtain $y'(0) = 0$ and

$$y''(x) = 2(xy^2 + x^2yy') \tag{13.215a}$$

Thus $y''(0) = 0$ and

$$y'''(x) = 2y^2 + 8xyy' + 2x^2y'^2 + 2x^2yy'' \qquad (13.215b)$$

from which $y'''(0) = 18$. Therefore, the MacLaurin expansion of y is

$$y(x) \simeq 3 + 0 + 0 + \frac{18}{3!}x^3 + \cdots = 3(1 + x^3) + \cdots \qquad (13.216a)$$

Using these first two nonzero terms in the MacLaurin expansion to represent $y(x)$, we find that

$$y(0.5) \simeq 3.3750 \qquad (13.216b)$$

which is not very accurate. □

If the value of x at which y is to be estimated is close to x_0, a small number of terms in the Taylor expansion will, in principle, be sufficient to achieve the required accuracy. However, if $x - x_0$ is relatively large, a small number of terms will not suffice. In the example above, we have obviously not achieved four-decimal-place accuracy by keeping terms up to x^3.

One method of improving the accuracy is to take a large enough number of terms that the truncated series yields the desired accuracy. Of course, this requires a large number of derivatives of y at x_0. In the example above, if we expand the series out to x^6, we obtain

$$\text{(a)} \quad y(x) \simeq 3 + \frac{18}{3!}x^3 + \frac{2160}{6!}x^6 = 3(1 + x^3 + x^6) \quad \Rightarrow \quad \text{(b)} \quad y(0.5) \simeq 3.4219 \qquad (13.217)$$

This is in better agreement with the exact result than the estimate obtained truncating the series after x^3. However, it is still not accurate to four decimal places. We note that equations 13.216a and 13.217a are truncated MacLaurin series expansions of the exact solution given in equation 13.214a. Thus it is expected that the method will yield accurate results if a large number of terms in the Taylor series are generated.

However, with a requirement of a high degree of accuracy, it may be necessary to use such a large number of terms that this approach becomes impractical. An iterative scheme can be employed in which the desired accuracy can be achieved using a small number of terms in the truncated Taylor series. The method involves using a series with a small number of terms to determine y at intermediate values of x, and repeating the method at increasing values of x until y is obtained at the specified x-value.

That is, we assume that we are trying to estimate $y(x_N)$ with $y(x_0)$ specified, and with $x_N - x_0$ too large to yield accurate results with the first few terms of the Taylor series. We take x_1 to be a point close enough to x_0 that we can estimate $y(x_1)$ accurately with a small number of terms. $y'(x_1)$, $y''(x_1)$, and so on, can be evaluated from the differential equation. Therefore, as an intermediate step, we determine $y(x_1)$ by

$$y(x_1) \simeq y(x_0) + y'(x_0)(x_1 - x_0) + \frac{1}{2!}y''(x_0)(x_1 - x_0)^2 + \cdots \qquad (13.218a)$$

Using this value of $y(x_1)$, and the differential equation to determine $y'(x_1)$, $y''(x_1)$, and so on, we then use the small series to estimate $y(x_2)$ from

$$y(x_2) \simeq y(x_1) + y'(x_1)(x_2 - x_1) + \frac{1}{2!}y''(x_1)(x_2 - x_1)^2 + \cdots \qquad (13.218b)$$

with acceptable accuracy. This process is carried out at several successively larger intermediate x-values until $y(x_N)$ is determined.

Example 13.26

For the differential equation of Example 13.25, if we use a two-step iteration, we would first estimate $y(0.25)$ from a series expansion around $x = 0$. The first two nonzero terms in the estimate of $y(x)$ expanded around $x = 0$ arise by keeping terms up to the third derivative of $y(x)$. Referring to equation 13.216a, we obtain

$$y(0.25) \simeq 3\left(1 + (0.25)^3\right) = 3.0469 \tag{13.219a}$$

From the differential equation, we obtain

$$y'(0.25) \simeq (0.25)^2(3.0469)^2 = 0.5802 \tag{13.219b}$$

Referring to equations 13.215 for the second and third derivatives, we have

$$\text{(c)} \quad y''(0.25) \simeq 4.8628 \qquad \text{(d)} \quad y'''(0.25) = 23.9970 \tag{13.219}$$

$$\Rightarrow \quad y(0.5) \simeq y(0.25) + y'(0.25)(0.5 - 0.25) + \frac{1}{2!}y''(0.25)(0.5 - 0.25)^2 + \frac{1}{3!}y'''(0.25)(0.5 - 0.25)^3 = 3.4064 \tag{13.220}$$

Although this two-step iteration still does not yield four-decimal-place accuracy, the result is more accurate than the single-step Taylor series truncated at x^3. The iteration is very straightforward to code, so the iterative process can be performed easily with a computer. For example, using a step size of 0.1 and five iterations, the author's computer estimate is

$$y(0.5) \simeq 3.4260 \quad \square \tag{13.221}$$

Picard's method for first-order equations

A Picard method (like that introduced in Chapter 12 to estimate the solution to various integral equations) can be developed for a first-order differential equation. Starting from the integrated form of equation 13.210a,

$$y(x) = y(x_0) + \int_{x_0}^{x} f(x, y)\, dx \tag{13.222}$$

we choose a step size Δx, such that $x_{n+1} = x_n + \Delta x$ and generate $y(x_1)$ from $y(x_0)$ by

$$\text{(a)} \quad y(x_1) = y(x_0) + \int_{x_0}^{x_1} f(x, y_0)\, dy \quad \Rightarrow \quad \text{(b)} \quad y(x_2) = y(x_1) + \int_{x_1}^{x_2} f(x, y_1)\, dy \tag{13.223}$$

and so on. In general, then,

$$y(x_{n+1}) = y(x_n) + \int_{x_n}^{x_{n+1}} f(x, y_n)\, dx \tag{13.224}$$

That is, for each iteration, the integral is evaluated over the interval Δx from x_n to x_{n+1}. If Δx is chosen small enough, it is assumed that the value of y does not vary much over this interval. As such, at each step, we approximate y in the argument of $f(x, y)$ by the constant y_n obtained from the previous iteration. Therefore, the explicit y-dependence (and thus the implicit x-dependence) has been removed from $f(x, y)$ and the intergrand in equation 13.224 contains only the explicit x-dependence. Therefore, the integral can, in principle, be evaluated.

Example 13.27

Returning to the equation of example 13.25, we estimate the value of $y(0.5)$ for

$$y' = x^2 y^2 \tag{13.213a}$$

by Picard's method. Equation 13.224 applied to this example becomes

$$y(x_{n+1}) \simeq y(x_n) + [y(x_n)]^2 \int_{x_n}^{x_{n+1}} x^2\, dx = y(x_n) + [y(x_n)]^2 \frac{(x_{n+1}^3 - x_n^3)}{3} \tag{13.225}$$

With $x_0 = 0$, $y(0) = 3$, and a step size $\Delta x = 0.1$, we obtain

$$y(0.1) \simeq y(0) + [y(0)]^2 \frac{(0.1^2 - 0^2)}{3} = 3.0300 \tag{13.226a}$$

$$\Rightarrow \quad y(0.2) \simeq y(0.1) + [y(0.1)]^2 \frac{(0.2^2 - 0.1^2)}{3} = 3.1218 \tag{13.226b}$$

and so on. Continuing this procedure, we find that

$$y(0.5) = 3.4072 \tag{13.226c}$$

With $\Delta x = 0.05$, we obtain

$$y(0.5) = 3.4172 \tag{13.227}$$

Both of these compare reasonably well with the exact result,

$$y(0.5) = 3.4286 \quad \square \tag{13.214b}$$

Runge – Kutta method for first-order equations

A commonly used approach for first-order equations is the Runge–Kutta method. Let x be the point at which we want the solution of equation 13.210a. The solution at a specified x is obtained in terms of four parameters. With $x - x_0 \equiv \Delta x$, we define

$$\text{(a)} \quad R_1 \equiv f(x_0, y_0)\,\Delta x \qquad \text{(b)} \quad R_2 \equiv f\left(x_0 + \tfrac{1}{2}\Delta x, y_0 + \tfrac{1}{2}R_1\right)\Delta x$$

$$\text{(c)} \quad R_3 \equiv f\left(x_0 + \tfrac{1}{2}\Delta x, y_0 + \tfrac{1}{2}R_2\right)\Delta x \qquad \text{(d)} \quad R_4 \equiv f(x_0 + \Delta x, y_0 + R_3)\,\Delta x \tag{13.228}$$

The Runge–Kutta solution for y in terms of these parameters is

$$y(x) = y(x_0) + \tfrac{1}{6}(R_1 + 2R_2 + 2R_3 + R_4) \tag{13.229}$$

The Runge–Kutta approach has the advantage over the Taylor expansion in that unlike the expansion in a Taylor series, the Runge–Kutta R-parameters do not require the determination of the derivatives of y.

To describe the Runge–Kutta process in geometric terms, we note that since

$$y'(x) = f[x, y(x)] \tag{13.210a}$$

is the slope of y at x, each $R/\Delta x$ represents the slope of y at a specified value of x. For example, $R_1/\Delta x$ is the slope of y at x_0. $R_4/\Delta x$ is the slope of y at $x_0 + \Delta x$, with y approximated by the constant $y_0 + R_3$ at this point. Similarly, $R_2/\Delta x$ and $R_3/\Delta x$ approximate the slope of y at the midpoint, $x_0 + \frac{1}{2}\Delta x$.

The relative weights of these slope parameters are chosen so that the Runge–Kutta solution is the same as the Taylor series expansion about x_0 up to order Δx^3. To see this, we expand the R-parameters in a Taylor series. Since the R-parameters have two arguments, we must develop the Taylor series for a function of two variables.

Let $F \equiv F(x, y)$, $\Delta x = x - x_0$, and $\Delta y = y - y_0$. We first expand F around x_0 holding y constant.

$$F(x, y) = \left(F(x, y) + \frac{\partial F}{\partial x}\Delta x + \frac{1}{2!}\frac{\partial^2 F}{\partial x^2}\Delta x + \cdots \right)_{x=x_0} \tag{13.230}$$

Each term in this expansion is then expanded around y_0. The terms are

$$F(x_0, y) = F(x_0, y_0) + \left.\frac{\partial F}{\partial y}\right|_{x_0 y_0}\Delta y + \left.\frac{\partial^2 F}{\partial y^2}\right|_{x_0 y_0}\Delta y^2 + \cdots \tag{13.231a}$$

$$\Delta x \left.\frac{\partial F}{\partial x}\right|_{x_0} = \Delta x \frac{\partial}{\partial x}\left(F(x, y) + \frac{\partial F}{\partial y}\Delta y + \frac{\partial^2 F}{\partial y^2}\Delta y^2 + \cdots \right)_{x_0 y_0} \tag{13.231b}$$

$$\Delta x^2 \left.\frac{\partial^2 F}{\partial x^2}\right|_{x_0} = \Delta x^2 \frac{\partial^2}{\partial x^2}\left(F(x, y) + \frac{\partial F}{\partial y}\Delta y + \frac{\partial^2 F}{\partial y^2}\Delta y^2 + \cdots \right)_{x_0 y_0} \tag{13.231c}$$

and so on. Therefore, the Taylor series expansion of a function of two variables is

$$\begin{aligned} F(x, y) = F(x_0, y_0) &+ \left.\frac{\partial F}{\partial x}\right|_{x_0 y_0}\Delta x + \left.\frac{\partial F}{\partial y}\right|_{x_0 y_0}\Delta y + \frac{1}{2!}\left.\frac{\partial^2 F}{\partial x^2}\right|_{x_0 y_0}\Delta x^2 + \frac{1}{2!}\left.\frac{\partial^2 F}{\partial y^2}\right|_{x_0 y_0}\Delta y^2 \\ &+ \frac{1}{1!1!}\left.\frac{\partial^2 F}{\partial x \partial y}\right|_{x_0 y_0}\Delta x\,\Delta y + \frac{1}{3!}\left.\frac{\partial^3 F}{\partial x^3}\right|_{x_0 y_0}\Delta x^3 + \frac{1}{2!\,1!}\left.\frac{\partial^3 F}{\partial x^2 \partial y}\right|_{x_0 y_0}\Delta x^2\,\Delta y \\ &+ \frac{1}{1!\,2!}\left.\frac{\partial^3 F}{\partial x\,\partial y^2}\right|_{x_0 y_0}\Delta x\,\Delta y^2 + \frac{1}{3!}\left.\frac{\partial^3 F}{\partial y^3}\right|_{x_0 y_0}\Delta y^3 + \cdots \end{aligned} \tag{13.232}$$

Using this result, we expand the R-parameters of equations 13.228 in a Taylor series. Although the algebra is quite cumbersome, when the expansions of the R-parameters are substituted into the Runge–Kutta solution of equation 13.229, it is straightforward to show that the Runge–Kutta solution is identical to the Taylor series expansion of $y(x)$ up to Δx^3. Thus the error in the Runge–Kutta solution is proportional to Δx^4.

Example 13.28

We apply the Runge–Kutta method to the differential equation

$$y' = x^2 y^2 \tag{13.213a}$$

of Example 13.25. With $x_0 = 0$ and $x = 0.5$ (and therefore, $\Delta x = 0.5$), the R-parameters are

$$R_1 = \Delta x f(x_0, y_0) = 0.5(0)^2(3)^2 = 0 \tag{13.233a}$$

$$R_2 = \Delta x f(x_0 + \tfrac{1}{2}\Delta x, y_0 + \tfrac{1}{2}R_1) = 0.5(0.25)^2(3)^2 = 0.2813 \tag{13.233b}$$

$$R_3 = \Delta x f(x_0 + \tfrac{1}{2}\Delta x, y_0 + \tfrac{1}{2}R_2) = 0.3082 \tag{13.233c}$$

$$R_4 = \Delta x f(x_0 + \Delta x, y_0 + R_3) = 1.3681 \tag{13.233d}$$

$$\Rightarrow \quad y(0.5) \simeq 3 + \tfrac{1}{6}(0 + 2 \cdot 0.2813 + 2 \cdot 0.3082 + 1.3681) = 3.4245 \tag{13.234}$$

in good agreement with the exact result. □

If Δx is too large to yield a result of the required accuracy with a single step from x_0 to x, it is very straightforward to generate a multistep iterative Runge–Kutta method. Starting at x_0, we can generate $y(x_1)$ using equations 13.228 and 13.229. Once $y(x_1)$ is known, the Runge–Kutta method is again used to determine $y(x_2)$. This procedure is repeated until $y(x_N)$ is generated.

Example 13.29

With $\Delta x = 0.25$, a two-step Runge–Kutta iteration for

$$y' = x^2 y^2 \tag{13.213a}$$

with $x_0 = 0$ yields

$$R_1 = (0.25)f(0,3) = (0.25)[0^2 \cdot 3^2] = 0 \tag{13.235a}$$

$$R_2 = (0.25)\left[\tfrac{1}{2}(0.25)\right]^2 \cdot (3)^2 = 0.0352 \tag{13.235b}$$

$$R_3 = (0.25)\left[\tfrac{1}{2}(0.25)\right]^2 \cdot \left[3 + \tfrac{1}{2}(0.0352)\right]^2 = 0.0356 \tag{13.235c}$$

$$R_4 = (0.25)(0.25)^2 \cdot [3 + 0.0356]^2 = 0.1440 \tag{13.235d}$$

$$\Rightarrow \quad y(0.25) = 3.0476 \tag{13.236}$$

Then with $x_0 = 0.25$ and $y_0 = 3.0476$, we obtain

$$R_1 = (0.25)f(0.25, 3.0476) = (0.25)(0.25)^2 \cdot (3.0476)^2 = 0.1451 \tag{13.237a}$$

$$R_2 = (0.25)\left[0.25 + \tfrac{1}{2}(0.25)\right]^2 \cdot \left[3.0476 + \tfrac{1}{2}(0.1451)\right]^2 = 0.3423 \tag{13.237b}$$

$$R_3 = (0.25)\left[0.25 + \tfrac{1}{2}(0.25)\right]^2 \cdot \left[3.0476 + \tfrac{1}{2}(0.3423)\right]^2 = 0.3642 \tag{13.237c}$$

$$R_4 = (0.25)[0.25 + 0.25]^2 \cdot [3.0476 + 0.3642]^2 = 0.7275 \tag{13.237d}$$

$$\Rightarrow \quad y(0.5) = 3.4285 \quad \square \tag{13.238}$$

Consider the Runge–Kutta method when $f(x, y) = f(x)$. Then

$$\begin{aligned} &\text{(a)} \quad R_1 = f(x_0)\,\Delta x \qquad \text{(b)} \quad R_2 = f\left(x_0 + \tfrac{1}{2}\Delta x\right)\Delta x \\ &\text{(c)} \quad R_3 = f\left(x_0 + \tfrac{1}{2}\Delta x\right)\Delta x = R_2 \qquad \text{(d)} \quad R_4 = f(x_0 + \Delta x)\,\Delta x \end{aligned} \tag{13.239}$$

$$y(x) \simeq y(x_0) + \tfrac{1}{6}\left[f(x_0) + 4f\left(x_0 + \tfrac{1}{2}\Delta x\right) + f(x_0 + \Delta x)\right] \tag{13.240}$$

But if $f(x, y) = f(x)$, integration of

$$\text{(a)} \quad y'(x) = f(x) \quad \Rightarrow \quad \text{(b)} \quad y(x) = y(x_0) + \int_{x_0}^{x} f(x)\,dx \tag{13.241}$$

If we use Simpson's $\frac{1}{3}$ rule, we must divide $[x_0, x]$ into an even number of segments. Dividing Δx into two segments, the width of each is $\frac{1}{2}(x - x_0) = \frac{1}{2}\Delta x$. Then

$$\text{(a)} \quad x_1 = x_0 \qquad \text{(b)} \quad x_2 = x_0 + \tfrac{1}{2}\Delta x \qquad \text{(c)} \quad x_3 = x_0 + \Delta x \tag{13.242}$$

Therefore, Simpson's $\frac{1}{3}$ rule yields

$$y(x) = y(x_0) + \int_{x_0}^{x} f(x)\,dx \simeq y(x_0) + \tfrac{1}{3}\left[f(x_0) + 4f\left(x_0 + \tfrac{1}{2}\Delta x\right) + f(x_0 + \Delta x)\right] \tag{13.243}$$

Comparing this to equation 13.240, we see that when $f(x, y) = f(x)$, the single-step Runge–Kutta solution of equation 13.241a is identical to an integration solution of the differential equation using a two-segment Simpson's $\frac{1}{3}$ rule.

Method of finite differences for first-order equations

As noted in equation 13.40, when the x-values are equally spaced, the derivative operator, D, can be written in terms of the finite difference operator Δ as

$$D = \frac{1}{h}\log(1+\Delta) \simeq \frac{1}{h}\Delta \tag{13.244}$$

where

$$h = x_{n+1} - x_n \tag{13.245}$$

$$\Rightarrow \quad Dy = f(x, y) \tag{13.246}$$

can be approximated by

$$\Delta y \simeq hf(x, y) \tag{13.25a}$$

Setting $x = x_k$, with $y_k \equiv y(x_k)$, and using

$$\Delta y_k = y_{k+1} - y_k \tag{13.24b}$$

equation 13.25a becomes the recurrence relation

$$y_{k+1} \simeq y_k + hf(x_k, y_k) \qquad 0 \leqslant k \leqslant N \tag{13.247}$$

The value of $y_0 = y(x_0)$ is given by the initial condition. Therefore,

$$\text{(a)} \quad y_1 = y_0 + hf(x_0, y_0) \qquad \Rightarrow \qquad \text{(b)} \quad y_2 = y_1 + hf(x_1, y_1) \tag{13.248}$$

and so on.

Example 13.30

To illustrate, we use the equation of example 13.25

$$\text{(a)} \quad Dy = x^2y^2 \qquad \text{with} \qquad \text{(b)} \quad y(0) = 3 \tag{13.213}$$

and determine $y(0.5)$ taking $h = 0.1$. As noted earlier, the analytic solution is

$$y(x) = \frac{3}{(1 - x^3)} \tag{13.214a}$$

from which $y(0.5) = 3.4286$.

With $x_0 = 0$, $y_0 = 3$, we obtain

$$y_1 = y(0.1) = y_0 + hf(x_0, y_0) = 3 + 0.1\left[(0^2)(3^2)\right] = 3.0000 \tag{13.249a}$$

$$y_2 = y(0.2) = y_1 + hf(x_1, y_1) = 3 + 0.1\left[(0.1^2)(3^2)\right] = 3.0090 \tag{13.249b}$$

$$y_3 = y(0.3) = y_2 + hf(x_2, y_2) = 3.0090 + 0.1\left[(0.2^2)(3.0090^2)\right] = 3.0452 \tag{13.249c}$$

$$y_4 = y(0.4) = y_3 + hf(x_3, y_3) = 3.0452 + 0.1\left[(0.3^2)(3.0452^2)\right] = 3.1287 \tag{13.249d}$$

$$y_5 = y(0.5) = y_4 + hf(x_4, y_4) = 3.1287 + 0.1\left[(0.4^2)(3.1287^2)\right] = 3.2853 \tag{13.249e}$$

We note that this result is not as accurate as the result obtained using other methods we have developed. □

Consider the Taylor expansion

$$y_{k+1} = y(x_k + h) = y_k + hy'(x_k) + \frac{1}{2!}h^2y''(x_k) + \cdots \tag{13.250}$$

From this, we obtain

$$\frac{y_{k+1} - y_k}{h} = \frac{1}{h}\,\Delta y_k = y'(x_k) + \frac{1}{2!}hy''(x_k) + \cdots \tag{13.251}$$

or, written in terms of the differential operator,

$$\frac{1}{h}\,\Delta y_k = Dy|_{x_k} + \frac{1}{2!}hD^2y|_{x_k} + \cdots \tag{13.252}$$

We see that by taking

$$Dy|_{x_k} \simeq \frac{1}{h}\,\Delta y_k \tag{13.253}$$

all terms in the Taylor series of order h^2 and higher are being ignored. This approximation of $y'(x_k)$ is known as the Euler approximation. It is often noted as having an accuracy up to order h.

Now consider the Taylor series

$$y_{k-1} = y(x_k - h) = y(x_k) - hy'(x_k) + \frac{1}{2!}h^2y''(x_k) - \cdots \tag{13.254}$$

Combining this with equation 13.250, we obtain

$$y_{k+1} - y_{k-1} = 2hy'(x_k) + \frac{2}{3!}h^3y'''(x_k) + \cdots \tag{13.255}$$

$$\Rightarrow \quad y'(x_k) = \frac{x_{k+1} - y_{k-1}}{2h} - \frac{2}{3!}h^2y'''(x_k) - \cdots \tag{13.256}$$

Thus, an approximation of the first derivative by

$$y'(x_k) \simeq \frac{y_{k+1} - y_{k-1}}{2h} \tag{13.257a}$$

is accurate to order h^2 whereas the Euler approximation is only accurate to order h. Thus this approximation, known as the Milne approximation of $y'(x_k)$, is more accurate than the Euler approximation.

Using the Milne estimate, a first-order differential equation is approximated by the finite difference equation

$$y_{k+1} = y_{k-1} + 2hf(x_k, y_k) \tag{13.257b}$$

for $1 \leqslant k \leqslant N$. Setting $k = 1$, this becomes

$$y_2 = y_0 + 2hf(x_1, y_1) \tag{13.258}$$

y_0 is given by the initial condition, but there is no specified value of y_1. In order to determine the value of y_2 an estimate of y_1 must be made that is independent of the differential equation. One such estimate can be made from the Euler approximation of the first derivative and the initial condition. That is, setting $k = 0$ in equation 13.247, we obtain

$$y_1 \simeq y_0 + hf(x_0, y_0) \tag{13.248a}$$

As we have seen, this Euler approximation of y_1 is a truncated Taylor series, accurate to order h. However, the Milne approximation is accurate to order h^2. Therefore an estimate of y_1 of order h^2 is more appropriate for the Milne approximation. To accomplish this, y_1 should be approximated by the truncated Taylor series

$$y_1 \simeq y_0 + hy_0' + \frac{1}{2}h^2y_0'' \tag{13.259}$$

where

$$y_0'' = \left[\frac{\partial f}{\partial x} + f\frac{\partial f}{\partial y}\right]_{x_0} \tag{13.212b}$$

It is not worthwhile estimating y_1 by a Taylor series with higher powers of h since the Milne approximation is accurate only to order h^2.

With this estimate of y_1, y_2 and all higher index y-values are obtained using the Milne approximation as specified in equation 13.257b.

Example 13.31

To illustrate the use of the Milne approximation, we again solve the differential equation of example 13.30

$$\text{(a)} \quad Dy = x^2y^2 \qquad \text{with} \qquad \text{(b)} \quad y(0) = y_0 = 3 \tag{13.213}$$

We use equation 13.259 to estimate y_1 rather than the less accurate Euler approximation. Taking $h = 0.1$, we obtain $y_1 = 3$. For this example y_1 has the same value as that obtained with the Euler estimate. From this, using the Milne approximation, the values of y_k are

$$y_2 = y(0.2) = y_0 + 2h(x_1)^2(y_1)^2 = 3 + 2(0.1)(0.1)^2(3)^2 = 3.0180 \tag{13.260a}$$

$$y_3 = y(0.3) = y_1 + 2h(x_2)^2(y_2)^2 = 3 + 2(0.1)(0.2)^2(30.180)^2 = 3.0729 \tag{13.260b}$$

$$y_4 = y(0.4) = y_2 + 2h(x_3)^2(x_3)^2 = 3.0180 + 2(0.1)(0.3)^2(3.0729)^2 = 3.1880 \tag{13.260c}$$

$$y_5 = y(0.5) = y_3 + 2h(x_4)^2(y_4)^2 = 3.0729 + 2(0.1)(0.4)^2(3.1880)^2 = 3.3981 \tag{13.260d}$$

This result is noticably more accurate than that obtained using the Euler approximation. □

Taylor series method for second-order linear equations

In expanding $y(x)$ in a Taylor series around x_0, it is necessary to know or be able to determine $y(x)$ and its derivatives at x_0. A knowledge of $y(x)$ at two different values of x is not needed for the Taylor series. Therefore, it is possible to apply the Taylor series approach to a second-order equation only if initial conditions $y(x_0) \equiv y_0$ and $y'(x_0) \equiv y_0'$ are specified. Equations for which boundary conditions $y(a) = y_a$ and $y(b) = y_b$ are imposed cannot be solved by this approach.

If initial conditions are given, the development of the method is identical to that for first-order equations. As before, the solution is approximated by a truncated Taylor series around $x = x_0$.

$$y(x) = y(x_0) + y'(x_0)(x - x_0) + \frac{1}{2!}y''(x_0)(x - x_0)^2 + \cdots \tag{13.261}$$

Since $y(x_0)$ and $y'(x_0)$ are known from the initial conditions, $y''(x_0)$ is obtained from the differential equation. Then by taking derivatives of the differential equation, $y'''(x_0)$, $y''''(x_0)$, and so on, can be determined.

Example 13.32

As a numerical example, we consider a Taylor series for

$$(1 - x^2)y'' - 2xy' + 6y = 0 \tag{13.262a}$$

which will be recognized as the Legendre differential equation for $l = 2$. If we take the initial conditions to be

$$\text{(b)}\quad y(1) = 1 \qquad \text{(c)}\quad y'(1) = 3 \tag{13.262}$$

the solution will be

$$y = P_2(x) = \tfrac{1}{2}(3x^2 - 1) \tag{13.262d}$$

These initial conditions are inconsistent with, and thus exclude, $Q_2(x)$.

Since the initial conditions are specified at 1, we take $x_0 = 1$. Thus the Taylor series for $y(x)$ is

$$y(x) = y(1) + y'(1)(x-1) + \frac{1}{2!}y''(1)(x-1)^2 + \frac{1}{3!}y'''(1)(x-1)^3 + \cdots \tag{13.263}$$

From equation 13.262a,

$$y''(x) = \frac{(2xy' - 6y)}{(1 - x^2)} \tag{13.264}$$

We see that in this form, $y''(1)$ is indeterminate. Therefore, using L'Hospital's rule,

$$\text{(a)}\quad y''(1) \doteq \lim_{x\to 1} \frac{\dfrac{d}{dx}(2xy' - 6y)}{\dfrac{d}{dx}(1 - x^2)} = 2y'(1) - y''(1)$$

$$\Rightarrow \qquad \text{(b)}\quad y''(1) = y'(1) = 3 \tag{13.265}$$

To evaluate $y'''(1)$, we consider

$$\frac{d}{dx}\left[(1 - x^2)y''\right] = -2xy'' + (1 - x^2)y''' \tag{13.266}$$

But from the Legendre differential equation,

$$\frac{d}{dx}\left[(1 - x^2)y''\right] = \frac{d}{dx}(2xy' - 6y) = 2xy'' - 4y' \tag{13.267}$$

Therefore,

$$y'''(1) = \lim_{x\to 1} \frac{(4xy'' - 4y')}{(1 - x^2)} = \frac{0}{0} \tag{13.268}$$

which by L'Hospital's rule becomes

$$\text{(a)}\quad y''' = -2y''' \qquad \Rightarrow \qquad \text{(b)}\quad y''' = 0 \tag{13.269}$$

Therefore, the truncated Taylor series becomes

$$y(x) = 1 + 3(x-1) - \tfrac{3}{2}(x-1)^2 = \tfrac{1}{2}(3x^2 - 1) = P_2(x) \tag{13.270}$$

It is not surprising that the truncated Taylor solution is the exact solution to the Legendre equation since that solution is a finite polynomial, $P_2(x)$. □

Example 13.33

As a second example, for which the solution is not a polynomial, consider the differential equation for the Bessel function of order $\frac{1}{2}$.

$$x^2y'' + xy' + \left(x^2 - \tfrac{1}{4}\right)y = 0 \tag{5.449}$$

The solution that is finite at $x = 0$ is

$$J_{1/2}(x) = \left(\frac{2}{\pi x}\right)^{1/2} \sin x \tag{6.572}$$

We can impose the initial conditions

$$\text{(a)} \quad y\left(\frac{\pi}{2}\right) = \frac{2}{\pi} \qquad \text{(b)} \quad y'\left(\frac{\pi}{2}\right) = -\frac{2}{\pi^2} \tag{13.271}$$

which will exclude all solutions other than $J_{1/2}(x)$.

The Taylor series expansion about $\pi/2$ up to the $(x - \pi/2)^3$ term is

$$y(x) \simeq y\left(\frac{\pi}{2}\right) + y'\left(\frac{\pi}{2}\right)\left(x - \frac{\pi}{2}\right) + \frac{1}{2!}y''\left(\frac{\pi}{2}\right)\left(x - \frac{\pi}{2}\right)^2 + \frac{1}{3!}y'''\left(\frac{\pi}{2}\right)\left(x - \frac{\pi}{2}\right)^2 \tag{13.272}$$

From the differential equation and one derivative of the differential equation, we obtain

$$\text{(a)} \quad y''\left(\frac{\pi}{2}\right) = \frac{6}{\pi^3} - \frac{2}{\pi} \qquad \text{(b)} \quad y'''\left(\frac{\pi}{2}\right) = \frac{6}{\pi^2} - \frac{30}{\pi^4} \tag{13.273}$$

Therefore, from equation 13.272 we obtain

$$y\left(\frac{\pi}{4}\right) \simeq 0.6349 \tag{13.274}$$

Referring to equation 6.572, the exact result is

$$y\left(\frac{\pi}{4}\right) = 0.6366 \tag{13.275}$$

So the third-order truncated Taylor series yields a reasonably accurate result. □

As we did with a first-order equation, we can achieve greater accuracy without increasing the number of terms in the series by iterating in steps from x_0 to the desired value of x. For example, if we estimate the value of $y(x_{n+1})$ using a series truncated after four terms, then

$$y(x_{n+1}) \simeq y(x_n) + y'(x_n)(x_{n+1} - x_n) + \frac{1}{2!}y''(x_n)(x_{n+1} - x_n)^2 + \frac{1}{3!}y'''(x_n)(x_{n+1} - x_n)^3 \tag{13.276}$$

For a first-order equation, once an estimate of $y(x_n)$ was made, the differential equation was used to obtain estimates of the required derivatives of y at x_n. For a second-order equation, we need an estimate of $y'(x_n)$ as well as $y(x_n)$. This can also be obtained by approximating $y'(x)$ by a truncated Taylor series. For example, we can estimate $y'(x_{n+1})$ by

$$y'(x_{n+1}) \simeq y'(x_n) + y''(x_n)(x_{n+1} - x_n) + \frac{1}{2!}y'''(x_n)(x_{n+1} - x_n)^2 + \frac{1}{3!}y''''(x_n)(x_{n+1} - x_n)^3 \tag{13.277}$$

Example 13.34

To illustrate, we again consider the example of the half-order Bessel equation, and use a two-step iteration from $\pi/2$ to $\pi/4$ using a step size of $\pi/8$ for each iteration. We approximate both $y(x_n)$ and $y'(x_n)$ by a four-term truncated Taylor series.

Since they will be needed for intermediate calculations, we differentiate the differential equation to obtain

$$y'''(x) = -\frac{y''}{x} + \left(\frac{5}{4x^2} - 1\right)y' - \frac{y}{2x^3} \tag{13.278a}$$

$$y''''(x) = -\frac{y'''}{x}\left(\frac{9}{4x^2} - 1\right)y'' - \frac{3y'}{x^3} + \frac{3y}{2x^4} \tag{13.278b}$$

$$\Rightarrow \quad y''''\left(\frac{\pi}{2}\right) = \frac{210}{\pi^5} - \frac{36}{\pi^3} + \frac{2}{\pi} \tag{13.279}$$

From this result, along with those expressed in equations 13.271 and 13.273, we obtain

$$y\left(\frac{\pi}{2} - \frac{\pi}{8}\right) = y\left(\frac{3\pi}{8}\right) \simeq y\left(\frac{\pi}{2}\right) + y'\left(\frac{\pi}{2}\right)\left(-\frac{\pi}{8}\right) + \frac{1}{2!}y''\left(\frac{\pi}{2}\right)\left(-\frac{\pi}{8}\right)^2$$

$$+ \frac{1}{3!}y'''\left(\frac{\pi}{2}\right)\left(-\frac{\pi}{8}\right)^3 = 0.6790 \tag{13.280a}$$

$$y'\left(\frac{3\pi}{8}\right) \simeq y'\left(\frac{\pi}{2}\right) + y''\left(\frac{\pi}{2}\right)\left(-\frac{\pi}{8}\right) + \frac{1}{2!}y'''\left(\frac{\pi}{2}\right)\left(-\frac{\pi}{8}\right)^2 + \frac{1}{3!}y''''\left(\frac{\pi}{2}\right)\left(-\frac{\pi}{8}\right)^3$$

$$= -0.0071 \tag{13.280b}$$

Then, with

$$y\left(\frac{3\pi}{8} - \frac{\pi}{8}\right) = y\left(\frac{\pi}{4}\right) \simeq y\left(\frac{3\pi}{8}\right) + y'\left(\frac{3\pi}{8}\right)\left(-\frac{\pi}{8}\right) + \frac{1}{2!}y''\left(\frac{3\pi}{8}\right)\left(-\frac{\pi}{8}\right)^2$$

$$+ \frac{1}{3!}y'''\left(\frac{3\pi}{8}\right)\left(-\frac{\pi}{8}\right)^3 \tag{13.281}$$

$$\Rightarrow \quad y\left(\frac{\pi}{4}\right) \simeq 0.6367 \tag{13.282}$$

which is quite accurate. □

If the initial conditions are specified as $y(x_0) = y_0$ and $y'(x_1) = y_1'$, evaluated at different points, the Taylor series cannot be developed directly from these conditions since the series requires knowing $y(x_0)$ and $y'(x_0)$. However, as discussed above, $y'(x_0)$ can be obtained from $y'(x_1)$ by iteration as in equation 13.280b. Thus an estimate of $y'(x_0)$ can be made. Then an iterated solution for $y(x_n)$ can be obtained as described above.

Runge – Kutta method for second-order equations

The Runge–Kutta method can also be applied to second-order differential equations with specified initial conditions. Consider

$$y'' + P(x)y' + Q(x)y = f(x) \tag{13.283}$$

with $y(x_0)$ and $y'(x_0)$ specified. This can be separated into two first-order equations by defining $w(x) = y'(x)$. The two first-order equations are

$$\text{(a)} \quad y'(x) = w(x) \equiv F(x, y, w) \qquad \text{with} \qquad \text{(b)} \quad y(x_0) \equiv y_0 \tag{13.284}$$

and

$$w'(x) = y''(x) = f(x) - P(x)w(x) - Q(x)y(x) \equiv G(x, y, w) \tag{13.285a}$$

subject to

$$w(x_0) = y'(x_0) \equiv w_0 \tag{13.285b}$$

We note that equation 13.268a is a first-order equation for $y(x)$. The Runge–Kutta solution for $y(x)$ is obtained from equation 13.285a in terms of the R-parameters. To compute the R-parameters, and to determine $y'(x)$, we require parameters which we denote by S. The R and S-parameters are

$$R_1 = F(x_0, y_0, w_0)\,\Delta x = w_0\,\Delta x \tag{13.286a}$$

$$S_1 = G(x_0, y_0, w_0)\,\Delta x = \left[f(x_0) - P(x_0)w_0 - Q(x_0)y_0\right]\Delta x \tag{13.286b}$$

$$\Rightarrow \quad R_2 = F\left(x + \tfrac{1}{2}\,\Delta x, y_0 + \tfrac{1}{2}R_1, w_0 + \tfrac{1}{2}S_1\right)\Delta x = \left(w_0 + \tfrac{1}{2}S_1\right)\Delta x \tag{13.286c}$$

$$S_2 = G\left(x_0 + \tfrac{1}{2}\,\Delta x, y_0 + \tfrac{1}{2}R_1, w_0 + \tfrac{1}{2}S_1\right)\Delta x$$
$$= \left[f\left(x_0 + \tfrac{1}{2}\,\Delta x\right) - P\left(x_0 + \tfrac{1}{2}\,\Delta x\right)\left(w_0 + \tfrac{1}{2}S_1\right) - Q\left(x_0 + \tfrac{1}{2}\,\Delta x\right)\left(y_0 + \tfrac{1}{2}R_1\right)\right]\Delta x \tag{13.286d}$$

$$\Rightarrow \quad R_3 = F\left(x_0 + \tfrac{1}{2}\,\Delta x, y_0 + \tfrac{1}{2}R_2, w_0 + \tfrac{1}{2}S_2\right)\Delta x = \left(w_0 + \tfrac{1}{2}S_2\right)\Delta x \tag{13.286e}$$

$$S_3 = G\left(x_0 + \tfrac{1}{2}\,\Delta x, y_0 + \tfrac{1}{2}R_2, w_0 + \tfrac{1}{2}S_2\right)\Delta x$$
$$= \left[f\left(x_0 + \tfrac{1}{2}\,\Delta x\right) - P\left(x_0 + \tfrac{1}{2}\,\Delta x\right)\left(w_0 + \tfrac{1}{2}S_2\right) - Q\left(x_0 + \tfrac{1}{2}\,\Delta x\right)\left(y_0 + \tfrac{1}{2}R_2\right)\right]\Delta x \tag{13.286f}$$

$$\Rightarrow \quad R_4 = F(x_0 + \Delta x, y_0 + R_3, w_0 + S_3)\,\Delta x = (w_0 + S_3)\,\Delta x \tag{13.286g}$$

$$S_4 = G(x_0 + \Delta x, y_0 + R_3, w_0 + S_3)\,\Delta x$$
$$= \left[f(x_0 + \Delta x) - P(x_0 + \Delta x)(w_0 + S_3) - Q(x_0 + \Delta x)(y_0 + R_3)\right]\Delta x \tag{13.286h}$$

As before, the solution to the first-order differential equation expressed in equation 13.284a is

$$y(x) = y(x_0) + \tfrac{1}{6}(R_1 + 2R_2 + 2R_3 + R_4) \tag{13.287a}$$

or since each R-parameter is linearly related to the corresponding S-parameter, this can be written

$$y(x) = y(x_0) + \frac{\Delta x}{6}(6w_0 + S_1 + S_2 + S_3) \tag{13.287b}$$

The derivative, $y'(x)$ is obtained from the Runge–Kutta expression of equation 13.285a involving the S-parameters. That is

$$y'(x) = y'(x_0) + \tfrac{1}{6}(S_1 + 2S_2 + 2S_3 + S_4) \tag{13.288}$$

Example 13.35

As an example, we will apply this to the Bessel equation of half-order, which we write in the form

$$y'' + \frac{1}{x}y' + \left(1 - \frac{1}{4x^2}\right)y = 0 \tag{13.289}$$

subject to $y(\pi/2) = 2/\pi$, and $y'(\pi/2) = w(\pi/2) = -2/\pi^2$. Then the coupled equations for this example are

$$\text{(a)}\quad y'(x) = w(x) \qquad \text{(b)}\quad w'(x) = -\frac{w}{x} - \left(1 - \frac{1}{4x^2}\right)y \equiv G(x, y, w) \tag{13.290}$$

To estimate $y(\pi/4)$, it is a straightforward problem to compute the numerical values of the

R and S parameters of equations 13.286. With $\Delta x = -\pi/4$ we obtain

$$\text{(a)} \quad R_1 = w_0\,\Delta x = 0.1592 \qquad \text{(b)} \quad S_1 = \left[-\frac{w_0}{x_0} - y_0\left(1 - \frac{1}{4x_0^2}\right)\right]\Delta x = 0.3480$$

$$\Rightarrow \quad \text{(c)} \quad R_2 = \left(w_0 + \tfrac{1}{2}S_1\right)\Delta x = 0.0225$$

$$\text{(d)} \quad S_2 = \left[-\frac{\left(w_0 + \frac{1}{2}S_1\right)}{\left(x_0 + \frac{1}{2}\Delta x\right)} - \left(y_0 + \tfrac{1}{2}R_1\right)\left(1 - \frac{1}{4\left(x_0 + \frac{1}{2}\Delta x\right)^2}\right)\right] = 0.4421 \tag{13.291}$$

$$\Rightarrow \quad \text{(e)} \quad R_3 = -0.0145 \qquad \text{(f)} \quad S_3 = 0.4295$$

$$\Rightarrow \quad \text{(g)} \quad R_4 = -0.1781 \qquad \text{(h)} \quad S_4 = 0.5174$$

The R-parameters yield the estimate

$$y\left(\frac{\pi}{4}\right) \simeq 0.6348 \tag{13.292}$$

which compares reasonably well with the exact result of 0.6366. From the S-parameters, we find that

$$y'\left(\frac{\pi}{4}\right) \simeq 0.2321 \tag{13.293a}$$

The exact value is

$$y'\left(\frac{\pi}{4}\right) = 0.2313 \quad \square \tag{13.293b}$$

Using the value of $y(x_1)$ can be found from the R-parameters and the values of $w(x_1) = y'(x_1)$ can be determined from the S-parameters. Thus it is straightforward to develop an iterative Runge–Kutta scheme for solving second-order equations. The reader is referred to problem 16 for such a development.

Method of finite differences for second-order equations with initial conditions

The method of finite differences was developed earlier for first-order differential equations by approximating the differential operator by

$$D \simeq \frac{1}{h}\Delta \tag{13.244}$$

for x-values equally spaced by $h = x_{n+1} - x_n$. Second-order equations in which the initial values $y(x_0) \equiv y_0$ and $y'(x_0) \equiv y_0'$ are given, can also be solved by finite difference methods.

Squaring the operator, we have

$$D^2 y \simeq \frac{1}{h^2}\Delta^2 y \tag{13.294}$$

$$\Rightarrow \quad y''(x_k) \simeq \frac{1}{h^2}\Delta(y_{k+1} - y_k) = \frac{1}{h^2}(y_{k+2} - 2y_{k+1} + y_k) \tag{13.295}$$

Then, with the Euler estimate of the first derivative, the equation

$$D^2 y + P(x)Dy + Q(x)y = f(x, y, y') \tag{13.296}$$

becomes

$$\frac{y_{k+2} - 2y_{k+1} + y_k}{h^2} + P(x_k)\frac{y_{k+1} - y_k}{h} + Q(x_k)y_k = f\left(x_k, y_k, \frac{y_{k+1} - y_k}{h}\right) \tag{13.297a}$$

with $k \geqslant 0$. For the purpose of writing computer code, that is more conveniently written as

$$\frac{1}{h^2}y_{k+2} + \left[-\frac{2}{h^2} + \frac{1}{h}P_k\right]y_{k+1} + \left[\frac{1}{h^2} - \frac{1}{h}P_k + Q_k\right]y_k = f\left(x_k, y_k, \frac{y_{k+1} - y_k}{h}\right) \tag{13.297b}$$

The initial conditions can be expressed as

$$y(x_0) = y_0 \tag{13.298a}$$

and, with the Euler approximation,

$$y'(x_0) = y_0' \simeq \frac{1}{h}\Delta y_0 = \frac{1}{h}(y_1 - y_0) \tag{13.298b}$$

From the second initial condition, we obtain

$$y_1 \simeq y_0 + hy_0' \tag{13.299}$$

Therefore, since y_0 and y_1 can be determined from the initial conditions, y_2 and all y-values of higher index values can be obtained from the recurrence relation of equation 13.297b.

Example 13.36

As an example, consider

$$\text{(a)}\quad y'' + xy' + y = (y')^2 \qquad \text{with} \qquad \text{(b)}\quad y(0) = -\frac{3}{2} \qquad \text{(c)}\quad y'(0) = 0 \tag{13.300}$$

It is straightforward to see by substitution that the solution to this equation is

$$y = \tfrac{3}{4}x^2 - \tfrac{3}{2} \tag{13.301}$$

Referring to equation 13.297b, with $x_0 = 0$ and a step size of $h = 0.1$, we determine $y_5 = y(0.5)$ from

$$\frac{1}{h^2}y_{k+2} + \left[-\frac{2}{h^2} + \frac{x_k}{h}\right]y_{k+1} + \left[\frac{1}{h^2} - \frac{x_k}{h} + 1\right]y_k = \frac{(y_{k+1} - y_k)^2}{h^2} \tag{13.302}$$

Using the initial conditions, equation 13.299 yields

$$y_1 = y(0.1) = y_0 + hy_0' = -1.5 = y_0 \tag{13.303}$$

We then set $k = 0$ in equation 13.302 to obtain

$$\frac{1}{(0.1)^2}y_2 + \left[\frac{-2}{(0.1)^2} + \frac{0}{0.1}\right](-1.5) + \left[\frac{1}{(0.1)^2} - \frac{0}{0.1} + 1\right](-1.5) = 0 \tag{13.304}$$

$$\Rightarrow \quad y_2 = y(0.2) = -1.48500 \tag{13.305}$$

Setting $k = 1$, with $x_1 = 0.1$, equation 13.302 becomes

$$\frac{1}{(0.1)^2}y_3 + \left[\frac{-2}{(0.1)^2} + \frac{0.1}{0.1}\right](-1.48500) + \left[\frac{1}{(0.1)^2} - \frac{0.1}{0.1} + 1\right](-1.5)$$

$$= \frac{(-1.485 + 1.5)^2}{(0.1)^2} \tag{13.306}$$

$$\Rightarrow \quad y_3 = y(0.3) = -1.45493 \tag{13.307}$$

Using this result, we set $k = 2$ to obtain

$$y_4 = y(0.4) = -1.40970 \tag{13.308}$$

and from this, with $k = 3$, we find

$$y_5 = y(0.5) = -1.36378 \tag{13.309}$$

This compares reasonably well with the exact result

$$y(0.5) = \tfrac{3}{4}(0.5)^2 - \tfrac{3}{2} = -1.31250 \quad \square \tag{13.310}$$

To use the Milne approximation, we note that the Milne estimate of the first derivative is symmetric about y_k and contains the y-values y_{k+1} and y_{k-1}. If we use the finite difference approximation of D^2y expressed in equation 13.295, the Milne finite difference equation will contain y_{k+2}, y_{k+1}, y_k, and y_{k-1}. The presence of four y-values in one equation makes the computation difficult, if not impossible, in most problems.

To avoid having four different y-values in the recurrence equation, we change the index notation of the second derivative and write it in the symmetric form

$$D^2 y|_{x_k} \simeq \frac{y_{k+1} - 2y_k + y_{k-1}}{h^2} \tag{13.311}$$

Using this symmetric form of D^2y, the Milne finite difference equation for a second-order equation becomes

$$\frac{y_{k+1} - 2y_k + y_{k-1}}{h^2} - P(x_k)\frac{y_{k+1} - y_{k-1}}{2h} + Q(x_k)y_k = f(x_k, y_k, y_k') \tag{13.312}$$

with $k \geqslant 1$. This equation only involves three y-values, y_{k+1}, y_k, and y_{k-1}.

We note that this symmetric form of the second derivative, when used with the Euler estimate of Dy, will also yield a finite difference equation with three y-values. As such, it is appropriate for both the Euler and Milne finite differences.

Setting $k = 1$, equation 13.312 becomes

$$\frac{y_2 - 2y_1 + y_0}{h^2} + P_1\frac{y_2 - y_0}{2h} + Q_1 y_1 = f(x_1, y_1, y_1') \tag{13.313}$$

y_0 is specified by one of the initial conditions. Therefore, to determine y_2 for example, y_1 must be determined from the second initial condition or some independent method. Alternatively, if y_2 can be found from some condition, y_1 can then be determined from equation 13.313.

But with $k = 1$, the first derivative, expressed in terms of the Milne approximation, is

$$y'(x_1) = y_1' = \frac{y_2 - y_0}{2h} \tag{13.314}$$

Since the value of y_1' is not given by the initial conditions, this cannot be used to determine y_2.

We can estimate y_1 by approximating y_0' by the Euler finite difference. Then

$$y_1 \simeq y_0 + hy_0' \tag{13.315a}$$

This is a truncated Taylor series for y_1 of order h. But, as noted earlier, the Milne estimate is accurate to order h^2. Therefore a preferable choice for approximating a quantity by a truncated Taylor series is to take a sum of order h^2. As such, when using the Milne approach a more appropriate estimate of y_1 is

$$y_1 \simeq y_0 + hy'_0 + \frac{1}{2}h^2 y''_0 \tag{13.315b}$$

Using a truncated Taylor series containing powers higher than h^2 is unnecessary since the Milne estimate is only accurate to order h^2. With this estimate of y_1, y_2 can then be determined from equation 13.313. All higher index y-values then follow using equation 13.312.

If $f(x, y, y')$ is independent of y' or if it depends linearly on y', then no further manipulation is required to use equation 13.312. In particular, if $f(x, y, y')$ depends linearly on y', the differential equation can be written

$$y'' + P(x)y' + Q(x)y = y'g_1(x, y) + g_2(x, y) \tag{13.316a}$$

which can be rewritten

$$y'' + R(x)y' + Q(x)y = g_2(x, y) \tag{13.316b}$$

This is the form of the differential equation with $f(x, y, y')$ independent of y'. As such, we need only consider the case for which

$$f(x, y, y') = f(x, y) \tag{13.317}$$

Then the finite difference equation becomes

$$\frac{y_{k+1} - 2y_k + y_{k-1}}{h^2} + P(x_k)\frac{y_{k+1} - y_{k-1}}{2h} + Q(x_k)y_k = f(x_k, y_k) \tag{13.318}$$

for $k \geqslant 1$. With y_1 determined as described in either of equations 13.315 (with equation 13.315b preferred), all y-values with index 2 and higher can then be found.

Example 13.37

To illustrate, we consider the equation

$$\text{(a)} \quad y'' + xy' + y = 2xy^4 \quad \text{with} \quad \text{(b)} \quad y(1) = 1 \quad \text{(c)} \quad y'(1) = -1 \tag{13.319}$$

The solution to this equation is

$$y(x) = \frac{1}{x} \tag{13.320}$$

The Milne finite difference estimate for this differential equation, with the second derivative written in symmetric form, is

$$\frac{y_{k+1} - 2y_k + y_{k-1}}{h^2} + x_k\frac{y_{k+1} - y_{k-1}}{2h} + y_k = 2x_k y_k^4 \tag{13.321}$$

with $k \geqslant 1$. Taking $h = 0.1$, we determine the value of $y(1.5)$, the exact value of which is $\frac{2}{3}$.

For $k = 1$, we have

$$\frac{y_2 - 2y_1 + y_0}{h^2} + x_1\frac{y_2 - y_0}{2h} + y_1 = 2x_1 y_1^4 \tag{13.322}$$

We estimate y_1 from the Taylor series of equation 13.315b, with $y'(1) = y'_0$ and $y''(1) = y''_0$ evaluated from the differential equation. We find

$$y_1 = y(1.1) \simeq y_0 + hy'_0 + \tfrac{1}{2}h^2 y''_0 = 1 + 0.1(-1) + \tfrac{1}{2}(0.01)(2) = 0.91 \tag{13.323}$$

$$\Rightarrow \quad y_2 = y(1.2) \simeq 0.83506 \tag{13.324}$$

Setting $k = 2$, equation 13.321 becomes

$$\frac{y_3 - 2y_2 + y_1}{h^2} + x_2\frac{y_3 - y_1}{2h} + y_2 = 2x_2y_2^4 \tag{13.325}$$

$$\Rightarrow \quad y_3 = y(1.3) \simeq 0.77173 \tag{13.326}$$

Continuing in this way, we eventually obtain

$$y_5 = y(1.5) \simeq 0.67069 \tag{13.327a}$$

which is a reasonable result. Using $h = 0.05$, we obtain

$$y_5 = y(1.5) \simeq 0.66768 \quad \square \tag{13.327b}$$

If $f(x, y, y')$ depends on y' in a nonlinear way, the Milne finite difference equation with the symmetric approximation to the second derivative is

$$\frac{y_{k+1} - 2y_k + y_{k-1}}{h^2} + P(x_k)\frac{y_{k+1} - y_{k-1}}{2h} + Q(x_k)y_k = f(x_k, y_k, y_k') \tag{13.312}$$

If y_k' is expressed in the Milne form in $f(x_k, y_k, y_k')$, y_{k+1} becomes the solution to a transcendental equation, or is one of the roots of a polynomial (as will be noted in example 13.38). In such cases, the procedure may be too cumbersome to be practical, and/or there may be multiple solutions (like several roots of a polynomial) with no prescription for deciding which solution to choose. As such, one must find an unambiguous estimate of y_k' at each step, which is then used in $f(x_k, y_k, y_k')$ to determine y_{k+1}. For example, writing

$$y_k' = y'(x_{k-1} + h) \tag{13.328}$$

y_k' can be approximated by the Euler estimate

$$y_k' \simeq y_{k-1}' + hy_{k-1}'' \tag{13.329a}$$

or the more accurate truncated Taylor sum

$$y_k' \simeq y_{k-1}' + hy_{k-1}'' + \tfrac{1}{2}h^2y_{k-1}''' \tag{13.329b}$$

where y_{k-1}'' is evaluated by setting $x = x_{k-1}$ in the differential equation and y_{k-1}''' is found by taking one derivative of the differential equation and setting $x = x_{k-1}$.

If $f(x, y, y')$ depends on y' and if that dependence is not linear, the additional manipulation that is required may not justify the improvement in accuracy provided by the Milne approach. Use of the Euler finite difference with the nonsymmetric second derivative, using a smaller step size may be preferable.

Example 13.38

An illustration of the effort required is provided by the equation of Example 13.36,

$$\text{(a)}\quad y'' + xy' + y = (y')^2 \quad \text{with} \quad \text{(b)}\quad y(0) = -1.5 \quad \text{(c)}\quad y'(0) = 0 \tag{13.300}$$

the solution to which is

$$y(x) = \tfrac{3}{4}x^2 - \tfrac{3}{2} \tag{13.301}$$

The finite difference approximation given in equation 13.312 is

$$\frac{y_{k+1} - 2y_k + y_{k-1}}{h^2} + x_k\frac{y_{k+1} - y_{k-1}}{2h} + y_k = (y_k')^2 \tag{13.330}$$

If $(y_k')^2$ were approximated by either the Euler or Milne estimates, then

$$\text{(a)} \quad (y_k')^2 \simeq \left(\frac{y_{k+1} - y_k}{h}\right)^2 \quad \text{or} \quad \text{(b)} \quad (y_k')^2 \simeq \left(\frac{y_{k+1} - y_{k-1}}{2h}\right)^2 \tag{13.331}$$

then y_{k+1} would be one of the roots of a quadratic equation, requiring a prescription for determining which root to select. As such, it is preferable to estimate $(y_k')^2$ as

$$(y_k')^2 = \left(y_{k-1}' + hy_{k-1}'' + \tfrac{1}{2}h^2 y_{k-1}'''\right)^2 \tag{13.332}$$

From the differential equation, we have

$$\text{(a)} \quad y_{k-1}'' = (y_{k-1}')^2 - x_{k-1}y_{k-1}' - y_{k-1} \qquad \text{(b)} \quad y_{k-1}''' = (2y_{k-1}' - x_{k-1} - 2)y_{k-1}'' \tag{13.333}$$

Thus setting $k = 1$, we have

$$\frac{y_2 - 2y_1 + y_0}{h^2} + x_1\frac{y_2 - y_0}{2h} + y_1 = \left(y_0' + hy_0'' + \tfrac{1}{2}h^2 y_0'''\right)^2 \tag{13.334}$$

We take $h = 0.1$ and use equations 13.333 to determine

$$y_1 = y(0.1) \simeq y_0 + hy_0' + \tfrac{1}{2}h^2 y_0'' = -1.49250 \tag{13.335}$$

$$\Rightarrow \quad y_2 = y(0.2) \simeq -1.47138 \tag{13.336}$$

Using equation 13.329b with $k = 1$ we obtain

$$y_1' \simeq y_0' + hy_0'' + \tfrac{1}{2}h^2 y_0''' = 0.13500 \tag{13.337}$$

Setting $k = 2$ in equations 13.333, we obtain values for y_1'' and y_1'''. These results then allow us to generate a value for y_3. Continuing this process, we finally obtain

$$y_5 = y(0.5) \simeq -1.31445 \tag{13.338}$$

which agrees well with the exact value of -1.31250. □

Method of finite differences for second-order equations with boundary conditions

The method of finite differences is also applicable to second-order equations subject to boundary conditions. Consider

$$\begin{aligned} &\text{(a)} \quad D^2y + P(x)Dy + Q(x)y = f(x, y, y') \\ &\text{with} \quad \text{(b)} \quad y(x_0) \equiv y_0 \qquad \text{(c)} \quad y(x_N) \equiv y_N \end{aligned} \tag{13.339}$$

The method of solution using a Milne approach very closely parallels the method with the Euler estimate. As such, we only present the details of the method using Euler finite difference methods. Using

$$\text{(a)} \quad y_k' \simeq \frac{y_{k+1} - y_k}{h} \quad \text{and} \quad \text{(b)} \quad y_k'' \simeq \frac{y_{k+2} - 2y_{k+1} + y_k}{h^2} \tag{13.340}$$

equation 13.339a becomes

$$\frac{y_{k+2} - 2y_{k+1} + y_k}{h^2} + P_k\frac{y_{k+1} - y_k}{h} + Q_k y_k = f(x_k, y_k, y_k') \tag{13.341}$$

with $0 \leqslant k \leqslant N - 2$.

Since y_0 and y_N are known from the boundary conditions, setting k to each of its values results in a set of $N-1$ simultaneous equations for the unknowns $y_1, y_2, \ldots, y_{N-1}$. If $f(x, y, y')$ contains any y-dependence or any y'-dependence (except perhaps depending linearly on one of these variables), the resulting simultaneous equations are nonlinear, and as such difficult to solve. Thus these finite difference methods are straightforwardly applicable to boundary value equations for which $f(x, y, y')$ is of the form

$$f(x, y, y') = g(x) + y g_1(x) + y' g_2(x) \tag{13.342}$$

where $g_1(x) = 0$ if the inhomogeneous term contains no y dependence, and $g_2(x) = 0$ if f is independent of y'. By combining the terms containing y and y' with the appropriate terms on the left-hand side of the differential equation, we obtain an equation in which the inhomogeneous term depends only on x. Then we can write equation 13.341 in the form

$$\frac{1}{h^2} y_{k+2} + \left[\frac{-2}{h^2} + \frac{P_k}{h}\right] y_{k+1} + \left[\frac{1}{h^2} - \frac{P_k}{h} + Q_k\right] y_k = g(x_k) \tag{13.343}$$

We introduce the notation

$$\text{(a)}\quad \alpha_k \equiv \left[\frac{-2}{h^2} + \frac{P_k}{h}\right] \qquad \text{(b)}\quad \beta_k \equiv \left[\frac{1}{h^2} - \frac{P_k}{h} + Q_k\right] \qquad \text{(c)}\quad g_k \equiv g(x_k) \tag{13.344}$$

Then, setting $k = 0, 1, 2, \ldots, N-2$, and using the fact that y_0 and y_N are known, we obtain the equations

$$\text{(a)}\quad \frac{1}{h^2} y_2 + \alpha_0 y_1 = g_0 - \beta_0 y_0 \qquad \text{(b)}\quad \frac{1}{h^2} y_3 + \alpha_1 y_2 + \beta_1 y_1 = g_1$$

$$\text{(c)}\quad \frac{1}{h^2} y_4 + \alpha_2 y_3 + \beta_2 y_2 = g_2 \qquad \cdots \qquad \text{(d)}\quad \alpha_{N-2} y_{N-1} + \beta_{N-2} y_{N-2} = g_N - \frac{1}{h^2} y_N \tag{13.345}$$

This set of equations, written in matrix form, is

$$\begin{bmatrix} \alpha_0 & \frac{1}{h^2} & 0 & 0 & 0 & \cdots & 0 \\ \beta_1 & \alpha_1 & \frac{1}{h^2} & 0 & 0 & \cdots & 0 \\ 0 & \beta_2 & \alpha_2 & \frac{1}{h^2} & 0 & \cdots & 0 \\ 0 & 0 & \beta_3 & \alpha_3 & \frac{1}{h^2} & \cdots & 0 \\ \vdots & & & & & & \\ 0 & 0 & 0 & \cdots & \cdots & \beta_{N-2} & \alpha_{N-2} \end{bmatrix} \begin{bmatrix} y_1 \\ y_2 \\ y_3 \\ \vdots \\ \\ y_{N-1} \end{bmatrix} = \begin{bmatrix} g_0 - \beta_0 y_0 \\ g_1 \\ g_2 \\ \vdots \\ g_{N-1} \\ g_N - \frac{1}{h^2} y_N \end{bmatrix} \tag{13.346}$$

which is then solved by matrix inversion, for example, as described in Chapter 8.

We note the pattern of entries in the coefficient matrix. Except in the first and last rows, all the elements in the lth row are zero except those in columns $l-1$, l, and $l+1$. The nonzero elements are β_{l-1}, α_{l-1} and $1/\mathrm{h}^2$ respectively. Recognizing this pattern will help in writing computer code to solve a problem of this type.

If the Milne estimate is used, with the symmetric approximation for the second derivative, the finite difference equation is

$$\frac{y_{k+1} - 2y_k + y_{k-1}}{h^2} + P(x_k)\frac{y_{k+1} - y_{k-1}}{2h} + Q(x_k)y_k = g(x_k) \tag{13.347a}$$

or

$$\left(\frac{1}{h^2} + \frac{P_k}{2h}\right)y_{k+1} + \left(Q_k - \frac{2}{h^2}\right)y_k + \left(\frac{1}{h^2} + Q_k\right)y_{k-1} = g_k \tag{13.347b}$$

With

$$\text{(a)}\quad a_k \equiv \left(\frac{1}{h^2} + \frac{P_k}{2h}\right) \qquad \text{(b)}\quad b_k \equiv \left(Q_k - \frac{2}{h^2}\right) \qquad \text{(c)}\quad c_k \equiv \left(\frac{1}{h^2} + Q_k\right) \tag{13.348}$$

the matrix form of the simultaneous equations is

$$\begin{bmatrix} b_1 & a_1 & 0 & 0 & 0 & \cdots & 0 \\ c_2 & b_2 & a_2 & 0 & 0 & \cdots & 0 \\ 0 & c_3 & b_3 & a_3 & 0 & \cdots & 0 \\ 0 & 0 & c_4 & b_4 & a_4 & \cdots & 0 \\ \vdots & & & & & & \\ 0 & 0 & 0 & \cdots & c_{N-2} & b_{N-2} & a_{N-2} \\ 0 & 0 & 0 & \cdots & \cdots & c_{N-1} & b_{N-1} \end{bmatrix} \begin{bmatrix} y_1 \\ y_2 \\ y_2 \\ \vdots \\ \vdots \\ y_{N-1} \end{bmatrix} = \begin{bmatrix} g_1 - c_1 y_0 \\ g_2 \\ g_3 \\ \vdots \\ g_{N-2} \\ g_{N-1} - a_{N-1}y_N \end{bmatrix} \tag{13.349}$$

Example 13.39

To illustrate, consider the differential equation

$$\text{(a)}\quad y'' - 2xy' - 2y = 0 \qquad \text{with} \qquad \text{(b)}\quad y(0) = 1 \qquad \text{(c)}\quad y(1) = e \tag{13.350}$$

The solution is

$$y(x) = e^{x^2} \tag{13.351}$$

Using the Euler finite difference approximation with the nonsymmetric second derivative, this equation is

$$\frac{y_{k+2} - 2y_{k+1} + y_k}{h^2} - \frac{2x_k}{h}(y_{k+1} - y_k) - 2y_k = 0 \tag{13.352}$$

$0 \leqslant k \leqslant N-2$. We take $h = 0.25$ and $x_0 = 0$, so that $x_4 = 1$. For $k = 0$, with $y_0 = 1$, we obtain

$$\frac{1}{(0.25)^2}y_2 + \left[\frac{-2}{(0.25)^2} - \frac{(2)(0)}{0.25}\right]y_1 = \left[\frac{-1}{(0.25)^2} - \frac{(2)(0)}{0.25} + 2\right]y_0 \tag{13.353a}$$

$$\Rightarrow \quad 16y_2 - 32y_1 = -14 \tag{13.353b}$$

The equation for $k = 1$ is

$$\frac{1}{(0.25)^2}y_3 + \left[\frac{-2}{(0.25)^2} - \frac{(2)(0.25)}{0.25}\right]y_2 + \left[\frac{1}{(0.25)^2} + \frac{(2)(0.25)}{0.25} - 2\right]y_1 = 0 \tag{13.354a}$$

$$\Rightarrow \quad 16y_3 - 34y_2 + 16y_1 = 0 \tag{13.354b}$$

For $k = 2$ we obtain

$$\left[\frac{-2}{(0.25)^2} - \frac{(2)(0.5)}{0.25}\right] y_3 + \left[\frac{1}{(0.25)^2} + \frac{(2)(0.5)}{0.25} - 2\right] y_2 = \frac{-1}{(0.25)^2} y_4 \tag{13.355a}$$

$$\Rightarrow \quad -36 y_3 + 18 y_2 = -43.49251 \tag{13.355b}$$

In matrix form, these equations are

$$\begin{bmatrix} -32 & 16 & 0 \\ 16 & -34 & 16 \\ 0 & 18 & -36 \end{bmatrix} \begin{bmatrix} y_1 \\ y_2 \\ y_3 \end{bmatrix} = \begin{bmatrix} -14.00000 \\ 0 \\ -43.49251 \end{bmatrix} \tag{13.356}$$

$$\Rightarrow \begin{bmatrix} y_1 \\ y_2 \\ y_3 \end{bmatrix} = \begin{bmatrix} y(0.25) \\ y(0.50) \\ y(0.75) \end{bmatrix} = \begin{bmatrix} 1.16889 \\ 1.46278 \\ 1.93952 \end{bmatrix} \tag{13.357}$$

The exact values are

$$\begin{bmatrix} y(0.25) \\ y(0.50) \\ y(0.75) \end{bmatrix} = \begin{bmatrix} 1.06449 \\ 1.28403 \\ 1.75505 \end{bmatrix} \tag{13.358}$$

As can be seen, the numerical results agree somewhat with the exact values, but are not very accurate. The accuracy is improved by taking $h = 0.1$. Then we obtain a set of nine simultaneous equations for $y_1 = y(0.1), \ldots, y_9 = y(0.9)$. The boundary conditions are given by $y(0) = y_0$ and $y_{10} = y(1)$. The numerical results are

$$\begin{bmatrix} y(0.1) \\ y(0.2) \\ y(0.3) \\ y(0.4) \\ y(0.5) \\ y(0.6) \\ y(0.7) \\ y(0.8) \\ y(0.9) \end{bmatrix} = \begin{bmatrix} 1.02836 \\ 1.07673 \\ 1.14663 \\ 1.24086 \\ 1.36367 \\ 1.52113 \\ 1.72161 \\ 1.97657 \\ 2.30165 \end{bmatrix} \tag{13.359}$$

The exact values are

$$\begin{bmatrix} y(0.1) \\ y(0.2) \\ y(0.3) \\ y(0.4) \\ y(0.5) \\ y(0.6) \\ y(0.7) \\ y(0.8) \\ y(0.9) \end{bmatrix} = \begin{bmatrix} 1.01001 \\ 1.04081 \\ 1.09417 \\ 1.17351 \\ 1.28403 \\ 1.43333 \\ 1.63232 \\ 1.89648 \\ 2.24791 \end{bmatrix} \tag{13.360}$$

The accuracy is improved with $h = 0.1$, but the results are still noticably inaccurate.

With the Milne finite difference and the symmetric second derivative, using $h = 0.25$, we obtain the matrix equation

$$\begin{bmatrix} -34 & 15 & 0 \\ 17 & -34 & 14 \\ 0 & 19 & -34 \end{bmatrix} \begin{bmatrix} y_1 \\ y_2 \\ y_3 \end{bmatrix} = \begin{bmatrix} -17 \\ 0 \\ -13e \end{bmatrix} \tag{13.361}$$

$$\Rightarrow \begin{bmatrix} y(0.25) \\ y(0.50) \\ y(0.75) \end{bmatrix} = \begin{bmatrix} 1.04451 \\ 1.23422 \\ 1.72905 \end{bmatrix} \tag{13.362}$$

which is in reasonable agreement with the exact values given in equation 13.358. With $h = 0.1$, the solution vector is found to be

$$\begin{bmatrix} y(0.1) \\ y(0.2) \\ y(0.3) \\ y(0.4) \\ y(0.5) \\ y(0.6) \\ y(0.7) \\ y(0.8) \\ y(0.9) \end{bmatrix} = \begin{bmatrix} 1.01533 \\ 1.05118 \\ 1.11022 \\ 1.19360 \\ 1.30675 \\ 1.45655 \\ 1.65279 \\ 1.90992 \\ 2.24535 \end{bmatrix} \tag{13.363}$$

which compares well with the exact results displayed in equation 13.360, and is more accurate than the analogous results obtained using the Euler approximation. □

Eigenvalue equations and finite differences

The eigenvalue equation

$$\begin{aligned} &\text{(a)} \quad P(x)y'' + Q(x)y' + R(x)y = \lambda\rho(x)y \\ &\text{with} \quad \text{(b)} \quad y(x_0) = 0 \qquad \text{(c)} \quad y(x_N) = 0 \end{aligned} \tag{13.364}$$

is an equation with $f(x, y, y')$ in the form given in equation 13.342 with $g(x) = g_2(x) = 0$. If we approximate the equation using the symmetric second derivative and the more accurate Milne estimate, the finite difference approximation of the eigenvalue equation is

$$\frac{P_k}{h^2}[y_{k+1} - 2y_k + y_{k-1}] + \frac{Q_k}{2h}[y_{k+1} - y_{k-1}] + R_k y_k = \lambda\rho_k y_k \tag{13.365a}$$

or

$$\frac{1}{\rho_k}\left[\frac{P_k}{h^2} + \frac{Q_k}{2h}\right]y_{k+1} + \frac{1}{\rho_k}\left[R_k - \frac{2P_k}{h^2}\right]y_k + \frac{1}{\rho_k}\left[\frac{P_k}{h^2} - \frac{Q_k}{2h}\right]y_{k-1} = \lambda y_k \tag{13.365b}$$

for $1 \leqslant k \leqslant N - 1$. We must require that $\rho(x_k) \neq 0$ so that division by ρ_k does not create a numerical problem.

With $y_0 = y_N = 0$, the set of equations for the unknown y-values $y_1, y_2, \ldots, y_{N-1}$ can be expressed in matrix form. Defining

$$\text{(a)} \quad \alpha_k \equiv \frac{1}{\rho_k}\left[\frac{P_k}{h^2} + \frac{Q_k}{2h}\right] \qquad \text{(b)} \quad \beta_k \equiv \frac{1}{\rho_k}\left[R_k - \frac{2P_k}{h^2}\right] \qquad \text{(c)} \quad \gamma_k \equiv \frac{1}{\rho_k}\left[\frac{P_k}{h^2} - \frac{Q_k}{2h}\right] \tag{13.366}$$

the equations, in matrix form, are

$$\begin{bmatrix} \beta_1 & \alpha_1 & 0 & 0 & 0 & \cdots & \cdots & 0 \\ \gamma_2 & \beta_2 & \alpha_2 & 0 & 0 & \cdots & \cdots & 0 \\ 0 & \gamma_3 & \beta_3 & \alpha_3 & 0 & \cdots & \cdots & 0 \\ \vdots & & & & & & & \\ 0 & 0 & \cdots & 0 & 0 & \gamma_{N-2} & \beta_{N-2} & \alpha_{N-2} \\ 0 & 0 & \cdots & 0 & 0 & 0 & \gamma_{N-1} & \beta_{N-1} \end{bmatrix} \begin{bmatrix} y_1 \\ y_2 \\ y_3 \\ \vdots \\ y_{N-2} \\ y_{N-1} \end{bmatrix} = \lambda \begin{bmatrix} y_1 \\ y_2 \\ y_3 \\ \vdots \\ y_{N-2} \\ y_{N-1} \end{bmatrix} \tag{13.367a}$$

or

$$\mathbf{MY} = \lambda \mathbf{Y} \tag{13.367b}$$

This is a standard form of an eigenvalue equation. The methods for determining the eigenvalues and eigenvectors are described in detail in Chapter 8.

We note that if the boundary conditions are any other than the Dirichlet conditions of equations 13.364b and 13.364c, the above method cannot be used to determine the eigenvalues and eigenvectors. For example, if either

$$\text{(a)} \quad y_0 = c_0 \neq 0 \qquad \text{or} \qquad \text{(b)} \quad y_N = c_N \neq 0 \tag{13.368}$$

the $k = 1$ equation is

$$\text{(a)} \quad \alpha_1 y_2 + \beta_1 y_1 + \gamma_1 c_0 = \lambda y_1 \qquad \text{or} \qquad \text{(b)} \quad \alpha_{N-1} c_N + \beta_{N-1} y_{N-1} + \gamma_{N-1} y_{N-2} = \lambda y_{N-1} \tag{13.369}$$

for $k = N - 1$. The presence of the nonzero constant terms $\gamma_1 c_0$ or $\alpha_{N-1} c_N$ yields a set of simultaneous equations that cannot be expressed in matrix form as an eigenvalue equation.

Example 13.40

To illustrate this procedure, consider the eigenvalue equation

$$\text{(a)} \quad -D^2 y = \lambda y \qquad \text{with} \qquad \text{(b)} \quad y(0) = 0 \qquad \text{(c)} \quad y(1) = 0 \tag{13.370}$$

This equation describes the distortion of a thin beam of unit length when a force is applied along its axis.

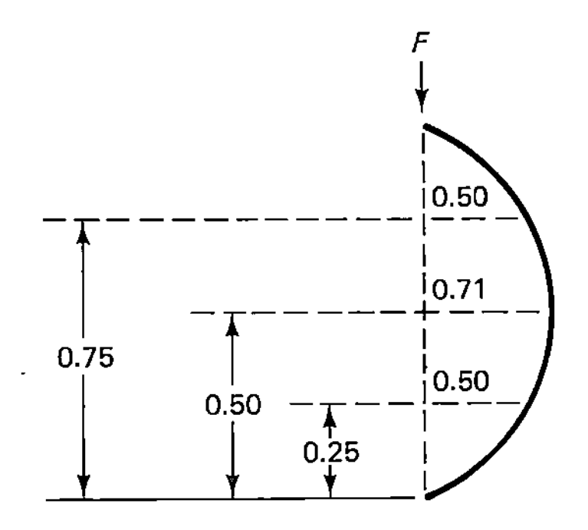

Figure 13.18 Horizontal distortion of a thin beam of unit length under axial force.

The eigenvalues of the system are related to the applied force and the elasticity of the material. Referring to equation 13.370a, we see that

$$\text{(a)} \quad P(x) = -1 \qquad \text{(b)} \quad Q(x) = 0 \qquad \text{(c)} \quad R(x) = 0 \qquad \text{(d)} \quad \rho(x) = 1 \tag{13.371}$$

$$\Rightarrow \quad \text{(a)} \quad \alpha_k = -\frac{1}{h^2} \qquad \text{(b)} \quad \beta_k = \frac{2}{h^2} \qquad \text{(c)} \quad \gamma_k = -\frac{1}{h^2} \tag{13.372}$$

Therefore, the coefficient matrix **M** is

$$\mathbf{M} = \begin{bmatrix} \frac{2}{h^2} & -\frac{1}{h^2} & 0 & \cdots & \cdots & \cdots & 0 \\ -\frac{1}{h^2} & \frac{2}{h^2} & -\frac{1}{h^2} & 0 & \cdots & \cdots & 0 \\ 0 & -\frac{1}{h^2} & \frac{2}{h^2} & -\frac{1}{h^2} & 0 & \cdots & 0 \\ \vdots & & & & & & \\ 0 & \cdots & 0 & & -\frac{1}{h^2} & \frac{2}{h^2} & -\frac{1}{h^2} \end{bmatrix} \tag{13.373}$$

With $h = 0.25$, **M** is the 3×3, real, symmetric (and thus hermitian) matrix

$$\mathbf{M} = 16\begin{bmatrix} 2 & -1 & 0 \\ -1 & 2 & -1 \\ 0 & -1 & 2 \end{bmatrix} \tag{13.374}$$

which has real eigenvalues, $\lambda = 32, 32(1 \pm 1/\sqrt{2})$, or

$$\text{(a)} \quad \lambda_1 = 9.37 \qquad \text{(b)} \quad \lambda_2 = 32 \qquad \text{(c)} \quad \lambda_3 = 54.63 \tag{13.375}$$

The exact eigenvalues of the differential equation are $\lambda_n = (n\pi)^2$, so

$$\text{(a)} \quad \lambda_1 = 9.87 \qquad \text{(b)} \quad \lambda_2 = 39.48 \qquad \text{(c)} \quad \lambda_3 = 88.83 \tag{13.376}$$

We note that the smallest eigenvalue is in best agreement with the corresponding exact value. Smaller eigenvalues correspond to smaller distorting forces.

For $\lambda = \lambda_{\min} = 9.37$, the eigenvector elements are

$$\text{(a)} \quad y_1 = y(0.25) = 0.50 \qquad \text{(b)} \quad y_2 = y(0.50) = 0.71 \qquad \text{(c)} \quad y_3 = y(0.75) = 0.50 \tag{13.377}$$

which are the distances of distortion (relative to a length of 1) as indicated in Figure 13.18. □

Finite difference methods for partial differential equations

The finite difference approximations for the derivatives of single variable functions can easily be extended to the partial derivatives of functions of two or more variables. To develop the approach, we will concentrate on partial differential equations containing first and second derivatives. The smaller the number of independent variables, the easier the numerical solution is to display. For this reason, we will consider examples involving a dependent function of two variables.

The Euler approximation for the first derivatives of $f(x, y)$ are

$$\left.\frac{\partial f}{\partial x}\right|_{x_k, y_l} \simeq \frac{f(x_{k+1}, y_l) - f(x_k, y_l)}{h_x} \equiv \frac{f_{k+1,l} - f_{k,l}}{h_x} \tag{13.378a}$$

$$\left.\frac{\partial f}{\partial y}\right|_{x_k, y_l} \simeq \frac{f(x_k, y_{l+1}) - f(x_k, y_l)}{h_y} \equiv \frac{f_{k,l+1} - f_{k,l}}{h_y} \tag{13.378b}$$

where the fixed differences are

$$\text{(a)} \quad h_x \equiv x_{k+1} - x_k \qquad \text{(b)} \quad h_y \equiv y_{l+1} - y_l \tag{13.379}$$

The Milne approximations of these derivatives are

$$\left.\frac{\partial f}{\partial x}\right|_{x_k, y_l} \simeq \frac{f(x_{k+1}, y_l) - f(x_{k-1}, y_l)}{2h_x} \equiv \frac{f_{k+1,l} - f_{k-1,l}}{2h_x} \tag{13.380a}$$

$$\left.\frac{\partial f}{\partial y}\right|_{x_k, y_l} \simeq \frac{f(x_k, y_{l+1}) - f(x_k, y_{l-1})}{2h_y} \equiv \frac{f_{k,l+1} - f_{k,l-1}}{2h_y} \tag{13.380b}$$

In nonsymmetric form, the second derivatives are approximated by

$$\text{(a)} \quad \left.\frac{\partial^2 f}{\partial x^2}\right|_{x_k, y_l} \simeq \frac{f_{k+2,l} - 2f_{k+1,l} + f_{k,l}}{h_x^2} \qquad \text{(b)} \quad \left.\frac{\partial^2 f}{\partial y^2}\right|_{x_k, y_l} \simeq \frac{f_{k,l+2} - 2f_{k,l+1} + f_{k,l}}{h_y^2} \tag{13.381}$$

$$\left.\frac{\partial^2 f}{\partial x \, \partial y}\right|_{x_k, y_l} = \left.\frac{\partial}{\partial y}\left(\frac{\partial f}{\partial x}\right)\right|_{x_k, y_l} \simeq \frac{\partial}{\partial y}\left[\frac{f(x_{k+1}, y) - f(x_k, y)}{h_x}\right]_{y_l}$$

$$\simeq \frac{1}{h_x}\left[\left(\frac{f_{k+1,l+1} - f_{k+1,l}}{h_y}\right) - \left(\frac{f_{k,l+1} - f_{k,l}}{h_y}\right)\right] = \frac{1}{h_x h_y}[f_{k+1,l+1} - f_{k+1,l} - f_{k,l+1} + f_{k,l}] \tag{13.381c}$$

The symmetric forms of these derivatives are

$$\text{(a)} \quad \left.\frac{\partial^2 f}{\partial x^2}\right|_{x_k, y_l} \simeq \frac{f_{k+1,l} - 2f_{k,l} + f_{k-1,l}}{h_x^2} \qquad \text{(b)} \quad \left.\frac{\partial^2 f}{\partial y^2}\right|_{x_k, y_l} \simeq \frac{f_{k,l+1} + 2f_{k,l} + f_{k,l-1}}{h_y^2} \tag{13.382}$$

$$\text{(c)} \quad \left.\frac{\partial^2 f}{\partial x \, \partial y}\right|_{x_k, y_l} = \frac{1}{4h_x h_y}[f_{k+1,l+1} - f_{k+1,l-1} - f_{k-1,l+1} + f_{k-1,l-1}]$$

Example 13.41

As a first example, we consider the diffusion equation

$$\frac{\partial^2 T}{\partial x^2} = \frac{1}{c^2}\frac{\partial T}{\partial t} \tag{13.383a}$$

with $x \in [0, 1]$, $t > 0$. We subject $T(x, t)$ to the initial and boundary conditions

$$\text{(b)} \quad T(x,0) = F(x) \equiv x(1 - x) \qquad \text{(c)} \quad T(0,t) = G_1(t) = 0 \qquad \text{(d)} \quad T(1,t) = G_2(t) = t^2 \tag{13.383}$$

If we use the Euler approximation to estimate the first derivative, and the nonsymmetric estimate of the second derivative, the finite difference approximation to equation 13.383a is

$$\frac{T_{k+2,l} - 2T_{k+1,l} + T_{k,l}}{h_x^2} = \frac{1}{c^2}\frac{T_{k,l+1} - T_{k,l}}{h_t} \tag{13.384a}$$

$$\Rightarrow \quad T_{k,l+1} = T_{k,l} + \frac{c^2 h_t}{h_x^2}[T_{k+2,l} - 2T_{k+1,l} + T_{k,l}] \tag{13.384b}$$

If the Milne estimate of $\partial T/\partial t$ is used along with the symmetric form of $\partial^2 T/\partial x^2$, the difference equation becomes

$$\frac{T_{k+1,l} - 2T_{k,l} + T_{k-1,l}}{h_x^2} = \frac{1}{c^2}\frac{T_{k,l+1} - T_{k,l-1}}{2h_t} \tag{13.385a}$$

$$\Rightarrow \quad T_{k,l+1} = T_{k,l-1} + \frac{2c^2 h_t}{h_x^2}[T_{k+1,l} - 2T_{k,l} + T_{k-1,l}] \tag{13.385b}$$

For this example, we take $c = 0.2$, $h_x = 0.25$ and $h_t = 0.5$. We therefore estimate the values of $T(x, t)$ at $t_0 = 0$, $t_1 = 0.5$, $t_2 = 1.0$, $t_3 = 1.5$, $t_4 = 2.0$, and $t_5 = 2.5$, for x-values $x_0 = 0$, $x_1 = 0.25$, $x_2 = 0.5$, $x_3 = 0.75$, and $x_4 = 1.0$. When the problem involves two independent variables, a most convenient method for presenting the results is in the form of a table. We refer to this as the *solution table.*

Referring to Table 13.14 for this example, we note that the values of $T(x, t)$ in the first row of the table are obtained from the initial condition;

$$T(x_k, 0) = x_k(1 - x_k) \tag{13.386}$$

The values in the first and last columns are found from the boundary conditions

$$\text{(a)} \quad T(0, t_l) = 0 \qquad \text{(b)} \quad T(1, t_l^2) = t_l^2 \tag{13.387}$$

Thus the solution table, partially filled from the initial and boundary conditions, is

TABLE 13.14 PARTIAL SOLUTION TABLE FOR THE DIFFUSION EQUATION FOR $c = 0.4$

		x_0 0.00	x_1 0.25	x_2 0.50	x_3 0.75	x_4 1.00
t_0	0.0	0.000000	0.187500	0.250000	0.187500	0.000000
t_1	0.5	0.000000				0.250000
t_2	1.0	0.000000				1.000000
t_3	1.5	0.000000				2.250000
t_4	2.0	0.000000				4.000000
t_5	2.5	0.000000				6.250000

The remaining entries in the table are calculated from the finite difference equation.

For example, with $l = 0$, equation 13.384b becomes

$$\begin{aligned} T_{k,1} &= T(x_k, t_1) = T(x_k, 0.5) \\ &= T(x_k, 0) + \frac{(0.2)^2(0.5)}{(0.25)^2}[T(x_{k+2}, 0) - 2T(x_{k+1}, 0) + T(x_k, 0)] \end{aligned} \tag{13.388}$$

Setting $k = 3$, we see that in order to determine $T(x_3, t_1) = T(0.75, 0.5)$ requires the value of $T(x_5, 0)$. Referring to Table 13.14, we note that this value is beyond the range of the table. Thus, since the nonsymmetric approximation to the second derivative contains the term $T(x_{k+2}, t_l)$, we cannot determine the entries in the next to last column in the solution table.

The finite difference relation of equation 13.385b, using the Milne estimate of $\partial T/\partial t$ and the symmetric approximation of $\partial^2 T/\partial x^2$ requires that T be known at two prior times, t_{l-1} and t_l in order to determine T at t_{l+1}. Since the initial condition specifies T at only one instant, t_0, an independent value of t_1 (such as a truncated Taylor sum) would be needed to find T at t_2 and thus all larger t-values.

It is straightforward to avoid the need for additional computations required by using the Milne first derivative by approximating $\partial T/\partial t$ by the Euler finite difference. The inability to calculate the next to last column in the solution table inherent in the nonsymmetric estimate of the second derivative is overcome by using the symmetric approximation to $\partial^2 T/\partial x^2$.

With the Euler first derivative and the symmetric form of the second derivative, the finite difference equation is

$$\frac{T_{k+1,l} - 2T_{k,l} + T_{k-1,l}}{h_x^2} = \frac{1}{c^2}\frac{T_{k,l+1} - T_{k,l}}{h_t} \tag{13.389a}$$

$$\Rightarrow \quad T_{k,l+1} = T_{k,l} + \frac{c^2 h_t}{h_x^2}[T_{k+1,l} - 2T_{k,l} + T_{k-1,l}] \tag{13.389b}$$

Taking $l = 0$, the entries in the second row of Table 13.14 for $x_1 = 0.25$, $x_2 = 0.50$, and $x_3 = 0.75$ can be determined. With $l = 0$, equation 13.389b becomes

$$T(x_k, t_1) = T(x_k, 0.5) = T(x_k, 0) + 0.32[T(x_{k+1}, 0) - 2T(x_k, 0) + T(x_{k-1}, 0)] \tag{13.390}$$

With $k = 1, 2$, and 3, the values of $T(x_{k+1}, 0)$, $T(x_k, 0)$ and $T(x_{k-1}, 0)$ are the known entries in the first row of Table 13.14. Thus, the missing entries in the second row can be determined from equation 13.390.

Once the entries for $t = 0.5$ have been computed, setting $l = 2$ in equation 13.389b yields the entries in the third row of the solution table. This process is continued with successively larger l-values until the table is completed. The result of these computations is presented in Table 13.15. □

TABLE 13.15 COMPLETE SOLUTION TABLE FOR THE DIFFUSION EQUATION FOR $c = 0.4$

		x_0 0.00	x_1 0.25	x_2 0.50	x_3 0.75	x_4 1.00
t_0	0.0	0.000000	0.187500	0.250000	0.187500	0.000000
t_1	0.5	0.000000	0.147500	0.210000	0.147500	0.250000
t_2	1.0	0.000000	0.120300	0.170000	0.200300	1.000000
t_3	1.5	0.000000	0.097708	0.163792	0.446508	2.250000
t_4	2.0	0.000000	0.087588	0.233114	0.933156	4.000000
t_5	2.5	0.000000	0.106128	0.410559	1.690533	6.250000

Example 13.42

As a second example, consider the wave equation

$$\frac{\partial^2 U}{\partial x^2} = \frac{1}{c^2}\frac{\partial^2 U}{\partial t^2} \tag{13.391a}$$

In this example, the time derivative of the unknown function is second order. Thus, two initial conditions are required. We take them to be

$$\text{(b)} \quad U(x, 0) = G_1(x) = x^3 \qquad \text{(c)} \quad \left.\frac{\partial U}{\partial t}\right|_{t=0} = G_2(x) = 0 \tag{13.391}$$

along with two boundary conditions which we specify to be

$$\text{(d)} \quad U(0, t) = F_1(t) = 0 \qquad \text{(e)} \quad U(1, t) = F_2(t) = 1 \tag{13.391}$$

For this example, we take $c^2 = 0.4$, $h_x = 0.25$ and $h_t = 1.0$.

From experience with Example 13.41, we express the second derivatives in the symmetric form. With these approximations, the wave equation becomes

$$\left[\frac{U(x_{k+1}, t_l) - 2U(x_k, t_l) + U(x_{k-1}, t_l)}{h_x^2}\right] = \frac{1}{0.4}\left[\frac{U(x_k, t_{l+1}) - 2U(x_k, t_l) + U(x_k, t_{l-1})}{h_t^2}\right] \tag{13.392}$$

In order to determine $U(x, t)$ at t_{l+1}, we will need the value of $U(x, t)$ at two earlier times t_l and t_{l-1}. Once $U(x, t_0)$ and $U(x, t_1)$ are known, $U(x, t_l)$ at all larger t-values can be found from equation 13.392. Thus, in this example, we cannot avoid the need for an independent determination of $U(x_k, t_1)$. For example, estimating $\partial U/\partial t$ by the Euler approximation, with $t_0 = 0$, the initial condition of equation 13.391c can be written

$$\text{(a)} \quad \left.\frac{\partial U}{\partial t}\right|_{x_k, t_0} \simeq \frac{U(x_k, t_1) - U(x_k, 0)}{h_t} = 0 \quad \Rightarrow \quad U(x_k, t_1) = U(x_k, t_0) = x_k^3 \tag{13.393}$$

Again the results are displayed in a solution table.

As with the solution table for the diffusion equation, the first row is determined from the initial condition of equation 13.391b. The entries in the first and last columns are determined from the boundary conditions of equations 13.391d and 13.391e. Using equation 13.393b, the second initial condition yields the entries in the second row of the solution table. That is

$$U(x_k, t_1) = U(x_k, 1.0) = x_k^3 \tag{13.394}$$

The rows corresponding to times $t \geqslant t_2 = 2.0$ are obtained from the difference equation.

With $l = 1$, equation 13.392 becomes

$$\left[\frac{U(x_{k+1}, t_1) - 2U(x_k, t_1) + U(x_{k-1}, t_1)}{h_x^2}\right] = \frac{1}{0.4}\left[\frac{U(x_k, t_2) - 2U(x_k, t_1) + U(x_k, t_0)}{h_t^2}\right] \tag{13.395}$$

Since the values of U in the first and second rows (for t_0 and t_1) have been determined as described above, $U(x_k, t_2) = U(x_k, 2.0)$ can be obtained from equation 13.395. For example, with $k = 1$,

$$\begin{aligned}&\left[\frac{U(0.50, 1.0) - 2U(0.25, 1.0) + U(0.00, 1.0)}{(0.25)^2}\right] \\ &\quad = \frac{1}{0.4}\left[\frac{U(0.25, 2.0) - 2U(0.25, 1.0) + U(0.25, 0.0)}{(1.0)^2}\right]\end{aligned} \tag{13.396}$$

From the values in the rows corresponding to $t = 1.0$ and $t = 0.0$, we obtain

$$U(0.25, 2.0) = 0.615625 \tag{13.397}$$

In an identical way, with $k = 2$ and $k = 3$, the remainder of the row for $t = 2.0$ can be completed. Then, with $l = 2$, and $k = 1, 2, 3$, the entries in the last row of Table 13.16 can be determined.

TABLE 13.16 SOLUTION TABLE FOR THE WAVE EQUATION FOR $c^2 = 0.4$

		x_0 0.00	x_1 0.25	x_2 0.50	x_3 0.75	x_4 1.00
t_0	0.0	0.000000	0.015625	0.125000	0.421875	1.000000
t_1	1.0	0.000000	0.015625	0.125000	0.421875	1.000000
t_2	2.0	0.000000	0.615625	1.325000	2.221875	1.000000
t_3	3.0	0.000000	1.815625	3.725000	5.821875	1.000000

□

Example 13.43

The solution for Laplace's equation, subject only to boundary conditions, is obtained by solving a set of simultaneous equations. To illustrate, consider the equation

$$\text{(a)} \quad \frac{\partial^2 V}{\partial x^2} + \frac{\partial^2 V}{\partial y^2} = 0$$

with (13.398)

$$\text{(b)} \quad V(x,0) = x \qquad \text{(c)} \quad V(x,1) = \tfrac{1}{2}x^2 \qquad \text{(d)} \quad V(0,y) = 0 \qquad \text{(e)} \quad V(2,y) = 2$$

We will determine the solution at $x = 0.0$, 0.5, 1.0, 1.5, and 2.0, and at y-values of 0.00, 0.25, 0.50, 0.75, and 1.0. Labelling the columns by the x-values and the rows by the y-values, the first and last rows of the solution table are determined by the boundary conditions on y given in equations 13.398b and 13.398c. The boundary conditions of equations 13.398d and 13.398e yield the first and last columns of the table. To facilitate the discussion, we denote all other entries in the solution table by V_i. That is, as a preliminary step, we construct the solution table as

TABLE 13.17 PRELIMINARY SOLUTION TABLE FOR LAPLACE'S EQUATION

		x_0 0.0	x_1 0.5	x_2 1.0	x_3 1.5	x_4 2.0
y_0	0.00	0.0000	0.5000	1.0000	1.5000	2.0000
y_1	0.25	0.0000	V_1	V_2	V_3	2.0000
y_2	0.50	0.0000	V_4	V_5	V_6	2.0000
y_3	0.75	0.0000	V_7	V_8	V_9	2.0000
y_4	1.00	0.0000	0.1250	0.5000	1.1250	2.0000

As in the previous examples, we use the symmetric form of the approximated second derivatives. Then the finite difference approximation to Laplace's equation is

$$\left[\frac{V(x_{k+1}, y_l) - 2V(x_k, y_l) + V(x_{k-1}, y_l)}{h_x^2}\right] = -\left[\frac{V(x_k, y_{l+1}) - 2V(x_k, y_l) + V(x_k, y_{l-1})}{h_y^2}\right] \tag{13.399}$$

We have denoted the undetermined entries in the solution table by equalities such as

$$\text{(a)} \quad V_1 = V(x_1, y_1) = V(0.5, 0.25) \qquad \text{(b)} \quad V_6 = V(x_3, y_2) = V(1.5, 0.5)$$
$$\text{(c)} \quad V_8 = V(x_2, y_3) = V(1.0, 0.75) \tag{13.400}$$

and so on. Thus, for example, with $k = 1$, $l = 1$, the difference equation becomes

$$\left[\frac{V(x_2, y_1) - 2V(x_1, y_1) + V(x_0, y_1)}{(0.5)^2}\right] = -\left[\frac{V(x_1, y_2) - 2V(x_1, y_1) + V(x_1, y_0)}{(0.25)^2}\right] \tag{13.401a}$$

or, referring to the entries in Table 13.17

$$\left[\frac{V_2 - 2V_1 + 0}{(0.5)^2}\right] = -\left[\frac{V_4 - 2V_1 + 0.5}{(0.25)^2}\right] \tag{13.401b}$$

Simplifying, this can be written

$$10V_1 - V_2 - 4V_4 = 2 \tag{13.402a}$$

By an identical analysis, we obtain eight more equations involving the V_i. They are

(b) $V_1 - 10V_2 + V_3 + 4V_5 = -4$ (c) $V_2 - 10V_3 + 4V_6 = -8$

(d) $4V_1 - 10V_4 + V_5 + 4V_7 = 0$ (e) $4V_2 + V_4 - 10V_5 + V_6 + 4V_8 = 0$

(f) $4V_3 + V_5 - 10V_6 + 4V_9 = -2$ (g) $4V_4 - 10V_7 + V_8 = -0.5$

(h) $4V_5 + V_7 - 10V_8 + V_9 = -2$ (i) $4V_6 + V_8 - 10V_9 = -6.5$ (13.402)

It is straightforward to solve this set of simultaneous equations (by matrix inversion, for example). Thus, we can complete the solution table. It is

TABLE 13.18 COMPLETE SOLUTION TABLE FOR LAPLACE'S EQUATION

		x_0 0.0	x_1 0.5	x_2 1.0	x_3 1.5	x_4 2.0
y_0	0.00	0.0000	0.5000	1.0000	1.5000	2.0000
y_1	0.25	0.0000	0.4343	0.9088	1.4343	2.0000
y_2	0.50	0.0000	0.3585	0.8049	1.3585	2.0000
y_3	0.75	0.0000	0.2608	0.6741	1.2608	2.0000
y_4	1.00	0.0000	0.1250	0.5000	1.1250	2.0000

□

13.5 Numerical Solutions to Integral Equations

In Chapter 12 we discussed methods of obtaining an analytic solution of Fredholm integral equations of the form

$$\Psi(x) = f(x) + \lambda \int_a^b K(x, y)\Psi(y)\, dy \tag{12.1c}$$

The inhomogeneous term $f(x)$ and the kernel $K(x, y)$ are known, and $\Psi(x)$ is the function to be determined. The most common approach to solving such equations numerically is by matrix inversion methods.

Unlike differential equations, a Taylor series solution for $\Psi(x)$ cannot be generated in a natural way. When solving a differential equation with initial conditions, the point about which the unknown function can be expanded is specified. In addition, the differential equation is used in a straightforward way to determine the various higher derivatives required for the Taylor series. For a Fredholm integral equation, the point about which $\Psi(x)$ is expanded can be taken to be any point $x \in [a, b]$. To obtain the necessary derivatives at that point, one can differentiate equation 12.1c and evaluate the derivative at the expansion point. However, the derivative of $\Psi(x)$ involves an integral containing the unknown function. That is,

$$\Psi'(x_0) = f'(x_0) + \lambda \int_a^b \left[\frac{\partial}{\partial x} K(x, y)|_{x=x_0}\right] \Psi(y)\, dy \tag{13.403}$$

involves the unknown function under the integral. Thus $\Psi'(x_0)$ cannot be determined without knowing $\Psi(x)$.

The most direct approach for numerically solving equation 12.1c is to approximate the integral by a quadrature sum. Then the integral equation can be written as

$$\Psi(x) \simeq f(x) + \lambda \sum_{l=1}^{N} w_l K(x, x_l)\Psi(x_l) \tag{13.404}$$

Letting $x = x_m$, one of the quadrature abscissae, equation 13.404 becomes

$$\Psi(x_m) \simeq f(x_m) + \lambda \sum_{l=1}^{N} w_l K(x_m, x_l)\Psi(x_l) \tag{13.405}$$

Sequentially setting $m = 1$, $m = 2, \ldots, m = N$, we obtain a set of N equations in N unknowns. Defining column vectors $\mathbf{\Psi}$ and $\mathbf{f}$, the elements of which are the $\Psi(x_l)$ and $f(x_l)$, and matrix elements

$$M_{ml} = w_l K(x_m, x_l) \tag{13.406a}$$

the solution for the elements of the vector $\mathbf{\Psi}$ are obtained from

$$\mathbf{\Psi} = (\mathbf{1} - \lambda \mathbf{M})^{-1}\mathbf{f} \tag{13.406b}$$

The elements of the resulting $\mathbf{\Psi}$ are then substituted into the sum in equation 13.404 to obtain an approximation of $\Psi(x)$ at any x.

To get a sense of whether this method is stable, one should solve the equation using at least two different sets of quadrature points. If the results obtained with the smaller sets of points agree well with the results using the larger set, one has a basis for accepting the results as valid.

Example 13.44

As an example, we consider the equation

$$\Psi(x) = e^x - x\int_{-1}^{1} e^{y(1-x)}\Psi(y)\,dy \tag{13.407}$$

which has solution

$$\Psi(x) = e^{-x} \tag{13.408}$$

To obtain a numerical solution, we approximate the integral of equation 13.407 by 10-point and 20-point Legendre quadratures rules. Then, at any x,

$$\int_{-1}^{1} e^{y(1-x)}\Psi(y)\,dy \simeq \sum_{l=1}^{N} w_l e^{y_l(1-x)}\Psi(y_l) \tag{13.409}$$

With $x = y_k$, the integral equation then becomes the set of linear equations

$$\Psi_k = e^{y_k} - y_k \sum_{l=1}^{N} w_l e^{y_l(1-y_k)}\Psi_l \tag{13.410}$$

The values of Ψ_l, obtained by matrix inversion, are then substituted into

$$\Psi(x) = e^x - x \sum_{l=1}^{N} w_l e^{y_l(1-x)}\Psi_l \tag{13.411}$$

TABLE 13.19 COMPARISON OF 10- AND 20-POINT APPROXIMATE SOLUTIONS TO INTEGRAL EQUATION USING LEGENDRE QUADRATURE

x	Approximate solution: 10 Points	Approximate solution: 20 Points	Exact solution
−0.9	2.459718	2.459603	2.459603
−0.5	1.648690	1.648721	1.648721
−0.1	1.105151	1.105171	1.105171
0	1.000000	1.000000	1.000000
0.3	0.740908	0.740818	0.740818
0.7	0.496847	0.496585	0.496585

The results obtained at various values of x are listed in Table 13.19. Clearly, the results using the 10-point set are almost the same as those using 20 points. Thus we would accept the results from the larger set as valid, even without an independent knowledge of the exact results to verify the accuracy. □

A method that avoids approximating the integral by a quadrature sum involves approximating $\Psi(y)$ under the integral in some prescribed way. For example, we can approximate $\Psi(y)$ by an interpolating polynomial

$$\Psi(y) \simeq \sum_{l=1}^{N} \mu_l(y)\Psi(x_l) \tag{13.412}$$

where $\mu_l(x_k) = \delta_{lk}$. Then equation 13.404 becomes

$$\Psi(x) = f(x) + \lambda \sum_{l=1}^{N} \Psi(x_l) \int_{-1}^{1} K(x, y)\mu_l(y)\, dy \tag{13.413}$$

Defining

$$M_l(x) \equiv \int_{-1}^{1} K(x, y)\mu_l(y)\, dy \tag{13.414}$$

$$\Rightarrow \quad \Psi(x) = f(x) + \lambda \sum_{l=1}^{N} \Psi(x_l) M_l(x) \tag{13.415a}$$

Evaluating x at each abscissa, we obtain the set of N linear equations of the form

$$\Psi(x_k) = f(x_k) + \lambda \sum_{l=1}^{N} M_l(x_k)\Psi(x_l) \tag{13.415b}$$

Defining the matrix $\mathbf{M}$ with elements $\mathbf{M}_{kl} = M_l(x_k)$ the column of elements $\Psi(x_k)$ is again obtained by inverting the matrix $(\mathbf{1} - \lambda\mathbf{M})$. Substituting these elements into equation 13.415a yields a solution to $\Psi(x)$ at any x.

One facet of this approach that requires more discussion is the integral that results from the interpolation; the integral of equation 13.414. Often it will be necessary to evaluate such an integral numerically. If $\mu_l(y)$ is a polynomial of order $N - 1$, the integrand of equation 13.414 will have at least $N - 1$ zeros for a kernel that is analytic for all x and y. A representative description of such an integrand is shown in Figure 13.19.

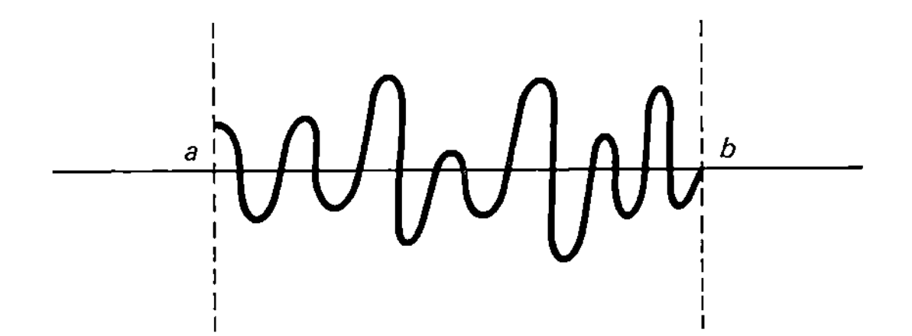

Figure 13.19 Representative integrand containing a polynomial with large number of zeros.

If the integral is evaluated numerically, it is necessary to use a large enough quadrature set that the integrand will be evaluated at several quadrature points between each zero. That is, a numerical evaluation of the integral will be in the form

$$\int_{a}^{b} K(x, y)\mu_l(y)\, dy \simeq \sum_{k=1}^{M} w_k K(x, z_k)\mu_l(z_k) \tag{13.416}$$

where w_k and z_k are the weights and abscissae of some relatively large quadrature set. By using a quadrature set that is much larger than the order of the polynomial, $\mu_l(y)$ is evaluated at several quadrature points between each of its zeros. In that way, a valid numerical sampling of $\mu_l(y)$ is obtained. The author has found that as a rule of thumb, if the interpolating polynomial $\mu_l(y)$ is of order N, one should use a quadrature set with

$10N$ points or more. In that way, the integrand will, on the average, be evaluated at a minimum of 10 points between each zero of $\mu_l(y)$.

Example 13.45

Using the example

$$\Psi(x) = e^x - x\int_{-1}^{1} e^{y(1-x)}\Psi(y)\,dy \tag{13.407}$$

we interpolate $\Psi(y)$ with a polynomial of the form

$$\mu_l(y) = \frac{P_N(y)}{(y - x_l)P_N'(x_l)} \tag{13.417}$$

where $P_N(y)$ is the Nth Legendre polynomial, and x_l is one of its roots. The Legendre polynomial interpolation is used because x and y range from -1 to 1. Then

$$\Psi(x) \simeq e^x - x\sum_{l=1}^{N} \Psi(x_l)\int_{-1}^{1} e^{y(1-x)}\mu_l(y)\,dy \tag{13.418}$$

The resulting integral

$$\lambda M_l(x) = -x\int_{-1}^{1} e^{y(1-x)}\mu_l(y)\,dy \tag{13.419}$$

can be evaluated analytically with great difficulty. We have chosen to evaluate it numerically. Using tenth- and twentieth-order Legendre polynomial interpolations ($N = 10$ and $N = 20$), we have integrated equation 13.418 using a 256-point Legendre quadrature. The resulting values for $\Psi(x)$, obtained from matrix inversion, are displayed in Table 13.20. We note that for both interpolating polynomials, evaluating the integral by a 256-point quadrature rule leads to results that are accurate to six decimal places.

TABLE 13.20 COMPARISON OF SOLUTIONS TO INTEGRAL EQUATION OBTAINED BY INTERPOLATING UNKNOWN FUNCTION

x	Approximate solution $N = 10$	Approximate solution $N = 20$	Exact solution
−0.9	2.459603	2.459603	2.459603
−0.5	1.648721	1.648721	1.648721
−0.1	1.105171	1.105171	1.105171
0	1.000000	1.000000	1.000000
0.3	0.740818	0.740818	0.740818
0.7	0.496585	0.496585	0.496585

□

This second approach still involves an approximation to the unknown function which cannot be validated without an independent way of verifying the resulting values of $\Psi(x)$.

Obtaining stable results by approximating the integral of equation 13.404 by a quadrature sum, or interpolating $\Psi(x)$ as in equation 13.412 does not guarantee that the approximation used is a valid one. It is possible for inaccurate results to be stable. The only way we can be assured that the methods above lead to accurate estimates of $\Psi(x)$ is to have an independent solution for $\Psi(x)$ to which we can compare the result. In practice, of course, the existence of such an independent solution negates the need for a numerical approximation.

A third approach that avoids any approximation of the unknown function is obtained by interpolating the known kernel $K(x, y)$. That is, we write

$$K(x,y) \approx \sum_{l=1}^{N} K(x,x_l)\mu_l(y) \tag{13.420}$$

where x_l is one of the roots of the polynomial used in defining $\mu_l(y)$. This is the only approximation made (other than possibly numerical integration). Therefore, before continuing with the procedure, we can test whether $K(x, y)$ is well described by the interpolation. If it is, the results can be expected to be as accurate as the interpolation and we do not need further independent verification of the solution.

With equation 13.420, the Fredholm equation becomes

$$\Psi(x) = f(x) + \lambda \sum_{l=1}^{N} K(x,x_l) \int_a^b \mu_l(y)\Psi(y)\,dy \tag{13.421}$$

The integrals in the sum contain no x-dependence and are therefore constants. We define

$$C_l \equiv \int_a^b \mu_l(y)\Psi(y)\,dy \tag{13.422}$$

$$\Rightarrow \quad \Psi(x) = f(x) + \lambda \sum_{l=1}^{N} K(x,x_l)C_l \tag{13.423}$$

Substituting this expression for $\Psi(x)$ into the definition of C_l leads to a set of linear equations for the constants C_l.

$$C_l = \int_a^b \mu_l(y)f(y)\,dy + \lambda \sum_{k=1}^{N} C_k \int_a^b \mu_l(y)K(y,x_k)\,dy \tag{13.424}$$

Defining

$$\text{(a)}\quad D_l \equiv \int_a^b \mu_l(y)f(y)\,dy \qquad \text{(b)}\quad M_{lk} \equiv \lambda \int_a^b \mu_l(y)K(y,x_k)\,dy \tag{13.425}$$

$$\Rightarrow \quad C_l = D_l + \sum_{k=1}^{N} M_{lk}C_k \tag{13.426}$$

$$\Rightarrow \quad \mathbf{C} = (\mathbf{1} - \mathbf{M})^{-1}\mathbf{D} \tag{13.427}$$

Substitution of the C_l into equation 13.423 yields $\Psi(x)$ at any value of x.

As with the previous interpolative approach, if integrals involving $\mu_l(y)$ are evaluated numerically, a quadrature set should be used that has at least 10 times as many points as the order of this interpolating polynomial.

Example 13.46

Applying this method to

$$\Psi(x) = e^x - x\int_{-1}^{1} e^{y(1-x)}\Psi(y)\,dy \tag{13.407}$$

the elements of the **M** and **D** arrays are

$$\text{(a)}\quad M_{lk} = -x_k \int_{-1}^{1} e^{y(1-x_k)}\mu_l(y)\,dy \qquad \text{(b)}\quad D_l = \int_{-1}^{1} e^{y}\mu_l(y)\,dy \tag{13.428}$$

As mentioned, with this method, we can test the approximation of the kernel as expressed in equation 13.420 before attempting a solution. Table 13.21 displays a sample of the values of the approximation

$$xe^{y(1-x)} \simeq \sum_{l=1}^{10} e^{x_l(1-x)}\mu_l(y) \tag{13.429}$$

using the interpolating polynomial

$$\mu_l(y) = \frac{P_{10}(y)}{(y - x_l)P'_{10}(x_l)} \tag{13.430}$$

TABLE 13.21 APPROXIMATE VALUES OF A KERNEL AT SELECTED POINTS USING A POLYNOMIAL INTERPOLATION

	y			
x	0.7	0.1	−0.1	−0.5
0.9	−0.965280	−0.909127	−0.891125	−0.856155
0.3	−0.489707	−0.321781	−0.279743	−0.211418
−0.5	1.428861	0.580968	0.430391	0.236195
−0.7	2.301014	0.829785	0.590616	0.299204

TABLE 13.22 EXACT VALUES OF THE KERNEL AT SELECTED POINTS, AND VALUES USING EXPONENTIAL FUNCTIONAL INTERPOLATION

	y			
x	0.7	0.1	−0.1	−0.5
0.9	−0.965257	−0.909045	−0.891045	−0.856106
0.3	−0.489695	−0.321752	−0.279718	−0.211406
−0.5	1.428826	0.580917	0.430354	0.236183
−0.7	2.300957	0.829714	0.590566	0.299191

where $P_{10}(y)$ is the tenth-order Legendre polynomial. In Table 13.22 we list the exact values of $xe^{y(1-x)}$. As can be seen, the polynomial interpolation is reasonably accurate. If we use a tenth-order Lagrange-like interpolation over the function $q(y) = e^y$ as in equation 13.10a instead of using a polynomial interpolation, the interpolated approximations are identical to the exact values of $e^{y(1-x)}$ shown in Table 13.22. Therefore, we use an interpolation over the function e^y.

The values of $\Psi(x)$ obtained by the interpolation method, using interpolation over the function e^y are listed in Table 13.23 at selected values of x. As can be seen, these approximate values are identical to the exact values to six decimal places. □

TABLE 13.23 COMPARISON OF APPROXIMATE AND EXACT VALUES OF SOLUTION TO AN INTEGRAL EQUATION OBTAINED BY FUNCTIONAL INTERPOLATION OF THE KERNEL

x	Approximate Solution	Exact Solution
−0.9	2.459603	2.459603
−0.5	1.648721	1.648721
−0.1	1.105171	1.105171
0	1.000000	1.000000
0.3	0.740818	0.740818
0.7	0.496585	0.496585

Interpolative method for weakly singular Fredholm equations

A kernel is defined to be weakly singular if it satisfies

$$\text{(a)} \quad \lim_{x \to y} K(x, y) = \infty \qquad \text{and} \qquad \text{(b)} \quad \lim_{x \to y} (x - y) K(x, y) = 0 \tag{13.431}$$

Using a spline-like interpolation method, it is possible to obtain a numerical solution to a Fredholm equation with such a kernel. The success of the method is based on the property that the integral of a function with a weak singularity is finite. That is, choosing an interval around the singularity, if $K(x, y)$ satisfies equations 13.431, then

$$I(x) \equiv \int_{x-\alpha_1}^{x+\alpha_2} K(x, y)\, dy \tag{13.432}$$

is finite.

For the Fredholm equation

$$\Psi(x) g(x) + \lambda \int_a^b K(x, y) \Psi(y)\, dy \tag{12.1c}$$

we begin by dividing the interval $[a, b]$ into several small segments, writing the equation

$$\Psi(x) = g(x) + \lambda \sum_{l=1}^{N} \int_{\alpha_l}^{\alpha_{l+1}} K(x, y) \Psi(y)\, dy \tag{13.433}$$

With x_l being some point in the segment $[\alpha_l, \alpha_{l+1}]$, we fit $\Psi(y)$ to some function $\mu_l(y)$ by a spline-like interpolation

$$\Psi(y) = \mu_l(y) \Psi(x_l) \qquad \alpha_l \leqslant y \leqslant \alpha_{l+1} \tag{13.434}$$

Unless there is some information (graphical, for example) on the behavior of $\Psi(y)$ over each interval, one has no idea about the functional form of $\mu_l(y)$. As such, the simplest function, $\mu_l(y) = 1$ might be reasonably satisfactory if the widths of the segments are taken to be small. With $\mu_l(y) = 1$, equation 13.433 becomes

$$\Psi(x) \simeq g(x) + \lambda \sum_{l=1}^{N} \Psi(x_l) \int_{\alpha_l}^{\alpha_{l+1}} K(x, y)\, dy \tag{13.435}$$

Setting $x = x_n$ and defining

$$M_{nl} \equiv \int_{\alpha_l}^{\alpha_{l+1}} K(x_n, y)\, dy \tag{13.436}$$

$$\Rightarrow \quad \Psi(x_n) = g(x_n) + \lambda \sum_{l=1}^{N} M_{nl} \Psi(x_l) \tag{13.437}$$

The solution for the elements $\Psi(x_n)$ of the column vector $\mathbf{\Psi}$ is then obtained by matrix inversion. As noted above, since the singularity of $K(x, y)$ is weak, all matrix elements M_{nl} are finite.

Example 13.47

As an example, we consider an integral equation with a weakly singular kernel derived by Kirkwood and Riseman for describing the intrinsic viscosity of a fluid. It is

$$\Psi(x) = g(x) - \lambda \int_{-1}^{1} \frac{\Psi(y)}{|x-y|^{\alpha}}\, dy \qquad 0 < \alpha < 1 \tag{13.438}$$

The reader interested in this derivation is referred to J. Kirkwood and J. Riseman's article in the *Journal of Chemical Physics* (vol. 16, 1948, pp. 565–573). For large values of λ, it is possible to obtain accurate estimates of $\Psi(x)$ analytically. [The interested reader is referred to a paper by the author and J. Ickovic in the *Journal of Computational Physics* (vol. 16, 1974, pp. 371–382) and references listed there.]

Since $x \in [-1, 1]$, we take the abscissae of the 20-point Gauss–Legendre quadrature set as the set $\{x_l\}$. Thus the first segments are taken to be

$$\text{(a)}\quad [\alpha_1, \alpha_2] = [-1, \tfrac{1}{2}(x_1 + x_2)] \qquad \text{(b)}\quad [\alpha_2, \alpha_3] = [\tfrac{1}{2}(x_1 + x_2), \tfrac{1}{2}(x_2 + x_3)] \tag{13.439}$$

and so on. The last (twentieth) segment is

$$[\alpha_{19}, \alpha_{20}] = [\tfrac{1}{2}(x_{19} + x_{20}), 1] \tag{13.439c}$$

$\Psi(x)$ is obtained by the approach outlined above with

$$\text{(a)}\quad g(x) = x^2 \qquad \text{(b)}\quad \alpha = \tfrac{1}{2} \qquad \text{(c)}\quad \lambda = 200 \tag{13.440}$$

The values of $\Psi(x)$ at selected values of x are listed in Table 13.24 along with corresponding best estimates of the analytic values. It will be noted that the values of $\Psi(x)$ are presented for $x \in [0, 1]$. By replacing y by $-y$, in the integral, it is straightforward to show that $\Psi(-x)$ satisfies the same Fredholm equation as $\Psi(x)$. Therefore, $[\Psi(x) - \Psi(-x)]$ satisfies the homogeneous Fredholm equation. By the Fredholm alternative, if λ is not an eigenvalue of the kernel, the homogeneous solution has only the trivial solution. Then $\Psi(-x) = \Psi(x)$. □

TABLE 13.24 NUMERICAL SOLUTION TO KIRKWOOD – RISEMAN EQUATION FOR $g(x) = x^2$, $\alpha = \frac{1}{2}$, $\lambda = 200$

	$\Psi(x)$	
x	This Method	Analytic
0.99313	0.00655	0.00639
0.96397	0.00382	0.00393
0.74633	0.00111	0.00113
0.51087	0.00003	0.00005
0.07653	−0.00074	−0.00073

Numerical solution to the Lippmann – Schwinger equation by interpolation

The Lippmann–Schwinger equation, as described in Chapter 12, has a kernel with a pole singularity. For a particle with initial and final momenta p and p', in a potential field described by $V(p^2, p'^2)$, in an angular momentum state designated by l, the Lippmann–Schwinger equation is

$$\Psi_l(p^2, p'^2) = V(p^2, p'^2) - \frac{2}{\pi} \int_0^{\infty} \frac{V(p^2, k^2)}{(k^2 - k_0^2 - i\varepsilon)} \Psi(k^2) k^2\, dk \tag{12.246}$$

The kernel has a pole at a fixed value $k = k_0$.

Using equation 2.299b, we write the denominator as

$$\frac{1}{(k^2 - k_0^2 - i\varepsilon)} = P\frac{1}{(k^2 - k_0^2)} + i\pi\delta(k^2 - k_0^2) \tag{13.441a}$$

We note that since both $k > 0$ and $k_0 > 0$, $\delta(k + k_0) = 0$. Therefore, referring to the identity expressed in equation 6.754, we can write

$$\frac{1}{(k^2 - k_0^2 - i\varepsilon)} = P\frac{1}{(k^2 - k_0^2)} + \frac{i\pi}{2k_0}\delta(k - k_0) \tag{13.441b}$$

Taking the final momentum to be $p' = k_0$, suppressing the angular momentum signature, and integrating over the resulting δ-function, the Lippmann–Schwinger equation can be written

$$\begin{aligned}\Psi(p^2, k_0^2) &= V(p^2, k_0^2) - ik_0V(p^2, k_0^2)\Psi(k_0^2, k_0^2)\\ &\quad - \frac{2}{\pi}P\int_0^\infty \frac{V(p^2, k^2)\Psi(k^2, k_0^2)}{(k^2 - k_0^2)}k^2\,dk\end{aligned} \tag{13.442a}$$

To solve for $\Psi(k_0^2, k_0^2)$, we set $p^2 = k_0^2$ to obtain

$$\begin{aligned}\Psi(k_0^2, k_0^2) &= \frac{V(k_0^2, k_0^2)}{[1 + ik_0V(k_0^2, k_0^2)]}\\ &\quad - \frac{2}{\pi}\frac{1}{[1 + ik_0V(k_0^2, k_0^2)]}P\int_0^\infty \frac{V(k_0^2, k^2)\Psi(k^2, k_0^2)}{(k^2 - k_0^2)}k^2\,dk\end{aligned} \tag{13.442b}$$

We substitute this back into equation 13.442a to obtain

$$\begin{aligned}\Psi(p^2, k_0^2) &= \frac{V(p^2, k_0^2)}{[1 + ik_0V(k_0^2, k_0^2)]}\\ &\quad - \frac{2}{\pi}P\int_0^\infty \left\{V(p^2, k^2) - ik_0\frac{V(p^2, k_0^2)V(k_0^2, k^2)}{[1 + ik_0V(k_0^2, k_0^2)]}\right\}\frac{\Psi(k^2, k_0^2)}{(k^2 - k_0^2)}k^2\,dk\end{aligned} \tag{13.443}$$

This is the equation we will use to obtain numerical estimates of $\Psi(p^2, k_0^2)$.

Using the transformations

$$\text{(a)}\quad x = \frac{p - 1}{p + 1} \qquad \text{(b)}\quad y = \frac{k - 1}{k + 1} \qquad \text{(c)}\quad E = \frac{k_0 - 1}{k_0 + 1} \tag{13.444}$$

we replace the variables p, k, and $k_0 \in [0, \infty]$ by x, y, and $E \in [-1, 1]$. This casts the Lippmann–Schwinger equation into the form

$$\Psi(x; E) = g(x; E) - \frac{2}{\pi}P\int_{-1}^1 \frac{U(x, y; E)}{(y - E)}\Psi(y; E)\,dy \tag{13.445}$$

where, with $k_0 = (1 + E)/(1 - E)$,

$$g(x; E) = \frac{V(x, E)}{1 + ik_0V(E, E)} \tag{13.446a}$$

$$U(x, y; E) = \frac{1}{2}\left(\frac{1 + y}{1 - y}\right)^2\frac{(1 - E)^2}{(1 - Ey)}\left\{V(x, y) - ik_0\frac{V(x, E)V(E, y)}{[1 + ik_0V(E, E)]}\right\} \tag{13.446b}$$

We note that the quantity $1/(1-y)^2$ in this expression for $U(x, y; E)$ is a reflection of k^2 in the kernel of the original Lippmann–Schwinger equation. Since that term is infinite when $y \to 1$ (or $k \to \infty$), it will be necessary for the potential to satisfy

$$\text{(a)} \quad \lim_{y \to 1} \frac{V(x,y)}{(1-y)^2} = 0 \qquad \Rightarrow \qquad \text{(b)} \quad \lim_{k \to \infty} k^2 V(p^2, k^2) = 0 \tag{13.447}$$

The term $1/(1-Ey)$ looks as if it might become infinite when $Ey = 1$. E and y have values between -1 and $+1$. Thus $Ey = 1$ only if $E = y = \pm 1$. But when $E = y = 1$, the term $(1-E)^2$ and the term in the potential will cause $U(x, 1; 1)$ to be zero. When $E = y = -1$, the term $(1+y)^2$ will make $U(x, -1; -1)$ zero. Thus there will not be a problem of an infinity arising from $1/(1-Ey)$.

Writing equation 13.445 as

$$\Psi(x,E) = g(x,E) - \frac{2}{\pi}\int_{-1}^{1} \frac{[U(x,y;E) - U(x,E;E)]}{(y-E)}\Psi(y,E)\,dy$$
$$- \frac{2}{\pi}U(x,E;E)P\int_{-1}^{1}\frac{\Psi(x,E)}{(y-E)}\,dy \tag{13.448}$$

we note that the integrand of the first integral no longer contains the pole singularity. Thus this integral can be approximated directly by an N-point Gauss–Legendre quadrature sum:

$$\int_{-1}^{1} \frac{[U(x,y;E) - U(x,E;E)]}{(y-E)}\Psi(y,E)\,dy \simeq \sum_{l=1}^{N} w_l \left[\frac{U(x,x_l;E) - U(x,E;E)}{(x_l - E)}\right] \tag{13.449}$$

Clearly, if E has a value that is very close (or equal) to one of the quadrature points, the quantity in brackets for that term is replaced by the derivative, $\partial U(x, y; E)/\partial y|_{y=E}$.

To obtain a solution by matrix inversion, the second integral of equation 13.448 must be approximated numerically in such a way that Ψ is evaluated at the same points x_l as those used in the quadrature sum for the first integral. As such, we approximate $\Psi(y, E)$ in the second integral by the interpolated function

$$\Psi(y,E) \simeq \sum_{l=1}^{N} \mu_l(y)\Psi(x_l,E) \tag{13.450}$$

where

$$\mu_l(y) = \frac{P_N(y)}{(y-x_l)P_N'(x_l)} \tag{13.417}$$

The polynomial $P_N(y)$ is the Legendre polynomial for which these required x_l are the roots. Then the second integral becomes

$$P\int_{-1}^{1}\frac{\Psi(x,E)}{(y-E)}\,dy \simeq \sum_{l=1}^{N}\frac{\Psi(x_l,E)}{P_N'(x_l)}\int_{-1}^{1}\frac{P_N(y)}{(y-E)_P(y-x_l)}\,dy \tag{13.451}$$

Writing

$$\frac{1}{(y-E)_P(y-x_l)} = \frac{1}{(E-x_l)}\left(\frac{1}{(y-E)_P} - \frac{1}{(y-x_l)}\right) \quad (13.452)$$

and using the Neumann identity for the second Legendre function

$$Q_N(x) = Re[Q_N(z)] = \frac{1}{2}\int_{-1}^{1} \frac{P_N(y)}{(x-y)_P}\,dy \quad (6.247)$$

the second integral can be written

$$P\int_{-1}^{1} \frac{\Psi(x,E)}{(y-E)}\,dy \simeq -2\sum_{l=1}^{N} \frac{\Psi(x_l,E)}{P_N'(x_l)}\left[\frac{Q_N(E)-Q_N(x_l)}{(E-x_l)}\right] \quad (13.453)$$

As with the approximate form of the first integral, if E is close to one of the quadrature points, the quantity in brackets for that term is replaced by the derivative, $Q_N'(E)$.

Example 13.48

As an example, we apply the method described above to the Lippmann–Schwinger equation for the Yamaguchi potential (see Problem 30 in Chapter 12).

$$\begin{aligned} V(p^2,k_0^2) &= \frac{\alpha}{(p^2+\beta^2)(k_0^2+\beta^2)} \\ &= \frac{\alpha(1-y)^2(1-E)^2}{\left[(1+y)^2+\beta^2(1-y)^2\right]\left[(1+E)^2+\beta^2(1-E)^2\right]} \end{aligned} \quad (13.454)$$

As noted in Chapter 12, since this potential is separable, an analytic solution can be found. Using the values $\alpha = -8.110$ and $\beta = 1.444$, which have meaning for a physical problem, we solve the Lippmann–Schwinger equation using a 20-point Legendre quadrature set. We present a comparison of the results to the exact values obtained from the analytic solution in Table 13.25. It will be noted that the numerical results are quite accurate. □

TABLE 13.25 NUMERICAL SOLUTION TO LIPPMANN – SCHWINGER EQUATION FOR YAMAGUCHI POTENTIAL, USING INTERPOLATION METHOD

	x	Numerical Solution	Analytic Solution
$E = -0.5195$	0.9931	$0.0067 - 0.0216i$	$0.0067 - 0.0216i$
$(k_0^2 = 0.1)$	0.3737	$0.4589 - 1.4720i$	$0.4590 - 1.4719i$
	-0.7463	$0.8803 - 2.8236i$	$0.8805 - 2.8235i$
	-0.9931	$0.9401 - 3.0153i$	$0.9403 - 3.0152i$
$E = +0.5195$	0.9931	$0.0019 - 0.0003i$	$0.0019 - 0.0003i$
$(k_0^2 = 10)$	0.3737	$0.1330 - 0.0203i$	$0.1330 - 0.0203i$
	-0.7463	$-0.2552 - 0.0389i$	$-0.2552 - 0.0389i$
	-0.9931	$-0.2725 - 0.0415i$	$-0.2725 - 0.0415i$

Interpolation method for solving the Muskhelishvili-Omnes equation

The interpolative approach used to obtain a numerical solution to the Lippmann–Schwinger equation can be applied to the Muskhelishvili-Omnes (MO) equation, which has a variable (movable) pole. It was noted in Chapter 12 that the MO equation has a unique solution if the phase shift parameter has a certain energy dependence. If the phase shift has that energy dependence, the numerical approach will approximate that unique solution.

The development of the interpolative approach for the MO equation very closely parallels the development for the Lippmann–Schwinger equation. We begin by writing

$$\frac{1}{(z' - z - i\varepsilon)} = P\frac{1}{(z' - z)} + i\pi\delta(z' - z) \qquad (2.299b)$$

so that the MO equation

$$\Phi(z) = \Phi_0(z) + \frac{1}{\pi}\int_1^{\infty} \frac{\sin\delta(z')e^{-i\delta(z')}}{(z' - z - i\varepsilon)}\Phi(z')\,dz' \qquad (13.455a)$$

can be cast into the form

$$\chi(z) = \chi_0(z) + \frac{1}{\pi}P\int_1^{\infty} \frac{\tan\delta(z')}{(z' - z)}\chi(z')\,dz' \qquad (13.455b)$$

where

$$\chi(z) \equiv \cos\delta(z)e^{-i\delta(z)}\Phi(z) \qquad (13.456)$$

Using the mapping

$$z = \frac{2}{(1 + x)} \qquad (13.457)$$

and defining

$$\text{(a)}\quad \Psi(x) \equiv \frac{\chi[z(x)]}{(1 + x)} \qquad \text{(b)}\quad g(x) \equiv \frac{\chi_0[z(x)]}{(1 + x)} \qquad (13.458)$$

the MO equation can be written as

$$\Psi(x) = g(x) + \frac{1}{\pi}P\int_{-1}^{1} \frac{\tan\delta(y)}{(x - y)}\Psi(y)\,dy$$

$$= g(x) - \frac{1}{\pi}\int_{-1}^{1} \frac{\tan\delta(x) - \tan\delta(y)}{(x - y)}\Psi(y)\,dy + \frac{1}{\pi}\tan\delta(x)P\int_{-1}^{1} \frac{\Psi(y)}{(x - y)}\,dy \qquad (13.459)$$

As with the Lippmann–Schwinger, the integrand of the first integral is no longer singular, and the first integral is approximated by a Gauss–Legendre quadrature sum,

$$\int_{-1}^{1} \frac{\tan\delta(x) - \tan\delta(y)}{(x - y)}\Psi(y)\,dy \simeq \sum_{l=1}^{N} w_l\left[\frac{\tan\delta(x) - \tan\delta(x_l)}{(x - x_l)}\right] \qquad (13.460)$$

To ensure that $\Psi(x)$ in the second integral is evaluated at the same quadrature points x_l as are used for the first integral, $\Psi(y)$ is interpolated over the Legendre polynomials as

$$\Psi(y) \simeq \sum_{l=1}^{N} \mu_l(y)\Psi(x_l) \qquad (13.412)$$

where

$$\mu_l(y) = \frac{P_N(y)}{(y - x_l)P_N'(x_l)} \qquad (13.417)$$

The second integral then becomes

$$P\int_{-1}^{1}\frac{\Psi(y)}{(x-y)}\,dy \simeq \sum_{l=1}^{N}\frac{\Psi(x_l)}{P_N'(x_l)}\int_{-1}^{1}\frac{P_N(y)}{(y-x)_P(y-x_l)}\,dy$$

$$= \sum_{l=1}^{N}\frac{\Psi(x_l)}{P_N'(x_l)}\frac{1}{(x-x_l)}\int_{-1}^{1}\left[\frac{1}{(x_l-y)}-\frac{1}{(x-y)_P}\right]P_N(y)\,dy \qquad (13.461)$$

We again use the integral representation

$$Q_N(x) = \frac{1}{2}\int_{-1}^{1}\frac{P_N(y)}{(x-y)_P}\,dy \qquad (6.247)$$

to evaluate the second integral as

$$P\int_{-1}^{1}\frac{\Psi(y)}{(x-y)}\,dy \simeq -2\sum_{l=1}^{N}\frac{\Psi(x_l)}{P_N'(x_l)}\left[\frac{Q_N(x)-Q_N(x_l)}{(x-x_l)}\right] \qquad (13.462)$$

Then, setting x to each of the quadrature points, we obtain a set of equations of the form

$$\Psi(x_k) = g(x_k) - \frac{1}{\pi}\sum_{l=1}^{N}M_{kl}\Psi(x_l) \qquad (13.463)$$

where

$$M_{kl} = w_l\left[\frac{\tan\delta(x_k)-\tan\delta(x_l)}{(x_k-x_l)}\right] + 2\frac{\tan\delta(x_k)}{P_N'(x_l)}\left[\frac{Q_N(x_k)-Q_N(x_l)}{(x_k-x_l)}\right] \qquad k\neq l \qquad (13.464a)$$

$$M_{kk} = w_k\frac{d}{dx}[\tan\delta(x)]_{x=x_k} + 2\frac{\tan\delta(x_k)}{P_N'(x_l)}Q_N'(x_k) \qquad k=l \qquad (13.464b)$$

As discussed before, the solution for the N quantities $\Psi(x_k)$ are obtained from equation 13.463 by matrix inversion.

Example 13.49

As a concrete example, we use

$$\text{(a)}\quad \delta(z) = \pi\frac{(z-1)}{z^2} \quad\Rightarrow\quad \text{(b)}\quad \delta(x) = \frac{\pi}{4}(1-x^2) \qquad (13.465)$$

$$\text{(a)}\quad \Phi_0(z) = \frac{1}{2z} \quad\Rightarrow\quad \text{(b)}\quad g(x) = \tfrac{1}{4} \qquad (13.466)$$

With this phase shift, the analytic solution is unique. It is

$$\Phi(z) = e^{i\delta(z)}\left[\Phi_0(z)\cos\delta(z) + \frac{1}{\pi}P\int_{1}^{\infty}\frac{\Phi_0(z')\sin\delta(z')e^{-\rho(z')}}{(z'-z)}\,dz'\right] \qquad (13.467)$$

where

$$\rho(z) \equiv \frac{1}{\pi}P\int_{1}^{\infty}\frac{\delta(z')}{(z'-z)}\,dz' \qquad (13.468)$$

We express the results of this numerical solution to the MO equation in terms of $\Psi(x)$, which is real. The analytic expression for $\Psi(x)$ is obtained from equation 13.467. With z expressed as

$$z = \frac{2}{(1+x)} \qquad (13.457)$$

the analytic form of $\Psi(x)$ is

$$\Psi(x) = \frac{\cos\delta(x)}{(1+x)}\left[\Phi_0(z)\cos\delta(z) + \frac{1}{\pi}P\int_1^\infty \frac{\Phi_0(z')\sin\delta(z')e^{-\rho(z')}}{(z'-z)}\,dz'\right] \tag{13.469}$$

To obtain the "exact" values of $\Psi(x)$, the integral of equation 13.467 must be evaluated numerically. We use a 20-point Legendre quadrature rule. The points at which $\Psi(x)$ are evaluated in equation 13.463 are also those from the 20-point Legendre quadrature set.

The results are displayed in Table 13.26. The differences between the numerical determination of $\Psi(x)$ using equations 13.463 and 13.464, and the "exact" solution of equation 13.469 are very small. Therefore, we display the values of $\Psi(x)$ and the percent difference between the numerical estimates and the "exact" values. □

TABLE 13.26 COMPARISON OF INTERPOLATIVE NUMERICAL METHOD WITH "EXACT" SOLUTION OF MO EQUATION

x	$z = \frac{2}{(1+x)}$	$\Psi(x)$	Percent Difference
0.9931	1.003	0.4188	6.9×10^{-4}
−0.0765	2.166	0.1645	1.5×10^{-3}
−0.5109	4.089	0.1315	1.2×10^{-3}
−0.7463	7.884	0.1307	1.9×10^{-4}
−0.9122	22.788	0.1380	4.3×10^{-4}
−0.9931	291.061	0.1492	2.9×10^{-3}

Taylor series for the Volterra equation

Unlike the Fredholm equation, it is possible to develop an approximate Taylor series solution to the Volterra equation

$$\Psi(x) = g(x) + \lambda\int_a^x K(x,y)\Psi(y)\,dy \tag{12.2c}$$

The reason for this is that unlike the Fredholm integral $\int_a^b K(x,y)\Psi(y)\,dy$, Volterra integrals such as $\int_a^x K(x,y)\Psi(y)\,dy$ are zero at $x = a$. Thus in an attempt to develop a Taylor series for Fredholm equations, the derivatives of Ψ and x_0 contain integrals involving the unknown function $\Psi(y)$, whereas the Taylor series for the Volterra function does not contain such an integral.

Clearly, the key to developing the Taylor expansion

$$\Psi(x) = \Psi(a) + \Psi'(a)(x-a) + \frac{1}{2!}\Psi''(a)(x-a)^2 + \frac{1}{3!}\Psi'''(a)(x-a)^3 + \cdots \tag{13.470}$$

is the determination of the various derivatives $\Psi^{(n)}(a)$. Since, at $x = a$, the integral in equation 12.2c is zero

$$\Psi(a) = g(a) \tag{13.471}$$

The first derivative of equation 12.2c is

$$\Psi'(x) = g'(x) + \lambda\int_a^x \frac{\partial K(x,y)}{\partial x}\Psi(y)\,dy + \lambda K(x,x)\Psi(x) \tag{13.472}$$

The term $K(x, x)\Psi(x)$ arises from differentiating the x-dependence in the upper limit of the integral. That is, if

$$G(x, y) \equiv \int K(x, y)\Psi(y)\, dy \tag{13.473}$$

then the derivative of the upper limit x-dependence is

$$\frac{\partial}{\partial x}\int_a^x K(x, y)\Psi(y)\, dy = \frac{\partial}{\partial y}[G(x, y) - G(x, a)]_{y=x} = K(x, y)\Psi(y)|_{y=x} \tag{13.474}$$

Thus from equation 13.472 with the integral at $x = a$ being zero, we obtain

$$\Psi'(a) = g'(a) + \lambda K(a, a)\Psi(a) = g'(a) + \lambda K(a, a)g(a) \tag{13.475}$$

Using the same arguments, the second derivative of $\Psi(x)$ is

$$\begin{aligned}\Psi''(x) = {} & g''(x) + \lambda \int_a^x \frac{\partial^2 K(x, y)}{\partial x^2}\Psi(y)\, dy + \lambda \frac{\partial K(x, y)}{\partial x}\Psi(y)\Bigg|_{y=x} \\ & + \lambda \frac{\partial K(x, y)}{\partial y}\Psi(y)\Bigg|_{y=x} + \lambda K(x, x)\Psi'(x)\end{aligned} \tag{13.476}$$

Using the shorthand notation

$$K_x(x, y) \equiv \frac{\partial K(x, y)}{\partial x} \tag{13.477}$$

and again setting the remaining integral to zero at $x = a$, we obtain

$$\Psi''(a) = g''(a) + \lambda\left[K_x(a, a) + K_y(a, a)\right]\Psi(a) + \lambda K(a, a)\Psi'(a) \tag{13.478a}$$

Using equations 13.471 and 13.475, this becomes

$$\begin{aligned}\Psi''(a) = {} & g''(a) + \lambda\left[K_x(a, a) + K_y(a, a)\right]g(a) \\ & + \lambda K(a, a)\left[g'(a) + \lambda K(a, a)g(a)\right]\end{aligned} \tag{13.478b}$$

Obviously, in order for the Taylor series to exist, $g(x)$ and $K(x, y)$ and all derivatives must be finite at $x = y = a$. That is, $g(x)$ and $K(x, y)$ must be analytic at $x = y = a$.

It is clear that this process quickly becomes computationally complicated. It is therefore advisable to use a Taylor series truncated after a small number of terms. If $(x - a)$ is too large to yield accurate results with a truncated series containing a small number of terms, we might consider a multistep iterative process such as that used to solve differential equations, generating a Taylor series with a small number of terms to determine $\Psi(a + \Delta)$. Then, from that result, we would attempt to determine $\Psi(a + 2\Delta)$, and so on.

However, to develop the Taylor series at $a + 2\Delta$, we would expand

$$\Psi(a + 2\Delta) = \Psi(a + \Delta) + \Psi'(a + \Delta)\Delta + \cdots \tag{13.479}$$

This requires being able to determine $\Psi(x)$ and its derivatives at $a + 2\Delta$ from $\Psi(a + \Delta)$. But, for example,

$$\Psi(a + \Delta) = g(a + \Delta) + \lambda \int_a^{a+\Delta} K(a + \Delta, y)\Psi(y)\, dy \tag{13.480}$$

Unlike the determination of $\Psi(a)$, this integral, which involves the unknown function, is not zero. Thus, such a multistep process is not possible, and one can only develop a truncated Taylor series around $x = a$.

Example 13.50

As an example, we consider

$$\Psi(x) = 1 + x + \int_0^x e^{(x-y)}\Psi(y)\, dy \tag{13.481}$$

Using the techniques developed in Chapter 12, the solution can be shown to be

$$\Psi(x) = \tfrac{1}{4} + \tfrac{1}{2}x + \tfrac{3}{4}e^{2x} \tag{13.482}$$

Using

$$\text{(a)}\quad g(x) = 1 + x \qquad \text{(b)}\quad K(x, y) = e^{(x-y)} \tag{13.483}$$

$$\text{(a)}\quad \Psi(0) = g(0) = 1 \qquad \text{(b)}\quad \Psi'(0) = g'(0) + g(0)e^0 = 2$$

$$\text{(c)}\quad \Psi''(0) = g''(0) + 2g(0)\frac{\partial e^{(x-y)}}{\partial x}\bigg|_{x=y=0} + g(0)\frac{\partial e^{(x-y)}}{\partial y}\bigg|_{x=y=0} + \left[g'(0) + e^0 g(0)\right]e^0 = 3 \tag{13.484}$$

Therefore, the MacLaurin series, truncated after three terms, is

$$\Psi(x) \simeq 1 + 2x + \tfrac{3}{2}x^2 \tag{13.485}$$

In Table 13.27 we present a comparison of the exact and approximate values of $\Psi(x)$ at various values of x near zero. As expected, the closer x is to zero, the more accurate the results. □

TABLE 13.27 SOLUTION OF VOLTERRA EQUATION BY TAYLOR SERIES METHOD

x	Three-term Series	Exact Value
0.1	1.2150	1.2161
0.2	1.4600	1.4689
0.3	1.7350	1.7666
0.4	2.0400	2.1192
0.5	2.3750	2.5387

Iterative solution to the Volterra equation

The reader will recall that by spline interpolation of the integrand by second- or third-order polynomials, we developed the Simpson $\frac{1}{3}$ and Simpson $\frac{3}{8}$ Newton–Coates quadrature rules. These rules required that the interval be divided into a number of segments that was a multiple of 2 (for the $\frac{1}{3}$ rule) or a multiple of 3 (for the $\frac{3}{8}$ rule). However, using the cardinal or linear spline interpolations, we developed the rectangular and trapezoidal quadrature schemes in which the number of segments was not restricted to be a multiple of some number greater than 1. Thus on the most crude scale, one could use the entire range of integration as one segment. Then, referring to equation 13.104a, we have

$$\int_a^b f(x)\, dx \simeq (b - a)f(a) \tag{13.486a}$$

if the rectangular rule is used, and from equation 13.109,

$$\int_a^b f(x)\, dx \simeq \frac{(b-a)}{2}[f(a) + f(b)] \tag{13.486b}$$

for the simplest trapezoidal rule. Since the entire range of integration can be taken as a single segment, we will refer to the two integration schemes described in equations 13.486 as *single-segment* integration rules. Using either of them, it is possible to develop an iterative approach to solve a Volterra equation.

Let x_n be the point at which Ψ is to be determined, and let

$$\text{(a)} \quad x_1 \equiv a + \Delta x \qquad \text{(b)} \quad x_2 \equiv a + 2\Delta x = x_1 + \Delta x \tag{13.487}$$

and so on, be points between $x = a$ and $x = x_n$. Then with Δx small, we evaluate the integral in

$$\Psi(x_1) = g(x_1) + \lambda \int_a^{x_1} K(x_1, y)\Psi(y)\, dy \tag{13.488}$$

by a single-segment integration rule. That is, if we use the rectangular rule, we approximate $\Psi(x_1)$ by

$$\Psi(x_1) \simeq g(x_1) + \lambda\, \Delta x K(x_1, a)\Psi(a) \tag{13.489}$$

and with the single-segment trapezoidal rule,

$$\Psi(x_1) \simeq g(x_1) + \lambda \frac{\Delta x}{2}[K(x_1, a)\Psi(a) + K(x_1, x_1)\Psi(x_1)] \tag{13.490a}$$

Since at $x = a$ the Volterra integral is zero, we obtain

$$\Psi(a) = g(a) \tag{13.471}$$

from the Volterra equation. Then, equation 13.490a yields

$$\Psi(x_1) = \frac{g(x_1) + \frac{1}{2}\lambda\, \Delta x K(x_1, a) g(a)}{1 - \frac{1}{2}\lambda\, \Delta x K(x_1, x_1)} \tag{13.490b}$$

Therefore $\Psi(x_1)$ can be determined using either single-segment integration.

The process is then repeated to determine

$$\begin{aligned}\Psi(x_2) &= g(x_2) + \lambda \int_a^{x_2} K(x_2, y)\Psi(y)\, dy \\ &= g(x_2) + \lambda \int_a^{x_1} K(x_2, y)\Psi(y)\, dy + \lambda \int_{x_1}^{x_2} K(x_2, y)\Psi(y)\, dy \end{aligned} \tag{13.491}$$

Again, each integral is approximated by the single-segment integration rule being used. The rectangular rule yields

$$\Psi(x_2) \simeq g(x_2) + \lambda\, \Delta x K(x_2, a)\Psi(a) + \lambda\, \Delta x K(x_2, x_1)\Psi(x_1) \tag{13.492a}$$

Using the trapezoidal rule, we obtain

$$\begin{aligned}\Psi(x_2) \simeq g(x_2) &+ \lambda \frac{\Delta x}{2}[K(x_2, a)\Psi(a) + K(x_2, x_1)\Psi(x_1)] \\ &+ \lambda \frac{\Delta x}{2}[K(x_2, x_1)\Psi(x_1) + K(x_2, x_2)\Psi(x_2)]\end{aligned} \tag{13.492b}$$

(The reader will recognize equation 13.492b as the trapezoidal rule for the integral over $[a, x_2]$.) Since $\Psi(a)$ and $\Psi(x_1)$ have been determined from previous iterations, a value of $\Psi(x_2)$ can be determined from either of equations 13.492 (see Problem 26).

Example 13.51

We apply this approach to the example

$$\Psi(x) = 1 + x + \int_0^x e^{(x-y)}\Psi(y)\,dy \tag{13.481}$$

to determine an approximate value of $\Psi(0.3)$, using the rectangular rule and iterating in steps of $\Delta x = 0.1$. With $\Psi(0) = g(0) = 1$,

$$\Psi(0.1) = 1 + 0.1 + \int_0^{0.1} e^{(0.1-y)}\Psi(y)\,dy \simeq 1.1 + (0.1)e^{(0.1-0)}\Psi(0) = 1.2105 \tag{13.493a}$$

$$\Rightarrow \quad \Psi(0.2) = 1 + 0.2 + \int_0^{0.1} e^{(0.2-y)}\Psi(y)\,dy + \int_{0.1}^{0.2} e^{(0.2-y)}\Psi(y)\,dy$$

$$\simeq 1.2 + (0.1)e^{(0.2-0)}\Psi(0) + (0.1)e^{(0.2-0.1)}\Psi(0.1) = 1.4559 \tag{13.493b}$$

$$\Rightarrow \quad \Psi(0.3) = 1 + 0.3 + \int_0^{0.1} e^{(0.3-y)}\Psi(y)\,dy + \int_{0.1}^{0.2} e^{(0.3-y)}\Psi(y)\,dy + \int_{0.2}^{0.3} e^{(0.3-y)}\Psi(y)\,dy$$

$$\simeq 1.3 + (0.1)e^{0.3}\Psi(0) + (0.1)e^{0.2}\Psi(0.1) + (0.1)e^{0.1}\Psi(0.2) = 1.7437 \tag{13.493c}$$

Comparing these results to those listed in Table 13.27, we see that this iterative approach yields values that are similar to those obtained with the Taylor series approach without the need to calculate derivatives of $\Psi(x)$. They are reasonable approximations to the exact values. □

Interpolative solution to the Volterra equation

Using a Lagrange like interpolation of $\Psi(y)$, the Volterra equation can be solved using matrix inversion methods similar to the method developed for the Fredholm equation.

By using the approximation

$$\Psi(y) \simeq \sum_{l=1}^{N} \mu_l(y)\Psi(x_l) \tag{13.412}$$

the Volterra equation becomes

$$\Psi(x) \simeq g(x) + \lambda \sum_{l=1}^{N} \Psi(x_l) \int_a^x K(x, y)\mu_l(y)\,dy \tag{13.494}$$

Since both $K(x, y)$ and $\mu_l(y)$ are known functions, the integrals can, in principle, be evaluated. We define

$$C_l(x) \equiv \int_a^x K(x, y)\mu_l(y)\,dy \tag{13.495}$$

Setting x to each of the interpolation points in sequence, equation 13.494 results in a set of N equations of the form

$$\Psi(x_k) \simeq g(x_k) + \lambda \sum_{l=1}^{N} M_{kl}\Psi(x_l) \tag{13.496}$$

where

$$M_{kl} \equiv C_l(x_k) \tag{13.497}$$

As seen before, such a set of equations is soluble by matrix inversion.

We note from equation 13.495 that $C_l(a) = 0$. Thus if we use $x = a$ as one of the interpolation points, every element of one of the rows of the matrix **M** will be zero and **M** will be singular. In that case the solution cannot be determined. But since $\Psi(a) = g(a)$ is known, it is not necessary to include $x = a$ in the set of interpolation points, and to avoid making **M** singular, it is important that $x = a$ be excluded from the set of points.

Example 13.52

To illustrate this approach, we will again use the example

$$\Psi(x) = 1 + x + \int_0^x e^{(x-y)}\Psi(y)\,dy \tag{13.481}$$

We will take the interpolation points to be $x_1 = 0.1$, $x_2 = 0.2$, and $x_3 = 0.3$. As such, the interpolation must be via a second-order polynomial. That is,

$$\Psi(y) \simeq \sum_{l=1}^{3} \mu_l(y)\Psi(x_l) \tag{13.498}$$

with

$$\text{(a)}\quad \mu_1(y) = \frac{(y-0.2)(y-0.3)}{(-0.1)(-0.2)} \qquad \text{(b)}\quad \mu_2(y) = \frac{(y-0.1)(y-0.3)}{(0.1)(-0.1)} \qquad \text{(c)}\quad \mu_3(y) = \frac{(y-0.1)(y-0.2)}{(0.2)(0.1)} \tag{13.499}$$

Then

$$\Psi(x) = 1 + x + \sum_{l=1}^{3} \psi(x_l)\int_0^x e^{(x-y)}\mu_l(y)\,dy \tag{13.500}$$

For the integrals involved, we find

$$C_1(x) = \frac{e^x}{0.02}\left[1.56 - e^{-x}(x^2 + 1.50x + 1.56)\right] \tag{13.501a}$$

$$C_2(x) = -\frac{e^x}{0.01}\left[1.63 - e^{-x}(x^2 + 1.60x + 1.63)\right] \tag{13.501b}$$

$$C_3(x) = \frac{e^x}{0.02}\left[1.72 - e^{-x}(x^2 + 1.70x + 1.72)\right] \tag{13.501c}$$

The matrix equation becomes

$$\begin{bmatrix}\Psi(0.1)\\ \Psi(0.2)\\ \Psi(0.3)\end{bmatrix} = \begin{bmatrix}1.1\\ 1.2\\ 1.3\end{bmatrix} + \begin{bmatrix}0.2033 & -0.1429 & 0.0447\\ 0.2694 & -0.0886 & 0.0406\\ 0.2890 & -0.0270 & 0.0879\end{bmatrix}\begin{bmatrix}\Psi(0.1)\\ \Psi(0.2)\\ \Psi(0.3)\end{bmatrix} \tag{13.502}$$

$$\Rightarrow \begin{bmatrix}\Psi(0.1)\\ \Psi(0.2)\\ \Psi(0.3)\end{bmatrix} = \begin{bmatrix}1.2164\\ 1.4693\\ 1.7671\end{bmatrix} \tag{13.503}$$

As can be seen by comparing this to the results displayed in Table 13.27 these results are reasonably accurate. □

TABLE 13.28 TABULATED FUNCTION THAT CONTAINS AT LEAST ONE ZERO

x	$F(x)$
0	0.5
1	0.3
2	−0.1
3	−0.4

13.6 Zeros of a Function

Another problem that arises in scientific investigations is that of determining the value(s) of x for which a function $F(x) = 0$.

Interpolation

The method of interpolation is particularly well suited to the problem of finding the zeros of a function that is only known in tabulated form. To illustrate the method by example, consider the function described in tabulated form in Table 13.28. Unless the function behaves in a bizarre way between the tabulated points, it is clear that $F(x)$ has a single zero, and it occurs at a value of x between 1 and 2. For a smoothly varying function, as this tabulated function seems to be, we would interpolate $F(x)$ by a polynomial in x. Setting $F(x) = 0$ in the example above would then require that we solve for the roots of a third-order polynomial. If this function has only one zero, the other two roots of the polynomial must be complex.

The solution for the roots of a high-order polynomial is a rather complicated process. An equivalent and arithmetically much simpler approach involves inverting the equation. We rewrite

$$\text{(a)} \quad y = F(x) \qquad \text{as} \qquad \text{(b)} \quad x = F^{-1}(y) \equiv G(y) \tag{13.504}$$

In this example, the tabulated values of y are given at four values of x. Thus x can be interpolated as a third-order polynomial in y. That is, for this example,

$$x = \sum_{l=1}^{4} \mu_l(y) G(y_l) = \sum_{l=1}^{4} \mu_l(y) x_l \tag{13.505}$$

Since the problem is to determine the value of x for which $y = 0$, we simply set $y = 0$ in equation 13.505 to obtain

$$x = \sum_{l=1}^{4} \mu_l(0) G(y_l) = \sum_{l=1}^{4} \mu_l(0) x_l \tag{13.506}$$

As described before, $\mu_l(y)$ does not have to be a polynomial. If it is known that $F(x)$ has behavior similar to the function $p(x)$, (exponential, sinusoidal, etc.), it is reasonable to expect that $G(y)$ will have behavior similar to $p^{-1}(y) \equiv q(y)$, and interpolating $G(y)$ over the function $q(y)$ as in equation 13.11, will yield results that are more accurate than those obtained with a polynomial interpolation. Of course, if tabulated data are all that is available, it is possible that there will be no indication of the functional behavior of $F(x)$. In such cases, polynomial interpolation is the most commonly used interpolation approximation of $G(y)$.

Example 13.53

For the data in Table 13.28, interpolation by a third-order polynomial yields

$$x = \sum_{l=1}^{4} \prod_{k \neq l} \frac{(y - y_k)}{(y_l - y_k)} x_l = \frac{(y - 0.3)(y + 0.1)(y + 0.4)}{(0.5 - 0.3)(0.5 + 0.1)(0.5 + 0.4)}(0)$$

$$+ \frac{(y - 0.5)(y + 0.1)(y + 0.4)}{(0.3 - 0.5)(0.3 + 0.1)(0.3 + 0.4)}(1) + \frac{(y - 0.5)(y - 0.3)(y + 0.4)}{(-0.1 - 0.5)(-0.1 - 0.3)(-0.1 + 0.4)}(2)$$

$$+ \frac{(y - 0.5)(y - 0.3)(y + 0.1)}{(-0.4 - 0.5)(-0.4 - 0.3)(-0.4 + 0.1)}(3) \tag{13.507}$$

Setting $y = 0$ yields $x = 1.78571$. Of course, since the function is known only in tabulated form, there is no way to check the accuracy of this estimate. □

Regula falsi method

The method of false position, or regula falsi, is an iterative procedure for locating a zero of $F(x)$ when $F(x)$ is known in functional form. The first step in the process is to search $F(x)$ for two values of x, which will be denoted by a and b, such that $F(a)$ and $F(b)$ are of opposite sign. Then at least one zero of $F(x)$ will lie between a and b. Figure 13.20 illustrates such a situation. $F(x)$ is now approximated by a straight line between a and b, with the constants of the linear fit determined by $F(a)$ and $F(b)$. That is, $F(x)$ is written

$$F(x) = \alpha x + \beta \tag{13.508}$$

with

$$\alpha = \frac{F(b) - F(a)}{(b - a)} \tag{13.509}$$

$$\beta = \frac{bF(a) - aF(b)}{(b - a)} \tag{13.510}$$

Setting this linear approximation to zero yields a first approximation to the zero of $F(x)$. It is

$$x_0 = -\frac{\beta}{\alpha} = \frac{aF(b) - bF(a)}{F(b) - F(a)} \tag{13.511}$$

The next step is to determine the sign of $F(x_0)$. In our example, $F(x_0)$ is negative, as indicated in Figure 13.20. Since $F(a)$ is positive and $F(x_0)$ is negative, the procedure is repeated taking a linear approximation to $F(x)$ between $x = a$ and $x = x_0$. Had

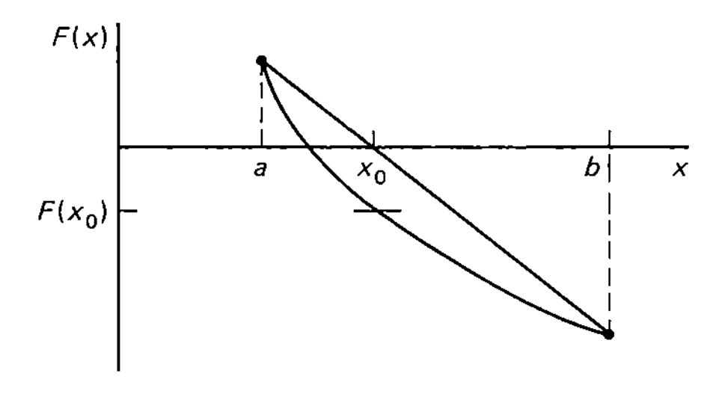

Figure 13.20 First step in the location of an approximate zero of $F(x)$ by regula falsi method.

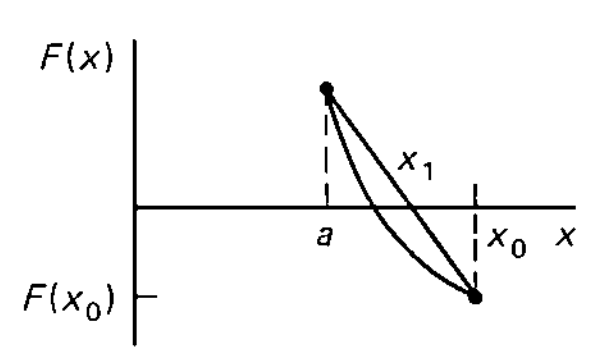

Figure 13.21 Subsequent step in the location of an approximation zero of $F(x)$ by regula falsi method.

$F(x_0)$ been positive, the linear fit would have been made between x_0 and b. For the function depicted in Figure 13.20, the second step is illustrated in Figure 13.21. From this a second estimate is found as

$$x_1 = \frac{aF(x_0) - x_0 F(a)}{F(x_0) - F(a)} \tag{13.512}$$

The process is repeated generating estimates of x that are progressively closer to the point where $F(x)$ passes through zero.

Example 13.54

As an example, we estimate the zero of

$$F(x) = e^x - 3 = 0 \tag{13.513}$$

The exact result is

$$x = \log(3) = 1.09861 \tag{13.514}$$

We note that

$$\text{(a)} \quad F(1) = -0.28172 < 0 \qquad \text{(b)} \quad F(1.5) = 1.48169 > 0 \tag{13.515}$$

Therefore, the first estimate of the zero is

$$x_0 = \frac{1 \cdot F(1.5) - 1.5 \cdot F(1)}{F(1.5) - F(1)} = 1.07988 \tag{13.516}$$

Since

$$F(1.07988) = -0.05568 < 0 \tag{13.517}$$

the procedure is repeated between $x = 1.07988$ and $x = 1.5$. From this, the second estimate is

$$x_1 = \frac{1.07988 \cdot F(1.5) - 1.5 \cdot F(1.07988)}{F(1.5) - F(1.07988)} = 1.09510 \tag{13.518}$$

With

$$F(1.09510) = -0.01053 \tag{13.519}$$

we repeat the process between $x = 1.09510$ and $x = 1.5$ to obtain

$$x_2 = 1.09796 \tag{13.520}$$

It is evident that the method is converging to the correct result. Therefore, with sufficient iterations, we will achieve stability to a required number of decimal places. (That is, x_{n+1} and x_n will have the same value to a specified number of decimal places.) □

Newton – Raphson method

Using the regula falsi method, $F(x)$ is approximated by the linear fit

$$F(x) = \alpha x + \beta \tag{13.508}$$

where α is the slope of the line, given by

$$\alpha = \frac{F(b) - F(a)}{(b - a)} \tag{13.509}$$

This is actually an approximation to the derivative of $F(x)$ over the interval $[a, b]$. The Newton–Raphson method also approximates $F(x)$ by a linear function, but uses a truncated Taylor series.

Let x_n be an estimate of the zero of $F(x)$. Then a Taylor series expansion of $F(x)$ around x_n, truncated after two terms, is

$$F(x) \simeq F(x_n) + F'(x_n)(x - x_n) \tag{13.521}$$

With $x = x_{n+1}$, we set $F(x_{n+1}) = 0$ to obtain the $(n+1)$th estimate of the zero of $F(x)$. It is

$$x_{n+1} = x_n - \frac{F(x_n)}{F'(x_n)} \tag{13.522}$$

The iterative process is begun by choosing a first estimate of the zero which we designate x_0. The next estimate is then obtained from

$$\text{(a)} \quad x_1 = x_0 - \frac{F(x_0)}{F'(x_0)} \quad \text{and} \quad \text{(b)} \quad x_2 = x_1 - \frac{F(x_1)}{F'(x_1)} \tag{13.523}$$

The process is continued until $|x_{n+1} - x_n|$ is smaller than some required limit.

Example 13.55

Applying this to the example of finding the zero of

$$F(x) = e^x - 3 \tag{13.513}$$

$$\Rightarrow \quad F'(x) = e^x \tag{13.524}$$

we choose $x = 1$ as our initial estimate. Then

$$\text{(a)} \quad x_1 = 1 - \frac{e-3}{e} = 1.10364 \quad \Rightarrow \quad \text{(b)} \quad x_2 = 1.10364 - \frac{e^{1.10364} - 3}{e^{1.10364}} = 1.09863$$

$$\Rightarrow \quad \text{(c)} \quad x_3 = 1.09863 - \frac{e^{1.09863} - 3}{e^{1.09863}} = 1.09861 \tag{13.525}$$

Thus, after two iterations, the Newton–Raphson approach has determined the zero of $F(x)$ accurately to five decimal places. □

In general, as in the example above, the Newton–Raphson method converges more rapidly than the regula falsi approach.

One of the potential drawbacks of the Newton–Raphson technique is that $F(x)$ may have a very small or zero slope at one of the iteration estimates. This can occur if $F(x)$ has a zero near an extremum, such as the function illustrated in Figure 13.22. As x_n approaches one of the zeros of Figure 13.22, it will also approach the minimum. Near this minimum, $F'(x_n)$ would be very small (or zero), and $F(x_n)/F'(x_n)$ would be very large (or infinite). That would destroy the convergence of the iterative process.

A somewhat less rapidly converging Newton–Raphson process can be developed that avoids this potential hazard. It requires choosing x_0 such that $F'(x_0)$ is not small or zero, and generating the estimates of the zero from

$$x_{n+1} = x_n - \frac{F(x_n)}{F'(x_0)} \tag{13.526}$$

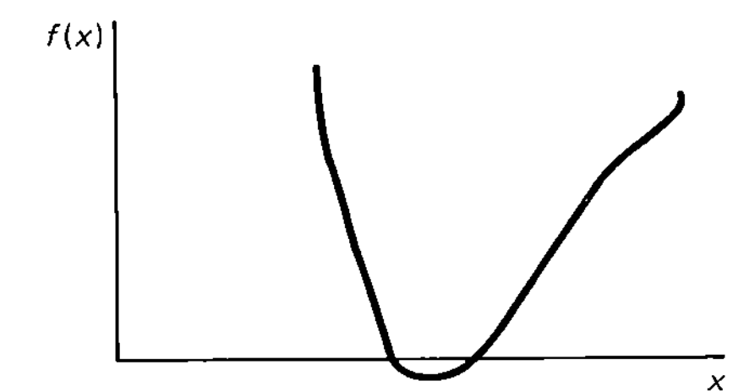

Figure 13.22 Function with an extremum close to two of its zeros.

Example 13.56

For

$$F(x) = e^x - 3 \tag{13.513}$$

taking $x_0 = 1$ leads to

$$\text{(a)}\quad x_1 = 1 - \frac{F(1)}{F'(1)} = 1.10364 \qquad \text{(b)}\quad x_2 = 1.10364 - \frac{F(1.10364)}{F'(1)} = 1.09808$$

$$\text{(c)}\quad x_3 = 1.09808 - \frac{F(1.09808)}{F'(1)} = 1.09867 \qquad \text{(d)}\quad x_4 = 1.09867 - \frac{F(1.09867)}{F'(1)} = 1.09861 \tag{13.527}$$

Like the results obtained in equations 13.525, x_4 is exact to five decimal places. □

We note that the exact value obtained in equations 13.525 requires three iterations, whereas to obtain the results expressed in equations 13.527, we need four iterations.

If $F(x)$ is a polynomial, both the regula falsi and the Newton–Raphson methods can be used to determine the roots. Since it is possible for the roots of a polynomial to be complex, one should take the initial estimate x_0 to be complex. The reader is referred to Problem 29.

Graeff root squaring method for the real roots of a polynomial

The Graeff technique is an iterative method for finding the real roots of a polynomial with real coefficients. We designate the roots of $F(x)$ as $x_1, x_2, \ldots, x_N$ and assume that the roots are ordered according to magnitude. We take x_1 to have largest magnitude, x_2 has the second largest magnitude, and so on. We assume for the present that none of the roots is a multiple or repeated root. The polynomial can be written either in the form

$$F(x) = (x - x_1)(x - x_2) \cdots (x - x_N) \tag{13.528a}$$

or as

$$F(x) = x^N + a_1 x^{N-1} + a_2 x^{N-2} + \cdots + a_N \tag{13.528b}$$

When the factors are multiplied to put equation 13.528a in the form expressed in equation 13.528b it is clear that the coefficient of x^N is 1. The terms in x^{N-1} arise from $N - 1$ products of x multiplied by one of the roots. Since there are N such terms, the coefficient of x^{N-1} is given by

$$-a_1 = \sum_{l=1}^{N} x_l \tag{13.529a}$$

By similar analysis, one can determine that

$$\text{(b)}\quad a_2 = (x_1 x_2 + x_1 x_3 + \cdots + x_2 x_3 + x_2 x_4 + \cdots) = \sum_{\substack{l,m \\ l>m}} x_l x_m$$

$$\text{(c)}\quad -a_3 = \sum_{\substack{l,m,n \\ l>m>n}} x_l x_m x_n \cdots \qquad \text{(d)}\quad a_N = (-1)^N \prod_{l=1}^{N} x_l \tag{13.529}$$

Let $G(x)$ be a polynomial with roots $-x_l$. Then $G(x)$ can be written

$$G(x) = (x + x_1)(x + x_2) \cdots (x + x_N) \tag{13.530a}$$

Since the roots of $G(x)$ are the negatives of the roots of $F(x)$, we can use the analysis that leads to equations 13.529 to write

$$G(x) = x^N - a_1 x^{N-1} + a_2 x^{N-2} - a_3 x^{N-3} + \cdots \tag{13.530b}$$

Multiplying $F(x)$ by $G(x)$ in the forms of equations 13.528a and 13.530a, we obtain

$$R(x^2) = F(x)G(x) = (x^2 - x_1^2)(x^2 - x_2^2) \cdots (x^2 - x_N^2) \tag{13.531a}$$

or with $y \equiv x^2$,

$$R(y) = (y - x_1^2)(y - x_2^2) \cdots (y - x_N^2) \tag{13.531b}$$

That is, the roots of $R(y)$ are the squares of the roots of $F(x)$. Multiplying $F(x)$ by $G(x)$ using equations 13.528b and 13.530b yields

$$\begin{aligned} R(x^2) = R(y) &= (x^N + a_1 x^{N-1} + \cdots)(x^N - a_1 x^{N-1} + \cdots) \\ &= (x^2)^N + (2a_2 - a_1^2)(x^2)^{N-1} + (a_2^2 - 2a_1 a_3 + 2a_4)(x^2)^{N-2} + \cdots \\ &= y^N + (2a_2 - a_1^2) y^{N-1} + (a_2^2 - 2a_1 a_3 + 2a_4) y^{N-2} + \cdots \\ &\equiv y^N + b_1 y^{N-1} + b_2 y^{N-2} + \cdots \end{aligned} \tag{13.532}$$

In this way, we obtain expressions relating the a-coefficients to the new b-coefficients. That is,

$$\text{(a)} \quad -b_1 = a_1^2 - 2a_2 \qquad \text{(b)} \quad b_2 = a_2^2 - 2a_1 a_3 + 2a_4 \qquad \text{etc.} \tag{13.533}$$

The roots of $R(y)$ are related to the b-coefficients in the same way the roots of $F(x)$ are related to the a-coefficients of equations 13.529. Therefore, since the roots of $R(y)$ are the squares of the roots of $F(x)$,

$$\begin{aligned} &\text{(a)} \quad \sum_{l=1}^{N} x_l^2 = -b_1 \qquad \text{(b)} \quad \sum_{\substack{l,m=1 \\ l>m}}^{N} x_l^2 x_m^2 = b_2 \\ &\text{(c)} \quad \sum_{\substack{l,m,n=1 \\ l>m>n}}^{N} x_l^2 x_m^2 x_n^2 = -b_3 \cdots \qquad \text{(d)} \quad \prod_{l=1} x_l^2 = (-1)^N b_N \end{aligned} \tag{13.534}$$

If the magnitude of the largest root is significantly larger than the magnitude of each of the other roots, its square will dominate the sum of equation 13.534a. That is, with x_1 designated as the dominant root, equation 13.534a can be written

$$\sum_{l=1}^{N} x_l^2 = -b_1 \simeq x_1^2 \tag{13.535}$$

Thus, x_1 can be approximated from the first coefficient of the $R(y)$ polynomial b_1, which is obtained from the a-coefficients of $F(x)$.

If x_1 is not large enough so that its square dominates the sum of equation 13.534a, this process can be repeated, calculating c-coefficients, which are the coefficients of a polynomial in x^4. Then, with

$$\text{(a)} \quad -c_1 \equiv b_1^2 - 2b_2 \qquad \text{(b)} \quad c_2 \equiv b_2^2 - 2b_1 b_3 + 2b_4 \qquad \text{etc.,} \tag{13.536}$$

the roots of $F(x)$ satisfy

$$\text{(a)} \quad \sum_{l=1}^{N} x_l^4 = -c_1 \qquad \text{(b)} \quad \sum_{\substack{l,m=1 \\ l>m}}^{N} x_l^4 x_m^4 = b_2^2 - 2b_1 b_3 + 2b_4 \equiv c_2 \qquad \text{etc.} \tag{13.537}$$

If the magnitude of x_1 is such that x_1^4 dominates the sum of equation 13.537a, then

$$\sum_{l=1}^{N} x_l^4 \simeq x_1^4 = -c_1 \tag{13.538}$$

This process can be repeated until the value of x_1 converges to the value of the largest root to a desired level of accuracy. For example, the next iteration will result in the approximation

$$\sum_{l=1}^{N} x_l^8 = c_1^2 - 2c_2 \equiv -d_1 \simeq x_1^8 \tag{13.539}$$

If another iteration were required, the coefficients $d_2, d_3, \ldots$ would be analogously determined so that the e-coefficients could be calculated for the estimate of x_1^{16}.

In general, after k iterations, we have

$$\sum_{l=1}^{N} x_l^{(2^k)} = -(\text{new first coefficient}) \simeq x_1^{(2^k)} \tag{13.540}$$

Therefore, to develop this approach, all that is needed is a prescription for generating the new coefficients (the b-coefficients from the a-coefficients, the c-coefficients from the b's, etc.).

We designate the set of coefficients used in the last iteration by $\{\alpha_l\}$, and define $\{\beta_l\}$ to be the set of coefficients to be determined for the next iteration. The prescription for generating the β-coefficients from the α-coefficients is

$$\begin{gathered} \text{(a)} \quad \beta_0 = 1 \qquad \text{(b)} \quad -\beta_1 = \alpha_1^2 - 2\alpha_2 \\ \text{(c)} \quad \beta_2 = \alpha_2^2 - 2\alpha_1\alpha_3 + 2\alpha_4 \qquad \text{(d)} \quad -\beta_3 = \alpha_3^2 - 2\alpha_2\alpha_4 + 2\alpha_1\alpha_5 - 2\alpha_6 \end{gathered} \tag{13.541}$$

or, in general, the nth coefficient for the next iteration is

$$\beta_n = (-1)^n\left(\alpha_n^2 - 2\alpha_{n-1}\alpha_{n+1} + 2\alpha_{n-2}\alpha_{n+2} - \cdots + (-1)^n 2\alpha_{2n}\right) \tag{13.542}$$

Clearly, if $n > N/2$, α_{2n} has no meaning. The most straightforward way to automate this is to define $\alpha_0 \equiv 1$, and $\alpha_{n+l} = 0$ for $n + l > N$.

The second largest root, x_2, can be determined in a similar way. If x_2 is large enough, the product $x_1^2 x_2^2$ will dominate the sum of equation 13.534b, and x_2 can be estimated from

$$\sum_{\substack{l,m=1 \\ l>m}}^{N} x_l^2 x_m^2 = b_2 \simeq x_1^2 x_2^2 \tag{13.543a}$$

Taking

$$x_1^2 \simeq -b_1 \qquad (13.535)$$

$$\Rightarrow \quad x_2^2 \simeq \frac{b_2}{-b_1} \qquad (13.543b)$$

As with the largest root, if x_2 is not accurately given by this estimate, the next iteration will be

$$\text{(a)} \quad \sum_{\substack{l,m=1 \\ l>m}} x_l^4 x_m^4 = c_2 \simeq x_1^4 x_2^4 \quad \Rightarrow \quad \text{(b)} \quad x_2^4 \simeq \frac{c_2}{-c_1} \qquad (13.544)$$

and so on. With sufficient iterations, $x_1^{(2^k)} x^{(2^k)}$ will dominate the sum of the product of the roots taken two at a time. Thus, in general, after k iterations, we have

$$\text{(a)} \quad x_1^{(2^k)} \simeq -\alpha_1 \qquad \text{(b)} \quad x_1^{(2^k)} x_2^{(2^k)} \simeq \alpha_2 \qquad (13.545)$$

$$\Rightarrow \quad x_2^{(2^k)} \simeq -\frac{\alpha_2}{\alpha_1} \qquad (13.546)$$

An analysis similar to this can be carried out to find the third largest root from the sum of the products of the roots taken three at a time, and so on.

By approximating the product of the roots three at a time by the third coefficient, the kth iteration yields

$$x_1^{(2^k)} x_2^{(2^k)} x_3^{(2^k)} = -\alpha_3 \qquad (13.547a)$$

Referring to equation 13.546a, we obtain

$$x_3^{(2^k)} \simeq -\frac{\alpha_3}{\alpha_2} \qquad (13.547b)$$

The generalization of this leads to the estimate of the lth root by

$$x_l^{(2^k)} \simeq -\frac{\alpha_l}{\alpha_{l-1}} \qquad (13.548)$$

Example 13.57

As an illustrative example, we apply the Graeff root squaring method to determine the roots of

$$x^3 - 6x^2 + 11x - 6 = 0 \qquad (13.549)$$

the roots of which are 1, 2, and 3. With

$$\text{(a)} \quad a_1 = -6 \qquad \text{(b)} \quad a_2 = 11 \qquad \text{(c)} \quad a_3 = -6 \qquad (13.550)$$

$$\Rightarrow \quad \sum_{l=1}^{3} x_l \simeq x_1 = -a_1 = 6 \qquad (13.551)$$

For the first iteration of x_1, we compute

$$\text{(a)} \quad -b_1 = a_1^2 - 2a_2 = 14 \qquad \text{(b)} \quad b_2 = a_2^2 - 2a_1a_3 = 49 \qquad \text{(c)} \quad -b_3 = a_3^2 = 36 \qquad (13.552)$$

$$\Rightarrow \quad \text{(a)} \quad \sum_{l=1}^{3} x_l^2 \simeq x_1^2 = -b_1 = 14 \quad \text{or} \quad \text{(b)} \quad x_1 \simeq 3.74166 \qquad (13.553)$$

Since this differs significantly from the value obtained from the previous iteration ($x_1 = 6$), we continue the process, calculating

$$(a) \quad -c_1 = b_1^2 - 2b_2 = 98 \qquad (b) \quad c_2 = b_2^2 - 2b_1b_3 = 1393 \qquad (c) \quad -c_3 = b_3^2 = 1296 \tag{13.554}$$

$$\Rightarrow \quad (a) \quad \sum_{l=1}^{3} x_l^4 = 98 \simeq x_1^4 \qquad \Rightarrow \qquad (b) \quad x_1 \simeq 3.14635 \tag{13.555}$$

The third iteration, with coefficients

$$(a) \quad -d_1 = 6818 \qquad (b) \quad d_2 = 1686433 \qquad (c) \quad -d_3 = 1679616 \tag{13.556}$$

$$\Rightarrow \quad (a) \quad \sum_{l=1}^{3} x_l^8 = 6816 \simeq x_1^8 \qquad \Rightarrow \qquad (b) \quad x_1 \simeq 3.01444 \tag{13.557}$$

As can be seen the results are converging to the correct value.

From the coefficients we have calculated, we also obtain the successive estimates of the second largest root. The first iteration yields

$$(a) \quad x_2^2 \simeq -\frac{b_2}{b_1} = \frac{49}{14} \qquad \text{so} \qquad (b) \quad x_2 \simeq 1.87083 \tag{13.558}$$

From the second iteration, we obtain

$$x_2 \simeq \left(-\frac{c_2}{c_1}\right)^{1/4} = \left(\frac{1393}{98}\right)^{1/4} = 1.94170 \tag{13.559}$$

and from the third iteration,

$$x_2 \simeq \left(-\frac{d_2}{d_1}\right)^{1/8} = \left(\frac{1{,}686{,}433}{6818}\right)^{1/8} = 1.99143 \tag{13.560}$$

which is clearly converging to $x_2 = 2$.

Referring to equation 13.549, after three iterations, the third largest root of our example is estimated to be

$$x_3 \simeq 0.99949 \quad \square \tag{13.561}$$

In the Graeff root squaring method, the roots of one iteration are the square of the roots of the previous iteration. Therefore, the technique is insensitive to the signs of the roots. That is, with successive iterations, the estimated values converge to the magnitude of the roots. Therefore, after obtaining an accurate estimate of a root, one must substitute both positive and negative values of that root into the original polynomial to determine which sign is correct.

The Graeff method can easily be adapted to determine the smallest roots without determining any of the largest roots. If we divide the original polynomial equation

$$x^N + a_1x^{N-1} + \cdots + a_N = 0 \tag{13.562}$$

by $a_N x^N$, and define $1/x \equiv y$, we obtain

$$\frac{1}{a_N} + \frac{a_1}{a_N}\frac{1}{x} + \cdots + \frac{a_{N-1}}{a_N}\frac{1}{x^{N-1}} + \frac{1}{x^N} = \frac{1}{a_N} + \frac{a_1}{a_N}y + \cdots + \frac{a_{N-1}}{a_N}y^{N-1} + y^N = 0 \tag{13.563a}$$

Defining $a'_N \equiv 1/a_N$ and $a'_{N-k} \equiv a_k/a_N$, this equation becomes

$$y^N + a'_1y^{N-1} + \cdots + a'_{N-1}y + a'_N = 0 \tag{13.563b}$$

This is identical in form to the original polynomial.

The root with the largest magnitude y_1 is then determined by the Graeff method. But if y_1 is the largest root, $1/y_1$ is the smallest root x_N of the original polynomial.

Thus far, we have assumed that all the roots are distinct. If one of the roots is repeated, the Graeff procedure will still converge to the correct root. However, the convergence will be extremely slow. The rate of convergence will depend on the multiplicity of the root. The higher the multiplicity, the slower will be the rate of convergence.

For example, if the largest root is a double root, then at each iteration

$$2x_1^{(2^k)} \simeq -\alpha_1 \tag{13.564}$$

$$\Rightarrow \quad x_1 \simeq \left(-\frac{\alpha_1}{2}\right)^{(1/2)^k} \tag{13.565}$$

Since $2^{((1/2)^{17})} = 1$ to five decimal places, it takes 17 iterations before the stability of the largest root to five decimal places is unaffected factor of 2. That is a very slow rate of convergence.

If x_1 were a triply repeated root, the kth iteration of x_1 would be

$$x_1 \simeq \left(-\frac{\alpha_2}{3}\right)^{(1/2)^k} \tag{13.566}$$

Twenty-one iterations are required before the first five decimal places of the estimate of the root are unaffected by the factor of 3.

If x_1 is a doubly repeated root, then $x_2 = x_1$ and the estimate of the second largest root becomes

$$x_1^{(2^k)} x_2^{(2^k)} = x_1^{2(2^k)} = -\frac{\alpha_2}{\alpha_1} \tag{13.567}$$

This estimate will converge to the same value x_1 as the procedure for the largest root, and the convergence will also be at the same slow rate. From this, it will become evident that x_1 is a multiply repeated root.

Example 13.58

For example, the polynomial

$$x^3 - 8x^2 + 21x - 18 = 0 \tag{13.568}$$

has roots 3, 3, 2. The first Graeff iteration yields

$$\text{(a)} \quad -b_1 = 22 \qquad \text{(b)} \quad b_2 = 153 \qquad \text{(c)} \quad -b_3 = 324 \tag{13.569}$$

$$\Rightarrow \quad \text{(a)} \quad x_1 \simeq 4.69042 \qquad \text{(b)} \quad x_2 \simeq 2.63715 \tag{13.570}$$

The second iteration yields

$$\text{(a)} \quad -c_1 = 178 \qquad \text{(b)} \quad c_2 = 9153 \qquad \text{(c)} \quad -c_3 = 104976 \tag{13.571}$$

$$\Rightarrow \quad \text{(a)} \quad x_1 = 3.65262 \qquad \text{(b)} \quad x_2 = 2.87322 \tag{13.572}$$

We can see that both roots are converging very slowly to 3 (but only because we know the correct result). Clearly, we cannot discern the correct value of x_1 from these first few iterations since none of the succeeding iterates is close enough to the previous one. □

The slow rate of the convergence is the clue that there is either a multiplicity to the root being sought (or that two or more roots are nearly equal). When this occurs, convergence will be achieved much more rapidly by taking the ratio of successive iterates of a given root.

To see this, let the multiplicity of a root be M. For the sake of arithmetic simplicity, let the largest root have this multiplicity. The kth iteration of the Graeff process for the largest root will yield

$$Mx_1^{(2^k)} \simeq -\alpha_k \tag{13.573a}$$

and for the $(k+1)$th iteration we have

$$Mx_1^{(2^{k+1})} \simeq -\beta_1 \tag{13.573b}$$

Taking the ratio of these equations, we eliminate M and obtain

$$x_1^{(2^k)} \simeq \frac{\beta_1}{\alpha_1} \tag{13.574}$$

Since this ratio of two successive iterations is independent of the multiplicity of a root, the method can be applied even if the roots are not repeated, and rapid convergence will be obtained.

Example 13.59

Applying this technique to

$$x^3 - 8x^2 + 21x - 8 = 0 \tag{13.568}$$

of Example 13.58, we have

$$\text{(a)}\quad \frac{Mx_1^4}{Mx_1^2} = x_1^2 \simeq \frac{-c_1}{-b_1} = \frac{178}{22} = 8.09091 \qquad \Rightarrow \qquad \text{(b)}\quad x_1 \simeq 2.84445 \tag{13.575}$$

From the ratio of the coefficient of the third iteration to the second, we have

$$\text{(a)}\quad \frac{Mx_1^8}{Mx_1^4} = x_1^4 \simeq \frac{-d_1}{-c_1} = \frac{13378}{178} = 75.15730 \qquad \Rightarrow \qquad \text{(b)}\quad x_1 \simeq 2.94437 \tag{13.576}$$

Taking the ratio of the first coefficient from the fourth iteration to that from the third iteration, we obtain

$$x_1 \simeq 2.99305 \tag{13.577}$$

We note that the convergence to $x_1 = 3$ is relatively rapid.

Once the value of the repeated root is obtained by this method, its multiplicity can also be obtained. If the kth iteration is the final iteration, then

$$Mx_1^{(2^k)} \simeq -\alpha_1 \tag{13.578}$$

from which the multiplicity of the root can be obtained.

In the example above, using the fourth iteration,

$$Mx_1^8 \simeq 13387 \tag{13.579}$$

Therefore, since

$$x_1 \simeq 2.99305 \tag{13.577}$$

$$\Rightarrow \quad M = 2.07722 \tag{13.580}$$

Clearly, the value of M is converging to 2. □

As mentioned earlier, the Graeff method is insensitive to the sign of a root, and thus one must substitute both the positive and negative values of the determined root into the polynomial. If two distinct roots have the same magnitude but have opposite signs ($\pm x_i$), after squaring, these will behave like a single repeated root of multiplicity 2 (or higher). Therefore, when x_i is determined to have a multiplicity of 2 or higher, both $+x_i$ and $-x_i$ should be tested in the polynomial.

PROBLEMS

1. Selected tabulated values of $x^2 - 2e^{-x}$ are given below.

x	$x^2 - 2e^{-x}$
0.50	−0.963
0.75	−0.382
1.00	0.264

Estimate the value of $x^2 - 2e^{-x}$ at $x = 0.65$:
(a) Using a Lagrange polynomial interpolation
(b) Using a linear spline interpolation
(c) Using an operator expansion

2. Using the data

x	$x^2 - 2e^{-x}$
0.25	−1.495
0.50	−0.963
0.75	−0.382
1.00	0.264
1.25	0.989

determine the approximate values of $x^2 - 2e^{-x}$ at $x = 0.35$ and $x = 1.10$ using a quadratic spline interpolation.

3. The values of $\sin(0.7x)$ at selected values of x (in radians) are listed below.

x	$\sin(0.7x)$
0.5	0.34290
0.7	0.47963
0.9	0.58914

Use a Lagrange-like interpolation over the function $q(x) = \sin x$ to estimate $\sin(0.595)$.

4. (a) Find the best straight line to describe the following data:

x	F
0.2	7.0
0.4	9.5
0.6	12.5
0.8	14.7
1.0	16.2

and determine the rms error to the fit. From that straight line, determine the values of $F(x)$ at the values of x listed in the table.
(b) Find the best parabolic fit to the data in the table. Determine the rms error for this curve and the values of $F(x)$ at each x listed in the table.

5. The data

x	F
−0.50	0.298
−0.25	−1.116
0.00	2.302
0.25	−1.113
0.50	0.301

represents an even, periodic function of period $\lambda = 2\pi/k = 0.4$. Find the best fit to these data by the function

$$F(x) = \sum_{m=0}^{2} \alpha_m \cos(mkx)$$

6. The data

x	F
1.0	1.30
1.5	0.73
2.0	0.41
2.5	0.19
3.0	0.08

decrease with increasing x. Find the best three-parameter curve to fit the data in the form:
(a) $F(x) = \alpha_0 + \alpha_1 \frac{1}{x} + \alpha_2 \frac{1}{x^2}$
(b) $F(x) = \alpha_0 + \alpha_1 e^{-x} + \alpha_2 e^{-2x}$
(c) $F(x) = \alpha_0 + \alpha_1 e^{-x^2} + \alpha_2 e^{-x^4}$
For each curve, determine the rms error and find $F(x)$ at each value of x in the table. Determine which of the three curves gives the best least squares fit.

7. Find the best approximation to $F(x) = e^{-x}$ in the range $x \in [-1, 1]$:
(a) By a straight line
(b) By a parabola
Present the comparisons in a table, listing the exact and approximate values at $x = -1.0, -0.5, 0.0, 0.5,$ and 1.0.

8. Evaluate

$$\int_1^3 e^{\sqrt{x}}\, dx$$

(a) Using a four-segment trapezoidal rule
(b) Using a four-segment Simpson's $\frac{1}{3}$ rule
(c) Using a four-point Gauss–Legendre quadrature

9. Evaluate

$$\int_1^{\infty} e^{-\sqrt{x}}\, dx$$

(a) Using a three-point Laguerre quadrature
(b) Using a five-point Laguerre quadrature
(c) Using a three-point Legendre quadrature
By substituting $x = (1 + z)^2$ this integral can be evaluated exactly. The result is $4e^{-1}$. Compare each result to the exact value of the integral.

10. Evaluate

$$\int_{-\infty}^{\infty} e^{-x^4}\, dx$$

(a) Using a three-point Hermite quadrature
(b) Using a three-point Legendre quadrature

11. The Chebyshev (Tshebychev) polynomials of the second kind have the form

$$U_N(\theta) = \frac{\sin[(N+1)\theta]}{\sin\theta}$$

with $x = \cos\theta$, they satisfy the orthonormality condition

$$\int_{-1}^{1} (1 - x^2) U_N(x) U_M\, dx = \frac{\pi}{8}\delta_{MN}$$

(a) Deduce the expressions for the weights and abscissae for a Gauss–Chebyshev quadrature that approximates

$$\int_{-1}^{1} (1 - x^2) f(x)\, dx \simeq \sum_{l=1}^{N} w_l f(x_l)$$

(b) Find the numerical values of the abscissae of a three-point Chebyshev quadrature and show that the weights are given by

$$w_l = -\frac{\sin^2\theta_l}{4\cos(4\theta_l)} \int_0^{\pi} \frac{\sin(4\theta)\sin^2\theta}{(\cos\theta - \cos\theta_l)}\, d\theta$$

12. Evaluate

$$P\int_{-3}^{3} \frac{1}{(1+x)(1+x^2)}\, dx$$

(a) By adding and subtracting the nonsingular part of the integrand and evaluating the integral of the resulting nonsingular integrand by a three-point Legendre quadrature rule.
(b) By interpolating the nonsingular part of the integrand over the Legendre polynomial $P_3(x)$ and using the identity

$$Q_N(x) = \frac{1}{2}\int_{-1}^{1} \frac{P_N(y)}{(x - y)}\, dy$$

(c) By dividing the interval $[-3, 3]$ into three regions, two of which are symmetric about the position of the pole. Evaluate each integral by a three-point Legendre quadrature rule. The correct result is $\frac{1}{2}[\log(2) - 2\tan^{-1}(\frac{1}{3}) + \pi] = 1.59562$.

13. Find an approximate value of $y(x)$ at $x = 0.2$, where y satisfies

$$y'(x) = e^{-xy} \qquad \text{subject to } y(0) = 0.$$

(a) Use a noniterated Taylor series truncated after the x^3 term.
(b) Use a two-step iteration, approximating $y(x)$ by a Taylor series truncated after the $(x - x_0)^3$ term at each step. Take $\Delta x = 0.1$ at each step.

14. Find an approximate solution for $y(x)$ at $x = 0.2$, where y satisfies

$$y'(x) = e^{-xy} \qquad \text{subject to } y(0) = 0$$

(a) Use a noniterated Runge–Kutta method.
(b) Use a two-step Runge–Kutta iteration. Find $y(0.1)$ from the first Runge–Kutta iteration. Use that to find $y(0.2)$ by the Runge–Kutta technique.

15. Find an approximate value of $y(0.2)$ using a two-step Picard iteration where $y(x)$ satisfies

$$y'(x) = e^{-xy} \qquad \text{subject to } y(0) = 0$$

16. The differential equation

$$(1 - x^2)y'' - 2xy' + 12y = 0$$

has solutions

$$P_3(x) = \tfrac{1}{2}(5x^3 - 3x)$$

and

$$Q_3(x) = P_3(x)\log\left(\frac{1+x}{1-x}\right) - \tfrac{5}{2}x + \tfrac{2}{3}$$

The initial conditions $y(0) = 0$ and $y'(0) = -\frac{3}{2}$ will select the polynomial solution, $P_3(x)$. The initial conditions $y(0) = \frac{2}{3}$ and $y'(0) = 0$ will select the second Legendre function $Q_3(x)$.
(a) Develop a two-step iteration of a Taylor series solution for $Q_3(0.2)$ taking $\Delta x = 0.1$ as the step intervals.
(b) Determine the value of $Q_3(0.2)$ by a two-step iteration of the Runge–Kutta method taking $\Delta x = 0.1$ at each step.

17. Use both the Euler and Milne finite difference approximations to determine $y(0.25)$, $y(0.50)$, $y(0.75)$,

and $y(1.00)$ where y satisfies

$$\frac{dy}{dx} = x^2 + y \qquad \text{with } y(0) = -2$$

The solution to this equation is

$$y(x) = (x^2 + 2x + 2)$$

Express the results in the form of a table comparing the result of each of the methods to the exact values of y at the four specified values of x.

18. Use the easiest method of finite differences to solve the initial value equation

$$y'' = -\frac{y'}{y^2} \qquad \text{subject to } y(2) = 2,\ y'(2) = 0.5$$

The analytic solution is $y(x) = \sqrt{2x}$. Take the step size to be $h = 0.2$ to determine $y(2.2)$, $y(2.4)$, $y(2.6)$, $y(2.8)$, and $y(3.0)$. Express the result in a table comparing the approximate and exact values of y at the specified values of x.

19. The solution to

$$x^2 y'' + 4y' - 2y = 0 \quad \text{subject to } y(0) = 8,\ y(1) = 13$$

is $y(x) = x^2 + 4x + 8$. Determine the values of $y(0.25)$, $y(0.50)$, and $y(0.75)$ by a finite difference method. Take the derivatives to be approximated by

(a) the Milne estimate of y' and the symmetric form of y''

(b) the Euler estimate of y' and the nonsymmetric form of y''

20. Use the finite differences with a step size of $h = \frac{2}{3}$ to find the two smallest eigenvalues of the Sturm-Liouville equation

$$(1 + x^2)y'' + 2xy' + 2\lambda y = 0 \text{ with } y(0) = y(2) = 0.$$

Determine the corresponding normalized eigenvectors.

21. Use the method of finite differences to find the solution to

$$\frac{\partial^2 T}{\partial x^2} - \frac{\partial T}{\partial t} = 0$$

with $T(x, 0) = 1 - x^2$, $T(-1, t) = 0$, $T(1, t) = t$.

Present the results in a solution table for $t = 0.0$, 0.5, and 1.0 at $x = -1.0$, -0.5, 0.0, 0.5, and 1.0.

22. Consider Laplace's equation

$$\frac{\partial^2 V}{\partial x^2} + \frac{\partial^2 V}{\partial y^2} = 0$$

with the boundary conditions

$$V(x, 0) = 0,\ V(x, 1) = 1,\ V(0, y) = y^2,\ V(1, y) = y^3$$

Using the method of finite differences, construct a solution table containing the values of $V(x, y)$ obtainable from the boundary conditions, and undetermined parameters V_i for those values not determined from the boundary conditions. Taking $x = 0, \frac{1}{3}, \frac{2}{3}, 1$ and $y = 0, \frac{1}{3}, \frac{2}{3}, 1$, determine a set of simultaneous equations for the unknown parameters V_i.

23. The Fredholm integral equation

$$\Psi(x) = 1 + \tfrac{1}{2}\int_{-1}^{1} e^{-(x+y)}\Psi(y)\, dy$$

can be solved exactly using the fact that the kernel is separable. Develop the numerical estimate of $\Psi(x)$ where x is not a root of the Legendre polynomial $P_N(x)$

(a) By approximating the integral by an N-point Legendre quadrature rule

(b) By approximating $\Psi(y)$ under the integral by an N-term interpolation over the Legendre polynomial $P_N(x)$ using

$$\mu_l(y) = \frac{P_N(y)}{(y - x_l)P_N'(x_l)}$$

(c) By approximating the kernel by an N-term interpolation over the Legendre polynomial $P_N(y)$

24. The exact solution to the Fredholm integral equation

$$\Psi(x) = \frac{1}{1 + x^2} + \frac{1}{2}\int_{-\infty}^{\infty} s^{-(x^2+y^2)}\Psi(y)\, dy$$

can be determined using the fact that the kernel is separable.

(a) Develop the numerical estimate of $\Psi(x)$ where x is not a root of the Hermite polynomial $H_N(x)$ by approximating the integral by an N-point Hermite quadrature rule.

(b) Develop the numerical estimate of $\Psi(x)$ where x is not a root of the Hermite polynomial $H_N(x)$ by approximating $\Psi(y)$ under the integral by an N-term interpolation over the Hermite polynomial $H_N(x)$ using

$$\mu_l(y) = \frac{H_N(y)}{(y - x_l)H_N'(x_l)}$$

(c) For

$$\mu_l(y) = \frac{H_N(y)}{(y - x_l)H_N'(x_l)}$$

determine why the numerical approximation

$$e^{-(x^2+y^2)} \simeq \sum_{l=1}^{N} e^{-(x^2+x_l^2)}\mu_l(y)$$

cannot be used to estimate $\Psi(x)$ for any $N > 1$.

25. Develop a three-term Taylor series for $\Psi(x)$ which satisfies

$$\Psi(x) = x + \int_1^x \sin(xy)\Psi(y)\,dy$$

Estimate $\Psi(1.2)$ from the series.

26. Using a single-segment trapezoidal integration rule, determine the value of $\Psi(1.2)$ using segment widths of $\Delta x = 0.1$. $\Psi(x)$ satisfies

$$\Psi(x) = x + \int_1^x \sin(xy)\Psi(y)\,dy$$

27. Using

$$\mu_l(y) = \frac{P_2(y)}{(y - x_l)P_2'(x_l)}$$

estimate the value of $\Psi(1.2)$ by interpolating $\Psi(y)$ in the integral of

$$\Psi(x) = x + \int_1^x \sin(xy)\Psi(y)\,dy$$

by

$$\Psi(y) \approx \sum_{l=1}^{2} \mu_l(y)\Psi(x_l)$$

28. A function $F(x)$ has the tabulated values

x	$F(x)$
0	−0.50000
$\frac{1}{3}$	−0.16701
$\frac{2}{3}$	0.15571
1	0.41732

(a) Estimate the zero of this function by a Lagrange interpolation of $x = F^{-1}(y)$.

(b) The table lists selected values of the function $F(x) \equiv \sqrt{\sin(x^2)} - \frac{1}{2}$. Estimate the zero of $F(x)$ using

(a) Three iterations of the regula falsi method

(b) Three iterations of the full Newton–Raphson method as expressed in equation 13.522

(c) Three iterations of the modified Newton–Raphson method as expressed in equations 13.526.

29. Use three iterations of the full Newton-Raphson method expressed in equations 13.522 to find one of the complex roots of

$$x^2 + 2x + 5 = 0$$

The two roots are $x = -1 \pm 2i$. Thus, starting with $x_0 = 1 + i$, which has a positive imaginary part, the iterations should converge on the root with the positive imaginary part, $-1 + 2i$.

30. For the polynomial

$$x^3 + x^2 - 17x + 15$$

(a) Using four iterations of the Graeff root squaring method, find the largest and second largest roots of the polynomial.

(b) Form a polynomial in $1/x$ and find the smallest root of the polynomial, using four iterations of the Graeff method.

31. For the polynomial

$$x^5 + 11x^4 + 34x^3 - 8x^2 - 160x - 128$$

use four iterations of the Graeff method to find the value of the root with the largest magnitude and determine its multiplicity.

INDEX

C

D

E

F

G

H

I

J

K

L

M

N

O

P

Q

R

S

T

U

V

W